Process Dynamics and Control

AN INDIAN ADAPTATION

Process Dynamics and Control

Fourth Edition

AN INDIAN ADAPTATION

Dale E. Seborg
University of California, Santa Barbara

Thomas F. Edgar
University of Texas at Austin

Duncan A. Mellichamp
University of California, Santa Barbara

Francis J. Doyle III
Harvard University

Content Contributions for Indian Adaptation by

Dr. Hare Krishna Mohanta
Associate Professor
Department of Chemical Engineering, Birla Institute of Technology & Science, Pilani, Rajasthan

Dr. Ajaya Kumar Pani
Assistant Professor,
Department of Chemical Engineering, Birla Institute of Technology & Science, Pilani, Rajasthan

WILEY

Process Dynamics and Control

Fourth Edition

WILEY INDIA ADAPTATION

Dale E. Seborg, Thomas F. Edgar, Duncan A. Mellichamp, and Francis J. Doyle III

Published by Wiley India Pvt. Ltd.,
4436/7, Ansari Road, Daryaganj, New Delhi-110002.

Printed at: Yash Printographics

Edition: 2021

ISBN: 978-93-5424-842-9

ISBN: 978-93-5424-628-9 (ebk)

Other Wiley Editorial Offices:
John Wiley & Sons, Inc. 111 River Street, Hoboken, NJ 07030, USA
Wiley-VCH Verlag GmbH, Pappellaee 3, D-69469 Weinheim, Germany
John Wiley & Sons Australia Ltd, 42 McDougall Street, Milton, Queensland 4064, Australia
John Wiley & Sons (Asia) Pte Ltd, 1 Fusionpolis Walk #07-01 Solaris, South Tower Singapore 138628
John Wiley & Sons Canada Ltd, 22 Worcester Road, Etobicoke, Ontario, Canada, M9W ILI

www.wileyindia.com

ABOUT THE AUTHORS

To our families

Dale E. Seborg is a Professor Emeritus and Research Professor in the Department of Chemical Engineering at the University of California, Santa Barbara. He received his B.S. degree from the University of Wisconsin and his Ph.D. degree from Princeton University. Before joining UCSB, he taught at the University of Alberta for nine years. Dr. Seborg has published over 230 articles and co-edited three books on process control and related topics. He has received the American Statistical Association's *Statistics in Chemistry Award,* the American Automatic Control Council's *Education Award,* and the ASEE *Meriam-Wiley Award*. He was elected to the *Process Automation Hall of Fame* in 2008. Dr. Seborg has served on the Editorial Advisory Boards for several journals and a book series. He has also been a co-organizer of several major national and international control engineering conferences.

Thomas F. Edgar holds the Abell Chair in chemical engineering at the University of Texas at Austin and is Director of the UT Energy Institute. He earned a B.S. degree in chemical engineering from the University of Kansas and his Ph.D. from Princeton University. Before receiving his doctorate, he was employed by Continental Oil Company. His professional honors include the AIChE Colburn and Lewis Awards, ASEE Meriam-Wiley and Chemical Engineering Division Awards, ISA and AACC Education Awards, AACC Bellman Control Heritage Award, and AIChE Computing in Chemical Engineering Award. He has published over 500 papers in the field of process control, optimization, and mathematical modeling of processes such as separations, combustion, microelectronics processing, and energy systems. He is a co-author of *Optimization of Chemical Processes*, published by McGraw-Hill in 2001. Dr. Edgar was the president of AIChE in 1997, President of the American Automatic Control Council in 1989–1991 and is a member of the National Academy of Engineering.

Duncan A. Mellichamp is a founding faculty member of the Department of Chemical Engineering of the University of California, Santa Barbara. He is editor of an early book on data acquisition and control computing and has published more than 100 papers on process modeling, large scale/plantwide systems analysis, and computer control. He earned a B.S. degree from Georgia Tech and a Ph.D. from Purdue University with intermediate studies at the Technische Universität Stuttgart (Germany). He worked for four years with the Textile Fibers Department of the DuPont Company before joining UCSB. Dr. Mellichamp has headed several organizations, including the CACHE Corporation (1977), the UCSB Academic Senate (1990–1992), and the University of California Systemwide Academic Senate (1995–1997), where he served on the UC Board of Regents. He presently serves on the governing boards of several nonprofit organizations and as president of Opera Santa Barbara. Emeritus Professor since 2003, he still guest lectures and publishes in the areas of process profitability and plantwide control.

Francis J. Doyle III is the Dean of the Harvard Paulson School of Engineering and Applied Sciences. He is also the John A. & Elizabeth S. Armstrong Professor of Engineering & Applied Sciences at Harvard University. He received his B.S.E. from Princeton, C.P.G.S. from Cambridge, and Ph.D. from Caltech, all in Chemical Engineering. Prior to his appointment at Harvard, Dr. Doyle held faculty appointments at Purdue University, the University of Delaware, and UCSB. He also held visiting positions at DuPont, Weyerhaeuser, and Stuttgart University. He is a Fellow of IEEE, IFAC, AAAS, and AIMBE; he is also the recipient of multiple research awards (including the AIChE Computing in Chemical Engineering Award) as well as teaching awards (including the ASEE Ray Fahien Award). He is the Vice President of the Technical Board of IFAC and is the President of the IEEE Control Systems Society in 2016.

PREFACE TO THE ADAPTED EDITION

Process Dynamics and Control provides full topical coverage for process control courses in the chemical engineering curriculum, emphasizing how process control and its related fields of process modeling and optimization are essential to the development of high-value products. A principal objective of this new edition is to describe modern techniques for control processes, with an emphasis on complex systems necessary to the development, design, and operation of modern processing plant. This Indian adaptation of the fourth edition of the book, has been developed based on the detailed feedback obtained from established readers of the book. Building on the conceptual strength of the previous editions, the focus of this adaptation has been on insertion and reorganization of topics to make it a better-fit textbook for Indian Universities with complete curriculum coverage. Internationally accepted SI system of units has been used in all solved examples and end of chapter problems. The end-of-chapter material now includes Multiple-choice Questions based on various competitive examinations. The chapters that are beyond the scope of coverage in the curriculum are now set as appendices for optional read.

The book aims to serve as a complete textbook for the basic as well as advanced process control course offered at undergraduate and post graduate Chemical Engineering curriculum in Indian universities. Chapters 1 to 11 are more oriented towards the syllabus of basic process control taught at undergraduate level dealing with need of process control, modeling aspects, dynamic behavior of open loop processes and feedback control systems. Chapter 12 through 19 along with the material covered in a few appendices are intended for students pursuing course in advanced process control. The book can also be a useful reference book for students of other disciplines pursuing course in process control as well as for practicing engineers.

The following are the salient features of this edition:

Updated Contents

- Chapter 1 on Introduction to Process Control includes sections on Need for Control Systems and their Design Aspects.
- Chapter 2 on The Theoretical Modeling has section on Degrees of Freedom Analysis revised and new section on Solving Differential Equations using MATLAB added.
- Chapter 3 on Laplace Transforms includes a section on Solving Laplace Transform Problems using MATLAB.
- Chapter 4 on Transfer Functions Models is enhanced with State Space Models and Interconversion of Different Forms of Models using MATLAB.
- Chapter 5 on Dynamic Behavior of First-Order and Second-Order Processes includes new topics on Zero Order Processes, Characteristics of First-Order processes, First-Order Processes with Variable Time Constants and Gain, First-Order Processes with Numerator Dynamics, Determining Overshoot, Decay Ratio and Settling Time using MATLAB.
- Chapter 8 on Control System Instrumentation includes a new section on Piping and Instrumentation Diagrams.
- Chapter 9 on Dynamic Behavior and Stability of Closed-Loop Control Systems is enhanced by including Block Diagram Reduction techniques and Generating Root Locus Diagram using MATLAB.
- Chapter 10 on Frequency Response Analysis and Control System Design is enhanced with the sections on Control System Design based on Bode Stability Criteria and Controller Design Based on Gain- and Phase Margins.
- Chapter 11 on PID Controller Design, Tuning, and Troubleshooting includes two new sections – Controller Tuning Based on One Quarter Decay Ratio criteria and Cohen-Coon Technique.
- Chapter 12 on Enhanced Single-Loop Control Strategies has reorganization of topics
- Chapter 13 on Digital Sampling, Filtering, and Control has reorganization of topics and enhanced coverage under section z-Transform Analysis for Digital Control.
- Chapter 14 on Multivariable Control includes more equations and explanations related to relative gain array analysis and input-output pairing.
- In Chapter 15 on Model Predictive Control, more explanation added to dynamic matrix control and

section on MPC with Inequality Constraints is thoroughly revised.

- Chapter 16 on Empirical Modeling has enhanced coverage of topics on Linear Regression, Neural Network Models and Soft Sensors.
- Chapter 17 on Process Monitoring has new content on principal component analysis (PCA) based monitoring technique and recent advances.
- Chapter 18 on Batch Process Control, has a new table on Boolean algebra and more material on programmable logic controller.
- A Chapter on Digital Process Control System: Hardware and Software has been brought from Appendix and made Chapter 19 in view of increasing importance of digital control.

Revised Pedagogy

All chapters have solved examples with detailed explanation of MATLAB application in problem solving.

- Complete SI version.
- More than one-third of the examples and end-of-chapter problems are new or revised.
- Objective questions, including questions asked in GATE are covered at the end of each chapter.

About the Contributors

Hare Krishna Mohanta is working as an Associate Professor in the Department of Chemical Engineering, BITS Pilani, Pilani Campus, Rajasthan (India). Prof. Mohanta has done his graduation in Chemical Engineering from NIT Rourkela, post-graduation in Chemical Engineering from IIT Kanpur and PhD from BITS Pilani. He has over 20 publications in the international and national journals of repute and served as a reviewer for a number of international journals. His areas of research include process identification, process monitoring, modeling, simulation and process control.

Ajaya Kumar Pani is working as Assistant Professor in the Department of Chemical Engineering at BITS Pilani, Pilani Campus, Rajasthan (India). Dr. Pani has done his graduation in Chemical Engineering from NIT Rourkela, post-graduation in Chemical Engineering from IIT BHU Varanasi and PhD from BITS Pilani. He has published over 12 research/review articles in reputed journals and served as a Reviewer for a number of international journals. His areas of research include modeling, simulation, process monitoring and process control.

PREFACE

Global competition, rapidly changing economic conditions, faster product development, and more stringent environmental and safety regulations have made process control increasingly important in the process industries. Process control and its allied fields of process modeling and optimization are critical in the development of more flexible and complex processes for manufacturing high-value-added products. Furthermore, the continuing development of improved and less-expensive digital technology has enabled high-performance measurement and control systems to become an essential part of industrial plants.

Overall, it is clear that the scope and importance of process control technology will continue to expand during the 21st century. Consequently, chemical engineers need to master this subject in order to be able to develop, design, and operate modern processing plants. The concepts of dynamic behavior, feedback, and stability are important for understanding many complex systems of interest to chemical engineers, such as bioengineering and advanced materials. An introductory process control course should provide an appropriate balance of theory and practice. In particular, the course should emphasize dynamic behavior, physical and empirical modeling, computer simulation, measurement and control technology, fundamental control concepts, and advanced control strategies. We have organized this book so that the instructor can cover the basic material while having the flexibility to include advanced topics on an individual basis. The textbook provides the basis for 10–30 weeks of instruction for a single course or a sequence of courses at either the undergraduate or first-year graduate levels. It is also suitable for self-study by engineers in industry. The book is divided into reasonably short chapters to make it more readable and modular. This organization allows some chapters to be omitted without a loss of continuity.

The mathematical level of the book is oriented toward a junior or senior student in chemical engineering who has taken at least one course in differential equations. Additional mathematical tools required for the analysis of control systems are introduced as needed. We emphasize process control techniques that are used in practice and provide detailed mathematical analysis only when it is essential for understanding the material. Key theoretical concepts are illustrated with numerous examples, exercises, and simulations.

Initially, the textbook material was developed for an industrial short course. But over the past 40 years, it has significantly evolved at the University of California, Santa Barbara, and the University of Texas at Austin. The first edition was published in 1989 and adopted by over 80 universities worldwide. In the second edition (2004), we added new chapters on the important topics of process monitoring, batch process control, and plantwide control. For the third edition (2011), we were very pleased to add a fourth coauthor, Professor Frank Doyle (then at UCSB) and made major changes that reflect the evolving field of chemical and biological engineering. These previous editions have been very successful and translated into Japanese, Chinese, Korean, and Turkish.

General revisions for the fourth edition include reducing the emphasis on lengthy theoretical derivations and increasing the emphasis on analysis using widely available software: MATLAB®, Simulink®, and Mathematica. We have also significantly revised material on major topics including control system design, instrumentation, and troubleshooting to include new developments. In addition, the references at the end of each chapter have been updated and new exercises have been added.

Exercises in several chapters are based on MATLAB® simulations of two physical models, a distillation column and a furnace. Both the book and the MATLAB simulations are available on the book's website (*www.wiley.com/college/seborg*). National Instruments has provided multimedia modules for a number of examples in the book based on their LabVIEW™ software.

Revisions to the five parts of the book can be summarized as follows. Part I provides an introduction to process control and an in-depth discussion of process modeling. It is an important topic because control system design and analysis are greatly enhanced by the availability of a process model.

Steady-state and unsteady-state behavior of processes are considered in Part II (Chapters 3 through 7). Transfer functions and state-space models are used to characterize the dynamic behavior of linear and nonlinear systems. However, we have kept derivations using classical analytical methods (e.g., Laplace transforms) to a minimum and prefer the use of computer simulation to determine dynamic responses. In addition, the important topics of empirical models and their development from experimental data are considered.

Part III (Chapters 8 through 15) addresses the fundamental concepts of feedback and feedforward control. Topics include an overview of process instrumentation (Chapter 9) and control hardware and software that are necessary to implement process control (Chapter 8 and Appendix A). Chapters 8–10 have been extensively revised to include new developments and recent references, especially in the area of process safety. The design and analysis of feedback control systems is a major topic with emphasis on industry-proven methods for controller design, tuning, and troubleshooting. Frequency response analysis (Chapter 14) provides important insights into closed-loop stability and why control loops can oscillate. Part III concludes with a chapter on feedforward and ratio control.

Part IV (Chapters 16 through 22) is concerned with advanced process control techniques. The topics include digital control, multivariable control, process monitoring, batch process control, and enhancements of PID control, such as cascade control, selective control, and gain scheduling. Up-to-date chapters on real-time optimization and model predictive control (MPC) emphasize the significant impact these powerful techniques have had on industrial practice. Material on Plantwide Control (Appendices G–I) and other important appendices are located on the book's website: *www.wiley.com/college/seborg*.

The website contains errata for current and previous editions that are available to both students and instructors. In addition, there are resources that are available for instructors (only): the Solutions Manual, lecture slides, figures from the book, and a link to the authors' websites. In order to access these password-protected resources, instructors need to register on the website.

We gratefully acknowledge the very helpful suggestions and reviews provided by many colleagues in academia and industry: Joe Alford, Anand Asthagiri, Karl Åström, Tom Badgwell, Michael Baldea, Max Barolo, Noel Bell, Larry Biegler, Don Bartusiak, Terry Blevins, Dominique Bonvin, Richard Braatz, Dave Camp, Jarrett Campbell, I-Lung Chien, Will Cluett, Oscar Crisalle, Patrick Daugherty, Bob Deshotels, Rainer Dittmar, Jim Downs, Ricardo Dunia, David Ender, Stacy Firth, Rudiyanto Gunawan, Juergen Hahn, Sandra Harris, John Hedengren, Karlene Hoo, Biao Huang, Babu Joseph, Derrick Kozub, Jietae Lee, Bernt Lie, Cheng Ling, Sam Mannan, Tom McAvoy, Greg McMillan, Randy Miller, Samir Mitragotri, Manfred Morari, Duane Morningred, Kenneth Muske, Mark Nixon, Srinivas Palanki, Bob Parker, Michel Perrier, Mike Piovoso, Joe Qin, Larry Ricker, Dan Rivera, Derrick Rollins, Alan Schneider, Sirish Shah, Mikhail Skliar, Sigurd Skogestad, Tyler Soderstrom, Ron Sorensen, Dirk Thiele, John Tsing, Ernie Vogel, Doug White, Willy Wojsznis, and Robert Young.

We also gratefully acknowledge the many current and recent students and postdocs at UCSB and UT-Austin who have provided careful reviews and simulation results: Ivan Castillo, Marco Castellani, David Castineira, Dan Chen, Jeremy Cobbs, Jeremy Conner, Eyal Dassau, Doug French, Scott Harrison, Xiaojiang Jiang, Ben Juricek, Fred Loquasto III, Lauren Huyett, Doron Ronon, Lina Rueda, Ashish Singhal, Jeff Ward, Dan Weber, and Yang Zhang. Eyal Dassau was instrumental in converting the old PCM modules to the version posted on this book's Website. The Solution Manual has been revised with the able assistance of two PhD students, Lauren Huyett (UCSB) and Shu Xu (UT-Austin). The Solution Manuals for earlier editions were prepared by Mukul Agarwal and David Castineira, with the help of Yang Zhang. We greatly appreciate their careful attention to detail. We commend Kristine Poland for her word processing skill during the numerous revisions for the fourth edition. Finally, we are deeply grateful for the support and patience of our long-suffering wives (Judy, Donna, Suzanne, and Diana) during the revisions of the book. We were saddened by the loss of Donna Edgar due to cancer, which occurred during the final revisions of this edition.

In the spirit of this continuous improvement, we are interested in receiving feedback from students, faculty, and practitioners who use this book. We hope you find it to be useful.

Dale E. Seborg
Thomas F. Edgar
Duncan A. Mellichamp
Francis J. Doyle III

CONTENTS

PART TWO:
DYNAMIC BEHAVIOR OF PROCESSES

PART FOUR: ADVANCED PROCESS CONTROL

*Appendices I to N are available on our website, https://www.wileyindia.com/student-resources and https://www.wileyindia.com/Instructor_Manuals/Register/login.php/.

CHAPTER 1

Introduction to Process Control

CHAPTER CONTENTS

In recent years the performance requirements for process plants have become increasingly difficult to satisfy. Stronger competition, tougher environmental and safety regulations, and rapidly changing economic conditions have been key factors. Consequently, product quality specifications have been tightened and increased emphasis has been placed on more profitable plant operation. A further complication is that modern plants have become more difficult to operate because of the trend toward complex and highly integrated processes. Thus, it is difficult to prevent disturbances from propagating from one unit to other interconnected units.

In view of the increased emphasis placed on safe and efficient plant operation, it is only natural that the subject of *process control* has become increasingly important in recent years. Without computer-based process control systems, it would be impossible to operate modern plants safely and profitably while satisfying product quality and environmental requirements. Thus, it is important for chemical engineers to have an understanding of both the theory and practice of process control.

The two main subjects of this book are *process dynamics* and *process control*. The term *process dynamics* refers to unsteady-state (or transient) process behavior. By contrast, most of the chemical engineering curricula emphasize steady-state and equilibrium conditions in such courses as material and energy balances, thermodynamics, and transport phenomena. But the topic of process dynamics is also very important. Transient operation occurs during important situations such as start-ups and shutdowns, unusual process disturbances, and planned transitions from one product grade to another. Consequently, the first part of this book is concerned with process dynamics.

The primary objective of process control is to maintain a process at the desired operating conditions, safely and economically, while satisfying environmental and product quality requirements. The subject of process control is concerned with how to achieve these goals. In large-scale, integrated processing plants such as oil refineries or ethylene plants, thousands of process variables such as compositions, temperatures, and pressures are measured and must be controlled. Fortunately, thousands of process variables (mainly flow rates) can usually be manipulated for this purpose. Feedback control systems compare measurements with their desired values and then adjust the manipulated variables accordingly.

Feedback control is a fundamental concept that is absolutely critical for both biological and manmade systems. Without feedback control, it would be very difficult, if not impossible, to keep complicated systems at the desired conditions. Feedback control is embedded in many modern devices that we take for granted: computers, cell phones, consumer electronics, air conditioning, automobiles, airplanes, as well as automatic control systems for industrial processes. The scope and history of feedback control and automatic control systems have been well described elsewhere (Mayr, 1970; Åström and Murray, 2008; Blevins and Nixon, 2011).

For living organisms, feedback control is essential to achieve a stable balance of physiological variables, a condition that is referred to as *homeostasis*. In fact, homeostasis is considered to be a defining feature of physiology (Widmaier *et al.*, 2011). In biology, feedback control occurs at many different levels including gene, cellular, metabolic pathways, organs, and even entire ecosystems. For the human body, feedback is essential to regulate critical physiological variables (e.g., temperature, blood pressure, and glucose concentration) and processes (e.g., blood circulation, respiration, and digestion). Feedback is also an important concept in education and the social sciences, especially economics (Rao, 2013) and psychology (Carver and Scheier, 1998).

1.1 NEED FOR CONTROL SYSTEMS

A control system commands, regulates, directs, or manages the behavior of other systems or devices using control loops. It is a rational arrangement of equipment (measuring devices, final control elements, controllers, computers, etc.) and operators to accomplish certain objectives. A chemical plant needs control systems to accomplish the following requirements:

(a) ***Safety.*** The safety of the persons working is the prime concern in a process plant. If the process conditions like temperature, pressure, concentration of the chemicals etc. exceed their allowable limits then hazardous situations may arise. This leads to accidents with loss of manpower and property. If a control system is in place, then it will bring back the process parameters to their allowable limits whenever it exceeds. Thus, safety can be ensured.

(b) ***Productivity.*** After safety, the next concern for a process plant is its productivity, i.e. the rate of production and the quality of products. Both are related to customer demand and satisfaction. The product quality is to be maintained for customer satisfaction. Without the control system it is very difficult to maintain the product quality.

(c) ***Environmental requirements.*** The process plant has to meet the restrictions imposed from time to time by various Government laws and regulations on the allowable limits of its effluents (gases, solids or liquids). This is again an impossible task without the control systems in place.

(d) ***Operational constraints.*** There always remain certain operational constraints in a process plant. For example, a distillation column should not be flooded, a storage tank should not overflow or run dry, the temperature must not exceed the allowable upper limit in a catalytic reactor, etc. These can be ensured by use of control systems.

(e) ***Profitability.*** Maximizing profit and minimizing operating cost is the primary goal of a process plant. For this, it is required that the operating conditions of the plants are to be controlled at certain optimum levels. Hence, control systems are required.

The above mentioned requirements dictate the need of control systems. Without control systems we cannot ensure safety, productivity and profitability in a process plant. Also, it will not be possible to meet the environmental and the operational requirements without control systems.

Process control is a subset of control systems where the main focus is on the control of process parameters like temperature, pressure, level, flow rate, concentration, etc. It enables automation. Without automation, plant operators must physically monitor the process parameters.

Process control can be defined as that branch of engineering science which deals with the technique of maintaining the outputs of a process or plant within desired limits. Many a time the phrases control systems and process control are used interchangeably.

1.2 CHARACTERISTICS OF PROCESS CONTROL PROBLEMS

The foundation of process control is *process understanding*. Thus, we begin this section with a basic question: what is a process? For our purposes, a brief definition is appropriate:

> ***Process:*** *The conversion of feed materials to products using chemical and physical operations. In practice, the term* process *tends to be used for both the processing operation and the processing equipment.*

There are three broad categories of processes: continuous, batch, and semibatch. Next, we consider representative processes and briefly summarize key control issues.

1.2.1 Continuous Processes

Four continuous processes are shown schematically in Fig. 1.1:

(a) ***Tubular heat exchanger.*** A process fluid on the tube side is cooled by cooling water on the shell side. Typically, the exit temperature of the process fluid is controlled by manipulating the cooling water flow rate. Variations in the inlet temperatures and the process fluid flow rate affect the

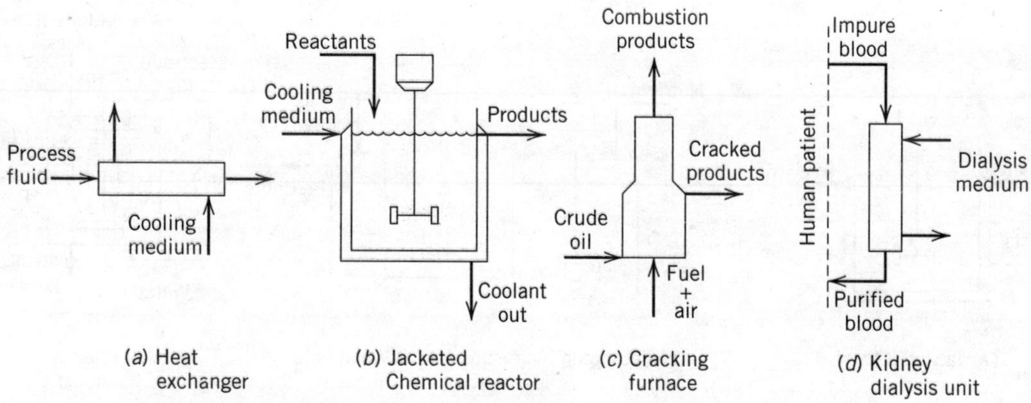

Figure 1.1 Some typical continuous processes.

heat exchanger operation. Consequently, these variables are considered to be disturbance variables.

(b) ***Continuous stirred-tank reactor (CSTR).*** If the reaction is highly exothermic, it is necessary to control the reactor temperature by manipulating the flow rate of coolant in a jacket or cooling coil. The feed conditions (composition, flow rate, and temperature) can be manipulated variables or disturbance variables.

(c) ***Thermal cracking furnace.*** Crude oil is broken down ("cracked") into a number of lighter petroleum fractions by the heat transferred from a burning fuel/air mixture. The furnace temperature and amount of excess air in the flue gas can be controlled by manipulating the fuel flow rate and the fuel/air ratio. The crude oil composition and the heating quality of the fuel are common disturbance variables.

(d) ***Kidney dialysis unit.*** This medical equipment is used to remove waste products from the blood of human patients whose own kidneys are failing or have failed. The blood flow rate is maintained by a pump, and "ambient conditions," such as temperature in the unit, are controlled by adjusting a flow rate. The dialysis is continued long enough to reduce waste concentrations to acceptable levels.

The process control problems are characterized by identifying the type of variables (temperature, pressure, flow rate, concentration, level etc.) associated with the process. These process variables are broadly divided into two groups:

- Input variables, which affect the process or system.
- Output variables, which get affected by the process or system.

The input variables are further classified as Manipulated Variables (MVs) and Disturbance Variables (DVs).

The output variables are also classified as Measured Outputs and Unmeasured Outputs. The Controlled Variables (CVs) can be either measured outputs or unmeasured outputs.

For each of these four examples, the process control problem has been characterized by identifying these three important types of process variables.

- *Controlled variables (CVs):* The process variables that are controlled. The desired value of a controlled variable is referred to as its *set point.*
- *Manipulated variables (MVs):* The process variables that can be adjusted in order to keep the controlled variables at or near their set points. Typically, the manipulated variables are flow rates.
- *Disturbance variables (DVs):* Process variables that affect the controlled variables but cannot be manipulated. Disturbances generally are related to changes in the operating environment of the process: for example, its feed conditions or ambient temperature. Some disturbance variables can be measured on-line, but many cannot such as the crude oil composition for Process (c), a thermal cracking furnace. The DVs are also called as the load variables.

The specification of CVs, MVs, and DVs is a critical step in developing a control system. The selections should be based on process knowledge, experience, and control objectives.

1.2.2 Batch and Semibatch Processes

Batch and semibatch processes are used in many process industries, including microelectronics, pharmaceu-

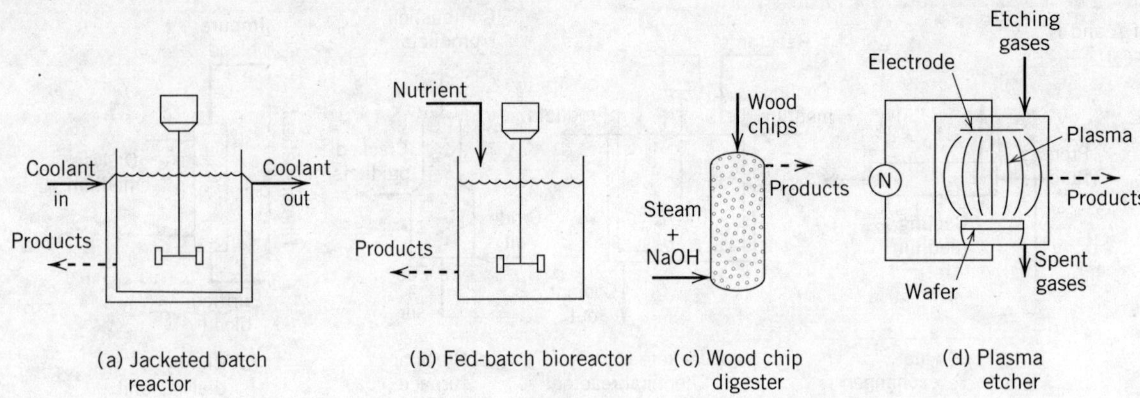

Figure 1.2 Some typical processes whose operation is noncontinuous. (Dashed lines indicate product removal after the operation is complete.)

ticals, specialty chemicals, and fermentation. Batch and semibatch processes provide needed flexibility for multiproduct plants, especially when products change frequently and production quantities are small. Figure 1.2 shows four representative batch and semibatch processes:

(a) ***Jacketed batch reactor.*** In a batch reactor, an initial charge (e.g., reactants and catalyst) is placed in the reactor, agitated, and brought to the desired starting conditions. For exothermic reactions, cooling jackets are used to keep the reactor temperature at or near the desired set point. Typically, the reactor temperature is regulated by adjusting the coolant flow rate. The endpoint composition of the batch can be controlled by adjusting the temperature set point and/or the *cycle time*, the time period for reactor operation. At the end of the batch, the reactor contents are removed and either stored or transferred to another process unit such as a separation process.

(b) ***Semibatch bioreactor.*** For a semibatch reactor, one of the two alternative operations is used: (i) a reactant is gradually added as the batch proceeds or (ii) a product stream is withdrawn during the reaction. The first configuration can be used to reduce the side reactions while the second configuration allows the reaction equilibrium to be changed by withdrawing one of the products (Fogler, 2010).

For bioreactors, the first type of semibatch operation is referred to as a *fed-batch operation*; it is shown in Fig. 1.2(*b*). In order to better regulate the growth of the desired microorganisms, a nutrient is slowly added in a predetermined manner.

(c) ***Semibatch digester in a pulp mill.*** Both continuous and semibatch digesters are used in paper manufacturing to break down wood chips in order to extract the cellulosic fibers. The end point of the chemical reaction is indicated by the kappa number, a measure of lignin content. It is controlled to a desired value by adjusting the digester temperature, pressure, and/or cycle time.

(d) ***Plasma etcher in semiconductor processing.*** A single wafer containing hundreds of printed circuits is subjected to a mixture of etching gases under conditions suitable to establish and maintain a plasma (a high voltage applied at high temperature and extremely low pressure). The unwanted material on a layer of a microelectronics circuit is selectively removed by chemical reactions. The temperature, pressure, and flow rates of etching gases to the reactor are controlled by adjusting electrical heaters and control valves.

Next, we consider an illustrative example in more detail.

1.3 DESIGNING CONTROL SYSTEMS FOR A PROCESS

A simple blending process is used to introduce some important issues in control system design. Blending operations are commonly used in many industries to ensure that final products meet customer specifications.

A continuous, stirred-tank blending system is shown in Fig. 1.3. The control objective is to blend the two inlet streams to produce an outlet stream that has the desired composition. Stream 1 is a mixture of two chemical species, A and B. We assume that its mass flow rate w_1 is constant, but the mass fraction of A, x_1, varies with time. Stream 2 consists of pure A and thus $x_2 = 1$. The mass flow rate of Stream 2, w_2, can be manipulated using a control valve. The mass fraction of A in the outlet stream is denoted by x and the desired value (set point) by x_{sp}. Thus for this control problem, the controlled variable is x, the manipulated variable is w_2, and the disturbance variable is x_1.

Next we consider two questions.

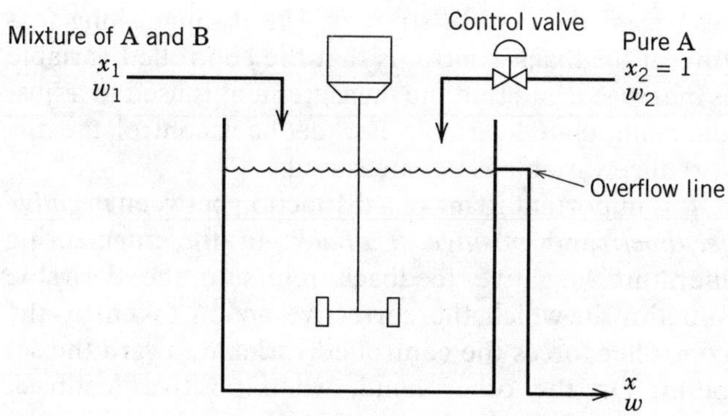

Figure 1.3 Stirred-tank blending system.

Design Question. *If the nominal value of x_1 is $\bar{x}_1$, what nominal flow rate $\bar{w}_2$ is required to produce the desired outlet concentration, x_{sp}?*

To answer this question, we consider the steady-state material balances:

Overall balance:

$$0 = \bar{w}_1 + \bar{w}_2 - \bar{w} \tag{1-1}$$

Component A balance:

$$0 = \bar{w}_1\bar{x}_1 + \bar{w}_2\bar{x}_2 - \bar{w}\bar{x} \tag{1-2}$$

The overbar over a symbol denotes its nominal steady-state value, for example, the value used in the process design. According to the process description, $\bar{x}_2 = 1$ and $\bar{x} = x_{sp}$. Solving Eq. 1-1 for $\bar{w}$, substituting these values into Eq. 1-2, and rearranging gives

$$\bar{w}_2 = \bar{w}_1 \frac{x_{sp} - \bar{x}_1}{1 - x_{sp}} \tag{1-3}$$

Equation 1-3 is the design equation for the blending system. If our assumptions are correct and if $x_1 = \bar{x}_1$, then this value of w_2 will produce the desired result, $x = x_{sp}$. But what happens if conditions change?

Control Question. *Suppose that inlet concentration x_1 varies with time. How can we ensure that the outlet composition x remains at or near its desired value, x_{sp}?*

As a specific example, assume that x_1 increases to a constant value that is larger than its nominal value, $\bar{x}_1$. It is clear that the outlet composition will also increase due to the increase in inlet composition. Consequently, at this new steady state, $x > x_{sp}$.

Next we consider several strategies for reducing the effects of x_1 disturbances on x.

Method 1. Measure x and adjust w_2. It is reasonable to measure controlled variable x and then adjust w_2 accordingly. For example, if x is too high, w_2 should be reduced; if x is too low, w_2 should be increased. This control strategy could be implemented by a person (*manual control*). However, it would normally be more convenient and economical to automate this simple task (*automatic control*).

Method 1 can be implemented as a simple control algorithm (or control law),

$$w_2(t) = \bar{w}_2 + K_c[x_{sp} - x(t)] \tag{1-4}$$

where K_c is a constant called the *controller gain*. The symbols, $w_2(t)$ and $x(t)$, indicate that w_2 and x change with time. Equation 1-4 is an example of *proportional control*, because the change in the flow rate, $w_2(t) - \bar{w}_2$, is proportional to the deviation from the set point, $x_{sp} - x(t)$. Consequently, a large deviation from set point produces a large corrective action, while a small deviation results in a small corrective action. Note that we require K_c to be positive because w_2 must increase when x decreases, and vice versa. However, in other control applications, negative values of K_c are appropriate, as discussed in Chapter 7.

A schematic diagram of Method 1 is shown in Fig. 1.4. The outlet concentration is measured and transmitted to the controller as an electrical signal. (Electrical signals are shown as dashed lines in Fig. 1.4.) The controller executes the control law and sends an appropriate

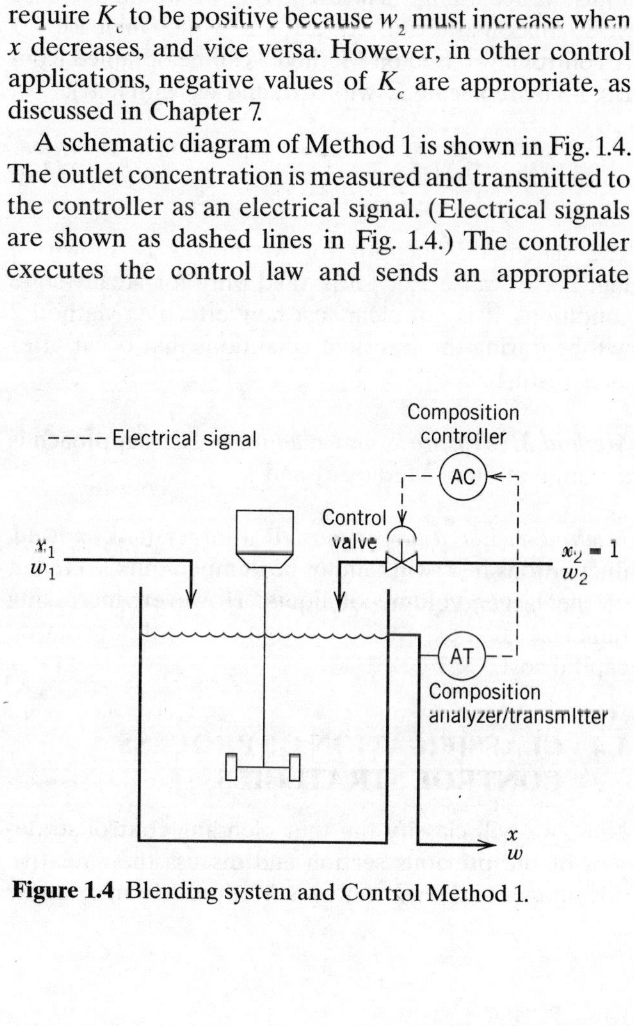

Figure 1.4 Blending system and Control Method 1.

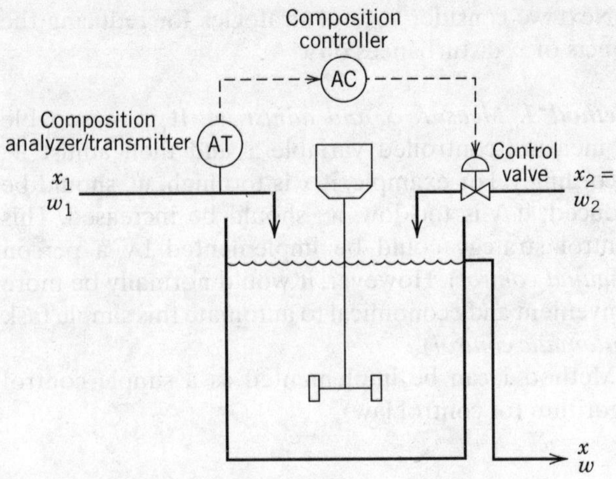

Figure 1.5 Blending system and Control Method 2.

electrical signal to the control valve. The control valve opens or closes accordingly. In Chapters 8 and 9, we consider process instrumentation and control hardware in more detail.

Method 2. *Measure x_1, adjust w_2.* As an alternative to Method 1, we could measure disturbance variable x_1 and adjust w_2 accordingly. Thus, if $x_1 > \bar{x}_1$, we would decrease w_2 so that $w_2 < \bar{w}_2$. If $x_1 < \bar{x}_1$, we would increase w_2. A control law based on Method 2 can be obtained from Eq. 1-3 by replacing $\bar{x}_1$ with $x_1(t)$ and $\bar{w}_2$ with $w_2(t)$:

$$w_2(t) = \bar{w}_1 \frac{x_{sp} - x_1(t)}{1 - x_{sp}} \tag{1-5}$$

The schematic diagram for Method 2 is shown in Fig. 1.5. Because Eq. 1-3 is valid only for steady-state conditions, it is not clear just how effective Method 2 will be during the transient conditions that occur after an x_1 disturbance.

Method 3. *Measure x_1 and x, adjust w_2.* This approach is a combination of Methods 1 and 2.

Method 4. *Use a larger tank.* If a larger tank is used, fluctuations in x_1 will tend to be damped out as a result of the larger volume of liquid. However, increasing tank size is an expensive solution due to the increased capital cost.

1.4 CLASSIFICATION OF PROCESS CONTROL STRATEGIES

Next, we will classify the four blending control strategies of the previous section and discuss their relative advantages and disadvantages. Method 1 is an example of a *feedback control* strategy. The distinguishing feature of feedback control is that the controlled variable is measured, and that the measurement is used to adjust the manipulated variable. For feedback control, the disturbance variable is *not* measured.

It is important to make a distinction between *negative feedback* and *positive feedback*. In the engineering literature, negative feedback refers to the desirable situation in which the corrective action taken by the controller forces the controlled variable toward the set point. On the other hand, when positive feedback occurs, the controller makes things worse by forcing the controlled variable farther away from the set point. For example, in the blending control problem, positive feedback takes place if $K_c < 0$, because w_2 will increase when x increases.[1] Clearly, it is of paramount importance to ensure that a feedback control system incorporates negative feedback rather than positive feedback.

An important advantage of feedback control is that corrective action occurs regardless of the source of the disturbance. For example, in the blending process, the feedback control law in Eq. 1-4 can accommodate disturbances in w_1, as well as x_1. Its ability to handle disturbances of unknown origin is a major reason why feedback control is the dominant process control strategy. Another important advantage is that feedback control reduces the sensitivity of the controlled variable to unmeasured disturbances and process changes. However, feedback control does have a fundamental limitation: no corrective action is taken until after the disturbance has upset the process, that is, until after the controlled variable deviates from the set point. This shortcoming is evident from the control law of Eq. 1-4.

Method 2 is an example of a *feedforward control strategy*. The distinguishing feature of feedforward control is that the disturbance variable is measured, but the controlled variable is not. The important advantage of feedforward control is that corrective action is taken *before* the controlled variable deviates from the set point. Ideally, the corrective action will cancel the effects of the disturbance so that the controlled variable is not affected by the disturbance. Although ideal cancelation is generally not possible, feedforward control can significantly reduce the effects of measured disturbances, as discussed in Chapter 12.

Feedforward control has three significant disadvantages: (i) the disturbance variable must be measured (or accurately estimated), (ii) no corrective action is

[1] Note that social scientists use the terms negative feedback and positive feedback in a very different way. For example, they would say that teachers provide "positive feedback" when they compliment students who correctly do assignments. Criticism of a poor performance would be an example of "negative feedback."

taken for unmeasured disturbances, and (iii) a process model is required. For example, the feedforward control strategy for the blending system (Method 2) does not take any corrective action for unmeasured w_1 disturbances. In principle, we could deal with this situation by measuring both x_1 and w_1 and then adjusting w_2 accordingly. However, in industrial applications, it is generally uneconomical to attempt to measure all potential disturbance variables. A more practical approach is to use a combined feedforward–feedback control system, in which feedback control provides corrective action for unmeasured disturbances, while feedforward control reacts to measured disturbances before the controlled variable is upset. Consequently, in industrial applications, feedforward control is normally used in combination with feedback control. This approach is illustrated by Method 3, a combined feedforward–feedback control strategy because both x and x_1 are measured.

Finally, Method 4 consists of a process design change and thus is not really a control strategy. The four strategies for the stirred-tank blending system are summarized in Table 1.1.

Table 1.1 Concentration Control Strategies for the Blending System

Method	Measured Variable	Manipulated Variable	Category
1	x	w_2	FB
2	x_1	w_2	FF
3	x_1 and x	w_2	FF/FB
4	—	—	Design change

FB = feedback control; FF = feedforward control; FF/FB = feedforward control and feedback control.

1.4.1 Process Control Diagrams

Next we consider the equipment that is used to implement control strategies. For the stirred-tank mixing system under feedback control (Method 1) in Fig. 1.4, the exit concentration x is controlled and the flow rate w_2 of pure species A is adjusted using proportional control. To consider how this feedback control strategy could be implemented, a block diagram for the stirred-tank control system is shown in Fig. 1.6.

The operation of the feedback control system can be summarized as follows:

1. *Analyzer and transmitter:* The tank exit concentration is measured by an analyzer and then the measurement is converted to a corresponding electrical current signal by a transmitter.
2. *Feedback controller:* The controller performs three distinct calculations. First, it converts the actual set point x_{sp} into an equivalent internal signal $\tilde{x}_{sp}$. Second, it calculates an error signal $e(t)$ by subtracting the measured value $x_m(t)$ from the set point $\tilde{x}_{sp}$, that is, $e(t) = \tilde{x}_{sp} - \tilde{x}_m(t)$. Third, controller output $p(t)$ is calculated from the proportional control law similar to Eq. 1-4.
3. *Control valve:* The controller output $p(t)$ in this case is a DC current signal that is sent to the control valve to adjust the valve stem position, which in turn affects flow rate $w_2(t)$. (The controller output signal is traditionally denoted by p because early controllers were pneumatic devices with pneumatic (pressure) signals as inputs and outputs.)

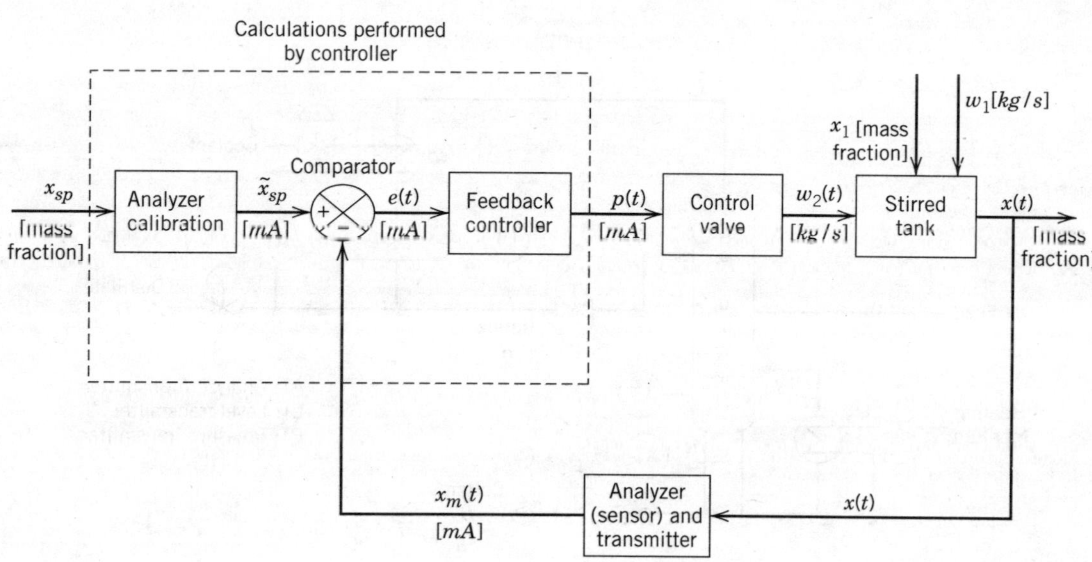

Figure 1.6 Block diagram for the outlet composition feedback control system in Fig. 1.4.

The block diagram in Fig. 1.6 provides a convenient starting point for analyzing process control problems. The physical units for each input and output signal are also shown. Note that the schematic diagram in Fig. 1.4 shows the *physical connections* between the components of the control system, while the block diagram shows the *flow of information* within the control system. The block labeled "control valve" has $p(t)$ as its input signal and $w_2(t)$ as its output signal, which illustrates that the signals on a block diagram can represent either a physical variable such as $w_2(t)$ or an instrument signal such as $p(t)$.

Each component in Fig. 1.6 exhibits behavior that can be described by a differential or algebraic equation. One of the tasks facing a control engineer is to develop suitable mathematical descriptions for each block; the development and analysis of such dynamic models are considered in Chapters 2–6.

The elements of the block diagram (Fig. 1.6) are discussed in detail in future chapters. Sensors, transmitters, and control valves are presented in Chapter 8, and the feedback controllers are considered in Chapter 7.

The feedback control system in Fig. 1.6 is shown as a single, standalone controller. However, for industrial applications, it is more economical to have a digital computer implement multiple feedback control loops. In particular, networks of digital computers can be used to implement thousands of feedback and feedforward control loops. Computer control systems are the subject of Chapters 13 and 19.

1.5 MULTILOOP VERSUS MULTIVARIABLE CONTROL

The blending control system in the previous section is quite simple, because there is only one controlled variable and one manipulated variable. For most practical applications, there are multiple controlled variables and multiple manipulated variables. As a representative example, we consider the distillation column in Fig. 1.7, with five controlled variables and five manipulated variables. The controlled variables are product compositions, x_D and x_B, column pressure, P, and the liquid levels in the reflux drum and column base, h_D and h_B. The five manipulated variables are product flow rates, D and B, reflux flow, R, and the heat duties for the condenser and reboiler, Q_D and Q_B. The heat duties are adjusted via the control valves on the coolant and heating medium lines. The feed stream is assumed to come from an upstream unit. Thus, the feed flow rate cannot be manipulated, but it can be measured and used for feedforward control.

A conventional *multiloop control* strategy for this distillation column would consist of five feedback control loops. Each control loop uses a single manipulated variable to control a single controlled variable. But how should the controlled and manipulated variables be paired? The total number of different multiloop control configurations that could be considered is 5!, or 120. Many of these control configurations are impractical or unworkable, such as any configuration that attempts to control the base level h_B by manipulating distillate flow D or condenser heat duty Q_D. However, even after the infeasible control configurations are eliminated, there are still many reasonable configurations left. Thus, there is a need for systematic techniques that can identify the most promising multiloop configurations. Fortunately, such tools are available and are discussed in Chapter 14.

In control applications, for which conventional multiloop control systems are not satisfactory, an alternative approach, *multivariable control*, can be advantageous.

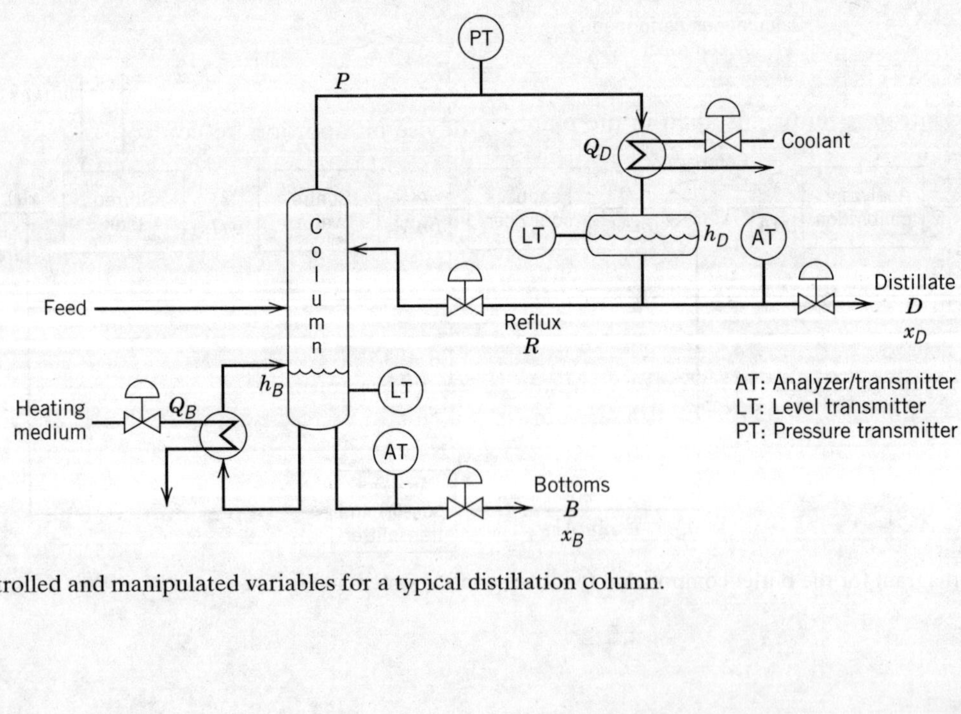

Figure 1.7 Controlled and manipulated variables for a typical distillation column.

In multivariable control, each manipulated variable is adjusted based on the measurements of at least two controlled variables rather than only a single controlled variable, as in multiloop control. The adjustments are based on a dynamic model of the process that indicates how the manipulated variables affect the controlled variables. Consequently, the performance of multivariable control, or any model-based control technique, will depend heavily on the accuracy of the process model. A specific type of multivariable control, *model predictive control*, has had a major impact on industrial practice, as discussed in Chapter 15.

1.6 DESIGN ASPECTS OF CONTROL SYSTEMS

In this section, we introduce some important aspects of control system design. However, it is appropriate first to describe the relationship between process design and process control.

Historically, process design and control system design have been separate engineering activities. Thus, in the traditional approach, control system design is not initiated until after plant design is well underway, and major pieces of equipment may even have been ordered. This approach has serious limitations because the plant design determines the process dynamics as well as the operability of the plant. In extreme situations, the process may be uncontrollable, even though the design appears satisfactory from a steady-state perspective. A better approach is to consider process dynamics and control issues early in the process design.

Next, we consider two general approaches to control system design:

1. ***Traditional Approach.*** The control strategy and control system hardware are selected based on knowledge of the process, experience, and insight. After the control system is installed in the plant, the controller settings (such as controller gain K_c in Eq. 1-4) are adjusted. This activity is referred to as *controller tuning*.
2. ***Model-Based Approach.*** A dynamic model of the process is first developed that can be helpful in at least three ways: (i) it can be used as the basis for model-based controller design methods (Chapters 10 and 11), (ii) the dynamic model can be incorporated directly in the control law (e.g., model predictive control), and (iii) the model can be used in a computer simulation to evaluate alternative control strategies and to determine preliminary values of the controller settings.

In this book, we advocate the philosophy that for complex processes, a dynamic model of the process should be developed so that the control system can be properly designed. Of course, for many simple process control problems, controller specification is relatively straightforward and a detailed analysis or an explicit model is not required. For complex processes, however, a process model is invaluable both for control system design and for an improved understanding of the process. As mentioned earlier, process control should be based on process understanding.

The major steps involved in designing and installing a control system using the model-based approach are shown in the flow chart of Fig. 1.8. The first step, formulation of the control objectives, is a critical decision. The formulation is based on the operating objectives for the plants and the process constraints. For example, in the distillation column control problem, the objective might be to regulate a key component in the distillate stream, the bottoms stream, or key components in both streams. An alternative would be to minimize energy consumption (e.g., reboiler heat duty) while meeting product quality specifications on one or both product streams. The inequality constraints should include upper and lower limits on manipulated variables, conditions that lead to flooding or weeping in the column, and product impurity levels.

After the control objectives have been formulated, a dynamic model of the process is developed. The dynamic model can have a theoretical basis, for example, physical and chemical principles such as conservation laws and rates of reactions (Chapter 2), or the model can be developed empirically from experimental data (Chapter 16). If experimental data are available, the dynamic model should be validated, and the model accuracy is characterized. This latter information is useful for control system design and tuning.

The next step in the control system design is to devise an appropriate control strategy that will meet the control objectives while satisfying process constraints. As indicated in Fig. 1.8, this design activity is both an art and a science. Process understanding and the experience and preferences of the design team are key factors. Computer simulation of the controlled process is used to screen alternative control strategies and to provide preliminary estimates of appropriate controller settings.

Finally, the control system hardware and instrumentation are selected, ordered, and installed in the plant. Then the control system is tuned in the plant using the preliminary estimates from the design step as a starting point. Controller tuning usually involves trial-and-error procedures, as described in Chapter 11.

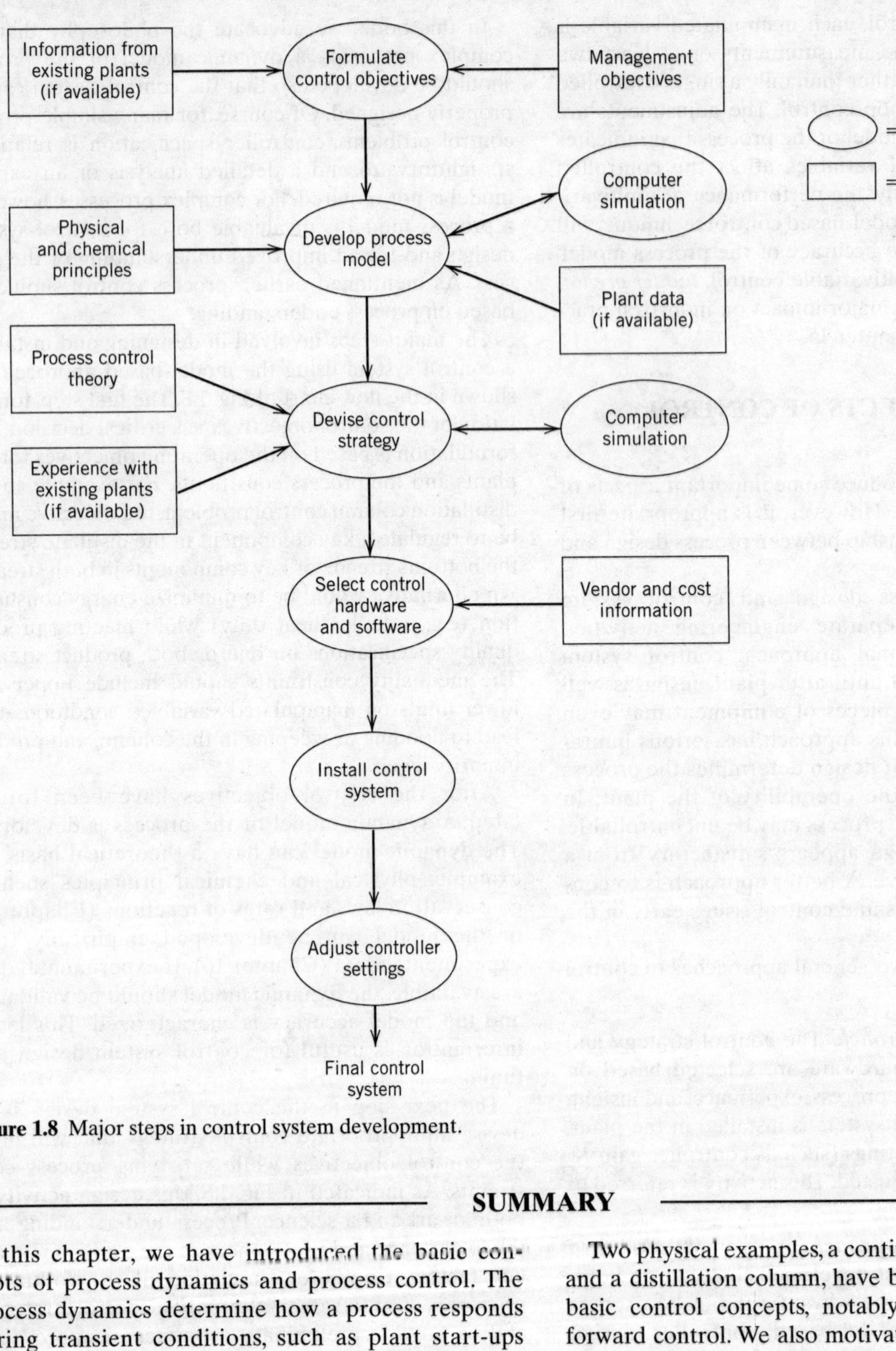

Figure 1.8 Major steps in control system development.

SUMMARY

In this chapter, we have introduced the basic concepts of process dynamics and process control. The process dynamics determine how a process responds during transient conditions, such as plant start-ups and shutdowns, grade changes, and unusual disturbances. Process control enables the process to be maintained at the desired operating conditions, safely and economically, while satisfying environmental and product quality requirements. Without effective process control, it would be impossible to operate large-scale industrial plants.

Two physical examples, a continuous blending system and a distillation column, have been used to introduce basic control concepts, notably, feedback and feedforward control. We also motivated the need for a systematic approach for the design of control systems for complex processes. Control system development consists of a number of separate activities that are shown in Fig. 1.8. In this book, we advocate the design philosophy that for complex processes, a dynamic model of the process should be developed so that the control system can be properly designed.

REFERENCES

Åström, K. J., and R. M. Murray, *Feedback Systems: An Introduction for Scientists and Engineers*, Princeton University Press, Princeton, NJ, 2008.

Blevins T., and M. Nixon, *Control Loop Foundation—Batch and Continuous Processes*, ISA, Research Triangle Park, NC, 2011.

Carver, C. S., and M. F. Scheier, *On the Self-Regulation of Behavior*, Cambridge University Press, Cambridge, UK, 1998.

Fogler, H. S., *Essentials of Chemical Reaction Engineering*, Prentice Hall, Upper Saddle River, NJ, 2010.

Mayr, O., *The Origins of Feedback Control*, MIT Press, Cambridge, MA, 1970.

Rao, C. V., Exploiting Market Fluctuations and Price Volatility through Feedback Control, *Comput. Chem. Eng.*, **51**, 181–186, 2013.

EXERCISES

1.1 Which of the following statements are true? For the false statements, explain why you think they are false:

(a) Feedforward and feedback control require a measured variable.

(b) For feedforward control, the measured variable is the variable to be controlled.

(c) Feedback control theoretically can provide perfect control (i.e., no deviations from set point) if the process model used to design the control system is perfect.

(d) Feedback control takes corrective action for all types of process disturbances, both known and unknown.

(e) Feedback control is superior to feedforward control.

1.2 Consider a home heating system consisting of a natural gas-fired furnace and a thermostat. In this case, the *process* consists of the interior space to be heated. The thermostat contains both the temperature sensor and the controller. The furnace is either on (heating) or off. Draw a schematic diagram for this control system. On your diagram, identify the controlled variables, manipulated variables, and disturbance variables. Be sure to include several possible sources of disturbances that can affect room temperature.

1.3 In addition to a thermostatically operated home heating system, identify two other feedback control systems that can be found in most residences. Describe briefly how each of them works; include sensor, actuator, and controller information.

1.4 Does a typical microwave oven utilize feedback control to set the cooking temperature or to determine if the food is "cooked"? If not, what technique is used? Can you think of any disadvantages to this approach, for example, in thawing and cooking foods?

1.5 Driving an automobile safely requires considerable skill. Even if not generally recognized, the driver needs an intuitive ability to utilize feedforward and feedback control methods.

(a) In the process of steering a car, one objective is to keep the vehicle generally centered in the proper traffic lane. Thus, the controlled variable is some measure of that distance. If so, how is feedback control used to accomplish this objective? Identify the sensor(s), the actuator, how the appropriate control action is determined, and some likely disturbances.

(b) The process of braking or accelerating an automobile is highly complex, requiring the skillful use of both feedback and feedforward mechanisms to drive safely. For feedback control, the driver normally uses distance to the vehicle ahead as the measured variable. This "set point" is often recommended to be some distance related to speed, for example, one car length separation for each 16 km/h. If this recommendation is used, how does feedforward control come into the accelerating/braking process when one is attempting to drive in traffic at a constant speed? In other words, what other information—in addition to distance separating the two vehicles—does the driver utilize to avoid colliding with the car ahead?

1.6 The human body contains numerous feedback control loops that are essential for regulating key physiological variables. For example, body temperature in a healthy person must be closely regulated within a narrow range.

(a) Briefly describe one or more ways in which body temperature is regulated by the body using feedback control.

(b) Briefly describe a feedback control system for the regulation of another important physiological variable.

1.7 The distillation column shown in Fig. E1.7 is used to distill a binary mixture. Symbols x, y, and z denote mole fractions of the more volatile component, while

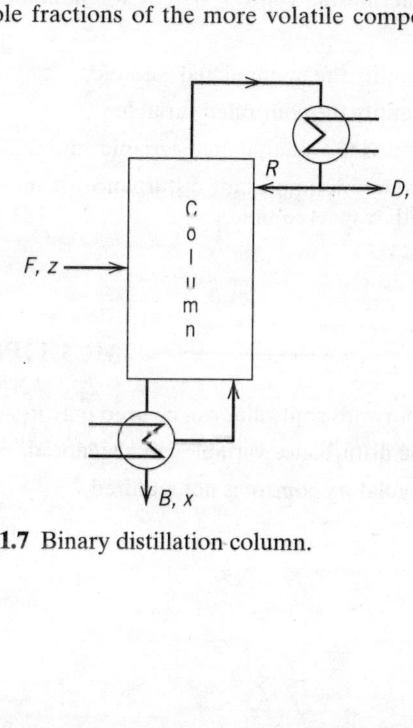

Figure E1.7 Binary distillation column.

B, D, R, and F represent molar flow rates. It is desired to control distillate composition y despite disturbances in feed flow rate F. All flow rates can be measured and manipulated with the exception of F, which can only be measured. A composition analyzer provides measurements of y.

(a) Propose a feedback control method and sketch the schematic diagram.

(b) Suggest a feedforward control method and sketch the schematic diagram.

1.8 Describe how a bicycle rider utilizes concepts from both feedforward control and feedback control while riding a bicycle.

1.9 Two flow control loops are shown in Fig. E1.9. Indicate whether each system is either a feedback or a feedforward control system. Justify your answer. It can be assumed that the distance between the flow transmitter (FT) and the control valve is quite small in each system.

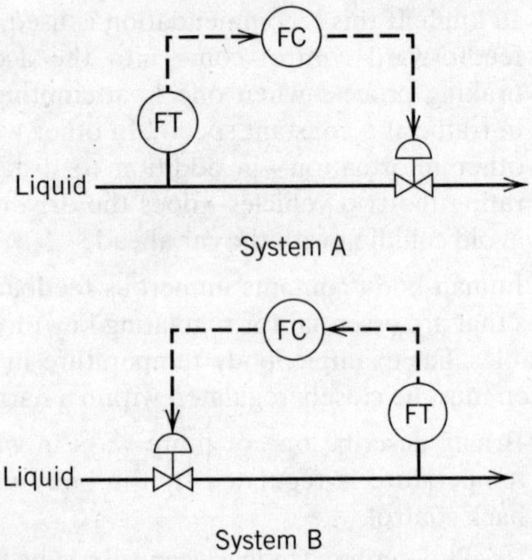

Figure E1.9 Flow control loops.

1.10 In a thermostat control system for a home heating system,

(a) Identify the manipulated variable

(b) Identify the controlled variable

(c) How is the manipulated variable adjusted?

(d) Name one important disturbance (it must change with respect to time).

1.11 Identify and describe three automatic control systems in a modern automobile (besides cruise control).

1.12 In Figure 1.1(d), identify the controlled, manipulated, and disturbance variables (there may be more than one of each type). How does the length of time for the dialysis treatment affect the waste concentration?

1.13 Find all the variables associated with the jacketed CSTR (Fig. E1.13), where a first-order exothermic reaction $A \rightarrow B$ takes place. The cooling water flows in the jacket.

Draw the feedback control loops for the following control objectives and categorize the variables as CVs, DVs or MVs.

(i) To maintain the temperature inside the reactor at T_{sp}.

(ii) To keep the liquid level in the reactor at h_{sp}.

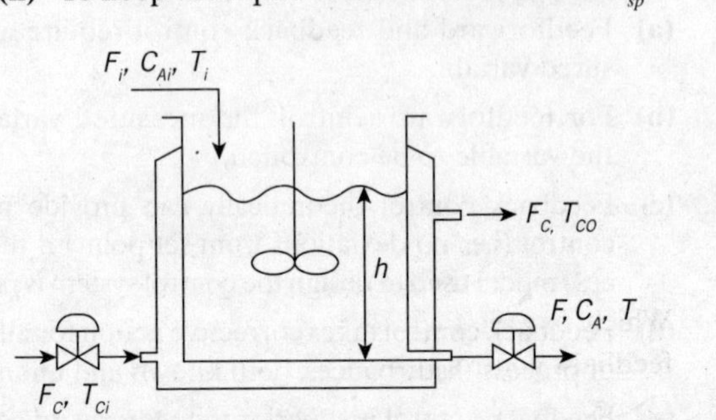

Figure E1.13 Jacketed CSTR.

1.14 In the heat exchanger process shown in Fig. E1.14, the control objective is to keep the temperature of the process stream T at a desired value T_{sp}. Saturated steam is used to heat the process stream.

(i) Identify the variables as CV, DVs and MV.

(ii) Draw the feedback and feedforward control loops for the same.

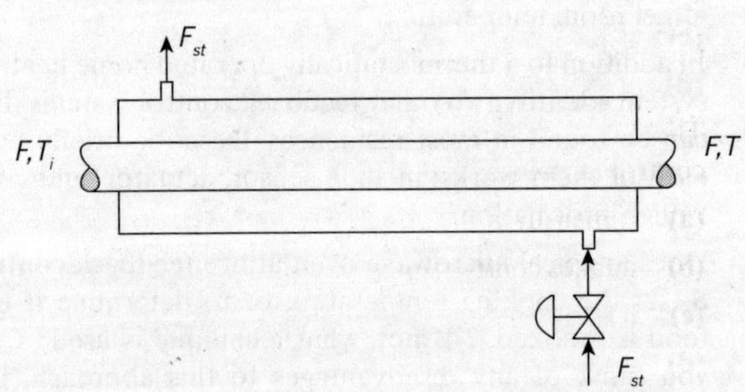

Figure E1.14 Heat exchanger process.

MULTIPLE CHOICE QUESTIONS

1. A feedforward controller can be used only if

(a) the disturbance variable can be ignored.

(b) regulatory control is not required.

(c) the disturbance variable can be manipulated.

(d) the disturbance variable can be measured.

(GATE 2021)

2. It is required to control the volume of the contents in the jacketed reactor shown in the figure.

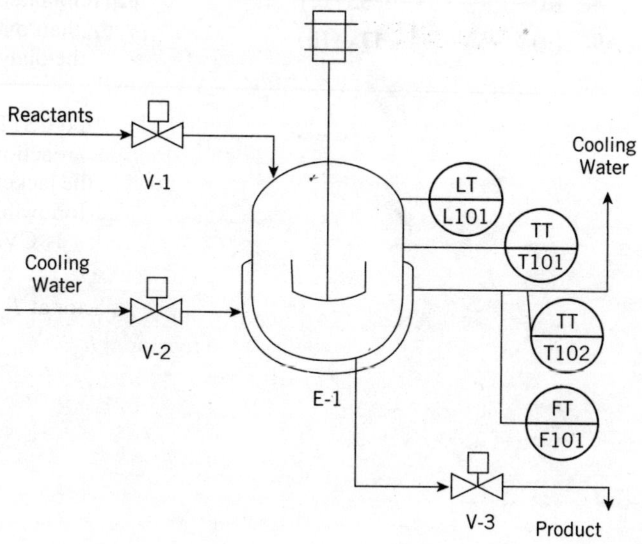

Which one of the following schemes can be used for feedback control?

(a) Measure L101 and manipulate valve V-2.

(b) Measure T101 and manipulate valve V-1.

(c) Measure F101 and manipulate valve V-1.

(d) Measure L101 and manipulate valve V-3.

(GATE 2021)

3. The feedforward control is superior to the feedback control in the sense that

(a) it measures the disturbance variable.

(b) it takes control action before the process is upset.

(c) it measures the controlled variable.

(d) it depends upon the model of the process.

4. The feedback control is superior to the feedforward control in the sense that

(a) it measures the controlled variable.

(b) it takes control action once the process is upset.

(c) it measures the disturbance variable.

(d) it takes care of the unmeasured disturbances.

5. In process control, the manipulated variables are generally

(a) temperatures.

(b) pressures.

(c) flow rates.

(d) liquid levels.

6. The desired value of the controlled variable is called as

(a) desired point. **(b)** measured point.

(c) set point. **(d)** unmeasured point.

7. Controlled variable is a/an

(a) input variable.

(b) output variable.

(c) measured input.

(d) unmeasured input.

8. Controlled variable

(a) is always a measured variable.

(b) is always an unmeasured variable.

(c) can be a measured or unmeasured variable.

(d) is not a process variable.

9. Disturbance variables are

(a) measured output variables.

(b) unmeasured output variables.

(c) either measured or unmeasured output variables.

(d) either measured or unmeasured input variables.

10. For sensitive processes

(a) feedback control is preferred.

(b) feedforward control is preferred.

(c) inferential control is preferred.

(d) combined feedforward-feedback control is preferred.

11. Process dynamics refers to

(a) steady state process behavior.

(b) unsteady state or transient process behavior.

(c) process behavior at either steady state or unsteady state condition.

(d) process behavior during normal operation.

12. The variables which affect the process are called as

(a) input variables.

(b) output variables.

(c) controlled variables.

(d) measured outputs.

13. The variables which get affected by the process are called as

(a) input variables.

(b) output variables.

(c) manipulated variables.

(d) disturbance variables.

14. In positive feedback

(a) the corrective action taken by the controller forces the CV towards the setpoint.

(b) the corrective action taken by the controller forces the CV away from the setpoint.

(c) no corrective action is taken by the controller..

(d) corrective action taken by the controller does not affect the CV.

ANSWER KEY

1. (d)	**2.** (d)	**3.** (b)	**4.** (d)	**5.** (c)	**6.** (c)
7. (b)	**8.** (c)	**9.** (d)	**10.** (d)	**11.** (b)	**12.** (a)
13. (b)	**14.** (b)				

CHAPTER 2

Theoretical Models of Chemical Processes

CHAPTER CONTENTS

In this chapter we consider the derivation of unsteady-state models of chemical processes from physical and chemical principles. Unsteady-state models are also referred to as *dynamic models*. We first consider the rationale for dynamic models and then present a general strategy for deriving them from first principles such as conservation laws. Then dynamic models are developed for several representative processes. Finally, we describe how dynamic models that consist of sets of ordinary differential equations and algebraic relations can be solved numerically using computer simulation.

2.1 DYNAMIC PROCESS MODELS – THEIR STRENGTHS AND LIMITATIONS

Dynamic models play a central role in the subject of process dynamics and control. The models can be used to:

1. ***Improve understanding of the process.*** Dynamic models and computer simulation allow transient process behavior to be investigated without having to disturb the process. Computer simulation allows valuable information about dynamic and steady-state process behavior to be acquired, even before the plant is constructed.
2. ***Train plant operating personnel.*** Process simulators play a critical role in training plant operators to run complex units and to deal with dangerous situations or emergency scenarios. By interfacing a process simulator to standard process control equipment, a realistic training environment is created. This role is analogous to flight training simulators used in the aerospace industry.
3. ***Develop a control strategy for a new process.*** A dynamic model of the process allows alternative control strategies to be evaluated. For example, a dynamic model can help identify the process variables that should be controlled and those that should be manipulated. Preliminary controller tuning may be derived using a model, prior to plant start-up using empirical models

(Chapter 11). For model-based control strategies (Chapters 11, 12 and 15), the process model is an explicit element of the control law.

4. ***Optimize process operating conditions.*** It can be advantageous to recalculate the optimum operating conditions periodically in order to maximize profit or minimize cost. A steady-state process model and economic information can be used to determine the most profitable operating conditions (see Appendix G).

For many of the examples cited above—particularly where new, hazardous, or difficult-to-operate processes are involved—development of a suitable process model can be crucial to success. Models can be classified based on how they are obtained:

(a) *Theoretical models* are developed using the principles of chemistry, physics, and biology.

(b) *Empirical models* are obtained by fitting experimental data (more in Chapter 16).

(c) *Semi-empirical models* are a combination of the models in categories (a) and (b); the numerical values of one or more of the parameters in a theoretical model are calculated from experimental data.

Theoretical models offer two very important advantages: they provide physical insight into process behavior, and they are applicable over wide ranges of conditions. However, there are disadvantages associated with theoretical models. They tend to be expensive and time-consuming to develop. In addition, theoretical models of complex processes typically include some model parameters that are not readily available, such as reaction rate coefficients, physical properties, or heat transfer coefficients.

Although empirical models are easier to develop and to use in controller design than theoretical models, they have a serious disadvantage: *empirical models typically do not extrapolate well.* More specifically, empirical models should be used with caution for operating conditions that were not included in the experimental data used to fit the model. The range of the data is typically quite small compared to the full range of process operating conditions.

Semi-empirical models have three inherent advantages: (i) they incorporate theoretical knowledge, (ii) they can be extrapolated over a wider range of operating conditions than purely empirical models, and (iii) they require less development effort than theoretical models. Consequently, semi-empirical models are widely used in industry.

This chapter is concerned with the development of theoretical models from first principles such as conservation laws.

2.1.1 An Illustrative Example: A Blending Process

In Chapter 1 we developed a steady-state model for a stirred-tank blending system based on mass and component balances. Now we develop an unsteady-state model that will allow us to analyze the more general situation where process variables vary with time and accumulation terms must be included.

As an illustrative example, we consider the isothermal stirred-tank blending system in Fig. 2.1. It is a more general version of the blending system in Fig. 1.3 because the overflow line has been omitted and inlet stream 2 is not necessarily pure A (that is, $x_2 \neq 1$). Now the volume of liquid in the tank V can vary with time, and the exit flow rate is not necessarily equal to the sum of the inlet flow rates. An unsteady-state mass balance for the blending system in Fig. 2.1 has the form

$$\begin{Bmatrix}\text{rate of accumulation}\\ \text{of mass in the tank}\end{Bmatrix} = \begin{Bmatrix}\text{rate of}\\ \text{mass in}\end{Bmatrix} - \begin{Bmatrix}\text{rate of}\\ \text{mass out}\end{Bmatrix} \quad (2\text{-}1)$$

The mass of liquid in the tank can be expressed as the product of the liquid volume V and the density ρ. Consequently, the rate of mass accumulation is simply $d(V\rho)/dt$, and Eq. 2-1 can be written as

$$\frac{d(V\rho)}{dt} = w_1 + w_2 - w \quad (2\text{-}2)$$

where w_1, w_2, and w are mass flow rates.

The unsteady-state material balance for component A can be derived in an analogous manner. We assume that the blending tank is perfectly mixed. This assumption has two important implications: (i) there are no concentration gradients in the tank contents

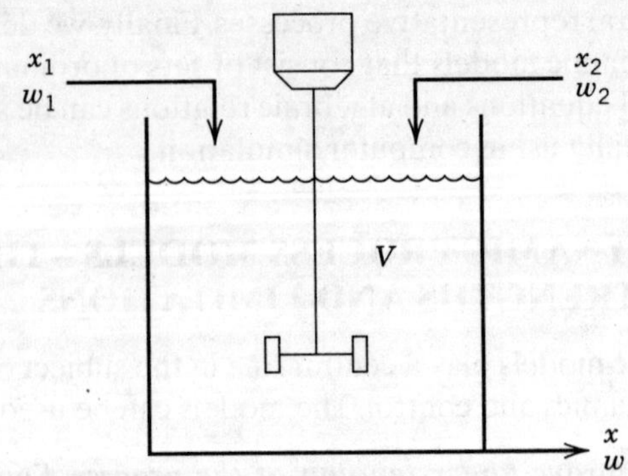

Figure 2.1 Stirred-tank blending process.

and (ii) the composition of the exit stream is equal to the tank composition. The perfect mixing assumption is valid for low-viscosity liquids that receive an adequate degree of agitation. In contrast, the assumption is less likely to be valid for high-viscosity liquids such as polymers or molten metals. Nonideal mixing is modeled in books on reactor analysis (e.g., Fogler, 2006).

For the perfect mixing assumption, the rate of accumulation of component A is $d(V\rho x)/dt$, where x is the mass fraction of A. The unsteady-state component balance is

$$\frac{d(V\rho x)}{dt} = w_1 x_1 + w_2 x_2 - wx \tag{2-3}$$

Equations 2-2 and 2-3 provide an unsteady-state model for the blending system. The corresponding steady-state model was derived in Chapter 1 (cf. Eqs. 1-1 and 1-2). It also can be obtained by setting the accumulation terms in Eqs. 2-2 and 2-3 equal to zero,

$$0 = \overline{w}_1 + \overline{w}_2 - \overline{w} \tag{2-4}$$

$$0 = \overline{w}_1 \overline{x}_1 + \overline{w}_2 \overline{x}_2 - \overline{w}\overline{x} \tag{2-5}$$

where the nominal steady-state conditions are denoted by $\overline{x}$ and $\overline{w}$ and so on. In general, a steady-state model is a special case of an unsteady-state model that can be derived by setting accumulation terms equal to zero.

A dynamic model can be used to characterize the transient behavior of a process for a wide variety of conditions. For example, some relevant concerns for the blending process: How would the exit composition change after a sudden increase in an inlet flow rate or after a gradual decrease in an inlet composition? Would these transient responses be very different if the volume of liquid in the tank is quite small, or quite large, when an inlet change begins? These questions can be answered by solving the ordinary differential equations (ODE) in Eqs. 2-2 and 2-3 for specific initial conditions and for particular changes in inlet flow rates or compositions. The solution of dynamic models is considered further in this chapter and in Chapters 3–6.

Before exploring the blending example in more detail, we first present general principles for the development of dynamic models.

2.2 GENERAL MODELING PRINCIPLES

It is important to remember that a process model is nothing more than a mathematical abstraction of a real process. The model equations are at best an approximation to the real process as expressed by the adage that "all models are wrong, but some are useful." Consequently, the model cannot incorporate all of the features, whether macroscopic or microscopic, of the real process. Modeling inherently involves a compromise between model accuracy and complexity on one hand, and the cost and effort required to develop the model and verify it on the other hand. The required compromise should consider a number of factors, including the modeling objectives, the expected benefits from use of the model, and the background of the intended users of the model (e.g., research chemists versus plant engineers).

Process modeling is both an art and a science. Creativity is required to make simplifying assumptions that result in an appropriate model. Consequently, careful enumeration of all the assumptions that are invoked in building a model is crucial for its final evaluation. The model should incorporate all of the important dynamic behavior while being no more complex than is necessary. Thus, less important phenomena are omitted in order to keep the number of model equations, variables, and parameters at reasonable levels. The failure to choose an appropriate set of simplifying assumptions invariably leads to either (1) rigorous but excessively complicated models or (2) overly simplistic models. Both extremes should be avoided. Fortunately, modeling is also a science, and predictions of process behavior from alternative models can be compared, both qualitatively and quantitatively. This chapter provides an introduction to the subject of theoretical dynamic models and shows how they can be developed from first principles such as conservation laws. Additional information is available in the books by Bequette (1998), Aris (1999), Elnashaie and Garhyan (2003), and Cameron and Gani (2011).

A systematic procedure for developing dynamic models from first principles is summarized in Table 2.1. Most of the steps in Table 2.1 are self-explanatory, with the possible exception of Step 7. The *degrees of freedom analysis* in Step 7 is required in model development for complex processes. Because these models typically contain large numbers of variables and equations, it is not obvious whether the model can be solved, or whether it has a unique solution. Consequently, we consider the degrees of freedom analysis in Sections 2.3 and 2.4.

Table 2.1 A Systematic Approach for Developing Dynamic Models

1. State the modeling objectives and the end use of the model. Then determine the required levels of model detail and model accuracy.
2. Draw a schematic diagram of the process and label all process variables.
3. List all of the assumptions involved in developing the model. Try to be parsimonious: the model should be no more complicated than necessary to meet the modeling objectives.
4. Determine whether spatial variations of process variables are important. If so, a partial differential equation model will be required.
5. Write appropriate conservation equations (mass, component, energy, and so forth).
6. Introduce equilibrium relations and other algebraic equations (from thermodynamics, transport phenomena, chemical kinetics, equipment geometry, etc.).
7. Perform a degrees of freedom analysis (Section 2.3) to ensure that the model equations can be solved.
8. Simplify the model. It is often possible to arrange the equations so that the output variables appear on the left side and the input variables appear on the right side. This model form is convenient for computer simulation and subsequent analysis.
9. Classify inputs as disturbance variables or as manipulated variables.

Dynamic models of chemical processes consist of ODE and/or partial differential equations (PDE), plus related algebraic equations. In this book we will restrict our discussion to ODE models. Additional details about PDE models for reaction engineering can be found in Fogler (2006) and numerical procedures for solving such models are available in, for example, Chapra and Canale (2014). For process control problems, dynamic models are derived using unsteady-state conservation laws. In this section, we first review general modeling principles, emphasizing the importance of the mass and energy conservation laws. Force–momentum balances are employed less often. For processes with momentum effects that cannot be neglected (e.g., some fluid and solid transport systems), such balances should be considered. The process model often also includes algebraic relations that arise from thermodynamics, transport phenomena, physical properties, and chemical kinetics. Vapor–liquid equilibria, heat transfer correlations, and reaction rate expressions are typical examples of such algebraic equations.

2.2.1 Conservation Laws

Theoretical models of chemical processes are based on conservation laws such as the conservation of mass and energy. Consequently, we now consider important conservation laws and use them to develop dynamic models for representative processes.

Conservation of Mass

$$\begin{Bmatrix}\text{rate of mass}\\ \text{accumulation}\end{Bmatrix} = \begin{Bmatrix}\text{rate of}\\ \text{mass in}\end{Bmatrix} - \begin{Bmatrix}\text{rate of}\\ \text{mass out}\end{Bmatrix} \quad (2\text{-}6)$$

Conservation of Component i

$$\begin{Bmatrix}\text{rate of component } i\\ \text{accumulation}\end{Bmatrix} = \begin{Bmatrix}\text{rate of component } i\\ \text{in}\end{Bmatrix} - \begin{Bmatrix}\text{rate of component } i\\ \text{out}\end{Bmatrix} + \begin{Bmatrix}\text{rate of component } i\\ \text{produced}\end{Bmatrix} \quad (2\text{-}7)$$

The last term on the right-hand side of Eq. 2-7 represents the rate of generation (or consumption) of component i as a result of chemical reactions. Conservation equations can also be written in terms of molar quantities, atomic species, and molecular species (Felder and Rousseau, 2015).

Conservation of Energy

The general law of energy conservation is also called the First Law of Thermodynamics (Sandler, 2006). It can be expressed as

$$\begin{Bmatrix}\text{rate of energy}\\ \text{accumulation}\end{Bmatrix} = \begin{Bmatrix}\text{rate of energy in}\\ \text{by convection}\end{Bmatrix} - \begin{Bmatrix}\text{rate of energy out}\\ \text{by convection}\end{Bmatrix} + \begin{Bmatrix}\text{net rate of heat addition}\\ \text{to the system from}\\ \text{the surroundings}\end{Bmatrix} + \begin{Bmatrix}\text{net rate of work}\\ \text{performed on the system}\\ \text{by the surroundings}\end{Bmatrix} \quad (2\text{-}8)$$

The total energy of a thermodynamic system, U_{tot}, is the sum of its internal energy, kinetic energy, and potential energy:

$$U_{tot} = U_{int} + U_{KE} + U_{PE} \quad (2\text{-}9)$$

For the processes and examples considered in this book, it is appropriate to make two assumptions:

1. Changes in potential energy and kinetic energy can be neglected, because they are small in comparison with changes in internal energy.
2. The net rate of work can be neglected, because it is small compared to the rates of heat transfer and convection.

For these reasonable assumptions, the energy balance in Eq. 2-8 can be written as (Bird et al., 2002)

$$\frac{dU_{\text{int}}}{dt} = -\Delta(w\hat{H}) + Q \tag{2-10}$$

where U_{int} is the internal energy of the system, $\hat{H}$ is the enthalpy per unit mass, w is the mass flow rate, and Q is the rate of heat transfer to the system. The Δ operator denotes the difference between outlet conditions and inlet conditions of the flowing streams. Consequently, the $-\Delta(w\hat{H})$ term represents the enthalpy of the inlet stream(s) minus the enthalpy of the outlet stream(s). The analogous equation for molar quantities is

$$\frac{dU_{\text{int}}}{dt} = -\Delta(\tilde{w}\tilde{H}) + Q \tag{2-11}$$

where $\tilde{H}$ is the enthalpy per mole and $\tilde{w}$ is the molar flow rate.

Note that the conservation laws of this section are valid for batch and semibatch processes, as well as for continuous processes. For example, in batch processes, there are no inlet and outlet flow rates. Thus, $w = 0$ and $\tilde{w} = 0$ in Eqs. 2-10 and 2-11.

In order to derive dynamic models of processes from the general energy balances in Eqs. 2-10 and 2-11, expressions for U_{int} and $\hat{H}$ or $\tilde{H}$ are required, which can be derived from thermodynamics. These derivations and a review of related thermodynamics concepts are included in Appendix A.

2.2.2 The Blending Process Revisited

Next, we show that the dynamic model of the blending process in Eqs. 2-2 and 2-3 can be simplified and expressed in a more appropriate form for computer simulation. For this analysis, we introduce the additional assumption that the density of the liquid, ρ, is a constant. This assumption is reasonable because often the density has only a weak dependence on composition. For constant ρ, Eqs. 2-2 and 2-3 become

$$\rho\frac{dV}{dt} = w_1 + w_2 - w \tag{2-12}$$

$$\rho\frac{d(Vx)}{dt} = w_1x_1 + w_2x_2 - wx \tag{2-13}$$

Equation 2-13 can be further simplified by expanding the accumulation term using the "chain rule" for differentiation of a product:

$$\rho\frac{d(Vx)}{dt} = \rho V\frac{dx}{dt} + \rho x\frac{dV}{dt} \tag{2-14}$$

Substitution of Eq. 2-14 into Eq. 2-13 gives

$$\rho V\frac{dx}{dt} + \rho x\frac{dV}{dt} = w_1x_1 + w_2x_2 - wx \tag{2-15}$$

Substitution of the mass balance in Eq. 2-12 for $\rho dV/dt$ in Eq. 2-15 gives

$$\rho V\frac{dx}{dt} + x(w_1 + w_2 - w) = w_1x_1 + w_2x_2 - wx \tag{2-16}$$

After canceling common terms and rearranging Eqs. 2-12 and 2-16, a more convenient model form (the so-called "state-space" form) is obtained:

$$\frac{dV}{dt} = \frac{1}{\rho}(w_1 + w_2 - w) \tag{2-17}$$

$$\frac{dx}{dt} = \frac{w_1}{V\rho}(x_1 - x) + \frac{w_2}{V\rho}(x_2 - x) \tag{2-18}$$

The dynamic model in Eqs. 2-17 and 2-18 is quite general and is based on only two assumptions: perfect mixing and constant density. For special situations, the liquid volume V is constant (i.e., $dV/dt = 0$), and the exit flow rate equals the sum of the inlet flow rates, $w = w_1 + w_2$. These conditions might occur when

1. An overflow line is used in the tank as shown in Fig. 1.3.
2. The tank is closed and filled to capacity.
3. A liquid-level controller keeps V essentially constant by adjusting a flow rate.

In all three cases, Eq. 2-17 reduces to the same form as Eq. 2-4, not because each flow rate is constant, but because $w = w_1 + w_2$ at all times.

The dynamic model in Eqs. 2-17 and 2-18 is in a convenient form for subsequent investigation based on analytical or numerical techniques. In order to obtain a solution to the ODE model, we must specify the inlet compositions (x_1 and x_2) and the flow rates (w_1, w_2, and w) as functions of time. After specifying initial conditions for the dependent variables, $V(0)$ and $x(0)$, we can determine the transient responses, $V(t)$ and $x(t)$. The derivation of an analytical expression for $x(t)$ when V is constant is illustrated in Example 2.1.

EXAMPLE 2.1

A stirred-tank blending process with a constant liquid holdup of 2 m³ is used to blend two streams whose densities are both approximately 900 kg/m³. The density does not change during mixing.

(a) Assume that the process has been operating for a long period of time with flow rates of $w_1 = 500$ kg/min and $w_2 = 200$ kg/min, and feed compositions (mass fractions) of $x_1 = 0.4$ and $x_2 = 0.75$. What is the steady-state value of x?

(b) Suppose that w_1 changes suddenly from 500 to 400 kg/min and remains at the new value. Determine an expression for $x(t)$ and plot it.

(c) Repeat part **(b)** for the case where w_2 (instead of w_1) changes suddenly from 200 to 100 kg/min and remains there.

(d) Repeat part **(c)** for the case where x_1 suddenly changes from 0.4 to 0.6 (in addition to the change in w_2).

(e) For parts **(b)** through **(d)**, plot the normalized response $x_N(t)$,

$$x_N(t) = \frac{x(t) - x(0)}{x(\infty) - x(0)}$$

where $x(0)$ is the initial steady-state value of $x(t)$ and $x(\infty)$ represents the final steady-state value, which is different for each part.

Solution

(a) Denote the initial steady-state conditions by $\bar{x}$, $\bar{w}$, and so on. For the initial steady state, Eqs. 2-4 and 2-5 are applicable. Solve Eq. 2-5 for $\bar{x}$:

$$\bar{x} = \frac{\bar{w}_1\bar{x}_1 + \bar{w}_2\bar{x}_2}{\bar{w}} = \frac{(500)(0.4) + (200)(0.75)}{700} = 0.5$$

(b) The component balance in Eq. 2-3 can be rearranged (for constant V and ρ) as

$$\tau\frac{dx}{dt} + x = \frac{w_1 x_1 + w_2 x_2}{w}$$

$$x(0) = \bar{x} = 0.5 \tag{2-19}$$

where $\tau \triangleq \bar{V}\rho/\bar{w}$. In each of the three parts, **(b)**–**(d)**, $\tau = 3$ min and the right side of Eq. 2-19 is constant for this example. Thus, Eq. 2-19 can be written as

$$3\frac{dx}{dt} + x = C^*$$

$$x(0) = 0.5 \tag{2-20}$$

where

$$C^* \triangleq \frac{\bar{w}_1\bar{x}_1 + \bar{w}_2\bar{x}_2}{\bar{w}} \tag{2-21}$$

The solution to Eq. 2-20 can be obtained by applying standard solution methods (Kreyszig, 2011):

$$x(t) = 0.5e^{-t/3} + C^*(1 - e^{-t/3}) \tag{2-22}$$

For case **(b)**,

$$C^* = \frac{(400\,\text{kg/min})(0.4) + (200\,\text{kg/min})(0.75)}{600\,\text{kg/min}} = 0.517$$

Substituting C^* into Eq. 2-22 gives the desired solution for the step change in w_1:

$$x(t) = 0.5e^{-t/3} + 0.517(1 - e^{-t/3}) \tag{2-23}$$

(c) For the step change in w_2,

$$C^* = \frac{(500\ \text{kg/min})(0.4) + (100\ \text{kg/min})(0.75)}{600\ \text{kg/min}} = 0.458$$

and the solution is

$$x(t) = 0.5e^{-t/3} + 0.458(1 - e^{-t/3}) \qquad (2\text{-}24)$$

(d) Similarly, for the simultaneous changes in x_1 and w_2, Eq. 2-21 gives $C^* = 0.625$. Thus, the solution is

$$x(t) = 0.5e^{-t/3} + 0.625(1 - e^{-t/3}) \qquad (2\text{-}25)$$

(e) The individual responses in Eqs. 2-22–2-24 have the same normalized response:

$$\frac{x(t) - x(0)}{x(\infty) - x(0)} = 1 - e^{-t/3} \qquad (2\text{-}26)$$

The responses of **(b)**–**(c)** are shown in Fig. 2.2.

The individual responses and normalized response have the same time dependence for cases **(b)**–**(d)** because $\tau = \bar{V}\rho/\bar{w} = 3\ \text{min}$ for each part. Note that τ is the mean residence time of the liquid in the blending tank. If $\bar{w}$ changes, then τ and the time dependence of the solution also change. This situation would occur, for example, if $\bar{w}_1$ changed from 500 kg/min to 600 kg/min. These more general situations will be addressed in Chapter 4.

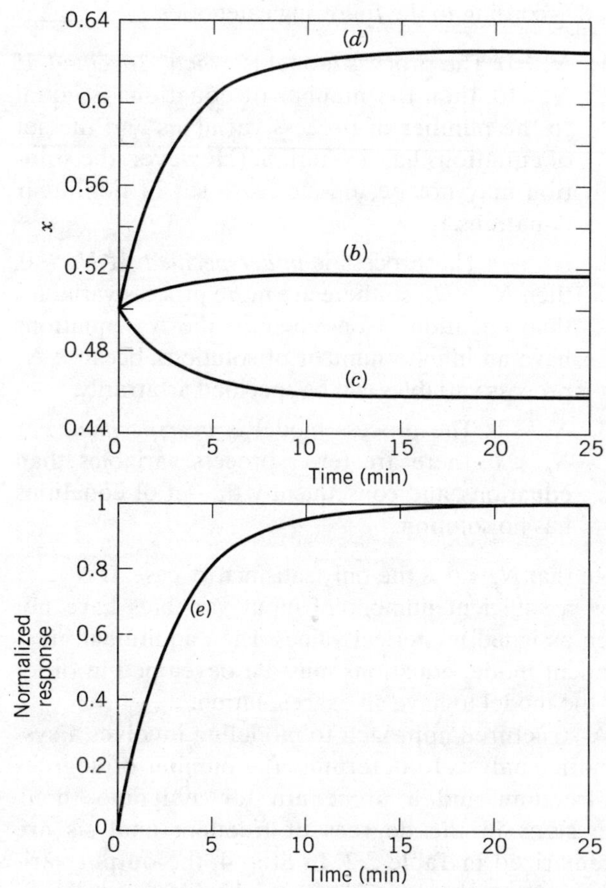

Figure 2.2 Exit composition responses of a stirred-tank blending process to step changes in
(b) Flow rate w_1
(c) Flow rate w_2
(d) Flow rate w_2 and inlet composition x_1
(e) Normalized response for parts (b)–(d).

2.3 DEGREES OF FREEDOM ANALYSIS

To simulate a process, we must first ensure that its model equations (differential and algebraic) constitute a solvable set of relations. In other words, the output variables, typically the variables on the left side of the equations, can be solved in terms of the input variables on the right side of the equations. For example, consider a set of linear algebraic equations, $\mathbf{y} = \mathbf{A}\mathbf{x}$. In order for these equations to have a unique solution for $\mathbf{x}$, vectors $\mathbf{x}$ and $\mathbf{y}$ must contain the same number of elements and matrix $\mathbf{A}$ must be nonsingular (that is, have a nonzero determinant).

It is not easy to make a similar evaluation for a large, complicated steady-state or dynamic model. However, there is one general requirement. In order for the model to have a unique solution, the number of unknown variables must equal the number of independent model equations. An equivalent statement is that all of the available *degrees of freedom* must be utilized. The number of degrees of freedom, N_F, can be calculated from the expression

$$N_F = N_V - N_E \qquad (2\text{-}27)$$

where N_V is the total number of process variables (distinct from constant process parameters) and N_E is the number of independent equations. A degrees of freedom analysis allows modeling problems to be classified according to the following categories:

1. $N_F = 0$: The process model is *exactly specified*. If $N_F = 0$, then the number of equations is equal to the number of process variables and the set of equations has a solution. (However, the solution may not be unique for a set of nonlinear equations.)
2. $N_F > 0$: The process is *underspecified*. If $N_F > 0$, then $N_V > N_E$, so there are more process variables than equations. Consequently, the N_E equations have an infinite number of solutions, because N_F process variables can be specified arbitrarily.
3. $N_F < 0$: The process model is *overspecified*. For $N_F < 0$, there are fewer process variables than equations, and consequently the set of equations has no solution.

Note that $N_F = 0$ is the only satisfactory case. If $N_F > 0$, then a sufficient number of input variables have not been assigned numerical values. Then additional independent model equations must be developed in order for the model to have an exact solution.

A structured approach to modeling involves a systematic analysis to determine the number of degrees of freedom and a procedure for assigning them. The steps in the degrees of freedom analysis are summarized in Table 2.2. In Step 4, the output variables include the dependent variables in the ordinary differential equations.

For Step 5, the N_F degrees of freedom are assigned by specifying a total of N_F input variables to be either *disturbance variables* or *manipulated variables*. In general, disturbance variables are determined by other process units or by the environment. Ambient temperature and feed conditions determined by the operation of upstream processes are typical examples of disturbance variables. By definition, a disturbance variable d varies with time and is independent of the other $N_V - 1$ process variables. Thus, we can express the transient behavior of the disturbance variable as

$$d(t) = f(t) \tag{2-28}$$

where $f(t)$ is an arbitrary function of time that must be specified if the model equations are to be solved. Thus, specifying a process variable to be a disturbance variable increases N_E by one and reduces N_F by one, as indicated by Eq. 2-27.

In general, a degree of freedom is also utilized when a process variable is specified to be a manipulated variable that is adjusted by a controller. In this situation, a new equation is introduced, namely the control law that indicates how the manipulated variable is adjusted (cf. Eqs. 1-4 or 1-5). Consequently, N_E increases by one and N_F decreases by one, again utilizing a degree of freedom.

Table 2.2 Degrees of Freedom Analysis

1. List all quantities in the model that are *known* constants (or parameters that can be specified) on the basis of equipment dimensions, kncwn physical properties, and so on.
2. Determine the number of equations N_E and the number of process variables, N_V. Note that time t is *not* considered to be a process variable, because it is neither a process input nor a process output.
3. Calculate the number of degrees of freedom, $N_F = N_V - N_E$.
4. Identify the N_E output variables that will be obtained by solving the process model.
5. Identify the N_F input variables that must be specified as either disturbance variables or manipulated variables, in order to utilize the N_F degrees of freedom.
6. Identify N_D disturbance variables which are generally determined by upstream conditions.
7. Calculate the control degrees of freedom (N_{FC}) as:
$$N_{FC} = N_F - N_D = N_{MV} = N_{CV}$$
8. Identify the MVs and make CV–MV pairing for the system from the process knowledge.
9. Define the control objectives (N_{CO}) and calculate the net degrees of freedom (N_{FN}) as:
$$N_{FN} = N_{FC} - N_{CO}.$$
If $N_{FN} = 0$, the system is completely specified.

2.4 DEGREES OF FREEDOM ANALYSIS FOR PROCESS CONTROL

For process control applications, it is very important to determine the maximum number of process variables that can be independently controlled, that is, to determine the *control degrees of freedom*, N_{FC}:

Definition. *The control degrees of freedom, N_{FC}, is the number of process variables that can be controlled independently.*

In order to make a clear distinction between N_F and N_{FC}, we refer to N_F as the model degrees of freedom and to N_{FC} as the control degrees of freedom. They are related by,

$$N_F = N_{FC} + N_D \tag{2-29}$$

where N_D is the number of DVs.

2.4.1 Control Degrees of Freedom

The control degrees of freedom N_{FC} is closely related to the number of independent MVs that are available:

General Rule. *For most practical control problems, the control degrees of freedom N_{FC} is equal to the number of independent input variables that can be manipulated.*

It is important that the manipulated inputs be independent. For example, if a process stream splits, or if two process streams merge to form a third stream, it is not possible to adjust all three flow rates independently. These situations are shown in Fig. 2.3.

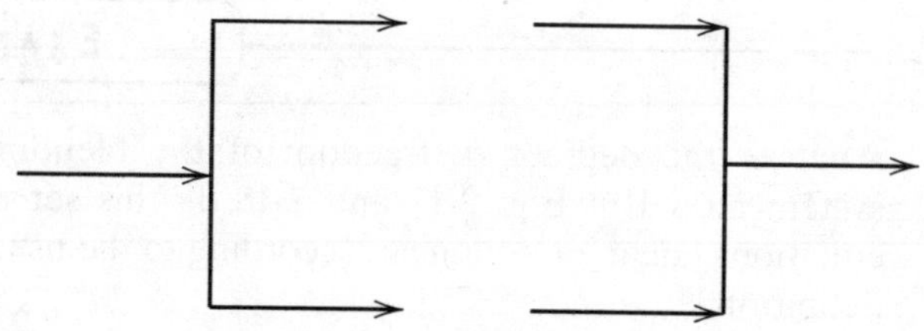

Figure 2.3 Two examples where all three streams cannot be manipulated independently.

2.4.2 Net Degrees of Freedom

Disturbance variables (DVs) are generally determined from the upstream process conditions. Once the DVs are identified one can make CV–MV pairing from the process knowledge. For a given CV, the input variable which affects it the most should be chosen as its MV. Then it will be easy to define the control objectives. The net number of degrees of freedom (N_{FN}) is given by

$$N_{FN} = N_{FC} - N_{CO} \tag{2-30}$$

where N_{CO} is the number of control objectives.

N_{FN} must be brought to zero for the system to be completely specified.

The addition of a feedback controller can change the control degrees of freedom, N_{FC}. In general, adding a feedback controller utilizes a control degree of freedom, because an MV is now adjusted by the controller. However, if the controller set-point is adjusted by a higher-level (or supervisory) control system, neither N_F nor N_{FC} changes. The reason is as follows. Adding a controller introduces a new equation, the control law, and a new variable, the set point. Thus both N_E and N_V increase by one. But Eqs. 2-27 and 2-29 indicate that N_F and N_{FC} do not change.

We illustrate the degrees of freedom analysis by considering the following examples:

EXAMPLE 2.2

Analyze the degrees of freedom for the blending model of Eq. 2-3 for the special condition where volume V is constant.

Solution

For this example, there are

2 parameters: V, ρ
4 variables ($N_V = 4$): x, x_1, w_1, w_2
1 equation ($N_E = 1$): Eq. 2-3

The degrees of freedom are calculated as $N_F = 4 - 1 = 3$. Thus, we must identify three input variables that can be specified as known functions of time in order for the equation to have a unique solution. The dependent variable x is an obvious choice for the output variable in this simple example. Consequently, we have

1 output: x
3 inputs: x_1, w_1, w_2

The three degrees of freedom can be utilized by specifying the inputs as

2 disturbance variables: x_1, w_1
1 manipulated variable: w_2

Because all of the degrees of freedom have been utilized, the single equation is exactly specified and can be solved.

EXAMPLE 2.3

Analyze the degrees of freedom of the blending system model in Eqs. 2-17 and 2-18. Is this set of equations linear, or nonlinear, according to the usual definition?[1]

Solution

In this case, volume is now considered to be a variable rather than a constant parameter. Consequently, for the degrees of freedom analysis, we have

1 parameter: ρ
7 variables ($N_V = 7$): $V, x, x_1, x_2, w, w_1, w_2$
2 equations ($N_E = 2$): Eqs. 2-17 and 2-18

Thus, $N_F = 7 - 2 = 5$. The dependent variables on the left side of the differential equations, V and x, are the model outputs. The remaining five variables must be chosen as inputs. Note that a physical output, effluent flow rate w, is classified as a mathematical input, because it can be specified arbitrarily. Any process variable that can be specified arbitrarily should be identified as an input. Thus, we have

2 outputs: V, x
5 inputs: w, w_1, w_2, x_1, x_2

Because the two outputs are the only variables to be determined in solving the system of two equations, no degrees of freedom are left. The system of equations is exactly specified and hence solvable.

To utilize the degrees of freedom, the five inputs are classified as either disturbance variables or manipulated variables. A reasonable classification is

3 disturbance variables: w_1, x_1, x_2
2 manipulated variables: w, w_2

For example, w could be used to control V and w_2 to control x.

Note that Eq. 2-17 is a linear ODE, while Eq. 2-18 is a nonlinear ODE as a result of the products and quotients.

EXAMPLE 2.4

Determine N_F and N_{FC} for the steam-heated, stirred-tank system model in Eqs. 2-54 through 2-56 in Chapter 2. Assume that only steam pressure P_s can be manipulated.

Solution

To calculate N_F from Eq. 2-27, we need to determine N_V and N_E. The dynamic model contains three equations ($N_E = 3$) and six process variables ($N_V = 6$): T_s, P_s, w, T_i, T, and T_w. Thus, $N_F = 6 - 3 = 3$. If feed temperature T_i and mass flow rate w are considered to be disturbance variables, $N_D = 2$ and thus $N_{FC} = 1$ from Eq. 2-29. This single degree of freedom could be used to control temperature T by manipulating steam pressure, P_s.

EXAMPLE 2.5

A conventional distillation column with a single feed stream and two product streams is shown in Fig. 2.4. The feed conditions are disturbance variables. Determine the control degrees of freedom N_{FC} and identify potential MVs and CVs.

Solution

For a typical distillation column, five input variables can be manipulated: product flow rates, B and D, reflux flow rate R, coolant flow rate q_c, and heating medium flow rate q_h. Thus, according to the General

[1]A linear model cannot contain any nonlinear combinations of variables (e.g., a product of two variables) or any variable raised to a power other than one.

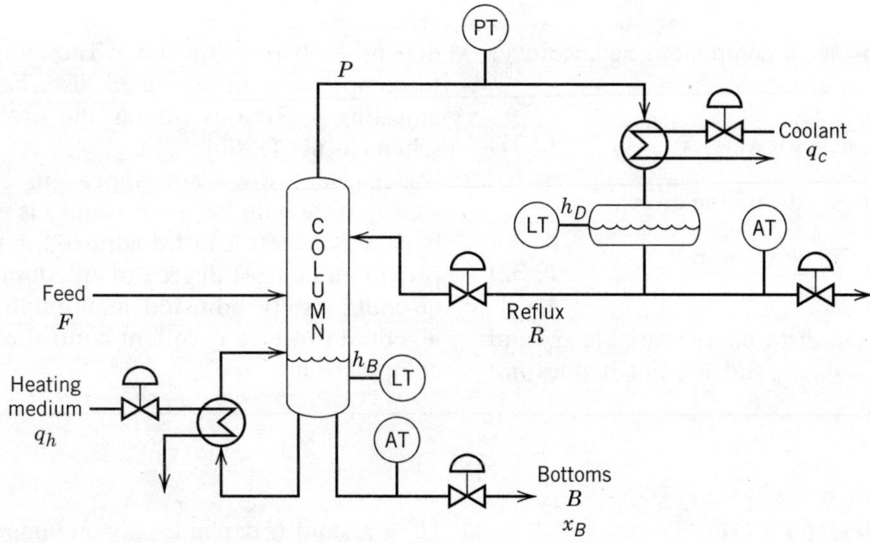

Figure 2.4 Schematic diagram of a distillation column.

Rule, $N_{FC} = 5$. This result can also be obtained from Eqs. 2-27 and 2-29, but considerable effort is required to develop the required dynamic model. Although five output variables could be selected as CVs, x_D, x_B, h_B, h_D, and P, for many distillation control problems, it is not necessary to control all five. Also, if it not feasible to measure the product compositions on-line, tray temperatures near the top and bottom of the column are often controlled instead, as discussed in the next section.

Although the General Rule is simple and widely applicable, there are exceptions where it is not valid. For example, N_{FC} should be reduced by 1 when a MV does not have a significant steady-state effect on any of the CVs, that is, when these steady-state gains are very small. This situation is illustrated in Example 2.6.

EXAMPLE 2.6

The blending system in Fig. 2.5 has a bypass stream that allows a fraction f of inlet stream w_2 to bypass the stirred tank. It is proposed that product composition x be controlled by adjusting f via the control valve. Analyze the feasibility of this control scheme by considering its steady-state and dynamic characteristics. In your analysis, assume that x_1 is the principal disturbance variable and that x_2, w_1, and w_2 are constant. Variations in liquid volume V can be neglected because $w_2 \ll w_1$.

Solution

The dynamic characteristics of the proposed control scheme are quite favorable because product composition x responds rapidly to changes in the bypass flow rate. In order to evaluate the steady-state

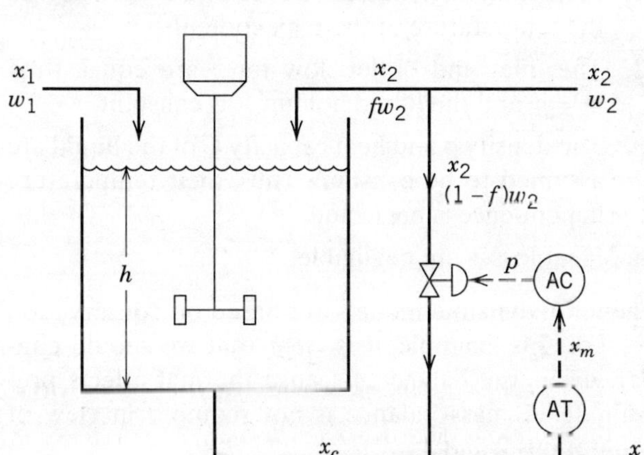

Figure 2.5 Blending system with bypass stream.

characteristics, consider a component balance over the entire system:

$$w_1\bar{x}_1 + w_2 x_2 = w\bar{x} \tag{2-31}$$

Solving for the controlled variable gives,

$$\bar{x} = \frac{w_1\bar{x}_1 + w_2 x_2}{w} \tag{2-32}$$

Thus $\bar{x}$ depends on disturbance variable $\bar{x}_1$ and four constants (w_1, w_2, x_2, and w), but it does *not* depend on bypass fraction, *f*. Thus, it is not possible to compensate for sustained disturbances in x_1 by adjusting *f*. For this reason, the proposed control scheme is not feasible.

Because *f* does not appear in Eq. 2-32, the steady-state gain between *x* and *f* is zero. Thus the bypass flow rate can be adjusted, but it does not provide a control degree of freedom. However, if w_2 could also be adjusted, manipulating both *f* and w_2 could produce excellent control of the product composition.

2.5 DYNAMIC MODELS OF REPRESENTATIVE PROCESSES

For the simple process discussed so far, the stirred-tank blending system, energy effects were not considered due to the assumed isothermal operation. Next, we illustrate how dynamic models can be developed for processes where energy balances are important.

2.5.1 Stirred-Tank Heating Process: Constant Holdup

Consider the stirred-tank heating system shown in Fig. 2.6. The liquid inlet stream consists of a single component with a mass flow rate w_i and an inlet temperature T_i. The tank contents are agitated and heated using an electrical heater that provides a heating rate, *Q*. A dynamic model will be developed based on the following assumptions:

1. Perfect mixing; thus, the exit temperature *T* is also the temperature of the tank contents.
2. The inlet and outlet flow rates are equal; thus, $w_i = w$ and the liquid holdup *V* is constant.
3. The density ρ and heat capacity *C* of the liquid are assumed to be constant. Thus, their temperature dependence is neglected.
4. Heat losses are negligible.

In general, dynamic models are based on conservation laws. For this example, it is clear that we should consider an energy balance, because thermal effects predominate. A mass balance is not required in view of Assumptions 2 and 3.

Next, we show how the general energy balance in Eq. 2-10 can be simplified for this particular example. For a pure liquid at low or moderate pressures, the internal energy is approximately equal to the enthalpy, $U_{int} \approx H$, and H depends only on temperature (Sandler, 2006). Consequently, in the subsequent development, we assume that $U_{int} = H$ and $\hat{U}_{int} = \hat{H}$ where the caret (ˆ) means per unit mass. As shown in Appendix A, a differential change in temperature, dT, produces a corresponding change in the internal energy per unit mass, $d\hat{U}_{int}$,

$$d\hat{U}_{int} = d\hat{H} = C\,dT \tag{2-33}$$

where *C* is the constant pressure heat capacity (assumed to be constant). The total internal energy of the liquid in the tank can be expressed as the product of $\hat{U}_{int}$ and the mass in the tank, ρV:

$$U_{int} = \rho V \hat{U}_{int} \tag{2-34}$$

An expression for the rate of internal energy accumulation can be derived from Eqs. 2-33 and 2-34:

$$\frac{d\hat{U}_{int}}{dt} = \rho V C \frac{dT}{dt} \tag{2-35}$$

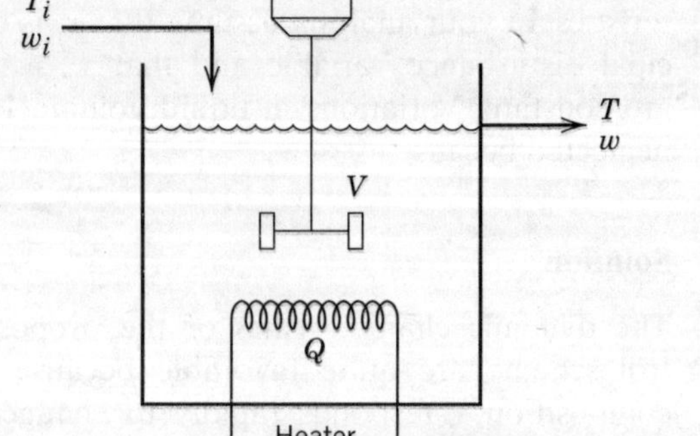

Figure 2.6 Stirred-tank heating process with constant holdup, *V*.

Note that this term appears in the general energy balance of Eq. 2-10.

Next, we derive an expression for the enthalpy term that appears on the right-hand side of Eq. 2-10. Suppose that the liquid in the tank is at a temperature T and has an enthalpy, $\hat{H}$. Integrating Eq. 2-33 from a reference temperature T_{ref} to T gives

$$\hat{H} - \hat{H}_{ref} = C(T - T_{ref}) \tag{2-36}$$

where $\hat{H}_{ref}$ is the value of $\hat{H}$ at T_{ref}. Without loss of generality, we assume that $\hat{H}_{ref} = 0$ (see Appendix A). Thus, Eq. 2-36 can be written as

$$\hat{H} = C(T - T_{ref}) \tag{2-37}$$

Similarly, for the inlet stream,

$$\hat{H}_i = C(T_i - T_{ref}) \tag{2-38}$$

Substituting Eq. 2-37 and Eq. 2-38 into the convection term of Eq. 2-10 gives

$$-\Delta(w\hat{H}) = w[C(T_i - T_{ref})] - w[C(T - T_{ref})] \tag{2-39}$$

Finally, substitution of Eq. 2-35 and Eq. 2-39 into Eq. 2-10 gives the desired dynamic model of the stirred-tank heating system:

$$V\rho C\frac{dT}{dt} = wC(T_i - T) + Q \tag{2-40}$$

Note that the T_{ref} terms have canceled, because C was assumed to be constant, and thus independent of temperature.

A degrees of freedom analysis for this model gives

3 parameters: V, ρ, C
4 variables: T, T_i, w, Q
1 equation: Eq. 2-40

Thus, the degrees of freedom are $N_F = 4 - 1 = 3$. The process variables are classified as

1 output variable: T
3 input variables: T_i, w, Q

For control purposes, it is reasonable to classify the three inputs as

2 disturbance variables: T_i, w
1 manipulated variable: Q

2.5.2 Stirred-Tank Heating Process: Variable Holdup

Now we consider the more general situation in which the tank holdup can vary with time. This analysis also is based on Assumptions 1, 3, and 4 of the previous section. Now an overall mass balance is needed, because the holdup is not constant. The overall mass balance is

$$\frac{d(V\rho)}{dt} = w_i - w \tag{2-41}$$

The energy balance for the current stirred-tank heating system can be derived from Eq. 2-10 in analogy with the derivation of Eq. 2-40. We again assume that $U_{int} = H$ for the liquid in the tank. Thus, for constant ρ,

$$\frac{dU_{int}}{dt} = \frac{dH}{dt} = \frac{d(\rho V\hat{H})}{dt} = \rho\frac{d(V\hat{H})}{dt} \tag{2-42}$$

From the definition of $-\Delta(w\hat{H})$ and Eqs. 2-37 and 2-38, it follows that

$$-\Delta(w\hat{H}) = w_i\hat{H}_i - w\hat{H} = w_iC(T_i - T_{ref}) - wC(T - T_{ref}) \tag{2-43}$$

where w_i and w are the mass flow rates of the inlet and outlet streams, respectively. Substituting Eq. 2-42 and Eq. 2-43 into Eq. 2-10 gives

$$\rho\frac{d(V\hat{H})}{dt} = w_iC(T_i - T_{ref}) - wC(T - T_{ref}) + Q \tag{2-44}$$

Next we simplify the dynamic model. Because ρ is constant, Eq. 2-41 can be written as

$$\rho\frac{dV}{dt} = w_i - w \tag{2-45}$$

The chain rule can be applied to expand the left side of Eq. 2-44 as:

$$\rho\frac{d(V\hat{H})}{dt} = \rho V\frac{d\hat{H}}{dt} + \rho\hat{H}\frac{dV}{dt} \tag{2-46}$$

From Eq. 2-33 or 2-37, it follows that $d\hat{H}/dt = CdT/dt$. Substituting this expression and Eqs. 2-37 and 2-45 into Eq. 2-46 gives

$$\rho\frac{d(V\hat{H})}{dt} = C(T - T_{ref})(w_i - w) + \rho CV\frac{dT}{dt} \tag{2-47}$$

Substituting Eq. 2-47 into Eq. 2-44 and rearranging gives

$$C(T - T_{\text{ref}})(w_i - w) + \rho CV \frac{dT}{dt}$$

$$= w_i C(T_i - T_{\text{ref}}) - wC(T - T_{\text{ref}}) + Q \qquad (2\text{-}48)$$

Rearranging Eq. 2-45 and Eq. 2-48 provides a simpler form for the dynamic model:

$$\frac{dV}{dt} = \frac{1}{\rho}(w_i - w) \qquad (2\text{-}49)$$

$$\frac{dT}{dt} = \frac{w_i}{V\rho}(T_i - T) + \frac{Q}{\rho CV} \qquad (2\text{-}50)$$

This example and the blending example in Section 2.2.2 have demonstrated that process models with variable holdups can be simplified by substituting the overall mass balance into the other conservation equations.

Equations 2-49 and 2-50 provide a model that can be solved for the two outputs (V and T) if the two parameters (ρ and C) are known and the four inputs (w_i, w, T_i, and Q) are known functions of time (i.e., there are four remaining degrees of freedom).

2.5.3 Electrically Heated Stirred Tank

Now we again consider the stirred-tank heating system with constant holdup (Section 2.5.1), but we relax the assumption that energy is transferred instantaneously from the heating element to the contents of the tank. Suppose that the metal heating element has a significant thermal capacitance and that the electrical heating rate Q directly affects the temperature of the element rather than the liquid contents. For simplicity, we neglect the temperature gradients in the heating element that result from heat conduction and assume that the element has a uniform temperature, T_e. This temperature can be interpreted as the average temperature for the heating element.

Based on this new assumption, and the previous assumptions of Section 2.5.1, the unsteady-state energy balances for the tank and the heating element can be written as

$$mC\frac{dT}{dt} = wC(T_i - T) + h_e A_e (T_e - T) \qquad (2\text{-}51)$$

$$m_e C_e \frac{dT_e}{dt} = Q - h_e A_e (T_e - T) \qquad (2\text{-}52)$$

where $m = V\rho$ and $m_e C_e$ is the product of the mass of metal in the heating element and its specific heat. The term $h_e A_e$ is the product of the heat transfer coefficient and area available for heat transfer. Note that mC and $m_e C_e$ are the thermal capacitances of the tank contents and the heating element, respectively. Q is an input variable, the thermal equivalent of the instantaneous electrical power dissipation in the heating element.

Is the model given by Eqs. 2-51 and 2-52 in suitable form for calculation of the unknown output variables T_e and T? There are two output variables and two differential equations. All of the other quantities must be either model parameters (constants) or inputs (known functions of time). For a specific process, m, C, m_e, C_e, h_e, and A_e are known parameters determined by the design of the process, its materials of construction, and its operating conditions. Input variables w, T_i, and Q must be specified as functions of time for the model to be completely determined—that is, to utilize the available degrees of freedom. The dynamic model can then be solved for T and T_e as functions of time by integration after initial conditions are specified for T and T_e.

If flow rate w is constant, Eqs. 2-51 and 2-52 can be converted into a single second-order differential equation. First, solve Eq. 2-51 for T_e and then differentiate to find dT_e/dt. Substituting the expressions for T_e and dT_e/dt into Eq. 2-52 yields

$$\frac{mm_eC_e}{wh_eA_e}\frac{d^2T}{dt^2} + \left(\frac{m_eC_e}{h_eA_e} + \frac{m_eC_e}{wC} + \frac{m}{w}\right)\frac{dT}{dt} + T$$

$$= \frac{m_eC_e}{h_eA_e}\frac{dT_i}{dt} + T_i + \frac{1}{wC}Q \qquad (2\text{-}53)$$

The reader should verify that the dimensions of each term in the equation are consistent and have units of temperature. In addition, the reader should consider the steady-state versions of Eqs. 2-40 and 2-53. They are identical, which is to be expected. Analyzing limiting cases is one way to check the consistency of a more complicated model. Confirming that dimensional units match on both sides of the equation is another check for consistency.

The model in Eq. 2-53 can be simplified when m_eC_e, the thermal capacitance of the heating element, is very small compared to mC. When $m_eC_e = 0$, Eq. 2-53 reverts to the first-order model, Eq. 2-40, which was derived for the case where the heating element has a negligible thermal capacitance.

It is important to note that the model of Eq. 2-53 consists of only a single equation and a single output variable, T. The intermediate variable, T_e, is less important than T and has been eliminated from the earlier model (Eqs. 2-51 and 2-52). Both models are exactly specified; that is, they have no unassigned degrees of freedom. To integrate Eq. 2-53, we require initial conditions for both T and dT/dt at $t = 0$, because it is a second-order differential equation. The initial condition for dT/dt can be found by evaluating the right side of Eq. 2-51 when $t = 0$, using the values of $T_e(0)$ and $T(0)$. For both models, the inputs (w, T_i, Q) must be specified as functions of time.

EXAMPLE 2.7

An electrically heated stirred-tank process can be modeled by Eqs. 2-51 and 2-52 or, equivalently, by Eq. 2-53 alone. Process design and operating conditions are characterized by the following four parameter groups:

$$\frac{m}{w} = 10\,\text{min} \qquad \frac{m_eC_e}{h_eA_e} = 1.0\,\text{min}$$

$$\frac{m_eC_e}{wC} = 1.0\,\text{min} \qquad \frac{1}{wC} = 0.05\,°\text{C min/ kcal}$$

The nominal values of Q and T_i are

$$\bar{Q} = 5000\,\text{kcal/min} \qquad \bar{T}_i = 100\,°\text{C}$$

(a) Calculate the nominal steady-state temperature, $\bar{T}$.

(b) Assume that the process is initially at the steady-state determined in part (**a**). Calculate the response, $T(t)$, to a sudden change in Q from 5000 to 5400 kcal/min using Eq. 2-53. Plot the temperature response.

(c) Suppose that it can be assumed that the term m_eC_e/h_eA_e is small relative to other terms in Eq. 2-53. Calculate the response $T(t)$ for the conditions of part (**b**), using a first-order differential equation approximation to Eq. 2-53. Plot $T(t)$ on the graph for part (**b**).

(d) What can we conclude about the accuracy of the approximation for part (**c**)?

Solution

(a) The steady-state form of Eq. 2-53 is

$$\bar{T} = \bar{T}_i + \frac{1}{wC}\bar{Q}$$

Substituting parameter values gives $\bar{T} = 350\,°\text{C}$.

(b) Substitution of the parameter values in Eq. 2-53 gives

$$10\frac{d^2T}{dt^2} + 12\frac{dT}{dt} + T = 370$$

The following solution can be derived using standard solution methods (Kreyszig, 2011):

$$T(t) = 350 + 20[1 - 1.089e^{-t/11.099} + 0.0884e^{-t/0.901}]$$

This response is plotted in Fig. 2.7 as the dashed curve (a).

(c) If we assume that m_eC_e is small relative to other terms, then Eq. 2-53 can be approximated by the first-order differential equation:

$$12\frac{dT}{dt} + T = 370, \quad T(0) = 350\,°\text{C}$$

The solution is

$$T(t) = 350 + 20(1 - e^{-t/12})$$

(d) Figure 2.7 shows that the approximate solution (b) is quite good, matching the exact solution very well over the entire response. For purposes of process control, this approximate model is likely to be as useful as the more complicated, exact model.

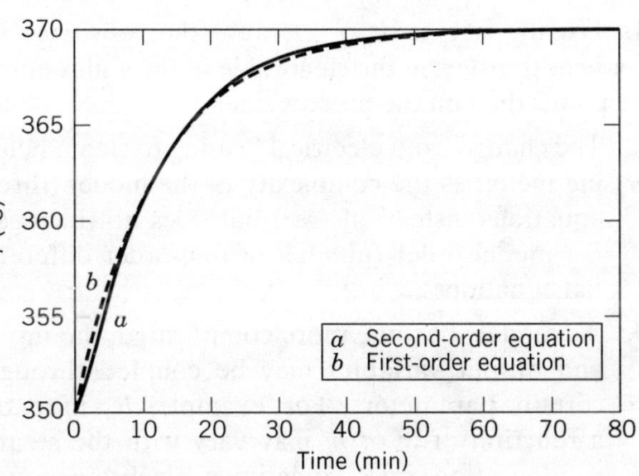

Figure 2.7 Responses of an electrically heated stirred-tank process to a sudden change in the heater input.

2.5.4 Steam-Heated Stirred Tank

Steam (or some other heating medium) can be condensed within a coil or jacket to heat liquid in a stirred tank, and the inlet steam pressure can be varied by adjusting a control valve. The condensation pressure P_s then fixes the steam temperature T_s through an appropriate thermodynamic relation or from tabular information such as the steam tables (ASME, 2014):

$$T_s = f(P_s) \tag{2-54}$$

Consider the stirred-tank heating system of Section 2.5.1 with constant holdup and a steam heating coil.

We assume that the thermal capacitance of the liquid condensate is negligible compared to the thermal capacitances of the tank liquid and the wall of the heating coil. This assumption is reasonable when a steam trap is used to remove the condensate from the coil as it is produced. As a result of this assumption, the dynamic model consists of energy balances on the liquid and the heating coil wall:

$$mC\frac{dT}{dt} = wC(T_i - T) + h_pA_p(T_w - T) \tag{2-55}$$

$$m_wC_w\frac{dT_w}{dt} = h_sA_s(T_s - T_w) - h_pA_p(T_w - T) \tag{2-56}$$

where the subscripts w, s, and p refer, respectively, to the wall of the heating coil and to its steam and process sides. Note that these energy balances are similar to Eqs. 2-51 and 2-52 for the electrically heated example.

The dynamic model contains three output variables (T_s, T, and T_w) and three equations: an algebraic equation with T_s related to P_s (a thermodynamic equation) and two differential equations. Thus, Eqs. 2-54 through 2-56 constitute an exactly specified model with three input variables: P_s, T_i, and w. Several important features are noted.

1. Usually $h_sA_s \gg h_pA_p$, because the resistance to heat transfer on the steam side of the coil is much lower than on the process side.
2. The change from electrical heating to steam heating increases the complexity of the model (three equations instead of two) but does not increase the model order (number of first-order differential equations).
3. As models become more complicated, the input and output variables may be coupled through certain parameters. For example, h_p may be a function of w or h_s may vary with the steam condensation rate; sometimes algebraic equations cannot be solved explicitly for a key variable. In this situation, numerical solution techniques have to be used. Usually, implicit algebraic equations must be solved by iterative methods at each time step in the numerical integration.

We now consider some simple models for liquid storage systems utilizing a single tank. In the event that two or more tanks are connected in series (cascaded), the single-tank models developed here can be easily extended, as shown in Chapter 5.

2.5.5 Liquid Storage Systems

A typical liquid storage process is shown in Fig. 2.8 where q_i and q are volumetric flow rates. A mass balance yields

$$\frac{d(\rho V)}{dt} = \rho q_i - \rho q \tag{2-57}$$

Assume that liquid density ρ is constant and the tank is cylindrical with cross-sectional area, A. Then the volume of liquid in the tank can be expressed as $V = Ah$, where h is the liquid level (or *head*). Thus, Eq. 2-57 becomes

$$A\frac{dh}{dt} = q_i - q \tag{2-58}$$

Note that Eq. 2-58 appears to be a *volume balance*. However, in general, volume is *not* conserved for fluids. This result occurs in this example due to the constant density assumption. We refer to volume as a "state" of the system, which is a dependent variable that correlates with a fundamentally conserved quantity (in this case, mass). Similarly, energy balances often yield differential equations for temperature, which is a state variable that correlates with energy.

There are three important variations of the liquid storage process:

1. The inlet or outlet flow rates might be constant; for example, exit flow rate q might be kept constant by a constant-speed, fixed-volume (metering) pump. An important consequence of this configuration is that the exit flow rate is then completely independent of liquid level over a wide range of conditions. Consequently, $q = \bar{q}$ where $\bar{q}$ is the steady-state value. For this situation, the tank operates essentially as a flow *integrator*. We will return to this case in Section 5.5.
2. The tank exit line may function simply as a resistance to flow from the tank (distributed along the entire line), or it may contain a valve that provides significant resistance to flow at a single point. In the simplest case, the flow may be assumed to be

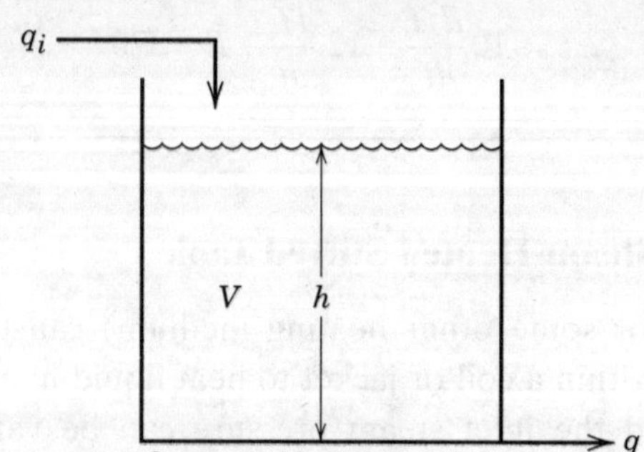

Figure 2.8 A liquid-level storage process.

linearly related to the driving force, the liquid level, in analogy to Ohm's law for electrical circuits ($E = IR$)

$$h = qR_v \tag{2-59}$$

where R_v is the resistance of the line or valve. Rearranging Eq. 2-59 gives the following *flow-head equation*:

$$q = \frac{1}{R_v}h \tag{2-60}$$

Substituting Eq. 2-60 into Eq. 2-56 gives a first-order differential equation:

$$A\frac{dh}{dt} = q_i - \frac{1}{R_v}h \tag{2-61}$$

This model of the liquid storage system exhibits dynamic behavior similar to that of the stirred-tank heating system of Eq. 2-40.

3. A more realistic expression for flow rate q can be obtained when a fixed valve has been placed in the exit line and turbulent flow can be assumed. The driving force for flow through the valve is the pressure drop ΔP:

$$\Delta P = P - P_a \tag{2-62}$$

where P is the pressure at the bottom of the tank and P_a is the pressure at the end of the exit line. We assume that P_a is the ambient pressure. If the valve is considered to be an orifice, a mechanical energy balance, or *Bernoulli equation* (Bird et al., 2002), can be used to derive the relation

$$q = C_v^*\sqrt{\frac{P - P_a}{\rho}} \tag{2-63}$$

where C_v^* is a constant. The value of C_v^* depends on the particular valve and the valve setting (how much it is open). See Chapter 8 for more information about control valves.

The pressure P at the bottom of the tank is related to liquid level h by a force balance

$$P = P_a + \frac{\rho g}{g_c}h \tag{2-64}$$

where the acceleration of gravity g is constant. Substituting Eqs. 2-63 and 2-64 into Eq. 2-58 yields the dynamic model

$$A\frac{dh}{dt} = q_i - C_v\sqrt{h} \tag{2-65}$$

where $C_v \triangleq C_v^*\sqrt{g/g_c}$. This model is nonlinear due to the square root term.

The liquid storage processes discussed above could be operated by controlling the liquid level in the tank or by allowing the level to fluctuate without attempting to control it. For the latter case (operation as a surge tank), it may be of interest to predict whether the tank would overflow or run dry for particular variations in the inlet and outlet flow rates. Thus, the dynamics of the process may be important even when automatic control is not utilized.

2.5.6 The Continuous Stirred-Tank Reactor (CSTR)

Continuous stirred-tank reactors (CSTR) have widespread application in industry and embody many features of other types of reactors. CSTR models tend to be simpler than models for other types of continuous reactors such as tubular reactors and packed-bed reactors. Consequently, a CSTR model provides a convenient way of illustrating modeling principles for chemical reactors.

Consider a simple liquid-phase, irreversible chemical reaction where chemical species A reacts to form species B. The reaction can be written as A → B. We assume that the rate of reaction is first-order with respect to component A,

$$r = kc_A \tag{2-66}$$

where r is the rate of reaction of A per unit volume, k is the reaction rate constant (with units of reciprocal time), and c_A is the molar concentration of species A. For single-phase reactions, the rate constant is typically a strong function of reaction temperature given by the Arrhenius relation,

$$k = k_0 e^{-E/RT} \tag{2-67}$$

where k_0 is the frequency factor, E is the activation energy, and R is the gas constant. The expressions in Eqs. 2-66 and 2-67 are based on theoretical considerations, but model parameters k_0 and E are usually determined by fitting experimental data. Thus, these two equations can be considered to be *semi-empirical* relations, according to the definition in Section 2.2.

The schematic diagram of the CSTR is shown in Fig. 2.9. The inlet stream consists of pure component A with molar concentration, c_{Ai}. A cooling coil is used to maintain the

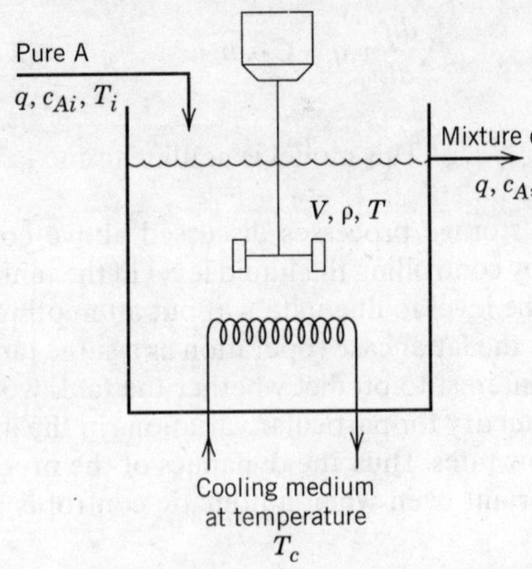

Figure 2.9 A nonisothermal continuous stirred-tank reactor.

reaction mixture at the desired operating temperature by removing heat that is released in the exothermic reaction. Our initial CSTR model development is based on three assumptions:

1. The CSTR is perfectly mixed.
2. The mass densities of the feed and product streams are equal and constant. They are denoted by ρ.
3. The liquid volume V in the reactor is kept constant by an overflow line.

For these assumptions, the unsteady-state mass balance for the CSTR is

$$\frac{d(\rho V)}{dt} = \rho q_i - \rho q \qquad (2\text{-}68)$$

Because V and ρ are constant, Eq. 2-68 reduces to

$$q = q_i \qquad (2\text{-}69)$$

Thus, even though the inlet and outlet flow rates may change due to upstream or downstream conditions, Eq. 2-69 must be satisfied at all times. In Fig. 2.9, both flow rates are denoted by the symbol q.

For the stated assumptions, the unsteady-state component balances for species A (in molar concentration units) is

$$V\frac{dc_A}{dt} = q(c_{Ai} - c_A) - Vkc_A \qquad (2\text{-}70)$$

This balance is a special case of the general component balance in Eq. 2-7.

Next, we consider an unsteady-state energy balance for the CSTR. But first we make five additional assumptions:

4. The thermal capacitances of the coolant and the cooling coil wall are negligible compared to the thermal capacitance of the liquid in the tank.
5. All of the coolant is at a uniform temperature, T_c. (That is, the increase in coolant temperature as the coolant passes through the coil is neglected.)
6. The rate of heat transfer from the reactor contents to the coolant is given by

$$Q = UA(T_c - T) \qquad (2\text{-}71)$$

where U is the overall heat transfer coefficient and A is the heat transfer area. Both of these model parameters are assumed to be constant.

7. The enthalpy change associated with the mixing of the feed and the liquid in the tank is negligible compared with the enthalpy change for the chemical reaction. In other words, the heat of mixing is negligible compared to the heat of reaction.
8. Shaft work and heat losses to the ambient can be neglected.

The following form of the CSTR energy balance is convenient for analysis and can be derived from Eqs. 2-66 and 2-67 and Assumptions 1–8 (Fogler, 2006; Russell and Denn, 1972),

$$V\rho C\frac{dT}{dt} = wC(T_i - T) + (-\Delta H_R)Vkc_A + UA(T_c - T) \qquad (2\text{-}72)$$

where ΔH_R is the heat of reaction per mole of A that is reacted.

In summary, the dynamic model of the CSTR consists of Eqs. 2-66 to 2-68, 2-70, 2-71, and 2-72. This model is nonlinear as a result of the many product terms and the exponential temperature dependence of k in Eq. 2-67. Consequently, it must be solved by numerical integration techniques (Fogler, 2006). The CSTR model will become considerably more complex if

1. More complicated rate expressions are considered. For example, a mass action kinetics model for a second-order, irreversible reaction, 2A → B, is given by

$$r = k_2 c_A^2 \qquad (2\text{-}73)$$

2. Additional species or chemical reactions are involved. If the reaction mechanism involved production of an intermediate species, 2A → B* → B, then unsteady-state component balances for both A and B* would be necessary (to calculate c_A and c_B^*), or balances for both A and B could be written (to calculate c_A and c_B). Information concerning the reaction mechanisms would also be required.

Reactions involving multiple species are described by high-order, highly coupled, nonlinear reaction models, because several component balances must be written.

EXAMPLE 2.8

To illustrate how the CSTR can exhibit nonlinear dynamic behavior, we simulate the effect of a step change in the coolant temperature T_c in positive and negative directions. Table 2.3 shows the parameters and nominal operating condition for the CSTR based on Eqs. 2-70 and 2-72 for the exothermic, irreversible first-order reaction A → B. The two state variables of the ODEs are the concentration of A (c_A) and the reactor temperature T. The manipulated input variable is the jacket water temperature, T_c.

Two cases are simulated, one based on increased cooling by changing T_c from 300 to 290 K and one reducing the cooling rate by increasing T_c from 300 to 305 K.

These model equations are solved in MATLAB with a numerical integrator (ode15s) over a 10-min horizon. The decrease in T_c results in an increase in c_A. The results are displayed in two plots of the temperature and reactor concentration as a function of time (Figs. 2.10 and 2.11).

At a jacket temperature of 305 K, the reactor model has an oscillatory response. The oscillations are characterized by apparent reaction run away with a temperature spike. However, when the concentration drops to a low value, the reactor then cools until the concentration builds, then there is another temperature rise. It is not unusual for chemical reactors to exhibit such widely different behaviors for different directional changes in the operating conditions.

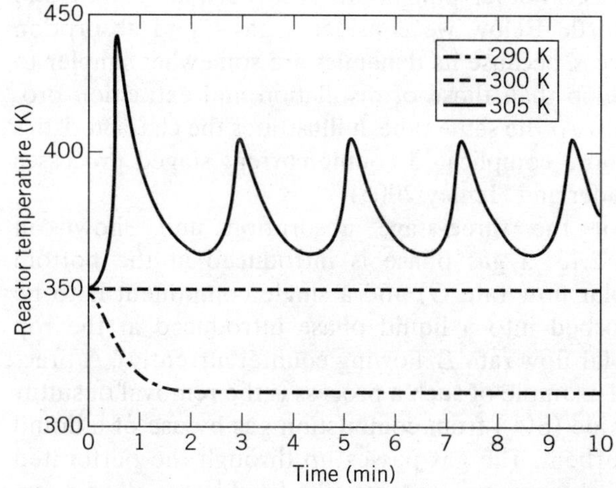

Figure 2.10 Reactor temperature variation with step changes in cooling water temperature from 300 to 305 K and from 300 to 290 K.

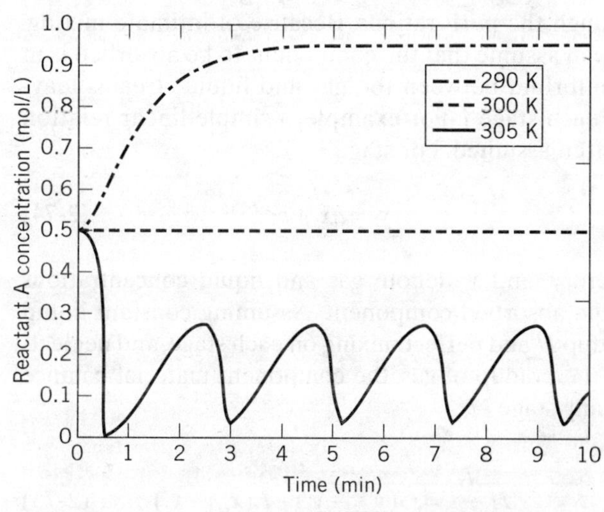

Figure 2.11 Reactant A concentration variation with step changes in cooling water temperature to 305 and 290 K.

Table 2.3 Nominal Operating Conditions for the CSTR

Parameter	Value	Parameter	Value
q	100 L/min	E/R	8750 K
c_{Ai}	1 mol/L	k_0	7.2×10^{10} min^{-1}
T_i	350 K	UA	5×10^4 J/min K
V	100 L	$T_c(0)$	300 K
ρ	1000 g/L	$c_A(0)$	0.5 mol/L
C	0.239 J/g K	$T(0)$	350 K
$-\Delta H_R$	$5 \cdot 10^4$ J/mol		

Although the modeling task becomes much more complex, the same principles illustrated above can be extended and applied. We will return to the simple CSTR model again in Chapter 4.

2.5.7 Staged Systems (a Three-Stage Absorber)

Chemical processes, particularly separation processes, often consist of a sequence of stages. In each stage, materials are brought into intimate contact to obtain (or approach) equilibrium between the individual phases. The most important examples of staged processes include distillation, absorption, and extraction. The stages are usually arranged as a cascade with immiscible or partially miscible materials (the separate phases) flowing either cocurrently or countercurrently. Countercurrent contacting, shown in Fig. 2.12, usually permits the highest degree of separation to be attained in a fixed number of stages and is considered here.

The feeds to staged systems may be introduced at each end of the process, as in absorption units, or a single feed may be introduced at a middle stage, as is usually the case with distillation. The stages may be physically connected in either a vertical or horizontal configuration, depending on how the materials are transported, that is, whether pumps are used between stages, and so forth. Below we consider a gas–liquid absorption process, because its dynamics are somewhat simpler to develop than those of distillation and extraction processes. At the same time, it illustrates the characteristics of more complicated countercurrent staged processes (Seader and Henley, 2005).

For the three-stage absorption unit shown in Fig. 2.13, a gas phase is introduced at the bottom (molar flow rate G) and a single component is to be absorbed into a liquid phase introduced at the top (molar flow rate L, flowing countercurrently). A practical example of such a process is the removal of sulfur dioxide (SO_2) from combustion gas by use of a liquid absorbent. The gas passes up through the perforated (sieve) trays and contacts the liquid cascading down through them. A series of weirs and downcomers typically are used to retain a significant holdup of liquid on each stage while forcing the gas to flow upward through the perforations. Because of intimate mixing, we can assume that the component to be absorbed is in equilibrium between the gas and liquid streams leaving each stage i. For example, a simple linear relation is often assumed. For stage i

$$y_i = ax_i + b \tag{2-74}$$

where y_i and x_i denote gas and liquid concentrations of the absorbed component. Assuming constant liquid holdup H and perfect mixing on each stage, and neglecting the holdup of gas, the component material balance for any stage i is

$$H\frac{dx_i}{dt} = G(y_{i-1} - y_i) + L(x_{i+1} - x_i) \tag{2-75}$$

In Eq. 2-75, we also assume that molar liquid and gas flow rates L and G are unaffected by the absorption, because changes in concentration of the absorbed

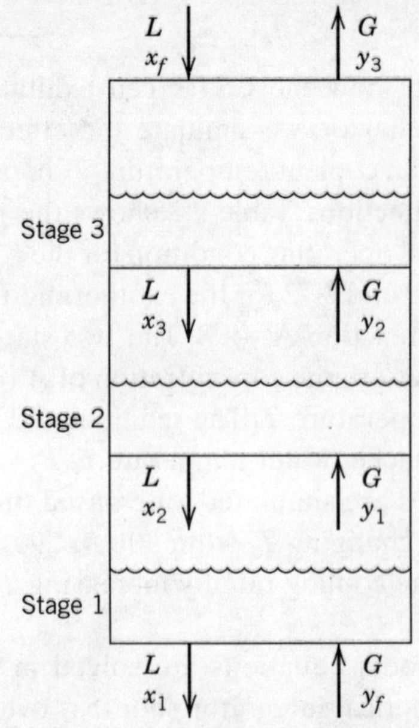

Figure 2.13 A three-stage absorption unit.

component are small, and L and G are approximately constant. Substituting Eq. 2-74 into Eq. 2-75 yields

$$H\frac{dx_i}{dt} = aGx_{i-1} - (L + aG)x_i + Lx_{i+1} \tag{2-76}$$

Dividing by L and substituting $\tau = H/L$ (the stage liquid residence time), $\mathcal{S} = aG/L$ (the *stripping factor*), and $K = G/L$ (the gas-to-liquid ratio), the following model is obtained for the three-stage absorber:

$$\tau\frac{dx_1}{dt} = K(y_f - b) - (1+\mathcal{S})x_1 + x_2 \tag{2-77}$$

$$\tau\frac{dx_2}{dt} = \mathcal{S}x_1 - (1+\mathcal{S})x_2 + x_3 \tag{2-78}$$

$$\tau\frac{dx_3}{dt} = \mathcal{S}x_2 - (1+\mathcal{S})x_3 + x_f \tag{2-79}$$

In the model of Eqs. 2-77 to 2-79, note that the individual equations are linear but also coupled, meaning that each output variable—x_1, x_2, x_3—appears in more than one equation. This feature can make it difficult to

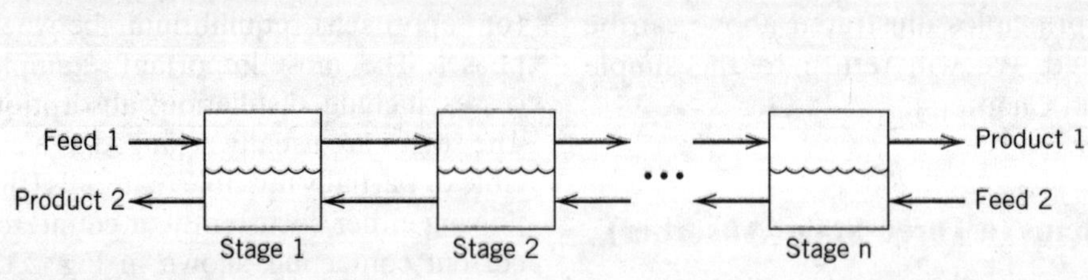

Figure 2.12 A countercurrent-flow staged process.

convert these three equations into a single higher-order equation in one of the outputs, as was done in Eq. 2-53.

2.5.8 Fed-Batch Bioreactor

Biological reactions that involve microorganisms and enzyme catalysts are pervasive and play a crucial role in the natural world. Without such bioreactions, plant and animal life, as we know it, simply could not exist. Bioreactions also provide the basis for production of a wide variety of pharmaceuticals and healthcare and food products. Other important industrial processes that involve bioreactions include fermentation and wastewater treatment. Chemical engineers are heavily involved with biochemical and biomedical processes. In this section we present a dynamic model for a representative process, a bioreactor operated in a semi-batch mode. Additional biochemical and biomedical applications appear in other chapters.

In general, bioreactions are characterized by the conversion of feed material (or *substrate*) into products and cell mass (or *biomass*). The reactions are typically catalyzed by enzymes (Bailey and Ollis, 1986; Fogler, 2006). When the objective is to produce cells, a small amount of cells (*inoculum*) is added to initiate subsequent cell growth. A broad class of bioreactions can be represented in simplified form as

$$\text{cells} + \text{substrate} \rightarrow \text{more cells} + \text{products} \tag{2-80}$$

The stoichiometry of bioreactions can be very complex and depends on many factors that include the environmental conditions in the vicinity of the cells. For simplicity we consider the class of bioreactions where the substrate contains a single limiting nutrient and only one product results. The following *yield coefficients* are based on the reaction stoichiometry:

$$Y_{X/S} = \frac{\textit{mass of new cells formed}}{\textit{mass of substrate consumed to form new cells}} \tag{2-81}$$

$$Y_{P/S} = \frac{\textit{mass of product formed}}{\textit{mass of substrate consumed to form product}} \tag{2-82}$$

$$Y_{P/X} = \frac{\textit{mass of product formed}}{\textit{mass of new cells formed}} \tag{2-83}$$

Many important bioreactors are operated in a semi-continuous manner that is referred to as *fed-batch* operation, which is illustrated in Fig. 2.14. A feed stream containing substrate is introduced to the fed-batch reactor continually. The mass flow rate

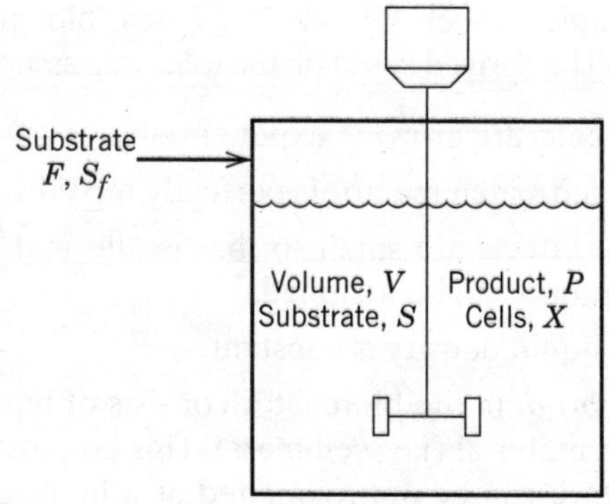

Figure 2.14 Fed-batch reactor for a bioreaction.

is denoted by F and the substrate mass concentration by S_f. Because there is no exit stream, the volume V of the bioreactor contents increases during the batch. The advantage of fed-batch operation is that it allows the substrate concentration to be maintained at a desired level, in contrast to batch reactors where the substrate concentration varies continually throughout the batch (Shuler and Kargi, 2002).

Fed-batch operation is used to manufacture many important industrial products, including antibiotics and protein pharmaceuticals (Appendix I) to be checked. In batch and fed-batch reactors, cell growth occurs in different stages after the inoculum is introduced. We will consider only the exponential growth stage where the cell growth rate is autocatalytic and is assumed to be proportional to the cell concentration. A standard reaction rate expression to describe the rate of cell growth with a single limiting substrate is given by (Bailey and Ollis, 1986; Fogler, 2006)

$$r_g = \mu X \tag{2-84}$$

where r_g is the rate of cell growth per unit volume, X is the cell mass, and μ is the *specific growth rate*, which is well described by the *Monod equation*:

$$\mu = \mu_{max} \frac{S}{K_S + S} \tag{2-85}$$

Note that μ has units of reciprocal time—for example, h^{-1}. Model parameter μ_{max} is referred to as the *maximum growth rate,* because μ has a maximum value of μ_{max} when $S \gg K_S$. The second model parameter, K_S, is called the *Monod constant*. The Monod equation has the same form as the Michaelis–Menten equation, a standard rate expression for enzyme reactions (Bailey and Ollis, 1986; Fogler, 2006). More complex versions of the specific growth rate are possible including, for example, product inhibition.

A dynamic model for the fed-batch bioreactor in Fig. 2.14 will be derived based on the following assumptions:

1. The cells are growing exponentially.
2. The fed-batch reactor is perfectly mixed.
3. Heat effects are small so that isothermal reactor operation can be assumed.
4. The liquid density is constant.
5. The *broth* in the bioreactor consists of liquid plus solid material (i.e., cell mass). This heterogeneous mixture can be approximated as a homogeneous liquid.
6. The rate of cell growth r_g is given by Eqs. 2-84 and 2-85.
7. The rate of product formation per unit volume r_p can be expressed as

$$r_P = Y_{P/X} r_g \tag{2-86}$$

8. The feed stream is sterile and thus contains no cells.

The dynamic model of the fed-batch reactor consists of individual balances for substrate, cell mass, and product, plus an overall mass balance. The general form of each balance is

$$\{Rate\ of\ accumulation\} = \{rate\ in\} + \{rate\ of\ formation\} \tag{2-87}$$

The individual component balances are

$$\text{Cells:} \quad \frac{d(XV)}{dt} = Vr_p \tag{2-88}$$

$$\text{Product:} \quad \frac{d(PV)}{dt} = Vr_p \tag{2-89}$$

$$\text{Substrate:} \quad \frac{d(SV)}{dt} = FS_f - \frac{1}{Y_{X/S}} Vr_g - \frac{1}{Y_{P/S}} Vr_p \tag{2-90}$$

where P is the mass concentration of the product and V is reactor volume. Reaction rates r_g and r_p and yield coefficients were defined in Eqs. 2-81 through 2-86. The overall mass balance (assuming constant density) is

$$\text{Mass:} \quad \frac{dV}{dt} = F \tag{2-91}$$

The dynamic model is simulated for two different feed rates (0.02 and 0.05 L/h). Figure 2.15 shows the profile of cell, product, and substrate concentration, together with liquid volume in the reactor. The model parameters and simulation conditions are given in Table 2.4. For different feed rates, the bioreactor gives different responses; thus, the product can be maximized by varying F.

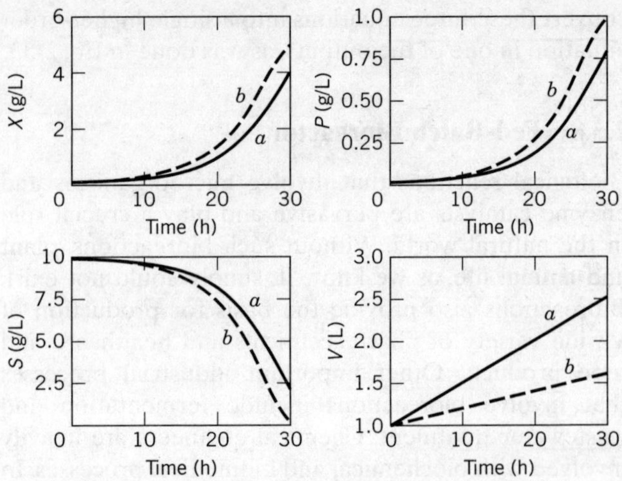

Figure 2.15 Fed-batch reaction profile (*a* (solid): $F = 0.05$ L/h; b (dashed): $F = 0.02$ L/h).

Table 2.4 Model Parameters and Simulation Conditions for Bioreactor

Model Parameters			Simulation Conditions		
μ_{max}	0.20	h^{-1}	S_f	10.0	g/L
K_S	1.0	g/L	$X(0)$	0.05	g/L
$Y_{X/S}$	0.5	g/g	$S(0)$	10.0	g/L
$Y_{P/X}$	0.2	g/g	$P(0)$	0.0	g/L
			$V(0)$	1.0	L

2.6 SOLVING DIFFERENTIAL EQUATIONS USING MATLAB

The mathematical model developed using differential equations can be solved through the symbolic math toolbox in MATLAB using the command 'dsolve'.

The terms used in the symbolic differential equation can be expressed in two ways – using differential operator 'D' or symbolic operator 'diff'. Table 2.5 gives the MATLAB equivalence for various differential equation terms.

The following examples will explain the used of 'dsolve' command in various cases for solving differential equations.

Table 2.5 MATLAB equivalence for differential equation terms

Terms in Differential Equation	Equivalence using 'D'	Equivalence using 'diff'
y	y	Y,
$\frac{dy}{dt}$	Dy	diff(y,t)
$\frac{d^2y}{dt^2}$	D2y	diff(y,t,2)
$\frac{d^ny}{dt^n}$	Dny	diff(y,t,n)

EXAMPLE 2.9

Solve the following differential equation using MATLAB.

$$\frac{dy}{dt} + ay = b$$

Solution

Method 1 (Using differential operator 'D')

```
>> syms y t a b
>> ys=dsolve('Dy+a*y=b')

ys = (b - C1*exp(-a*t))/a
```

Method 2 (Using symbolic operator 'diff')

```
>> syms y(t) a b
>> eqn = diff(y,t) + a*y==b;
>> s = dsolve(eqn)

s =(b - c1*exp(-a*t))/a
```

EXAMPLE 2.10

Solve the following differential equation with initial condition using MATLAB.

$$\frac{dy}{dt} + 3y = 10 \qquad y(0) = 5$$

Solution

Method 1 (Using differential operator 'D')

```
>> syms y t
>> ys=dsolve('Dy+3*y=10','y(0)=5')

ys = (5*exp(-3*t))/3 + 10/3
```

Method 2 (Using symbolic operator 'diff')

```
>> syms y(t)
>> eqn = diff(y,t)+3*y==10;
>> cond = y(0)==5;

>> S = dsolve(eqn,cond)

S = (5*exp(-3*t))/3 + 10/3
```

EXAMPLE 2.11

Solve the following second-order differential equation using MATLAB.

$$\frac{d^2y}{dt^2} = ay + b$$

Solution

Method 1

```
>> syms y t a b
>> ys=dsolve('D2y=a*y+b')

ys = C1*exp(a^(1/2)*t)
     - b/a + C2*exp(-a^(1/2)*t)
```

Method 2

```
>> syms y(t)a b
>> eqn = diff(y,t,2)==a*y+b;
>> ySol(t) = dsolve(eqn)

ySol(t) = C1*exp(a^(1/2)*t)
          - b/a + C2*exp(-a^(1/2)*t)
```

EXAMPLE 2.12

Solve the following second-order differential equation using MATLAB.

$$\frac{d^2y}{dt^2} - a^2y = 0 \text{ with initial conditions}$$

$$y(0) = 1 \,\&\, y'(0) = 0.$$

Solution

Method 1

```
>> syms y t a
>> ys=dsolve('D2y-a^2*y=0','y(0)=1',
'Dy(0)=0')
ys = exp(a*t)/2 + exp(-a*t)/2
```

Method 2

```
>> syms y(t) a
>> eqn = diff(y,t,2)==a^2*y;
>> Dy=diff(y,t);
>> cond = [y(0)==1, Dy(0)==0];
>> S = dsolve(eqn,cond)

S = exp(a*t)/2 + exp(-a*t)/2
```

EXAMPLE 2.13

Solve the following system of differential equations without and with the initial conditions $y(0) = 0$ and $z(0) = 1$.

$$\frac{dy}{dt} = 2y + 3z$$

$$\frac{dz}{dt} = -3y + 2z$$

Solution

Method 1

```
>> syms y z t
>> [y,z]=dsolve('Dy=2*y+3*z',
'Dz=-3*y+2*z')

y = C2*cos(3*t)*exp(2*t)
    + C1*sin(3*t)*exp(2*t)
z = C1*cos(3*t)*exp(2*t)
    - C2*sin(3*t)*exp(2*t)

>> % With initial conditions
>> syms y z t
>> [y,z]=dsolve('Dy=2*y+3*z',
   'Dz= -3*y+2*z','y(0)=0', 'z(0)=1')
```

Method 2

```
>> syms y(t)z(t)
>> eqns = [diff(y,t)==2*y+3*z,
diff(z,t)==-3*y+2*z];
>> S = dsolve(eqns);
>> ysol(t)=S.y
>> zsol(t)=S.z

ysol(t) = C2*cos(3*t)*exp(2*t)
          + C1*sin(3*t)*exp(2*t)
zsol(t) = C1*cos(3*t)*exp(2*t)
          - C2*sin(3*t)*exp(2*t)

>> % With initial conditions
>> syms y(t)z(t)
>> eqns = [diff(y,t)==2*y+3*z,
           diff(z,t)==-3*y+2*z];
>> cond = [y(0)==0, z(0)==1];
>> S = dsolve(eqns,cond);
>> ysol(t)=S.y
>> zsol(t)=S.z

ysol(t) = sin(3*t)*exp(2*t)
zsol(t) = cos(3*t)*exp(2*t)
```

EXAMPLE 2.14

Solve the following second-order differential equation using MATLAB.

$$2\frac{d^2T}{dt^2}+5\frac{dT}{dt}+3T=48 \quad T(0)=10^\circ\mathrm{C} \quad T'(0)=0^\circ\mathrm{C/min}$$

Also plot the temperature response for time 0 to 6 minutes.

Solution

```
>> syms T t
>> Tsol=dsolve('2*D2T+5*DT+3*T=48',
'T(0)=10','DT(0)=0')
>> ezplot(Tsol,[0 6])
>> xlabel('Time(min)')
>> ylabel('Temperature(^°C)')

Tsol = 12*exp(-(3*t)/2) - 18*exp(-t)
      + 16
```

The plot for temperature response is shown in Fig. 2.16.

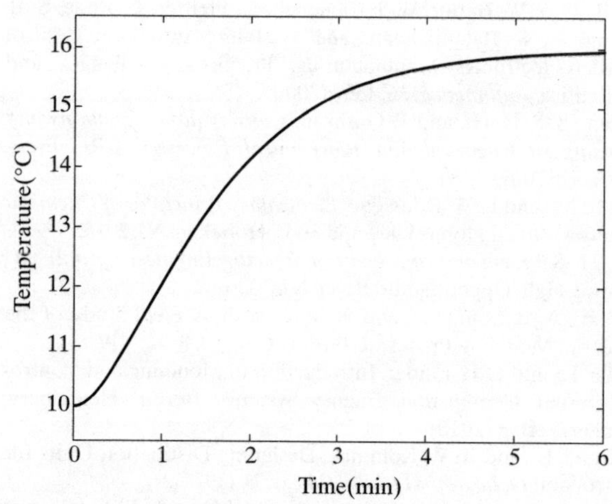

Figure 2.16 Temperature response.

SUMMARY

In this chapter, we have considered the derivation of dynamic models from first principles, especially conservation equations. Model development is an art as well as a science. It requires making assumptions and simplifications that are consistent with the modeling objectives and the end use of the model. A systematic approach for developing dynamic models is summarized in Table 2.1. This approach has been illustrated by deriving models for representative processes. Although these illustrative examples are rather simple, they demonstrate fundamental concepts that are also valid for more complicated processes. Finally, we have discussed how the development and solution of dynamic models continues to be made easier by commercial simulation software MATLAB® which is a product from MathWorks (www.mathworks.com).

In Chapters 3–6, we will consider analytical solutions of linear dynamic models using Laplace transforms and transfer functions. These useful techniques allow dynamic response characteristics to be analyzed on a more systematic basis. They also provide considerable insight into common characteristics shared by complex processes.

REFERENCES

Aris, R., *Mathematical Modeling: A Chemical Engineer's Perspective*, Academic Press, New York, 1999.

ASME *International Steam Tables for Industrial Use*, ASME, New York, 2014.

Bailey, J. E., and D. F. Ollis, *Biochemical Engineering Fundamentals*, 2nd ed., McGraw-Hill, New York, 1986.

Bequette, B. W., *Process Dynamics: Modeling, Analysis, and Simulation*, Prentice Hall, Upper Saddle River, NJ, 1998.

Bird, R. B., W. E. Stewart, and E. N. Lightfoot, *Transport Phenomena*, 2nd ed., John Wiley and Sons, Hoboken, NJ, 2002.

Braunschweig, B. L., C. C. Pantelides, H. I. Britt, and S. Sama, Process Modeling: The Promise of Open Software Architectures, *Chem Eng. Progr.*, **96**(9), 65 (2000).

Cameron, I. T., and R. Gani, *Product and Process Modelling: A Case Study Approach*, Elsevier, Oxford, 2011.

Chapra, S. C., and R. P. Canale, *Numerical Methods for Engineers*, 7th ed., McGraw-Hill, New York, 2014.

Coughanowr, D. R., *Process Systems Analysis and Control*, 2nd ed. McGraw-Hill, New York, 1991.

Doyle, F. J. III, E. P. Gatzke, and R. S. Parker, Practical Case Studies for Undergraduate Process Dynamics and Control Using Process Control Modules, *Comput. Appl. Eng. Educ.* **6**, 181–191 (1998).

Edgar, T. F., S. W. Butler, W. J. Campbell, C. Pfeiffer, C. Bode, S. B. Hwang, K. S. Balakrishnan, and J. Hahn, Automatic Control in Microelectronics Manufacturing: Practices, Challenges and Possibilities, *Automatica*, **36**, 1567 (2000).

Elnashaie, S. S. E. H., and P. Garhyan, *Conservation Equations and Modeling of Chemical and Biochemical Processes*, CRC Press, New York, 2003.

Felder, R. M., and R. W. Rousseau, *Elementary Principles of Chemical Processes*, 4th ed., John Wiley and Sons, Hoboken, NJ, 2015.

Fogler, H. S., *Elements of Chemical Reactor Engineering*, 4th ed., Prentice-Hall, Upper Saddle River, NJ, 2006.

Foss, B. A., A. B. Lohmann, and W. Marquardt, A Field Study of the Industrial Modeling Process, *J. Process Control*, **8**, 325 (1998).

Guzzella, L., and C. H. Onder, Introduction to Modeling and Control of Internal Combustion Engine Systems, Berlin, Heidelberg: Springer-Verlag (2010).

Hughart, C. L., and K. W. Kominek, Designing Distillation Units for Controllability, *Instrum. Technol.* **24** (5), 71 (1977).

Kiwitz, P., C. Onder, and L. Guzzella, Control-Oriented Modeling of a Three-Way Catalytic Converter with Observation of the Relative Oxygen Level Profile, *J. Process Control*, **22**, 984 (2012).

Kreyszig, E., *Advanced Engineering Mathematics*, 10th ed., John Wiley and Sons, Hoboken, NJ, 2011.

Lipták, B. (Ed.), *Instrument Engineers' Handbook*, 4th ed., Section 8.19, Furnace and Reformer Controls, Chilton Pub. Co., Radnor, PA, 2003.

Luyben, W. L. (Ed.), *Practical Distillation Control*, Van Nostrand Reinhold, NY, 1992.

Luyben, W. L., *Plantwide Dynamic Simulators in Chemical Processing and Control*, Marcel Dekker, New York, 2002.

Marquardt, W., Trends in Computer-Aided Modeling, *Comput. Chem. Eng.*, **20**, 591 (1996).

Newell, R. B., and P. L. Lee, *Applied Process Control*, Prentice-Hall of Australia, Brookvale, NSW, Australia, 1989.

Russell, T. F. W., and M. M. Denn, *Introduction to Chemical Engineering Analysis*, John Wiley and Sons, Hoboken, NJ, 1972.

Sandler, S. I., *Chemical and Engineering Thermodynamics*, 4th ed., John Wiley and Sons, Hoboken, NJ, 2006.

Seader, J. D., and E. J. Henley, *Separation Process Principles*, 2nd ed., John Wiley and Sons, Hoboken, NJ, 2005.

Schmidt, L. D., *The Engineering of Chemical Reactions*, 2nd ed., pp. 293–295, Oxford University Press, NY, 2005.

Shinskey, F. G., *Process Control Systems*, 4th ed., Chapter 9, McGraw-Hill, NY, 1996.

Shuler, M. L., and F. Kargi, *Bioprocess Engineering*, 2nd ed., Prentice-Hall, Upper Saddle River, NJ, 2002.

Skogestad, S., Dynamics and Control of Distillation Columns–A Tutorial Introduction, *Trans IChemE, Part A, Chem. Eng. Res. Des.*, **75**, 539–562 (1997).

Skogestad, S., The Dos and Don'ts of Distillation Columns Control, *Trans IChemE, Part A, Chem. Eng. Res. Des.*, **85**(A1), 13 (2007).

Simulation Software: Web Sites

Advanced Continuous Simulation Language (ACSL), Mitchell and Gauthier Associates Inc., www.mga.com.

Aspen Custom Modeler, www.aspentech.com.

CHEMCAD, www.chemstations.com.

Daesim, www.daesim.com.

Global Cape Open, www.global-cape-open.org.

GNU-Octave, www.octave.org.

gPROMS, www.psenterprise.com.

HYSYS, www.aspentech.com.

IMSL, www.vni.com.

LabVIEW, www.ni.com/labview.

Mathcad, www.mathcad.com.

Mathematica, www.wolfram.com.

MATLAB, www.mathworks.com.

Modelica, www.modelica.org.

POLYMATH, www.cache.org/polymath.html.

PRO II, www.simsci.com.

Scilab, www.scilab.org.

Unisim, www.honeywell.com/ps.

EXERCISES

2.1 A perfectly stirred, constant-volume tank has two input streams, both consisting of the same liquid. The temperature and flow rate of each of the streams can vary with time.

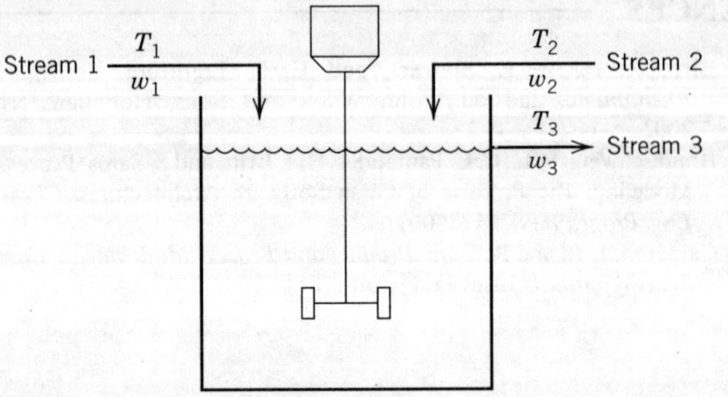

Figure E2.1

(a) Derive a dynamic model that will describe transient operation. Make a degrees of freedom analysis assuming that both Streams 1 and 2 come from upstream units (i.e., their flow rates and temperatures are known functions of time).

(b) Simplify your model, if possible, to one or more differential equations by eliminating any algebraic equations. Also, simplify any derivatives of products of variables.

Notes:

w_i denotes mass flow rate for stream *i*.
Liquid properties are constant (not functions of temperature).

2.2 A completely enclosed stirred-tank heating process is used to heat an incoming stream whose flow rate varies.

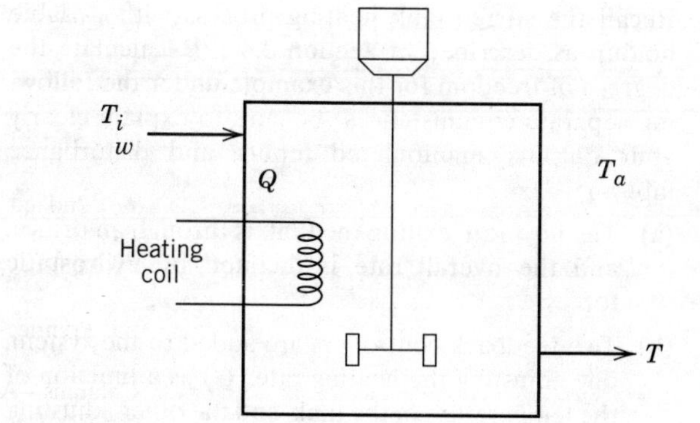

Figure E2.2

The heating rate from this coil and the volume are both constant.

(a) Develop a mathematical model (differential and algebraic equations) that describes the exit temperature if heat losses to the ambient occur and if the ambient temperature (T_a) and the incoming stream's temperature (T_i) both can vary.

(b) Discuss qualitatively what you expect to happen as T_i and w increase (or decrease). Justify by reference to your model.

Notes:

ρ and C_p are constants.
U, the overall heat transfer coefficient, is constant.
A_s is the surface area for heat losses to ambient.
$T_i > T_a$ (inlet temperature is higher than ambient temperature).

2.3 Two tanks are connected together in the following unusual way in Fig. E2.3.

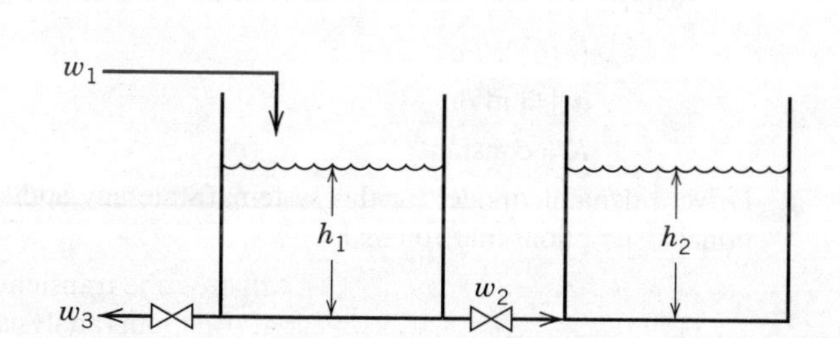

Figure E2.3

(a) Develop a model for this system that can be used to find h_1, h_2, w_2, and w_3 as functions of time for any given variations in inputs.

(b) Perform a degrees of freedom analysis. Identify all input and output variables.

Notes:

The density of the incoming liquid, ρ, is constant.
The cross-sectional areas of the two tanks are A_1 and A_2.
w_2 is positive for flow from Tank 1 to Tank 2.
The two valves are linear with resistances R_2 and R_3.

2.4 Consider a liquid flow system consisting of a sealed tank with noncondensible gas above the liquid as shown in Fig. E2.4. Derive an unsteady-state model relating the liquid level h to the input flow rate q_i. Is operation of this system independent of the ambient pressure P_a? What about for a system open to the atmosphere?

You may make the following assumptions:

(i) The gas obeys the ideal gas law. A constant amount of m_g/M moles of gas is present in the tank.

(ii) The operation is isothermal.

(iii) A square root relation holds for flow through the valve.

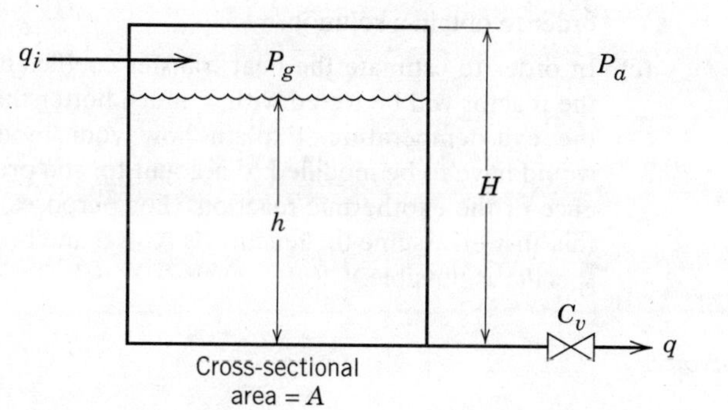

Figure E2.4

2.5 Two surge tanks are used to dampen pressure fluctuations caused by erratic operations of a large air compressor. (See Fig. E2.5.)

(a) If the discharge pressure of the compressor is $P_d(t)$ and the operating pressure of the furnace is P_f (constant), develop a dynamic model for the pressures in the two surge tanks as well as for the air mass flows at points a, b, and c. You may assume that the valve resistances are constant, that the valve flow characteristics are linear, for example, $w_b = (P_1 - P_2)/R_b$, that the surge processes operate isothermally, and that the ideal gas law holds.

(b) How would you modify your model if the surge tanks operated adiabatically? What if the ideal gas law were not a good approximation?

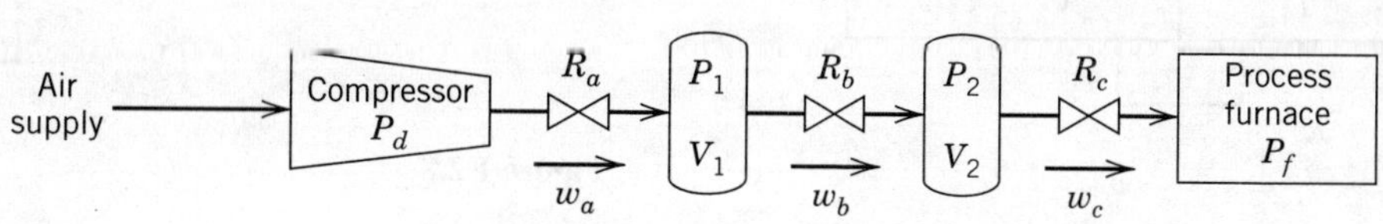

Figure E2.5

2.6 A closed stirred-tank reactor with two compartments is shown in Fig. E2.6. The basic idea is to feed the reactants continuously into the first compartment, where they will be preheated by energy liberated in the exothermic reaction, which is anticipated to occur primarily in the second compartment. The wall separating the two compartments is quite thin, thus allowing heat transfer; the outside of the reactor is well insulated; and a cooling coil is built into the second compartment to remove excess energy liberated in the reaction. Both tanks are completely full at all times.

Tests are to be conducted initially with a single-component feed (i.e., no reaction) to evaluate the reactor's thermal characteristics.

(a) Develop a dynamic model for this process under the conditions of no reaction. Assume that q_0, T_i, and T_c all may vary.

(b) Make a degrees of freedom analysis for your model—identifying all parameters, outputs, and inputs that must be known functions of time in order to obtain a solution.

(c) In order to estimate the heat transfer coefficients, the reactor will be tested with T_i much hotter than the exit temperature. Explain how your model would have to be modified to account for the presence of the exothermic reaction. (For purposes of this answer, assume the reaction is A → B and be as specific as possible.)

Notes:

U_t, A_t: Overall heat transfer coefficient and surface area between compartments.

U_c, A_c: Overall heat transfer coefficient and surface area of cooling tube.

V_1: Volume of Compartment 1.

V_2: Volume of Compartment 2.

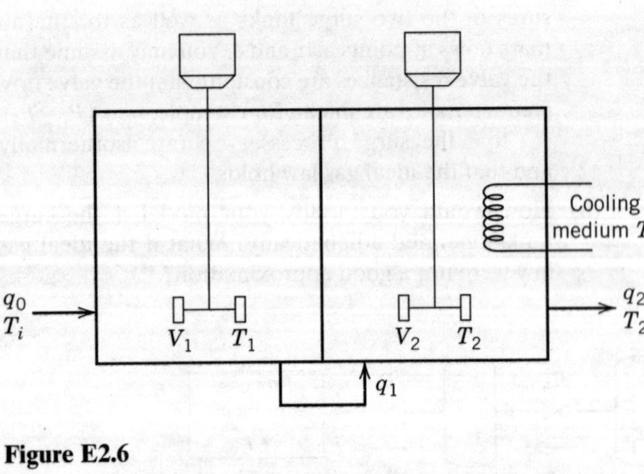

Figure E2.6

2.7 Recall the stirred-tank heating process with variable holdup as described in Section 2.4.2. Recalculate the degrees of freedom for this example under the following separate circumstances (be sure to explain clearly your outputs, manipulated inputs, and disturbance inputs):

(a) The flow rate exiting the tank is through an orifice, and the overall rate is dictated by hydrostatic forces.

(b) Two feedback controllers are added to the system, one adjusting the heating rate (Q) as a function of the temperature of the tank, and the other adjusting the exit flow rate (w) as a function of the volume of the tank. (*Hint:* the exact forms of the controllers are not required to solve this part.)

2.8 A jacketed vessel is used to cool a process stream as shown in Fig. E2.8. The following information is available:

(i) The volume of liquid in the tank V and the volume of coolant in the jacket V_J remain constant. Volumetric flow rate q_F is constant, but q_J varies with time.

(ii) Heat losses from the jacketed vessel are negligible.

(iii) Both the tank contents and the jacket contents are well mixed and have significant thermal capacitances.

(iv) The thermal capacitances of the tank wall and the jacket wall are negligible.

(v) The overall heat transfer coefficient for transfer between the tank liquid and the coolant varies with coolant flow rate:

$$U = Kq_J^{0.8}$$

where

U [=] W/m² °C

q_J [=] m³/h

K = constant

Drive a dynamic model for this system. (State any additional assumptions that you make.)

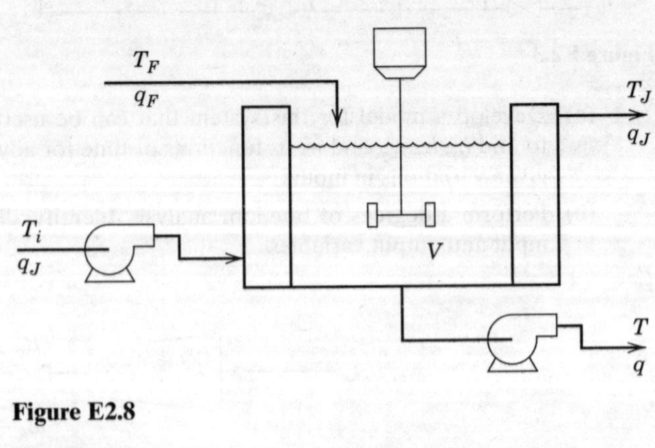

Figure E2.8

2.9 Irreversible consecutive reactions $A \xrightarrow{k_1} B \xrightarrow{k_2} C$ occur in a jacketed, stirred-tank reactor as shown in Fig. E2.9. Derive a dynamic model based on the following assumptions:

(i) The contents of the tank and cooling jacket are well mixed. The volumes of material in the jacket and in the tank do not vary with time.

(ii) The reaction rates are given by

$$r_1 = k_1 e^{-E_1/RT} c_A \,[=]\, \text{mol A / h L}$$
$$r_2 = k_2 e^{-E_2/RT} c_B \,[=]\, \text{mol B / h L}$$

(iii) The thermal capacitances of the tank contents and the jacket contents are significant relative to the thermal capacitances of the jacket and tank walls, which can be neglected.

(iv) Constant physical properties and heat transfer coefficients can be assumed.

Note:

All flow rates are volumetric flow rates in L/h. The concentrations have units of mol/L. The heats of reaction are ΔH_1 and ΔH_2.

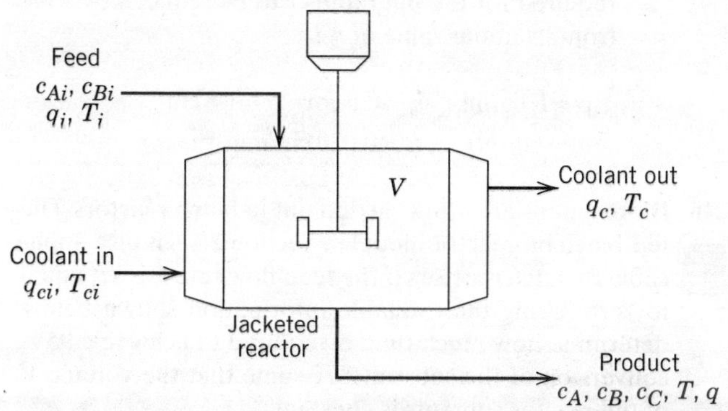

Figure E2.9

2.10 Example 2.1 plots responses for changes in input flows for the stirred tank blending system. Repeat part (b) and plot it. Next, relax the assumption that V is constant, and plot the response of $x(t)$ and $V(t)$ for the change in w_1 for $t = 0$ to 15 minutes. Assume that w_2 and w remain constant.

2.11 A process tank has two input streams—Stream 1 at mass flow rate w_1 and Stream 2 at mass flow rate w_2. The tank's effluent stream, at flow rate w, discharges through a fixed valve to atmospheric pressure. Pressure drop across the valve is proportional to the flow rate squared. The cross-sectional area of the tank, A, is 5 m², and the mass density of all streams is 950 kg/m³.

(a) Draw a schematic diagram of the process and write an appropriate dynamic model for the tank level. What is the corresponding steady-state model?

(b) At initial steady-state conditions, with $w_1 = 2.2$ kg/s and $w_2 = 1.4$ kg/s, the tank level is 2.56 m. What is the value of the valve constant (give units)?

(c) A process control engineer decides to use a feed forward controller to hold the level approximately constant at the set-point value ($h_{sp} = 2.56$ m) by measuring w_1 and manipulating w_2. What is the mathematical relation that will be used in the controller? If the w_1 measurement is not very accurate and always supplies a value that is 1.1 times the actual flow rate, what can you conclude about the resulting level control? (*Hint:* Consider the process initially at the desired steady-state level and with the feedforward controller turned on. Because the controller output is slightly in error, $w_2 \neq 1.4$, so the process will come to a new steady state. What is it?) What conclusions can you draw concerning the need for accuracy in a steady-state model? For the accuracy of the measurement device? For the accuracy of the control valve? Consider all of these with respect to their use in a feedforward control system.

2.12 The liquid storage tank shown in Fig. E2.12 has two inlet streams with mass flow rates w_1 and w_2 and an exit stream with flow rate w_3. The cylindrical tank is 3 m tall and 1.4 m in diameter. The liquid has a density of 900 kg/m³. Normal operating procedure is to fill the tank until the liquid level reaches a nominal value of 1.75 m using constant flow rates: w_1 – 130 kg/min, w_2 – 110 kg/min, and $w_3 = 210$ kg/min. At that point, inlet flow rate w_1 is adjusted so that the level remains constant. However, on this particular day, corrosion of the tank has opened up a hole in the wall at a height of 1 m, producing a leak whose volumetric flow rate q_4 (m³/min) can be approximated by

$$q_4 = 0.025\sqrt{h-1}$$

where h is height in meters.

(a) If the tank was initially empty, how long did it take for the liquid level to reach the corrosion point?

(b) If mass flow rates w_1, w_2, and w_3 are kept constant indefinitely, will the tank eventually overflow? Justify your answer.

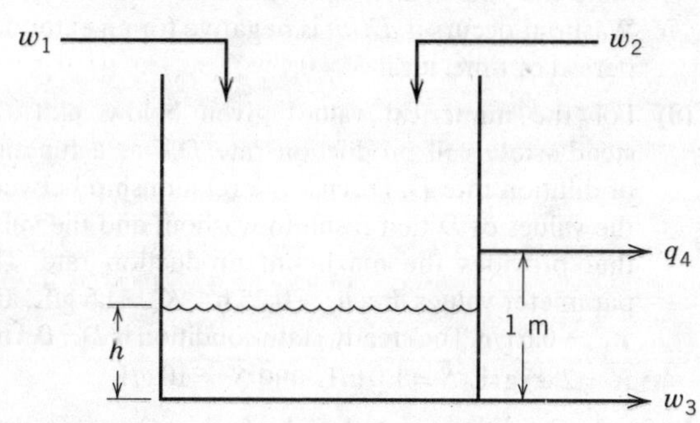

Figure E2.12

2.13 Consider a blending tank that has the same dimensions and nominal flow rates as the storage tank in Exercise 2.12 but that incorporates a valve on the outflow line that is used to establish flow rate w_3. (For this exercise, there is no leak in the tank as in Exercise 2.12.) In addition, the nominal inlet stream mass fractions of component A are $x_1 = x_2 = 0.5$.

The process has been operating for a long time with constant flow rates and inlet concentrations. Under these conditions, it has come to steady state with exit mass fraction $x = 0.5$ and level $h = 1.75$ m. Using the information below, answer the following questions:

(a) What is the value of w_3? The constant, C_v?

(b) If x_1 is suddenly changed from 0.5 to 0.7 without changing the inlet flow rates (of course, x_2 must change as well), what is the final value of x_3? How long does it take to come within 1% of this final value?

(c) If w_1 is changed from 130 kg/min to 110 kg/min without changing the inlet concentrations, what will be the final value of the tank level? How long will it take to come within 1% of this final value?

(d) Would it have made any difference in part (c) if the concentrations had changed at the same time the flow rate was changed?

Useful information: The tank is perfectly stirred.

$$w_3 = C_v\sqrt{h}$$

2.14 Suppose that the fed-batch bioreactor in Fig. 2.14 is converted to a continuous, stirred-tank bioreactor (also called a *chemostat*) by adding an exit stream. Assume that the inlet and exit streams have the same mass flow rate F and thus the volume of liquid V in the chemostat is constant.

(a) Derive a dynamic model for this chemostat by modifying the fed-batch reactor model in Section 2.4.8.

(b) Derive the steady-state relationship between growth rate μ in Eq. 2-80 and *dilution rate D* where by definition, $D = F/V$. Suggest a simple control strategy for controlling the growth rate based on this result.

(c) An undesirable situation called *washout* occurs when all of the cells are washed out of the bioreactor and thus cell mass X becomes zero. Determine the values of D that result in washout. (*Hint:* Washout occurs if dX/dt is negative for an extended period of time, until $X = 0$.)

(d) For the numerical values given below, plot the steady-state cell production rate DX as a function of dilution rate D. Discuss the relationship between the values of D that result in washout and the value that provides the maximum production rate. The parameter values are: $\mu_m = 0.25\ \text{h}^{-1}$; $K_S = 1.5$ g/L, and $Y_{X/S} = 0.6$ g/g. The steady-state condition is $\bar{D} = 0.1\ \text{h}^{-1}$, $\bar{X} = 2.25$ g/L, $\bar{S} = 1.0$ g/L, and $\bar{S}_f = 10$ g/L.

2.15 In medical applications the chief objectives for drug delivery are (i) to deliver the drug to the correct location in the patient's body, and (ii) to obtain a specified drug concentration profile in the body through a controlled release of the drug over time. Drugs are often administered as pills. In order to derive a simple dynamic model of pill dissolution, assume that the rate of dissolution r_d of the pill in a patient is proportional to the product of the pill surface area and the concentration driving force:

$$r_d = kA(c_s - c_{aq})$$

where c_{aq} is the concentration of the dissolved drug in the aqueous medium, c_s is the saturation value, A is the surface area of the pill, and k is the mass transfer coefficient. Because $c_s \gg c_{aq}$, even if the pill dissolves completely, the rate of dissolution reduces to $r_d = kAc_s$.

(a) Derive a dynamic model that can be used to calculate pill mass M as a function of time. You can make the following simplifying assumptions:

(i) The rate of dissolution of the pill is given by $r_d = kAc_s$.

(ii) The pill can be approximated as a cylinder with radius r and height h. It can be assumed that $h/r \gg 1$. Thus the pill surface area can be approximated as $A = 2\pi rh$.

(b) For the conditions given below, how much time is required for the pill radius r to be reduced by 95% from its initial value of r_0?

$$\rho = 1.1\ \text{g/ml} \qquad r_0 = 0.5\ \text{cm} \qquad h = 1.8\ \text{cm}$$
$$c_s = 600\ \text{g/L} \qquad k = 0.020\ \text{cm/min}$$

2.16 Bioreactions are often carried out in batch reactors. The fed-batch bioreactor model in Section 2.4.8 is also applicable to batch reactors if the feed flow rate F is set equal to zero. Using the available information shown below, determine how much time is required to achieve a 95% conversion of the substrate. Assume that the volume V of the reactor contents is constant.

Available information:

(i) *Initial conditions:*

$$X(0) = 0.05\ \text{g/L}, \quad S(0) = 10\ \text{g/L}, \quad P(0) = 0\ \text{g/L}.$$

(ii) *Parameter values:*

$$V = 1\ \text{L}, \quad \mu_m = 0.25\ \text{h}^{-1}, \quad K_S = 1.5\ \text{g/L},$$
$$Y_{X/S} = 0.6\ \text{g/g}, \quad Y_{P/X} = 0.3\ \text{g/g}.$$

2.17 Sketch the level response for a bathtub with cross-sectional area of 0.75 m² as a function of time for the following sequence of events; assume an initial level of 0.15 m with the drain open. The inflow and outflow are initially equal to 0.05 m³/min.

(a) The drain is suddenly closed, and the inflow remains constant for 3 min ($0 \le t \le 3$).

(b) The drain is opened for 15 min; assume a time constant in a linear transfer function of 3 min, so a steady state is essentially reached ($3 \le t \le 18$).

(c) The inflow rate is doubled for 6 min ($18 \le t \le 24$).

(d) The inflow rate is returned to its original value for 16 min ($24 \le t \le 40$).

2.18 Perform a degrees of freedom analysis for the model in Eqs. 2-64 through 2-68. Identify parameters, output variables, and inputs (manipulated and disturbance variables).

2.19 PCM Consider the PCM distillation column module of Appendix C, in which a 50%–50% mixture of methanol (MeOH) and ethanol is separated.

Do the following sequence of simulations:

(a) Change the *Vapor Flow Rate* from the initialized value to a new value of 0.045 m^3/s, and start the simulation. Generate plots for the *Overhead MeOH Composition* and the *Bottom MeOH Composition.* Record the starting values and final (steady-state) values for these variables.

(b) Once the system has reached a new steady state (and you have collected the information requested in part (a)), increase the *Feed MeOH Composition* to 0.55. Generate plots for the *Overhead MeOH Composition* and the *Bottom MeOH Composition.* Record the starting values and final (steady-state) values for these variables.

(c) Comment on the relative differences between these two input changes (strength of effect, timing of response).

2.20 PCM Consider the PCM Furnace module of Appendix C, which is used to preheat a high-molecular-weight hydrocarbon feed (C_{16}–C_{26}) to a cracking unit at a petroleum refinery.

Do the following sequence of simulations:

(a) Change the *Fuel Gas Purity* from the initialized value of 1.0 to a new value of 0.95, and start the simulation. Generate plots for the *Oxygen Exit Concentration* and the *Hydrocarbon Outlet Temperature.* Record the starting values and final (steady-state) values for these variables.

(b) Once the system has reached a new steady state (and you have collected the information requested in part (a)), increase the *Hydrocarbon Flow Rate* by 10% over the current value. Generate plots for the *Oxygen Exit Concentration* and the *Hydrocarbon Outlet Temperature.* Record the starting values and final (steady-state) values for these variables.

(c) Comment on the relative differences between these two input changes (strength of effect, timing of response).

2.21 Plot the level response for a tank with constant cross-sectional area of 0.37 m^2 as a function of time for the following sequence of events; assume an initial level of 0.3 m with the drain open, and that level and outflow rate are linearly related. The steady-state inflow and outflow are initially equal to 55 L/min. The graph should show numerical values of level vs. time.

(a) The drain is suddenly closed, and the inflow remains constant for 3 min ($0 \le t \le 3$).

(b) The drain is opened for 15 min, keeping the inflow at 55 L/min, where a steady state is essentially reached ($3 \le t \le 18$).

(c) The inflow rate is doubled to 110 L/min for 15 min ($18 \le t \le 33$).

(d) The inflow rate is returned to its original value of 55 L/min for 17 min ($33 \le t \le 50$).

2.22 Consider the unusual piping diagram for the four tanks in Fig. E2.22 in which both the flow rates F_1 and F_2 are split between two streams entering the upper and lower tanks (denoted by the fractions in the diagram). For the exit lines leaving the bottom of each of the four tanks through an orifice, you may assume that flow through that orifice obeys the square root dependence on the height of liquid in the tanks, as described in Section 2.4.5.

(a) Derive the mass balances for each of the four tanks, and express them as simple equations (four total) with one derivative term on the left hand side.

(b) For the case of $\gamma_1 = 0.5$ and $\gamma_2 = 0.5$ (i.e., equal splitting of each stream), what is the resulting form of the equations? Can the levels be solved independently? Can the flow rates be used independently to adjust the heights in the tanks?

(c) For the extreme case of $\gamma_1 = 0$ and $\gamma_2 = 0$, what is the resulting form of the equations? Does this make sense in terms of the process schematic?

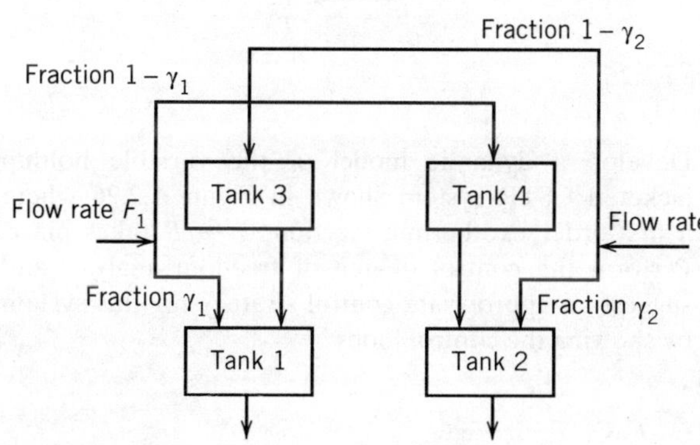

Figure E2.22

2.23 While solving part (c) of Problem 2.13, the differential equation obtained is

$$800\pi \frac{dh}{dt} = 200 - 166.3\sqrt{h} \qquad h(0) = 1.75$$

Plot the curve h vs. t using `ezplot` or `fplot` command of MATLAB and show time (t) when level $h = 1.46$ m.

2.24 The following simultaneous differential equations are obtained while solving for Problem 2.16.

$$\frac{dX}{dt} = 0.2\frac{SX}{1+S}$$

$$\frac{dP}{dt} = (0.2)(0.2)\frac{SX}{1+S}$$

$$\frac{dS}{dt} = 0.2\frac{SX}{1+S}\left[-\frac{1}{0.5} - \frac{0.2}{0.1}\right]$$

Numerically simulate the system using Euler's method. Plot the graph of X, P and S vs. t and show time when conversion of S is 90%. The initial conditions are $X(0) =$ 005 g/L, $P(0)$= 0 g/L, $S(0)$ = 10 g/L.

2.25 Consider the jacketed continuous stirred tank heater (CSTH) shown in Figure E2.25. The liquid gets heated by the steam condensing in the jacket and leaves at a higher temperature. The liquid holdup in the tank (V) is constant. The heat transfer area is A, and the overall heat transfer coefficient is U. Specific heat of the liquid is C_p and the density is ρ. Develop the dynamic model of the system and perform the degrees of freedom analysis. Neglect the thermal capacities of jacket content, jacket wall and the tank wall.

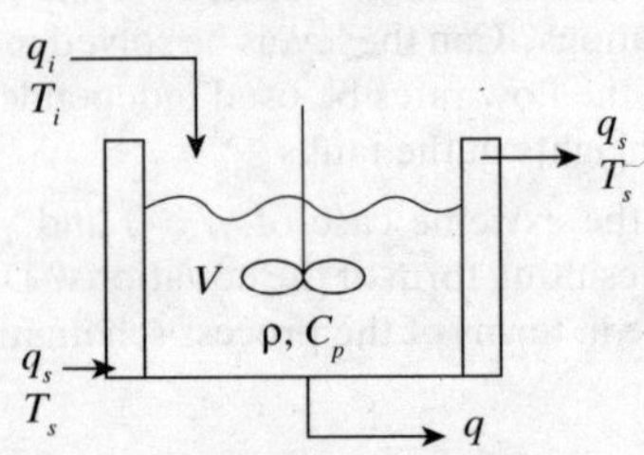

Figure E2.25

2.26 Develop a dynamic model of the variable holdup jacketed CSTR system shown in Figure E2.26 where a first-order exothermic reaction $A \rightarrow B$ takes place. Perform the control degree of freedom analysis and suggest an appropriate control strategy for this system by showing the control loops.

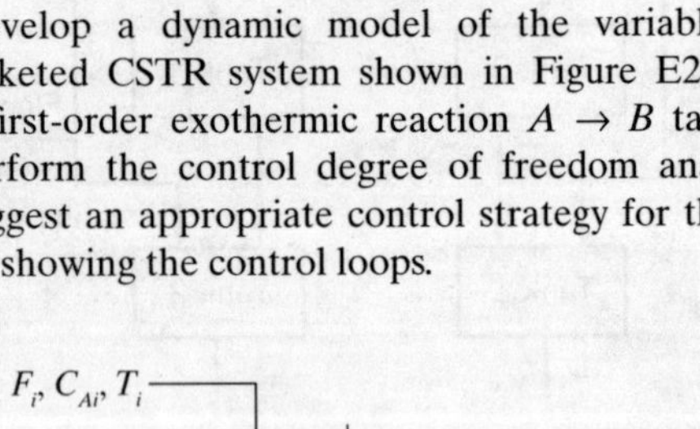

Figure E2.26

2.27 Develop a dynamic model of the stirred tank heating system with variable holdup shown in Fig. E2.27. Perform the control degree of freedom analysis and suggest an appropriate control strategy for this system by showing the control loops.

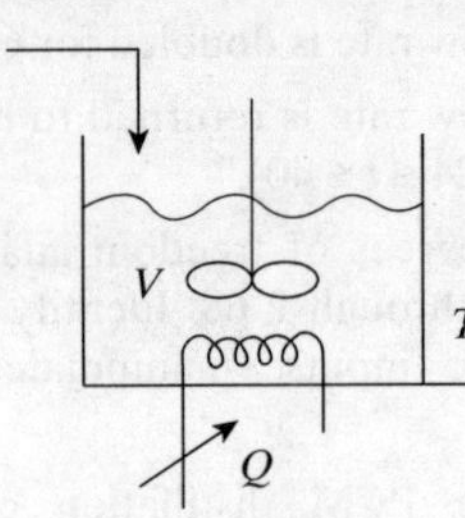

Figure E2.27

2.28 A binary distillation column (see Fig. E2.28) having 20 trays, a total condenser and a reboiler is used to separate a binary mixture of components A and B as top and bottom products, respectively.

(a) Make a dynamic modeling of the process.

(b) Perform the control degrees of freedom analysis and show the control loops.

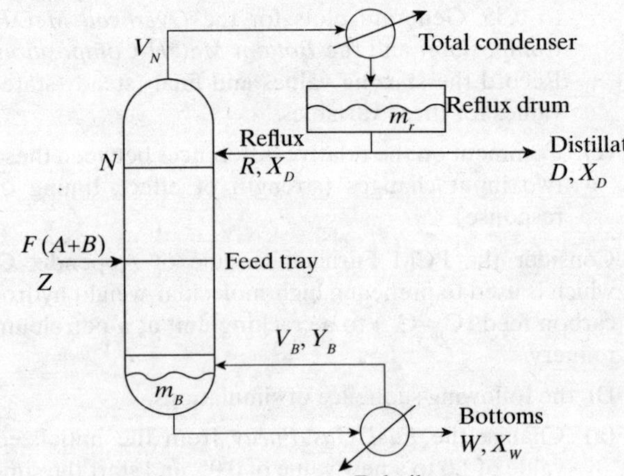

Figure E2.28

2.29 A stirred-tank blending system with a bypass stream is shown in Fig. E2.29. The control objective is to control the composition of a key component in the exit stream, x_4. The chief disturbance variables are the mass fractions of the key component in the inlet streams, x_1 and x_2. Using the following information, discuss which flow rate should be selected as the manipulated variable: (i) inlet flow rate w_2, (ii) the bypass fraction f, or (iii) exit flow rate, w_4. Your choice should reflect both steady-state and dynamic considerations.

Available Information:

(a) The tank is perfectly mixed.

(b) Constant physical properties can be assumed because the composition changes are quite small.

(c) Because the variations in liquid level are small, h does not have to be controlled.

(d) The bypass piping results in a negligible time delay.

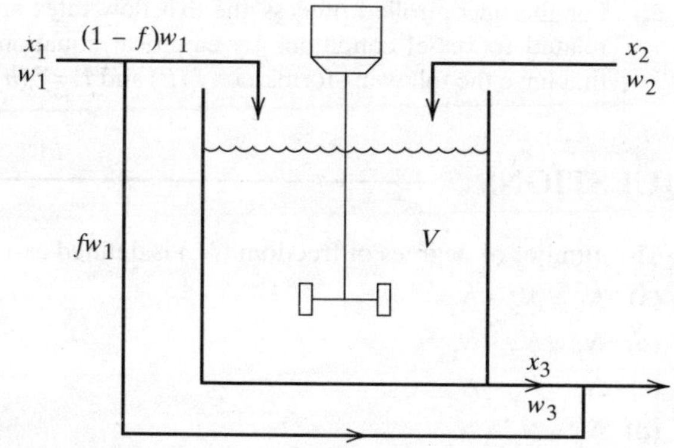

Figure E2.29

2.30 Consider the liquid storage system shown in Fig. E2.30. Only volumetric flow rates, q_1 and q_2, can be manipulated. The volumetric flow rate between the two tanks q_4 is related to the tank liquid levels by a relationship of the form, $q_4 = f(h_1, h_2)$. Determine the model degrees of freedom, N_F, and the control degrees of freedom, N_{FC}.

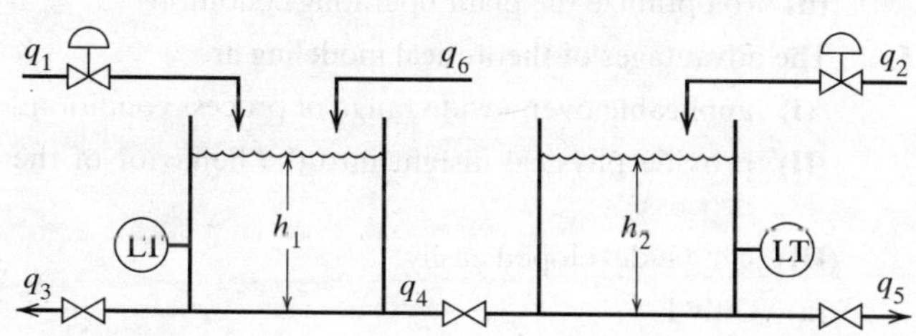

Figure E2.30

2.31 A double pipe heat exchanger with a partial bypass for the cold stream is shown in Fig. E2.31. The mass flow rate of the hot stream, w_h, and the bypass fraction, f, can be manipulated. Heat losses can be neglected.

(a) Determine the model degrees of freedom, N_F, and control degrees of freedom, N_{FC}, based on a steady-state analysis.

(b) Determine the number of disturbance variables, N_D, and specify reasonable choices for the disturbance variables.

(c) Would N_F or N_{FC} change if a cocurrent heat exchanger configuration is analyzed instead of the countercurrent configuration? Justify your answer.

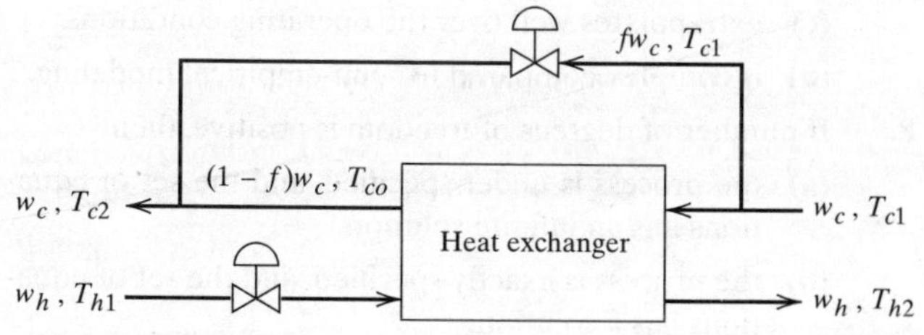

Figure E2.31

2.32 Consider the blending system of Exercise 2.29. Inlet flow rate, w_2, and the bypass fraction, f, can be manipulated. Determine the model degrees of freedom, N_F, and the control degrees of freedom, N_{FC}.

2.33 A stirred-tank heating system is shown in Fig. E2.33. Briefly critique these two control strategies.

(a) It is proposed that h and T be controlled by manipulating w_h and w_c using two PI controllers.

(b) Suppose that two PI controllers are to be used, with h controlled by manipulating w_h and T controlled by manipulating w.

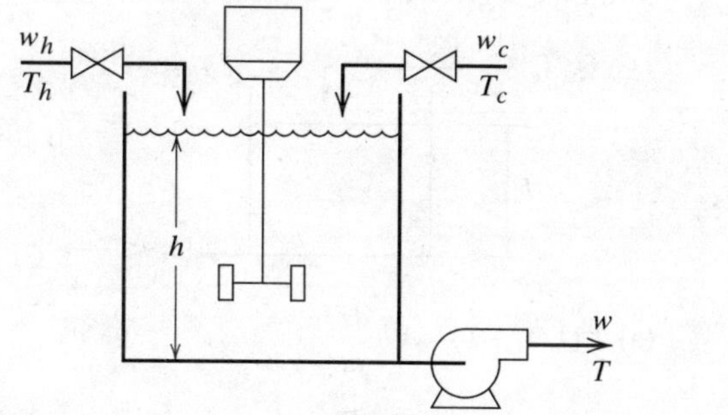

Figure E2.33

2.34 A two-phase feed to the gas-liquid separator (or flash drum), shown in Fig. E2.34, consists of a mixture of two hydrocarbons. Because the vessel pressure P is lower than the feed pressure, the feed flashes as it enters the separator. Using the following information and a degrees of freedom analysis, do the following:

(a) Determine the model degrees of freedom, N_F.

(b) Determine the control degrees of freedom, N_{FC}.

(c) Specify reasonable CVs and MVs. Justify your answers.

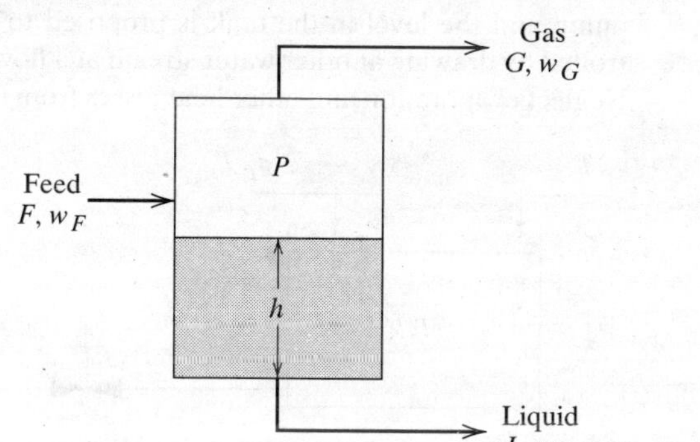

Figure E2.34

Available Information:

1. The flash drum operates isothermally with the two phases in equilibrium.
2. Each phase is perfectly mixed.
3. The mass flow rates are in units of kg/min and the compositions (e.g., w_F) are expressed as mass fractions.

4. Each flow rate can be adjusted by a control valve (not shown).

5. For the uncontrolled process, the exit flow rates are related to vessel conditions by empirical equations that have the following forms: $G = f_1(P)$ and $L = f_2(h)$.

MULTIPLE CHOICE QUESTIONS

1. Water is entering a storage tank at a temperature T_0 and flow rate Q_0 and leaving at a flow rate Q and temperature T. There are negligible heat losses in the tank. The area of cross-section of the tank is A_c. The model that describes the dynamic variation of water temperature in the tank with time is given as:

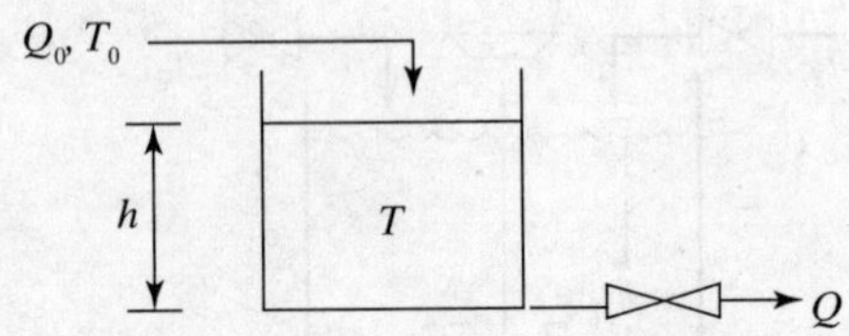

(a) $Q_0(T_0 - T) = A_c h \dfrac{dT}{dt}$

(b) $Q_0 T_0 - QT = A_c h \dfrac{dT}{dt}$

(c) $Q(T_0 - T) = A_c h \dfrac{dT}{dt}$

(d) $Q(T_0 - T) = A_c h \dfrac{d(Th)}{dt}$ **(GATE 2003)**

2. The dynamic model for a mixing tank open to atmosphere at its top as shown below is to be written. The objective of mixing is to cool the hot water stream entering the tank at a flow rate q_2 and feed temperature of T_s with a cold water feed stream entering the tank at a flow rate q_1 and feed temperature of T_0. A water stream is drawn from the tank bottom at a flow rate of q_4 by a pump and the level in the tank is proposed to be controlled by drawing another water stream at a flow rate q_3. Neglect evaporation and other heat losses from the tank.

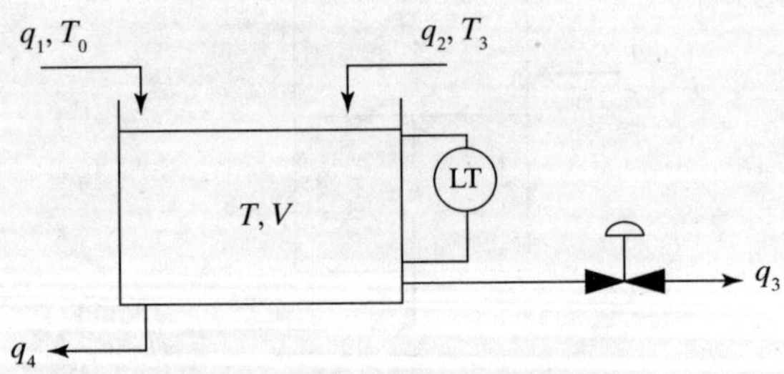

The dynamic model for the tank is given as

(a) $\dfrac{dV}{dt} = q_1 + q_2 - q_3$, $\quad V\dfrac{dT}{dt} = q_1 T_0 + q_2 T_s - q_3 T$

(b) $\dfrac{dV}{dt} = q_1 - q_4$, $\quad \dfrac{d(VT)}{dt} = q_1 T_s - q_4 T$

(c) $\dfrac{dV}{dt} = q_1 + q_2 - q_4$, $\quad \dfrac{d(VT)}{dt} = q_1 T_0 + q_2 T_s - q_4 T$

(d) $\dfrac{dV}{dt} = q_1 + q_2 - q_3 - q_4$, $\quad V\dfrac{dT}{dt} = q_1(T_0 - T) + q_2(T_s - T)$

(GATE 2007)

3. The number of degrees of freedom (N_F) is defined as

(a) $N_F = N_V + N_E$

(b) $N_F = N_E - N_V$

(c) $N_F = N_V - N_E$

(d) $N_F = N_E + N_V$

where, N_V = Number of variables and N_E = Number of equations

4. Which of the following statements is not correct regarding the need for a mathematical model of the system?

(a) To understand the process and to train the plant operators

(b) To perform fault-tree analysis of the system

(c) To develop the control system for the process

(d) To optimize the plant operating conditions

5. The advantages of theoretical modeling are

(I) applicable over a wide range of process conditions.

(II) provide physical insight into the behavior of the process.

(III) can be developed easily.

(a) Only I

(b) Both I and II

(c) Only II

(d) Both I nor III

6. The limitations of the theoretical modeling are

(I) time consuming to develop and expensive.

(II) not applicable for wider range of process conditions.

(III) models for complex process typically include some model parameters that are not readily available.

(a) Only I

(b) Only II

(c) Both I nor II

(d) Both I and III

7. Advantage of empirical modeling is that it

(a) provides physical insight into the behavior of the process.

(b) is easier to develop than theoretical modeling.

(c) extrapolates well over the operating conditions.

(d) is complex compared to semi-empirical modeling.

8. If number of degrees of freedom is positive, then

(a) the process is under-specified, and the set of equations has an infinite solution.

(b) the process is exactly specified, and the set of equations has a solution.

(c) the process is over specified and the set of equations has an infinite solution.

(d) nothing can be said about the process.

9. If number of degrees of freedom is negative, then

(a) the process is under-specified, and the set of equations has an infinite solution.

(b) the process is exactly specified, and the set of equations has a unique solution.

(c) the process is over specified, and the set of equations has no solution.

(d) nothing can be said about the process.

10. If number of degrees of freedom is zero, then

(a) the process is under-specified, and the set of equations has an infinite solution.

(b) the process is exactly specified, and the set of equations has a unique solution.

(c) the process is over specified, and the set of equations has no solution.

(d) nothing can be said about the process.

ANSWER KEY

1. (b)	**2.** (d)	**3.** (c)	**4.** (b)	**5.** (b)	**6.** (d)
7. (b)	**8.** (a)	**9.** (c)	**10.** (b)		

CHAPTER 3

Laplace Transforms

CHAPTER CONTENTS

In Chapter 2 we developed a number of mathematical models that describe the dynamic operation of selected processes. Solving such models—that is, finding the output variables as functions of time for some change in the input variable(s)—requires either analytical or numerical integration of the differential equations. Sometimes considerable effort is involved in obtaining the solutions. One important class of models includes systems described by linear ordinary differential equations (ODEs). Such *linear systems* represent the starting point for many analysis techniques in process control.

Both linear and nonlinear ODE problems can be solved numerically by using standard numerical integration methods (Chapra and Canale, 2014) and computer software. For linear ODEs, it is also possible to generate analytical (or "closed-form") solutions. The important advantage of an analytical solution is that the effects of changing the model structure or the model parameters are readily apparent. By contrast, for a numerical solution, this information can only be obtained by recalculating the solution for each situation of interest. As indicated in Appendix B, the analytical solutions to ODEs can also be generated using software for symbolic problems.

In this chapter we introduce a mathematical tool, the *Laplace transform*, which can significantly reduce the effort required to solve and analyze linear differential equation models. A major benefit is that this transformation converts ordinary differential equations to algebraic equations, which can simplify the mathematical manipulations required to obtain a solution or perform an analysis.

First, we define the Laplace transform and show how it can be used to derive the Laplace transforms of simple functions. Then we show that linear ODEs can be solved using Laplace transforms, and a technique called *partial fraction expansion*. Some important general properties of Laplace transforms are presented, and we illustrate the use of these techniques with a series of examples. Software solutions for Laplace transform problems are also demonstrated.

3.1 LAPLACE TRANSFORMS OF REPRESENTATIVE FUNCTIONS

The Laplace transform of a function $f(t)$ is defined as

$$F(s) = \mathcal{L}[f(t)] = \int_0^\infty f(t)e^{-st}dt \tag{3-1}$$

where $F(s)$ is the symbol for the Laplace transform, s is a complex independent variable, $f(t)$ is some function of time to be transformed, and $\mathcal{L}$ is an operator, defined by the integral. The function $f(t)$ must satisfy mild conditions that include being piecewise continuous for $0 < t < \infty$ (Churchill, 1971; Schiff, 1999); this requirement almost always holds for functions that are useful in process modeling and control. When the integration is performed, the transform becomes a function of the Laplace transform variable s. The *inverse Laplace transform* ($\mathcal{L}^{-1}$) operates on the function $F(s)$ and converts it to $f(t)$. Notice that $F(s)$ contains no information about $f(t)$ for $t < 0$. Hence, $f(t) = \mathcal{L}^{-1}\{F(s)\}$ is not defined for $t < 0$ (Schiff, 1999).

One of the important properties of the Laplace transform and the inverse Laplace transform is that they are linear operators; a linear operator satisfies the *Principle of Superposition*:

$$\mathcal{F}(ax(t)+by(t)) = a\,\mathcal{F}(x(t)) + b\,\mathcal{F}(y(t)) \tag{3-2}$$

where $\mathcal{F}$ denotes a particular operation to be performed, such as differentiation or integration with respect to time. If $\mathcal{F} \equiv \mathcal{L}$, then Eq. 3-2 becomes

$$\mathcal{L}(ax(t)+by(t)) = aX(s)+bY(s) \tag{3-3}$$

Therefore, the Laplace transform of a sum of functions $x(t)$ and $y(t)$ is the sum of the individual Laplace transforms, $X(s)$ and $Y(s)$; in addition, multiplicative constants can be factored out of the operator, as shown in Eq. 3-3.

In this book we are more concerned with operational aspects of Laplace transforms—that is, using them to obtain solutions or the properties of solutions of *linear differential equations*. For more details on mathematical aspects of the Laplace transform, the texts by Churchill (1971), Dyke (1999), and Schiff (1999) are recommended.

Before we consider solution techniques, the application of Eq. 3-1 should be discussed. The Laplace transform can be derived easily for most simple functions, as shown below.

Constant Function. For $f(t) = a$ (a constant),

$$\mathcal{L}(a) = \int_0^\infty ae^{-st}\,dt = -\frac{a}{s}e^{-st}\Big|_0^\infty$$

$$= 0 - \left(-\frac{a}{s}\right) = \frac{a}{s} \tag{3-4}$$

Step Function. The unit step function, defined as

$$S(t) = \begin{cases} 0 & t < 0 \\ 1 & t \geq 0 \end{cases} \tag{3-5}$$

is an important input that is used frequently in process dynamics and control. The Laplace transform of the unit step function is the same as that obtained for the constant above when $a = 1$:

$$\mathcal{L}[S(t)] = \frac{1}{s} \tag{3-6}$$

If the step magnitude is a, the Laplace transform is a/s. The step function incorporates the idea of *initial time*, *zero time*, or *time zero* for the function, which refers to the time at which $S(t)$ changes from 0 to 1. To avoid any ambiguity concerning the value of the step function at $t = 0$ (it is discontinuous), we will consider $S(t = 0)$ to be the value of the function approached from the positive side, $t = 0^+$.

Shifted Unit Step Function. When the (unit) step function is shifted by θ time units, it can be expressed as

$$S(t-\theta) = \begin{cases} 0 & t < \theta \\ 1 & t \geq \theta \end{cases} \tag{3-7}$$

The Laplace transform of the shifted (or delayed) unit step function is given by

$$\mathcal{L}[S(t-\theta)] = \int_0^\infty S(t-\theta)\,e^{-st}\,dt$$

$$= \int_0^\theta S(t-\theta)\,e^{-st}\,dt + \int_\theta^\infty S(t-\theta)\,e^{-st}\,dt$$

$$= \int_0^\theta 0.e^{-st}\,dt + \int_\theta^\infty 1.e^{-st}\,dt = 0 + \frac{e^{-st}}{-s}\Big|_\theta^\infty$$

$$= \frac{e^{-\theta s}}{s} \tag{3-8}$$

The unit step function and shifted unit step function are shown in Fig. 3.1.

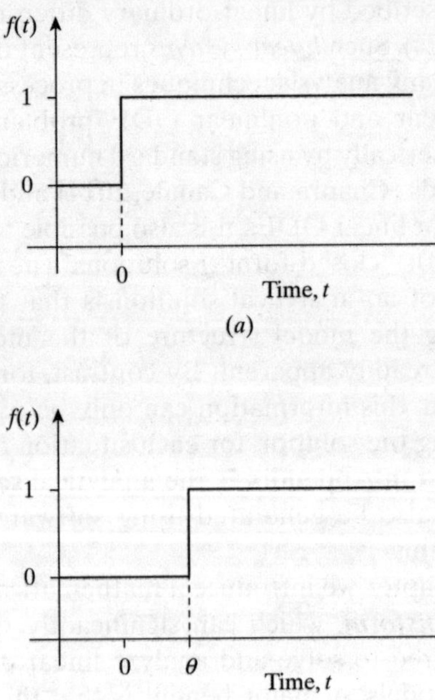

Figure 3.1 (*a*) Unit step function, $f(t) = S(t)$; (*b*) Shifted unit step function, $f(t) = S(t-\theta)$.

Derivatives. The transform of a first derivative of f is important because such derivatives appear in dynamic models:

$$\mathcal{L}(df/dt) = \int_0^\infty (df/dt)e^{-st}dt \tag{3-9}$$

Integrating by parts,

$$\mathcal{L}(df/dt) = \int_0^\infty f(t)e^{-st}s\,dt + f(t)e^{-st}\Big|_0^\infty \tag{3-10}$$

$$= s\mathcal{L}(f(t)) - f(0) = sF(s) - f(0) \tag{3-11}$$

where $F(s)$ is the Laplace transform of $f(t)$. Generally, the point at which we start keeping time for a solution is arbitrary. Model solutions are most easily obtained assuming that time *starts* (i.e., $t = 0$) at the moment the process model is first perturbed. For example, if the process initially is assumed to be at steady state and an input undergoes a unit step change, *zero time* is taken to be the moment at which the input changes in magnitude. In many process modeling applications, functions are defined so that they are zero at initial time—that is, $f(0) = 0$. In these cases, Eq. 3-11 simplifies to $\mathcal{L}(df/dt) = sF(s)$.

The Laplace transform for higher-order derivatives can be found by repeated application of Eq. 3-11:

$$\mathcal{L}\left(\frac{d^n f}{dt^n}\right) = s^n F(s) - s^{n-1}f(0) - s^{n-2}f^{(1)}(0) - \cdots$$
$$- sf^{(n-2)}(0) - f^{(n-1)}(0) \tag{3-12}$$

where $f^{(i)}(0)$ is the ith derivative evaluated at $t = 0$.

Exponential Functions. The Laplace transform of an exponential function is important because exponential functions appear in the solution to most linear differential equations. For an exponential, e^{-bt}, with $b > 0$,

$$\mathcal{L}(e^{-bt}) = \int_0^\infty e^{-bt}e^{-st}dt = \int_0^\infty e^{-(b+s)t}dt \tag{3-13}$$

$$= \frac{1}{b+s}[-e^{-(b+s)t}]\Big|_0^\infty = \frac{1}{s+b} \tag{3-14}$$

The Laplace transform for $b < 0$ is unbounded if $s < b$; therefore, the real part of s must be restricted to be larger than $-b$ for the integral to be finite. This condition is satisfied for all problems we consider in this book.

Table 3.1 lists some important Laplace transform pairs that occur in the solution of linear differential equations. For a more extensive list of transforms, see Dyke (1999).

Table 3.1 Laplace Transforms for Various Time-Domain Functions*

$f(t)$	$F(s)$
1. $\delta(t)$ (unit impulse)	1
2. $S(t)$ (unit step)	$\frac{1}{s}$
3. t (ramp)	$\frac{1}{s^2}$
4. t^{n-1}	$\frac{(n-1)!}{s^n}$
5. e^{-bt}	$\frac{1}{s+b}$
6. $\frac{1}{\tau}e^{-t/\tau}$	$\frac{1}{\tau s+1}$
7. $\frac{t^{n-1}e^{-bt}}{(n-1)!}(n>0)$	$\frac{1}{(s+b)^n}$
8. $\frac{1}{\tau^n(n-1)!}t^{n-1}e^{-t/\tau}$	$\frac{1}{(\tau s+1)^n}$
9. $\frac{1}{b_1-b_2}(e^{-b_2t}-e^{-b_1t})$	$\frac{1}{(s+b_1)(s+b_2)}$

(*Continued*)

Table 3.1 (Continued)

$f(t)$	$F(s)$
10. $\dfrac{1}{\tau_1-\tau_2}(e^{-t/\tau_1}-e^{-t/\tau_2})$	$\dfrac{1}{(\tau_1 s+1)(\tau_2 s+1)}$
11. $\dfrac{b_3-b_1}{b_2-b_1}e^{-b_1 t}+\dfrac{b_3-b_2}{b_1-b_2}e^{-b_2 t}$	$\dfrac{s+b_3}{(s+b_1)(s+b_2)}$
12. $\dfrac{1}{\tau_1}\dfrac{\tau_1-\tau_3}{\tau_1-\tau_2}e^{-t/\tau_1}+\dfrac{1}{\tau_2}\dfrac{\tau_2-\tau_3}{\tau_2-\tau_1}e^{-t/\tau_2}$	$\dfrac{\tau_3 s+1}{(\tau_1 s+1)(\tau_2 s+1)}$
13. $1-e^{-t/\tau}$	$\dfrac{1}{s(\tau s+1)}$
14. $\sin \omega t$	$\dfrac{\omega}{s^2+\omega^2}$
15. $\cos \omega t$	$\dfrac{s}{s^2+\omega^2}$
16. $\sin(\omega t+\phi)$	$\dfrac{\omega\cos\phi+s\sin\phi}{s^2+\omega^2}$
17. $e^{-bt}\sin \omega t$ (b, ω real)	$\dfrac{\omega}{(s+b)^2+\omega^2}$
18. $e^{-bt}\sin \omega t$ (b, ω real)	$\dfrac{s+b}{(s+b)^2+\omega^2}$
19. $\dfrac{1}{\tau\sqrt{1-\zeta^2}}e^{-\zeta t/\tau}\sin(\sqrt{1-\zeta^2}\,t/\tau)$ $(0\le\lvert\zeta\rvert<1)$	$\dfrac{1}{\tau^2 s^2+2\zeta\tau s+1}$
20. $1+\dfrac{1}{\tau_2-\tau_1}(\tau_1 e^{-t/\tau_1}-\tau_2 e^{-t/\tau_2})$ $(\tau_1\ne\tau_2)$	$\dfrac{1}{s(\tau_1 s+1)(\tau_2 s+1)}$
21. $1-\dfrac{1}{\sqrt{1-\zeta^2}}e^{-\zeta t/\tau}\sin\left[\sqrt{1-\zeta^2}\,t/\tau+\psi\right]$ $\psi=\tan^{-1}\dfrac{\sqrt{1-\zeta^2}}{\zeta},\ (0\le\lvert\zeta\rvert<1)$	$\dfrac{1}{s(\tau^2 s^2+2\zeta\tau s+1)}$
22. $1-e^{-\zeta t/\tau}\left[\cos(\sqrt{1-\zeta^2}\,t/\tau)+\dfrac{\zeta}{\sqrt{1-\zeta^2}}\sin(\sqrt{1-\zeta^2}\,t/\tau)\right]$ $(0\le\lvert\zeta\rvert<1)$	$\dfrac{1}{s(\tau^2 s^2+2\zeta\tau s+1)}$
23. $1+\dfrac{\tau_3-\tau_1}{\tau_1-\tau_2}e^{-t/\tau_1}+\dfrac{\tau_3-\tau_2}{\tau_2-\tau_1}e^{-t/\tau_2}$ $(\tau_1\ne\tau_2)$	$\dfrac{\tau_3 s+1}{s(\tau_1 s+1)(\tau_2 s+1)}$
24. $\dfrac{df}{dt}$	$sF(s)-f(0)$

Table 3.1 (Continued)

$f(t)$	$F(s)$
25. $\dfrac{d^n f}{dt^n}$	$s^nF(s) - s^{n-1}f(0) - s^{n-2}f^{(1)}(0) - \cdots - sf^{(n-2)}(0) - f^{(n-1)}(0)$
26. $f(t-\theta)S(t-\theta)$	$e^{-\theta s}F(s)$
27. $\int_0^t f(\tau)d\tau$	$\dfrac{1}{s}F(s)$

*Note that $f(t)$ and $F(s)$ are defined for $t \geq 0$ only.

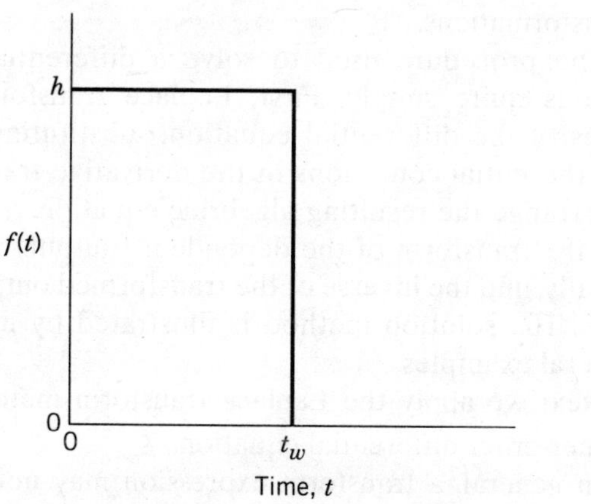

Figure 3.2 The rectangular pulse function.

Note that in most of the transform cases derived above, $F(s)$ is a ratio of polynomials in s, that is, a *rational form.* There are some important cases when nonpolynomial (irrational) forms occur, such as item 26 in Table 3.1. Another case is discussed next.

The Rectangular Pulse Function. An illustration of the rectangular pulse is shown in Fig. 3.2. The pulse has height h and width t_w. This type of signal might be used to depict the opening and closing of a valve regulating flow into a tank. The flow rate would be held at h for a duration of t_w units of time. The area under the curve in Fig. 3.2 could be interpreted as the amount of material delivered to the tank ($= ht_w$). Mathematically, the function $f(t)$ is defined as

$$f(t) = \begin{cases} 0 & t < 0 \\ h & 0 \leq t < t_w \\ 0 & t \geq t_w \end{cases} \tag{3-15}$$

The Laplace transform of the rectangular pulse can be derived by evaluating the integral (3-1) between $t = 0$ and $t = t_w$ because $f(t)$ is zero everywhere else:

$$F(s) = \int_0^\infty f(t)e^{-st}dt = \int_0^{t_w} he^{-st}dt \tag{3-16}$$

$$F(s) = -\frac{h}{s}e^{-st}\Big|_0^{t_w} = \frac{h}{s}(1-e^{-t_w s}) \tag{3-17}$$

Note that an exponential term in $F(s)$ results. For a *unit rectangular pulse*, $h = 1/t_w$ and the area under the pulse is unity.

Impulse Function. A limiting case of the unit rectangular pulse is the *impulse* or *Dirac delta function*, which has the symbol $\delta(t)$. This function is obtained when $t_w \to 0$ while keeping the area under the pulse equal to unity. A pulse of infinite height and infinitesimal width results. Mathematically, this can be accomplished by substituting $h = 1/t_w$ into Eq. 3-17; the Laplace transform of $\delta(t)$ is

$$\mathcal{L}(\delta(t)) = \lim_{t_w \to 0} \frac{1}{t_w s}(1-e^{-t_w s}) \tag{3-18}$$

Equation 3-18 is an indeterminate form that can be evaluated by application of L'Hospital's rule (also spelled L'Hôpital), which involves taking derivatives of both numerator and denominator with respect to t_w:

$$\mathcal{L}(\delta(t)) = \lim_{t_w \to 0} \frac{se^{-t_w s}}{s} = 1 \tag{3-19}$$

If the impulse magnitude (i.e., area $t_w h$) is a constant a rather than unity, then

$$\mathcal{L}(a\delta(t)) = a \tag{3-20}$$

The unit impulse function may also be interpreted as the time derivative of the unit step function $S(t)$. The response of a process to a unit impulse is called its *impulse response*.

A physical example of an impulse function is the rapid injection of dye or tracer into a fluid stream, where $f(t)$ corresponds to the concentration or the flow rate of the tracer. This type of signal is sometimes used in process testing, for example, to obtain the residence time distribution of a piece of equipment, as illustrated in Section 3.5.

Composite Functions. Many composite functions can be obtained from different combinations of step function, ramp function and their shifted versions. The composite function (car function) shown in Fig. 3.3 can be expressed as the combination of ramp and shifted ramp functions. It is given by

$$f(t) = 0.5t - 0.5(t-2)S(t-2) - 0.5(t-4)S(t-4) + 0.5(t-6)S(t-6)$$

The following Laplace transform can be obtained for this composite (car) function.

$$F(s) = \frac{1}{2s^2} - \frac{e^{-2s}}{2s^2} - \frac{e^{-4s}}{2s^2} + \frac{e^{-6s}}{2s^2}$$

$$f(t) = 0.5t - 0.5(t-2)S(t-2) - 0.5(t-4)S(t-4) + 0.5(t-6)S(t-6)$$

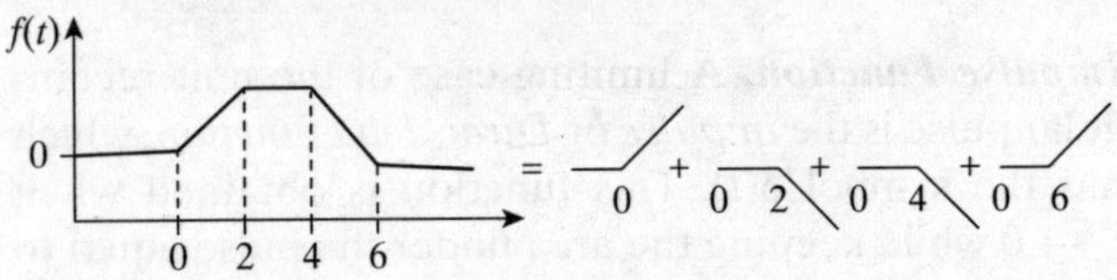

Figure 3.3 Composite function.

3.2 SOLUTION OF DIFFERENTIAL EQUATIONS BY LAPLACE TRANSFORM TECHNIQUES

In the previous section we developed the techniques required to obtain the Laplace transform of each term in a linear ordinary differential equation. Table 3.1 lists important functions of time, including derivatives, and their Laplace transform equivalents. Because the Laplace transform converts any function $f(t)$ to $F(s)$ and the inverse Laplace transform converts $F(s)$ back to $f(t)$, the table provides an organized way to carry out these transformations.

The procedure used to solve a differential equation is quite simple. First, Laplace transform both sides of the differential equation, substituting values for the initial conditions in the derivative transforms. Rearrange the resulting algebraic equation, and solve for the transform of the dependent (output) variable. Finally, find the inverse of the transformed output variable. The solution method is illustrated by means of several examples.

Next we apply the Laplace transform method to a higher-order differential equation.

In general, a transform expression may not exactly match any of the entries in Table 3.1. This problem always arises for higher-order differential equations,

EXAMPLE 3.1

Solve the differential equation,

$$5\frac{dy}{dt} + 4y = 2 \quad y(0) = 1 \tag{3-21}$$

using Laplace transforms.

Solution

First, take the Laplace transform of both sides of Eq. 3-21:

$$\mathcal{L}\left(5\frac{dy}{dt} + 4y\right) = \mathcal{L}(2) \tag{3-22}$$

Using the Principle of Superposition, each term can be transformed individually:

$$\mathcal{L}\left(5\frac{dy}{dt}\right) + \mathcal{L}(4y) = \mathcal{L}(2) \tag{3-23}$$

$$\mathcal{L}\left(5\frac{dy}{dt}\right) = 5\mathcal{L}\left(\frac{dy}{dt}\right) = 5(sY(s) - 1) = 5sY(s) - 5 \tag{3-24}$$

$$\mathcal{L}(4y) = 4\mathcal{L}(y) = 4Y(s) \tag{3-25}$$

$$\mathcal{L}(2) = \frac{2}{s} \tag{3-26}$$

Substitute the individual terms:

$$5sY(s) - 5 + 4Y(s) = \frac{2}{s} \tag{3-27}$$

Rearrange (3-27) and factor out $Y(s)$:

$$Y(s)(5s+4) = 5 + \frac{2}{s} \tag{3-28}$$

or

$$Y(s) = \frac{5s+2}{s(5s+4)} \tag{3-29}$$

Take the inverse Laplace transform of both sides of Eq. 3-29:

$$\mathcal{L}^{-1}[Y(s)] = \mathcal{L}^{-1}\left[\frac{5s+2}{s(5s+4)}\right] \tag{3-30}$$

The inverse Laplace transform of the right side of Eq. 3-30 can be found by using Table 3.1. First, divide the numerator and denominator by 5 to put all factors in the $s + b$ form corresponding to the table entries:

$$y(t) = \mathcal{L}^{-1}\left(\frac{s+0.4}{s(s+0.8)}\right) \tag{3-31}$$

Because entry 11 in the table, $(s+b_3)/[(s+b_1)(s+b_2)]$, matches Eq. 3-31 with $b_1 = 0.8, b_2 = 0$, and $b_3 = 0.4$, the solution can be written immediately:

$$y(t) = 0.5 + 0.5e^{-0.8t} \tag{3-32}$$

Note that in solving (3-21) both the forcing function (the constant 2 on the right side) and the initial condition have been incorporated easily and directly. As for any differential equation solution, Eq. 3-32 should be checked to make sure it satisfies the initial condition and the original differential equation for $t \geq 0$.

EXAMPLE 3.2

Solve the ordinary differential equation

$$\frac{d^3y}{dt^3} + 6\frac{d^2y}{dt^2} + 11\frac{dy}{dt} + 6y = 1 \tag{3-33}$$

with initial conditions $y(0) = y'(0) = y''(0) = 0$ where the primes denote derivatives.

Solution

Take Laplace transforms, term by term, using Table 3.1:

$$\mathcal{L}\left(\frac{d^3y}{dt^3}\right) = s^3Y(s)$$

$$\mathcal{L}\left(6\frac{d^2y}{dt^2}\right) = 6s^2Y(s)$$

$$\mathcal{L}\left(11\frac{dy}{dt}\right) = 11sY(s)$$

$$\mathcal{L}(6y) = 6Y(s)$$

$$\mathcal{L}(1) = \frac{1}{s}$$

Rearranging and factoring $Y(s)$, we obtain

$$Y(s)(s^3 + 6s^2 + 11s + 6) = \frac{1}{s} \tag{3-34}$$

$$Y(s) = \frac{1}{s(s^3 + 6s^2 + 11s + 6)} \tag{3-35}$$

To invert (3-35) to find $y(t)$, we must find a similar expression in Table 3.1. Unfortunately, no formula in the table has a fourth-order polynomial in the denominator. This example will be continued later, after we develop the techniques necessary to generalize the solution method in Section 3.3.

First, factor the denominator of Eq. 3-35 into a product of first-order terms ($n = 4$ in Eq. 3-38). Simple integer factors, as in this case, rarely occur in actual applications.

$$s(s^3 + 6s^2 + 11s + 6) = s(s+1)(s+2)(s+3) \tag{3-36}$$

This result determines the four terms that will appear in the PFE—namely,

$$\begin{aligned} Y(s) &= \frac{1}{s(s+1)(s+2)(s+3)} \\ &= \frac{\alpha_1}{s} + \frac{\alpha_2}{s+1} + \frac{\alpha_3}{s+2} + \frac{\alpha_4}{s+3} \end{aligned} \tag{3-37}$$

The Heaviside expansion method gives $\alpha_1 = 1/6$, $\alpha_2 = -1/2$, $\alpha_3 = 1/2$, $\alpha_4 = -1/6$.

After the transform has been expanded into a sum of first-order terms, invert each term individually using Table 3.1:

$$\begin{aligned} y(t) &= \mathcal{L}^{-1}[Y(s)] \\ &= \mathcal{L}^{-1}\left(\frac{1/6}{s} - \frac{1/2}{s+1} + \frac{1/2}{s+2} - \frac{1/6}{s+3}\right) \\ &= \frac{1}{6}\mathcal{L}^{-1}\left(\frac{1}{s}\right) - \frac{1}{2}\mathcal{L}^{-1}\left(\frac{1}{s+1}\right) \\ &\quad + \frac{1}{2}\mathcal{L}^{-1}\left(\frac{1}{s+2}\right) - \frac{1}{6}\mathcal{L}^{-1}\left(\frac{1}{s+3}\right) \\ &= \frac{1}{6} - \frac{1}{2}e^{-t} + \frac{1}{2}e^{-2t} - \frac{1}{6}e^{-3t} \end{aligned} \tag{3-38}$$

Equation 3-38 is thus the solution $y(t)$ to the differential equation 3-33. The α_i's are simply the coefficients of the solution. Equation 3-38 also satisfies the three initial conditions of the differential equation. The reader should verify the result.

In Eq. 3-36, the cubic polynomial has three roots that are all integer numbers. This special situation is used for the purposes of illustration. In practical problems, the roots are usually not integers.

because the order of the denominator polynomial (characteristic polynomial) of the transform is equal to the order of the original differential equation, and no table entries are higher than third order in the denominator. It is simply not practical to expand the number of entries in the table ad infinitum. Instead, we use a procedure based on elementary transform building blocks. This procedure, called *partial fraction expansion*, is presented in the next section.

3.3 PARTIAL FRACTION EXPANSION (PFE)

The high-order denominator polynomial in a Laplace transform solution arises from the differential equation terms (its *characteristic polynomial*) plus terms contributed by the inputs. The factors of the characteristic polynomial correspond to the roots of the characteristic polynomial set equal to zero. The input factors may be quite simple. Once the factors are obtained, the Laplace transform is then expanded into *partial fractions*. As an example, consider

$$Y(s) = \frac{s+5}{s^2+5s+4} \tag{3-39}$$

The denominator can be factored into a product of first-order terms, $(s + 1)(s + 4)$. This transform can be expanded into the sum of two partial fractions:

$$\frac{s+5}{(s+1)(s+4)} = \frac{\alpha_1}{s+1} + \frac{\alpha_2}{s+4} \tag{3-40}$$

where α_1 and α_2 are unspecified coefficients that must satisfy Eq. 3-40. The expansion in Eq. 3-40 indicates that the original denominator polynomial has been factored into a product of first-order terms. In general, for every partial fraction expansion (PFE), there will be a unique set of α_i that satisfy the equation.

One of the fastest and most popular methods for calculating the values of coefficients such as α_1 and α_2 is the *Heaviside expansion*. In this method multiply both sides of the equation by one of the denominator terms $(s + b_i)$ and then set $s = -b_i$, which causes all terms except one to be multiplied by zero. Multiplying Eq. 3-40 by $s + 1$ and then letting $s = -1$ gives

$$\alpha_1 = \left.\frac{s+5}{s+4}\right|_{s=-1} = \frac{4}{3}$$

Similarly, after multiplying by $(s + 4)$ and letting $s = -4$, the expansion gives

$$\alpha_2 = \left.\frac{s+5}{s+1}\right|_{s=-4} = -\frac{1}{3}$$

Thus, the coefficients of the PFE can be found by simple calculations.

For a more general transform, where the factors are real and distinct (no complex or repeated factors appear), the following expansion formula can be used:

$$Y(s) = \frac{N(s)}{D(s)} = \frac{N(s)}{\prod_{i=1}^{n}(s+b_i)} = \sum_{i=1}^{n}\frac{\alpha_i}{s+b_i} \tag{3-41}$$

where $D(s)$, an nth-order polynomial, is the denominator of the transform. $D(s)$ is the *characteristic polynomial*. The numerator $N(s)$ has a maximum order of $n - 1$. The ith coefficient can be calculated using the Heaviside expansion

$$\alpha_i = (s+b_i)\left.\frac{N(s)}{D(s)}\right|_{s=-b_i} \tag{3-42}$$

Alternatively, an expansion for real, distinct factors may be written as

$$Y(s) = \frac{N'(s)}{D'(s)} = \frac{N'(s)}{\prod_{i=1}^{n}(\tau_i s+1)} = \sum_{i=1}^{n}\frac{\alpha_i'}{\tau_i s+1} \tag{3-43}$$

Then we can calculate the coefficients by

$$\alpha_i' = (\tau_i s + 1)\frac{N'(s)}{D'(s)}\bigg|_{s=-\frac{1}{\tau_i}} \qquad (3\text{-}44)$$

Note that several entries in Table 3.1 have the $\tau s + 1$ format.

We now can use the Heaviside expansion to complete the solution of Example 3.2.

3.3.1 General Procedure for Solving Differential Equations

We now state a general procedure to solve ordinary differential equations using Laplace transforms. The procedure consists of four steps, as shown in Fig. 3.4.

Note that solution involves use of Laplace transforms as an intermediate step. Step 3 can be bypassed if the transform found in Step 2 matches an entry in Table 3.1. In order to factor $D(s)$ in Step 3, software such as MATLAB, Mathematica, or Mathcad can be utilized (Chapra and Canale, 2014).

So far, we have assumed that the denominator of the characteristic polynomial $D(s)$ in Eq. 3-43 has factors that are real and distinct. However, a factor can also be

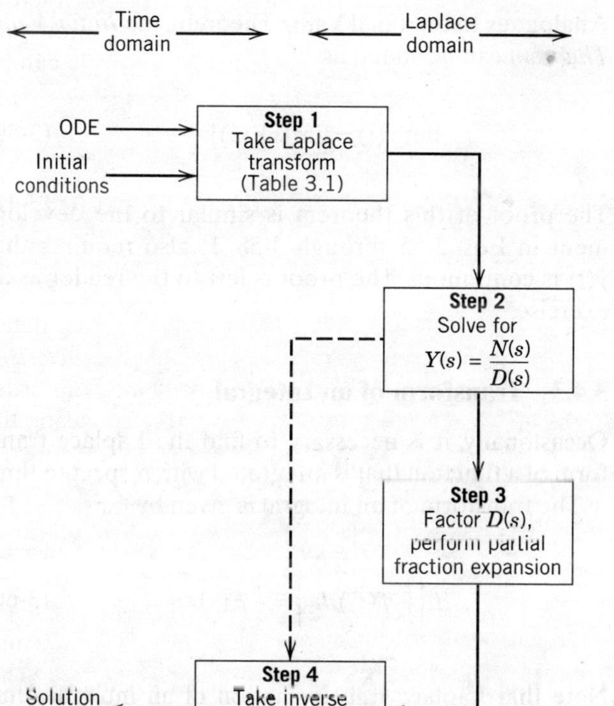

Figure 3.4 The general procedure for solving an ordinary differential equation using Laplace transforms.

repeated or a complex number. These two special cases require modification to the PFE in Eq. 3-43. The modifications are summarized below and considered in more detail in Appendix N.

Repeated Real Factors

Suppose that a function $Y(s)$ has the form

$$Y(s) = \frac{1}{(s+b)^n} \qquad (3\text{-}45)$$

where n is a positive integer. Item 7 in Table 3.1 indicates that the inverse Laplace transform of Eq. 3-45 is

$$y(t) = \frac{t^{n-1}e^{-bt}}{(n-1)!} \qquad (n > 0) \qquad (3\text{-}46)$$

The appropriate PFE for Eq. 3-45 is

$$Y(s) = \frac{\alpha_n}{(s+b)^n} + \frac{\alpha_{n-1}}{(s+b)^{n-1}} + \cdots + \frac{\alpha_1}{s+b} \qquad (3\text{-}47)$$

In general, if a function $X(s)$ has the form

$$X(s) = \frac{N(s)}{D(s)}\frac{1}{(s+b)^n} \qquad (3\text{-}48)$$

The required PFE has the form

$$\frac{N(s)}{D(s)}\frac{1}{(s+b)^n} = \cdots \frac{\alpha_n}{(s+b)^n} + \frac{\alpha_{n-1}}{(s+b)^{n-1}} + \cdots + \frac{\alpha_1}{s+b}$$
$$+ \text{terms based on the factors of } D(s) \qquad (3\text{-}49)$$

The procedures for determining the α_i coefficients are presented in Appendix N.

Equations 3-45 and 3-46 provide an important insight:

> Repeated, real factors for $Y(s)$ result in non-oscillatory terms in the time-domain expression, $y(t)$. If $b < 0$, these terms become unbounded as $t \to \infty$.

Complex Factors

An important special case for PFE occurs when a function $Y(s)$ is factored and has a factor of the form

$$\frac{c_1 s + c_0}{s^2 + d_1 s + d_0} \tag{3-50}$$

with

$$\frac{d_1^2}{4} < d_0 \tag{3-51}$$

Then the two roots of the denominator of Eq. 3-51 are complex and can be written as

$$r_1, r_2 = R \pm Ij \tag{3-52}$$

where R is the real part and I is the imaginary part of the two roots.

As shown in Appendix N, Eq. 3-50 can be converted to the *standard form* on the right side of Eq. 3-53 by a procedure known as *completing the square*:

$$\frac{c_1 s + c_0}{s^2 + d_1 s + d_0} = \frac{a_1(s+b) + a_2}{(s+b)^2 + \omega^2} \tag{3-53}$$

Items 17 and 18 in Table 3.1 indicate that the inverse Laplace transform of the right side of Eq. 3-53 is

$$a_1 e^{-bt} \cos \omega t + \frac{a_2}{\omega} e^{-bt} \sin \omega t \tag{3-54}$$

Thus Eqs. 3-50, 3-53, and 3-54 lead to an important insight:

> Complex factors of $Y(s)$ result in oscillatory terms in the time-domain expression, $y(t)$.

3.4 OTHER LAPLACE TRANSFORM PROPERTIES

In this section, we consider several Laplace transform properties that are useful in process dynamics and control.

3.4.1 Final Value Theorem

The asymptotic value of $y(t)$ for large values of time $y(\infty)$ can be found from Eq. 3-55, providing that $\lim_{s\to 0}[sY(s)]$ exists for all $Re(s) \geq 0$:

$$\lim_{t\to\infty} y(t) = \lim_{s\to 0} [sY(s)] \tag{3-55}$$

The *Final Value Theorem* in Eq. 3-55 can be proved using the Laplace transform of a derivative (Eq. 3-11):

$$\int_0^\infty \frac{dy}{dt} e^{-st} dt = sY(s) - y(0) \tag{3-56}$$

Taking the limit as $s \to 0$ and assuming that dy/dt is continuous and that $sY(s)$ has a limit for all $Re(s) \geq 0$,

$$\int_0^\infty \frac{dy}{dt} dt = \lim_{s\to 0} [sY(s)] - y(0) \tag{3-57}$$

Integrating the left side and simplifying yields

$$\lim_{t\to\infty} y(t) = \lim_{s\to 0} [sY(s)] \tag{3-58}$$

If $y(t)$ is unbounded for $t \to \infty$, Eq. 3-58 gives erroneous results. For example, if $Y(s) = 1/(s-5)$, Eq. 3-58 predicts $y(\infty) = 0$. But in this case, $y(t) = e^{5t}$, which is unbounded for $t \to \infty$. However, Eq. 3-55 does not apply here because $sY(s) = s/(s-5)$ does not have a limit for a real value of $s \geq 0$, namely, $s = 5$.

3.4.2 Initial Value Theorem

Analogous to the Final Value Theorem, the *Initial Value Theorem* can be stated as

$$\lim_{t\to 0} y(t) = \lim_{s\to\infty} [sY(s)] \tag{3-59}$$

The proof of this theorem is similar to the development in Eqs. 3-55 through 3-58. It also requires that $y(t)$ is continuous. The proof is left to the reader as an exercise.

3.4.3 Transform of an Integral

Occasionally, it is necessary to find the Laplace transform of a function that is integrated with respect to time.

The transform of an integral is given by

$$\mathcal{L}\left\{\int_0^t f(t^*)dt^*\right\} = \frac{1}{s}F(s) \tag{3-60}$$

Note that Laplace transformation of an integral function of time leads to division of the transformed function by s. We have already seen in Eq. 3-11 that transformation of a time derivative leads to an inverse relation—that is, multiplication of the transform by s.

EXAMPLE 3.3

Apply the Initial and Final Value Theorems to the transform derived in Example 3.1:

$$Y(s) = \frac{5s+2}{s(5s+4)} \tag{3-61}$$

Solution

Initial Value

$$y(0) = \lim_{s\to\infty}[sY(s)] = \lim_{s\to\infty}\frac{5s+2}{5s+4} = 1 \tag{3-62a}$$

Final Value

$$y(\infty) = \lim_{s\to 0}[sY(s)] = \lim_{s\to 0}\frac{5s+2}{5s+4} = 0.5 \tag{3-62b}$$

The initial value of one corresponds to the initial condition given in Eq. 3-21. The final value of 0.5 agrees with the time-domain solution in Eq. 3-33. Both theorems are useful for checking mathematical errors that may occur in obtaining Laplace transform solutions.

EXAMPLE 3.4

A process is described by a third-order ODE:

$$\frac{d^3y}{dt^3} + 6\frac{d^2y}{dt^2} + 11\frac{dy}{dt} + 6y = 4u \tag{3-63}$$

with all initial conditions on y, dy/dt, and dy^2/dt^2 equal to zero. Show that for a step change in u of two units, the steady-state result in the time domain is the same as applying the Final Value Theorem.

Solution

If $u = 2$, the steady-state result for y can be found by setting all derivatives equal to zero and substituting for u. Therefore,

$$6y = 8 \quad \text{or} \quad y = 4/3 \tag{3-64}$$

The Laplace transform of Eq. 3-63 is

$$(s^3 + 6s^2 + 11s + 6)Y(s) = 8/s \tag{3-65}$$

$$Y(s) = \frac{8}{s^4 + 6s^3 + 11s^2 + 6s} \tag{3-66}$$

One of the benefits of the Final Value Theorem is that we do not have to solve for the analytical solution of $y(t)$. Instead, simply apply Eq. 3-58 to the transform $Y(s)$ as follows:

$$\lim_{s\to 0} sY(s) = \lim_{s\to 0}\frac{8}{s^3 + 6s^2 + 11s + 6} = \frac{8}{6} = \frac{4}{3} \tag{3-67}$$

This is the same answer as obtained in Eq. 3-64. The time-domain solution obtained from a PFE is

$$y = 4/3 - 2e^{-t} + 2e^{-2t} - 2/3e^{-3t} \tag{3-68}$$

As $t \to \infty$, only the first term remains, which is the same result as Eq. 3-67 (from the Final Value Theorem).

3.4.4 Time Delay (Translation in Time)

Time delays play an important role in process dynamics and control. They commonly occur as a result of the transport time required for a fluid to flow through a pipe. Consider the stirred tank heating system example of Chapter 2. Suppose one thermocouple is located at the outflow point of the stirred tank, and a second thermocouple is immersed in the liquid a short distance ($L = 10$ m) downstream. The heating system is off initially and turned on at $t = 0$. If there is no mixing in the pipe (the liquid is in plug flow) and if no heat losses occur from the pipe, the shapes of the two temperature responses should be identical. However, the second sensor response will be translated in time; that is, it will exhibit a *time delay*. If the liquid velocity is $v = 1$ m/s, the time delay ($\theta = L/v$) is 10 s. If we denote $f(t)$ as the transient temperature response at the first sensor and $f_d(t)$ as the temperature response at the second sensor,

Fig. 3.5 shows how they are related. The function $f_d = 0$ for $t < t_0$. Therefore, f_d and f are related by

$$f_d(t) = f(t-\theta)S(t-\theta) \tag{3-69}$$

Note that f_d is the function $f(t)$ delayed by θ time units. The unit step function $S(t - \theta)$ is included to denote explicitly that $f_d(t) = 0$ for all values of $t < \theta$. If $\mathcal{L}(f(t)) = F(s)$, then the Laplace transform of a time-delayed function can be derived as follows:

$$\begin{aligned}\mathcal{L}(f_d(t)) &= \mathcal{L}(f(t-\theta)S(t-\theta)) \\ &= \int_0^\infty f(t-\theta)S(t-\theta)e^{-st}dt \\ &= \int_0^\theta f(t-\theta)(0)e^{-st}dt + \int_\theta^\infty f(t-\theta)e^{-st}dt \\ &= \int_\theta^\infty f(t-\theta)e^{-s(t-\theta)}e^{-s\theta}d(t-\theta)\end{aligned} \tag{3-70}$$

Because $(t - \theta)$ is now the artificial variable of integration, it can be replaced by t^*.

$$\mathcal{L}(f(t)) = e^{-s\theta}\int_0^\infty f(t^*)e^{-st^*}dt^* \tag{3-71}$$

yielding the *Real Translation Theorem*:

$$F_d(s) = \mathcal{L}(f(t-\theta)S(t-\theta)) = e^{-s\theta}F(s) \tag{3-72}$$

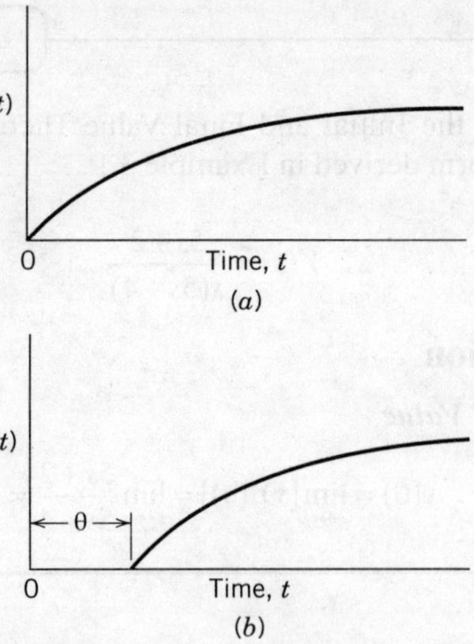

Figure 3.5 A time function with and without time delay. (*a*) Original function (no delay); (*b*) function with delay θ.

In inverting a transform that contains an $e^{-s\theta}$ element (time-delay term), the following procedure will easily yield results and also avoid the pitfalls of dealing with translated (shifted) time arguments. Starting with the Laplace transform

$$Y(s) = e^{-s\theta}F(s) \tag{3-73}$$

1. Invert $F(s)$ in the usual manner; that is, perform PFE and find $f(t)$.
2. Find $y(t) = f(t - \theta)S(t - \theta)$ by replacing the argument t, wherever it appears in $f(t)$, by $(t - \theta)$; then multiply the function by the shifted unit step function, $S(t - \theta)$.

EXAMPLE 3.5

Find the inverse transform of

$$Y(s) = \frac{1+e^{-2s}}{(4s+1)(3s+1)} \tag{3-74}$$

Solution

Equation 3-74 can be split into two terms:

$$Y(s) = Y_1(s) + Y_2(s) \tag{3-75}$$

$$= \frac{1}{(4s+1)(3s+1)} + \frac{e^{-2s}}{(4s+1)(3s+1)} \tag{3-76}$$

The inverse transform of $Y_1(s)$ can be obtained directly from Table 3.1:

$$y_1(t) = e^{-t/4} - e^{-t/3} \tag{3-77}$$

Because $Y_2(s) = e^{-2s}Y_1(s)$, its inverse transform can be written immediately by replacing t by $(t - 2)$ in (3-77), and then multiplying by the shifted step function:

$$y_2(t) = [e^{-(t-2)/4} - e^{-(t-2)/3}]S(t-2) \tag{3-78}$$

Thus, the complete inverse transform is

$$y(t) = e^{-t/4} - e^{-t/3} + [e^{-(t-2)/4} - e^{-(t-2)/3}]S(t-2)$$
$$= y_1(t) + y_2(t) \tag{3-79}$$

Equation 3-79 can be numerically evaluated without difficulty for particular values of t, noting that the term in brackets is multiplied by 0 (the value of the unit step function) for $t < 2$, and by 1 for $t \geq 2$. An equivalent and simpler method is to evaluate the contributions from the bracketed time functions only when the time arguments are nonnegative. An alternative way of writing Eq. 3-79 is as two equations, each one applicable over a particular interval of time:

$$0 \leq t < 2: \quad y(t) = e^{-t/4} - e^{-t/3} \tag{3-80}$$

and

$$t \geq 2: \quad y(t) = e^{-t/4} - e^{-t/3} + [e^{-(t-2)/4} - e^{-(t-2)/3}]$$
$$= e^{-t/4}(1 + e^{2/4}) - e^{-t/3}(1 + e^{2/3})$$
$$= 2.6487e^{-t/4} - 2.9477e^{-t/3} \tag{3-81}$$

Note that Eqs. 3-80 and 3-81 give equivalent results for $t = 2$, because in this case, $y(t)$ is continuous at $t = 2$.

3.5 A TRANSIENT RESPONSE EXAMPLE

In Chapter 4, we will develop a standardized approach for using Laplace transforms to calculate transient responses. However, we already have the tools to analyze an example of a transient response situation in some detail. Example 3.6 illustrates many features of Laplace transform methods in investigating the dynamic characteristics of a physical process.

EXAMPLE 3.6

The Ideal Gas Company has a fixed-volume, stirred-tank reactor that is shown in Fig. 3.6. The three IGC engineers who are responsible for reactor operations—Kim Ng, Casey Gain, and Tim Delay—are concerned about the adequacy of mixing in the tank and want to run a tracer test on the system to determine whether dead zones and/or channeling exist.

Their idea is to operate the reactor at a temperature low enough that reaction will not occur and then apply a rectangular pulse in the feed concentration for test purposes. In this way, available instrumentation on the exit line can be used to measure the exit concentration.

Before performing the test, the engineers would like to have a good idea of the results that should be expected if perfect mixing actually is accomplished in the reactor. A rectangular pulse input for the change in feed concentration will be used with the restriction that the resulting exit concentration changes must be large enough to be measured accurately.

The operating conditions for the reactor test are given in Table 3.2. Figure 3.6 shows the proposed pulse change of 0.25 min duration that can be made while maintaining the total reactor input flow rate constant. As part of the theoretical analysis, Kim, Casey, and Tim would like to know how closely the exit concentration response $c(t)$ to the rectangular pulse can be approximated by the response to an impulse of equivalent magnitude. Based on these considerations, they need to obtain the following information:

(a) The magnitude of an impulse input equivalent to the rectangular pulse of Fig. 3.7.

(b) The impulse and rectangular pulse responses of the exit reactant concentration.

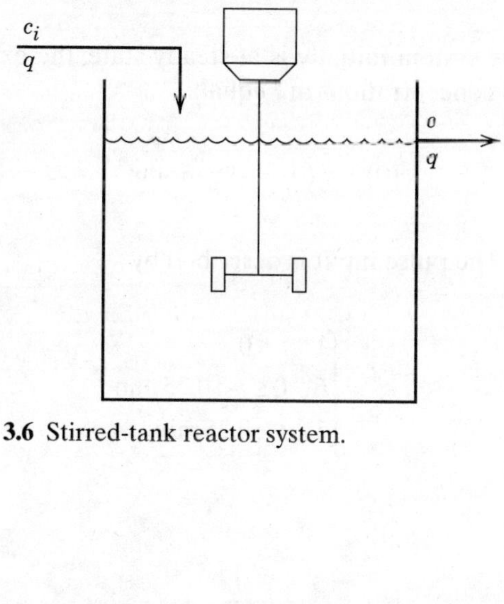

Figure 3.6 Stirred-tank reactor system.

Table 3.2 Stirred-Tank Reactor System and Operating Conditions

Volume	$= 4\ \text{m}^3$
Flow rate q	$= 2\ \text{m}^3/\text{min}$
Nominal feed concentration c_i	$= 1\ \text{kg mol/m}^3$

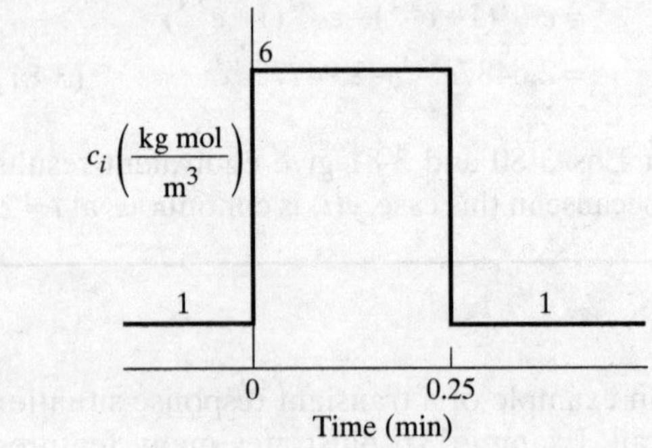

Figure 3.7 Proposed input pulse in reactant concentration.

Solution

The reactor model for a single-stage CSTR was given in Eq. 2-70 as

$$V\frac{dc}{dt} = q(c_i - c) - Vkc \tag{2-70}$$

where c is the reactant concentration of component A. Because the reaction term can be neglected in this example ($k = 0$), the reactor is merely a continuous-flow mixer. Substituting numerical values gives

$$4\frac{dc}{dt} + 2c = 2c_i \tag{3-82}$$

If the system initially is at steady state, the exit and feed concentrations are equal:

$$c(0) = c_i(0) = 1\ \text{kg mol/m}^3 \tag{3-83}$$

(a) The pulse input is described by

$$c_i^p = \begin{cases} 1 & t < 0 \\ 6 & 0 \le t < 0.25\ \text{min} \\ 1 & t \ge 0.25\ \text{min} \end{cases} \tag{3-84}$$

A convenient way to interpret Eq. 3-84 is as a constant input of 1 added to a rectangular pulse of height = 5 kg mol/m³:

$$c_i^p = 6 \quad \text{for} \quad 0 \le t < 0.25\ \text{min} \tag{3-85}$$

The magnitude of an impulse input that is equivalent to the time-varying portion of Eq. 3-85 is simply the integral of the rectangular pulse:

$$M = 5\frac{\text{kg mol}}{\text{m}^3} \times 0.25\ \text{min} = 1.25\frac{\text{kg mol}\cdot\text{min}}{\text{m}^3}$$

Therefore, the equivalent impulse input is

$$c_i^\delta(t) = 1 + 1.25\delta(t) \tag{3-86}$$

Although the units of M have little physical meaning, the product

$$qM = 2\frac{\text{m}^3}{\text{min}} \times 1.25\frac{\text{kg mol}\cdot\text{min}}{\text{m}^3} = 2.5\ \text{kg mol}$$

can be interpreted as the amount of additional reactant fed into the reactor as either the rectangular pulse or the impulse.

(b) The impulse response $c(t)$ is obtained by Laplace transforming Eq. 3-82, using $c(0) = 1$:

$$4sC(s) - 4(1) + 2C_1(s) = 2C_i(s) \tag{3-87}$$

By rearranging Eq. 3-87, we obtain $C(s)$:

$$C(s) = \frac{4}{4s+2} + \frac{2}{4s+2}C_i(s) \tag{3-88}$$

The transform of the impulse input in feed concentration in Eq. 3-86 is

$$C^\delta(s) = \frac{1}{s} + 1.25 \tag{3-89}$$

Substituting Eq. 3-89 into 3-88 gives

$$C^\delta(s) = \frac{2}{s(4s+2)} + \frac{6.5}{4s+2} \tag{3-90}$$

Equation 3-90 does not correspond exactly to any entries in Table 3.1. However, putting the denominator in $\tau s + 1$ form yields

$$C^{\delta}(s)=\frac{1}{s(2s+1)}+\frac{3.25}{2s+1} \tag{3-91}$$

which can be directly inverted using the table, yielding

$$c^{\delta}(t)=1-e^{-t/2}+1.625e^{-t/2}=1+0.625e^{-t/2} \tag{3-92}$$

The rectangular pulse response is obtained in the same way. The transform of the rectangular input pulse Eq. 3-84 is given by Eq. 3-17, so that

$$c_i^p(s)=\frac{1}{s}+\frac{5(1-e^{-0.25s})}{s} \tag{3-93}$$

Substituting (3-93) into (3-88) and solving for $C^p(s)$ yields

$$C^p(s)=\frac{4}{4s+2}+\frac{12}{s(4s+2)}-\frac{10e^{-0.25s}}{s(4s+2)} \tag{3-94}$$

Again, we have to put (3-94) into a form suitable for inversion:

$$C^p(s)=\frac{2}{2s+1}+\frac{6}{s(2s+1)}-\frac{5e^{-0.25s}}{s(2s+1)} \tag{3-95}$$

Before inverting Eq. 3-95, note that the term containing $e^{-0.25s}$ will involve a translation in time. Utilizing the procedure discussed above, we obtain the following inverse transform:

$$c^p(t)=e^{-t/2}+6(1-e^{-t/2})-5[1-e^{-(t-0.25)/2}]S(t-0.25) \tag{3-96}$$

where $S(t-0.25)$ is a delayed unit step function. Note that there are two solutions; for $t<0.25$ min (or t_w) the rightmost term, including the time delay, is zero in the time solution. Thus, the solution is

$$t<0.25\text{ min:}\quad c^p(t)=e^{-t/2}+6(1-e^{-t/2})=6-5e^{-t/2} \tag{3-97}$$

$$\begin{aligned} t\geq 0.25\text{ min:}\quad c^p(t)&=e^{-t/2}+6(1-e^{-t/2})\\ &\quad-5(1-e^{-(t-0.25)/2})\\ &=1-5e^{-t/2}+5e^{-(t-0.25)/2} \end{aligned} \tag{3-98}$$

Plots of Eqs. 3-92, 3-97, and 3-98 are shown in Fig. 3.8. Note that the rectangular pulse response approximates the impulse response fairly well for $t>0.25$ min. Obviously, the approximation cannot be very good for $t<0.25$ min, because the full effect of the rectangular pulse is not felt until that time, while the full effect of the hypothetical impulse begins immediately at $t=0$.

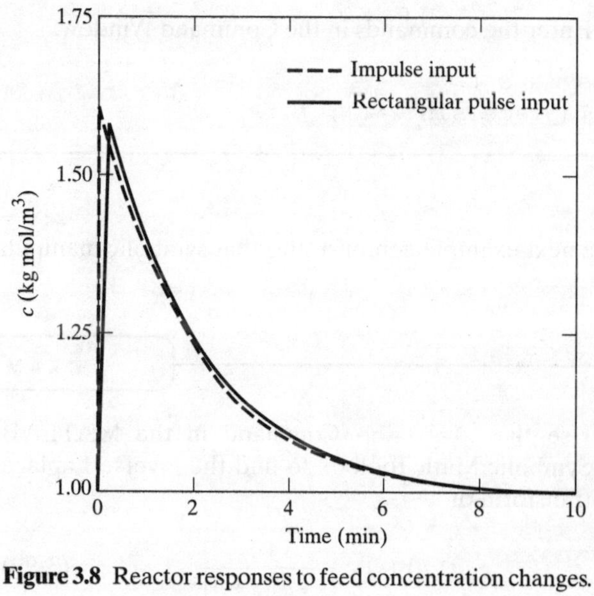

Figure 3.8 Reactor responses to feed concentration changes.

3.6 SOLVING LAPLACE TRANSFORM PROBLEMS USING MATLAB

This chapter has considered analytical solutions of dynamic model responses based on derivations using the Laplace transform method. In contrast to solutions based on numerical methods, analytical solutions provide greater insight into the structure of the solution and its sensitivity to changes in model parameters. However, the disadvantage of analytical methods is that they are restricted to linear models and only a few special nonlinear models. By contrast, solutions of most nonlinear dynamic models can only be obtained using numerical methods.

Analytical solutions to linear dynamic models can also be obtained using software for symbolic mathematical problems such as MATLAB (Mathworks, 2016), Mathematica (Wolfram, 2016), and Maple (Maplesoft, 2016).

In this section, we include five examples to illustrate the use of the MATLAB Symbolic Math Toolbox to generate analytical solutions for Laplace transform problems. The MATLAB commands used for Laplace transform, Inverse Laplace transform and Unit step function are *laplace*, *ilaplace* and *heaviside*, respectively.

The following steps are involved while getting the Laplace transform of an ordinary differential equation using MATLAB:

1. Define the symbolic variables.
2. Write the equation in terms of the symbolic variables and operators.
3. Get the Laplace transform of the equation using command '*laplace*'.
4. Substitute the Laplace transform of the dependent variables (e.g., $\mathcal{L}[y(t)]$) by their corresponding s-domain variables (e.g, $Y(s)$).
5. Substitute the initial conditions for the variables and the derivatives.
6. Solve the Laplace transform equation for the s-domain dependent variable.

EXAMPLE 3.7

Find the Laplace transform of $f(t) = \sin(2t)$ using the `laplace` command in the MATLAB Symbolic Toolbox.

Solution

Enter the commands in the Command Window:

```
syms s t
laplace(sin(2*t), t, s)
```

The `syms` command indicates that `s` and `t` are symbols that do not have numerical values. The MATLAB solution for $F(s)$ is displayed as:

```
ans = 2/(s^2+4)
```

Note that this solution for $F(s)$ is consistent with item #14 in the table of Laplace transform pairs, Table 3.1.

The next example demonstrates that symbolic manipulations can involve non-numeric symbols.

EXAMPLE 3.8

Use the `ilaplace` Command in the MATLAB Symbolic Math Toolbox to find the inverse Laplace transform of

$$Y(s) = \frac{s+1}{(s+a)(s+b)} \tag{3-99}$$

where a and b do not have numerical values.

Solution

Enter the following commands in the Command Window:

```
syms s b a
ilaplace((s+1)/((s+a)*(s+b)))
```

Note that the `syms` command indicates that these three symbols do not have numerical values.

The MATLAB solution for $y(t)$ is displayed as:

```
ans = (exp(-a*t)*(a - 1))/(a - b)
      -(exp(-b*t)*(b - 1))/(a - b)
```

This solution can be verified by considering item #11 in Table 3.1.

Next, we consider an example where the denominator of the Laplace domain function contains a pair of complex conjugate roots.

EXAMPLE 3.9

Use the ilaplace Command in the MATLAB Symbolic Math Toolbox to find the inverse Laplace transform of a function:

$$Y(s) = \frac{s+2}{s^2+s+1} \quad (3\text{-}100)$$

Note that the roots of the denominator are a complex conjugate pair, $r_1, r_2 = -0.5 \pm j\frac{\sqrt{3}}{2}$.

Solution

Enter the commands in the Command Window:

```
sym s
ilaplace((s+2)/(s^2+s+1))
```

The MATLAB solution for $y(t)$ is displayed as

```
ans = exp(-t/2)*(cos((3^(1/2)*t)/2)
      +3^(1/2)*sin((3^(1/2)*t)/2
```

This solution can be verified by comparison with the derivation in Appendix N.

The next example contains a pair of repeated roots.

EXAMPLE 3.10

Use the ilaplace Command in the MATLAB Symbolic Math Toolbox to find the inverse Laplace transform of a function:

$$Y(s) = \frac{s+1}{s^2+4s+4} = \frac{s+1}{(s+2)^2} \quad (3\text{-}101)$$

Solution

Enter the commands in the Command Window:

```
sym s
ilaplace((s+1)/(s^2+4s+4))
```

The MATLAB solution for $y(t)$ is displayed as:

```
ans = exp(-2*t) - t*exp(-2*t)
```

This solution can be verified by comparison with the derivation in Appendix N.

Complex symbolic expressions in MATLAB can be simplified by using commands such as collect, expand, and simplify. For example, consider the expression

$$y = 2(x+1)^2 + 4(x+2)^3 + 7$$

Terms on the right side can be collected by using the expand command:

```
syms x
y = 2*(x+1)^2+4*(x+2)^3+7
expand (y)
ans = 4*^3+26*x^2+41
```

EXAMPLE 3.11

For the CSTR model in Example 3.6, use the MATLAB Symbolic Math Toolbox to generate solutions for the impulse and rectangular pulse inputs.

Solution

(a) ***Impulse response***

Rearrange Eq. 3-91 as

$$C^{\delta}(s) = \frac{1+3.25s}{s(2s+1)} \quad (3\text{-}102)$$

Using the MATLAB Command window, enter the commands

```
sim s
ilaplace((1+3.25*s)/(s*(2*s+1)))
```

The MATLAB solution for $c^{\delta}(t)$ is displayed as:

```
ans = (5*exp(-t/2))/8+1
```

Thus, the impulse response, $c^\delta(t)$, can be written as

$$c^\delta(t) = 1 + 0.625e^{-t/2} \qquad (3\text{-}103)$$

This result is also given in Eq. 3-92.

(b) ***Rectangular pulse***
Rearrange Eq. 3-95 as

$$C^p(s) = \left(\frac{1}{2s+1}\right)\left(2 + \frac{6}{s} - \frac{5e^{-0.25s}}{s}\right) \qquad (3\text{-}104)$$

In the MATLAB Command window, enter the commands

```
sym s
ilaplace((1/(2*s+1))*(2+6/s —
        -(5*exp(-0.25*s))/s))
```

The MATLAB solution for $c^p(t)$ is displayed as

```
ans = 10*heaviside(t - 1/4)
      *(exp(1/8 - t/2)/2 - 1/2)
      - 5*exp(-t/2) + 6
```

where "heaviside(t - 1/4)" denotes the delayed unit step function, $S(t - 1/4)$. After some simplification, an equivalent expression is obtained:

$$c^p(t) = 6 - 5e^{-t/2} \quad \text{for} \quad t < 0.25\,\text{min} \qquad (3\text{-}105)$$

$$c^p(t) = 1 + 5e^{-(t-0.25)/2} - 5e^{-t/2} \quad \text{for } t \geq 0.25\,\text{min} \quad (3\text{-}106)$$

This result is also given in Eqs. 3-97 and 3-98.

EXAMPLE 3.12

Obtain the Laplace transform $X(s)$ and its inverse $x(t)$ from the following differential equation and plot the curve.

$$\frac{d^2x}{dt^2} + 6\frac{dx}{dt} + 5x = 20 \qquad x(0) = 5 \qquad x'(0) = 3 \quad (3\text{-}107)$$

Solution

The MATLAB code using symbolic math toolbox can be written as:

```
clear all; close all; clc
syms x(t) s F X
Dx = diff(x,t);
D2x = diff(x,t,2);
F = laplace(D2x+6*Dx+5*x == 20);
F=subs(F,{'laplace(x(t),t,s)'},{X});
F=subs(F,{'x(0),D(x)(0)'},{5,3});
X=solve(F,'X')
x=ilaplace(X)
ezplot(x,[0 10])
xlabel('Time,t')
ylabel('x(t)')
```

The program outputs are as follows:

```
X = (5*s+20/s+33)/(s^2+6*s+5)
x = 2*exp(-t) - exp(-5*t)+4
```

The simulation is shown in Fig. 3.9.

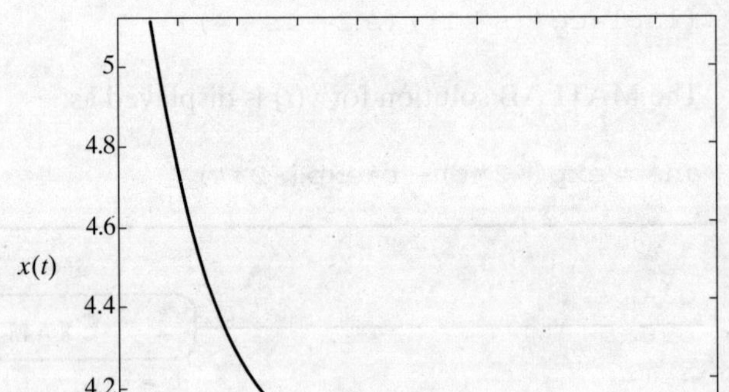

Figure 3.9 Simulation of $x(t)$.

EXAMPLE 3.13

If X and Y are the Laplace Transforms of $x(t)$ and $y(t)$, respectively, obtain the ratio Y/X from the following differential equation using MATLAB commands.

$$\frac{d^2y}{dt^2}+6\frac{dy}{dt}+2y=2x$$
$$y(0)=0$$
$$y'(0)=0 \qquad (3\text{-}108)$$
$$x(0)=0$$

Solution

```
clear all; close all; clc
syms y(t) x(t) s F Y X
Dy = diff(y,t);
D2y = diff(y,t,2);
ode = D2y+6*Dy+5*y == 20*x;
F = laplace(ode);
F=subs(F,{'laplace(y(t),t,s)'},{Y});
F=subs(F,{'laplace(x(t),t,s)'},{X});
F=subs(F,{'y(0),D(y)(0),x(0)'},{0,0,0});
Y=solve(F,'Y')
Ratio=Y/X
```

Outputs from the program:

```
Y = (20*X)/(s^2+6*s+5)
Ratio = 20/(s^2+6*s+5)
```

So, the desired ratio is

```
Y/X = 20/(s^2+6*s+5)
```

EXAMPLE 3.14

Using MATLAB symbolic math toolbox, find the Laplace transform of the following differential equation in terms of $Y(s)$ and its Laplace inverse $y(t)$.

$$\frac{d^2y}{dt^2}+3\frac{dy}{dt}+2y=g(t) \quad y(0)=0 \quad y'(0)=1 \qquad (3\text{-}109)$$

where $g(t)=\begin{cases}1 & t<1\\ 0 & t\geq 1\end{cases}$

Solution

Using unit step and shifted step functions, the discontinuous function $g(t)$ can be written as

$$g(t)=S(t)-S(t\text{–}1).$$

Now we can make the following code to obtain the desired outputs.

```
clear all; close all; clc
syms y(t) s F Y
Dy = diff(y,t);
D2y = diff(y,t,2);
ode = D2y+3*Dy+2*y == heaviside(t)-
      heaviside(t-1);
F = laplace(ode);
F=subs(F,{'laplace(y(t),t,s)'},{Y});
F=subs(F,{'y(0),D(y)(0)'},{0,1});
Y=solve(F,'Y')
y=ilaplace(Y)
ezplot(y,[0 10])
xlabel('Time,t')
ylabel('y(t)')
```

The outputs are as follows:

```
Y = (1/s - exp(-s)/s+1)/(s^2+3*s+2)
y = 1/2 - heaviside(t - 1)*(exp(2 - 2*t)/
    2 - exp(1 - t)+1/2) - exp(-2*t)/2
```

The simulation is shown in Fig. 3.10.

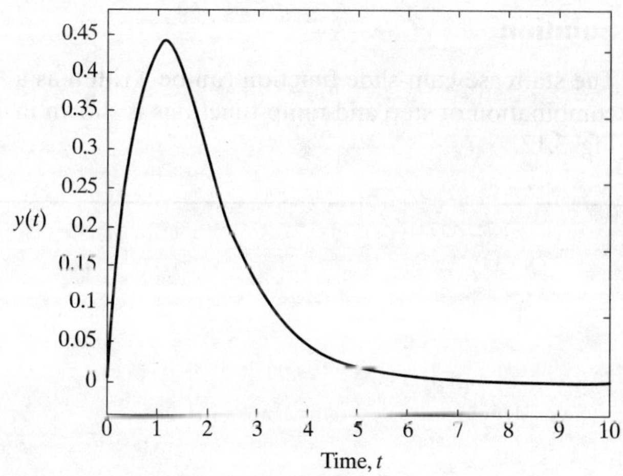

Figure 3.10 Simulation of $y(t)$.

EXAMPLE 3.15

Obtain the Laplace transform of the following 3rd order ordinary differential equation in terms of $Y(s)$ using MATLAB.

$$\frac{d^3y}{dt^3}+6\frac{dy^2}{dt^2}+28\frac{dy}{dt}+40y=\cos(4t)$$

$$y(0)=0 \quad y'(0)=1 \quad y''(0)=0 \qquad (3\text{-}110)$$

Solution

The following MATLAB code can be written

```
clear all; close all; clc
syms y(t) s F Y
Dy = diff(y,t);
D2y = diff(y,t,2);
D3y = diff(y,t,3);
ode = D3y+6*D2y+28*Dy+40*y ==
    cos(4*t);
F = laplace(ode);
F = subs(F,{'laplace(y(t),t,s)'},{Y});
F = subs(F,{'y(0)', 'D(y)(0)', 'D(D(y))
    (0)'},{0,1,0});
Y = solve(F,'Y')
```

The output obtained from this program is given by

```
Y =(s+s/(s^2+16)+6)/
    (s^3+6*s^2+28*s+40)
```

EXAMPLE 3.16

Find the Laplace transform for the composite (staircase-cum-slide) function shown in Fig. 3.11.

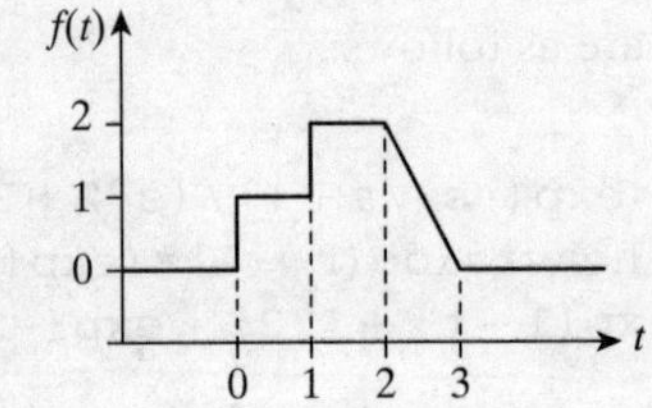

Figure 3.11 Composite (Staircase-cum-slide) function.

Solution

The staircase-cum-slide function can be written as a combination of step and ramp functions is shown in Fig. 3.12.

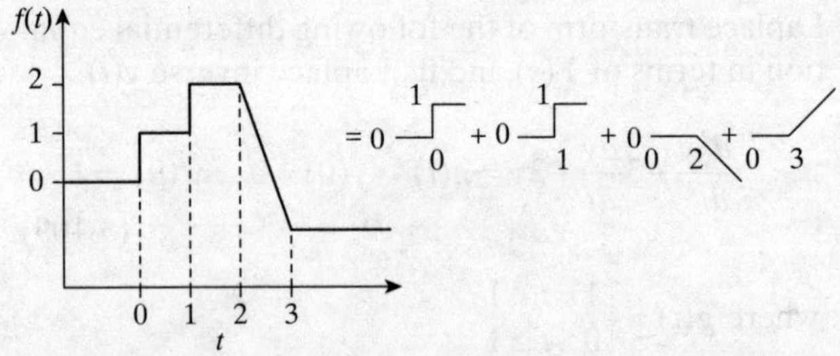

Figure 3.12 Staircase-cum-slide function as a sum of step and ramp functions.

So, mathematically it can be expresses as

$$f(t)=S(t)-S(t-1)-2(t-2)S(t-2)+2(t-3)S(t-3)$$

Its Laplace transform is given by

$$F(s)=\frac{1}{s}+\frac{e^{-s}}{s}-2\frac{e^{-s2}}{s}+2\frac{e^{-3s}}{s}.$$

EXAMPLE 3.17

Sketch the following composite function (for time $t = 0$ to $t = 3$) and get its Laplace Transform.

$$f(t) = 3t\,S(t) - 3\,S(t-1) - S(t-2)$$

Solution

The sketch of the composite function is shown in Fig. 3.13.

The Laplace transform if $f(t)$ is given by

```
F = 3/s^2 - exp(-2*s)/s - (3*exp(-s))/s.
```

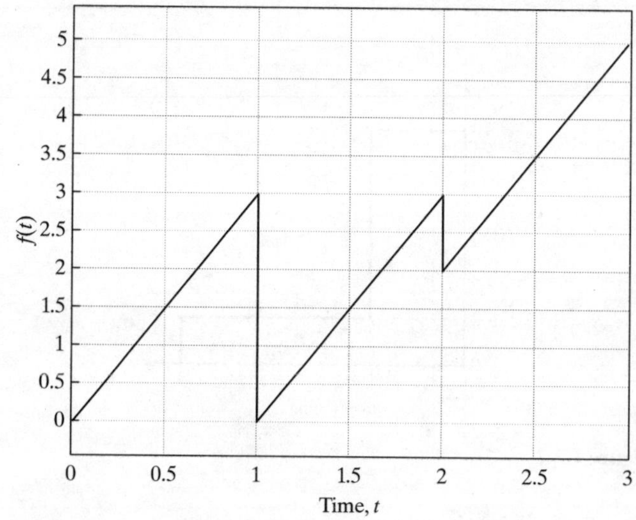

Figure 3.13 Sketch of the composite function, $f(t) = 3t\,S(t) - 3\,S(t-1) - S(t-2)$.

SUMMARY

In this chapter we have considered the application of Laplace transform techniques to solve linear differential equations. Although this material may be a review for some readers, an attempt has been made to concentrate on the important properties of the Laplace transform and its inverse, and to point out the techniques that make manipulation of transforms easier and less prone to error. MATLAB solutions for Laplace transform problems have also been demonstrated.

The use of Laplace transforms can be extended to obtain solutions for models consisting of simultaneous differential equations. However, before addressing such extensions, we introduce the concept of input–output models described by transfer functions. The conversion of differential equation models into transfer function models, covered in the next chapter, represents an important simplification in the methodology, one that can be exploited extensively in process modeling and control system design.

REFERENCES

Churchill, R. V., *Operational Mathematics*, 3rd ed., McGraw-Hill, New York, 1971.

Chapra, S. C., and R. P. Canale, *Numerical Methods for Engineers*, 7th ed., McGraw-Hill, New York, 2014.

Dyke, P. R. G., *An Introduction to Laplace Transforms and Fourier Series*, Springer-Verlag, New York, 1999.

Schiff, J. L., *The Laplace Transform: Theory and Application*, Springer, New York, 1999.

www.maplesoft.com (2016).

www.mathworks.com (2016).

www.wolfram.com (2016).

EXERCISES

3.1 Find the Laplace transforms of the following functions, using the information in Table 3.1. (However, some of the individual terms in these functions may *not* have Laplace transforms.)

(a) $f(t) = 10 + e^{-2t} + te^{-3t}$

(b) $f(t) = \sin(2t) + t - 2 + e^{-(t-2)} + 4/t$

(c) $f(t) = e^{-t}\cos(2t) + t/4$

(d) $f(t) = S(t-2)\cos(3(t-2)) + t^2$

3.2 Derive Laplace transforms of the input signals shown in Figs. E3.2 and E3.3 by summing component functions found in Table 3.1.

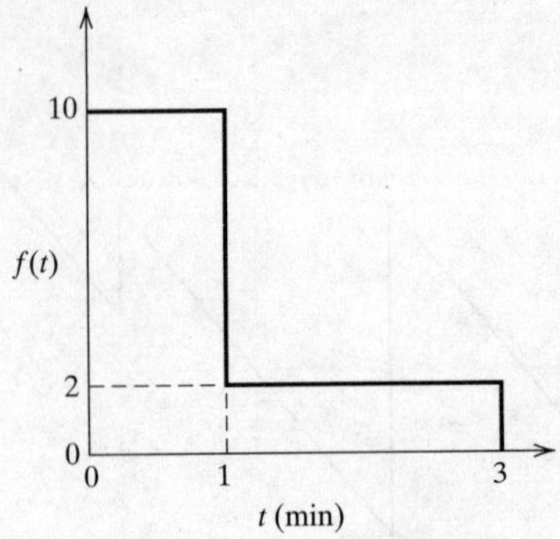

Figure E3.2

3.3 Figure E3.3 shows a pulse function, $u(t)$.

(a) From the information shown in Fig. 3.3, calculate the pulse width, t_w.

(b) Express $u(t)$ as the sum of simpler functions (some perhaps translated in time), whose transforms can be obtained from Table 3.1.

(c) Find $U(s)$.

(d) What is the area under the pulse?

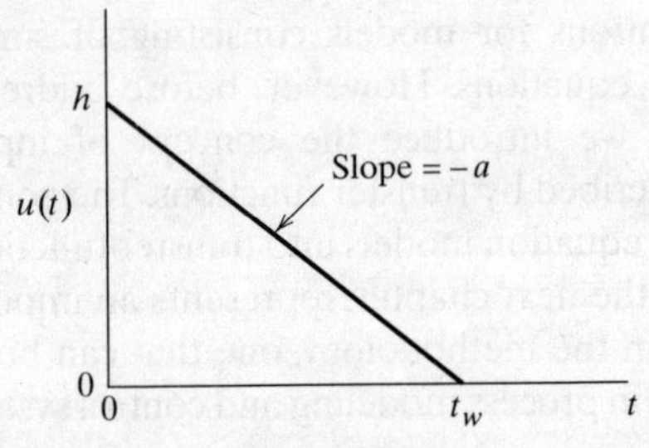

Figure E3.3 Triangular pulse function.

3.4 The dynamic model for a process is given by

$$\frac{d^2y}{dt^2} + 7\frac{dy}{dt} + 12y = 5u(t)$$

where $u(t)$ is the input function and $y(0)$ and $dy/dt\,(0)$ are both zero.

What are the functions of the time (e.g., $e^{-t/\tau}$) in the solution for each of the following cases?

(a) $u(t) = be^{-2t}$

(b) $u(t) = ct$

b and c are constants.

Note: You do not have to find $y(t)$ in these cases. Just determine the functions of time that will appear in $y(t)$.

3.5 The start-up procedure for a batch reactor includes a heating step where the reactor temperature is gradually heated to the nominal operating temperature of 75 °C. The desired temperature profile $T(t)$ is shown in Fig. E3.5. What is $T(s)$?

Hint: Expand $T(t)$ as a sum of individual functions in Table 3.1, some of which may be delayed in time (e.g., $f(t - \theta)$).

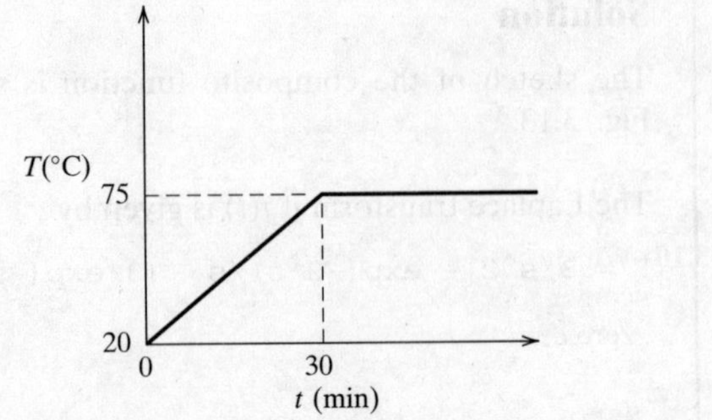

Figure E3.5

3.6 Using PFE where required, find $x(t)$ for

(a) $X(s) = \dfrac{(s+1)(s+2)}{(s+3)(s+4)(s+5)}$

(b) $X(s) = \dfrac{s+2}{(s+1)^2}$

(c) $X(s) = \dfrac{1}{s^2+s+1}$

(d) $X(s) = \dfrac{s+1}{s(s+4)(s+3)}e^{-0.5s}$

3.7 Expand each of the following s-domain functions into partial fractions:

(a) $Y(s) = \dfrac{6(s+1)}{s^2(s+1)}$

(b) $Y(s) = \dfrac{5(s+2)}{s(s^2+4)}$

(c) $Y(s) = \dfrac{(s+2)(s+3)}{(s+4)(s+5)(s+6)}$

(d) $Y(s) = \dfrac{1}{[(s+1)^2+1]^2(s+2)}$

3.8 For the integro-differential equation,

$$\frac{d^2x}{dt^2} + 5\frac{dx}{dt} + 6x = 3\int_0^t e^{-\tau}d\tau$$

$$\text{with } \frac{dx}{dt}(0) = x(0) = 0.$$

(a) Find $x(t)$.

(b) What is the limiting value of t as $t \to \infty$?

3.9 For each of the following cases, determine what functions of time, e.g., sin 3t, e^{-8t}, will appear in $y(t)$.

(Note that you do not have to find $y(t)$!) Which $y(t)$ are oscillatory? Which have a constant value of $y(t)$ for large values of t?

(i) $Y(s)=\dfrac{5}{s(s^2+4s)}$

(ii) $Y(s)=\dfrac{5}{s(s^2+4s+3)}$

(iii) $Y(s)=\dfrac{5}{s(s^2+4s+4)}$

(iv) $Y(s)=\dfrac{5}{s(s^2+4s+8)}$

(v) $Y(s)=\dfrac{2(s+1)}{s(s^2+4)}$

3.10 Which solutions of the following equations will exhibit convergent behavior? Which are oscillatory? Assume zero initial conditions for y and its derivatives.

(a) $\dfrac{d^3x}{dt^3}+2\dfrac{d^2x}{dt^2}+2\dfrac{dx}{dt}+x=4$

(b) $\dfrac{d^2x}{dt^2}-x=3e^t$

(c) $\dfrac{d^3x}{dt^3}+x=2\sin t$

(d) $\dfrac{d^2x}{dt^2}+\dfrac{dx}{dt}=5$

Note: All of the above differential equations have one common factor in their characteristic equations.

3.11 The differential equation model for a particular chemical process has been obtained from experimental test data,

$$\tau_1\tau_2\frac{d^2y}{dt^2}+(\tau_1+\tau_2)\frac{dy}{dt}+y=Ku(t)$$

where τ_1 and τ_2 are constants and $u(t)$ is the input function.

What are the functions of time (e.g., e^{-t}) in the solution for output $y(t)$ for the following cases? (Optional: find the solutions for $y(t)$.)

(a) $u(t)=aS(t)$ step change of magnitude, A

(b) $u(t)=be^{t/\tau}$ $\tau\neq\tau_1\neq\tau_2$

(c) $u(t)=ce^{-t/\tau}$ $\tau=\tau_1\neq\tau_2$

(d) $u(t)=d\sin\omega t$ $\tau_1\neq\tau_2$

3.12 Find the complete time-domain solutions for the following differential equations using Laplace transforms:

(a) $\dfrac{d^3x}{dt^3}+4x=e^t$ with $x(0)=0,\ \dfrac{dx(0)}{dt}=0,\ \dfrac{d^2x(0)}{dt^2}=0$

(b) $\dfrac{dx}{dt}-12x=\sin 3t\ \ x(0)=0$

(c) $\dfrac{d^2x}{dt^2}+6\dfrac{dx}{dt}+25x=e^{-t}\ \ x(0)=0,\ \dfrac{dx(0)}{dt}=0$

3.13 The dynamic model between an output variable y and an input variable u can be expressed by

$$\frac{d^2y(t)}{dt^2}+3\frac{dy(t)}{dt}+y(t)=4\frac{du(t-2)}{dt}-u(t-2)$$

(a) Does this system exhibit an oscillatory response after an arbitrary change in u?

(b) What is the steady-state value of y when $u(t-2)=1$?

(c) For a step change in u of magnitude 1.5, what is $y(t)$?

3.14 For the equation,

$$\ddot{y}+5\dot{y}+6y(t)=7,\quad \dot{y}(0)=0,\quad y(0)=1$$

Use the partial fraction method, to find $y(t)$.

3.15 Find the solution of

$$\frac{dx}{dt}+4x=f(t)$$

where $f(t)=\begin{cases}0 & t<0\\ h & 0\le t<1/h\\ 0 & t\ge 1/h\end{cases}$

$$x(0)=0$$

Plots the solutions for $0\le t\le 2$ and values of $h=1$, 10, 100 and the limiting case of $h\to\infty$ (i.e., $h(t)=\delta(t)$). Place all four plots on a single figure.

3.16 (a) The differential equation

$$\frac{d^2y}{dt^2}+6\frac{dy}{dt}+9y=\cos t$$

has initial conditions, $y(0)=1$, $y'(0)=2$. Find $Y(s)$ and, without finding $y(t)$, determine what functions of time will appear in the solution.

(b) If $Y(s)=\dfrac{s+1}{s(s^2+4s+8)}$, find $y(t)$.

3.17 Use Laplace transforms to find the solution to the following set of equations

$$\frac{dy_1}{dt}+y_2=x$$

$$\frac{dy_2}{dt}+3y_2=2y_1$$

for $x=e^{-t}$ and zero initial conditions.

3.18 A continuous, stirred-tank reactor is initially full of water with the inlet and exit volumetric flow rates of water having the same numerical value. At a particular time, an operator shuts off the water flow and adds caustic solution at the same volumetric flow rate q, but with concentration $\bar{c}_i$. If the liquid volume V is constant, the dynamic model for this process is

$$V\frac{dc}{dt}+qc=q\bar{c}_i \qquad c(0)=0$$

where $c(t)$ is the exit concentration. Calculate $c(t)$ and plot it as a function of time.

Data: $V = 3$ m^3; $q = 0.6$ m^3/min; $\bar{c}_i = 100$ kg/m^3

3.19 A liquid storage facility can be modeled by

$$\ddot{y}+5\dot{y}+y(t)=8\dot{u}+u(t)$$

where y is the liquid level (m) and u is an inlet flow rate (m^3/s). Both are defined as deviations from the nominal steady-state values. Thus, $y = u = 0$ at the nominal steady state. Also, the initial values of all the derivatives are zero.

(a) If $u(t)$ suddenly changes from 0 to 1 m^3/s at $t = 0$, determine the liquid level response, $y(t)$.

(b) If the tank height is 2.5 m, will the tank overflow?

(c) Based on your results for **(b)**, what is the maximum flow change, u_{max}, that can occur without the tank overflowing? (*Hint:* Consider the Principle of Superposition.)

3.20 Three stirred-tanks in series are used in a reactor train (see Fig. E3.20). The flow rate into the system of some inert species is maintained constant while tracer tests are conducted. Assuming that mixing in each tank is perfect and the volumes are constant:

(a) Derive expressions for the concentration of tracer leaving each tank; c_i is the concentration of tracer entering the first tank.

(b) If c_i has been constant and equal to zero for a long time and an operator suddenly injects a large amount of tracer material in the inlet to tank 1, what will be the form of $c_3(t)$ (i.e., what types of time functions will be involved) if

1. $V_1 = V_2 = V_3$

2. $V_1 \neq V_2 \neq V_3$.

(c) If the amount of tracer injected is unknown, is it possible to back-calculate it from experimental data? How would you do it?

3.21 For the three stirred-tank system of Exercise 3.20 (Part **(b)1**), use symbolic system software to find the exit concentration of tank 3, $c_3(t)$, after a rectangular pulse in $c_i(t)$ occurs at $t = 0$. The pulse magnitude is A and the pulse width is t_w.

3.22 Solve this ODE using a Symbolic software program:

$$\frac{d^4y}{dt^4}+16\frac{d^3y}{dt^3}+86\frac{d^2y}{dt^2}+176\frac{dy}{dt}+105y=1$$

All initial conditions for y and its derivatives are zero.

3.23 Obtain the Laplace transform $Y(s)$ and its inverse $y(t)$ from the following differential equation and plot the curve.

$$\frac{d^2y}{dt^2}+6\frac{dy}{dt}+2y=2$$
$$y(0)=0$$
$$y'(0)=0$$

3.24 The composite M-function $f(t)$ is shown below in Fig. E3.24. Obtain a single time-domain equation of this discontinuous function and get its Laplace transform.

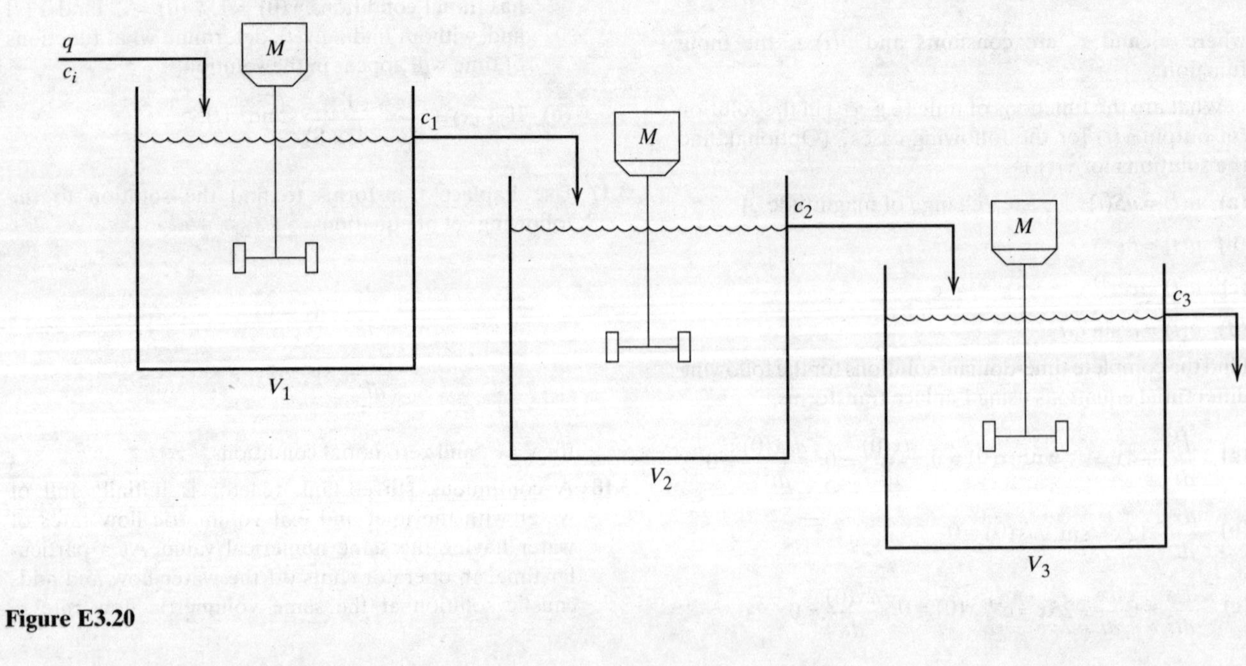

Figure E3.20

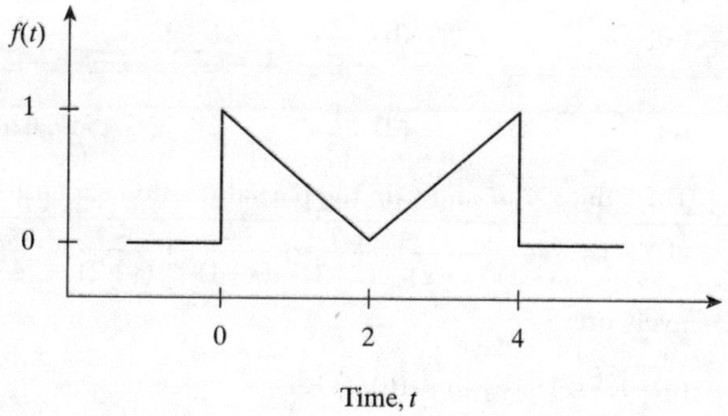

Figure E3.24

3.25 Solve Problem 3.17 using MATLAB.

3.26 Solve the following differential equation using Laplace transforms. Assume zero initial conditions for $y(t)$ and $y'(t)$.

$$2\frac{d^2y}{dt^2}+3\frac{dy}{dt}+y=\delta(t)$$

3.27 Solve the following differential equation using Laplace Transforms. Assume zero initial conditions for the dependent variable and its derivative.

$$\frac{d^2y}{dt^2}+4y=2e^{-t}$$

3.28 Express the function $f(t)$ given in Fig. E3.28 in the t-domain and obtain its Laplace transform.

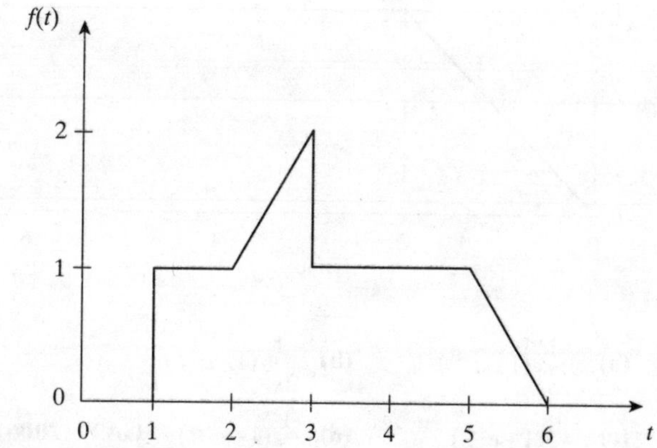

Figure E3.28

3.29 Solve the following differential equation using Laplace transforms.

$$\frac{d^3x}{dt^3}+2\frac{d^2x}{dt^2}-\frac{dx}{dt}-2x=4+e^{2t}$$
$$x(0)=1$$
$$x'(0)=0$$
$$x''(0)=-1$$

3.30 Solve the following differential equation using Laplace transforms.

$$\frac{d^4x}{dt^4}+\frac{d^3x}{dt^3}=\cos t$$
$$x(0)=x'(0)=x'''(0)=0$$
$$x''(0)=1$$

MULTIPLE CHOICE QUESTIONS

1. For the ordinary differential equation

$$\frac{d^3y}{dt^3}+6\frac{d^2y}{dt^2}+11\frac{dy}{dt}+6y=1$$

with initial conditions $y(0)=y'(0)=y''(0)=y'''=(0)=0$, the value of $\lim_{t\to\infty} y(t)=$ __________ (round off to 3 decimal places).

(a) 0.436 **(b)** 0.167
(c) 0.248 **(d)** 0.176 **(GATE 2021)**

2. The Laplace transform of a staircase function $S(b, t)$ formed by successive addition of unit step functions at $0, b, 2b, 3b$, etc. is

(a) $\frac{1}{s}$ **(b)** $\frac{1}{bs}$
(c) $s(1-e^{-bs})$ **(d)** $\frac{1}{s(1-e^{-bs})}$ **(GATE 1996)**

3. For an input forcing function, $X(t)=2t^2$, the Laplace transform of this function is

(a) $2/s^2$ **(b)** $4/s^2$
(c) $2/s^3$ **(d)** $4/s^3$ **(GATE 1997)**

4. The Laplace transform of e^{at}, where $a>0$, is defined only for the Laplace parameter, $s>a$, since

(a) the function is exponential.
(b) the Laplace transform of integral of exp(at) has finite values only for $s>a$.
(c) the Laplace transform of integral of exp(at) has initial values only for $s>a$.
(d) the function exp(at) is piece-wise continuous only for $s>a$. **(GATE 1999)**

5. The Laplace transform of the input function, $X(t)$, given in the figure below, is given by

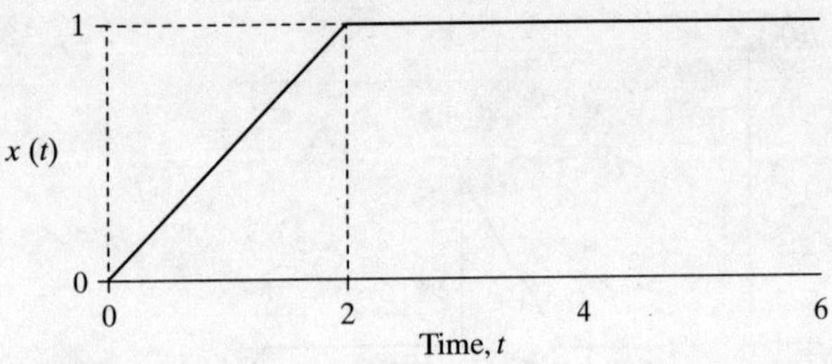

(a) $\frac{1}{2s^2}(1-e^{-2s})$ **(b)** $\frac{1}{2s^2}(1+e^{-2s})$

(c) $\frac{1}{s^2}(1+e^{+2s})$ **(d)** $\frac{1}{s^2}(1-e^{-2s})$ **(GATE 2006)**

6. The inverse Laplace transform of $\frac{1}{2s^2+3s+1}$ is

(a) $e^{-t/2}-e^{-t}$ **(b)** $2e^{-t/2}-e^{-t}$

(c) $e^{-t}-2e^{-t/2}$ **(d)** $e^{-t}-e^{-t/2}$ **(GATE 2009)**

7. The Laplace transform of a function is $\frac{s}{s(s+2)}$. The initial and final values, respectively, of the function are

(a) 0 and 1 **(b)** 1 and $\frac{1}{2}$

(c) $\frac{1}{2}$ and 1 **(d)** $\frac{1}{2}$ and 0 **(GATE 2017)**

8. Consider the following unit step function. The Laplace transform of this function is

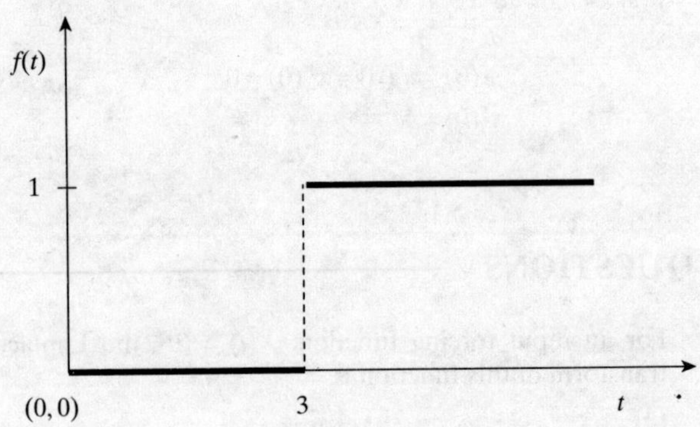

(a) $\frac{e^{-3s}}{s}$ **(b)** $\frac{e^{-3s}}{s^2}$

(c) $\frac{e^{-3s}}{3s}$ **(d)** $\frac{e^{-6s}}{s}$ **(GATE 2020)**

9. The values A, B and C in the partial fraction expansion of $X(s)=\frac{1}{(s+1)^2(s+2)}=\frac{A}{(s+1)}+\frac{B}{(s+1)^2}+\frac{C}{(s+2)}$ respectively, are

(a) 1, –1, 1 **(b)** 1, 1, 1

(c) 1, 1, –1 **(d)** –1, 1, 1

10. The Laplace transform of the dynamic equation $\left\{5\frac{d^2x}{dt^2}+3\frac{dx}{dt}+2x(t)=S(t);\ x(0)=1,\ x'(0)=-1\right\}$ is given by

(a) $X(s)=\frac{1}{s(5s^2+3s+2)}+\frac{(5s+1)}{(5s^2+3s+2)}$

(b) $X(s)=\frac{1}{s(5s^2+3s+2)}+\frac{(5s+2)}{(5s^2+3s+2)}$

(c) $X(s)=\frac{1}{s(5s^2+3s+2)}+\frac{(5s-1)}{(5s^2+3s+2)}$

(d) $X(s)=\frac{1}{s(5s^2+3s+2)}+\frac{(5s-2)}{(5s^2+3s+2)}$

11. The solution of the differential equation $\left\{\frac{d^2y}{dt^2}-2=0;\ y'(0)=y(0)=0.\right\}$ using Laplace transform is

(a) $y(t)=t^3$ **(b)** $y(t)=2t^3$

(c) $y(t)=t^2$ **(d)** $y(t)=3t^2$

12. The inverse Laplace transform of the function $X(s)=\frac{s+1}{s^2-2s+5}$ is

(a) $x(t)=e^{-t}[\cos(2t)+\sin(2t)]$

(b) $x(t)=e^{t}[\cos(2t)+\sin(2t)]$

(c) $x(t)=e^{-t/2}[\cos(t)+\sin(t)]$

(d) $x(t)=e^{-2t}[\cos(t)+\sin(t)]$

ANSWER KEY

1. (b) **2.** (d) **3.** (d) **4.** (b) **5.** (a) **6.** (a)

7. (b) **8.** (a) **9.** (c) **10.** (d) **11.** (c) **12.** (b)

CHAPTER 4

Transfer Function and State-Space Models

CHAPTER CONTENTS

Chapters 2 and 3 have considered dynamic models in the form of ordinary differential equations (ODE). In this chapter, we introduce an alternative model form based on Laplace transforms: the *transfer function model*. Both types of models can be used to determine the dynamic behavior of a process after changes in input variables. The transfer function also plays a key role in the design and analysis of control systems, as will be considered in later chapters.

A transfer function model characterizes the dynamic relationship of two process variables, a dependent variable (or *output variable*) and an independent variable (or *input variable*). For example, in a continuous chemical reactor, the output variable could be the exit concentration and the input variable, a feed flow rate. Thus the input can be considered to be a "cause" and the output an "effect." Transfer function models are only directly applicable to processes that exhibit linear dynamic behavior, such as a process that can be modeled by a linear ODE. If the process is nonlinear, a transfer function can provide an approximate linear model, as described in Section 4.3.

The state-space and transfer function matrix models are discussed in Section 4.4. In Section 4.5, poles and zeros and their effects on the process response are described.

4.1 INTRODUCTION TO TRANSFER FUNCTION MODELS

Figure 4.1 shows a dynamic model with a single input variable u and a single output variable y.

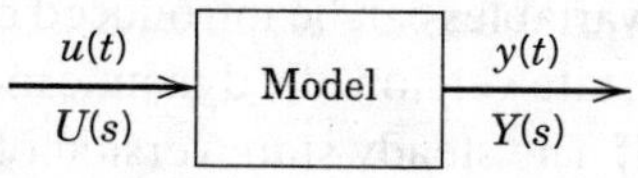

Figure 4.1 Input–output model.

The time-domain model that relates u and y is an ODE. For a linear ODE, there is an equivalent model in the Laplace domain, the transfer function model. By definition, the transfer function between u and y is

$$G(s) = \frac{Y(s)}{U(s)} \tag{4-1}$$

where $Y(s) = \mathcal{L}\{y(t)\}$ and $U(s) = \mathcal{L}\{u(t)\}$.

The transfer function in Eq. 4-1 is expressed in terms of the physical variables, y and u. However, it is usually more convenient to express transfer functions in terms of *deviation variables* that are deviations from a nominal steady state. For example, the deviation variables for y and u are defined as,

$$y' \triangleq y - \bar{y} \quad \text{and} \quad u' \triangleq u - \bar{u} \tag{4-2}$$

where y' and u' denote the deviation variables and $\bar{y}$ and $\bar{u}$ are the nominal steady-state values.

Example 4.1 illustrates how a transfer function model can be derived easily from an ODE model.

EXAMPLE 4.1

Consider the liquid storage process of Section 2.5.5. The dynamic model consists of the linear first-order ODE in Eq. 2-61:

$$A\frac{dh}{dt} = q_i - \frac{1}{R_v}h \tag{2-61}$$

where h is the liquid level, q_i is the inlet flow rate, and A and R_v are constants.

(a) Derive a transfer function model between h and q_i (in deviation variables).

(b) Use the transfer function model to determine the response $h(t)$ to a step change in q_i of magnitude M at $t = 0$.

Solution

(a) For this example, there are two deviation variables, h' and q_i':

$$h' \triangleq h - \bar{h} \quad \text{and} \quad q_i' \triangleq q_i - \bar{q}_i \tag{4-3}$$

Deviation variables can be introduced by subtracting the steady-state version of a dynamic model from the model itself. The steady-state version of Eq. 2-61 is

$$0 = \bar{q}_i - \frac{1}{R_v}\bar{h} \tag{4-4}$$

Subtracting Eq. 4-4 from Eq. 2-61 gives

$$A\frac{dh}{dt} = (q_i - \bar{q}_i) - \frac{1}{R_v}(h - \bar{h}) \tag{4-5}$$

Because $\bar{h}$ is a constant, it follows from Eq. 4-3 that

$$\frac{dh'}{dt} = \frac{d(h - \bar{h})}{dt} = \frac{dh}{dt} \tag{4-6}$$

Substituting Eq. 4-6 into the left side of Eq. 4-5, and then substituting Eq. 4-3 gives the dynamic model in deviation variable form:

$$A\frac{dh'}{dt} = q_i' - \frac{1}{R_v}h' \tag{4-7}$$

Next, we assume that the liquid storage system is initially at the nominal steady state. Thus, $h(0) = \bar{h}$, and from Eq. 4-3, it follows that $h'(0) = h(0) - \bar{h} = h(0) - h(0) = 0$. Taking the Laplace transform of both sides of Eq. 4-7 gives

$$A\left[sH'(s) - \cancel{h'(0)}\right] = Q_i'(s) - \frac{1}{R_v}H'(s) \tag{4-8}$$

Rearranging gives the desired transfer function, $G(s)$:

$$G(s) = \frac{H'(s)}{Q_i'(s)} = \frac{R_v}{AR_v s + 1} \tag{4-9}$$

(b) In order to derive the step response, first rearrange Eq. 4-9:

$$H'(s) = \left(\frac{R_v}{AR_v s + 1}\right)Q_i'(s) \tag{4-10}$$

For a step change of magnitude M at $t = 0$, $u_i(t) = M$, and $Q_i(s) = M/s$. Substituting in Eq. 4-10 gives

$$H'(s) = \left(\frac{R_v}{AR_v s + 1}\right)\frac{M}{s} \tag{4-11}$$

From the table of Laplace transforms (Table 3.1), taking the inverse Laplace transform of Eq. 4-11 gives the liquid level response:

$$h'(t) = R_v M(1 - e^{-t/AR_v}) \tag{4-12}$$

Finally, substituting Eq. 4-3 and rearranging gives the desired result:

$$h(t) = \bar{h} + h'(t) = \bar{h} + R_v M(1 - e^{-t/AR_v}) \tag{4-13}$$

This example has demonstrated that the transfer function model could be used to derive the liquid level response $h(t)$ for a step change in the input, q_i. Furthermore, the response to any arbitrary input change $q_i(t)$ can be calculated from Eq. 4-10 after substituting $Q_i(s)$. Thus we have the following general result.

Important Property of Transfer Functions

For any transfer function,

$$\frac{Y(s)}{U(s)} = G(s) \tag{4-14}$$

the output response $y(t)$ can be determined for any input change $u(t)$ that has a Laplace transform.

The derivation in Example 4.1 leads to two questions concerning individual steps:

1. Why were deviation variables introduced? Why not simply consider the physical variables rather than deviation variables?
2. The derivation assumed that the process was initially at steady state. Can transfer function analysis be used for other initial conditions?

The next two examples provide the answers.

EXAMPLE 4.2

Derive a transfer function model for the liquid storage system of Example 4.1 using physical variables instead of deviation variables.

Solution

Taking the Laplace transform of Eq. 2-61 gives

$$A[sH(s) - h(0)] = Q_i(s) - \frac{1}{R_v}H(s) \tag{4-15}$$

Rearranging gives

$$\frac{H(s)}{Q_i(s)} = \frac{R_v}{AR_v s + 1} + \left(\frac{AR_v}{AR_v s + 1}\right)\left(\frac{h(0)}{Q_i(s)}\right) \tag{4-16}$$

While Eq. 4-16 is correct, it is unwieldy because $Q_i(s)$ appears on the right side and $h(0) \neq 0$ unless the tank is initially empty, a very restrictive condition. Transfer function derivations based on deviation variables rely on the much less restrictive assumption of steady-state initial conditions (cf. Example 4.1). Thus it is more convenient to use deviation variables rather than physical variables for transfer function analysis.

EXAMPLE 4.3

Show that transfer function analysis is also applicable for situations where the process is not initially at steady state. As an example, consider the liquid storage system of Example 4.1.

Solution

The transfer function of Eq. 4-9 does not contain initial conditions. However they can be introduced by transforming the model into the equivalent ODE model. Rearrange Eq. 4-9 as

$$H'(s)(AR_v s + 1) = R_v Q_i'(s) \tag{4-17}$$

Taking inverse Laplace transforms of both sides and rearranging gives

$$A\frac{dh'}{dt} = q_i' - \frac{1}{R_v}h' \tag{4-18}$$

Taking the Laplace transform of Eq. 4-18 assuming that $h'(0) \neq 0$ results in Eq. 4-15 and then Eq. 4-16. In analogy with Example 4.1, the response $h(t)$ for a specified change in q_i can be derived. Thus transform analysis based on deviation variables is applicable to problems with arbitrary initial conditions.

EXAMPLE 4.4

Consider the continuous blending process of Section 2.1.1. For simplicity, we make the following assumptions:

1. Liquid density ρ and volume V are constant.
2. Mass flow rates w_1, w_2, and w are constant.

Then the component balance in Eq. 2-3 becomes

$$\rho V \frac{dx}{dt} = w_1 x_1 + w_2 x_2 - wx \tag{4-19}$$

Derive transfer function models for two cases:

(a) Inlet concentration x_1 varies while x_2 is constant.
(b) Inlet concentrations x_1 and x_2 vary.

Solution

Case (a): Inlet concentration x_1 varies while x_2 is constant We will derive a transfer function model between exit composition x (the output variable) and inlet composition x_1 (the input variable), starting with Eq. 4-19. The steady-state version of Eq. 4-19 is

$$0 = w_1\bar{x}_1 + w_2 x_2 - w\bar{x} \tag{4-20}$$

where the bar over a symbol denotes a nominal steady-state value. Subtracting Eq. 4-20 from Eq. 4-19 gives

$$\rho V \frac{dx}{dt} = w_1 x_1' - wx' \tag{4-21}$$

where the two deviation variables are defined as

$$x_1' \triangleq x_1 - \bar{x}_1$$
$$x' \triangleq x - \bar{x} \tag{4-22}$$

Because $\bar{x}$ is a constant, it follows that

$$\frac{dx}{dt} = \frac{d(x-\bar{x})}{dt} = \frac{dx'}{dt} \tag{4-23}$$

Substituting Eq. 4-23 into Eq. 4-21 gives the solute component balance in deviation variable form:

$$\rho V \frac{dx'}{dt} = w_1 x_1' - wx' \tag{4-24}$$

Assume that the blending system is initially at the nominal steady state. Thus, $x(0) = \bar{x}$ and $x'(0) = 0$. Taking the Laplace transform of Eq. 4-24 gives

$$\rho V s\left[X'(s) - \cancel{x'(0)}^{\,0}\right] = w_1 X_1'(s) - wX'(s) \tag{4-25}$$

where $X'(s) \triangleq \mathcal{L}[X'(t)]$ and $X_1'(s) \triangleq \mathcal{L}[X'(t)]$. Rearranging gives the transfer function $G(s)$ between the exit and inlet and compositions:

$$\frac{X'(s)}{X_1'(s)} = \frac{w_1}{\rho V s + w} \tag{4-26}$$

It is useful to place the transfer function in a standard form by dividing both the numerator and the denominator by w:

$$\frac{X'(s)}{X_1'(s)} = G(s) \triangleq \frac{K_1}{\tau s + 1} \tag{4-27}$$

where constants K_1 and τ are defined as

$$K_1 \triangleq \frac{w_1}{w} \tag{4-28}$$

$$\tau \triangleq \frac{\rho V}{w} \tag{4-29}$$

Useful physical interpretations of K and τ are provided in Section 5.4.

Case (b): Both inlet concentrations, x_1 and x_2, vary For the case of two input variables, x_1 and x_2, two transfer functions are needed to describe their effects on output variable x. Their derivation is analogous to the derivation for Case (a).

For this case, the steady-state version of Eq. 4-19 can be written as

$$0 = w_1\bar{x}_1 + w_2\bar{x}_2 - w\bar{x} \tag{4-30}$$

Subtracting Eq. 4-30 from Eq. 4-19 and introducing deviation variables gives

$$\rho V \frac{dx'}{dt} = w_1 x_1' + w_2 x_2' - wx' \tag{4-31}$$

where $x_2' \triangleq x_2 - \bar{x}_2$. Again assuming that the blending system is initially at the nominal steady state, taking the Laplace transform of Eq. 4-31 gives

$$\rho V s[X'(s) - 0] = w_1 X_1'(s) + w_2 X_2'(s) - wX'(s) \tag{4-32}$$

which can be rearranged as

$$X'(s) = \frac{K_1}{\tau s + 1} X_1'(s) + \frac{K_2}{\tau s + 1} X_2'(s) \tag{4-33}$$

where K_2 is defined as

$$K_2 \triangleq \frac{w_2}{w} \tag{4-34}$$

In order to derive the transfer function between x and x_1, assume that x_2 is constant at its nominal steady-state value, $x_2 = \bar{x}_2$. Therefore, $x_2'(t) = 0$, $X_2'(s) = 0$, and Eq. 4-33 reduces to the previous transfer function relating x and x_1 (see Eq. 4-27).

$$\frac{X'(s)}{X_1'(s)} \triangleq G_1(s) \triangleq \frac{K_1}{\tau s+1} \qquad (4\text{-}35)$$

Similarly, the transfer function between x and x_2 can be derived from Eq. 4-33 and the assumption that x_1 is constant at its nominal steady-state value, $x_1 = \bar{x}_1$:

$$\frac{X'(s)}{X_2'(s)} \triangleq G_2(s) \triangleq \frac{K_2}{\tau s+1} \qquad (4\text{-}36)$$

Example 4.4 demonstrates that transfer function analysis can also be used when there is more than one input variable.

The models in Eqs. 4-35 and 4-36 are referred to as *first-order transfer functions*, because the denominators are first order in the Laplace variable s.

Three important aspects of these derivations are

1. A comparison of Eq. 4-33 to Eq. 4-36 shows that the effects of the individual input variables on the output variable are additive. This result is a consequence of the Principle of Superposition for linear models (see Section 3.1).
2. The assumption of an input being constant in the derivations of Eqs. 4-35 and 4-36 seems restrictive but actually is not, for the following reason. Because a transfer function concerns the effect of a *single input* on an output, it is not restrictive to assume that the other independent inputs are constant for purposes of the derivation. Simultaneous changes in both inputs can be analyzed, as indicated by Eq. 4-33.
3. A transfer function model allows the output response to be calculated for a specified input change. For example, Eq. 4-35 can be rearranged as

$$X'(s) = G_1(s)\,X_1'(s) \qquad (4\text{-}37)$$

After specifying $x_1'(t)$, its Laplace transform $X_1'(s)$ can be determined using Table 3.1. Then the output response $x'(t)$ can be derived from Eq. 4-37, as illustrated in Example 4.5.

EXAMPLE 4.5

Consider the stirred-tank blending process for Example 4.4 and Eqs. 4-19 and 4-20 with $w_1 = 600$ kg/min, $w_2 = 2$ kg/min, $x_1 = 0.050$, and $x_2 = 1$ (for pure solute). The liquid volume and density are constant: $V = 2\ \text{m}^3$ and $\rho = 900\ \text{kg/m}^3$, respectively.

(a) Calculate the nominal exit concentration, $\bar{x}$.

(b) Derive an expression for the response, $x(t)$, to a sudden change in x_1 from 0.050 to 0.075 that occurs at time, $t = 0$. Assume that the process is initially at the nominal steady state.

Solution

(a) Exit flow rate w can be calculated from a steady-state mass balance

$$w = w_1 + w_2 = 600 + 2 = 602\ \text{kg/min}$$

and weight fraction $\bar{x}$ can be determined from Eq. 4-30:

$$\bar{x} = \frac{w_1\bar{x}_1 + w_2 x_2}{w} = \frac{(600)(0.05)+(2)(1)}{602} = 0.053$$

(b) To determine $x(t)$ for a sudden change in x_1, we first derive an expression for $x_1'(t)$ and then obtain $X_1'(s)$. Thus, the appropriate starting point for the derivation is the transfer function in Eq. 4-27 where $K_1 = 600/602 = 0.997$ and $\tau = \rho V/w = (900)(2)/(602) = 2.99$ min. The sudden change in x_1 can be expressed in deviation variable form as

$$x_1'(t) = x_1 - \bar{x}_1 = 0.075 - 0.050$$
$$= 0.025 \ \Rightarrow\ X_1'(s) = \frac{0.025}{s}$$

Rearranging Eq. 4-27 and substituting numerical values gives

$$X'(s) = \left(\frac{K_1}{\tau s+1}\right) X_1'(s) = \left(\frac{0.997}{2.99s+1}\right)\left(\frac{0.025}{s}\right)$$
$$= \frac{0.0249}{s(2.99s+1)} \qquad (4\text{-}38)$$

Using Item 13 in Table 3.1, the inverse Laplace transform is

$$x'(t) = 0.0249\,(1 - e^{-t/2.99}) \qquad (4\text{-}39)$$

From Eq. 4-22,

$$x(t) = \bar{x} + x'(t) = 0.053 + 0.0249(1 - e^{-t/2.99}) \qquad (4\text{-}40)$$

Example 4.5 has shown how an expression can be derived for the response $x(t)$ to a step change in x_1. Analogous derivations could be made for other types of x_1 or x_2 changes, such as a sinusoidal change, and for simultaneous changes in x_1 and x_2. The starting point for the latter derivation would be Eq. 4-33.

4.2 PROPERTIES OF TRANSFER FUNCTIONS

One important property of the transfer function is that the steady-state output change for a sustained input change can be calculated directly. The *steady-state gain* is the ratio of the output variable change to an input variable change when the input is changed to a new value and held there, thus allowing the process to reach a new steady state. Stated another way, the steady-state gain K of a process corresponds to the ratio:

$$K = \frac{\bar{y}_2 - \bar{y}_1}{\bar{u}_2 - \bar{u}_1} \tag{4-41}$$

where subscripts 1 and 2 indicate different steady states and $(\bar{y}, \bar{u})$ denote the corresponding steady-state values of the output and input variables. The steady-state gain is constant for linear processes regardless of the operating conditions. This is not true for a nonlinear process, as discussed in Section 4.3.

Setting $s = 0$ in $G(s)$ gives the steady-state gain of a process if the gain exists.[1] For example, in Eqs. 4-35 and 4-36, K_1 and K_2 are steady-state gains. This feature is a consequence of the Final Value Theorem presented in Chapter 3. If a unit step change in input is assumed, the corresponding output change for $t \to \infty$ is $\lim G(s)$ as $s \to 0$.

Another important property of the transfer function is that the order of the denominator polynomial (in s) is the same as the order of the corresponding differential equation. A general linear nth-order differential equation has the form

$$a_n \frac{d^n y}{dt^n} + a_{n-1}\frac{d^{n-1} y}{dt^{n-1}} + \cdots + a_1 \frac{dy}{dt} + a_0 y = b_m \frac{d^m u}{dt^m} + b_{m-1}\frac{d^{m-1} u}{dt^{m-1}} + \cdots + b_1 \frac{du}{dt} + b_0 u \tag{4-42}$$

where u and y are input and output deviation variables, respectively. The transfer function obtained by Laplace transformation of (4-42) with $y(0) = 0$ and all initial conditions for the derivatives of u and y set equal to zero is

$$G(s) = \frac{Y(s)}{U(s)} = \frac{\sum_{i=0}^{m} b_i s^i}{\sum_{i=0}^{n} a_i s^i} = \frac{b_m s^m + b_{m-1}s^{m-1} + \cdots + b_0}{a_n s^n + a_{n-1}s^{n-1} + \cdots + a_0} \tag{4-43}$$

Note that the numerator and denominator polynomials of the transfer function have the same orders (m and n, respectively) as the ODE in (4-42).

The steady-state gain of $G(s)$ is b_o/a_o, obtained by setting $s = 0$. If both the numerator and denominator of Eq. 4-43 are divided by a_o, the characteristic (or denominator) polynomial can be factored into a product $\Pi(\tau_i s + 1)$ where τ_i denotes a time constant.

$$G(s) = \frac{Y(s)}{U(s)} = \frac{KB(s)}{(\tau_1 s+1)(\tau_2 s+1)\cdots(\tau_n s+1)} \tag{4-44}$$

where gain K and m-th order polynomial $B(s)$ are obtained from the numerator of Eq. 4-43.

In this *time constant form*, inspection of the individual time constants provides information about the speed, and qualitative features, of the system response. This important point is discussed in detail in Chapter 5, after some additional mathematical tools have been developed.

The orders of the numerator and denominator polynomials in Eq. 4-43 are restricted by physical considerations so that $n \geq m$. Suppose that a process was modeled by

$$a_0 y = b_1 \frac{du}{dt} + b_0 u \tag{4-45}$$

That is, $n = 0$ and $m = 1$ in Eq. 4-42. This system will respond to a step change in $u(t)$ with an impulse at time zero, because du/dt is infinite at the time the step change occurs. The ability to respond infinitely quickly to a sudden change in input is impossible to achieve with any real (physical) process, although it can be approximated in some instances—for example, in an explosion. Therefore, we refer to the restriction $n \geq m$ as a *physical realizability* condition. It provides a diagnostic check on transfer functions derived from a high-order differential equation or from a set of first-order differential equations. Those transfer functions where $m > 0$, are said to exhibit *numerator dynamics*. There are, however, many important cases where m is zero.

[1] Some processes do not exhibit a steady-state gain, for example, the integrating process in Chapter 5.

4.2.1 Additive Property of Transfer Functions

An output variable can be affected by more than one input variable. Because transfer functions are models of *linear systems*, the Principal of Superposition applies (cf. Section 3.1). Thus, the effects of the individual inputs on the output are additive, as shown in Fig. 4.2 for a two-input, single-output system:

$$Y(s) = G_1(s)U_1(s) + G_2(s)U_2(s) \tag{4-46}$$

This additive property is also illustrated in the transfer function model for the blending system in Eq. 4-33.

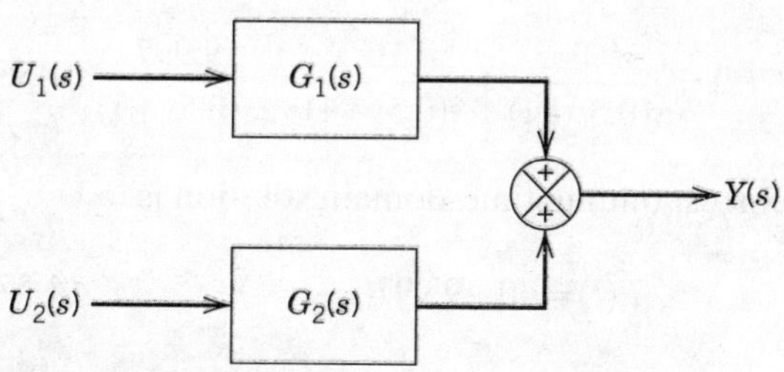

Figure 4.2 Block diagram of an additive transfer function model.

4.2.2 Multiplicative Property of Transfer Functions

Transfer function models also exhibit a *multiplicative property* for a sequence of individual models. For the series configuration of two transfer functions shown in Fig. 4.3, it follows that

$$Y_1(s) = G_1(s)U(s) \tag{4-47}$$

$$Y_2(s) = G_2(s)Y_1(s) = G_2(s)G_1(s)U(s) \tag{4-48}$$

Thus, the overall transfer function between Y_2 and U is the product of the two individual transfer functions, G_1 and G_2.

The additive and multiplicative properties of transfer function models are illustrated by Examples 4.6 and 4.7, respectively.

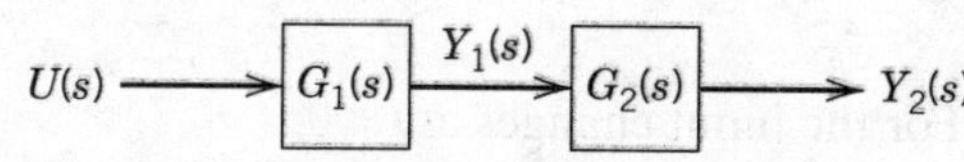

Figure 4.3 Block diagram of multiplicative (series) transfer function model.

EXAMPLE 4.6

The dynamic model of a stirred-tank heating process with constant volume in Eq. 2-40 was derived in Section 2.5.1:

$$V\rho C\frac{dT}{dt} = wC(T_i - T) + Q \tag{2-40}$$

At the nominal steady state, the inlet stream temperature is $\bar{T}_i = 22\,°\text{C}$ and the heater input is $\bar{Q} = 36$ kW. Additional information:

$w = 90$ kg/min $\qquad \rho = 1000\ \text{kg/m}^3$

$C = 1340$ J/kg °C $\qquad V = 45$ L

Substituting numerical values in the steady-state version of Eq. 2-40 gives $\bar{T} = 40\,°\text{C}$. At $t = 0$, inlet temperature T_i is changed from 22 to 32 °C and the heater input Q is changed from 36 to 30 kW.

(a) Derive a transfer function model for this system.

(b) Use the model to calculate the exit stream temperature response, $T(t)$.

Solution

(a) The heating process has two inputs (T_i and Q) and one output (T). A dynamic model was derived in Section 2.5.1:

$$V\rho C\frac{dT}{dt} = wC(T_i - T) + Q \tag{2-40}$$

The steady-state version of Eq. 2-40 is

$$0 = wC(\bar{T}_i - \bar{T}) + \bar{Q} \tag{4-49}$$

Introduce deviation variables by subtracting Eq. 4-49 from Eq. 2-40 and noting that $dT/dt = dT'/dt$:

$$V\rho C\frac{dT'}{dt} = wC[(T_i - \bar{T}_i) - (T - \bar{T})] + (Q - \bar{Q}) \tag{4-50}$$

or

$$V\rho C\frac{dT'}{dt} = wC(T_i' - T') + Q' \tag{4-51}$$

Taking Laplace transforms and assuming that the process is initially at the nominal steady state gives $T'(t=0)=0$:

$$V\rho C s T'(s) = wC[T_i'(s) - T'(s)] + Q'(s) \quad (4\text{-}52)$$

Rearranging gives the desired transfer function model:

$$T'(s) = \frac{K_2}{\tau s+1}Q'(s) + \frac{K_1}{\tau s+1}T_i'(s) \quad (4\text{-}53)$$

where

$$\tau \triangleq \frac{\rho V}{w}, \quad K_1 \triangleq 1, \quad K_2 \triangleq \frac{1}{wC} \quad (4\text{-}54)$$

(b) For the input changes at $t=0$,

$$T_i'(s) = \frac{32-22}{s} = \frac{10}{s}$$

$$Q'(s) = \frac{30-36}{s} = -\frac{6}{s}$$

The time constant and gains are

$$\tau = \frac{(1000)(45\times10^{-3})}{90} = 0.5\,\text{min}, \quad K_1 = 1,$$

$$K_2 = \frac{1}{(90)(1340)/60} = 5\times10^{-4}\,\text{W}$$

Substituting in Eq. 4-53 yields

$$T'(s) = \frac{5\times10^{-4}}{0.5s+1}\left(-\frac{6}{s}\right) + \frac{1}{0.5s+1}\left(\frac{10}{s}\right) \quad (4\text{-}55)$$

After simplification,

$$T'(s) = \frac{-0.003}{s(0.5s+1)} + \frac{10}{s(0.5s+1)} = \frac{9.997}{s(0.5s+1)} \quad (4\text{-}56)$$

The corresponding time-domain solution is

$$T(t) = 40 + 9.997(1-e^{-2t}) \quad (4\text{-}57)$$

Equation 4-56 shows the individual effects of the two input changes. At steady state, the reduction in the heater input lowers the temperature by 0.003 °C, while the inlet temperature change increases it by 10 °C, for a net increase of 9.997 °C.

EXAMPLE 4.7

Two surge tanks are placed in series so that the exit flow from the top tank flows into the lower tank, as shown in Fig. 4.4. If an exit flow rate is proportional to liquid level (or *head*) in that tank, derive the transfer function that relates changes in the exit flow rate q_2 of the lower tank to changes in the inlet flow rate to the top tank q_i. Show how this overall transfer function, $Q_2'(s)/Q_1'(s)$, is related to the individual transfer functions, $H_1'(s)/Q_i'(s)$, $Q_1'(s)/H_1'(s)$, $H_2'(s)/Q_1'(s)$, and $Q_2'(s)/H_2'(s)$. $H_1'(s)$ and $H_2'(s)$ denote the Laplace transforms of the deviations in Tank 1 and Tank 2 levels, respectively. Assume that the two tanks have cross-sectional areas, A_1 and A_2, and valve resistances, R_1 and R_2, respectively.

Solution

Equations 2-56 and 2-57 are valid for each tank; for Tank 1,

$$A_1\frac{dh_1}{dt} = q_i - q_1 \quad (4\text{-}58)$$

$$q_1 = \frac{1}{R_1}h_1 \quad (4\text{-}59)$$

Substituting Eq. 4-59 into Eq. 4-58 eliminates q_1:

$$A_1\frac{dh_1}{dt} = q_i - \frac{1}{R_1}h_1 \quad (4\text{-}60)$$

Putting Eq. 4-60 and Eq. 4-59 into deviation variable form gives

$$A_1\frac{dh_1'}{dt} = q_i' - \frac{1}{R_1}h_1' \quad (4\text{-}61)$$

$$q_1' = \frac{1}{R_1}h_1' \quad (4\text{-}62)$$

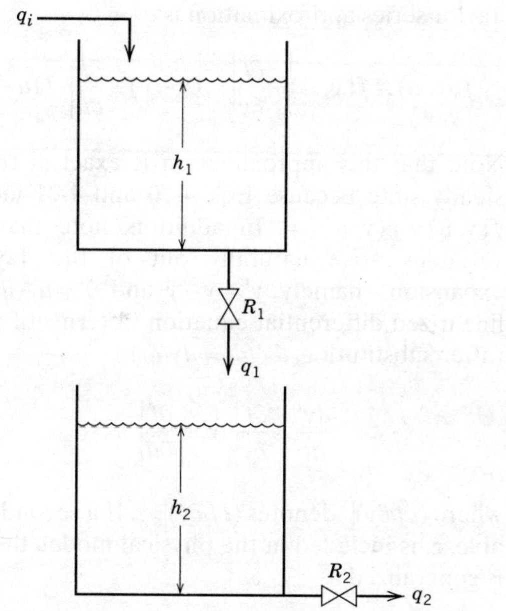

Figure 4.4 Schematic diagram of two surge tanks in series.

Transfer function $H_1'(s)/Q_{1i}'(s)$ is obtained by Laplace transforming Eq. 4-61 and rearranging to obtain

$$\frac{H_1'(s)}{Q_i'(s)} = \frac{R_1}{A_1R_1s+1} = \frac{K_1}{\tau_1 s+1} \tag{4-63}$$

where $K_1 \triangleq R_1$ and $\tau_1 \triangleq A_1R_1$. Similarly, the transfer function relating $Q_1'(s)$ to $H_1'(s)$ is obtained by Laplace transforming Eq. 4-62.

$$\frac{Q_1'(s)}{H_1'(s)} = \frac{1}{R_1} = \frac{1}{K_1} \tag{4-64}$$

The same procedure leads to the corresponding transfer functions for Tank 2,

$$\frac{H_2'(s)}{H_1'(s)} = \frac{R_2}{A_2R_2s+1} = \frac{K_2}{\tau_2 s+1} \tag{4-65}$$

$$\frac{Q_2'(s)}{H_2'(s)} = \frac{1}{R_2} = \frac{1}{K_2} \tag{4-66}$$

where $K_2 \triangleq R_2$ and $\tau_2 \triangleq A_2R_2$. The overall transfer function relating the outflow from Tank 2 to the inflow to Tank 1 can be derived by forming the product of Eq. 4-63 through Eq. 4-66.

$$\frac{Q_2'(s)}{Q_i'(s)} = \frac{Q_2'(s)}{H_2'(s)}\frac{H_2'(s)}{Q_1'(s)}\frac{Q_1'(s)}{H_1'(s)}\frac{H_1'(s)}{Q_i'(s)} \tag{4-67}$$

or

$$\frac{Q_2'(s)}{Q_i'(s)} = \frac{1}{K_2}\frac{K_2}{\tau_2 s+1}\frac{1}{K_1}\frac{K_1}{\tau_1 s+1} \tag{4-68}$$

which can be simplified to yield

$$\frac{Q_2'(s)}{Q_i'(s)} = \frac{1}{(\tau_1 s+1)(\tau_2 s+1)} \tag{4-69}$$

which is a second-order transfer function (does the unity gain make sense on physical grounds?). Figure 4.5 is a block diagram showing the *information flow* for this system.

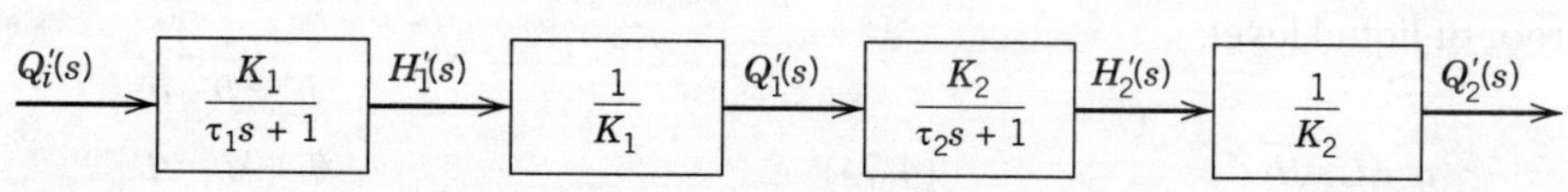

Figure 4.5 Block diagram model for two liquid surge tanks in series.

The multiplicative property of transfer functions proves to be quite valuable in designing process control systems because of the series manner in which process units are connected.

4.3 LINEARIZATION OF NONLINEAR MODELS

In the previous sections, the discussion has been limited to those processes that can be modeled by linear ordinary differential equations. However, for a wide variety of processes, the dynamic behavior depends on process variables in a nonlinear manner. Prominent examples include the exponential dependence of reaction rate on temperature (considered in Chapter 2), the nonlinear behavior of pH with flow rate of an acid or a base stream, and the asymmetric responses of distillate and bottoms compositions in a distillation column to increases and decreases in feed flow.

Classical process control methodology has been developed based on linear dynamic models such as linear

ODEs and transfer-function models. Thus, for highly nonlinear processes, the corresponding nonlinear models must be linearized in order to use classical process control techniques. As discussed later in this section, the linearization is typically based on the Taylor series expansion of the nonlinear model.

For continuous processes, the linearized model is based on a nominal steady-state condition. The model is relatively accurate in a region near the nominal conditions but becomes less accurate for large changes in operating conditions or large disturbances. Linearization is less often used for batch and semibatch processes that inherently operate during transient conditions.

In many instances, nonlinear processes remain in the vicinity of the nominal conditions for long periods of time. For these situations, a linearized model may be sufficiently accurate. Suppose a nonlinear dynamic model has been derived from first principles (e.g., material, energy, or momentum balances):

$$\frac{dy}{dt} = f(y,u) \tag{4-70}$$

where y is the output and u is the input. A linear approximation of this equation can be obtained by using a Taylor series expansion and truncating after the first-order terms. The reference point for linearization is the nominal steady-state operating point $(\bar{y}, \bar{u})$. Thus, the Taylor series approximation is:

$$f(y,u) \cong f(\bar{y},\bar{u}) + \left.\frac{\partial f}{\partial y}\right|_{\bar{y},\bar{u}} (y-\bar{y}) + \left.\frac{\partial f}{\partial u}\right|_{\bar{y},\bar{u}} (u-\bar{u}) \tag{4-71}$$

Note that this approximation is exact at the nominal steady state because Eqs. 4-70 and 4-71 indicate that $f(y,u) = f(\bar{y},\bar{u}) = 0$. In addition, note that deviation variables arise naturally out of the Taylor series expansion—namely, $y' = y - \bar{y}$ and $u' = u - \bar{u}$. Thus, the linearized differential equation in terms of y' and u' is (after substituting, $dy'/dt = dy/dt$)

$$\frac{dy'}{dt} = \left.\frac{\partial f}{\partial y}\right|_s y' + \left.\frac{\partial f}{\partial u}\right|_s u' \tag{4-72}$$

where $(\partial f/\partial y)|_s$ denotes $(\partial f/\partial y)|_{\bar{y},\bar{u}}$. If a second input variable, z, is included in the physical model, then Eq. 4-71 is generalized as

$$\frac{dy'}{dt} = \left.\frac{\partial f}{\partial y}\right|_s y' + \left.\frac{\partial f}{\partial u}\right|_s u' + \left.\frac{\partial f}{\partial z}\right|_s z' \tag{4-73}$$

where $z' = z - \bar{z}$.

A general procedure for developing a transfer function model from a nonlinear process model is summarized in Fig. 4.6, and demonstrated in the following example.

EXAMPLE 4.8

Consider a single tank liquid-level system where the outflow passes through a valve. Recalling Eq. 2-60, assume now that the valve discharge rate is related to the square root of liquid level:

$$q = C_v\sqrt{h} \tag{4-74}$$

where C_v depends on the size of the valve (see Chapter 8). Derive an approximate dynamic model for this process by linearization and compare with the results in Example 4.1.

Solution

The material balance for the process (Eq. 2-58) after substituting Eq. 4-74 is

$$A\frac{dh}{dt} = q_i - C_v\sqrt{h} \tag{4-75}$$

To obtain the system transfer function, linearize Eq. 4-75 about the steady-state conditions $(\bar{h}, \bar{q}_i)$. The deviation variables are

$$h' = h - \bar{h}$$

$$q_i' = q_i - \bar{q}_i$$

Applying Eq. 4-72 where $y = h$ and $x = q_i$, and $f(h, q_i)$ is the right side of Eq. 4-75, the linearized differential equation is

$$A\frac{dh'}{dt} = q_i' - \frac{C_v}{2\sqrt{\bar{h}}}h' \tag{4-76}$$

If we define the valve resistance R using the relation

$$\frac{1}{R} = \frac{C_v}{2\sqrt{\bar{h}}} \tag{4-77}$$

the resulting dynamic equation is analogous to the linear model presented earlier in Eq. 4-61:

$$A\frac{dh'}{dt} = q_i' - \frac{1}{R}h' \qquad (4\text{-}78)$$

The transfer function corresponding to Eq. 4-78 was derived earlier as Eq. 4-63.

Note that the linearized model in Eq. 4-78 has the same form as the corresponding linear model in Eq. 4-7 of Example 4.1. But the valve resistance parameter R_1 for the linearized model depends on the nominal steady-state liquid level $\bar{h}$, while the corresponding parameter R for the linear model was a constant.

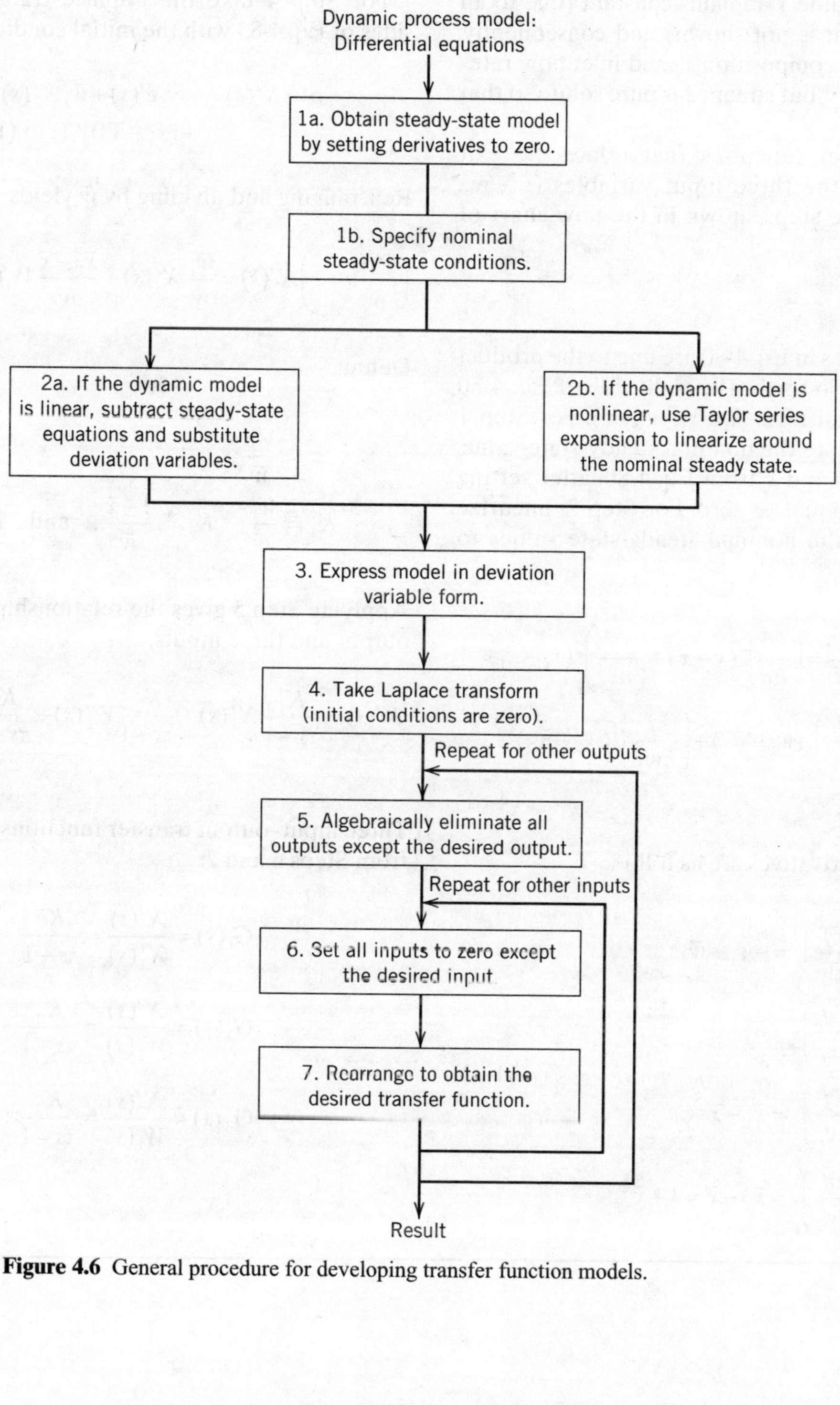

Figure 4.6 General procedure for developing transfer function models.

EXAMPLE 4.9

Again consider the stirred-tank blending system in Eqs. 2-17 and 2-18, written as

$$\rho\frac{dV}{dt} = w_1 + w_2 - w \quad (4\text{-}79)$$

$$\rho V\frac{dx}{dt} = w_1(x_1 - x) + w_2(x_2 - x) \quad (4\text{-}80)$$

Assume that volume V remains constant (due to an overflow line that is not shown) and consequently, $w = w_1 + w_2$. Inlet composition x_1 and inlet flow rates w_1 and w_2 can vary, but stream 2 is pure solute so that $x_2 = 1$.

Derive transfer functions that relate the exit composition to the three input variables (w_1, w_2, and x_1) using the steps shown in the flow chart of Fig. 4.6.

Solution

The nonlinearities in Eq. 4-80 are due to the product terms, w_1x_1, and so forth. The right side of Eq. 4-80 has the functional form $f(x, x_1, w_1, w_2)$. For Step 1 of Fig. 4.6, calculate the nominal steady-state value, $\bar{w}$, from Eq. 4-79 and $\bar{x}$ from Eq. 4-80, after setting the derivatives equal to zero. For Step 2, linearize Eq. 4-80 about the nominal steady-state values to obtain

$$\rho V\frac{dx}{dt} = \rho V\frac{dx'}{dt} = \left(\frac{\partial f}{\partial x}\right)_s (x - \bar{x}) + \left(\frac{\partial f}{\partial x_1}\right)_s (x_1 - \bar{x}_1) + \left(\frac{\partial f}{\partial w_1}\right)_s (w_1 - \bar{w}_1) + \left(\frac{\partial f}{\partial w_2}\right)_s (w_2 - \bar{w}_2) \quad (4\text{-}81)$$

The partial derivatives are as follow:

$$\left(\frac{\partial f}{\partial x}\right)_s = -\bar{w}_1 - \bar{w}_2$$
$$\left(\frac{\partial f}{x_1}\right)_s = \bar{w}_1$$
$$\left(\frac{\partial f}{w_1}\right)_s = \bar{x}_1 - \bar{x} \quad (4\text{-}82)$$
$$\left(\frac{\partial f}{w_2}\right)_s = \bar{x}_2 - \bar{x} = 1 - \bar{x}$$

Substitute Eq. 4-82 into Eq. 4-81 and introduce deviation variables (Step 3):

$$\rho V\frac{dx'}{dt} = -\bar{w}x' + \bar{w}_1x_1' + (\bar{x}_1 - \bar{x})w_1' + (1 - \bar{x})w_2' \quad (4\text{-}83)$$

The above equation is general in that it applies to any specified operating point.

For Step 4 take the Laplace transform of both sides of Eq. 4-83 with the initial condition, $x'(0) = 0$:

$$\rho VsX'(s) = -\bar{w}X'(s) + \bar{w}_1X_1'(s) + (\bar{x}_1 - \bar{x})W_1'(s) + (1 - \bar{x})W_2'(s)$$

Rearranging and dividing by $\bar{w}$ yields

$$\left(\frac{V\rho}{\bar{w}}s + 1\right)X'(s) = \frac{\bar{w}_1}{\bar{w}}X_1'(s) + \frac{\bar{x}_1 - \bar{x}}{\bar{w}}W_1'(s) + \frac{1 - \bar{x}}{\bar{w}}W_2'(s)$$

Define

$$\tau \triangleq \frac{V\rho}{\bar{w}}$$

$$K_1 \triangleq \frac{\bar{w}_1}{\bar{w}}, \quad K_2 \triangleq \frac{1 - \bar{x}}{\bar{w}}, \quad \text{and} \quad K_3 \triangleq \frac{\bar{x}_1 - \bar{x}}{\bar{w}}$$

Applying Step 5 gives the relationship for the single output and three inputs:

$$X'(s) = \frac{K_1}{\tau s + 1}X_1'(s) + \frac{K_2}{\tau s + 1}W_2'(s) + \frac{K_3}{\tau s + 1}W_1'(s) \quad (4\text{-}84)$$

Three input–output transfer functions can be derived from Steps 6 and 7:

$$G_1(s) = \frac{X'(s)}{X_1'(s)} = \frac{K_1}{\tau s + 1} \quad (4\text{-}85)$$

$$G_2(s) = \frac{X'(s)}{W_2'(s)} = \frac{K_2}{\tau s + 1} \quad (4\text{-}86)$$

$$G_3(s) = \frac{X'(s)}{W_1'(s)} = \frac{K_3}{\tau s + 1} \quad (4\text{-}87)$$

This example shows that individual transfer functions for a model with several inputs can be obtained by linearization of the nonlinear differential equation model. Note that all three transfer functions have the same time constant τ but different gains.

$$K_1 = \frac{\bar{w}_1}{\bar{w}} > 0$$

$$K_2 = \frac{1-\bar{x}}{\bar{w}} > 0$$

$$K_3 = \frac{\bar{x}_1 - \bar{x}}{\bar{w}} < 0$$

If a gain is positive, a steady-state increase in the input produces a steady-state increase in the output. A negative gain (e.g., K_3) has just the opposite effect.

Note that the gains of this nonlinear process depend on the nominal steady-state conditions. Thus, if these conditions were changed (e.g., to improve process performance), the numerical values of the gains and time constant would also change.

Finally, we consider the application of linearization methods when the model involves more than one nonlinear equation.

EXAMPLE 4.10

As shown in Chapter 2, a continuous stirred-tank reactor with a single first-order chemical reaction has the following material and energy balances:

$$V\frac{dc_A}{dt} = q(c_{Ai} - c_A) - Vkc_A \tag{2-66}$$

$$V\rho C\frac{dT}{dt} = wC(T_i - T) + (-\Delta H_R)Vkc_A + UA(T_c - T) \tag{2-68}$$

If the reaction rate coefficient k is given by the Arrhenius equation,

$$k = k_0 e^{-E/RT} \tag{2-63}$$

this model is nonlinear. However, it is possible to find approximate transfer functions relating the inputs and outputs. For the case where the flow rate (q or w) and inlet conditions (c_{Ai} and T_i) are assumed to be constant, derive the transfer function relating changes in the reactor concentration c_A to changes in the coolant temperature T_c.

Solution

For this situation, there is a single input variable T_c and two output variables c_A and T. First, the steady-state operating point must be determined (Step 1 in Fig. 4.6). Note that such a determination will require iterative solution of two nonlinear algebraic equations; this can be done using a Newton–Raphson method or similar algorithm (Chapra and Canale, 2014). Normally, we would specify $\bar{T}_i$, $\bar{c}_{Ai}$, and $\bar{c}_A$ and then determine $\bar{T}$ and $\bar{T}_c$ that satisfy Eqs. 2-66 and 2-68 at steady state. Then we can proceed with the linearization of Eqs. 2-70 and 2-72. Defining deviation variables c_A', T', and T_c', gives the following equations:

$$\frac{dc_A'}{dt} = a_{11}c_A' + a_{12}T' \tag{4-88}$$

$$\frac{dT'}{dt} = a_{21}c_A' + a_{22}T' + b_2 T_c' \tag{4-89}$$

where

$$a_{11} = -\frac{q}{V} - k_0 e^{-E/R\bar{T}}$$

$$a_{12} = -k_0 e^{-E/R\bar{T}}\bar{c}_A\left(\frac{E}{R\bar{T}^2}\right)$$

$$a_{21} = \frac{(-\Delta H_R)k_0 e^{-E/R\bar{T}}}{\rho C}$$

$$a_{22} = \frac{1}{V\rho C}\left[-(wC + UA) + (-\Delta H_R)V\bar{c}_A k_0 e^{-E/R\bar{T}}\left(\frac{E}{R\bar{T}^2}\right)\right]$$

$$b_2 = \frac{UA}{V\rho C}$$

Note that Eq. 2-70 does not contain input variable T_c, so no T_c' term appears in Eq. 4-88. We can convert Eqs. 4-88 and 4-89 into a transfer function between the coolant temperature $T_c'(s)$ and the tank outlet concentration $C_A'(s)$ via Laplace transformation:

$$(s - a_{11})C_A'(s) = a_{12}T'(s) \tag{4-90}$$

$$(s - a_{22})T'(s) = a_{21}C_A'(s) + b_2 T_c'(s) \tag{4-91}$$

Substituting for $T'(s)$, Eq. 4-90 becomes

$$(s - a_{11})(s - a_{22})C_A'(s) = a_{12}a_{21}C_A'(s) + a_{12}b_2 T_c'(s) \tag{4-92}$$

Yielding

$$\frac{C_A'(s)}{T_c'(s)} = \frac{a_{12}b_2}{s^2 - (a_{11} + a_{22})s + a_{11}a_{22} - a_{12}a_{21}} \tag{4-93}$$

which is a second-order transfer function. The a and b coefficients can be evaluated for specified nominal conditions.

4.4 STATE-SPACE AND TRANSFER FUNCTION MATRIX MODELS

Dynamic models derived from physical principles typically consist of one or more ordinary differential equations (ODEs). In this section, we consider a general class of ODE models referred to as *state-space models*, that provide a compact and useful representation of dynamic systems. Although we limit our discussion to linear state-space models, nonlinear state-space models are also very useful and provide the theoretical basis for the analysis of nonlinear processes (Henson and Seborg, 1997; Khalil, 2002).

Consider a linear state-space model,

$$\dot{\boldsymbol{x}} = \boldsymbol{Ax} + \boldsymbol{Bu} + \boldsymbol{Ed} \tag{4-94}$$

$$\boldsymbol{y} = \boldsymbol{Cx} \tag{4-95}$$

where $\boldsymbol{x}$ is the *state vector*; $\boldsymbol{u}$ is the input vector of manipulated variables (also called *control variables*); $\boldsymbol{d}$ is the disturbance vector; and $\boldsymbol{y}$ is the output vector of measured variables. The time derivative of $\boldsymbol{x}$ is denoted by $\dot{\boldsymbol{x}}(= d\boldsymbol{x}/dt)$; it is also a vector. (**Boldface symbols** are used to denote vectors and matrices, and plain text to represent scalars.) The elements of $\boldsymbol{x}$ are referred to as *state variables*. The elements of $\boldsymbol{y}$ are typically a subset of $\boldsymbol{x}$, namely, the state variables that are measured. In general, $\boldsymbol{x}$, $\boldsymbol{u}$, $\boldsymbol{d}$, and $\boldsymbol{y}$ are functions of time. Matrices $\boldsymbol{A}, \boldsymbol{B}, \boldsymbol{C}$, and $\boldsymbol{E}$ are constant matrices. The vectors in Eq. 4-94 can have different dimensions (or "lengths") and are usually written as deviation variables.

Because the state-space model in Eqs. 4-94 and 4-95 may seem rather abstract, it is helpful to consider a physical example.

EXAMPLE 4.11

Show that the linearized CSTR model of Example 4.9 can be written in the state-space form of Eqs. 4-94 and 4-95. Derive state-space models for two cases:

(a) Both c_A and T are measured
(b) Only T is measured

Solution

The linearized CSTR model in Eqs. 4-88 and 4-89 can be written in vector-matrix form using deviation variables:

$$\begin{bmatrix} \dfrac{dc'_A}{dt} \\ \dfrac{dT'}{dt} \end{bmatrix} = \begin{bmatrix} a_{11} & a_{12} \\ a_{21} & a_{22} \end{bmatrix} \begin{bmatrix} c'_A \\ T' \end{bmatrix} + \begin{bmatrix} 0 \\ b_2 \end{bmatrix} T'_c \tag{4-96}$$

Let $x_1 \triangleq c'_A$ and $x_2 \triangleq T'$, and denote their time derivatives by $\dot{x}_1$ and $\dot{x}_2$. In Eq. 4-96 the coolant temperature T_c is considered to be a manipulated variable. For this example, there is a single control variable, $u \triangleq T'_c$, and no disturbance variable. Substituting these definitions into Eq. 4-96 gives

$$\begin{bmatrix} \dot{x}_1 \\ \dot{x}_2 \end{bmatrix} = \underbrace{\begin{bmatrix} a_{11} & a_{12} \\ a_{21} & a_{22} \end{bmatrix}}_{\boldsymbol{A}} \begin{bmatrix} x_1 \\ x_2 \end{bmatrix} + \underbrace{\begin{bmatrix} 0 \\ b_2 \end{bmatrix}}_{\boldsymbol{B}} u \tag{4-97}$$

which is in the form of Eq. 4-94 with $\boldsymbol{x} = \text{col}[x_1, x_2]$. ("col" denotes a column vector.)

(a) If both T and c_A are measured, then $\boldsymbol{y} = \boldsymbol{x}$ and $\boldsymbol{C} = \boldsymbol{I}$ in Eq. 6-61, where $\boldsymbol{I}$ denotes the 2×2 identity matrix. $\boldsymbol{A}$ and $\boldsymbol{B}$ are defined in Eq. 4-97.

(b) When only T is measured, output vector $\boldsymbol{y}$ is a scalar, $y = T'$, and $\boldsymbol{C}$ is a row vector, $\boldsymbol{C} = [0, 1]$.

Note that the state-space model for Example 6.5 has $\boldsymbol{d} = \boldsymbol{0}$, because disturbance variables were not included in Eq. 4-96. By contrast, suppose that the feed composition and feed temperature are considered to be disturbance variables in the original nonlinear CSTR model in Eqs. 2-66 and 2-68. Then the linearized model would include two additional deviation variables c'_{Ai} and T'_i, which would also be included in Eq. 4-96. As a result, Eq. 4-97 would be modified to include two disturbance variables, $d_1 \triangleq c'_{Ai}$ and $d_2 \triangleq T'_i$.

The state-space model in Eq. 4-94 contains both dependent variables, the elements of $\boldsymbol{x}$, and independent variables, the elements of $\boldsymbol{u}$ and $\boldsymbol{d}$. But why is $\boldsymbol{x}$ referred to as the "state vector"? This term is used because $\boldsymbol{x}(t)$ uniquely determines the state of the system at any time, t. Suppose that at time t, the initial value $\boldsymbol{x}(0)$ is specified and $\boldsymbol{u}(t)$ and $\boldsymbol{d}(t)$ are known over the time period $[0, t]$. Then $\boldsymbol{x}(t)$ is unique and can be determined from the analytical solution or by numerical integration. Analytical techniques are described in control engineering textbooks (e.g., Franklin et al., 2005; Ogata, 2008), while numerical and analytical solutions can be readily obtained using software packages such as *MATLAB* or *Mathematica*.

4.4.1 Stability of State-Space Models

A detailed analysis of state-space models is beyond the scope of this book but is available elsewhere (e.g., Franklin et al., 2005; Ogata, 2008). One important property of state-space models is *stability*. A state-space model is said to be stable if the response $\boldsymbol{x}(t)$ is bounded for all $\boldsymbol{u}(t)$ and $\boldsymbol{d}(t)$ that are bounded. The stability characteristics of a state-space model can be determined from a necessary and sufficient condition:

Stability Criterion for State-Space Models

The state-space model in Eq. 4-94 will exhibit a bounded response $\mathbf{x}(t)$ *for all bounded* $\mathbf{u}(t)$ *and* $\mathbf{d}(t)$ *if and only if all of the eigenvalues of* $\mathbf{A}$ *have negative real parts.*

Note that stability is solely determined by $\boldsymbol{A}$; the $\boldsymbol{B}, \boldsymbol{C}$, and $\boldsymbol{E}$ matrices have no effect.

Next, we review concepts from linear algebra that are used in stability analysis. Suppose that $\boldsymbol{A}$ is an $n \times n$ matrix where n is the dimension of the state vector, $\boldsymbol{x}$. Let λ denote an *eigenvalue* of $\boldsymbol{A}$. By definition, the eigenvalues are the n values of λ that satisfy the equation $\lambda \boldsymbol{x} = \boldsymbol{A}\boldsymbol{x}$ (Strang, 2005). The corresponding values of $\boldsymbol{x}$ are the *eigenvectors* of $\boldsymbol{A}$. The eigenvalues are the roots of the characteristic equation.

$$|\lambda \boldsymbol{I} - \boldsymbol{A}| = 0 \tag{4-98}$$

where $\boldsymbol{I}$ is the $n \times n$ identity matrix and $|\lambda \boldsymbol{I} - \boldsymbol{A}|$ denotes the determinant of the matrix $\lambda \boldsymbol{I} - \boldsymbol{A}$.

EXAMPLE 4.12

Determine the stability of the state-space model with the following $\mathbf{A}$ matrix:

$$\mathbf{A} = \begin{bmatrix} -4.0 & 0.3 & 1.5 \\ 1.2 & -2.0 & 1.0 \\ -0.5 & 2.0 & -3.5 \end{bmatrix}$$

Solution

The stability criterion for state-space models indicates that stability is determined by the eigenvalues of $\mathbf{A}$. They can be calculated using the MATLAB command, *eig*, after defining $\mathbf{A}$:

$$\mathbf{A} = [-4.0 \;\; 0.3 \;\; 1.5; \;\; 1.2 \;\; -2 \;\; 1.0; \;\; -0.5 \;\; 2.0 \;\; -3.5]$$
$$eig(A)$$

The eigenvalues of $\mathbf{A}$ are -0.83, $-4.33 + 1.18j$, and $-4.33 - 1.18j$ where $j \triangleq \sqrt{-1}$. Because all three eigenvalues have negative real parts, the state-space model is stable, although the dynamic behavior will exhibit oscillation due to the presence of imaginary components in the eigenvalues.

4.4.2 The Relationship between State-Space and Transfer Function Models

State-space models can be converted to equivalent transfer function models. Consider again the CSTR model in Eq. 4-97, which can be expanded as

$$\dot{x}_1 = a_{11}x_1 + a_{12}x_2 \tag{4-99}$$

$$\dot{x}_2 = a_{21}x_1 + a_{22}x_2 + b_2 u \tag{4-100}$$

Apply the Laplace transform to each equation (assuming zero initial conditions for each deviation variable, x_1 and x_2):

$$sX_1(s) = a_{11}X_1(s) + a_{12}X_2(s) \tag{4-101}$$

$$sX_2(s) = a_{21}X_1(s) + a_{22}X_2(s) + b_2U(s) \tag{4-102}$$

Solving Eq. 4-101 for $X_2(s)$ and substituting into Eq. 4-102 gives the equivalent transfer function model relating X_1 and U:

$$\frac{X_1(s)}{U(s)} = \frac{a_{12}b_2}{s^2 - (a_{11} + a_{22})s + a_{11}a_{22} - a_{12}a_{21}} \tag{4-103}$$

Equation 4-101 can be used to derive the transfer function relating X_2 and U:

$$\frac{X_2(s)}{U(s)} = \frac{b_2(s - a_{11})}{s^2 - (a_{11} + a_{22})s + a_{11}a_{22} - a_{12}a_{21}} \tag{4-104}$$

Note that these two transfer functions are also the transfer functions for $C'_A(s)/T'_A(s)$ and $T'(s)/T'_c(s)$, respectively, as a result of the definitions for x_1, x_2, and u. Furthermore, the roots of the denominator of Eqs. 4-103 and 4-104 are also the eigenvalues of $\boldsymbol{A}$ in Eq. 4-97.

EXAMPLE 4.13

To illustrate the relationships between state-space models and transfer functions, again consider the electrically heated, stirred tank model in Section 2.4.3. First, equations (2-51) and (2-52) are converted into state-space form by dividing (2-51) by mC and (2-52) by m_eC_e, respectively:

$$\frac{dT}{dt} = \frac{w}{m}(T_i - T) + \frac{h_eA_e}{mC}(T_e - T) \tag{4-105}$$

$$\frac{dT_e}{dt} = \frac{Q}{m_eC_e} - \frac{h_eA_e}{m_eC_e}(T_e - T) \tag{4-106}$$

The nominal parameter values are the same as in Example 2.7:

$$\frac{m}{w} = 10\,\text{min} \quad \frac{m_eC_e}{h_eA_e} = 1.0\,\text{min}$$

$$\frac{m_eC_e}{wC} = 1.0\,\text{min} \quad \frac{1}{wC} = 0.05\,°\text{C min/kcal}$$

Consequently,

$$m_eC_e = 20\,\text{kcal/°C}$$
$$mC = 200\,\text{kcal/°C}$$
$$h_eA_e = 20\,\text{kcal/°C min}$$

(a) Using deviation variables (T', T_e', Q') determine the transfer function between temperature T' and heat input Q'. Consider the conditions used in Example 2.7: $\bar{Q} = 5000$ kcal/min and $\bar{T}_i = 100$ °C; at $t = 0$, Q is changed to 5400 kcal/min. Compare the expression for $T'(s)$ with the time domain solution obtained in Example 2.7.

(b) Calculate the eigenvalues of the state-space model.

Solution

(a) Substituting numerical values for the parameters gives

$$\frac{dT}{dt} = 0.1(T_i - T) + 0.1(T_e - T) \tag{4-107}$$

$$\frac{dT_e}{dt} = 0.05Q - (T_e - T) \tag{4-108}$$

The model can be written in deviation variable form (note that the steady-state values can be calculated to be $\bar{T} = 350$ °C and $\bar{T}_e = 640$ °C):

$$\frac{dT'}{dt} = 0.1(T_i' - T') + 0.1(T_e' - T') \tag{4-109}$$

$$\frac{dT_e'}{dt} = 0.05Q' - (T_e' - T') \tag{4-110}$$

$T_i' = 0$ because the inlet temperature is assumed to be constant. Taking the Laplace transform gives

$$sT'(s) = -0.2T'(s) + 0.1T_e'(s) \tag{4-111}$$

$$sT_e'(s) = 0.05Q'(s) - T_e'(s) + T'(s) \tag{4-112}$$

Using the result derived earlier in Eq. 4-103, the transfer function is

$$\frac{T'(s)}{Q'(s)} = \frac{0.05}{10s^2 + 12s + 1} = \frac{0.005}{s^2 + 1.2s + 0.1} \tag{4-113}$$

For the step change of 400 kcal/min, $Q'(s) = \frac{400}{s}$ kcal/min, then

$$T'(s) = \frac{2}{s(s^2 + 12s + 0.1)}$$

The reader can verify that the inverse Laplace transform is

$$T'(t) = 20[1 - 1.089e^{-0.09t} + 0.0884e^{-1.11t}] \tag{4-114}$$

which is the same solution as obtained in Example 2.4.

(b) The state-space model in 4-109 and 4-110 can be written as

$$\begin{bmatrix} \dot{T}' \\ \dot{T}_e' \end{bmatrix} = \begin{bmatrix} -0.2 & 0.1 \\ 1 & -1 \end{bmatrix} \begin{bmatrix} T' \\ T_e' \end{bmatrix} + \begin{bmatrix} 0 \\ 0.5 \end{bmatrix} Q' \tag{4-115}$$

The 2 × 2 state matrix for this linear model is the same when either deviation variables (T', T_e', Q') or the original physical variables (T, T_e, Q) are employed. The eigenvalues λ_i of the state matrix can be calculated from setting the determinant of $\boldsymbol{A} - \lambda\boldsymbol{I}$ equals zero.

$$\det\begin{bmatrix}-0.2-\lambda & 0.1\\ 1 & -1-\lambda\end{bmatrix}=0$$

$$(-0.2-\lambda)(-1-\lambda)-0.1=0$$

$$\lambda^2+1.2\lambda+0.1=0$$

Solving for λ using the quadratic formula gives

$$\lambda_{1,2}=\frac{-1.2+\sqrt{1.44-0.4}}{2}=-1.11,\ -0.09$$

which is the same result that was obtained using the transfer function approach. Because both eigenvalues are real, the response is non-oscillating, as shown in Figure. 2.7.

The state-space form of the dynamic system is not unique. If we are principally interested in modeling the dynamics of the temperature T, the state variables of the process model can be defined as

$$x=\begin{bmatrix}T'\\ \dfrac{dT'}{dt}\end{bmatrix}\ \text{(note } T'_e \text{ is not an explicit variable).}$$

The resulting state-space description analogous to Eqs. 4-107 and 4-108 would be

$$\frac{dx_1}{dt}=x_2 \tag{4-116}$$

$$\frac{dx_2}{dt}=-1.2x_2-0.1x_1+0.05u \tag{4-117}$$

Note that if Eq. 4-116 is differentiated once and we substitute the right-hand side of Eq. 4-117 for dx_2/dt, then the same second-order model for T' is obtained. This is left as an exercise for the reader to verify. In addition, it is possible to derive other state-space descriptions of the same second-order ODE, because the state-space form is not unique.

A general expression for the conversion of a state-space model to the corresponding transfer function model will now be derived. The starting point for the derivation is the standard state-space model in Eqs. 4-94 and 4–95. Taking the Laplace transform of these two equations gives

$$s\boldsymbol{X}(s)=\boldsymbol{AX}(s)+\boldsymbol{BU}(s)+\boldsymbol{ED}(s) \tag{4-118}$$

$$\boldsymbol{Y}(s)=\boldsymbol{CX}(s) \tag{4-119}$$

where $\boldsymbol{Y}(s)$ is a column vector that is the Laplace transform of $\boldsymbol{y}(t)$. The other vectors are defined in an analogous manner. After applying linear algebra and rearranging, a transfer function representation can be derived (Franklin et al., 2005):

$$\boldsymbol{Y}(s)=\boldsymbol{G}_p(s)\boldsymbol{U}(s)+\boldsymbol{G}_d(s)\boldsymbol{D}(s) \tag{4-120}$$

where the *process transfer function matrix*, $\boldsymbol{G}_p(s)$ is defined as

$$\boldsymbol{G}_p(s)\triangleq\boldsymbol{C}[s\boldsymbol{I}-\boldsymbol{A}]^{-1}\boldsymbol{B} \tag{4-121}$$

and the *disturbance transfer function matrix* $\boldsymbol{G}_d(s)$ is defined as

$$\boldsymbol{G}_d(s)\triangleq\boldsymbol{C}[s\boldsymbol{I}-\boldsymbol{A}]^{-1}\boldsymbol{E} \tag{4-122}$$

Note that the dimensions of the transfer function matrices depend on the dimensions of **Y**, **U**, and **D**.

Fortunately, we do not have to perform tedious evaluations of expressions such as Eqs. 4-121 and 4–122 by hand. State-space models can be converted to transfer function form using the MATLAB command *ss2tf*.

EXAMPLE 4.14

Determine $G_p(s)$ for temperature T' and input Q' for Example 4.13 using Eqs. 4-121 and 4-122 and the MATLAB command *ss2tf*.

Solution

For part (a) of Example 4.13, $Y(s) = X_1(s)$, and there is one manipulated variable and no disturbance variable. Consequently, Eq. 4-120 reduces to

$$Y(s) = C(sI - A)^{-1}BU(s) \tag{4-123}$$

where $G_p(s)$ is now a scalar transfer function.

To find the multivariable transfer function for $T'(s)/Q'(s)$, we use the following matrices from the state-space model Eq. 4-115:

$$A = \begin{bmatrix} -0.2 & 0.1 \\ 1 & -1 \end{bmatrix} \quad B = \begin{bmatrix} 0 \\ 0.05 \end{bmatrix} \quad C = [1 \ \ 0]$$

Using the MATLAB command *ss2tf* yields

$$G_p(s) = \frac{0.005}{s^2 + 1.2s + 0.1} \tag{4-124}$$

which is the same result as in Eq. 4-114. The reader can also derive $G_d(s)$ relating $T'(s)$ and $T_i'(s)$ using the matrix-based approach in Eq. 4-122.

It is also possible to convert a general transfer function matrix in the form of Eq. 4-121 to a state-space model, and vice versa, using a single command in MATLAB. Using such software is recommended when the state matrix is larger than 2 × 2.

4.5 POLES AND ZEROS AND THEIR EFFECT ON PROCESS RESPONSE

An important feature of the simple process models discussed in Chapter 5 is that their response characteristics are determined by the factors of the transfer function denominator. For example, consider a transfer function,

$$G(s) = \frac{K}{s(\tau_1 s + 1)(\tau_2^2 s^2 + 2\zeta\tau_2 s + 1)} \tag{4-125}$$

where $0 \le \zeta < 1$. Using partial fraction expansion followed by the inverse transformation operation, we know that the response of system (Eq. 4-125) to *any* input will contain the following functions of time:

- A constant term resulting from the s factor
- An e^{-t/τ_1} term resulting from the $(\tau_1 s + 1)$ factor
- $e^{-\zeta t/\tau_2} \sin \frac{\sqrt{1-\zeta^2}}{\tau_2} t$ and
- $e^{-\zeta t/\tau_2} \cos \frac{\sqrt{1-\zeta^2}}{\tau_2} t$

  terms resulting from the $(\tau_2^2 s^2 + 2\zeta\tau_2 s + 1)$ factor if $0 \le \zeta < 1$

Additional terms determined by the specific input forcing will also appear in the response, but the intrinsic dynamic features of the model, the so-called *response modes* or *natural modes,* are determined by the model itself. Each of the above response modes is determined from the factors of the denominator polynomial, which is also called the characteristic polynomial (cf. Section 3.3). The roots of these factors for Eq. 4-125 are

$$\begin{aligned} s_1 &= 0 \\ s_2 &= -\frac{1}{\tau_1} \\ s_3 &= -\frac{\zeta}{\tau_2} + j\frac{\sqrt{1-\zeta^2}}{\tau_2} \\ s_4 &= -\frac{\zeta}{\tau_2} - j\frac{\sqrt{1-\zeta^2}}{\tau_2} \end{aligned} \tag{4-126}$$

Roots s_3 and s_4 are obtained by applying the quadratic formula.

Control engineers refer to the values of s that are roots of the denominator polynomial as the *poles* of transfer function $G(s)$. Sometimes it is useful to plot the roots (poles) and to discuss process response characteristics in terms of root locations in the complex s plane. In Fig. 4.7, the ordinate expresses the imaginary part of each root; the abscissa expresses the real part. Figure 4.7 is based on Eq. 4-126 and indicates the presence of four poles: an integrating element (pole at the origin), one real pole (at $-1/\tau_1$), and a pair of complex conjugate poles, s_3 and s_4. The real pole is closer to the imaginary axis than the complex pair, indicating a slower response mode (e^{-t/τ_1} decays slower than $e^{-\zeta t/\tau_2}$). In general, the speed of response for a given mode increases as the pole location moves farther away from the imaginary axis.

Historically, plots such as Fig. 4.7 have played an important role in the design of mechanical and electrical control systems, but they are rarely used in designing process control systems. However, it is helpful to develop some intuitive feeling for the influence of pole locations. A pole to the right of the imaginary axis (called a *right-half plane pole*), for example, $s = +1/\tau$, indicates that one of the system response modes is $e^{t/\tau}$. This mode grows without bound as t

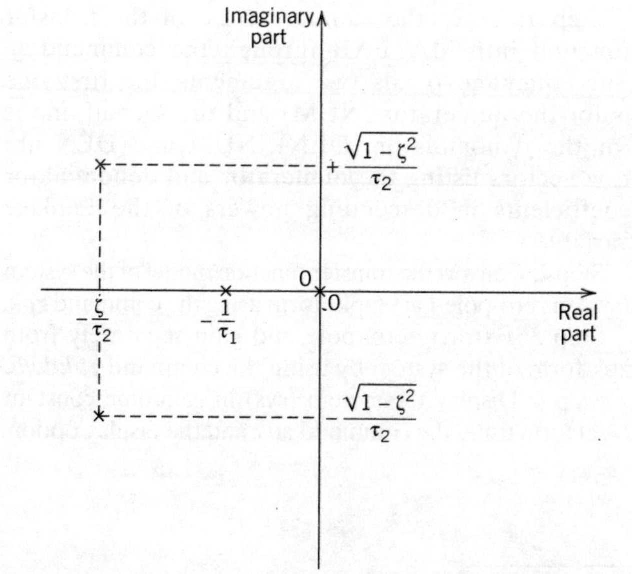

Figure 4.7 Poles of $G(s)$ (Eq. 4-125) plotted in the complex s plane (× denotes a pole location).

becomes large, a characteristic of unstable systems. As a second example, a complex pole always appears as part of a conjugate pair, such as s_3 and s_4 in Eq. 4-126. The complex conjugate poles indicate that the response will contain sine and cosine terms; that is, it will exhibit oscillatory modes.

All of the transfer functions discussed so far can be extended to represent more complex process dynamics simply by adding numerator terms. For example, some control systems contain a *lead–lag element*. The differential equation for this element is

$$\tau_1 \frac{dy}{dt} + y = K\left(\tau_a \frac{du}{dt} + u\right) \tag{4-127}$$

In Eq. 4-127, the standard first-order dynamics have been modified by the addition of the *du*/*dt* term multiplied by a time constant τ_a. The corresponding transfer function is

$$G(s) = \frac{K(\tau_a s + 1)}{\tau_1 s + 1} \tag{4-128}$$

Transfer functions with numerator terms such as $\tau_a s + 1$ above are said to exhibit *numerator dynamics*. Suppose that the integral of u is included in the input terms:

$$\tau_1 \frac{dy}{dt} + y = K\left(u + \frac{1}{\tau_a}\int_0^t u(t^*)dt^*\right) \tag{4-129}$$

The transfer function for Eq. 4-129, assuming zero initial conditions, is

$$G(s) = \frac{K(\tau_a s + 1)}{\tau_a s(\tau_1 s + 1)} \tag{4-130}$$

In this example, integration of the input introduces a pole at the origin (the $\tau_a s$ term in the denominator), an important point that will be discussed later.

The dynamics of a process are affected not only by the poles of $G(s)$, but also by the values of s that cause the numerator of $G(s)$ to become zero. These values are called the *zeros* of $G(s)$.

Before discussing zeros, it is useful to show several equivalent ways in which transfer functions can be written.

$$G(s) = \frac{\sum_{i=0}^{m} b_i s^i}{\sum_{i=0}^{n} a_i s^i} = \frac{b_m s^m + b_{m-1}s^{m-1} + \cdots + b_0}{a_n s^n + a_{n-1}s^{n-1} + \cdots + a_0} \tag{4-43}$$

which can also be written as

$$G(s) = \frac{b_m(s-z_1)(s-z_2)\ldots(s-z_m)}{a_n(s-p_1)(s-p_2)\ldots(s-p_n)} \tag{4-131}$$

where the z_i and p_i are zeros and poles, respectively. Note that the poles of $G(s)$ are also the roots of the characteristic equation. This equation is obtained by setting the denominator of $G(s)$, the characteristic polynomial, equal to zero.

It is convenient to express transfer functions in *gain/time constant form*; that is, b_0 is factored out of the numerator of Eq. 4-43 and a_0 out of the denominator to show the steady-state gain explicitly ($K = b_0/a_0 = G(0)$). Then the resulting expressions are factored to give

$$G(s) = K\frac{(\tau_a s + 1)(\tau_b s + 1)\cdots}{(\tau_1 s + 1)(\tau_2 s + 1)\cdots} \tag{4-132}$$

for the case where all factors represent real roots. Thus, the relationships between poles and zeros and the time constants are

$$z_1 = -1/\tau_a, \quad z_2 = -1/\tau_b, \cdots \tag{4-133}$$

$$p_1 = -1/\tau_1, \quad p_2 = -1/\tau_2, \cdots \tag{4-134}$$

The presence or absence of system zeros in Eq. 4-131 has no effect on the number and location of the poles and their associated response modes unless there is an exact cancellation of a pole by a zero with the same numerical value. However, the zeros exert a profound effect on the coefficients of the response modes (i.e., how they are weighted) in the system response. Such coefficients are found by partial fraction expansion. For practical control systems the number of zeros in

Eq. 4-131 is less than or equal to the number of poles ($m \le n$). When $m = n$, the output response is discontinuous after a step input change.

4.6 CONVERTING ONE FORM OF MODEL TO ANOTHER USING MATLAB

Different forms of transfer functions are discussed in section 4.5. In MATLAB it is easier to convert the standard form (or polynomial form) of transfer function to zero-pole-gain (zpk) form using the command *zpk*. However, to convert to gain/time constant (k/τ) form, we have to use the commands *zpk* and *zpkdata*. The following steps may be followed:

Step 1: Enter the standard from of the transfer function into MATLAB through the command *tf*. The command *tf* has two arguments, the first one is for the numerator (NUM) and the second one is for the denominator (DEN). NUM and DEN are row vectors listing the numerator and denominator coefficients in descending powers of the Laplace variable *s*.

Step 2: Convert the transfer function model of the system (sys) to zero-pole-gain (zpk) form using the command *zpk*.

Step 3: Extract zero, pole and gain separately from zpk form of the system by using the command *zpkdata*.

Step 4: Display the system (sys) in gain/time constant (k/τ) form using the command *zpk* and the display option.

EXAMPLE 4.15

Obtain Zero-Pole-Gain form and Gain/Time constant form from the following transfer function manually as well as through MATLAB commands.

$$G(s) = \frac{3s+2}{2s^2+3s+1}$$

Solution

$$G(s) = \frac{3s+2}{2s^2+3s+1} = \frac{2(1.5s+1)}{(2s+1)(s+1)}$$

(Gain/Time constant form)

$$= \frac{2(1.5s+1)}{(2s+1)(s+1)} = \frac{3(s+2/3)}{2(s+1/2)(s+1)}$$

$$= \frac{3(s+0.67)}{2(s+0.5)(s+1)} \quad \text{(zpk form)}$$

We can obtain the same result in MATLAB.

```
% sys is the given transfer function,
  G(s) in standard form
num = [3 2];
den = [2 3 1];
sys =tf(num,den);
% convert G(s) to zpk form
sys_zpk=zpk(sys)
%extract z,p and k
[z,p,k]=zpkdata(sys_zpk);
% display G(s) in time-constant-format
  as sys_ktau
sys_ktau=zpk(z,p,k,'DisplayFormat',
  'time constant')
```

The output obtained is shown below:

```
sys_zpk =

1.5 (s+0.6667)
--------------
(s+1) (s+0.5)

Continuous-time zero/pole/gain model.
sys_ktau =
  2 (1+1.5s)
 ------------
(1+s) (1+2s)

Continuous-time zero/pole/gain model.
```

4.6.1 Interconversion of Models

The state-space (SS) model of the dynamic system is discussed in Section 4.4. In general, we have three types of models – transfer function model, zero-pole-gain model and state-space model. These models can be converted from one form to another using the commands *tf2zp, tf2ss, zp2tf, zp2ss, ss2tf* and *ss2zp*.

Table 4.1 Conversion of dynamic models from one form to another

Purpose (Conversion from)	MATLAB statement
transfer function model to zero-pole-gain model	>>[Z,P,K]=tf2zp(NUM,DEN)
transfer function model to state-space model	>>[A,B,C,D]=tf2ss(NUM,DEN)
zero-pole-gain model to transfer function model	>>[NUM,DEN]=zp2tf(Z,P,K)
zero-pole-gain model to state-space model	>>[A,B,C,D]=zp2ss(Z,P,K)
state-space model to transfer function model	>>[NUM,DEN]=ss2tf(A,B,C,D,iu)
state-space model to zero-pole-gain model	>>[Z,P,K]=ss2zp(A,B,C,D,iu)

where iu = i^{th} input in the state-space model (default value is 1)

EXAMPLE 4.16

Convert the following transfer function model to (a) zero-pole-gain model and (b) state-space model using MATLAB.

$$G(s)=\frac{s+4}{s^3+6s^2+11s+6}$$

Solution

The following MATLAB code can be written

```
>>num=[1 4];
>>den=[1 6 11 6];
>>[Z,P,K]=tf2zp(num,den)
>>[A,B,C,D]=tf2ss(num,den)
```

From the outputs we can construct the ZPK model and SS model as follows

$$G(s)=\frac{s+4}{s^3+6s^2+11s+6}=\frac{s+4}{(s+1)(s+2)(s+3)} \quad \text{(ZPK model)}$$

$$\dot{x}=Ax+Bu=\begin{bmatrix}-6 & -11 & -6\\ 1 & 0 & 0\\ 0 & 1 & 0\end{bmatrix}x+\begin{bmatrix}1\\0\\0\end{bmatrix}u \quad \text{(SS Model)}$$

$$y=Cx+Du=\begin{bmatrix}0 & 1 & 4\end{bmatrix}x$$

EXAMPLE 4.17

The state space model of a system is given by

$$\dot{x}=Ax+Bu=\begin{bmatrix}0 & 1 & 0\\ 0 & 0 & 1\\ -10 & -11 & -6\end{bmatrix}x+\begin{bmatrix}0\\0\\8\end{bmatrix}u$$

$$y=Cx+Du=\begin{bmatrix}0 & 0 & 1\end{bmatrix}x$$

Obtain the transfer function model of the system using MATLAB command *ss2tf*.

Solution

The MATLAB code can be written as

```
>>A=[0 1 0;0 0 1;-10 -11 -6];
>>B=[0;0;8];
>>C=[1 0 0];
>>D=0;
>>[NUM,DEN]=ss2tf(A,B,C,D,1)
```

The output is given by

```
NUM =
    0 0 0 8
DEN =
    1.0000 6.0000 11.0000 10.0000
```

So, the transfer function model of the system can be expressed as

$$G(s)=\frac{8}{s^3+6s^2+11s+10}$$

EXAMPLE 4.18

A system is described by the differential equation

$$\frac{dx_1}{dt} = 2x_2 + 5u$$

$$\frac{dx_2}{dt} = -x_1 - 3x_2$$

where x_1 is the output and u is the input to the system. Obtain the transfer function representation of the system using the MATLAB command ss2tf.

Solution

The state space representation of the system is given by

$$\dot{x} = Ax + Bu$$
$$y = Cx + Du$$

where

$$A = \begin{bmatrix} 0 & 2 \\ -1 & -3 \end{bmatrix} \quad B = \begin{bmatrix} 5 \\ 0 \end{bmatrix}$$

$$C = \begin{bmatrix} 1 & 0 \end{bmatrix} \quad D = 0 \quad x = \begin{bmatrix} x_1 \\ x_2 \end{bmatrix}$$

MATLAB code

```
>>A= [0 2;-1 -3];
>>B= [5;0];
>>C= [1 0];
>>D=0;
>> [NUM,DEN] =ss2tf(A,B,C,D)
```

From the output we can get the transfer function model of the system as

$$G(s) = \frac{5s+15}{s^2+3s+2}$$

SUMMARY

In this chapter, we have introduced two important concepts, the transfer function model and state-space model, which are widely used in control system design and analysis. The transfer function model relates changes in a process output to changes in a process input and can be derived from a linear differential equation model using Laplace transformation methods. The transfer function contains key information about the steady-state and dynamic relations between input and output variables, namely, the process gain and time constants, respectively. Transfer functions are usually expressed in terms of deviation variables, that is, deviations from nominal steady-state conditions.

State-space models provide a convenient representation of dynamic models that can be expressed as a set of first-order, ordinary differential equations. State-space models can be derived from first principles models (e.g., material and energy balances) and used to describe both linear and nonlinear dynamic systems.

REFERENCES

Buckley, P. S., W. L. Luyben, and J. P. Shunta, *Design of Distillation Column Control Systems*, Instrument Society of America, Research Triangle Park, NC, 1985.

Chapra, S. C., and R. P. Canale, *Numerical Methods for Engineers*, 7th ed., McGraw-Hill, New York, 2014.

Franklin, G. F., J. D. Powell, and A. Emani-Naeini, *Feedback Control of Dynamic Systems*, 5th ed., Prentice-Hall, Upper Saddle River, NJ, 2005.

Henson, M. A., and D. E. Seborg (Eds.), *Nonlinear Process Control*, Prentice-Hall, Upper Saddle River, NJ, 1997.

Ogata, K., *Modern Control Engineering*, 5th ed., Prentice-Hall, Upper Saddle River, NJ, 2008.

Strang, G., *Linear Algebra and Its Applications*, 4th ed., Brooks Cole, Florence, KY, 2005.

EXERCISES

4.1 Consider a transfer function:

$$\frac{Y(s)}{U(s)} = \frac{d}{bs+c}$$

(a) What is the steady-state gain?

(b) For a step change of magnitude *M* in the input, will the output response be bounded for all values of constants *b*, *c* and *d*? Briefly justify your answer.

4.2 Consider the following transfer function:

$$G(s) = \frac{Y(s)}{U(s)} = \frac{2e^{-s}}{5s+1}$$

(a) What is the steady-state gain?

(b) What is the time constant?

(c) If $U(s) = 4/s$, what is the value of the output $y(t)$ when $t \to \infty$?

(d) For the same input, what is the value of the output when $t = 10$? What is the output when expressed as a fraction of the new steady-state value?

(e) If $U(s) = (1 - e^{-s})/s$, the unit rectangular pulse, what is the output when $t \to \infty$?

(f) If $u(t) = \delta(t)$, the unit impulse at $t = 0$, what is the output when $t \to \infty$?

(g) If $u(t) = 5 \sin 2t$, what is the value of the output when $t \to \infty$?

4.3 The dynamic behavior of a pressure sensor/transmitter can be expressed as a first-order transfer function (in deviation variables) that relates the measured value P_m to the actual pressure, P:

$$\frac{P_m'(s)}{P'(s)} = \frac{1}{10s+1}$$

Both P_m and P have units of kPa and the time constant has units of seconds. Suppose that an alarm will sound if P_m exceeds 310 kPa. If the process is initially at steady state $(\bar{P}_m = \bar{P} = 240\text{ kPa})$, and then P suddenly changes from 240 to 345 kPa at 1:30 PM, at what time will the alarm sound?

4.4 Consider the transfer function model in Exercise 4.2. For an initial condition of $y(0) = 4$ and a step change in u of magnitude 2 (at $t = 0$), calculate the response, $y(t)$.

Hint: First determine the corresponding differential equation model.

4.5 For the process modeled by

$$2\frac{dy_1}{dt} = -2y_1 - 3y_2 + 2u_1$$

$$\frac{dy_2}{dt} = 4y_1 - 6y_2 + 2u_1 + 4u_2$$

Find the four transfer functions relating two outputs (y_1, y_2) to two inputs (u_1, u_2). The u_i and y_i are deviation variables.

4.6 A stirred-tank blending system can be described by a first-order transfer function between the exit composition x and the inlet composition x_1 (both are mass fractions of solute):

$$\frac{X'(s)}{X_i'(s)} = \frac{K}{\tau s+1}$$

where $K = 0.6$ (dimensionless) and $\tau = 10$ min. When the blending system is at steady state $(\bar{x} = 0.3)$, the dynamic behavior is tested by quickly adding a large amount of a radioactive tracer, thus approximating an impulse function with magnitude 1.5.

(a) Calculate the exit composition response $x(t)$ using Laplace transforms and sketch $x(t)$. Based on this analytical expression, what is the value of $x(0)$?

(b) Using the Initial Value Theorem of Section 3.4, determine the value of $x(0)$.

(c) If the process is initially at a steady state with $\bar{x} = 0.3$, what is the value of $x(0)$?

(d) Compare your answer for parts (a)–(c) and briefly discuss any differences.

4.7 A single equilibrium stage in a distillation column is shown in Fig. E4.7. The model that describes this stage is

$$\frac{dH}{dt} = L_0 + V_2 - (L_1 + V_1)$$

$$\frac{dHx_1}{dt} = L_0x_0 + V_2y_2 - (L_1x_1 + V_1y_1)$$

$$y_1 = a_0 + a_1x_1 + a_2x_1^2 + a_3x_1^3$$

Assume that the molar holdup H in the stage is constant and that constant molal overflow occurs. Thus, $L_1 = L_0$ and $V_2 = V_1$.

(a) Linearize the resulting model and introduce deviation variables.

(b) For constant liquid and vapor flow rates, derive the four transfer functions relating outputs x_1 and y_1 to inputs x_0 and y_2. Place it in standard form.

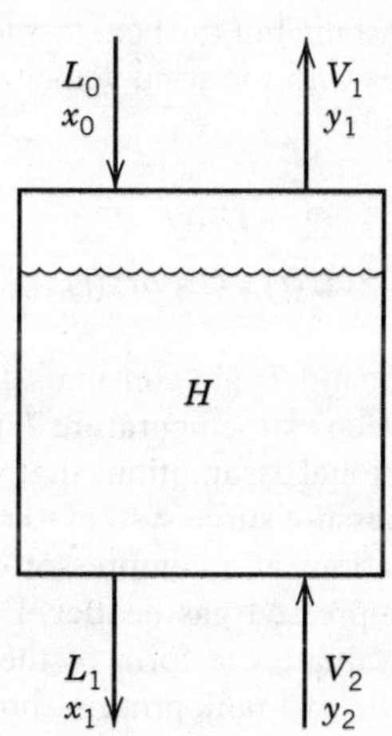

Figure E4.7

4.8 A surge tank in Fig. E4.8 is designed with a slotted weir so that the outflow rate w is related to the liquid level h to the 1.5 power; that is,

$$w = Rh^{1.5}$$

where R is a constant. If a single stream enters the tank with flow rate w_i, find the transfer function, $H'(s)/W_i'(s)$. Identify the gain and all time constants. Verify units.

The cross-sectional area of the tank is A. Liquid density ρ is constant.

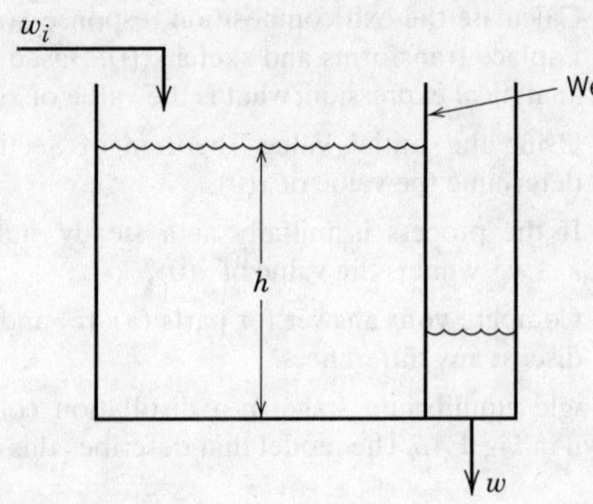

Figure E4.8

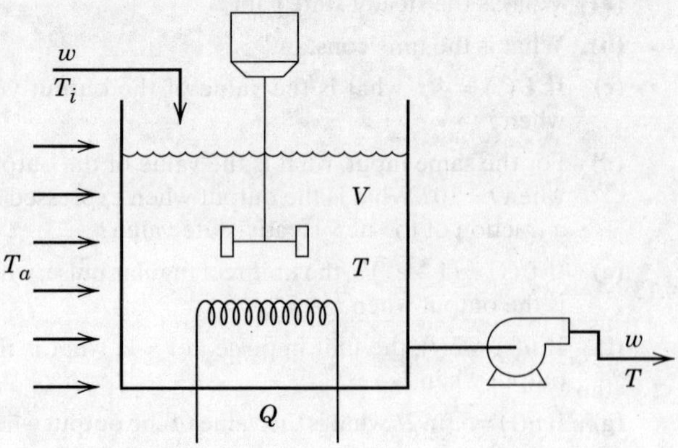

Figure E4.10

4.9 For the steam-heated stirred-tank system model in Eqs. 2-55 and 2-56, assume that the steam temperature T_s is constant.

(a) Find a transfer function relating tank temperature T to inlet liquid temperature T_i.

(b) What is the steady-state gain for this transfer function?

(c) Based on physical arguments only, should the gain be unity? Justify your answer.

4.10 The contents of the stirred-tank heating system shown in Fig. E4.10 are heated at a constant rate of Q(kW) using a gas-fired heater. The flow rate w(kg/h) and volume V(m^3) are constant, but the heat loss to the surroundings Q_L(kW) varies with the wind velocity v (m/s) according to the expressions

$$Q_L = UA(T - T_a)$$
$$U(t) = \bar{U} + bv^2(t)$$

where $\bar{U}$, A, b, and T_a are constants. Derive the transfer function between exit temperature T and wind velocity v. List any additional assumptions that you make.

4.11 Consider a pressure surge system to reduce the effect of pressure variations at a compressor outlet on the pressure in a compressed gas header. Develop a dynamic model and evaluate the form of the resulting transfer function for the two-tank process shown in Fig. E4.11.

(a) Develop a dynamic model that can be used to solve for the gas flow rate, $w_3(t)$, to the header given known pressures at the compressor, $P_c(t)$, and in the header, $P_h(t)$. Determine the degrees of freedom.

Available Information:

(i) The three valves operate linearly with resistances R_1, R_2, R_3. E.g., $w_1 = (P_c - P_1)R_1$.

(ii) The tank volumes (V_1 and V_2) are constant.

(iii) The Ideal Gas Law holds.

(iv) The molecular weight of the gas is M.

(v) Operation is isothermal.

(b) Develop the model (linearize, Laplace transform, etc.) just to the point where you can identify the following characteristics of the transfer function

$$\frac{W_3'(s)}{P_c'(s)}$$

(i) What is the order of the denominator?

(ii) What is the order of the numerator?

(iii) Does the gain equal one?

Note: There is no need to derive the actual transfer function. On the other hand, you should justify your answer to each question.

4.12 Consider the model of the electrically heated stirred-tank system in Section 2.4.3. Subscript e refers to the heating element:

$$mC\frac{dT}{dt} = wC(T_i - T) + h_eA_e(T_e - T)$$
$$m_eC_e\frac{dT_e}{dt} = Q - h_eA_e(T_e - T)$$

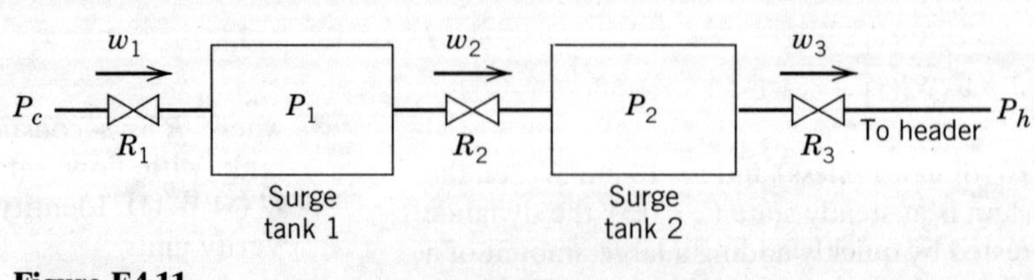

Figure E4.11

(a) Derive transfer functions relating changes in outlet temperature T to changes in the two input variables: heater input Q (assuming no change in inlet temperature), and inlet temperature T_i (for no change in heater input).

(b) Show how these transfer functions are simplified when negligible thermal capacitance of the heating element ($m_eC_e \to 0$) is assumed.

4.13 A horizontal cylindrical tank shown in Fig. E4.13a is used to slow the propagation of liquid flow surges in a processing line. Figure E4.13b illustrates an end view of the tank and w_t is the width of the liquid surface, which is a function of its height, both of which can vary with time. Develop a model for the height of liquid h in the tank at any time with the inlet and outlet volumetric flow rates as model inputs. Linearize the model assuming that the process initially is at steady state and that the liquid density ρ is constant.

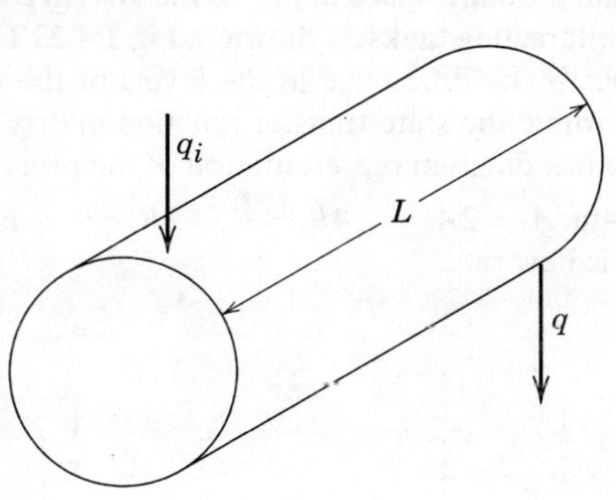

Figure E4.13a

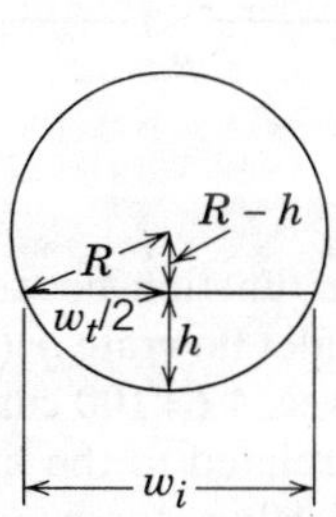

Figure E4.13b

4.14 An exothermic reaction, A → 2B, takes place adiabatically in a stirred-tank reactor. This liquid reaction occurs at constant volume in a 3785 L reactor. The reaction can be considered to be first-order and irreversible with the rate constant given by

$$k = 2.4 \times 10^{15} e^{-36{,}000/T} \text{ (min}^{-1}\text{)}$$

where T is in K (Kelvin).

(a) Using the information below, derive a transfer function relating the exit temperature T to the inlet concentration c_{Ai}. State all assumptions that you make.

(b) How sensitive is the transfer function gain K to the operating conditions? Find an expression for the gain in terms of $\bar{q}$, $\bar{T}$, and $\bar{c}_{Ai}$, and evaluate the sensitivities (that is, $\partial K/\partial \bar{q}$, etc.).

Available Information:

(i) Nominal steady-state conditions are

$$\bar{T} = 65\,°\text{C}, \quad \bar{c}_{Ai} = 0.028 \text{ mol/L}$$

$$\bar{q} = 76 \text{ L/min} = \text{flow in and out of the reactor}$$

(ii) Physical property data for the mixture at the nominal steady state:

$$C = 3.35 \text{ kJ/kg°C}, \quad \rho = 833 \text{ kg/m}^3, \quad -\Delta H_R = 500 \text{ kJ/mol}$$

4.15 A chemostat is a continuous stirred tank bioreactor that can carry out fermentation of a plant cell culture. Its dynamic behavior can be described by the following model:

$$\dot{X} = \mu(S)X - DX$$

$$\dot{S} = -\mu(S)X/Y_{X/S} + D(S_f - S)$$

X and S are the cell and substrate concentrations, respectively, and S_f is the substrate feed concentration. The dilution rate D is defined as the feed flow rate divided by the bioreactor volume. D is the input, while the cell concentration X and substrate concentration S are the output variables. Typically, the rate of reaction is referred to as the specific growth rate μ and is modeled by a Monod equation,

$$\mu(S) = \frac{\mu_m S}{K_s + S}$$

Assume $\mu_m = 0.20 \text{ h}^{-1}$, $K_s = 1.0$ g/L, and $Y_{X/S} = 0.5$ g/g. Use a steady-state operating point of $\bar{D} = 0.1 \text{ h}^{-1}$, $\bar{X} = 4.5$ g/L, $\bar{S} = 1.0$ g/L, and $\bar{S}_f = 10$ g/L.

Using linearization, derive a transfer function relating the deviation variables for the cell concentration $(X - \bar{X})$ to the dilution ratio $(D - \bar{D})$.

4.16 Consider the transfer function:

$$G(s) = \frac{0.7(s^2 + 2s + 2)}{s^5 + 5s^4 + 2s^3 + 4s^2 + 6}$$

(a) Plot its poles and zeros in the complex plane. A computer program that calculates the roots of the polynomial (such as the command *roots* in MATLAB) can help you factor the denominator polynomial.

(b) From the pole locations in the complex plane, what can be concluded about the output modes for any input change?

(c) Plot the response of the output to a unit step input. Does the form of your response agree with your analysis for part (b)? Explain.

4.17 For a lead–lag unit,

$$\frac{Y(s)}{X(s)} = \frac{K(\tau_a s + 1)}{\tau_1 s + 1}$$

show that for a step input of magnitude M:

(a) The value of y at $t = 0^+$ is given by $y(0^+) = KM\tau_a/\tau_1$.

(b) Overshoot occurs only for $\tau_a > \tau_1$.

(c) Inverse response occurs only for $\tau_a < 0$.

4.18 A process consists of an integrating element operating in parallel with a first-order element (Fig. E4.18).

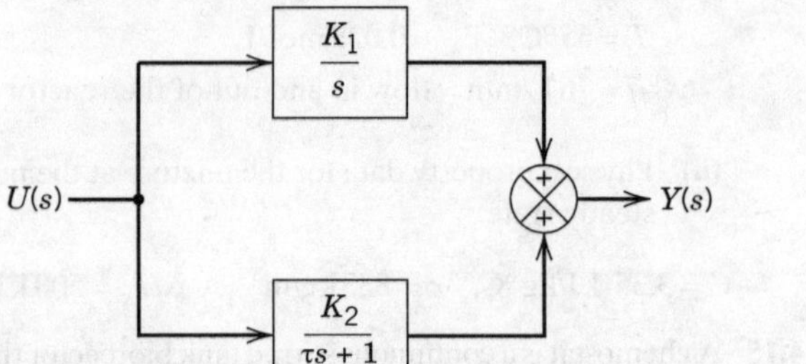

Figure E4.18

(a) What is the order of the overall transfer function, $G(s) = Y(s)/U(s)$?

(b) What is the gain of $G(s)$?

(c) What are the poles of $G(s)$? Where are they located in the complex s-plane?

(d) What are the zeros of $G(s)$? Where are they located? Under what condition(s) will one or more of the zeros be located in the right-half s-plane?

(e) For what conditions, will this process exhibit both a negative gain and a right-half plane zero?

(f) For any input change, what functions of time (response modes) will be included in the response, $y(t)$?

(g) Is the output bounded for any bounded input change, for example, $u(t) = M$?

4.19 The system equations for two liquid surge tanks in series are

$$A_1\frac{dh_1'}{dt} = q_i' - \frac{1}{R_1}h_1', \qquad q_1' = \frac{1}{R_1}h_1'$$

$$A_2\frac{dh_2'}{dt} = \frac{1}{R_1}h_1' - \frac{1}{R_2}h_2', \quad q_2' = \frac{1}{R_2}h_2'$$

Using state-space notation, determine the matrices A, B, C, and E, assuming that the level deviations is h_1' and h_2' are the state variables, the input variable is q_1', and the output variable is the flow rate deviation, q_2'.

4.20 The staged system model for a three-stage absorber is presented in Eqs. 2-77–2-79, which are in state-space form. A numerical example of the absorber model suggested by Wong and Luus[2] has the following parameters: $H = 34.34$ kg, $L = 18.5$ kg/min, $G = 30.25$ kg/min, $a = 0.72$, and $b = 0.0$. Using the MATLAB function *ss2tf*, calculate the three transfer functions ($Y_1'/Y_f', Y_2'/Y_f', Y_3'/Y_f'$) for the three state variables and the feed composition deviation Y_f' as the input.

4.21 A two-input two-output system is given by

$$\frac{dx_1}{dt} = 2x_1 + 4x_2 - 3u_1 + 5u_2$$

$$\frac{dx_2}{dt} = 6x_1 + 8x_2 + 12u_1 - 10u_2$$

All the variables are in their deviation forms. Obtain the state transfer function matrix. Make a block-diagram representation of the system. Is the system stable?

4.22 Obtain the state-space model of the system consisting of two interacting tanks as shown in Fig. E4.22 The desired output is the difference in the levels of the two tanks. Determine the state transfer function matrix and make the block diagram representation of the process.

Data: $A_1 = 2A_2 = 1$; $4R_1 = R_2 = 2R_3 = 2$; q_i is the volumetric flow rate.

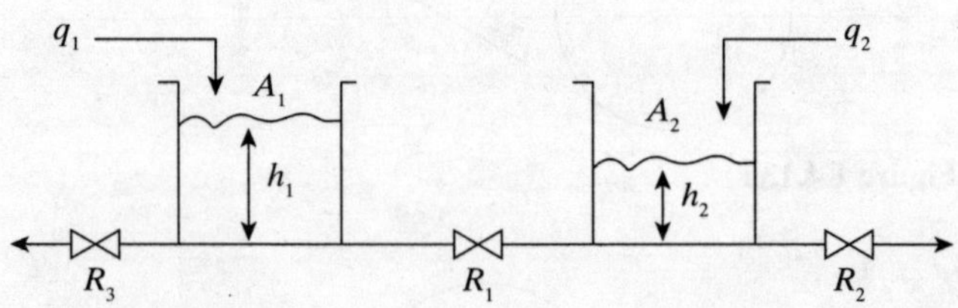

Figure E4.22

4.23 Find the transfer function between the liquid level h (in cm) and the inlet flowrate q_i (in cm³/s) for a tank of cross-sectional area A (= 100 cm²). The outlet flowrate q_{out} (in cm³/s) is related to the liquid level as $q = 3\sqrt{h}$. The steady state values for q_i and h are 18 cm³/s and 36 cm, respectively. Also find the time (in seconds) at which the tank will begin to overflow when a step change of 4 cm³/s is given in the inlet flow rate. The tank capacity is 5 L.

MULTIPLE CHOICE QUESTIONS

1. A unit step input is given to a process that is represented by the transfer function $\frac{s+2}{s+5}$. The initial value ($t = 0^+$) of the response of the process to the step input is

(a) 0 **(b)** 2/5 **(c)** 1 **(d)** ∞

(GATE 2008)

2. A tank of volume 0.25 m³ and height 1 m has water flowing in at 0.05 m³/min. The outlet flow rate is governed by the relation $q_{out} = 0.1$ h, where h is the height of the water in the tank in m and q_{out} is the outlet flow rate in m³/min. The inlet flow rate changed suddenly from its nominal value of 0.05 m³/min to 0.15 m³/min and remains there.

[2] Wong, K. T., and R Luus, Model Reduction of High-order Multistage Systems by the Method of Orethogonal Collocation, *Can. J. Chem. Eng.* **58**, 382 (1980).

The time (in minutes) at which the tank will begin to overflow is given by

(a) 0.28 **(b)** 1.01 **(c)** 1.73 **(d)** ∞

(GATE 2008)

3. Which ONE of the following transfer functions corresponds to an inverse response process with a positive gain?

(a) $\dfrac{1}{2s+1}-\dfrac{1}{3s+1}$ **(b)** $\dfrac{2}{s+1}-\dfrac{5}{s+10}$

(c) $\dfrac{3(0.5s-1)}{(2s+1)(s+1)}$ **(d)** $\dfrac{5}{s+1}-\dfrac{3}{2s+1}$

(GATE 2008)

4. A 2-input, 2-output process can be described in the Laplace domain as given below:

$$(\tau_1 s+1)Y_1(s)=K_1U_1(s)+K_2U_2(s)$$
$$(\tau_{21} s+1)Y_2(s)=K_3U_2(s)+K_4U_1(s)$$

where U_1 and U_2 are the inputs and Y_1 and Y_2 are the outputs. The gains of the transfer functions $Y_1(s)/U_2(s)$ and $Y_2(s)/U_2(s)$, respectively, are

(a) K_2 and K_3 **(b)** K_1 and $K_3+K_2K_4$

(c) K_2 and $K_3+K_1K_4$ **(d)** K_2 and $K_3+K_2K_4$

(GATE 2006)

5. A process is initially at steady state with its output $y-1$ for an input $u=1$. The input is suddenly changed to 2 at time $t=0$. The output response is $y(t)=1+2t$. The transfer function of the process is

(a) $\dfrac{2}{s}$ **(b)** $1+\dfrac{2}{s^2}$

(c) $1+\dfrac{2}{s}$ **(d)** $\dfrac{1}{s}\left(1+\dfrac{2}{s}\right)$ **(GATE 2001)**

6. Consider two non-interacting tank-in series as shown in figure. Water enters TANK 1 at q cm³/s and frains down to TANK 2 by gravity at a rate $k\sqrt{h_1}\,(\text{cm}^3/\text{s})$. Similarly, water drains from TANK 2 by gravity at a rate of $k\sqrt{h_2}\,(\text{cm}^3/\text{s})$ where h_1 and h_2 represents levels of TANK 1 and TANK 2, respectively (see figure). Drain valve constant $k=4\ \text{cm}^{2.5}/\text{s}$ and cross-sectional areas of the two tanks are $A_1=A_2=28\,\text{cm}^2$.

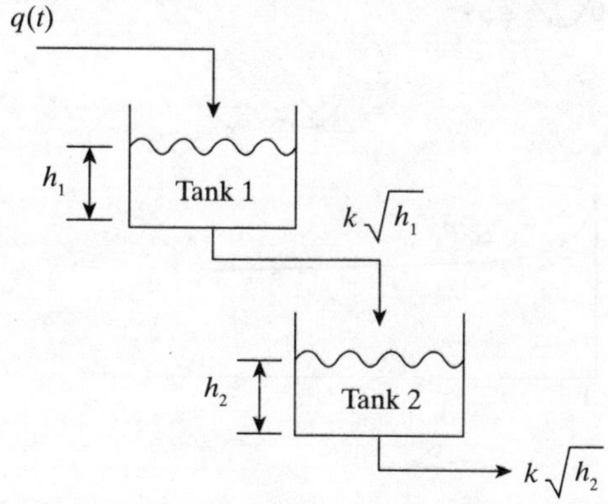

At steady state operation, the water inlet flow rate is $q_{ss}=16\,\text{cm}^3/\text{s}$. The transfer function relating the deviation variables $\tilde{h}_2$ (cm) to flow rate $\tilde{q}$(cm/s) is,

(a) $\dfrac{2}{(56s+1)^2}$ **(b)** $\dfrac{2}{(62s+1)^2}$

(c) $\dfrac{2}{(36s+1)^2}$ **(d)** $\dfrac{2}{(49s+1)^2}$

(GATE 2019)

7. Choose the option that correctly matches the step response curves in GROUP – I with the appropriate transfer functions in GROUP – II. The step input change occurs at time $t=0$.

GROUP – I (Step Response)

(P)

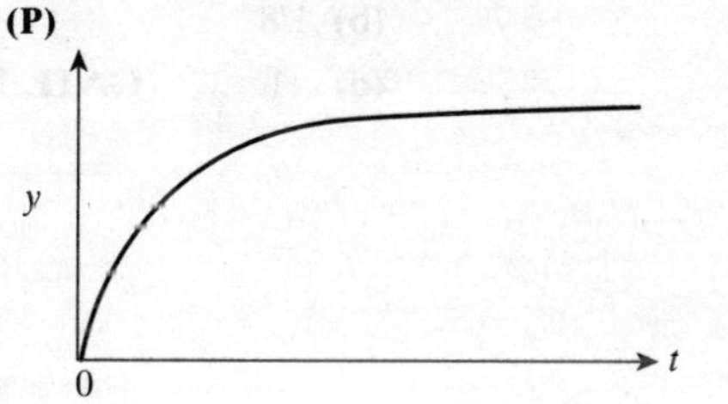

(Q)

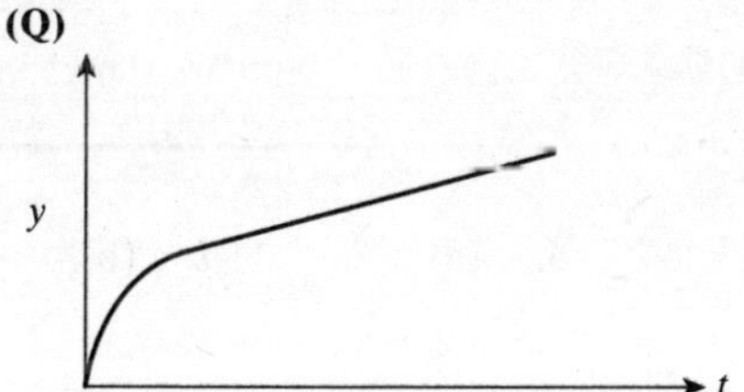

GROUP – II (Transfer Function)

(I) $\dfrac{K(\zeta s+1)}{s(\tau s+1)}$ $K>0,\ \xi>0,\tau>0$

(II) $\dfrac{K(\xi s+1)}{(\tau s+1)}$ $K>0,\ \xi>\tau>0$

(R)

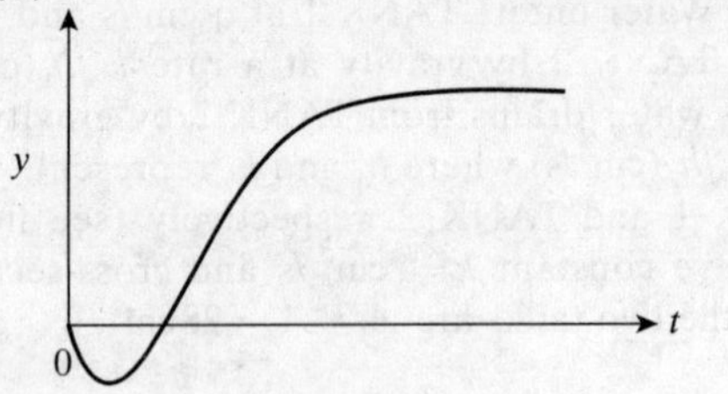

(III) $\frac{K}{(\tau s+1)}$ $K>0, \tau>0$

(S)

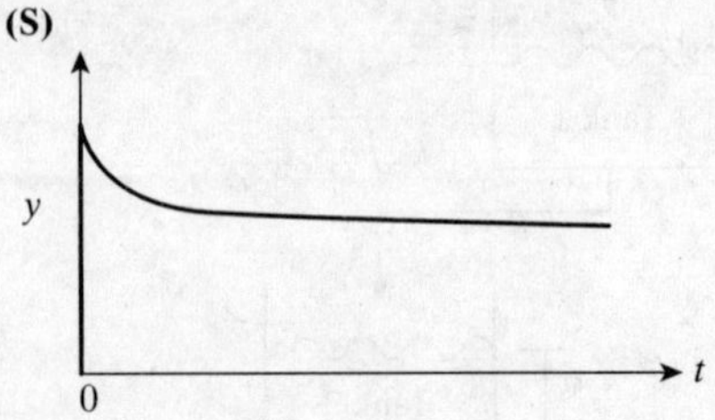

(IV) $\frac{K(\xi s+1)}{(\tau_1 s+1)(\tau_2 s+1)}$ $K>0, \xi<0$

$\tau_1>0, \tau_2>0$

(a) P–III, Q–IV; R–II, S–I **(b)** P–III, Q–I; R–IV, S–II **(c)** P–IV, Q–III; R–II, S–I **(d)** P–III, Q–II; R–IV, S–I **(GATE 2019)**

8. A system exhibits inverse response for a unit step change in the input. Which one of the following statements must necessarily be satisfied?

(a) The transfer function of the system has at least one negative pole

(b) The transfer function of the system has at least one positive pole

(c) The transfer function of the system has at least one negative zero

(d) The transfer function of the system has at least one positive zero **(GATE 2016)**

9. Match the location of the poles/zeros in the *s*-plane listed in GROUP – I, with the system response characteristics in GROUP – II.

GROUP – I	GROUP – II
(P) Pole in the right half plane	**(I)** Stable response
(Q) Pole at origin	**(II)** Integrating response
(R) Zero in the right half plane	**(III)** Unstable response
	(IV) Inverse response

(a) P–I, Q–II; R–III **(b)** P–III, Q–IV; R–I

(c) P–III, Q–II; R–IV **(d)** P–L, Q–IV, R–III **(GATE 2010)**

10. A system with a double pole at the origin is unstable since the corresponding term in the time domain

(a) is a constant.

(b) grows exponentially with time.

(c) grows linearly with time.

(d) decays linearly with time. **(GATE 1999)**

11. A control system has the following transfer function.

$$F(s)=\frac{(s-1)(s+1)}{s(s-2)(s+4)}$$

The initial value of the corresponding time function is

(a) 1 **(b)** 1/8

(c) 7/8 **(d)** –1 **(GATE 1999)**

ANSWER KEY

1. (c) **2.** (c) **3.** (c) **4.** (d) **5.** (a) **6.** (a) **7.** (b)

8. (d) **9.** (c) **10.** (c) **11.** (a)

CHAPTER 5

Dynamic Behavior of First-Order and Second-Order Processes

CHAPTER CONTENTS

In Chapter 2, we derived dynamic models for several typical processes, and in Chapter 4, we showed how these models can be put into standard transfer function form. Now we investigate how processes respond to typical changes in their environment, that is, to changes in their inputs. We have already seen in Chapter 1 that process inputs fall into two categories:

1. **1.** Inputs that can be manipulated to control the process.
2. **2.** Inputs that are not manipulated, classified as disturbance variables.

The transfer function representation makes it easy to compare the effects of different inputs. For example, the dynamic model for the constant-flow stirred-tank blending system was derived in Section 4.1. Rewriting Eq. 4-33 in terms of process parameters yields

$$X'(s) = \frac{w_1/w}{\frac{V\rho}{w}s+1}X_1'(s) + \frac{w_2/w}{\frac{V\rho}{w}s+1}X_2'(s) \tag{5-1}$$

The resulting first-order transfer functions,

$$\frac{X'(s)}{X_1'(s)} = \frac{w_1/w}{\frac{V\rho}{w}s+1} \quad \text{and} \quad \frac{X'(s)}{X_2'(s)} = \frac{w_2/w}{\frac{V\rho}{w}s+1} \tag{5-2}$$

relate changes in outlet mass fraction $X'(s)$ to changes in inlet mass fractions $X_1'(s)$ and $X_2'(s)$, respectively.

A second advantage of the transfer function representation is that the dynamic behavior of a given process can be generalized easily. Once we analyze the response of the process to an input change, the response of any other process described by the same generic transfer function is then known.

For a general first-order transfer function with output $Y(s)$ and input $U(s)$,

$$Y(s) = \frac{K}{\tau s+1}U(s) \tag{5-3}$$

an analytical time-domain solution can be found once the nature of the input change is specified (e.g., step or impulse change). This solution applies to any other process with a first-order transfer function, for example, the liquid surge tanks of Eqs. 4-63 and 4-65. Another benefit of transfer function form (e.g., Eq. 5-3) is that it is not necessary to re-solve the ODE when K, τ, or $U(s)$ changes.

We will exploit this ability to develop general process dynamic formulas as much as possible, concentrating on transfer functions that commonly arise in describing the dynamic behavior of industrial processes. This chapter covers the simplest transfer functions: zero-order processes, first-order processes, integrating units, and second-order processes. In Chapter 6, the responses of more complicated transfer functions will be discussed. To keep the results as general as possible, we now consider several standard process inputs that are used to characterize the behavior of many actual processes.

5.1 STANDARD PROCESS INPUTS

We have previously discussed outputs and inputs for process models; we now introduce more precise working definitions. The word *output* generally refers to a dependent variable, usually a controlled variable in a process, which is a process variable to be maintained at a desired value (set point). For example, the output from the stirred blending tank just discussed is the mass fraction x of the effluent stream. The word *input* refers to any variable that influences the process output, such as the flow rate of the stream flowing into the stirred blending tank. The characteristic feature of all inputs, whether they are disturbance variables or manipulated variables, is that they influence the output variables.

In analyzing process dynamics and in designing control systems, it is important to know how the process outputs will respond to changes in the process inputs. There are six important types of input changes used in industrial practice for the purposes of modeling and control.

1. ***Step Input.*** One characteristic of industrial processes is that they can be subjected to sudden and sustained input changes; for example, a reactor feedstock may be changed quickly from one supply to another, causing a corresponding change in important input variables such as feed concentration and feed temperature. Such a change can be approximated by the step change

$$u_S(t) = \begin{cases} 0 & t < 0 \\ M & t \geq 0 \end{cases} \tag{5-4}$$

where *zero time*, as noted earlier, is taken to be the time at which the sudden change of magnitude M occurs. Note that $u_S(t)$ is defined as a deviation variable—that is, the change from the normal steady state. Suppose the heat input to a stirred-tank heating unit suddenly is changed from 8000 to 10,000 kcal/h, by changing the electrical heater input. Then

$$Q(t) = 8000 + 2000\,S(t) \tag{5-5a}$$

$$Q'(t) = 2000\,S(t) \tag{5-5b}$$

where $S(t)$ is the unit step function. The Laplace transform of a step of magnitude M (cf. Eq. 3-4) is

$$u_S(s) = \frac{M}{s} \tag{5-6}$$

2. ***Ramp Input.*** Industrial processes often are subjected to inputs that *drift*—that is, they gradually change upward or downward for some period of time with a roughly constant slope. For example, ambient conditions (air temperature and relative humidity) can change slowly during the day so that the plant cooling tower temperature also changes slowly. Set points are sometimes ramped from one value to another rather than making a step change. We can approximate such a gradual change in an input variable by means of the ramp function:

$$u_R(t) = \begin{cases} 0 & t < 0 \\ at & t \geq 0 \end{cases} \tag{5-7}$$

where $u_R(t)$ is a deviation variable. The Laplace transform of a ramp input with a slope of 1 is given in Table 3.1 as $1/s^2$. Hence, transforming Eq. 5-7 yields

$$u_R(s) = \frac{a}{s^2} \tag{5-8}$$

3. ***Rectangular Pulse.*** Processes sometimes are subjected to a sudden step change that then returns to its original value. Suppose that a feed to a reactor is shut off for a certain period of time or a natural-gas-fired furnace experiences a brief interruption in fuel gas. We might approximate this type of input change as a rectangular pulse:

$$u_{RP}(t) = \begin{cases} 0 & t < 0 \\ h & 0 \leq t < t_w \\ 0 & t \geq t_w \end{cases} \tag{5-9}$$

where the pulse width t_w can range from very short (approximation to an impulse) to very long.

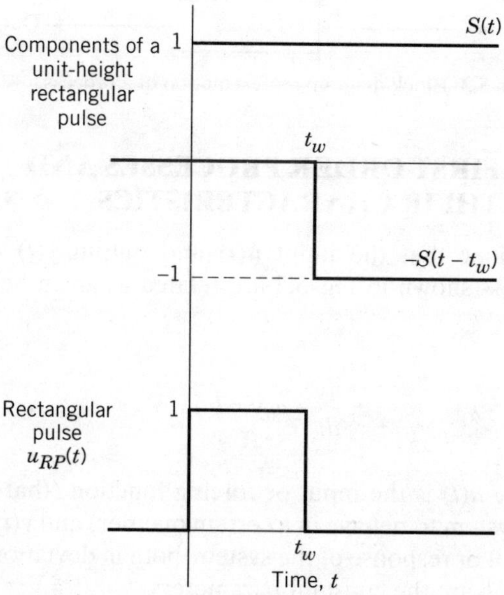

Figure 5.1 How two step inputs can be combined to form a rectangular pulse.

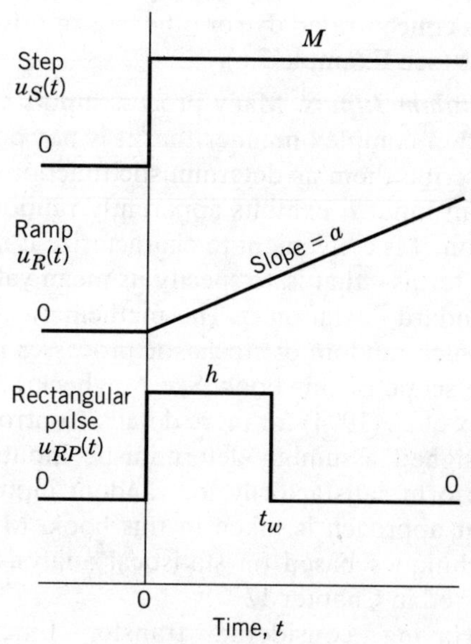

Figure 5.2 Three important examples of deterministic inputs.

An alternative way of expressing Eq. 5-9 utilizes the shifted unit step input $S(t - t_w)$, which is equal to unity for $t \geq t_w$ and equal to zero for $t < t_w$. Equation 5-9 is depicted in Fig. 5.1 as the sum of two steps, one step of magnitude equal to 1 occurring at $t = 0$ combined with a second step of magnitude equal to −1 occurring at $t = t_w$. Mathematically, this combination can be expressed as

$$u_{RP}(t) = h[S(t) - S(t - t_w)]$$

Because the Laplace transform is only defined for $t \geq 0$, this expression can be simplified to

$$u_{RP}(t) = h[1 - S(t - t_w)] \qquad t \geq 0 \tag{5-10}$$

which can be Laplace transformed to yield

$$u_{RP}(s) = \frac{h}{s}(1 - e^{-t_w s}) \tag{5-11}$$

which is the same result given in Eq. 3-15.

The three important inputs discussed above—step, ramp, rectangular pulse—are depicted in Fig. 5.2.

4. ***Sinusoidal Input.*** Processes are also subjected to inputs that vary periodically. As an example, the drift in cooling water temperature discussed earlier can often be closely tied to diurnal (day-to-night-to-day) fluctuations in ambient conditions. Cyclic process changes within a 24-h period often are caused by variations in cooling water temperature that can be approximated as a sinusoidal function:

$$u_{\sin}(t) = \begin{cases} 0 & t < 0 \\ A \sin \omega t & t \geq 0 \end{cases} \tag{5-12}$$

The amplitude of the sinusoidal function is A, while the period P is related to the angular frequency ω by $P = 2\pi/\omega$ (ω in radians/time). On a shorter time scale, high-frequency disturbances are associated with mixing and pumping operations, and with 60-Hz electrical noise arising from AC electrical equipment and instrumentation.

Sinusoidal inputs are particularly important, because they play a central role in frequency response analysis, which is discussed in Chapter 10. The Laplace transform of the sine function in Eq. 5-12 can be obtained by multiplying entry 14 in Table 3.1 by the amplitude A to obtain

$$u_{\sin}(s) = \frac{A\omega}{s^2 + \omega^2} \tag{5-13}$$

5. ***Impulse Input.*** The unit impulse function discussed in Chapter 3 has the simplest Laplace transform, $U(s) = 1$ (Eq. 3-19). However, exact impulse functions are not encountered in normal plant operations. To obtain an impulse input, it is necessary to inject a finite amount of energy or material into a process in an infinitesimal length of time, which is not possible. However, this type of input can be approximated through the injection

of a concentrated dye or other tracer into the process (see Example 3.6).

6. ***Random Inputs.*** Many process inputs change in such a complex manner that it is not possible to describe them as deterministic functions of time. If an input u exhibits apparently random fluctuation, it is convenient to characterize it in statistical terms—that is, to specify its mean value μ and standard deviation σ. The mathematical analysis of such random or stochastic processes is beyond the scope of this book. See Maybeck (1997) and Box et al. (1994) for more details. Control systems designed assuming deterministic inputs usually perform satisfactorily for random inputs; hence that approach is taken in this book. Monitoring techniques based on statistical analysis are discussed in Chapter 17.

Having considered transfer functions in Chapter 4 and important types of forcing functions (process inputs) here, we now can discuss the dynamic behavior of processes in an organized way.

5.2 ZERO-ORDER SYSTEMS (INSTANTANEOUS PROCESSES)

Consider the process shown in Fig. 5.3. If the output $y(t)$ can be expressed as a factor of input $u(t)$, as shown in Eq. 5-14, then the process is called as a zero-order process or instantaneous process.

$$y(t) = Ku(t) \tag{5-14}$$

Eq. 5-14 represents a zero-order process because there is no derivative describing the dynamics of the process. If y and u are deviation variables, we can take the Laplace transform on both sides and get its transfer function as follows:

$$\mathscr{L}[y(t)] = \mathscr{L}[Ku(t)] = K\,\mathscr{L}[u(t)]$$

or
$$Y(s) = KU(s) \tag{5-15}$$

or
$$G(s) = \frac{Y(s)}{U(s)} = K \tag{5-16}$$

This is called instantaneous process because the output follows the changes in the input (Eq. 5-15) so quickly that, the process, for all practical purposes, remains at a steady state. Here, K is the static gain (sensitivity) of the system.

An example of zero-order process is the ideal spring, where output (force, F) and input (elongation, x) are related as $F = Kx$.

Figure 5.3 Block diagram representation of a process.

5.3 FIRST-ORDER PROCESSES AND THEIR CHARACTERISTICS

Consider that the input $u(t)$ and output $y(t)$ of the process shown in Fig. 5.3 are related as given below in Eq. 5-17.

$$a_1\frac{dy}{dt} + a_0 y = bu \tag{5-17}$$

where $u(t)$ is the input or forcing function (that forces the system to behave in a certain manner) and $y(t)$ is the output or response of the system, both in deviation form a_0, a_1, b are the constant parameters.

Case 1: If $a_0 \neq 0$

We can write Eq. 5-17 as:

$$\frac{a_1}{a_0}\frac{dy}{dt} + y = \frac{b}{a_0}u$$

or
$$\tau\frac{dy}{dt} + a_0 y = Ku \tag{5-18}$$

where
$$\tau = \frac{a_1}{a_0},\ K = \frac{b}{a_0}$$

Taking Laplace transform of Eq. 5-18 and after rearranging the terms, we get:

$$G(s) = \frac{Y(s)}{U(s)} = \frac{K}{\tau s + 1} \tag{5-19}$$

Eq. 5-19 represents a first-order lag process or linear lag system or exponential transfer lag or self-regulating system. Here, K is called as the steady state gain of the system and τ is the time-constant of the system.

Case 2: If $a_0 = 0$.

We can write Eq. 5-17 as:

$$a_1\frac{dy}{dt} = bu$$

or
$$\frac{dy}{dt} = \frac{b}{a_1}u \tag{5-20}$$

Taking Laplace transforms on both sides of Eq. (5-20) we get:

$$G(s)=\frac{Y(s)}{U(s)}=\frac{K'}{s} \tag{5-21}$$

where $K'=\frac{b}{a_1}$

Eq. 5-21 represents a first-order integrating process or pure capacitive process or pure integrator process. The self-regulating process and integrating process don't have a same steady state gain, i.e., $K \neq K'$.

The first-order lag processes are characterized as having capacity to store material, energy or momentum and they have resistance associated with the flow of mass, energy, or momentum in reaching the capacity. For example, the tank system (capacity to store liquids or gases) has resistances associated with the pumps, valves, weirs and pipe attached to inflow or outflow. Similarly, solids, liquids or gases have capacity to store thermal energy (in the form of specific heat, C_p) and resistances associated with transfer of heat through them. In other words, any system that has capacity to store mass or energy and acts as a buffer between inlet and outlet streams can be modeled as a first-order process.

In case of a first-order lag process, the time constant (τ) can be expressed as the product of storage capacity (matter or energy) and the resistance to flow. This indicated the speed of response. Higher the value of time constant, lower is the speed and vice-versa. The steady-stage gain (K) is an indication of the sensitivity of the system. Higher K indicates highly sensitive systems.

5.4 RESPONSE OF FIRST-ORDER PROCESSES

The general first-order transfer function is given by

$$\frac{Y(s)}{U(s)}=\frac{K}{\tau s+1} \tag{5-22}$$

where K is the steady-state gain and τ is the time constant. The transfer function was useful in describing the dynamics of the blending system in Section 4.4. Now we investigate some particular forms of input $U(s)$, deriving expressions for the response, $y(t)$.

5.4.1 Step Response

For a step input of magnitude M, $U(s) = M/s$, and Eq. 5-22 becomes

$$Y(s)=\frac{KM}{s(\tau s+1)} \tag{5-23}$$

Using Table 3.1, the time-domain response is

$$y(t)=KM(1-e^{-t/\tau}) \tag{5-24}$$

The plot of this equation in Fig. 5.4 shows that a first-order process does not respond instantaneously to a sudden change in its input. In fact, after a time interval equal to the time constant ($t = \tau$), the process response is still only 63.2% complete. Theoretically, the process output never reaches the new steady-state value except as $t \to \infty$; it does approximate the final steady-state value when $t \approx 5\tau$, as shown in Table 5.1. Notice that Fig. 5.4 has been drawn in dimensionless or normalized form, with time divided by the time constant and the output change divided by the product of the process gain and magnitude of the input change. Now we consider a more specific example.

Table 5.1 Normalized Response of a First-Order Process to a Step Input

t	$y(t)/KM = 1 - e^{-t/\tau}$
0	0
τ	0.6321
2τ	0.8647
3τ	0.9502
4τ	0.9817
5τ	0.9933

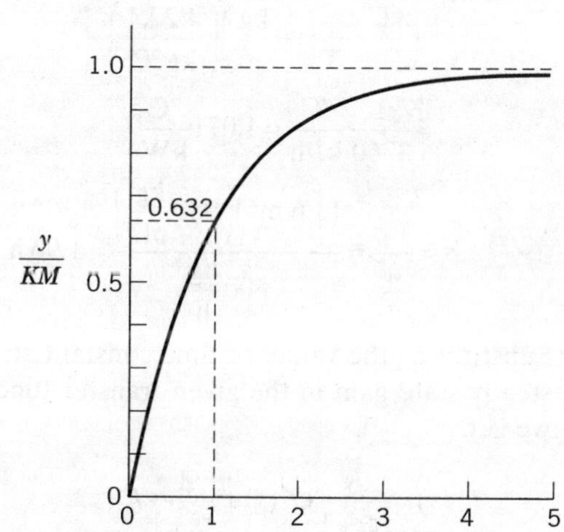

Figure 5.4 Step response of a first-order process.

EXAMPLE 5.1

A stirred-tank heating system is used to preheat a reactant containing a suspended solid catalyst at a constant flow rate of 800 kg/h. The volume in the tank is 1.6 m³, and the density and specific heat of the suspended mixture are, respectively, 980 kg/m³ and 4.2 kJ/kg °C. The process initially is operating with inlet and outlet temperatures of 110 and 140 °C, respectively. The steady-state energy input is 34.89 kW. The transfer function is

$$T'(s) = \frac{1/wC}{\frac{m}{w}s+1}Q'(s) + \frac{1}{\frac{m}{w}s+1}T_i'(s) \quad (5\text{-}25)$$

The following questions concerning process operations are posed:

(a) What are the values of K and τ?

(b) If the heater input is suddenly increased by +30%, how long will it take for the tank temperature to achieve 99% of the final temperature change? What will be the outlet temperature at the final steady state?

(c) If the inlet temperature is increased suddenly from 110 to 130 °C, how long will it take before the outlet temperature changes from 140 to 145 °C? What will be the outlet temperature at the final steady state?

Solution

(a) Using Eq. 5-25, the gain and time constants can be determined

$$K = \frac{1}{wC} = \frac{1}{\left(800\frac{\text{kg}}{\text{h}}\right)\left(\frac{4.2\text{ kJ}}{\text{kg}\,°\text{C}}\right)}$$

$$= \frac{1}{3360\text{ kJ/h}}\frac{°\text{C}}{} = 1.071\frac{°\text{C}}{\text{kW}}$$

$$\tau = \frac{V\rho}{w} = \frac{(1.6\text{ m}^3)\left(980\frac{\text{kg}}{\text{m}^3}\right)}{800\frac{\text{kg}}{\text{h}}} = 1.96\text{ h}$$

(b) Substituting the values of time constant and the steady state gain in the given transfer function, we get:

$$T'(s) = \frac{K}{\tau s+1}Q'(s) + \frac{1}{\tau s+1}T_i'(s)$$

$$= \frac{1.071}{1.96\,s+1}Q'(s) + \frac{1}{1.96\,s+1}T_i'(s)$$

When the heater input is increased by +30% (keeping the inlet temperature constant), the above equation can be written as

$$T'(s) = \frac{1.071}{1.96\,s+1}Q'(s) = \frac{1.071}{1.96\,s+1}\times\frac{(0.30)(34.89)}{s}$$

$$= \frac{11.21}{s(1.96\,s+1)}$$

Applying Eq 5-24 we get:

$$T'(t) = 11.21(1-e^{-t/1.96})$$

Final steady state value in deviation form

$$T'(t\to\infty) = 11.21\,°\text{C}.$$

Time for 99% of the final response is given by

$$(0.99)(11.21) = 11.21(1-e^{-t/1.96})$$

Solving the above equation, we get time for 99% response as

$$t = 9.02\text{ h}$$

Table 5.1 indicates that the time required to attain the 99.33% response following a step change of any magnitude in heater input will be 5τ that is, $5\times1.96 = 9.8$ h, which is a close approximation to the value obtained.

Also, the steady-state change in temperature due to a change of +30% in Q (10.467 kW) can be calculated from the Final Value Theorem, Eq. 3-53:

$$T'(t\to\infty) = \lim_{s\to0} sT'(s) = \lim_{s\to0} s\frac{11.21}{s(1.96s+1)}$$

$$= 11.21\,°\text{C}$$

So, the actual outlet temperature at the final steady state = 140 + 11.21 = 151.21 °C.

(c) When the inlet temperature is increased from 110 to 130 °C (keeping the heater input constant), we get

$$T'(s) = \frac{1}{\tau s+1}T_i'(s) = \frac{1}{1.96\,s+1}T_i'(s)$$

$$= \frac{1}{1.96\,s+1}\times\frac{130-110}{s} = \frac{20}{s(1.96\,s+1)}$$

Applying Eq 5-24 we get:

$$T'(t) = 20(1 - e^{-t/1.96})$$

The final steady state value in deviation form

$$T'(t \to \infty) = 20°\text{C}.$$

Actual outlet temperature at the final steady state = 140 + 20 = 160°C.

When actual temperature is 145°C, $T'(t)$ = 145–140 = 5°C

Therefore,

$T'(t) = 20(1 - e^{-t/1.96})$ or, $5 = 20(1 - e^{-t/1.96})$

Solving the above equation, we get time, t = 0.56 h.

5.4.2 Ramp Response

We now evaluate the response of a first-order system to a ramp input, $U(s) = a/s^2$ of Eq. 5-8. Performing a partial fraction expansion yields

$$Y(s) = \frac{Ka}{(\tau s+1)s^2} = \frac{\alpha_1}{\tau s+1} + \frac{\alpha_2}{s} + \frac{\alpha_3}{s^2} \tag{5-26}$$

The Heaviside expansion (Chapter 3) gives

$$Y(s) = \frac{Ka\tau^2}{\tau s+1} - \frac{Ka\tau}{s} + \frac{Ka}{s^2} \tag{5-27}$$

Using Table 3.1,

$$y(t) = Ka\tau(e^{-t/\tau} - 1) + Kat \tag{5-28}$$

The above expression has the interesting property that for large values of time ($t \gg \tau$)

$$y(t) \approx Ka(t - \tau) \tag{5-29}$$

Equation 5-29 implies that after an initial transient period, the ramp input yields a ramp output with slope equal to Ka, but shifted in time by the process time constant τ (see Fig. 5.5). An unbounded ramp input will ultimately cause some process component to saturate, so the duration of the ramp input ordinarily is limited. A process input frequently will be *ramped* from one value to another in a fixed amount of time so as to avoid the sudden change associated with a step change. Ramp inputs of this type are particularly useful during the start-up of a continuous process or in operating a batch process.

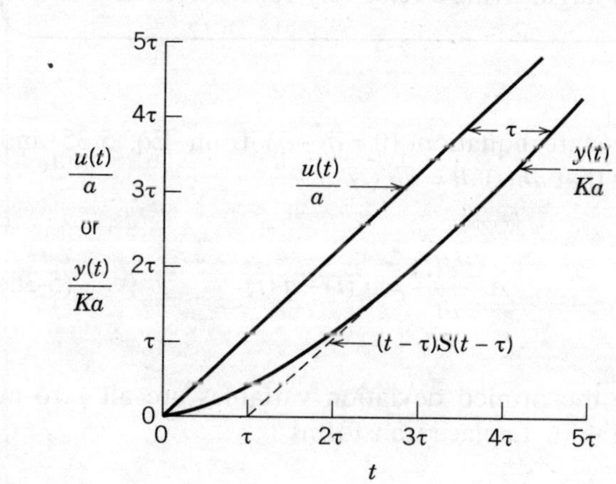

Figure 5.5 Ramp response of a first-order process (comparison of input and output).

5.4.3 Sinusoidal Response

As a final example of the response of first-order processes, consider a sinusoidal input $u_{\sin}(t) = A \sin \omega t$, with The Laplace transform is given by Eq. 5-13:

$$\begin{aligned} Y(s) &= \frac{KA\omega}{(\tau_s+1)(s^2+\omega^2)} \\ &= \frac{KA}{\omega^2\tau^2+1}\left(\frac{\omega\tau^2}{\tau s+1} - \frac{s\omega\tau}{s^2+\omega^2} + \frac{\omega}{s^2+\omega^2}\right) \end{aligned} \tag{5-30}$$

Taking the inverse Laplace transform gives

$$y(t) = \frac{KA}{\omega^2\tau^2+1}(\omega\tau e^{-t/\tau} - \omega\tau\cos\omega t + \sin\omega t) \tag{5-31}$$

or, by using trigonometric identities,

$$y(t) = \frac{KA\omega\tau}{\omega^2\tau^2+1}e^{-t/\tau} + \frac{KA}{\sqrt{\omega^2\tau^2+1}}\sin(\omega t + \phi) \tag{5-32}$$

where

$$\phi = -\tan^{-1}(\omega\tau) \tag{5-33}$$

Notice that in both Eqs. 5-31 and 5-32, the exponential term goes to zero as $t \to \infty$, leaving a pure sinusoidal response. This property is exploited in Chapter 10 for frequency response analysis.

Students often have difficulty imagining how a real process variable might change sinusoidally. How can the flow rate into a reactor be negative as well as positive? Remember that we have defined the input u and output

y in these relations to be deviation variables. An actual input might be

$$q(t) = 0.4\frac{\text{m}^3}{\text{s}} + \left(0.1\frac{\text{m}^3}{\text{s}}\right)\sin \omega t \tag{5-34}$$

where the amplitude of the deviation input signal A is 0.1 m³/s. After a long period of time, the output response Eq. 5-32 also will be a sinusoidal deviation, similar to that given in Eq. 5-34.

EXAMPLE 5.2

A liquid surge tank similar to the one described by Eq. 4-60 has the transfer function model form of Eq. 4-63:

$$\frac{H'(s)}{Q_i'(s)} = \frac{10}{50s+1}$$

where h is the tank level (m), q_i is the flow rate (m³/s), the gain is 10 m/m³/s, or 10 s/m², and the time constant is 50 s. The system is operating at steady state with $\bar{q} = 0.4\,\text{m}^3/\text{s}$ and $\bar{h} = 4\,\text{m}$ when a sinusoidal perturbation in inlet flow rate begins with amplitude = 0.1 m³/s and a cyclic frequency of 0.002 cycles/s. What are the maximum and minimum values of the tank level after the flow rate disturbance has occurred for 6 min or more? What are the largest level perturbations expected as a result of this sinusoidal variation in flow rate? What is the effect of high-frequency variations, say, 0.2 cycles/s?

Solution

Note that the actual input signal $q(t)$ is given by Eq. 5-34, but only the amplitude of the input deviation (0.1 m³/s) is required. From Eq. 5-32, the value of the exponential term 6 min after the start of sinusoidal forcing is $e^{-360/50} = e^{-7.2} < 10^{-3}$. Thus, the effect of the exponential transient term is less than 0.1% of the disturbance amplitude and can be safely neglected.

Consequently, from Eq. 5-32 the amplitude of the output (level) perturbation is

$$\frac{KA}{\sqrt{\omega^2\tau^2+1}}$$

where A is the input amplitude and ω is the frequency (in radians) = (2π) (cyclic frequency) = (6.28) (0.002) radians/s. The amplitude of the perturbation in the liquid level is

$$\frac{10(\text{s/m}^2)(0.1\,\text{m}^3/\text{s})}{\sqrt{[(6.28\,\text{rad/cycles})(0.002\,\text{cycles/s})(50\,\text{s})]^2+1}}$$

or 0.85 m. Hence, the actual tank level varies from a minimum of 3.15 m to a maximum of 4.85 m.

The largest output deviations that can result from sinusoidal variations of amplitude 0.1 m³/s occur for $\omega \to 0$—that is, for very low frequencies. In this case, the deviations would be $\pm KA = \pm(10\,\text{s/m}^2)$ $(0.1\,\text{m}^3/\text{s}) = \pm 1$ m. Hence, the minimum and maximum values of liquid level would be 3 and 5 m, respectively.

For high-frequency variations (0.2 cycles/s), the output amplitude approaches zero. This occurs because the rapid variations of flow rate are averaged in the tank when the residence time is sufficiently large, giving a relatively constant level.

5.5 RESPONSE OF FIRST-ORDER INTEGRATING PROCESSES

In Section 2.5 we briefly considered a liquid-level system with a pump attached to the outflow line. Assuming that the outflow rate q can be set at any time by adjusting the speed of the pump, Eq. 2-58 becomes

$$A\frac{dh(t)}{dt} = q_i(t) - q(t) \tag{5-35}$$

Suppose at $t = 0$, the process is at the nominal steady state where $q_i = \bar{q}$ and $h = \bar{h}$. After subtracting the steady-state equation $(0 = \bar{q}_i - \bar{q})$ from Eq. 5-35 and noting that $dh(t)/dt = dh'(t)/dt$,

$$A\frac{dh'(t)}{dt} = q_i'(t) - q'(t) \tag{5-36}$$

where the primed deviation variables are all zero at $t = 0$. Taking Laplace transforms

$$sAH'(s) = Q_i'(s) - Q'(s) \tag{5-37}$$

and rearranging gives

$$H'(s) = \frac{1}{As}[Q_i'(s) - Q'(s)] \tag{5-38}$$

Both transfer functions, $H'(s)/Q_i'(s) = 1/As$ and $H'(s)/Q'(s) = -1/As$, represent *integrating models*, characterized by the term $1/s$. The integral of Eq. 5-35 is

$$h(t) - \bar{h} = \frac{1}{A}\int_0^t [q_i(t^*) - q(t^*)]dt^* \tag{5-39}$$

hence the term, *integrating process*. Integrating processes do not have a steady-state gain in the usual sense. For such a process operating at steady state, any positive step change in q_i (increase in q_i above q) will cause the tank level to increase linearly with time in proportion to the difference, $q_i(t) - q(t)$, while a positive step change in q will cause the tank level to decrease linearly. Thus, no new steady state will be attained, unless the tank overflows or empties. In contrast, a tank with an exit line valve, rather than a pump, will reach a steady state when the outflow rate becomes equal to the inflow rate. This process is described by a

EXAMPLE 5.3

A vented cylindrical tank is used for storage between a tank car unloading facility and a continuous reactor that uses the tank car contents as feedstock (Fig. 5.6). The reactor feed exits the storage tank at a constant flow rate of 0.02 m^3/s. During some periods of operation, feedstock is simultaneously transferred from the tank car to the feed tank and from the tank to the reactor. The operators have to be particularly careful not to let the feed tank overflow or empty. The feed tank is 5 m high (distance to the vent) and has an internal cross-sectional area of 4 m^2.

(a) Suppose that after a long period of operation, the storage tank level is 2 m at the time the tank car empties. How long can the reactor be operated before the feed tank is depleted?

(b) Another tank car is moved into place and connected to the tank, while flow continues into the reactor at 0.02 m^3/s. If flow is introduced into the feed tank just as the tank level reaches 1 m, how long can the transfer pump from the tank car be operated? Assume that it pumps at a constant rate of 0.1 m^3/s when switched on.

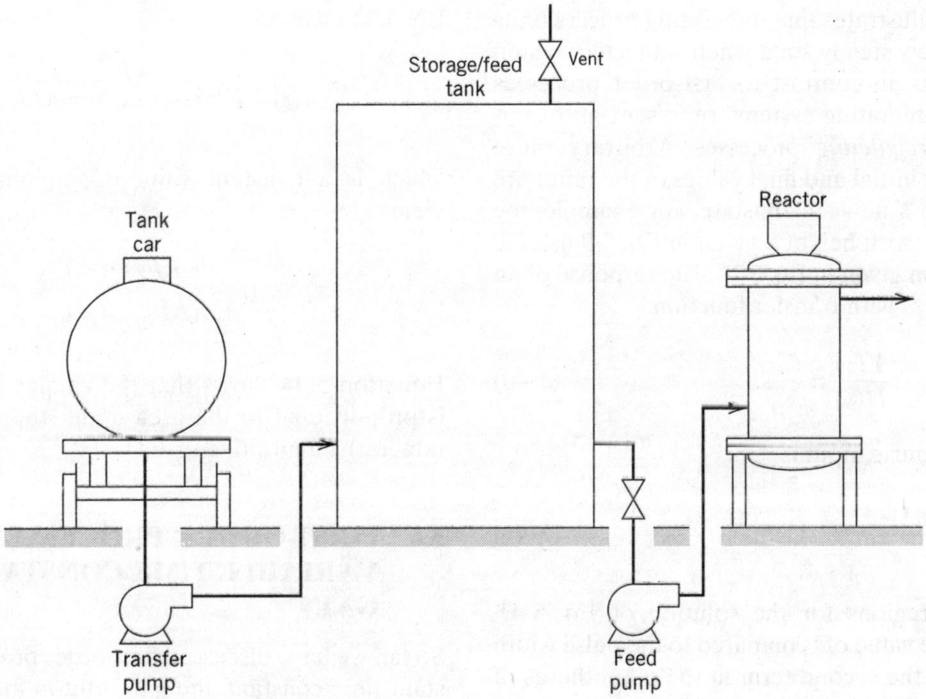

Figure 5.6 Unloading and storage facility for a continuous reactor.

Solution

(a) For this system, there is no unique steady-state level corresponding to a particular value of input and output flow rate. Suppose the initial level is $h = 2$ m and the constant flow rate from the feed pump to the reactor, $q = 0.02$ m³/s, is the basis for defining deviation variables for h, q, and q_i. Then

$$\bar{h} = 2\,\text{m}$$
$$\bar{q}_i = \bar{q} = 0.02\,\text{m}^3/\text{s}$$

and, from Eq. 5-30, the process model for the tank is

$$H'(s) = \frac{1}{4s}[Q_i'(s) - Q'(s)]$$

At the time the tank car empties

$$q_i = 0 \Rightarrow q_i' = -0.02 \Rightarrow Q_i'(s) = -\frac{0.02}{s}$$
$$q = 0.02 \Rightarrow q' = 0 \Rightarrow Q'(s) = 0$$

Thus

$$H'(s) = \frac{1}{4s}\left(-\frac{0.02}{s} - 0\right) = -\frac{0.005}{s^2}$$

Inversion to the time domain gives $h'(t) = -0.005t$ and $h(t) = 2 - 0.005t$. The length of time for $h(t)$ to go to zero is $t = 2/0.005 = 400$ s.

(b) For the tank-filling period and using the same basis for deviation variables,

$$q_i = 0.1 \Rightarrow q_i' = +0.08 \Rightarrow Q_i'(s) = \frac{0.08}{s}$$
$$q = 0.02 \Rightarrow q' = 0 \Rightarrow Q'(s) = 0$$

Consequently, from Eq. 5-38, the tank model is

$$H'(s) = \frac{1}{4s}\left(\frac{0.08}{s} - 0\right) = \frac{0.02}{s^2}$$

Inversion to the time domain yields $h(t) = 1 + 0.02t$. Thus, the transfer pump can be operated for 200 s until $h(t) = 5$ m, when the tank would overflow. Note that this time (as well as the time to empty the tank in (a)) can be calculated without using Laplace transforms, simply by using the constant rate of inflow (or outflow) and the tank volume.

first-order transfer function rather than an integrator (cf. Example 4.8).

This example illustrates that integrating process units do not reach a new steady state when subjected to step changes in inputs, in contrast to first-order processes (cf. Eq. 5-24). Integrating systems represent an example of *non-self-regulating* processes. Arbitrary pulse inputs, where the initial and final values of the input are equal, do lead to a new steady state. For example, the rectangular pulse with height h given in Eq. 5-9 has the Laplace transform given in Eq. 5-10. The response of an integrating process with transfer function

$$\frac{Y(s)}{U(s)} = \frac{K}{s} \tag{5-40}$$

to a rectangular pulse input is

$$Y(s) = \frac{Kh(1 - e^{-t_w s})}{s^2} = Kh\left(\frac{1}{s^2} - \frac{e^{-t_w s}}{s^2}\right) \tag{5-41}$$

There are two regions for the solution of Eq. 5-41, depending on the value of t compared to the pulse width t_w. For $0 \le t < t_w$, the second term in the parentheses of Eq. 5-41 is 0, hence

$$y(t) = Kht \tag{5-42}$$

corresponding to a linear increase with respect to time. For $t \ge t_w$, taking the inverse Laplace transform of Eq. 5-41 gives

$$y(t) = Kh[t - (t - t_w)] = Kht_w \tag{5-43}$$

which is a constant value. Combining the solutions yields

$$y(t) = \begin{cases} Kht & t < t_w \\ Kht_w & t \ge t_w \end{cases} \tag{5-44}$$

Equation 5-44 shows that the change in y at any time is proportional to the area under the input pulse (the integral), an intuitive result.

5.6 FIRST-ORDER PROCESSES WITH VARIABLE TIME CONSTANT AND GAIN

So far, we have discussed first-order processed with constant time constant and gain. But in most of the chemical processes the time constants and the gains are not constant. The following examples will illustrate this fact.

EXAMPLE 5.4

Consider the liquid storage system shown in Fig. 5.7 with inlet flow rate q_i and cross-sectional area A. The resistance to flow is nonlinear and the outlet volumetric flow rate q is given by, $q = \beta\sqrt{h}$. Obtain the transfer function of the system between liquid level (h) and inlet flow rate.

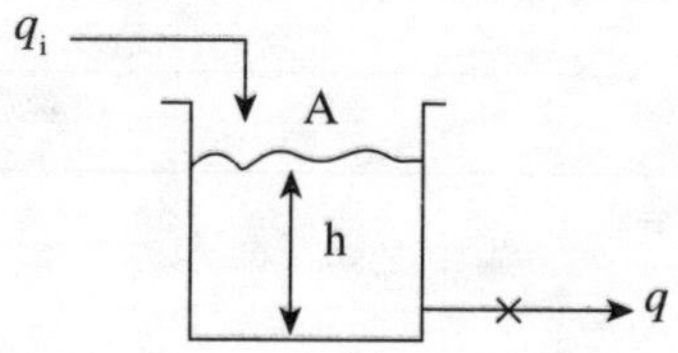

Figure 5.7 Liquid level system with nonlinear resistance.

Solution

Let ρ denote density of the liquid. Overall mass balance will give

Rate of mass accumulation = Rate of mass flow in – Rate of mass flow out

$$\frac{d}{dt}(\rho A h) = q_i\rho - q\rho$$

or $$\rho A\frac{dh}{dt} = q_i\rho - q\rho$$

or $$A\frac{dh}{dt} = q_i - q \qquad (5\text{-}45)$$

or $$A\frac{dh}{dt} = q_i - \beta\sqrt{h}$$

Eq 5-45 is nonlinear and we need to linearize for getting its Laplace transform.

$$A\frac{dh}{dt} = q_i - \beta\sqrt{h} = f(q_i, h)$$

$$A\frac{dh'}{dt} = \left.\frac{\partial f}{\partial q_i}\right|_{ss} q_i' + \left.\frac{\partial f}{\partial h}\right|_{ss} h' = q_i' - \frac{\beta}{2\sqrt{\bar{h}}}h' \qquad (5\text{-}46)$$

Taking Laplace transforms of both sides of Eq. 5-46, we get

$$AsH'(s) = Q_i'(s) - \frac{\beta}{2\sqrt{\bar{h}}}H'(s)$$

$$AsH'(s) + \frac{\beta}{2\sqrt{\bar{h}}}H'(s) = Q_i'(s)$$

$$\left(\frac{2A\sqrt{\bar{h}}}{\beta}s + 1\right)H'(s) = \frac{2\sqrt{\bar{h}}}{\beta}Q_i'(s)$$

$$G(s) = \frac{H'(s)}{Q_i'(s)} = \frac{K}{\tau s + 1}$$

where $\tau = \frac{2A\sqrt{\bar{h}}}{\beta}$ and $K = \frac{2\sqrt{\bar{h}}}{\beta}$. We can see that both time constant (τ) and gain (K) are dependent upon the steady-state value of the liquid level ($\bar{h}$). As $\bar{h}$ can be varied by varying the steady-stage value of the inlet flow rate ($\bar{q}_i$), we can say that the process has variable time constant and gain.

5.7 FIRST-ORDER PROCESSES WITH NUMERATOR DYNAMICS

The transfer function of the first order lag system with numerator dynamics can be represented as follows:

$$G(s) = \frac{Y(s)}{U(s)} = \frac{K(\tau_a s + 1)}{\tau s + 1} \qquad (5\text{-}47)$$

For a step change of magnitude M in the input, $U(s) = M/s$.

We can obtain the step response from Eq. 5-47 as

$$Y(s) = G(s)U(s) = \frac{K(\tau_a s + 1)}{\tau s + 1}\frac{M}{s} = \frac{KM(\tau_a s + 1)}{s(\tau s + 1)}$$

$$= \frac{A}{s} + \frac{B}{\tau s + 1} \qquad (5\text{-}48)$$

$$A = \lim_{s\to 0} sY(s) = \left.\frac{KM(\tau_a s + 1)}{\tau s + 1}\right|_{s=0} = KM$$

$$B = \lim_{s\to 0}(\tau s + 1)Y(s) = \left.\frac{KM(\tau_a s + 1)}{s}\right|_{s=-1/\tau}$$

$$= \frac{KM(1 - \tau_a/\tau)}{-1/\tau} = \frac{KM(\tau_a - \tau)/\tau}{1/\tau} = KM(\tau_a - \tau)$$

Hence, $$Y(s) = \frac{KM}{s} + \frac{KM(\tau_a - \tau)}{\tau s + 1}$$

Therefore, $$y(t) = KM + \frac{KM}{\tau}(\tau_a - \tau)e^{-t/\tau}$$

$$= KM\left(1 - \frac{\tau - \tau_a}{\tau}e^{-t/\tau}\right) \qquad (5\text{-}49)$$

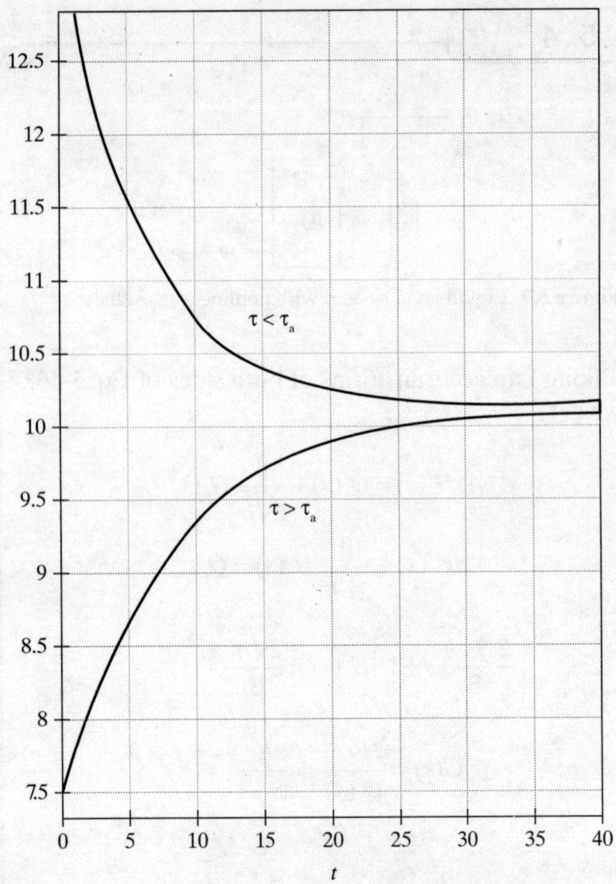

Figure 5.8 Step response of first-order system with numerator dynamics.

For $\tau_a = 0$, Eq. 5-49 and Eq. 5-24 will become identical. Step response of first-order system with numerator dynamics is shown in Fig. 5.8 for two cases..

5.8 SECOND-ORDER PROCESSES AND THEIR TYPES

Consider that the input $u(t)$ and output $y(t)$ of the process shown in Fig. 5.3 are related as given below in Eq. 5-50.

$$a_2 \frac{d^2y}{dt^2} + a_1 \frac{dy}{dt} + a_0 y = bu(t) \tag{5-50}$$

For $a_0 \neq 0$

$$\frac{a_2}{a_0}\frac{d^2y}{dt^2} + \frac{a_1}{a_0}\frac{dy}{dt} + y = \frac{b}{a_0}u(t)$$

$$\tau^2 \frac{d^2y}{dt^2} + 2\zeta\tau \frac{dy}{dt} + y = Ku \tag{5-51}$$

where $\tau^2 = \dfrac{a_2}{a_0}$, $2\zeta\tau = \dfrac{a_1}{a_0}, K = \dfrac{b}{a_0}$

Assuming the variables $u(t)$ and $y(t)$ to be in deviation form and taking Laplace transform on both sides of Eq. 5-51 we can get the transfer function of the second-order system as:

$$G(s) = \frac{Y(s)}{U(s)} = \frac{K}{\tau^2 s^2 + 2\zeta\tau s + 1}$$

There are basically three types of second-order processes. Process that consists of two capacities (first-order systems) in series, through which a material or energy must flow is of one kind. Processes possessing inertia and subject to acceleration are of second type. Inertia to motion is the characteristic of inherently second-order processes. The example of such processes are manometers, externally mounted level indicator, variable capacitance differential pressure transducer, pneumatic valve, etc. A processing system with its controller may exhibit second (or higher) order dynamics. These are third type of second-order processes.

5.9 RESPONSE OF SECOND-ORDER PROCESSES

As noted in Chapter 4, a second-order transfer function can arise physically whenever two first-order processes are connected in series. For example, two stirred-tank blending processes, each with a first-order transfer function relating inlet to outlet mass fraction, might be physically connected so that the outflow stream of the first tank is used as the inflow stream of the second tank. Figure 5.9 illustrates the signal flow relation for such a process. Here

$$G(s) = \frac{Y(s)}{U(s)} = \frac{K_1K_2}{(\tau_1 s+1)(\tau_2 s+1)} = \frac{K}{(\tau_1 s+1)(\tau_2 s+1)} \tag{5-52}$$

where $K = K_1K_2$. Alternatively, a second-order process transfer function will arise upon transforming either a second-order differential equation process model such as the two coupled first-order differential equations, for the CSTR (Eqs. 4-88 and 4-89). In this chapter we consider second-order transfer functions in the standard form

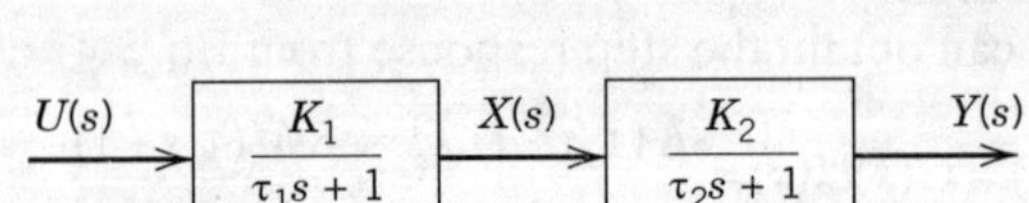

Figure 5.9 Two first-order systems in series yield an overall second-order system.

$$G(s) = \frac{K}{\tau^2 s^2 + 2\zeta\tau s + 1} \quad (5\text{-}53)$$

We defer discussion of the more general models where a time-delay term in the numerator or other numerator dynamics are present until Chapter 6.

In Eq. 5-53, K and τ have the same importance as for a first-order transfer function. K is the steady-state gain, and τ determines the speed of response (or, equivalently, the response time) of the system. The damping coefficient ζ (zeta) is dimensionless. It provides a measure of the amount of *damping* in the system—that is, the degree of oscillation in a process response after an input change. Small values of ζ imply little damping and a large amount of oscillation, as, for example, in an automobile suspension system with ineffective shock absorbers. Hitting a bump causes a vehicle to bounce up and down dangerously. In some textbooks, Eq. 5-53 is written in terms of $\omega_n = 1/\tau$, the undamped natural frequency of the system. This name arises because it represents the frequency of oscillation of the system when there is no damping ($\zeta = 0$).

There are three important classes of second-order systems, as shown in Table 5.2 The case where $\zeta < 0$ is omitted here because it corresponds to an unstable second-order system that has an unbounded response to any input (effects of instability are covered in Chapter 9). The overdamped and critically damped forms of the second-order transfer function most often appear when two first-order systems occur in series (see Fig. 5.9). The transfer functions given by Eqs. 5-52 and 5-53 differ only in the form of the denominators. Equating the denominators yields the relation between the two alternative forms for the *overdamped* second-order system:

$$\tau^2 s^2 + 2\zeta\tau s + 1 = (\tau_1 s + 1)(\tau_2 s + 1) \quad (5\text{-}54)$$

Table 5.2 The Three Classes of Second-Order Transfer Functions

Damping Coefficient	Characterization of Response	Roots of Characteristic Equation*
$\zeta > 1$	Overdamped	Real and unequal
$\zeta = 1$	Critically damped	Real and equal
$0 < \zeta < 1$	Underdamped	Complex conjugates (of the form $a + jb$ and $a - jb$)

*This equation is $\tau^2 s^2 + 2\zeta\tau s + 1 = 0$.

Expanding the right side of Eq. 5-54 and equating coefficients of the s terms,

$$\tau^2 = \tau_1 \tau_2 \quad (5\text{-}55)$$

$$2\zeta\tau = \tau_1 + \tau_2 \quad (5\text{-}56)$$

from which we obtain

$$\tau = \sqrt{\tau_1 \tau_2} \quad (5\text{-}57)$$

$$\zeta = \frac{\tau_1 + \tau_2}{2\sqrt{\tau_1 \tau_2}} \quad (5\text{-}58)$$

Solving for τ_1 and τ_2 gives

$$\tau_1 = \frac{\tau}{\zeta - \sqrt{\zeta^2 - 1}} \quad (\zeta \geq 1) \quad (5\text{-}59)$$

$$\tau_2 = \frac{\tau}{\zeta + \sqrt{\zeta^2 - 1}} \quad (\zeta \geq 1) \quad (5\text{-}60)$$

EXAMPLE 5.5

An overdamped system consists of two first-order processes operating in series ($\tau_1 = 4$, $\tau_2 = 1$). Find the equivalent values of τ and ζ for this system.

Solution

From Eqs. 5-57 and 5-58,

$$\tau = \sqrt{(4)(1)} = 2$$

$$\zeta = \frac{4+1}{(2)(2)} = 1.25$$

Equations 5-59 and 5-60 provide a check on these results:

$$\tau_1 = \frac{2}{1.25 - \sqrt{(1.25)^2 - 1}} = \frac{2}{1.25 - 0.75} = 4$$

$$\tau_2 = \frac{2}{1.25 + \sqrt{(1.25)^2 - 1}} = \frac{2}{1.25 + 0.75} = 1$$

The underdamped form of Eq. 5-53 can arise from some mechanical systems, from flow or other processes such as a pneumatic (air) instrument line with too little line capacity, or from a mercury manometer, where inertial effects are important.

For process control problems the underdamped form is frequently encountered in investigating the properties of processes under feedback control. Next we develop the relations for the step responses of all three classes of second-order processes.

5.9.1 Step Response

For the step input ($U(s) = M/s$) to a process described by Eq. 5-53,

$$Y(s) = \frac{KM}{s(\tau^2 s^2 + 2\zeta\tau s + 1)} \tag{5-61}$$

After inverting to the time domain, the responses can be categorized into three classes:

Overdamped ($\zeta > 1$)

If the denominator of Eq. 5-61 can be factored using Eqs. 5-59 and 5-60, then the response can be written

$$y(t) = KM\left(1 - \frac{\tau_1 e^{-t/\tau_1} - \tau_2 e^{-t/\tau_2}}{\tau_1 - \tau_2}\right) \tag{5-62}$$

Critically Damped ($\zeta = 1$)

$$y(t) = KM\left[1 - \left(1 + \frac{t}{\tau}\right)e^{-t/\tau}\right] \tag{5-63}$$

Underdamped ($0 \le \zeta < 1$)

$$y(t) = KM\left\{1 - e^{-\zeta t/\tau}\left[\cos\left(\frac{\sqrt{1-\zeta^2}}{\tau}t\right) + \frac{\zeta}{\sqrt{1-\zeta^2}}\sin\left(\frac{\sqrt{1-\zeta^2}}{\tau}t\right)\right]\right\} \tag{5-64}$$

Plots of the step responses for different values of ζ are shown in Figs. 5.10 and 5.11, where the time axis is normalized with respect to τ. Thus, when τ is small, a rapid response is signified, implying a large value for the undamped natural frequency, $\omega_n = 1/\tau$.

Several general remarks can be made concerning the responses shown in Figs. 5.10 and 5.11:

1. Responses exhibit a higher degree of oscillation and overshoot ($y/KM > 1$) as ζ approaches zero.
2. Large values of ζ yield a sluggish (slow) response.
3. The fastest response without overshoot is obtained for the critically damped case ($\zeta = 1$).

Control system designers sometimes attempt to make the response of the controlled variable to a set-point change approximate the ideal step response of an underdamped second-order system, that is, make it exhibit a prescribed amount of overshoot and oscillation as it settles at the new operating point. When damped oscillation is desirable, values of ζ in the range 0.4 to 0.8 may be chosen. In this range, the controlled variable y reaches the new operating point faster than with $\zeta = 1.0$ or 1.5, but the response is much less oscillatory (settles faster) than with $\zeta = 0.2$.

Figure 5.12 illustrates the characteristics of the step response of a second-order underdamped process. The following terms are used to describe the dynamics of underdamped processes:

1. ***Rise Time.*** t_r is the time the process output takes to first reach the new steady-state value.
2. ***Time to First Peak.*** t_p is the time required for the output to reach its first maximum value.
3. ***Settling Time.*** t_s is the time required for the process output to reach and remain inside a band whose width is equal to ±5% of the total change in y (±1% is also used for some applications).
4. ***Overshoot.*** OS = a/b (% overshoot is 100 a/b).
5. ***Decay Ratio.*** DR = c/a (where c is the height of the second peak).
6. ***Period of Oscillation.*** P is the time between two successive peaks or two successive valleys of the response.

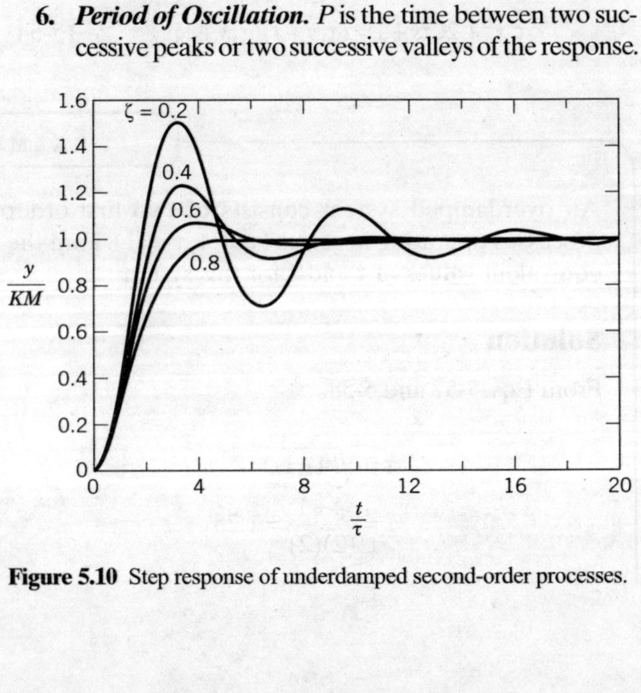

Figure 5.10 Step response of underdamped second-order processes.

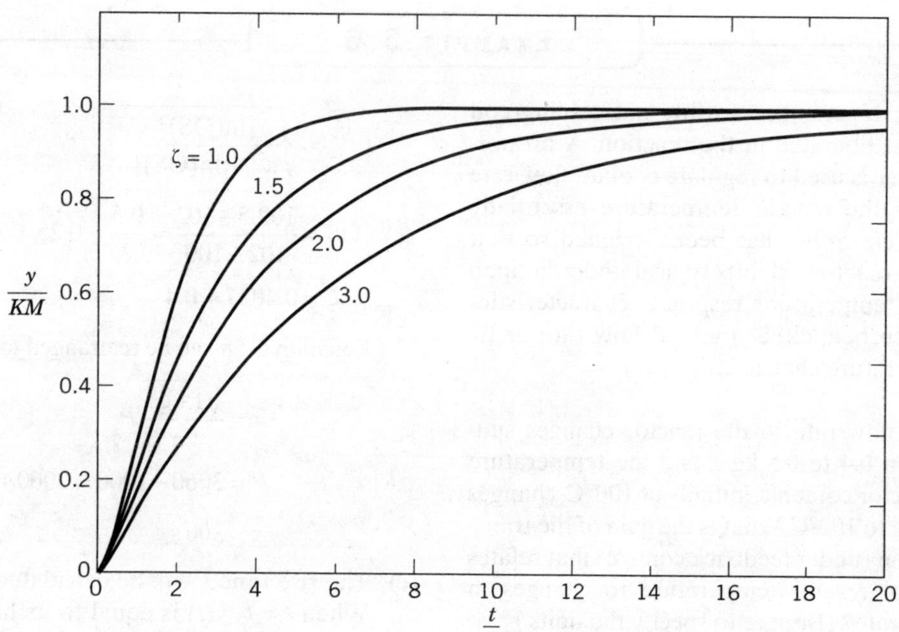

Figure 5.11 Step response of critically damped and overdamped second-order processes.

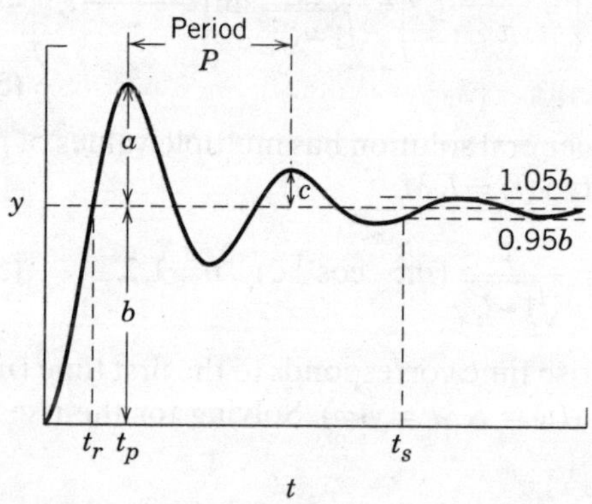

Figure 5.12 Performance characteristics for the step response of an underdamped process.

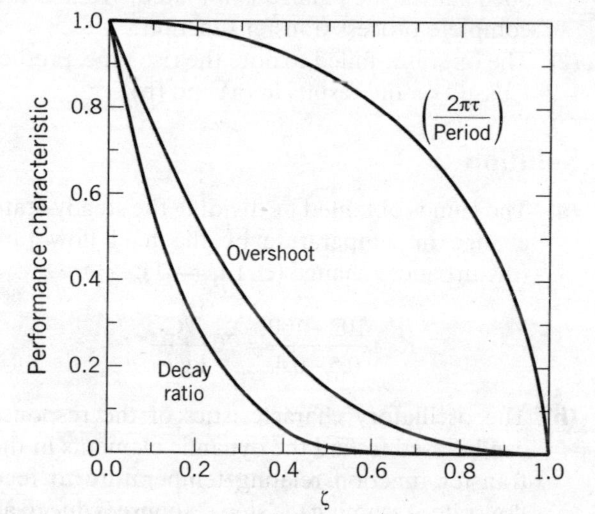

Figure 5.13 Relation between some performance characteristics of an underdamped second-order process and the process damping coefficient.

Note that the above definitions apply more generally to the step response of any underdamped process. If the process does not exhibit overshoot, the rise time definition is modified to be the time required to go from 10% to 90% of the steady-state response (Åström and Hägglund, 2006). For the particular case of an underdamped second-order process, we can develop analytical expressions for some of these characteristics. Using Eq. 5-64

Time to first peak: $$t_p = \pi\tau/\sqrt{1-\zeta^2} \tag{5-65}$$

Overshoot: $$\text{OS} = \exp\left(-\pi\zeta/\sqrt{1-\zeta^2}\right) \tag{5-66}$$

Decay ratio: $$\text{DR} = (\text{OS})^2 = \exp\left(-2\pi\zeta/\sqrt{1-\zeta^2}\right) \tag{5-67}$$

Period: $$P = \frac{2\pi\tau}{\sqrt{1-\zeta^2}} \tag{5-68}$$

Note that OS and DR are functions of ζ only. For a second-order system, the decay ratio is constant for each successive pair of peaks. Figure 5.12 illustrates the dependence of overshoot and decay ratio on damping coefficient.

For an underdamped second-order transfer function, Figs. 5.10 and 5.13 and Eq. 5-68 can be used to obtain estimates of ζ and τ based on step response characteristics.

EXAMPLE 5.6

A stirred-tank reactor has an internal cooling coil to remove heat liberated in the reaction. A proportional controller is used to regulate coolant flow rate so as to keep the reactor temperature essentially constant. The controller has been designed so that the controlled reactor exhibits typical underdamped second-order temperature response characteristics when it is disturbed, either by feed flow rate or by coolant temperature changes.

(a) The feed flow rate to the reactor changes suddenly from 0.4 to 0.5 kg/s, and the temperature of the reactor contents, initially at 100°C, changes eventually to 102°C. What is the gain of the transfer function (under feedback control) that relates changes in reactor temperature to changes in feed flow rate? (Be sure to specify the units.)

(b) The operator notes that the resulting response is slightly oscillatory with maxima estimated to be 102.5 and 102.0°C occurring at times 1000 and 3060 s after the change is initiated. What is the complete process transfer function?

(c) The operator failed to note the rise time. Predict t_r based on the results in (a) and (b).

Solution

(a) The gain is obtained by dividing the steady-state change in temperature by the feed flow rate (disturbance) change (cf. Eq. 4-41):

$$K = \frac{102-100}{0.5-0.4} = 20\,\frac{°\text{C}}{\text{kg/s}}$$

(b) The oscillatory characteristics of the response can be used to find the dynamic elements in the transfer function relating temperature to feed flow rate. Assuming the step response is due to an underdamped second-order process, Figs. 5.10 and 5.13 can be used to obtain estimates of ζ and τ. Alternatively, analytical expressions can be used, which is the approach taken here. Either Eq. 5-66 or 5-67 can be employed to find ζ independently of τ. Because the second peak value of temperature (102.0°C) is essentially the final value (102°C), the calculated value of peak height c will be subject to appreciable measurement error. Instead, we use the relation for overshoot (rather than decay ratio) to take advantage of the greater precision of the first peak measurement. Rearranging Eq. 5-66 gives

$$\zeta = \sqrt{\frac{[\ln(\text{OS})]^2}{\pi^2 + [\ln(\text{OS})]^2}}$$

$$\text{OS} = \frac{102.5-102}{102-100} = \frac{0.5}{2} = 0.25 \text{ (i.e.,25\%)} \qquad (5\text{-}69)$$

$$\zeta = 0.4037 \approx 0.4$$

Equation 5-68 can be rearranged to find τ:

$$\tau = \frac{\sqrt{1-\zeta^2}}{2\pi}P$$

$$P = 3060 - 1000 = 2060 \text{ s} \qquad (5\text{-}70)$$

$$\tau = 300 \text{ s}$$

(c) The rise time t_r can be calculated from Eq. 5-64. When $t = t_r$, $y(t)$ is equal to its final steady-state value, KM. In other words, the bracketed quantity is identically zero at $t = t_r$:

$$\cos\left(\frac{\sqrt{1-\zeta^2}}{\tau}t_r\right) + \frac{\zeta}{\sqrt{1-\zeta^2}}\sin\left(\frac{\sqrt{1-\zeta^2}}{\tau}t_r\right) = 0 \qquad (5\text{-}71)$$

The general solution has multiple values of t that satisfy $y(t) = KM$:

$$t = \frac{\tau}{\sqrt{1-\zeta^2}}(n\pi - \cos^{-1}\zeta) \quad n = 1,2,\ldots \qquad (5\text{-}72)$$

The rise time corresponds to the first time ($n = 1$) that $y(t) = KM = y(\infty)$. Solving for the rise time gives

$$t_r = \frac{\tau}{\sqrt{1-\zeta^2}}(\pi - \cos^{-1}\zeta) \qquad (5\text{-}73)$$

where the result of the inverse cosine computation must be in radians. Because τ = 300 s and $\zeta = 0.40$

$$t_r = 649 \text{ s}$$

In summary, the disturbance transfer function between feed flow rate and outlet temperature while under feedback control is

$$\frac{T'(s)}{W'(s)} = \frac{20}{(300)^2 s^2 + 2(0.4)(300)s + 1} = \frac{20}{90{,}000s^2 + 240s + 1}$$

where the process gain has units of °C/kg/s.

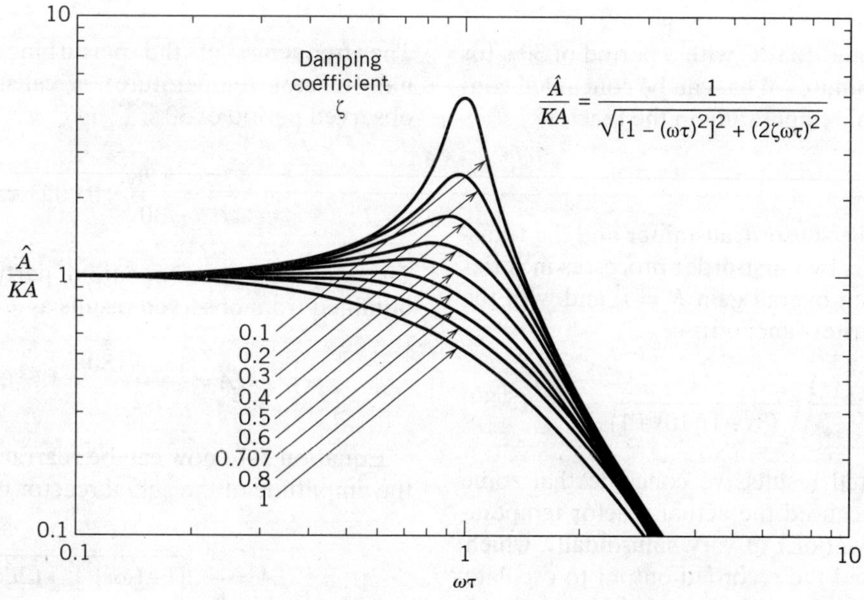

Figure 5.14 Sinusoidal response amplitude of a second-order system after exponential terms have become negligible.

5.9.2 Sinusoidal Response

When a linear second-order system is forced by a sinusoidal input $A \sin \omega t$, the output for large values of time (after exponential terms have become negligible) is also a sinusoidal signal given by

$$y(t) = \frac{KA}{\sqrt{[1-(\omega\tau)^2]^2 + (2\zeta\omega\tau)^2}} \sin(\omega t + \phi) \tag{5-74}$$

where

$$\phi = -\tan^{-1}\left[\frac{2\zeta\omega\tau}{1-(\omega\tau)^2}\right] \tag{5-75}$$

The output amplitude $\hat{A}$ is obtained directly from Eq. 5-74:

$$\hat{A} = \frac{KA}{\sqrt{[1-(\omega\tau)^2]^2 + (2\zeta\omega\tau)^2}} \tag{5-76}$$

The ratio of output to input amplitude is the *amplitude ratio* AR ($= \hat{A}/A$). When normalized by the process gain, it is called the *normalized amplitude ratio* AR_N

$$AR_N = \frac{\hat{A}}{KA} = \frac{1}{\sqrt{[1-(\omega\tau)^2]^2 + (2\zeta\omega\tau)^2}} \tag{5-77}$$

AR_N represents the effect of the dynamic model parameters (ζ, τ) on the sinusoidal response; that is, AR_N is independent of steady-state gain K and the amplitude of the forcing function, A. The maximum value of AR_N (if it exists) can be found by differentiating Eq. 5-77 with respect to ω and setting the derivative to zero. Solving for ω_{max} gives

$$\omega_{max} = \frac{\sqrt{1-2\zeta^2}}{\tau} \quad \text{for } 0 < \zeta < 0.707 \tag{5-78}$$

For $\zeta \geq 0.707$, there is no maximum, as Fig. 5.14 illustrates. Substituting Eq. 5-78 into Eq. 5-77 yields an expression for the maximum value of AR_N:

$$AR_N\big|_{max} = \frac{\hat{A}}{KA}\bigg|_{max} = \frac{1}{2\zeta\sqrt{1-\zeta^2}} \tag{5-79}$$
$$\text{for } 0 < \zeta < 0.707$$

Equation 5-79 indicates that the maximum output amplitude for a second-order process that has no damping ($\zeta = 0$) is undefined. Small values of ζ are invariably avoided in the design of processes and control systems. Equation 5-79 indicates that a process with little damping can exhibit very large output oscillations if it is perturbed by periodic signals with ω near ω_{max}.

EXAMPLE 5.7

An engineer uses a temperature sensor mounted in a thermowell to measure the temperature in a CSTR. The temperature sensor/transmitter combination operates approximately as a first-order system with a time constant of 3 s. The thermowell behaves like a first-order system with a time constant of 10 s. The engineer notes that the measured reactor temperature has been cycling approximately sinusoidally between 180 °C and 183 °C with a period of 30 s for at least several minutes. What can be concluded concerning the actual temperature in the reactor?

Solution

First, note that the sensor/transmitter and the transmission line act as two first-order processes in series (Eq. 5-52) with an overall gain K = 1, and with the approximate transfer function

$$\frac{T'_{\text{meas}}(s)}{T'_{\text{reactor}}} = \frac{1}{(3s+1)(10s+1)} \tag{5-80}$$

From the reported results, we conclude that some disturbance has caused the actual reactor temperature (and its deviation) to vary sinusoidally, which, in turn, has caused the recorded output to oscillate. The cycling has continued for a period of time that is much longer than the time constants of the process—that is, the instrumentation system. Hence, the transients have died out and we can infer the conditions in the reactor from the measured results, using Eq. 5-76 for the sinusoidal response of a second-order system. From Eq. 5-80, τ_1 = 3 s and τ_2 = 10 s; τ and ζ are calculated from Eqs. 5-55 and 5-56:

$$\tau = \sqrt{(3)(10)} = 5.48\text{ s}$$

$$\zeta = \frac{13}{2(5.477)} = 1.19$$

The frequency of the perturbing sinusoidal signal (reactor temperature) is calculated from the observed period of 30 s:

$$\omega = \frac{2\pi}{P} = \frac{6.28}{30} = 0.2093\text{ rad/s}$$

The amplitude of the output perturbation also is obtained from observed results as

$$\hat{A} = \frac{183-180}{2} = 1.5\,°\text{C}$$

Equation 5-76 now can be rearranged to calculate the amplitude of the actual reactor temperature

$$A = \frac{\hat{A}}{K}\sqrt{[1-(\omega\tau)^2]^2 + (2\zeta\omega\tau)^2}$$

from which A = 4.12 °C. Thus, the actual reactor temperature is varying between 181.5 – 4.12 = 177.38 °C and 181.5 + 4.12 = 185.62 °C, nearly three times the variation indicated by the measured value.

Because the second-order process in this example is overdamped (ζ = 1.19), we expect that sinusoidal perturbations in the reactor temperature always will be attenuated (reduced in amplitude) in the measurement system regardless of the frequency of the perturbation. Further discussion of sinusoidal forcing is contained in Chapter 14 on frequency response techniques.

5.10 SECOND-ORDER PROCESSES WITH NUMERATOR DYNAMICS

The presence of a zero in the first-order system (cf. Eq. 4-128) causes a jump discontinuity in $y(t)$ at t = 0 when a step input is applied. Such an instantaneous step response is possible only when the numerator and denominator polynomials have the same order, which includes the case, $G(s) = K$. Industrial processes have higher-order dynamics in the denominator, causing them to exhibit some degree of *inertia*. This feature prevents them from responding instantaneously to any input, including an impulse input. Thus, we say that $m \leq n$ for a system to be *physically realizable*.

EXAMPLE 5.8

For the case of a single zero in an overdamped second-order transfer function,

$$G(s) = \frac{K(\tau_a s+1)}{(\tau_1 s+1)(\tau_2 s+1)} \tag{5-81}$$

calculate the response to a step input of magnitude M and plot the results for τ_1 = 4, τ_2 = 1 and several values of τ_a.

Solution

The response of this system to a step change in input is (see Table 3.1)

$$y(t) = KM\left(1 + \frac{\tau_a-\tau_1}{\tau_1-\tau_2}e^{-t/\tau_1} + \frac{\tau_a-\tau_2}{\tau_2-\tau_1}e^{-t/\tau_2}\right) \tag{5-82}$$

Note that $y(t \to \infty) = KM$ as expected; thus, the effect of including the single zero does not change the final value, nor does it change the number or locations of the poles. But the zero does affect how the response modes (exponential terms) are weighted in the solution, Eq. 5-82.

Mathematical analysis shows that three types of responses can occur, as illustrated for eight values of τ_a in Fig. 5.15:

$$\textbf{Case i}: \quad \tau_a > \tau_1 \qquad (\tau_a = 8, 16)$$
$$\textbf{Case ii}: \quad 0 < \tau_a \le \tau_1 \qquad (\tau_a = 0.5, 1, 2, 4)$$
$$\textbf{Case iii}: \quad \tau_a < 0 \qquad (\tau_a = -1, -4)$$

where $\tau_1 > \tau_2$ is arbitrarily chosen. Case (i) shows that *overshoot* can occur if τ_a is sufficiently large. Case (ii) is similar to a first-order process response. Case (iii), which has a *positive zero*, also called a *right-half plane zero*, exhibits an *inverse response*, an infrequently encountered yet important dynamic characteristic. An inverse response occurs when the initial response to a step input is in one direction but the final steady state is in the opposite direction. For example, for case (iii), the initial response is in the negative direction while the new steady state $y(\infty)$ is in the positive direction in the sense that $y(\infty) > y(0)$. Inverse responses are associated with right-half plane zeros.

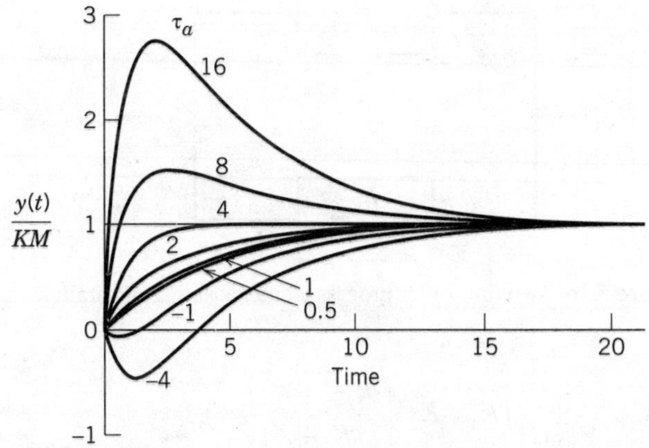

Figure 5.15 Step response of an overdamped second-order system (Eq. 5-81) for different values of τ_a ($\tau_1 = 4$, $\tau_2 = 1$).

The phenomenon of overshoot or inverse response results from the zero in the above example and will not occur for an overdamped second-order transfer function containing two poles but no zero. Inverse responses arise from competing dynamic effects that operate on two different time scales (τ_1 and τ_2 in Example 5.8). For example, an inverse response can occur in a distillation column when the steam pressure to the reboiler is suddenly changed. An increase in steam pressure ultimately will decrease the reboiler level (in the absence of level control) by boiling off more of the liquid. However, the initial effect usually is to increase the amount of frothing on the trays immediately above the reboiler, causing a rapid spillover of liquid from these trays into the reboiler below. This initial increase in reboiler liquid level is later overwhelmed by a decrease due to the increased vapor boil-up. See Buckley et al. (1985) for a detailed analysis of this phenomenon.

As a second example, tubular catalytic reactors with exothermic chemical reactions exhibit an inverse response in exit temperature when the feed temperature is increased. Initially, increased conversion in the entrance region of the bed momentarily depletes reactants at the exit end of the bed, causing less heat generation there and decreasing the exit temperature. Subsequently, higher reaction rates occur, leading to a higher exit temperature, as would be expected. Conversely, if the feed temperature is decreased, the inverse response initially yields a higher exit temperature.

Inverse response or overshoot can be expected whenever two physical effects act on the process output variable in different ways and with different time scales. For the case of reboiler level mentioned above, the *fast* effect of a steam pressure increase is to spill liquid off the trays above the reboiler immediately as the vapor flow increases. The *slow* effect is to remove significant amounts of the liquid mixture from the reboiler through increased boiling. Hence, the relationship between reboiler level and reboiler steam pressure can be represented approximately as an overdamped second-order transfer function with a right-half plane zero.

Next, we show that inverse responses can occur for two first-order transfer functions in a parallel arrangement, as shown in Fig. 5.16. The relationship between $Y(s)$ and $U(s)$ can be expressed as

$$\frac{Y(s)}{U(s)} = \frac{K_1}{\tau_1 s+1} + \frac{K_2}{\tau_2 s+1} = \frac{K_1(\tau_2 s+1)+K_2(\tau_1 s+1)}{(\tau_1 s+1)(\tau_2 s+1)} \tag{5-83}$$

or, after rearranging the numerator into standard gain/time constant form, we have

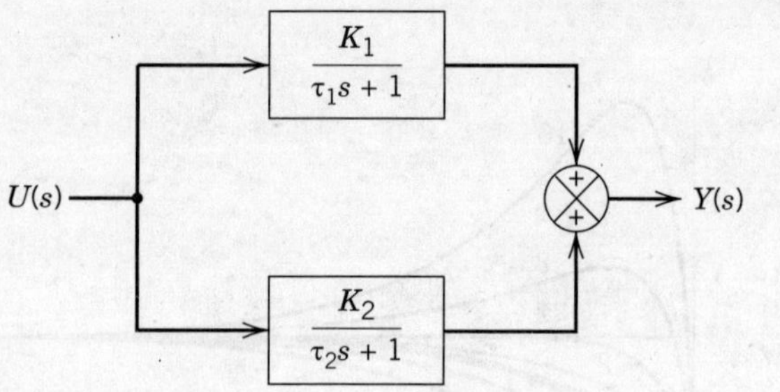

Figure 5.16 Two first-order process elements acting in parallel.

$$\frac{Y(s)}{U(s)} = \frac{(K_1 + K_2)\left(\dfrac{K_1\tau_2 + K_2\tau_1}{K_1 + K_2}s + 1\right)}{(\tau_1 s + 1)(\tau_2 s + 1)} \tag{5-84}$$

$$K = K_1 + K_2 \tag{5-85}$$

and

$$\tau_a = \frac{K_1\tau_2 + K_2\tau_1}{K_1 + K_2} \tag{5-86}$$

$$= \frac{K_1\tau_2 + K_2\tau_1}{K} \tag{5-87}$$

The condition for an inverse response to exist is $\tau_a < 0$, or

$$\frac{K_1\tau_2 + K_2\tau_1}{K} < 0 \tag{5-88}$$

For $K > 0$, Eq. 5-88 can be rearranged to the more convenient form

$$-\frac{K_2}{K_1} > \frac{\tau_2}{\tau_1} \tag{5-89}$$

For $K < 0$, the inequality in Eq. 5-89 is reversed. Note that Eq. 5-88 indicates that K_1 and K_2 have opposite signs, because $\tau_1 > 0$ and $\tau_2 > 0$. It is left to the reader to show that $K > 0$ when $K_1 > 0$ and that $K < 0$ when $K_1 < 0$. In other words, the sign of the overall transfer function gain is the same as that of the slower process.

The step response of the process described by Eq. 5-81 will have a negative slope initially (at $t = 0$) if the product of the gain and step change magnitude is positive ($KM > 0$), τ_a is negative, and τ_1 and τ_2 are both positive.

5.11 DETERMINING STEP RESPONSE CHARACTERISTICS USING MATLAB

The step response characteristics for a dynamic system model can be computed by using the MATLAB command `stepinfo`. The function returns the characteristics in a structure containing the following fields:

Rise Time. Time it takes for the response to rise from 10% to 90% of the steady-state response.

Settling Time. Time it takes for the error $|y(t) - y_{ss}|$ between the response $y(t)$ and the steady-state response y_{ss} to fall to within 2% of y_{ss}.

Settling Min. Minimum value of $y(t)$ once the response has risen.

Settling Max. Maximum value of $y(t)$ once the response has risen.

Overshoot. Percentage overshoot, relative to y_{ss}.

Undershoot. Percentage undershoot.

Peak. Peak absolute value of $y(t)$.

Peak Time. Time at which the peak value occurs.

By default, `stepinfo` defines settling time as the time taken by the error between the response $y(t)$ and the steady-state response y_{ss} to come within 2% of y_{ss}. Also, `stepinfo` defines the rise time as the time it takes for the response to rise from 10% of y_{ss} to 90% of y_{ss}. These definitions can be changed using `SettlingTimeThreshold` and `RiseTimeThreshold`.

It is also possible to display the system characteristics such as rise time, peak time, overshoot, settling time etc. on the step response plot. Simply by right-clicking anywhere in the figure we can see select "Characteristics" from the field and choose any one characteristic from the list to view it.

EXAMPLE 5.9

Find the step response characteristics of the following system using the MATLAB command 'stepinfo':

$$G(s) = \frac{5}{18s^2 + 2s + 1}$$

Solution

The following MATLAB code can be written for the given system.

```
clc;
clear all;
```

```
close all;
sys=tf([5],[18 2 1])
step(sys);
S=stepinfo(sys, 'SettlingTime
Threshold',0.005,'RiseTimeThreshold',
[0 1])
```

We can get the characteristics from the output of this programme as follows:

```
S =
       RiseTime: 7.9057
   SettlingTime: 87.2663
    SettlingMin: 3.9109
    SettlingMax: 7.3247
      Overshoot: 46.4942
     Undershoot: 0
           Peak: 7.3247
       PeakTime: 14.0918
```

EXAMPLE 5.10

Obtain the characteristics of the following system from the step response plot:

$$G(s) = \frac{s^2+5s+5}{s^4+8s^3+18s^2+2s+1}.$$

Solution

The following MATLAB code can be written to display the plot.

```
clc;
clear all;
close all;
sys = tf([1 5 5],[1 8 18 2 1]);
step(sys);
```

We can obtain the step response characteristics by right clicking on the plot area and choosing required items one by one. These are shown in Fig. 5.17.

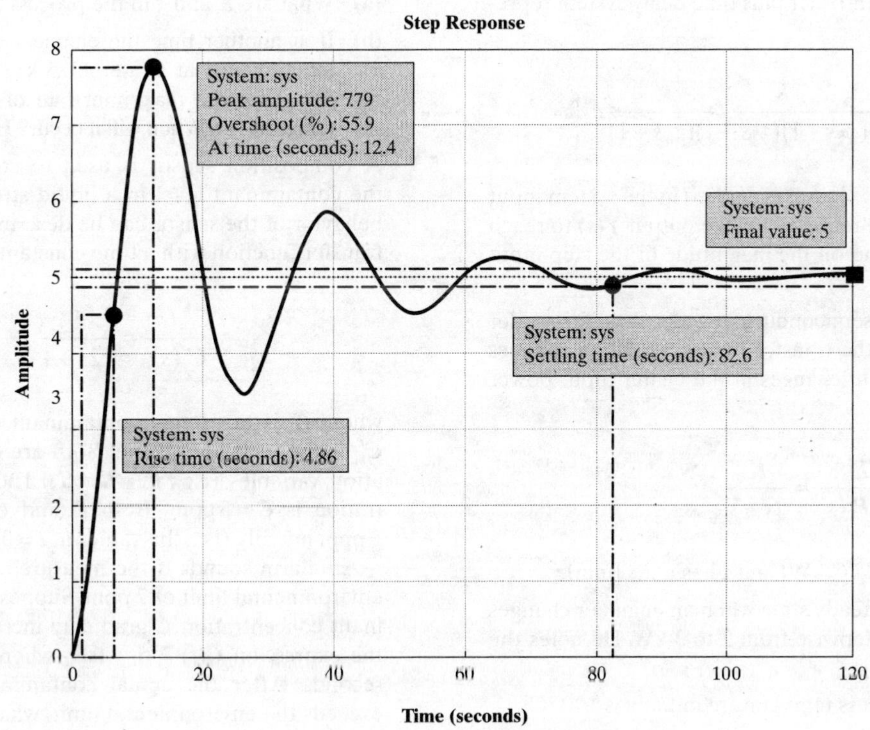

Figure 5.17 Step response characteristics of the system given in Example 5.10.

SUMMARY

Transfer functions can be used to obtain output responses to any type of input change. In this chapter we have focused on first- and second-order transfer functions and integrating processes. Because a relatively small number of input changes have industrial or analytical significance, we have considered in detail the responses of these basic process transfer functions to important types of inputs, such as step, ramp, impulse, and sine inputs.

If a process can be modeled as a first-order or second-order transfer function, the process response to any standard input change can be found analytically or using computer software. When a theoretical model is not available, as occurs in many plant situations, data can be used to obtain an approximate process transfer function if the input is measured. A model permits predictions of how a process will react to other types of disturbances or input changes.

Unfortunately, not all processes can be modeled by such simple transfer functions. Hence, in Chapter 6, several additional transfer function elements are introduced in order to construct more complicated transfer functions. However, the emphasis there is to show how complex process behavior can be explained with combinations of simple transfer function elements.

REFERENCES

Åström, K. J., and T. Hägglund, *Advanced PID Control*, 3rd ed., Instrument Society of America, Research Triangle Park, NC, 2006.

Box, G. E. P., G. M. Jenkins, and G. C. Reinsel, *Time Series Analysis, Forecasting, and Control*, 3rd ed., Prentice-Hall, Englewood Cliffs, NJ, 1994.

Buckley, P. S., W. L. Luyben, and J. P. Shunta, *Design of Distillation Column Control Systems*, Instrument Society of America, Research Triangle Park, NC, 1985.

Maybeck, P. S., *Stochastic Models, Estimation, and Control*, 2nd ed., Academic Press, New York, 1997.

EXERCISES

5.1 Consider the fourth-order plus time delay system represented below:

$$G(s) = \frac{K}{(\tau_1 s+1)(\tau_2 s+1)(\tau_3 s+1)(\tau_4 s+1)} e^{-\tau s}$$

Assume that a step change is applied in the system input $U(s)$. Will the time required for the output $Y(s)$ to reach steady state depend on the magnitude of the step input in U? Explain.

5.2 A heater for a semiconductor wafer has first-order dynamics, that is, the transfer function relating changes in temperature T to changes in the heater input power level P is

$$\frac{T'(s)}{P'(s)} = \frac{K}{\tau s+1}$$

where K has units [°C/kW] and τ has units [min].

The process is at steady state when an engineer changes the power input stepwise from 2 to 3 kW. He notes the following:

(i) The process temperature initially is 100 °C.

(ii) Four minutes after changing the power input, the temperature is 400 °C.

(iii) Thirty minutes later the temperature is 500 °C.

(a) What are K and τ in the process transfer function?

(b) If at another time the engineer changes the power input linearly at a rate of 0.5 kW/min, what can you say about the maximum rate of change of process temperature: When will it occur? How large will it be?

5.3 A composition sensor is used to continually monitor the contaminant level in a liquid stream. The dynamic behavior of the sensor can be described by a first-order transfer function with a time constant of 12 s,

$$\frac{C'_m(s)}{C'(s)} = \frac{1}{12s+1}$$

where C' is the actual contaminant concentration and C'_m is the measured value. Both are expressed as deviation variables (e.g., $C' = C - \bar{C}$). The nominal concentration is $\bar{C} = 5$ ppm. Both C and C_m have values of 5 ppm initially (i.e., the values at $t = 0$).

An alarm sounds if the measured value exceeds the environmental limit of 7 ppm. Suppose that the contaminant concentration C gradually increases according to the expression $C(t) = 5 + 0.2t$, where t is expressed in seconds. After the actual contaminant concentration exceeds the environmental limit, what is the time interval, Δt, until the alarm sounds?

5.4 The dynamic response of a stirred-tank bioreactor can be represented by the transfer function

$$\frac{C'(s)}{C'_F(s)} = \frac{2}{4s+1}$$

where C' is the exit substrate concentration, mol/L, and C'_F is the feed substrate concentration, mol/L.

(a) Derive an expression for $c'(t)$ if $c_F(t)$ is a rectangular pulse (Fig. 5.2) with the following characteristics:

$$c_F(t) = \begin{cases} 2 & t < 0 \\ 4 & 0 \le t < 2 \\ 2 & 2 \le t < \infty \end{cases}$$

(b) What is the maximum value of $c'(t)$? When does it occur? What is the final value of $c'(t)$?

(c) If the initial value is $c(0) = 1$, how long does it take for $c(t)$ to return to a value of 1.05 after it has reached its maximum value?

5.5 A thermocouple has the following characteristics when it is immersed in a stirred bath:

Mass of thermocouple = 1 g

Heat capacity of thermocouple = 0.25 cal/g °C

Heat transfer coefficient = 20 cal/cm²h °C (for thermocouple and bath)

Surface area of thermocouple = 3 cm²

(a) Derive a transfer function model for the thermocouple relating the change in its indicated output T to the change in the temperature of its surroundings T_s assuming uniform temperature (no gradients in the thermocouple bead), no conduction in the leads, constant physical properties, and conversion of the millivolt-level output directly to a °C reading by a very fast sensor.

(b) If the thermocouple is initially out of the bath and at room temperature (23 °C), what is the maximum temperature that it will register if it is suddenly plunged into the bath (80 °C) and held there for 20 s?

(c) Verify the result in (b) using computer simulation.

5.6 Consider the transfer function

$$G(s) = \frac{Y(s)}{U(s)} = \frac{20}{(5s+1)(3s+1)}$$

What is $y(t \to \infty)$ for the following inputs:

(a) step input of height M

(b) unit impulse input $\delta(t)$

(c) $\sin t$

(d) unit rectangular pulse (Eq. 3-15, $h = 1$)

5.7 Appelpolscher has just left a meeting with Stella J. Smarly, IGC's vice-president for process operations and development. Smarly is concerned about an upcoming extended plant test of a method intended to improve the yields of a large packed-bed reactor. The basic idea, which came from IGC's university consultant and was recently tested for feasibility in a brief run, involves operating the reactor cyclically so that nonlinearities in the system cause the time-average yield at the exit to exceed the steady-state value. Smarly is worried about the possibility of sintering the catalyst during an extended run, particularly in the region of the "hotspot" (axially about one-third of the way down the bed and at the centerline) where temperatures invariably peak. Appelpolscher, who plans to leave the next day on a two-week big game photo safari, doesn't want to cancel his vacation. On the other hand, Smarly has told him he faces early, unexpected retirement in Botswana if the measurement device (located near the hot spot) fails to alert operating people and the reactor catalyst sinters. Appelpolscher likes Botswana but doesn't want to retire there. He manages to pull together the following data and assumptions before heading for the airport and leaves them with you for analysis with the offer of the use of his swimming pool while he is gone. What do you report to Smarly?

Data:

Frequency of cyclic operation = 0.2 cycles/min

Amplitude of thermal wave (temperature) at the measurement point obtained experimentally in the recent brief run = 15 °C

Average operating temperature at the measurement point, T_{meas} = 350 °C

Time constant of temperature sensor and thermowell = 1.5 min

Temperature at the reactor wall = 200 °C

Temperature at which the catalyst sinters if operated for several hours = 740 °C

Temperature at which the catalyst sinters instantaneously = 750 °C

Assumptions:

The reactor operational cycle is approximately sinusoidal at the measurement point.

The thermowell is located near the reactor wall so as to measure a "radial average" temperature rather than the centerline temperature.

The approximate relation is

$$T = \frac{T_{center} + 2T_{wall}}{3}$$

which also holds during transient operation.

5.8 A liquid storage system is shown below. The normal operating conditions are $\bar{q}_1 = 284$ L/min, $\bar{q}_2 = 142$ L/min, $\bar{h} = 1.22$ m. The tank is 1.83 m in diameter, and the density of each stream is 96 kg/m³. Suppose that a pulse change in q_1 occurs as shown in Fig. E5.8.

(a) What is the transfer function relating H' to Q'_1?

(b) Derive an expression for $h(t)$ for this input change.

(c) What is the new steady-state value of liquid level $\bar{h}$?

(d) Repeat (b) and (c) for a change in q_1 from 284 to 568 (at $t = 0$) to 426 L/min at $t = 12$.

5.9 Two liquid storage systems are shown in Fig. E5.9. Each tank is 1.22 m in diameter. For System I, the valve acts as a linear resistance with the flow–head relation $q = 104\,h$, where q is in L/min and h is in m. For System II, variations in liquid level h do not affect exit flow rate q. Suppose that each system is initially at steady state with $\bar{h} = 1.83$ m and $\bar{q}_i = 189$ L/min and that at time $t = 0$ the inlet flow rate suddenly changes from 189 to 265 L/min. For each system, determine the following information:

(a) The transfer function $H'(s)/Q_i'(s)$ where the primes denote deviation variables.

(b) The transient response $h(t)$.

(c) The new steady-state levels.

(d) If each tank is 2.44 m tall, which tank overflows first? When?

5.10 The dynamic behavior of the liquid level in a leg of a manometer tube, responding to a change in pressure, is given by

$$\frac{d^2h'}{dt^2} + \frac{6\mu}{R^2\rho}\frac{dh'}{dt} + \frac{3}{2}\frac{g}{L}h' = \frac{3}{4\rho L}p'(t)$$

where $h'(t)$ is the level of fluid measured with respect to the initial steady-state value, $p'(t)$ is the pressure change, and R, L, g, ρ, and μ are constants.

(a) Rearrange this equation into standard gain-time constant form and find expressions for K, τ, ζ in terms of the physical constants.

(b) For what values of the physical constants does the manometer response oscillate?

(c) Would changing the manometer fluid so that ρ (density) is larger make its response more oscillatory, or less? Repeat the analysis for an increase in μ (viscosity).

5.11 A process is described by the following transfer function:

$$\frac{Y(s)}{U(s)} = \frac{K}{s(\tau s + 1)}$$

Thus, it exhibits characteristics of both first-order and integrating processes.

How could you utilize a step change in the input of magnitude M to find quickly the two parameters K and τ? (Be sure to show all work and sketch the anticipated process response.)

5.12 For the equation

$$\frac{d^2y}{dt^2} + K\frac{dy}{dt} + 4y = u$$

(a) Find the transfer function and put it in standard gain/time constant form.

(b) Discuss the qualitative form of the response of this system (independent of the input forcing) over the range $-10 \le K \le 10$.

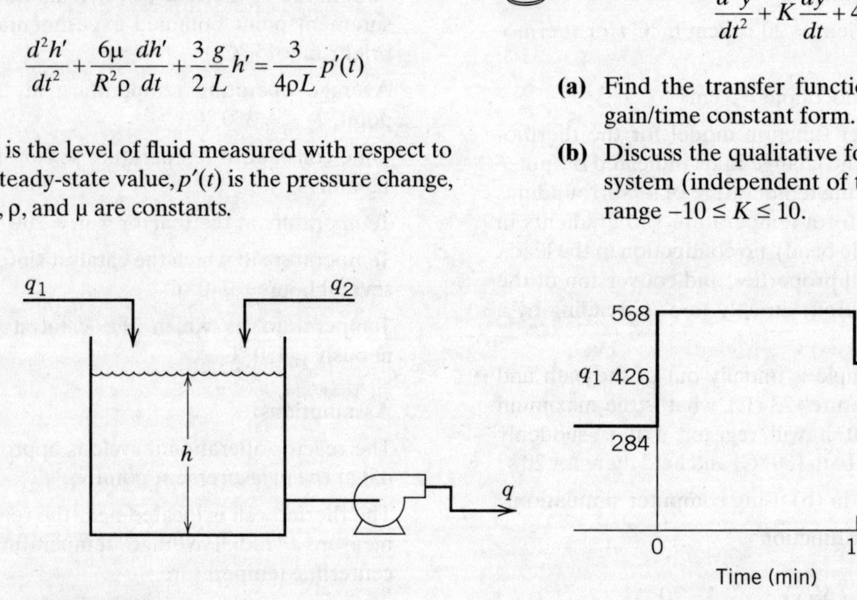

Figure E5.8

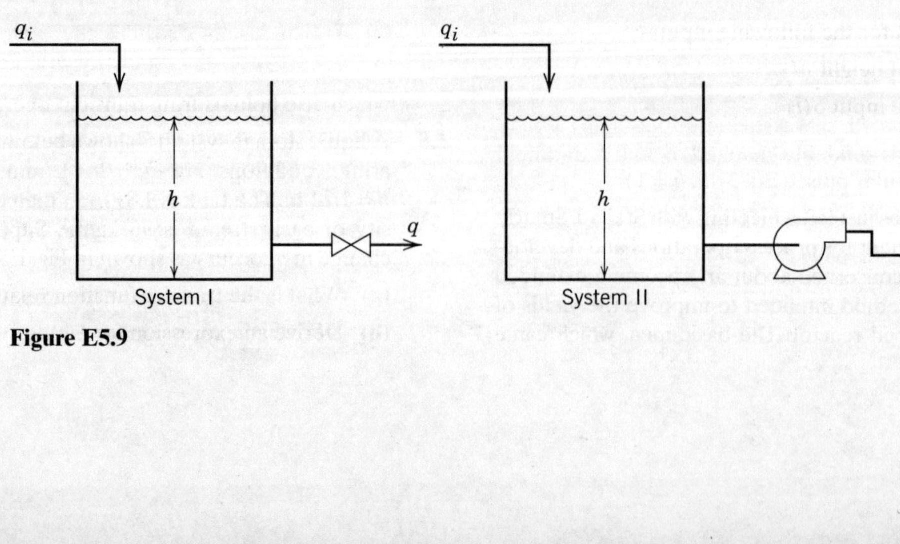

Figure E5.9

Specify values of K where the response will converge and where it will not. Write the form of the response without evaluating any coefficients.

5.13 A second-order critically damped process has the transfer function

$$\frac{Y(s)}{U(s)} = \frac{K}{(\tau s + 1)^2}$$

(a) For a step change in input of magnitude M, what is the time (t_s) required for this process to settle to within 5% of the total change in the output?

(b) For $K = 1$ and a ramp change in input, $u(t) = at$, by what time period does $y(t)$ lag behind $u(t)$ once the output is changing linearly with time?

(c) Verify the results in parts (a) and (b) using computer simulation and $M = \tau = a = 1$.

5.14 A step change from 104 to 214 kPa in actual pressure results in the measured response from a pressure-indicating element shown in Fig. E5.14.

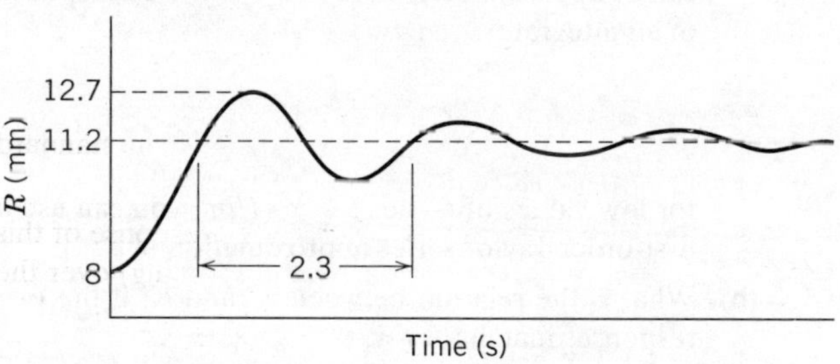

Figure E5.14

(a) Assuming second-order dynamics, calculate all important parameters and write an approximate transfer function in the form

$$\frac{R'(s)}{P'(s)} = \frac{K}{\tau^2 s^2 + 2\zeta\tau s + 1}$$

where R' is the instrument output deviation (mm), P' is the actual pressure deviation (psi).

(b) Write an equivalent differential equation model in terms of actual (not deviation) variables.

5.15 An electrically heated process is known to exhibit second-order dynamics with the following parameter values: $K = 3$ °C/kW, $\tau = 3$ min, $\zeta = 0.7$. If the process initially is at steady state at 70 °C with heater input of 30 kW and the heater input is suddenly changed to 36 kW and held there,

(a) Derive an expression for the measured temperature as a function of time.

(b) What will be the maximum observed temperature? When will it occur?

5.16 Starting with Eq. 5-64, derive expressions for the following response characteristics of the underdamped second-order system.

(a) The time to first peak t_p (Eq. 5-65).

(b) The fraction overshoot (Eq. 5-66).

(c) The decay ratio (Eq. 5-67).

(d) The settling time (t_s, defined in Fig. 5.12). Can a *single* expression be used for t_s over the full range of ζ, $0 < \zeta < 1$?

5.17 A tank used to dampen liquid flow rate surges is known to exhibit second-order dynamics. The input flow rate changes suddenly from 681 to 795 L/min. An operator notes that the tank level changes as follows:

Before input change: level = 3.05 m and steady
Four minutes later: level = 5.03 m
Forty minutes later: level = 4.6 m and steady

(a) Find a transfer function model that describes this process, at least approximately. Evaluate all parameters in your model, including units.

(b) Is your model unique? Why or why not?

5.18 A process has the transfer function

$$G(s) = \frac{10}{s^2 + s + 1} = \frac{Y(s)}{U(s)}$$

(a) For a step change in the input $U(s) = 2/s$, sketch the response $y(t)$ (you do not need to solve the differential equation). Show as much detail as possible, including the steady-state value of $y(t)$, and whether there is oscillation.

(b) What is the decay ratio?

5.19 A surge tank system is to be installed as part of a pilot plant facility. The initial proposal calls for the configuration shown in Fig. 4.4. Each tank is 1.5 m high and 0.91 m in diameter. The design flow rate is $q_i = 378.5$ L/min. It has been suggested that an improved design will result if the two-tank system is replaced by a single tank that is 1.22 m in diameter and has the same total volume (i.e., $V = V_1 + V_2$).

(a) Which surge system (original or modified) can handle larger step disturbances in q_i? Justify your answer.

(b) Which system provides the best damping of step disturbances in q_i? (Justify your answer).

In your analysis you may assume that:

(i) The valves on the exit lines act as linear resistances.

(ii) The valves are adjusted so that each tank is half full at the nominal design condition of $q_i = 378.5$ L/min.

(c) Use computer simulation to verify your answers in (a) and (b).

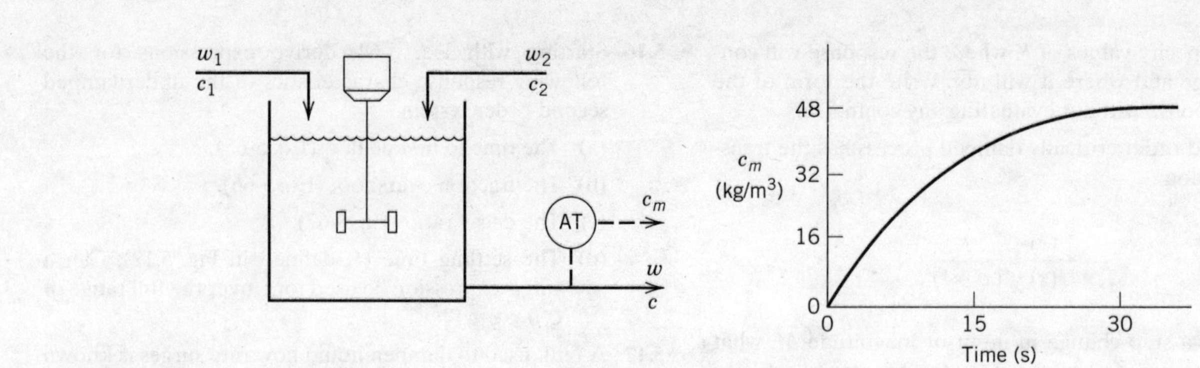

Figure E5.20

5.20 The caustic concentration of the mixing tank shown in Fig. E5.20 is measured using a conductivity cell. The total volume of solution in the tank is constant at 0.2 m³ and the density (ρ = 1122 kg/m³) can be considered to be independent of concentration. Let c_m denote the caustic concentration measured by the conductivity cell. The dynamic response of the conductivity cell to a step change (at $t = 0$) of 48 kg/m³ in the actual concentration (passing through the cell) is also shown in Fig. E5.20.

(a) Determine the transfer function $C_m'(s)/C_1'(s)$ assuming that the flow rates are equal and constant: ($w_1 = w_2 = 80$ kg/m³).

(b) Find the response for a step change in c_1 from 224 to 272 kg/m³.

(c) If the transfer function $C_m'(s)/C'(s)$ were approximated by 1 (unity), what would be the step response of the system for the same input change?

(d) By comparison of (b) and (c), what can you say about the dynamics of the conductivity cell? Plot both responses, if necessary.

5.21 An exothermic reaction, A → 2B, takes place adiabatically in a stirred-tank system. This liquid phase reaction occurs at constant volume in a 378.5 L reactor. The reaction can be considered to be first order and irreversible with the rate constant given by

$$k = 2.4 \times 10^{15} e^{-11111/T} (\text{min}^{-1})$$

where T is in K. Using the information below, derive a transfer function relating the exit temperature T to the inlet concentration c_{Ai}. State any assumptions that you make. Simplify the transfer function by making a first-order approximation and show that the approximation is valid by comparing the step responses of both the original and the approximate models.

Available Information

(i) Nominal steady-state conditions are:

$$\bar{T} = 66\ °\text{C},\ \bar{c}_{Ai} = 12.8\ \text{kmol/m}^3$$

$$q = 76\ \text{L/min} = \text{flow rate in and out of the reactor}$$

(ii) Physical property data for the mixture at the nominal steady state: $C_p = 3350$ J/kg-K,

$$\rho = 833\ \text{kg/m}^3,\quad -\Delta H_R = 1101\ \text{J/mol}$$

5.22 Using the step responses of (1) an integrating element and (2) a first-order process to an input change of magnitude M.

(a) Show that the step response for an input change M of a first-order process

$$G_1(s) = \frac{K_1}{\tau s + 1}$$

can be approximately modeled by the step response of an integrator.

$$G_0(s) = \frac{K_0}{s}$$

for low values of t—i.e., $t \ll \tau$. (*Hint:* you can use a first-order Taylor series approximation of $e^{-t/\tau}$.)

(b) What is the relation between K_0 and K_1 if the two responses match for $t \ll \tau$?

(c) This relationship motivates the use of an integrator model to approximate a first-order process by means of a single-parameter model. Explain how you would analyze a single step test to find K_0 and a time delay (if one exists).

5.23 For a stirred-tank heater, assume the transfer function between the heater input change $u(t)$ (cal/sec) and the tank temperature change $y(t)$ (°C) can be modeled as

$$G(s) = \frac{5}{3s + 1}$$

(a) Using the Final Value Theorem, find the steady-state response for a unit rectangular pulse change in the heating rate $\left(U(s) = \dfrac{1 - e^{-s}}{s}\right)$.

(b) Repeat the calculation in (a) for a unit ramp $\left(U(s) = \dfrac{1}{s^2}\right)$.

(c) For both cases (a) and (b), explain your answer physically. Is there a physical limitation on the ramping of the heating rate?

5.24 An additive process model is depicted in the figure below. For $G_1 = \dfrac{1}{s}, G_2 = \dfrac{4}{2s+1}, G_3 = \dfrac{-3}{s+1}, U(s) = 1$ (unit impulse)

(a) Derive the response $Y(s)$ and describe $y(t)$ quantitatively.

(b) Simulate the response and identify its major characteristics.

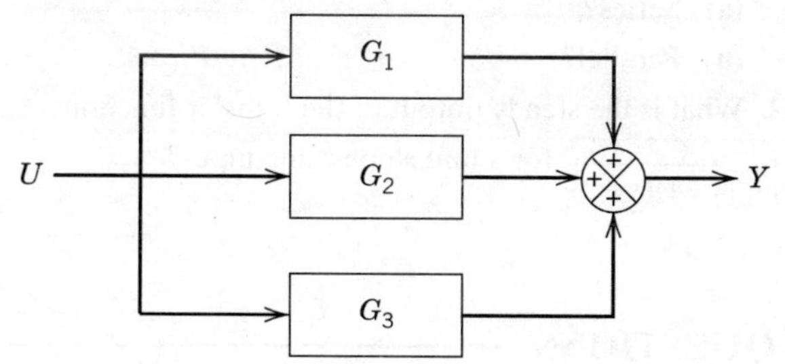

Figure E5.24

5.25 Can a tank with the outflow rate fixed by a constant speed pump reach a steady state if the inlet flow rate undergoes a step change? Why, or why not? If the transfer function is $G(s) = K/s$, is it possible to calculate a steady-state gain?

5.26 A thermometer with first-order time constant = 0.2 min and gain = 1.0 is placed in a temperature bath (25°C). After the thermometer comes to equilibrium with the bath, the temperature of the bath is increased linearly at a rate of 1°/min.

(a) What is the difference between the measured temperature T_m and the bath temperature T at $t = 0.2$ min and $t = 1.0$ min after the change in temperature?

(b) What is the maximum deviation between $T_m(t)$ and $T(t)$? When does it occur?

(c) Plot both $T(t)$ and $T_m(t)$ to 3 mins. For large values of t, determine the time lag between T_m and T.

5.27 A thermometer has first-order dynamics with a time constant of 1 sec and is placed in a temperature bath at 49°C. After the thermometer reaches steady state, it is suddenly placed in a bath at 60°C for $0 \le t \le 10$ sec. Then it is returned to the bath at 38°C.

(a) Sketch the variation of the measured temperature $T_m(t)$ with time.

(b) Calculate $T_m(t)$ at $t = 0.5$ sec and at $t = 15.0$ sec.

5.28 Show that when $K = 1$, the ramp response of the third-order system

$$G(s) = \frac{K}{(\tau s + 1)^3}$$

lags behind the input signal ($u = at$) by three time constants, once the output is changing linearly in time.

5.29 A vertical, cylindrical tank is filled with water at 20°C. The tank is insulated at the top and bottom, with diameter of 0.5 m and height of 1.0 m. The overall heat transfer coefficient is $U = 120$ W/m²-K. The density of water is $\rho = 1000$ kg/m³, the heat capacity $C_p = 4180$ J/kg-K, the melting point is 0°C and the heat of fusion is $\lambda = 334$ kJ/kg.

(a) In the evening, the tank is suddenly exposed to air at −15°C. Calculate how many minutes it will take for the first crystal of ice to form in the tank. Model this process by assuming thermal equilibrium between the tank and the environment at an initial steady state, followed by a sudden drop in the outside temperature to −15°C.

(b) How long will it take to completely freeze the water in the tank? You many neglect any volume expansion associated with freezing and assume that the tank is well-mixed, that is, the temperature is uniform within the tank and there are no radial temperature gradients.

5.30 An operator tests the dynamic behavior of a furnace in order to identify the transfer function relating the furnace temperature (output) to the heat input. The operator performs a series of step increases/decreases in the amount of heat input and records the temperature changes in the furnace. The system is initially at steady state, and the test takes 10 h (see Fig. E5.30). Data are recorded for another 5 h after changes in the input are no longer made.

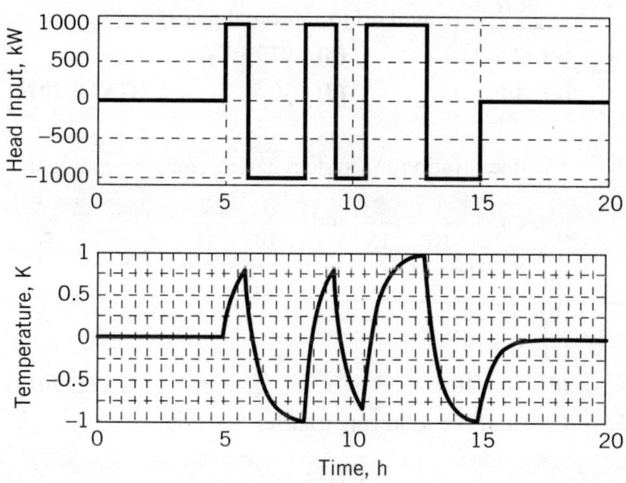

Figure E5.30

(a) Does a visual inspection of the plots above provide enough information to determine

(i) the order of the system transfer function

(ii) the gain

(iii) the time constant(s)

If yes, please provide the corresponding values and justify your answers. If an exact determination is not possible for some of the above, justify why and discuss what a good estimate would be. Please give the units where necessary.

(b) How should the test procedure be modified to improve the quality of the data and make it easier (or possible) to identify the transfer function of the system from the plot?

(c) What can you qualitatively infer about the following physical parameters of the furnace (ignore the air inside):

(i) mass

(ii) density

(iii) heat capacity

(iv) furnace height

Do not perform any calculation; use words such as "large," "small," and "significant." Please justify your answer.

5.31 The transfer of a reactor is given by $\dfrac{K}{(\tau s+1)}$

What is the transfer function of the system with two of these reactors if they are connected in,

(a) Series?

(b) Parallel?

5.32 What is the steady output of the transfer function $\dfrac{s}{(s+1)(s+2)}$ for a unit slope ramp input?

MULTIPLE CHOICE QUESTIONS

1. A process has a transfer function $G(s)=\dfrac{Y(s)}{X(s)}=\dfrac{20}{90000s^2+240s+1}$. Initially the process is at steady state with $x(t=0)=0.4$ and $y(t=0)=100$. If a step change in x is given from 0.4 to 0.5, the maximum value of y that will be observed before it reaches the new steady state is ________ (round off to 1 decimal place).

(a) 2.5 **(b)** 97.5

(c) 102.5 **(d)** 10.5 **(GATE 2021)**

2. The decay ratio for a system having complex conjugate poles as $\left(-\dfrac{1}{10}+j\dfrac{2}{15}\right)$ and $\left(-\dfrac{1}{10}-j\dfrac{2}{15}\right)$

(a) 7×10^{-1}

(b) 8×10^{-2}

(c) 9×10^{-3}

(d) 10×10^{-4} **(GATE 2018)**

3. Consider the following transfer function:

$$G(s)=\frac{3}{(5s+1)^2}$$

The damping coefficient and the fractional overshoot for a step change in its input are given respectively by

(a) 0 and 0.5

(b) 1 and 0.0

(c) 1 and 0.6

(d) 0.5 and 0.0 **(GATE 2018)**

4. The transfer fraction of a system is $\dfrac{1}{4s^2+1.2s+1}$. For a unit step increase in the input, the fractional overshoot, rounded to 2 decimal place, is

(a) 0.368 **(b)** 0.632

(c) 0.372 **(d)** 0.864 **(GATE 2017)**

5. What is the order of response exhibited by; a U-tube manometer?

(a) Zero-order

(b) First-order

(c) Second-order

(d) Third-order **(GATE 2016)**

6. For a unit step, the response of a second-order system is

$$y(t)=K_p\left[1-\frac{1}{\sqrt{1-\zeta^2}}e^{-\frac{\zeta}{\tau}t}\sin\left(\frac{\sqrt{1-\zeta^2}}{\tau}t+\phi\right)\right]$$

where K_p is the steady state gain, ζ is the damping coefficient, τ is the natural period of oscillation and ϕ is the phase lag. The overshoot of the system is $\exp\left(-\dfrac{\pi\zeta}{\sqrt{1-\zeta^2}}\right)$. For a unit step input, the response of the system from an initial steady state condition at $t=0$ is shown in the figure below.

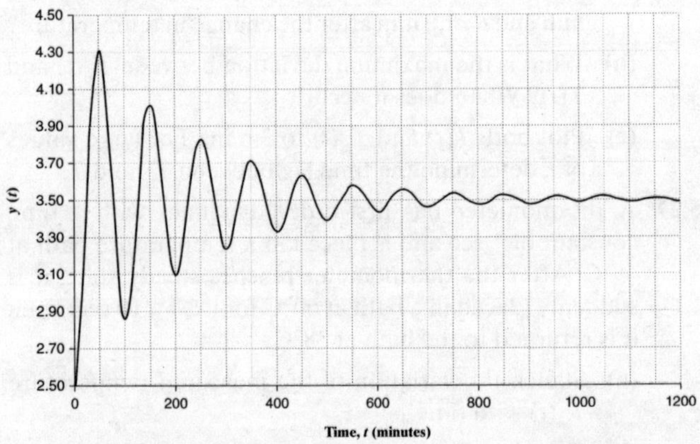

What is the natural period of oscillation (in seconds) of the system?

(a) 15.9 **(b)** 50

(c) 63.2 **(d)** 100 **(GATE 2016)**

7. A unit IMPULSE response of a first order system with time constant τ and steady state gain K_p is given by

(a) $\dfrac{1}{K_p\tau}e^{t/\tau}$

(b) $\dfrac{1}{K_p\tau}e^{-t/\tau}$

(c) $K_p\tau e^{-t/\tau}$

(d) $\dfrac{K_p}{\tau}e^{-t/\tau}$ **(GATE 2014)**

8. Assume that an ordinary mercury-in-glass thermometer follows first order dynamics with a time constant of 10 s. It is at a steady state temperature of 0 °C. At time $t = 0$, the thermometer is suddenly immersed in a constant temperature bath at 100 °C. The time required (in s) for the thermometer to read 95 °C, approximately is

(a) 60

(b) 40

(c) 30

(d) 20 **(GATE 2014)**

9. A step change of magnitude M is introduced into a system having the following transfer function

$$G(s) = \frac{2}{s^2 + 2s + 4}$$

The percent overshoot is

(a) 6.3

(b) 16.3

(c) 26.3

(d) 36.3 **(GATE 2014)**

10. A unit gain second order underdamped process has a period of oscillation 1 second and decay ratio 0.25. The transfer function of the process is

(a) $\dfrac{1}{0.024s^2 + 0.0675s + 1}$

(b) $\dfrac{1}{0.067s^2 + 0.024s + 1}$

(c) $\dfrac{1}{0.021s^2 + 0.1176s + 1}$

(d) $\dfrac{1}{0.1176s^2 + 0.021s + 1}$ **(GATE 2013)**

11. The unit impulse response of a first order process is given by $2e^{-0.5t}$. The gain and time constant of the process are, respectively,

(a) 4 and 2.

(b) 2 and 2.

(c) 2 and 0.5.

(d) 1 and 0.5. **(GATE 2008)**

12. A unit step input is given to a process that is represented by the transfer function $\dfrac{(s+2)}{(s+5)}$. The initial value ($t = 0^+$) of the response of the process to the step input is

(a) 0

(b) 2/5

(c) 1

(d) ∞ **(GATE 2008)**

ANSWER KEY

1. (c)	**2.** (c)	**3.** (b)	**4.** (c)	**5.** (d)	**6.** (a)
7. (d)	**8.** (c)	**9.** (b)	**10.** (a)	**11.** (a)	**12.** (c)

CHAPTER 6

Dynamic Behavior of Higher-Order Processes

CHAPTER CONTENTS

Chapter 5 discussed the dynamics of relatively simple processes, those that can be modeled as either first- or second-order transfer functions or as an integrator. Now we consider more complex transfer function models that include additional time constants in the denominator and/or functions of s in the numerator. We show that the forms of the numerator and denominator of the transfer function model influence the dynamic behavior of the process. We also consider a very important concept, the *time delay,* and present the approximation of complicated transfer function models by simpler, low-order models. Additional topics in this chapter include interacting processes, processes with multiple inputs and outputs, and fitting first- and second-order models using step tests.

6.1 PROCESSES WITH TIME DELAYS

Whenever material or energy is physically moved in a process or plant, there is a *time delay* associated with the movement. For example, if a fluid is transported through a pipe in plug flow, as shown in Fig. 6.1, then the transportation time between points 1 and 2 is given by

$$\theta = \frac{\text{length of pipe}}{\text{fluid velocity}} \tag{6-1}$$

or equivalently, by

$$= \frac{\text{volume of pipe}}{\text{volumetric flowrate}}$$

where length and volume both refer to the pipe segment between 1 and 2. The first relation in Eq. 6-1 indicates why a time delay sometimes is referred to as a *distance–velocity lag*. Other synonyms are *transportation lag, transport delay,* and *dead time*. If the plug flow assumption does not hold, for example, with laminar flow or for non-Newtonian liquids, approximation of the bulk transport dynamics using a time delay still may be useful, as discussed below.

Suppose that x is some fluid property at point 1, such as concentration, and y is the same property at point 2. Both x and y are deviation variables. Then they are related by a time delay θ

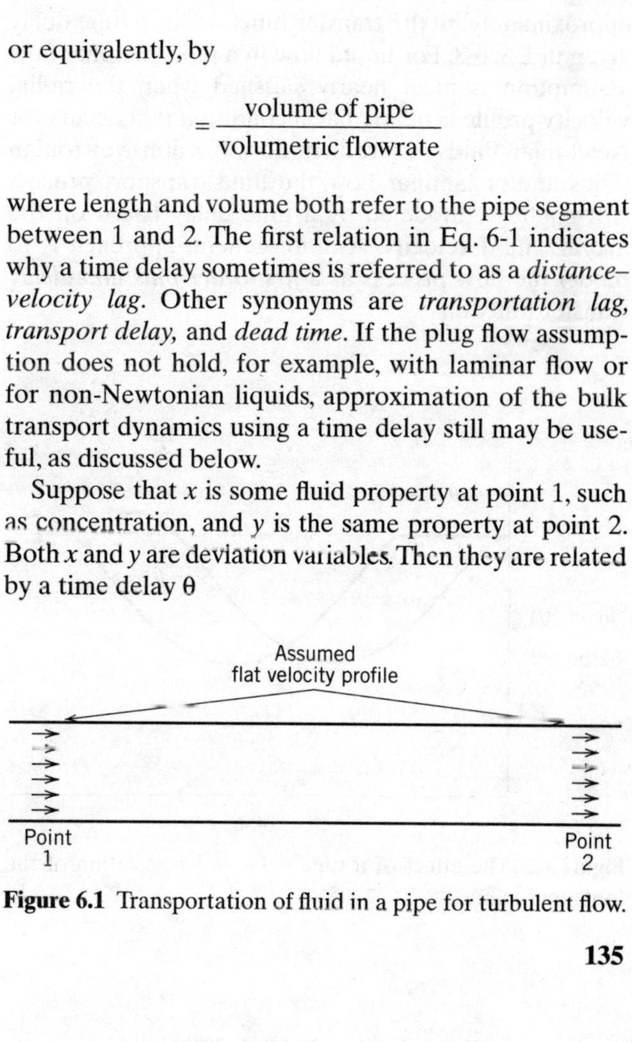

Figure 6.1 Transportation of fluid in a pipe for turbulent flow.

$$y(t)=\begin{cases}0 & t<\theta\\ x(t-\theta) & t\geq\theta\end{cases} \tag{6-2}$$

Thus, the output $y(t)$ is simply the same input function shifted backward in time by θ. Figure 6.2 shows this *translation in time* for an arbitrary $x(t)$.

Equation 3-73 shows that the Laplace transform of a function shifted in time by t_0 units is simply $e^{-t_0 s}$. Thus, the transfer function of a time delay of magnitude θ is given by

$$\frac{Y(s)}{X(s)}=G(s)=e^{-\theta s} \tag{6-3}$$

Besides the physical movement of liquid and solid materials, there are other sources of time delays in process control problems. For example, the use of a chromatograph to measure concentration in liquid or gas stream samples taken from a process introduces a time delay, the analysis time. One distinctive characteristic of chemical processes is the common occurrence of time delays.

Even when the plug flow assumption is not valid, transportation processes usually can be modeled approximately by the transfer function for a time delay given in Eq. 6-3. For liquid flow in a pipe, the plug flow assumption is most nearly satisfied when the radial velocity profile is nearly flat, a condition that occurs for Newtonian fluids in turbulent flow. For non-Newtonian fluids and/or laminar flow, the fluid transport process still might be modeled by a time delay based on the average fluid velocity. A more general approach is to model the flow process as a *first-order plus time-delay* transfer function

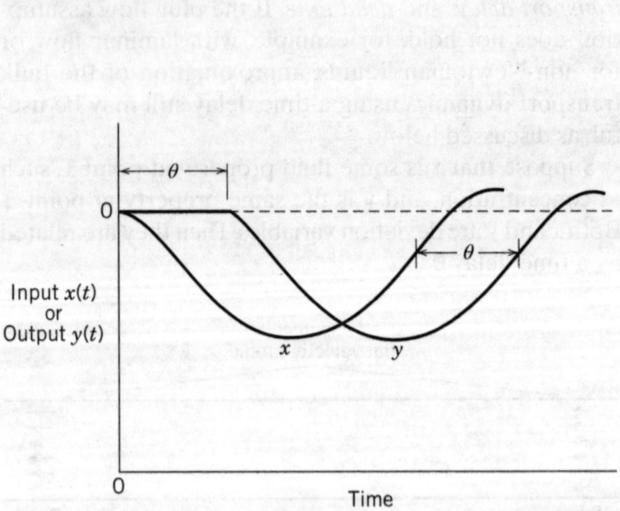

Figure 6.2 The effect of a time delay is a translation of the function in time.

$$G(s)=\frac{e^{-\theta s}}{\tau s+1} \tag{6-4}$$

where τ is a time constant associated with the degree of mixing in the pipe or channel. Both τ and θ may have to be determined from empirical relations or by experiment. Note that the process gain in Eq. 6-4 is unity when y and x are material properties such as composition.

Next we demonstrate that analytical expressions for time delays can be derived from the application of conservation equations. In Fig. 6.1, suppose that a very small cell of liquid passes point 1 at time t. It contains $Vc_1(t)$ units of the chemical species of interest where V is the volume of material in the cell and c_1 is the concentration of the species. At time $t + \theta$, the cell passes point 2 and contains $Vc_2(t + \theta)$ units of the species. If the material moves in plug flow, not mixing at all with adjacent material, then the amount of species in the cell is constant:

$$Vc_2(t+\theta)=Vc_1(t) \tag{6-5}$$

or

$$c_2(t+\theta)=c_1(t) \tag{6-6}$$

An equivalent way of writing Eq. 6-6 is

$$c_2(t)=c_1(t-\theta) \tag{6-7}$$

if the flow rate is constant. Putting Eq. 6-7 in deviation form (by subtracting the steady-state version of Eq. 6-7) and taking Laplace transforms gives

$$\frac{C_2'(s)}{C_1'(s)}=e^{-\theta s} \tag{6-8}$$

When the fluid is incompressible, flow rate changes at point 1 propagate instantaneously to any other point in the pipe. For compressible fluids such as gases, the simple expression of Eq. 6-8 may not be accurate. Note that use of a constant time delay implies constant flow rate.

6.1.1 Polynomial Approximations to $e^{-\theta s}$

The exponential form of Eq. 6-3 is a irrational transfer function that cannot be expressed as a rational function, a ratio of two polynominals in s. Consequently, Eq. 6-3 cannot be factored into poles and zeros, a convenient form for analysis, as discussed in Section 4.5. However, it is possible to approximate $e^{-\theta s}$ by polynomials using either a Taylor series expansion or a Padé approximation.

The Taylor series expansion for $e^{-\theta s}$ is

$$e^{-\theta s} = 1 - \theta s + \frac{\theta^2 s^2}{2!} - \frac{\theta^3 s^3}{3!} + \frac{\theta^4 s^4}{4!} - \frac{\theta^5 s^5}{5!} + \cdots \qquad (6\text{-}9)$$

The Padé approximation for a time delay is a ratio of two polynomials in s with coefficients selected to match the terms of a truncated Taylor series expansion of $e^{-\theta s}$. The simplest pole–zero approximation is the 1/1 Padé approximation:

$$e^{-\theta s} \approx G_1(s) = \frac{1 - \frac{\theta}{2}s}{1 + \frac{\theta}{2}s} \qquad (6\text{-}10)$$

Equation 6-10 is called the 1/1 Padé approximation because it is first-order in both numerator and denominator.

Performing the indicated long division in Eq. 6-10 gives

$$G_1(s) = 1 - \theta s + \frac{\theta^2 s^2}{2} - \frac{\theta^3 s^3}{4} + \cdots \qquad (6\text{-}11)$$

A comparison of Eqs. 6-9 and 6-11 indicates that $G_1(s)$ is correct through the first three terms.

EXAMPLE 6.1

The trickle-bed catalytic reactor shown in Fig. 6.3 utilizes product recycle to obtain satisfactory operating conditions for temperature and conversion. Use of a high recycle rate eliminates the need for mechanical agitation. The concentration of the reactant is measured at a point in the recycle line where the product stream is removed. A liquid phase first-order reaction is assumed.

Under normal operating conditions, the following assumptions may be made:

(i) The reactor operates isothermally with a reaction rate given by $r = kc$, where $-r$ denotes the rate of disappearance of reactant per unit volume, c is the concentration of reactant, and k is the rate constant.

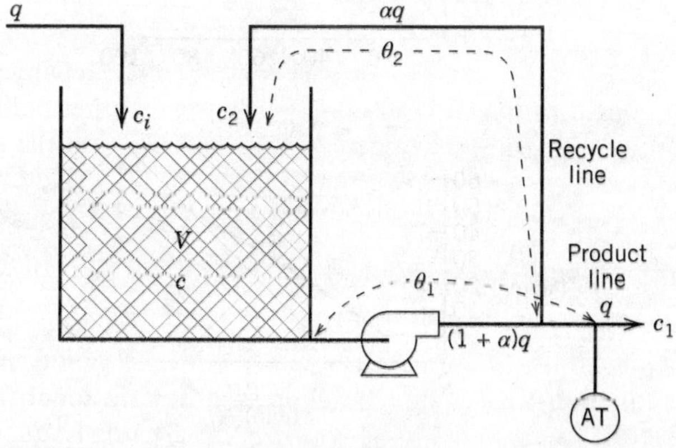

Figure 6.3 Schematic diagram of a trickle-bed reactor with recycle line. (AT: analyzer/transmitter; θ_1: time delay associated with material flow from reactor outlet to the composition analyzer; θ_2: time delay associated with material flow from analyzer to reactor inlet.)

(ii) All flow rates and the liquid volume V are constant.

(iii) No reaction occurs in the piping. The dynamics of the exit and recycle lines can be approximated as constant time delays, θ_1 and θ_2, as indicated in the figure. Let c_1 denote the reactant concentration at the measurement point.

(iv) Because of the high recycle flow rate, the reactor is completely mixed.

(a) Derive an expression for the transfer function $C_1'(s)/C_i'(s)$.

(b) Using the following information, calculate $c_1'(t)$ for a step change in $c_1'(t)$ of 2000 kg/m^3.

Parameter Values

$V = 5\,\text{m}^3$	$\alpha = 12$
$q = 0.05\,\text{m}^3/\text{min}$	$\theta_1 = 0.9\,\text{min}$
$k = 0.04\,\text{min}^{-1}$	$\theta_2 = 1.1\,\text{min}$

Solution

(a) In analogy with Eq. 2-70; the component balance around the reactor is,

$$V\frac{dc}{dt} = qc_i + \alpha q c_2 - (1+\alpha)qc - Vkc \qquad (6\text{-}12)$$

where the concentration of the reactant is denoted by c. Equation 6-12 is a linear ODE with constant coefficients. Subtracting the steady-state equation and substituting deviation variables yields

$$V\frac{dc'}{dt} = qc'_i + \alpha qc'_2 - (1+\alpha)qc' - Vkc' \quad (6\text{-}13)$$

Additional relations are needed for $c'_2(t)$ and $c'_1(t)$. Because the exit and recycle lines can be modeled as time delays,

$$c'_1(t) = c'(t-\theta_1) \quad (6\text{-}14)$$

$$c'_2(t) = c'_1(t-\theta_2) \quad (6\text{-}15)$$

Equations (6-13) through (6-15) provide the process model for the isothermal reactor with recycle. Taking the Laplace transform of each equation yields

$$sVC'(s) = qC'_i(s) + \alpha qC'_2(s) - (1+\alpha)qC'(s) - VkC'(s) \quad (6\text{-}16)$$

$$C'_1(s) = e^{-\theta_1 s}C'(s) \quad (6\text{-}17)$$

$$C'_2(s) = e^{-\theta_2 s}C'_1(s) = e^{-(\theta_1+\theta_2)s}C'(s) = e^{-\theta_3 s}C'(s) \quad (6\text{-}18)$$

where $\theta_3 \triangleq \theta_1 + \theta_2$. Substitute Eq. 6-18 into Eq. 6-16 and solve for $C'(s)$:

$$C'(s) = \frac{q}{sV - \alpha qe^{-\theta_3 s} + (1+\alpha)q + Vk}C'_i(s) \quad (6\text{-}19)$$

Equation 6-19 can be rearranged to the following form:

$$C'(s) = \frac{K}{\tau s + 1 + \alpha K(1 - e^{-\theta_3 s})}C'_i(s) \quad (6\text{-}20)$$

where

$$K = \frac{q}{q+Vk} \quad (6\text{-}21)$$

$$\tau = \frac{V}{q+Vk} \quad (6\text{-}22)$$

Note that, in the limit as $\theta_3 \to 0$, $e^{-\theta_3 s} \to 1$ and

$$C'(s) = \frac{K}{\tau s+1}C'_i(s) \quad (6\text{-}23)$$

So K and τ can be interpreted as the process gain and time constant, respectively, of a recycle reactor with no time delay in the recycle line, which is equivalent to a stirred isothermal reactor with no recycle.

The desired transfer function $C'_1(s)/C'_i(s)$ is obtained by combining Eqs. 6-20 and 6-17 to obtain

$$\frac{C'_1(s)}{C'_i(s)} = \frac{Ke^{-\theta_1 s}}{\tau s + 1 + \alpha K(1 - e^{-\theta_3 s})} \quad (6\text{-}24)$$

The numerical parameters in Eq. 6-24 are

$$K = \frac{q}{q+Vk} = \frac{0.05}{0.05+(5)(0.04)} = 0.2$$

$$\tau = \frac{V}{q+Vk} = 20 \text{ min}$$

(b) The easiest way to simulate the step response is to construct a Simulink diagram (see Appendix B) corresponding to Fig. 6.3, after substituting the numerical values into the transfer functions and transforms. The recycle loop can be treated graphically in Simulink instead of solving the model Eq. 6-12. Figure 6.4 shows the measured composition at the analyzer location for two different time scales as a result of the step change in inlet composition.

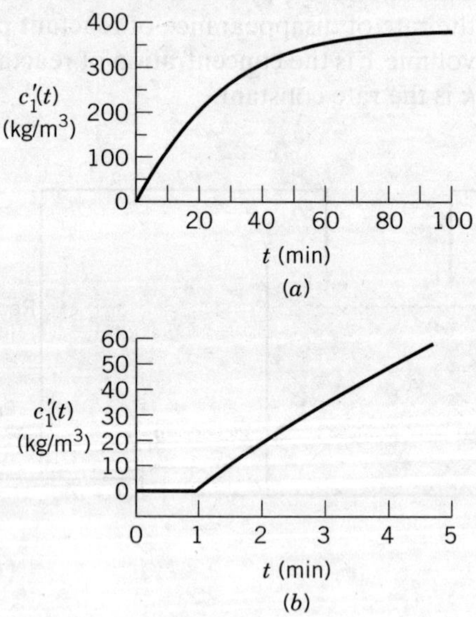

Figure 6.4 Recycle reactor composition measured at analyzer: (*a*) complete response; (*b*) detailed view of short-term response.

6.2 APPROXIMATION OF HIGHER-ORDER TRANSFER FUNCTIONS

In this section, we present a general approach for approximating higher-order transfer function models with lower-order models that have similar dynamic and steady-state characteristics. The low-order models are more convenient for control system design and analysis, as discussed in Chapter 11.

In Section 6.1.1, we showed that the transfer function for a time delay can be expressed as a Taylor series expansion. For small values of s, truncating the expansion after the first-order term provides a suitable approximation:

$$e^{-\theta s} \approx 1 - \theta s \tag{6-25}$$

Note that this time-delay approximation is a right-half plane (RHP) zero at $s = +\theta$. An alternative first-order approximation consists of the transfer function,

$$e^{-\theta s} = \frac{1}{e^{\theta s}} \approx \frac{1}{1 + \theta s} \tag{6-26}$$

which is based on the approximation, $e^{\theta s} \approx 1 + \theta s$. Note that the time constant has a value of θ.

Equations 6-25 and 6-26 were derived to approximate time-delay terms. However, these expressions can be reversed to approximate the pole or zero on the right-hand side of the equation by the time-delay term on the left side. These pole and zero approximations will be demonstrated in Example 6.2.

6.2.1 Skogestad's "Half Rule"

Skogestad (2003) has proposed a related approximation method for higher-order models that contain multiple time constants. He approximates the largest neglected time constant in the denominator in the following manner. One-half of its value is added to the existing time delay (if any), and the other half is added to the smallest retained time constant. Time constants that are smaller than the largest neglected time constant are approximated as time delays using Eq. 6-26. A right-half plane zero is approximated by Eq. 6-25. The motivation for this "half rule" is to derive approximate low-order models that are more appropriate for control system design. Examples 6.2 and 6.3 illustrate Skogestad's half rule.

EXAMPLE 6.2

Consider a transfer function:

$$G(s) = \frac{K(-0.1s + 1)}{(5s + 1)(3s + 1)(0.5s + 1)} \tag{6-27}$$

Derive an approximate first-order-plus-time-delay model,

$$\tilde{G}(s) = \frac{Ke^{-\theta s}}{\tau s + 1} \tag{6-28}$$

using two methods:

(a) The Taylor series expansions of Eqs. 6-25 and 6-26.
(b) Skogestad's half rule.

Compare the normalized responses of $G(s)$ and the approximate models for a unit step input.

Solution

(a) The dominant time constant (5) is retained. Applying the approximations in Eqs. 6-25 and 6-26 gives

$$-0.1s + 1 \approx e^{-0.1s} \tag{6-29}$$

and

$$\frac{1}{3s + 1} \approx e^{-3s} \qquad \frac{1}{0.5s + 1} \approx e^{-0.5s} \tag{6-30}$$

Substitution into Eq. 6-27 gives the Taylor series approximation, $\tilde{G}_{TS}(s)$:

$$\tilde{G}_{TS}(s) = \frac{Ke^{-0.1s}e^{-3s}e^{-0.5s}}{5s + 1} = \frac{Ke^{-3.6s}}{5s + 1} \tag{6-31}$$

(b) To apply Skogestad's method, we note that the largest neglected time constant in Eq. 6-27 has a value of 3. According to his "half rule," half of this value is added to the next largest time constant to generate a new time constant, $\tau = 5 + 0.5(3) = 6.5$. The other half provides a new time delay of $0.5(3) = 1.5$. The approximation of the RHP zero in Eq. 6-29 provides an additional time delay of 0.1. Approximating the smallest time constant of 0.5 in Eq. 6-27 by Eq. 6-26 produces an additional time delay of 0.5. Thus, the total time delay in Eq. 6-28 is

$$\theta = 1.5 + 0.1 + 0.5 = 2.1$$

and $G(s)$ can be approximated as

$$\tilde{G}_{SK}(s) = \frac{Ke^{-2.1s}}{6.5s+1} \tag{6-32}$$

The normalized step responses for $G(s)$ and the two approximate models are shown in Fig. 6.5. Skogestad's method provides better agreement with the actual response.

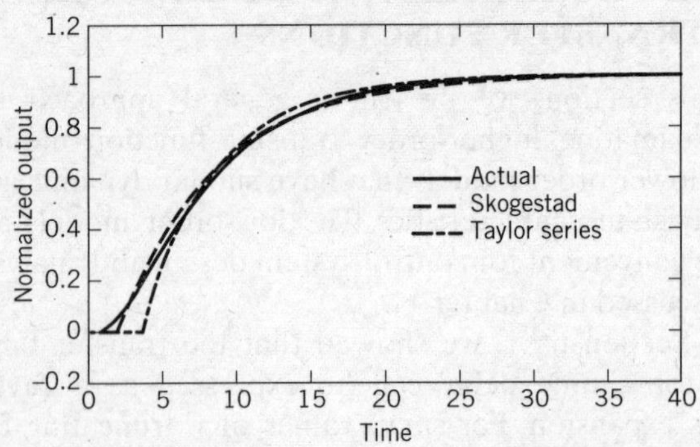

Figure 6.5 Comparison of the actual and approximate models for Example 6.2.

EXAMPLE 6.3

Consider the following transfer function:

$$G(s) = \frac{K(1-s)e^{-s}}{(12s+1)(3s+1)(0.2s+1)(0.05s+1)} \tag{6-33}$$

Use Skogestad's method to derive two approximate models:

(a) A first-order-plus-time-delay model in the form of Eq. 6-28.

(b) A second-order-plus-time-delay model in the form

$$\tilde{G}(s) = \frac{Ke^{-\theta s}}{(\tau_1 s+1)(\tau_2 s+1)} \tag{6-34}$$

Compare the normalized output responses for $G(s)$ and the approximate models to a unit step input.

Solution

(a) For the first-order-plus-time-delay model, the dominant time constant (12) is retained. One-half of the largest neglected time constant (3) is allocated to the retained time constant and one-half to the approximate time delay. Also, the small time constants (0.2 and 0.05) and the zero (1) are added to the original time delay. Thus, the model parameters in Eq. 6-28 are

$$\theta = 1 + \frac{3}{2} + 0.2 + 0.05 + 1 = 3.75$$

$$\tau = 12 + \frac{3}{2} = 13.5$$

(b) An analogous derivation for the second-order-plus-time-delay model gives

$$\theta = 1 + \frac{0.2}{2} + 0.05 + 1 = 2.15$$

$$\tau_1 = 12, \quad \tau_2 = 3 + 0.1 = 3.1$$

In this case, the half rule is applied to the third largest time constant, 0.2.

The normalized step responses of the original and approximate transfer functions are shown in Fig. 6.6. The second-order model provides an excellent approximation, because the neglected time constants are much smaller than the retained time constants. The first-order-plus-time-delay model is not as accurate, but it does provide a suitable approximation of the actual response.

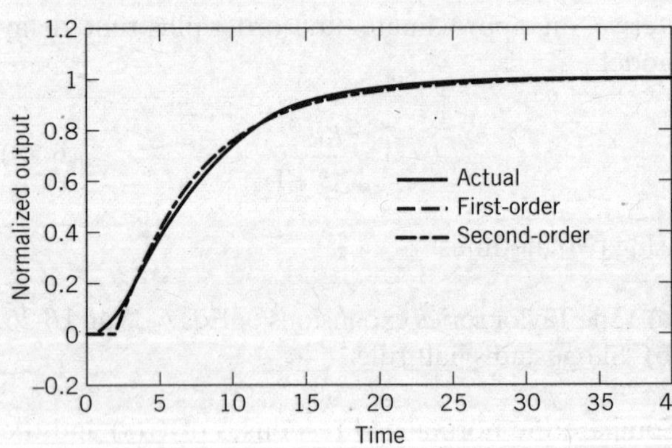

Figure 6.6 Comparison of the actual model and approximate models for Example 6.3. The actual and second-order model responses are almost indistinguishable.

Skogestad (2003) has also proposed approximations for left-half plane zeros of the form, $\tau_a s + 1$, where $\tau_a > 0$. However, these approximations are more complicated and beyond the scope of this book.

6.3 INTERACTING AND NONINTERACTING PROCESSES

Many processes consist of individual units that are connected in various configurations that include series and parallel structures, as well the recycle of material or energy. It is convenient to classify process configurations as being either *interacting* or *noninteracting.* The distinguishing feature of a *noninteracting process* is that changes in a downstream unit have no effect on upstream units. By contrast, for an *interacting process*, downstream units affect upstream units, and *vice versa.* For example, suppose that the exit stream from a chemical reactor serves as the feed stream to a distillation column used to separate product from unreacted feed. Changes in the reactor affect column operation but not *vice versa*—a noninteracting process. But suppose that the distillate stream from the column contains largely unreacted feed; then, it could be beneficial to increase the reactor yield by recycling the distillate to the reactor where it would be added to the fresh feed. Now, changes in the column affect the reactor, and *vice versa*—an interacting process.

An example of a system that does *not* exhibit interaction was discussed in Example 4.7. The two storage tanks were connected in series in such a way that liquid level in the second tank did not influence the level in the first tank (Fig. 6.7). The following transfer functions relating tank levels and flows were derived:

$$\frac{H_1'(s)}{Q_i'(s)} = \frac{K_1}{\tau_1 s+1} \tag{4-63}$$

$$\frac{Q_1'(s)}{H_1'(s)} = \frac{1}{K_1} \tag{4-64}$$

$$\frac{H_2'(s)}{Q_1'(s)} = \frac{K_2}{\tau_2 s+1} \tag{4-65}$$

$$\frac{Q_2'(s)}{H_2'(s)} = \frac{1}{K_2} \tag{4-66}$$

where $K_1 = R_1$, $K_2 = R_2$, $\tau_1 = A_1R_1$, and $\tau_2 = A_2R_2$. Each tank level has first-order dynamics with respect to its inlet flow rate. Tank 2 level h_2 is related to q_i by a second-order transfer function that can be obtained by simple multiplication:

$$\frac{H_2'(s)}{Q_i'(s)} = \frac{H_2'(s)}{Q_1'(s)}\frac{Q_1'(s)}{H_1'(s)}\frac{H_1'(s)}{Q_i'(s)} = \frac{K_2}{(\tau_1 s+1)(\tau_2 s+1)} \tag{6-35}$$

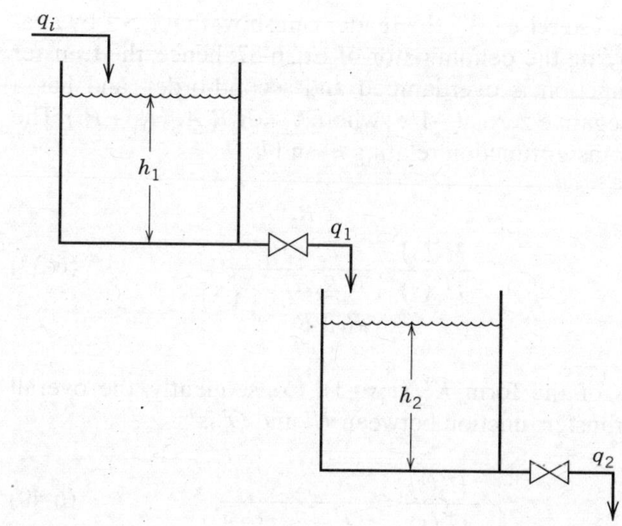

Figure 6.7 A series configuration of two noninteracting tanks.

Next consider an example of an interacting process that is similar to the two-tank process in Chapter 4. The process shown in Fig. 6.8 is called an *interacting system* because h_1 depends on h_2 (and vice versa) as a result of the interconnecting stream with flow rate q_1. Therefore, the equation for flow from Tank 1 to Tank 2 must be written to reflect that physical feature:

$$q_1 = \frac{1}{R_1}(h_1 - h_2) \tag{6-36}$$

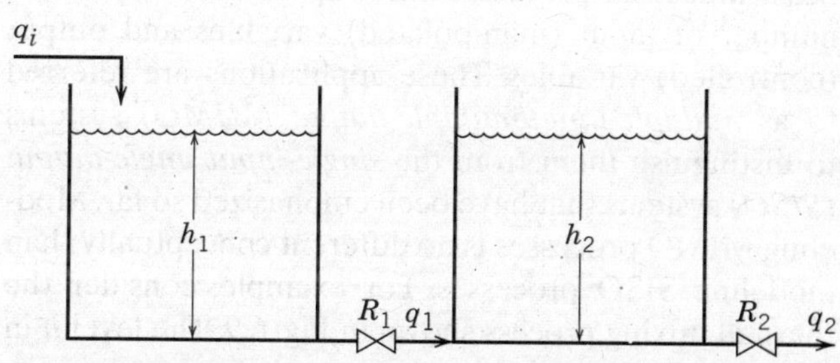

Figure 6.8 Two tanks in series whose liquid levels interact.

For the Tank 1 level transfer function, a much more complicated expression than (4-63) results:

$$\frac{H_1'(s)}{Q_i'(s)} = \frac{(R_1+R_2)\left(\dfrac{R_1R_2A_2}{R_1+R_2}s+1\right)}{R_1R_2A_1A_2s^2+(R_2A_2+R_1A_1+R_2A_1)s+1} \tag{6-37}$$

It is of the form

$$\frac{H_1'(s)}{Q_i'(s)} = \frac{K_1'(\tau_a s+1)}{\tau^2 s^2+2\zeta\tau s+1} \tag{6-38}$$

In Exercise 6.12, the reader can show that $\zeta > 1$ by analyzing the denominator of Eq. 6-37; hence, the transfer function is overdamped and second-order, and has a negative zero at $-1/\tau_a$, where $\tau_a = R_1R_2A_2/(R_1 + R_2)$. The transfer function relating h_1 and h_2,

$$\frac{H_2'(s)}{H_1'(s)} = \frac{\dfrac{R_2}{R_1+R_2}}{\dfrac{R_1R_2A_2}{R_1+R_2}s+1} \tag{6-39}$$

is of the form $K_2'/(\tau_a s+1)$. Consequently, the overall transfer function between H_2' and Q_i' is

$$\frac{H_2'(s)}{Q_i'(s)} = \frac{R_2}{\tau^2 s^2 + 2\zeta\tau s + 1} \tag{6-40}$$

The above analysis of the interacting two-tank system is more complicated than that for the noninteracting system of Example 4.7. The denominator polynomial can no longer be factored into two first-order terms, each associated with a single tank. Also, the numerator of the first tank transfer function in Eq. 6-38 contains a zero that modifies the dynamic behavior along the lines suggested in Section 4.5.

6.4 MULTIPLE-INPUT, MULTIPLE-OUTPUT (MIMO) PROCESSES

Most industrial process control applications involve a number of input (manipulated) variables and output (controlled) variables. These applications are referred to as *multiple-input/multiple-output (MIMO)* systems to distinguish them from the *single-input/single-output (SISO)* systems that have been emphasized so far. Modeling MIMO processes is no different conceptually than modeling SISO processes. For example, consider the thermal mixing process shown in Fig. 6.9. The level h in the stirred tank and the temperature T are to be controlled by adjusting the flow rates of the hot and cold streams, w_h and w_c, respectively. The temperatures of the inlet streams T_h and T_c are considered to be disturbance variables. The outlet flow rate w is maintained constant by the pump, and the liquid properties are assumed to be constant (not affected by temperature) in the following derivation.

Noting that the liquid volume can vary with time, the energy and mass balances for this process are

$$\rho C\frac{d[V(T-T_{\text{ref}})]}{dt} = w_hC(T_h - T_{\text{ref}}) + w_cC(T_c - T_{\text{ref}}) - wC(T - T_{\text{ref}}) \tag{6-41}$$

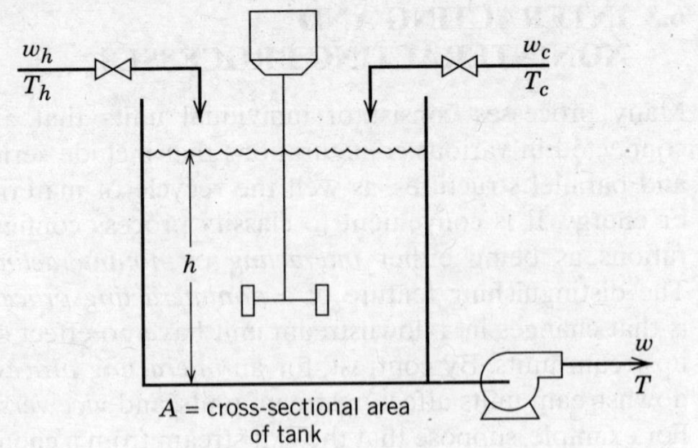

Figure 6.9 A multi-input, multi-output thermal mixing process.

$$\rho\frac{dV}{dt} = w_h + w_c - w \tag{6-42}$$

The energy balance includes a thermodynamic reference temperature T_{ref} (see Section 2.5.1). Expanding the derivative gives

$$\frac{d[V(T-T_{\text{ref}})]}{dt} = (T - T_{\text{ref}})\frac{dV}{dt} + V\frac{dT}{dt} \tag{6-43}$$

Equation 6-43 can be substituted for the left side of Eq. 6-41. Following substitution of the mass balance Eq. 6-42, a simpler set of equations results with $V = Ah$

$$\frac{dT}{dt} = \frac{1}{\rho Ah}[w_hT_h + w_cT_c - (w_h + w_c)T] \tag{6-44}$$

$$\frac{dh}{dt} = \frac{1}{\rho A}(w_h + w_c - w) \tag{6-45}$$

After linearizing Eqs. 6-44 and 6-45, putting them in deviation form, and taking Laplace transforms, we obtain a set of eight transfer functions that describe the effect of each input variable (w_h', w_c', T_h', T_c') on each output variable (T' and h'):

$$\frac{T'(s)}{W_h'(s)} = \frac{(\bar{T}_h - \bar{T})/\bar{w}}{\tau s+1} \tag{6-46}$$

$$\frac{T'(s)}{W_c'(s)} = \frac{(\bar{T}_c - \bar{T})/\bar{w}}{\tau s+1} \tag{6-47}$$

$$\frac{T'(s)}{T_h'(s)} = \frac{\bar{w}_h/\bar{w}}{\tau s+1} \tag{6-48}$$

$$\frac{T'(s)}{T_c'(s)} = \frac{\bar{w}_c/\bar{w}}{\tau s+1} \tag{6-49}$$

$$\frac{H'(s)}{W_h'(s)} = \frac{1/A\rho}{s} \tag{6-50}$$

$$\frac{H'(s)}{W_c'(s)} = \frac{1/A\rho}{s} \tag{6-51}$$

$$\frac{H'(s)}{T_h'(s)} = 0 \tag{6-52}$$

$$\frac{H'(s)}{T_c'(s)} = 0 \tag{6-53}$$

where $\tau = \rho A\bar{h}/\bar{w}$ is the average residence time in the tank and an overbar denotes a nominal steady-state value.

Equations 6-46 through 6-49 indicate that all four inputs affect the tank temperature through first-order transfer functions and a single time constant τ. Equations 6-50 and 6-51 show that the inlet flow rates affect level through integrating transfer functions that result from the pump on the exit line. Finally, it is clear from Eqs. 6-52 and 6-53, as well as from physical intuition, that inlet temperature changes have no effect on liquid level.

A very compact way of expressing Eqs. 6-46 through 6-53 is by means of the *transfer function matrix*:

$$\begin{bmatrix} T'(s) \\ H'(s) \end{bmatrix} = \begin{bmatrix} \dfrac{(\bar{T}_h-\bar{T})/\bar{w}}{\tau s+1} & \dfrac{(\bar{T}_c-\bar{T})/\bar{w}}{\tau s+1} & \dfrac{\bar{w}_h/\bar{w}}{\tau s+1} & \dfrac{\bar{w}_c/\bar{w}}{\tau s+1} \\ \dfrac{1/A\rho}{s} & \dfrac{1/A\rho}{s} & 0 & 0 \end{bmatrix} \begin{bmatrix} W_h'(s) \\ W_c'(s) \\ T_h'(s) \\ T_c'(s) \end{bmatrix} \tag{6-54}$$

Equivalently, two transfer function matrices can be used to separate the manipulated variables, w_h and w_c, from the disturbance variables, T_h and T_c:

$$\begin{bmatrix} T'(s) \\ H'(s) \end{bmatrix} = \begin{bmatrix} \dfrac{(\bar{T}_h-\bar{T})/\bar{w}}{\tau s+1} & \dfrac{(\bar{T}_c-\bar{T})/\bar{w}}{\tau s+1} \\ \dfrac{1/A\rho}{s} & \dfrac{1/A\rho}{s} \end{bmatrix} \begin{bmatrix} W_h'(s) \\ W_c'(s) \end{bmatrix} + \begin{bmatrix} \dfrac{\bar{w}_h/\bar{w}}{\tau s+1} & \dfrac{\bar{w}_c/\bar{w}}{\tau s+1} \\ 0 & 0 \end{bmatrix} \begin{bmatrix} T_h'(s) \\ T_c'(s) \end{bmatrix} \tag{6-55}$$

The block diagram in Fig. 6.10 illustrates how the four input variables affect the two output variables.

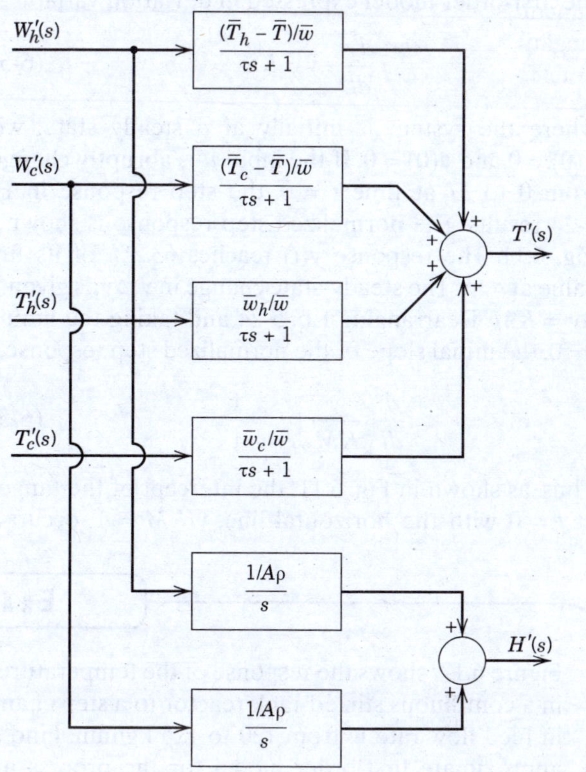

Figure 6.10 Block diagram of the MIMO thermal mixing system with variable liquid level.

Two points are worth mentioning in conclusion:

1. A transfer function matrix, or, equivalently, the set of individual transfer functions, facilitates the design of control systems that deal with the interactions between inputs and outputs. For example, for the thermal mixing process in this section, control strategies can be devised that minimize or eliminate the effect of flow changes on temperature and level. This type of multivariable control system is considered in Chapters 14 and 15.
2. The development of physically based MIMO models can require a significant effort. Thus, empirical models rather than theoretical models often must be used for complicated processes. Empirical modeling is the subject of Chapter 16.

6.5 FITTING FIRST- AND SECOND-ORDER MODELS USING STEP TESTS

A plot of the output response of a process to a step change in input is sometimes referred to as the *process reaction curve*. If the process of interest can be approximated by a first-order or second-order linear model, the model parameters can be obtained by inspection of the process reaction curve. For example, recall the first-order model expressed in deviation variables,

$$\tau \frac{dy}{dt} + y = Ku \tag{6-56}$$

where the system is initially at a steady state, with $u(0) = 0$ and $y(0) = 0$. If the input u is abruptly changed from 0 to M at time $t = 0$, the step response in Eq. 5-24 results. The normalized step response is shown in Fig. 6.11. The response $y(t)$ reaches 63.2% of its final value at $t = \tau$. The steady-state change in y, Δy, is given by $\Delta y = KM$. Rearranging Eq. 5-24 and taking the limit at $t = 0$, the initial slope of the normalized step response is

$$\frac{d}{dt}\left(\frac{y}{KM}\right)_{t=0} = \frac{1}{\tau} \tag{6-57}$$

Thus, as shown in Fig. 6.11, the intercept of the tangent at $t = 0$ with the horizontal line, $y/KM = 1$, occurs at

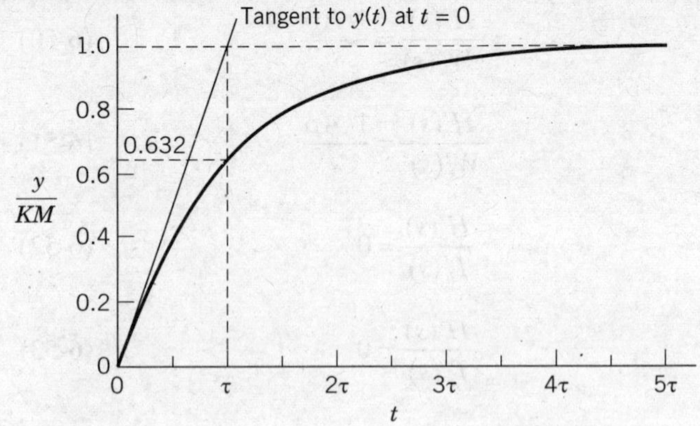

Figure 6.11 Step response of a first-order system and graphical constructions used to estimate the time constant, τ.

$t = \tau$. As an alternative, τ can be estimated from a step response plot using the value of t at which the response is 63.2% complete, as shown in the following example.

EXAMPLE 6.4

Figure 6.12 shows the response of the temperature T in a continuous stirred-tank reactor to a step change in feed flow rate w from 120 to 125 kg/min. Find an approximate first-order model for the process and these operating conditions.

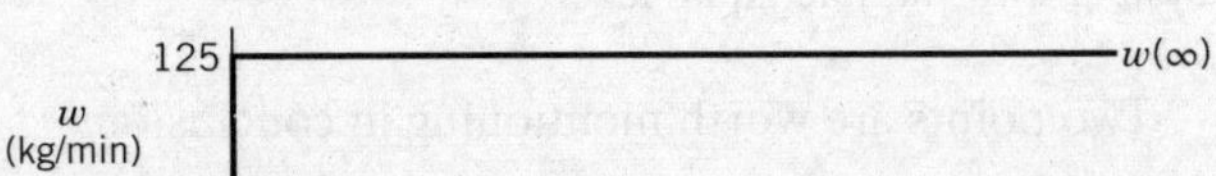

Figure 6.12 Temperature response of a stirred-tank reactor for a step change in feed flow rate.

Solution

First note that $\Delta w = M = 125 - 120 = 5$ kg/min. Because $\Delta T = T(\infty) - T(0) = 160 - 140 = 20°\text{C}$, the steady-state gain is

$$K = \frac{\Delta T}{\Delta w} = \frac{20\,°\text{C}}{5\text{ kg/min}} = 4\frac{°\text{C}}{\text{kg/min}}$$

The time constant obtained from the tangent construction shown in Fig. 6.12 is $\tau = 5$ min. Note that this result is consistent with the "63.2% method," because

$$T = 140 + 0.632(20) = 152.6\,°\text{C}$$

Consequently, the resulting process model is

$$\frac{T'(s)}{W'(s)} = \frac{4}{5s+1}$$

where the steady-state gain is 4°C/kg/min.

Very few experimental step responses exhibit exactly first-order behavior, because

1. The true process model is usually *neither first-order nor linear*. Only the simplest processes exhibit such ideal dynamics.
2. The output data are usually corrupted with noise; that is, the measurements contain a random component. Noise can arise from normal operation of the process, for example, inadequate mixing that produces eddies of higher and lower concentration (or temperature), or from electronic instrumentation. If noise is completely random (i.e., uncorrelated), a first-order response plot may still be drawn that fits the output data well in a time-averaged sense. However, autocorrelated random noise, such as in drifting disturbances, can cause problems in the analysis.

3. Another process input (disturbance) may change in an unknown manner during the duration of the step test. In the CSTR example, undetected changes in inlet composition or temperature are examples of such disturbances.
4. It can be difficult to generate a perfect step input. Process equipment, such as the pumps and control valves discussed in Chapter 8, cannot be changed instantaneously from one setting to another but must be ramped over a period of time. However, if the ramp time is small compared to the process time constant, a reasonably good approximation to a step input may be obtained.

In summary, departures from the ideal response curve in Fig. 6.11 are common.

In order to account for higher-order dynamics that are neglected in a first-order model, a time-delay term can be included. This modification can improve the agreement between model and experimental responses. The fitting of a first-order-plus-time-delay model (FOPTD),

$$G(s) = \frac{Ke^{-\theta s}}{\tau s + 1} \tag{6-58}$$

to the actual step response requires the following steps, as shown in Fig. 6.13:

1. The process gain K is found by calculating the ratio of the steady-state change in y to the size of the input step change, M.
2. A tangent is drawn at the point of inflection of the step response; the intersection of the tangent line and the time axis (where $y = 0$) is the time delay.
3. If the tangent is extended to intersect the steady-state response line (where $y = KM$), the point of intersection corresponds to time $t = \theta + \tau$. Therefore, τ can be found by subtracting θ from the point of intersection.

The tangent method presented here for obtaining the time constant suffers from using only a single point to estimate the time constant. Use of several points from the response may provide a better estimate. For a FOPTD model presented in Eq. 6-58, the dynamic response can be obtained from Eq. 5-24 by replacing time t by time shift $t - \theta$ as follows:

$$y(t_i) = KM[1 - e^{-(t_i-\theta)/\tau}] = y(\infty)[1 - e^{-(t_i-\theta)/\tau}]$$

Take logarithm on both sides and rearrange to give the expression

$$\ln\left[\frac{y(\infty) - y_i}{y(\infty)}\right] = -\frac{t_i - \theta}{\tau} \tag{6-59}$$

where $y_i = y(t_i)$.

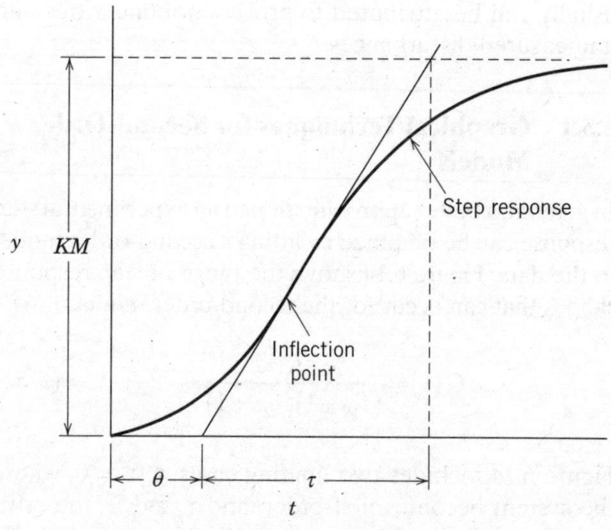

Figure 6.13 Graphical analysis of the process reaction curve to obtain parameters of a first-order-plus-time-delay model.

The final steady-state value, $y(\infty)$, equals KM. In Eq. 6-59, $y(\infty) - y_i$ can be interpreted as the *incomplete response* for data point i; dividing by $y(\infty)$ yields the *fraction incomplete response*: a semilog plot of $[y(\infty) - y_i]/y(\infty)$ vs. $(t_i - \theta)$ will then yield a straight line with a slope of $-1/\tau$, from which an average value of τ is obtained. An equation equivalent to Eq. 6-59 for the variables of Example 6.4 is

$$\ln\left[\frac{T(\infty) - T(t)}{T(\infty) - T(0)}\right] = -\frac{t - \theta}{\lambda\tau} \tag{6-60}$$

The major disadvantage of the time-delay estimation method in Fig. 6.13 is that it is difficult to find the point of inflection, as a result of measurement noise and small-scale recorder charts or computer displays. The method of Sundaresan and Krishnaswamy (1978) avoids use of the point of inflection construction entirely to estimate the time delay. They proposed that two times, t_1 and t_2, be estimated from a step response curve. These times correspond to the 35.3% and 85.3% response times, respectively. The time delay and time constant are then calculated from the following equations:

$$\begin{aligned} \theta &= 1.3t_1 - 0.29t_2 \\ \tau &= 0.67(t_2 - t_1) \end{aligned} \tag{6-61}$$

These values of θ and τ approximately minimize the difference between the measured response and the model response, based on a correlation for many data sets. By using actual step response data, model parameters K, θ, and τ can vary considerably, depending on the operating conditions of the process, the size of the input step change, and the direction of the change. These variations usually can be attributed to process nonlinearities and unmeasured disturbances.

6.5.1 Graphical Techniques for Second-Order Models

In general, a better approximation to an experimental step response can be obtained by fitting a second-order model to the data. Figure 6.14 shows the range of step response shapes that can occur for the second-order model,

$$G(s) = \frac{K}{(\tau_1 s+1)(\tau_2 s+1)} \tag{5-52}$$

Figure 6.14 includes two limiting cases: $\tau_2/\tau_1 = 0$, where the system becomes first-order, and $\tau_2/\tau_1 = 1$, the critically damped case. The larger of the two time constants, τ_1, is called the *dominant time constant*. The S-shaped response becomes more pronounced as the ratio of τ_2/τ_1 becomes closer to one.

Model parameters for second-order systems which include time delays can be estimated using graphical or numerical methods. A method due to Smith (1972) utilizes a model of the form

$$G(s) = \frac{Ke^{-\theta s}}{\tau^2 s^2 + 2\zeta\tau s + 1} \tag{6-62}$$

which includes both overdamped and underdamped cases. Smith's method requires the times (with apparent time delay removed) at which the normalized response reaches 20% and 60%, respectively. Using Fig. 6.15, the ratio of t_{20}/t_{60} gives the value of ζ. An estimate of τ can be obtained from the plot of t_{60}/τ vs. t_{20}/t_{60}.

When graphically fitting second-order models, some caution must be exercised in estimating θ. A second-order model with no time delay exhibits a point-of-inflection (see

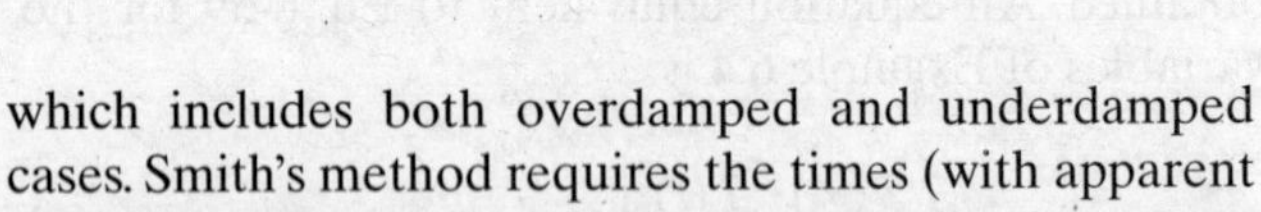

Figure 6.14 Step response for several overdamped second-order systems.

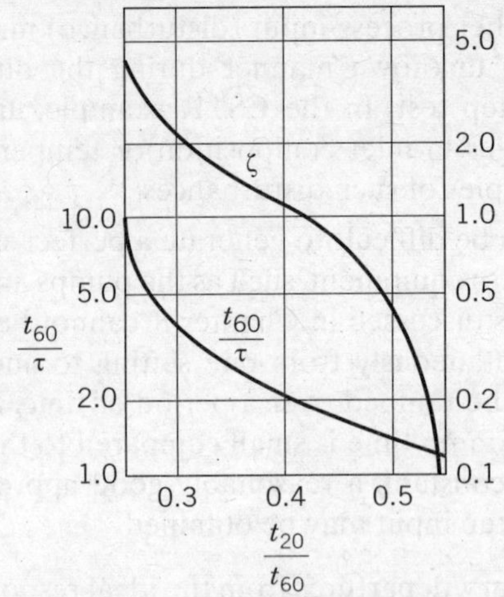

Figure 6.15 Smith's method: relationship of ζ and τ to τ_{20} and τ_{60}.

Fig. 6.14 when $\tau_1 \approx \tau_2$). If the tangent to the point-of-inflection shown in Fig. 6.13 is applied to this case, however, a nonzero time delay is indicated. To avoid this conflict, visual determination of θ is recommended for graphical estimation, but estimation of θ by trial and error may be required to obtain a good fit. In the folowing examples, the time delay is subtracted from the actual time value; then the adjusted time, $t' = t - \theta$, is employed for the actual graphical analysis. An alternative approach to fitting the parameters of the second-order model utilizes three points in the step response. Rangaiah and Krishnaswamy (1994, 1996).

6.5.2 Nonlinear Regression of Step Response Data

Model parameters of transfer function models can be estimated using nonlinear regression and standard software such as Excel and MATLAB. To use Excel, the measured data must be placed in one column. The model predictions to be compared with the measured data are placed in a second column. The sum of squares of the errors is calculated and put into a cell, called the target cell. The target cell value can be minimized using the built-in *Solver* in the *Tools* menu. The window of the Solver allows the user to select the cell to minimize/maximize, the range of cells to be adjusted (the model parameters), and the restrictions, if any, that apply. Clicking on ⟨solve⟩ will calculate the parameter values that minimize the sum of squares. The optimization method used by Excel is based on the generalized reduced gradient technique (Edgar et al., 2001).

In order to use MATLAB, it is necessary to write an M-file that defines the sum of squares of errors. Then the command *fminunc* is used to calculate the minimum. The default algorithm in MATLAB is the BFGS quasi-Newton method (Ljung, 2007).

EXAMPLE 6.5

Step test data have been obtained for the off-gas CO_2 concentration response obtained from changing the feed rate to a bioreactor. Use Smith's method as well as nonlinear regression based on Excel and MATLAB to estimate parameters in a second-order model from experimental step response data shown in Fig. 6.16. For all three methods, assume $\theta = 0$ because the response curve becomes nonzero immediately after $t = 0$. Compare the results with an FOPTD model that is fit using the 63.2% response method to estimate the time constant.

Solution

Smith's Method

The two points of interest are the 20% response time, $t_{20} = 1.85$ min, and the 60% response time, $t_{60} = 5.0$ min. Hence, $t_{20}/t_{60} = 0.37$. From Fig. 6.15, $\zeta = 1.3$ and $t_{60}/\tau = 2.8$; thus, $\tau = 5.0/2.8 = 1.79$ min. Because the model is overdamped, the two time constants can be calculated from the following expressions modified from Eqs. 5-59 and 5-60:

$$\tau_1 = \tau\zeta + \tau\sqrt{\zeta^2 - 1}, \quad \tau_2 = \tau\zeta - \tau\sqrt{\zeta^2 - 1}$$

Solving gives $\tau_1 = 3.81$ min and $\tau_2 = 0.84$ min.

For Fig. 6.16 the 63.2% response is estimated to occur at $t = 5.3$ min. Using the slope at the point of inflection, we can estimate the time delay to be $\theta = 0.7$ min. Note that $\tau = 4.6$ min, which is approximately equal to the sum of τ_1 and τ_2 for the second-order model.

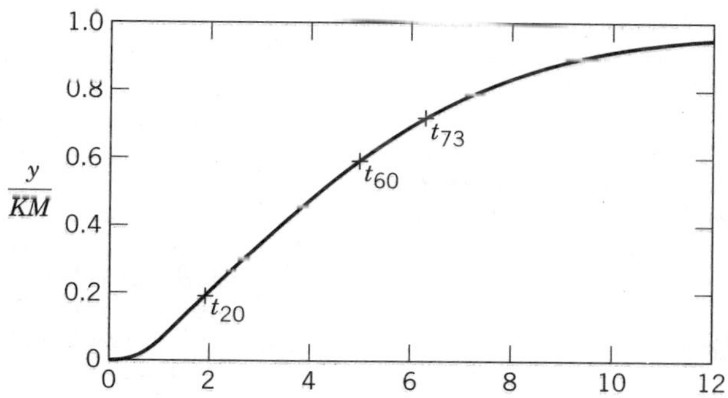

Figure 6.16 Normalized experimental step response.

Nonlinear Regression

Using Excel and MATLAB, we calculate the time constants in Eq. 5-62 that minimize the sum of the squares of the errors between experimental data and model predictions (see Eq. 6-16). The data consisted of 25 points between $t = 0$ and $t = 12$ min with a sampling period of 0.5 min. A comparison of the model parameters and the sum of squared errors for each method is shown below; the time delay is set to zero for the three second-order methods.

	τ_1 (min)	τ_2 (min)	Sum of Squares
Smith	3.81	0.84	0.0769
First order ($\theta = 0.7$ min)	4.60	—	0.0323
Excel and MATLAB	3.34	1.86	0.0057

Clearly, the nonlinear regression method is superior in terms of the goodness of fit, as measured by the sum of squares of the prediction error, but the required calculations are more complicated. Note that the nonlinear regression methods employed by Excel and MATLAB produce identical results.

The step responses are plotted in Fig. 6.17; all three calculated models give an acceptable fit to the original step response curve. In fact, the nonlinear regression model is indistinguishable from the experimental response. Nonlinear regression does not depend on graphical correlations and always provides a better fit to the data. It also permits the experimental step test to be terminated before the final steady state is reached; however, sufficient response data must be obtained for the regression method to be effective.

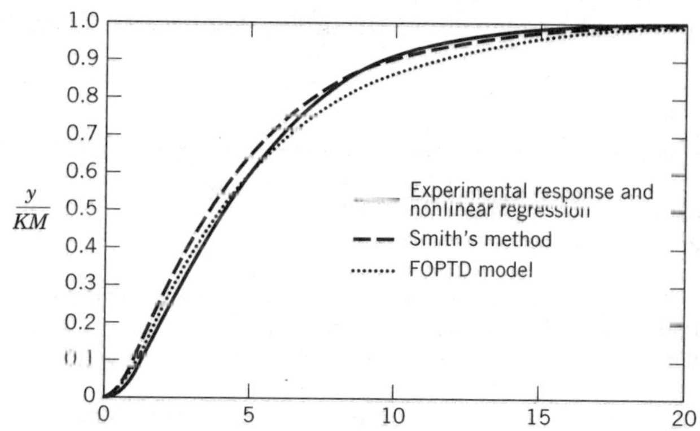

Figure 6.17 Comparison of step responses of fitted models with the original data.

6.5.3 Other Types of Input Excitation

Sometimes a step change in a process input is not permissible owing to safety considerations or the possibility of producing off-specification (*off-spec*) material as a result of the process output deviating significantly from the desired value. In these situations, other types of input changes that do not move the process to a new steady state can be selected. They include rectangular pulses (see Fig. 6.18), pulses of arbitrary shape, or even white (Gaussian) noise. Such "plant-friendly" inputs should be as short as possible, stay within actuator limits, and cause minimum disruption to the controlled variables (Rivera and Jun, 2000). For pulse forcing, the input is suddenly changed, left at its new value for a period of time, and then returned to its original value. Consequently, the process output also returns to its initial steady state, unless the process has an integrating mode.

Random Binary Sequence (RBS) forcing involves a series of pulses of fixed height and random duration. At each sampling instant, a random number generator determines whether the input signal is set at its maximum or minimum value. However, it is more convenient to implement a *pseudo random binary sequence (PRBS)*, which is a two-level, periodic, deterministic signal of a specified length, shown in Fig. 6.18. The actual sequence of inputs can be repeated multiple times. The term *pseudo random* indicates the input is a repeating sequence that has the spectral characteristics of a random signal (Godfrey, 1993). The advantage of a PRBS is that the input excitation can be concentrated in particular frequency ranges that correspond to the process dynamics and that are important for control system design (Rivera and Jun, 2000). See Chapter 10 for more information on frequency response analysis.

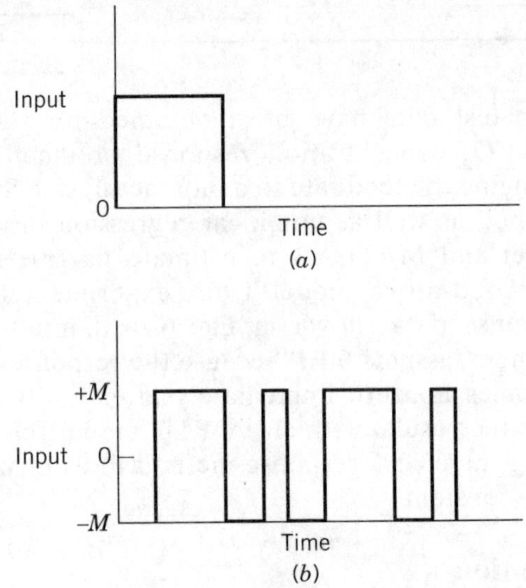

Figure 6.18 (*a*) Pulse and (*b*) PRBS inputs (one cycle).

SUMMARY

In this chapter, we have considered the dynamics of processes that cannot be described by simple transfer functions. Models for these processes include numerator dynamics such as time delays or process zeros. An observed time delay is often a manifestation of higher-order dynamics; consequently, a time-delay term in a transfer function model provides a way of approximating high-order dynamics (e.g., one or more small time constants). Important industrial processes typically have several input variables and several output variables. Fortunately, the transfer function methodology for single-input, single-output processes is also applicable to such multiple-input, multiple-output processes. In Chapter 16, we show that empirical transfer function models can be easily obtained from experimental input–output data.

REFERENCES

Edgar, T. F., D. M. Himmeblau, and L. S. Lasdon, *Optimization of Chemical Processes*, 2nd ed., McGraw-Hill, New York, 2001.

Godfrey, K. (Ed.), *Perturbation Signals for System Identification*, Prentice-Hall, Englewood Cliffs, NJ (1993).

Khalil, H. K., *Nonlinear Systems*, 3rd ed., Prentice-Hall, Upper Saddle River, NJ, 2002.

Ljung, L., *System Identification Toolbox for Use with Matlab. Version 7.* MathWorks, Inc, Natick, MA, 2007.

Rangaiah, G. P., and P. R. Krishnaswamy, Estimating Second-Order plus Dead Time Model Parameters, *Ind. Eng. Chem. Res.*, **33**, 1867 (1994).

Rangaiah, G. P., and P. R. Krishnaswamy, Estimating Second-Order Dead Time Parameters from Underdamped Process Transients, *Chem. Engr. Sci.*, **51**, 1149 (1996).

Rivera, D. E., and K. S. Jun, An Integrated Identification and Control Design Methodology for Multivariable Process System Applications, *IEEE Control Systems Magazine*, **20**(3), 25 (2000).

Smith, C. L., *Digital Computer Process Control*, Intext, Scranton, PA, 1972.

Skogestad, S., Simple Analytic Rules for Model Reduction and PID Controller Tuning, *J. Process Control*, **13**, 291 (2003).

Sundaresan, K. R., and R. R. Krishnaswamy, Estimation of Time Delay, Time Constant Parameters in Time, Frequency, and Laplace Domains, *Can. J. Chem. Eng.*, **56**, 257 (1978).

EXERCISES

6.1 The following transfer function is not written in a standard form:

$$G(s) = \frac{3(s+4)}{(0.25s+1)(3s+1)}e^{-3s}$$

(a) Put it in standard gain/time constant form.

(b) Determine the gain, poles and zeros.

(c) If the time-delay term is replaced by a 1/1 Padé approximation, repeat part (b).

6.2 Consider a process model:

$$\frac{Y(s)}{X(s)} = \frac{K(\tau_a s+1)}{(\tau_1 s+1)(\tau_2 s+1)} \quad (\tau_1 > \tau_2)$$

For a step input, show that:

(a) $y(t)$ can exhibit an extremum (maximum or minimum value) in the step response only if

$$\frac{1-\tau_a/\tau_2}{1-\tau_a/\tau_1} > 1$$

(b) Overshoot occurs only for $\tau_a/\tau_1 > 1$.

(c) Inverse response occurs only for $\tau_a < 0$.

(d) If an extremum in y exists, the time at which it occurs can be found analytically. What is it?

6.3 The transfer function relating the blood pressure of a patient to the infusion rate of a blood pressure drug is given by

$$G_p(s) = \frac{e^{-\theta_1 s}(1+0.4e^{-\theta_2 s})}{\tau_1 s+1}$$

where $\theta_1 = 30$ s, $\theta_2 = 50$ s, and $\tau_1 = 40$ s. Sketch the unit step response of this system. Show an approximate time axis and all the relevant values on the y-axis.

6.4 A pressure-measuring device has been analyzed and can be described by a model with the structure shown in Fig. E6.4a. In other words, the device responds to pressure changes as if it were a first-order process in parallel with a second-order process. Preliminary tests have shown that the gain of the first-order process is –3 and the time constant equals 20 min, as shown in Fig. E6.4*a*. An additional test is made on the overall system. The actual output P_m (not P_m') resulting from a step change in P from 30 to 40 kPa is plotted in Fig. E6.4*b*.

(a) Determine $Q'(t)$.

(b) What are the values of K, τ, and ζ?

(c) What is the overall transfer function for the measurement device $P_m'(s)/P'(s)$?

(d) Find an expression for the overall process gain.

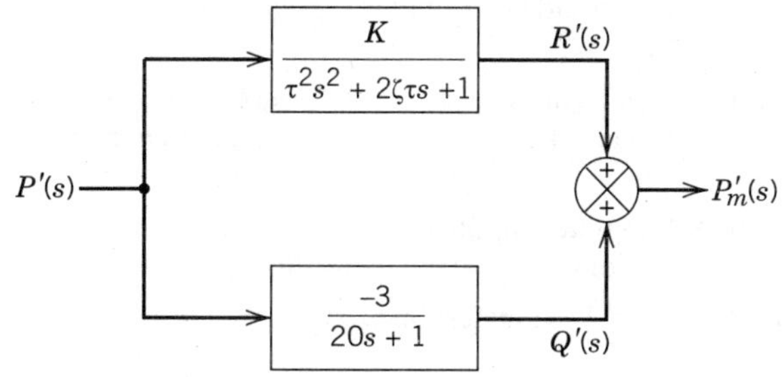

Figure E6.4*a*

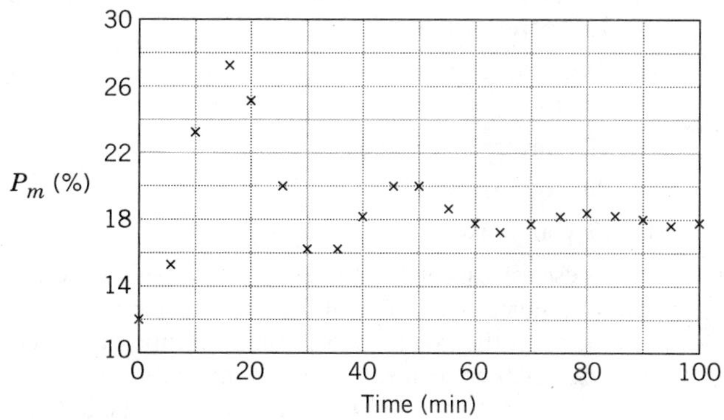

Figure E6.4*b*

6.5 A blending tank that provides nearly perfect mixing is connected to a downstream unit by a long transfer pipe. The blending tank operates dynamically like a first-order process.

The mixing characteristics of the transfer pipe, on the other hand, are somewhere between plug flow (no mixing) and perfectly mixed. A test is made on the transfer pipe that shows that it operates as if the total volume of the pipe were split into five equal-sized perfectly stirred tanks connected in series.

The process (tank plus transfer pipe) has the following characteristics:

$$V_{\text{tank}} = 2\text{ m}^3$$
$$V_{\text{pipe}} = 0.1\text{ m}^3$$
$$q_{\text{total}} = 1\text{ m}^3/\text{min}$$

where q_{total} represents the sum of all flow rates into the process.

(a) Using the information provided above, what would be the most accurate transfer function $C_{\text{out}}'(s)/C_{\text{in}}'(s)$ for the process (tank plus transfer pipe) that you can develop? *Note:* c_{in} and c_{out} are inlet and exit concentrations.

(b) For these particular values of volumes and flow rate, what approximate (low-order) transfer function could be used to represent the dynamics of this process?

(c) What can you conclude concerning the need to model the transfer pipe's mixing characteristics very accurately for this particular process?

(d) Simulate approximate and full-order model responses to a step change in c_{in}.

6.6 Consider the transfer function

$$G(s) = \frac{240(1-4s)e^{-3s}}{18s^2+21s+3}$$

(a) What are the gain, time delay, time constants, poles, and zeros of $G(s)$?

(b) Will the step response of this transfer function exhibit

(i) inverse response or

(ii) oscillations?

Justify your answer.

6.7 A process consists of five perfectly stirred tanks in series. The volume in each tank is 30 L, and the volumetric flow rate through the system is 5 L/min. At some particular time, the inlet concentration of a nonreacting species is changed from 0.60 to 0.45 (mass fraction) and held there.

(a) Write an expression for c_5 (the concentration leaving the fifth tank) as a function of time.

(b) Simulate and plot $c_1, c_2, \ldots, c_5$. Compare c_5 at $t = 30$ min with the corresponding value for the expression in part (a).

6.8 A composition analyzer is used to measure the concentration of a pollutant in a wastewater stream. The relationship between the measured composition C_m and the actual composition C is given by the following transfer function (in deviation variable form):

$$\frac{C'_m(s)}{C'(s)} = \frac{e^{-\theta s}}{\tau s+1}$$

where $\theta = 2$ min and $\tau = 4$ min. The nominal value of the pollutant is $\bar{C} = 5$ ppm. A warning light on the analyzer turns on whenever the measured concentration exceeds 25 ppm.

Suppose that at time $t = 0$, the actual concentration begins to drift higher, $C(t) = 5 + 2t$, where C has units of ppm and t has units of minutes. At what time will the warning light turn on?

6.9 For the process described by the exact transfer function

$$G(s) = \frac{5}{(9s+1)(6s+1)(0.15s+1)(0.05s+1)}$$

(a) Find an approximate first-order-plus-time-delay transfer function model.

(b) Simulate and plot the response $y(t)$ of both the approximate model and the exact model on the same graph for a unit step change in input $x(t)$.

(c) What is the maximum error between the two responses? When does it occur?

6.10 Find the transfer functions $P'_1(s)/P'_d(s)$ and $P'_2(s)/P'_d(s)$ for the compressor-surge tank system of Exercise 2.5 when it is operated isothermally. Put the results in standard (gain/time constant) form. For the second-order model, determine whether the system is overdamped or underdamped.

6.11 A process has the block diagram

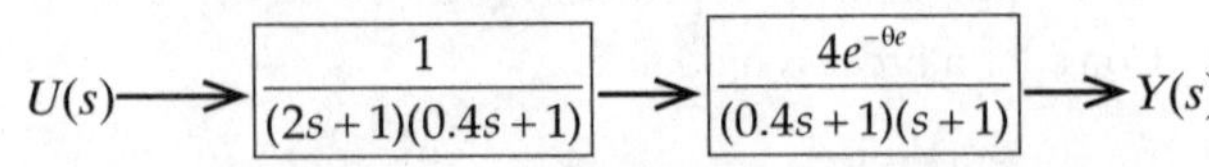

Derive an approximate first-order-plus-time-delay transfer function model.

6.12 Show that the liquid-level system consisting of two interacting tanks (Fig. 6.8) exhibits overdamped dynamics; that is, show that the damping coefficient in Eq. 6-38 is larger than one.

6.13 A liquid surge system (ρ = constant) is designed with a side tank that normally is isolated from the flowing material as shown in Fig. E6.13.

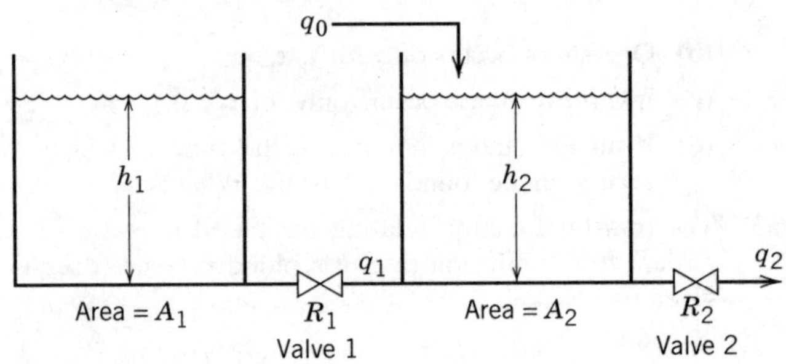

Figure E6.13

(a) In normal operation, Valve 1 is closed ($R_1 \to \infty$) and $q_1 = 0$. What is the transfer function relating changes in q_0 to changes in outflow rate q_2 for these conditions?

(b) Someone inadvertently leaves Valve 1 partially open ($0 < R_1 < \infty$). What is the differential equation model for this system?

(c) What do you know about the form of the transfer function $Q'_2(s)/Q'_0(s)$ for Valve 1 partially open? Discuss but do not derive.

(d) Is the response to changes in q_0 faster or slower for Case (b) compared to Case (a)? Explain why but do not derive an expression for the response.

6.14 The dynamic behavior of a packed-bed reactor can be approximated by a transfer function model

$$\frac{T'(s)}{T'_i(s)} = \frac{-4.5(2-s)}{(10s+1)(5s+1)}$$

where T_i' is the inlet temperature, T is the outlet temperature (°C), and the time constants are in hours. The inlet temperature varies in a cyclic fashion due to the changes in ambient temperature from day to night.

As an approximation, assume that T_i' varies sinusoidally with a period of 24 hours and amplitude of 8 °C. What is the maximum variation in the outlet temperature, T?

6.15 Example 5.1 derives the gain and time constant for a first-order model of a stirred-tank heating process.

(a) Simulate the response of the tank temperature to a step change in heat input from 34.89 kW to 58.15 kW.

(b) Suppose there are dynamics associated with changing the heat input to the system. If the dynamics of the heater itself can be expressed by a first-order transfer function with a gain of one and a time constant of 10 min, what is the overall transfer function for the heating system (tank plus heater)?

(c) For the process in (b), simulate the step increase in heat input.

6.16 Consider the following cascade connection of isothermal reactors. A first-order reaction A → B occurs in both reactors, with reaction rate constant k. The volumes of liquid in the reactors, V, are constant and equal; the flow rates F_0, F_1, F_2 and R are constant. Assume constant physical properties and a negligible time delay for the recycle line.

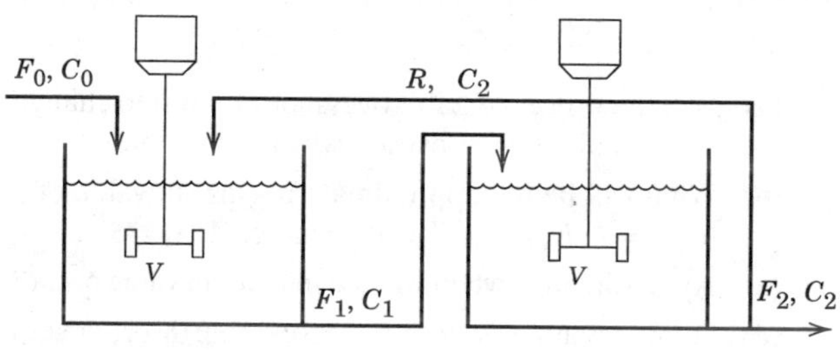

Figure E6.16

(a) Write a mathematical model for this process.

(b) Derive a transfer function model relating the output concentration of A, C_2 to the inlet concentration of A, C_0.

(c) Verify that, in the limit of no recycle ($R \to 0$), the transfer function derived in (b) is equivalent to the transfer function of the two tanks connected in series.

(d) Show that when $k = 0$ and a very large recycle flow rate is used (i.e., the limit as $R \to \infty$), the transfer function derived in (b) becomes the transfer function of a single tank that has the volume equal to $2V$ and a gain of one.

Hint: Recognize that $F_1 = R + F_2$ and $F_0 = F_2$.

6.17 A two-input/two-output process involving simultaneous heating and liquid-level changes is illustrated in Fig. E6.17. Find the transfer function models and expressions for the gains and the time constant τ for this process. What is the output response for a unit step change in Q for a unit step change in w? *Note:* Transfer function models for a somewhat similar process depicted in Fig. 6.13 are given in Eqs. 6-96 through 6-103. They can be compared with your results. For this exercise, T and h are the outputs and Q and w are the inputs.

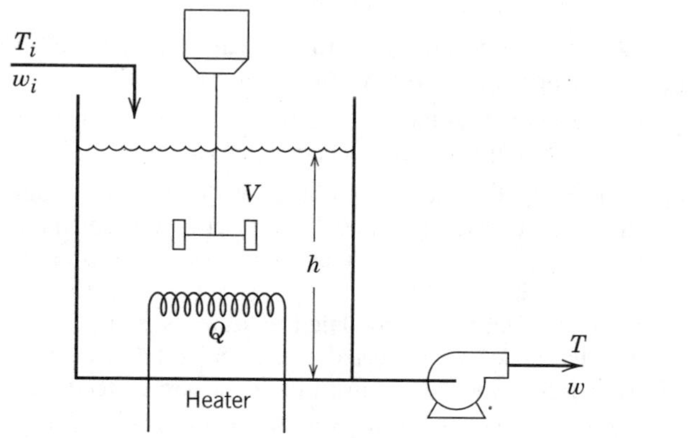

Figure E6.17

6.18 The jacketed vessel in Fig. E6.18 is used to heat a liquid by means of condensing steam. The following information is available:

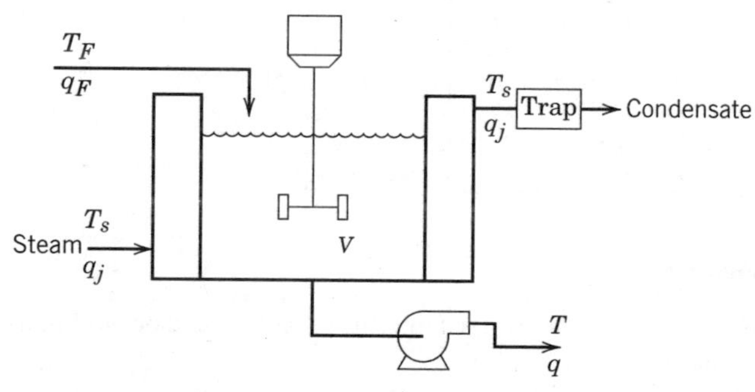

Figure E6.18

(i) The volume of liquid within the tank may vary, thus changing the area available for heat transfer.

(ii) Heat losses are negligible.

(iii) The tank contents are well mixed. Steam condensate is removed from the jacket by a steam trap as soon as it has formed.

(iv) The thermal capacitances of the tank and jacket walls are negligible.

(v) The steam condensation pressure P_s is set by a control valve and is not necessarily constant.

(vi) The overall heat transfer coefficient U for this system is constant.

(vii) Flow rates q_F and q are independently set and may vary. Derive a dynamic model for this process. The model should be simplified as much as possible. State any additional assumptions that you make.

(a) Find transfer functions relating the two primary output variables h (level) and T (liquid temperature) to inputs q_F, q, and T_S. You should obtain six separate transfer functions.

(b) Briefly interpret the form of each transfer function using physical arguments as much as possible.

(c) Discuss qualitatively the form of the response of each output to a step change in each input.

6.19 Your company is having problems with the feed stream to a reactor. The feed must be kept at a constant mass flow rate ($\overline{w}$) even though the supply from the upstream process unit varies with time, $w_i(t)$. Your boss feels that an available tank can be modified to serve as a surge unit, with the tank level expected to vary up and down somewhat as the supply fluctuates around the desired feed rate. She wants you to consider whether (1) the available tank should be used, or (2) the tank should be modified by inserting an interior wall, thus effectively providing two tanks in series to smooth the flow fluctuations.

The available tank would be piped as shown in Fig. E6.19*a*:

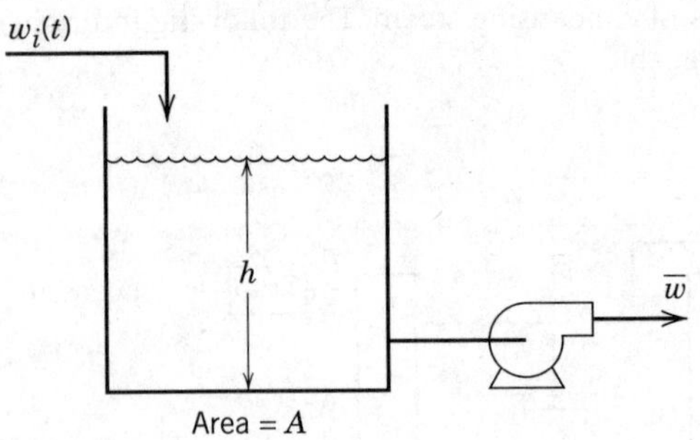

Figure E6.19*a*

In the second proposed scheme, the process would be modified as shown in Fig. E6.19*b*:

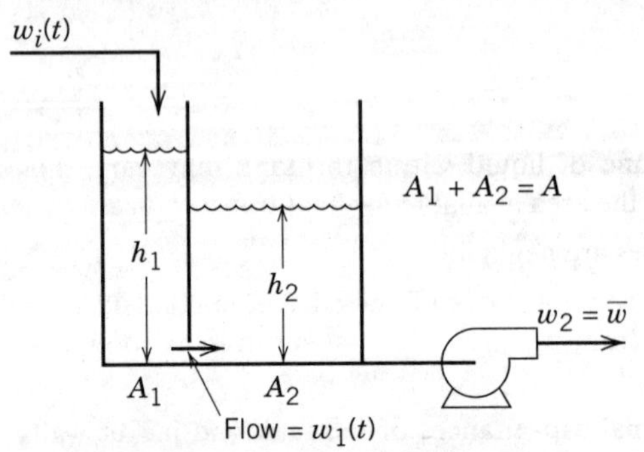

Figure E6.19*b*

In this case, an opening placed at the bottom of the interior wall permits flow between the two partitions. You can assume that the resistance to flow $w_1(t)$ is linear and constant (R).

(a) Derive a transfer function model for the two-tank process $[H_2'(s)/W_i'(s)]$ and compare it to the one-tank process $[H'(s)/W_i'(s)]$. In particular, for each transfer function indicate its order, presence of any zeros, gain, time constants, presence or absence of an integrating element, and whether it is interacting or noninteracting.

(b) Evaluate how well the two-tank surge process would work relative to the one-tank process for the case $A_1 = A_2 = A/2$ where A is the cross-sectional area of the single tank. Your analysis should consider whether h_2 will vary more or less rapidly than h for the same input flow rate change, for example, a step input change.

(c) Determine the best way to partition the volume in the two-tank system to smooth inlet flow changes. In other words, should the first tank contain a larger portion of the volume than the second?

(d) Plot typical responses to corroborate your analysis. For this purpose, you should size the opening in the two-tank interior wall (i.e., choose R) such that the tank levels will be the same at whatever nominal flow rate you choose.

6.20 A process has the following block diagram representation:

$$U(s) \longrightarrow \boxed{\frac{K_1}{(0.1s+1)^2}} \longrightarrow \boxed{\frac{K_2}{4s^2+2s+1}} \longrightarrow Y(s)$$

(a) Will the process exhibit overshoot for a step change in u? Explain/demonstrate why or why not.

(b) What will be the approximate maximum value of y for $K = K_1K_2 = 1$ and a step change, $U(s) = 3/s$?

(c) Approximately when will the maximum value occur?

(d) Simulate and plot both the actual fourth-order step response and the response for a second-order-plus-time delay model that approximates the fourth order system. Using the same second-order-plus-time delay model, what happens when the time constant in the first block is changed to 1 and then 5? Compare the step responses graphically.

6.21 An operator introduces a step change in the flow rate q_i to a particular process at 3:05 A.M., changing the flow from 1900 to 1970 lpm. The first change in the process temperature T (initially at 50°C) occurs at 3:08 A.M. After that, the response in T is quite rapid, slowing down gradually until it appears to reach a steady-state value of 51.5°C. The operator notes in the logbook that there is no change after 3:34 A.M. What approximate transfer function might be used to relate temperature to flow rate for this process in the absence of more accurate information? What should the operator do next time to obtain a better estimate?

6.22 Assume that step response data obtained from an FOPTD system with $K = \tau = 1$ are available. Determine the accuracy of the FOPTD approximate model derived from these data using the Sundaresan and Krishnaswamy method, and consider three cases: $\theta/\tau = 0.1, 1.0, 10.0$. Plot the results and calculate the sum of squared errors.

6.23 For the unit step response shown in Fig. E6.23, estimate the following models using graphical methods:

(a) First-order plus time delay.

(b) Second-order using Smith's method and nonlinear regression.

Plot all three predicted model responses on the same graph.

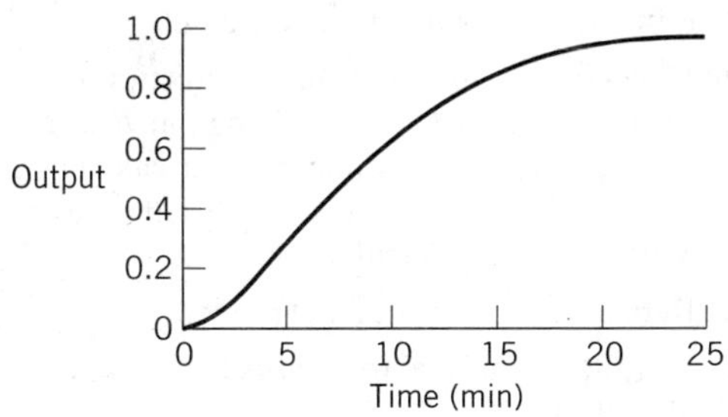

Figure E6.23

6.24 A heat exchanger used to heat a glycol Solution with a hot oil is known to exhibit FOPTD behavior, $G_1(s) = T'(s)/Q'(s)$, where T' is the outlet temperature deviation and Q' is the hot oil flow rate deviation. A thermocouple is placed 3 m downstream from the outlet of the heat exchanger. The average velocity of the glycol in the outlet pipe is 0.5 m/s. The thermocouple also is known to exhibit first-order behavior; however, its time constant is expected to be considerably smaller than the heat exchanger time constant.

(a) Data from a unit step test in Q' on the complete system are shown in Fig. E6.24. Using a method of your choice, calculate the time constants of this process from the step response.

(b) From your empirical model, find transfer functions for the heat exchanger, pipe, and thermocouple. Think of the model as the product of three transfer functions: process, pipe flow, and sensor. What assumptions do you have to make to obtain these individual transfer functions from the overall transfer function?

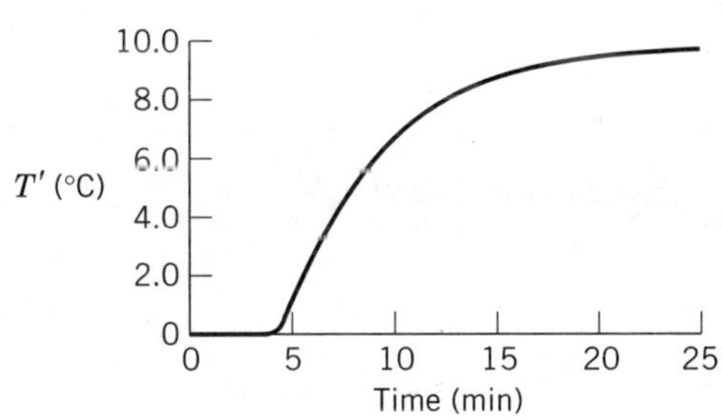

Figure E6.24

6.25 The level in a tank responds as a first-order system to changes in its inlet flow. The data shown below were gathered after the inlet flow was increased quickly from 5.68 to 18.7 L/min.

(a) Determine the transfer function by estimating the time constant using one of the methods of Section 6.5. Be sure to use deviation variables and include units for the model parameters.

(b) Repeat part (a) using nonlinear regression (e.g., Excel) and the liquid level data.

(c) Graphically compare the two model responses with the data. Which model is superior? (Justify your answer)

Table E6.25

Time (min)	Level (m)	Time (min)	Level (m)
0.00	3.17	1.75	6.19
0.25	3.66	2.00	6.55
0.50	4.12	2.25	6.74
0.75	4.60	2.50	6.98
1.00	5.12	2.75	7.22
1.25	5.52	...	...
1.50	5.85	15.00	9.36 (steady state)

6.26 The output response data y shown in Table E6.26 were generated from a step change in input u from 1 to 5 at time $t = 0$. Develop a transfer function model of the form

$$\frac{Y(s)}{U(s)} = \frac{Ke^{-\theta s}}{(\tau_1 s + 1)(\tau_2 s + 1)}$$

Table E6.26

t	y	t	y
0	0	7	1.8
1	0	8	2.4
2	0	9	2.7
3	0.3	10	2.8
4	0.6	11	2.9
5	0.9	12	3.0
6	1.3	13	3.0

6.27 Noisy data for the step response of a boiler temperature T to a decrease in air flow rate q from 28.3 to 26.9 m^3/min. are shown below. Develop a FOPTD model using a method from Chapter 6. Be sure to use deviation variables and report units for the model parameters.

Table E6.27

t (min)	q (m³/min)	T (°C)
0	28.3	849
1	28.3	851
2	28.3	850
3	26.9	851
4	26.9	849
5	26.9	860
6	26.9	867
7	26.9	873
8	26.9	878
9	26.9	882
10	26.9	886
11	26.9	888
12	26.9	890
13	26.9	890

6.28 Consider the step response data for a furnace shown in Fig. E6.28. Assume that hydrocarbon temperature T_{HC} is the output variable and that air flow rate F_A is the input variable.

(a) Develop an FOPTD model from response data for a step change in F_A at $t = 14$ min from 17.9 to 20.0 m³/min. Summarize your calculated model parameters in a table and briefly describe the method used to calculate them.

(b) Repeat (a) for an SOPTD model.

(c) Plot the two model responses for the F_A step change of part (a).

6.29 Consider the step response data for methyl alcohol distillation column shown in Fig. E6.29. Assume that distillate MeOH composition x_D is the output variable and that reflux ratio R is the input variable.

(a) Develop an FOPTD transfer function model from response data for a step change in R at $t = 3950$ s from 1.75 to 2.0. Summarize your calculated model parameters in a table and briefly describe the method used to calculate them.

(b) Repeat (a) for an SOPTD model.

(c) Plot the two model responses for the R step change of part (a).

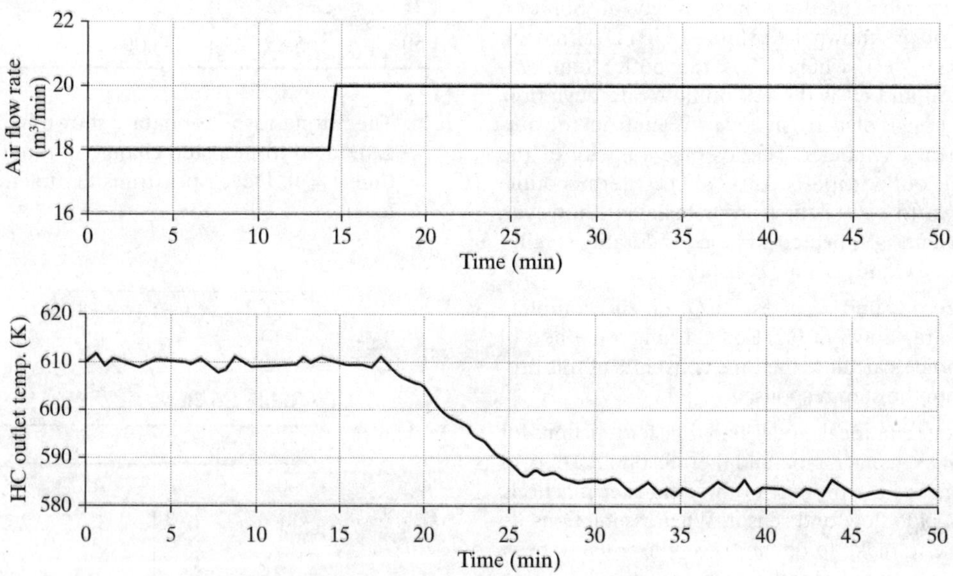

Figure E6.28 Step response data for the furnace module.

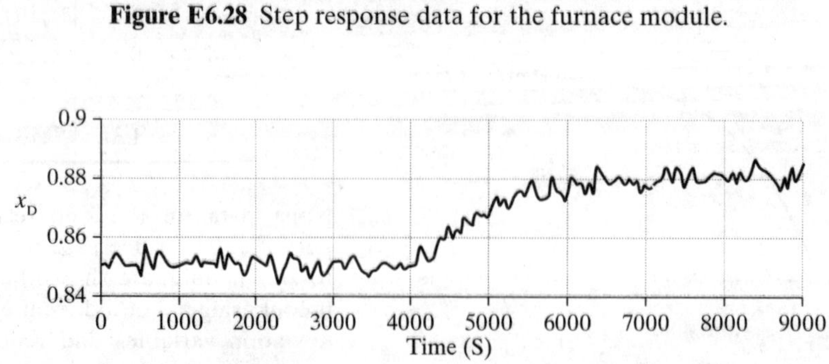

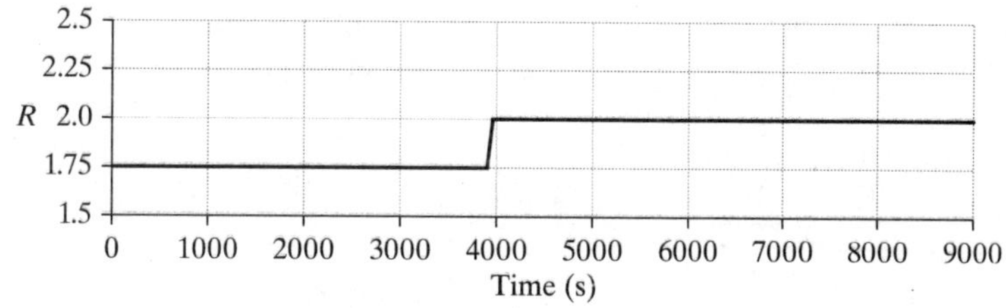

Figure E6.29 Step response data for MeOH distillation column.

MULTIPLE CHOICE QUESTIONS

1. A system has a transfer function $G(s) = \dfrac{3e^{-4s}}{12s+1}$. When a step change of magnitude M is given to the system input, the final value of the system output is measured to be 120. The value of M is ________.

   (a) 20 **(b)** 30
   (c) 40 **(d)** 50 **(GATE 2021)**

2. Match the output signals as obtained from four measuring devices in response to a unit step change in the input signal.

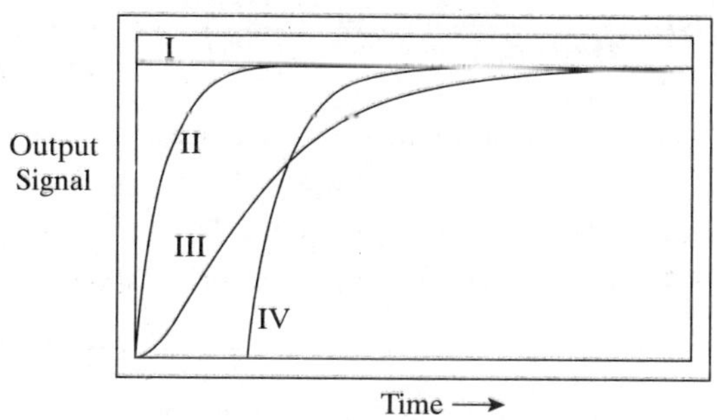

P: Gas chromatograph, with a long capillary tube
Q: Venturi tube
R: Thermocouple with first order dynamics
S: Pressure transducer with second order dynamics

   (a) P-IV, Q-III, R-II, S-I **(b)** P-III, Q-I, R-II, S-IV
   (c) P-IV, Q-I, R-II, S-III **(d)** P-II, Q-IV, R-III, S-I **(GATE 2015)**

3. Which one of the following transfer functions corresponds to an inverse response process with a positive gain?

   (a) $\dfrac{1}{2s+1} - \dfrac{1}{3s+1}$ **(b)** $\dfrac{2}{s+1} - \dfrac{5}{s+10}$

   (c) $\dfrac{3(0.5s-1)}{(2s+1)(s+1)}$ **(d)** $\dfrac{5}{s+1} - \dfrac{3}{2s+1}$ **(GATE 2008)**

4. Match the transfer functions with the responses to a unit step input shown in the figure.

   (I) $\dfrac{-2.5(-4s+1)}{4s^2+4s+1}$ **(II)** $\dfrac{-2e^{-10s}}{10s+1}$

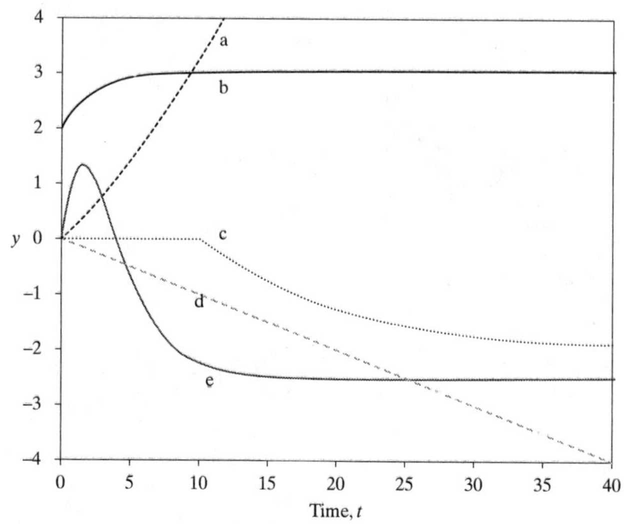

   (III) $\dfrac{-5}{-20s+1}$ **(IV)** $\dfrac{-0.1s}{s}$ **(V)** $\dfrac{4s+3}{2s+1}$

   (a) I-e, II-c, III-a, IV-d, V-b
   (b) I-a, II-b, III-c, IV-d, V-e
   (c) I-b, II-a, III-c, IV-e, V-d
   (d) I-e, II-a, III-c, IV-b, V-d **(GATE 2007)**

5. The experimental response of the controlled variable $y(t)$ for a step change of magnitude P in the manipulated variable $x(t)$ is shown below.

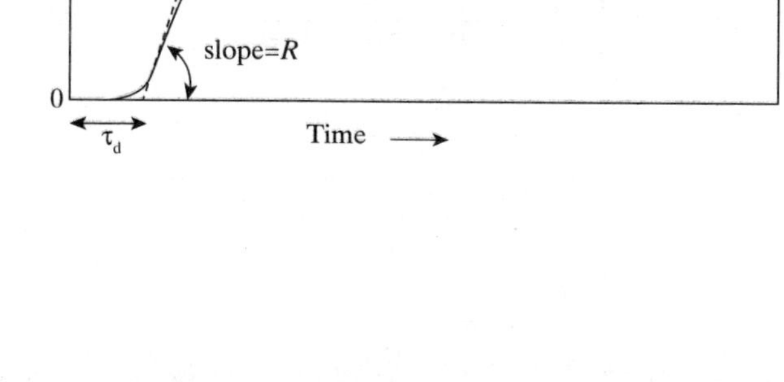

The appropriate transfer function of the process is

(a) $\dfrac{(Q/P)e^{-(Q/P)s}}{\tau_d s+1}$ **(c)** $\dfrac{(Q/P)e^{-\tau_d s}}{(Q/P)s+1}$

(c) $\dfrac{(Q/R)e^{-\tau_d s}}{(Q/R)s+1}$ **(d)** $\dfrac{(Q/R)e^{-(Q)s}}{\tau_d s+1}$

(GATE 2004)

6. Consider a system of two tanks in series as shown below:

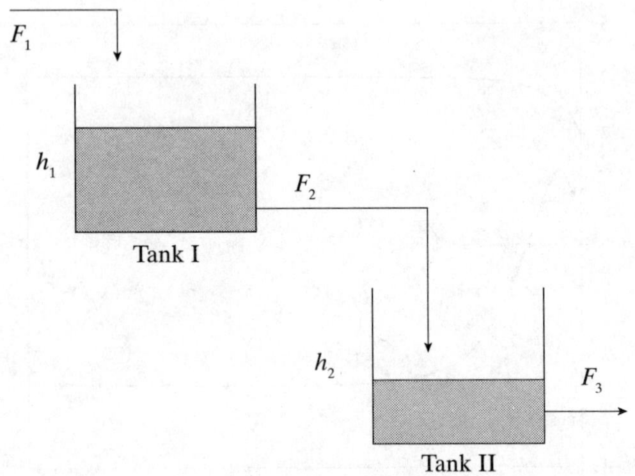

The level h_2 in Tank II is measured and has to be controlled by manipulating the flow rate F_1. It is given that $F_2 = 0.005h_1$ and $F_3 = 0.0025h_2$ m³/s, where h is in m. The cross-sectional areas of Tank I and II are both equals to 1 m².

The transfer function of $H_2(s)$ to $F_1(s)$ is given by:

(a) $G(s)=\dfrac{200}{(400s+1)(200s+1)}$

(b) $G(s)=\dfrac{400}{(400s+1)(200s+1)}$

(c) $G(s)=\dfrac{300}{(400s+1)(200s+1)}$

(d) $G(s)=\dfrac{100}{(400s+1)(200s+1)}$ **(GATE 2002)**

7. The transfer function of a system consisting of a thermal process, value and measuring system is given by

$$\frac{\Delta T(s)}{\Delta I(s)}=\frac{5}{(5s+1)(0.5s+1)(0.1s+1)}$$

FOPTD approximation of the system using Skogestard's half rule is

(a) $G(s)=\dfrac{5e^{-0.35s}}{5.5s+1}$ **(b)** $G(s)=\dfrac{5e^{-0.6s}}{5s+1}$

(c) $G(s)=\dfrac{5e^{-0.35s}}{5.25s+1}$ **(d)** $G(s)=\dfrac{5e^{-0.35s}}{5s+1}$

8. A control system has the following transfer function,

$$F(s)=\frac{(s-1)(s+1)}{s(s-2)(s+4)}.$$

The initial value of the corresponding time function is

(a) 1 **(b)** 1/8 **(c)** 7/8 **(d)** –1 **(GATE 1999)**

9. The transfer function for a first-order process with time delay is

(a) $\dfrac{e^{\tau_D s}}{(\tau s+1)}$ **(b)** $\dfrac{e^{-\tau_D s}}{(\tau s+1)}$

(c) $\dfrac{1}{(\tau s+1)(\tau_D s+1)}$ **(d)** $\dfrac{\tau_D s}{(\tau s+1)}$ **(GATE 1997)**

10. The response of two tanks of same size and resistance in series is

(a) under damped **(b)** critically damped
(c) over damped **(d)** none of the above

(GATE 1995)

11. Transfer function of a system is given by

$$G(s)=\frac{3}{(7s+1)(3s+1)(2s+1)(0.1s+1)}.$$

The SOPTD model of the system using Skogestad's half rule is

(a) $G(s)=\dfrac{3e^{-(1.1)s}}{(7s+1)(4s+1)}$

(b) $G(s)=\dfrac{3e^{-(1.1)s}}{(8s+1)(3s+1)}$

(c) $G(s)=\dfrac{3e^{-(2.1)s}}{(7s+1)(4s+1)}$

(d) $G(s)=\dfrac{3e^{-(2.1)s}}{(7s+1)(3s+1)}$

ANSWER KEY

1. (c)	**2.** (c)	**3.** (c)	**4.** (a)	**5.** (c)	**6.** (b)
7. (c)	**8.** (a)	**9.** (b)	**10.** (c)	**11.** (a)	

CHAPTER 7

Feedback Controllers

CHAPTER CONTENTS

In previous chapters we analyzed the dynamic behavior of representative processes and developed mathematical tools required to analyze process dynamics. We now consider the important topic of feedback control that was introduced in Chapter 1.

The standard feedback *control algorithms* (also called *control laws*) are presented, with emphasis on the two methods that are most widely used in the process industries: *proportional-integral-derivative* (PID) control and *on–off control*. Both analog (i.e., continuous) and digital (i.e., discrete) versions of PID control are considered. The remaining elements of a standard feedback control loop—sensors, transmitters, and control valves—will be considered in Chapter 8. Advanced feedback control strategies are considered in Chapters 12, 14, 15 and Appendices I and J.

7.1 INTRODUCTION

We introduce feedback control systems by again considering the stirred-tank blending process of Chapters 2 and 4.

7.1.1 Illustrative Example: The Continuous Blending Process

A schematic diagram of a stirred-tank blending process and control system is shown in Fig. 7.1. The control objective is to keep the tank exit composition x at the desired value set point by adjusting w_2, the flow rate of pure component A, via the control valve. The composition analyzer-transmitter (AT) measures the exit composition and transmits it as an electronic signal to the feedback controller (AC). The controller compares the measured value x_m to the desired value (set point) and calculates an appropriate output signal p, an electronic signal that is sent to a *current-to-pressure transducer* (I/P) where it is converted to an equivalent pneumatic (air) signal that is compatible with the pneumatically-actuated control valve. The symbols of Fig. 7.1 are examples of the standard instrumentation symbols published by the Instrumentation, Systems and Automation (ISA) Society. In particular, an electronic signal is denoted by a dashed line and a pneumatic signal by a solid line with crosshatches.

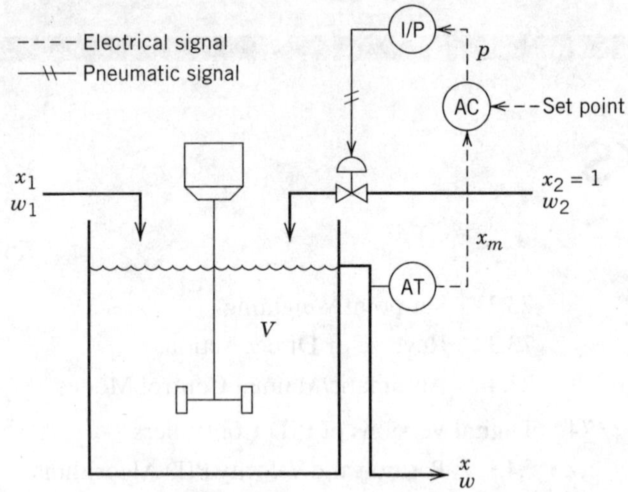

Figure 7.1 Schematic diagram for a stirred-tank blending process and control system.

This example illustrates that the basic components in a feedback control loop are:

- Process being controlled (blending system)
- Sensor-transmitter combination (AT)
- Feedback controller (AC)[1]
- Current-to-pressure transducer (I/P)
- Final control element (control valve)
- Transmission lines between the various instruments (electrical cables and pneumatic tubing)

A current-to-pressure (or voltage-to-pressure) transducer is required if the control loop contains both electronic instruments and a pneumatic control valve. The term *final control element* refers to the device that is used to adjust the manipulated variable. It is usually a control valve but could be some other type of device, such as a variable speed pump or an electrical heater. The operation of a similar blending control system (with an electronically actuated control valve) has been described in Section 1.2.

The blending system in Fig. 7.1 involves *analog* instrumentation. For an analog device, the input and output signals are continuous (analog) rather than discontinuous (digital or discrete time). Analog devices can be either electronic or pneumatic. For electronic devices such as sensors and controllers, the standard ranges for input and output signals are 4–20 mA and 1–5 V (DC). Pneumatic instruments continue to be used, particularly in older plants or hazardous areas where electronic instruments are not intrinsically safe. For a pneumatic instrument, the input and output signals are air pressures in the range of 20 to 100 kPa(g) (0.2–1 barg or 3–15 psig). Metal or plastic tubing (usually 6 or 10 mm OD) is used to interconnect the various pneumatic instruments. As indicated in Fig. 7.1, both electronic and pneumatic devices can be used in the same feedback control loop.

Most new control systems utilize digital technology with the control algorithms implemented via digital computers and digital signal pathways (networks) used for data transmission (see Chapter 19). Instrumentation for process control, including computer hardware and software, are considered in greater detail in Chapter 8 and Chapter 19.

Now we consider the heart of a feedback control system, the controller itself.

7.1.2 Historical Perspective

We tend to regard automatic control devices as a modern development. However, water clocks with ingenious feedback mechanisms were developed by the ancient Greeks as early as 250 B.C. (Mayr, 1970). A thousand years earlier, Polynesians sailed outrigger canoes over large portions of the Pacific Ocean. These remarkable feats were made possible by building canoes with a feedback device to ensure roll stability (Abramovitch, 2005). Finally, the fly-ball governor played a key role in the development of steam power in the nineteenth century.

During the 1930s, *three-mode controllers* with proportional, integral, and derivative (PID) feedback control action became commercially available (Ziegler, 1975). The first theoretical papers on process control were published during this same period. Pneumatic PID controllers gained widespread industrial acceptance during the 1940s, and their electronic counterparts entered the market in the 1950s. The first computer control applications in the process industries were reported in the late 1950s and early 1960s. Since the 1980s, digital hardware has been used on a routine basis and has had a tremendous impact on process control.

As a simple example of feedback control, consider the flow control loop in Fig. 7.2 where the flow rate of a process stream is measured and transmitted electronically to a flow controller. The controller compares the

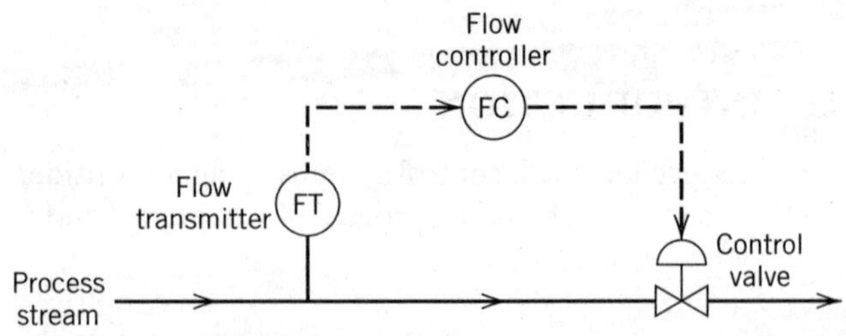

Figure 7.2 Flow control system.

[1] The letter "A" in symbols, AT and AC, refers to *Analysis*.

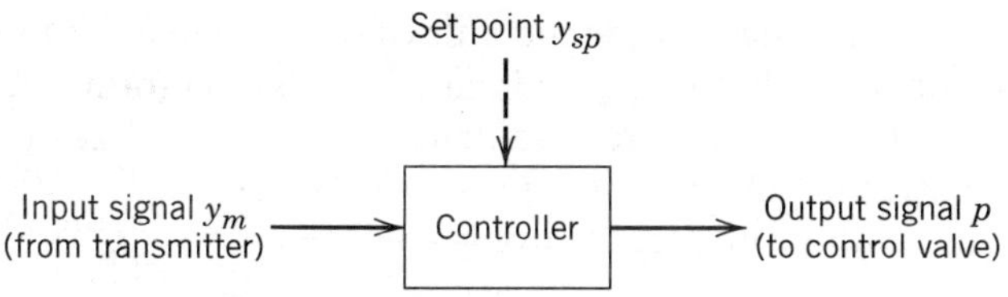

Figure 7.3 Simple diagram of a feedback controller.

measured value to the set point and takes the appropriate corrective action by calculating the controller output and transmitting it as an electronic signal to the control valve.

The block diagram for the feedback controller of Fig. 7.2 is shown in Fig. 7.3. The set point is shown as a dashed line. For digital control systems, the set point would be entered by an operator using a computer terminal. For an analog controller, the set point would be specified via a dial setting on the equipment. In addition to this local set point, some controllers have a remote set-point option that permits them to receive an external set-point from another controller or a computer. The input and output signals for analog controllers are continuous signals that are either electrical or pneumatic. For digital control systems, the input signals are first converted from analog to digital form prior to the control calculations. Then, the calculated value of the controller output is converted from a digital signal to an analog signal for transmission to the control valve (or some other type of final control element). These types of signal conversions are described in Chapter 19.

7.2 BASIC CONTROL MODES

Next we consider the three basic feedback control modes starting with the simplest mode, *proportional control.*

7.2.1 Proportional Control

In feedback control, the objective is to reduce the error signal to zero where

$$e(t) = y_{sp}(t) - y_m(t) \tag{7-1}$$

and

$e(t)$ = error signal
$y_{sp}(t)$ = set point
$y_m(t)$ = measured value of the controlled variable (or equivalent signal from the sensor/transmitter)

Although Eq. 7-1 indicates that the set point can be time-varying, in many process control problems it is kept constant for long periods of time.

For proportional control, the controller output is proportional to the error signal,

$$p(t) = \bar{p} + K_c e(t) \tag{7-2}$$

where

$p(t)$ = controller output
$\bar{p}$ = bias (steady-state) value
K_c = controller gain (usually dimensionless)

The key concepts behind proportional control are that (1) the controller gain can be adjusted to make the controller output changes as sensitive as desired to deviations between set point and controlled variable, and that (2) the sign of K_c can be chosen to make the controller output increase (or decrease) as the error signal increases. For example, for the blending process in Fig. 7.1, we want w_2 to decrease as x increases; hence, K_c should be a positive number.

For proportional controllers, bias $\bar{p}$ can be adjusted, a procedure referred to as *manual reset.* Because the controller output equals $\bar{p}$ when the error is zero, $\bar{p}$ is adjusted so that the controller output, and consequently the manipulated variable, are at their nominal steady-state values when the error is zero. For example, if the final control element is a control valve, $\bar{p}$ is adjusted so that the flow rate through the control valve is equal to the nominal, steady-state value when $e = 0$. The controller gain K_c is adjustable and is usually *tuned* (i.e., adjusted) after the controller has been installed.

For general-purpose controllers, K_c is dimensionless. This situation occurs when p and e in Eq. 7-2 have the same units. For example, the units could be associated with electronic or pneumatic instruments (mA, volts, psi, etc.). For digital implementation, p and e are often expressed as numbers between 0 and 100%. The latter representation is especially convenient for graphical displays using computer control software.

On the other hand, in analyzing control systems it can be more convenient to express the error signal in engineering units such as °C or mol/L. For these situations, K_c will not be dimensionless. As an example, consider the stirred-tank blending system. Suppose that e [=] mass fraction and p [=] mA; then Eq. 7-2 implies that K_c [=] mA because mass fraction is a dimensionless quantity. If a controller gain is not dimensionless, it includes the steady-state gain for another component of the control loop such as a transmitter or control valve. This situation is discussed in Chapter 9.

Some controllers have a proportional band setting instead of a controller gain. The *proportional band PB* (in %) is defined as

$$PB \triangleq \frac{100\%}{K_c} \tag{7-3}$$

This definition applies only if K_c *is* dimensionless. Note that a small (narrow) proportional band corresponds to a large controller gain, whereas a large (wide) *PB* value implies a small value of K_c.

The ideal proportional controller in Eq. 7-2 and Fig. 7.4 does not include physical limits on the controller output, *p*. A more realistic representation is shown in Fig. 7.5, where the controller *saturates* when its output reaches a physical limit, either p_{max} or p_{min}. In order to derive the transfer function for an ideal proportional controller (without saturation limits), define a deviation variable $p'(t)$ as

$$p'(t) \triangleq p(t) - \bar{p} \tag{7-4}$$

Then Eq. 7-2 can be written as

$$p'(t) = K_c\, e(t) \tag{7-5}$$

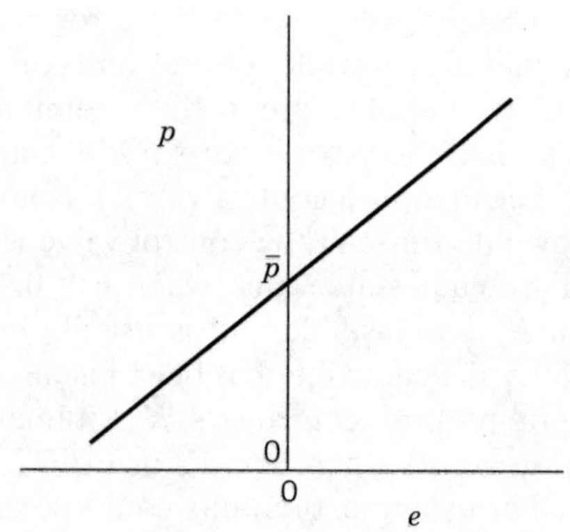

Figure 7.4 Proportional control: ideal behavior (slope of line = K_c).

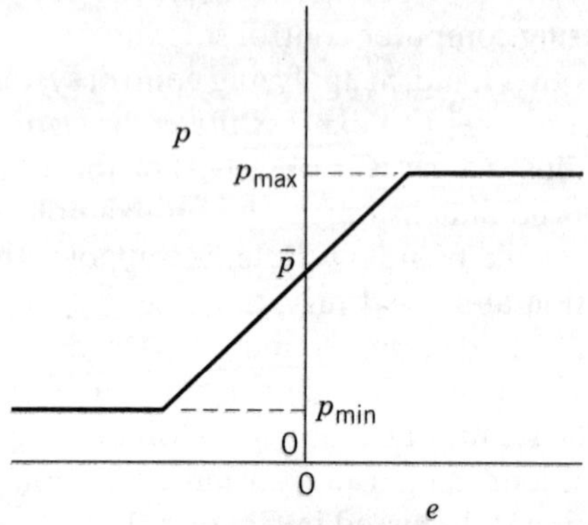

Figure 7.5 Proportional control: actual behavior.

It is unnecessary to define a deviation variable for the error signal, because *e* is already in deviation form, and its nominal steady-state value is $\bar{e} = 0$. Taking Laplace transforms and rearranging Eq. 7-5 gives the transfer function for proportional-only control:

$$\frac{P'(s)}{E(s)} = K_c \tag{7-6}$$

An inherent disadvantage of proportional-only control is that a steady-state error (or *offset*) occurs after a set-point change or a sustained disturbance. In Chapter 9 we demonstrate that offset will occur for proportional-only control regardless of the value of K_c that is employed. Fortunately, the addition of the integral control mode facilitates offset elimination, as discussed in the next section.

For control applications where offsets can be tolerated, proportional-only control is attractive because of its simplicity. For example, in some level control problems, maintaining the liquid level close to the set point is not as important as merely ensuring that the storage tank does not overflow or run dry.

7.2.2 Integral Control

For integral control action, the controller output depends on the integral of the error signal over time,

$$p(t) = \bar{p} + \frac{1}{\tau_I}\int_0^t e(t^*)dt^* \tag{7-7}$$

where τ_I, an adjustable parameter referred to as the *integral time* or *reset time*, has units of time. In the past, integral control action has been referred to as *reset* or *floating control*, but these terms are seldom used anymore.

Integral control action is widely used because it provides an important practical advantage, the elimination of offset. To understand why offset is eliminated, consider Eq. 7-7. In order for the controlled process to be at steady state, the controller output *p* must be constant so that the manipulated variable is also constant. Equation 7-7 implies that *p* changes with time unless $e(t^*) = 0$. Thus, when integral action is used, *p* automatically changes until it attains the value required to make the steady-state error zero. This desirable situation always occurs unless the controller output or final control element saturates and thus is unable to bring the controlled variable back to the set point. Controller saturation occurs whenever the disturbance or set-point change is so large that it is beyond the range of the manipulated variable.

Although elimination of offset is usually an important control objective, the integral controller in Eq. 7-7 is seldom used by itself, because little control action takes place until the error signal has persisted for some time. In contrast, proportional control action takes immediate corrective action as soon as an error is detected. Consequently, integral control action is normally used in conjunction with proportional control as the *proportional-integral (PI)* controller:

$$p(t) = \bar{p} + K_c\left(e(t) + \frac{1}{\tau_I}\int_0^t e(t^*)dt^*\right) \tag{7-8}$$

The corresponding transfer function for the PI controller in Eq. 7-8 is given by

$$\frac{P'(s)}{E(s)} = K_c\left(1 + \frac{1}{\tau_I s}\right) = K_c\left(\frac{\tau_I s + 1}{\tau_I s}\right) \tag{7-9}$$

The response of the PI controller to a unit step change in $e(t)$ is shown in Fig. 7.6. At time zero, the controller output changes instantaneously due to the proportional action. Integral action causes the ramp increase in $p(t)$ for $t > 0$. When $t = \tau_I$, the integral term has contributed the same amount to the controller output as the proportional term. Thus, the integral action has *repeated* the proportional action once. Some commercial controllers are calibrated in terms of $1/\tau_I$ (repeats per minute) rather than τ_I (minutes, or minutes per repeat). For example, if $\tau_I = 0.2$ min, this corresponds to $1/\tau_I$ having a value of 5 repeats/minute.

One disadvantage of using integral action is that it tends to produce oscillatory responses of the controlled variable and, as we will see in Chapter 9, it reduces the stability of the feedback control system. A limited amount of oscillation can usually be tolerated, because it often is associated with a faster response. The undesirable effects of too much integral action can be avoided by proper tuning of the controller or by including

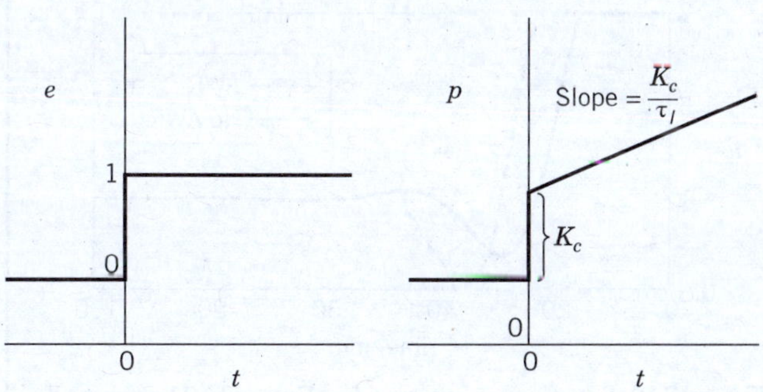

Figure 7.6 Response of proportional-integral controller to unit step change in $e(t)$.

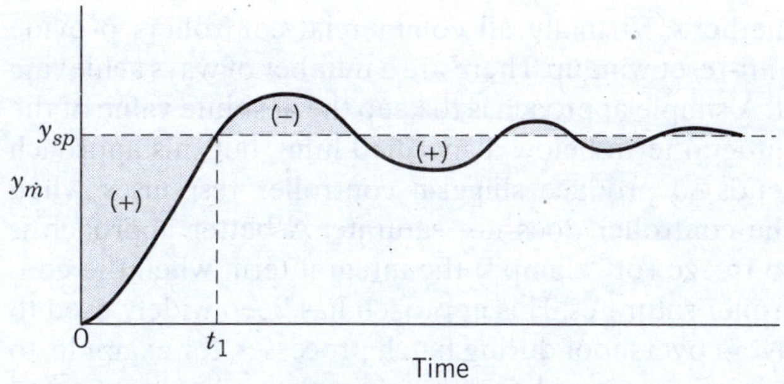

Figure 7.7 Reset windup during a set-point change.

derivative control action (Section 7.2.3), which tends to counteract the destabilizing effects.

Reset Windup

An inherent disadvantage of integral control action is a phenomenon known as *reset windup*. Recall that the integral mode causes the controller output to change as long as $e(t^*) \neq 0$ in Eq. 7-8. When a sustained error occurs, the integral term becomes quite large and the controller output eventually saturates. Further buildup of the integral term while the controller is saturated is referred to as reset windup or *integral windup*. Figure 7.7 shows a typical response to a step change in set point when a PI controller is used. Note that the indicated areas under the curve provide either positive or negative contributions to the integral term, depending on whether the measurement of the controlled variable y_m is below or above the set point, y_{sp}. The large overshoot in Fig. 7.7 occurs because the integral term continues to increase until the error signal changes sign at $t = t_1$. Only then, does the integral term begin to decrease. After the integral term becomes sufficiently small, the controller output moves away from the saturation limit and has the value determined by Eq. 7-8.

Reset windup occurs when a PI or PID controller experiences a sustained error, for example, during the start-up of a batch process or after a large set-point change. It can also occur as a consequence of a large sustained disturbance that is beyond the range of the manipulated variable. In this situation, a physical limitation (e.g., a control valve fully open or completely shut) prevents the controller from reducing the error signal to zero. Thus the feedback control is now essentially disabled because u is constant, regardless of the value of y.

Clearly, it is undesirable to have the integral term continue to build up after the controller output saturates, because the controller is already implementing its maximum control effort. Modifications to standard PI or PID control that limit the contribution of the integral term are referred to as *anti-reset windup (AWR)* methods. Virtually all commercial controllers provide anti-reset windup. There are a number of ways achieving it. A simple approach is to keep the absolute value of the integral term below a specified limit, but this approach tends to produce sluggish controller responses when the controller does not saturate. A better approach is to freeze (or "clamp") the integral term when the controller saturates. This approach has been widely used to avoid overshoot during batch processes, for example, to minimize or avoid overshooting when a batch is heated from room temperature to the desired temperature set point. This approach is equivalent to the *back calculation method* (Åström and Hägglund, 2006).

An effective and widely used anti-reset windup method is based on implementing integral control action using a positive feedback loop, as shown in Fig. 7.8 (Blevins et al., 2012). When this controller does not saturate, this implementation is mathematically equivalent to the standard PI control algorithm in Eq. 7-8. When the PI controller output is saturated for an extended period of time, u is constant and the output signal of the positive feedback loop X settles to a constant value after $t \approx 5\tau_I$. Thus, the integral term does not windup. By contrast, for the standard PI algorithm in Eq. 7-8, the integral term does windup because it continually grows during saturation conditions. The advantages of using anti-reset windup are illustrated in Example 7.1.

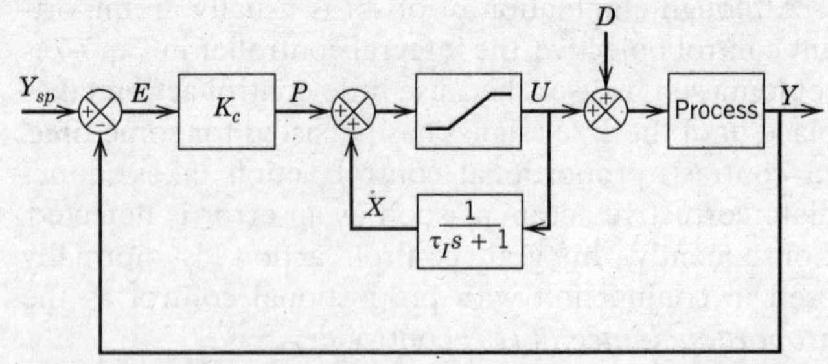

Figure 7.8 Block diagram for PI control with the integral control action implemented as a positive feedback loop with saturation limits.

EXAMPLE 7.1

A standard PI control system for a first-order-plus-time-delay process is shown in Fig. 7.9 with the following parameter values:

Model: $K = 1, \tau = 20$ min, $\theta = 4$ min

Controller: $K_c = 3, \tau_I = 11$ min

Compare the responses of this feedback control system, with and without anti-reset windup, for a unit step change in set-point at $t = 0$. Assume that the controller saturation limits are ± 2.0.

Solution

The set-point responses with and without anti-reset-windup (AWR) are shown in Fig. 7.10. With AWR, the set-point response is improved because y exhibits less overshoot and settles at the new set point more rapidly. Note that this improvement is due to the controller output u being saturated for a shorter period of time.

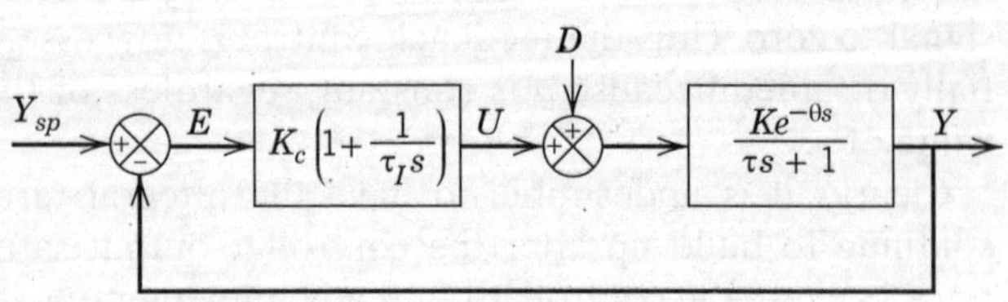

Figure 7.9 Block diagram for the PI control system without anti-reset windup.

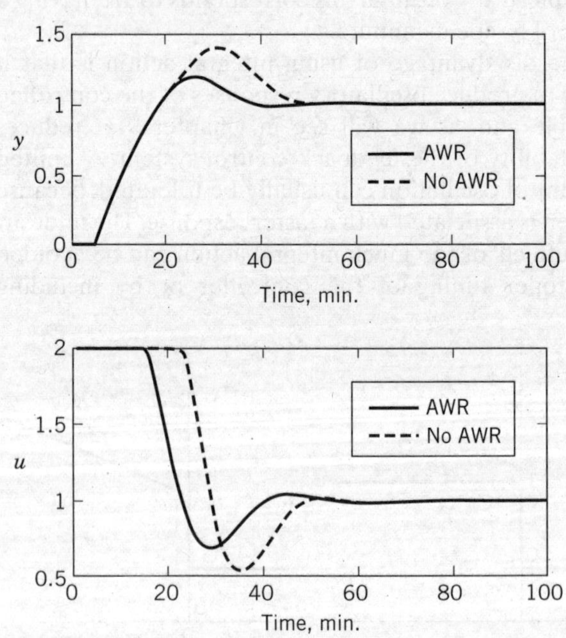

Figure 7.10 Set-point responses for Example 7.1 with and without anti-reset-windup.

7.2.3 Derivative Control

The function of derivative control action is to anticipate the future behavior of the error signal by considering its rate of change. In the past, derivative action was also referred to as *rate action*, *pre-act*, or *anticipatory control*. For example, suppose that a reactor temperature increases by 10 °C in a short period of time, say, 3 min. This clearly is a more rapid increase in temperature than a 10 °C rise in 30 min, and it could indicate a potential *runaway* situation for an exothermic reaction. If the reactor were under manual control, an experienced plant operator would anticipate the consequences and quickly take appropriate corrective action to reduce the temperature. Such a response would not be obtainable from the proportional and integral control modes discussed so far. Note that a proportional controller reacts to a deviation in temperature only, making no distinction as to the time period over which the deviation develops. Integral control action is also ineffective for a sudden deviation in temperature, because the corrective action depends on the duration of the deviation.

The anticipatory strategy used by the experienced operator can be incorporated in automatic controllers by making the controller output proportional to the rate of change of the error signal or the controlled variable. Thus, for *ideal* derivative action,

$$p(t) = \bar{p} + \tau_D \frac{de(t)}{dt} \tag{7-10}$$

where τ_D, the *derivative time*, has units of time. Note that the controller output is equal to the nominal value $\bar{p}$ as long as the error is constant (that is, as long as $de(t)/dt = 0$). Consequently, derivative action is never used alone; it is always used in conjunction with proportional or proportional-integral control. For example, an ideal PD controller has the transfer function:

$$\frac{P'(s)}{E(s)} = K_c(1 + \tau_D s) \tag{7-11}$$

By providing anticipatory control action, the derivative mode tends to stabilize the controlled process. Thus, it is often used to counteract the destabilizing tendency of the integral mode (see Chapters 9 and 10).

Derivative control action also tends to improve the dynamic response of the controlled variable by reducing the settling time, the time it takes to reach steady state. But if the process measurement is *noisy*, that is, if it contains high-frequency, random fluctuations, then the derivative of the measured variable will change wildly, and derivative action will amplify the noise unless the measurement is *filtered*, as discussed in Chapter 13. Consequently, derivative action is seldom used for flow control, because flow control loops respond quickly and flow measurements tend to be noisy.

Unfortunately, the ideal proportional-derivative control algorithm in Eq. 7-11 is *physically unrealizable* because it cannot be implemented exactly using either analog or digital controllers. For analog controllers, the transfer function in Eq. 7-11 can be approximated in several ways. For example, by

$$\frac{P'(s)}{E(s)} = K_c\left(1 + \frac{\tau_D s}{\alpha \tau_D s + 1}\right) \tag{7-12}$$

where the constant α typically has a value between 0.05 and 0.2, with 0.1 being a common choice. In Eq. 7-12 the denominator term serves as a *derivative mode filter* (or a *derivative filter*) that reduces the sensitivity of the control calculations to noisy measurements. Derivative filters are used in virtually all commercial PD and PID controllers.

7.2.4 Proportional-Integral-Derivative Control

Now we consider the combination of the proportional, integral, and derivative control modes as a PID controller. PI and PID control have been the dominant control techniques for process control for many decades. For example, a survey indicated that large-scale continuous processes typically have between 500 and 5,000 feedback controllers for individual process variables such as flow rate and liquid level (Desborough and Miller, 2002). Of these controllers, 97% utilize some form of PID control.

Many variations of PID control are used in practice; next, we consider the three most common forms.

Parallel Form of PID Control

The *parallel form* of the PID control algorithm (without a derivative filter) is given by

$$p(t) = \bar{p} + K_c\left[e(t) + \frac{1}{\tau_I}\int_0^t e(t^*)dt^* + \tau_D \frac{de(t)}{dt}\right] \tag{7-13}$$

The corresponding transfer function is

$$\frac{P'(s)}{E(s)} = K_c\left[1 + \frac{1}{\tau_I s} + \tau_D s\right] \tag{7-14}$$

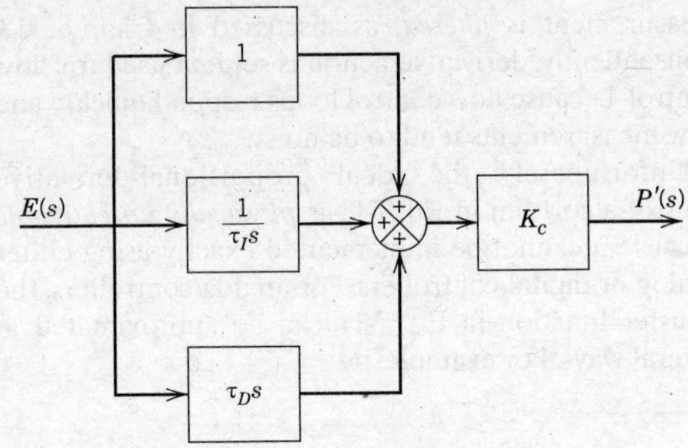

Figure 7.11 Block diagram of the parallel form of PID control (without a derivative filter).

Figure 7.11 illustrates that this controller can be viewed as three separate elements operating in parallel on $E(s)$. The parallel-form PID controller, with and without a derivative filter, is shown in Table 7.1.

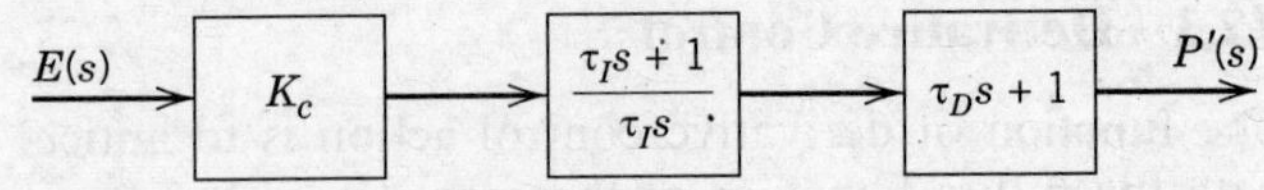

Figure 7.12 Block diagram of the series form of PID control (without a derivative filter).

Series Form of PID Control

Historically, it was convenient to construct early analog controllers (both electronic and pneumatic) so that a PI element and a PD element operated in series. The *series form* of PID control without a derivative filter is shown in Fig. 7.12. In principle, it makes no difference whether the PD element or the PI element comes first. Commercial versions of the series-form controller have a derivative filter that is applied to either the derivative term, as in Eq. 7-12, or to the PD term, as in Eq. 7-15:

$$\frac{P'(s)}{E(s)} = K_c\left(\frac{\tau_I s+1}{\tau_I s}\right)\left(\frac{\tau_D s+1}{\alpha\tau_D s+1}\right) \tag{7-15}$$

Table 7.1 Common PID Controllers

Controller Type	Other Names Used	Controller Equation	Transfer Function
Parallel	Ideal, additive, ISA form	$p(t) = \bar{p} + K_c\left(e(t) + \frac{1}{\tau_I}\int_0^t e(t^*)dt^* + \tau_D\frac{de(t)}{dt}\right)$	$\frac{P'(s)}{E(s)} = K_c\left(1 + \frac{1}{\tau_I s} + \tau_D s\right)$
Parallel with derivative filter	Ideal, realizable, ISA Standard	*See Exercise 7.10*	$\frac{P'(s)}{E(s)} = K_c\left(1 + \frac{1}{\tau_I s} + \frac{\tau_D s}{\alpha\tau_D s+1}\right)$
Series	Multiplicative, interacting	$p(t) = \bar{p} + \frac{K_c}{\tau_I}\left[(\tau_I + \tau_D)e(t) + \int_0^t e(t^*)dt^* + \tau_D\frac{de(t)}{dt}\right]$	$\frac{P'(s)}{E(s)} = K_c\left(\frac{\tau_I s+1}{\tau_I s}\right)(\tau_D s+1)$
Series with derivative filter	Physically realizable	*See Exercise 7.10(b)*	$\frac{P'(s)}{E(s)} = K_c\left(\frac{\tau_I s+1}{\tau_I s}\right)\left(\frac{\tau_D s+1}{\alpha\tau_D s+1}\right)$
Expanded	Noninteracting	$p(t) = \bar{p} + K_c e(t) + K_I\int_0^t e(t^*)dt^* + K_D\frac{de(t)}{dt}$	$\frac{P'(s)}{E(s)} = K_c + \frac{K_I}{s} + K_D s$
Parallel, with proportional and derivative weighting	Ideal β, γ controller	$p(t) = \bar{p} + K_c\left(e_P(t) + \frac{1}{\tau_I}\int_0^t e(t^*)dt^* + \tau_D\frac{de_D(t)}{dt}\right)$ where $e_P(t) = \beta y_{sp}(t) - y_m(t)$, $e(t) = y_{sp}(t) - y_m(t)$, $e_D(t) = \gamma y_{sp}(t) - y_m(t)$	$P'(s) = K_c\left(E_P(s) + \frac{1}{\tau_I s}E(s) + \tau_D s E_D(s)\right)$ where $E_P(s) = \beta Y_{sp}(s) - Y_m(s)$, $E(s) = Y_{sp}(s) - Y_m(s)$, $E_D(s) = \gamma Y_{sp}(s) - Y_m(s)$

Expanded Form of PID Control

The *expanded form* of PID control is:

$$p(t) = \bar{p} + K_c e(t) + K_I \int_0^t e(t^*)dt^* + K_D \frac{de(t)}{dt} \qquad (7\text{-}16)$$

Note that the controller parameters for the expanded form are three "gains," K_c, K_I, and K_D, rather than the standard parameters, K_c, τ_I, and τ_D. The expanded form of PID control is used in MATLAB. This form might appear to be well suited for controller tuning, because each gain independently influences only one control mode. But the well-established controller tuning relations presented in Chapters 10 and 11 were developed for the series and parallel forms. Thus, there is little advantage in using the expanded form in Eq. 7-16, except for simulation.

7.3 FEATURES OF PID CONTROLLERS

Next, we consider common extensions of the basic PID controllers that greatly enhance their performance.

7.3.1 Elimination of Derivative Kick

One disadvantage of the previous PID controllers is that a sudden change in set point (and hence the error, e) will cause the derivative term momentarily to become very large and thus provide a *derivative kick* to the final control element. This sudden "spike" is undesirable and can be avoided by basing the derivative action on the measurement, y_m, rather than on the error signal, e. To illustrate the elimination of derivative kick, consider the parallel form of PID control in Eq. 7-13. Replacing de/dt by $-dy_m/dt$ gives

$$p(t) = \bar{p} + K_c\left[e(t) + \frac{1}{\tau_I}\int_0^t e(t^*)dt^* - \tau_D \frac{dy_m(t)}{dt}\right] \qquad (7\text{-}17)$$

This method of eliminating derivative kick is a standard feature in most commercial controllers. For a series-form PID controller, it can be implemented quite easily by placing the PD element in the feedback path, as shown in Fig. 7.13. Note that the elimination of derivative kick for set-point changes does not affect the controller performance when y_{sp} is constant. Thus, Eqs. 7-13 and 7-17 provide identical responses to process disturbances when the set point is constant.

7.3.2 Set-point Weighting

A more flexible PID control algorithm can be obtained by weighting the set point in the proportional and

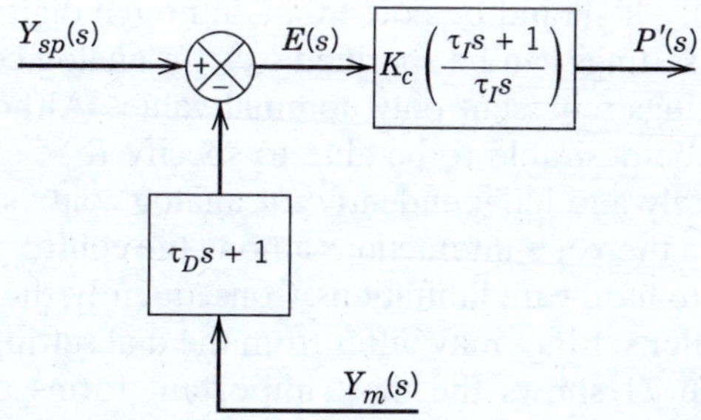

Figure 7.13 Block diagram of the series form of PID control that eliminates derivative kick.

derivative terms. These modifications allow adjustment of the control system response to set-point changes independently of the response to disturbances. The modified PID control algorithm with set-pointing weighting has the form

$$p(t) = \bar{p} + K_c\left(e_P(t) + \frac{1}{\tau_I}\int_0^t e(t^*)dt^* + \tau_D \frac{de_D(t)}{dt}\right) \qquad (7\text{-}18)$$

with different error terms defined for each mode:

$$e_P(t) \triangleq \beta y_{sp}(t) - y_m(t) \qquad (7\text{-}19)$$

$$e(t) \triangleq y_{sp}(t) - y_m(t) \qquad (7\text{-}20)$$

$$e_D(t) \triangleq \gamma y_{sp}(t) - y_m(t) \qquad (7\text{-}21)$$

This control algorithm is known as the parallel PID controller with proportional and derivative mode weighting, or the *beta-gamma controller*. In general, the values of β and γ are in the interval, [0, 1]. Often, γ is selected to be zero.

The modified PID control algorithm reduces to the standard algorithm in Eq. 7-13 when $\beta = \gamma = 1$. Setting $\gamma = 0$ eliminates derivative kick. After a step change in set point, the standard algorithm results in a *proportional kick,* a sudden change in controller output p due to the sudden increase in e which results in a corresponding change in the proportional term. The amount of proportional kick after a sudden set-point change can be reduced by setting $\beta < 1$.

The β weighting parameter can be used to tune this PID controller performance for set-point changes, as discussed in Chapter 11. Note that the definition of the integral mode error in Eq. 7-20 is the same as for the standard control law in Eq. 7-13; this error term is essential in order to eliminate offset after a set-point change or sustained disturbance.

Finally, it should be noted that, although digital controller settings can be specified exactly, analog controller settings represent only nominal values. Although it would be desirable to be able to specify K_c, τ_I, and τ_D accurately and independently for analog controllers, in practice there are interactions among the control modes owing to hardware limitations. Consequently, the actual controller settings may differ from the dial settings.

Table 7.1 shows the most important forms of PID controllers, controller equations, and transfer functions. The derivation of several controller equation forms is left as an exercise for the reader. The table is organized by the descriptive names used in this book, but common synonyms are also included. However, all these terms should be used with caution as a result of the inconsistent terminology that occurs in the literature. For example, referring to the parallel form (the first line of Table 7.1) as an "ideal controller" is misleading, because its derivative mode amplifies noise, an undesirable characteristic. In addition, the terms *interacting* and *noninteracting* can be quite confusing, because a controller's modes can be noninteracting in the time domain (controller equation) but interacting in the Laplace domain (transfer function) and vice versa. Some of these idiosyncrasies are evident from the exercises and from the frequency response analysis of Chapter 10.

Table 7.2 summarizes important characteristics of representative commercial PID controllers.

7.3.3 Reverse or Direct Action

The controller gain can be either negative or positive.[2] For proportional control, when $K_c > 0$, the controller output $p(t)$ increases as its input signal $y_m(t)$ decreases, as is apparent after combining Eqs. 7-2 and 7-1:

$$p(t) - \overline{p} = K_c[y_{sp}(t) - y_m(t)] \tag{7-22}$$

Thus if $K_c > 0$, the controller is called a *reverse-acting* controller. When $K_c < 0$, the controller is said to be *direct acting,* because p increases as y_m increases. Note that these definitions are based on the measurement, $y_m(t)$, rather than the error, $e(t)$. Direct-acting and reverse-acting proportional controllers are compared in Fig. 7.14.

To illustrate why both direct-acting and reverse-acting controllers are needed, again consider the flow control loop in Fig. 7.2. Suppose that the flow transmitter is designed to be direct-acting so that its output signal increases as the flow rate increases. Most transmitters are designed to be direct-acting. Also assume that the control valve is designed so that the flow rate through the valve increases as the signal to the valve, $p(t)$, increases. In this case the valve is designated as *air-to-open* (or *fail close*). The question is: should the flow controller have direct or reverse action? Clearly, when the measured flow rate is higher than the set point, we want to reduce the flow by closing the control

Table 7.2 Key Characteristics of Commercial PID Controllers

Controller Feature	Controller Parameter	Symbol	Units	Typical Range*
Proportional mode	*Controller gain*	K_c	Dimensionless [%/%, mA/mA]	0.1–100
	Proportional band	$PB = 100\%/K_c$	%	1–1000%
Integral mode	*Integral time (or reset time)*	τ_I	Time [min, s]	0.02–20 min 1–1000 s
	Reset rate	$1/\tau_I$	Repeats/time [min^{-1}, s^{-1}]	0.001–1 repeats/s 0.06–60 repeats/min
	Integral mode "gain"	K_I	Time^{-1} [min^{-1}, s^{-1}]	0.1–100
Derivative mode	*Derivative time*	τ_D	Time [min, s]	0.1–10 min 5–500 s
	Derivative mode "gain"	K_D	Time [min, s]	0.1–100
	Derivative filter parameter	α	Dimensionless	0.05–0.2
Control interval (Digital controllers)		Δt	Time [s, min]	0.1 s–10 min

*Based on McMillan (2015).

[2] For some computer control software, K_c must be positive. The user enters the designation of reverse or direct action as a separate binary parameter.

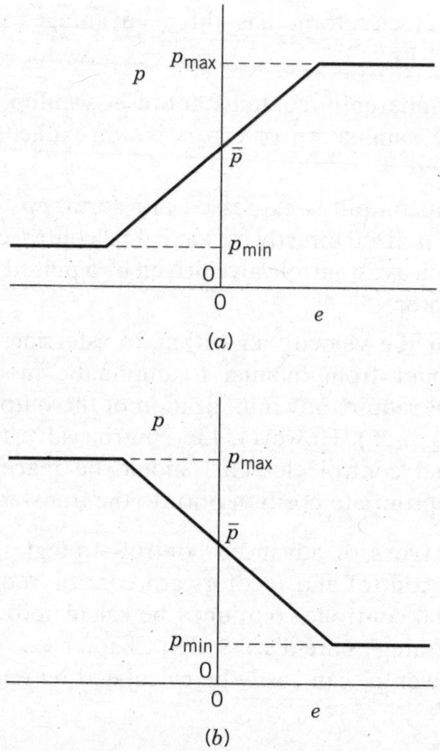

Figure 7.14 Reverse and direct-acting proportional controllers: (*a*) reverse acting ($K_c > 0$), (*b*) direct acting ($K_c < 0$).

valve. For an air-to-open valve, the controller output signal should be decreased. Thus, the controller should be *reverse-acting*.

But what if the control valve is air-to-close (or *fail open*) rather than air-to-open? Now when the flow rate is too high, the controller output should increase to further close the valve. Here, a *direct-acting* controller is required.

It is extremely important that the controller action be specified correctly, because an incorrect choice usually results in loss of control. For the flow control example, having the wrong controller action would force the control valve to stay fully open or fully closed (why?). Thus, the controller action must be carefully specified when a controller is installed or when a troublesome control loop is being analyzed. The following guideline is very useful and can be justified by the stability analysis techniques of Chapter 9

General Guideline for Specifying the Controller Action (Direct or Reverse): *The overall product of the gains for all of the components in the feedback control loop must be positive.*

For example, the blending control system in Fig. 7.1 has five components in the feedback control loop: the process, the sensor, the controller, the I/P transducer, and the control valve.

7.3.4 Automatic/Manual Control Modes

Equations 7-2 to 7-21 describe how controllers perform during the *automatic mode* of operation. However, in certain situations, the plant operator may decide to override the automatic mode and adjust the controller output manually.

This *manual mode* of controller operation is very useful during a plant start-up, shutdown, or emergency situation. A manual/automatic switch, or the software equivalent, is used to transfer the controller from the automatic mode to the manual mode, and vice versa. During these transfers, it is important that the controller output not change abruptly and "bump" the process. Consequently, most controllers facilitate *bumpless transfers*.

A controller may be left in manual for long periods of time (or indefinitely) if the operator is not satisfied with its performance in the automatic mode. Consequently, if a significant percentage of the controllers in a plant is in manual, it is an indication that the control systems are not performing well or that the plant operators do not have much confidence in them. The topic of troubleshooting poorly performing control loops is considered in Chapter 11.

7.4 DIGITAL VERSIONS OF PID CONTROLLERS

So far we have assumed that the input and output signals of the controller are continuous functions of time. However, digital control systems are much more common due to their flexibility, computational power, and cost effectiveness. In this section we briefly introduce digital control techniques by considering digital versions of PID control. A more complete discussion of digital computer control is presented in Chapter 13 and Chapter 19.

In recent years there has been increasing interest in using wireless technology in manufacturing operations. In particular, it is feasible to implement monitoring and control strategies wirelessly (Blevins et al., 2015).

When a feedback control strategy is implemented digitally, the controller input and output are digital (or discrete-time) signals rather than continuous (or analog) signals. Thus, the continuous signal from the measurement device (sensor/transmitter) is sampled and converted to a digital signal by an *analog-to-digital converter* (ADC). A digital control algorithm is then used to calculate the controller output, also a digital signal. Because most final control elements are analog devices, the digital output signal is usually converted to a corresponding analog signal by a *digital-to-analog converter* (DAC). However, some electronic final control elements can receive digital signals directly, as discussed in Chapter 8.

7.4.1 Position and Velocity PID Algorithms

There are two alternative forms of the digital PID control equation, the *position form* and the *velocity form*. A straightforward way of deriving a digital version of the parallel form of the PID controller (Eq. 7-13) is to replace the integral and derivative terms by finite difference approximations,

$$\int_0^t e(t^*)dt^* \approx \sum_{j=1}^{k} e_j \Delta t \tag{7-23}$$

$$\frac{de}{dt} \approx \frac{e_k - e_{k-1}}{\Delta t} \tag{7-24}$$

where

Δt = the sampling period (the time between successive measurements of the controlled variable)

e_k = error at the kth sampling instant for $k = 1, 2, \ldots$

Substituting Eqs. 7-23 and 7-24 into Eq. 7-13 gives the *position form*,

$$p_k = \bar{p} + K_c\left[e_k + \frac{\Delta t}{\tau_I}\sum_{j=1}^{k} e_j + \frac{\tau_D}{\Delta t}(e_k - e_{k-1})\right] \tag{7-25}$$

where p_k is the controller output at the kth sampling instant. The other symbols in Eq. 7-25 have the same meaning as in Eq. 7-13. Equation 7-25 is referred to as the position form, because the actual value of the controller output is calculated.

In the *velocity form*, the change in controller output is calculated. The velocity form can be derived by writing Eq. 7-25 for the $(k - 1)$ sampling instant:

$$p_{k-1} = \bar{p} + K_c\left[e_{k-1} + \frac{\Delta t}{\tau_I}\sum_{j=1}^{k-1} e_j + \frac{\tau_D}{\Delta t}(e_{k-1} - e_{k-2})\right] \tag{7-26}$$

Note that the summation still begins at $j = 1$, because it is assumed that the process is at the desired steady state for $j \le 0$, and thus $e_j = 0$ for $j \le 0$. Subtracting Eq. 7-26 from Eq. 7-25 gives the velocity form of the digital PID algorithm:

$$\Delta p_k = p_k - p_{k-1} = K_c\left[(e_k - e_{k-1}) + \frac{\Delta t}{\tau_I}e_k + \frac{\tau_D}{\Delta t}(e_k - 2e_{k-1} + e_{k-2})\right] \tag{7-27}$$

The velocity form has three advantages over the position form:

1. It inherently contains antireset windup, because the summation of errors is not explicitly calculated.
2. This output is expressed in a form, Δp_k, that can be utilized directly by some final control elements, such as a control valve driven by a pulsed stepping motor.
3. For the velocity algorithm, transferring the controller from manual to automatic model does not require any initialization of the output ($\bar{p}$ in Eq. 7-25). However, the control valve (or other final control element) should be placed in the appropriate position prior to the transfer.

Certain types of advanced control strategies, such as cascade control and feedforward control, require that the actual controller output p_k be calculated explicitly. These strategies are discussed in Chapter 12.

However, p_k can easily be calculated by rearranging Eq. 7-27:

$$p_k = p_{k-1} + K_c\left[(e_k - e_{k-1}) + \frac{\Delta t}{\tau_I}e_k + \frac{\tau_D}{\Delta t}(e_k - 2e_{k-1} + e_{k-2})\right] \tag{7-28}$$

A minor disadvantage of the velocity form is that the integral mode *must* be included. When the set point is constant, it cancels out in both the proportional and derivative error terms. Consequently, if the integral mode were omitted, the process response to a disturbance would tend to drift away from the set point.

The position form of the PID algorithm (Eq. 7-25) requires a value of $\bar{p}$ while the velocity form in Eq. 7-27 does not. Initialization of either algorithm is straightforward, because manual operation of the control system usually precedes the transfer to automatic control. Hence, $\bar{p}$ (or p_{k-1} for the velocity algorithm) is simply set equal to the signal to the final control element at the time of transfer. As noted previously, the velocity form is less prone to reset windup problems.

7.4.2 Modifications of the Basic PID Algorithm

We now consider several modifications of the basic PID control algorithms that are widely used in industry.

1. ***Elimination of Reset Windup.*** For digital controllers that contain integral control action, *reset windup* can occur when the error summation grows to a very large value. Suppose the controller

output saturates at an upper or lower limit, as the result of a large sustained error signal. Even though the measured variable eventually reaches its set point (where $e_k = 0$), the controller may be wound up because of the summation term. Until the error changes sign for a period of time, thereby reducing the value of the summation, the controller will remain at its saturation limit.

For the position algorithm, anti-reset windup can be achieved, using the methods of Section 7.2.2.

(a) When the controller saturates, suspend the summation.

(b) Implement the integral control action as a positive feedback loop, as shown in Fig. 7.8.

Experience and Example 7.1 indicate that approach (b) is superior to (a), although it is somewhat more complicated.

For the velocity form in Eq. 7-27, no summation appears, and thus the reset windup problem is avoided. However, the control algorithm must be implemented so that Δp_k is disregarded if p_k violates a saturation limit, implying that p_k should be monitored at all times. In general, the velocity form is preferred over the position form.

2. ***Elimination of Derivative Kick.*** When a sudden set-point change is made, the PID control algorithms in Eq. 7-25 or Eq. 7-27 will produce a large immediate change in the output due to the derivative control action. For digital control algorithms, several methods are available for eliminating derivative kick:

   (a) In analogy with Eq. 7-17, derivative action can be applied to the measurement, $y_{m,k}$, rather than the error signal. Thus, for the position form in Eq. 7-25, e_k is replaced by $-y_{m,k}$ in the derivative term:

$$p_k = \bar{p} + K_c\left[e_k + \frac{\Delta t}{\tau_I}\sum_{k}^{j=1} e_j - \frac{\tau_D}{\Delta t}(y_{m,k} - y_{m,k-1})\right] \tag{7-29}$$

   The velocity form in Eq. 7-27 can be modified in an analogous fashion.

   (b) Change the set point gradually by ramping it to the new value. This strategy limits the rate of change of the set point and thus reduces the derivative kick.

If measurement noise combined with a large derivative time to sampling period ratio ($\tau_D/\Delta t$) causes an overactive derivative mode, then the error signal must be filtered before calculating the derivative action (see Chapter 13).

3. ***Effect of Saturation on Controller Performance.*** Another difficulty that can occur for a digital controller equation such as Eq. 7-29 is that a small change in the error can cause the controller output to saturate for certain values of the controller settings. Suppose that $K_c\tau_D/\Delta t = 100$ due to a small sampling period, and that e_k and p_k are both scaled from 0 to 100%. A 1% change in $\Delta e_k = e_k - e_{k-1}$ will cause a 100% change in p_k, thus exceeding its upper limit. Therefore, the values of the controller settings and Δt should be checked to ensure that they do not cause such *overrange* problems. For the velocity algorithm, the change in the controller output can be constrained by using *rate limits* or *clamps*, that is, lower and upper bounds on the change, Δp_k.

4. ***Other Optional Features.*** For some control applications, it is desirable that the controller output signal not be changed when the error is small, within a specified tolerance. This optional feature is referred to as *gap action*. Finally, in *gain scheduling*, the numerical value of K_c depends on the value of the error signal. These controller options are discussed in more detail in Chapter 12.

For a more detailed discussion of digital control algorithms, see Chapter 13.

7.5 TYPICAL RESPONSES OF FEEDBACK CONTROL SYSTEMS

In this section, simulation results are used to demonstrate the typical behavior of a controlled process after step change in a disturbance variable occurs. The dynamic behavior of the simulated process is described by a first-order-plus-time-delay (FOPTD) model between a controlled variable y and a manipulated variable u,

$$\frac{Y(s)}{U(s)} = \frac{Ke^{-\theta s}}{\tau s + 1} \tag{7-30}$$

where $K = 0.5$, $\theta = 4$, and $\tau = 20$. Both u and y are deviation variables and the set point is $y_{sp} = 0$. The closed-loop responses to a step disturbance for P, PI and PID control are compared to the response with no feedback control in Fig. 7.15. The controller settings were calculated using the IMC tuning relations that will be introduced in Chapter 11.

In Fig. 7.15 when feedback control is not used, y slowly reaches a new steady state and has a large offset. Proportional control speeds up the response and reduces

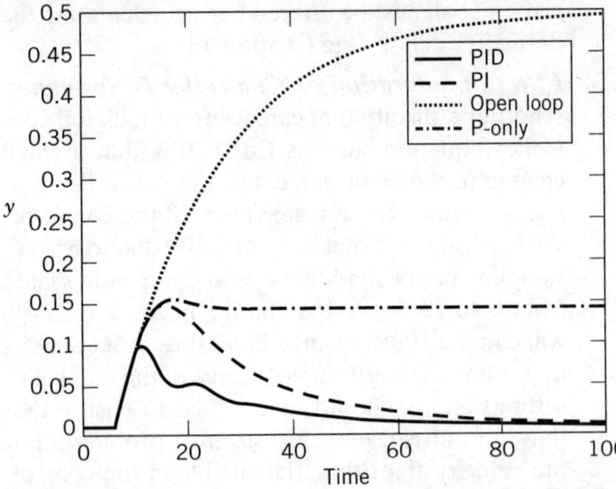

Figure 7.15 Closed-loop responses for a unit step disturbance and a FOPTD model.

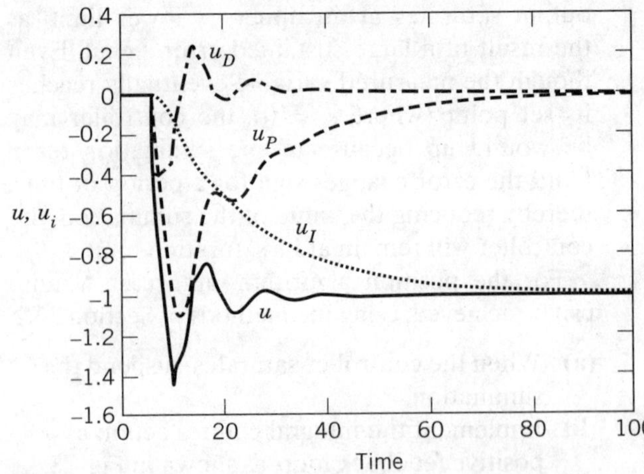

Figure 7.16 Manipulated variable u and its individual components for the PID control system of Fig. 7.15.

the offset, while the addition of integral control action completely eliminates offset. Adding derivative action reduces both the "peak error" (the maximum value of y) and the response time, while producing a small amount of oscillation. Note that PID controllers do *not* always result in oscillations. The closed-loop responses depend on the controller settings (K_c, τ_I, and τ_D) and of course, the process dynamics. Similarly, P and PI controllers do *not* necessarily result in overdamped responses, for the same reasons. But the responses in Fig. 7.15 are representative of what occurs in practice. The responses in Fig. 7.15 were generated using the FOPTD model in Eq. 7-30.

In evaluating control system performance, it is informative to consider the MV response, as well as the CV response. Figure 7.16 shows the MV response $u(t)$ for the PID control simulation of Fig. 7.15, as well as the individual contributions of the P, I, and D modes. By definition

$$u(t) = u_P(t) + u_I(t) + u_D(t) \tag{7-31}$$

where the right side of Eq. 7-31 corresponds to the last three terms on the right side of Eq. 7-17. The proportional contribution $u_P(t)$ tracks the controlled variable $y(t)$ in Fig. 7.15 but in the opposite direction (i.e., negative instead of positive). The integral contribution $u_I(t)$ monotonically decreases to the steady-state value of −1. The contribution term $u_D(t)$ is large when $y(t)$ is changing rapidly and then approaches zero as steady state is reached, as would be expected.

The effects of changing individual PID controller settings are shown in Figs. 7.17 to 7.19 for a unit step disturbance at $t = 2$. The "nominal values" are the same PID settings that were used in Fig. 7.15; they are compared with changes of ±50% from the nominal value.

As controller gain K_c increases in Fig. 7.17, $y(t)$ responds faster and becomes more oscillatory and almost unstable for the +50% K_c increase. Decreasing the integral time τ_I in Fig. 7.18 makes the response less sluggish. However, if τ_I is too small, oscillatory responses can result (not shown). Theoretically, offset will be eliminated for all positive values of τ_I. But for very large values of τ_I the controlled variable will return to the set point very slowly, as demonstrated by the −50% change. The trends for PID control in Figs. 7.17

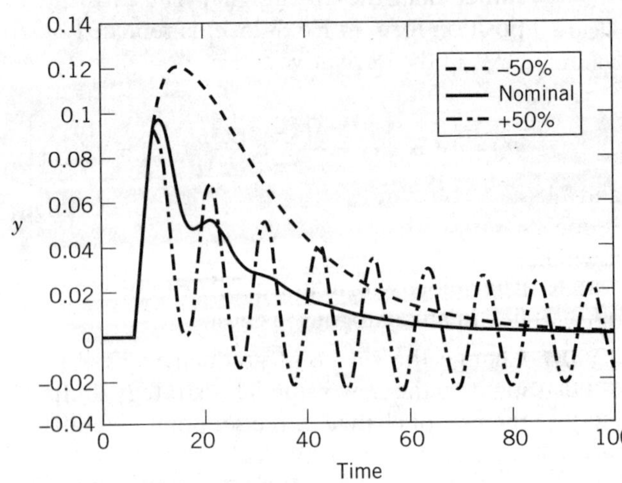

Figure 7.17 The effects of different K_c values for the PID control system of Fig. 7.15.

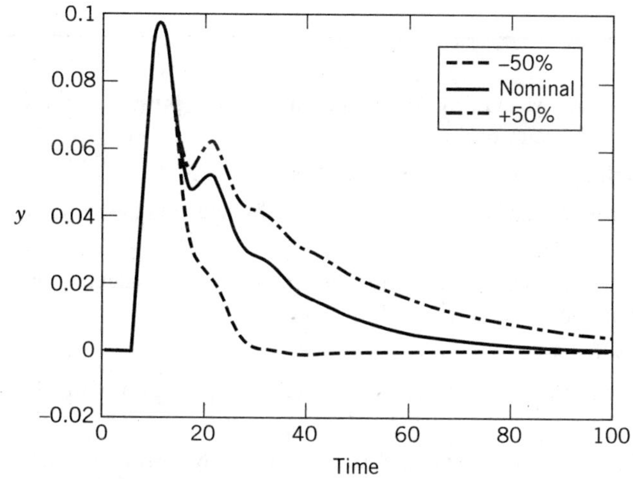

Figure 7.18 The effects of different τ_I values for the PID control system of Fig. 7.15.

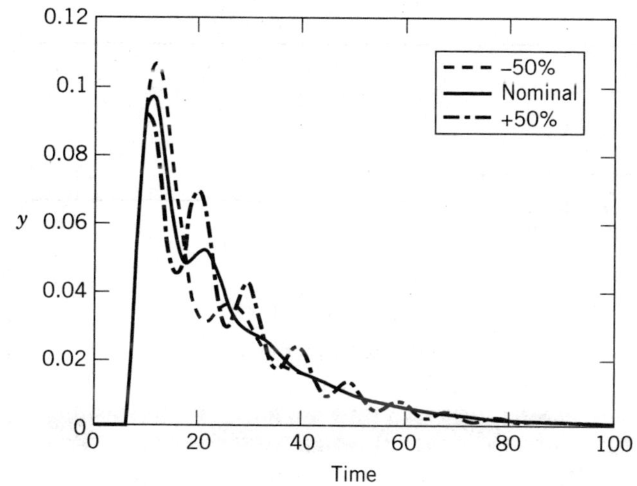

Figure 7.19 The effects of different τ_D values for the PID control system of Fig. 7.15.

and 7.18 also commonly occur for PI control. But there is one exception: the integrating process of Section 5.5; it will be considered again in Section 9.3.4.

It is difficult to generalize about the effect of the derivative time τ_D. For small values of τ_D, increasing it tends to improve the response by reducing the maximum deviation, response time, and degree of oscillation, as shown in Fig. 7.15. However, if τ_D is too large, measurement noise is amplified and $y(t)$ may become oscillatory. Thus, an intermediate value of τ_D is required.

More information on how PID controller settings should be specified are presented in Chapters 9, 10, and 11.

7.6 ON–OFF CONTROLLERS

On–off controllers are simple, inexpensive feedback controllers that are commonly used as thermostats in home heating systems and domestic refrigerators. They are also used in noncritical industrial applications such as some level control loops and heating systems. However, on–off controllers are less widely used than PID controllers, because they are not as versatile or as effective.

For ideal on–off control, the controller output has only two possible values:

$$p(t) = \begin{cases} p_{max} & \text{if } e \geq 0 \\ p_{min} & \text{if } e < 0 \end{cases} \tag{7-32}$$

where p_{max} and p_{min} denote the on and off values, respectively (for example, for a typical digital computer implementation, p_{max} = 100% and p_{min} = 0%; for a current-based electronic controller, p_{max} = 20 mA and p_{min} = 4 mA). On–off controllers can be modified to include a dead band for the error signal to reduce sensitivity to measurement noise (Shinskey, 1996). Equation 7-32 also indicates why on–off control is sometimes referred to as *two-position* or *bang-bang* control. Note that on–off control can be considered a special case of proportional control with a very high controller gain (see Fig. 7.4).

The disadvantages of on–off control are that it results in continual cycling of the controlled variable and produces excessive wear on the control valve (or other final control element). The latter disadvantage is significant if a control valve is used, but less of a factor for solenoid valves or solenoid switches that are normally employed with on–off controllers.

7.7 SIMULINK MODEL FOR A FEEDBACK CONTROL SYSTEM

SIMULINK is a GUI based simulation software comes along with MATLAB. The details about SIMULINK is given in section B.2 of Appendix B. The steps involved in making a Simulink model are given in Table 7.3.

Table 7.3 Steps for Development of Simulink Model

1. Open MATLAB.
2. Click SIMULINK button in MATLAB or write "***simulink***" in the MATLAB command window and press enter key. Simulink Library Function window will open.
3. Create new Simulink model workspace (.***mdl*** or .***slx***) file from the menu of the Library function window.
4. Drag and drop required icons from library to the workspace. The icons can be 'flipped' or 'rotated' by selecting the icon and going to the format "pull-down" menu and selecting "***flip block***" or "***rotate block***".
5. Connect the icons (output of one icon to the input of the other) as per the block diagram. This can be done by selecting the source icon and pressing ctrl key and then left-clicking the destination icon.
6. Change the labels below each icon by clicking and entering new names.
7. Change the ***Transfer function*** blocks by double-clicking and entering the coefficients of descending powers of *s* for the numerator and denominator polynomials.
8. Specify appropriate variable names to ***simout*** blocks. The default format is "structure". Change to "array" format for data storage in an appropriate form for plotting.
9. To have a visual display connect the variables to *Scope* block through ***Mux*** block. The number of inputs can be increased from two to any number by double clicking the ***Mux*** block.
10. Start the simulation by clicking the ***Run*** button.
11. Observe the responses by double-clicking the ***Scope*** icon or by giving the necessary plot command in the MATLAB command window.

The development of SIMULINK model is explained in Example 7.2.

EXAMPLE 7.2

Generate a PID controller block in SIMULINK.

Solution

The parallel form of PID controller is given in Eq. 7.13 and its transfer function is shown in Eq. 7.14. The corresponding SIMULINK model is made using the library functions *Gain, Constant, Sum/Add, Divide, Integrator, Derivative,* and is shown in Fig. 7.20. The complete set is further made into one icon (PID Controller Block) using *Subsystem* and its masking. On double clicking the icon, we can see the dialog box (Fig. 7.21) where we can enter the values for the parameters, K_c, τ_I and τ_D.

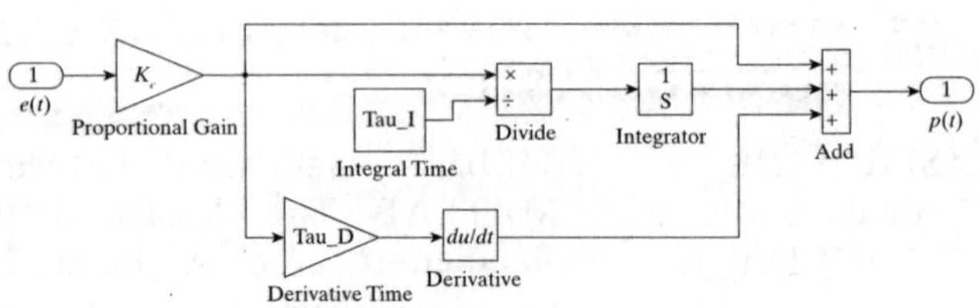

Figure 7.20 SIMULINK Model for a PID controller.

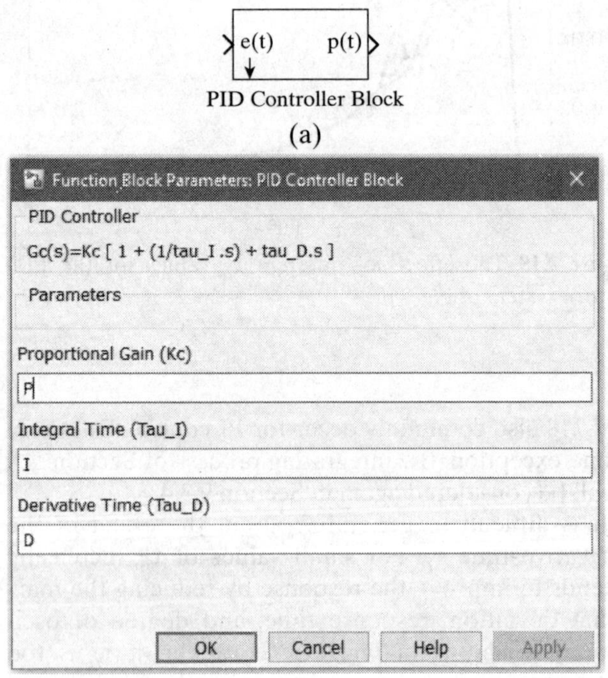

Figure 7.21 SIMULINK model: 1(*a*) PID controller block; (*b*) dialog box.

SUMMARY

In this chapter we have considered the most common types of feedback controllers. Although there are potentially many forms of feedback control, the process industries rely largely on variations of PID control and on–off control. The remaining important elements within the control loop—sensors, transmitters, and final control elements—are discussed in detail in the next chapter. Once the steady-state and dynamic characteristics of these elements are understood, we can investigate the dynamic characteristics of the controlled process (Chapter 9).

REFERENCES

Åström, K. J., and T. Hägglund, *Advanced PID Control*, ISA, Research Triangle Park, NC, 2006.

Blevins, T. L., D. Chen, M. J. Nixon, and W. Wojsznis, *Wireless Control Foundation: Continuous and Discrete Control for the Process Industry*, ISA, Research Triangle Park, NC, 2015.

Blevins, T. L., D. Coyne, W. K. Wojsznis, and M. J. Nixon, *Improving PID Recovery from Limit Conditions, Proc. IFAC Conference on Advances in PID Controllers*, Brescia, Italy, 2012.

Blevins, T. L., and M. J. Nixon, *Control Loop Foundation—Batch and Continuous Processes*, ISA, Research Triangle Park, NC 2011.

Desborough, L., and R. Miller, Increasing Customer Value of Industrial Control Performance Monitoring—Honeywell, Experience, *Proc. 6th Internat. Conf. on Chemical Process Control (CPC VI)*, p. 169, AIChE, NY (2002).

Instrumentation Symbols and Identification, Standard ANSI/ISA-5.1–2009, International Society of Automation (ISA), Research Triangle Park, NC 2009.

Mayr, O., *The Origins of Feedback Control*, MIT Press, Cambridge, MA, 1970.

McMillan, G. K., *Tuning and Control-Loop Performance*, 4th ed., Momentum Press, New York, 2015.

Shinskey, F. G., *Process Control Systems*, 4th ed., McGraw-Hill, New York, 1996.

Ziegler, J. G., Those Magnificent Men and Their Controlling Machines, *J. Dynamic Systems, Measurement and Control, Trans. ASME*, **97**, 279 (1975).

EXERCISES

7.1 An electronic PI temperature controller has an output p of 12 mA when the set point equals the nominal process temperature. The controller response to step change in the temperature set point of 2.5 mA (equivalent to a change of 3°C) is shown below:

t, s	p, mA
0–	8.00
0+	7.00
20	6.35
60	5.00
80	4.35
90	4.00

Determine the controller gain K_c (mA/mA) and the integral time, τ_I. Is the controller reverse-acting or direct-acting?

7.2 A physically realizable form of the ideal PD controller transfer function in Eq. 7-11 is given by

$$\frac{P'(s)}{E(s)} = \frac{K_c(\tau_D s+1)}{\alpha\tau_D s+1}$$

where $0.05 < \alpha < 0.2$.

(a) Show how to obtain this transfer function with a parallel arrangement of two much simpler functions in Fig. E7.2.

(b) Find expressions for K_1, K_2, and τ_1 that can be used to obtain desired values of K_c, τ_D, and α.

(c) Verify the relations for $K_c = 4$, $\tau_D = 1$, $\alpha = 0.1$.

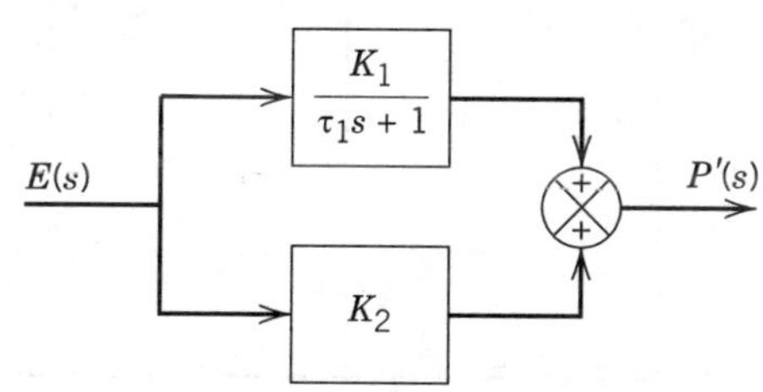

Figure E7.2

7.3 The parallel form of the PID controller has the transfer function given by Eq. 7-14. Many commercial analog controllers can be described by the series form given by Eq. 7-15.

(a) For the simplest case, $\alpha \to 0$, find the relations between the settings for the parallel form ($K_c^\dagger, \tau_I^\dagger, \tau_D^\dagger$) and the settings for the series form (K_c, τ_I, τ_D).

(b) Does the series form make each controller setting (K_c, τ_I, or τ_D) larger or smaller than would be expected for the parallel form?

(c) What are the magnitudes of these interaction effects for $K_c = 4$, $\tau_I = 5$ min, $\tau_D = 1$ min?

(d) What can you say about the effect of nonzero α on these relations? (Discuss only first-order effects.)

7.4 Exercise 1.9 shows two possible ways to design a feedback control loop to obtain a desired rate of liquid flow. Assume that in both Systems I and II, the flow transmitter is direct-acting (i.e., the output increases as the flow rate increases). The control valve in System I is "air-to-open," meaning that an increasing pressure signal from the controller will open the valve more, thus increasing the flow rate (See Chapter 8). On the other hand, the control valve in System II is "air-to-close." The dynamics for both of the valves are negligible.

(a) For each of these valves, what is the sign of its gain, K_v?

(b) Which controller must be direct-acting? Reverse-acting? Use physical arguments to support your answers.

(c) What sign should the controller gain have for each case?

7.5 A liquid-level control system can be configured in either of two ways: with a control valve manipulating flow of liquid into the holding tank (Fig. E7.5*a*), or with a control valve manipulating the flow of liquid from the tank (Fig. E7.5*b*). Assuming that the liquid-level transmitter always is direct-acting,

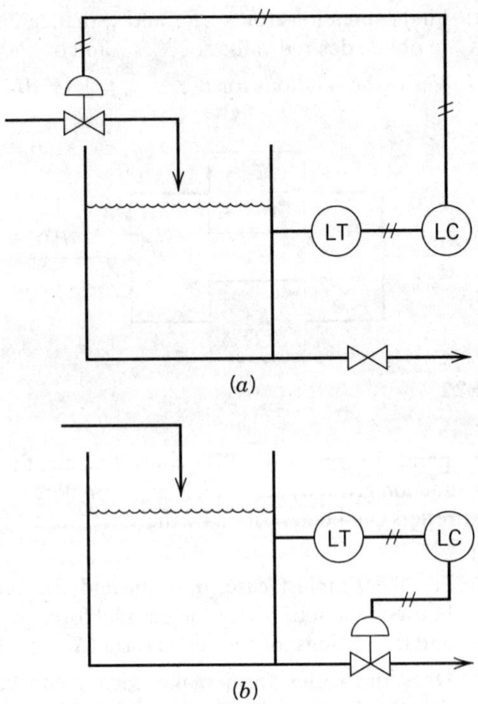

Figure E7.5

(a) For each configuration, what control action (direct or reverse) should a proportional pneumatic controller have if the control valve is air-to-close?

(b) If the control valve is air-to-open?

7.6 If the measured input to a PI controller is a step change ($Y_m(s) = 4/s$) and the controller output changes initially as shown in Fig. E7.6, what are the values of the controller gain and integral time?

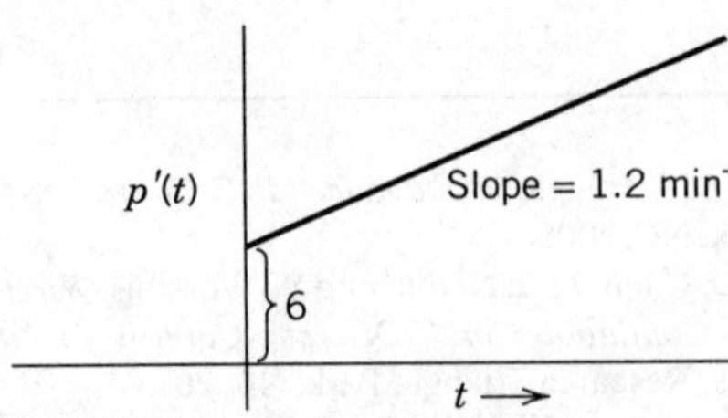

Figure E7.6

7.7 An electronic PID temperature controller is at steady state with an output of 12 mA. The set point equals the nominal process temperature initially. At $t = 0$, the error signal is increased at the rate of 0.6 mA/min (equivalent to a rate of 2°C/min). If the current settings are

$$K_c = 2 \text{ (dimensionless)}$$
$$\tau_I = 1.5 \text{ min}$$
$$\tau_D = 0.5 \text{ min}$$

(a) Derive an expression for the controller output $p(t)$.

(b) Repeat (a) for $\tau_D = 0$ (i.e., a PI controller).

(c) Plot the two controller outputs and qualitatively discuss their differences.

7.8 Find an expression for the amount of derivative kick that will be applied to the process for the position form of the PID digital algorithm (Eq. 7-25) if a set-point change of magnitude Δy_{sp} is made between the $k-1$ and k sampling instants.

(a) Repeat for the proportional kick, that is, the sudden change caused by the proportional mode.

(b) Plot the sequence of controller output values at the $k-1$, k, . . . sampling times for the case of a set-point change of Δy_{sp} magnitude made just after the $k-1$ sampling time if the controller receives a constant measurement $\bar{y}_m$ and the initial set point is $\bar{y}_{sp} = \bar{y}_m$. Assume that the controller output initially is $\bar{p}$.

(c) How can Eq. 7-25 be modified to eliminate derivative kick?

7.9 **(a)** For the digital velocity P and PD algorithms, show how the set point enters into calculation of Δp_k on the assumption that it is not changing, that is, y_{sp} is a constant.

(b) What do the results indicate about use of the velocity form of P and PD digital control algorithms?

(c) Are similar problems encountered if the integral mode is present, that is, with PI and PID forms of the velocity algorithm? Explain.

7.10 What differential equation model represents the parallel PID controller with a derivative filter? (*Hint:* Find a common denominator for the transfer function first.)

(a) Repeat for the series PID controller with a derivative filter.

(b) Simulate the time response of each controller for a unit step change in $e(t)$ and the following parameter values:

$$K_c = 2, \quad \tau_I = 3\,\text{min}, \quad \tau_D = 0.5\,\text{min}, \quad \alpha = 0.1, \quad M = 1$$

7.11 Consider the simulation results for PID control in Figs. 7.15 and 7.16. Note that the u_I response in Fig. 7.16 changes monotonically. Do you think this feature is typical of u_I responses for other PID controllers? Justify your answer.

7.12 Consider a standard feedback control system where each component is functioning properly. Briefly indicate whether you agree or disagree with the following statements:

(a) For proportional-only control, the controller output is always proportional to the error signal.

(b) A PID controller always eliminates offset after a sustained, unmeasured disturbance.

(c) In presence of derivative term, we can use more proportional and more integral actions.

7.13 Consider the liquid storage system in Fig. E7.13. Suppose that q_1 must be kept constant and h_2 is to be controlled either by adjusting q_2 or q_5.
Suppose that the q_2 control valve is fail-close. Should the level controller for h_2 be reverse-acting or direct-acting? Justify your answer.
If q_5 control valve is air-to-close, should the level controller for h_2 be direct-acting or reverse-acting? Justify.

7.14 A steam-heated evaporator used to concentrate a feed stream by evaporating water is shown in Fig. E7.14. The mass fraction of solute in the exit stream x is measured and controlled by adjusting the steam flow rate, S. The control valve is fail-open. Should the composition controller be direct-acting? Justify your answer.

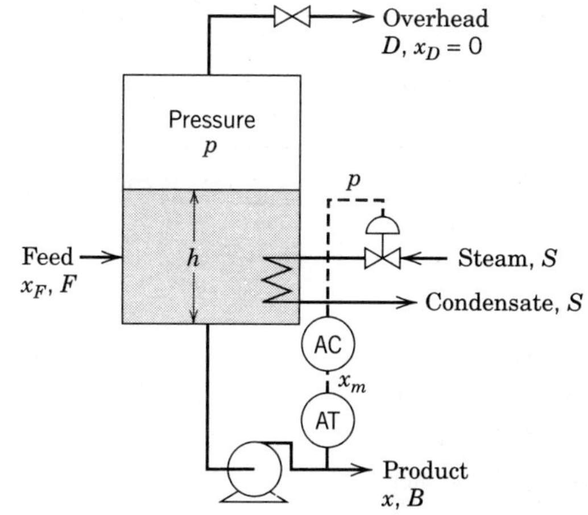

Figure E7.14

7.15 A hot liquid is cooled by cold water in a counter-current heat exchanger: shown in Fig. E7.15:

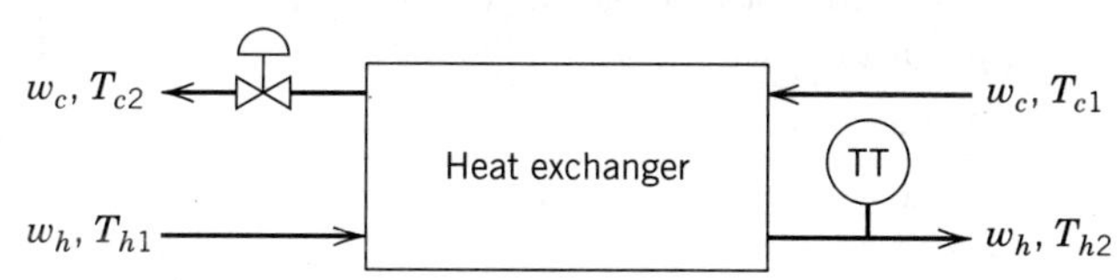

Figure E7.15

Temperature T_{h2} is to be controlled by adjusting flow rate, w_c. The temperature sensor/transmitter (TT) is direct-acting. Based on safety considerations, should the control valve be air-to-open or air-to-close? Should the feedback controller be direct-acting or reverse-acting?

7.16 Consider the schematic diagram of a controlled blending process shown in Fig. 7.1. The control objective is to control the mass fraction of the exit stream, x, by adjusting inlet flow rate, w_2, using a feedback controller. The mass fractions of a key chemical component in the inlet streams, x_1 and x_2, are constant, and mass flow rate w_1 is a disturbance variable. The liquid volume V is constant. The composition sensor/transmitter (AT) and the current-to-pressure transducer (I/P) are both direct-acting devices.

What is the *minimum* amount of information you would need in order to decide whether the feedback controller, AC, should be reverse-acting or direct-acting?

7.17 Using SIMULINK model obtain the response of a PI controller to unit step change in $e(t)$. Compare your result with that shown in Fig. 7.6.

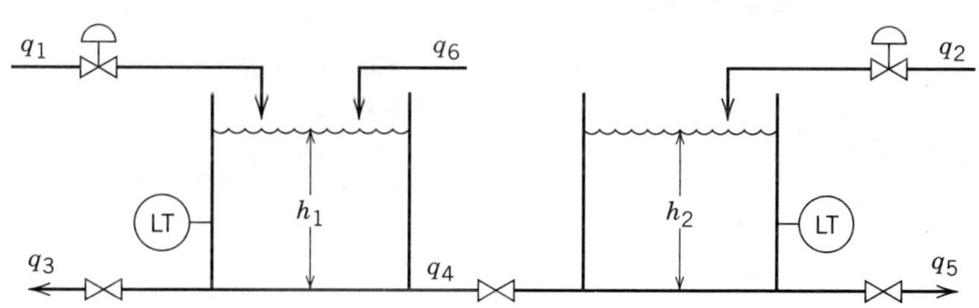

Figure E7.13

MULTIPLE CHOICE QUESTIONS

1. A microprocessor based P-only controller has an input range of 4 to 20 mA and an output range of 1 to 5 Volt. For a change of 5 mA in the input, what is the change in the output if the proportional band is 50%?
   (a) 2.5 V
   (b) 1.0 V
   (c) 2.0 V
   (d) 1.5 V
2. Offset is a characteristic of
   (a) PID controller.
   (b) PI controller.
   (c) P-only controller.
   (d) I-only controller.
3. The gain of a proportional controller is 20. What is the magnitude of disturbance if the controller output undergoes a variation of 10% due to this change?
   (a) 30%
   (b) 50%
   (c) 40%
   (d) 20%
4. The input and output ranges of a P-only controller are 50 to 500°C and 20 to 100 kPa(g), respectively. What is the controller gain (in %) if its output changes from 20 to 60 kPa for an input change of 50°C?
   (a) 0.8
   (b) 1.5
   (c) 4.5
   (d) 5.5
5. The actual or physically realizable PD controller is given by
   (a) $G_c(s) = K_c(1+\tau_D s)$
   (b) $G_c(s) = K_c\left(1+\dfrac{\tau_D s}{\alpha\tau_D s+1}\right)$
   (c) $G_c(s) = K_c\left(1+\dfrac{1}{\alpha\tau_D s+1}\right)$
   (d) $G_c(s) = K_c\left(1-\dfrac{\tau_D s}{\alpha\tau_D s+1}\right)$
6. The parallel-form of PID controller with derivative filter is given by
   (a) $G_c(s) = K_c\left(1+\dfrac{1}{\tau_I s}+\dfrac{\tau_D s}{\alpha\tau_D s+1}\right)$
   (b) $G_c(s) = K_c\left(1+\dfrac{1}{\tau_I s}+\tau_D s\right)$
   (c) $G_c(s) = K_c\left(\dfrac{\tau_I s+1}{\tau_I s}\right)\left(\dfrac{\tau_D s+1}{\alpha\tau_D s+1}\right)$
   (d) $G_c(s) = K_c\left(\dfrac{\tau_I s+1}{\tau_I s}\right)(\tau_D s+1)$
7. The series-form of PID controller with derivative filter is given by
   (a) $G_c(s) = K_c\left(1+\dfrac{1}{\tau_I s}+\dfrac{\tau_D s}{\alpha\tau_D s+1}\right)$
   (b) $G_c(s) = K_c\left(1+\dfrac{1}{\tau_I s}+\tau_D s\right)$
   (c) $G_c(s) = K_c\left(\dfrac{\tau_I s+1}{\tau_I s}\right)\left(\dfrac{\tau_D s+1}{\alpha\tau_D s+1}\right)$
   (d) $G_c(s) = K_c\left(\dfrac{\tau_I s+1}{\tau_I s}\right)(\tau_D s+1)$
8. The series-form of PID controller is given by
   (a) $G_c(s) = K_c\left(1+\dfrac{1}{\tau_I s}+\dfrac{\tau_D s}{\alpha\tau_D s+1}\right)$
   (b) $G_c(s) = K_c\left(1+\dfrac{1}{\tau_I s}+\tau_D s\right)$
   (c) $G_c(s) = K_c\left(\dfrac{\tau_I s+1}{\tau_I s}\right)\left(\dfrac{\tau_D s+1}{\alpha\tau_D s+1}\right)$
   (d) $G_c(s) = K_c\left(\dfrac{\tau_I s+1}{\tau_I s}\right)(\tau_D s+1)$
9. The ideal PD controller $G_c(s) = K_c(1+\tau_D s)$ is not physically realizable because
   (a) the order of denominator polynomial is higher than that of the numerator polynomial.
   (b) derivative time is a positive quantity.
   (c) the order of denominator polynomial is lower than that of the numerator polynomial.
   (d) proportional/controller gain is a positive quantity.
10. A PI controller has the transfer function

$$G_c(s) = \frac{6s+12}{s}$$

   The values of K_c and τ_I are, respectively
   (a) 6 and 2
   (b) 2 and 0.5
   (c) 0.2 and 0.5
   (d) 6 and 0.5

11. The transfer function of a PID controller is

(a) $K_c(1 + \tau_I s + \tau_D s)$

(b) $K_c\left(1+\dfrac{1}{\tau_I s}+\tau_D s\right)$

(c) $K_c\left(1+\tau_I s+\dfrac{1}{\tau_D s}\right)$

(d) $K_c\left(1+\dfrac{1}{\tau_I s}+\dfrac{1}{\tau_D s}\right)$ **(GATE 1997)**

where τ_I is the integral (reset) time and τ_D is the derivative time.

12. The transfer function for an ideal proportional plus reset controller (Reset time τ) is

(a) $K_c\left[1+\dfrac{1}{\tau s}\right]$ **(b)** $K_c[1 + \tau s]$

(c) $\dfrac{K_c}{1+\tau s}$ **(d)** $\dfrac{K_c}{1+\dfrac{s}{\tau}}$ **(GATE 1996)**

13. The transfer function for a PD controller is

(a) $K_c(1 + \tau_D s)$ **(b)** $K_c\left(1+\dfrac{1}{\tau_D s}\right)$

(c) $K_c \tau_D s$ **(d)** $\dfrac{K_c}{\tau_D s}$ **(GATE 1995)**

14. The offset introduced by proportional controller with gain K_c in response of first-order system can be reduced by

(a) reducing value of K_c

(b) inroducing integral control

(c) introducing derivative control

(d) none of the above **(GATE 1995)**

15. The type of process that most often can benefit from derivative control is

(a) Flow

(b) Level

(c) Temperature

(d) Pressure **(GATE 1991)**

ANSWER KEY

1. (a)	**2.** (c)	**3.** (b)	**4.** (c)	**5.** (b)	**6.** (a)	**7.** (c)
8. (d)	**9.** (c)	**10.** (d)	**11.** (b)	**12.** (a)	**13.** (a)	**14.** (b)
15. (c)						

CHAPTER 8

Control System Instrumentation

CHAPTER CONTENTS

Having considered PID controllers in Chapter 7, we now consider the other components of the feedback control loop. As an illustrative example, consider the stirred-tank heating system in Fig. 8.1. A thermocouple measures the liquid temperature and converts it to a millivolt-level electrical signal. This signal is then converted to a voltage level and transmitted to the electronic controller. The feedback controller performs the control calculations and sends the calculated value as an output signal to the final control element, an electrical heater that adjusts the rate of heat transfer to the liquid. This example illustrates the three important functions of a feedback control loop: (1) measurement of the *controlled variable* (CV), (2) adjustment of the *manipulated variable* (MV), and (3) signal transmission between components.

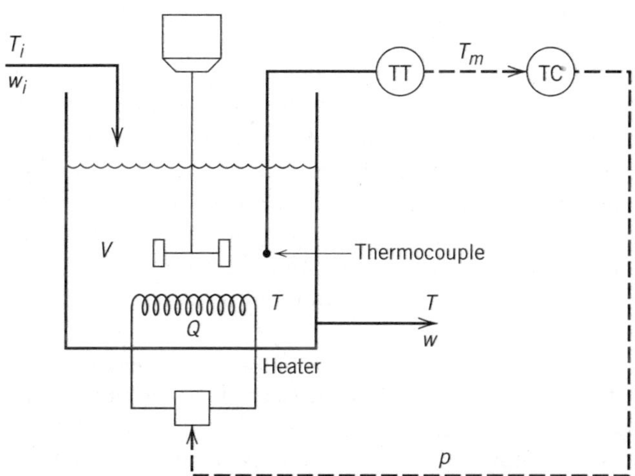

Figure 8.1 Schematic diagram for a stirred-tank heating control system.

The interconnection between the process and the controller in Fig. 8.1 can be considered to be an *interface*. The interconnection is required for a single controller or for a number of controllers in a computer control system (Fig. 8.2). In each case, the interface consists of all measurement, manipulation, and transmission instruments. The interface elements in Fig. 8.2 all contain a common feature. Each involves the conversion of a variable, for example, temperature to a voltage-level signal. Final control elements, or *actuators*, are used to manipulate process variables (usually flow rates).

This chapter introduces key instrumentation concepts and emphasizes how the choice of measurement and manipulation hardware affects the characteristics of the control system. Many of the assumptions that are commonly used to simplify the design of control systems—linear behavior of instruments and actuators, negligible instrumentation and signal transmission dynamics—depend on the proper design and specification of control loop instrumentation. A number of general references and handbooks for process instrumentation are available (e.g., Lipták, 2006a; Johnson, 2007; Edgar et al., 2016; Scott, 2007).

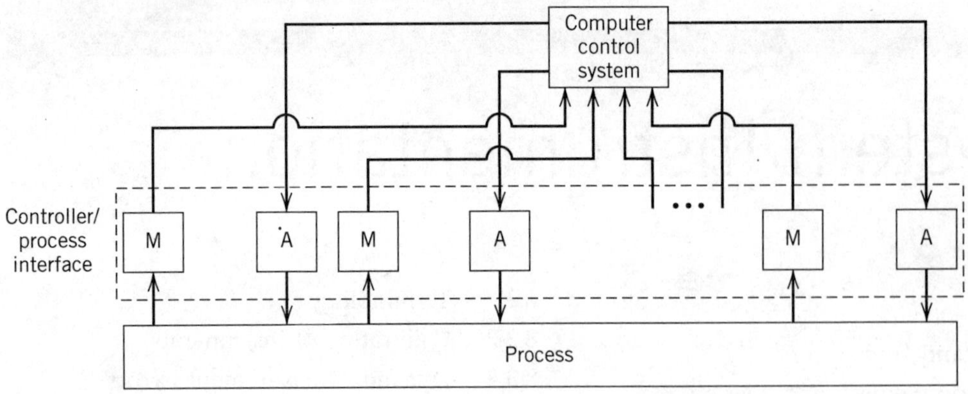

Figure 8.2 Computer control system with multiple measurements (M) and multiple actuators (A).

Chapter 19 describes digital computer control and digital instrumentation systems. A significant amount of instrumentation used today is based on digital technology, although traditional analog instrumentation is still used. Consequently, we consider both in this chapter.

8.1 SENSORS, TRANSMITTERS, AND TRANSDUCERS

The operation of complex industrial plants would be difficult, if not impossible, without the measurement and control of critical process variables. Large plants typically have thousands of process variables that are repetitively measured on-line every few seconds or minutes. In addition, important product properties are measured in quality control labs less frequently—e.g., once per hour, once an eight-hour shift, or daily. Consequently, the design and maintenance of accurate, reliable measurement systems is a critical aspect of process control. The lack of a reliable, cost-effective on-line sensor can be a key limitation on the effectiveness of a process control system.

A physical variable is measured by a *sensor* which produces a physical response (e.g., electrical or mechanical) that is related to the value of the process variable. For example, in the stirred-tank heating system, in Fig. 8.1, the thermocouple generates a millivolt electrical signal that increases as the temperature of the liquid increases. However, for this temperature measurement to be used in the control calculations, the millivolt-level signal must be converted to an appropriate voltage or current signal in a standard input range for the controller (see Section 8.1.1). This signal conversion is done by a *transmitter*.

In the process control literature, the terms *sensor*, *transmitter*, and *sensor-transmitter* are used more or less interchangeably; we follow suit in this book.

It is often necessary to convert an instrumentation signal from one form to another. A device that performs this conversion is referred to as a *transducer*. One common application is when the controller output signal is a current signal and the final control element is a pneumatic control valve (see Section 8.2.1). The required conversion is performed by a current-to-pressure (I/P) transducer. Voltage-to-pressure (E/P) transducers are also quite common.

8.1.1 Instrumentation Signals and Data Networks

Process control systems make extensive use of three types of instrument signals: pneumatic (air) signals, analog (continuous) electronic signals, and digital (discrete) signals. Before 1960, instrumentation in the process industries predominantly used pneumatic signals to transmit measurement and control information. These instruments were powered by compressed air and used mechanical force-balance elements to generate signals in the range of 20–100 kPa(g), an industry standard. Pneumatic pressure signals are transmitted by tubing, usually 6 or 10 mm diameter. Current-to-pressure (I/P) and voltage-to-pressure (E/P) transducers are widely used to convert electronic signals to equivalent pneumatic signals that are compatible with pneumatically-actuated control valves (see Section 8.2.1). The instrumentation symbols on process diagrams have been standardized and are available from professional organizations (ISA, 2009).

Since the 1950s, electronic instrumentation has become predominant due to the development of solid-state electronics and the need to transmit signals over longer distances (e.g., hundreds to thousands of feet) than were feasible with pneumatic devices. The standard signal ranges for electronic analog instruments are 4–20 mA and 1–5 V, direct current (VDC). Most transmitter analog signals are in the form of current rather than voltage, because voltage is affected by wire and connector resistances that change with wire length, temperature and aging. Voltage-level instrumentation signals (e.g., 1–5 VDC) are better suited to situations where short distances are involved.

Digital instruments and devices became increasingly important starting in the 1960s. A microprocessor in each instrument or controller is responsible for communicating periodically over a *data highway*, either receiving or requesting information from other devices. A major advantage of digital communication is the significant reduction in wiring and maintenance costs, compared to analog systems. Major advances in computing equipment (e.g., microprocessors and personal computers) and control technology during the 1970s and 1980s led to proprietary computer networks (i.e., distributed control systems) and smart sensors with digital memory and logic capabilities.

In recent years, instrumentation and control equipment have been interconnected by digital networks and open architecture. In particular, *fieldbuses* are a generic name for digital communication networks between instrumentation, devices and control equipment (Lipták, 2006a; Anon, 2013). Several fieldbus network protocols such as Foundation fieldbus (Fieldbus Foundation, 2015) and Profibus (Profibus & Profinet Int., 2015) have been developed. For new industrial plants, fieldbuses offer the advantages of lower installation and maintenance costs plus improved process diagnostics. But for existing plants, the integration of a fieldbus system with existing instrumentation and control systems may be difficult (Hebert, 2007).

In recent years, there has been considerable interest in using optical fiber communications (Agrawal, 2010) and wireless networks for process control applications (Blevins et al., 2014; Caro, 2014). The major incentives for wireless technologies is their increased flexibility and greatly reduced wiring and installation costs. However, there also well-known challenges that must be overcome, which include security concern, line-of-sight limitations, and how the networks will be powered (Blevins et al., 2014; Caro, 2014).

An informative history of early measurement and control devices up to the early 1990s has been reported by Strothman (1995). Chapter 19 provides additional information on the hardware and software for computer control systems.

8.1.2 Sensors

We now briefly discuss common sensors for the most important process variables. Additional information is available in general handbooks (Lipták, 2006a; Edgar et al., 2016), books (Shuler et al., 2014; Scott, 2007; Johnson, 2007; Nichols, 2010) and more specialized references in individual subsections.

The main types of measurements used in process control are temperature, pressure, flow rate, liquid level, and chemical composition. Table 8.1 lists sensor options for each category.

Selection Criteria. The selection of a measurement device should consider the following factors (Edgar et al., 2016):

1. ***Measurement range (span).*** The required measurement range for the process variable must lie entirely within the instrument's range of performance.
2. ***Performance and Cost.*** Depending on the application, accuracy, repeatability, or some other measure of performance is appropriate. For closed-loop control, speed of response is also important. And of course, cost is an important consideration (White, 2014).
3. ***Reliability.*** Manufacturers provide baseline conditions. Previous experience with the measurement device is also very important.
4. ***Materials of construction.*** An instrument may need to withstand high temperatures, high pressures, and corrosive and abrasive environments. For some applications, seals or purges may be necessary.
5. ***Prior use.*** For the first installation of a specific measurement device at a site, training of maintenance personnel and purchases of spare parts are essential.
6. ***Potential for releasing process materials to the environment.*** Preventing exposure to fugitive emissions for maintenance personnel is important when the process fluid is corrosive or toxic. Sterility in bioprocesses must be maintained.
7. ***Electrical classification.*** If the sensor is not inherently compatible with possible exposure to hazards, suitable enclosures must be provided.
8. ***Invasive or noninvasive.*** The insertion of a probe (invasive) can cause fouling, which leads to inaccurate measurements. Probe location must be selected carefully to ensure measurement accuracy and to minimize fouling.

Temperature. The most common temperature sensors are filled systems, thermocouples, resistance temperature detectors (RTDs), and pyrometers. Measurement principles are, respectively, based on measurement of volumetric expansion, electromotive force generated by two dissimilar metals, resistance as a function of temperature, and wavelength of radiated energy (Lipták, 2006a; Michalski et al., 2001). Thermocouples and RTDs can be used up to 1000 °C, although RTDs are

Table 8.1 On-Line Measurement Options for Process Control

Temperature	Flow	Pressure	Level	Composition
Thermocouple	Orifice	Liquid column	Float-activated	Gas-liquid chromatography (GLC)
Resistance temperature detector (RTD)	Venturi	Elastic element	—chain gauge, lever	Mass spectrometry (MS)
Filled-system thermometer	Rotameter	—Bourdon tube	—magnetically coupled	Magnetic resonance analysis (MRA)
Bimetal thermometer	Turbine	—bellows	Head devices	Infrared (IR) spectroscopy
Pyrometer	Vortex-shedding	—diaphragm	—bubble tube	Raman spectroscopy
—total radiation	Ultrasonic	Strain gauges	Electrical (conductivity)	Ultraviolet (UV) spectroscopy
—photoelectric	Magnetic	Piezoresistive transducers	Radiation	Thermal conductivity
—ratio	Thermal mass	Piezoelectric transducers	Radar	Refractive index (RI)
Laser	Coriolis	Optical fiber	Ultrasonic	Capacitance probe
Surface acoustic wave	Target			Surface acoustic wave
Semiconductor				Electrophoresis
				Electrochemical
				Paramagnetic
				Chemi/bioluminescence
				Tunable diode laser absorption

much more accurate. Pyrometry is typically used above 900 °C (e.g., in high-temperature applications such as combustion).

Differential Pressure and Pressure. Pressure measurement and transmission in the process industries is usually based on a small deflection of a diaphragm or membrane in response to a pressure. The deflection is produced by two pressure connections, one on each side of the diaphragm. The pressure sensor is designed so that the deflection is proportional to the pressure difference, or *differential pressure (DP)*, between the two connections. The deflection is then transduced to a corresponding electrical or pneumatic output signal that can be recorded and used for control. Gauge pressure can be measured by having an atmospheric pressure connection on one side of the diaphragm. Industrial DP sensors differ on the type of membrane material and construction, and how the deflections are measured (Lipták, 2006a; Gassman, 2014).

Other important sensors include

1. *Piezo devices*. As the pressure changes, the dimensions of a metal or crystal change, which produces a change in electric potential (the *piezoelectric effect*) or a change in electrical resistance (the *piezoresistive effect*). These short-lived changes can be measured and converted to sensor output signals.

   Piezo devices are well suited for measuring dynamic pressures where the sensor response must be very rapid (e.g., due to shocks, vibrations and explosions) but are less effective for measuring static pressures.

2. *Capacitive sensor*. Two parallel metal membranes are used; one is fixed while the other can move (i.e., deflect). The process fluid is in contact with one side of the moving membrane. As the fluid pressure increases, the distance between the two membranes changes, which changes the electrical capacitance between the plates. The capacitance change is detected by an electrical circuit and converted to a corresponding value of pressure.

Liquid or Gas Flow Rate. Accurate measurement of fluid flow rates is a critical activity in the process industries for many purposes that include process safety, process monitoring, and the calculation of material and energy balances. Selection of a flow transmitter should consider the following factors: nature of the flowing material (liquid/gas/solid/multiphase), corrosiveness, mass versus volumetric flow measurement, nature of the signal, cost, accuracy, space available, and required maintenance (Spitzer, 2001; Emerson Process Management, 2005; Lipták, 2006a).

Volumetric flow rate has traditionally been determined indirectly by measuring the pressure drop across an obstruction in a pipe such as an orifice, venturi, or pitot tube. A steady-state energy balance indicates that the volumetric flow rate is proportional to the square root of the pressure drop across the obstruction. The pressure drop can be measured using conventional DP instrumentation, as described in the previous section. The orifice plate is normally sized to provide a pressure drop in the range of 0.51–5.1 m of water. But this approach is less accurate and has a larger pressure drop than when a venturi replaces the orifice plate. For large pipe sizes, pressure drops are important and thus venturi meters are favored. DP flow measurements are inexpensive and tend to be used for clean liquid and gas streams. But due to their relatively poor accuracy, the use of DP flow sensors in the process industries is on the decline, compared to the newer types of sensors described below.

Volumetric flow rates can also be measured using turbine flowmeters or by measuring the deflection of a vane inserted in the pipe or channel (e.g., a target meter or a vortex shedding meter). Magnetic flow meters are more accurate than orifice plates but their application is limited to conducting fluids (see Fig. 8.3).

Coriolis meters use a vibrating flow loop that undergoes a twisting action due to the Coriolis effect. There are two basic configurations: the curved tube meter (see Fig. 8.4) and the straight tube meter. The amplitude of the deflection angle is converted to a voltage that is nearly proportional to the liquid mass flow rate.

A capacitance probes measure the dielectric constant of the fluid and is useful for flow slurries and other two-phase flows. Non-contact, ultrasonic meters that can be clamped to the outside of pipes are attractive because of their noninvasive nature and the absence of moving parts that can wear out (Baker, 2000).

Courtesy of Honeywell, Inc.

Figure 8.3 Magnetic flow meter.

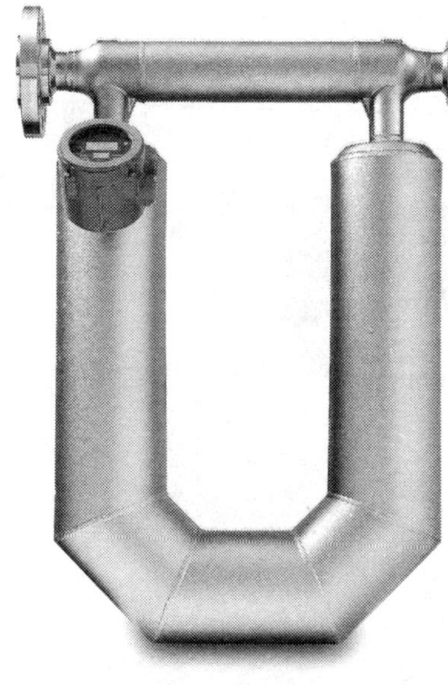

Courtesy of Emerson Process Management

Figure 8.4 Coriolis flow meter.

Mass flow meters that are independent of changes in pressure, temperature, viscosity, and density include thermal mass meters and Coriolis meters. Thermal mass meters are widely used in semiconductor manufacturing and bioprocessing for control of low flow rates (called *mass flow controllers*, or MFCs). MFCs measure the heat loss from a heated element that varies with flow rate; their accuracy is ±1%. Multivariable flow transmitters are available that determine the volumetric or mass flow rate of a fluid from standard process variables (see Fig. 8.5).

Liquid Level. Measurement of liquid level in process vessels is very important to prevent overflow and potential safety problems due to excessively high liquid levels and pump damage due to very low levels. For some process operations, the liquid level must be tightly regulated in order to have a specified residence time in the vessel (e.g., some chemical reactors). In this section, we consider several common level sensors.

Liquid level can be measured indirectly by using a differential pressure sensor with one connection at or near the bottom of the tank and the other connection at or near the top of the tank. Liquid level h is related to the DP measurement ΔP by the equation for static pressure,

$$\Delta P = \rho g h \tag{8-1}$$

where ρ is the density of the liquid and g is the gravitational constant. A disadvantage of this widely used sensor is that when significant density variations occur due

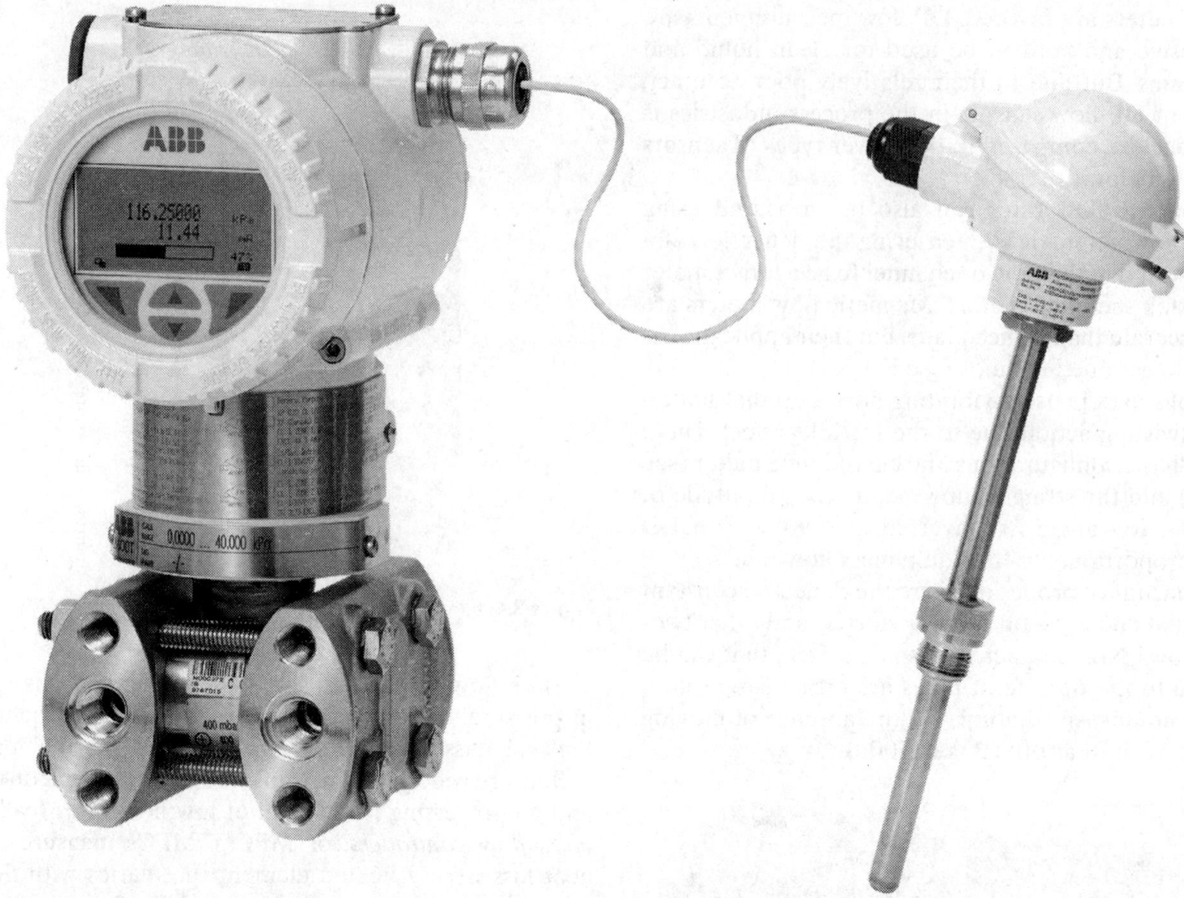

Courtesy of ABB, Inc.

Figure 8.5 This multivariable flow transmitter calculates the mass flow rate, or standard volumetric flow rate for gases, vapors and liquids.

to temperature or liquid composition variations, they need to be taken into account. An advantage of the D/P approach is that when the vessel is not completely full, the D/P measurement is not affected by the gas pressure in the space above the liquid because it has the same effect on each side of the diaphragm.

Capacitance sensors have a probe that is inserted into the liquid. The probe and the vessel act as two plates of a capacitor. The capacitance of the liquid changes with the liquid level. A radio frequency signal is applied to the probe; the capacitance is then measured and the level sensor converts it to an electronic equivalent of the liquid level (Lipták, 2006a).

Noncontact level sensors are available for applications where the sensor is not allowed to contact the liquid. The sensor sends an ultrasonic sound wave to the liquid surface; the level sensor detects the reflected sound wave (the echo) and uses this information to determine the distance to the surface. The calculated liquid level value is then transmitted as an electronic output signal. Noncontact radar and gamma radiation level sensors are available that operate in an analogous manner. A disadvantage of noncontact sensors is that factors such as dust, foam, and surface turbulence can affect performance.

Chemical Composition. Chemical composition is generally the most challenging on-line measurement. It is also very important because accurate composition measurements are essential for evaluating process performance. Before the era of on-line analyzers, samples were manually delivered to an on-site analytical laboratory. The resulting long time delays prevented process adjustments being made in timely manner, thus affecting product quality. The development of on-line analyzers has automated this activity and reduced analysis time. Because a chemical analyzer can be very expensive to purchase, install, and maintain, its specifications and selection require careful consideration (Lipták, 1994; Lipták, 2014; Nichols. 2014).

In order to obtain quantitative composition measurements, specific instruments must be chosen depending on the nature of the species to be analyzed. Measuring a specific concentration requires a unique chemical or physical attribute. Conductivity and pH analyzers are widely used in the process industries. In infrared (IR) spectroscopy, the vibrational frequency of specific molecules such as CO and CO_2 can be probed by absorbing electromagnetic radiation. Ultraviolet (UV) radiation analyzers operate similarly to IR analyzers in that the degree of absorption for specific compounds occurs at specific frequencies and can be measured. Turbidity, an indicator of cell mass in a bioreactor, can be measured by absorbance in a spectrophotometer. Magnetic resonance analysis (formerly called nuclear magnetic resonance) uses magnetic moments to discern molecular structure and concentrations for both chemical and biochemical systems.

Significant advances have occurred in recent years to lower analyzer costs, reduce analysis time, and introduce new technologies (Studebaker, 2014). Chemical sensors have been placed on microchips, even those requiring multiple, physical, chemical, and biochemical steps (such as electrophoresis). These devices have been called *lab-on-a-chip* (Chow, 2002; Daw and Finkelstein, 2006).

In gas chromatography (GC), a gas sample (or a vaporized liquid sample) is carried through the GC by an inert gas, and components of the sample are separated by a packed bed. Because each component has a different affinity for the column packing, it passes through the column at a different time during the sample analysis, allowing individual concentrations to be measured. The GC can simultaneously measure many components in a mixture, whereas most other analyzers can only detect a single component; hence, GC is widely used. A field-mounted gas chromatograph is shown in Fig. 8.6.

Figure 8.6 Field-mounted gas chromatograph.

Typically, all components can be analyzed in less than 10 min. A GC can measure concentrations ranging from parts per billion (ppb) to tens of percent, depending on the compound (Nichols, 2010). High-performance liquid chromatography (HPLC) can be used to measure dissolved solute levels, including proteins (Shuler et al., 2014).

Mass spectroscopy (MS) determines the partial pressures of gases in a mixture by directing ionized gases into a detector under a vacuum (10^{-6} torr); the gas phase composition is then monitored based on the molecular weight of the species (Nichols, 2010). Sometimes, a GC is combined with an MS in order to obtain a higher level of discrimination of the components present. As an example, complete analysis of a combustion gas might require multiple on-line analyzers as shown in Table 8.2.

Fiber-optic sensors are attractive options (but more expensive) for acquiring measurements in harsh environments such as high temperature or pressure. Because the transducing technique used by these sensors does not involve electrical signals, they are immune to electromagnetic interference. Raman spectroscopy uses fiber-optics and involves pulsed light scattering by molecules.

Many composition measurements are both difficult and expensive to obtain. Indirect means of measuring concentrations are often less expensive and faster, for example, relating the mole or mass fraction of a liquid component to pH or relating conductivity or the concentration of one component in a vapor stream to its IR or UV absorption. Often an indirect measurement is used to infer composition; for example, the liquid temperature on a tray near the top of a distillation column might be used to determine the distillate composition.

A related approach is to use a process model as a *soft sensor* (see Section 16.2) to estimate process variables that cannot be measured on-line. For example, *predictive emissions monitoring systems (PEMS)* relate trace pollutant concentrations to operating conditions such as temperature, pressure, and excess air (see Section 16.2).

Table 8.2 Analysis of a combustion gas

Component	Method
O_3	UV photometer
SO_2	UV fluorescence
NO_x	Chemiluminescence
CO, CO_2, SO_2	Infrared
O_2	Paramagnetic
Trace hydrocarbons	GC/MS

Physical Properties. This category includes such measurements such as density, moisture, turbidity, viscosity, refractive index, pH, dielectric constant, and thermal conductivity (Lipták, 2006a; Shuler et al., 2014).

8.1.3 Static and Dynamic Characteristics

As noted above, the output signal from a sensor-transmitter (or transmitter) must be compatible with the input range of the controller that receives the signal. Transmitters are generally designed to be *direct-acting*; that is, the output signal increases as the measured variable increases. In addition, most commercial transmitters have an adjustable input range. For example, a standard temperature transmitter with a 4–20 mA output signal might be adjusted so that the input range of a platinum resistance element (the sensor) is 50–150 °C. In this case, the following correspondence is obtained:

Input	Output
50 °C	4 mA
150 °C	20 mA

This instrument has a lower limit, or *zero*, of 50 °C and a range, or *span*, of 150 °C – 50 °C = 100 °C. The zero and span of a sensor/transmitter are adjustable. Figure 8.7 illustrates the concepts of zero and span. In this example, the relation between temperature and the transmitted (measured) signal is linear. But why is the zero of the temperature transmitter 4 mA, rather than 0 mA? If the transmitter is functioning properly, its output signal is the range of 4–20 mA. But if its power supply fails, the output signal will be 0 mA. Thus an output signal that is outside the 4–20 mA range is an indication that a malfunction has occurred.

For a linear transmitter, the steady-state relationship between the measured variable y and the transmitter output y_m is a linear equation,

$$y_m = K_m(y - y_0) + y_{m0} \tag{8-2}$$

where K_m is the transmitter steady-state gain. The general expression for K_m is

$$K_m = \frac{\text{output range}}{\text{transmitter span}} = \frac{y_2 - y_1}{y_{m2} - y_{m1}}$$

where subscripts 1 and 2 denote two arbitrary points on the calibration curve in Fig. 8.7.

For the temperature transmitter example, substituting $T_0 = 50$ °C and $T_{m0} = 4$ mA gives

$$K_m = \frac{20 - 4\,\text{mA}}{150 - 50\,°\text{C}} = 0.16\,\text{mA}/°\text{C}$$

Substituting numerical values into Eq. 8-1 gives the relation between the temperature transmitter input and output:

$$T_m = 0.16(T - 50) + 4 \tag{8-3}$$

For a nonlinear calibration curve, the gain at an operating point is the derivative of the characteristic input–output relation at the operating point. Figure 8.8 illustrates a typical case. Note that the gain changes whenever the operating point changes; hence, instruments that exhibit nearly linear behavior are preferable. Gain K_m changes when the span is changed but is invariant to changes in the zero.

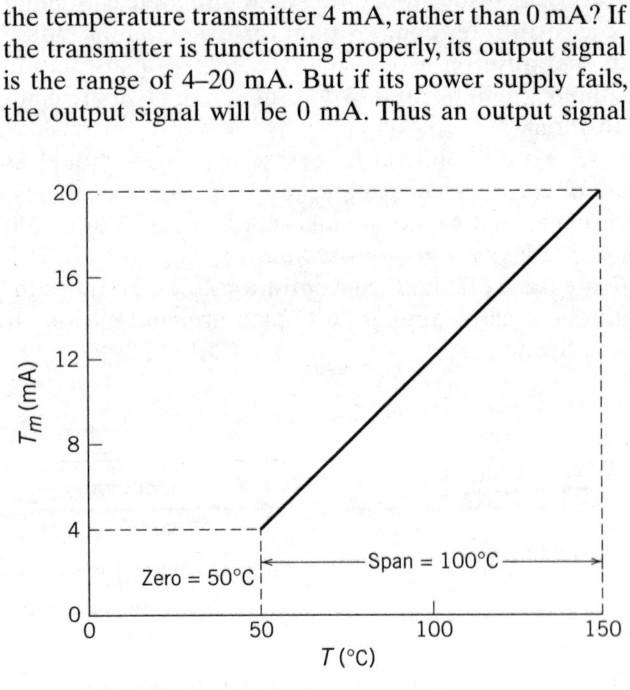

Figure 8.7 A linear instrument calibration showing its zero and span.

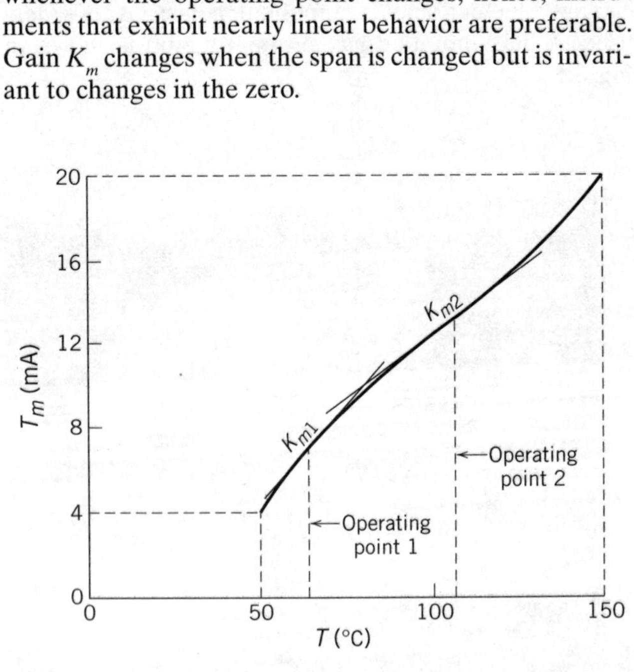

Figure 8.8 Gain of a nonlinear transmitter as a function of operating point.

Sensors and transmitters are normally designed to be *direct acting* ($K_m > 0$) so that the output signal increases as the measured variables increase. However, *reverse-acting* instruments ($K_m < 0$) are sometimes more appropriate, especially for some level measurement applications.

Dynamic Characteristics of Sensor-Transmitters. Many sensor-transmitters respond quickly and have measurement dynamics that are negligible compared to slower process dynamics. For other applications where measurement dynamics are not negligible, significant dynamic errors can occur, resulting in large differences between the actual values and the measured values. For example, a bare thermocouple has a rapid response to a changing fluid temperature. But a thermocouple placed in a protective thermowell with a large mass and large specific heat, can have a significant measurement time constant (see Section 8.3.3).

Many sensor-transmitters have overdamped dynamics and exhibit monotonic responses to a step change in the measured variable. Thus, it is reasonable to model this type of measurement dynamics as a first-order transfer function between the actual value y and the measured value y_m:

$$\frac{Y_m(s)}{Y(s)} = \frac{K_m}{\tau_m s + 1} \tag{8-4}$$

where K_m is the steady-state gain and τ_m is the measurement time constant. For the temperature transmitter example, the units of K_m are mA/°C.

Significant measurement dynamics can occur due to a poor sensor location or a long sampling line. For example, if a pH sensor for a continuous neutralization process is located in the exit line, a long distance from the process vessel, a significant time delay can arise due to the distance-velocity lag (see Section 6.1). Time delays can also result when an on-line composition measurement requires a long sample line to a remote analyzer in a protected environment. This common situation can produce a significant distance-velocity lag.

While manufacturers of sensors and transmitters often supply general information on the dynamic response characteristics of their devices, this information may be difficult to interpret for specific applications (Edgar et al., 2016). Representative values of sensor time constants and time delays have been reported by McMillan (2015).

8.2 FINAL CONTROL ELEMENTS

Every process control loop must contain a final control element that enables a process variable to be manipulated. For most chemical and petroleum processes, the final control element adjusts the flow rate of a material (solid, liquid, gas or multiphase) which indirectly adjusts the rates of material and energy transfer to and from the process. In these situations, the final control element is usually a control valve or a variable speed pump. Although control valves are more widely used, variable speed pumps are preferred for abrasive liquids such as slurries. They also avoid the energy loss associated with the pressure drop across the control valve.

In Fig. 8.1 an electrical resistance heater serves as the final control element. In this case, the controller output, a voltage signal, cannot be applied directly to the terminals of the heater, because the controller is not designed to supply the electrical energy requirements of the heater. Hence, a transducer must be placed between the controller and the heater (not shown in Fig. 8.1).

8.2.1 Control Valves

There are many ways to manipulate the flows of material and energy into or out of a process; for example, the speed of a pump drive, screw conveyer, or blower can be varied. However, a simple and widely used method of adjusting the flow rate of a fluid is to use a *control valve*. The main control valve components of a pneumatic control valve are shown schematically in Fig. 8.9; they are the valve body, stem, seat, and actuator. The *valve body* contains a variable orifice that adjusts the flow rate. The *stem* position determines the valve opening and consequently, the flow rate. The stem can be attached to a plug, ball, disk, or gate (Borden and Friedmann, 1998). The *seat* consists of a protective material (typically metal or soft polymer) inserted around the orifice to provide a tight shutoff and to increase the life of the valve when corrosive or solid materials pass through it. The *actuator* provides the force for opening and closing the valve. It is usually a pneumatically powered device, but electric and hydraulic powered actuators are also available (Emerson Process Management, 2005). Electric actuators are well suited for very large control valves where large forces are needed to open or close the valve.

Control valves are either linear (sliding stem) or rotary in design. Linear valves are often *globe valves* that open and close the valve by moving a plug vertically away from the orifice and seat. The movement changes the cross-sectional area available for fluid flow. *Rotary valves* (or *quarter-turn valves*) are completely closed by a 90° turn of the closing mechanism; they are used for both on-off operation and flow adjustment. In general, rotary valves are more compact, less expensive, and easier to maintain than sliding stem valves. The primary types of rotary valves are the *plug valve*, the *butterfly valve*, the *ball valve*, and the *rotary globe valve*.

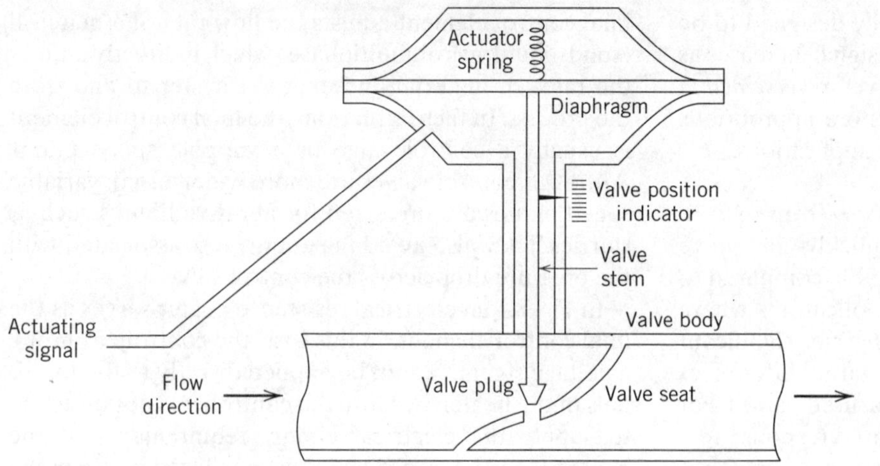

Figure 8.9 A control valve with a pneumatic actuator (air-to-open).

For more information concerning control valves, see the extensive literature on control valves (Borden and Friedmann, 1998; Lipták, 2006a,b; Emerson Process Management, 2005; Edgar et al., 2016).

Despite the growing use of motor-driven valves, most control applications utilize pneumatic actuators such as the rising stem type shown schematically in Fig. 8.9. Here the pneumatically-operated diaphragm device moves the stem vertically against the opposing force of a fixed spring. The actuating signal from the controller is transmitted to the control valve as a pneumatic signal. As this signal increases, the increased pressure on the diaphragm compresses the spring, thus pulling the stem out and opening the valve further. This type of control valve is referred to as *air-to-open* (A–O). By reversing either the plug/seat or the spring/air inlet orientation, the valve becomes *air-to-close* (A–C). For example, if the spring is located above the diaphragm and the air inlet is placed below the diaphragm, an A–O valve results, as shown in Fig. 8.9.

The choice of an A–O or A–C valve is based on safety considerations. The way that the control valve should fail (open vs. closed) is selected based on the desired valve response in an emergency situation, such as a power failure. Hence, A–C and A–O valves often are referred to as *fail-open* (F-O) and *fail-close* (F-C), respectively.

EXAMPLE 8.1

Pneumatic control valves are to be specified for the applications listed below. State whether an A–O or A–C valve should be specified for the following manipulated variables and give reason(s):

(a) Steam pressure in a reactor heating coil.
(b) Flow rate of reactants into a polymerization reactor.
(c) Flow of effluent from a wastewater treatment holding tank into a river.
(d) Flow of cooling water to a distillation condenser.

Solution

(a) A–O (fail-close) to make sure that a transmitter failure will not cause the reactor to overheat, which is usually more serious than having it operate at too low a temperature.
(b) A–O (fail-close) to prevent the reactor from being flooded with excessive reactants.
(c) A–O (fail-close) to prevent excessive and perhaps untreated waste from entering the stream.
(d) A–C (fail-open) to ensure that overhead vapor is completely condensed before it reaches the receiver. Otherwise, an excessively high pressure could develop.

8.2.2 Control Valve Dynamics

Control valve dynamics tend to be relatively fast compared to the dynamics of the process itself. But the overall behavior of pneumatic control valves can be complicated due to nonlinear phenomena such as *hysteresis*, *deadband*, and *stick-slip* (*stiction*) that are illustrated in Fig. 8.10 (Choudhury, 2005; McMillan, 2015).

The hysteresis diagram in Fig. 8.10(a) indicates that for a specified value of the control signal p, the valve stem position x can have two different values, depending on the past history of the stem movement, that is, whether x has been increasing or decreasing. The deadband in Fig. 8.10(b) is a time delay that is associated with reversing the direction of the valve stem position x, for example, the control signal p causes x to increase after it had been decreasing. Deadband is caused by the relative movement of mechanical parts in the valve assembly, that is, *backlash*.

Stick-slip phenomena is due to static friction between the control valve stem and the packing around the stem; the packing prevents leakage of the process fluid that is passing through the valve. Thus when the valve stem has been constant for a while it will not move ("it sticks") until the change in the control signal is greater than a threshold Δp (that is, it then "slips"). Stick-slip is detrimental because it can lead to control-loop oscillations. Note that all three phenomena (deadband, hysteresis, and stick-slip phenomena) can occur for a control valve.

The behavior of control valves also tends to be nonlinear due to the asymmetric responses to positive and negative changes in the signal to the control valve. But for purposes of control system analysis using transfer function models, the dynamic behavior of the control valve (and valve positioner) can be approximated by a first-order-plus-time-delay transfer function $G_v(s)$ between the signal to the control valve p and the manipulated flow rate u

$$\frac{U(s)}{P(s)} = G_v(s) = \frac{K_v e^{-\theta_v s}}{\tau_v s + 1} \tag{8-5}$$

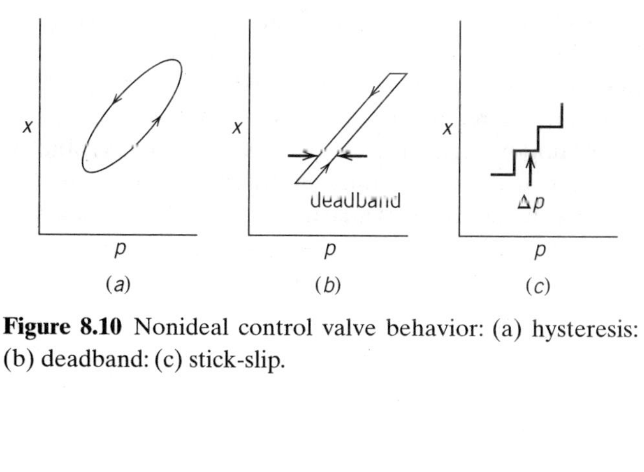

Figure 8.10 Nonideal control valve behavior: (a) hysteresis: (b) deadband: (c) stick-slip.

The valve time constant τ_v is typically much smaller than the largest process time constant τ_p, that is, $\tau_v \ll \tau_p$.

Emerson Process Management (2005, pg. 32) has reported experimental θ_v and τ_v values, for 0.1 m size control valves with different valve assemblies. The τ_v values varied from 0.3 to 7.8 s, while the θ_v/τ_v ratios varied from 0.3 to 0.8. The model parameters also depended on the size and direction of the step signal applied to the valve. Larger size control valves can have much slower dynamics.

Valve Positioners

For important control loops, pneumatic control valves should be equipped with a *valve positioner*, a type of mechanical or digital feedback controller that senses the actual stem position x, compares it to the desired position and adjusts the pneumatic control signal to the valve, accordingly. Valve positioners are used to increase the relatively small mechanical force that can be exerted by a 21-104 kPa psig pressure signal operating directly on the valve diaphragm. Because valve positioners can significantly reduce the effects of deadband, hysteresis and stick-slip, they are widely used. The valve positioner is usually mounted on the side of the valve actuator, as shown in Fig. 8.11. Additional information on valve positioners is available in manufacturers' documentation and handbooks (Emerson Process Management, 2005; Lipták, 2006a; Edgar et al., 2016).

Courtesy of Emerson Process Management

Figure 8.11 Modern control valves with a digital valve positioner.

8.2.3 Specifying and Sizing Control Valves

Control valves must be designed and installed very carefully. For example, a survey of 5000 control loops in the pulp and paper industry reported that 30% of the variability in control loop performance was attributed to the control valve (Ross and Bialkowski, 1996). Valve sizing can be particularly difficult because many of the published recommendations are ambiguous, conflicting, or not consistent with the control objectives. In sizing a control valve it is important to consider the other equipment on the pipeline, such as pumps, heat exchangers, restrictions (e.g., orifice plates), and filters that are placed in series with the control valve.

Control valves are specified by first considering both properties of the process fluid and the desired flow characteristics, in order to choose the valve body material and type. Then the desired characteristics for the actuator are considered. The choice of construction material depends on the corrosive properties of the process fluid at operating conditions. Commercial valves made of brass, carbon steel, and stainless steel can be ordered off the shelf, at least in smaller sizes. For large valves and more unusual materials of construction, special orders may be required.

The flow rate of a fluid through a control valve depends on several variables: the properties of the fluid, the valve design, the pressure drop across the valve, and the extent to which the valve is open or shut. As the fluid passes through the restriction of the valve, the fluid velocity increases and the pressure downstream of the valve decreases, according to the Bernoulli equation (Felder and Rousseau, 2005).

In this section we present the general features of control valve design and a simplified design method. More detailed descriptions of design methods are available in handbooks and general references (Borden and Friedmann, 1998; Emerson Process Management, 2005; Lipták, 2006a,b; Edgar et al., 2016).

Sizing Control Valves

The valve sizing equation for a turbulent, nonflashing liquid passing through a control valve has the general form:

$$q = C_v N f(\ell)\sqrt{\frac{\Delta P_v}{g_s}} \tag{8-6}$$

Here, q is the volumetric flow rate, ΔP_v is the pressure drop across the valve, and g_s is the specific gravity of the liquid. The symbol ℓ ($0 \le f(\ell) \le 1$) is the valve stem position[1] and indicates the extent to which the valve is full open ($\ell = 1$). The term $f(\ell)$ is called the *valve characteristic* and will be discussed later.

[1] For rotary control valves such as a ball valve, ℓ is the degree of rotation.

There are two constants in Eq. 8-6, C_v and N. The *valve sizing coefficient* C_v (or *valve coefficient*) is a dimensionless constant that determines the size and capacity of the control valve. Manufacturers tabulate C_v values for specific types and sizes of control valves.

Constant N is a *unit conversion factor* that depends on the units for q and ΔP_v. By definition, $N = 1$ gpm/(psi)$^{1/2}$ when q has units of gpm and ΔP_v has units of psi. This is because historically the formula for valve coefficient was derived using the units of flow rate and differential pressure in gpm (gallons/minute) and psi (pound per square inch), respectively. Based on the definition of N, when the valve is completely open ($f = 1$), a ΔP_v of 1 psi results in a flow rate of $q = C_v$ gpm. Thus, the C_v of a valve can be defined as the flow rate of water in gpm, which results in its fully open condition when the pressure drop across it is 1 psi. For metric units, $N = 0.0865$ m^3/h (kPa)$^{1/2}$ when q has units of m^3/h and ΔP_v has units of kPa.

Design equations analogous to Eq. 8-6 for other situations including mass flow rates, gases, vapors and two phase flow are available in handbooks (Emerson, 2005; Lipták, 2006a; Edgar et al., 2016) and vendor documentation.

An expression of C_v in metric systems for a fully open valve can be developed from Eq. 8.6 as:

$$C_v = 11.6\, q\sqrt{\frac{g_s}{\Delta P_v}} \tag{8-7}$$

where the flow rate q is in m^3/h and the differential pressure ΔP_v is in kPa.

Then substituting design conditions, q_d and ΔP_{vd}, and specifying that the control valve is half open ($f(\ell) = 0.5$) gives a control valve design equation:

$$C_v = 23\, q_d\sqrt{\frac{g_s}{\Delta P_{vd}}} \tag{8-8}$$

In control valve design, it is also important to consider the specified range of operating conditions and the control objectives. In particular, the values of q_{max} and q_{min} are important to ensure that the flow rate can be adjusted over a wide range of values. For example, for a feed flow rate, its value at any time may be unequal to q_d due to changes in the available feed supply or economic conditions. Similarly, if q is a manipulated variable such as cooling water flow rate, it is desirable to be able to adjust it over a wide values, in order to compensate for process disturbances. Thus, the *rangeability* R^* is a key concern where R^* is defined as

$$R^* = \frac{q_{max}}{q_{min}} \tag{8-9}$$

EXAMPLE 8.2

As a design example, consider the heat exchanger system in Fig. 8.12. The block labeled "equipment" represents all items (except the control valve) that reduce pressure in the pipeline (e.g., piping fittings and other obstructions). Denote the pressure drop for the "equipment" as the system pressure drop ΔP_s and its value at the design conditions as ΔP_{sd}. Similarly, the pressure drop across the control valve is denoted by ΔP_v and its design value as ΔP_{vd}.

Suppose that the nonflashing liquid has a specific gravity of $g_s = 1$ and the design flow rate is $q_d = 45$ m³/h. At this design flow rate, the equipment pressure drop is $\Delta P_{sd} = 275$ kPa. The pump discharge pressure P_2 is assumed to be constant and the discharge pressure at the pipe exit is atmospheric, $P_1 = 101.325$ kPa. Since P_1 and P_2 are constant, the total pressure drop ΔP is also constant. It can be expressed as

$$\Delta P = P_2 - P_1 = \Delta P_v + \Delta P_s \tag{8-10}$$

Assume that ΔP_s varies with square of the flow rate:

$$\Delta P_s = Kq^2 \tag{8-11}$$

The value of constant K can be determined from Eq. 8-11 and the design conditions

$$K = \frac{\Delta P_{sd}}{q_d^2} = \frac{275 \text{ kPa}}{(45 \text{ m}^3/\text{h})^2} = 0.1358 \text{ kPa/(m}^3/\text{h}^2) \tag{8-12}$$

(a) For five values of ΔP_{vd} (70, 140, 280, 700 and 1400 kPa) determine the corresponding values of C_v.

(b) Determine the maximum and minimum values of q (q_{max} and q_{min}) for each value of C_v.

Solution

(a) The values of C_v can be determined from Eq. 8-8 with $N = 0.0865$ m³/h(kPa)$^{1/2}$ and $g_s = 1$. The resulting C_v values are 123.7, 87.5, 61.8, 39.1 and 27.7, respectively.

(b) Find q_{max}:
Consider Eq. 8-6 with $N = 0.0865$, $q = q_{max}$ and $f = 1$ (because the maximum flow rate occurs when the control valve is full open):

$$q_{max} = C_v N \sqrt{\Delta P_v} \tag{8-13}$$

As noted earlier, ΔP_v changes with q. Thus $\Delta P_v \neq \Delta P_{vd}$ when $q \neq q_d$. Rearranging Eq. 8-10

$$\Delta P_v = \Delta P - \Delta P_s \tag{8-14}$$

and substituting Eqs. 8-11 and 8-12 into Eq. 8-13 with $q = q_{max}$ gives

$$q_{max} = C_v N \sqrt{\frac{\Delta P - Kq_{max}^2}{g_s}} \tag{8-15}$$

Solving Eq. 8-15 for q_{max} after substituting Eq. 8-12 gives the desired result:

$$q_{max} = K_1 \sqrt{\frac{\Delta P q_d^2}{q_d^2 + K_1^2 \Delta P_{sd}}} \tag{8-16}$$

where $K_1 = C_v N/\sqrt{g_s}$.

Find q_{min}:
A similar derivation with $f = 0.1$ (the minimum value of f for effective flow control) gives an expression for q_{min}:

$$q_{min} = K_2 \sqrt{\frac{\Delta P q_d^2}{q_d^2 + K_1^2 \Delta P_{sd}}} \tag{8-17}$$

where $K_2 = 0.1C_v N/\sqrt{g_s}$.

The results for this example are shown in Table 8.3.

Some important trends in Table 8.3 are

1. For small values of ΔP_{vd}, q_{max} is not much greater than $q_d = 45$ m³/h and R is small.

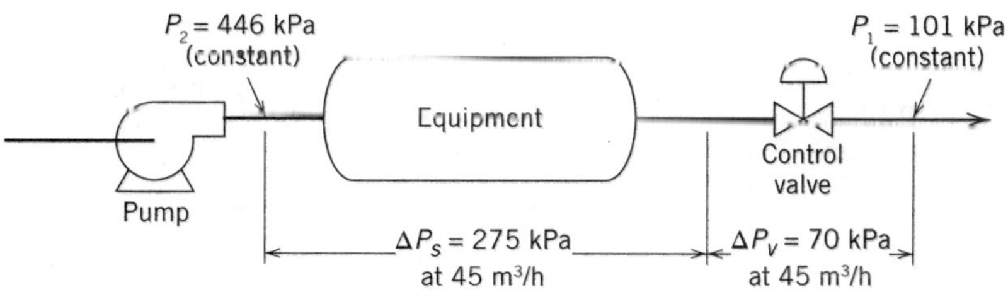

Figure 8.12 A control valve placed in series with a pump and other equipment. Pump discharge pressure is constant.

Table 8.3 Results for Five ΔP_{vd} Values (q_d = 45 m³/h, ΔP_{sd} = 275 kPa)

ΔP_{vd} (kPa)	70	140	280	700	1400
q_{max} (m³/h)	48.86	52.0	57.0	66.0	73.5
q_{min} (m³/h)	18.49	13.54	12.36	10.48	9.77
C_v	123.7	87.5	61.8	39.1	27.7
$\bar{R}$	2.64	3.84	4.61	6.30	7.52

2. As ΔP_{vd} increases, q_{max} increases, q_{min} decreases, and R increases. Thus from a control perspective, a large value of ΔP_{vd}, is desirable to provide effective flow control over a larger range.

3. But as ΔP_{vd} increases, ΔP also increases (see Eq. 8-10). Thus, a larger pump is required and the pump operating cost increases due to the cost of power for the pump motor. Thus from an economic perspective, a small value of ΔP_{vd} is desirable. Note that it is also possible to simultaneously calculate the required pump discharge pressure P_2 and the control valve sizing coefficient C_v that produce the desired values of q_{max} and q_{min}. Thus, Eqs. 8-16 and 8-17 can be solved simultaneously for the two unknowns, ΔP and C_v (Luyben and Luyben, 1996, pp. 76–81). Then P_2 can be determined from Eq. 8-10.

This example demonstrates that the choice of the ratio, $\Delta P_{vd}/\Delta P$, is a key decision in control valve design. It represents an engineering tradeoff between lower cost (small C_v) and more effective control (large C_v). Reported guidelines suggest values of ¼ to ½ for this ratio at the design flow rate. However, Example 8.2 demonstrates that the range of operating conditions and the control objectives are important and should be considered. Another rule of thumb is that the control valve C_v should correspond to a size that is one less than the size of the pipe diameter (Woods, 2007).

8.2.4 Selection of the Control Valve Trim

Three types of valve characteristic (or *valve trim*) are widely used. For a fixed pressure drop across the valve ΔP_v, the fractional flow rate, $f\,(0 \le f \le 1)$ is related to the fractional stem position $\ell(0 \le \ell \le 1)$ by the following valve trim relations:

$$\begin{aligned} &\text{Linear trim:} && f = \ell \\ &\text{Quick opening trim:} && f = \sqrt{\ell} \\ &\text{Equal percentage trim:} && f = R^{\ell-1} \end{aligned} \tag{8-18}$$

For equal percentage control valves, R is a design parameter that is usually in the range 20–50. Figure 8.13 illustrates the valve characteristics for these three valve trims. A quick-opening valve is also referred to as a *square root* valve. The *equal percentage valve* has this name because the slope of the f vs. ℓ curve, $df/d\ell$, is proportional to f, leading to an equal percentage change in flow for a change in ℓ over the entire range, $0 \le \ell \le 1$. Note that when the control valve is completely shut, $\ell = f = 0$ and when it is completely open, $\ell = f = 1$.

Because ΔP_v varies with flow rate, a nonlinear valve characteristic often will yield a more linear flow relation *after installation*, than will a linear characteristic. In particular, the equal percentage valve is designed to compensate, at least approximately, for changes in ΔP_v with flow rate. The objective is to obtain a nearly linear relation for q with respect to ℓ over the normal operating range of the valve.

Some general guidelines for the selection of control valve trim are

1. If the pump characteristic in Fig. 8.14 is fairly flat and the system pressure drop ΔP_s is quite small over the entire operating region, choose a linear trim. However, this situation occurs infrequently because it results from an overdesigned process (pump and piping capacity too large).

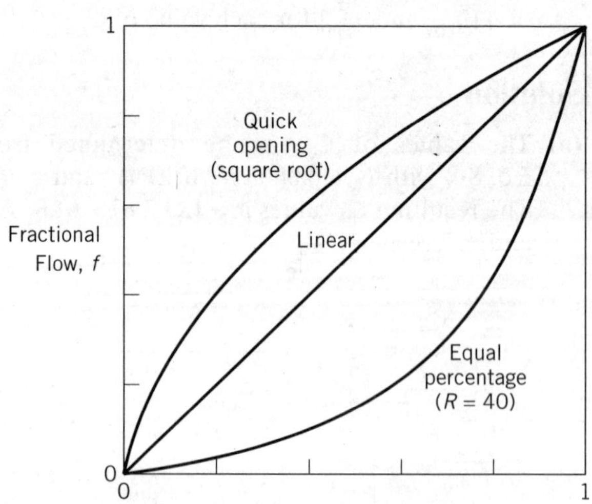

Figure 8.13 Inherent control valve characteristics.

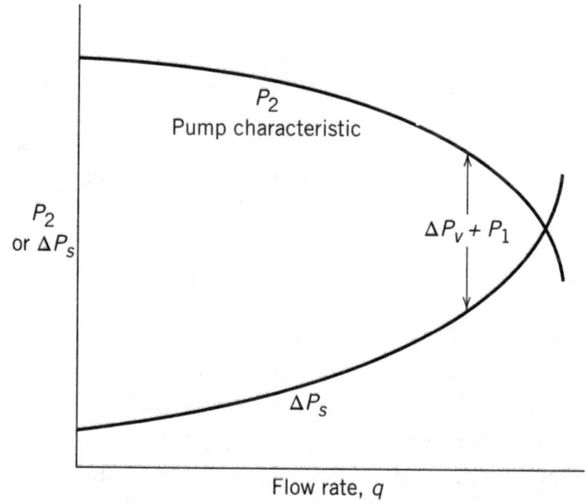

Figure 8.14 Calculation of the valve pressure drop (ΔP_v) from the pump characteristic curve and the system pressure drop without the valve (ΔP_s).

2. To select an equal percentage trim:
   (a) Plot the pump characteristic curve and ΔP_s, as shown in Fig. 8.14. The difference between these two curves is ΔP_v.
   (b) Calculate C_v for the design conditions, q_d and ΔP_{vd}.
   (c) Calculate q as a function of ℓ using Eq. 8-6 and ΔP_v from part (a). A plot of q vs. ℓ should be reasonably linear in the operating region of interest (or at least around the design flow rate). If it is not very linear, adjust C_v and repeat.

In some applications, it is advantageous to use two control valves, even though there is only a single controlled variable. For example, consider a jacketed batch reactor with an exothermic reaction. It may be necessary to initially heat the batch to a temperature at which the

EXAMPLE 8.3

Again consider the schematic diagram in Fig. 8.12. Assume that the pump furnishes a constant head of $\Delta P = 275$ kPa over the entire flow rate range of interest. The system pressure drop is $\Delta P_s = 206$ kPa at the design flow rate of $q_d = 45$ m³/h. The specific gravity of the liquid is $g_s = 1$. Assume that ΔP_s is proportional to q^2. Determine C_v and plot f versus ℓ for the installed control valve and four cases:

(a) A linear valve that is half open at the design flow rate.
(b) An equal percentage valve ($R = 50$ in Eq. 8-18) that is sized to be completely open at 110% of the design flow rate.
(c) Same as **(b)**, except with a C_v that is 20% higher than calculated.
(d) Same as **(b)**, except with a C_v that is 20% lower than calculated.

Solution

First, we write the expression for the system pressure drop,

$$\Delta P_s = Kq^2 \tag{8-19}$$

where

$$K = \frac{\Delta P_{sd}}{q_d^2} = \frac{206\ \text{kPa}}{45\ \text{m}^3/\text{h}} = 0.1017\ \text{kPa}/(\text{m}^3/\text{h})^2 \tag{8-20}$$

Thus

$$\Delta P_s = 0.1017q^2 \tag{8-21}$$

Because the pump head is constant at $\Delta P = 275$ kPa, the pressure drop available for the control valve is

$$\Delta P_v = \Delta P - \Delta P_s = 275 - 0.1017q^2 \tag{8-22}$$

Figure 8.15 illustrates this relationship.

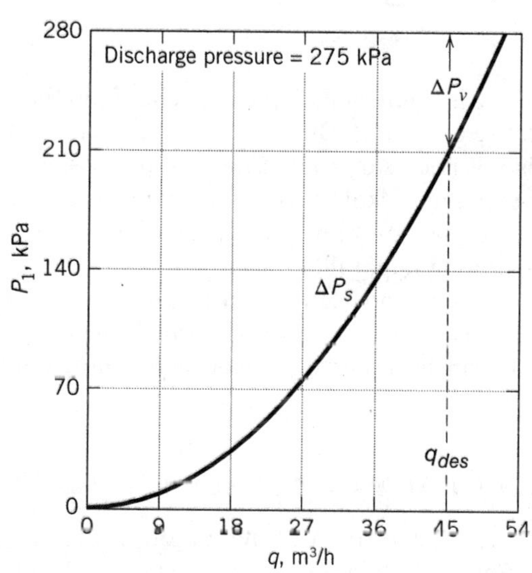

Figure 8.15 Pump characteristic and system pressure drop for Example 8.2.

(a) First, calculate the C_v at the design conditions using Eq. 8-8 with $f = 0.5$ and $N = 0.0865$,

$$C_v = \frac{45}{(0.5)(0.0865)\sqrt{275-206}} = 125.3 \quad (8\text{-}23)$$

Choose $C_v = 125$, a standard valve size in manufacturers' documentation. For a linear valve trim, substitute the relation between ℓ and q in Eq. 8-6:

$$\ell = \frac{q}{C_v N\sqrt{\Delta P_v}} \quad (8\text{-}24)$$

Using Eq. 8-24 and values of ΔP_v from Eq. 8-22, the installed valve characteristic curve can be plotted (Fig. 8.15).

(b) From Eqs. 8-22 and 8-23, $\Delta P_v = 275–(0.1017)(45 \times 1.1)^2 = 25.8$ kPa. The C_v value for 110% of q_d can be calculated.

$$C_v = \frac{(45 \times 1.1)/0.0865}{\sqrt{25.8}} = 112.66 \quad (8\text{-}25)$$

Use a value of $C_v = 115$. For the equal percentage valve, rearrange Eq. 8-6 as follows:

$$\frac{q}{C_v N\sqrt{\Delta P_v}} = R^{\ell-1} \quad (8\text{-}26)$$

Solving for ℓ gives

$$\ell = 1 + \ln\left(\frac{q}{C_v N\sqrt{\Delta P_v}}\right)/\ln R \quad (8\text{-}27)$$

Substituting $C_v = 115$, $R = 50$ and values for q and C_v yields the installed characteristic curve in Fig. 8.16.

(c) $C_v = 1.2(115) = 138$

(d) $C_v = 0.8(115) = 92$

Note that in all four cases, $\Delta P_{vd}/\Delta P_s = 70/206 = 33\%$. From the installed characteristics in Fig. 8.16, we conclude that an equal percentage valve with $C_v = 115$ would give a reasonably linear installed characteristic over a large range of flow rates and have sufficient capacity to accommodate flow rates as high as 110% of the design flow rate.

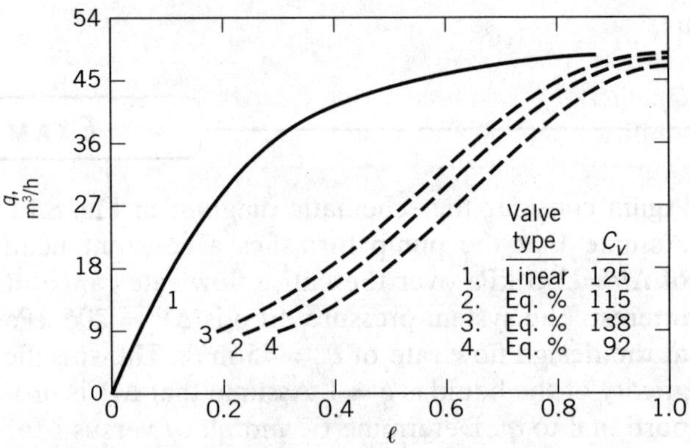

Figure 8.16 Installed valve characteristics for Example 8.2.

reaction rate is significant by passing a hot fluid through the jacket (e.g., steam). But as the exothermic reaction proceeds, it is necessary to remove heat by passing coolant through the jacket in order to control the batch temperature (Section 12.6) with the heating and cooling valves operating in parallel.

For flow control problems where the design flow rate is very large, two control valves in parallel, one large and one small, can be used to provide better flow control than is possible using a single control valve.

8.3 ACCURACY IN INSTRUMENTATION

The accuracy of control instrumentation is very important and inherently related to control system objectives. For example, cooling water flow errors on the order of 10% (of the measured flow rate) might be acceptable in a control loop regulating the temperature of a liquid leaving a condenser, as long as the measurements are simply biased from the true value by a constant amount. On the other hand, a 2% error in the feed flow rate to a process may be unacceptable because throughput/inventory calculations are based on mass balances.

8.3.1 Terminology

We now introduce several terms that are commonly associated with measurement accuracy. The *measurement error* (or *error*) is the difference between the true value and the measured value. Instrument vendors often express accuracy as a *percentage of full scale* (% FS) where the term *full scale* refers to the span of the instrument. Suppose that the % FS error of a temperature transmitter is reported as 1% and the zero and span are adjusted so that the instrument operates over the range of 10–70 °C. Since the span is 70 – 10 = 60 °C, the measurement error is 1% of 60 °C, or 0.6 °C. Consequently, the *relative error* (obtained by dividing the error by the value of the measurement) at 10 °C is 0.6/10 = 6%. Thus when instrument accuracy is expressed as % FS, the relative error can be quite large for small values of the measured variable.

Resolution refers to the smallest interval between two numerical values that can be distinguished. For example, if a temperature transmitter has a resolution of 0.1 °C, it is not possible for it to distinguish between actual temperatures of 21.62 °C and 21.67 °C.

Precision refers to the variability of a measurement for specified conditions and a particular instrument. It is usually expressed in terms of a *standard deviation* or *range*. For example, suppose that a composition sample is divided into four parts, and that the composition of each part is measured using an analyzer. If the four measurements of a key component are 21.3%, 22.7%, 20.6%, and 21.5%, the analyzer precision could be expressed as the range, 22.7 − 20.6 = 2.1%, or as the standard deviation, 0.87%.

Repeatability is similar to precision but refers to the variability of replicate measurements in a set of data. The variability is due to random errors from both known and unknown sources. The variability can be expressed as a range or standard deviation.

Bias refers to a constant error in the data due to a deterministic cause rather than random variations. A thermocouple measurement in a vessel could be consistently lower than the actual fluid temperature because of conduction heat losses, if the thermocouple is in contact with the vessel wall.

Figure 8.17 displays an error analysis for a dataset consisting of replicate measurements at known flow rates. The dataset contains both a systematic error (bias) and random errors.

Instrument manufacturers usually characterize the accuracy of an instrument at a set of standard conditions (e.g., temperature and pressure). But if the instrument is used at other conditions, its accuracy may be affected.

8.3.2 Calibration of Instruments

Any important instrument should be calibrated both initially (before commissioning) and periodically (during service). In recent years, the use of *smart sensors* has become more widespread. These devices incorporate a microcomputer as part of the sensor/transmitter, which can greatly reduce the need for in-service calibration and checkout. Their key features are

1. Checks on the internal electronics, such as verifying that the voltage levels of internal power supplies are within specifications.
2. Checks on environmental conditions (e.g., temperature) within the instruments.
3. Compensation of the measured value for conditions such as temperature and pressure at the measurement point.
4. Linearization of a transmitter output if necessary (e.g., square root of the differential pressure for a head-type flow transmitter) can be done within the instrument, instead of the control system.
5. Remote re-configuration the transmitter from a remote, such as changing its zero or span.
6. Automatic recalibration of the transmitter.

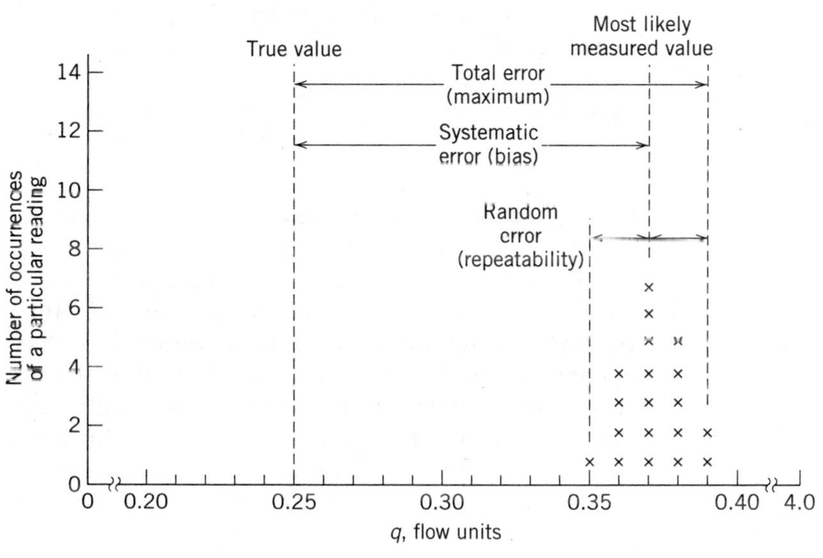

Figure 8.17 Analysis of types of error for a flow instrument whose range is 0 to 4 flow units.

8.3.3 Dynamic Measurement Errors

During transient conditions, the measured value of a process variable can differ significantly from the actual value due to the dynamic behavior of the sensor/transmitter. The concept of dynamic measurement error is illustrated for a temperature sensor. Figure 8.18 shows a thermocouple placed in a metal thermowell with mass m and specific heat C. The dynamic lag introduced by the thermowell/thermocouple combination can be estimated if simplifying assumptions are made. In particular, assume that the well and thermocouple are always at the same temperature T_m, which can be different from the surrounding fluid temperature T. Further assume that heat is transferred only between the fluid and the well (i.e., there are no ambient losses from the thermowell due to conduction along its length to the environment). An energy balance on the thermowell gives

$$mC\frac{dT_m}{dt} = UA(T - T_m) \tag{8-28}$$

where C is the heat capacity, U is the heat transfer coefficient, and A is the heat transfer area. Rearranging gives

$$\frac{mC}{UA}\frac{dT_m}{dt} + T_m = T \tag{8-29}$$

Introducing deviation variables and taking the Laplace transform gives

$$\frac{T'_m(s)}{T'(s)} = \frac{1}{\tau s + 1} \tag{8-30}$$

with $\tau \triangleq mC/UA$.

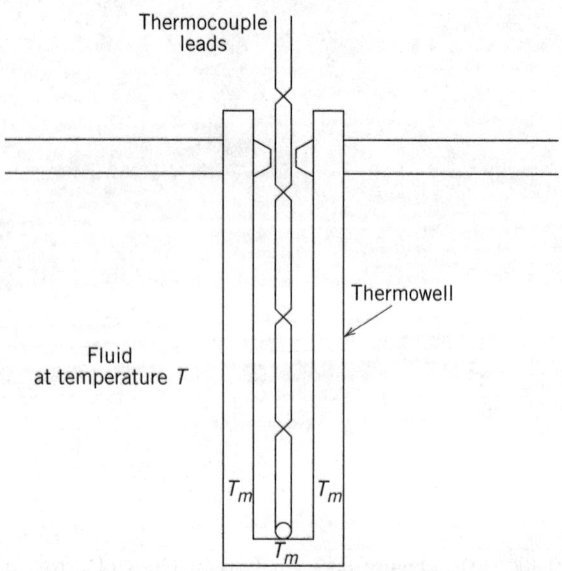

Figure 8.18 Schematic diagram of a thermowell/thermocouple.

The transfer function in Eq. 8-30 indicates that the sensor dynamics will be minimized if the thermal capacitance of the well (mC) is small and UA is large. The combined effect will be to make time constant τ small. Thus, the thermowell should be as thin as possible, consistent with maintaining isolation between the thermocouple and the process fluid. At the same time, because U will be strongly dependent on the fluid velocity, the thermowell should be placed in a region of maximum fluid velocity, near the centerline of a pipe or in the vicinity of a mixing impeller. The dynamic model indicates that materials such as a plastic, which have a lower specific heat C than a metal, will provide a faster response. However, such a material typically has low heat conductivity, which may invalidate the assumption that the entire thermowell is at the same temperature. In this case, a more rigorous model incorporating the effect of heat conduction in the thermowell could be used to evaluate the effect of heat capacitance/conduction tradeoffs.

Any measurement contains some dynamic error. An estimate of the error can be calculated if the sensor time constant τ and the maximum expected rate of change of the measured variable are known. For a ramp input, $x(t) = at$, and a first-order dynamic model (see Eq. 8-30), the sensor output y_m is related to the measured value y:

$$Y_m(s) = \frac{1}{\tau s + 1}Y(s) = \frac{1}{\tau s + 1}\frac{a}{s^2} \tag{8-31}$$

The ramp response $y(t)$ of a first-order system was obtained in Eqs. 5-18 through 5-20. The maximum deviation between input and output is $a\tau$ (obtained when $t \gg \tau$), as shown in Fig. 5.4. Hence, as a general result, we can say that the maximum dynamic error that can occur for any instrument with *first-order* dynamics is

$$\varepsilon_{\max} = \left|y_m(t) - y(t)\right|_{\max} = a\tau$$

Clearly, by reducing the sensor time constant, the dynamic error can be made negligibly small.

In general, measurement time constants should be less than one-tenth the largest process time constant, preferably much less, in order to reduce dynamic measurement errors. The dynamics of measurement and final control elements can significantly limit the speed of response of the controlled process (Chapter 9). Thus, it is important that the dynamics of these components be made as fast as is practical and economical.

8.4 PIPING AND INSTRUMENTATION DIAGRAMS (P&ID)

A piping and instrumentation diagram (P&ID) is the final diagram drawn by a process design engineer after drawing block flow diagram (BFD) and process flow diagram (PFD). It contains items that do not appear in PFD. P&ID shows the piping and process equipment together with the instrumentation and control devices; the positioning of all valves and the size, schedule and material of construction of piping. It is used by field technicians, engineers, and operators to better understand the process and to know how the instrumentation is inter connected.

8.4.1 Line Symbols

Various signal and material flow lines are given in Table 8.4.

Table 8.4 Line Symbols in Piping and Instrumentation Diagrams

Line	Symbol	Line	Symbol
Major process line		Minor process line	
Electric line		Pneumatic line	
Hydraulic line		Capilari tubing	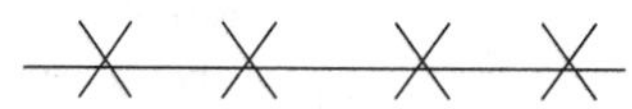
Mechanical link		Electromagnetic line	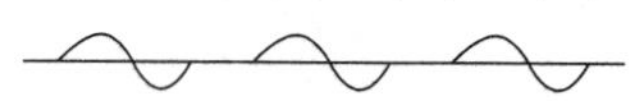
Software or data link		Connected line	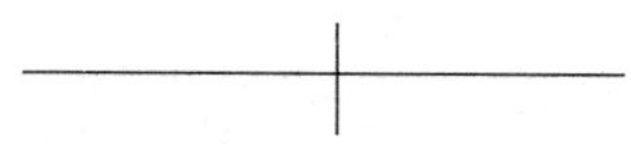
Jacketed line	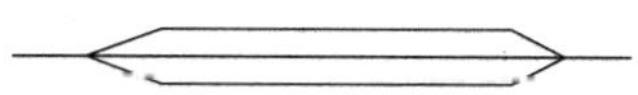	Non-connecting line	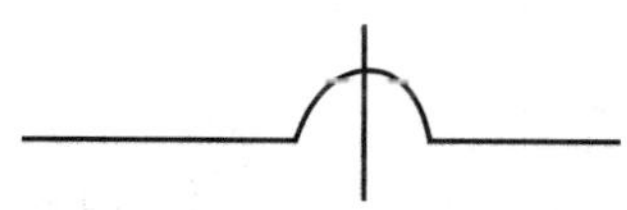
Electrically heated pipe		Insulated pipe	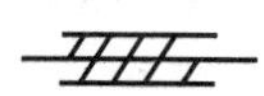
Jacketed pipe		Heated or cooled pipe	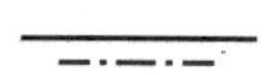
Future equipment		Pipe or line	
Flexible connection		Guided electromagnetic or sonic signal	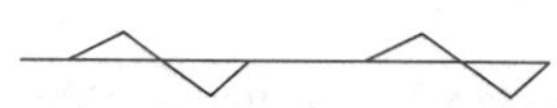

8.4.2 Valve Symbols

The commonly used valves are shown in Table 8.5 and the control valves are shown in Table 8.6.

8.4.3 General Instrument or Function Symbols

A circle indicates individual measuring instrument and a single horizontal line running across the center of the shape shows that the instrument is located in a primary location (e.g., a control room). A double line indicates that it is in an auxiliary location (e.g., an instrument rack) and absence of a line indicates that it is field mounted. The dotted line indicates that it is inaccessible whereas a square with a circle inside represents instruments that both display measurement readings and perform some control function. A square with a diamond inside represents PLCs. A hexagon represents computer functions, such as those carried out by a controller. The instrumentation or function symbols are shown in Table 8.7.

8.4.4 Identification Letters and Tag Numbers

In general there are two or more letters and a number are marked inside the instrument or function symbols.

Table 8.5 Symbols for Commonly Used Valves

Valve	Symbol	Valve	Symbol
Gate valve		Globe valve	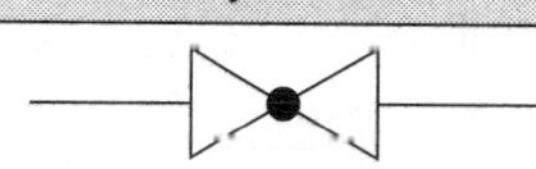
Check valve	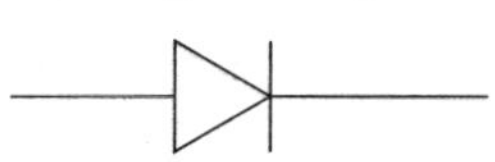	Ball valve	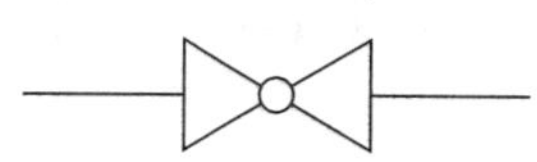

(Continued)

Table 8.5 (Continued)

Valve	Symbol	Valve	Symbol
Needle valve		Butterfly valve	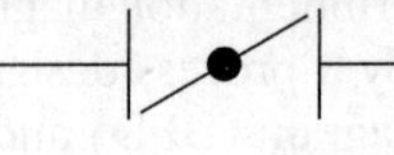
Angle valve	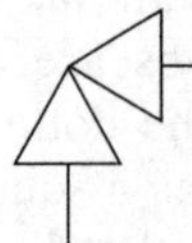	Relief valve	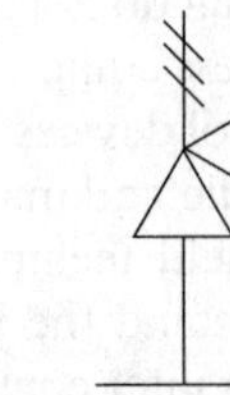

Table 8.6 Control Valves

Control Valve	Symbol	Control Valve	Symbol
Control valve (Fail closed)	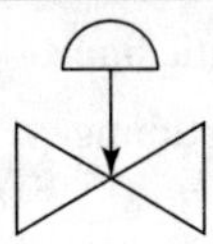	Motorized valve	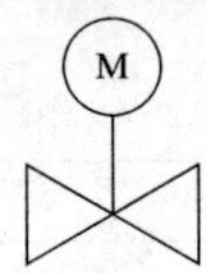
Control valve (Fail open)	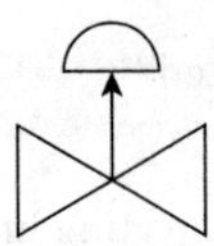	Solenoid valve	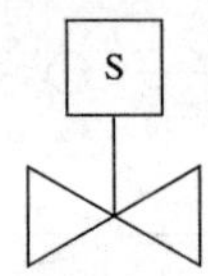
Manual valve	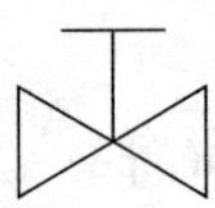	Electric valve	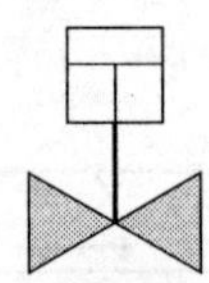

Table 8.7 Instrument or Function Symbols

Instrument or Function	Control Room	Auxiliary	Field	Not Accessible
Discrete Instruments	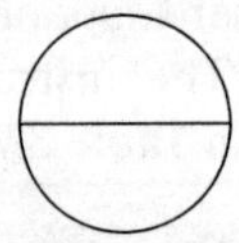	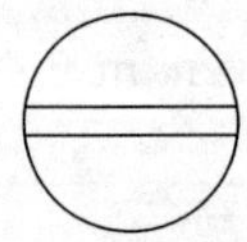	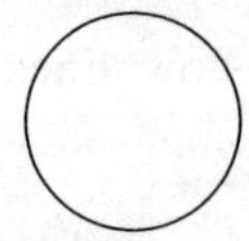	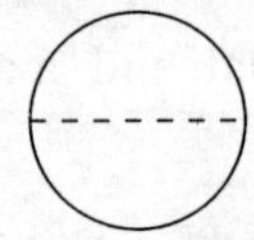
Shared Control/Display Elements	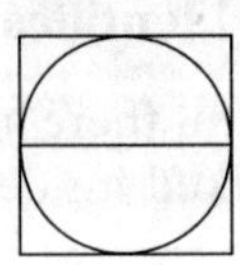	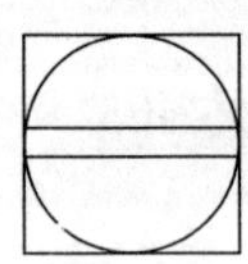	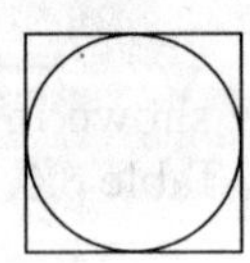	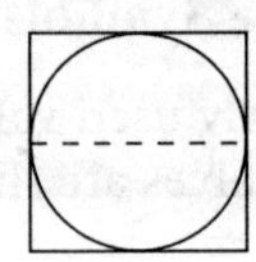
Computer Functions (Controllers)	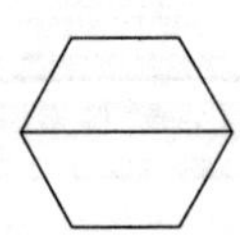	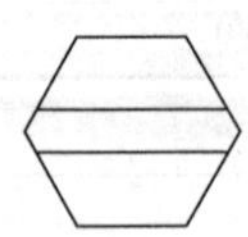	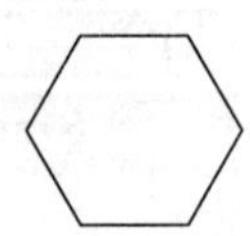	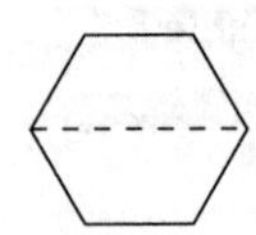
Programmable Logic Controllers (PLCs)	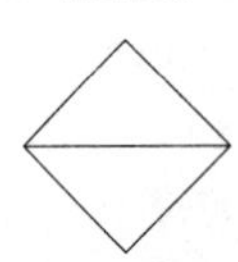	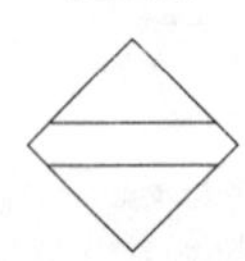	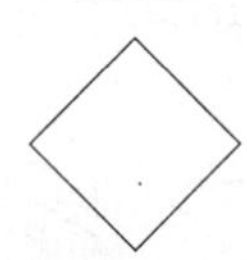	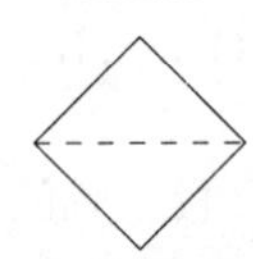

The letters and number are called as the identification letters and tag number, respectively. This is shown in Figs. 8.19 and 8.20. Some commonly used identification letters are given in Tables 8.8 and 8.9.

Figure 8.19 Identification Letters and the Tag Number.

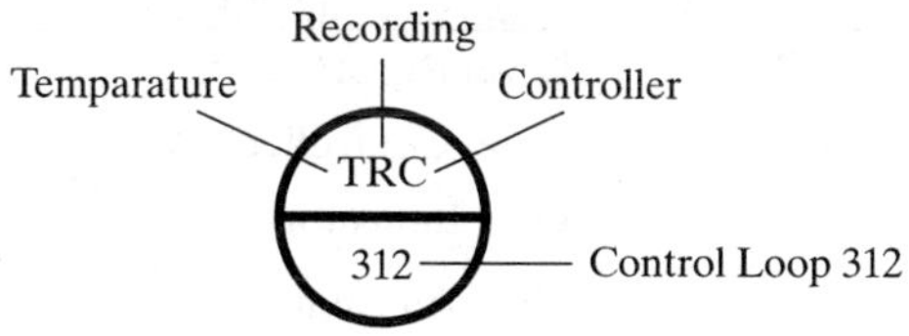

Figure 8.20 Meaning of the Identification Letter and the Tag No.

Table 8.8 Identification Letters

First Letter	Parameter controlled (Measured variable)	Second or Third Letter	Type of Conditioner
A	Analysis	A	Alarm
B	Burner, combustion	C	Controller
C	Conductivity	E	Sensor Element
D	Density	G	Gauge or Glass
E	Voltage	I	Indicate
F	Flow rate	K	Control station
H	Hand	L	Light or Low
I	Current	M	Middle or Intermediate
J	Power	O	Orifice, restriction
K	Time	P	Point
L	Level	R	Record
M	Moisture	S	Switch
N	Torque	T	Transmit
P	Pressure or Vacuum	V	Valve, damper or louver
Q	Quantity	Y	Relay, compute, convert
S	Speed or frequency	Z	Drive
T	Temperature		
V	Viscosity or Vibration		
W	Weight or Force		
Y	Event		
Z	Position		

Table 8.9 Some Typical Identification Codes

Identification Code	Meaning	Identification Code	Meaning
TI	Temperature Indicator	LA	Level Alarm
TT	Temperature Transmitter	LG	Level Gauge
TR	Temperature Recorder	TE	Temperature Element
TC	Temperature Controller	FE	Flow Element
LI	Level Indicator	I/P	Current to Pressure Transducer
LT	Level Transmitter	AT	Analyzer Transmitter

(*Continued*)

Table 8.9 (Continued)

Identification Code	Meaning	Identification Code	Meaning
LR	Level Recorder	PIC	Pressure Indicating Controller
LC	Level Controller	PRC	Pressure Recording Controller
FI	Flow Indicator	PSH	Pressure Switch High
FT	Flow Transmitter	PIT	Pressure Indicating Transmitter
FR	Flow Recorder	PAH	Pressure Alarm High
FC	Flow Controller	FSL	Flow Switch Low
PI	Pressure Indicator	LSH	Level Switch High
PT	Pressure Transmitter	LSL	Level Switch Low
PR	Pressure Recorder	LAH	Level Alarm High
PC	Pressure Controller	PDIR	Pressure Diff. Indicating Recorder
TY	Temperature Relay	RV	Relief Valve
TW	Temperature Well	PSV	Pressure Safety Valve
LG	Level Glass	SDV	Shut Down Valve

8.4.5 Symbols of Process Equipment

The P&ID symbols for some commonly used process equipment are given in Table 8.10.

The Piping and Instrumentation Diagram for an exothermic CSTR is given in Fig. 8.21.

Table 8.10 Symbols of Process Equipment

Equipment	Symbol	Equipment	Symbol
Vessel/Tank	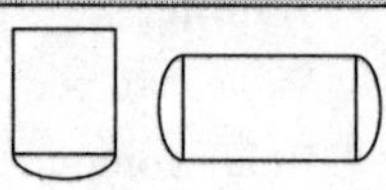	Pump	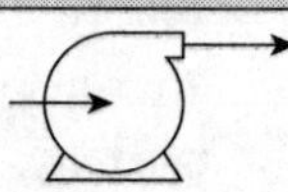
Heat Exchanger	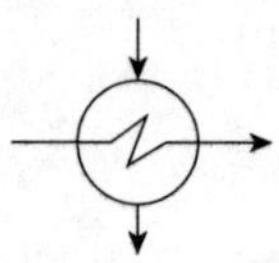	Dryer	
Plate/Tray Column	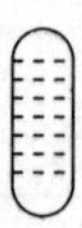	Pressure Vessel	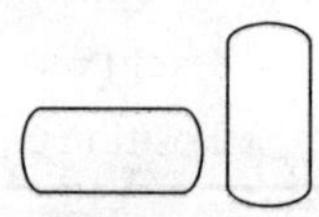
Packed Column		Jacketed Vessel	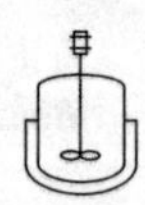
Furnace	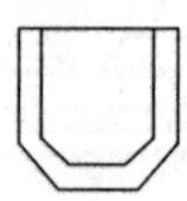	Cooling Tower	
Cooler		Fan	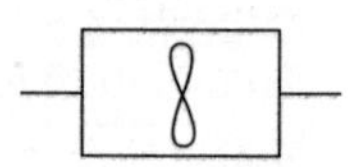
Ejector	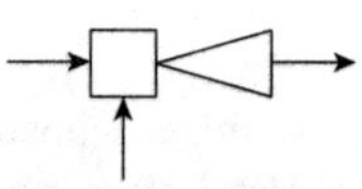	Compressor	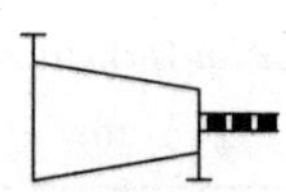

Table 8.10 (Continued)

Equipment	Symbol	Equipment	Symbol
Orifice plate	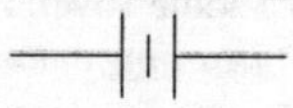	Pitot tube	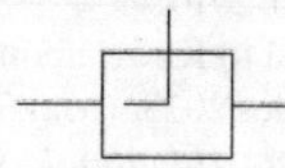
Venturi meter	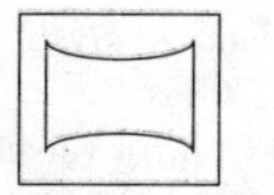	Pipe support	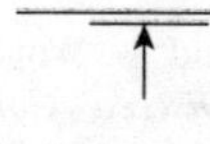

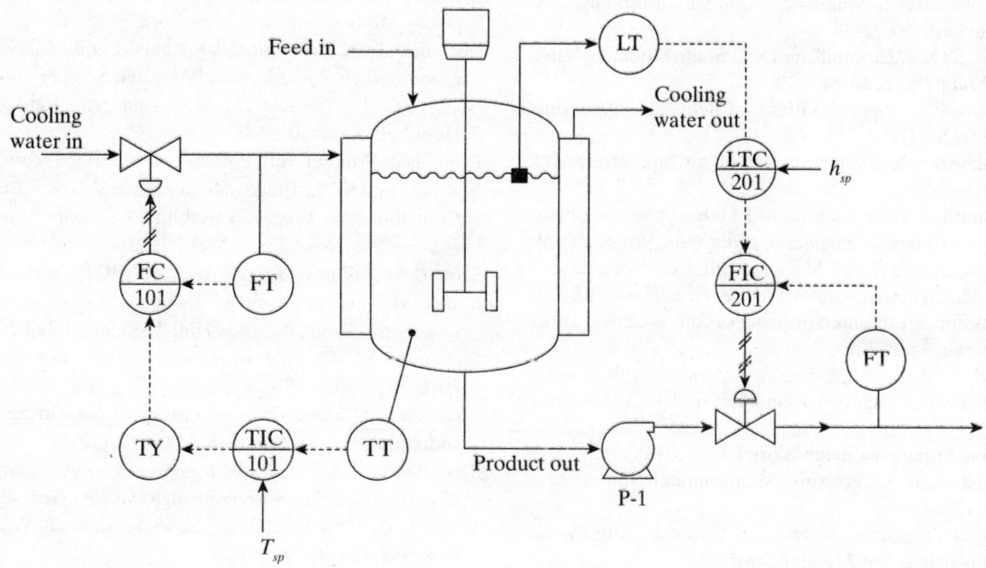

Figure 8.21 Piping and Instrumentation Diagram for a jacketed CSTR.

SUMMARY

This chapter has considered the instrumentation and final control elements required for process control applications. Sensors provide measurements of process variables, while transmitters convert the measurements into equivalent signals that can be transmitted to controllers. Accurate measurements of key process variables such as chemical composition, flow rate, and temperature are of critical importance for process safety and optimal process performance. Thus, sensors and transmitters should be carefully selected, installed, and maintained. An accurate instrument that is improperly installed or maintained loses much of, its inherent value.

Final control elements such as control valves and variable speed pumps adjust manipulated variables based on output signals from the controllers. As such, they are a critical component of a feedback control system and should be carefully designed, installed and maintained.

Two current technology trends are (1) the increasing use of smart sensors with digital logic and memory capabilities that facilitate internal calibration and diagnostic capabilities and (2) digital and wireless communication systems that provide greater flexibility and reduced installation and maintenance costs, compared to traditional wired networks. The diagnostics and flexibility of smart sensors and digital communication networks facilitate the development of improved control systems that can be readily updated for changes in process conditions or control objectives.

Piping and Instrumentation Diagrams (P&ID) are briefly discussed in this chapter.

REFERENCES

Agrawal, G. P., *Fiber-Optic Communication Systems,* 4th ed., John Wiley and Sons, Hoboken, NJ, 2010.

Anon., *Control*, What You Need to Know about Fieldbus Now, http://www.controlglobal.com/articles/2013/ (May 7, 2013).

Baker, R. C., *Flow Measurement Handbook*, Cambridge University Press, New York, 2000.

Blevins, T., D. Chen, M. Nixon, and W. Wojsznis, *Wireless Control Foundation: Continuous and Discrete Control for the Process Industry*, ISA, Research Triangle Park, NC, 2014.

Blevins, T., W. Wojsznis, and M. Nixon, *Advanced Control Foundation: Tools, Techniques, and Applications*, ISA, Research Triangle Park, NC, 2013.

Borden Jr., G., and P. G. Friedmann (Eds.), *Control Valves*, ISA, Research Triangle Park, NC, 1998.

Caro, D., *Wireless Networks for Industrial Automation* 4th ed., ISA, Research Triangle Park, NC, 2014.

Choudhury, M. A. A. S., N. F. Thornhill, and S. L. Shah, Modeling Valve Stiction, *Control Eng. Pract.*, **13**, 641 (2005).

Chow, A. W., Lab-on-a-Chip: Opportunities for Chemical Engineering, *AIChE J.*, **48**, 1590 (2002).

Daw, R., and J. Finkelstein, Introduction: Lab on a Chip, *Nature*, **442**, 367 (27 July 2006).

Edgar, T. F., C. L. Smith, B. W. Bequette, and J. Hahn, Process Control, Section 8 in *Perry's Chemical Engineers Handbook*, 9th ed., D. W. Green and M. Z. Southard (Eds.), McGraw-Hill, New York, 2016.

Emerson Process Management, *Control Valve Handbook*, 4th ed., http://www.documentation.emersonprocess.com/groups/public/documents/book/cvh99.pdf, 2005.

Felder, R. M., and R. W. Rousseau, *Elementary Principles of Chemical Processes, 3rd Ed. Update,* page 334, John Wiley and Sons, Hoboken, NJ 2005.

Fieldbus Foundation, http://www.fieldbus.org/(Feb., 2015).

Gassman, E., Introduction to Pressure Measurement, *Chem. Eng. Prog.*, **110** (6), 28 (2014).

Hebert, D., Digital Fieldbus Networks, *Control,* http://www.controlglobal.com/articles/2007 (Feb. 1, 2007).

Instrumentation Symbols and Identification, Standard ANSI/ISA-5.1–2009, ISA, Research Triangle Park, NC, 2009.

Johnson, C. D., *Process Control Instrumentation Technology*, 8th ed., Prentice-Hall, Upper Saddle River, NJ, 2007.

Lipták, B. G., *Analytical Instrumentation*, CRC Press, Boca Raton, FL, 1994.

Lipták, B. (Ed.), *Instrument Engineers' Handbook*, 4th ed., *Vol. I: Process Measurement and Analysis; Vol. 2: Process Control*, CRC Press, Boca Raton, FL, 2006a.

Lipták, B., How to Select Control Valves, Parts 1–3, *Control Magazine*, www.controlglobal.com (July 14, 2006b).

Lipták, B., How to Select an Analyzer, *Control Magazine*, www.controlglobal.com (Jan. 14, 2014).

Luyben, W. L., and M. L. Luyben, *Essentials of Process Control*, McGraw-Hill, New York, 1996.

McMillan, G. K., *Tuning and Control-Loop Performance*, 4th ed., p. 124, Momentum Press, New York, 2015.

Michalski, L., K. Eckersdorf, J. Kucharski, and J. McGhee, *Temperature Measurement,* 2nd ed., John Wiley and Sons, Hoboken, NJ, 2001.

Nichols, G. D., *On-Line Process Analyzers*, John Wiley and Sons, Hoboken, NJ, 2010.

Profibus & Profinet, Int., Profibus, www.Profibus.com. (Feb., 2015).

Ross, E., and W. L. Bialkowski, Control Valves—The Biggest Single Contributor to Process Variability, *Instrument Symposium for the Process Industries*, Texas A&M, January, 1996.

Scott, D. M., *Industrial Process Sensors*, CRC Press, Boca Raton, FL, 2007.

Shuler, M. P., F. Kargi, and M. DeLisa, *Bioprocess Engineering: Basic Concepts*, 3rd ed., Prentice-Hall, Upper Saddle River, NJ, 2014.

Spitzer, D. W., *Flow Measurement*, 2nd ed., ISA, Research Triangle Park, NC, 2001.

Strothman, J., More Than a Century of Measuring and Controlling Industrial Processes, *InTech*, **42** (6), 52 (1995).

Studebaker, P., 25 Years of Progress in Process Analyzers' Efficiency, *Control Magazine*, www.controlglobal.com (Feb. 19, 2014).

White, D. C., How Much Is a Good Measurement Worth?, *Chem. Eng. Progress*, **110** (1), 50 (2014).

Woods, D. R., *Rules of Thumb in Engineering Practice*, Wiley-VCH Verlag GmbH & Co. KGaA, Weinheim, Germany (2007).

EXERCISES

8.1 Several linear transmitters have been installed and calibrated as follows:

Flow rate:	92 m^3/h	→ 100 kPa(g)	Pneumatic transmitter
	0 m^3/h	→ 20 kPa(g)	
Pressure:	760 mm Hg	→ 20 mA	Current transmitter
	254 mm Hg	→ 4 mA	
Level:	10 m	→ 5 VDC	Voltage transmitter
	0.5 m	→ 1 VDC	
Concentration:	20 g/L	→ 10 VDC	Voltage transmitter
	3 g/L	→ 1 VDC	

(a) Develop an expression for the output of each transmitter as a function of its input. Be sure to include appropriate units.

(b) What is the gain of each transmitter? Zero? Span?

8.2 A process instrumentation diagram for a flash drum is shown in Fig. E8.2. Steam is condensed in a steam coil to vaporize a portion of the liquid feed, and the liquid product is removed by a pump. There are five control valves for the steam flow, vapor product, liquid product, feed flow, and steam chest (which allows the steam chest to be rapidly evacuated in emergency situations). Determine whether the five valves should be air-to-open (AO) or air-to-close (AC) for safe operation, for each of three cases:

(a) The safest conditions are achieved by the lowest temperature and pressure in the flash vessel.

(b) Vapor flow to downstream equipment can cause a hazardous situation.

(c) Liquid flow to downstream equipment can cause a hazardous situation.

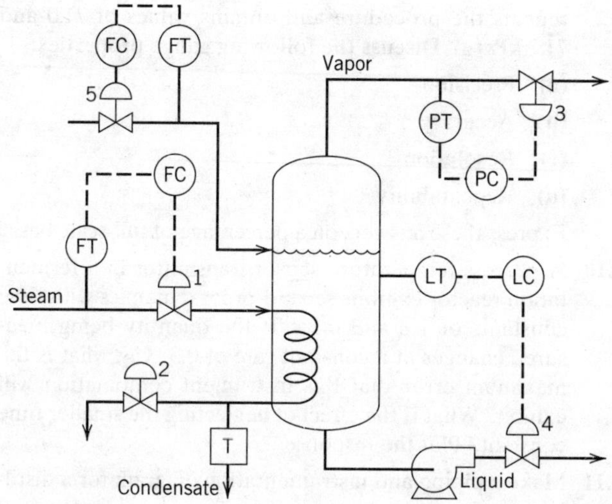

Figure E8.2

Discuss various scenarios of air failure (or power failure).

8.3 Suppose that the temperature in an exothermic continuous stirred-tank reactor is controlled by manipulating the coolant flow rate using a control valve. A PID controller is used and is well-tuned. Which of these changes could adversely affect the stability of the closed-loop system? Briefly justify your answers.

(a) The span of the temperature transmitter is increased from 25°C to 45°C.

(b) The zero of the temperature transmitter is increased from 20°C to 25°C.

(c) The control valve trim is changed from linear to equal percentage.

(d) The feed flow rate is doubled.

8.4 A coolant (specific gravity = 1.2) is pumped through a heat exchanger and control valve at a nominal (desired) flow rate of q_d = 0.6 m³/min. The total pressure drop over this heat exchanger system (including piping and fittings) is constant at 450 kPa. The pressure drop over the system without the control valve, ΔP_s, is proportional to the flow rate squared. Its nominal value at the desired flow rate is ΔP_{sd} = 200 kPa. The control valve is to be designed so that the nominal flow rate is 2/3 of the maximum flow rate when the valve is completely open. Based on this information, what value of the valve coefficient C_v should be specified?

8.5 A pneumatic control valve is used to adjust the flow rate of a petroleum fraction (specific gravity = 0.9) that is used as fuel in a cracking furnace. A centrifugal pump is used to supply the fuel, and an orifice meter/differential pressure transmitter is used to monitor flow rate. The nominal fuel rate to the furnace is q_n = 75 m³/h. Select an equal percentage valve that will be satisfactory to operate this system and $f = 1$ in Eq. 8-7: Use the following data (all pressures in kPa; all flow rates in m³/h):

(a) Pump characteristic (discharge pressure):

$$P = (1 - 4.61 \times 10^{-5} q^2) P_{de}$$

where P_{de} is the pump discharge pressure when the pump is dead ended (no flow).

(b) Pressure drop across the orifice:

$$\Delta P_0 = 0.0255 q^2$$

(c) Pressure drop across the furnace burners:

$$\Delta P_b = 275$$

(d) R = 50 for the control valve.

(e) Operating region of interest:

$$60 \le q \le 80$$

This design attempt should attempt to minimize pumping costs by keeping the pump capacity (related to P_{de}) as low as possible. In no case should $\Delta P_v / \Delta P_s$ be greater than 0.33 at the nominal flow rate. Show, by means of a plot of the installed valve characteristic (q vs. ℓ), just how linear the final design is.

8.6 Consider the evaporator and control system in Figure 13.6. Briefly answer the following questions:

(a) Should each control valve be air-to-open (AO) or air-to-close (AC)?

(b) Should each PI controller be direct-acting or reverse-acting?

8.7 It has been suggested that the liquid flow rate in a large diameter pipeline could be better regulated by using two control valves instead of one. Suppose that one control valve has a large C_v value, that the other has a small C_v value, and that the flow controller will primarily adjust the smaller valve while also making occasional adjustments to the large valve, as needed. Which of the two alternative configurations in Fig. E8.7 seems to be the more promising: placing the control valves in series (Configuration I), or in parallel (Configuration II)? Briefly justify your answer.

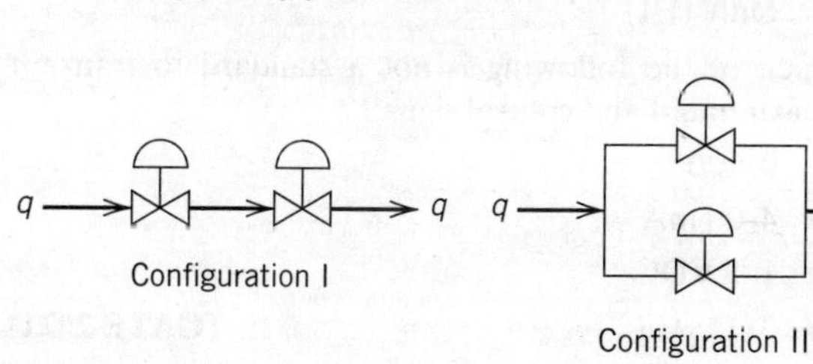

Figure E8.7

8.8 The dynamic behavior of a temperature sensor/transmitter can be modeled as a first-order transfer function (in deviation variables) that relates the measured value T_m to the actual temperature, T

$$\frac{T_m'(s)}{T'(s)} = \frac{1}{8s+1}$$

Both T_m and T' have units of °C and the time constant has units of seconds. An alarm will sound when T_m exceeds 30 °C. If the process is initially at steady state, and then T suddenly changes from 25 to 35 °C at 1:10 PM, at what time will the alarm sound?

8.9 An engineer sets the pressure in a supply tank using a very accurate manometer as a guide and then reads the output of a 140-kPa(g) pressure gauge attached to the tank as 70.3 kPa(g). Sometime later, the engineer repeats the procedure and obtains values of 71.0 and 71.7 kPa(g). Discuss the following gauge properties:

(a) Precision

(b) Accuracy

(c) Resolution

(d) Repeatability

Express these answers on a percentage of full-scale basis.

8.10 A process temperature sensor/transmitter in a fermentation reactor exhibits second-order dynamics with time constants of 1 s and 0.1 s. If the quantity being measured changes at a constant rate of 0.3 °C/s, what is the maximum error that this instrument combination will exhibit? What is the effect of neglecting the smaller time constant? Plot the response.

8.11 Make a piping and instrumentation diagram for a distillation column.

MULTIPLE CHOICE QUESTIONS

1. The inherent characteristics of three control values P, Q and R are shown in the figure.

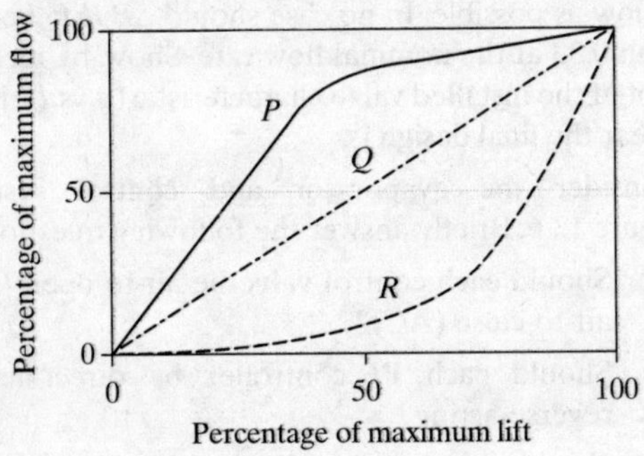

(I) P is a quick opening valve.

(II) P is an equal percentage valve.

(III) Q is a quick opening valve.

(IV) R is an equal percentage valve. **(GATE 2021)**

The correct option(s) is/are

(a) Only (I)

(b) (II) and (III)

(c) (I) and (IV)

(d) Only (III)

2. Which of the following is not a standard to transmit measurement and control signals?

(a) 0–100%

(b) 4–20 mA

(c) 1–5 VDC

(d) 3–15 psig **(GATE 2021)**

3. Which of the following is not a standard industrial range for pneumatic signals?

(a) 0.2 to 1 kPa(g)

(b) 0.2 to 1 kgf/cm²(g)

(c) 20 to 100 kPa(g)

(d) 3 to 15 psig

4. Hot-wire anemometer is used for the measurement of

(a) composition.

(b) flow.

(c) pressure.

(d) temperature. **(GATE 2020)**

5. A thermocouple senses temperature based on the

(a) Nernst effect

(b) Maxwell effect

(c) Seebeck effect

(d) Peltier effect **(GATE 2019)**

6. The liquid flow rate through an equal percentage control valve, when fully open, is 150 gal/min and the corresponding pressure drop is 50 psi. If the specific gravity of the liquid is 0.8, then the valve coefficient, C_v, in gal/(min psi$^{0.5}$) is ______ (rounded off to two decimal places).

(a) 20.08

(b) 16.75

(c) 18.92

(d) 18.97 **(GATE 2019)**

7. Match the variables in **Group – I** with the instruments in **Group – II**

Group – I	Group – II
(P) Temperature	(I) Capacitance probe
(Q) Liquid level	(II) McLeod gauge
(R) Vacuum	(III) Chromatograph
(S) Concentration	(IV) Thermistor

Choose the correct set of combinations.

(a) P-IV, Q-III, R-II, S-I

(b) P-I, Q-II, R-IV, S-III

(c) P-IV, Q-I, R-II, S-III

(d) P-III, Q-II, R-I, S-IV **(GATE 2017)**

8. An LVDT (Linear Variable Differential Transformer) is a transducer used for converting

(a) displacement to voltage.
(b) voltage to displacement.
(c) resistance to voltage.
(d) voltage to current. **(GATE 2017)**

9. Match the Instruments in Group – I with Process variables in Group – II.

Group – I	**Group – II**
(P) Conductivity meter	(I) Flow
(Q) Turbine meter	(II) Pressure
(R) Piezoresistivity elements	(III) Composition

(a) P-II, Q-I, R-III
(b) P-II, Q-III, R-I
(c) P-III, Q-II, R-I
(d) P-III, Q-I, R-II **(GATE 2016)**

10. A liquid flows through an "equal percentage" valve at a rate of 2 m^3/h when the valve is 10% open. When the valve opens to 20% the flowrate increases to 3 m^3/h. Assume that the pressure drop across the valve and the density of the liquid remain constant. When the valve opens to 50%, the flowrate (in m^3/h, rounded off to the second decimal place) is ______

(a) 9.12
(b) 8.13
(c) 11.13
(d) 10.12 **(GATE 2016)**

11. Match the following.

Group – I	**Group – II**
(P) Viscosity	(I) Pyrometer
(Q) Pressure	(II) Hot wire anemometer
(R) Velocity	(III) Rheometer
(S) Temperature	(IV) Piezoelectric element

(a) P-IV, Q-III, R-I, S-II
(b) P-III, Q-IV, R-II, S-I
(c) P-III, Q-IV, R-I, S-II
(d) P-IV, Q-III, R-II, S-I **(GATE 2013)**

12. A thermocouple having a linear relationship between 0°C and 350°C shows an emf of zero and 30.5 mV, respectively at these two temperatures. If the cold junction temperature is shifted form 0°C and 30°C, then the emf correction (in mV) is

(a) 3.13
(b) 2.92
(c) 2.61
(d) 2.02 **(GATE 2012)**

13. Match the process parameters in **Group – I**, with the measuring instruments in **Group – II.**

Group – I	**Group – II**
(P) Flame temperature	(I) Thermocouple
(Q) Composition of LPG	(II) Radiation pyrometer
(R) Liquid air temperature	(III) Gas chromatograph

(a) P-III, Q-I, R-II
(b) P-I, Q-III, R-II
(c) P-II, Q-III, R-I
(d) P-II, Q-I, R-III

14. The range of standard current signal in process instruments is 4 to 20 mA. Which ONE of the following is the reason for choosing the minimum signal as 4 mA instead of 0 mA?

(a) To minimize resistive heating in instruments
(b) To distinguish between signal failure and minimum signal condition
(c) To ensure a smaller difference between maximum and minimum signal
(d) To ensure compatibility with other instruments **(GATE 2011)**

15. Flow measuring instruments with different specifications (zero and span) are available for an application that requires flow rate measurements in the range of 300 litres/h to 400 L/h. The appropriate instruments for this application is the ONE whose specifications are

(a) zero = 175 L/h, span = 150 L/h
(b) zero = 375 L/h, span = 100 L/h
(c) zero = 275 L/h, span = 150 L/h
(d) zero = 475 L/h, span = 100 L/h **(GATE 2010)**

16. Which ONE of the following sensors is used for the measurement of temperature in a combustion process (T > 1800°C)?

(a) Type J thermocouple
(b) Thermistor
(c) Resistance temperature detector
(d) Pyrometer **(GATE 2009)**

17. Match the following.

Group – I	**Group – II**
(P) Temperature	(I) Hot wire anemometry
(Q) Pressure	(II) Strain Gauge
(R) Flow	(III) Chromatographic analyzer
	(IV) Pyrometer

(a) P-I, Q-II, R-III
(b) P-IV, Q-I, R-III
(c) P-I, Q-II, R-IV
(d) P-IV, Q-II, R-I **(GATE 2008)**

18. The control valve characteristics for three types of control valves (P, Q, and R) are given in the figure below. Match the control valve with its characteristics.

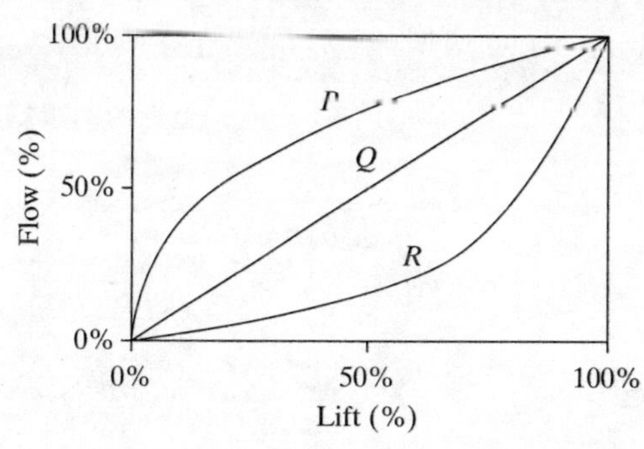

(a) P – Quick opening, Q – Linear, R – Equal percentage
(b) P – Linear, Q – Square root, R – Equal percentage
(c) P – Equal percentage, Q – Linear, R – Quick opening
(d) P – Square root, Q – Quick opening, R – Linear **(GATE 2006)**

19. The control valve characteristic is selected such that the product of process gain and the valve gain
(a) is a linearly increasing function of the manipulated variable.
(b) is a linearly decreasing function of the manipulated variable.
(c) remains constant as the value of the manipulated variable changes.
(d) is an exponentially increasing function of the manipulated variable. **(GATE 2005)**

20. Match the process variables **Group – I** given below with the measuring devices **Group – II**

Group – I	**Group – II**
(P) High Temperature	(I) Orifice meter
(Q) Flow	(II) Chromatograph
(R) Composition	(III) Radiation Pyrometer
	(IV) Bi-metallic Thermometer

(a) P-I, Q-II, R-III
(b) P-I, Q-III, R-II
(c) P-III, Q-I, R-II
(d) P-IV, Q-II, R-I **(GATE 2005)**

21. Match the measured process variables in **Group – I** with the list of measuring devices in **Group – II**:

Measured Process Variables	**Measuring Devices**
(P) Temperature	(I) Bourdon tube element
(Q) Pressure	(II) Orifice plates
(R) Flow	(III) Infrared analyzer
(S) Liquid level	(IV) Displacer devices
(T) Composition	(V) Pyrometer

(a) P-V, Q-I, R-II, S-IV, T-III
(b) P-III, Q-I, R-IV, S-II, T-V
(c) P-I, Q-III, R-IV, S-I, T-V
(d) P-III, Q-I, R-II, S-IV, T-V **(GATE 2003)**

22. The calibration data of a thermocouple with its cold junction at 0°C are given below.

Hot junction temperature (°C)	0	20	40	60	80	100
Thermo emf (mV)	0.00	0.80	1.61	2.43	3.26	4.10

The hot junction of the thermocouple is placed in a bath at 80°C while its cold junction is at 20°C. What is the emf of the thermocouple?
(a) 3.26 mV **(b)** 0.80 mV
(c) 2.46 mV **(d)** 2.43 mV **(GATE 2001)**

ANSWER KEY

1. (c)	**2.** (a)	**3.** (a)	**4.** (b)	**5.** (c)	**6.** (d)	**7.** (c)
8. (a)	**9.** (d)	**10.** (d)	**11.** (b)	**12.** (c)	**13.** (c)	**14.** (b)
15. (c)	**16.** (d)	**17.** (d)	**18.** (a)	**19.** (c)	**20.** (c)	**21.** (a)
22. (d)						

CHAPTER 9

Dynamic Behavior and Stability of Closed-Loop Control Systems

CHAPTER CONTENTS

In this chapter, we consider the dynamic behavior of processes that are operated using feedback control. This combination of the process, the feedback controller, and the instrumentation is referred to as a *feedback control loop* or a *closed-loop system*. Thus, the term *closed-loop system* is used to denote the controlled process. We begin by demonstrating that block diagrams and transfer functions provide a useful description of closed-loop systems. We then use block diagrams to analyze the dynamic behavior of several simple closed-loop systems.

Although feedback control yields many desirable characteristics, it also has one undesirable characteristic. If the controller is poorly designed or the process dynamic characteristics change after the controller is implemented, the resulting closed-loop system can be unstable. This means that the controller can produce a growing oscillation in the controlled variable rather than keeping it at the set point. Understanding the source of this unstable behavior, and how to prevent it, are important issues. In this chapter, several mathematical stability criteria are introduced, and practical methods for analyzing closed-loop stability are considered.

9.1 BLOCK DIAGRAM REPRESENTATION

In Chapters 1 and 7, we have shown that a block diagram provides a convenient representation of the flow of information around a feedback control loop. The previous discussion of block diagrams was qualitative rather than quantitative, because the blocks were labeled but did not indicate the relationships between process variables. However, quantitative information can also be included by showing the transfer function for each block.

To illustrate the development of a block diagram, we return to a previous example, the stirred-tank blending process considered in earlier chapters. The schematic diagram in Fig. 9.1 shows the blending tank with

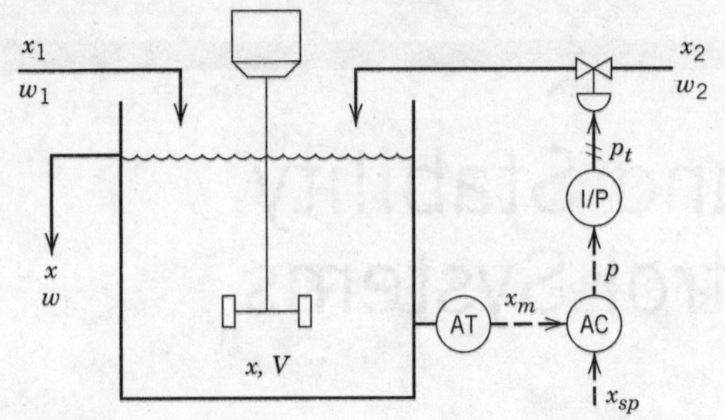

Figure 9.1 Composition control system for a stirred-tank blending process.

the flow rate of pure component A, w_2, as the manipulated variable. The control objective is to regulate the tank composition, x, by adjusting the mass flow rate w_2. The primary disturbance variable is assumed to be inlet composition x_1. The tank composition is measured by a sensor/transmitter whose output signal x_m is sent to an electronic controller. Because a pneumatic control valve is used, the controller output (an electrical signal in the range of 4 to 20 mA) must be converted to an equivalent pneumatic signal p_t (3–15 psig or 20–100 kPa(g) or 0.2–1 kgf/cm^2(g)) by a current-to-pressure transducer. The transducer output signal is then used to adjust the valve.

Next, we develop a transfer function for each of the five elements in the feedback control loop. For the sake of simplicity, flow rate w_1 is assumed to be constant, and the system is initially operating at the nominal steady rate. Later, we extend this analysis to more general situations.

9.1.1 Process

In Example 4.9, the approximate dynamic model of a stirred-tank blending system was developed:

$$X'(s) = \left(\frac{K_1}{\tau s+1}\right)X_1'(s) + \left(\frac{K_2}{\tau s+1}\right)W_2'(s) \qquad (9\text{-}1)$$

where

$$\tau = \frac{V\rho}{\overline{w}},\ K_1 = \frac{\overline{w}_1}{\overline{w}},\ \text{and}\ K_2 = \frac{1-\overline{x}}{\overline{w}} \qquad (9\text{-}2)$$

Figure 9.2 provides a block diagram representation of the information in Eqs. 9-1 and 9-2 and indicates the units for each variable. In the diagram, the deviation variable, $X_d'(s)$, denotes the change in exit composition due to a change in inlet composition $X_1'(s)$ (the disturbance). Similarly, $X_u'(s)$ is a deviation variable that denotes the change in $X'(s)$ due to a change in the manipulated variable (the flow rate of pure A, $W_2'(s)$).

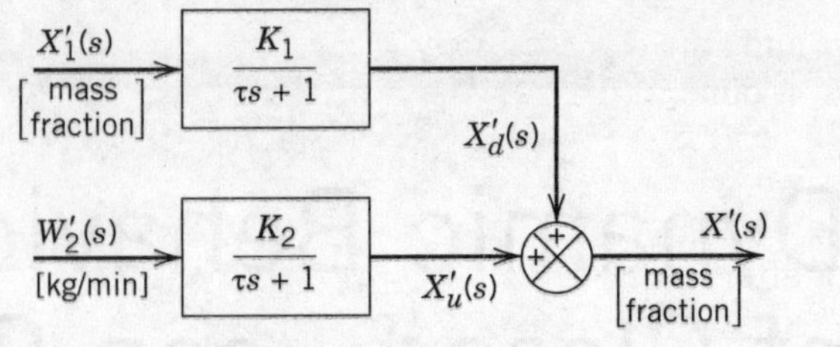

Figure 9.2 Block diagram of the blending process.

The effects of these changes are additive because $X'(s) = X_d'(s) + X_u'(s)$ as a direct consequence of the Superposition Principle for linear systems discussed in Chapter 3. Recall that this transfer function representation is valid only for linear systems and for nonlinear systems that have been linearized, as is the case for the blending process model.

9.1.2 Composition Sensor-Transmitter (Analyzer)

We assume that the dynamic behavior of the composition sensor-transmitter can be approximated by a first-order transfer function:

$$\frac{X_m'(s)}{X'(s)} = \frac{K_m}{\tau_m s+1} \qquad (9\text{-}3)$$

This instrument has negligible dynamics when $\tau \gg \tau_m$. For a change in one of the inputs, the measured composition $x_m'(t)$ rapidly follows the true composition $x'(t)$, even while $x'(t)$ is slowly changing with time constant τ. Hence, the dynamic error associated with the measurement can be neglected (cf. Section 8.3). A useful approximation is to set $\tau_m = 0$ in Eq. 9-3. The steady-state gain K_m depends on the input and output ranges of the composition sensor-transmitter combination, as indicated in Section 8.1.3. The block diagram for the sensor-transmitter is shown in Fig. 9.3.

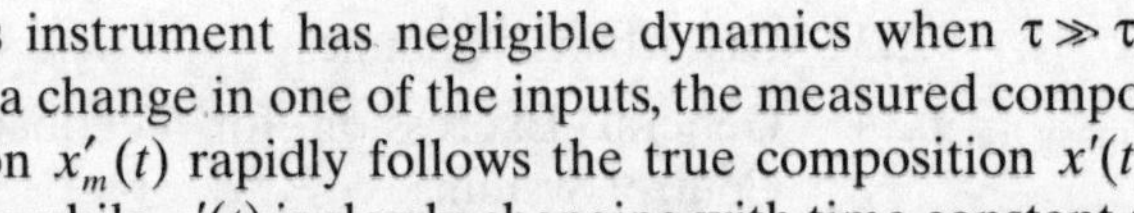

Figure 9.3 Block diagram for the composition sensor-transmitter (analyzer).

9.1.3 Controller

Suppose that an electronic proportional plus integral controller is used. From Chapter 7, the controller transfer function is

$$\frac{P'(s)}{E(s)} = K_c\left(1+\frac{1}{\tau_I s}\right) \qquad (9\text{-}4)$$

where $P'(s)$ and $E(s)$ are the Laplace transforms of the controller output $p'(t)$ and the error signal $e(t)$. Note that

p' and e are electrical signals that have units of mA, while K_c is dimensionless. The error signal is expressed as

$$e(t) = \tilde{x}'_{sp}(t) - x'_m(t) \tag{9-5}$$

or after taking Laplace transforms,

$$E(s) = \tilde{X}'_{sp}(s) - X'_m(s) \tag{9-6}$$

The symbol $\tilde{x}'_{sp}(t)$ denotes the *internal set-point* composition expressed as an equivalent electrical current signal. This signal is used internally by the controller. $\tilde{x}'_{sp}(t)$ is related to the actual composition set point $x'_{sp}(t)$ by the sensor-transmitter gain K_m (which is the slope of the calibration curve):

$$\tilde{x}'_{sp}(t) - K_m x'_{sp}(t) \tag{9-7}$$

Thus

$$\frac{\tilde{X}'_{sp}(s)}{X'_{sp}(s)} = K_m \tag{9-8}$$

The block diagram representing the controller in Eqs. 9-4 through 9-8 is shown in Fig. 9.4. The symbol for the subtraction operation is called a *comparator*.

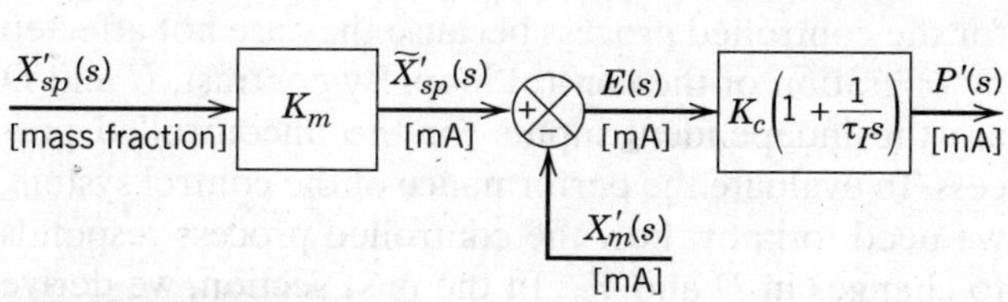

Figure 9.4 Block diagram for the controller.

In general, if a reported controller gain is not dimensionless, it includes the gain of at least one other device (such as an actuator) in addition to the dimensionless controller gain.

9.1.4 Current-to-Pressure (I/P) Transducer

Because transducers are usually designed to have linear characteristics and negligible (fast) dynamics, we assume that the transducer transfer function merely consists of a steady-state gain K_{IP}:

$$\frac{P'_t(s)}{P'(s)} = K_{IP} \tag{9-9}$$

In Eq. 9-9, $P'_t(s)$ denotes the output signal from the I/P transducer in deviation form. The corresponding block diagram is shown in Fig. 9.5.

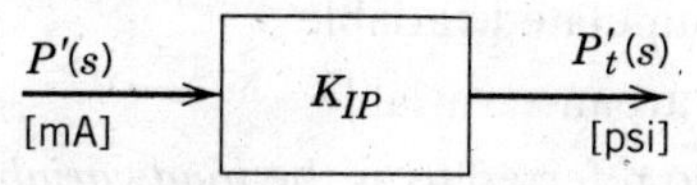

Figure 9.5 Block diagram for the I/P transducer.

9.1.5 Control Valve

As discussed in Section 8.2, control valves are usually designed so that the flow rate through the valve is a nearly linear function of the signal to the valve actuator. Therefore, a first-order transfer function usually provides an adequate model for operation of an installed valve in the vicinity of a nominal steady state. Thus, we assume that the control valve can be modeled as:

$$\frac{W'_2(s)}{P'_t(s)} = \frac{K_v}{\tau_v s + 1} \tag{9-10}$$

The block diagram for an I/P transducer plus pneumatic control valve is shown in Fig. 9.6.

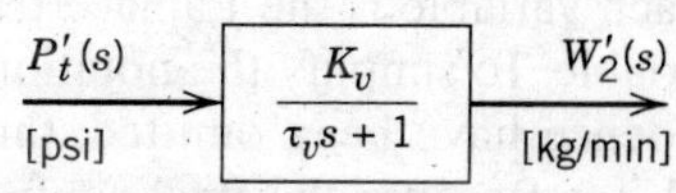

Figure 9.6 Block diagram for the control valve.

Now that transfer functions and block diagrams in Figs. 9.2–9.6 have been developed for the individual components of the feedback control system, we can combine this information to obtain the composite block diagram of the controlled system shown in Fig. 9.7.

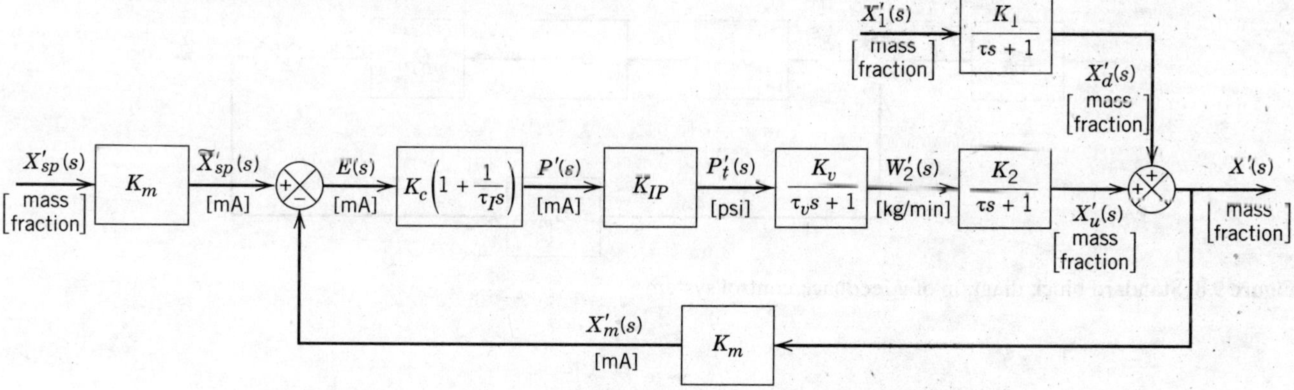

Figure 9.7 Block diagram for the entire blending process composition control system.

9.2 CLOSED-LOOP TRANSFER FUNCTIONS

The block diagrams considered so far have been specifically developed for the stirred-tank blending system. The more general block diagram in Fig. 9.8 contains the standard notation:

Y = controlled variable
U = manipulated variable
D = disturbance variable
(also referred to as the *load variable*)
P = controller output
E = error signal
Y_m = measured value of Y
Y_{sp} = set point
$\tilde{Y}_{sp}$ = internal set point (used by the controller)
Y_u = change in Y due to U
Y_d = change in Y due to D
G_c = controller transfer function
G_v = transfer function for the final control element
G_p = process transfer function
G_d = disturbance transfer function
G_m = transfer function for sensor-transmitter
K_m = steady-state gain for G_m

In Fig. 9.8, each variable is the Laplace transform of a deviation variable. To simplify the notation, the primes and s dependence have been omitted; thus, Y is used rather than $Y'(s)$. Because the final control element is often a control valve, its transfer function is denoted by G_v. Note that the process transfer function G_p indicates the effect of the manipulated variable on the controlled variable. The disturbance transfer function G_d represents the effect of the disturbance variable on the controlled variable. For the stirred-tank blending system, G_d and G_p are given in Eq. 9-1.

The standard block diagram in Fig. 9.8 can be used to represent a wide variety of practical control problems. Other blocks can be added to the standard diagram to represent additional elements in the feedback control loop such as the current-to-pressure transducer in Fig. 9.5. In Fig. 9.8, the signal path from E to Y through blocks G_c, G_v, and G_p is referred to as the *forward path*. The path from Y to the comparator through G_m is called the *feedback path*.

Figure 9.9 shows an alternative representation of the standard block diagram that is also used in the control literature. Because the disturbance transfer functions appear in different locations in Figs. 9.8 and 9.9, different symbols are used. For these two block diagrams to be equivalent, the relation between Y and D must be preserved. Thus, G_d and G_d^* must be related by the expression $G_d = G_p G_d^*$.

Note that Y_{sp} and D are the independent input signals for the controlled process because they are not affected by operation of the control loop. By contrast, U and D are the independent inputs for the uncontrolled process. To evaluate the performance of the control system, we need to know how the controlled process responds to changes in D and Y_{sp}. In the next section, we derive expressions for the *closed-loop transfer functions*, $Y(s)/Y_{sp}(s)$ and $Y(s)/D(s)$. But first, we review some block diagram algebra.

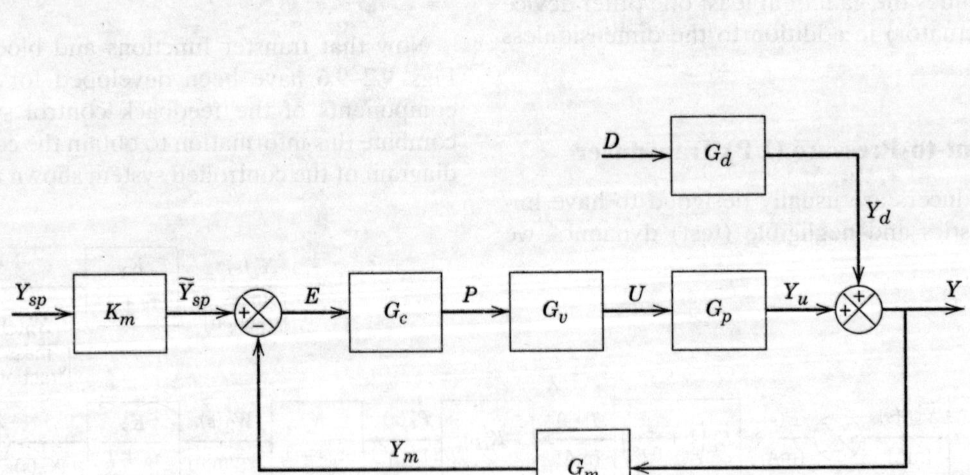

Figure 9.8 Standard block diagram of a feedback control system.

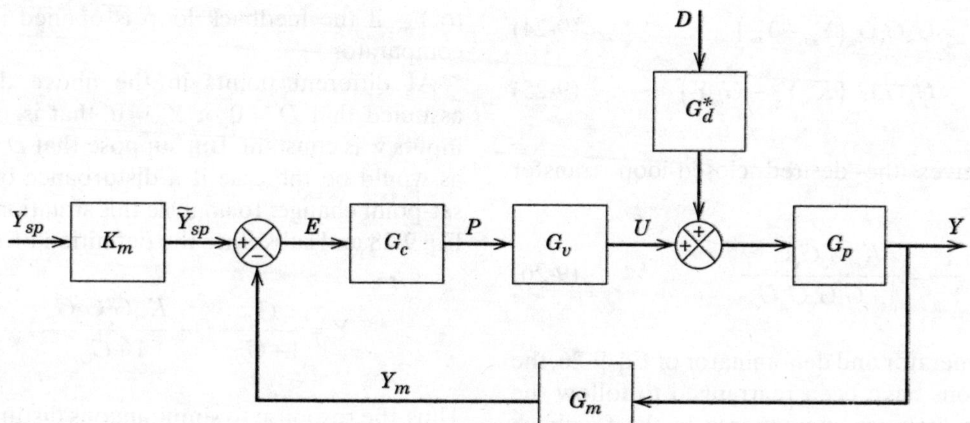

Figure 9.9 Alternative form of the standard block diagram of a feedback control system.

9.2.1 Block Diagram Reduction

In deriving closed-loop transfer functions, it is often convenient to combine several blocks into a single block. For example, consider the three blocks in series in Fig. 9.10. The block diagram indicates the following relations:

$$\begin{aligned} X_1 &= G_1 U \\ X_2 &= G_2 X_1 \\ X_3 &= G_3 X_2 \end{aligned} \tag{9-11}$$

By successive substitution,

$$X_3 = G_3 G_2 G_1 U \tag{9-12}$$

or

$$X_3 = GU \tag{9-13}$$

where $G \triangleq G_3 G_2 G_1$. Equation 9-13 indicates that the block diagram in Fig. 9.10 can be reduced to the equivalent block diagram in Fig. 9.11.

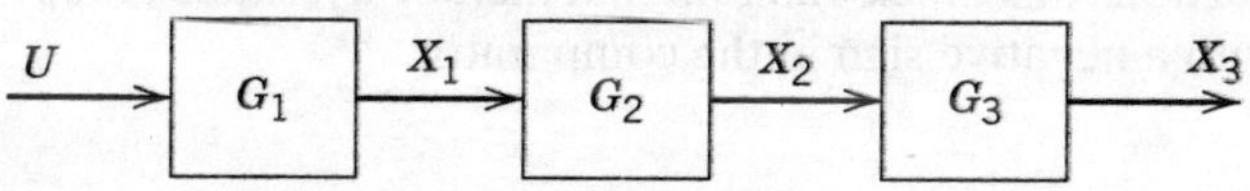

Figure 9.10 Three blocks in series.

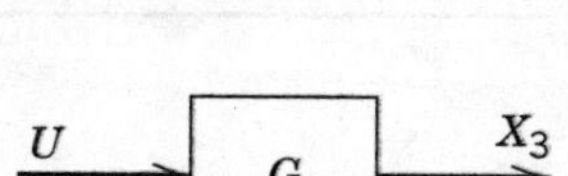

Figure 9.11 Equivalent block diagram.

9.2.2 Set-Point Changes

Next we derive the closed-loop transfer function for set-point changes. The closed-loop system behavior for set-point changes is also referred to as the servomechanism *(servo) problem* in the control literature, because early applications were concerned with positioning devices called servomechanisms. We assume for this case that no disturbance change occurs and thus $D = 0$. From Fig. 9.8, it follows that

$$Y = Y_d + Y_u \tag{9-14}$$

$$Y_d = G_d D = 0 \text{ (because } D = 0) \tag{9-15}$$

$$Y_u = G_p U \tag{9-16}$$

Combining gives

$$Y = G_p U \tag{9-17}$$

Figure 9.8 also indicates the following input/output relations for the individual blocks:

$$U = G_v P \tag{9-18}$$

$$P = G_c E \tag{9-19}$$

$$E = \tilde{Y}_{sp} - Y_m \tag{9-20}$$

$$\tilde{Y}_{sp} = K_m Y_{sp} \tag{9-21}$$

$$Y_m = G_m Y \tag{9-22}$$

Combining the above equations gives

$$Y = G_p G_v P = G_p G_v G_c E \tag{9-23}$$

$$= G_p G_v G_c \left(\tilde{Y}_{sp} - Y_m\right) \quad (9\text{-}24)$$

$$= G_p G_v G_c \left(K_m Y_{sp} - G_m Y\right) \quad (9\text{-}25)$$

Rearranging gives the desired closed-loop transfer function,

$$\frac{Y}{Y_{sp}} = \frac{K_m G_c G_v G_p}{1 + G_c G_v G_p G_m} \quad (9\text{-}26)$$

In both the numerator and denominator of Eq. 9-26, the transfer functions have been rearranged to follow the order in which they are encountered in the feedback control loop. This convention makes it easy to determine which transfer functions are present or missing in analyzing subsequent problems.

9.2.3 Disturbance Changes

Now consider the case of disturbance changes, which is also referred to as the *regulator problem* since the process is to be *regulated* at a constant set point. From Fig. 9.8,

$$Y = Y_d + Y_u = G_d D + G_p U \quad (9\text{-}27)$$

Substituting Eq. 9-18 through Eq. 9-22 gives

$$Y = G_d D + G_p U$$

$$= G_d D + G_p G_v G_c \left(K_m Y_{sp} - G_m Y\right) \quad (9\text{-}28)$$

Because $Y_{sp} = 0$ we can rearrange Eq. 9-28 to give the closed-loop transfer function for disturbance changes:

$$\frac{Y}{D} = \frac{G_d}{1 + G_c G_v G_p G_m} \quad (9\text{-}29)$$

A comparison of Eqs. 9-26 and 9-29 indicates that both closed-loop transfer functions have the same denominator, $1 + G_c G_v G_p G_m$. The denominator is often written as $1 + G_{OL}$ where G_{OL} is the *open-loop transfer function,* $G_{OL} \triangleq G_c G_v G_p G_m$. The term open-loop transfer function (or open-loop system) is used because G_{OL} relates Y_m to $\tilde{Y}_{sp}$ if the feedback loop is opened just before the comparator.

At different points in the above derivations, we assumed that $D = 0$ or $Y_{sp} = 0$, that is, one of the two inputs was constant. But suppose that $D \neq 0$ and $Y_{sp} \neq 0$. as would be the case if a disturbance occurs during a set-point change. To analyze this situation, we rearrange Eq. 9-28 and substitute the definition of G_{OL} to obtain

$$Y = \frac{G_d}{1 + G_{OL}} D + \frac{K_m G_c G_v G_p}{1 + G_{OL}} Y_{sp} \quad (9\text{-}30)$$

Thus, the response to simultaneous disturbance variable and set-point changes is merely the sum of the individual responses, as can be seen by comparing Eqs. 9-26, 9-29, and 9-30. This result is a consequence of the Superposition Principle for linear systems.

9.2.4 General Expression for Feedback Control Systems

Closed-loop transfer functions for more complicated block diagrams can be written in the general form

$$\frac{Z}{Z_i} = \frac{\Pi_f}{1 + \Pi_e} \quad (9\text{-}31)$$

where

Z is the output variable or any internal variable within the control loop
Z_i is an input variable (e.g., Y_{sp} or D)
Π_f = product of the transfer functions in the *forward* path from Z_i to Z
Π_e = product of *every* transfer function in the feedback loop

Thus, for the previous servo problem, we have $Z_i = Y_{sp}$, $Z = Y$, $\Pi_f = K_m G_c G_v G_p$, and $\Pi_e = G_{OL}$. For the regulator problem, $Z_i = D$, $Z = Y$, $\Pi_f = G_d$, and $\Pi_e = G_{OL}$. It is important to note that Eq. 9-31 is applicable only to portions of a block diagram that include a *feedback loop* with a negative sign in the comparator.

EXAMPLE 9.1

Find the closed-loop transfer function Y/Y_{sp} for the complex control system in Fig. 9.12. Notice that this block diagram has two feedback loops and two disturbance variables. This configuration arises when the cascade control scheme of Chapter 12 is employed.

Solution

Using the general rule in Eq. 9-31, we first reduce the inner loop to a single block as shown in Fig. 9.13. To solve the servo problem, set $D_1 = D_2 = 0$. Because

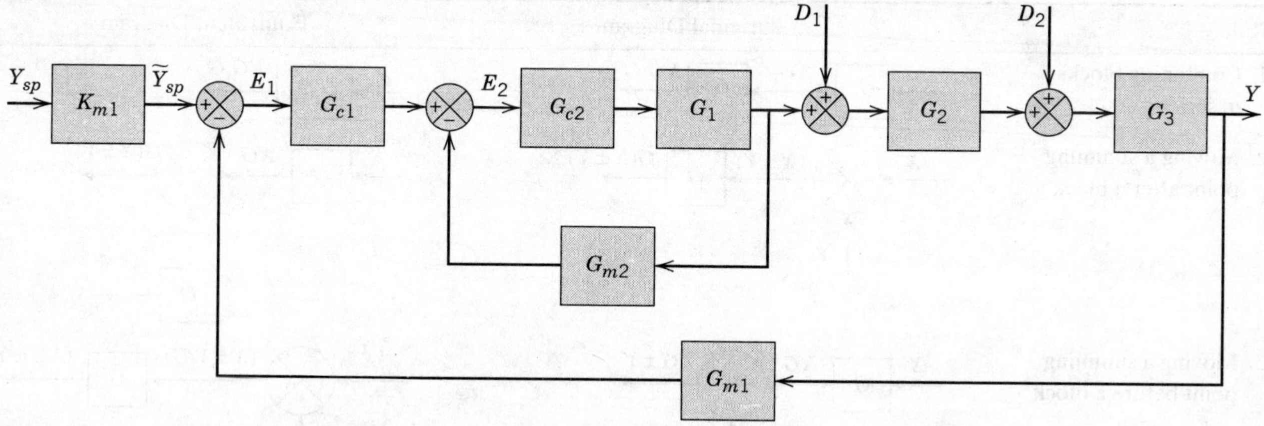

Figure 9.12 Complex control system.

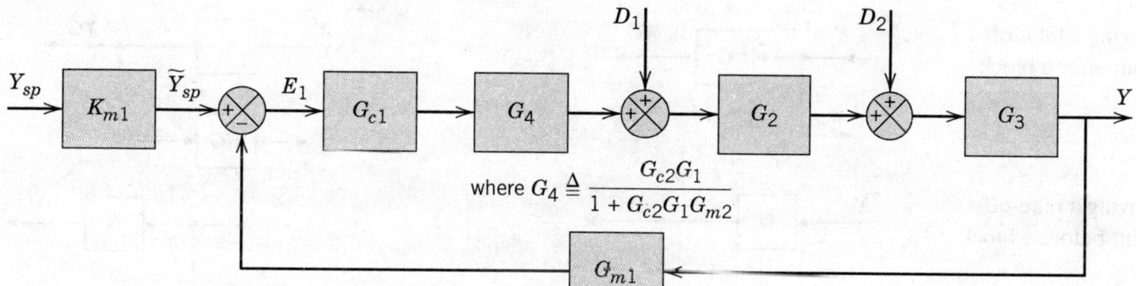

Figure 9.13 Block diagram for reduced system.

Fig. 9.13 contains a single feedback loop, use Eq. 9.31 to obtain Fig. 9.14*a*. The final block diagram is shown in Fig. 9.14*b* with $Y/Y_{sp} = K_{m1}G_5$. Substitution for G_4 and G_5 gives the desired closed-loop transfer function:

$$\frac{Y}{Y_{sp}} = \frac{K_{m1}G_{c1}G_{c2}G_1G_2G_3}{1 + G_{c2}G_1G_{m2} + G_{c1}G_2G_3G_{m1}G_{c2}G_1}$$

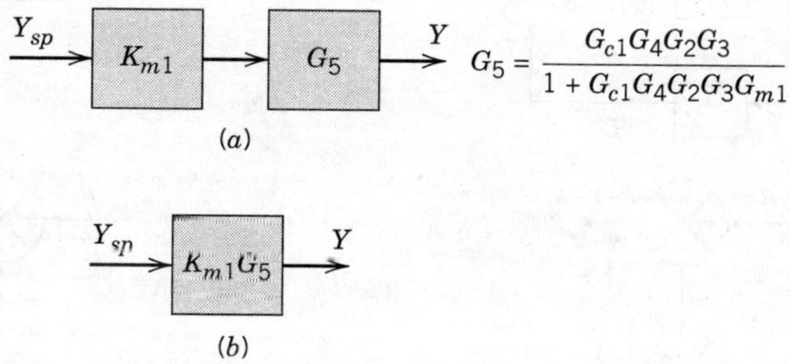

Figure 9.14 Final block diagrams for Example 9.1.

9.2.5 Rules for Block Diagram Reduction

The block diagram reduction method discussed in Section 9.2.1 is applicable for feedback loops. Table 9.1 shows block diagram algebra for reducing big and complicated block diagrams to smaller and simpler ones.

Table 9.1 The Block Diagram Algebra and the Associated Rules

<table>
<tr><th>Rule</th><th>Original Diagram</th><th>Equivalent Diagram</th></tr>
<tr><td>1. Combining blocks in series</td><td>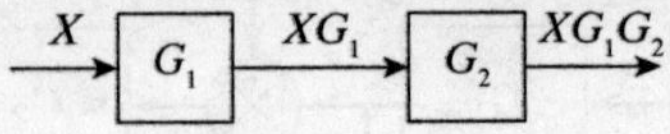
</td><td>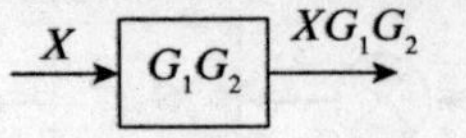
</td></tr>
<tr><td>2. Moving a summing point after a block</td><td>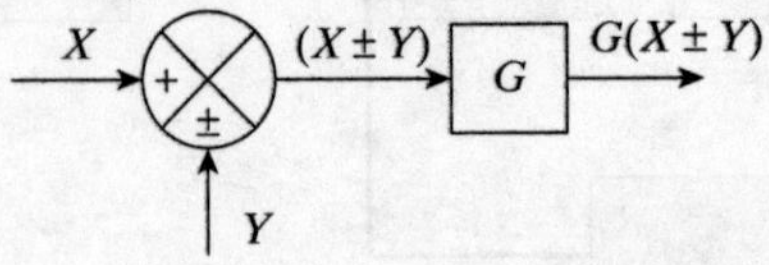
</td><td>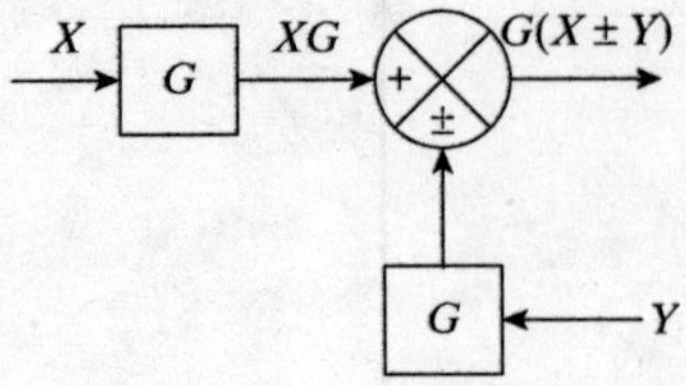
</td></tr>
<tr><td>3. Moving a summing point before a block</td><td>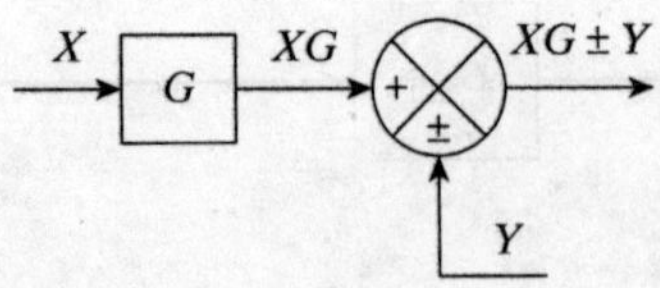
</td><td>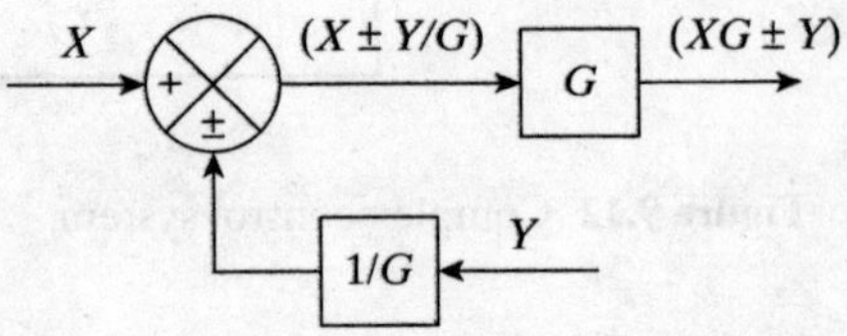
</td></tr>
<tr><td>4. Moving a take-off point after a block</td><td>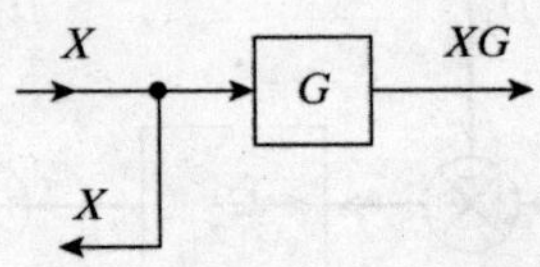
</td><td>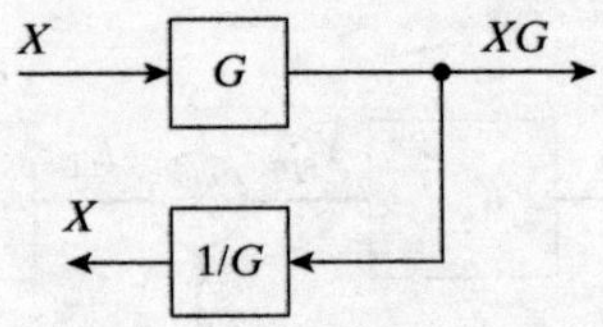
</td></tr>
<tr><td>5. Moving a take-off point before a block</td><td>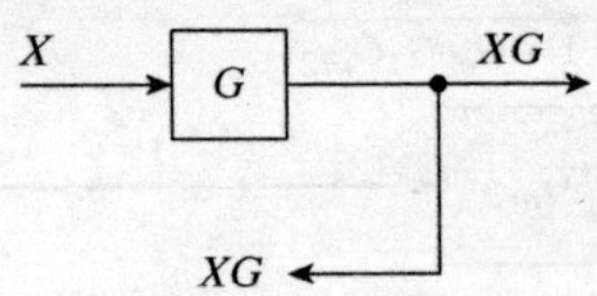
</td><td>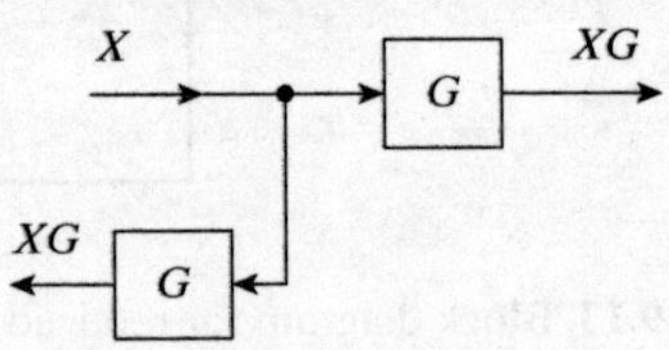
</td></tr>
<tr><td>6. Eliminating a feedback loop</td><td>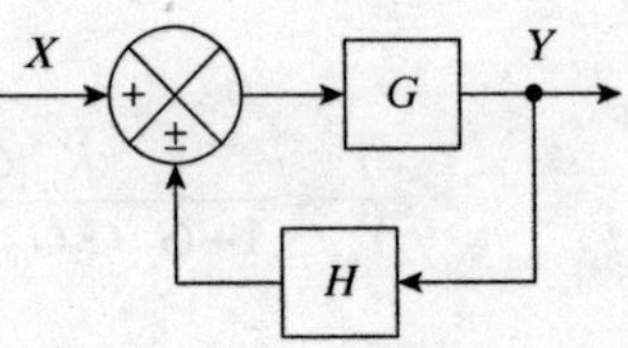
</td><td>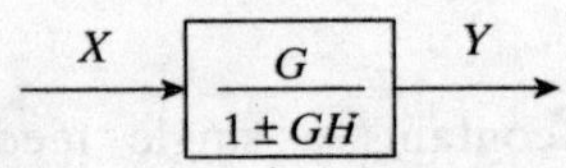
</td></tr>
<tr><td>7. Eliminating a parallel path</td><td>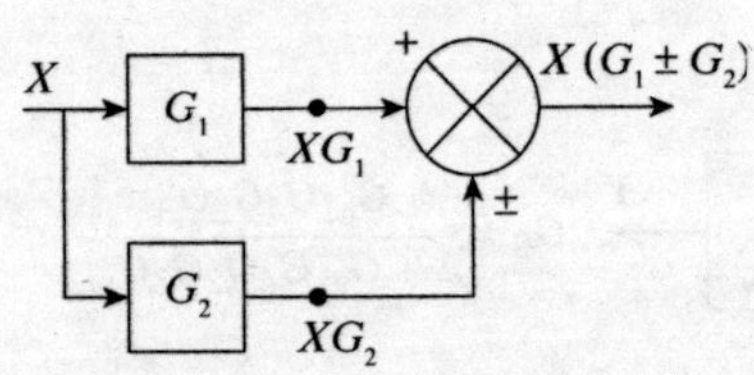
</td><td>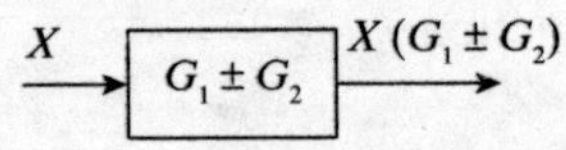
</td></tr>
<tr><td>8. Moving a summing point from feedback path to forward path</td><td>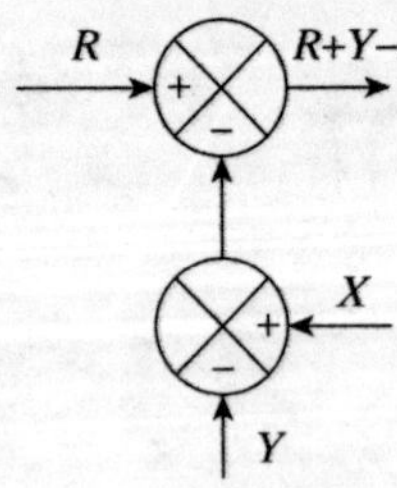
</td><td>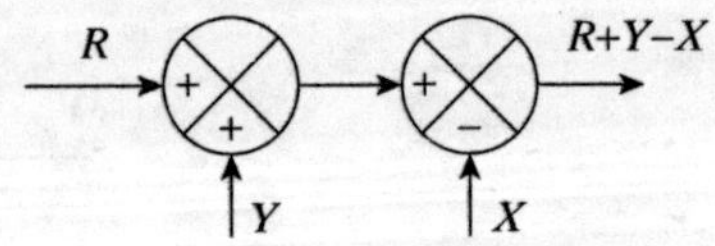
</td></tr>
</table>

Table 9.2 Steps for Block Diagram Reduction

The following steps are followed for reducing block diagrams containing a single input:

Step 1. Simplify the blocks connected in series (using Rule 1).

Step 2. Simplify the blocks connected in parallel (using Rule 7).

Step 3. Simplify the blocks connected in feedback loop (using Rule 6).

Step 4. Shift the take-off point towards right (using Rule 4) if there is difficulty in simplifying.

Step 5. Shift the summing point towards left (using Rule 3) if there is difficulty in simplifying.

Step 6. Repeat the above steps till you get a single block, which will give the transfer function of the overall block diagram.

The following steps are followed for reducing block diagrams containing multiple inputs:

Step 1. Find the transfer function of block diagram by considering one input at a time and make the remaining inputs as zero.

Step 2. Repeat Step 1 for remaining inputs.

Step 3. Obtain the overall transfer function by adding all the above transfer functions.

EXAMPLE 9.2

Determine the transfer function $Y(s)/X(s)$ for the block diagram shown in Fig. 9.15

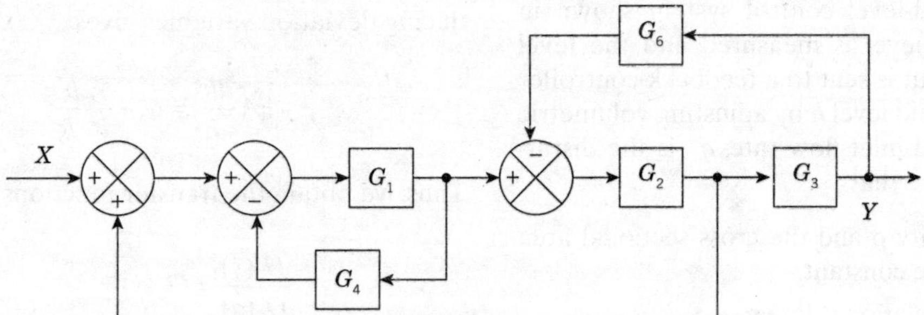

Figure 9.15 Block diagram for a feedback control system.

Solution

After simplifying the feedback loop consisting of G_1 and G_4 and moving the take-off point right of G_3, we can get the following diagram.

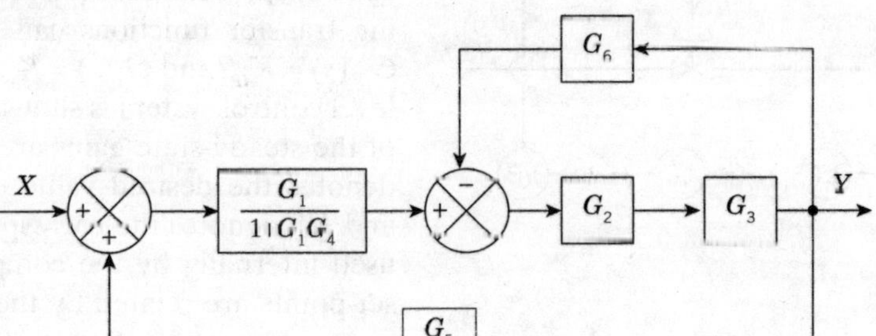

Next, the feedback loop consisting of G_2, G_3, and G_6 is simplified as follows:

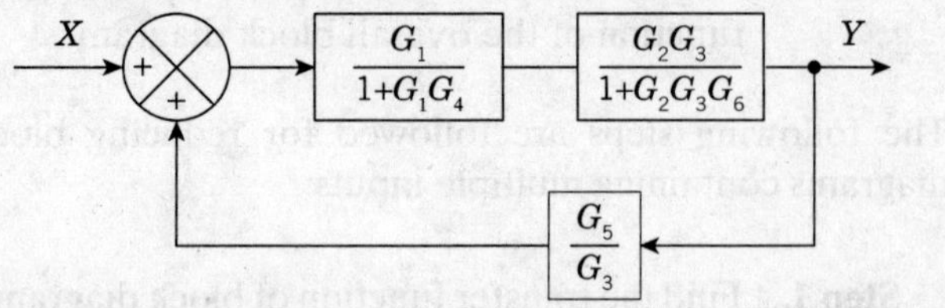

After eliminating the positive feedback loop, we get the simplified block diagram as given below.

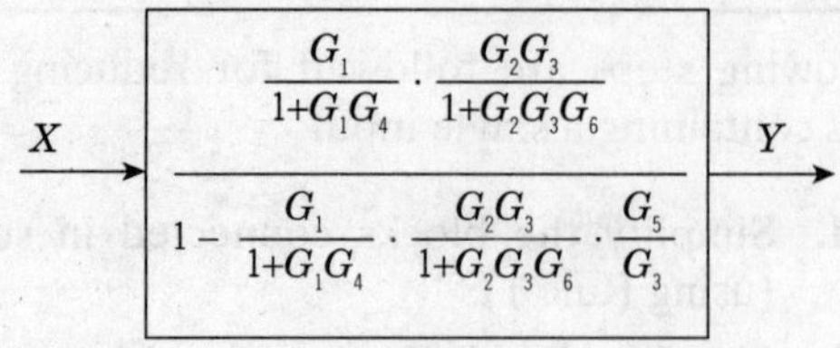

This can be further simplified as follows.

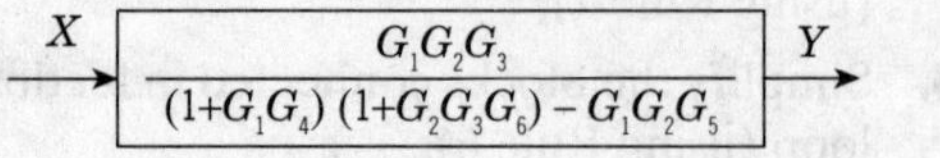

The transfer function Y/X is given by that is inside the simplified block.

9.3 CLOSED-LOOP RESPONSES OF SIMPLE CONTROL SYSTEMS

In this section, we consider the dynamic behavior of several elementary control problems for disturbance variable and set-point changes. The transient responses can be determined in a straightforward manner if the closed-loop transfer functions are available.

Consider the liquid-level control system shown in Fig. 9.16. The liquid level is measured and the level transmitter (LT) output is sent to a feedback controller (LC) that controls liquid level h by adjusting volumetric flow rate q_2. A second inlet flow rate, q_1, is the disturbance variable. Assume that

1. The liquid density ρ and the cross-sectional area A of the tank are constant.
2. The flow-head relation is linear, $q_3 = h/R$.
3. The level transmitter, I/P transducer, and pneumatic control valve have negligible dynamics.
4. An electronic controller with input and output in % is used (full scale = 100%).

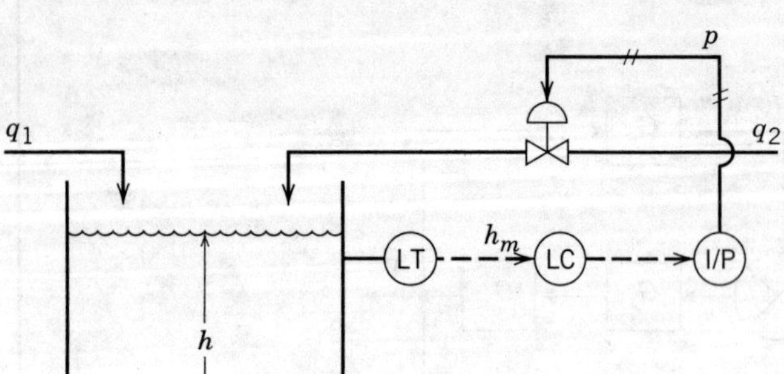

Figure 9.16 Liquid-level control system.

Derivation of the process and disturbance transfer functions directly follows Example 4.7. Consider the unsteady-state mass balance for the tank contents:

$$\rho A \frac{dh}{dt} = \rho q_1 + \rho q_2 \rho q_3 \tag{9-32}$$

Substituting the flow-head relation, $q_3 = h/R$, and introducing deviation variables gives

$$A \frac{dh'}{dt} = q_1' + q_2' \frac{h'}{R} \tag{9-33}$$

Thus, we obtain the transfer functions

$$\frac{H'(s)}{Q_2'(s)} = G_p(s) = \frac{K_p}{\tau s + 1} \tag{9-34}$$

$$\frac{H'(s)}{Q_1'(s)} = G_d(s) = \frac{K_p}{\tau s + 1} \tag{9-35}$$

where $K_p = R$ and $\tau = RA$. Note that $G_p(s)$ and $G_d(s)$ are identical, because q_1 and q_2 are both inlet flow rates and thus have the same effect on h.

Because the level transmitter, I/P transducer, and control valve have negligible dynamics, the corresponding transfer functions can be written as $G_m(s) = K_m$, $G_{IP}(s) = K_{IP}$, and $G_v(s) = K_v$. The block diagram for the level control system is shown in Fig.9.17, where the units of the steady-state gains are apparent. The symbol H'_{sp} denotes the desired value of liquid level (in meters), and $\tilde{H}'_{sp}$ denotes the corresponding value (in %) that is used internally by the computer. Note that these two set-points are related by the level transmitter gain K_m, as was discussed in Section 9.1.

The block diagram in Fig. 9.17 is in the alternative form of Fig. 9.9 with $G_d^*(s) = 1$.

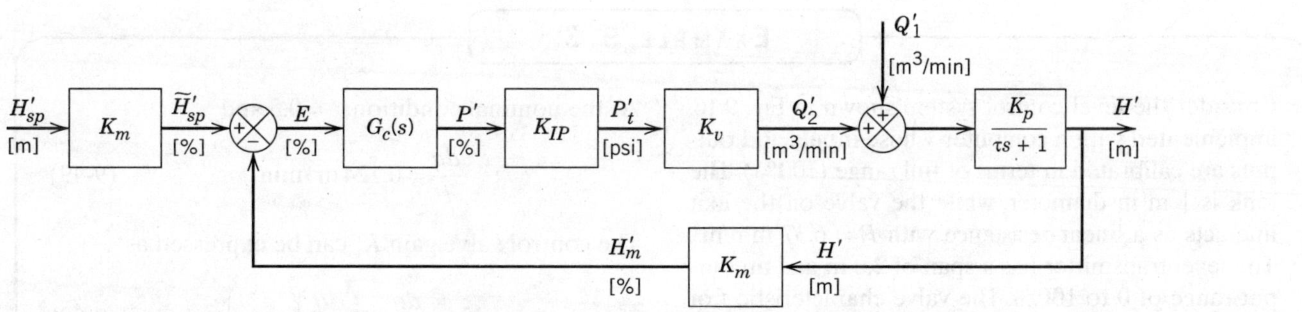

Figure 9.17 Block diagram for level control system.

9.3.1 Proportional Control and Set-Point Changes

If a proportional controller is used, $G_c(s) = K_c$. From Fig. 9.17 and the material in the previous section, it follows that the closed-loop transfer function for set-point changes is given by

$$\frac{H'(s)}{H'_{sp}(s)} = \frac{K_c K_{IP} K_v K_p K_m/(\tau s+1)}{1+K_c K_{IP} K_v K_p K_m/(\tau s+1)} \qquad (9\text{-}36)$$

This relation can be rearranged in the standard form for a first-order transfer function,

$$\frac{H'(s)}{H'_{sp}(s)} = \frac{K_1}{\tau_1 s+1} \qquad (9\text{-}37)$$

where

$$K_1 = \frac{K_{OL}}{1+K_{OL}} \qquad (9\text{-}38)$$

$$\tau_1 = \frac{\tau}{1+K_{OL}} \qquad (9\text{-}39)$$

and the *open-loop gain* K_{OL} is given by

$$K_{OL} = K_c K_{IP} K_v K_p K_m \qquad (9\text{-}40)$$

Equations 9-37 to 9-40 indicate that the closed-loop process has first-order dynamics with a time constant τ_1 that is smaller than the process time constant τ. We assume here that $K_{OL} > 0$; otherwise, the control system would not function properly, as will be apparent from the stability analysis later in this chapter. Because $\tau_1 < \tau$, the feedback controller enables the controlled process to respond more quickly than the uncontrolled process.

From Eq. 9-37, it follows that the closed-loop response to a unit step change of magnitude M in set point is given by

$$h'(t) = K_1 M(1-e^{-t/\tau_1}) \qquad (9\text{-}41)$$

This response is shown in Fig. 9.18. Note that a steady-state error or *offset* exists, because the new steady-state value is $K_1 M$ rather than the desired value $M (K_1 < 1)$. The offset is defined as

$$\text{offset} \triangleq h'_{sp}(\infty) - h'(\infty) \qquad (9\text{-}42)$$

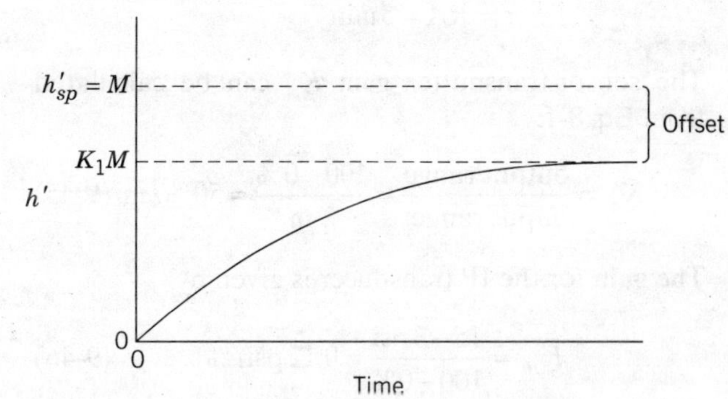

Figure 9.18 Step response for proportional control (set-point change).

For a step change of magnitude M in the set point, $h'_{sp}(\infty) = M$. From Eq. 9-41, it is clear that $h'(\infty) = K_1 M$. Substituting these values and Eq. 9-38 into Eq. 9-42 gives

$$\text{offset} = M - K_1 M = \frac{M}{1+K_{OL}} \qquad (9\text{-}43)$$

EXAMPLE 9.3

Consider the level control system shown in Fig. 9.16, implemented with a computer whose inputs and outputs are calibrated in terms of full range (100%). The tank is 1 m in diameter, while the valve on the exit line acts as a linear resistance with $R = 6.37$ min/m^2. The level transmitter has a span of 2.0 m and an output range of 0 to 100%. The valve characteristic f of the equal percentage control valve is related to the fraction of lift ℓ by the relation $f = (30)^{\ell-1}$. The air-to-open control valve receives a 3 to 15 psi signal from an I/P transducer, which, in turn, receives a 0–100% signal from the computer-implemented proportional-only controller. When the control valve is fully open ($\ell = 1$), the flow rate through the valve is 0.2 m^3/min. At the nominal operating condition, the control valve is half-open ($\ell = 0.5$). Using the dynamic model in the block diagram of Fig. 9.17, calculate the closed-loop responses to a step change in the set point of 0.3 m for three values of the controller gain: $K_c = 4, 8$, and 20.

Solution

From the given information, we can calculate the cross-sectional area of the tank A, the process gain K_p, and the time constant:

$$A = \pi(0.5\,\text{m})^2 = 0.785\,\text{m}^2$$

$$K_p = R = 6.37\,\text{min/m}^2 \tag{9-44}$$

$$\tau = RA = 5\,\text{min}$$

The sensor-transmitter gain K_m can be calculated from Eq. 8-1:

$$K_m = \frac{\text{output range}}{\text{input range}} = \frac{100-0\%}{2\text{ m}} = 50\%/\text{m} \tag{9-45}$$

The gain for the IP transducer is given by

$$K_{IP} = \frac{15-3\,\text{psi}}{100-0\%} = 0.12\,\text{psi}/\% \tag{9-46}$$

Next, we calculate the gain for the control valve K_v. The valve relation between flow rate q and fraction of lift ℓ can be expressed as (cf. Eqs. 8-6 and 8-18)

$$q = 0.2(30)^{\ell-1} \tag{9-47}$$

Thus

$$\frac{dq}{d\ell} = 0.2 \ln 30(30)^{\ell-1} \tag{9-48}$$

At the nominal condition, $\ell = 0.5$ and

$$\frac{dq}{d\ell} = 0.124\,\text{m}^3/\text{min} \tag{9-49}$$

The control valve gain K_v can be expressed as

$$K_v = \frac{dq}{dp_t} = \left(\frac{dq}{d\ell}\right)\left(\frac{d\ell}{dp_t}\right) \tag{9-50}$$

If the valve actuator is designed so that the fraction of lift ℓ varies linearly with the IP transducer output p_t, then

$$\frac{d\ell}{dp_t} = \frac{\Delta\ell}{\Delta p_t} = \frac{1-0}{15-3\,\text{psi}} = 0.0833\,\text{psi}^{-1} \tag{9-51}$$

Then, from Eqs. 9-48, 9-50, and 9-51,

$$K_v = 1.03\times10^{-2}\,\text{m}^3/\text{min psi} \tag{9-52}$$

An alternative method for estimating K_v is to use the tangent to the valve characteristic curve (see Chapter 8). Now that all of the gains and the time constant in Fig. 9.17, have been calculated, we can calculate the closed-loop gain K_1 and time constant τ_1 in Eq. 9-41. Substituting these numerical values into Eqs. 9-38 and 9-39 for the three values of K_c gives

K_c	τ_1 (min)	K_1
4	1.94	0.612
8	1.20	0.759
20	0.56	0.887

The closed-loop responses are shown in Fig. 9.19. Increasing K_c reduces both the offset and the time required to reach the new steady state.

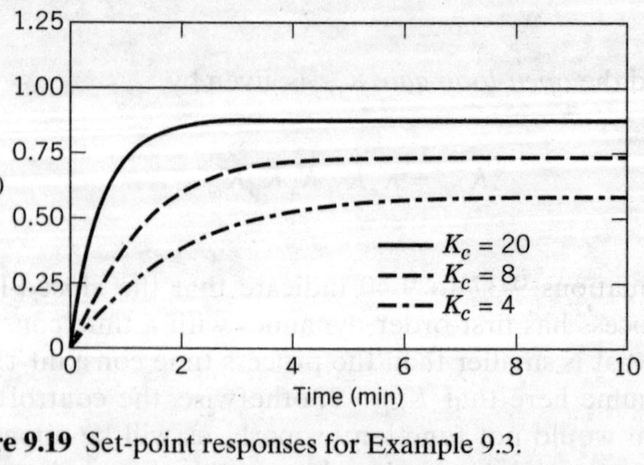

Figure 9.19 Set-point responses for Example 9.3.

Equation 9-43 suggests that offset can be reduced by increasing K_c. However, for most control problems, making K_c too large can result in oscillatory or unstable responses due to the effect of additional lags and time delays that have been neglected in the present analysis. For example, we have neglected the dynamics associated with the control valve, level transmitter, and pneumatic transmission line between the I/P transducer and control valve. A more rigorous analysis, including the dynamics of these components, would reveal the possibility of oscillations or instability. Stability problems associated with feedback control systems are analyzed in Section 9.4.

For many liquid-level control problems, a small offset can be tolerated, because the vessel serves as a surge capacity (or intermediate storage volume) between processing units.

9.3.2 Proportional Control and Disturbance Changes

From Fig. 9.17 and Eq. 9-29, the closed-loop transfer function for disturbance changes with proportional control is

$$\frac{H'(s)}{Q_1'(s)} = \frac{K_p/(\tau s+1)}{1+K_{OL}/(\tau s+1)} \tag{9-53}$$

Rearranging gives

$$\frac{H'(s)}{Q_1'(s)} = \frac{K_2}{\tau_1 s+1} \tag{9-54}$$

where τ_1 is defined in Eq. 9-39 and K_2 is given by

$$K_2 = \frac{K_p}{1+K_{OL}} \tag{9-55}$$

A comparison of Eqs. 9-54 and 9-37 indicates that both closed-loop transfer functions are first-order, and they have the same time constant. However, the steady-state gains, K_1 and K_2, are different.

From Eq. 9-54 it follows that the closed-loop response to a step change in disturbance of magnitude M is given by

$$h'(t) = K_2 M(1-e^{-t/\tau_1}) \tag{9-56}$$

The offset can be determined from Eq. 9-56. Now $h'_{sp}(\infty) = 0$, because we are considering disturbance changes and $h'(\infty) = K_2 M$ for a step change of magnitude M. Thus,

$$\text{offset} = 0 - h'(\infty) = -K_2 M = -\frac{K_p M}{1+K_{OL}} \tag{9-57}$$

As was the case for set-point changes, increasing K_c reduces the amount of offset for disturbance changes.

EXAMPLE 9.4

For the liquid-level control system and numerical parameter values of Example 9.3, calculate the closed-loop response to a step change in the disturbance variable of 0.05 m³/min. Calculate the offsets and plot the results for $K_c = 4$, 8, and 20.

Solution

The closed-loop responses in Fig. 9.20 indicate that increasing K_c reduces the offset and speeds up the closed-loop response. The offsets are

K_c	Offset
4	–0.124
8	–0.077
20	–0.036

The negative values of offset indicate that the controlled variable is greater than the set point. For $K_c = 0$ (no control), the offset is –0.318.

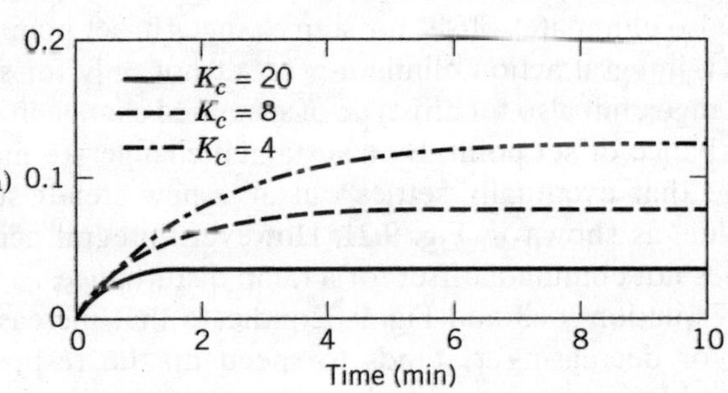

Figure 9.20 Disturbance responses for Example 9.4.

9.3.3 PI Control and Disturbance Changes

For PI control, $G_c(s) = K_c(1+1/\tau_I s)$. The closed-loop transfer function for disturbance changes can then be derived from Fig. 9.17.

$$\frac{H'(s)}{Q_1'(s)} = \frac{K_p/(\tau s+1)}{1+K_{OL}(1+1/\tau_I s)/(\tau s+1)} \quad (9\text{-}58)$$

Clearing terms in the denominator gives

$$\frac{H'(s)}{Q_1'(s)} = \frac{K_p \tau_I s}{\tau_I s(\tau s+1)+K_{OL}(\tau_I s+1)} \quad (9\text{-}59)$$

Further rearrangement allows the denominator to be placed in the standard form for a second-order transfer function:

$$\frac{H'(s)}{Q_1'(s)} = \frac{K_3 s}{\tau_3^2 s^2 + 2\zeta_3 \tau_3 s+1} \quad (9\text{-}60)$$

$$K_3 = \tau_I / K_c K_{IP} K_v K_m \quad (9\text{-}61)$$

$$\zeta_3 = \frac{1}{2}\left(\frac{1+K_{OL}}{\sqrt{K_{OL}}}\right)\sqrt{\frac{\tau_I}{\tau}} \quad (9\text{-}62)$$

$$\tau_3 = \sqrt{\tau\tau_I / K_{OL}} \quad (9\text{-}63)$$

For a unit step change in disturbance, $Q_1'(s) = 1/s$, and Eq. 9-59 becomes

$$H'(s) = \frac{K_3}{\tau_3^2 s^2 + 2\zeta_3 \tau_3 s+1} \quad (9\text{-}64)$$

For $0 < \zeta_3 < 1$, the response is a damped oscillation that can be described by

$$h'(t) = \frac{K_3}{\tau_3\sqrt{1-\zeta_3^2}} e^{-\zeta_3 t/\tau_3} \sin[\sqrt{1-\zeta_3^2}\, t/\tau_3] \quad (9\text{-}65)$$

It is clear from Eq. 9-65 that $h'(\infty) = 0$ because of the negative exponential term. Thus, the addition of integral action eliminates offset for a step change in disturbance. It also eliminates offset for step changes in set point. In fact, integral action eliminates offset not only for step changes, but also for any type of sustained change in disturbance or set point. By a sustained change, we mean one that eventually settles out at a new steady-state value, as shown in Fig. 9.21. However, integral action does not eliminate offset for a ramp disturbance.

Equation 9-63 and Fig. 9.22 indicate that increasing K_c or decreasing τ_I tends to speed up the response. In addition, the response becomes more oscillatory as either K_c or τ_I decreases. But in general, closed-loop responses become more oscillatory as K_c is *increased*

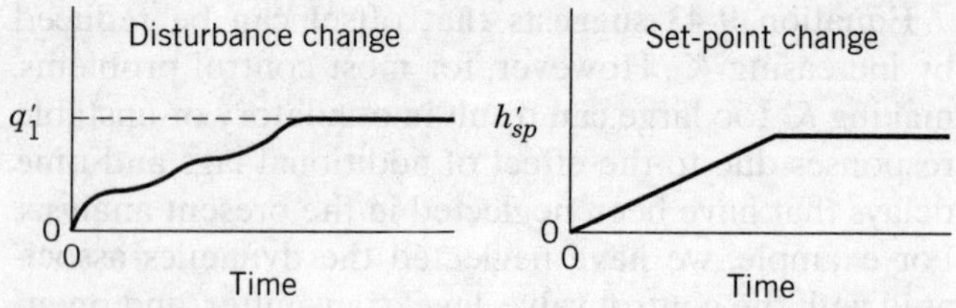

Figure 9.21 Sustained changes in disturbance and set point.

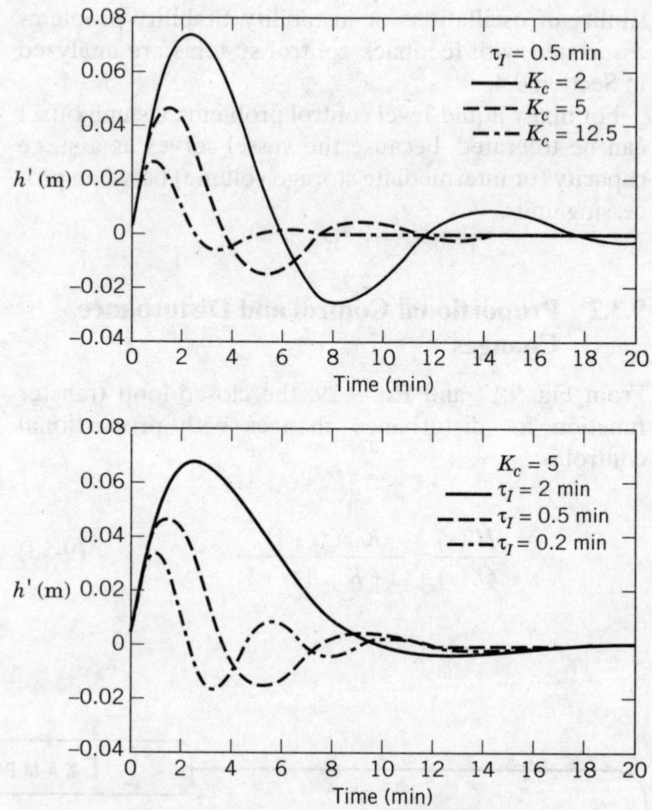

Figure 9.22 Effect of controller settings on disturbance responses.

(see Example 9.5) These anomalous results occur because the small dynamic lags associated with the control valve and transmitter were neglected. If these lags are included, the transfer function in Eq. 9-60 is no longer second-order, and then increasing K_c makes the response more oscillatory.

9.3.4 PI Control of an Integrating Process

Consider the electronic liquid-level control system shown in Fig. 9.23. This system differs from the previous example in two ways: (1) the exit line contains a pump, and (2) the manipulated variable is the exit flow rate rather than an inlet flow rate. In Section 5.5, we saw that a tank with a pump in the exit stream can act as an integrator with respect to flow rate changes, because

$$\frac{H'(s)}{Q_3'(s)} = G_p(s) = -\frac{1}{As} \tag{9-66}$$

$$\frac{H'(s)}{Q_1'(s)} = G_d(s) = \frac{1}{As} \tag{9-67}$$

If the level transmitter and control valve in Fig. 9.23 have negligible dynamics, then $G_m(s) = K_m$ and $G_v(s) = K_v$. For PI control, $G_c(s) = K_c(1+1/\tau_I s)$. Substituting these expressions into the closed-loop transfer function for disturbance changes

$$\frac{H'(s)}{Q_1'(s)} = \frac{G_d}{1+G_cG_vG_pG_m} \tag{9-68}$$

and rearranging gives

$$\frac{H'(s)}{Q_1'(s)} = \frac{K_4 s}{\tau_4^2 s^2 + 2\zeta_4\tau_4 s + 1} \tag{9-69}$$

where

$$K_4 = -\tau_I/K_cK_vK_m \tag{9-70}$$

$$\tau_4 = \sqrt{\tau_I/K_{OL}} \tag{9-71}$$

$$\zeta_4 = 0.5\sqrt{K_{OL}\tau_I} \tag{9-72}$$

and $K_{OL} = K_cK_vK_pK_m$ with $K_p = -1/A$. A comparison of Eqs. 9-67 and 9–69 indicates that feedback control significantly changes the relation between Q_1 and H. Note that Eq. 9-67 is the transfer function for the uncontrolled process, whereas Eq. 9-69 is the closed-loop transfer function for disturbance changes.

From the analysis of second-order transfer functions in Chapter 5, we know that the closed-loop response is oscillatory for $0 < \zeta_4 < 1$. Thus, Eq. 9-72 indicates that the degree of oscillation can be reduced by increasing either $K_c (K_c > 0)$ or τ_I. The effect of τ_I is familiar, because we have noted previously that increasing τ_I tends to make closed-loop responses less oscillatory. However, the effect of K_c is just the opposite of what normally is observed. In most control problems, increasing K_c tends to produce a more oscillatory response. However, Eq. 9-72 indicates that increasing K_c results in a less oscillatory response. This anomalous behavior is due to the integrating nature of the process (cf. Eq. 9-66).

This liquid-level system illustrates the insight that can be obtained from block diagram analysis. It also demonstrates the danger in blindly using a rule of thumb such as "decrease the controller gain to reduce the degree of oscillation."

The analysis of the level control system in Fig. 9.23 has neglected the small dynamic lags associated with

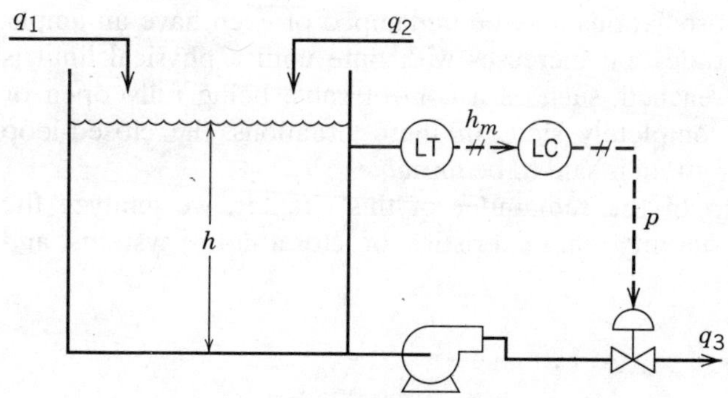

Figure 9.23 Liquid-level control system with pump in exit line.

the transmitter and control valve. If these lags were included, then for very large values of K_c the closed-loop response would indeed tend to become more oscillatory. Thus, if τ_I is held constant, the effect of K_c on the higher-order system can be summarized as follows:

Value of K_c	Closed-Loop Response
Small	Oscillatory
Moderate or large	Overdamped (nonoscillatory)
Very large	Oscillatory or unstable

Because the liquid-level system in Fig. 9.23 acts as an integrator, the question arises whether the controller must also contain integral action to eliminate offset. This question is considered further in Exercise 9.6.

In the previous examples, the denominator of the closed-loop transfer function was either a first- or second-order polynomial in s. Consequently, the transient responses to specified inputs were easily determined. In many control problems, the order of the denominator polynomial is three or higher, and the roots of the polynomial have to be determined numerically. In these cases, analysis of transfer functions becomes more difficult; thus computer simulation is the best tool for determining the nature of the dynamic response. Furthermore, for higher-order ($n > 2$) systems, feedback control can result in unstable responses if inappropriate values of the controller settings are employed, which is also best analyzed using computer simulation.

9.4 STABILITY OF CLOSED-LOOP CONTROL SYSTEMS

An important consequence of feedback control is that it can cause oscillatory responses. If the oscillation has a small amplitude and damps out quickly, then the control system performance is generally considered to be satisfactory. However, under certain circumstances, the oscillations may be undamped or even have an amplitude that increases with time until a physical limit is reached, such as a control valve being fully open or completely shut. In these situations, the closed-loop system is said to be *unstable*.

In the remainder of this chapter, we analyze the stability characteristics of closed-loop systems and present several useful criteria for determining whether a system will be stable. Additional stability criteria based on frequency response analysis are discussed in Chapter 10. But first we consider an illustrative example of a closed-loop system that can become unstable.

EXAMPLE 9.5

Consider the feedback control system shown in Fig. 9.8 with the following transfer functions:

$$G_c = K_c \qquad G_v = \frac{1}{2s+1} \tag{9-73}$$

$$G_p = G_d = \frac{1}{5s+1} \qquad G_m = \frac{1}{s+1} \tag{9-74}$$

Show that the closed-loop system produces unstable responses if controller gain K_c is too large.

Solution

To determine the effect of K_c on the closed-loop response $y(t)$, consider a unit step change in set point, $Y_{sp}(s) = 1/s$. In Section 9.2, we derived the closed-loop transfer function for set-point changes (cf. Eq. 9-26):

$$\frac{Y}{Y_{sp}} = \frac{K_m G_c G_v G_p}{1 + G_c G_v G_p G_m} \tag{9-75}$$

Substituting Eqs. 9-73 and 9-74 into Eq. 9-75 and rearranging gives

$$Y(s) = \frac{K_c(s+1)}{10s^3 + 17s^2 + 8s + 1 + K_c}\frac{1}{s} \tag{9-76}$$

After K_c is specified, $y(t)$ can be determined from the inverse Laplace transform of Eq. 9-76 or by computer simulation. The roots of the cubic polynomial in s must be determined using standard root-finding techniques before performing the partial fraction expansion (Chapra and Canale, 2014). Figure 9.24 demonstrates that as K_c increases, the response becomes more oscillatory and is unstable for $K_c = 15$. More details on the actual stability limit of this control system are given in Example 9.11.

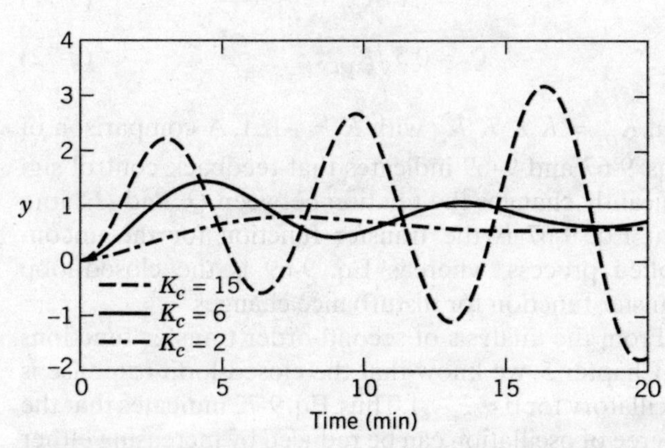

Figure 9.24 Effect of controller gains on closed-loop response to a unit step change in set point (Example 9.5).

The unstable response for Example 9.5 is oscillatory, with the amplitude growing in each successive cycle. In contrast, for an actual physical system, the amplitude will increase until a physical limit is reached or an equipment failure occurs. Because the final control element usually has saturation limits (see Chapter 8), the unstable response will manifest itself as a sustained oscillation with a constant amplitude instead of a continually increasing amplitude. Sustained oscillations can also occur without having the final control element *saturate*, as was mentioned in Section 9.3.

Clearly, a feedback control system must be stable as a prerequisite for satisfactory control. Consequently, it is of considerable practical importance to be able to determine under what conditions a control system becomes unstable. For example, for what values of the PID controller parameters K_c, τ_I, and τ_D is the controlled process stable?

9.4.1 General Stability Criterion

Most industrial processes are stable without feedback control. Thus, they are said to be *open-loop stable,* or *self-regulating*. An open-loop stable process will return to the original steady state after a transient disturbance (one that is not sustained) occurs. By contrast, there are a few processes, such as exothermic chemical reactors, that can be *open-loop unstable*. These processes are extremely difficult to operate without feedback control.

Before presenting various stability criteria, we introduce the following definition for unconstrained linear systems. We use the term *unconstrained* to refer to the ideal situation where there are no physical limits on the input and output variables.

> ***Definition of Stability.*** *An unconstrained linear system is said to be stable if the output response is bounded for all bounded inputs. Otherwise it is said to be unstable.*

By a bounded input, we mean an input variable that stays within upper and lower limits for all values of time. For example, consider a variable $u(t)$ that varies with time. If $u(t)$ is a step or sinusoidal function, then it is bounded. However, the functions $u(t) = t$ and $u(t) = e^{3t}$ are not bounded.

EXAMPLE 9.6

A liquid storage system is shown in Fig. 9.25. Show that this process is not self-regulating by considering its response to a step change in inlet flow rate.

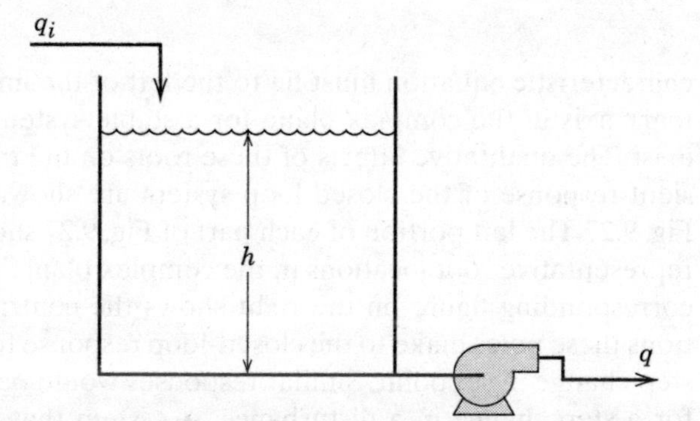

Figure 9.25 A liquid storage system that is not self-regulating.

Solution

The transfer function relating liquid level h to inlet flow rate q_i *was* derived in Section 5.5:

$$\frac{H'(s)}{Q_i'(s)} = \frac{1}{As} \tag{9-77}$$

where A is the cross-sectional area of the tank. For a step change of magnitude $M_0, Q_i'(s) = M_0/s$, and thus

$$H'(s) = \frac{M_0}{As^2} \tag{9-78}$$

Taking the inverse Laplace transform gives the transient response,

$$h'(t) = \frac{M_0}{A}t \tag{9-79}$$

We conclude that the liquid storage system is open-loop unstable (or *non-self-regulating*) because a bounded input has produced an unbounded response. However, if the pump in Fig. 9.25 were replaced by a valve, then the storage system would be self-regulating (cf. Example 4.7).

Characteristic Equation

As a starting point for the stability analysis, consider the block diagram in Fig.9.8. Using block diagram algebra that was developed earlier in this chapter, we obtain

$$Y = \frac{K_m G_c G_v G_p}{1+G_{OL}} Y_{sp} + \frac{G_d}{1+G_{OL}} D \tag{9-80}$$

where G_{OL} is the open-loop transfer function, $G_{OL} = G_c G_v G_p G_m$.

For the moment consider set-point changes only, in which case Eq. 9-80 reduces to the closed-loop transfer function

$$\frac{Y}{Y_{sp}} = \frac{K_m G_c G_v G_p}{1+G_{OL}} \tag{9-81}$$

If G_{OL} is a ratio of polynomials in s (i.e., a rational function), then the closed-loop transfer function in Eq. 9-81 is also a rational function. After rearrangement, it can be factored into poles (p_i) and zeroes (z_i) as

$$\frac{Y}{Y_{sp}} = K' \frac{(s-z_1)(s-z_2)\cdots(s-z_m)}{(s-p_1)(s-p_2)\cdots(s-p_n)} \tag{9-82}$$

where K' is a multiplicative constant that gives the correct steady-state gain. To have a physically realizable system, the number of poles must be greater than or equal to the number of zeroes; that is, $n \geq m$ (Kuo, 2002). Note that a *pole-zero cancellation* occurs if a zero and a pole have the same numerical value.

Comparing Eqs. 9-81 and 9-82 indicates that the poles are also the roots of the following equation, which is referred to as the *characteristic equation* of the closed-loop system:

$$1 + G_{OL} = 0 \tag{9-83}$$

The characteristic equation plays a decisive role in determining system stability, as discussed below.

For a unit change in set point, $Y_{sp}(s) = 1/s$, and Eq. 9-82 becomes

$$Y = \frac{K'}{s} \frac{(s-z_1)(s-z_2)\ldots(s-z_m)}{(s-p_1)(s-p_2)\ldots(s-p_n)} \tag{9-84}$$

If there are no repeated poles (i.e., if they are all *distinct poles*), then the partial fraction expansion of Eq. 9-84 has the form considered in Section 4.5,

$$Y(s) = \frac{A_0}{s} + \frac{A_1}{s-p_1} + \frac{A_2}{s-p_2} + \cdots + \frac{A_n}{s-p_n} \tag{9-85}$$

where the $\{A_i\}$ can be determined using the methods of Chapter 3. Taking the inverse Laplace transform of Eq. 9-85 gives

$$y(t) = A_0 + A_1 e^{p_1 t} + A_2 e^{p_2 t} + \cdots + A_n e^{p_n t} \tag{9-86}$$

Suppose that one of the poles is a positive real number; that is, $p_k > 0$. Then it is clear from Eq. 9-86 that $y(t)$ is unbounded, and thus the closed-loop system in Fig. 9.8 is unstable. If p_k is a complex number, $p_k = a_k + jb_k$, with a positive real part ($a_k > 0$), then the system is also unstable. By contrast, if all of the poles are negative (or have negative real parts), then the system is stable. These considerations can be summarized in the following stability criterion:

General Stability Criterion. *The feedback control system in Fig. 9.8 is stable if and only if all roots of the characteristic equation are negative or have negative real parts. Otherwise, the system is unstable.*

Figure 9.26 provides a graphical interpretation of this stability criterion. Note that all of the roots of the characteristic equation must lie to the left of the imaginary axis in the complex plane for a stable system to exist. The qualitative effects of these roots on the transient response of the closed-loop system are shown in Fig. 9.27. The left portion of each part of Fig. 9.27 shows representative root locations in the complex plane. The corresponding figure on the right shows the contributions these poles make to the closed-loop response for a step change in set point. Similar responses would occur for a step change in a disturbance. A system that has all negative real roots will have a stable, nonoscillatory response, as shown in Fig. 9.27*a*. On the other hand, if one of the real roots is positive, then the response is unbounded, as shown in Fig. 9.27*b*. A pair of complex conjugate roots results in oscillatory responses as shown in Figs. 9.27*c* and 9.27*d*. If the complex roots have negative real parts, the system is stable; otherwise it is unstable. Recall that complex roots always occur as complex conjugate pairs.

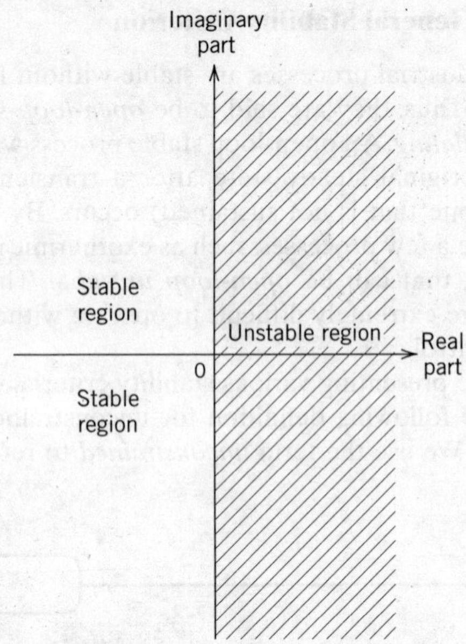

Figure 9.26 Stability regions in the complex plane for roots of the characteristic equation.

The root locations also provide an indication of how rapid the transient response will be. A real root at $s = -a$ corresponds to a closed-loop time constant of $\tau = 1/a$, as is evident from Eqs. 9-85 and 9-86. Thus, real roots close to the imaginary (vertical) axis result in slow responses. Similarly, complex roots near the imaginary axis correspond to slow response modes. The farther the complex roots are away from the real axis, the more oscillatory the transient response will be (see Example 9.5). However, the process zeros also influence the response, as discussed in Chapter 4.

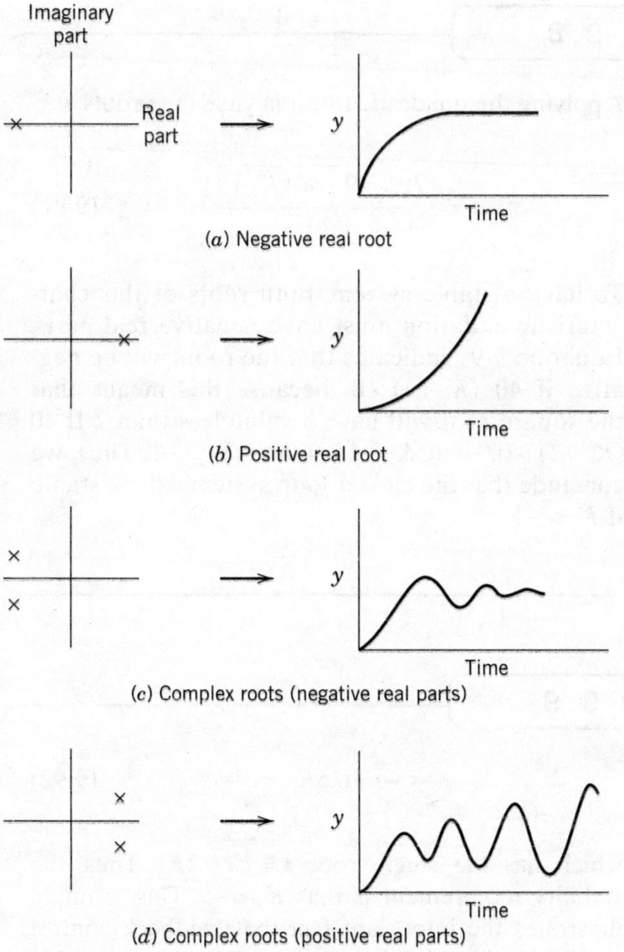

Figure 9.27 Contributions of characteristic equation roots to closed-loop response.

Note that the same characteristic equation occurs for both disturbance and set-point changes because the term, $1+G_{OL}$, appears in the denominator of both terms in Eq. 9-80. Thus, if the closed-loop system is stable for disturbances, it will also be stable for set-point changes.

The analysis in Eqs. 9-80 to 9-86 that led to the general stability criterion is based on a number of assumptions:

1. Set-point changes (rather than disturbance changes) are considered.
2. The closed-loop transfer function is a ratio of polynomials.
3. The poles in Eq. 9-82 are all distinct.

However, the general stability criterion is valid even if these assumptions are removed. In fact, this stability criterion is valid for *any* linear control system (comprised of linear elements described by transfer functions). By contrast, for nonlinear systems rigorous stability analyses tend to be considerably more complex and involve special techniques such as Liapunov and Popov stability criteria (Khalil, 2001). Fortunately, a stability analysis of a linearized system using the techniques presented in this chapter normally provides useful information for nonlinear systems operating near the point of linearization.

From a mathematical point of view, the general stability criterion presented above is a *necessary and sufficient condition*. Thus, linear system stability is completely determined by the roots of the characteristic equation.

EXAMPLE 9.7

Consider the feedback control system in Fig. 9.8 with $G_v = K_v$, $G_m = 1$, and $G_p = K_p/(\tau_p s+1)$. Determine the stability characteristics if a proportional controller is used, $G_c = K_c$.

Solution

Substituting the transfer functions into the characteristic equation in Eq. 9-83 gives

$$1+\frac{K_c K_v K_p}{\tau_p s+1}=0$$

which reduces to

$$\tau_p s+1+K_c K_v K_p=0 \tag{9-87}$$

This characteristic equation has a single root,

$$s=-\frac{1+K_c K_v K_p}{\tau_p} \tag{9-88}$$

The closed-loop system will be stable if this root is negative. Because time constants are always positive ($\tau_p > 0$), the feedback control system will be stable if $K_c K_v K_p > -1$. This means that as long as the controller has the correct control action (i.e., reverse- or direct-acting, as per Section 7.3), then the system will be stable. For example, if $K_p > 0$ and $K_v > 0$ then the controller must be made reverse-acting so that $K_c > 0$. By contrast, if $K_p < 0$, then a direct-acting controller ($K_c < 0$) is required.

EXAMPLE 9.8

Consider the feedback control system in Example 9.5, but now assume that $G_m = 1$. Determine the range of K_c values that result in a stable closed-loop system.

Solution

Substituting these transfer functions into Eq. 9-83 gives

$$1+\frac{K_c}{(2s+1)(5s+1)}=0 \tag{9-89}$$

which can be rearranged as

$$10s^2+7s+K_c+1=0 \tag{9-90}$$

Applying the quadratic formula yields the roots

$$s=\frac{-7\pm\sqrt{49-40(K_c+1)}}{20} \tag{9-91}$$

To have a stable system, both roots of this characteristic equation must have negative real parts. Equation 9-91 indicates that the roots will be negative if $40\ (K_c+1)>0$, because this means that the square root will have a value less than 7. If $40\ (K_c+1)>0$, then $K_c+1>0$, and $K_c>-1$. Thus, we conclude that the closed-loop system will be stable if $K_c>-1$.

EXAMPLE 9.9

Consider a process, $G_p = 0.2/(-s+1)$, that is open-loop unstable. If $G_v = G_m = 1$, determine whether a proportional controller can stabilize the closed-loop system.

Solution

The characteristic equation for this system is

$$-s+1+0.5K_c=0 \tag{9-92}$$

which has the single root $s=1+0.2K_c$. Thus, the stability requirement is that $K_c<-5$. This example illustrates the important fact that feedback control can be used to stabilize a process that is not stable without control.

The stability analyses for Examples 9.7 and 9.9 have indicated that these closed-loop systems will be stable for all positive values of K_c, no matter how large. However, this result is *not* typical, because it occurs only for the special case where the open-loop system is stable and the open-loop transfer function G_{OL} is first- or second-order with no time delay. In more typical problems, K_c must be below an upper limit to have a stable closed-loop system.[1] See the examples in the next section.

In Examples 9.7–9.9, the characteristic equations were either first- or second-order, and thus we could find the roots analytically. For higher-order polynomials, this is not possible, and numerical root-finding techniques (Chapra and Canale, 2014), also available in MATLAB and Mathematica, must be employed. An attractive alternative, the Routh stability criterion, can evaluate stability without requiring calculation of the roots of the characteristic equation.

9.4.2 Routh Stability Criterion

Routh (1905) published an analytical technique for determining whether any roots of a polynomial have positive real parts. According to their general stability criterion, a closed-loop system will be stable only if all of the roots of the characteristic equation have negative real parts. Thus, by applying Routh's technique to analyze the coefficients of the characteristic equation, we can determine whether the closed-loop system is stable. This approach is referred to as the *Routh stability criterion*. It can be applied only to systems whose

[1] If a direct-acting controller is used (i.e., $K_c<0$), then stability considerations place an upper limit on $-K_c$ rather than on K_c.

characteristic equations are polynomials in s. Thus, the Routh stability criterion is not directly applicable to systems containing time delays, which limits its applicability for realistic processes. If $e^{-\theta s}$ is replaced by a Padé approximation (see Section 6.1), then an *approximate* stability analysis can be performed using the Routh array (see Seborg et al., 2010). An exact stability analysis of systems containing time delays can be performed by direct root-finding or by using a frequency response analysis and the Bode or Nyquist stability criterion presented in Chapter 10.

The Routh stability criterion is based on a characteristic equation that has the form

$$a_n s^n + a_{n-1} s^{n-1} + \cdots + a_1 s + a_0 = 0$$

We arbitrarily assume that $a_n > 0$. If $a_n < 0$, simply multiply the above equation by −1 to generate a new equation that satisfies this condition. A *necessary* (but not sufficient) condition for stability is that all of the coefficients $(a_0, a_1, \ldots, a_n)$ in the characteristic equation be positive. If any coefficient is negative or zero, then at least one root of the characteristic equation lies to the right of, or on, the imaginary axis, and the system is unstable. If all of the coefficients are positive, we next construct the following Routh array:

Row				
1	a_n	a_{n-2}	a_{n-4}	...
2	a_{n-1}	a_{n-3}	a_{n-5}	...
3	b_1	b_2	b_3	...
4	c_1	c_2	...	
⋮	⋮			
$n+1$	z_1			

The Routh array has $n + 1$ rows, where n is the order of the characteristic equation (in polynomial form, shown in the beginning of this section). The Routh array has a roughly triangular structure with only a single element in the last row. The first two rows are merely the coefficients in the characteristic equation, arranged according to odd and even powers of s. The elements in the remaining rows are calculated from the formulas.

$$b_1 = \frac{a_{n-1}a_{n-2} - a_n a_{n-3}}{a_{n-1}}$$

$$b_2 = \frac{a_{n-1}a_{n-4} - a_n a_{n-5}}{a_{n-1}}$$

$$\vdots$$

$$c_1 = \frac{b_1 a_{n-3} - a_{n-1} b_2}{b_1}$$

$$c_2 = \frac{b_1 a_{n-5} - a_{n-1} b_3}{b_1}$$

$$\vdots$$

Note that the expressions in the numerators of the above equations are similar to the calculation of a determinant for a 2 × 2 matrix except that the order of subtraction is reversed. Having constructed the Routh array, we can now state the Routh stability criterion:

Routh Stability Criterion. *A necessary and sufficient condition for all roots of the characteristic equation* $(a_n s^n + a_{n-1} s^{n-1} + \cdots + a_1 s + a_0 = 0)$ *to have negative real parts is that all of the elements in the left column of the Routh array are positive.*

9.4.3 Direct Substitution Method

The imaginary axis divides the complex plane into stable and unstable regions for the roots of the characteristic equation, as indicated in Fig. 9.27. On the imaginary axis, the real part of s is zero, and thus $s = j\omega$. Substituting $s = j\omega$ into the characteristic equation provides a way to calculate the stability limit indicated by the maximum value of K_c (Luyben and Luyben, 1997), which is called K_{cm}. As the gain K_c (in absolute value) is increased, the roots of the characteristic equation cross the imaginary axis when $K_c = K_{cm}$.

EXAMPLE 9.10

Determine the stability of a system that has the characteristic equation

$$s^4 + 5s^2 + 3s^2 + 1 = 0 \quad (9\text{-}93)$$

Solution

Because the s term is missing, its coefficient is zero. Thus, the system is unstable. A necessary condition for stability is that all of the coefficients in the characteristic equation must be positive.

EXAMPLE 9.11

Find the values of controller gain K_c that make the feedback control system of Example 9.5 stable. Use the direct substitution method to determine K_{cm} for the system.

Solution

From Eq. 9-76, the characteristic equation is

$$10s^3+17s^2+8s+1+K_c=0 \qquad (9\text{-}94)$$

All coefficients are positive, provided that $1+K_c>0$ or $K_c>-1$. Substitute $s=j\omega$, $\omega=\omega_m$, and $K_c=K_{cm}$ into Eq. 9-94:

$$-10j\omega_m^3-17\omega_m^2+8j\omega_m+1+K_{cm}=0 \qquad (9\text{-}95)$$

or

$$(1+K_{cm}-17\omega_m^2)+j(8j\omega_m-10\omega_m^3)=0$$

Equation 9-95 is satisfied if both the real and imaginary parts are identically zero:

$$1+K_{cm}-17\omega_m^2=0 \qquad (9\text{-}96a)$$

$$8\omega_m-10\omega_m^3=\omega_m(8-10\omega_m^2)=0 \qquad (9\text{-}96b)$$

Therefore,

$$\omega_m^2=0.8 \Rightarrow \omega_m=\pm 0.894 \qquad (9\text{-}97)$$

and from (9-96a),

$$K_{cm}=12.6$$

Thus, we conclude that $K_c<12.6$ for stability. Equation 9-97 indicates that at the stability limit (where $K_c=K_{cm}=12.6$), a sustained oscillation occurs that has a frequency of $\omega_m=0.894$ radian/min if the time constants have units of minutes. (Recall that a pair of complex roots on the imaginary axis, $s=\pm j\omega$, results in an undamped oscillation of frequency ω.) The corresponding period P is $2\pi/0.894=7.03$ min.

EXAMPLE 9.12

Consider a feedback control system with $G_c=K_c$, $G_v=2$, $G_m=0.25$, and $G_p=4e^{-s}/(5s+1)$. Find K_{cm} using direct substitution.

Solution

The characteristic equation is

$$1+5s+2K_ce^{-s}=0 \qquad (9\text{-}98)$$

Set $s=j\omega$

$$1+5j\omega+2K_ce^{-j\omega}=0 \qquad (9\text{-}99)$$

$$1+5j\omega+2K_c(\cos\omega-j\sin\omega)=0 \qquad (9\text{-}100)$$

Re:

$$1+2K_c\cos\omega=0 \qquad (9\text{-}101)$$

Im:

$$5\omega-2K_c\sin\omega=0 \qquad (9\text{-}102)$$

Solve for K_c in Eq. 9-101 and substitute into Eq. 9-102:

$$5\omega+\frac{\sin\omega}{\cos\omega}=5\omega+\tan\omega=0 \qquad (9\text{-}103)$$

Solve for ω:

$$\omega_m=1.69 \text{ rad/min}(96.87°/\text{min})$$

From Eq. 9-101,

$$K_{cm}=4.25$$

The direct-substitution method is related to the Routh stability criterion in Section 9.4.2. If the characteristic equation has a pair of roots on the imaginary axis, equidistant from the origin, and all other roots are in the left-hand plane, the single element in the next-to-last row of the Routh array will be zero. Then the location of the two imaginary roots can be obtained from the solution of the equation.

$$Cs^2+D=0$$

where C and D are the two elements in the $(n-1)$ row of the Routh array, as read from left to right.

The direct-substitution method is also related to the frequency response approach of Chapter 10, because both techniques are based on the substitution $s=j\omega$.

EXAMPLE 9.13

A PI controller with integral time constant of 0.1 min is to be designed to control a process with transfer function,

$$G_p(s) = \frac{10}{s^2 + 2s + 100}$$

Given $G_m = 1, G_v = 1$.

Find the value of the controller gain that will constitute the critical condition for stability of the PI feedback control system. Use Routh stability criterion.

Solution

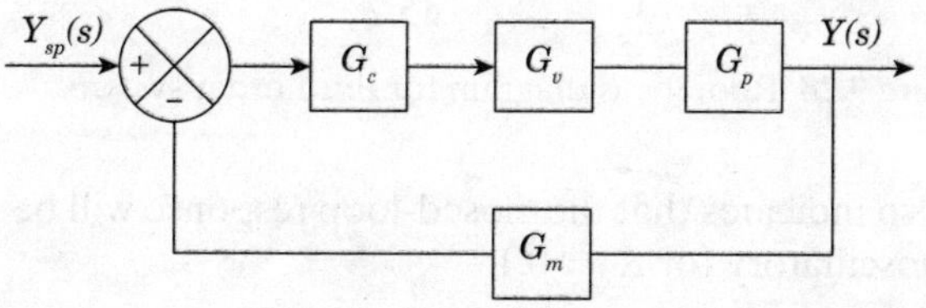

$$G_c(s) = K_c\left(1+\frac{1}{Y_I s}\right) = K_c\left(1+\frac{1}{0.1s}\right)$$

$$G_v(s) = 1$$

$$G_p(s) = \frac{10}{s^2 + 2s + 100}$$

$$G_m(s) = 1$$

Characteristic equation ($1 + G_{OL} = 0$)

$$1 + G_c G_v G_p G_m = 0$$

$$1+\left[K_c\left(1+\frac{1}{0.1s}\right)\cdot\left(\frac{10}{s^2+2s+100}\right)\right] = 0$$

or $$1+\left[\frac{K_c(0.1s+1)}{0.1s}\cdot\frac{10}{(s^2+2s+100)}\right] = 0$$

or $$1+\left[\frac{K_c(s+10)}{0.1s(s^2+2s+100)}\right] = 0$$

or $$1+\left[\frac{K_c(s+10)}{(0.1s^3+0.2s^2+10s)}\right] = 0$$

$$0.1s^3 + 0.2s^2 + 10s + sK_c + 10K_c = 0$$

or $$0.1s^3 + 0.2s^2 + s(10+K_c) + 10K_c = 0$$

Using Routh Stability Criteria

0.1	$(10+K_c)$
0.2	$10K_c$
$\dfrac{0.2(10+K_c) - 0.1\times 10K_c}{0.2}$	–

Critical or marginal condition of stability

$$0.2(10+K_c) - 0.1\times 10K_c = 0$$

or $$0.2(10+K_c) = K_c$$

or $$10+K_c = \frac{K_c}{0.2} = 5K_c$$

or $$4K_c = 10$$

or $$K_c = \frac{10}{4} = 2.5$$

or $$K_{cm} = 2.5$$

9.5 ROOT LOCUS DIAGRAMS

In the previous section we have seen that the roots of the characteristic equation play a crucial role in determining system stability and the nature of the closed-loop responses. In the design and analysis of control systems, it is instructive to know how the roots of the characteristic equation change when a particular system parameter such as a controller gain changes. A root locus diagram provides a convenient graphical display of this information, as indicated in the following example.

The MATLAB command *rlocus* can be used to generate a diagram such as Fig. 9.28.

EXAMPLE 9.14

Consider a feedback control system that has the open-loop transfer function,

$$G_{OL}(s) = \frac{4K_c}{(s+1)(s+2)(s+3)} \quad (9\text{-}104)$$

Plot the root locus diagram for $0 \le K_c \le 20$.

Solution

The characteristic equation is $1 + G_{OL} = 0$ or

$$(s+1)(s+2)(s+3) + 4K_c = 0 \qquad (9\text{-}105)$$

The root locus diagram in Fig. 9.28 shows how the three roots of this characteristic equation vary with K_c. When $K_c = 0$, the roots are merely the poles of the open-loop transfer function, –1, –2, and –3. These are designated by an × symbol in Fig. 9.28. As K_c increases, the root at –3 decreases monotonically. The other two roots converge and then form a complex conjugate pair when $K_c = 0.1$. When $K_c = K_{cm} = 15$, the complex roots cross the imaginary axis and enter the unstable region. This illustrates why the substitution of $s = j\omega$ (Section 9.3) determines the unstable controller gain. Thus, the root locus diagram indicates that the closed-loop system is unstable for $K_c > 15$.

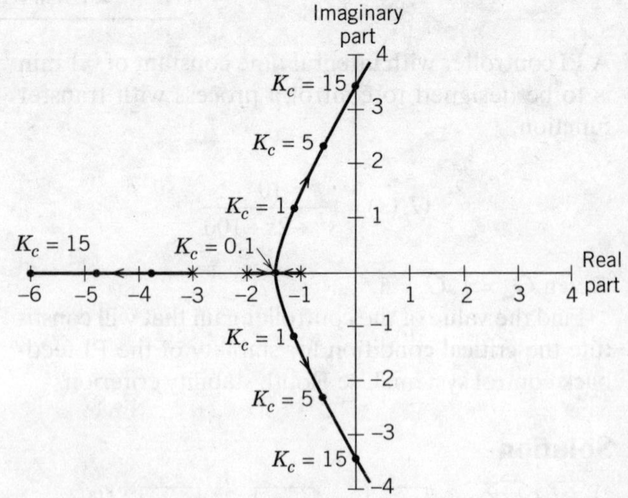

Figure 9.28 Root locus diagram for third-order system.

It also indicates that the closed-loop response will be nonoscillatory for $K_c < 0.1$.

The root locus diagram can be used to provide a quick estimate of the transient response of the closed-loop system. The roots closest to the imaginary axis correspond to the slowest response modes. If the two closest roots are a complex conjugate pair, then the closed-loop system can be approximated by an underdamped second-order system (cf. Section 5.9).

The utility of root locus diagrams has been illustrated by the third-order system of Example 9.4, the roots of which can be generated using a MATLAB command.

The major disadvantage of root locus analysis is that time delays cannot be handled conveniently, and they require iterative solution of the nonlinear and nonrational characteristic equation. Nor is it easy to display simultaneous changes in more than one parameter (e.g., controller parameters K_c and τ_I). For these reasons, the root locus technique is rarely used as a design tool in process control.

Root locus diagrams can be quickly generated by using software such as MATLAB.

9.6 RULES FOR DRAWING ROOT LOCUS DIAGRAM

Consider an open-loop transfer function as shown in Eq. 9-106.

$$G_{OL}(s) = K\frac{(s-z_1)(s-z_2)\cdots(s-z_n)}{(s-p_1)(s-p_2)\cdots(s-p_m)} = K\frac{N(s)}{D(s)} \qquad (9\text{-}106)$$

where $z_1, z_2, \ldots z_n$ are the zeros and $p_1, p_2, \ldots p_m$ are the poles of the system, m is the number of poles and n is the number of zeros of the G_{OL}.

The following rules are to be followed for constructing the root locus diagram:

1. Locate the open loop poles and zeros in a complex plane with real and imaginary axes.
2. Find the number of root loci.

   Number of separate root loci = greater of m and n. where m and n are number of poles and zeros, respectively.
3. Identify and draw root loci on the real axis.

The real axis is part of the root locus when the sum of the number of poles and zeros to the right of a point on the real axis is odd. An nth-order pole must be counted n times. Complex poles and zeros are not to be counted.

The root loci begin at poles (corresponding to $K_c = 0$) and terminate at zeros (corresponding to $K_c \to \infty$). Termination of $(m - n)$ number of root loci occur at the zeros at infinite long asymptotes.

4. Find the number of asymptotes and the center of gravity (centroid).

Number of asymptotes = $m - n$

$$\text{Centre of gravity } \sigma = \frac{\sum Poles - \sum zeros}{m-n} \quad (9\text{-}107)$$

5. Obtain the angle of asymptotes.

Angle of asymptote,

$$\theta_k = \frac{(2k+1)\pi}{m-n} \qquad k = 0, 1, 2, \ldots, m-n-1. \quad (9\text{-}108)$$

6. Find Break-away and Break-in points.

The break-away point is the point at which two root loci intersect and then leave the real axis. Whereas, the point at which two root loci meet and then enter the real axis is the break-in point. If there exists a real axis root locus branch between two open loop poles, then there will be a break-away point in between them. But if there exists a real axis root locus branch between two open loop zeros, then there will be a break-in point in between them.

Solve any of the following equations to find break-away and break-in points.

$$\frac{d}{ds}(G_{OL}) = 0 \text{ or } \frac{dK}{ds} = 0$$

or
$$N(s)D'(s) = D(s)N'(s)$$

or
$$\sum_{j=1}^{m} \frac{1}{s-p_j} = \sum_{i=1}^{n} \frac{1}{s-z_i} \quad (9\text{-}109)$$

where K is a function of s and is obtained from the characteristic equation, $1 + G_{OL}(s) = 0$.

This can be solved by trial and error method. It will be near to the center of gravity. When breakaway point is same as the center of gravity, the root loci coincide with the asymptotes.

7. Find the angle of departure and the angle of arrival. Angle of departure (or approach), θ, associated with the imaginary poles (if any) is given by

$$\sum(\text{Angles produced by zeros on positive real axis}) - \sum(\text{Angles produced by poles on positive real axis}) - \theta = 180^\circ \quad (9\text{-}110)$$

8. Find the intersection points of root loci with the imaginary axis.

Point of intersection of the root loci with the imaginary axis comes from the Auxiliary Equation in Routh Table for s^2-row with values of K_c for marginal stability (when all elements in a row are zeros).

Identify the row in such a way that if the first element is made zero, then the elements of the entire row are zeros. Find the value of K_c for this combination. Substitute this K_c value in the auxiliary equation and solve to get the desired intersection point.

EXAMPLE 9.15

Plot the root locus diagram for the system given in Example 9.14 using the rules outlined in section 9.6.

Solution

$$G_{OL} = \frac{4K_c}{(s+1)(s+2)(s+3)}$$

Here, there are three poles and no zeros. The poles are: −1, −2, −3.

So, $m = 3$ and $n = 0$.

The poles are marked as ∗ on the complex plane shown in Fig. 9.29. N indicates the absence of root locus on that portion of the real axis (Rule 3).

Number of root loci = greater of m and n = 3

Number of asymptotes = $m - n$ = 3

$$\text{Centre of gravity } \sigma = \frac{\sum \text{Poles} - \sum \text{Zeros}}{m-n} = \frac{(-1)+(-2)+(-3)}{3-0} = -2$$

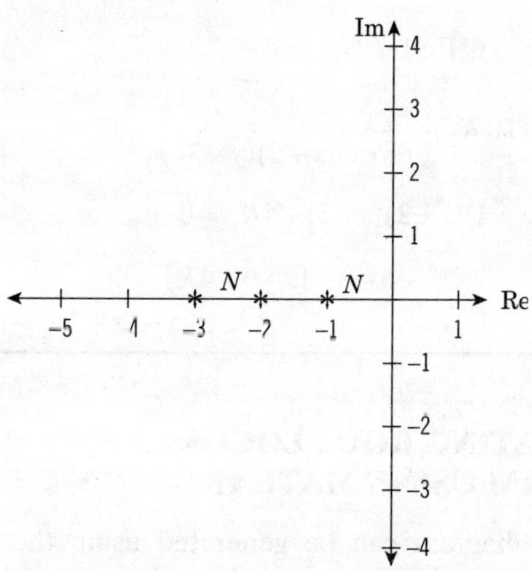

Figure 9.29 Location of poles of the system.

$$\text{Angle of asymptotes } \theta_k = \frac{(2k+1)\pi}{m-n} \qquad k = 0, 1, 2 \text{ (for 3 asymptotes)}$$

$$\theta_0 = \frac{\pi}{3} = 60° \quad \theta_1 = \frac{3\pi}{3} = \pi = 180° \quad \theta_2 = \frac{5\pi}{3} = 300°$$

Asymptotes are drawn at $s = -2$, with the above angles.

Break-away point:

$$\frac{d}{ds}(G_{OL}) = 0 \text{ or } \frac{d}{ds}\left(\frac{4K_c}{(s+1)(s+2)(s+3)}\right) = 0$$

or
$$\frac{-(3s^2 + 12s + 11)}{[(s+1)(s+2)(s+3)]^2} = 0$$

$$\Rightarrow (3s^2 + 12s + 11) = 0 \Rightarrow s = -1.423 \text{ and } -2.58.$$

As there is no root locus present between −3 and −2, $s = -2.58$ is not applicable for the break-away point.

So, the break-away point is $s = -1.423$.

Draw the asymptotes at the centre of gravity ($s = -2$) with angles 60°, 180°, and 300°. The root loci from poles −1 and −2 approach each other and depart from the real axis at the break-away point. They follow the asymptotes at 60° and 300°. The root locus starting from pole −3 coincides with the asymptote and goes to −∞. This is shown in Fig. 9.30.

The following procedure is followed (as per Rule 8) to obtain the intersection with the imaginary axis.

$$\text{C.E}: 1 + G_{OL} = 0$$

$$1 + (1)(1)(K_c)\left(\frac{4}{(s+1)(s+2)(s+3)}\right) = 0$$

$$(s+1)(s+2)(s+3) + 4K_c = 0$$

Or
$$s^3 + 6s^2 + 11s + 6 + 4K_c = 0$$

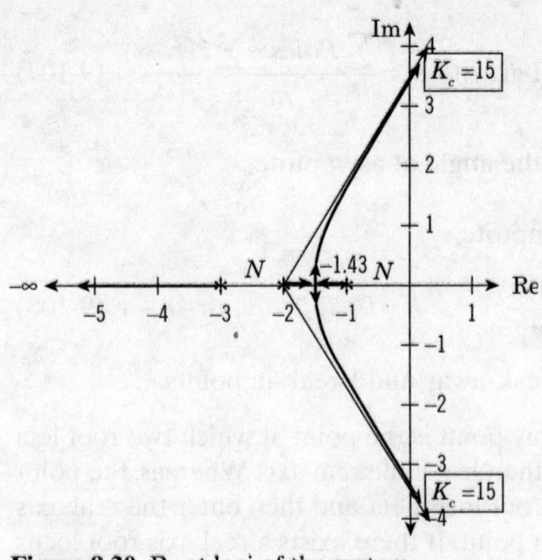

Figure 9.30 Root loci of the system.

Routh Array

s^3	6	11
s^2	1	$6 + 4K_c$
s^1	$\frac{66 - (6 + 4K_c)}{6}$	0
s^0	$6 + 4K_c$	

For marginal stability, all the elements in a row should be zero. The s^1-row has one element zero. For marginal stability we have to make the other element zero. So,

$$\frac{66 - (6 + 4K_c)}{6} = 0 \quad \text{or} \quad 60–4K_c - 0 \quad \text{or} \quad K_c = 15$$

Now, the auxiliary equation is to be made from s^2-row:

$$6s^2 + 6 + 4K_c = 0 \text{ or } 6s^2 + 6 + 60 = 0$$
$$\text{or } s^2 + 11 = 0 \text{ or, } s = \pm 3.316\,j$$

The root locus crosses the imaginary axis at these points corresponding to $K_c = 15$.

9.7 GENERATING ROOT LOCUS DIAGRAM USING MATLAB

The root locus diagram can be generated using the MATLAB command ***rlocus.***

Consider the following open loop transfer function.

$$G_{OL} = \frac{K_c}{(s+1)(s+2)(s+3)} = \frac{K_c}{s^3 + 6s^2 + 11s + 6}$$

The MATLAB code to obtain the root locus diagram is as follows:

```
>>num = [1];
>>den = [16 11 6];
>>sys = tf(num, den);
>>rlocus(sys)
```

The output of this program is shown in Fig 9.31.

The MATLAB code can also be written as follows:

```
>>clear all;
>>s = tf('s')
>>Kc = 1;
>>GoL = Kc/((s + 1)*(s + 2)*(s + 3));
>>rlocus(GoL)
```

The same root locus diagram will be obtained in both the cases.

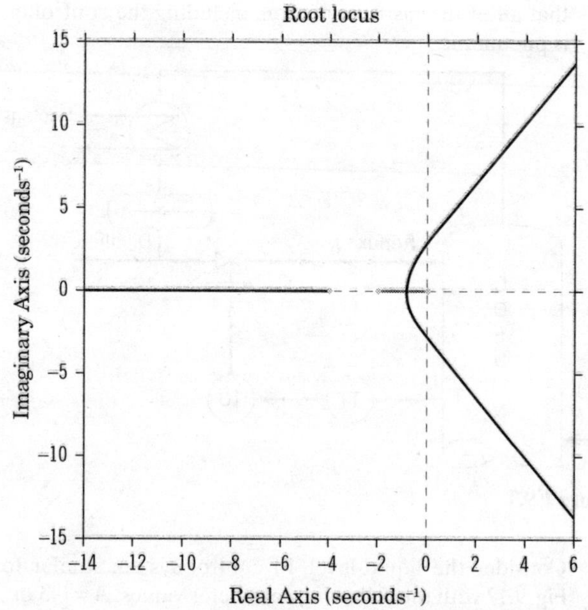

Figure 9.31 Root Locus Diagram Using MATLAB.

SUMMARY

This chapter has considered the dynamic behavior of processes that are operated under feedback control. A block diagram provides a convenient representation for analyzing control system performance. By using block diagram algebra, expressions for closed-loop transfer functions can be derived and used to calculate the closed-loop responses to disturbance and set-point changes. Several liquid-level control problems have been considered to illustrate key features of proportional and proportional-integral control. Proportional control results in offset for sustained disturbance or set-point changes; however, these offsets can be eliminated by including integral control action.

We have also considered several stability criteria for linear systems that can be described by transfer function models. If the process model is nonlinear, then advanced stability theory can be used (Khalil, 2001), or an approximate stability analysis can be performed based on a linearized transfer function model. If the transfer function model includes time delays, then an exact stability analysis can be performed using root-finding or, preferably, the frequency response methods of Chapter 10.

Having dealt with the stability of closed-loop systems, we can consider our next topic, the design of feedback control systems. This important subject is considered in Chapters 10 and 11. A number of prominent control system design and tuning techniques are based on stability criteria.

REFERENCES

Chapra, S. C., and R. P. Canale, *Numerical Methods for Engineers*, 7th ed., McGraw-Hill, New York, 2014.

Khalil, H. K., *Nonlinear Systems*, 3rd ed., Prentice Hall, Englewood Cliffs, NJ, 2001.

Kuo, B. C., *Automatic Control Systems*, 8th ed., Prentice Hall, Englewood Cliffs, NJ, 2002.

Luyben, W. L., and M. L. Luyben, *Essentials of Process Control*, McGraw-Hill, New York, 1997.

Routh, E. J., *Dynamics of a System of Rigid Bodies, Part 11*, Macmillan, London, 1905.

Seborg, D. E., T. F. Edgar, D. A. Mellichamp, and F. J. Doyle III, *Process Dynamics and Control*, 3rd ed., John Wiley and Sons, Hoboken, NJ, 2010.

EXERCISES

9.1 A temperature control system for a distillation column is shown in Fig. E9.1. The temperature T of a tray near the top of the column is controlled by adjusting the reflux flow rate R. Draw a block diagram for this feedback control system. You may assume that both feed flow rate F and feed composition x_F are disturbance variables and that all of the instrumentation, including the controller, is pneumatic.

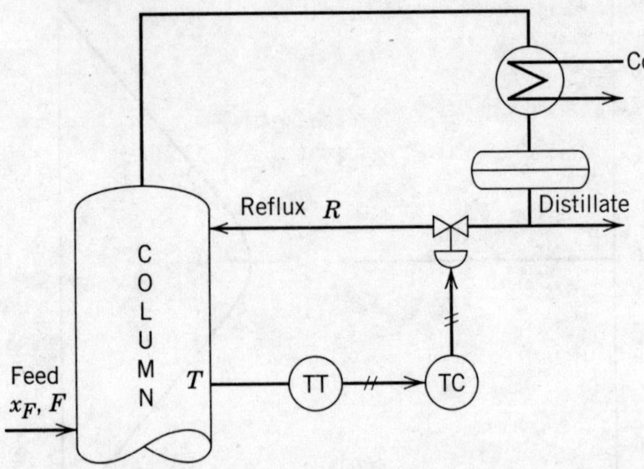

Figure E9.1

9.2 Consider the liquid-level, PI control system similar to Fig. 9.17 with the following parameter values: $A = 0.3\ \text{m}^2$, $R = 10\ \text{min/m}^2$, $K_v = 820\ \text{cm}^3/\text{min}\cdot\text{kPa}$, $K_m = 13\ \text{mA/m}$, $K_c = 5.33$, $K_{IP} = 5\ \text{kPa/mA}$, and $\tau_I = 3\ \text{min}$. Suppose that the system is initially at the nominal steady state with a liquid level of 60 cm. If the set point is suddenly changed from 60 to 100 cm, how long will it take the system to reach (a) 75 cm and (b) 100 cm?

9.3 Consider proportional-only control of the stirred-tank heater control system in Fig. E9.3. The temperature transmitter has a span of 50 °C and a zero of 55 °C. The nominal design conditions are $\bar{T} = 80$ °C and $\bar{T}_i = 65$ °C. The controller has a gain of 5, while the gains for the control valve and current-to-pressure transducer are $K_v = 1.2$ (dimensionless) and $K_{IP} = 5\ \text{kPa/mA}$, respectively. The time constant for the tank is $\tau = 5$ min. The control valve and transmitter dynamics are negligible. After the set point is changed from 80 to 85 °C, the tank temperature eventually reaches a new steady-state value of 84.14 °C, which was measured with a highly accurate thermometer.

(a) What is the offset?

(b) What is the process gain K_2?

(c) What is the pressure signal p_t to the control valve at the final steady state?

9.4 It is desired to control the exit concentration of c_3 of the liquid blending system shown in Fig. 9.4. Using the information given below, do the following:

(a) Draw a block diagram for the composition control scheme, using the symbols in Fig. E9.4.

(b) Derive an expression for each transfer function and substitute numerical values.

(c) Suppose that the PI controller has been tuned for the nominal set of operating conditions below. Indicate whether the controller should be retuned for each of the following situations. (Briefly justify your answers).

(i) The nominal value of c_2 changes to $\bar{c}_2 = 130$ kg solute/m².

(ii) The span of the composition transmitter is adjusted so that the transmitter output varies from 4 to 20 mA as c_3 varies from 50 to 225 kg solute/m³.

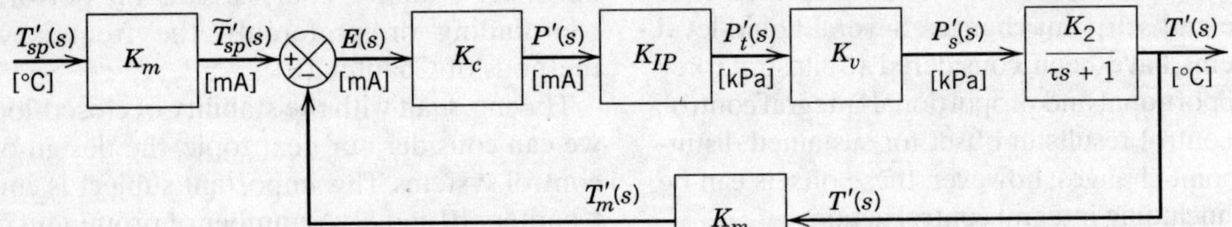

Figure E9.3

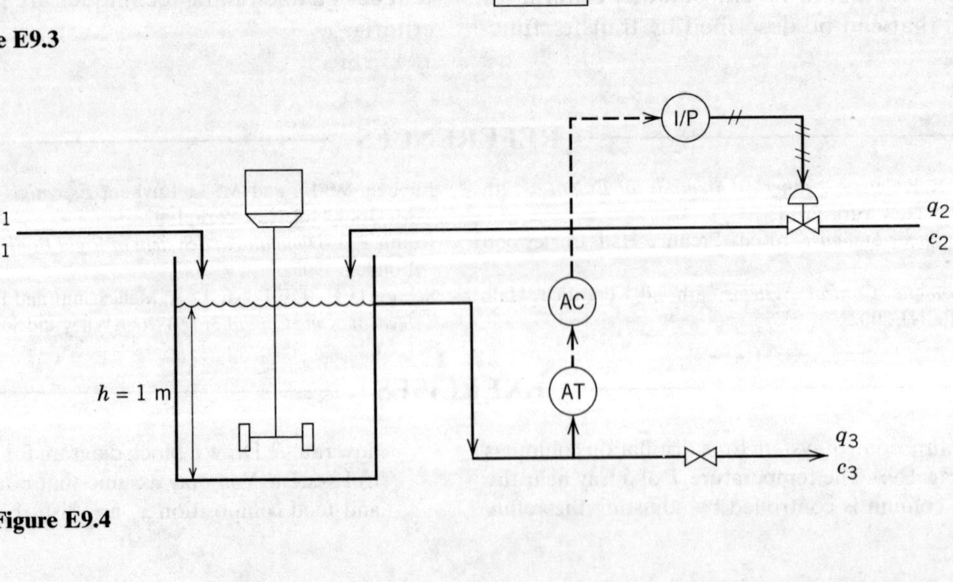

Figure E9.4

(iii) The zero of the composition transmitter is adjusted so that the transmitter output varies from 4 to 20 mA as c_3 varies from 60 to 160 kg solute/m³.

Available Information

1. The tank is perfectly mixed.
2. An overflow pipe is used to keep the mixture height at 1 m.
3. The volumetric flow rate and solute concentration of stream 2, q_2 and c_2, vary with time, whereas those of stream 1 are constant.
4. The density of all three streams are identical and do not vary with time.
5. A 2-min time delay is associated with the composition measurement. The transmission output signal varies linearly from 4 to 20 mA as c_3 varies from 50 to 150 kg solute/m³.
6. The pneumatic control valve has negligible dynamics. Its steady-state behavior is summarized below where p_t is the air pressure signal to the control valve from the I/P transducer.

p_t (kPa)	q_2 (L/min)
40	80
60	60
80	40

7. An electronic, direct-acting PI controller is used.
8. The current-to-pressure transducer has negligible dynamics and a gain of 2 kPa/mA.
9. The nominal operating conditions are:

$\rho = 1200$ kg/m^2 $\quad \bar{c}_3 = 80$ kg solute/m^3

$\bar{q}_1 = 40$ lpm $\quad \bar{c}_2 = 110$ kg solute/m^3

$\bar{q}_2 = 60$ lpm $\quad \bar{c}_1 = 30$ kg solute/m^3

D = tank diameter = 1 m.

9.5 A control system has the following transfer functions in its block diagram (see Fig. 9.8): $G_c = 1$, $G_v = 2$, $G_d - G_p = \dfrac{5}{s(s+4)}$, $G_m = 1$. For a unit step change in Y_{sp}, determine

(a) $Y(s)/Y_{sp}(s)$

(b) $y(\infty)$

(c) Offset (note proportional control is being used)

(d) $y(0.5)$

(e) if the closed-loop response is oscillatory

9.6 For a liquid-level control system similar to that in Fig. 9.23, Appelpolscher has argued that integral control action is not required because the process acts as an integrator (cf. Eq. 9-77). To evaluate his assertion, determine whether proportional-only control will eliminate offset for step changes in (a) set point and (b) disturbance variable.

9.7 A block diagram for internal model control, a control technique that is considered in Chapter 11, is shown in Fig. E9.7. Transfer function $\tilde{G}_p$ denotes the process model, while G_p denotes the actual process transfer function. It has been assumed that $G_v = G_m = 1$ for simplicity. Derive closed-loop transfer functions for both the servo and regulator problems.

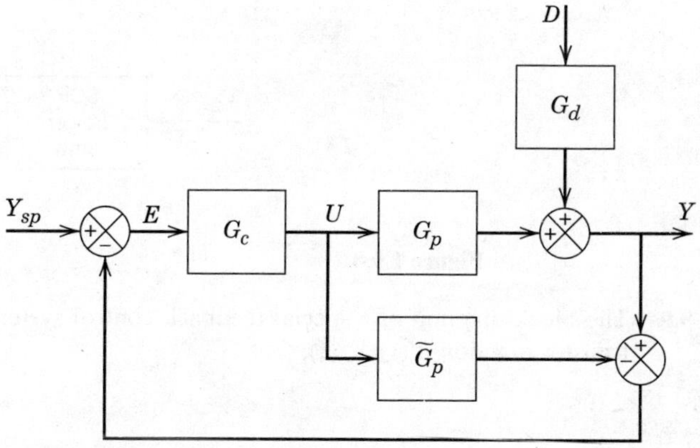

Figure E9.7

9.8 An electrically heated, stirred-tank system is shown in Fig. E9.8. Using the given information, do the following:

(a) Draw a block diagram for the case where T_3 is the controlled variable and voltage signal V_2 is the manipulated variable. Derive an expression for each transfer function.

(b) Repeat part (a) using V_1 as the manipulated variable.

(c) Which of these two control configurations would provide better control? Justify your answer.

Available Information

1. The volume of liquid in each tank is kept constant using an overflow line.
2. Temperature T_0 is constant.
3. A 3 L/m decrease in q_0 ultimately makes T_1 increase by 2 °C. Two-thirds of this total temperature change occurs in 12 min. This change in q_0 ultimately results in a 3 °C increase in T_3.
4. A change in V_1 from 10 to 12 volts ultimately causes T_1 to change from 20 to 25 °C. A similar test for V_2 causes T_3 to change from 30 to 33 °C. The apparent time constant for these tests is 10 min.
5. A step change in T_2 produces a transient response in T_3 that is essentially complete in 50 (= 5 τ) min.
6. The thermocouple output is amplified to give $V_3 = 0.15T_3 + 5$, where V_3[=] volts and T_3[=] °C.
7. The pipe connecting the two tanks has a mean residence time of 30 s.

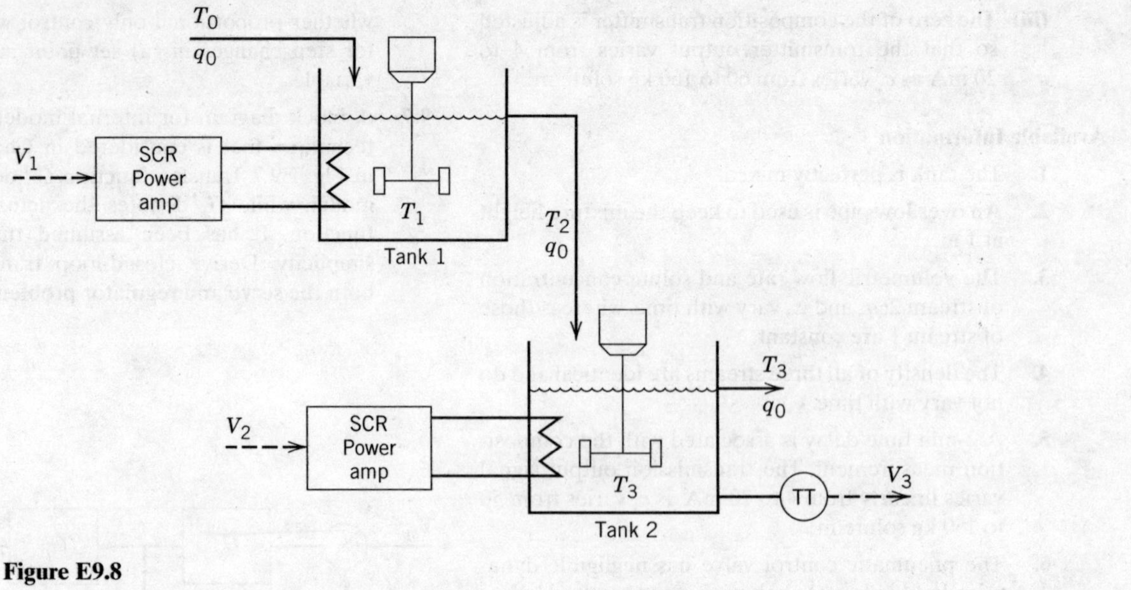

Figure E9.8

9.9 The block diagram of a special feedback control system is shown in Fig. E9.9. Derive an expression for the closed-loop transfer function, $Y(s)/Y_{sp}(s)$.

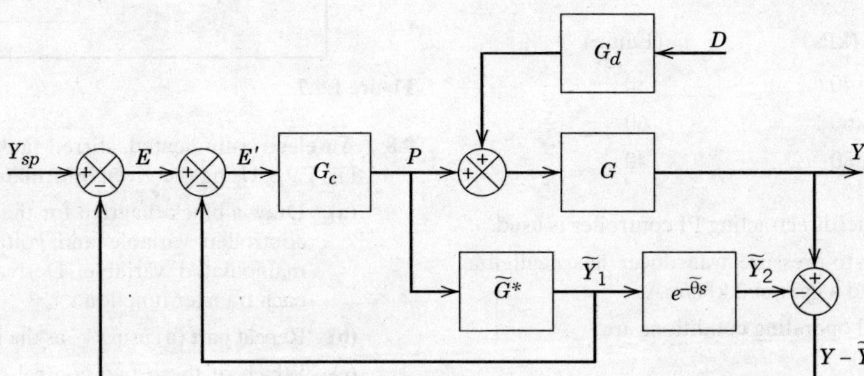

Figure E9.9

9.10 A block diagram of a closed-loop system is shown in Fig. E9.10

(a) Derive a closed-loop transfer function for set-point changes, $Y(s)/Y_{sp}(s)$.

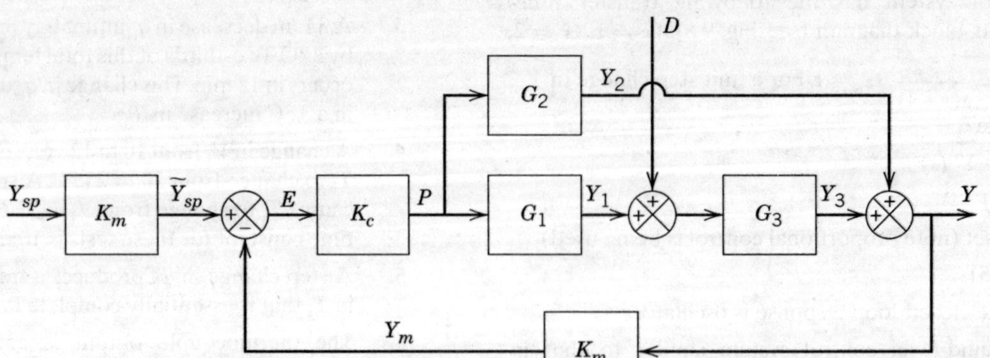

Figure E9.10

(b) For the following transfer functions, what values of K_c will result in a stable closed-loop system?

$$G_1(s) = 5 \quad G_2(s) = \frac{4}{2s+1}$$

$$K_m = 1 \quad G_3(s) = \frac{1}{s-1}$$

9.11 A mixing process consists of a single stirred-tank instrumented as shown in Fig. E9.11. The concentration of a single species A in the feed stream varies. The controller attempts to compensate for this by varying the flow rate of pure A through the control valve. The transmitter dynamics are negligible.

(a) Draw a block diagram for the controlled process.

(b) Derive a transfer function for each block in the block diagram.

Process

(i) The volume is constant (5 m^3).

(ii) The feed flow rate is constant ($\bar{q}_F = 7$ m^3/min).

(iii) The flow rate of the A stream varies but is small compared to $\bar{q}_F$ ($\bar{q}_A = 0.5$ m^3/min).

(iv) $\bar{c}_F = 50$ kg/m^3 and $\bar{c}_A = 800$ kg/m^3.

(v) All densities are constant and equal.

Transfer Line

(i) The transfer line is 20 m long and has 0.5 m inside diameter.

(ii) Pump volume can be neglected.

Composition Transmitter Data

c (kg/m^3)	c_m (mA)
0	4
200	20

Transmitter dynamics are negligible.

PID Controller

(i) Derivative on measurement only (cf. Eq. 7-17)

(ii) Direct or reverse acting, as required

(iii) Current (mA) input and output signals

I/P Transducer Data

p (mA)	p_t (kPa-g)
4	20
20	100

Control Valve

An equal percentage valve is used, which has the following relation:

$$q_A = 0.17 + 0.03(20)^{\frac{p_v - 3}{12}}$$

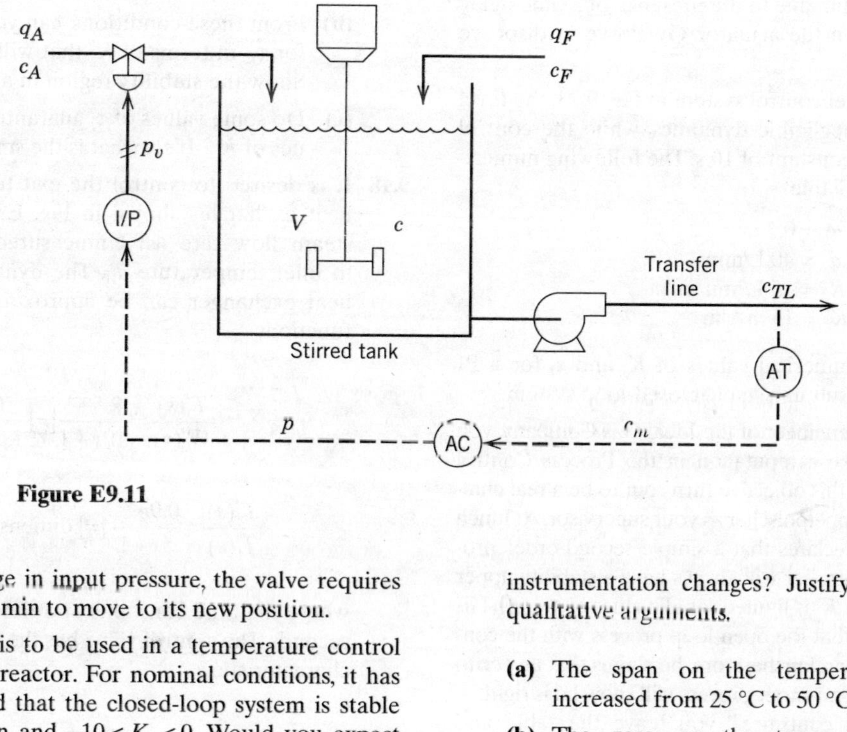

Figure E9.11

For a step change in input pressure, the valve requires approximately 1 min to move to its new position.

9.12 A PI controller is to be used in a temperature control system for a bioreactor. For nominal conditions, it has been determined that the closed-loop system is stable when $\tau_I = 10$ min and $-10 < K_c < 0$. Would you expect these stability limits to change for any of the following instrumentation changes? Justify your answers using qualitative arguments.

(a) The span on the temperature transmitter is increased from 25 °C to 50 °C.

(b) The zero on the temperature transmitter is decreased from 20 °C to 10 °C.

(c) The control valve "trim" is changed from linear to equal percentage.

9.13 A process is described by the transfer function

$$G(s) = \frac{1}{(5s+1)(s+1)}$$

Find the range of controller settings that yield stable closed-loop systems for:

(a) A proportional-only controller.

(b) A proportional-integral controller ($\tau_I = 0.1, 1.0, 10.0$).

(c) What can you say about the effect of adding larger amounts of the integral action on the stability of the controlled system; that is, does it tend to stabilize or destabilize the system relative to proportional-only control? Justify your answer.

9.14 The block diagram of a feedback control system is shown in Fig. E9.14. Determine the values of K_c that result in a stable closed-loop system.

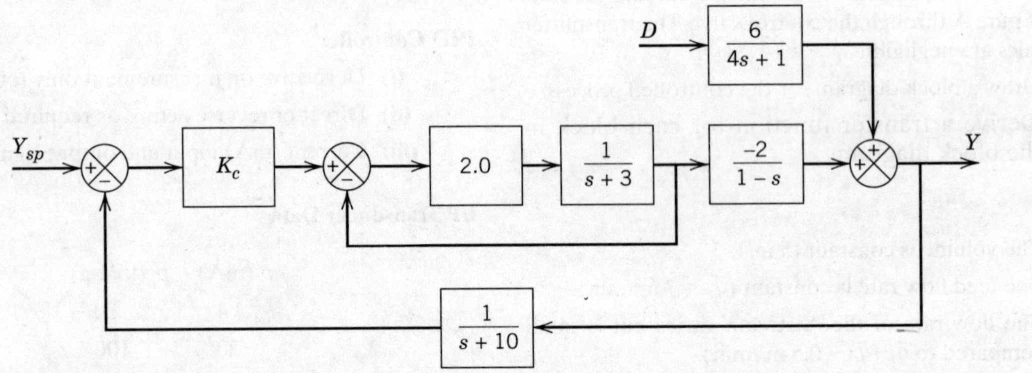

Figure E9.14

9.15 Consider the block diagram of a feedback control system in Fig. 9.8, where $K_m = 1$, $G_p = \frac{5}{s+1}$, $G_d = 0$, $G_v = K_v e^{-s}$, $G_c = 1$, and $G_m = 1$. A control engineer claims that the process will exhibit an inverse response to changes in set-point, due to the presence of a time delay in the dynamics of the actuator Gv. Prove or disprove this assertion.

9.16 For the liquid-level control system in Fig. 9.23, the level transmitter has negligible dynamics, while the control valve has a time constant of 10 s. The following numerical values are available:

$A = 0.3\ \text{m}^2$
$\bar{q}_3 = 40$ L/min
$K_v = -5$ L/min/mA
$K_m = 15$ mA/m

Determine the numerical values of K_c and τ_I for a PI controller that result in a stable closed-loop system.

9.17 As a newly hired engineer of the Ideal Gas Company, you are trying to make a reputation in the Process Control Group. However, this objective turns out to be a real challenge with I. M. Appelpolscher as your supervisor. At lunch one day, I.M.A. declares that a simple second-order process with a PI controller will always have a stability upper limit on K_c; that is, K_c is limited for all values of $\tau_I > 0$. His best argument is that the open-loop process with the controller is third order. Furthermore, he claims that any critically damped second-order process will show he is right.

Muttering "au contraire," you leave the table and quickly investigate the properties of

$$G_v G_p G_m = \frac{5}{(10s+1)^2}$$

(a) What are the conditions for closed-loop stability for a PI controller?

(b) From these conditions, can you find a relationship for τ_I in terms of K_c that will guarantee stability? Show the stability region in a plot of τ_I versus K_c.

(c) Do some values of τ_I guarantee stability for all values of K_c? If so, what is the smallest value?

9.18 It is desired to control the exit temperature T_2 of the heat exchanger shown in Fig. E9.18 by adjusting the steam flow rate w_s. Unmeasured disturbances occur in inlet temperature T_1. The dynamic behavior of the heat exchanger can be approximated by the transfer functions

$$\frac{T_2'(s)}{W_s'(s)} = \frac{2.5e^{-s}}{10s+1} \ [=] \ \frac{°\text{C}}{\text{kg/s}}$$

$$\frac{T_2'(s)}{T_1'(s)} = \frac{0.9e^{-2s}}{5s+1} \ [=] \ \text{dimensionless}$$

where the time constants and time delays have units of seconds. The control valve has the following steady-state characteristic:

$$w_s = 0.6\sqrt{p-4}$$

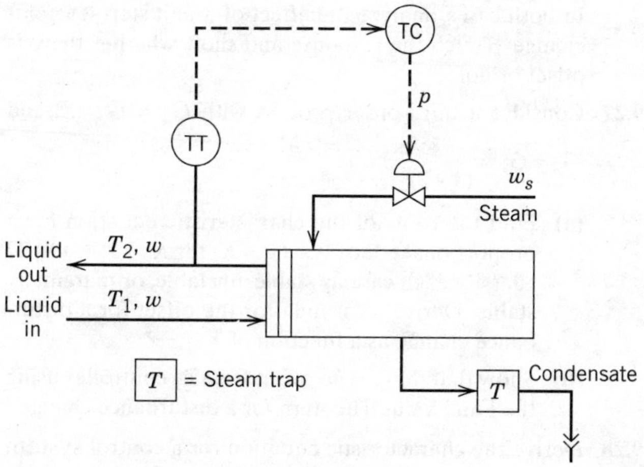

Figure E9.18

where p is the controller output expressed in mA. At the nominal operating condition, $p = 12$ mA. After a sudden change in the controller output, w_s reaches a new steady-state value in 20 s (assumed to take five time constants). The temperature transmitter has negligible dynamics and is designed so that its output signal varies linearly from 4 to 20 mA as T_2 varies from 120 to 160 °C.

(a) If a proportional-only feedback controller is used, what is K_{cm}? What is the frequency of the resulting oscillation when $K_c = K_{cm}$? (*Hint:* Use the direct-substitution method and Euler's identity.)

(b) Estimate K_{cm} using direct substitution and a 1/1 Padé approximation for the time-delay term. Does this analysis provide a satisfactory approximation?

9.19 A process model is described by the transfer function

$$G(s) = \frac{4(1-5s)}{(2s+1)(4s+1)(2s+1)}$$

which includes actuator and measurement dynamics. The process engineering group has the option of redesigning the process to eliminate the right-half plane zero. They need to determine whether this modification will yield a substantially better (faster) controlled process.

(a) For a proportional only controller, find the stability bounds for K_c for the existing process using direct substitution.

(b) Repeat part (a) for the case where the RHP zero has been eliminated.

(c) From analysis and/or your knowledge of closed-loop systems, what can you conclude about the potential speed of response of this controlled process if the zero can be eliminated?

9.20 A feedback control system hasthe open-loop transfer function, $G_{OL}(s) = 0.5K_c e^{-3s}/(10s+1)$. Determine the values of K_c for which the closed-loop system is stable using two approaches:

(a) An approximate analysis using a 1/1 Padé approximation for e^{-3s} and the direct substitution method.

(b) An exact analysis based on substitution of $s = j\omega$. (*Hint:* Recall Euler's identity.)

(c) Compare your answers for (a) and (b) using computer simulation.

9.21 A process control system contains the following transfer functions:

$$G_p(s) = \frac{2e^{-1.5s}}{(60s+1)(5s+1)}$$

$$G_v(s) = \frac{0.5e^{-0.3s}}{3s+1}$$

$$G_m(s) = \frac{3e^{-0.2s}}{2s+1}$$

$$G_c(s) = K_c$$

(a) Show how $G_{OL}(s)$ can be approximated by a FOPTD model;

$$G_p G_v G_m G_c = G_{OL}(s) \approx \frac{Ke^{-\theta s}}{\tau s+1}$$

Find K, τ, and θ for the open-loop process transfer function.

(b) Use direct substitution and your FOPTD model to find the range of K_c values that will yield a stable closed-loop system. Check your value of K_{cm} for the full-order model using simulation.

9.22 For the control system based on the standard feedback control configuration in Fig. 9.8

$$G_c = K_c G_p = \frac{4-as}{(s+1)^3} G_v = G_m = 1 \quad a > 0$$

determine whether the value of a affects the stability of the closed-loop system. Assume $K_c > 0$. You do not need to solve for the roots of the characteristic equation to answer the question.

9.23 Consider proportional-only control of the level control system in Fig. 9.17. The level transmitter has a span of 3 m and a zero of 5 m. Recall that the standard instrument ranges are 4 to 20 mA and 20 to 100 kPa-g. The nominal design conditions are $h = 6$ m. The controller has a gain of 5 while the gain for the control valve is K_v = 1.65 lpm/kPa, respectively. The time constant for the tank is $\tau = 5$ min. After the set-point is changed from 6 to 7 m, the tank level eventually reaches a new steady-state value of 6.58 m.

(a) What is the offset?

(b) What are the gains K_m and K_p in Fig. 9.17? (Give their units also)

(c) How could the controller be modified to obtain zero offset?

9.24 A control system has $G_v = G_m = 1$ and a second-order process G_p with $K_p = 2$, $\tau_1 = 4$ min, and $\tau_2 = 1$ min, which is to be controlled by a PI controller with $K_c = 2$ and $\tau_I = \tau_1 = 4$ min (i.e., the integral time of the controller is set equal to the dominant time constant). For a set-point change

(a) Determine the closed-loop transfer function.

(b) Derive the characteristic equation, which is a quadratic polynomial in s. Is it overdamped or underdamped?

(c) Can a large value of K_c make the closed-loop process unstable?

9.25 The set-point of the control system under proportional control ($K_c = 5$) undergoes a step change of magnitude 2. For

$$G_vG_p = \frac{2}{(s+1)(2s+1)} \text{ and } G_m = 1,$$

(a) Determine when the maximum value of y occurs.

(b) Determine the offset.

(c) Determine the period of oscillation.

(d) Sketch $y(t)$ as a function of time, showing key characteristics as determined in (a), (b), and (c).

9.26 A plasma etching batch process has a process gain of E Å/min (but no dynamics). The manipulated variable is the etch time, so the controlled variable is the film thickness. There are no time constants that need to be included. Assume $G_v = G_m = 1$. Derive the closed-loop transfer function for a set-point change for two different controllers:

(a) $G_c = K_c$

(b) $G_c = \dfrac{1}{\tau_I s}$

In both cases, analyze the effect of a unit step set-point change. Sketch the response and show whether there is offset or not.

9.27 Consider a third-order process with $G_m = G_v = 2$ and $G_p = G_d = \dfrac{2}{(s+2)^3}$.

(a) Find the roots of the characteristic equation for a proportional controller $G_c = K_c$ for $K_c = 1, 8$, and 27. Classify each case as stable, unstable, or marginally stable. Derive a formula for the offset for a disturbance change as a function of K_c.

(b) Show that there is no offset for a PI controller using the Final Value Theorem for a disturbance change.

9.28 Derive the characteristic equation for a control system with the following transfer functions: $G_c = K_c$, $G_vG_p = \dfrac{0.5}{(s+1)(0.5s+1)}$, $G_m = \dfrac{6}{s+3}$. Is the system stable for (a) $K_c = 9$, (b) $K_c = 11$, (c) $K_c = 13$? Check your answers using simulation and by direct substitution.

9.29 Suppose a control system is modeled by $G_vG_p = \dfrac{0.25}{(s+1)^3}$, $G_m = 4$, and $G_c = K_c$. Find the maximum value of K_c for a proportional controller for which the system is stable, using direct substitution, and verify your result using simulation. Replace the controller with a PD controller ($K_c = 10$). Determine the range of τ_D for which the system is stable. Check your results using simulation.

9.30 Obtain the transfer function $Y(s)/X(s)$ from the block diagram shown in Fig. E9.30 using block diagram algebra.

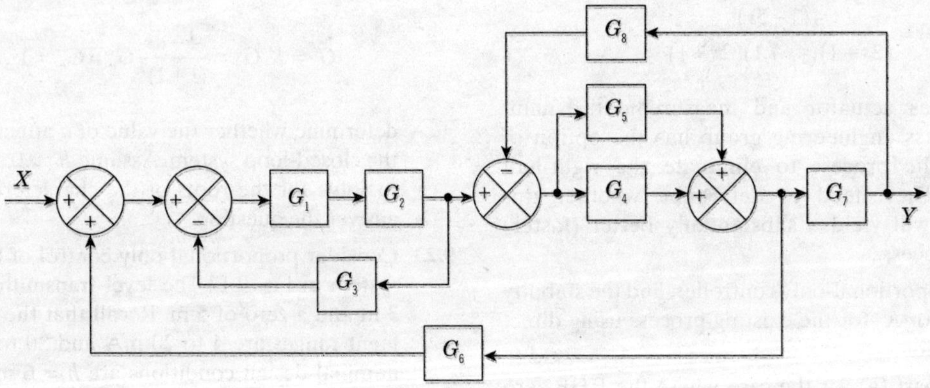

Figure E9.30 Block diagram of a chemical process.

MULTIPLE CHOICE QUESTIONS

1. Which of the following is NOT a necessary condition for a process under closed-loop control to be stable?

(a) Dead-time term(s) must be absent in the open-loop transfer function.

(b) Open-loop transfer function must have an amplitude ratio less than 1 at the critical frequency.

(c) Roots of the characteristic equation must have negative real part.

(d) All the elements in the left (first) column of the Routh array must have the same sign.

(GATE 2021)

2. A first-order process having a transfer function,

$$G_p = \frac{2}{7s+1}$$

is controlled by a proportional controller with gain of 3.2. The process time constant is in minutes. Addition of the integral control action with an integral time constant of 5 min leads to increase in

(a) offset.

(b) speed of response.

(c) order of the closed loop system.

(d) proportional band. **(GATE 2018)**

3. For a closed-loop system, consider the following transfer functions: process $G_p(s)$, controller $G_c(s)$, measuring device $G_m(s)$, and final control element $G_f(s)$

$$G_p(s) = \frac{2}{7s+1};\quad G_c(s) = 2;\quad G_m(s) = 1;\quad G_f(s) = 1$$

The offset in the closed loop response due to a unit step change introduced in the set point of the output variable is ____________.

(a) 0.4 **(b)** 0.1

(c) 0.3 **(d)** 0.2 **(GATE 2018)**

4. Consider the following closed loop system with G_p and G_c as the transfer functions of the process and the controller, respectively.

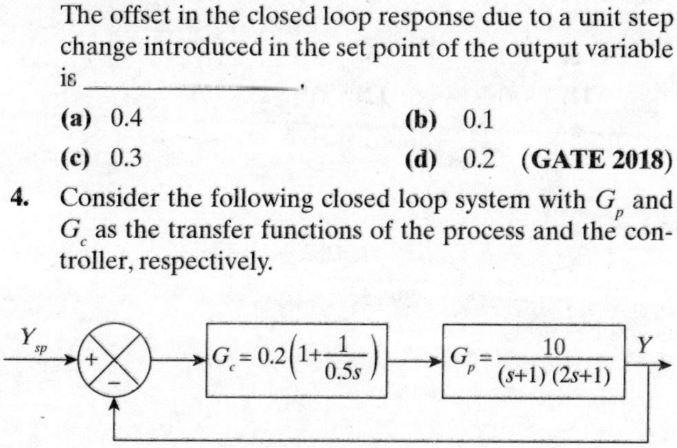

For a unit step change in the set-point (y_{sp}), the change in the value of the response (y) at steady state is ________ (round off to 1 decimal place).

(a) 1.0 **(b)** 2.1

(c) 3.2 **(d)** 4.1 **(GATE 2020)**

5. The characteristic equation of a closed-loop system is $6s^3 + 11s^2 + 6s + (1+K) = 0$, where $K > 0$. The value of K beyond which the system just becomes unstable, rounded to the nearest integer, is ____________.

(a) 8 **(b)** 9

(c) 10 **(d)** 11 **(GATE 2017)**

6. The block diagram for a process with feedback control for output deviation variable h is shown in the figure below. All transfer functions are given with pre-factor of s in minutes. A unit step change is made in the set-point at $t = 0$. The time required for h to reach 50% of its ultimate value, in minutes (up to two decimal places), is:

____

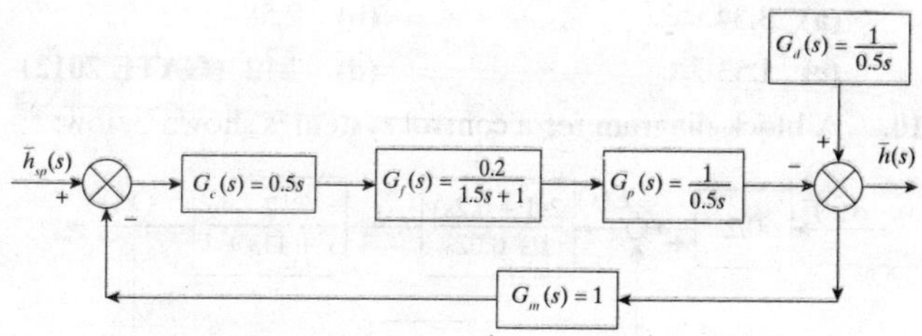

(a) 0.968 **(b)** 0.867

(c) 0.657 **(d)** 0.429

(GATE 2015)

7. A process with transfer function, $G_p = \frac{2}{s-1}$ is to be controlled by a feedback proportional controller with a gain K_c. If the transfer functions of all other elements in the control loop are unity, then which ONE of the following conditions produces a stable closed loop response?

(a) $K_c = 0.25$ **(b)** $0 < K_c < 0.25$

(c) $0.25 < K_c < 0.5$ **(d)** $K_c > 0.5$

(GATE 2014)

8. Consider the following block diagram for a closed-loop feedback control system

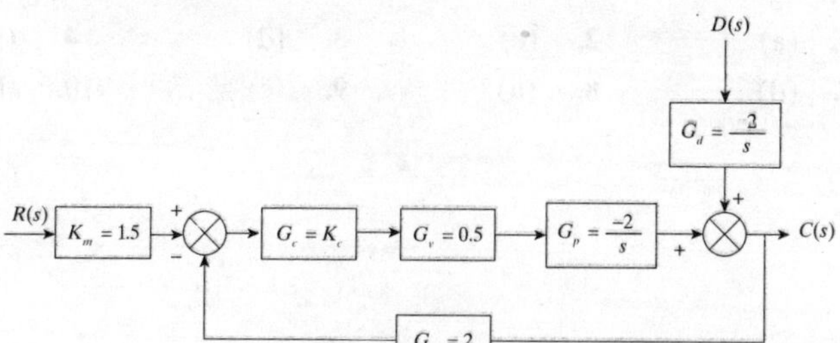

A proportional controller is being used with $K_c = -4$. If a step change in disturbance of magnitude 2 affects the system, then the value of the offset is

(a) 0.50 **(b)** 0.60

(c) 0.70 **(d)** 0.80 **(GATE 2014)**

9. The block diagram of a system with a proportional controller is shown below

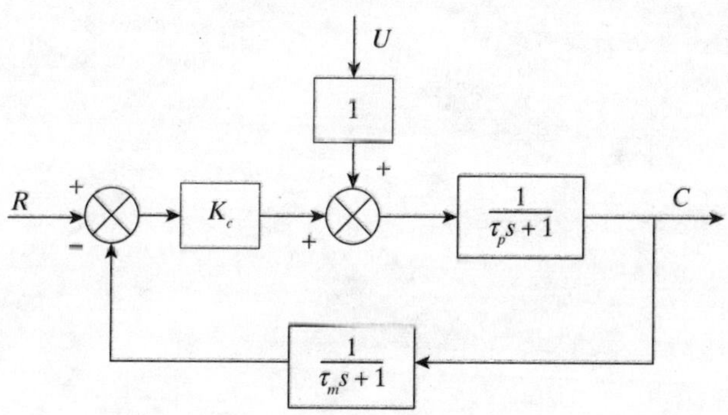

A unit step input is introduced in the set point. The value of K_c to provide a critically damped response for $U = 0$, $\tau_p = 8$ and $\tau_m = 1$ is

(a) 3.34 **(b)** 2.58

(c) 1.53 **(d)** 1.12 **(GATE 2012)**

10. A block diagram for a control system is shown below:

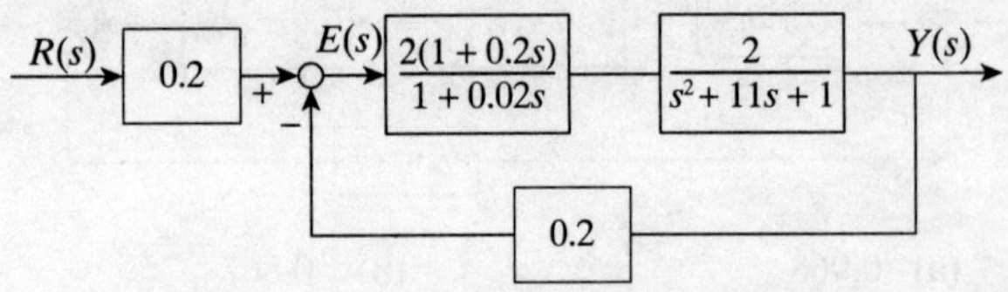

The steady state gain of the closed loop system, between output $Y(s)$ and set point $R(s)$, is

(a) 5/9 **(b)** 4/9

(c) 1/3 **(d)** 2/9 **(GATE 2010)**

11. The block diagram for a control system is shown below:

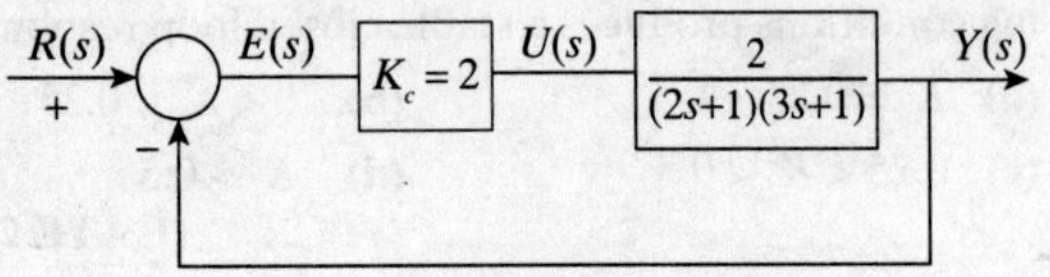

For a unit step change in the set point, $R(s)$, the steady state offset in the output $Y(s)$ is

(a) 0.2 **(b)** 0.3

(c) 0.4 **(d)** 0.5 **(GATE 2009)**

12. Given the process transfer function, $G_p = 20/(s - 2)$, and controller transfer function $G_c = K_c$, and assuming the transfer functions of all other elements in the control loop are unity, select the range of K_c for which the closed loop response will be stable

(a) $K_c < 1/10$ **(b)** $K_c < 1/100$

(c) $1/100 < K_c < 1/10$ **(d)** $K_c > 1/10$

(GATE 2005)

ANSWER KEY

1. (a)	**2.** (c)	**3.** (d)	**4.** (a)	**5.** (c)	**6.** (b)
7. (d)	**8.** (a)	**9.** (c)	**10.** (b)	**11.** (a)	**12.** (d)

CHAPTER 10

Frequency Response Analysis and Control System Design

CHAPTER CONTENTS

In previous chapters, Laplace transform techniques were used to calculate transient responses from transfer functions. This chapter focuses on an alternative way to analyze dynamic systems by using frequency response analysis. Frequency response concepts and techniques play an important role in stability analysis, control system design, and robustness assessment. Historically, frequency response techniques provided the conceptual framework for early control theory and important applications in the field of communications (MacFarlane, 1979).

We introduce a simplified procedure to calculate the frequency response characteristics from the transfer function model of any linear process. Two concepts, the Bode and Nyquist stability criteria, are generally applicable for feedback control systems and stability analysis. Next we introduce two useful metrics for relative stability, namely gain and phase margins. These metrics indicate how close a control system is to instability. A related issue is robustness, that is, the sensitivity of control system performance to process variations and to uncertainty in the process model.

The design of robust feedback control systems is considered in Appendix L.

10.1 SINUSOIDAL FORCING OF A FIRST-ORDER PROCESS

We start with the response properties of a first-order process when forced by a sinusoidal input and show how the output response characteristics depend on the frequency of the input signal. This is the origin of the term *frequency response*. The responses for first- and second-order processes forced by a sinusoidal input were presented in Chapter 5. Recall that these responses consisted of sine, cosine, and exponential terms. Specifically, for a first-order transfer function with gain K and time constant τ, the response to a general sinusoidal input, $x(t) = A \sin \omega t$, is

$$y(t) = \frac{KA}{\omega^2\tau^2 + 1}(\omega\tau e^{-t/\tau} - \omega\tau \cos \omega t + \sin \omega t) \qquad (5\text{-}31)$$

where y is in deviation form.

If the sinusoidal input is continued for a long time, the exponential term $(\omega\tau e^{-t/\tau})$ becomes negligible. The remaining sine and cosine terms can be combined via a trigonometric identity to yield

$$y_\ell(t) = \frac{KA}{\sqrt{\omega^2\tau^2+1}}\sin(\omega t+\phi) \qquad (10\text{-}1)$$

where $\phi = -\tan^{-1}(\omega\tau)$. The long-time response $y_\ell(t)$ is called the frequency response of the first-order system and has two distinctive features (see Fig. 10.1).

1. The output signal is a sine wave that has the same frequency, but its phase is shifted relative to the input sine wave by the angle ϕ (referred to as the *phase shift* or the *phase angle*); the amount of phase shift depends on the forcing frequency ω.
2. The sine wave has an amplitude $\hat{A}$ that is a function of the forcing frequency:

$$\hat{A} = \frac{KA}{\sqrt{\omega^2\tau^2+1}} \qquad (10\text{-}2)$$

Dividing both sides of Eq. 10-2 by the input signal amplitude A yields the *amplitude ratio* (AR)

$$AR = \frac{\hat{A}}{A} = \frac{K}{\sqrt{\omega^2\tau^2+1}} \qquad (10\text{-}3a)$$

which can, in turn, be divided by the process gain to yield the *normalized amplitude ratio* (AR_N):

$$AR_N = \frac{AR}{K} = \frac{1}{\sqrt{\omega^2\tau^2+1}} \qquad (10\text{-}3b)$$

Next we examine the physical significance of the preceding equations, with specific reference to the blending process example discussed earlier. In Chapter 4, the transfer function model for the stirred-tank blending system was derived as

$$X'(s) = \frac{K_1}{\tau s+1}X_1'(s) + \frac{K_2}{\tau s+1}W_2'(s) + \frac{K_3}{\tau s+1}W_1'(s) \qquad (4\text{-}84)$$

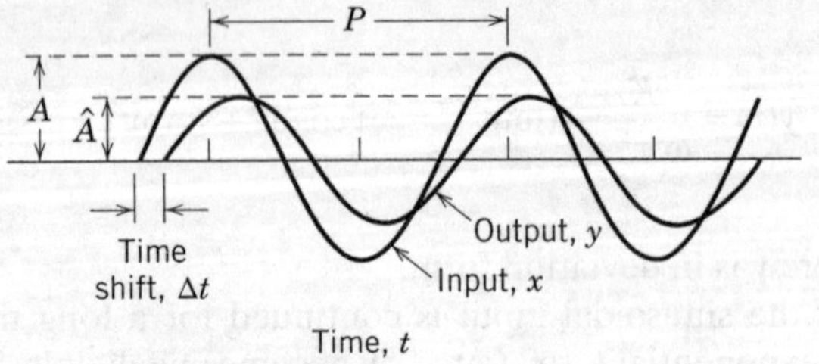

Figure 10.1 Attenuation and time shift between input and output sine waves. The phase angle ϕ of the output signal is given by $\phi = \Delta t/P \times 360°$, where Δt is the time shift and P is the period of oscillation.

Suppose flow rate w_2 is varied sinusoidally about a constant value, while the other inlet conditions are kept constant at their nominal values; that is, $w_1'(t) = x_1'(t) = 0$. Because $w_2(t)$ is sinusoidal, the output composition deviation $x'(t)$ eventually becomes sinusoidal according to Eq. 5-32. However, there is a phase shift in the output relative to the input, as shown in Fig. 10.1, owing to the material holdup of the tank. If the flow rate w_2 oscillates very slowly relative to the residence time $\tau(\omega \ll 1/\tau)$, the phase shift is very small, approaching 0°, whereas the normalized amplitude ratio $(\hat{A}/KA)$ is very nearly unity. For the case of a low-frequency input, the output is in phase with the input, tracking the sinusoidal input as if the process model were $G(s) = K$.

On the other hand, suppose that the flow rate is varied rapidly by increasing the input signal frequency. For $\omega \gg 1/\tau$, Eq. 10-1 indicates that the phase shift approaches a value of $-\pi/2$ radians (–90°). The presence of the negative sign indicates that the output lags behind the input by 90°; in other words, the phase lag is 90°. The amplitude ratio approaches zero as the frequency becomes large, indicating that the input signal is almost completely attenuated; namely, the sinusoidal deviation in the output signal is very small.

These results indicate that positive and negative deviations in w_2 are essentially canceled by the capacitance of the liquid in the blending tank if the frequency is high enough. High frequency implies $\omega \gg 1/\tau$. Most processes behave qualitatively similar to the stirred-tank blending system, when subjected to a sinusoidal input. For high-frequency input changes, the process output deviations are so completely attenuated that the corresponding periodic variation in the output is difficult (perhaps impossible) to detect or measure.

Input–output phase shift and attenuation (or amplification) occur for any stable transfer function, regardless of its complexity. In all cases, the phase shift and amplitude ratio are related to the frequency ω of the sinusoidal input signal. In developments up to this point, the expressions for the amplitude ratio and phase shift were derived using the process transfer function. However, the frequency response of a process can also be obtained experimentally. By performing a series of tests in which a sinusoidal input is applied to the process, the resulting amplitude ratio and phase shift can be measured for different frequencies. In this case, the frequency response is expressed as a table of measured amplitude ratios and phase shifts for selected values of ω. However, the method is very time-consuming because of the repeated experiments for different values of ω. Thus other methods, such as pulse testing (Ogunnaike and Ray, 1994), are utilized, because only a single test is required.

In this chapter, the focus is on developing a powerful analytical method to calculate the frequency response for any stable process transfer function. Later in this chapter, we show how this information can be used to design controllers and analyze the properties of the closed loop system responses.

10.2 SINUSOIDAL FORCING OF AN *n*TH-ORDER PROCESS

This section presents a general approach for deriving the frequency response of any stable transfer function. The physical interpretation of frequency response is not valid for unstable systems, because a sinusoidal input produces an unbounded output response instead of a sinusoidal response. A rather simple procedure can be employed to find the sinusoidal response.

After setting $s = j\omega$ in $G(s)$, by algebraic manipulation we can separate the expression into real (R) and imaginary (I) terms (j indicates an imaginary component):

$$G(j\omega) = R(\omega) + jI(\omega) \tag{10-4}$$

Similar to Eq. 10-1, we can express the long time response for a linear system (cf. Eq. 10-1) as

$$y_{\ell}(t) = \hat{A}\sin(\omega t + \phi) \tag{10-5}$$

$\hat{A}$ and ϕ are related to $I(\omega)$ and $R(\omega)$ by the following relations (Seborg et al., 2004):

$$\hat{A} = A\sqrt{R^2 + I^2} \tag{10-6a}$$

$$\phi = \tan^{-1}(I/R) \tag{10-6b}$$

Both $\hat{A}$ and ϕ are functions of frequency ω. A simple but elegant relation for the frequency response can be derived, where the amplitude ratio is given by

$$AR = \frac{\hat{A}}{A} = |G| = \sqrt{R^2 + I^2} \tag{10-7}$$

The absolute value denotes the *magnitude* of G, and the phase shift (also called the *phase angle* or *argument* of G, $\angle G$) between the sinusoidal output and input is given by

$$\phi = \angle G = \tan^{-1}(I/R) \tag{10-8}$$

Because $R(\omega)$ and $I(\omega)$ (and hence AR and ϕ) can be obtained without calculating the complete transient response $y(t)$, these characteristics provide a convenient shortcut method to determine the frequency response of transfer functions.

Equations 10-7 and 10-8 can calculate the frequency response characteristics of any stable $G(s)$, including those with time-delay terms.

The shortcut method can be summarized as follows:

Step 1. Substitute $s = j\omega$ in $G(s)$ to obtain $G(j\omega)$.

Step 2. Rationalize $G(j\omega)$, i.e., express $G(j\omega)$ as the sum of real (R) and imaginary (I) parts $R + jI$, where R and I are functions of ω, using complex conjugate multiplication.

Step 3. The output sine wave has amplitude $\hat{A} = A\sqrt{R^2 + I^2}$ and phase angle $\phi = \tan^{-1}(I/R)$. The amplitude ratio is $AR = \sqrt{R^2 + I^2}$ and is independent of the value of A.

EXAMPLE 10.1

Find the frequency response of a first-order system, with

$$G(s) = \frac{1}{\tau s + 1} \tag{10-9}$$

Solution

First substitute $s = j\omega$ in the transfer function

$$G(j\omega) = \frac{1}{\tau j\omega + 1} = \frac{1}{j\omega\tau + 1} \tag{10-10}$$

Then multiply both numerator and denominator by the complex conjugate of the denominator, that is, $-j\omega\tau + 1$

$$G(j\omega) = \frac{-j\omega\tau + 1}{(j\omega\tau + 1)(-j\omega\tau + 1)} = \frac{-j\omega\tau + 1}{\omega^2\tau^2 + 1}$$

$$= \frac{1}{\omega^2\tau^2 + 1} + j\frac{(-\omega\tau)}{\omega^2\tau^2 + 1} = R + jI \tag{10-11}$$

Where

$$R = \frac{1}{\omega^2\tau^2 + 1} \tag{10-12a}$$

and

$$I = \frac{-\omega\tau}{\omega^2\tau^2 + 1} \tag{10-12b}$$

From Eq. 10-7,

$$AR = |G(j\omega)| = \sqrt{\left(\frac{1}{\omega^2\tau^2+1}\right)^2 + \left(\frac{-\omega\tau}{\omega^2\tau^2+1}\right)^2}$$

Simplifying,

$$AR = \sqrt{\frac{(1+\omega^2\tau^2)}{(\omega^2\tau^2+1)^2}} = \frac{1}{\sqrt{\omega^2\tau^2+1}} \quad (10\text{-}13a)$$

$$\phi = \angle G(j\omega) = \tan^{-1}(-\omega\tau) = -\tan^{-1}(\omega\tau) \quad (10\text{-}13b)$$

If the process gain had been a positive value K instead of 1,

$$AR = \frac{K}{\sqrt{\omega^2\tau^2+1}} \quad (10\text{-}14)$$

and the phase angle would be unchanged (Eq. 10-13b). Both the amplitude ratio and phase angle are identical to those values calculated in Section 10.1 using the time-domain derivation.

From this example, we conclude that direct analysis of the complex transfer function $G(j\omega)$ is computationally easier than solving for the actual long-time output response $j_\ell(t)$. The computational advantages are even greater when dealing with more complicated processes, as shown in the following. Start with a general transfer function in factored form

$$G(s) = \frac{G_a(s)G_b(s)G_c(s)\cdots}{G_1(s)G_2(s)G_3(s)\cdots} \quad (10\text{-}15)$$

$G(s)$ is converted to the complex form $G(j\omega)$ by the substitution $s = j\omega$:

$$G(j\omega) = \frac{G_a(j\omega)G_b(j\omega)G_c(j\omega)\cdots}{G_1(j\omega)G_2(j\omega)G_3(j\omega)\cdots} \quad (10\text{-}16)$$

The magnitude and phase angle of $G(j\omega)$ are as follows:

$$|G(j\omega)| = \frac{|G_a(j\omega)||G_b(j\omega)||G_c(j\omega)|\cdots}{|G_1(j\omega)||G_2(j\omega)||G_3(j\omega)|\cdots} \quad (10\text{-}17a)$$

$$\angle G(j\omega) = \angle G_a(j\omega) + \angle G_b(j\omega) + \angle G_c(j\omega) + \cdots$$
$$-[\angle G_1(j\omega) + \angle G_2(j\omega) + \angle G_3(j\omega) + \cdots] \quad (10\text{-}17b)$$

Equations 10-17a and 10-17b greatly simplify the computation of $|G(j\omega)|$ and $\angle G(j\omega)$ and, consequently, AR and ϕ, for factored transfer functions. These expressions eliminate much of the complex algebra associated with the rationalization of complicated transfer functions. Hence, the factored form (Eq. 10-15) may be preferred for frequency response analysis. On the other hand, if the frequency response curves are generated using software such as MATLAB, there is no need to factor the numerator or denominator, as discussed in Section 10.3.

EXAMPLE 10.2

Calculate the amplitude ratio and phase angle for the over-damped second-order transfer function

$$G(s) = \frac{K}{(\tau_1 s+1)(\tau_2 s+1)}$$

Solution

Using Eq. 10-15, let

$$G_a = K$$

$$G_1 = \tau_1 s + 1$$

$$G_2 = \tau_2 s + 1$$

Substituting $s = j\omega$

$$G_a(j\omega) = K$$

$$G_1(j\omega) = j\omega\tau_1 + 1$$

$$G_2(j\omega) = j\omega\tau_2 + 1$$

The magnitudes and angles of each component of the complex transfer function are

$$|G_a| = K \qquad \angle G_a = 0$$

$$|G_1| = \sqrt{\omega^2\tau_1^2+1} \qquad \angle G_1 = \tan^{-1}(\omega\tau_1)$$

$$|G_2| = \sqrt{\omega^2\tau_2^2+1} \qquad \angle G_2 = \tan^{-1}(\omega\tau_2)$$

Combining these expressions via Eqs. 10-17a and 10-17b yields

$$AR = |G(j\omega)| = \frac{|G_a(j\omega)|}{|G_1(j\omega)||G_2(j\omega)|}$$

$$= \frac{K}{\sqrt{\omega^2\tau_1^2+1}\sqrt{\omega^2\tau_2^2+1}} \quad \text{(10-18a)}$$

$$\phi = \angle G(j\omega) = \angle G_a(j\omega) - (\angle G_1(j\omega) + \angle G_2(j\omega))$$

$$= -\tan^{-1}(\omega\tau_1) - \tan^{-1}(\omega\tau_2) \quad \text{(10-18b)}$$

10.3 BODE DIAGRAMS

The *Bode diagram* (or Bode plot) provides a convenient display of the frequency response characteristics in which AR and ϕ are each plotted as a function of ω. Ordinarily, ω is expressed in units of radians/time to simplify inverse tangent calculations (e.g., Eq. 10-18b) where the arguments must be dimensionless, that is, in radians. Occasionally, a cyclic frequency, $\omega/2\pi$, with units of cycles/time, is used. Phase angle ϕ is normally expressed in degrees rather than radians. For reasons that will become apparent in the following development, the Bode diagram consists of: (1) a log–log plot of AR versus ω and (2) a semilog plot of ϕ versus ω. These plots are particularly useful for rapid analysis of the response characteristics and stability of closed-loop systems.

10.3.1 First-Order Process

In the past, when frequency response plots had to be generated by hand, they were of limited utility. A much more practical approach now utilizes spreadsheets or control-oriented software such as MATLAB to simplify calculations and generate Bode plots. Although spreadsheet software can be used to generate Bode plots, it is much more convenient to use software designed specifically for control system analysis. Thus, after describing the qualitative features of Bode plots of simple transfer functions, we illustrate how the AR and ϕ components of such a plot are generated by a MATLAB program in Example 10.3.

For a first-order model, $K/(\tau s+1)$, Fig. 10.2 shows a general log–log plot of the normalized amplitude ratio versus $\omega\tau$, for positive K. For a negative valve of K, the phase angle is decreased by –180°. A semilog plot of ϕ versus $\omega\tau$ is also shown. In Fig. 10.2, the abscissa $\omega\tau$ has units of radians. If K and τ are known, AR_N (or AR) and ϕ can be plotted as a function of ω. Note that, at high frequencies, the amplitude ratio drops to an infinitesimal level, and the phase lag (the phase angle expressed as a positive value) approaches a maximum value of 90°.

Some books and software define AR differently, in terms of decibels. The amplitude ratio in decibels AR_{db} is defined as

$$AR_{db} = 20\log AR \quad \text{(10-19)}$$

The use of decibels merely results in a rescaling of the Bode plot AR axis. The decibel unit is employed in electrical communication and acoustic theory and is seldom used today in the process control field. Note that the MATLAB *bode* routine uses decibels as the default option; however, it can be modified to plot AR

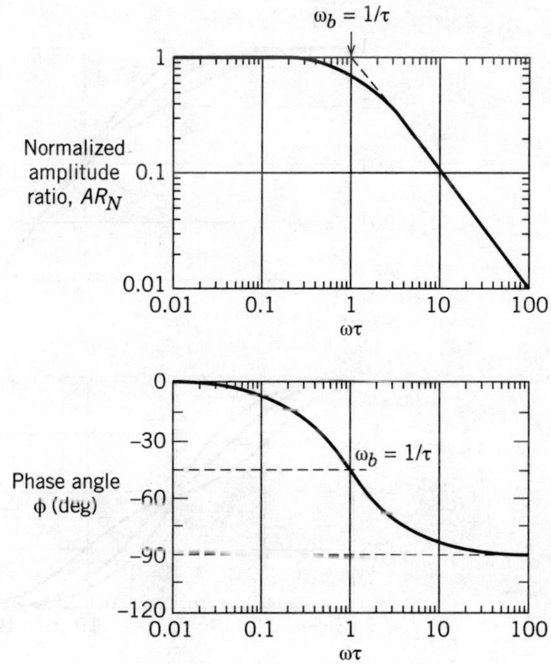

Figure 10.2 Bode diagram for a first-oreder process.

results, as shown in Fig 10.2. In the rest of this chapter, we only derive frequency responses for simple transfer functions (integrator, first-order, second-order, zeros, time delay). Software should be used for calculating frequency responses of more complicated transfer functions.

10.3.2 Integrating Process

The transfer function for an integrating process was given in Chapter 5.

$$G(s) = \frac{Y(s)}{U(s)} = \frac{K}{s} \tag{5-40}$$

Because of the single pole located at the origin, this transfer function represents a marginally stable process. The shortcut method of determining frequency response outlined in the preceding section was developed for stable processes, that is, those that converge to a bounded oscillatory response for a sinusoidal input. Because the output of an integrating process is bounded when forced by a sinusoidal input, the shortcut method does apply for this marginally stable process:

$$AR = |G(j\omega)| = \left|\frac{K}{j\omega}\right| = \frac{K}{\omega} \tag{10-20}$$

$$\phi = \angle G(j\omega) = \angle K - \tan^{-1}(j\infty) = -90^\circ \tag{10-21}$$

10.3.3 Second-Order Process

A general transfer function for a second-order system without numerator dynamics is

$$G(s) = \frac{K}{\tau^2 s^2 + 2\zeta\tau s + 1} \tag{10-22}$$

Substituting $s = j\omega$ and rearranging into real and imaginary parts (see Example 10.1) yields

$$AR = \frac{K}{\sqrt{(1-\omega^2\tau^2)^2 + (2\zeta\omega\tau)^2}} \tag{10-23a}$$

$$\phi = \tan^{-1}\left[\frac{-2\zeta\omega\tau}{1-\omega^2\tau^2}\right] \tag{10-23b}$$

Note that, in evaluating ϕ, multiple results are obtained because Eq. 10-23b has infinitely many solutions, each differing by $n180°$, where n is a positive integer. The appropriate solution of Eq. 10-23b for the second-order system yields $-180° < \phi < 0$.

Figure 10.3 shows the Bode plots for overdamped ($\xi > 1$), critically damped ($\xi = 1$), and underdamped ($0 < \xi < 1$) processes as a function of $\omega\tau$. The low-frequency limits of the second-order system are identical to those of the first-order system. However, the limits are different at high frequencies, $\omega\tau \gg 1$.

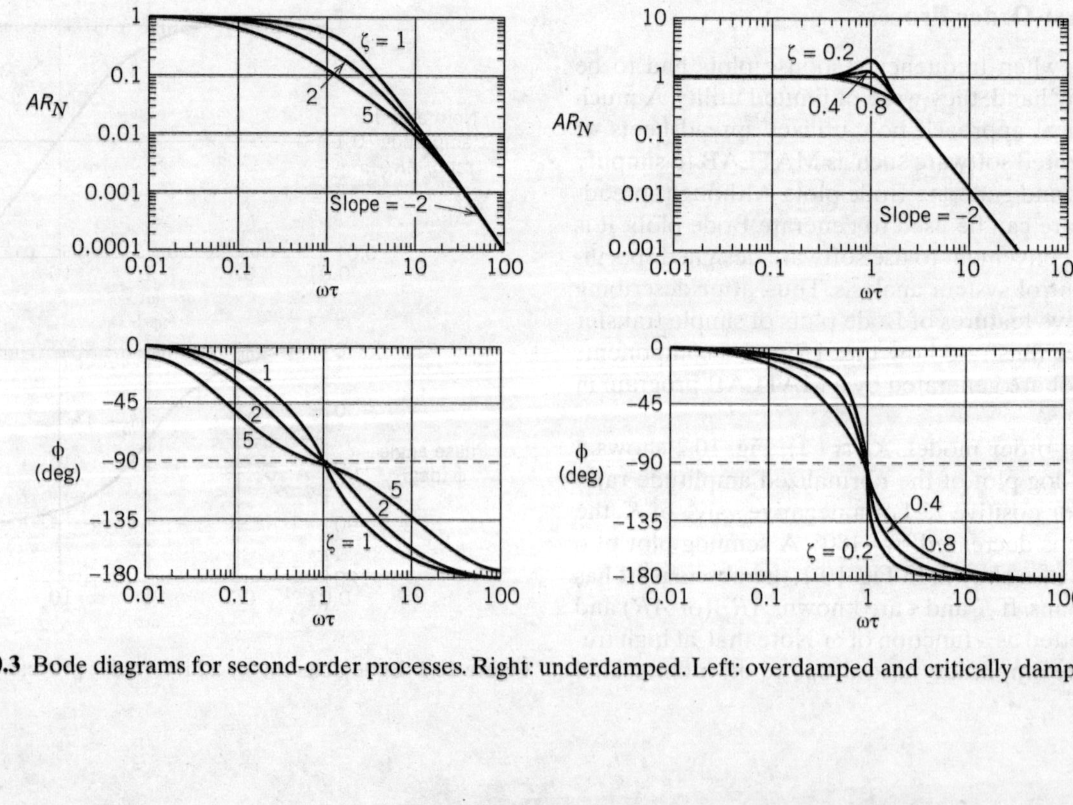

Figure 10.3 Bode diagrams for second-order processes. Right: underdamped. Left: overdamped and critically damped.

$$AR_N \approx 1/(\omega\tau)^2 \tag{10-24a}$$

$$\phi \approx -180^\circ \tag{10-24b}$$

For overdamped systems, the normalized amplitude ratio is attenuated ($\hat{A}/KA < 1$) for all ω. For underdamped systems, the amplitude ratio plot exhibits a maximum (for values of $0 < \xi < \sqrt{2}/2$) at the resonant frequency

$$\omega_r = \frac{\sqrt{1-2\zeta^2}}{\tau} \tag{10-25}$$

$$(AR_N)_{\max} = \frac{1}{2\zeta\sqrt{1-\zeta^2}} \tag{10-26}$$

These expressions can be derived by the interested reader. The *resonant frequency* ω_r is that frequency for which the sinusoidal output response has the maximum amplitude for a given sinusoidal input. Equations 10-25 and 10-26 indicate how ω_r and $(AR_N)_{\max}$ depend on ξ. This behavior is used in designing organ pipes to create sounds at specific frequencies. However, excessive resonance is undesirable, for example, in automobiles, where a particular vibration is noticeable only at a certain speed. For industrial processes operated without feedback control, resonance is seldom encountered, although some measurement devices are designed to exhibit a limited amount of resonant behavior. On the other hand, feedback controllers can be tuned to give the controlled process a slight amount of oscillatory or underdamped behavior in order to speed up the controlled system response (see Chapter 11).

10.3.4 Process Zero

A term of the form $\tau s + 1$ in the denominator of a transfer function is sometimes referred to as a process lag, because it causes the process output to lag the input (the phase angle is negative). Similarly, a process zero of the form $\tau s + 1$ ($\tau > 0$) in the numerator (see Section 4.5) causes the sinusoidal output of the process to lead the input ($\phi > 0$); hence, a left-half plane (LHP) zero often is referred to as a *process lead*. Next we consider the amplitude ratio and phase angle for this term.

Substituting $s = j\omega$ into $G(s) = \tau s + 1$ gives

$$G(j\omega) = j\omega\tau + 1 \tag{10-27}$$

from which

$$AR = |G(j\omega)| = \sqrt{\omega^2\tau^2 + 1} \tag{10-28a}$$

$$\phi = \angle G(j\omega) = \tan^{-1}(\omega\tau) \tag{10-28b}$$

Therefore, a process zero contributes a positive phase angle that varies between 0 and +90°. The output signal amplitude becomes very large at high frequencies (i.e., $AR \to \infty$ *as* $\omega \to \infty$), which is a physical impossibility. Consequently, in practice a process zero is always found in combination with one or more poles. The order of the numerator of the process transfer function must be less than or equal to the order of the denominator, as noted in Section 4.5.

Suppose that the numerator of a transfer function contains the term $1 - \tau s$, with $\tau > 0$. As shown in Section 4.5, a right-half plane (RHP) zero is associated with an inverse step response. The frequency response characteristics of $G(s) = 1 - \tau s$ are

$$AR = \sqrt{\omega^2\tau^2 + 1} \tag{10-29a}$$

$$\phi = -\tan^{-1}(\omega\tau) \tag{10-29b}$$

Hence, the amplitude ratios of LHP and RHP zeros are identical. However, an RHP zero contributes phase lag to the overall frequency response because of the negative sign. Processes that contain an RHP zero or time delay are sometimes referred to as *nonminimum phase systems* because they exhibit more phase lag than another transfer function that has the same AR characteristics (Franklin et al., 2014). Exercise 10.11 illustrates the importance of zero location on the phase angle.

10.3.5 Time Delay

The time delay $e^{-\theta s}$ is the remaining important process element to be analyzed. Its frequency response characteristics can be obtained by substituting $s = j\omega$:

$$G(j\omega) = e^{-j\omega\theta} \tag{10-30}$$

which can be written in rational form by substitution of the Euler identity

$$G(j\omega) = \cos\omega\theta - j\sin\omega\theta \tag{10-31}$$

From Eq. 10.6,

$$AR = |G(j\omega)| = \sqrt{\cos^2\omega\theta + \sin^2\omega\theta} = 1 \tag{10-32}$$

$$\phi = \angle G(j\omega) = \tan^{-1}\left(-\frac{\sin \omega\theta}{\cos \omega\theta}\right)$$

or

$$\phi = -\omega\theta \tag{10-33}$$

Because ω is expressed in radians/time, the phase angle in degrees is $-180\,\omega\theta/\pi$. Figure 10.4 illustrates the Bode plot for a time delay. The phase angle is unbounded, that is, it approaches $-\infty$ as ω becomes large. By contrast, the c This unbounded phase lag is an important attribute of a time delay and is detrimental to closed-loop system stability, as is discussed in Section 10.6

The frequency response characteristics of important process transfer functions are given in Table 10.1.

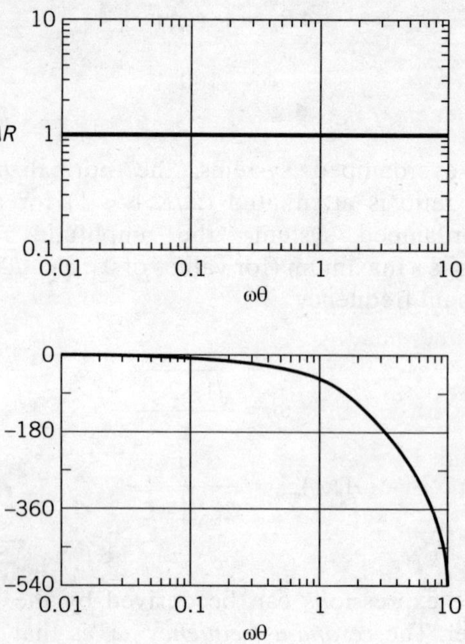

Figure 10.4 Bode diagram for a time delay, $e^{-\theta s}$.

Table 10.1 Frequency Response Characteristics of Important Process Transfer Functions

Transfer Function	$G(s)$	$AR = \|G(j\omega)\|$	Plot of log AR_N vs. log ω	$\phi = \angle G(j\omega)$	Plot of φ vs. log ω
1. First-order	$\frac{K}{\tau s+1}$	$\frac{K}{\sqrt{(\omega\tau)^2+1}}$	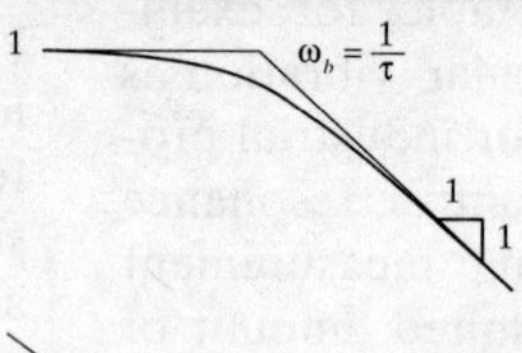	$-\tan^{-1}(\omega\tau)$	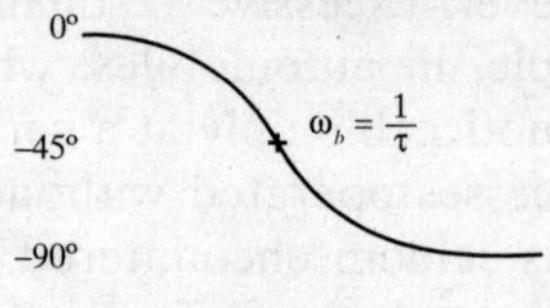
2. Integrator	$\frac{K}{s}$	$\frac{K}{\omega}$	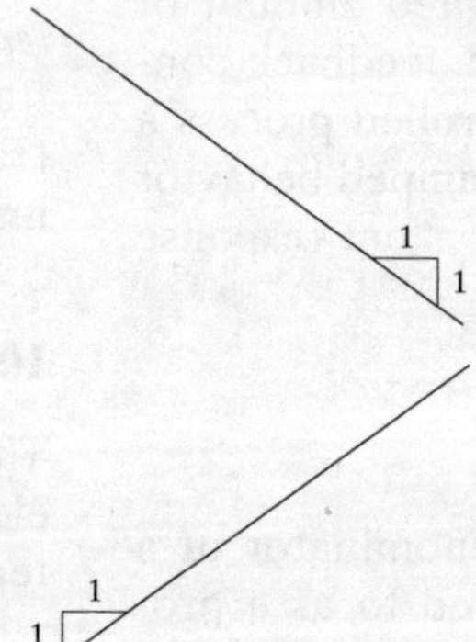	$-90°$	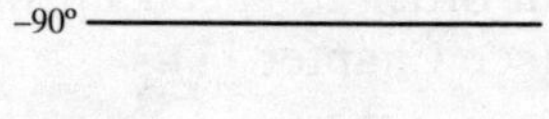
3. Derivative	Ks	$K\omega$		$+90°$	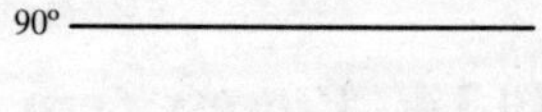
4. Overdamped second-order	$\frac{K}{(\tau_1 s+1)(\tau_2 s+1)}$	$\frac{K}{\sqrt{(\omega\tau_1)^2+1}\sqrt{(\omega\tau_2)^2+1}}$	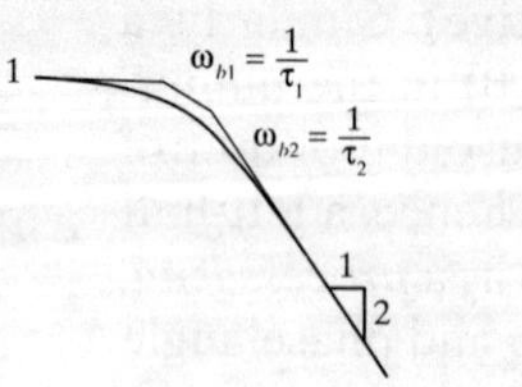	$-\tan^{-1}(\omega\tau_1) - \tan^{-1}(\omega\tau_2)$	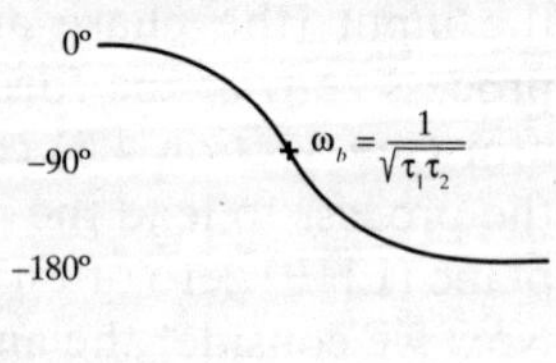
5. Critically damped second-order	$\frac{K}{(\tau s+1)^2}$	$\frac{K}{(\omega\tau)^2+1}$	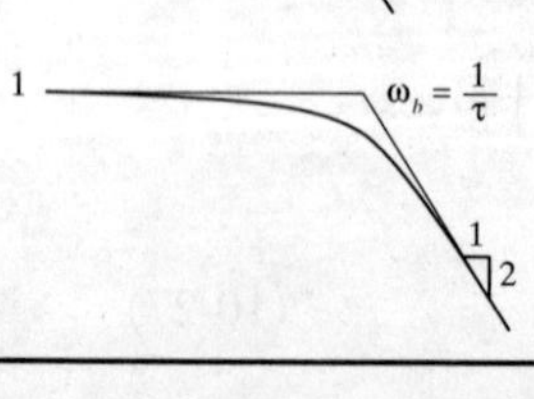	$-2\tan^{-1}(\omega\tau)$	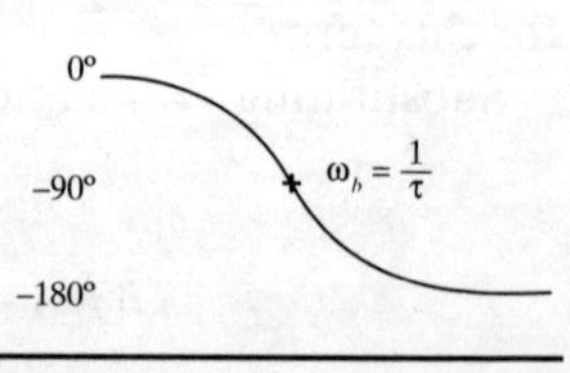

EXAMPLE 10.3

Generate the Bode plot for the transfer function

$$G(s) = \frac{5(0.5s+1)e^{-0.5s}}{(20s+1)(4s+1)}$$

where the time constants and time delay have units of minutes.

Solution

The Bode plot is shown in Fig. 10.5. The steady-state gain ($K = 5$) is the value of AR when $\omega \rightarrow 0$. The phase angle at high frequencies is dominated by the time delay. The MATLAB code for generating a Bode plot of the transfer function is shown in Table 10.2. In this code the normalized AR is used (AR_N).

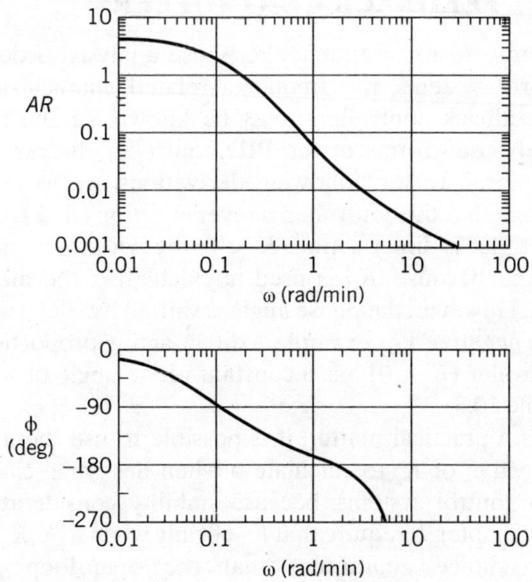

Figure 10.5 Bode plot of the transfer function in Example 10.3.

Table 10.2 MATLAB Program to Calculate and Plot the Frequency Response in Example 10.3

```
%Make a Bode plot for G = 5 (0.5s + 1)e^-0.5s/(20s + 1) %(4s +1)
close all
gain = 5;
tdead = 0.5;
num = [0.5 1];
den = [80 24 1];
G = tf (gain*num, den) %Define the system as a transfer %function
points = 500; %Define the number of points
ww = logspace (-2, 2, points); %Frequencies to be evaluated [mag, phase, ww] = bode
(G, ww); % Generate numerical %values for Bode plot
AR = zeros (points, 1); % Preallocate vectors for Amplitude %Ratio and Phase Angle
PA = zeros (points, 1);
for i = 1 : points
  AR(i) = mag (1,1,i)/gain; %Normalized AR
  PA(i) = phase (1,1,i) - ((180/pi) *tdead*ww(i));
end
fiqure
subplot (2,1,1)
loglog(ww, AR)
axis ([0.01 100 0.001 1])
title ('Frequency Response of a SOPTD with Zero')
ylabel('AR/K')
subplot (2,1,2)
semilogx(ww,PA)
axis ([0.01 100-270 0])
ylabel('Phase Angle (degrees)')
xlabel('Frequency (rad/time)')
```

10.4 FREQUENCY RESPONSE CHARACTERISTICS OF FEEDBACK CONTROLLERS

In order to use frequency response analysis to design control systems, the frequency-related characteristics of feedback controllers must be known for the most widely used forms of the PID controller discussed in Chapter 7. In the following derivations, we generally assume that the controller is reverse-acting ($K_c > 0$). If a controller is direct-acting ($K_c < 0$), the AR plot does not change, because $|K_c|$ is used in calculating the magnitude. However, the phase angle is shifted by −180° when K_c is negative. For example, a direct-acting proportional controller ($K_c < 0$) has a constant phase angle of −180° (Table 10.3)

As a practical matter, it is possible to use the absolute value of K_c to calculate ϕ when designing closed-loop control systems, because stability considerations (see Chapter 9) require that $K_c < 0$ only when $K_v K_p K_m < 0$. This choice guarantees that the open-loop gain ($K_{OL} = K_c K_v K_p K_m$) will always be positive. Use of this convention conveniently yields ϕ = 0° for any proportional controller and, in general, eliminates the need to consider the −180° phase shift contribution of the negative controller gain.

The frequency response characteristics of feedback controllers are given in Table 10.3.

Proportional Controller. Consider a proportional controller with positive gain

$$G_c(s) = K_c \tag{10-34}$$

In this case, $|G_c(j\omega)| = K_c$, which is independent of ω. Therefore,

$$AR = K_c \tag{10-35}$$

and

$$\phi = 0^\circ \tag{10-36}$$

Proportional-Integral Controller. A proportional-integral (PI) controller has the transfer function,

$$G_c(s) = K_c\left(1 + \frac{1}{\tau_I s}\right) = K_c\left(\frac{\tau_I s + 1}{\tau_I s}\right) \tag{10-37}$$

Substituting $s = j\omega$ gives

$$G_c(j\omega) = K_c\left(1 + \frac{1}{\tau_I j\omega}\right) = K_c\left(1 - \frac{j}{\omega\tau_I}\right) \tag{10-38}$$

Thus, the amplitude ratio and phase angle are

$$AR = |G_c(j\omega)| = K_c\sqrt{1 + \frac{1}{(\omega\tau_I)^2}} = K_c\frac{\sqrt{(\omega\tau_I)^2 + 1}}{\omega\tau_I} \tag{10-39}$$

$$\phi = \angle G_c(j\omega) = \tan^{-1}(-1/\omega\tau_I) = \tan^{-1}(\omega\tau_I) - 90^\circ \tag{10-40}$$

Based on Eqs. 10-39 and 10-40, at low frequencies, the integral action dominates. As $\omega \to 0$, $AR \to \infty$, and $\phi \to -90°$. At high frequencies, $AR = K_c$ and $\phi = 0°$; neither is a function of ω in this region (cf. the proportional controller).

Ideal Proportional-Derivative Controller. The ideal proportional-derivative (PD) controller (cf. Eq. 7-11) is rarely implemented in actual control systems but is a component of PID control and influences PID control at high frequency. Its transfer function is

$$G_c(s) = K_c(1 + \tau_D s) \tag{10-41}$$

The frequency response characteristics are similar to those of an LHP zero:

$$AR = K_c\sqrt{(\omega\tau_D)^2 + 1} \tag{10-42}$$

$$\phi = \tan^{-1}(\omega\tau_D) \tag{10-43}$$

Proportional-Derivative Controller with Filter. As indicated in Chapter 7, the PD controller is most often realized by the transfer function

$$G_c(s) = K_c\left(\frac{\tau_D s + 1}{\alpha\tau_D s + 1}\right) \tag{10-44}$$

where α has a value in the range 0.05–0.2. The frequency response for this controller is given by

$$AR = K_c\sqrt{\frac{(\omega\tau_D)^2 + 1}{(\alpha\omega\tau_D)^2 + 1}} \tag{10-45}$$

$$\phi = \tan^{-1}(\omega\tau_D) - \tan^{-1}(\alpha\omega\tau_D) \tag{10-46}$$

The pole in Eq. 10-44 bounds the high-frequency asymptote of the AR

$$\lim_{\omega\to\infty} AR = \lim_{\omega\to\infty} |G_c(j\omega)| = K_c/\alpha = 2/0.1 = 20 \tag{10-47}$$

Table 10.3 Frequency Response Characteristics of Important Controller Transfer Functions

Controller	$G_c(s)$	$AR = \|G_c(j\omega)\|$	Plot of log AR_N vs. log ω	$\phi = \angle G_c(j\omega)$	Plot of φ vs. log ω
1. PI	$K_c\left(1+\frac{1}{\tau_I s}\right) = K_c\left(\frac{\tau_I s+1}{\tau_I s}\right)$	$K_c\left(\frac{\sqrt{(\omega\tau_I)^2+1}}{\omega\tau_I}\right)$	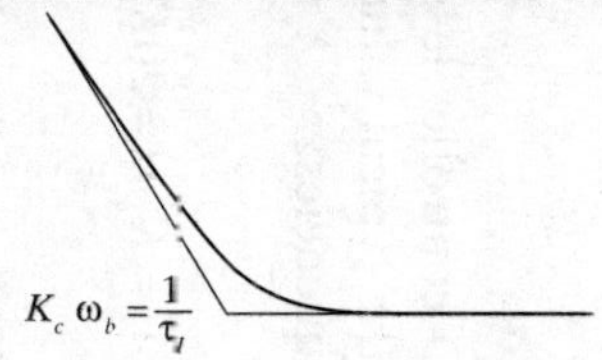	$\tan^{-1}(\omega\tau_I) - 90°$	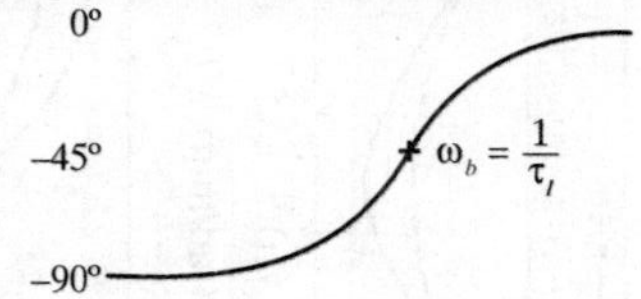
2. Ideal PD	$K_c(\tau_D s+1)$	$K_c\sqrt{(\omega\tau_D)^2+1}$	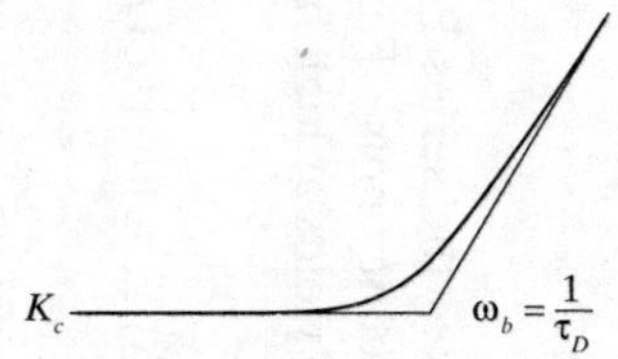	$\tan^{-1}(\omega\tau_D)$	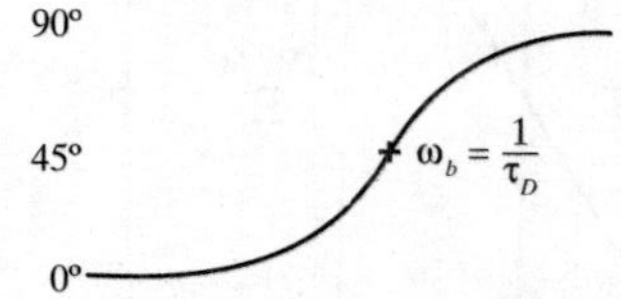
3. PD with Derivative Filter	$K_c\left(\frac{\tau_D s+1}{\alpha\tau_D s+1}\right)$	$K_c\sqrt{\frac{(\omega\tau_D)^2+1}{(\alpha\omega\tau_D)^2+1}}$	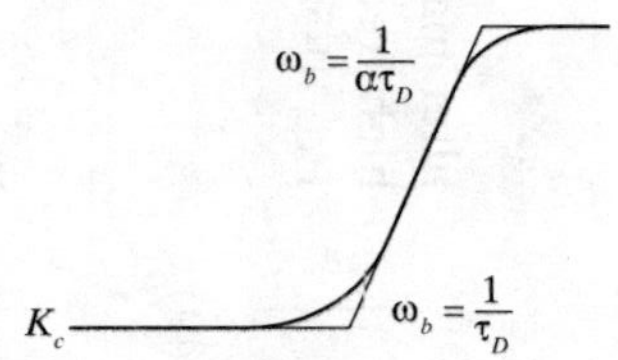	$\tan^{-1}(\omega\tau_D)$ $-\tan^{-1}(\alpha\omega\tau_D)$	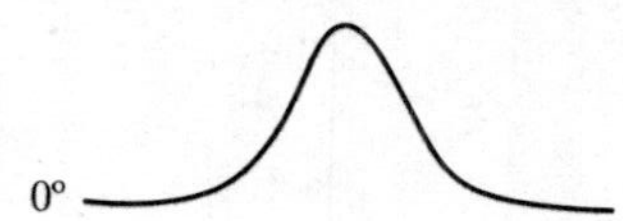
4. Parallel PID	$K_c\left(1+\frac{1}{\tau_I s}+\tau_D s\right)$	$K_c\sqrt{\left(\omega\tau_D - \frac{1}{\omega\tau_I}\right)^2+1}$	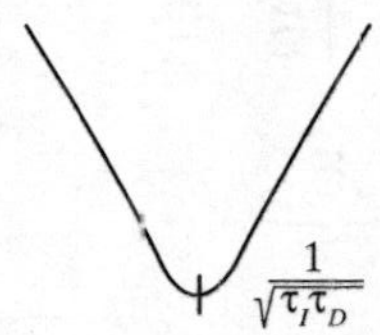	$-\tan^{-1}\left(\omega\tau_D - \frac{1}{\omega\tau_I}\right)$	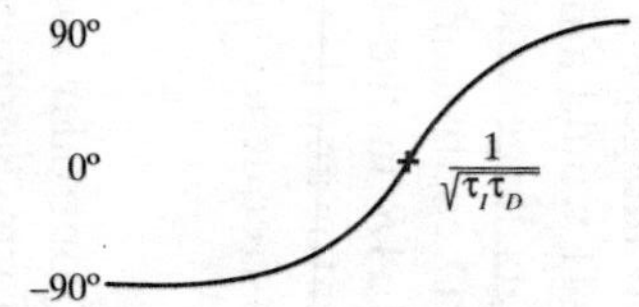
5. Series PID	$K_c\left(\frac{\tau_I s+1}{\tau_I s}\right)(\tau_D s+1)$	$K_c\frac{\sqrt{(\omega\tau_I)^2+1}}{\omega\tau_I}\sqrt{(\omega\tau_D)^2+1}$	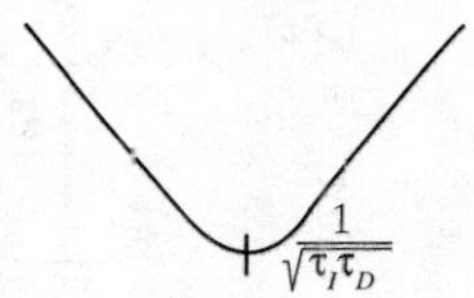	$\tan^{-1}(\omega\tau_I) + \tan^{-1}$ $(\omega\tau_D) - 90°$	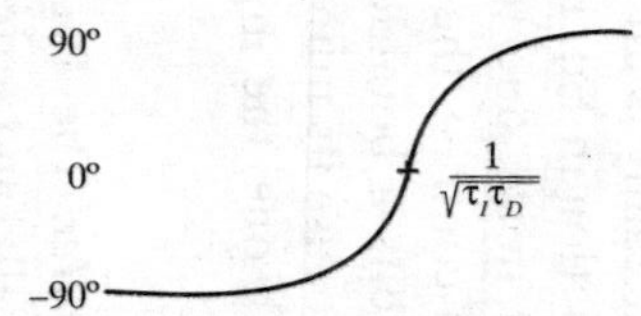
6. Series PID with Filter	$K_c\left(\frac{\tau_I s-1}{\tau_I s}\right)\left(\frac{\tau_D s+1}{\alpha\tau_D s+1}\right)$	$K_c\left(\frac{\sqrt{(\omega\tau_I)^2+1}}{\omega\tau_I}\right)\sqrt{\frac{(\omega\tau_D)^2+1}{(\alpha\omega\tau_D)^2+1}}$	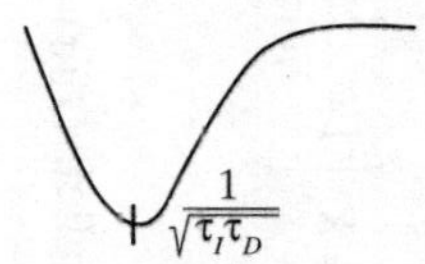	$\tan^{-1}(\omega\tau_I) + \tan^{-1}(\omega\tau_D)$ $-\tan^{-1}(\alpha\omega\tau_D) - 90°$	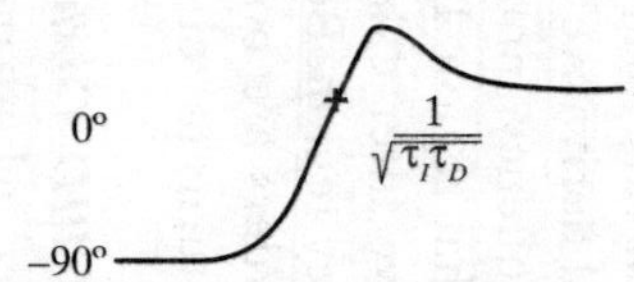

Note that this form actually is an advantage, because the ideal derivative action in Eq. 10-41 would amplify high-frequency input noise, due to its large value of AR in that region. In contrast, the PD controller with derivative filter exhibits a bounded AR in the high-frequency region. Because its numerator and denominator orders are both one, the high-frequency phase angle returns to zero.

Parallel PID Controller. The PID controller can be developed in both parallel and series forms, as discussed in Chapter 7. Either version exhibits features of both the PI and PD controllers. The simpler version is the following parallel form (cf. Eq. 7-14):

$$G_c(s) = K_c\left(1 + \frac{1}{\tau_I s} + \tau_D s\right) = K_c\left(\frac{1 + \tau_I s + \tau_I \tau_D s^2}{\tau_I s}\right) \tag{10-48}$$

Substituting $s = j\omega$ and rearranging gives

$$G_c(j\omega) = K_c\left(1 + \frac{1}{j\omega\tau_I} + j\omega\tau_D\right) = K_c\left[1 + j\left(\omega\tau_D - \frac{1}{\omega\tau_I}\right)\right] \tag{10-49}$$

Ideal parallel: $G_c(s) = 2\left(1 + \dfrac{1}{10s} + 4s\right)$

Parallel with derivative filter:

$G_c(s) = 2\left(1 + \dfrac{1}{10s} + \dfrac{4s}{0.4s + 1}\right)$

Parallel PID Controller with a Derivative Filter. The parallel controller with a derivative filter was described in Chapter 7 and Table 7.1.

$$G_c(s) = K_c\left(1 + \frac{1}{\tau_I s} + \frac{\tau_D s}{\alpha\tau_D s + 1}\right) \tag{10-50}$$

Figure 10.6 shows a Bode plot for an ideal PID controller, with and without a derivative filter (see Table 7.1). The controller settings are $K_c = 2$, $\tau_I = 10$ min, $\tau_D = 4$ min, and $\alpha = 0.1$. The phase angle varies from $-90°$ $(\omega \to 0)$ to $+90°$ $(\omega \to \infty)$.

A comparison of the amplitude ratios in Fig. 10.6 indicates that the AR for the controller without the derivative filter in Eq. 10-48 is unbounded at high frequencies, in contrast to the controller with the derivative filter (Eq. 10-50), which has a bounded AR at all frequencies. Consequently, the addition of the derivative filter makes the series PID controller less sensitive to high-frequency noise. For the typical value of $\alpha = 0.10$, Eq. 10-50 yields at high frequencies:

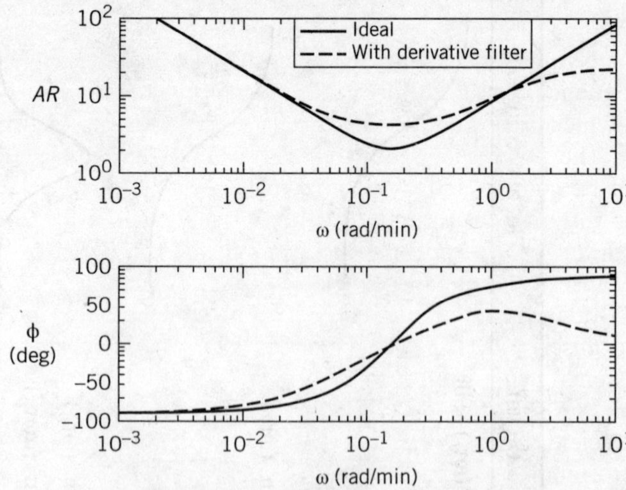

Figure 10.6 Bode plots of ideal parallel PID controller and ideal parallel PID controller with derivative filter ($\alpha = 0.1$).

$$AR_{\omega\to\infty} = \lim_{\omega\to\infty} |G_c(j\omega)| = K_c/\alpha = 20K_c \tag{10-51}$$

When $\tau_D = 0$, the parallel PID controller with filter is the same as the PI controller of Eq. 10-37.

By adjusting the values of τ_I and τ_D, one can prescribe the shape and location of the notch in the AR curve. Decreasing τ_I and increasing τ_D narrows the notch, whereas the opposite changes broaden it. Figure 10.6 indicates that the center of the notch is located at $\omega = 1/\sqrt{\tau_I\tau_D}$ where $\phi = 0°$ and $AR = K_c$. Varying K_c moves the amplitude ratio curve up or down, without affecting the width of the notch. Generally, the integral time τ_I is larger than τ_D, typically $\tau_I \approx 4\tau_D$.

Series PID Controller. The simplest version of the series PID controller is

$$G_c(s) = K_c\left(\frac{\tau_I s + 1}{\tau_I s}\right)(\tau_D s + 1) \tag{10-52}$$

This controller transfer function can be interpreted as the product of the transfer functions for PI and PD controllers. Because the transfer function in Eq. 10-52 is physically unrealizable and amplifies high-frequency noise, a more practical version includes a derivative filter.

10.5 NYQUIST DIAGRAMS

The Nyquist diagram is an alternative representation of frequency response information, a polar plot of $G(j\omega)$ in which frequency ω appears as an implicit parameter. The Nyquist diagram for a transfer function $G(s)$ can be constructed directly from $|G(j\omega)|$ and $\angle G(j\omega)$ for different values of ω. Alternatively, the Nyquist diagram can be constructed from the Bode diagram, because $AR = |G(j\omega)|$ and $\phi = \angle G(j\omega)$. The advantages of Bode plots are that frequency is plotted explicitly as the abscissa, and the log–log and semilog coordinate systems facilitate block multiplication. The Nyquist diagram, on the other hand, is more compact and is sufficient for many important analyses, for example, determining system stability (see Appendix L). Most of the recent interest in Nyquist diagrams has been in connection with designing multiloop controllers and for robustness (sensitivity) studies (Maciejowski, 1989; Skogestad and Postlethwaite, 2005). For single-loop controllers, Bode plots are used more often.

10.6 BODE STABILITY CRITERION

The Bode stability criterion has an important advantage in comparison with the alternative of calculating the roots of the characteristic equation in Chapter 9. It provides a measure of the relative stability rather than merely a yes or no answer to the question "Is the closed-loop system stable?"

Before considering the basis for the Bode stability criterion, it is useful to review the General Stability Criterion of Section 9.1: *A feedback control system is stable if and only if all roots of the characteristic equation lie to the left of the imaginary axis in the complex plane.*

Thus, the imaginary axis divides the complex plane into stable and unstable regions. Recall that the characteristic equation was defined in Chapter 9 as

$$1 + G_{OL}(s) = 0 \tag{10-53}$$

where the open-loop transfer function in Eq. 10-53 is $G_{OL}(s) = G_c(s)G_v(s)G_p(s)G_m(s)$.

Before stating the Bode stability criterion, we introduce two important definitions:

1. A *critical frequency* ω_c is a value of ω for which $\phi_{OL}(\omega) = -180^\circ$. This frequency is also referred to as a *phase crossover frequency*.
2. A *gain crossover frequency* ω_g is a value of ω for which $AR_{OL}(\omega) = 1$.

The Bode stability criterion allows the stability of a closed-loop system to be determined from the open-loop transfer function.

Bode Stability Criterion. *Consider an open-loop transfer function $G_{OL} = G_cG_vG_pG_m$ that is strictly proper (more poles than zeros) and has no poles located on or to the right of the imaginary axis, with the possible exception of a single pole at the origin. Assume that the open-loop frequency response has only a single critical frequency ω_c and a single gain crossover frequency ω_g. Then the closed-loop system is stable if the open-loop amplitude ratio $AR_{OL}(\omega_c) < 1$. Otherwise, it is unstable.*

The root locus diagrams of Section 9.5 (e.g., Fig. 9.28) show how the roots of the characteristic equation change as controller gain K_c changes. By definition, the roots of the characteristic equation are the numerical values of the complex variable, s, that satisfy Eq. 10-53. Thus, each point on the root locus also satisfies Eq. 10-54, which is a rearrangement of Eq. 10-53:

$$G_{OL}(s) = -1 \tag{10-54}$$

The corresponding magnitude and argument are

$$|G_{OL}(j\omega)| = 1 \quad \text{and} \quad \angle G_{OL}(j\omega) = -180^\circ \tag{10-55}$$

For a marginally stable system, $\omega_c = \omega_g$ and the frequency of the sustained oscillation, ω_c, is caused by a pair of roots on the imaginary axis at $s = \pm\omega_c j$. Substituting this expression for s into Eq. 10-55 gives the following expressions for a conditionally stable system:

$$AR_{OL}(\omega_c) = |G_{OL}(j\omega_c)| = 1 \tag{10-56}$$

$$\phi_{OL}(\omega_c) = \angle G_{OL}(j\omega_c) = -180^\circ \tag{10-57}$$

for some specific value of $\omega_c > 0$. Equations 10-56 and 10-57 provide the basis for the Bode stability criterion.

Some of the important properties of the Bode stability criterion are

1. It provides a necessary and sufficient condition for closed-loop stability, based on the properties of the open-loop transfer function.
2. The Bode stability criterion is applicable to systems that contain time delays.
3. The Bode stability criterion is very useful for a wide variety of process control problems. However, for any $G_{OL}(s)$ that does not satisfy the required conditions, the Nyquist stability criterion discussed in Appendix L can be applied.

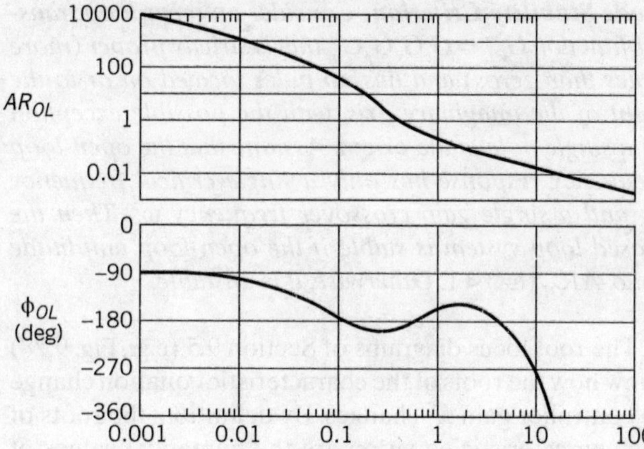

Figure 10.7 Bode plot exhibiting multiple critical frequencies.

For many control problems, there is only a single ω_c and a single ω_g. But multiple values for ω_c can occur, as shown in Fig. 10.7. In this somewhat unusual situation, the closed-loop system is stable for two different ranges of the controller gain (Luyben and Luyben, 1997). Consequently, increasing the absolute value of K_c can actually improve the stability of the closed-loop system for certain ranges of K_c. For systems with multiple ω_c or ω_g, the Bode stability criterion has been modified by Hahn et al. (2001) to provide a sufficient condition for stability.

As indicated in Chapter 9, when the closed-loop system is marginally stable, the closed-loop response exhibits a sustained oscillation after a set-point change or a disturbance. Thus, the amplitude neither increases nor decreases.

In order to gain physical insight into why a sustained oscillation occurs at the stability limit, consider the analogy of an adult pushing a child on a swing. The child swings in the same arc as long as the adult pushes at the right time and with the right amount of force. Thus the desired sustained oscillation places requirements on both timing (i.e., phase) and applied force (i.e., amplitude). By contrast, if either the force or the timing is not correct, the desired swinging motion ceases, as the child will quickly protest. A similar requirement occurs when a person bounces a ball.

To further illustrate why feedback control can produce sustained oscillations, consider the following thought experiment for the feedback control system shown in Fig. 10.8. Assume that the open-loop system is stable and that no disturbances occur ($D = 0$). Suppose that the set-point is varied sinusoidally at the critical frequency, $y_{sp}(t) = A \sin(\omega_c t)$, for a long period of time. Assume that during this period, the measured output, y_m, is disconnected, so that the feedback loop is broken before the comparator. After the initial transient dies out, y_m will oscillate at the excitation frequency ω_c, because the response of a linear system to a sinusoidal input is a sinusoidal output at the same frequency (see Section 10.2). Suppose the two events occur simultaneously: (i) the set-point is set to zero, and (ii) y_m is reconnected. If the feedback control system is marginally stable, the controlled variable y will then exhibit a sustained sinusoidal oscillation with amplitude A and frequency ω_c.

To analyze why this special type of oscillation occurs only when $\omega = \omega_c$, note that the sinusoidal signal E in Fig. 10.8 passes through transfer functions G_c, G_v, G_p, and G_m before returning to the comparator. In order to have a sustained oscillation after the feedback loop is reconnected, signal Y_m must have the same amplitude as E and a 180° phase shift relative to E. Note that the comparator also provides a −180° phase shift because of its negative sign. Consequently, after Y_m passes through the comparator, it is in phase with E and has the same amplitude, A. Thus, the closed-loop system oscillates indefinitely after the feedback loop is closed because the conditions in Eqs. 10-56 and 10-57 are both satisfied. But what happens if K_c is increased by a small amount? Then, $AR_{OL}(\omega_c)$ is greater than one, the oscillations grow, and the closed-loop system becomes unstable. In contrast, if K_c is reduced by a small amount, the oscillation is damped and eventually dies out.

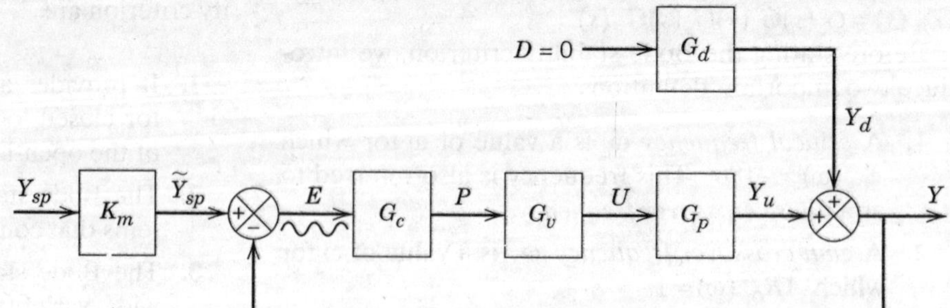

Figure 10.8 Sustained oscillation in a feedback control system.

EXAMPLE 10.4

A process has the third-order transfer function (time constant in minutes),

$$G_p(s) = \frac{2}{(0.5s+1)^3}$$

Also, $G_v = 0.1$ and $G_m = 10$. For a proportional controller, evaluate the stability of the closed-loop control system using the Bode stability criterion and three values of K_c: 1, 4, and 20.

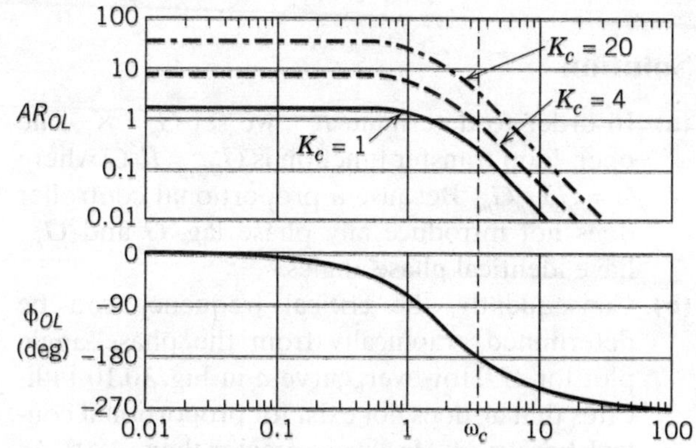

Figure 10.9 Bode plots for $G_{OL} = 2K_c/(0.5s+1)^3$.

Solution

For this example,

$$G_{OL} = G_c G_v G_p G_m = (K_c)(0.1)\frac{2}{(0.5s+1)^3}(10) = \frac{2K_c}{(0.5s+1)^3}$$

Figure 10.9 shows a Bode plot of G_{OL} for three values of K_c. Note that all three cases have the same phase angle plot, because the phase lag of a proportional controller is zero for $K_c > 0$.

From the phase angle plot, we observe that $\omega_c = 3.46$ rad/min. This is the frequency of the sustained oscillation that occurs at the stability limit, as discussed previously. Next, we consider the amplitude ratio AR_{OL} for each value of K_c. Based on Fig. 10.9, we make the following classifications:

K_c	AR_{OL} (for $\omega = \omega_c$)	Classification
1	0.25	Stable
4	1	Marginally Stable
20	5	Unstable

In Section 11.5.1, the concept of the ultimate gain was introduced. For proportional-only control, the ultimate gain K_{cu} was defined to be the largest value of K_c that results in a stable closed-loop system. The value of K_{cu} can be determined graphically from a Bode plot for transfer function $G = G_v G_p G_m$. For proportional-only control, $G_{OL} = K_c G$. Because a proportional controller has zero phase lag, ω_c is determined solely by G. Also,

$$AR_{OL}(\omega) = K_c AR_G(\omega) \tag{10-58}$$

where AR_G denotes the amplitude ratio of G. At the stability limit, $\omega = \omega_c$, $AR_{OL}(\omega_c) = 1$ and $K_c = K_{cu}$. Substituting these expressions into Eq. 10-58 and solving for K_{cu} gives an important result:

$$K_{cu} = \frac{1}{AR_G(\omega_c)} \tag{10-59}$$

The stability limit for K_c can also be calculated for PI and PID controllers and is denoted by K_{cm}, as demonstrated by Example 10.5.

EXAMPLE 10.5

Consider PI control of an overdamped second-order process (time constants in minutes),

$$G_p(s) = \frac{5}{(s+1)(10.5s+1)}$$

$$G_m = G_v = 1$$

(a) Determine the value of K_{cu}.

(b) Use a Bode plot to show that controller settings of $K_c = 0.4$ and $\tau_I = 0.2$ min produce an unstable closed-loop system.

(c) Find K_{cm}, the maximum value of K_c that can be used with $\tau_I = 0.2$ min and still have closed-loop stability.

(d) Show that $\tau_I = 1$ min results in a stable closed-loop system for all positive values of K_c.

Solution

(a) In order to determine K_{cu}, we set $G_c = K_c$. The open-loop transfer function is $G_{OL} = K_c G$ where $G = G_v G_p G_m$. Because a proportional controller does not introduce any phase lag, G and G_{OL} have identical phase angles.

(b) Consequently, the critical frequency can be determined graphically from the phase angle plot for G. However, curve *a* in Fig. 10.10 indicates that ω_c does not exist for proportional control, because ϕ_{OL} is always greater than –180°. As a result, K_{cu} does not exist, and thus K_c does not have a stability limit. Conversely, the addition of integral control action can produce closed-loop instability. Curve *b* in Fig. 10.10 indicates that an unstable closed-loop system occurs for $G_c(s) = 0.4(1 + 1/0.2s)$, because $AR_{OL} > 1$ when $\phi_{OL} = -180°$.

(c) To find K_{cm} for $\tau_I = 0.2$ min, we note that ω_c depends on τ_I but not on K_c, because K_c has no effect on ϕ_{OL}. For curve *b* in Fig. 10.10, $\omega_c = 2.2$ rad/min, and the corresponding amplitude ratio is $AR_{OL} = 1.38$. To find K_{cm}, multiply the current value of K_c by a factor, 1/1.38. Thus, $K_{cm} = 0.4/1.38 = 0.29$.

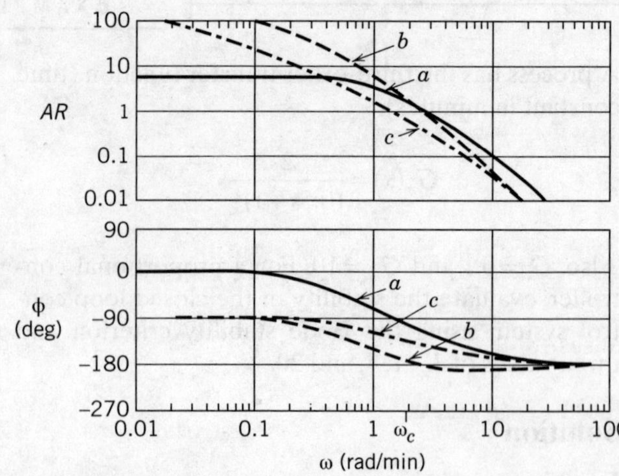

Figure 10.10 Bode plots for Example 10.5.

(d) When τ_I is increased to 1 min, curve *c* in Fig. 10.10 results. Because curve *c* does not have a critical frequency, the closed-loop system is stable for all positive values of K_c.

Curve a: $G(s)$

Curve b: $G_{OL}(s)$: $G_c(s) = 0.4\left(1 + \dfrac{1}{0.2s}\right)$

Curve c: $G_{OL}(s)$: $G_c(s) = 0.4\left(1 + \dfrac{1}{s}\right)$

EXAMPLE 10.6

Find the critical frequency for the following process and PID controller, assuming $G_v = G_m = 1$.

$$G_p(s) = \frac{e^{-0.3s}}{(9s+1)(11s+1)} \qquad G_c(s) = 20\left(1 + \frac{1}{2.5s} + s\right)$$

Solution

Figure 10.7 shows the open-loop amplitude ratio and phase angle plots for G_{OL}. Note that the phase angle crosses –180° at three points. Because there is more than one value of ω_c, the Bode stability criterion cannot be applied.

EXAMPLE 10.7

Evaluate the stability of the closed-loop system for:

$$G_p(s) = \frac{4e^{-s}}{5s+1}$$

The time constant and time delay have units of minutes and,

$$G_v = 2, \quad G_m = 0.25, \quad G_c = K_c$$

Obtain ω_c and K_{cu} from a Bode plot. Use an initial value of $K_c = 1$.

Solution

The Bode plot for G_{OL} and $K_c = 1$ is shown in Fig. 10.11. For $\omega_c = 1.69$ rad/min, $\phi_{OL} = -180°$, and $AR_{OL} = 0.235$. For $K_c = 1$, $AR_{OL} = AR_G$ and K_{cu} can be calculated from Eq. 10-59. Thus, $K_{cu} = 1/0.235 = 4.25$. Setting $K_c = 1.5K_{cu}$ gives $K_c = 6.38$. A larger value of K_c causes the closed-loop system to become unstable. Only values of K_c less than K_{cu} result in a stable closed-loop system.

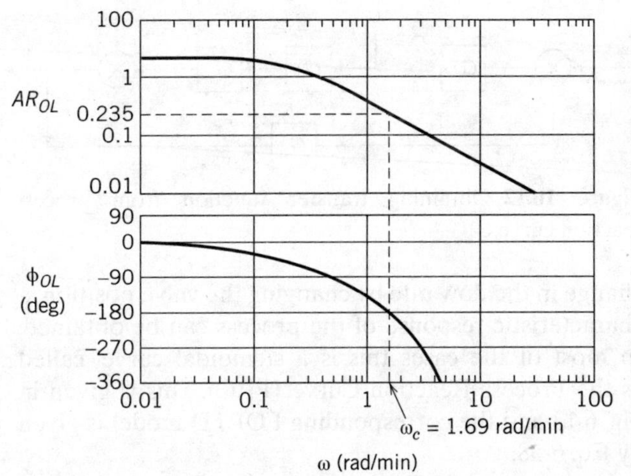

Figure 10.11 Bode plot for Example 10.7, $K_c = 1$.

10.7 CONTROLLER DESIGN BASED ON BODE STABILITY CRITERION

According to Bode Stability Criteria, a feedback control system is unstable if the amplitude ratio (AR_{OL}) of the corresponding open-loop transfer function (G_{OL}) is greater than unity at the phase crossover frequency or critical frequency (ω_c). So, the ultimate gain (K_{cu}) for a P-only controller can be obtained when $AR_{OL} = 1$ at the critical frequency. This is given in Eq. 10-59. The corresponding period of oscillation, called as ultimate period (P_u) can be obtained as:

$$P_u = \frac{2\pi}{\omega_c} \tag{10-60}$$

Using the values of K_{cu} and P_u, Ziegler and Nichols (1942) recommended the settings (popularly called as ZN settings) for classical feedback controllers. These are shown in Table 10.4.

The controller with ZN settings gives a very aggressive response. So, it is not useful for sensitive systems. For such systems the settings proposed by Tyreus and Luyben (1997) is used. It is called as Tyreus-Luyben settings (TL settings) or TLC settings, where TLC stands for Tender–Loving–Care. The TL settings give conservative response and so are useful for sensitive processes. These are shown in Table 10.5.

If the transfer functions, G_v, G_p, and G_m are not known precisely (as in the cases of complicated processes), the first-order-plus-time delay (FOPTD) model of the system is obtained using step test method or process reaction curve method as discussed in Section 6.5.

Consider the block diagram shown in Fig. 10.12. The control valve is disconnected from the controller so that the valve can be operated in manual mode. Giving a step

Table 10.4 Ziegler–Nichols Settings for Classical Feedback Controllers

Controllers	K_c	τ_I	τ_D
Proportional (P-only)	$K_{cu}/2$	-	-
Proportional-Integral (PI)	$K_{cu}/2.2$	$P_u/1.2$	-
Proportional-Integral-Derivative (PID)	$K_{cu}/1.7$	$P_u/2$	$P_u/8$

Table 10.5 Tyreus–Luyben Settings for Classical Feedback Controllers

Controllers	K_c	τ_I	τ_D
Proportional-Integral (PI)	$K_{cu}/3.2$	$2.2P_u$	-
Proportional-Integral-Derivative (PID)	$K_{cu}/2.2$	$2.2P_u$	$P_u/6.3$

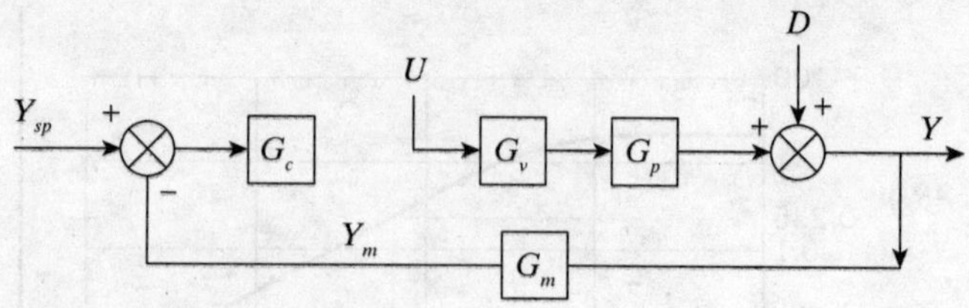

Figure 10.12 Obtaining transfer function from process reaction curve.

change in the flow rate by changing the valve position, a characteristic response of the process can be obtained. In most of the cases this is a sigmoidal curve, called as the Process Reaction Curve (PRC). This is given in Fig. 6.13 and the corresponding FOPTD model is given by Eq. 6-58.

$$G(s) = \frac{Ke^{-\theta s}}{\tau s+1} \quad (6\text{-}58)$$

The transfer function G in Eq. 6-58 is obtained from the process reaction curve (PRC). So, it is also called as G_{PRC}, which according to Fig.10.12 is given by

$$G = G_{PRC} = G_V G_P G_M \quad (10\text{-}61)$$

From Eqs. 6-58 and 10-61, we get

$$G_{PRC} = G_V G_P G_M = \frac{Ke^{-\theta s}}{\tau s+1} \quad (10\text{-}62)$$

The AR and phase lag are given by

$$AR_{PRC} = \frac{K}{\sqrt{\tau^2\omega^2+1}} \quad (10\text{-}63)$$

$$\varphi_{PRC} = -\tan^{-1}(\tau\omega) - \theta\omega \times \frac{180}{\pi} \text{ deg.} \quad (10\text{-}64)$$

For a P-only controller, we have $G_C = K_C$ and the open loop transfer function given by

$$G_{OL} = G_C G_V G_P G_M = G_C G_{PRC} = \frac{K_c Ke^{-\theta s}}{\tau s+1} \quad (10\text{-}65)$$

The frequency response characteristics are

$$AR_{OL} = \frac{KK_C}{\sqrt{\tau^2\omega^2+1}} = K_C AR_{PRC} \quad (10\text{-}66)$$

(Using AR_{PRC} from Eq. 10-63)

$$\varphi_{OL} = -\tan^{-1}(\tau\omega) - \theta\omega \times \frac{180}{\pi} \text{ deg.} \quad (10\text{-}67)$$

From Eq. (10.64) and Eq. (10.67), the phase lags are the same.

$$\phi_{OL} = \phi_{PRC} \text{ (For a P-only controller)} \quad (10\text{-}68)$$

Using Bode stability criteria [i.e., at crossover/critical frequency (ω_c), $\phi_{OL} = -180°$, and $AR_{OL} = 1$ for marginal stability] we can obtain the critical frequency and the ultimate controller gain from the following equations, respectively.

$$-180^\circ = -\tan^{-1}(\tau\omega_c) - \theta\omega_c \times \frac{180}{\pi} \quad (10\text{-}69)$$

$$AR_{OL}\big|_{\omega_c} = \frac{KK_{CU}}{\sqrt{\tau^2\omega_c^{\,2}+1}} = K_{CU} AR_{PRC} = 1 \quad (10\text{-}70)$$

Solving for K_{cu}, we get

$$K_{cu} = \frac{1}{AR_{PRC}\big|_{\omega_c}} = \frac{1}{AR_{PRC}(\omega_c)} \quad (10\text{-}71)$$

10.8 GAIN AND PHASE MARGINS

Rarely does the model of a chemical process stay unchanged for a variety of operating conditions and disturbances. When the process changes or the controller is poorly tuned, the closed-loop system can become unstable. Thus, it is useful to have quantitative measures of relative stability that indicate how close the system is to becoming unstable. The concepts of *gain margin (GM)* and *phase margin (PM)* provide useful metrics for relative stability.

Let AR_c be the value of the open-loop amplitude ratio at the critical frequency ω_c. Gain margin GM is defined as:

$$GM \triangleq \frac{1}{AR_c} \quad (10\text{-}72)$$

According to the Bode stability criterion, AR_c must be less than one for closed-loop stability. An equivalent stability requirement is that $GM > 1$. The gain margin provides a measure of relative stability, because it indicates how much any gain in the feedback loop component can increase before instability occurs. For example, if $GM = 2.1$, either process gain K_p or controller gain K_c could be doubled, and the closed-loop system would still be stable, although probably very oscillatory.

Next, we consider the phase margin. In Fig. 10.13, ϕ_g denotes the phase angle at the gain-crossover frequency ω_g where $AR_{OL} = 1$. Phase margin PM is defined as

$$PM \triangleq 180 + \phi_g \quad (10\text{-}73)$$

The phase margin also provides a measure of relative stability. In particular, it indicates how much additional time delay can be included in the feedback loop before instability will occur. Denote the additional time delay as $\Delta\theta_{max}$. For a time delay of $\Delta\theta_{max}$, the phase angle is $-\Delta\theta_{max}\omega$ (see Section 10.3.5). Thus, $\Delta\theta_{max}$ can be calculated from the following expression,

$$PM = \Delta\theta_{max}\omega_g\left(\frac{180^\circ}{\pi}\right) \quad (10\text{-}74)$$

or

$$\Delta\theta_{max} = \left(\frac{PM}{\omega_g}\right)\left(\frac{\pi}{180^\circ}\right) \quad (10\text{-}75)$$

where the ($\pi/180°$) factor converts PM from degrees to radians. Graphical representations of the gain and phase margins in a Bode plot are shown in Fig. 10.13.

The specification of phase and gain margins requires a compromise between performance and robustness. In general, large values of GM and PM correspond to sluggish closed-loop responses, whereas smaller values result in less sluggish, more oscillatory responses. The choices for GM and PM should also reflect model accuracy and the expected process variability.

Guideline. *In general, a well-tuned controller should have a gain margin between 1.7 and 4.0 and a phase margin between 30° and 45°.*

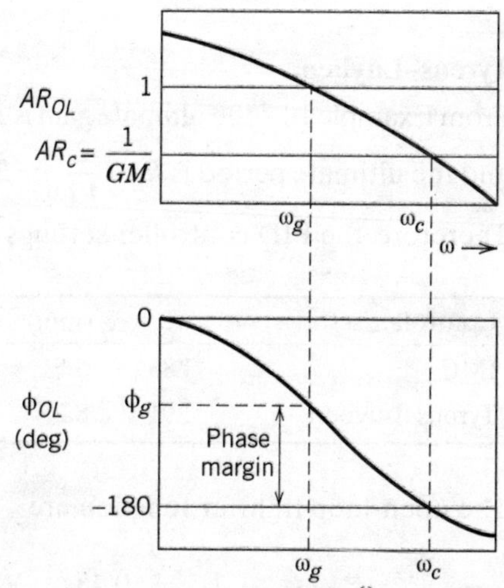

Figure 10.13 Gain and phase margins on a Bode plot.

Recognize that these ranges are approximate and that it may not be possible to choose PI or PID controller settings that result in specified GM and PM values. Tan et al. (1999) have developed graphical procedures for designing PI and PID controllers that satisfy GM and PM specifications. The GM and PM concepts are easily evaluated when the open-loop system does not have multiple values of ω_c or ω_g. However, for systems with multiple ω_g, gain margins can be determined from Nyquist plots (Doyle et al., 2009).

EXAMPLE 10.8

For the FOPTD model of Example 10.7, calculate the PID controller settings for the following approaches:

(a) IMC (Table 11.1 with $\tau_c = 1$)

(b) Continuous Cycling: Use the Tyreus–Luyben tuning relations (Luyben and Luyben, 1997), which are $K_c = 0.45K_{cu}$; $\tau_I = 2.2P_u$; $\tau_D = P_u/6.3$

Assume that the two PID controllers are implemented in the parallel form with a derivative filter ($\alpha = 0.1$) in Table 7.1. Plot the open-loop Bode diagram and determine the gain and phase margins for each controller.

For the Tyreus–Luyben settings, determine the maximum increase in the time delay $\Delta\theta_{max}$ that can occur while still maintaining closed-loop stability.

Solution

$$G_{OL} = G_cG_vG_pG_m = G_c\frac{2e^{-s}}{5s+1}$$

(a) IMC tuning:

Based on Table 11.1 (line H) for $\tau_c = 1$, we have

$$K_c = \frac{\tau + \frac{\theta}{2}}{\left(\tau_c + \frac{\theta}{2}\right)K} = \frac{5+0.5}{2(1+0.5)} = 1.83;$$

$$\tau_I = \tau + \frac{\theta}{2} = 5.5 \text{ min};$$

$$\tau_D = \frac{\tau\theta}{2\tau+\theta} = \frac{5}{10+1} = 0.45 \text{ min}$$

(b) Tyreus–Luyben:

From Example 10.7, the ultimate gain is $K_{cu} = 4.25$, and the ultimate period is $P_u = \dfrac{2\pi}{1.69} = 3.72$ min. Therefore, the PID controller settings are

Controller settings	K_c	τ_I (min)	τ_D (min)
IMC	1.83	5.5	0.45
Tyreus–Luyben	1.91	8.2	0.59

The open-loop transfer function are

$$G_{OL,\,IMC} = 1.83\left(1+\frac{1}{5.5s}+\frac{0.45s}{0.045s+1}\right)\left(\frac{2e^{-s}}{5s+1}\right)$$

$$G_{OL,\,T-L} = 1.91\left(1+\frac{1}{8.18s}+\frac{0.59s}{0.059s+1}\right)\left(\frac{2e^{-s}}{5s+1}\right)$$

Figure 10.14 shows the frequency response of G_{OL} for the two controllers. The gain and phase margins can be determined by inspection of the Bode diagram or by using the MATLAB command *margin*.

Controller	GM	PM	ω_c (rad/min)
IMC	2.2	68.5°	2.38
Tyreus–Luyben	1.8	76°	2.51

Based on Fig. 10.14, the Tyreus–Luyben controller settings are close to the IMC tuning result. The value of $\Delta\theta_{max}$ is calculated from Eq. 10-63, and the information in the preceding Table:

$$\Delta\theta_{max} = \frac{(76°)(\pi \text{ rad})}{(0.79 \text{ rad/min})(180°)} = 1.7 \text{ min}$$

Thus, time delay θ can increase by as much as 70% and still maintain closed-loop stability.

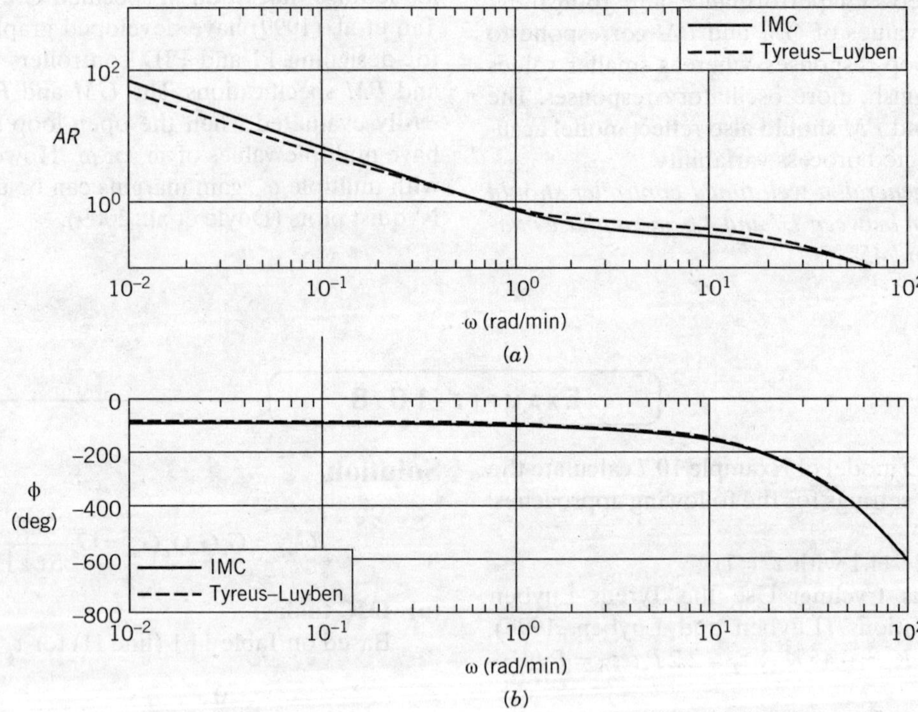

Figure 10.14 Comparison of G_{OL} Bode plots for Example 10.8.

10.8.1 Controller Design Based on Gain and Phase Margins

The concepts of *gain margin* (GM) and *phase margin* (PM) are two very important design considerations for ensuring stability of poorly known processes. Since process parameters such as time delays, steady state gains and time constants are almost never known exactly, a gain margin between 1.7 and 4.0 and/or a phase margin between 30° and 45° provide necessary safety factor.

The open-loop amplitude ratio at critical frequency (AR_c) is unity for a marginally stable system and for a stable system it must be a fractional value, preferably between 0.25 and 0.6 which provides necessary GM according to Eq. 10-72.

Similarly, the phase angle at the gain-crossover frequency (ϕ_g) is –180° for a marginally stable system and for a stable system it must be between –135° and –150°, which provides necessary PM according to Eq. 10-73.

EXAMPLE 10.9

A closed-loop system is having

$$G_v = 1, G_m = 1, G_p = \frac{e^{-s}}{5s+1}.$$

(a) Obtain the ZN settings for a PID controller.

(b) What are the values of GM and PM for the PID controller of part (a)? Modify the obtained PID controller for acceptable GM and PM.

Solution

(a) From Eq. 10.62, we get

$$G_{PRC} = G_V G_P G_M = \frac{e^{-s}}{5s+1}$$

$$AR_{PRC} = \frac{1}{\sqrt{25\omega^2+1}}$$

$$\varphi_{OL} = \varphi_{PRC} = -\tan^{-1}(5\omega) - \omega \times \frac{180}{\pi} \text{ deg}$$

At critical frequency:

$$-180^\circ = \tan^{-1}(-5\omega_c) - \omega_c \times \frac{180}{\pi} = f(\omega_c)$$

By trial and error approach:

ω_c	1	2	1.8	1.75	1.7	1.69
$f(\omega_c)$	–136	–199	–187	–184	–180.7	–180.08

So, the critical or crossover frequency, $\omega_c = 1.69$ radians.

$$AR_{PRC}|_{\omega_c} = \frac{1}{\sqrt{25(1.69)^2+1}} = 0.1175$$

From Eq. 10-71, we get

$$K_{cu} = \frac{1}{AR_{PRC}|_{\omega_c}} = \frac{1}{0.1175} = 8.5$$

$$P_u = \frac{2\pi}{\omega_c} = \frac{2\pi}{1.69} = 3.7$$

Using ZN settings from Table 10.4, we get for the PID Controller:

$$K_c = K_{CU}/1.7 = 5$$

$$\tau_I = P_U/2 = 1.85$$

$$\tau_D = P_U/8 = 0.46$$

$$G_C = 5\left(1 + \frac{1}{1.85s} + 0.46s\right).$$

(b) The open-loop transfer function is given by

$$G_{OL} = G_C G_V G_P G_M = 5\left(1 + \frac{1}{1.85s} + 0.46s\right)\left(\frac{e^{-s}}{5s+1}\right)$$

$$AR_{OL} = 5\sqrt{\left[\left(0.46\omega - \frac{1}{1.85\omega}\right)^2 + 1\right]}\left(\frac{1}{\sqrt{25\omega^2+1}}\right)$$

$$\phi_{OL} = \tan^{-1}\left(0.46\omega - \frac{1}{1.85\omega}\right) - \tan^{-1}(5\omega) - \frac{180\omega}{\pi}$$

At critical frequency, we have

$$-180 = \tan^{-1}\left(0.46\omega_c - \frac{1}{1.85\omega_c}\right) - \tan^{-1}(5\omega_c) - \frac{180\omega_c}{\pi} = f(\omega_c)$$

By trial and error approach:

ω_c	1	2	3	2.5	2.4	2.37
$f(\omega_c)$	–140.6	–165.9	–207.9	–185.6	–181.4	–180.2

So, $\omega_c = 2.37$

$$AR_{OL}(\omega_c) = AR_C = 5\sqrt{\left[\left(0.46 \times 1.37 - \frac{1}{1.85 \times 1.37}\right)^2 + 1\right]}\left(\frac{1}{\sqrt{25 \times 1.37^2 + 1}}\right) = 0.742$$

Gain Margin, $GM = 1/AR_C = 1/0.742 = 1.35$.

As this $GM < 1.7$, it is not acceptable.

To find the PM, we need to find the gain crossover frequency (ω_g):

At ω_g, $AR_{OL} = 1$.

$$1 = 5\sqrt{\left[\left(0.46\omega_g - \frac{1}{1.85\omega_g}\right)^2 + 1\right]}\left(\frac{1}{\sqrt{25\omega_g^2 + 1}}\right) = f(\omega_g)$$

By trial and error approach:

ω_g	1	2	0.9	0.95	0.98
$f(\omega_g)$	0.98	0.5933	1.10	1.04	1.005

So, $\omega_g = 0.98$

$$\varphi_{OL}(\omega_g) = \varphi_g = \tan^{-1}\left(0.46\omega_g - \frac{1}{1.85\omega_g}\right) - \tan^{-1}(5\omega_g) - \frac{180\omega_g}{\pi} = -140^\circ$$

Phase margin, PM = $180 + \phi_g = 40^\circ$.

The phase margin is between 30° to 45°. So, there is no need to correct for phase margin.

The GM can be changed by changing K_c value. For $GM = 2.0$, $AR_C = 1/GM = 0.5$.

Now,

$$AR_C = 0.5 = K_C\sqrt{\left[\left(0.46\times1.37 - \frac{1}{1.85\times1.37}\right)^2 + 1\right]}\left(\frac{1}{\sqrt{25\times1.37^2 + 1}}\right)$$

Solving, we get $K_c = 3.37$.

The designed PID controller is

$$G_C = 3.37\left(1 + \frac{1}{1.85s} + 0.46s\right).$$

SUMMARY

Frequency response techniques are powerful tools for the design and analysis of feedback control systems. The frequency response characteristics of a process, its amplitude ratio AR and phase angle, characterize the dynamic behavior of the process and can be plotted as functions of frequency in Bode diagrams. The Bode stability criterion provides exact stability results for a wide variety of control problems, including processes with time delays. It also provides a convenient measure of relative stability, such as gain and phase margins. Control system design involves trade-offs between control system performance and robustness (see Appendix L). Modern control systems are typically designed using a model-based technique, such as those described in Chapter 11.

REFERENCES

Doyle, J. C., B. A. Francis, and A. R. Tannenbaum, *Feedback Control Theory*, Macmillan, New York, 2009.

Franklin, G. F., J. D. Powell, and A. Emami-Naeini, *Feedback Control of Dynamic Systems*, 7th ed., Prentice Hall, Upper Saddle River, NJ, 2014.

Hahn, J., T. Edison, and T. F. Edgar, A Note on Stability Analysis Using Bode Plots, *Chem. Eng. Educ.* **35**(3), 208 (2001).

Luyben, W. L., and M. L. Luyben, *Essentials of Process Control*, McGraw-Hill, New York, 1997, Chapter 11.

MacFarlane, A. G. J., The Development of Frequency Response Methods in Automatic Control, *IEEE Trans. Auto. Control*, **AC-24**, 250 (1979).

Maciejowski, J. M., *Multivariable Feedback Design*, Addison-Wesley, New York, 1989.

Ogunnaike, B. A., and W. H. Ray, *Process Dynamics, Modeling, and Control*, Oxford University Press, New York, 1993.

Seborg, D. E., T. F. Edgar, and D. A., Mellichamp, *Process Dynamics and Cantrol*, 2nd ed., John Wiley and Sons, Hoboken, NJ, 2004.

Skogestad, S., and I. Postlethwaite, *Multivariable Feedback Design: Analysis and Design*, 2d ed., John Wiley and Sons, Hoboken, NJ, 2005.

Tan, K. K., Q.-G. Wang, C. C. Hang, and T. Hägglund, *Advances in PID Control*, Springer, New York, 1999.

EXERCISES

10.1 A heat transfer process has the following transfer function between a temperature T (in °C) and an inlet flow rate q where the time constants have units of minutes:

$$\frac{T'(s)}{Q'(s)} = \frac{4(1-s)}{s(2s+1)}$$

If the flow rate varies sinusoidally with an amplitude of 2 L/min and a period of 0.5 min, what is the amplitude of the temperature signal after the transients have died out?

10.2 Using frequency response arguments, discuss how well $e^{-\theta s}$ can be approximated by a two-term Taylor series expansion, $1 - \theta s$. Compare your results with those given in Section 6.1.1 for a 1/1 Padé approximation.

10.3 A data acquisition system for environmental monitoring is used to record the temperature of an air stream as measured by a thermocouple. It shows an essentially sinusoidal variation after about 15 s. The maximum recorded temperature is 53°C, and the minimum is 49°C at 1.8 cycles per min. It is estimated that the thermocouple has a time constant of 5 s. Estimate the actual maximum and minimum air temperatures.

10.4 A perfectly stirred tank is used to heat a flowing liquid. The dynamic model is shown in Fig. E10.4.

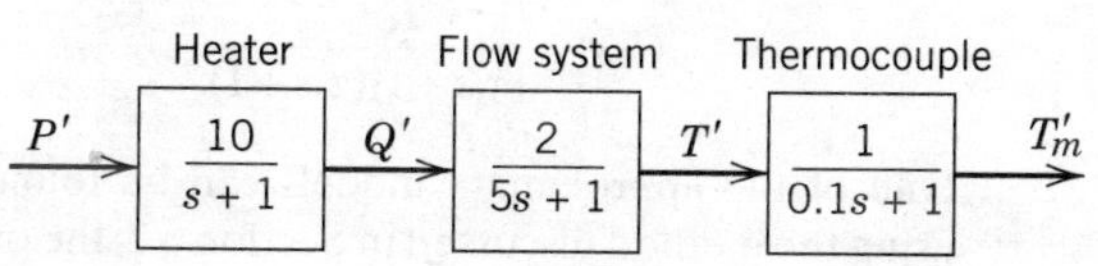

Figure E10.4

where:

P is the power applied to the heater

Q is the heating rate of the system

T is the actual temperature in the tank

T_m is the measured temperature time constants have units of min

A test has been made with P' varied sinusoidally as

$$P' = 0.5 \sin 0.2t$$

For these conditions, the measured temperature is

$$T'_m = 3.464 \sin(0.2t + \phi)$$

Find a value for the maximum error bound between T' and T'_m if the sinusoidal input has been applied for a long time.

10.5 Determine if the following processes can be made unstable by increasing the gain of a proportional controller K_c to a sufficiently large value using frequency response arguments:

(a) $G_v G_p G_m = \dfrac{3}{s+1}$

(b) $G_v G_p G_m = \dfrac{2}{(s+1)(2s+1)}$

(c) $G_v G_p G_m = \dfrac{1}{(s+1)(2s+1)(3s+1)}$

(d) $G_v G_p G_m = \dfrac{4e^{-s}}{2s+1}$

10.6 Two engineers are analyzing step-test data from a bioreactor. Engineer A says that the data indicate a second-order overdamped process, with time constants of 5 and 3 min but no time delay. Engineer B insists that the best fit is a FOPTD model, with $\tau = 7$ min and $\theta = 1$ min. Both engineers claim a proportional controller can be set at a large value for K_c to control the process and that stability is no problem. Based on their models, who is right, who is wrong, and why? Use a frequency-response argument.

10.7 Plot the Bode diagram ($0.1 \le \omega \le 100$) of the third-order transfer function,

$$G(s) = \frac{5}{(12s+1)(2s+1)(s+1)}$$

Find both the value of ω that yields a $-180°$ phase angle and the value of AR at that frequency.

10.8 Using MATLAB, plot the Bode diagram of the following transfer function:

$$G(s) = \frac{4(s+1)e^{-2s}}{(3s+1)(2s+1)}$$

Repeat for the situation where the time-delay term is replaced by a 1/1 Padé approximation. Discuss how the accuracy of the Padé approximation varies with frequency.

10.9 Two thermocouples, one of them a known standard, are placed in an air stream whose temperature is varying sinusoidally. The temperature responses of the two thermocouples are recorded at a number of frequencies, with the phase angle between the two measured temperatures as shown below. The standard is known to have first-order dynamics and a time constant of 0.15 min when operating in the air stream. From the data, show that the unknown thermocouple also is a first order and determine its time constant.

Frequency (cycles/min)	Phase Difference (deg)
0.05	4.5
0.1	8.7
0.2	16.0
0.1	24.5
0.8	26.5
1.0	25.0
2.0	16.7
4.0	9.2

10.10 Exercise 5.19 considered whether a two-tank liquid surge system provided better damping of step disturbances than a single-tank system with the same total volume. Reconsider this situation, this time with respect to sinusoidal disturbances; that is, determine which system better damps sinusoidal inputs of frequency ω. Does your answer depend on the value of ω?

10.11 A process has the transfer function of Eq. 5.84 with $K = 2$, $\tau_1 = 10$, $\tau_2 = 2$. If τ_a has the following values,

Case i: $\tau_a = 15$ Case iii: $\tau_a = 1$

Case ii: $\tau_a = 3$ Case iv: $\tau_a = -2$

Plot the composite amplitude ratio and phase angle curves on a single Bode plot for each of the four cases of numerator dynamics. What can you conclude concerning the importance of the zero location for the amplitude and phase characteristics of this second-order system?

10.12 Develop expressions for the amplitude ratio as a function of ω of each of the two forms of the PID controller:

(a) The parallel controller of Eq. 7-13.

(b) The series controller of Eq. 7-15.

Plot AR/K_c vs. $\omega\tau_D$ for each AR curve. Assume $\tau_1 = 4\tau_D$ and $\alpha = 0.1$.

For what region(s) of ω are the differences significant?

10.13 You are using proportional control ($G_c = K_c$) for a process with $G_v = \dfrac{3}{2s+1}$ and $G_p = \dfrac{0.5}{50s+1}$ (time constants in s). You have a choice of two measurements, both of which exhibit first-order dynamic behavior, $G_{m1} = \dfrac{2}{s+1}$ or $G_{m2} = \dfrac{2}{0.5s+1}$. Can G_c be made unstable for either process?
Which measurement is preferred for the best stability and performance properties? Why?

10.14 For the following statements, discuss whether they are always true, sometimes true, always false, or sometimes false. Cite evidence from this chapter.

(a) Increasing the controller gain speeds up the response for a set-point change.

(b) Increasing the controller gain always causes oscillation in the response to a setpoint change.

(c) Increasing the controller gain too much can cause instability in the control system.

(d) Selecting a large controller gain is a good idea in order to minimize offset.

10.15 Use arguments based on the phase angle in frequency response to determine if the following combinations of $G = G_vG_pG_m$ and G_c become unstable for some value of K_c.

(a) $G = \dfrac{1}{(4s+1)(2s+1)}$ $\quad G_c = K_c$

(b) $G = \dfrac{1}{(4s+1)(2s+1)}$ $\quad G_c = K_c\left(1 + \dfrac{1}{5s}\right)$

(c) $G = \dfrac{s+1}{(4s+1)(2s+1)}$ $\quad G_c = K_c\left(\dfrac{2s+1}{s}\right)$

(d) $G = \dfrac{1-s}{(4s+1)(2s+1)}$ $\quad G_c = K_c$

(e) $G = \dfrac{e^{-s}}{4s+1}$ $\quad G_c = K_c$

10.16 Plot the Bode diagram for a composite transfer function consisting of $G(s)$ in Exercise 10.8 multiplied by that of a parallel-form PID controller with $K_c = 0.21$, $\tau_I = 5$, and $\tau_D = 0.42$.

Repeat for a series PID controller with filter that employs the same settings. How different are these two diagrams? In particular, by how much do the two amplitude ratios differ when $\omega = \omega_c$?

10.17 For the process described by the transfer function

$$G(s) = \frac{12}{(8s+1)(2s+1)(0.4s+1)(0.1s+1)}$$

(a) Find second-order-plus-time-delay models that approximate $G(s)$ and are of the form

$$\hat{G}(s) = \frac{Ke^{-\theta s}}{(\tau_1 s+1)(\tau_2 s+1)}$$

One of the approximate models can be found by using the method discussed in Section 6.2; the other, by using a method from Chapter 6.5.1.

(b) Compare all three models (exact and approximate) in the frequency domain and a FOPTD model.

10.18 Obtain Bode plots for both the transfer function:

$$G(s) = \frac{10(2s+1)e^{-2s}}{(20s+1)(4s+1)(2s+1)}$$

and a FOPTD approximation obtained using the method discussed in Section 6.2. What do you conclude about the accuracy of the approximation relative to the original transfer function?

10.19 **(a)** Using the process, sensor, and valve transfer functions in Exercise 11.21, find the ultimate controller gain K_{cu} using a Bode plot. Using simulation, verify that values of $K_c > K_{cu}$ cause instability.

(b) Next fit a FOPTD model to G and tune a PI controller for a set-point change. What is the gain margin for the controller?

10.20 A process that can be modeled as a time delay (gain = 1) is controlled using a proportional feedback controller. The control valve and measurement device have negligible dynamics and steady-state gains of $K_v = 1$ and $K_m = 0.5$, respectively. After a small set-point change is made, a sustained oscillation occurs, which has a period of 10 min.

(a) What controller gain is being used? Explain.

(b) How large is the time-delay?

10.21 The block diagram of a conventional feedback control system contains the following transfer functions:

$$G_c = K_c\left(1+\frac{1}{5s}\right) \qquad G_v = 1$$

$$G_m = \frac{2}{s+1}$$

$$G_p = G_d = \frac{2.5e^{-2s}}{10s+1}$$

(a) Plot the Bode diagram for the open-loop transfer function.

(b) For what values of K_c is the system stable?

(c) If $K_c = 0.2$, what is the phase margin?

(d) What value of K_c will result in a gain margin of 1.7?

10.22 Consider the storage tank with sightglass in Fig. E10.22. The parameter values are $R_1 = 5$ min/m², $R_2 = 20$ min/m², $A_1 = 1$ m², $K_v = 0.07$ m³/mA, $A_2 = 0.08$ m², $K_m = 5$ mA/m, and $\tau_m = 0.5$ min.

(a) Suppose that R_2 is decreased to 5 min/m². Compare the old and new values of the ultimate gain and the critical frequency. Would you expect the control system performance to become better or worse? Justify your answer.

(b) If PI controller settings are calculated using the Ziegler-Nichols rules, what are the gain and phase margins? Assume $R_2 = 6.5$ min/m.

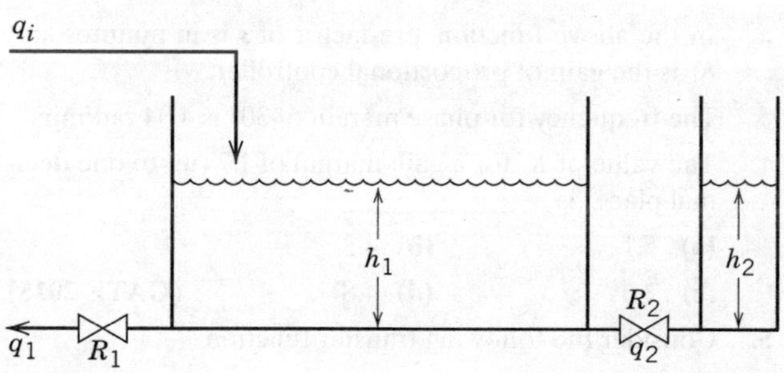

Figure E10.22

10.23 A process (including valve and sensor-transmitter) has the approximate transfer function, $G(s) = 2e^{-0.2s}/(s+1)$ with time constant and time delay in minutes. Determine PI controller settings and the corresponding gain margins by two methods:

(a) Direct synthesis ($\tau_c = 0.3$ min).

(b) Phase margin = 40° (assume $\tau_I = 0.5$ min).

(c) Simulate these two control systems for a unit step change in set-point. Which controller provides the better performance?

10.24 Consider the feedback control system in Fig. 10.8, and the following transfer functions:

$$G_c = K_c\left(\frac{2s+1}{0.1s+1}\right) \qquad G_v = \frac{2}{0.5s+1}$$

$$G_p = \frac{0.4}{s(5s+1)} \qquad G_d = \frac{3}{5s+1}$$

$$G_m = 1$$

(a) Plot a Bode diagram for the open-loop transfer function.

(b) Calculate the value of K_c that provides a phase margin of 30°.

(c) What is the gain margin when $K_c = 10$?

10.25 Hot and cold liquids are mixed at the junction of two pipes. The temperature of the resulting mixture is to be controlled using a control valve on the hot stream. The dynamics of the mixing process, control valve, and temperature sensor/transmitter are negligible and the sensor-transmitter gain is 6 mA/mA. Because the temperature sensor is located well downstream of the junction, an 8 s time delay occurs. There are no heat losses/gains for the downstream pipe.

(a) Draw a block diagram for the closed-loop system.

(b) Determine the Ziegler–Nichols settings (continuous cycling method) for both PI and PID controllers.

(c) For each controller, simulate the closed-loop responses for a unit step change in set point.

(d) Does the addition of derivative control action provide a significant improvement? Justify your answer.

10.26 For the process in Exercise 10.23, the measurement is to be filtered using a noise filter with transfer function $G_F(s) = 1/(0.2s+1)$. Would you expect this change to result in better or worse control system performance? Compare the ultimate gains and critical frequencies with and without the filter. Justify your answer.

10.27 The dynamic behavior of the heat exchanger shown in Fig. E10.27 can be described by the following transfer functions (H. S. Wilson and L. M. Zoss, *ISA J.*, 9, 59 (1962)):

Process:

$$\frac{T'}{W_s'} = \frac{2\text{°C/kg/min}}{(0.432s+1)(0.017s+1)}$$

Control valve:

$$\frac{X'}{P'} = \frac{0.047\ \text{m/kPa}}{0.083s+1} \qquad \frac{W_s'}{X'} = 112\frac{\text{kg}}{\text{min m}}$$

Temperature sensor-transmitter:

$$\frac{P'}{T'} = \frac{0.12\ \text{kPa/°C}}{0.024s+1}$$

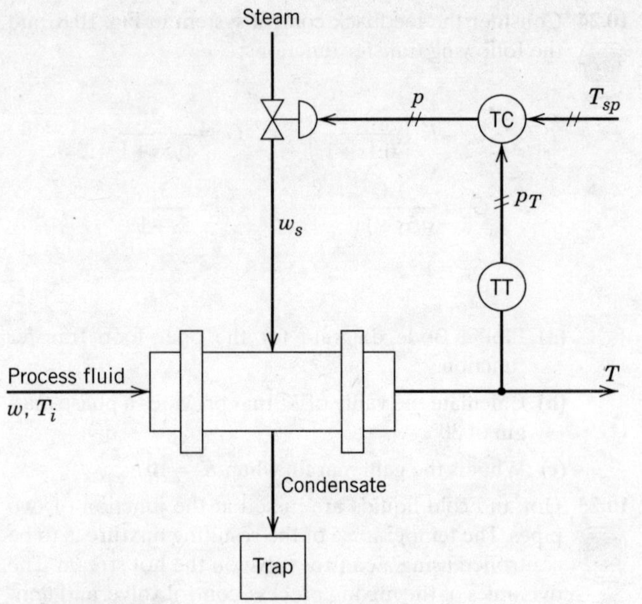

Figure E10.27

The valve lift x is measured in inches. Other symbols are defined in Fig. E10.27.

(a) Find the Ziegler–Nichols settings for a PI controller.

(b) Calculate the corresponding gain and phase margins.

10.28 Consider the control problem of Exercise 10.27 and a PI controller with $K_c = 5$ and $\tau_I = 0.3$ min.

(a) Plot the Nyquist diagram for the open-loop system.

(b) Determine the gain margin from the Nyquist plot.

MULTIPLE CHOICE QUESTIONS

1. For the closed loop system shown in figure, the phase margin (in degrees) is ________ (rounded off to one decimal place).

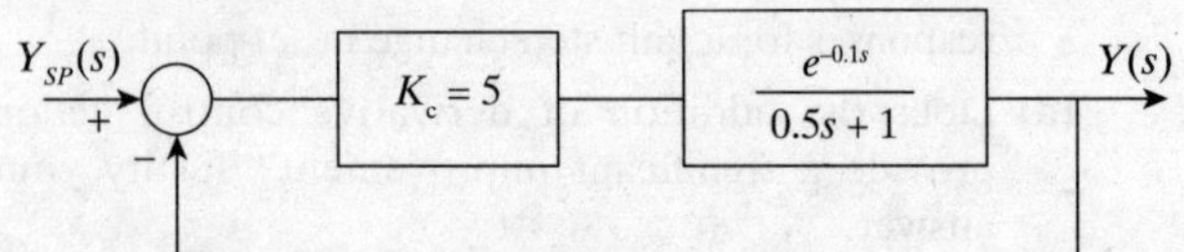

(a) 44.8 **(b)** 45.4
(c) 30.5 **(d)** 38.2 **(GATE 2019)**

2. Consider the following transfer function:

$$G(s) = \frac{3}{(5s+1)^2}$$

where the natural period of oscillation is in min. The amplitude ratio at a frequency of 0.5 rad/min is ________ (rounded off to second decimal place).

(a) 0.41 **(b)** 0.51
(c) 0.31 **(d)** 0.38 **(GATE 2018)**

3. The open loop transfer function of a process with a pro-proportional controller (gain K_C) is

$$G_{OL} = K_C \frac{e^{-2s}}{s}$$

Based on the Bode criterion for closed-loop stability, the ultimate gain of the controller, rounded to 2 decimal places, is ________.

(a) 0.87 **(b)** 0.75
(c) 0.78 **(d)** 0.72 **(GATE 2017)**

4. Consider a control system with the open loop transfer function given by:

$$G_{OL}(s) = \frac{K_c e^{-0.3c}}{1.5s+1}$$

In the above function, pre-factor of s is in minutes and K_c is the gain of proportional controller.

The frequency for phase margin of 30° is 4.04 rad/min.

The value of K_c for a gain margin of 1.7 (up to one decimal place) is ________

(a) 5.1 **(b)** 4.2
(c) 3.6 **(d)** 4.8 **(GATE 2015)**

5. Consider the following transfer function

$$G_p(s) = \frac{5}{(2s+1)^4}$$

(Note: The unit of the process time constant is in seconds)
The crossover frequency (in rad/s) of the process is

(a) 20 **(b)** 0.1
(c) 0.5 **(d)** 0.05 **(GATE 2013)**

6. Consider the following transfer function

$$G_p(s) = \frac{5}{(2s+1)^4}$$

(Note: The unit of the process time constant is in seconds) For the computation of Ziegler–Nichols settings, the ultimate period (in s/cycle) and the ultimate gain are

(a) π and 0.8, respectively.

(b) 4π and 0.8, respectively.

(c) 4π and 1.25, respectively.

(d) π and 1.25, respectively. **(GATE 2013)**

7. The Bode stability criterion is applicable when

(a) Gain and phase curves decrease continuously with frequency.

(b) Gain curve increases and phase curve decreases with frequency.

(c) Gain curve and phase curve both increase with frequency.

(d) Gain curve decreases and phase curve increases with frequency. **(GATE 2013)**

8. The transfer function, $G(s)$, whose asymptotic Bode diagram is shown below, is

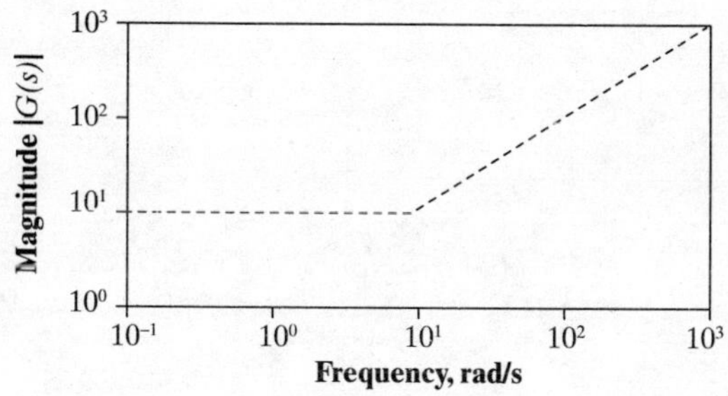

(a) $10s + 1$ **(b)** $s - 10$

(c) $s + 10$ **(d)** $10s - 1$ **(GATE 2010)**

9. The cross-over frequency associated with a feedback loop employing a proportional controller to control the process represented by the transfer function

$$G_p(s) = \frac{2e^{-z}}{(\tau s + 1)^2}$$

(units of time is minutes) is found to be 0.6 rad/min. Assume that the measurement and value transfer functions are unity. The time constant, τ (in minutes) is

(a) 1.14 **(b)** 1.92

(c) 3.223 **(d)** 5.39 **(GATE 2008)**

10. The cross-over frequency associated with a feedback loop employing a proportional controller to control the process represented by the transfer function

$$G_p(s) = \frac{2e^{-z}}{(\tau s + 1)^2}$$

(units of time is minutes) is found to be 0.6 rad/min. Assume that the measurement and value transfer functions are unity.

If the control loop is to operate at a gain margin of 2.0, the gain of the proportional controller must equal

(a) 0.85 **(b)** 2.87

(c) 3.39 **(d)** 11.50 **(GATE 2008)**

11. For the system shown below,

$$G_1(s) = \frac{1}{\tau_1 s + 1},\ G_2(s) = \frac{1}{\tau_2 s + 1} \text{ and } \tau_2 = 2\tau_1$$

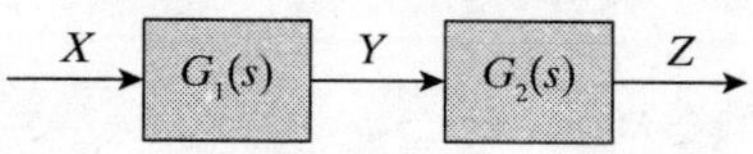

When the system is excited by the sinusoidal input $X(t) = \sin \omega t$, the intermediate response Y is given by $Y = A \sin(\omega t + \phi)$.

If the response of Y lags behind the input X by 45° and $\tau_1 = 1$, then the input frequency ω is

(a) 1 **(b)** $\pi/4$

(c) zero **(d)** –1 **(GATE 2006)**

12. For the system shown below,

$$G_1(s) = \frac{1}{\tau_1 s + 1},\ G_2(s) = \frac{1}{\tau_2 s + 1} \text{ and } \tau_2 = 2\tau_1$$

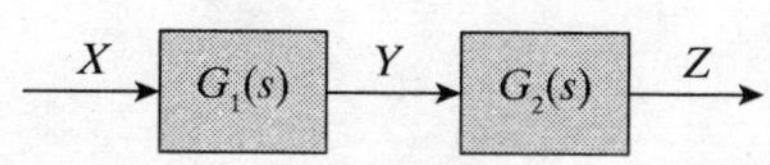

When the system is excited by the sinusoidal input $X(t) = \sin \omega t$, the intermediate response Y is given by $Y = A \sin(\omega t + \phi)$.

For the same input, the amplitude of the output Z will be

(a) 1.00 **(b)** 0.62

(c) 0.42 **(d)** 0.32 **(GATE 2006)**

ANSWER KEY

1. (b)	**2.** (a)	**3.** (c)	**4.** (c)	**5.** (c)	**6.** (b)
7. (a)	**8.** (c)	**9.** (d)	**10.** (b)	**11.** (a)	**12.** (d)

CHAPTER 11

PID Controller Design, Tuning, and Troubleshooting

CHAPTER CONTENTS

Several examples in Chapter 9 demonstrated that the controller settings have a major impact on closed-loop stability and performance. For most control problems, the closed-loop system is stable for a wide range of controller settings. Consequently, there is an opportunity to specify controller settings so that the desired control system performance is achieved.

To further illustrate the influence of controller settings, we consider a simple closed-loop system that consists of a first-order-plus-time-delay (FOPTD) model and a PI controller. The simulation results in Fig. 11.1 show the disturbance responses for nine combinations of the controller gain K_c and integral time τ_I. As K_c increases or τ_I decreases, the response to the step disturbance becomes more aggressive. Controller 1 produces an unstable response, while Controller 5 arguably provides the best response. This example demonstrates that controller settings can be adjusted to achieve the desired closed-loop system performance, a procedure referred to as *controller tuning*.

It is appropriate to devote an entire chapter to PID control for several reasons. First, it is the predominant control technique used in the process industries. Several surveys have reported that at least 95% of the control loops in industrial plants use PID control, and most of these controllers are actually PI controllers. Second, using feedback controllers such as PID can be worse than using no control if the controller is poorly designed or tuned. Third, there is a voluminous literature concerning PID controller design and tuning.[1] Thus it is important to condense this huge amount of material to key concepts, important methods, and insightful analyses. Four books (Shinskey, 1994; Åström and Hägglund, 2006; Blevins et al., 2013; McMillan, 2015) and four influential journal articles (Ziegler and Nichols, 1942; Rivera et al., 1986; Skogestad, 2003; Åström and

[1] An internet survey for "PID control" in July, 2015, resulted in more than 8100 publications. Tabulations of 79 patents and 45 software packages for PID control are also available (Ang et al., 2005).

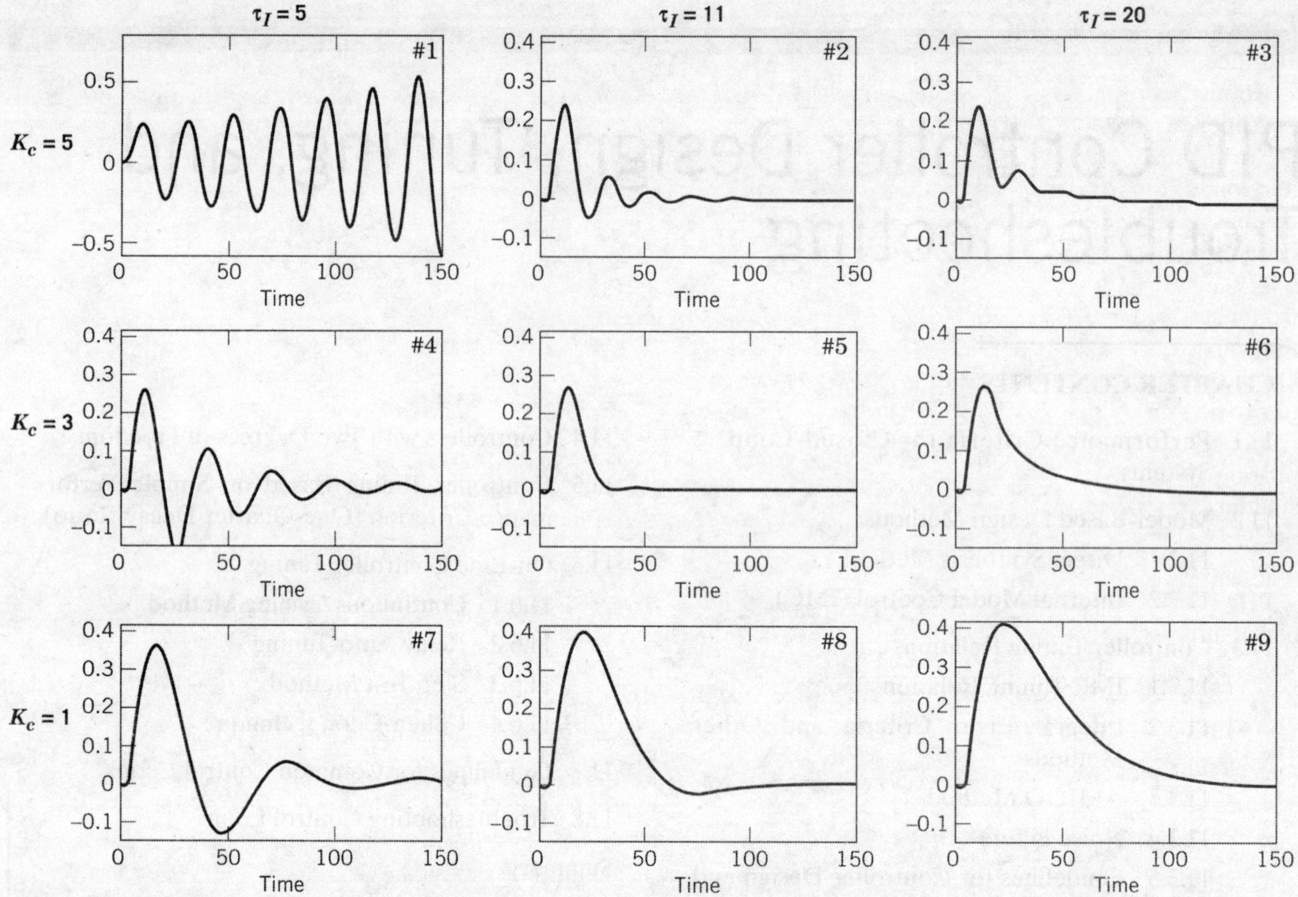

Figure 11.1 Unit-step disturbance responses for the candidate controllers (FOPTD Model: $K = 1, \theta = 4, \tau = 20$) and $G = G_d$.

Hägglund, 1984) are important references. They are cited often in this chapter.

This chapter considers PID controllers based on transfer function models and transient response criteria. Alternative frequency response design methods are considered in Chapter 10. While PID control is of tremendous importance and widely used, there are critical process control problems where it is inadequate (Bauer and Craig, 2008). For these situations, the advanced control strategies in Chapters 10 and 12–15 should be considered.

11.1 PERFORMANCE CRITERIA FOR CLOSED-LOOP SYSTEMS

The function of a feedback control system is to ensure that the closed-loop system has desirable dynamic and steady-state response characteristics. Ideally, we would like the closed-loop system to satisfy the following performance criteria:

1. The closed-loop system must be stable
2. The effects of disturbances are minimized, providing good *disturbance rejection*
3. Rapid, smooth responses to set-point changes are obtained, that is, good *set-point tracking*
4. Steady-state error (offset) is eliminated
5. Excessive control action is avoided
6. The control system is robust, that is, insensitive to changes in process conditions and to inaccuracies in the process model

In typical control applications, it is not possible to achieve all of these goals simultaneously, because they involve inherent conflicts and tradeoffs. The tradeoffs must balance two important objectives, *performance* and *robustness*. A control system exhibits a high degree of performance if it provides rapid and smooth responses to disturbances and set-point changes with little, if any, oscillation. A control system is *robust* if it provides satisfactory performance for a wide range of process conditions and for a reasonable degree of model inaccuracy. Robustness can be achieved by choosing conservative controller settings (typically, small values of K_c and large values of τ_I), but this choice tends to result in poor performance.[2] Thus, conservative controller settings sacrifice performance in order to achieve robustness.

Robustness analysis for PID controllers is considered in Appendix L and in the control engineering literature (Åström and Hägglund, 2006; Kristiansson and Lennartson, 2006). Detailed analyses of performance-robustness tradeoffs for PID controllers have recently been published (Garpinger and Hägglund, 2015).

A second type of tradeoff occurs because PID controller settings that provide excellent disturbance rejection can produce large overshoots for set-point changes. On the other hand, if the controller settings are specified to provide excellent set-point tracking, the disturbance responses can be very sluggish. Thus, a tradeoff between set-point tracking and disturbance rejection occurs for standard PID controllers. Fortunately, this tradeoff can be avoided by using a *controller with two degrees of freedom*, as shown in Section 11.4.

PID controller settings can be determined by a number of alternative techniques:

1. Direct Synthesis (DS) method
2. Internal Model Control (IMC) method
3. Controller tuning relations
4. Frequency response techniques
5. Computer simulation
6. On-line tuning after the control system is installed

Because Methods 1–5 are based on process models, they can be used to specify controller settings before the control system is installed. However, for important control loops, these initial controller settings are typically adjusted after the control system is installed. This *on-line tuning* is based on simple experimental tests that are often required because the process models used to calculate the preliminary controller settings are not exact. Consequently, the objective for Methods 1–5 is to provide good initial controller settings that can subsequently be fine-tuned on-line, if necessary. Because on-line tuning can be time-consuming, it is very useful to have good initial controller settings in order to minimize the required time and effort.

Methods 1 and 2 are based on simple transfer function models and will be considered in Section 11.2. The PID controller tuning relations of Method 3 are analytical expressions and correlations. They are considered in Section 11.3. Design techniques based on frequency response analysis (Method 4) are the subject of Chapter 10. Computer simulation of the controlled process (Method 5) can provide considerable insight into dynamic behavior and control system performance. In particular, software such as MATLAB and LabVIEW facilitates the comparison of alternative control strategies and different controller settings. (See Appendices B and C and Doyle III (2000).) On-line controller tuning (Method 6) is considered in Section 11.6.

A comparison of PID tuning relations in Section 11.7 and an introduction to the important practical problem of *troubleshooting control loops* in Section 11.8 conclude this chapter.

11.2 MODEL-BASED DESIGN METHODS

If a reasonably accurate dynamic model of the process is available, it is advantageous to base the controller design on the process model. A wide variety of model-based design strategies are available for designing PID controllers. In this section, we consider two important model-based design methods that are especially useful in process control. Model-based techniques can also be used to design feedforward controllers (Chapter 12) and advanced control systems (Chapters 12, 13, and 15).

11.2.1 Direct Synthesis Method

In the Direct Synthesis (DS) method, the feedback controller design is based on a process model and a desired closed-loop transfer function. The latter is usually specified for set-point changes, but closed-loop disturbance transfer functions can also be utilized (Chen and Seborg, 2002). In the DS method, an analytical expression for the controller transfer function is derived. Although the resulting feedback controller does not always have a PID structure, the DS method produces PI or PID controllers for many common process models, as will be demonstrated in this section and Section 11.2.2. An important advantage of the DS approach is that it provides valuable insight into the relationship between the process model and the resulting controller.

The DS method was developed in the 1950s. A special case, the Lambda tuning method, was introduced into the process control literature in the mid-1960s. The Internal Model Control (IMC) method was developed in the 1980s (Garcia and Morari, 1982) and is closely related to the DS method. The IMC approach has generated considerable interest in the process control community, especially after IMC tuning relations for PID control were published (Rivera et al., 1986). Today, the IMC and the Lambda tuning methods are widely used for process control applications and are described in

[2] As indicated in Chapter 7, values of K_c can be either positive or negative. Thus statements like "small (or large) values of K_c" should actually include the additional phrase, "in absolute value." For convenience in this and later chapters, we assume that $K_c > 0$ and omit this qualifying phrase.

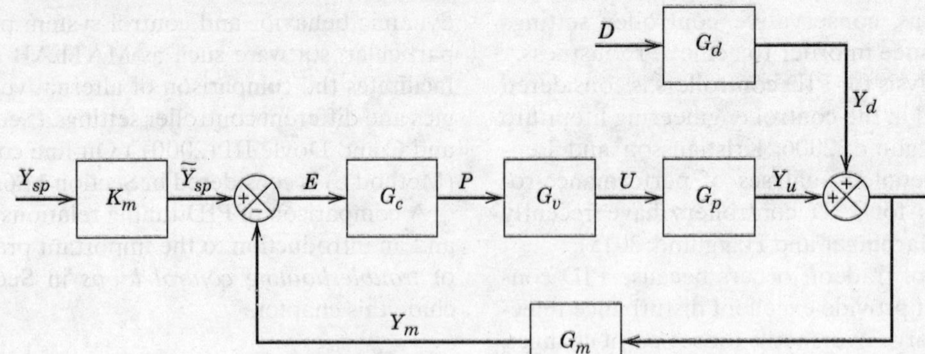

Figure 11.2 Block diagram for a standard feedback control system.

Section 11.3.3. An informative history of the development of the DS approach and related methods has been summarized in Chapter 6 of the book by Åström and Hägglund (2006).

As a starting point for the analysis, consider the block diagram of a feedback control system in Fig. 11.2. The closed-loop transfer function for set-point changes was derived in Section 9.2:

$$\frac{Y}{Y_{sp}} = \frac{K_m G_c G_v G_p}{1 + G_c G_v G_p G_m} \tag{11-1}$$

For simplicity, let $G \triangleq G_v G_p G_m$ and assume that $G_m = K_m$. Then Eq. 11-1 reduces to[3]

$$\frac{Y}{Y_{sp}} = \frac{G_c G}{1 + G_c G} \tag{11-2}$$

Rearranging and solving for G_c gives an expression for the ideal feedback controller:

$$G_c = \frac{1}{G}\left(\frac{Y/Y_{sp}}{1 - Y/Y_{sp}}\right) \tag{11-3a}$$

Equation 11-3a cannot be used for controller design, because the closed-loop transfer function Y/Y_{sp} is not known *a priori*. Also, it is useful to distinguish between the actual process G and the model, $\tilde{G}$, that provides an approximation of the process behavior. A practical design equation can be derived by replacing the unknown G by $\tilde{G}$, and Y/Y_{sp} by a *desired closed-loop transfer function*, $(Y/Y_{sp})_d$:

$$G_c = \frac{1}{\tilde{G}}\left[\frac{(Y/Y_{sp})_d}{1 - (Y/Y_{sp})_d}\right] \tag{11-3b}$$

The specification of $(Y/Y_{sp})_d$ is the key design decision and will be considered later in this section. Note that the controller transfer function in Eq. 11-3b contains the inverse of the process model due to the $1/\tilde{G}$ term. This feature is a distinguishing characteristic of model-based control.

Desired Closed-Loop Transfer Function

The performance of the DS controller in Eq. 11-3b strongly depends on the specification of the desired closed-loop transfer function, $(Y/Y_{sp})_d$. Ideally, we would like to have $(Y/Y_{sp})_d = 1$ so that the controlled variable tracks set-point changes instantaneously without any error. However, this ideal situation, called *perfect control*, cannot be achieved by feedback control because the controller does not respond until $e \neq 0$. For processes without time delays, the first-order model in Eq. 11-4 is a more reasonable choice

$$\left(\frac{Y}{Y_{sp}}\right)_d = \frac{1}{\tau_c s + 1} \tag{11-4}$$

where τ_c is the *desired closed-loop time constant*. This model has a settling time of ~$5\tau_c$, as shown in Section 5.4. Because the steady-state gain is one, no offset occurs for set-point changes. By substituting Eq. 11-4 into Eq. 11-3b and solving for G_c, the controller design equation becomes

$$G_c = \frac{1}{\tilde{G}}\frac{1}{\tau_c s} \tag{11-5}$$

The $1/\tau_c s$ term provides integral control action and thus eliminates offset. Design parameter τ_c provides a convenient controller tuning parameter that can be used to make the controller more aggressive (small τ_c) or less aggressive (large τ_c).

[3]We use the symbols G and G_c to denote $G(s)$ and $G_c(s)$, for the sake of simplicity.

If the process transfer function contains a known time delay θ, a reasonable choice for the desired closed-loop transfer function is

$$\left(\frac{Y}{Y_{sp}}\right)_d = \frac{e^{-\theta s}}{\tau_c s+1} \tag{11-6}$$

The time-delay term in Eq. 11-6 is essential, because it is physically impossible for the controlled variable to respond to a set-point change, before $t = \theta$. If the time delay is unknown, θ must be replaced by an estimate. Combining Eqs. 11-6 and 11-3b gives

$$G_c = \frac{1}{\widetilde{G}}\frac{e^{-\theta s}}{\tau_c s+1-e^{-\theta s}} \tag{11-7}$$

Although this controller is not in a standard PID form, it is physically realizable.

Next, we show that the design equation in Eq. 11-7 can be used to derive PID controllers for simple process models. The derivation is based on approximating the time-delay term in the denominator of Eq. 11-7 with a truncated Taylor series expansion:

$$e^{-\theta s} \approx 1-\theta s \tag{11-8}$$

Substituting Eq. 11-8 into the denominator of Eq. 11-7 and rearranging gives

$$G_c = \frac{1}{\widetilde{G}}\frac{e^{-\theta s}}{(\tau_c+\theta)s} \tag{11-9}$$

Note that this controller also contains integral control action.

Time-delay approximations are less accurate when the time delay is relatively large compared to the dominant time constant of the process. Note that it is not necessary to approximate the time-delay term in the numerator, because it is canceled by the identical term in $\widetilde{G}$, when the time delay is known exactly.

Next, we derive controllers for two important process models. For each derivation, we assume that the model is perfect ($\widetilde{G} = G$).

First-Order-Plus-Time-Delay (FOPTD) Model. Consider the standard FOPTD model,

$$\widetilde{G}(s) = \frac{Ke^{-\theta s}}{\tau s+1} \tag{11-10}$$

Substituting Eq. 11-10 into Eq. 11-9 and rearranging gives a PI controller, $G_c = K_c(1 + 1/\tau_I s)$, with

$$K_c = \frac{1}{K}\frac{\tau}{\tau_c+\theta}, \quad \tau_I = \tau \tag{11-11}$$

The expressions for the PI controller settings in Eq. 11-11 provide considerable insight. Controller gain K_c depends inversely on model gain K, which is reasonable based on the stability analysis in Chapter 9. In particular, if the product K_cK is constant, the characteristic equation and stability characteristics of the closed-loop system do not change. It is also reasonable that $\tau_I = \tau$, because slow processes have large values of τ, and thus τ_I should also be large for satisfactory control. As τ_c decreases, K_c increases, because a faster set-point response requires more strenuous control action and thus a larger value of K_c. The time delay θ imposes an upper: limit on K_c, even for the limiting case where $\tau_c \to 0$. By contrast, K_c becomes unbounded when $\theta = 0$ and $\tau_c \to 0$.

Second-Order-Plus-Time-Delay (SOPTD) Model. Consider a second-order-plus-time-delay (SOPTD) model,

$$\widetilde{G}(s) = \frac{Ke^{-\theta s}}{(\tau_1 s+1)(\tau_2 s+1)} \tag{11-12}$$

Substitution into Eq. 11-9 and rearrangement gives a PID controller in parallel form (cf. Chapter 7),

$$G_c = K_c\left(1+\frac{1}{\tau_I s}+\tau_D s\right) \tag{11-13}$$

where

$$K_c = \frac{1}{K}\frac{\tau_1+\tau_2}{\tau_c+\theta}, \quad \tau_I = \tau_1+\tau_2, \quad \tau_D = \frac{\tau_1\tau_2}{\tau_1+\tau_2} \tag{11-14}$$

The tuning relations in Eq. 11-14 indicate that for large values of θ, K_c decreases, but τ_I and τ_D do not. Again, the time delay imposes an upper limit on K_c as $\tau_c \to 0$.

The controller settings in Eqs. 11-11 and 11-14 become more conservative (smaller K_c) as τ_c increases. If θ is relatively large (e.g., $\theta/\tau_1 > 0.5$), a conservative choice of τ_c is prudent, because the controller design equations are based on the time-delay approximation in Eq. 11-8.

A number of guidelines for choosing τ_c that are applicable to both the Direct Synthesis method and the Internal Model Control method are presented in Section 11.2.2.

EXAMPLE 11.1

Use the DS design method to calculate PID controller settings for the process:

$$G = \frac{2e^{-s}}{(10s+1)(5s+1)}$$

Consider three values of the desired closed-loop time constant: $\tau_c = 1, 3,$ and 10. Evaluate the controllers for unit step changes in both the set point and the disturbance, assuming that $G_d = G$. Repeat the evaluation for two cases:

(a) The process model is perfect ($\widetilde{G} = G$).

(b) The model gain is incorrect, $\widetilde{K} = 0.9$, instead of the actual value, $K = 2$. Thus,

$$\widetilde{G} = \frac{0.9e^{-s}}{(10s+1)(5s+1)}$$

Solution

The controller settings from Eq. 11-14 are

	$\tau_c = 1$	$\tau_c = 3$	$\tau_c = 10$
$K_c(\widetilde{K} = 2)$	3.75	1.88	0.68
$K_c(\widetilde{K} = 0.9)$	8.33	4.17	1.51
τ_I	15	15	15
τ_D	3.33	3.33	3.33

The values of K_c decrease as τ_c increases, but the values of τ_I and τ_D do not change, as indicated by Eq. 11-14. Although some of these controller settings have been reported with three significant figures for purposes of comparison, calculating a third digit is not necessary in practice. For example, controllers with K_c values of 8.33 and 8.3 would produce essentially the same closed-loop responses.

Figures 11.3 and 11.4 compare the closed-loop responses for the three values of τ_c. As τ_c increases, the responses become more sluggish, and the maximum deviation is larger after the disturbance occurs at $\tau = 80$. For case (b), when the model gain is 0.9, about 50% too low, the closed-loop response for $\tau_c = 1$ in Fig. 11.4 is excessively oscillatory and would even become unstable if $\widetilde{K} = 0.8$ had been considered. The disturbance responses for $\tau_c = 3$ and $\tau_c = 10$ in Fig. 11.4 are actually better than the corresponding responses in Fig. 11.3 because the former have shorter settling times and smaller maximum deviations. This improvement is due to the larger values of K_c for case (b).

The Simulink diagram for this example is quite simple, as shown in Fig. 11.5. (See Appendix B for a Simulink tutorial.) However, the simulation results for Figs. 11.3 and 11.4 were generated using a modified controller that eliminated derivative kick (see Chapter 7).

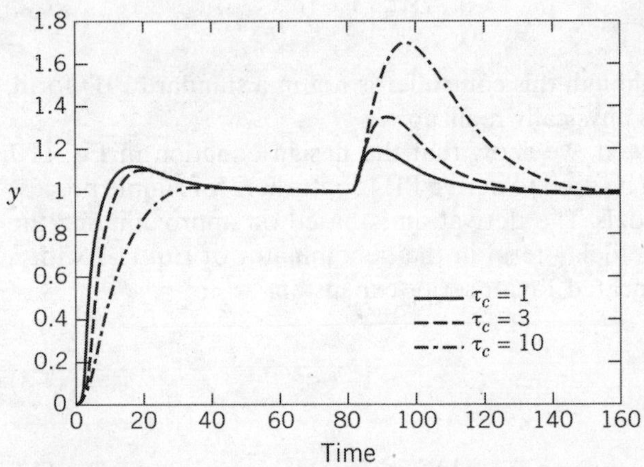

Figure 11.3 Simulation results for Example 11.1 (a): correct model gain.

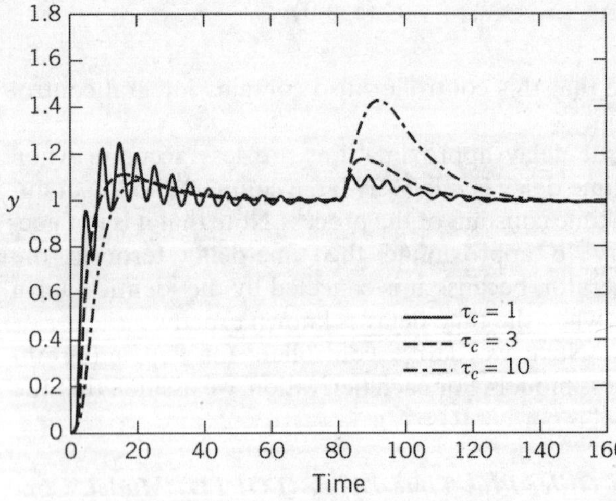

Figure 11.4 Simulation results for Example 11.1 (b): incorrect model gain.

The specification of the desired closed-loop transfer function, $(Y/Y_{sp})_d$, should be based on the assumed process model, as well as the desired set-point response. The FOPTD model in Eq. 11-6 is a reasonable choice for many processes but not all. For example, suppose that the process model contains a right-half plane zero term denoted by $(1 - \tau_a s)$ where $\tau_a > 0$ (cf. Section 4.5). Then if Eq. 11-6 is selected, the DS controller will have the $(1 - \tau_a s)$ term in its denominator and thus be unstable, a very undesirable feature. This problem can be avoided by replacing Eq. 11-6 with Eq. 11-15:

$$\left(\frac{Y}{Y_{sp}}\right)_d = \frac{(1-\tau_a s)e^{-\theta s}}{\tau_c s+1} \tag{11-15}$$

The DS approach should not be used directly for process models with unstable poles. However, it can be applied if the model is first stabilized by an additional feedback control loop.

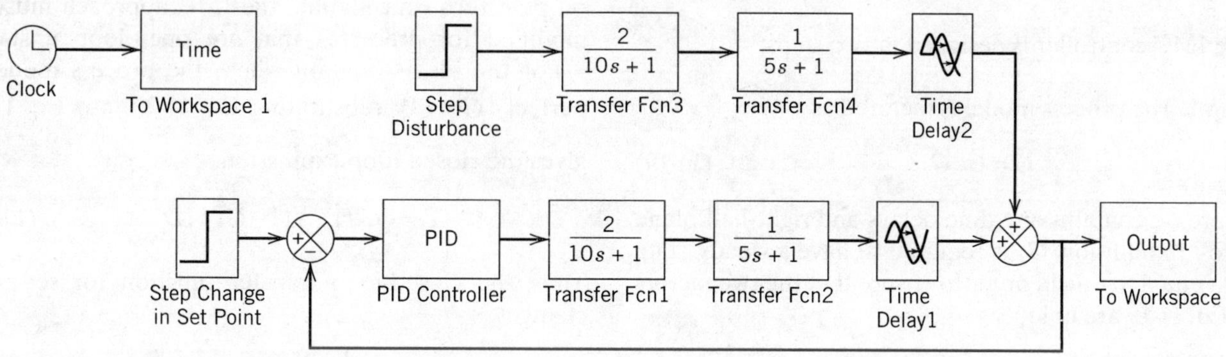

Figure 11.5 Simulink diagram for Example 11.1.

11.2.2 Internal Model Control (IMC)

A more comprehensive model-based design method, *Internal Model Control* (*IMC*), was developed by Morari and coworkers (Garcia and Morari, 1982; Rivera et al., 1986). The IMC method, like the DS method, is based on an assumed process model and leads to analytical expressions for the controller settings. These two design methods are closely related and produce identical controllers if the design parameters are specified in a consistent manner. However, the IMC approach has the advantage that it allows model uncertainty and tradeoffs between performance and robustness to be considered in a more systematic fashion.

Simplified block diagrams for conventional feedback control and IMC are compared in Fig. 11.6. The IMC method is based on the simplified block diagram shown in Fig. 11.6*b*. A process model $\tilde{G}$ and the controller output P are used to calculate the model response, $\tilde{Y}$. The model response is subtracted from the actual response Y, and the difference, $Y - \tilde{Y}$, is used as the input signal to the IMC controller, G_c^*. In general, $Y \# \tilde{Y}$ due to modeling errors $(\tilde{G} \# G)$ and unknown disturbances $(D \neq 0)$ that are not accounted for in the model.

It can be shown that the two block diagrams are equivalent if controllers G_c and G_c^* satisfy the relation

$$G_c = \frac{G_c^*}{1 - G_c^* \tilde{G}} \tag{11-16}$$

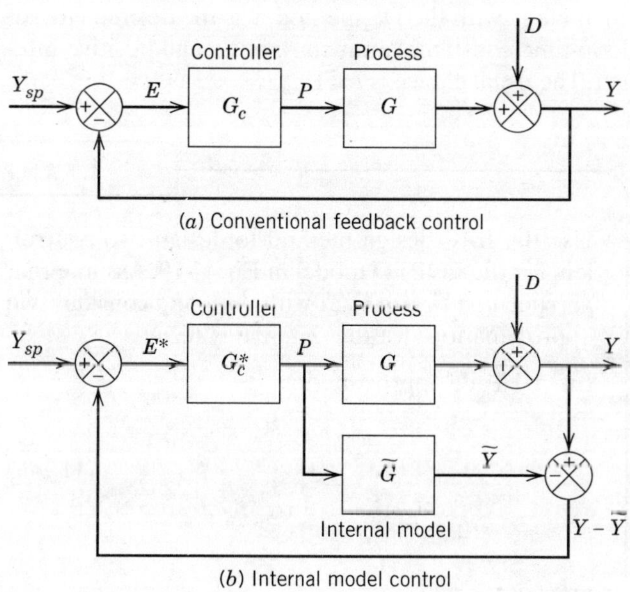

Figure 11.6 Feedback control strategies.

Thus, any IMC controller G_c^* is equivalent to a conventional feedback controller G_c, and vice versa.

The following closed-loop relation for IMC can be derived from Fig. 11.6*b* using the block diagram algebra of Chapter 9:

$$Y = \frac{G_c^* G}{1+G_c^*(G-\widetilde{G})} Y_{sp} + \frac{1-G_c^*\widetilde{G}}{1+G_c^*(G-\widetilde{G})} D \qquad (11\text{-}17)$$

For the special case of a perfect model, $\widetilde{G} = G$, Eq. 11-17 reduces to

$$Y = G_c^* G Y_{sp} + (1-G_c^* G) D \qquad (11\text{-}18)$$

The IMC controller is designed in two steps:

Step 1. The process model is factored as

$$\widetilde{G} = \widetilde{G}_+ \widetilde{G}_- \qquad (11\text{-}19)$$

where $\widetilde{G}_+$ contains any time delays and right-half plane zeros. In addition, $\widetilde{G}_+$ is required to have a steady-state gain equal to one in order to ensure that the two factors in Eq. 11-19 are unique.

Step 2. The IMC controller is specified as

$$G_c^* = \frac{1}{\widetilde{G}_-} f \qquad (11\text{-}20)$$

where f is a *low-pass filter* with a steady-state gain of one.[4] It typically has the form

$$f = \frac{1}{(\tau_c s+1)^r} \qquad (11\text{-}21)$$

In analogy with the DS method, τ_c is the desired closed-loop time constant. Parameter r is a small positive integer. The usual choice is $r = 1$.

Note that the IMC controller in Eq. 11-20 is based on the invertible part of the process model, $\widetilde{G}_-$, rather than the entire model, $\widetilde{G}$. If $\widetilde{G}$ had been used, the controller could contain a prediction term $e^{+\theta s}$ (if $\widetilde{G}_+$ contains a time delay θ), or an unstable pole (if $\widetilde{G}_+$ contains a right-half plane zero). Thus, by employing the factorization of Eq. 11-19 and the design equations in Eqs. 11-20 and 11-21, the resulting IMC controller G_c^* is guaranteed to be physically realizable and stable. In general, the noninvertible part of the model, $\widetilde{G}_+$, places limitations on the performance that can be achieved by any control system. Because the standard IMC design method is based on pole-zero cancellation, the IMC approach must be modified for processes that are open-loop unstable.

For the ideal situation where the process model is perfect $(\widetilde{G} = G)$, substituting Eq. 11-20 into Eq. 11-18 gives the closed-loop expression

$$Y = \widetilde{G}_+ f Y_{sp} + (1-\widetilde{G}_+ f) D \qquad (11\text{-}22)$$

Thus, the closed-loop transfer function for set-point changes is

$$\frac{Y}{Y_{sp}} = \widetilde{G}_+ f \qquad (11\text{-}23)$$

The IMC and Direct Synthesis (DS) design methods can produce equivalent controllers and identical closed-loop responses, even when modeling errors are present. This equivalence occurs if the desired transfer function $(Y/Y_{sp})_d$ in Eq. 11-3b is set equal to Y/Y_{sp} in Eq. 11-23. Recall that Eq. 11-16 shows how to convert G_c^* to the equivalent G_c.

The IMC design method is illustrated in the following example.

Example 11.2

Use the IMC design method to design two controllers for the FOPTD model in Eq. 11-10. Assume that f is specified by Eq. 11-21 with $r = 1$, and consider two approximations for the time-delay term:

(a) 1/1 Padé approximation:

$$e^{-\theta s} \cong \frac{1-\frac{\theta}{2}s}{1+\frac{\theta}{2}s} \qquad (11\text{-}24a)$$

(b) First-order Taylor series approximation:

$$e^{-\theta s} \cong 1-\theta s \qquad (11\text{-}24b)$$

Solution

(a) Substituting Eq. 11-24a into Eq. 11-10 gives

$$\widetilde{G}(s) = \frac{K\left(1-\frac{\theta}{2}s\right)}{\left(1+\frac{\theta}{2}s\right)(\tau s+1)} \qquad (11\text{-}25)$$

[4]The term *low-pass filter* is a frequency response concept that will be explained in Chapter 10.

Factor this model as $\widetilde{G} = \widetilde{G}_+ \widetilde{G}_-$ where

$$\widetilde{G}_+ = 1 - \frac{\theta}{2}s \qquad (11\text{-}26)$$

and

$$\widetilde{G}_- = \frac{K}{\left(1+\frac{\theta}{2}s\right)(\tau s+1)} \qquad (11\text{-}27)$$

Note that $\widetilde{G}_+$ has a steady-state gain of one, as required in the IMC design procedure.

Substituting Eqs. 11-27 and 11-21 into Eq. 11-20 and setting $r = 1$ gives

$$G_c^* = \frac{\left(1+\frac{\theta}{2}s\right)(\tau s+1)}{K(\tau_c s+1)} \qquad (11\text{-}28)$$

The equivalent controller G_c can be obtained from Eq. 11-16,

$$G_c = \frac{\left(1+\frac{\theta}{2}s\right)(\tau s+1)}{K\left(\tau_c+\frac{\theta}{2}\right)s} \qquad (11\text{-}29)$$

and rearranged into the PID controller of Eq. 11-13 with

$$K_c = \frac{1}{K}\frac{2\left(\frac{\tau}{\theta}\right)+1}{2\left(\frac{\tau_c}{\theta}\right)+1}, \quad \tau_I = \frac{\theta}{2}+\tau, \quad \tau_D = \frac{\tau}{2\left(\frac{\tau}{\theta}\right)+1} \qquad (11\text{-}30)$$

(b) Repeating this derivation for the Taylor series approximation gives a standard PI controller for G_c:

$$K_c = \frac{1}{K}\frac{\tau}{\tau_c+\theta}, \quad \tau_I = \tau \qquad (11\text{-}31)$$

A comparison of Eqs. 11-30 and 11-31 indicates that the type of controller that is designed depends on the time-delay approximation. Furthermore, the IMC controller in Eq. 11-31 is identical to the DS controller in Eq. 11-11 for an FOPTD model.

For more general process models with a dominant time constant, τ_{dom}, the guideline can be generalized by replacing τ by τ_{dom}. For example, setting $\tau_c = \tau_{dom}/3$ means that the desired closed-loop response is three times faster than the open-loop response.

11.3 CONTROLLER TUNING RELATIONS

In the last section, model-based design methods such as DS and IMC produced PI or PID controllers for certain classes of process models. Analytical expressions for PID controller settings have been developed from other perspectives as well. These expressions are referred to as *controller tuning relations*, or just *tuning relations*. In this section, we present some of the most widely used tuning relations as well as some promising new ones.

11.3.1 IMC Tuning Relations

Different IMC tuning relations can be derived depending on the type of low-pass filter f and time-delay approximation that are selected (Rivera et al., 1986; Chien and Fruehauf, 1990; Skogestad, 2003).

Table 11.1 presents the PID controller tuning relations for the parallel form that were derived by Chien and Fruehauf (1990) for common types of process models. The derivations are analogous to those for Example 11.2.

The IMC filter f was selected according to Eq. 11-21 with $r = 1$ for first-order and second-order models. For models with integrating elements, the following expression was employed:

$$f = \frac{(2\tau_c - C)s+1}{(\tau_c s+1)^2} \quad \text{where } C = \left.\frac{d\widetilde{G}_+}{ds}\right|_{s=0} \qquad (11\text{-}32)$$

In Table 11.1, two controllers are listed for some process models (cf. controllers G and H, and M and N).

For these models, the PI controller in the first row was derived based on the time-delay approximation in Eq. 11-24b, while the PID controller in the next row was derived based on Eq. 11-24a.

Chien and Fruehauf (1990) have reported the equivalent tuning relations for the series form of the PID controller in Chapter 7. The controller settings for the parallel form can easily be converted to the corresponding settings for the series form, and *vice versa*, as shown in Table 11.2.

The following example illustrates the use of the tuning relations in Table 11.1.

Table 11.1 IMC Controller Settings for Parallel-From PID Controller

Case	Model	PID Controller, $K_c\left[1+\frac{1}{\tau_I s}+\tau_D s\right]$
A	$\frac{K}{\tau s+1}$	$\frac{\tau}{\tau_c K}\left[1+\frac{1}{\tau s}\right]$
B	$\frac{K}{(\tau_1 s+1)(\tau_2 s+1)}$	$\frac{\tau_1+\tau_2}{\tau_c K}\left[1+\frac{1}{(\tau_1+\tau_2)s}+\frac{\tau_1\tau_2}{\tau_1+\tau_2}s\right]$
C	$\frac{K}{\tau^2 s^2+2\zeta\tau s+1}$	$\frac{2\zeta\tau}{\tau_c K}\left[1+\frac{1}{2\zeta\tau s}+\frac{\tau}{2\zeta}s\right]$
D	$\frac{K(-\beta s+1)}{\tau^2 s^2+2\zeta\tau s+1},\ \beta>0$	$\frac{2\zeta\tau}{(\tau_c+\beta)K}\left[1+\frac{1}{2\zeta\tau s}+\frac{\tau}{2\zeta}s\right]$
E	$\frac{K}{s}$	$\frac{2}{\tau_c K}\left[1+\frac{1}{2\tau_c s}\right]$
F	$\frac{K}{s(\tau s+1)}$	$\frac{2\tau_c+\tau}{\tau_c^2 K}\left[1+\frac{1}{(2\tau_c+\tau)s}+\frac{2\tau_c\tau}{2\tau_c+\tau}s\right]$
G	$\frac{Ke^{-\theta s}}{\tau s+1}$	$\frac{\tau}{(\tau_c+\theta)K}\left[1+\frac{1}{\tau s}\right]$
H	$\frac{Ke^{-\theta s}}{\tau s+1}$	$\frac{\tau+\frac{\theta}{2}}{\left(\tau_c+\frac{\theta}{2}\right)K}\left[1+\frac{1}{\left(\tau+\frac{\theta}{2}\right)s}+\frac{\tau\theta}{2\tau+\theta}s\right]$
I	$\frac{Ke^{-\theta s}}{s}$	$\frac{2\tau_c+\theta}{(\tau_c+\theta)^2 K}\left[1+\frac{1}{(2\tau_c+\theta)s}\right]$
J	$\frac{Ke^{-\theta s}}{s}$	$\frac{2\tau_c+\theta}{\left(\tau_c+\frac{\theta}{2}\right)^2 K}\left[1+\frac{1}{(2\tau_c+\theta)s}+\frac{\tau_c\theta+\frac{\theta^2}{4}}{2\tau_c+\theta}s\right]$
K	$\frac{Ke^{-\theta s}}{s(\tau s+1)}$	$\frac{2\tau_c+\tau+\theta}{(\tau_c+\theta)^2 K}\left[1+\frac{1}{(2\tau_c+\tau+\theta)s}+\frac{(2\tau_c+\theta)\tau}{2\tau_c+\tau+\theta}s\right]$

Table 11.2 Equivalent PID Controller Settings for the Parallel and Series Forms

Parallel Form	Series Form
$G_c(s) = K_c\left(1 + \frac{1}{\tau_I s} + \tau_D s\right)$	$G_c(s) = K'_c\left(1 + \frac{1}{\tau'_I s}\right)(1 + \tau'_D s)^*$
$K_c = K'_c\left(1 + \frac{\tau'_D}{\tau'_I}\right)$	$K'_c = \frac{K_c}{2}\left(1 + \sqrt{1 - 4\tau_D/\tau_I}\right)$
$\tau_I = \tau'_I + \tau'_D$	$\tau'_I = \frac{\tau_I}{2}\left(1 + \sqrt{1 - 4\tau_D/\tau_I}\right)$
$\tau_D = \frac{\tau'_D \tau'_I}{\tau'_I + \tau'_D}$	$\tau'_D = \frac{\tau_I}{2}\left(1 - \sqrt{1 - 4\tau_D/\tau_I}\right)$

* These conversion equations are only valid if $\tau_D/\tau_I \leq 0.25$.

EXAMPLE 11.3

A process model for a liquid storage system is given by Chien and Fruehauf (1990):

$$\widetilde{G}(s) = \frac{Ke^{-7.4s}}{s}$$

Use Table 11.1 to calculate PI and PID controller settings for $K = 0.2$ and $\tau_c = 8$. Repeat for $\tau_c = 15$ and do the following:

(a) Compare the four controllers for unit step changes in the set point and disturbance, assuming that $G_d = \widetilde{G}$.

(b) In order to characterize the robustness of each controller of part (a), determine K_{max}, the largest value of K that results in a stable closed-loop system for each controller.

Solution

(a) For this integrating process, $\widetilde{G}_+ = e^{-\theta s}$, and thus $C = -\theta$ in Eq. 11-32. The IMC controller settings for controllers M and N in Table 11.1 are

	K_c	τ_I	τ_D
PI ($\tau_c = 8$)	0.493	23.4	–
PI ($\tau_c = 15$)	0.373	37.4	–
PID ($\tau_c = 8$)	0.857	23.4	3.12
PID ($\tau_c = 15$)	0.535	37.4	3.33

The closed-loop responses in Fig. 11.7 are more sluggish and less oscillatory for $\tau_c = 15$ than they are for $\tau_c = 8$. Also, for $\tau_c = 15$ the overshoot is smaller for the set-point change, and the maximum deviation is larger after the disturbance.

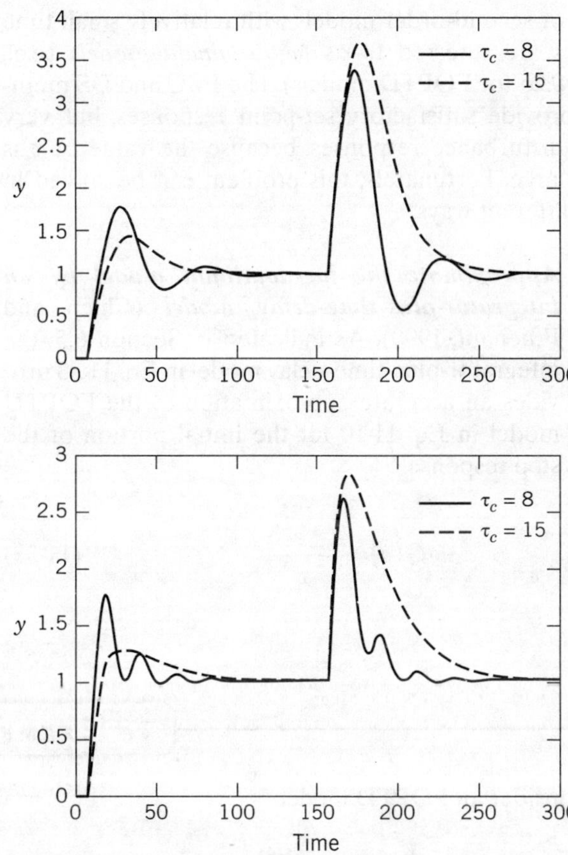

Figure 11.7 Simulation results for Example 11.3: PI control (top) and PID control (bottom).

The PID controller provides a better disturbance response than the PI controller with a smaller maximum deviation. In addition, the PID controller has a very short settling time for $\tau_c = 8$, which gives it the best performance of the four controllers considered.

(b) The numerical value of K_{max} can be obtained from a stability analysis using the Direct Substitution Method of Chapter 9.

The numerical results shown in the following table indicate that K can increase significantly from its nominal value of 0.2 before the closed-loop system becomes unstable. Thus, these IMC controllers are quite robust and become even more so as τ_c increases. The approximate values of K_{max} were obtained by using the time-delay approximation in Eq. 11-24b.

		K_{max}	
Controller	τ_c	Approximate	Exact
PI	8	0.274	0.356
PI	15	0.363	0.515
PID	8	0.376	0.277
PID	15	0.561	0.425

Lag-Dominant Models **($\theta/\tau \ll 1$)**

First-or second-order models with relatively small time delays are referred to as *lag-dominant models* (e.g., $\theta/\tau < 0.25$ for FOPTD models). The IMC and DS methods provide satisfactory set-point responses, but very slow disturbance responses, because the value of τ_I is very large. Fortunately, this problem can be solved in two different ways.

1. ***Approximate the lag-dominant model by an integrator-plus-time-delay model*** (Chien and Fruehauf, 1990). As indicated in Section 6.5, the integrator-plus-time-delay model in Eq. 11-33 provides an accurate approximation to the FOPTD model in Eq. 11-10 for the initial portion of the step response:

$$G(s) = \frac{K^* e^{-\theta s}}{s} \tag{11-33}$$

In Eq. 11-33, $K^* \triangleq K/\tau$. Then the IMC tuning relations in Table 11.1 for either controller M or N can be applied.

2. ***Skogestad IMC (SIMC) Method.*** For lag-dominant models, the standard IMC controllers for first-and second-order models provide sluggish disturbance responses because τ_I is very large. For example, controller G in Table 11.1 has $\tau_I = \tau$ where τ is very large. As a remedy, Skogestad (2003) proposed modifying the IMC tuning relations by limiting the value of τ_I:

$$\tau_I = \min\{\tau, 4(\tau_c + \theta)\} \tag{11-34}$$

For PID control and an FOPTD model, Grimholt and Skogestad (2013) suggested using τ_I from Eq. 11-34, $\tau_D = \theta/3$, $\tau_c = 0.5$, and $K_c = (1/K)(\tau + \theta/2)/(\tau_c + \theta/2)$ from Case H in Table 11.1.

Methods 1 and 2 for lag-dominant models are compared in Example 11.4.

EXAMPLE 11.4

Consider an FOPTD model:

$$\tilde{G}(s) = \frac{100}{100s+1} e^{-s}$$

This model is lag-dominant because $\theta/\tau = 0.01$. Specify three PI controllers using the following methods:

(a) IMC ($\tau_c = 1$)

(b) IMC ($\tau_c = 2$) and the integrator approximation method (Eq. 11-33)

(c) IMC ($\tau_c = 1$) and SIMC (Eq. 11-34)

Evaluate the three controllers by comparing their performance for unit step changes in both set point and disturbance. Assume that the model is perfect and that $G_d(s) = G(s)$.

Solution

The PI controller settings are:

Controller	K_c	τ_I
(a) IMC	0.5	100
(b) Integrator approximation	0.57	5
(c) SIMC	0.5	8

The simulation results in Fig. 11.8 indicate that the IMC controller provides an excellent set-point response. The two special controllers provide significantly improved disturbance rejection with much smaller settling times.

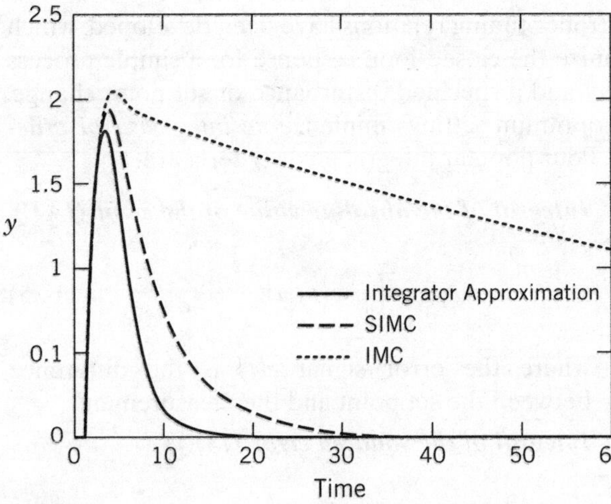

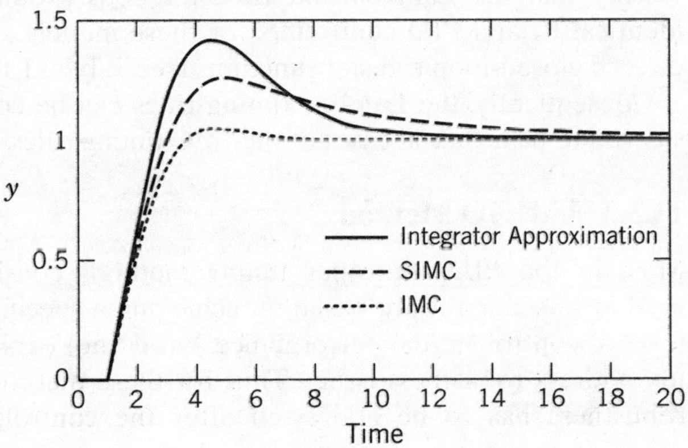

Figure 11.8 Comparison of set-point responses (top) and disturbance responses (bottom) for Example 11.4.

Thus, although the standard IMC tuning rules produce very sluggish disturbance responses for FOPTD models with small θ/τ ratios, simple remedies are available as demonstrated in this example. Also, the set-point overshoots can be reduced without affecting the disturbance responses by adjusting the set-point weighting coefficient, as considered in Section 11.4.

Table 11.3 Guidelines for Choosing τ_c for the IMC and SIMC Tuning Methods and an FOPTD Model

Recommendations	Conditions	References
$\tau_c = \begin{cases} 0.5\theta \\ \theta \\ 1.5\theta \end{cases}$	aggressive control default value smooth responses	Grimholt and Skogestad (2013)
$\tau_c = 3\theta$	robust control	Åström and Hägglund (2006, p. 187), McMillan (2015, p. 29)
$\left.\begin{array}{l} \tau_c > \tau \\ \tau_c = 2-3\tau \end{array}\right\}$	default value robust control}	Blevins et al. (2013, p.102)

Choice of τ_c

The IMC and SIMC tuning methods have a single design parameter, τ_c. Successful controller operation depends on specifying an appropriate value for τ_c. In general, increasing τ_c makes the controller more conservative while reducing τ_c makes the controller aggressive. There are many published guidelines for specifying τ_c based on theoretical analysis and/or practical experience. Table 11.3 provides representative recommendations for an FOPTD model, based on both research and industrial perspectives. For other model structures, these guidelines can be applied after the model is reduced to an FOPTD model (see Section 6.2).

11.3.2 Integral Error Criteria and Other Methods

Controller tuning relations have been developed, which optimize the closed-loop response for a simple process model and a specified disturbance or set-point change. The optimum settings minimize an *integral error criterion*. Four popular integral error criteria are

1. ***Integral of the absolute value of the error (IAE)***

$$\text{IAE} = \int_0^{\infty} |e(t)|\, dt \tag{11-35}$$

where the error signal $e(t)$ is the difference between the set point and the measurement.

2. ***Integral of the squared error (ISE)***

$$\text{ISE} = \int_0^{\infty} e^2(t)\, dt \tag{11-36}$$

3. ***Integral of the time-weighted absolute error (ITAE)***

$$\text{ITAE} = \int_0^{\infty} t\,|e(t)|\, dt \tag{11-37}$$

4. ***Integral of the error (IE)***

$$\text{IE} = \int_0^{\infty} e(t)\, dt \tag{11-38}$$

When the error signal $e(t)$ does not change sign, IE = IAE, and its value can be calculated analytically (Åström and Hägglund, 2006):

$$IE = \frac{\tau_I}{K_c} \quad (\text{when } e(t) \text{ does not change signs}) \tag{11-39}$$

For these situations, IE is a useful metric for evaluating disturbance responses with small values preferred. But if $e(t)$ *does* change signs (e.g., for an oscillatory response), the IE value can be very misleading and should not be used for evaluations.

The ISE criterion penalizes large errors, while the ITAE criterion penalizes errors that persist for long periods of time. In general, the ITAE is the preferred criterion, because it usually results in the most conservative controller settings. By contrast, the ISE criterion provides the most aggressive settings, while the IAE criterion tends to produce controller settings that are between those for the ITAE and ISE criteria. A graphical interpretation of the IAE performance index is shown in Fig. 11.9.

Two early controller tuning relations were published by Ziegler and Nichols (1942) and Cohen and Coon (1953). These well-known tuning relations were developed to provide closed-loop responses that have a 1/4 decay ratio (see Section 5.9). Because a response with a 1/4 decay ratio is considered to be excessively oscillatory for most process control applications, these tuning relations are not recommended.

Other PID design methods and tuning relations are available in books (Åström and Hägglund, 2006; Visioli, 2006; McMillan, 2015) and in an extensive compilation (O'Dwyer, 2009).

Lambda Method

The Lambda method is a popular tuning technique, especially in the pulp and paper industry. Its PI or PID controller tuning is based on a simple transfer function model and the desired closed-loop transfer function in Eq. 11-6. The traditional approach is to tune a PI controller based on an FOPTD model or an integrator-plus-time-delay model. However, the Lambda method can also be used to tune PID controllers for these models (Åström and Hägglund, 2006). It is noteworthy that the Lambda and IMC methods produce identical PI and PID controllers for these models and desired closed-loop transfer functions (see Table 11.1). Consequently, the Lambda tuning rules can be considered to be a special cases of the IMC tuning rules.[5]

11.3.3 AMIGO Method

Most of the PID controller tuning methods considered in this chapter are based on achieving a specified level of control system performance, but do not explicitly address robustness issues. Thus for these methods, robustness has to be addressed after the controller

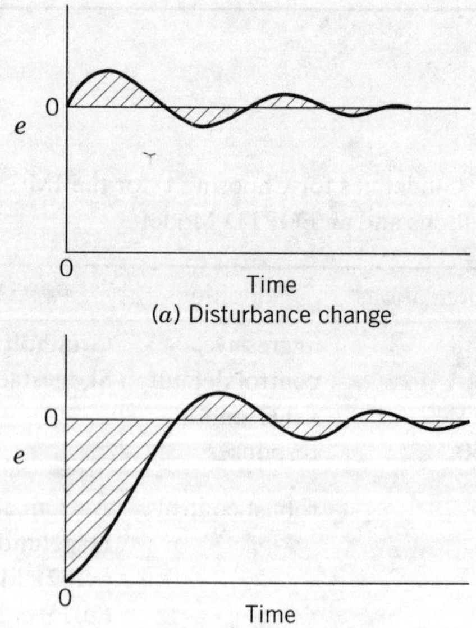

Figure 11.9 Graphical interpretation of IAE. The shaded area is the IAE value.

design has been completed (e.g., by detuning the controller). But both the IMC method and the AMIGO method (Hägglund and Åström, 2004; Åström and Hägglund, 2006) do consider robustness and performance in their development. The AMIGO method optimizes a performance metric while satisfying constraints on robustness and noise rejection.

The AMIGO (Approximate M-constrained Integral Gain Optimization) tuning relations are based on the expanded form of the PID controller in Eq. 11-40:

$$u(t) = K_c\left[y_{sp}(t) - y_m(t)\right] + \\ + K_I\int_0^t\left(y_{sp}(t^*) - y_m(t^*)\right)dt^* + K_D\left(-\frac{dy_m}{dt}\right) \quad (11\text{-}40)$$

The AMIGO controller settings are calculated by maximizing the *integral gain*, K_I, subject to a robustness constraint. The choice of K_I as a performance metric is based on disturbance rejection considerations: the degree of attenuation of low frequency disturbances (common in process control) by the closed-loop system is typically inversely proportional to K_I (cf. Eq. 11-39). Note that K_I is related to the PID parameters in the parallel PID form (Eq. 11-13) by

$$K_I = \frac{K_c}{\tau_I} \quad (11\text{-}41)$$

The AMIGO robustness constraint consists of a specified value for a key frequency-domain robustness metric, *M*, the maximum value of the *sensitivity function* (see Appendix L).

The AMIGO tuning relations were developed as follows. First, the optimum PI and PID controller settings were determined for each model in a test set of 134 process models, which consisted of 10 types of transfer function models that included higher-order models, integrating processes, and wide ranges of θ/τ values. Except for models with an integrator, all the test models had monotonic, or approximately monotonic, step responses. This is only a mild restriction because most process models satisfy this condition. However, the monotonicity requirement means that the AMIGO tuning relations are not recommended for underdamped processes or processes whose step responses exhibit significant overshoots or inverse responses (cf. Chapter 5).

For each of the 134 test models, optimum controller parameters were calculated by constrained optimization (with $M = 1.4$) and a simple approximate model (e.g., first and second order models with time delays) was developed using a model reduction technique. Then AMIGO tuning relations were developed by regression of the optimum controller parameters with the approximate model parameters.

11.3.4 Noise Filters

High-frequency noise is often associated with process measurements. The noise can be produced by a number of sources that include the sensor, electrical equipment, or the process itself. Process-induced noise can occur due to incomplete mixing, turbulence near the sensor, and nonuniform multiphase flows. High noise levels in measurements can have detrimental effects on closed-loop performance because the feedback controller amplifies the noise and re-inserts it into the process via the controller output. Consequently, noise can cause excessive wear on control valves and other final control elements.

The effects of high-frequency noise can be reduced by a type of signal processing, *filtering* the measurements. By contrast, low and medium frequency noise can be interpreted as disturbances that are dealt with by the integral control mode. In this section we briefly consider filter design and its effects on controller design and closed-loop performance. Chapter 13 provides a more detailed analysis of noise filters and digital control techniques.

Commercial computer-control software includes filtering options for each measured variable. The filters are often expressed as transfer functions. For example, the standard *first-order filter* G_F is given by

$$\frac{Y_F(s)}{Y(s)} = G_F(s) = \frac{1}{\tau_F s + 1} \quad (11\text{-}42)$$

where y_F is the filter output, y is the measured variable, and τ_F is the filter time constant. For the standard block diagram in Fig. 11.2, G_F would be included as an additional block in the feedback path, after G_m.

The choice of τ_F involves an inherent tradeoff: Large values of τ_F provide better smoothing of the measurements but add a dynamic lag to the closed-loop system. This additional time constant can have adverse effects on closed-loop stability, performance, and robustness. Its impact can be evaluated by replacing G by GG_F in the design and tuning methods of Sections 11.3 and 11.4.

[5]The term "Lambda method" was used because λ was the symbol for the reciprocal of the desired closed-loop time constant in early publications (e.g., Dahlin, 1968).

Guidelines for first-order filters indicate that τ_F should be specified as either a small fraction of the dominant process time constant τ_p or a fraction of a controller setting, either τ_I or τ_D. A typical guideline is τ_F should be less than $0.1\tau_p$. Simple tuning rules for τ_F and PI/PID control are available based on the tradeoff between measurement smoothing and closed-loop performance/robustness (Segovia et al., 2014; Garpinger and Hagglund, 2015).

11.3.5 Guidelines for Controller Design and Tuning

Although the design and tuning relations of the previous sections are based on different performance criteria, several generalizations and guidelines can be made for PI/PID controllers. These guidelines are expressed in terms of FOPTD models but are also valid for other models if they are first approximated by FOPTD models, as per Section 6.2.

Some important guidelines are:

1. Controller design and tuning involve tradeoffs between control system performance and robustness. High performance controllers tend not to be very robust, while very robust controllers tend to provide very sluggish responses. The required compromise depends largely on control objectives, control loop importance, and model accuracy.
2. Commercial software includes many variations of PID control algorithms (cf. Chapter 7). For example, series versus parallel forms, the use of PB versus K_c. Thus in designing or tuning a PID controller, it is important to specify which version of PID control is being considered.
3. The controller gain K_c should be inversely proportional to the product of the other gains in the feedback loop: $K_c \propto 1/K$ where $K = K_v K_p K_m$. For closed-loop stability, $K_c K > 0$.
4. K_c should decrease as θ/τ increases. In general, control system performance decreases as θ/τ increases, due to longer settling times and larger deviations from set point.
5. As stated earlier, the vast majority of feedback controllers used in industry are PI rather than PID. Derivative action is largely used for slow processes to speed up the closed-loop response. Derivative action is most effective for lag-dominant models (e.g., $\theta/\tau < 0.25$).
6. Both τ_I and τ_D should increase as θ/τ increases. For many controller tuning relations, the ratio, τ_D/τ_I is between 0.1 and 0.5. As a rule of thumb, use $\tau_D/\tau_I = 0.25$ as a first guess.
7. For set-point weighting, a conservative rule is to set $\beta = 1$ when $\theta/\tau > 0.43$ (Åström and Hägglund, 2006, p. 230).
8. Tuning relations (and these guidelines) should be viewed as providing a first estimate of satisfactory controller settings. Subsequent fine tuning is often necessary, especially when the process model is not very accurate, as is often the case.

Although the tuning relations in the previous sections were developed for the parallel form of PID control, they can be converted to the series form by using Table 11.2.

EXAMPLE 11.5

A blending system with a measurement time delay can be modeled as

$$G(s) = \frac{1.54e^{-1.07s}}{5.93s + 1}$$

Calculate PI controller settings using the following tuning relations:

(a) IMC ($\tau_c = \tau/3$)
(b) IMC ($\tau_c = \theta$)
(c) ITAE (disturbance)
(d) AMIGO

Solution

The calculated PI controller settings are

	K_c	τ_I
IMC ($\tau_c = \tau/3 = 1.97$)	1.27	5.93
IMC ($\tau_c = \theta = 1.07$)	1.80	5.93
ITAE (disturbance)	2.97	2.75
AMIGO	0.905	4.47

It appears that the ITAE (disturbance) settings are the most aggressive, and the AMIGO settings are the least aggressive. But this assertion should be checked by simulation.

11.4 CONTROLLERS WITH TWO DEGREES OF FREEDOM

The specification of controller settings for a standard PID controller typically requires a tradeoff between set-point tracking and disturbance rejection. For most single-loop controllers in the process industries, disturbance rejection is more important than set-point tracking, although an exception occurs for cascade control where the set point is calculated by another controller (see Section 12.3). Thus, it is reasonable to tune the controller for satisfactory disturbance rejection, especially if it can be achieved without sacrificing set-point tracking. Fortunately, two simple strategies can be used to adjust the set-point and disturbance responses independently. These strategies are referred to as *controllers with two degrees of freedom.*

The first strategy is very simple. Set-point changes are introduced gradually rather than as abrupt step changes. For example, the set point can be ramped to the new value as shown in Fig. 11.10 (or "filtered") by passing it through a first-order transfer function,

$$\frac{Y_{sp}^*}{Y_{sp}} = \frac{1}{\tau_{sp}+1} \tag{11-43}$$

where Y_{sp}^* denotes the *filtered set point* that is used in the control calculations. The filter time constant, τ_{sp}, determines how quickly the filtered set point will attain the new value, as shown in Fig. 11.10. This strategy can significantly reduce, or even eliminate, overshoot for set-point changes.

A second strategy for independently adjusting the set-point response is to specify the set-point weighting factor β in Eq. 11-40 so that $0 \leq \beta \leq 1$. In general, as β increases, the set-point response becomes faster but exhibits more overshoot.

We have considered two simple but effective strategies for adjusting the set-point response without affecting the disturbance response: set-point filtering (or ramping) and the use of a set-point weighting factor β.

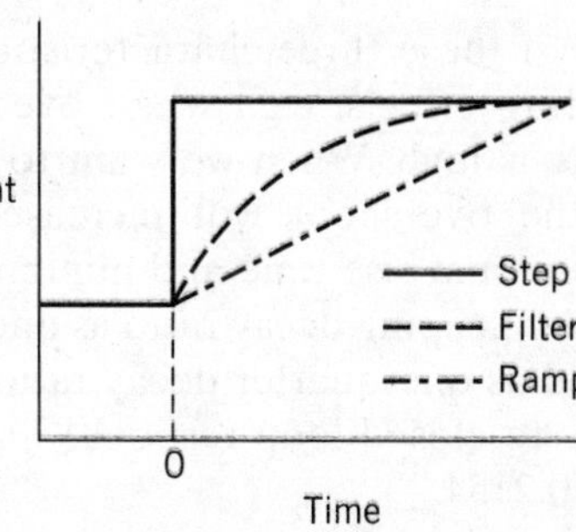

Figure 11.10 Implementation of set-point changes.

EXAMPLE 11.6

For the FOPTD model of Example 11.4, the PI controller for case (b) provided the best disturbance response. However, its set-point response had a significant overshoot. Can set-point weighting significantly reduce the overshoot without adversely affecting the settling time?

Solution

Figure 11.11 compares the set-point responses for a PI controller with and without set-point weighting. Set point weighting with β = 0.5 provides a significant improvement, because the overshoot is greatly reduced and the settling time is significantly decreased. Because the disturbance response in Fig. 11.8 is independent of the value of β, the stated goal is achieved.

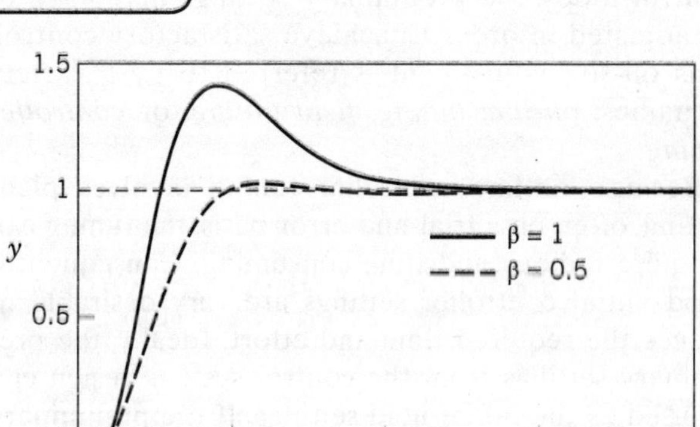

Figure 11.11 Influence of set-point weighting on closed-loop responses for Example 11.6.

11.5 CONTROLLER TUNING BASED ON SIMPLE PERFORMANCE CRITERION (ONE-QUARTER DECAY RATIO)

The three most important simple performance criteria for closed-loop response of the system are:

(a) Minimum overshoot

(b) Minimum rise time (i.e., the time needed for the response to reach the desired value for the first time)

(c) Minimum settling/response time (i.e., the time needed for the response to settle within ±5% of the desired value)

We know that these three characteristics have mutually conflicting effects, e.g., when overshoot is low, the rise time is high. When we want to minimize the rise time, the overshoot will increase. A trade-off between minimum rise time and minimum overshoot is achieved by keeping decay ratio as one-fourth (1/4). This is called as one-quarter decay ratio criteria. This corresponds to closed-loop response damping coefficient (ζ) of 0.2154.

11.6 ON-LINE CONTROLLER TUNING

The control systems for modern industrial plants typically include thousands of individual control loops. During control system design, preliminary controller settings are specified based on process knowledge, control objectives, and prior experience. After a controller is installed, the preliminary settings often prove to be satisfactory. But for critical control loops, the preliminary settings may have to be adjusted in order to achieve satisfactory control. This on-site adjustment is referred to by a variety of names: *on-line tuning, field tuning*, or *controller tuning*.

Because on-line controller tuning involves plant testing, often on a trial-and-error basis, the tuning can be quite tedious and time consuming. Consequently, good initial controller settings are very desirable to reduce the required time and effort. Idealiy, the preliminary settings from the control system design can be used as the initial field settings. If the preliminary settings are not satisfactory, alternative settings can be obtained from simple experimental tests. If necessary, the settings can be fine-tuned by a modest amount of trial and error.

In this chapter, we present three important methods for on-line controller tuning. More detailed analysis and a wealth of practical experience are available in books by industrial practitioners (McMillan, 2015; Shinskey, 1994) and university researchers (Åström and Hägglund, 2006; Yu, 1999). Software for controller tuning is also widely available.

Next, we make a few general observations:

1. ***Controller tuning inevitably involves a tradeoff between performance and robustness.*** The performance goals of excellent set-point tracking and disturbance rejection should be balanced against the robustness goal of stable operation over a wide range of conditions.
2. ***Controller settings do not have to be precisely determined.*** In general, a small change in a controller setting from its best value (e.g., ±10%) has little effect on closed-loop responses.
3. ***For most plants, it is not feasible to manually tune each controller.*** Tuning is usually done by a control specialist (engineer or technician) or by a plant operator. Because each person is typically responsible for 300 to 1000 control loops, it is not feasible to manually tune every controller. Instead, only the control loops that are perceived to be the most important or the most troublesome receive detailed attention. The other controllers typically operate using the preliminary settings from the control system design.
4. ***Diagnostic techniques for monitoring control system performance are available.*** This topic is introduced in Chapter 17.

Next we consider three important on-line tuning methods.

11.6.1 Continuous Cycling Method

In 1942, Ziegler and Nichols published a classic paper that introduced their *continuous cycling method* for on-line controller tuning (Ziegler and Nichols, 1942). This paper and its Ziegler–Nichols (Z-N) tuning relations have had an enormous impact on both process control research and practice. Despite serious shortcomings, the Z-N relations and (various modifications) have been widely used as a benchmark in comparisons of tuning relations. Consequently, we briefly describe the Z-N continuous cycling method in this section before considering a more recent on-line tuning method, relay auto-tuning, in Section 11.6.2.

The Z-N continuous cycling method is based on experimentally determining the closed-loop stability limit for proportional-only control. The following trial and error procedure is used:

Step 1. After the process has reached steady state (at least approximately), introduce proportional-only control by eliminating the integral and derivative modes (i.e., set $\tau_D = 0$ and $\tau_I = \infty$).

Step 2. Introduce a small, momentary set-point change so that the controlled variable moves away from the set point. Gradually increase K_c (in absolute value) in small increments until continuous cycling occurs. The term *continuous cycling* refers to a sustained oscillation with a constant amplitude. The value of K_c that produces continuous cycling (for proportional-only control) is defined to be the *ultimate gain*, K_{cu}. The period of this sustained oscillation is the *ultimate period*, P_u.

After K_{cu} and P_u have been determined, the controller settings can be calculated using the Z-N tuning relations in Table 11.4.

The Z-N tuning relations reported were determined empirically to provide closed-loop responses that have a 1/4 decay ratio. For proportional control, the Z-N settings in Table 11.4 provide a safety margin of two, because K_c is one-half of the stability limit, K_{cu}. When integral action is added for PI control, K_c is reduced from $0.5K_{cu}$ to $0.45K_{cu}$. The stabilizing effect of derivative action allows K_c to be increased to $0.6K_{cu}$ for PID control.

The continuous cycling method has two major disadvantages:

1. The trial-and-error determination of K_{cu} and P_u can be quite time consuming if the process dynamics are slow.
2. For many applications, continuous cycling is objectionable because the process is pushed to a stability limit. Consequently, if external disturbances or process changes occur during the test, unstable operation or a hazardous situation could result (e.g., a "runaway" chemical reaction).

The Z-N controller settings have been widely used as a benchmark for evaluating different tuning methods

Table 11.4 Controller Settings based on the Continuous Cycling Method of Ziegler and Nichols (1942)

Controller	K_c	τ_I	τ_D
P	$0.5K_{cu}$	—	—
PI	$0.45K_{cu}$	$P_u/1.2$	—
PID	$0.6K_{cu}$	$P_u/2$	$P_u/8$

and control strategies. Because they are based on a 1/4 decay ratio, the Z-N settings tend to produce oscillatory responses and large overshoots for set-point changes.

Despite their prominence in the process control literature, it is not certain whether the famous Z-N tuning relations for PID control were developed for the series or parallel form of the controller (Åström et al., 2001). Because the parallel form results in a more conservative controller (Skogestad, 2003), it is reasonable to calculate the Z-N settings for the parallel form and then convert them to series form, if necessary, using Table 11.2.

Calculation of FOPTD Model Parameters from K_{cu} and P_u

Model parameters for simple transfer functions can be calculated from K_{cu} and P_u. For example, consider the FOPTD model in Eq. 11-10. If time delay θ is known or can be estimated, then K and τ can be calculated (Rivera et al., 1986):

$$\tau = \frac{P_u}{2\pi} \tan\left[\frac{\pi(P_u - 2\theta)}{P_u}\right] \tag{11-44}$$

$$K = \frac{1}{K_{cu}}\left(1 + \left[\frac{2\pi\tau}{P_u}\right]^2\right)^{1/2} \tag{11-45}$$

After the FOPTD model is calculated, the model-based design methods in Sections 11.2 and 11.3 can be used to calculate PID controller settings.

Controller tuning relations are compared in Examples 11.7 and 11.8.

EXAMPLE 11.7

For the FOPTD model,

$$G = \frac{e^{-s}}{s+1}$$

compare four PI controllers:

(a) IMC settings (Table 11.1) with $\tau_c = 1$
(b) Z-N settings (Table 11.4)

Evaluate these controllers for unit step changes in set point (at $t = 0$) and a step disturbance of magnitude 1 at $t = 20$. Assume that $G_d = G$.

Solution

The ultimate gain and ultimate period can be determined by trial and error to be $K_{cu} = 2.26$ and $P_u = 3.10$.

Method	K_c	τ_I
IMC	0.5	1.0
Z-N	1.02	2.58

These controller settings and the closed-loop responses in Fig. 11.12 indicate that the ITAE and the Z-N settings are aggressive and produce damped oscillatory responses.

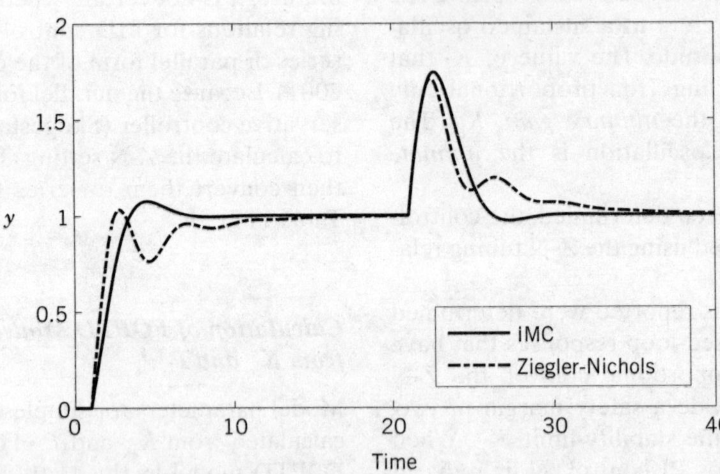

Figure 11.12 Comparison of two PI controllers for Example 11.7.

EXAMPLE 11.8

Consider a process model,

$$G = \frac{2e^{0.633s}}{(s+1)(0.2s+1)(0.04s+1)(0.008s+1)}$$

and compare four PID controllers that are designed using an FOPTD model:

(a) IMC settings (Table 11.1) with $\tau_c = \theta = 0.73$
(b) Z-N settings (Table 11.4)
(c) SIMC settings with $\tau_c = 0.37$ (Eq. 11-34)

Evaluate these controllers for a unit step change in set point (at $t = 0$) and a disturbance of magnitude 0.5 at $t = 10$. Assume that $G_d = G$.

Solution

In order use methods (a), (b), and (d), the fourth-order model is reduced to an FOPTD model:

$$G = \frac{Ke^{-\theta s}}{\tau s+1}$$

Applying Skogestad's model reduction technique (Section 6.2) to the fourth-order model gives:

$$K = 2$$
$$\theta = 0.633 + (0.5)(0.2) = 0.733$$
$$\tau = 1 + (0.5)(0.2) = 1.1$$

For methods (a) and (c), the τ_c design parameter is specified based on the guidelines reported earlier: For IMC, $\tau_c = \theta = 0.733$; for SIMC, $\tau_c = 0.5\theta = 0.37$. The ultimate gain and ultimate period for the Z-N settings are $K_{cu} = 5.01$ and $P_u = 0.331$. The PID controller settings are shown below:

Method	K_c	τ_I	τ_D
IMC	1.00	1.47	0.37
Z-N	1.51	1.51	0.38
SIMC	1.00	1.10	0.24

The closed-loop responses in Fig. 11.13 indicate that all four controllers produce satisfactory closed-loop responses. The set-point overshoots can be reduced without affecting the disturbance responses by adjusting the set-point weighting factor β (Section 11.4). The Z-N controller is the most aggressive and produces slightly oscillatory responses. The IMC controller is the least aggressive and produces a sluggish disturbance response. The AMIGO and SIMC controllers are very satisfactory and provide similar responses.

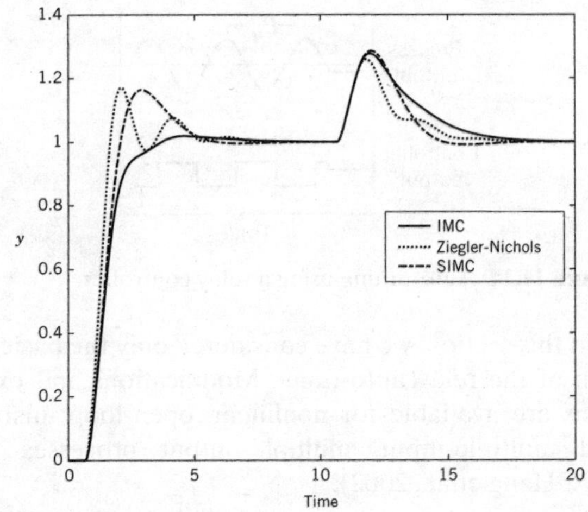

Figure 11.13 Comparison of three PID controllers for Example 11.8.

11.6.2 Relay Auto-Tuning

Åström and Hägglund (1984) developed an attractive alternative to the continuous cycling method. In their *relay auto-tuning* method, a simple experimental test is used to determine K_{cu} and P_u. For this test, the feedback controller is temporarily replaced by an on–off controller (or *relay*). After the control loop is closed, the controlled variable exhibits a sustained oscillation that is characteristic of on–off control (cf. Section 7.6). The operation of the relay auto-tuner includes a *dead zone* as shown in Fig. 11.14. The dead band is used to avoid frequent switching caused by measurement noise.

The ultimate gain and the ultimate period can easily be obtained from Fig. 11.14. The ultimate period P_u is equal to the period of oscillation for the process output. Åström and Hägglund (1984) derived an approximate expression for the ultimate gain,

$$K_{cu} = \frac{4d}{\pi a} \tag{11-46}$$

where d is the relay amplitude (set by the user) and a is the measured amplitude of the process oscillation. PID controller settings can then be calculated from Table 11.4 or from the model parameters in Eqs. 11-40 to 11-45. The relay auto-tuning method has several important advantages compared to the continuous cycling method:

1. Only a single experimental test is required instead of a trial-and-error procedure.
2. The amplitude of the process output a can be restricted by adjusting relay amplitude d.
3. The process is not forced to a stability limit.
4. The experimental test is easily automated.

The relay auto-tuning method also has a disadvantage. For slow processes, it may not be acceptable to subject the process to the two to four cycles of oscillation required to complete the test. Thus, the Step Test Method of Section 11.6.3 may be preferred for slow processes.

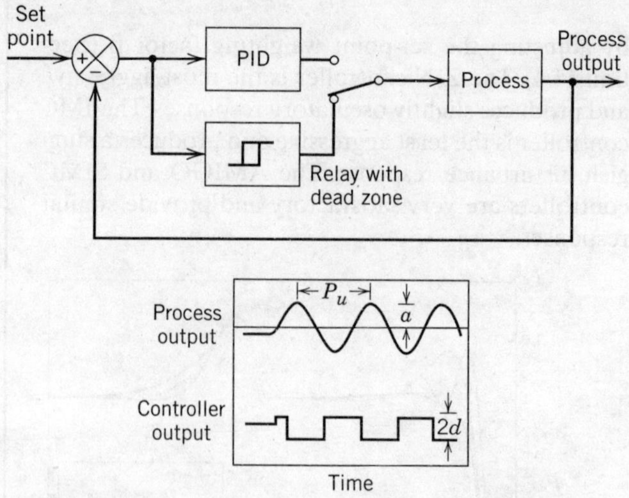

Figure 11.14 Auto-tuning using a relay controller.

In this section, we have considered only the basic version of the relay auto-tuner. Modifications and extensions are available for nonlinear, open-loop unstable, and multiple-input, multiple-output processes (Yu, 1999; Hang et al., 2002).

11.6.3 Step Test Method

In their classic paper, Ziegler and Nichols (1942) proposed a second on-line tuning technique based on a single step test. The experimental procedure is quite simple. After the process has reached steady state (at least approximately), the controller is placed in the manual mode. Then a small step change in the controller output (e.g., 3% to 5%) is introduced. The controller settings are based on the *process reaction curve* (Section 6.5), the open-loop step response. Consequently, this on-line tuning technique is referred to as *the step test method or the process reaction curve method.*

Two types of process reaction curves are shown in Fig. 11.15 for step changes occurring at $t = 0$. After an initial transient, the measured response y_m for Case (a) increases at a constant rate, indicating that the process appears to act as an integrating element and thus is *not* self-regulating. In contrast, the hypothetical process considered in Case (b) is self-regulating because the step response reaches a new steady state. Both step responses are characterized by two parameters: S, the slope of the tangent through the inflection point, and θ, the apparent time delay. The graphical determination of S and θ was described in Section 6.5. For Case (b), the slope S is equal to K/τ for an FOPTD model.

An appropriate transfer function model can be obtained from the step response by using the parameter estimation methods of Chapter 16. For processes that have monotonically increasing step responses, such as the responses in Fig. 11.15, the models in Eqs. 11-40 and 11-43 are appropriate. Then, any of the model-based tuning relations in Sections 11.2 and 11.3 can be employed.

The chief advantage of the step test method is that only a single experimental test is necessary. But the method does have four disadvantages:

1. The experimental test is performed under open-loop conditions. Thus, if a significant disturbance occurs during the test, no corrective action is taken. Consequently, the process can be upset and the test results may be misleading.
2. For a nonlinear process, the test results can be sensitive to the magnitude and direction of the step change. If the magnitude of the step change is too large, process nonlinearities can influence the result. But if the step magnitude is too small, the step response may be difficult to distinguish from the usual fluctuations due to noise and disturbances. The direction of the step

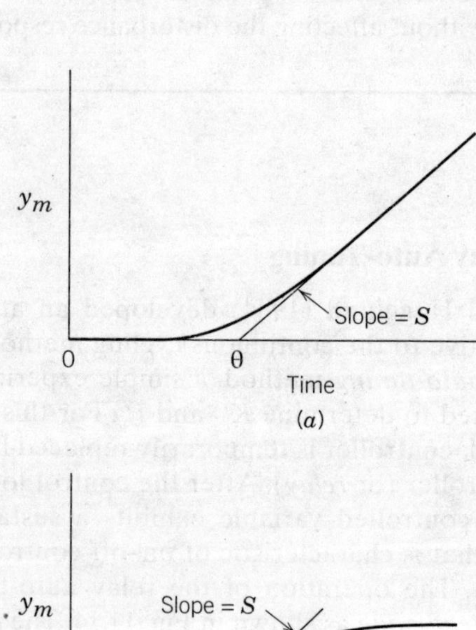

Figure 11.15 Typical process reaction curves: (a) non-self-regulating process, (b) self-regulating process.

change (positive or negative) should be chosen so that the controlled variable will not violate a constraint.

3. The method is not applicable to open-loop unstable processes.
4. For continuous (analog) controllers, the method tends to be sensitive to controller calibration errors. By contrast, the continuous cycling method is less sensitive to calibration errors in K_c, because it is adjusted during the experimental test.

Closed-loop versions of the step test method have also been proposed as a partial remedy for the first disadvantage (Lee et al., 1990).[6] Typically, a step change in the set point is introduced while the process is controlled using proportional-only or PI control. Then the parameters in simple transfer function models can be estimated from the closed-loop step response.

The second disadvantage can be avoided by making multiple step changes. For example, if a series of both positive and negative changes is made, the effects of disturbances, nonlinearities, and control valve hysteresis will become apparent.

EXAMPLE 11.9

Consider the feedback control system for the stirred-tank blending process shown in Fig. 11.1 and the following step test. The controller was initially in the manual mode, and then its output was suddenly changed from 30% to 43%. The resulting process reaction curve is shown in Fig. 11.16. Thus, after the step change occurred at $t = 0$, the measured exit composition changed from 35% to 55% (expressed as a percentage of the measurement span), which is equivalent to the mole fraction changing from 0.10 to 0.30. Determine an appropriate process model for $G \triangleq G_{IP}G_vG_pG_m$.

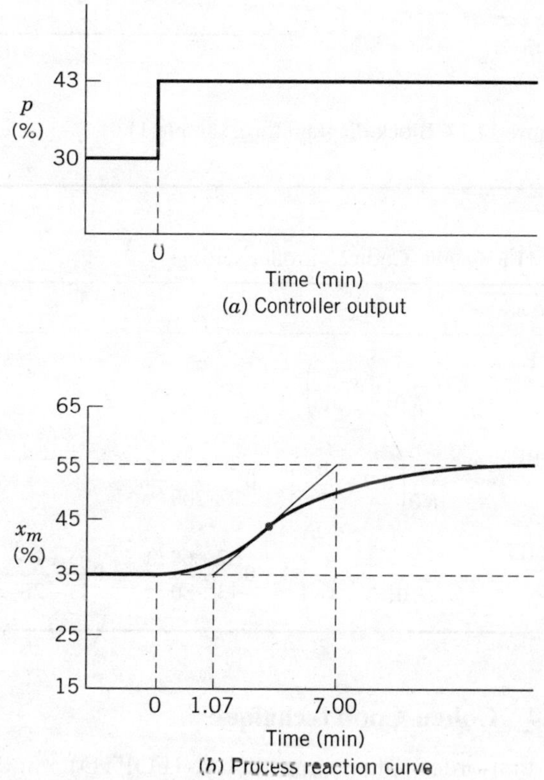

Figure 11.16 Process reaction curve for Example 11.8.

Solution

A block diagram for the closed-loop system is shown in Fig. 11.17. This block diagram is similar to Fig. 11.7, but the feedback loop has been opened between the controller and the current-to-pressure (I/P) transducer. An FOPTD model can be developed from the process reaction curve in Fig. 11.16 using the graphical method of Section 6.5. The tangent line through the inflection point intersects the horizontal lines for the initial and final composition values at 1.07 min and 7.00 min, respectively. The slope of the line is

$$S = \left(\frac{55\% - 35\%}{7.00 - 1.07\,\text{min}} \right) = 3.37\%/\text{min}$$

The model parameters can be calculated as

$$K = \frac{\Delta x_m}{\Delta p} = \frac{55\% - 35\%}{43\% - 30\%} = 1.54\,(\text{dimensionless})$$

$$\theta = 1.07\,\text{min}$$

$$\tau = 7.00 - 1.07\,\text{min} = 5.93\,\text{min}$$

[6]The first disadvantage can be minimized by using a series of small changes such as the pseudo random binary sequences (PRBS) of Chapter 16.

The apparent time delay of 1.07 min is subtracted from the intercept value of 7.00 min for the τ calculation.

The resulting empirical process model can be expressed as

$$\frac{X'_m(s)}{P'(s)} = G(s) = \frac{1.54e^{-1.07s}}{5.93s+1}$$

Example 11.5 provided a comparison of PI controller settings for this model that were calculated using different tuning relations.

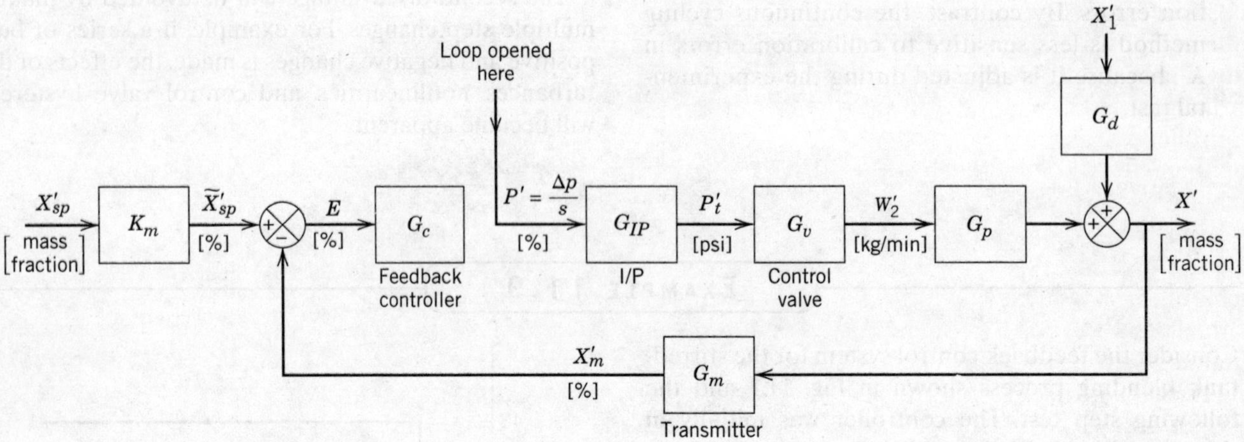

Figure 11.17 Block diagram for Example 11.9.

Table 11.5 Cohen-Coon controller settings

Controllers	K_c	τ_I	τ_D
P	$\frac{\tau}{K\theta}\left(1+\frac{\theta}{3\tau}\right)$	–	–
PI	$\frac{\tau}{K\theta}\left(0.9+\frac{\theta}{12\tau}\right)$	$\theta\frac{30+3\theta/\tau}{9+20\theta/\tau}$	–
PID	$\frac{\tau}{K\theta}\left(\frac{4}{3}+\frac{\theta}{4\tau}\right)$	$\theta\frac{32+6\theta/\tau}{13+8\theta/\tau}$	$\theta\frac{4}{11+2\theta/\tau}$

11.6.4 Cohen-Coon technique

The first-order plus time delay (FOPTD) model obtained from step test method is utilized by Cohen and Coon (1953) to obtain the classical PID controller settings as shown in Table 11.5.

11.7 GUIDELINES FOR COMMON CONTROL LOOPS

General guidelines for selection of controller type (P, PI, etc.) and controller settings are available for common process variables such as flow rate, liquid level, gas pressure, temperature, and composition. The general guidelines presented below are useful but they should be used with caution, because exceptions do occur.

Flow Rate

Flow control loops are widely used in the process industries. For example, they typically are about half of all of the control loops in large, continuous processing plants like oil refineries and petrochemical plants. Flow and pressure control loops are characterized by fast responses (on the order of seconds), with essentially no time delay. The process dynamics result from compressibility (in a gas stream) or inertial effects (in a liquid) plus control valve dynamics for large-diameter pipelines. Disturbances in flow control systems tend to be frequent but generally small. Most of the disturbances are high-frequency noise (periodic or random) due to upstream turbulence, valve changes, and pump vibration.

For flow control loops, PI control is generally used with intermediate values of the controller gain. Fruehauf et al. (1994) recommend the following controller settings: $0.5 < K_c < 0.7$ and $0.2 < \tau_I < 0.3$ min. The presence of recurring high-frequency noise discourages the use of derivative action, because it amplifies the noise. Furthermore, because flow control loops usually have relatively small settling times (compared to other control loops), there is little incentive to use derivative action to make the loop respond even faster.

Liquid Level

Liquid storage vessels typically have a pump on the exit line and act as integrating processes, which have been considered in Chapters 2, 5, and 9. Standard P or PI controllers are widely used for level control. However, as shown in Section 9.4, these level control problems have an unusual characteristic: increasing the gain of a PI controller can *increase* stability, while reducing the gain can increase the degree of oscillation and thus *reduce* stability. Of course, if K_c becomes too large, oscillations or even instability can result. Integral control action is often used for level control but can be omitted if small offsets can be tolerated. Derivative action is not normally used for level control because the level measurements are often noisy as a result of the splashing and turbulence of the liquid entering the tank.

There are two broad classes of level control problems with different control objectives:

1. Tight level control (control at or near a set point)
2. Averaging level control (control between high and low limits)

For some important process operations, tight level control is important. For example, constant liquid level is desirable for some chemical reactors and bioreactors in order to keep the residence time constant. In these situations, the level controller can be specified using standard tuning methods for integrating processes. If level control also involves heat transfer and a phase change (e.g., a vaporizer or evaporator), the controller design becomes much more complicated. In such situations, special control methods can be advantageous (Shinskey, 1994).

Averaging level control is very widely applied when a liquid storage vessel is used as a surge tank in order to damp out fluctuations in the inlet streams. The control objectives are that (i) the exit flow rate from the tank change should change gradually, rather than abruptly, in order to avoid upsetting downstream process units; (ii) the liquid level should be maintained within specified upper and lower limits; and (iii) the steady-state mass balance must be satisfied so that the inlet and outlet flows are equal. These three goals can be achieved by allowing the liquid level to rise or fall in response to inlet flow disturbances. In particular, little or no control action is taken when the level is within the high and low limits. But much stronger control is required as the level is close to a high or low limit.

Because offset is not important in averaging level control, it is reasonable to use a proportional-only controller. But if integral control action is desired, St. Clair (1993) recommends the following PI controller settings for averaging level control:

$$K_c = \frac{100\%}{\Delta h} \tag{11-47}$$

$$\tau_I = \frac{4V}{K_c Q_{max}} \tag{11-48}$$

where

$$\Delta h \triangleq \min\left(h_{max} - h_{sp}, h_{sp} - h_{min}\right) \tag{11-49}$$

In Eq. 11-49, h_{max} and h_{min} are the maximum and minimum allowable liquid levels, and h_{sp} is the set point. Each is expressed as a percentage of the level transmitter range. In Eq. 11-48, V is the tank volume, and Q_{max} is the maximum flow rate through the control valve. Equations 11-47 and 11-49 ensure that the controller output will be at a saturation limit (0% or 100%) when the absolute value of the controller error is larger than Δh.

Nonlinear versions of PI control are sometimes used for averaging level control, especially *error-squared controllers* where the controller gain is proportional to the error signal (see Chapter 12). Error-squared controllers offer the advantage of a large control action when the controlled variable is far from the set point, and a small control action when it is near. However, they must be tuned carefully in order to guarantee stable responses over the entire operating range (Shinskey, 1994).

Gas Pressure

The control of gas pressure is analogous to the control of liquid level in the sense that some applications use averaging control while others require tight control around a set point. However, high and low limits are usually a more serious concern for pressure control than for level control, because of safety and operational issues. For self-regulating processes, pressure is relatively easy to control, except when the gas is in equilibrium with a liquid. Gas pressure is self-regulating when the vessel (or pipeline) admits more feed when the pressure is low, and reduces the intake when the pressure becomes high. Integrating processes occur when the exit pressure is determined by a compressor, in analogy to liquid level when there is a pump for the exit stream. For pressure control, PI controllers are normally used with only a small amount of integral control action (i.e., τ_I is large). Usually the pressure vessel is not large, leading to relatively small residence times and time constants. Derivative action is normally not needed because the process response times are usually quite small compared to those of other process operations.

Temperature

General guidelines for temperature control loops are difficult to state because of the wide variety of processes and equipment involving heat transfer and their different time scales. For example, the temperature control problems are quite different for heat exchangers, distillation columns, chemical reactors, and evaporators. The presence of time delays and/or multiple thermal capacitances will usually place a stability limit on the controller gain. PID controllers are commonly employed to provide quicker responses than can be obtained with PI controllers.

Composition

Composition control loops generally have characteristics similar to temperature loops but with certain differences:

1. Measurement (instrument) noise is a more significant problem in composition loops.
2. The time delay associated with the analyzer and its sampling system may be a significant factor.

These two factors can limit the effectiveness of derivative action. Because of their importance and the difficulty of control, composition and temperature loops often are prime candidates for the advanced control strategies discussed in Chapters 12, 14, and 15.

11.8 TROUBLESHOOTING CONTROL LOOPS

If a control loop is not performing satisfactorily, then troubleshooting is necessary to identify the source of the problem. Ideally, it would be desirable to evaluate the control loop over the full range of process operating conditions during the commissioning of the plant. In practice, this is seldom feasible. Furthermore, the process characteristics can vary with time for a variety of reasons, including changes in equipment and instrumentation, different operating conditions, new feedstocks or products, and large disturbances. Surveys of thousands of control loops have confirmed that a large fraction of industrial control loops perform poorly. For example, surveys have reported that about one-third of the industrial control loops were in the manual mode, and another one-third actually increased process variability over manual control, a result of poor controller tuning (Ender, 1993; Desborough and Miller, 2002). Clearly, controller tuning and control loop troubleshooting are important activities.

This section provides a brief introduction to the basic principles and strategies that are useful in monitoring and troubleshooting control loops. More detailed analyses and useful insights are available elsewhere, from both an industrial perspective (Ender, 1992; Lieberman, 2008; Smuts, 2011) and a research perspective (Section 17.5). A survey of control system monitoring software is also available (Bacci di Capaci and Scali, 2015).

Troubleshooting begins with the realization that the control loop consists of a number of individual components: sensor/transmitter, controller, final control element, instrument lines, computer–process interface, and the process itself. Serious control problems can result from a malfunction of any single component. On the other hand, even if all the individual components are functioning properly, there is no guarantee that the overall system will perform properly. Thus, a *systems approach* is required.

In industrial applications, it is tempting to view controller re-tuning as a cure-all for solving control loop problems. But after a control loop has performed satisfactorily for a significant amount of time, it can still become excessively oscillatory or overly sluggish for a variety of reasons that include

(a) Changing process conditions or set points (e.g., changes in throughput rate),

(b) A malfunctioning control valve (e.g., stick-slip or hysteresis),

(c) A sensor problem (e.g., drift, plugged sampling line),

(d) A cavitating pump (usually due to a low suction pressure).

Note that only Item (a) provides a valid reason for retuning the controller.

Required Background Information

The starting point for control loop troubleshooting is to obtain enough background information to clearly define the problem. Many questions need to be answered:

1. What is the process being controlled?
2. What is the controlled variable?
3. What are the control objectives?
4. Are closed-loop response data available?
5. Is the controller in the manual or automatic mode? Is it reverse-or direct-acting?
6. If the process is cycling, what is the cycling frequency?
7. Is the process response too sluggish compared to expectations or previous experience?
8. What control algorithm is used? What are the controller settings?
9. Is the process self-regulating?
10. What documentation is available, such as control loop summary sheets, piping and instrumentation diagrams, operator logs, and maintenance records?

The answers to the first three questions should be obvious to the plant operators and engineers who run the process. But the answers may not be readily apparent if the control loops are monitored remotely or if the monitoring software handles many control loops (e.g., hundreds to thousands).

Preliminary Analysis

After acquiring the available background information (which is seldom complete), the next step is to check out the individual components in the control loop. In particular, determine whether or not the process, measurement device (sensor/transmitter), final control element, and signal transmission equipment are all working properly. Typically, sensors and control valves located in the field require more maintenance than control equipment located in a central control room. Any recent change in equipment, instrumentation, or maintenance could be the source of the problem. For example, cleaning heat exchanger tubes, using a new shipment of feed materials or catalyst, or changing a transmitter span or control valve trim could adversely affect the control-loop performance. Sensor problems are often associated with the small-diameter lines that transport process fluids to the sensors (e.g., DP cells, on-line gas chromatographs).

The presence of solid material, ice, or bubbles in a line can result in erroneous measurements.

Common Problems and Solutions

Next, we consider four common control-loop problems and simple troubleshooting strategies for their resolution.

1. *Dead or unresponsive sensor.* An abnormally small amount of variability (e.g., sample variance) in a set of consecutive measurements can indicate a "dead" sensor (see Appendix E), while a break in an electrical instrument line can be detected by a *rate-of-change* or *noise-spike filter* (see Chapter 13).
2. *A sticking control valve.* This problem can be diagnosed by performing a simple experimental test: the controller is placed in manual and a small step change is made in the output signal. If the control valve is working properly, the flow rate through the control valve should change accordingly. But if the control valve is stuck or sticking badly, the flow rate will not change. Then a series of step changes should be made to characterize the problem and find a solution. Techniques for stick-slip and hysteresis characterization and detection are available (Choudhury et al., 2008; Blevins et al., 2013).
3. *Excessively sluggish responses to disturbances.* In general, industrial control loops tend to be tuned very conservatively to provide robustness over a wide range of conditions. There is also a psychological motivation for conservative tuning: a control loop that is too conservatively tuned is often not obvious, while a control loop with excessive oscillation is readily apparent, can have serious consequences and can attract unwanted attention from plant management.

   Excessively sluggish control loops can affect adversely affect process performance due to the production of off-spec product and slow recovery from disturbances and set-point changes. These considerations have motivated the development of systematic methods for the detection of very sluggish control loops (Hägglund, 1999; Kuehl and Horch, 2005).
4. *Excessively oscillatory control loops.* Excessive oscillations can produce offset product, destabilize the process, and adversely affect other control loops. Thus, their detection and resolution is a major motivation for control-loop troubleshooting An extensive literature on the automatic detection of control loop oscillations has been published (see Section 17.5).

   A simple diagnostic procedure for control-loop oscillations is illustrated by two examples.

EXAMPLE 11.10

A control loop exhibits excessively oscillatory behavior even though there have been no recent equipment, instrument, or personnel changes. Suggest a general troubleshooting strategy to diagnose the problem.

Solution

Oscillatory control loops are often the result of (i) a cyclic process disturbance, (ii) a sticking control valve, or (iii) a poorly tuned controller. A simple test can be used to distinguish between case (i), and cases (ii) and (iii). If the controller is placed in the manual mode for a short period of time and the oscillations die out, then the oscillation is caused by the control loop, rather than by an external disturbance. In order to distinguish between cases (ii) and (iii), the controller should be placed in the manual mode and one or more small step changes made in the controller output. The test results can be used to characterize the degrees of stick-slip and hysteresis of the control valve. If the valve is functioning properly, then the controller may need to be re-tuned.

EXAMPLE 11.11

The bottom composition of a pilot-scale, methanol–water distillation column is measured using an on-line gas chromatograph (GC). The composition measurement is sent to a PI controller that adjusts the steam flow rate to the reboiler. Recently, the control-loop performance has deteriorated: the closed-loop composition response is more oscillatory and the period of oscillation is much larger than usual. The troubleshooting strategy used in Example 11.9 has indicated that neither a cyclic disturbance nor a sticking control valve is the source of the problem. According to the maintenance records, the filter in the sample line has been replaced recently. Suggest additional diagnostic tests that should be considered before controller re-tuning is performed as a last resort.

Solution

The combination of a more oscillatory response and a larger period of oscillation could occur if the time delay associated with the composition measurement has increased. For example, the transport delay associated with the sampling line to the GC would increase if the flow rate in the sampling line decreased. A decrease could occur due to a partial blockage in the line, or perhaps due to the new filter. Thus, the filter and the sample line should be inspected.

If the old filter had inadvertently been replaced with a new filter that had a smaller sieve size, a larger pressure drop would occur, and the downstream pressure would decrease. Consequently, the liquid velocity in the sample line would decrease, and the transport delay would increase. As a result, the composition control loop would become more oscillatory and exhibit a longer period of oscillation.

SUMMARY

In this chapter, we have considered three important issues associated with feedback control systems: design, tuning, and troubleshooting. Model-based design and tuning methods are recommended because they provide considerable insight and usually have only a single adjustable parameter. However, a reasonably accurate process model must be available. After a control system is installed, on-line tuning is commonly used to improve the performance of key control loops. For poorly performing control loops or control systems, systematic troubleshooting methods can be very useful.

This chapter has considered a variety of controller design and tuning methods. Consequently, it is appropriate to summarize conclusions and recommendations:

1. Control system design and tuning inevitably require tradeoffs between performance and

robustness. The tradeoffs should be carefully evaluated (Garpinger and Hägglund, 2015).

2. Model-based techniques are recommended for control system design, especially the Internal Model Control/Direct Synthesis and AMIGO methods. If the process is "lag dominant" (very small θ/τ ratio), the SIMC modification in Section 11.3.1 should be considered.
3. For most process control applications, the controller should be tuned for disturbances rather than set-point changes. Set-point tracking can be adjusted without affecting disturbance response by using a set-point filter or a set-point weighting factor (Section 11.4).
4. Many controller tuning relations exhibit the same general features, as noted in Section 11.3.5.
5. The recommended on-line tuning techniques are relay auto-tuning (Section 11.6.2) and step tests (Section 11.6.3). But subsequent fine tuning may be required.
6. Tuning relations based on a one-quarter decay ratio, such as the Ziegler–Nichols and Cohen–Coon methods, are not recommended.
7. Tuning relations based on integral error criteria (Section 11.3.2) provide useful benchmarks, but the resulting controllers are usually not very robust.

If a control loop is not performing satisfactorily, then troubleshooting is necessary to identify the source of the problem. This diagnostic activity should be based on a "systems approach" that considers the overall performance of the control loop as well as the performance of individual components. Retuning the controller is not a cure-all, especially if a sensor or a control valve is the source of the problem. Automated monitoring techniques for control loops are currently being developed and are commercially available (see Chapter 17).

REFERENCES

Ang, K. H., G. Chong, and Y. Li, PID Control System Analysis, Design, and Technology, *IEEE Trans. Control Systems Technol.*, **13**, 559 (2005).

Åström, K. J., D. B. Ender, S. Skogestad, and R. C. Sorensen, personal communications, 2001.

Åström, K. J., and T. Hägglund, Automatic Tuning of Simple Regulators with Specification on the Gain and Phase Margins, *Automatica*, **20**, 645 (1984).

Åström, K. J., and T. Hägglund, *Advanced PID Control*, ISA, Research Triangle Park, NC, 2006.

Bacci di Capaci, R., and C. Scali, Review on Valve Stiction. Part I: From Modeling to Smart Diagnosis, *Processes*, **3** (2), 422–451 (2015).

Bauer, M., and I. K. Craig, Economic Assessment of Advanced Process Control – A Survey and Framework, *J. Process Control*, **18**, 2 (2008).

Blevins, T., W. K. Wojsznis, and M. Nixon, *Advanced Control Foundation*, ISA, Research Triangle Park, N.C., 2013.

Chen, D., and D. E. Seborg, PI/PID Controller Design Based on Direct Synthesis and Disturbance Rejection, *Ind. Eng. Chem. Res.*, **41**, 4807 (2002).

Chien, I-L., and P. S. Fruehauf, Consider IMC Tuning to Improve Controller Performance, *Chem. Eng. Prog.*, **86** (10), 33 (1990).

Choudhury, M. A. A. S., M. Jain, and S. L. Shah, Stiction—Definition, Modelling, Detection and Quantification, *J. Process Control*, **18**, 232 (2008).

Cohen, G. H., and G. A. Coon, Theoretical Considerations of Retarded Control, *Trans. ASME*, **75**, 827 (1953).

Dahlin, E. B., Designing and Tuning Digital Controllers, *Instr. Control Syst.*, **42** (6) 77 (1968).

Desborough, L., and R. Miller, Increasing Customer Value of Industrial Control Performance Monitoring, *Proc. of the 6th Internat. Conf. on Chemical Process Control*, CPC-VII, AIChE Sympos. Series, **98**, 169 (2002).

Doyle, F. J., III, *Process Control Modules: A Software Laboratory for Control System Design*, Prentice Hall PTR, Upper Saddle River, NJ, 2000.

Ender, D. B., Troubleshooting Your PID Control Loop, *InTech*, **39** (5), 35 (1992).

Ender, D. B., Process Control Performance: Not as Good as You Think, *Control Eng.*, **40** (10), 189 (1993).

Fruehauf, P. S., I-L. Chien, and M. D. Lauritsen, Simplified IMCPID Tuning Rules, *ISA Trans.*, **33**, 43 (1994).

Garcia, C. E., and M. Morari, Internal Model Control I. A Unifying Review and Some New Results, *Ind. Eng. Chem. Process Des. Dev.*, **21**, 308 (1982).

Garpinger, O., and T. Hägglund, Software-Based Optimal PID Design with Robustness and Noise Sensitivity Consraints, *J. Process Control*, **33**, 90 (2015).

Grimholt, C., and S. Skogestad, Optimal PID-Control on First Order Plus Time Delay Systems & Verification of the SIMC Rules, *Proc. 10th IFAC DYCOPS Sympos.*, Mumbai, India (December, 2013).

Hägglund, T., Automatic Detection of Sluggish Control Loops, *Control Eng. Practice*, **7**, 1505 (1999).

Hägglund, T., and K. J. Åström, Revisiting the Ziegler–Nichols Step Response Method for PID Control, *J. Process Control*, **14**, 635 (2004).

Hang, C. C., K. J. Åström, and Q. G. Wang, Relay Feedback Auto-Tuning of Process Controllers—A Tutorial Review, *J. Process Control*, **12**, 143 (2002).

Kristiansson, B., and B. Lennartson, Robust Tuning of PI and PID Controllers, *IEEE Control Syst. Mag.*, **26** (1), 55 (2006).

Kuehl, P., and A. Horch, Detection of Sluggish Control Loops-Experiences and Improvements, *Control Eng. Prac.*, **13**, 1019 (2005).

Lee, J., W. Cho, and T. F. Edgar, An Improved Technique for PID Controller Tuning from Closed-Loop Tests, *AIChE J.*, **36**, 1891 (1990). Lieberman, N., *Troubleshooting Process Plant Control*, John Wiley and Sons, Hoboken, NJ, 2008.

Lipták, B. (Ed.), *Instrument Engineers' Handbook*, 4th ed., *Vol. 2: Process Control*, Section 8, CRC Press, Boca Raton, FL, 2006.

Madhuranthakam, C. R., A. Elkamel, and H. Budman, Optimal Tuning of PID controllers for FOPTD, SOPTD and SOPTD with Lead Processes, *Chem. Eng. Process.*, **47**, 251 (2008).

McMillan, G. K., *Tuning and Control-Loop Performance*, 4th ed., Momentum Press, New York, 2015.

O'Dwyer, A., *Handbook of PI and PID Controller Tuning Rules*, 3rd ed., World Scientific Press, Hackensack, NJ, 2009.

Rivera, D. E., M. Morari, and S. Skogestad, Internal Model Control, 4. PID Controller Design, *Ind. Eng. Process Des. Dev.*, **25**, 252 (1986).

Segovia, V. R., T. Hägglund, and K. J. Åström, Measurement Noise Filtering for Common PID Tuning Rules, *Control Eng. Practice*, **32**, 43 (2014).

Shinskey, F. G., *Feedback Controllers for the Process Industries*, McGraw-Hill, New York, 1994.

Skogestad, S., Simple Analytic Rules for Model Reduction and PID Controller Tuning, *J. Process Control* **13**, 291 (2003).

Smuts, J. F., *Process Control for Practitioners*, OptiControls Inc., (2011).

St. Clair, D. W., *Controller Tuning and Control Loop Performance: "PID without the Math,"* 2nd ed., Straight-Line Control Co. Inc., Newark, DE, 1993.

Visioli, A., *Practical PID Control*, Springer, London, 2006.

Yu, C.-C., *Autotuning of PID Controllers*, Springer-Verlag, New York, 1999.

Ziegler, J. G., and N. B. Nichols, Optimum Settings for Automatic Controllers, *Trans. ASME*, **64**, 759 (1942)

EXERCISES

11.1 A process has the transfer function,

$$G(s) = \frac{K}{(12s+1)(3s+1)}$$

where K has a nominal value of $K = 1$. PID controller settings are to be calculated using the Direct Synthesis approach with $\tau_c = 5$ min. Suppose that these controller constants are employed and that K changes unexpectedly from 1 to $1 + \alpha$.

(a) For what values of α will the closed-loop system be stable?

(b) Suppose that the PID controller constants are calculated using the nominal value of $K = 1$ but it is desired that the resulting closed-loop system be stable for $|\alpha| \le 0.2$. What is the smallest value of τ_c that can be used?

(c) What conclusions can be made concerning the effect that the choice of τ_c has on the robustness of the closed-loop system to changes in steady-state gain K?

11.2 Consider the two feedback control strategies shown in Fig. 11.6 (with $G = G_v G_p G_m$) and the following transfer functions:

$$G_p(s) = G_d(s) = \frac{4(1-s)}{s}, \quad G_v(s) = 1 \quad G_m(s) = 1$$

(a) Design an IMC controller, G_c^*, using a filter, $f = 1/(\tau_c s + 1)$.

(b) Suppose that the IMC controller will be implemented as a controller G_c in the classical feedback control configuration of Fig. 11.6(a). Derive an expression for G_c and report it in a standard form (e.g., P, PI, or PID).

11.3 A process has the transfer function, $G(s) = 2e^{-2s}/(s + 1)$ Compare the PI controller settings for the following design approaches:

(a) IMC method ($\tau_c = 0.2$)

(b) IMC method ($\tau_c = 1.0$)

(c) ITAE performance index (disturbance)

(d) ITAE performance index (set point)

(e) Which controller has the most conservative settings? Which has the least conservative?

(f) For the two controllers of part (e), simulate the closed-loop responses to a unit step disturbance, assuming that $G_d(s) = G(s)$.

11.4 A process, including the sensor and control valve, can be modeled by the transfer function:

$$G(s) = \frac{2e^{-3s}}{s}$$

(a) If a proportional-only controller is used, what is the maximum value of controller gain K_c that will result in a stable closed-loop system? (Determine the *exact* value of K_c, not an approximate value.)

(b) Specify a PI controller using the Z-N controller settings.

11.5 A process stream is heated using a shell and tube heat exchanger. The exit temperature is controlled by adjusting the steam control valve shown in Fig. E11.5. During an open-loop experimental test, the steam pressure P_s was suddenly changed from 120 to 140 kPa and the temperature data shown below

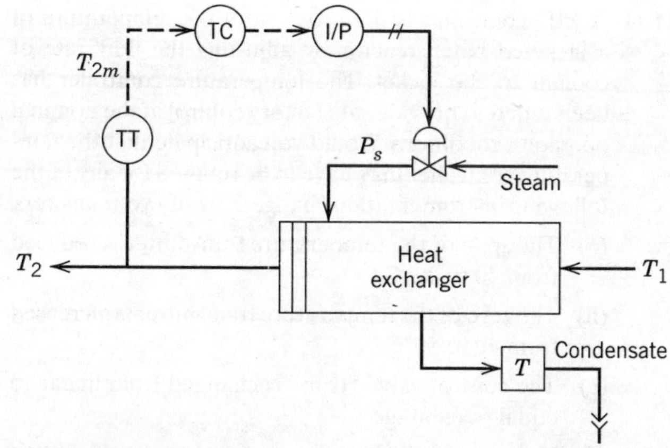

Figure E11.5

were obtained. At the nominal conditions, the control valve and current-to-pressure transducers have gains of $K_v = 0.9$ kPa/kPa and $K_{IP} = 0.75$ kPa/mA, respectively. Determine appropriate PID controller settings using the following approaches:

(a) Internal Model Control (select a reasonable value of τ_c)

(b) Cohen-Coon method

(c) SIMC method

Which controller is the most aggressive?

t (min)	T_{2m} (mA)
0	12.0
1	12.0
2	12.5
3	13.1
4	14.0
5	14.8
6	15.4
7	16.1
8	16.4
9	16.8
10	16.9
11	17.0
12	16.9

11.6 Consider the FOPTD model in Eq. 11-10 with $K = 6$, $\tau = 4$, and $\theta = 3$. Design PI and PID controllers using the IMC tuning method with $\tau_c = \theta = 3$. Simulate the closed-loop systems for a unit step change in set point. Does the addition of derivative action result in significant improvement? Justify your answer.

11.7 A process including sensor and control valve can be modeled by a fourth-order transfer function:

$$G(s) = \frac{e^{-0.002s}}{(s+1)(0.2s+1)(0.04s+1)(0.006s+1)}$$

(a) Design a PID controllers with a SOPTD model using the model reduction approach proposed by Skogestad (Section 6.2) and the IMC tuning relation in Table 11.1.

(b) Evaluate the controller by simulating the closed-loop responses to a unit step change in a disturbance, assuming that $G_d(s) = G(s)$.

11.8 Consider the level control problem in Example 11.3. Attempt to reduce the set-point overshoot of the PI controller ($\tau_c = 15$) by using set-point weighting (cf. Eq. 11-39). Which value of β gives the best results? Does the value of β affect the disturbance response?

11.9 Consider the PID controller of Example 11.3 for $\tau_c = 8$.

(a) Suppose that the PID controller is implemented as the series form in Table 11.2, rather than as the parallel form that was employed in Fig. 11.7. Are the simulated closed-loop responses significantly different from the ones in Fig. 11.7?

(b) Suppose that τ_D is varied, whereas K_c and τ_I are kept constant at the design values. For what values of τ_D is the closed-loop system stable? Does the closed-loop system become more or less oscillatory as τ_D increases?

11.10 Consider the blending system shown in Fig. E11.10. A feedback control system is used to reduce the effect of disturbances in feed composition x_1 on the controlled variable, product composition, x. Inlet flow rate w_2 can be manipulated. Do the following:

(a) Draw a block diagram of the feedback control system.

(b) Using the information shown below, derive a transfer function for each block.

(c) Simulate the closed-loop response for the PI controller settings given below and a step disturbance of +0.2 in x_1.

(d) Repeat part (c) for a set-point change of −0.1. Attempt to obtain better closed-loop responses by tuning the PI controller. Which controller settings give the best results?

(e) Attempt to obtain improved control by adding derivative action to your best PI controller of part (d). Try several values of derivative time, τ_D. Which one gives the best results?

(f) Suppose that the sampling line to the composition analyzer becomes partially plugged so that the measurement time delay is now three minutes. Using your best controller settings of part (d), simulate the closed-loop response for the same set-point change and the new time-delay value. Explain your new simulation results. Does the larger time delay have a major effect on control system performance?

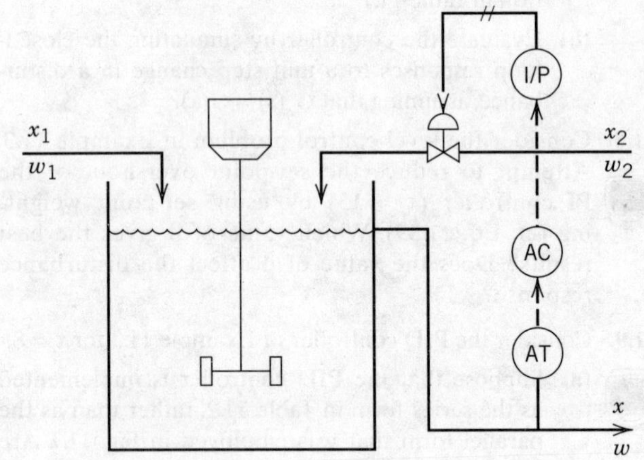

Figure E11.10

Process Information

The pilot-scale blending tank has an internal diameter of 2 m and a height of 3 m. Inlet flow rate w_1 and inlet composition x_2 are constant. The nominal steady-state operating conditions are as follows:

$\bar{w}_1 = 650$ kg/min $\quad \bar{x}_1 = 0.2 \quad \bar{h} = 1.5$ m

$\bar{w}_2 = 350$ kg/min $\quad \bar{x}_2 = 0.6$

$\rho = 1$ g/cm^3 $\quad \bar{x} = 0.34$

The overflow line maintains a constant liquid volume in the tank.

Instrumentation: The range for all of the electronic signals is 4 to 20 mA.

Current-to-pressure transducer: The I/P transducer acts as a linear device with negligible dynamics. The output signal changes from 20 to 100 kPa when the input signal changes full-scale from 4 to 20 mA.

Control valve: The behavior of the control valve can be approximated by a first-order transfer function with a time constant of 5 s (i.e., 0.0833 min). A 8 kPa change in the signal to the control valve produces a 300 kg/min change in w_2.

Composition measurement: The zero and span of the composition transmitter for the exit composition are 0 and 0.50 (mass fraction), respectively. A one-minute time delay is associated with the measurement.

Feedback controller: Initially, consider a standard PI controller tuned using the IMC relations in Table 11.1. Justify your choice of τ_c.

11.11 A PID controller is used to control the temperature of a jacketed batch reactor by adjusting the flow rate of coolant to the jacket. The temperature controller has been tuned to provide satisfactory control at the nominal operating conditions. Would you anticipate that the temperature controller may have to be retuned for any of the following instrumentation changes? Justify your answers.

(a) The span of the temperature transmitter is reduced from 40 to 20°C.

(b) The zero of the temperature transmitter is increased from 40 to 50°C.

(c) The control valve "trim" is changed from linear to equal percentage.

(d) The temperature of the coolant leaving the jacket is used as the controlled variable instead of the temperature in the reactor.

11.12 Suppose that a process can be adequately modeled by the FOPTD model in Eq. 11-10.

(a) Calculate PI controller settings using the IMC tuning relations.

(b) Compare the PI settings calculated from Cohen-Coon settings (Table 11.5) to the controller settings of part (a). Which would you expect to be more conservative?

(c) Simulate the controllers of parts (a) and (b) for a unit step change in set point, followed by a unit step disturbance at $t = 40$. Assume that the $G_d(s) = G(s)$ and that the model parameters are $K = 2$, $\tau = 3$, and $\theta = 1$. Which controller provides better control?

11.13 Consider the experimental step response data for the heat exchanger of Exercise 11.5. Determine the PI controller settings using the step response method and two controller tuning relations:

(a) IMC method with $\tau_c = \tau/3$

(b) Z-N settings in Table 11.4

Which controller provides the more conservative controller settings?

11.14 IGC's operations area personnel are experiencing problems with a particular feedback control loop on an interstage cooler. Appelpolscher has asked you to assess the situation and report back what remedies, if any, are available. The control loop is exhibiting an undesirable sustained oscillation that the operations people are sure is caused by the feedback loop itself (e.g., poor controller tuning). They want assistance in retuning the loop. Appelpolscher thinks that the oscillations are caused by external disturbances (e.g., cyclic disturbances such as cycling of the cooling water temperature); he wants the operations people to deal with the problem themselves. Suggest a simple procedure that will allow you to determine quickly what is causing the oscillations. How will you explain your logic to Appelpolscher?

11.15 A problem has arisen in the level control loop for the flash separation unit shown in Fig. E11.15. The level control loop had functioned in a satisfactory manner for a long period of time. However, the liquid level is gradually increasing with time, even though the PI level controller output has saturated. Furthermore, the liquid flow rate is well above the nominal value, while the feed flow rate is at the nominal value, according to the recorded measurements from the two flow transmitters. The accuracy of the level transmitter measurement has been confirmed by comparison with sight glass readings for the separator. The two flow measurements are obtained via orifice plates and differential pressure transmitters, as described in Chapter 8. Suggest possible causes for this problem and describe how you would troubleshoot this situation.

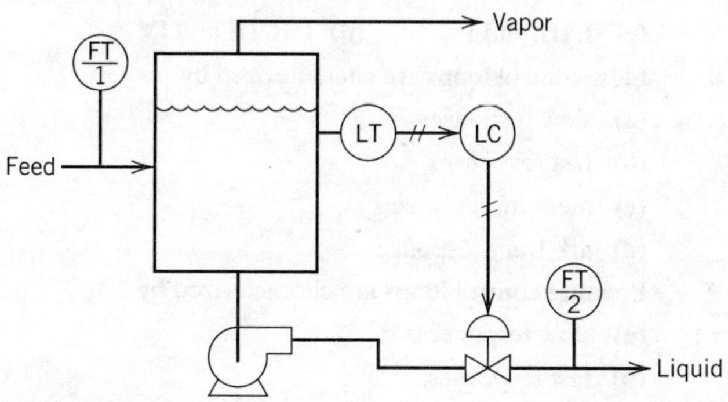

Figure E11.15

11.16 Consider the PCM furnace module of Appendix C. Assume that hydrocarbon temperature T_{HC} is the CV, that fuel gas flow rate F_{FG} is the MV, and that they are related by the following transfer function model:

$$\frac{T_{HC}}{F_{FG}} = G_p = \frac{200e^{-2s}}{6.5s+1}, \qquad G_v = G_m = 1$$

(a) Design a PID controller based on the IMC tuning relations and a reasonable choice for τ_c.

(b) Design a PID controller based on the relay auto-tuning feature of the PCM, and the Z-N settings.

(c) Simulate each controller for a sudden change in an unmeasured disturbance, the hydrocarbon flow rate F_{HC}, at $t = 10$ min, from 0.035 to 0.040 m³/min. Which controller is superior? Justify your answer.

(d) Attempt to obtain improved control system performance by fine tuning your better controller for part (c). For the performance criteria, consider maximum deviation from set point and settling time.

11.17 Consider the PCM distillation column module of Appendix C. Assume that distillate MeOH composition x_D is the CV, that reflux ratio R is the MV, and that they are related by the following transfer function model:

$$\frac{X_D(s)}{R(s)} = G_p(s) = \frac{0.126e^{-138s}}{762s+1}, \quad G_v = G_m = 1$$

(a) Design a PID controller based on the IMC tuning relations and a reasonable choice for τ_c.

(b) Design a PID controller based on the relay auto-tuning feature of the PCM and the Z-N settings.

(c) Simulate each controller for a sudden change in an unmeasured disturbance, feed composition x_F, at $t - 10$ min, from 0.50 to 0.55 (mole fraction). Which controller is superior? Justify your answer.

(d) Attempt to obtain improved control system performance by fine-tuning your better controller for part (c). For the performance criteria, consider maximum deviation from set point and settling time.

MULTIPLE CHOICE QUESTIONS

1. Integral of the time-weighted absolute error (ITAE) is expressed as

(a) $\int_0^\infty \frac{|\varepsilon(t)|}{t^2}dt$ **(b)** $\int_0^\infty \frac{|\varepsilon(t)|}{t}dt$

(c) $\int_0^\infty t|\varepsilon(t)|dt$ **(d)** $\int_0^\infty t^2|\varepsilon(t)|dt$ **(GATE 2014)**

2. Match the following

Group–I	Group–II
(P) Ziegler-Nichols	(I) Process Reaction Curve
(Q) Under damped response	(II) Decay ratio
(R) Feed forward control	(III) Frequency response
	(IV) Disturbance measurement

(a) P-III, Q-II, R-IV
(b) P-I, Q-II, R-III
(c) P-III, Q-IV, R-II
(d) P-I, Q-IV, R-II **(GATE 2008)**

3. The value of ultimate period of oscillation P_u is 183 minutes, and that of the ultimate controller gain K_{cu} is 2. Select the correct set of tuning parameters (controller gain K_c, the derivative time constant τ_D in minutes, and the integral time constant Z_1 in minutes) for a PID controller using Ziegler-Nichols controller settings.

(a) $K_c = 1.1; Z_1 = 2.1, \tau_D = 1.31$
(b) $K_c = 1.5; Z_1 = 1.8, \tau_D = 0.51$
(c) $K_c = 15; Z_1 = 1.8, \tau_D = 0.51$
(d) $K_c = 1.2; Z_1 = 1.5, \tau_D = 0.38$ **(GATE 2005)**

4. The experimental response of the controlled variable $y(t)$ for a step change of magnitude P in the manipulated variable $x(t)$ is shown below.

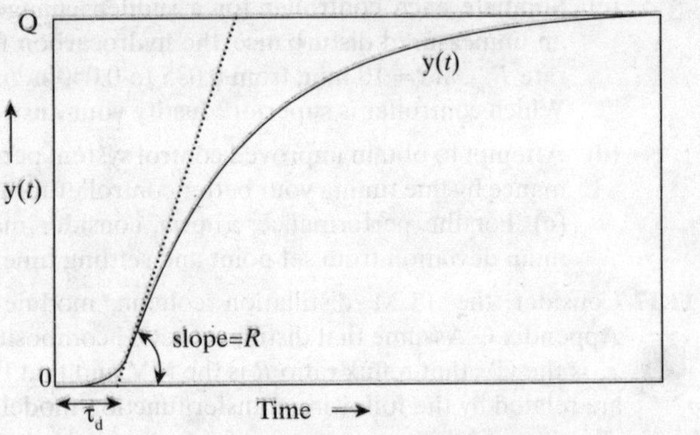

The appropriate transfer function of the process is

(a) $\dfrac{(Q/P)e^{-(Q/R)s}}{\tau_d s+1}$ **(b)** $\dfrac{(Q/R)e^{-\tau_d s}}{(Q/P)s+1}$

(c) $\dfrac{(Q/P)e^{-\tau_d s}}{(Q/R)s+1}$ **(d)** $\dfrac{(Q/R)e^{-(Q)s}}{\tau_d s+1}$

(GATE 2004)

5. Match the type of controller given in **Group–II** that is most suitable for each application given in **Group–I**

Group–I	Group–II
(P) Distillation column bottoms level to be controlled with bottom flow	(I) P control
(Q) Distillation column pressure to be controlled by manipulating vapor flow from the top plate	(II) P-I control
(R) Flow control of a liquid from a pump by positioning the valve in the line	(III) P-I-D control
(S) Control of temperature of a CSTR with coolant flow in the jacket.	

(a) P – I, Q – I, R – II, S – III
(b) P – II, Q – II, R – III, S – III
(c) P – II, Q – II, R – I, S – I
(d) P – II, Q – III, R – II, S – III **(GATE 2004)**

6. The type of process that most often can benefit from derivative control is

(a) flow. **(b)** level.
(c) temperature. **(d)** pressure. **(GATE 1991)**

7. The process characteristics can vary with time due to

I. changes in equipment and instrumentation
II. different operating conditions
III. new feed stocks or products
IV. large disturbances

(a) I, II, and III **(b)** II, III, and IV
(c) I, III, and IV **(d)** I, II, III and IV

8. Flow control loops are characterized by

(a) slow responses.
(b) fast responses.
(c) medium responses.
(d) arbitrary responses.

9. Pressure control loops are characterized by

(a) slow responses.
(b) fast responses.
(c) medium responses.
(d) arbitrary responses.

10. For flow control loops ____________ is generally used.

(a) P-only control **(b)** PI control
(c) PID control **(d)** PD control

11. For level control loops ____________ is/are generally used.

(a) P-only control
(b) P or PI control
(c) PI and PID control
(d) PD or PID control

12. For pressure control, ____________ controllers are normally used.

(a) P-only **(b)** PI
(c) PID **(d)** PD

13. For temperature control, ____________ controllers are generally used.

(a) P-only **(b)** PI
(c) PID **(d)** PD

ANSWER KEY

1. (c) **2.** (a) **3.** (d) **4.** (c) **5.** (d) **6.** (d) **7.** (d)
8. (b) **9.** (b) **10.** (b) **11.** (b) **12.** (b) **13.** (c)

CHAPTER 12

Enhanced Single-Loop Control Strategies

CHAPTER CONTENTS

In Chapter 7 it was emphasized that feedback control is an important technique that is widely used in the process industries. Its main advantages are.

1. Corrective action occurs as soon as the controlled variable deviates from the set point, regardless of the source and type of disturbance.
2. Feedback control requires minimal knowledge about the process to be controlled; in particular, a mathematical model of the process is *not* required, although it can be very useful for control system design.
3. The ubiquitous PID controller is both versatile and robust. If process conditions change, re-tuning the controller usually produces satisfactory control.

However, feedback control also has certain inherent disadvantages:

1. No corrective action is taken until after a deviation in the controlled variable occurs. Thus, *perfect control*, where the controlled variable does not deviate from the set point during disturbance or set-point changes, is theoretically impossible.
2. It does not provide predictive control action to compensate for the effects of known or measurable disturbances.
3. It may not be satisfactory for processes with large time constants and/or long time delays. If large and frequent disturbances occur, the process may operate continuously in a transient state and never attain the desired steady state.
4. In some situations, the controlled variable cannot be measured on-line, so feedback control is not feasible.

In this chapter, we introduce several specialized strategies that provide enhanced process control beyond what can be obtained with conventional single-loop PID controllers. Because processing plants have become more complex in order to increase efficiency or reduce costs, there are incentives for using such enhancements, which also fall under the general classification of advanced control. Although new methods are continually evolving and being field-tested (Henson and Badgwell, 2006; Rawlings et al., 2002; Daoutidis and Bartusiak, 2013; Åström and Kumar, 2014), this chapter emphasizes eight different strategies that have been proven commercially:

1. Feedforward control
2. Ratio control
3. Cascade control
4. Time-delay compensation
5. Inferential control
6. Selective and override control
7. Nonlinear control
8. Adaptive control

These techniques have gained increased industrial acceptance over the past 20 years, and in many cases they utilize the principles of single-loop PID feedback controller design. These strategies can incorporate additional measurements, controlled variables, or manipulated variables, and they can also incorporate alternative block diagram structures.

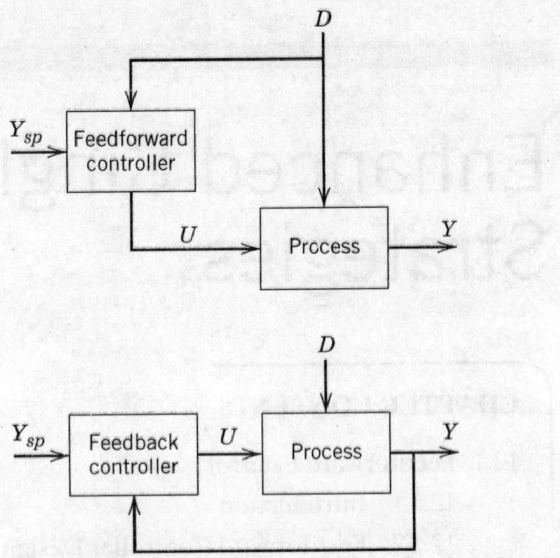

Figure 12.1 Simplified block diagrams for feedforward and feedback control.

12.1 FEEDFORWARD CONTROL

For situations in which feedback control by itself is not satisfactory, significant improvement can be achieved by adding feedforward control.

12.1.1 Introduction

The basic concept of feedforward control is to measure important disturbance variables and take corrective action before they upset the process. In contrast, a feedback controller does not take corrective action until after the disturbance has upset the process and generated a nonzero error signal. Simplified block diagrams for feedforward and feedback control are shown in Fig. 12.1.

Feedforward control has several disadvantages:

1. If only feedforward technique is used and no other control scheme is used, perfect control requires identification and continuous measurement of all possible disturbance variables. In many applications, this is not feasible.
2. To make effective use of feedforward control, at least an approximate process model should be available. In particular, we need to know how the controlled variable responds to changes in both the disturbance variable and the manipulated variable. The quality of feedforward control depends on the accuracy of the process model.
3. Ideal feedforward controllers that are theoretically capable of achieving perfect control may not be physically realizable. Fortunately, practical approximations of these ideal controllers can provide very effective control.
4. In some scenarios, the process model is quite accurate during initial stage. However, when process parameters gradually change over a period of time (e.g. catalyst deactivation, change in heat transfer coefficient due to fouling etc.), the accuracy of the process model gradually decreases. This results in poor performance of the feedforward controller since the impact of process parameters cannot be detected properly.

Feedforward control was not widely used in the process industries until the 1960s (Shinskey, 1996). Since then, it has been applied to a wide variety of industrial processes. However, the basic concept is much older and was applied as early as 1925 in the three-element level control system for boiler drums. We will use this control application to illustrate the use of feedforward control.

A boiler drum with a conventional feedback control system is shown in Fig. 12.2. The level of the boiling liquid is measured and used to adjust the feedwater flow rate. This control system tends to be quite sensitive to rapid changes in the disturbance variable, steam flow rate, as a result of the small liquid capacity of the boiler drum. Rapid disturbance changes are produced by steam demands made by downstream processing units. Another difficulty is that large controller gains cannot

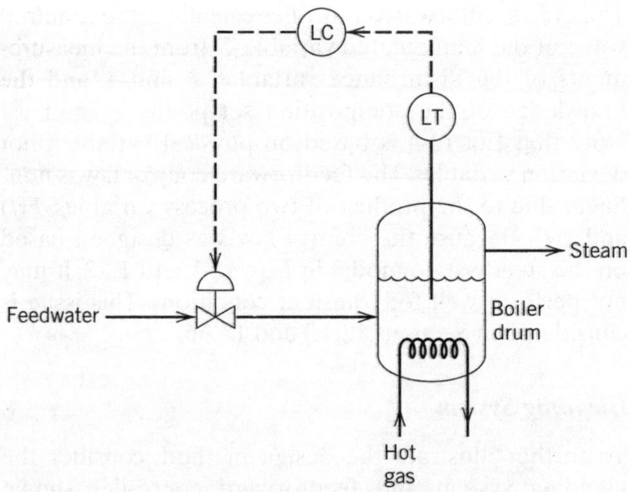

Figure 12.2 Feedback control of the liquid level in a boiler drum.

be used because level measurements exhibit rapid fluctuations for boiling liquids. Thus a high controller gain would tend to amplify the measurement noise and produce unacceptable variations in the feedwater flow rate.

The feedforward control scheme in Fig. 12.3 can provide better control of the liquid level. The steam flow rate is measured, and the feedforward controller adjusts the feedwater flow rate so as to balance the steam demand. Note that the controlled variable, liquid level, is not measured. As an alternative, steam pressure could be measured instead of steam flow rate. Feedforward control can also be used advantageously for level control problems where the objective is *surge control* (or *averaging control*), rather than tight level control. For example, the input streams to a surge tank will be intermittent if they are effluent streams from batch operations, but the tank exit stream can be continuous. Special feedforward control methods have been developed for these batch-to-continuous transitions to balance the surge capacity requirement for the measured inlet flow rates with the surge control objective of gradual changes in the tank exit stream (Blevins et al., 2003). In practical applications, feedforward control is normally used in combination with feedback control.

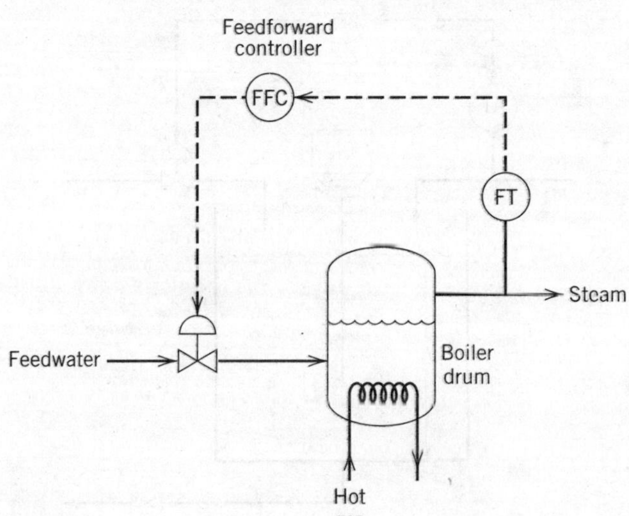

Figure 12.3 Feedforward control of the liquid level in a boiler drum.

Feedforward control is used to reduce the effects of measurable disturbances, while *feedback trim* compensates for inaccuracies in the process model, measurement errors, and unmeasured disturbances. The feedforward and feedback controllers can be combined in several different ways, as will be discussed in Section 12.1.5. A typical configuration is shown in Fig. 12.4, where the outputs of the feedforward and feedback controllers are added together and the combined signal is sent to the control valve.

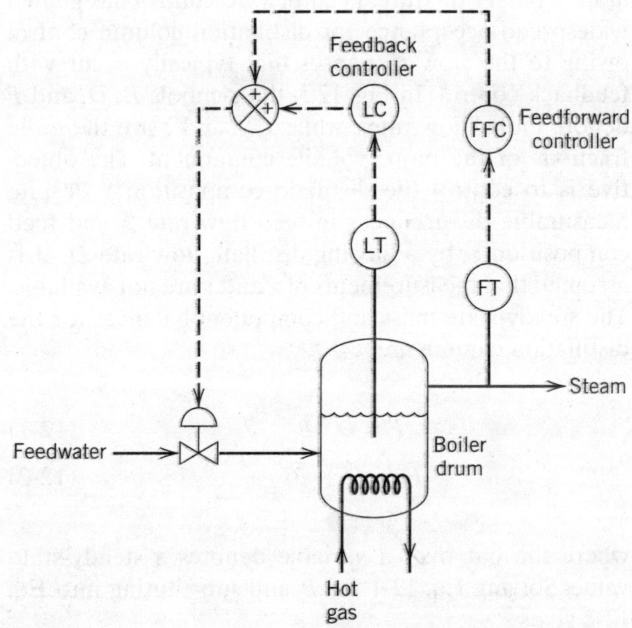

Figure 12.4 Feedforward–feedback control of the boiler drum level.

12.1.2 Feedforward Controller Design Based On Steady-State Models

A useful interpretation of feedforward control is that it continually attempts to balance the material or energy that must be delivered to the process against the demands of the disturbance (Shinskey, 1996). For example, the level control system in Fig. 12.3 adjusts the

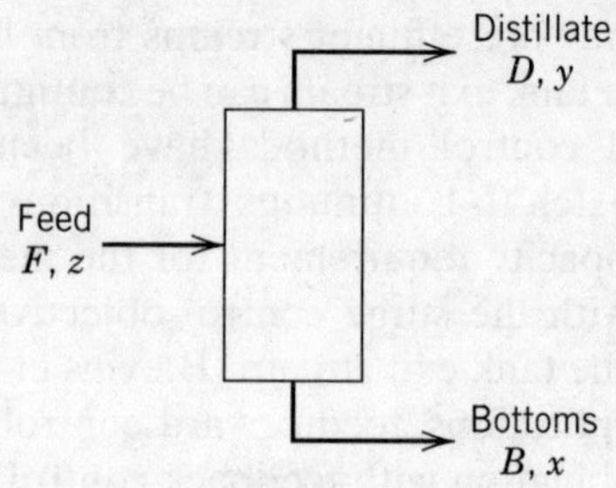

Figure 12.5 A simplified schematic diagram of a distillation column.

feedwater flow so that it balances the steam demand. Thus, it is natural to base the feedforward control calculations on material and energy balances. For simplicity, we will first consider designs based on steady-state balances using physical variables rather than deviation variables. Design methods based on dynamic models are considered in Section 12.1.3.

To illustrate the design procedure, consider the distillation column shown in Fig. 12.5, which is used to separate a binary mixture. Feedforward control has gained widespread acceptance for distillation column control owing to the slow responses that typically occur with feedback control. In Fig. 12.5, the symbols B, D, and F denote molar flow rates, while x, y, and z are the mole fractions of the more volatile component. The objective is to control the distillate composition y despite measurable disturbances in feed flow rate F and feed composition z, by adjusting distillate flow rate D. It is assumed that measurements of x and y are not available. The steady-state mass and component balances for the distillation column are

$$\bar{F} = \bar{D} + \bar{B} \tag{12-1}$$

$$\bar{F}\bar{z} = \bar{D}\bar{y} + \bar{B}\bar{x} \tag{12-2}$$

where the bar over a variable denotes a steady-state value. Solving Eq. 12-1 for $\bar{B}$ and substituting into Eq. 12-2 gives

$$\bar{D} = \frac{\bar{F}(\bar{z} - \bar{x})}{\bar{y} - \bar{x}} \tag{12-3}$$

Because x and y are not measured, we replace x and y by their set points and replace $\bar{D}$, $\bar{F}$, and $\bar{z}$ by $D(t)$, $F(t)$, and $z(t)$, respectively. These substitutions yield a feedforward control law:

$$D(t) = \frac{F(t)\left[z(t) - x_{sp}\right]}{y_{sp} - x_{sp}} \tag{12-4}$$

Thus, the feedforward controller calculates the required value of the manipulated variable D from the measurements of the disturbance variables, F and z, and the knowledge of the composition set points x_{sp} and y_{sp}. Note that Eq. 12-4 is based on physical variables, not deviation variables. The feedforward control law is nonlinear due to the product of two process variables, $F(t)$ and $z(t)$. Because the control law was designed based on the steady-state model in Eqs. 12-1 and 12-2, it may not perform well for transient conditions. This issue is considered in Sections 12.1.3 and 12.1.6.

Blending System

To further illustrate the design method, consider the blending system and feedforward controller shown in Fig. 12.6. We wish to design a feedforward control scheme to maintain exit composition x at a constant set point x_{sp}, despite disturbances in inlet composition, x_1. Suppose that inlet flow rate w_1 and the composition of the other inlet stream x_2 are constant. It is assumed that x_1 is measured but that x is not. (If x were measured, then feedback control would also be possible.) The manipulated variable is inlet flow rate w_2. The flow-head relation for the valve on the exit line is given by $w = C_v\sqrt{h}$. Note that the feedforward controller has two input signals: the x_1 measurement x_{1m}, and the set point for the exit composition x_{sp}.

The starting point for the feedforward controller design is the steady-state mass and component balances that were considered in Chapter 1,

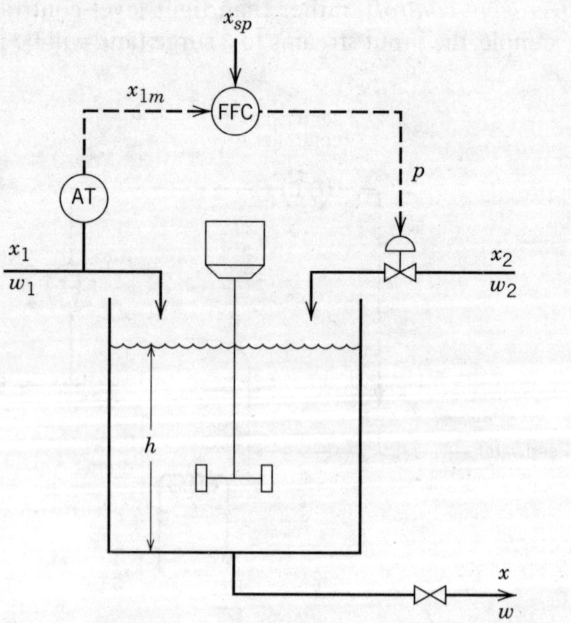

Figure 12.6 Feedforward control of exit composition in the blending system.

$$\bar{w} = \bar{w}_1 + \bar{w}_2 \quad (12\text{-}5)$$

$$\bar{w}\bar{x} = \bar{w}_1\bar{x}_1 + \bar{w}_2\bar{x}_2 \quad (12\text{-}6)$$

These equations are the steady-state version of the dynamic model in Eqs. 2-12 and 2-13. Substituting Eq. 12-5 into Eq. 12-6 and solving for $\bar{w}_2$ gives

$$\bar{w}_2 = \frac{\bar{w}_1(\bar{x} - \bar{x}_1)}{\bar{x}_2 - \bar{x}} \quad (12\text{-}7)$$

In order to derive a feedforward control law, we replace $\bar{x}$ by x_{sp} and $\bar{w}_2$ and $\bar{x}_1$ by $w_2(t)$ and $x_1(t)$, respectively:

$$w_2(t) = \frac{\bar{w}_1\left[x_{sp} - x_1(t)\right]}{\bar{x}_2 - x_{sp}} \quad (12\text{-}8)$$

Note that this feedforward control law is also based on physical variables rather than deviation variables. The feedforward control law in Eq. 12-8 is not in the final form required for actual implementation, because it ignores two important instrumentation considerations: First, the actual value of x_1 is not available, but its measured value x_{1m} is. Second, the controller output signal is p rather than inlet flow rate, w_2. Thus, the feedforward control law should be expressed in terms of x_{1m} and p, rather than x_1 and w_2. Consequently, a more realistic feedforward control law should incorporate the appropriate steady-state instrument relations for the w_2 flow transmitter and the control valve, as shown below.

Composition Measurement for x_1

Suppose that the sensor/transmitter for x_1 is an electronic instrument with negligible dynamics and a standard output range of 4–20 mA. In analogy with Section 8.1, if the calibration relation is linear, it can be written as

$$x_{1m}(t) = K_t\left[x_1(t) - (x_1)_0\right] + 4 \quad (12\text{-}9)$$

where $(x_1)_0$ is the zero of this instrument and K_t is its gain. From Eq. 8.2,

$$K_t = \frac{\text{output range}}{\text{input range}} = \frac{20 - 4\,\text{mA}}{S_t} \quad (12\text{-}10)$$

where S_t is the span of the instrument.

Control Valve and Current-to-Pressure Transducer

Suppose that the current-to-pressure transducer and the control valve are designed to have linear input–output relationships with negligible dynamics. Their input ranges (spans) are 4–20 mA and 20–100 kPa (3–15 psi), respectively. Then in analogy with Eq. 8-2, the relationship between the controller output signal $p(t)$ and inlet flow rate $w_2(t)$ can be written as

$$w_2(t) = K_v K_{IP}\left[p(t) - 4\right] + (w_2)_0 \quad (12\text{-}11)$$

where K_v and K_{IP} are the steady-state gains for the control valve and I/P transducer, respectively, and $(w_2)_0$ is the minimum value of the w_2 flow rate that corresponds to the minimum controller output value of 4 mA. Note that all of the symbols in Eqs. 12-5 through 12-11 denote physical variables rather than deviation variables. Rearranging Eq. 12-9 gives

$$x_1(t) = \frac{x_{1m}(t) - 4}{K_t} + (x_1)_0 \quad (12\text{-}12)$$

Substituting Eqs. 12-11 and 12-12 into Eq. 12-8 and rearranging the resulting equation provides a feedforward control law that is suitable for implementation:

$$p(t) = C_1 + C_2\left[\frac{K_t x_{sp} - x_{1m}(t) + C_3}{\bar{x}_2 - x_{sp}}\right] \quad (12\text{-}13)$$

where

$$C_1 \triangleq 4 - \frac{(w_2)_0}{K_v K_{IP}} \quad (12\text{-}14)$$

$$C_2 \triangleq \frac{\bar{w}_1}{K_v K_{IP} K_t} \quad (12\text{-}15)$$

$$C_3 \triangleq 4 + K_t(x_1)_0 \quad (12\text{-}16)$$

An alternative feedforward control scheme for the blending system is shown in Fig. 12.7. Here the feedforward controller output signal serves as a set point to a feedback controller for flow rate w_2. The advantage of this configuration is that it is less sensitive to valve sticking and upstream pressure fluctuations. Because the feedforward controller calculates the w_2 set point rather than the signal to the control valve p, it would not be necessary to incorporate Eq. 12-11 into the feedforward control law. The blending and distillation column examples illustrate that feedforward controllers can be designed using steady-state mass and energy balances. The advantages of this approach are that the required calculations are quite simple, and a detailed process model is not required. However, a disadvantage is that

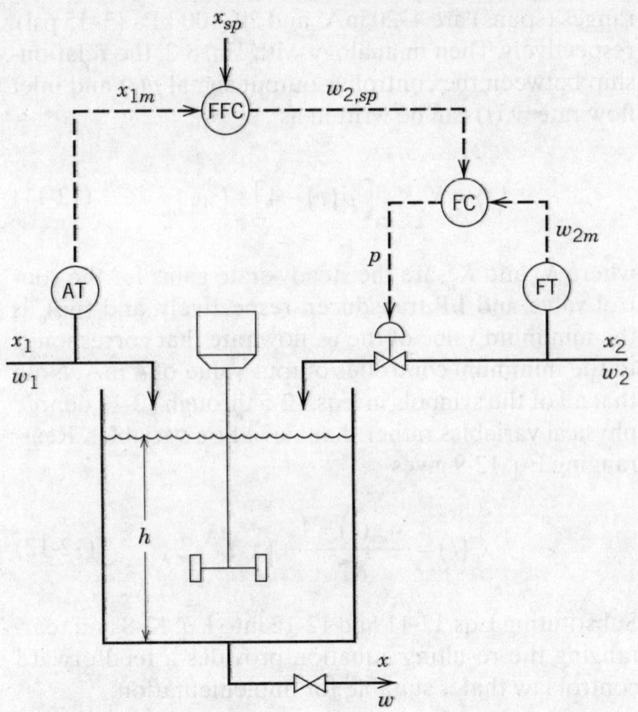

Figure 12.7 Feedforward control of exit composition using an additional flow control loop.

process dynamics are neglected, and consequently the control system may not perform well during transient conditions. The feedforward controllers can be improved by adding dynamic compensation, usually in the form of a lead–lag unit. This topic is discussed in Section 12.1.6. An alternative approach is to base the controller design on a dynamic model of the process, as discussed in the next section.

In many feedforward control applications (e.g., the two previous examples), the controller output is the desired value of a flow rate through a control valve. Because control valves tend to exhibit hysteresis and other nonlinear behavior (Chapter 8), the controller output is usually the set-point for the flow control loop, rather than the signal to the control valve. This strategy provides more assurance that the calculated flow rate is actually implemented.

12.1.3 Feedforward Controller Design Based On Dynamic Models

In this section, we consider the design of feedforward control systems based on dynamic, rather than steady-state, process models. We will restrict our attention to design techniques based on linear dynamic models. But nonlinear process models can also be used (Smith and Corripio, 2006).

As the starting point, consider the block diagram in Fig. 12.8. This diagram is similar to Fig. 9.8 for feedback control, but an additional signal path through transfer functions, G_t and G_f, has been added. The disturbance transmitter with transfer function G_t sends a measurement of the disturbance variable to the feedforward controller G_f. The outputs of the feedforward and feedback controllers are then added together, and the sum is sent to the control valve. In contrast to the steady-state design methods of Section 12.1.2, the block diagram in Fig. 12.8 is based on deviation variables. The closed-loop transfer function for disturbance changes in Eq. 12-17 can be derived using the block diagram algebra that was introduced in Chapter 9:

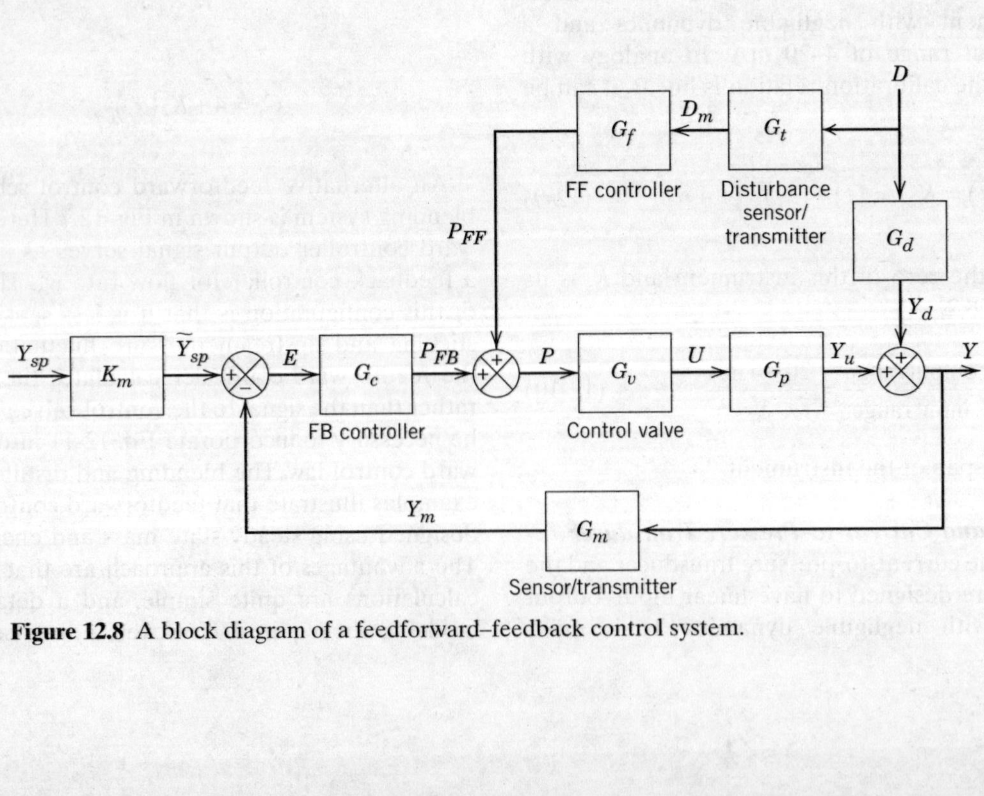

Figure 12.8 A block diagram of a feedforward–feedback control system.

$$\frac{Y(s)}{D(s)} = \frac{G_d + G_t G_f G_v G_p}{1 + G_c G_v G_p G_m} \quad (12\text{-}17)$$

Ideally, we would like the control system to produce *perfect control*, where the controlled variable remains exactly at the set point despite arbitrary changes in the disturbance variable, D. Thus, if the set point is constant ($Y_{sp}(s) = 0$), we want $Y(s) = 0$, even though $D(s) \neq 0$. This condition can be satisfied by setting the numerator of Eq. 12-17 equal to zero and solving for G_f:

$$G_f = -\frac{G_d}{G_t G_v G_p} \quad (12\text{-}18)$$

Figure 12.8 and Eq. 12-18 provide a useful interpretation of the ideal feedforward controller. Figure 12.8 indicates that a disturbance has two effects: it upsets the process via the disturbance transfer function G_d; however, a corrective action is generated via the path through $G_t G_f G_v G_p$. Ideally, the corrective action compensates exactly for the upset so that signals Y_d and Y_u cancel each other and $Y(s) = 0$. Next, we consider three examples in which feedforward controllers are derived for various types of process models. For simplicity, it is assumed that the disturbance transmitters and control valves have negligible dynamics, that is, $G_t(s) = K_t$ and $G_v(s) = K_v$, where K_t and K_v denote steady-state gains.

EXAMPLE 12.1

Suppose that

$$G_d = \frac{K_d}{\tau_d s+1} \quad G_p = \frac{K_p}{\tau_p s+1} \quad (12\text{-}19)$$

Then, from Eq. 12-18, the ideal feedforward controller is

$$G_f = -\left(\frac{K_d}{K_t K_v K_p}\right)\left(\frac{\tau_p s+1}{\tau_d s+1}\right) \quad (12\text{-}20)$$

This controller is a lead–lag unit with a gain given by $K_f = -K_d/K_t K_v K_p$. The dynamic response characteristics of lead–lag units were considered in Example 6.1 of Chapter 6.

EXAMPLE 12.2

Now consider

$$G_d = \frac{K_d}{\tau_d s+1} \quad G_p = \frac{K_p e^{-\theta s}}{\tau_p s+1} \quad (12\text{-}21)$$

From Eq. 12-18

$$G_f = -\left(\frac{K_d}{K_t K_v K_p}\right)\left(\frac{\tau_p s+1}{\tau_d s+1}\right)e^{+\theta s} \quad (12\text{-}22)$$

Because the term $e^{+\theta s}$ represents a *negative* time delay, implying a predictive element, the ideal feedforward controller in Eq. 12-22 is *physically unrealizable*. However, we can approximate the $e^{+\theta s}$ term by increasing the value of the lead time constant from τ_p to $\tau_p + \theta$.

EXAMPLE 12.3

Finally, if

$$G_d = \frac{K_d}{\tau_d s+1} \quad G_p = \frac{K_p}{(\tau_{p1} s+1)(\tau_{p2} s+1)} \quad (12\text{-}23)$$

then the ideal feedforward controller,

$$G_f = -\left(\frac{K_d}{K_t K_v K_p}\right)\frac{(\tau_{p1} s+1)(\tau_{p2} s+1)}{\tau_d s+1} \quad (12\text{-}24)$$

is physically unrealizable, because the numerator is a higher-order polynomial in s than the denominator (cf. Section 3.3). Again, we could approximate this controller by a physically realizable transfer function such as a lead–lag unit, where the lead time constant is the sum of the two time constants, $\tau_{p1} + \tau_{p2}$

Stability Considerations

To analyze the stability of the closed-loop system in Fig. 12.8, we consider the closed-loop transfer function in Eq. 12-17. Setting the denominator equal to zero gives the characteristic equation,

$$1 + G_c G_v G_p G_m = 0 \tag{12-25}$$

In Chapter 9, it was shown that the roots of the characteristic equation completely determine the stability of the closed-loop system. Because G_f does not appear in the characteristic equation, we have an important theoretical result: *the feedforward controller has no effect on the stability of the feedback control system*. This is a desirable situation that allows the feedback and feedforward controllers to be tuned individually.

Lead–Lag Units

The three examples in the previous section have demonstrated that lead–lag units can provide reasonable approximations to ideal feedforward controllers. Thus, if the feedforward controller consists of a lead–lag unit with gain K_f, its transfer function is:

$$G_f(s) = \frac{U(s)}{D(s)} = \frac{K_f(\tau_1 s + 1)}{\tau_2 s + 1} \tag{12-26}$$

where K_f, τ_1, and τ_2 are adjustable controller parameters. In Section 12.1.6, tuning techniques for this type of feedforward controller are considered.

EXAMPLE 12.4

Consider the blending system of Section 12.1.2, but now assume that a pneumatic control valve and an I/P transducer are used. A feedforward–feedback control system is to be designed to reduce the effect of disturbances in feed composition x_1 on the controlled variable, product composition x. Inlet flow rate w_2 can be manipulated. Using the information given below, design the following control systems and compare the closed-loop responses for a +0.2 step change in x_1.

(a) A feedforward controller based on a steady-state model of the process.

(b) Static and dynamic feedforward controllers based on a linearized, dynamic model.

(c) A PI feedback controller based on the Ziegler–Nichols settings for the continuous cycling method.

(d) The combined feedback–feedforward control system that consists of the feedforward controller of part (a) and the PI controller of part (c). Use the configuration in Fig. 12.8.

Process Information

The pilot-scale blending tank has an internal diameter of 2 m and a height of 3 m. Inlet flow rate w_1 and inlet composition x_2 are constant. The nominal steady-state operating conditions are

$$\bar{w}_1 = 650\,\text{kg/min} \quad \bar{x}_1 = 0.2 \quad \bar{h} = 1.5\,\text{m}$$

$$\bar{w}_2 = 350\,\text{kg/min} \quad \bar{x}_2 = 0.6 \quad \rho = 1\,\text{g/cm}^3 \quad \bar{x} = 0.34$$

The flow-head relation for the valve on the exit line is given by $w = C_v\sqrt{h}$.

Instrumentation (The range for each electronic signal is 4 to 20 mA.)

Current-to-pressure transducer: The I/P transducer acts as a linear device with negligible dynamics. The output signal changes from 20–100 kPa when the input signal changes full-scale from 4 to 20 mA.

Control valve: The behavior of the control valve can be approximated by a first-order transfer function with a time constant of 5 s (0.0833 min). A 20–100 kPa change in the signal to the control valve produces a 300-kg/min change in w_2.

Composition measurement: The zero and span of each composition transmitter are 0 and 0.50 (mass fraction), respectively. The output range is 4–20 mA. A one-minute time delay is associated with each composition measurement.

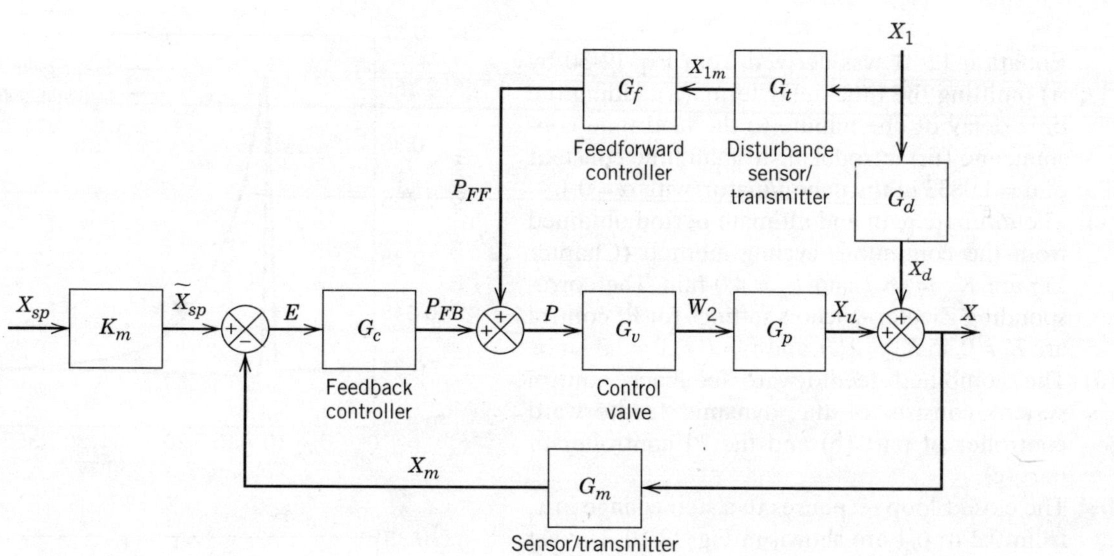

Figure 12.9 Block diagram for feedforward–feedback control of the blending system.

Solution

A block diagram for the feedforward–feedback control system is shown in Fig. 12.9.

(a) Using the given information, we can calculate the following steady-state gains:

$$K_{IP} = (100-20)/(20-4) = 5\,\text{kPa/mA}$$
$$K_v = 300/80 = 3.75\,\text{kg/min kPa}$$
$$K_t = (20-4)/0.5 = 32\,\text{mA}$$

Substitution into Eqs. 12-13 to 12-16 with $(w_2)_0 = 0$ and $(x_1)_0 = 0$ gives the following feedforward control law:

$$p(t) = 4 + 1.083\left[\frac{32x_{sp} - x_{1m}(t) + 4}{0.6 - x_{sp}}\right] \qquad (12\text{-}27)$$

(b) The following expression for the ideal feedforward controller can be derived in analogy with the derivation of Eq. 12-18:

$$G_f = -\frac{G_d}{K_{IP} G_t G_v G_p} \qquad (12\text{-}28)$$

The process and disturbance transfer functions are similar to the ones derived in Example 4.1:

$$\frac{X'(s)}{W_2'(s)} = G_p(s) = \frac{K_p}{\tau s+1}$$

$$\frac{X'(s)}{X_1'(s)} = G_d(s) = \frac{K_d}{\tau s+1}$$

where

$$K_p = \frac{\bar{x}_2 - \bar{x}}{\bar{w}},\ K_d = \frac{\bar{w}_1}{\bar{w}},\ \tau = V\rho/\bar{w},\ V = \pi R^2 h$$

Substituting numerical values gives

$$\frac{X'(s)}{W_2'(s)} = \frac{2.6\times10^{-4}}{4.71s+1},\ \frac{X'(s)}{X_1'(s)} = \frac{0.65}{4.71s+1} \qquad (12\text{-}29)$$

The transfer functions for the instrumentation can be determined from the given information:

$$G_{IP} = K_{IP} = 5.17\,\text{kPa/mA},$$
$$G_t(s) = G_m(s) = K_t e^{-\theta s} = 32e^{-s}$$
$$G_v(s) = \frac{K_v}{\tau_v s+1} = \frac{25}{0.0833s+1}$$

Substituting the individual transfer functions into Eq. 12-28 gives the ideal dynamic feedforward controller:

$$G_f(s) = -4.17(0.0833s+1)e^{+s} \qquad (12\text{-}30)$$

Note that $G_f(s)$ is physically unrealizable. The static (or steady-state) version of the controller is simply a gain, $G_f(s) = -4.17$. In order to derive a physically realizable dynamic controller, we approximate the unrealizable controller in Eq. 12-30 by a lead–lag unit:

$$G_f(s) = -4.17\frac{1.0833s+1}{\alpha(1.0833)s+1} \qquad (12\text{-}31)$$

Equation 12-31 was derived from Eq. 12-30 by (i) omitting the time-delay term, (ii) adding the time delay of one minute to the lead time constant, and (iii) introducing a small time constant of $\alpha = 1.0833$ in the denominator, with $\alpha = 0.1$.

(c) The ultimate gain and ultimate period obtained from the continuous cycling method (Chapter 11) are $K_{cu} = 48.7$, and $P_u = 4.0$ min. The corresponding Ziegler–Nichols settings for PI control are $K_c = 0.45\ K_{cu} = 21.9$, and $\tau_1 = P_u/1.2 = 3.33$ min.

(d) The combined feedforward–feedback control system consists of the dynamic feedforward controller of part (b) and the PI controller of part (c).

The closed-loop responses to a step change in x_1 from 0.2 to 0.4 are shown in Fig. 12.10. The set point is the nominal value, $x_{sp} = 0.34$. The static feedforward controllers for cases (a) and (b) are equivalent and thus produce identical responses. The comparison in part (a) of Fig. 12.10 shows that the dynamic feedforward controller is superior to the static feedforward controller, because it provides a better approximation to the ideal feedforward controller of Eq. 12-30. The PI controller in part (b) of Fig. 12.10 produces a larger maximum deviation than the dynamic feedforward controller. The combined feedforward–feedback control system of part (d) results in better performance than the PI controller, because it has a much smaller maximum deviation and a smaller IAE value. The peak in the response at approximately $t = 13$ min in Fig. 12.10*b* is a consequence of the x_1 measurement time delay.

For this example, feedforward control with dynamic compensation provides a better response to the measured x_1 disturbance than does combined feedforward–feedback control. However, feedback control is essential to cope with unmeasured disturbances and modeling errors. Thus, a combined feedforward–feedback control system is preferred in practice.

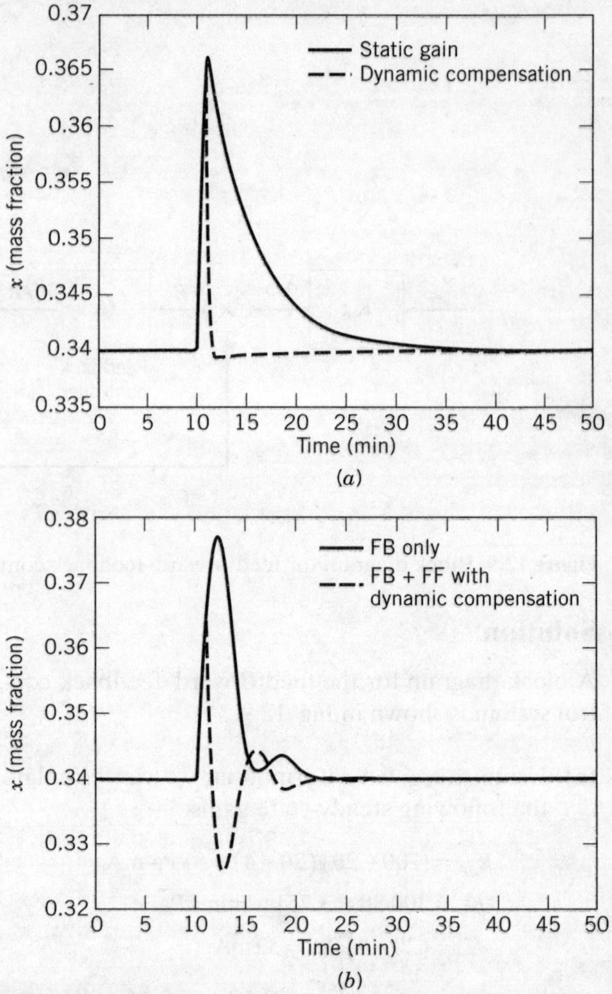

Figure 12.10 Comparison of closed-loop responses: (a) feedforward controllers with and without dynamic compensation; (b) feedback control and feedforward–feedback control.

12.1.4 The Relationship Between The Steady-State And Dynamic Design Methods

In the previous two sections, we considered two design methods for feedforward control. The design method of Section 12.1.2 was based on a nonlinear steady-state process model, while the design method of Section 12.1.3 was based on a transfer function model and block diagram analysis. Next, we show how the two design methods are related.

Dynamic Design Method

The block diagram of Fig. 12.8 indicates that the manipulated variable is related to the disturbance variable by

$$\frac{U(s)}{D(s)} = G_v(s)G_f(s)G_t(s) \tag{12-32}$$

Let the steady-state gain for this transfer function be denoted by K. Thus, as shown in Chapter 4:

$$K = \lim_{s\to 0} G_v(s)G_f(s)G_t(s) \tag{12-33}$$

Suppose that the disturbance changes from a nominal value, $\bar{d}$, to a new value, d_1. Denote the change as $\Delta d = d_1 - \bar{d}$. Let the corresponding steady-state change in the manipulated variable be denoted by $\Delta u = u_1 - \bar{u}$. Then, from Eqs. 12-32 and 12-33 and the definition of a steady-state gain in Chapter 4, we have

$$K = \frac{\Delta u}{\Delta d} \tag{12-34}$$

Steady-State Design Methods

The steady-state design method of Section 12.1.2 produces a feedforward control law that has the general nonlinear form:

$$u = f(d, y_{sp}) \tag{12-35}$$

Let K_{loc} denote the local derivative of u with respect to d at the nominal value $\bar{d}$:

$$K_{loc} = \left(\frac{\partial u}{\partial d}\right)_{\bar{d}} \tag{12-36}$$

A comparison of Eqs. 12-34 and 12-36 indicates that if Δd is small, $K_{loc} \approx K$. If the steady-state feedforward control law of Eq. 12-35 is indeed linear, then $K_{loc} = K$ and the gains for the two design methods are equivalent.

Steady-State Controller Design Based on Transfer Function Models

For some feedforward control applications, dynamic compensation is not necessarily based on physical or economic considerations, or controller simplicity. In these situations, the feedforward controller is simply a gain that can be tuned or adapted for changing process conditions. If a transfer function model is available, the feedforward controller gain can be calculated from the steady-state version of Eq. 12-18:

$$G_f = K_f = \frac{-K_d}{K_v K_t K_p} \tag{12-37}$$

12.1.5 Configurations for Feedforward–Feedback Control

As mentioned in Section 12.1.1 and illustrated in Example 12.4, *feedback trim* is normally used in conjunction with feedforward control to compensate for Feedforward and feedback controllers can be combined in several different ways. In a typical control configuration, the outputs of the feedforward and feedback

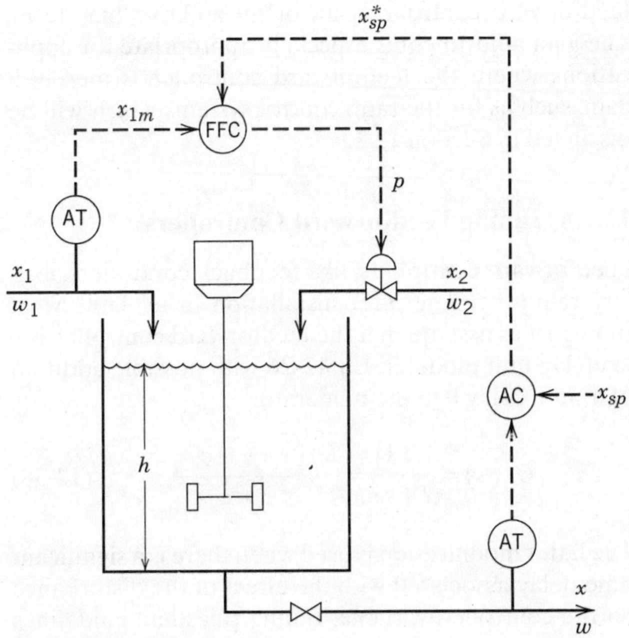

Figure 12.11 Feedforward–feedback control of exit composition in the blending system.

controllers are added together, and the sum is sent to the final control element. This configuration was introduced in Figs. 12.4 and 12.8. Its chief advantage is that the feedforward controller theoretically does not affect the stability of the feedback control loop. Recall that the feedforward controller transfer function $G_f(s)$ does not appear in the characteristic equation of Eq. 12-25.

An alternative configuration for feedforward–feedback control is to have the feedback controller output serve as the set point for the feedforward controller. It is especially convenient when the feedforward control law is designed using steady-state material and energy balances. For example, a feedforward–feedback control system for the blending system is shown in Fig. 12.11. Note that this control system is similar to the feedforward scheme in Fig. 12.6 except that the feedforward controller set point is now denoted as x_{sp}^*. It is generated as the output signal from the feedback controller. The actual set point x_{sp} is used as the set point for the feedback controller. In this configuration, the feedforward controller can affect the stability of the feedback control system, because it is now an element in the feedback loop. If dynamic compensation is included, it should be introduced outside of the feedback loop (e.g., applied to X_{1m}, not p). Otherwise, it will interfere with the operation of the feedback loop, especially when the controller is placed in the manual model (Shinskey, 1996).

Alternative ways of incorporating feedback trim into a feedforward control system include having the feedback controller output signal adjust either the feedforward controller gain or an additive bias term. The gain adjustment is especially appropriate for applications where the feedforward controller is merely a gain, such as for the ratio control systems which will be discussed in Section 12.2.

12.1.6 Tuning Feedforward Controllers

Feedforward controllers, like feedback controllers, usually require tuning after installation in a plant. Most tuning rules assume that the feedforward controller is a lead–lag unit model in Eq. 12-26 with possible addition of a time delay θ in the numerator.

$$G_f(s) = \frac{U(s)}{D(s)} = \frac{K_f(\tau_1 s+1)e^{-\theta s}}{\tau_2 s+1} \quad (12\text{-}38)$$

The latter modification is used when there is a significant time delay associated with the effect of the disturbance on the controlled variable. Tuning rules and guidelines for the feedforward controllers in Eqs. 12-26 and 12-38 are available (Guzmán and Hägglund, 2011; Hast and Hägglund, 2014;McMillan, 2015). Comparisons of alternative FF–FB configurations have also been reported (Guzmán et al., 2015; McMillan, 2015).

Next, we consider a simple tuning procedure for the lead–lag unit, feedforward controller in Eq. 12-26 with K_f, τ_1, and τ_2 as adjustable controller parameters.

Step 1. *Adjust* K_f. The effort required to tune a controller is greatly reduced if good initial estimates of the controller parameters are available. An initial estimate of K_f can be obtained from a steady-state model of the process or from steady-state data. For example, suppose that the open-loop responses to step changes in d and u are available, as shown in Fig. 12.12. After K_p and K_d have been determined, the feedforward controller gain can be calculated from Eq. 12-37. Gains K_t and K_v are available from the steady-state characteristics of the transmitter and control valve.

To tune the controller gain, K_f is set equal to an initial value and a small step change (3–5%) in the disturbance variable d is introduced, if this is feasible. If an offset results, then K_f is adjusted until the offset is eliminated. While K_f is being tuned, τ_1 and τ_2 should be set equal to their minimum values, ideally zero.

Step 2. *Determine initial values for* τ_1 *and* τ_2. Theoretical values for τ_1 and τ_2 can be calculated if a dynamic model of the process is available. Alternatively, initial estimates can be determined from open-loop response data. For example, if the step responses have the shapes shown in Fig. 12.12, a reasonable process model is

$$G_p(s) = \frac{K_p}{\tau_p s+1} \qquad G_d(s) = \frac{K_d}{\tau_d s+1} \quad (12\text{-}39)$$

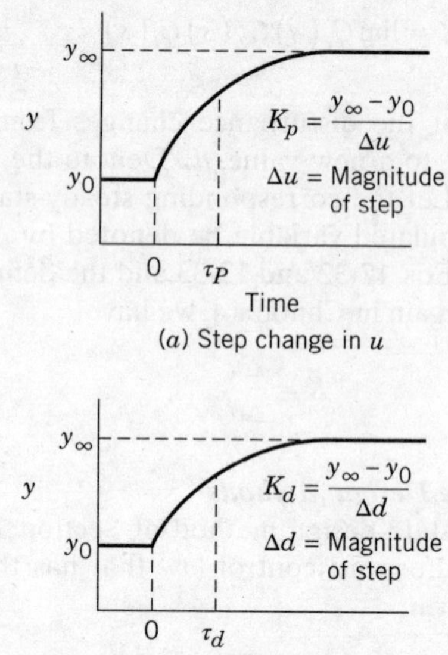

(*a*) Step change in u

(*b*) Step change in d ($\Delta d < 0$)

Figure 12.12 The open-loop responses to step changes in u and d.

where τ_p and τ_d can be calculated using one of the methods of Section 6.5. A comparison of Eqs. 12-20 and 12-26 leads to the following expressions for τ_1 and τ_2:

$$\tau_1 = \tau_p \quad (12\text{-}40)$$

$$\tau_2 = \tau_d \quad (12\text{-}41)$$

These values can then be used as initial estimates for the fine tuning of τ_1 and τ_2 in Step 3. If neither a process model nor experimental data are available, the relations $\tau_1/\tau_2 = 2$ or $\tau_1/\tau_2 = 0.5$ may be used, depending on whether the controlled variable responds faster to the disturbance variable or to the manipulated variable.

Step 3. *Fine-tune* τ_1 *and* τ_2. The final step is a trial-and-error procedure to fine-tune τ_1 and τ_2 by making small step changes in d, if feasible. The desired step response consists of small deviations in the controlled variable with equal areas above and below the set point (Shinskey, 1996), as shown in Fig. 12.13. For simple process models, it can be shown theoretically that equal areas above and below the set point imply that the difference, $\tau_1 - \tau_2$, is correct. In subsequent tuning to reduce the size of the areas, τ_1 and τ_2 should be adjusted so that $\tau_1 - \tau_2$ remains constant.

As a hypothetical illustration of this trial-and-error tuning procedure, consider the set of responses shown in Fig. 12.14 for positive step changes in disturbance variable d. It is assumed that $K_p > 0$, $K_d < 0$, and that controller gain K_f has already been adjusted so that offset is eliminated. For the initial values of τ_1 and τ_2 in Fig. 12.14*a*, the controlled variable is below the set point, which implies that τ_1 should be increased

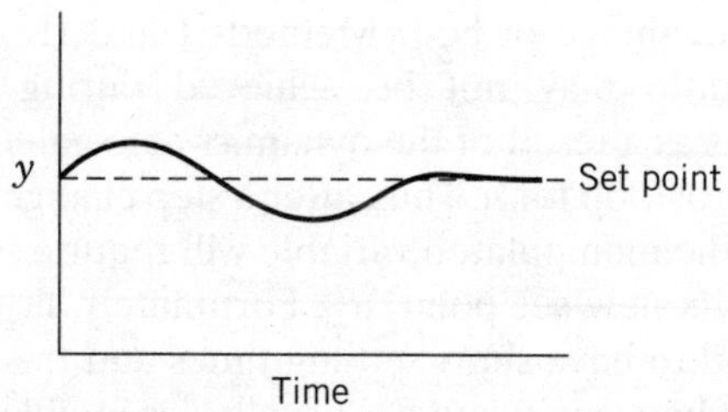

Figure 12.13 The desired response for a well-tuned feedforward controller. Note approximately equal areas above and below the set point.

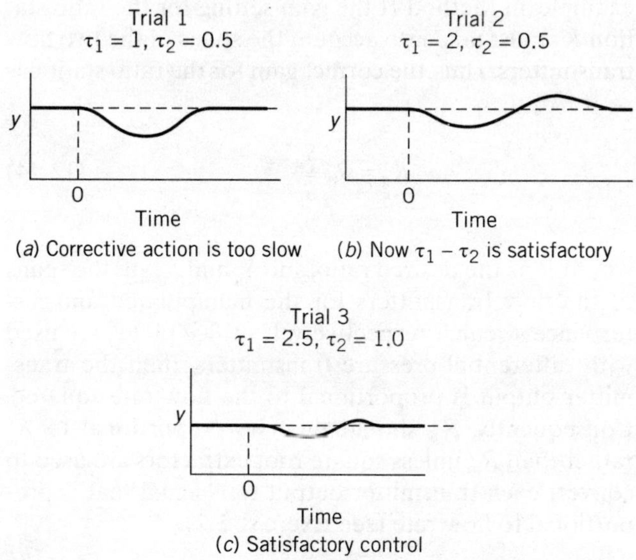

Figure 12.14 An example of feedforward controller tuning.

to speed up the corrective action. (Recall that $K_p > 0$, $K_d < 0$, and that positive step changes in d are introduced.) Increasing τ_1 from 1 to 2 gives the response in Fig. 12.14*b*, which has equal areas above and below the set point. Thus, in subsequent tuning to reduce the size of each area, $\tau_1 - \tau_2$ should be kept constant. Increasing both τ_1 and τ_2 by 0.5 reduces the size of each area, as shown in Fig. 12.14*c*. Because this response is considered to be satisfactory, no further controller tuning is required.

12.2 RATIO CONTROL

Ratio control is a special type of feedforward control that has had widespread application in the process industries. Its objective is to maintain the ratio of two process variables at a specified value. The two variables are usually flow rates, a manipulated variable u and a disturbance variable d. Thus, the ratio

$$R \triangleq \frac{u}{d} \tag{12-42}$$

is controlled rather than the individual variables. In Eq. 12-42, u and d are physical variables, not deviation variables.

Typical applications of ratio control include (1) specifying the relative amounts of components in blending operations, (2) maintaining a stoichiometric ratio of reactants to a reactor, (3) keeping a specified reflux ratio for a distillation column, and (4) holding the fuel-air ratio to a furnace at the optimum value.

Ratio control can be implemented in two basic schemes. For Method I in Fig. 12.15, the flow rates for both the disturbance stream and the manipulated stream are measured, and the measured ratio, $R_m = u_m/d_m$, is calculated. The output of the divider element is sent to a ratio controller (RC) that compares the calculated ratio R_m to the desired ratio R_d and adjusts the manipulated flow rate u accordingly. The ratio controller is typically a PI controller with the desired ratio as its set point.

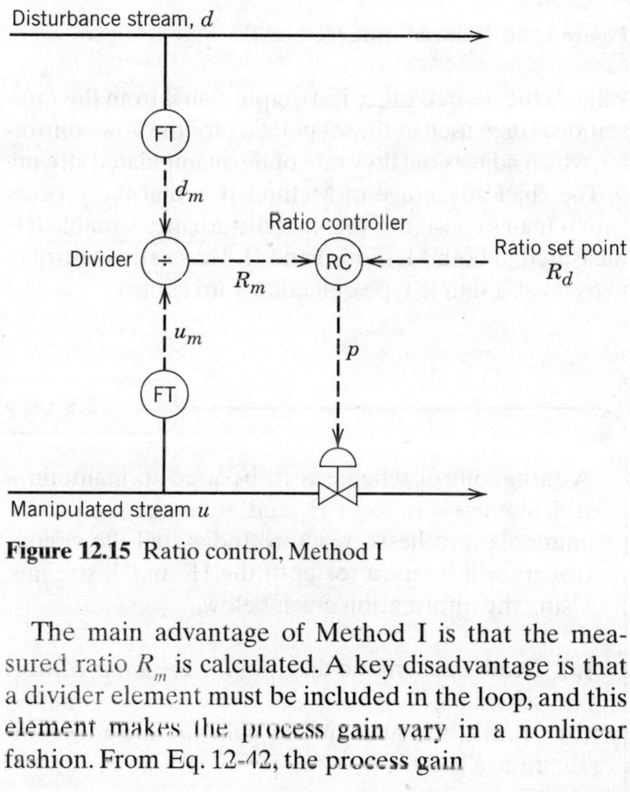

Figure 12.15 Ratio control, Method I

The main advantage of Method I is that the measured ratio R_m is calculated. A key disadvantage is that a divider element must be included in the loop, and this element makes the process gain vary in a nonlinear fashion. From Eq. 12-42, the process gain

$$K_p = \left(\frac{\partial R}{\partial u}\right)_d = \frac{1}{d} \tag{12-43}$$

is inversely related to the disturbance flow rate d. Because of this significant disadvantage, the preferred scheme for implementing ratio control is Method II, which is shown in Fig. 12.16.

In Method II, the flow rate of the disturbance stream is measured and transmitted to the *ratio station* (RS), which multiplies this signal by an adjustable gain, K_R, whose

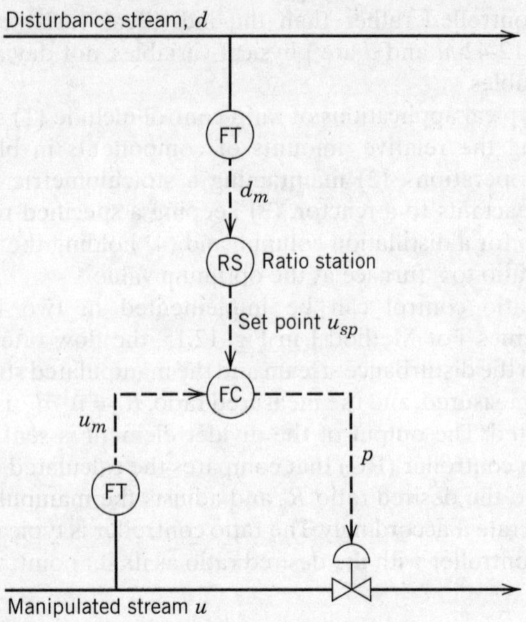

Figure 12.16 Ratio control, Method II.

value is the desired ratio. The output signal from the ratio station is then used as the set point u_{sp} for the flow controller, which adjusts the flow rate of the manipulated stream, *u*. The chief advantage of Method II is that the process gain remains constant. Note that disturbance variable *d* is measured in both Methods I and II. Thus, ratio control is, in essence, a simple type of feedforward control.

A disadvantage of both Methods I and II is that the desired ratio may not be achieved during transient conditions as a result of the dynamics associated with the flow control loop for *u*. Thus, after a step change in disturbance *d*, the manipulated variable will require some time to reach its new set point, u_{sp}. Fortunately, flow control loops tend to have short settling times and this transient mismatch between *u* and *d* is usually acceptable. For situations where it is not, modified versions of Method II have been proposed by Hägglund (2001) and Visioli (2005a,b).

Regardless of how ratio control is implemented, the process variables must be scaled appropriately. For example, in Method II the gain setting for the ratio station K_R must take into account the spans of the two flow transmitters. Thus, the correct gain for the ratio station is

$$K_R = R_d \frac{S_d}{S_u} \tag{12-44}$$

where R_d is the desired ratio, and S_u and S_d are the spans of the flow transmitters for the manipulated and disturbance streams, respectively. If orifice plates are used with differential-pressure transmitters, then the transmitter output is proportional to the flow rate squared. Consequently, K_R should then be proportional to R_d^2 rather than R_d, unless square root extractors are used to convert each transmitter output to a signal that is proportional to flow rate (see Exercise 5.2).

EXAMPLE 12.5

A ratio control scheme is to be used to maintain a stoichiometric ratio of H_2 and N_2 as the feed to an ammonia synthesis reactor. Individual flow controllers will be used for both the H_2 and N_2 streams. Using the information given below,

(a) Draw a schematic diagram for the ratio control scheme.
(b) Specify the appropriate gain for the ratio station, K_R.

Available information:

(i) The electronic flow transmitters have built-in square root extractors. The spans of the flow transmitters are 30 L/min for H_2 and 15 L/min for N_2.
(ii) The control valves have pneumatic actuators.
(iii) Each required current-to-pressure (I/P) transducer has a gain of 5.17 kPa/mA.
(iv) The ratio station is an electronic instrument with 4–20 mA input and output signals.

Solution

The stoichiometric equation for the ammonia synthesis reaction is

$$3H_2 + N_2 \rightleftarrows 2NH_3$$

In order to introduce a feed mixture in stoichiometric proportions, the ratio of the molar flow rates (H_2/N_2) should be 3:1. For the sake of simplicity, we assume that the ratio of the molar flow rates is equal to the ratio of the volumetric flow rates. But, in general, the volumetric flow rates also depend on the temperature and pressure of each stream (cf. the ideal gas law).

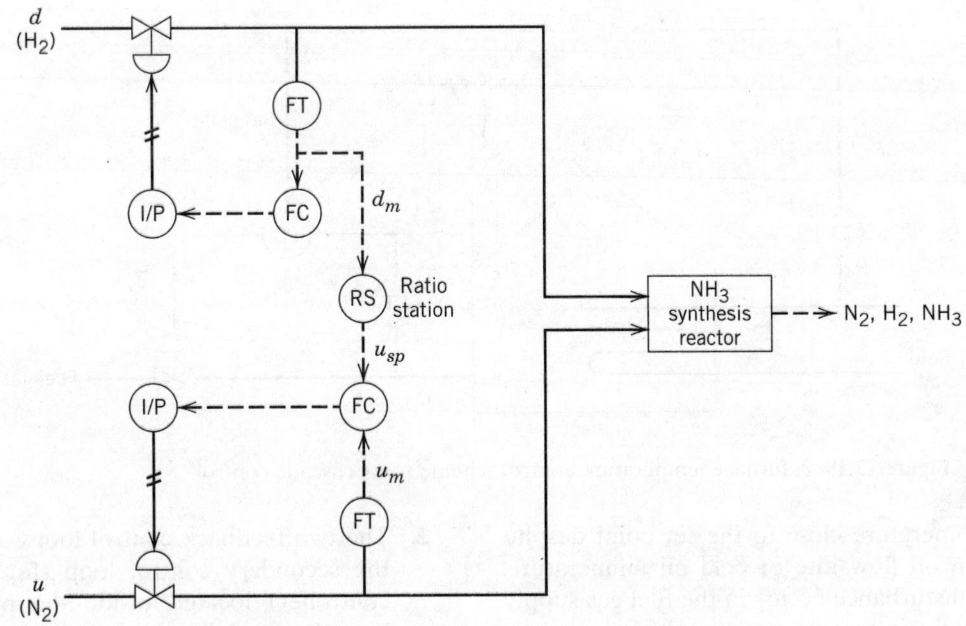

Figure 12.17 Ratio control scheme for an ammonia synthesis reactor of Example 12.5.

(a) The schematic diagram for the ammonia synthesis reaction is shown in Fig. 12.17. The H_2 flow rate is considered to be the disturbance variable, although this choice is arbitrary, because both the H_2 and N_2 flow rates are controlled.

Note that the ratio station is merely a device with an adjustable gain. The input signal to the ratio station is d_m, the measured H_2 flow rate. Its output signal u_{sp} serves as the set point for the N_2 flow control loop. It is calculated as $u_{sp} = K_R d_m$.

(b) From the stoichiometric equation, it follows that the desired ratio is $R_d = u/d = 1/3$. Substitution into Eq. 12-44 gives

$$K_R = \left(\frac{1}{3}\right)\left(\frac{30\,\text{L/min}}{15\,\text{L/min}}\right) = \frac{2}{3}$$

12.3 CASCADE CONTROL

A disadvantage of conventional feedback control is that corrective action for disturbances does not begin until after the controlled variable deviates from the set point. As discussed in Section 12.1, feedforward control offers large improvements over feedback control for processes that have large time constants or time delays. However, feedforward control requires that the disturbances be measured explicitly, and that a steady-state or dynamic model be available to calculate the controller output. An alternative approach that can significantly improve the dynamic response to disturbances employs a secondary measured variable and a secondary feedback controller. The secondary measured variable is located so that it recognizes the upset condition sooner than the controlled variable, but possible disturbances are not necessarily measured. This approach, called *cascade control*, is widely used in the process industries and is particularly useful when the disturbances are associated with the manipulated variable or when the final control element exhibits nonlinear behavior (Shinskey, 1996).

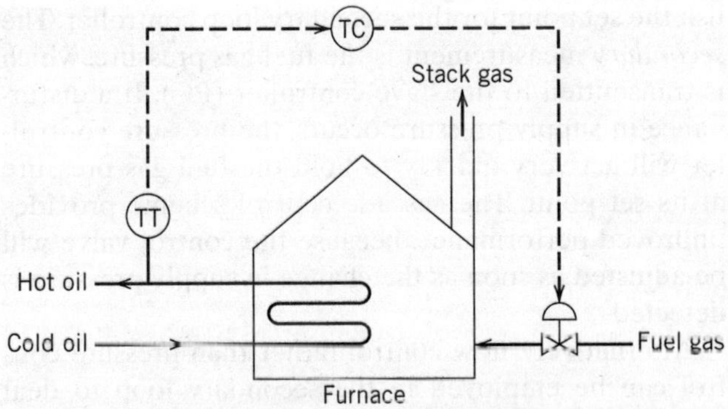

Figure 12.18 A furnace temperature control scheme that uses conventional feedback control.

As an example of where cascade control may be advantageous, consider the natural draft furnace temperature control system shown in Fig. 12.18. The conventional feedback control system in Fig. 12.18 may keep

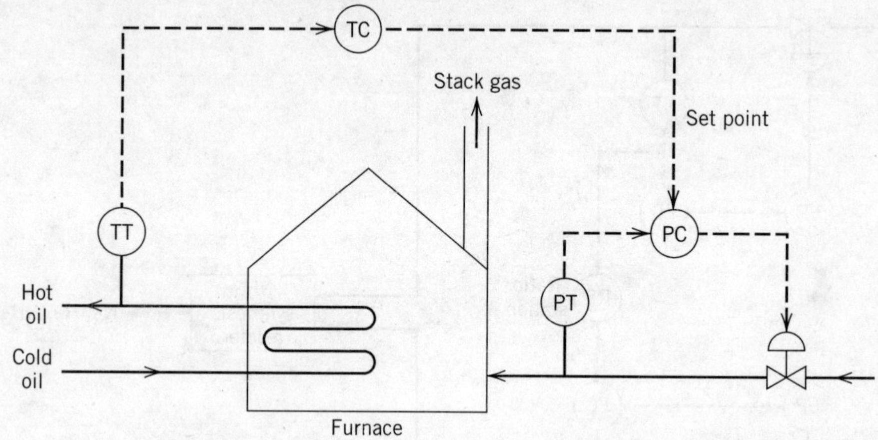

Figure 12.19 A furnace temperature control scheme using cascade control.

the hot oil temperature close to the set point despite disturbances in oil flow rate or cold oil temperature. However, if a disturbance occurs in the fuel gas supply pressure, the fuel gas flow will change, which upsets the furnace operation and changes the hot oil temperature. Only then can the temperature controller (TC) begin to take corrective action by adjusting the fuel gas flow based on the error from the setpoint, which can result in very sluggish responses to changes in fuel gas supply pressure. This disturbance is clearly associated with the manipulated variable.

Figure 12.19 shows a cascade control configuration for the furnace, which consists of a primary control loop (utilizing TT and TC) and a secondary control loop that controls the fuel gas pressure via PT and PC. The primary measurement is the hot oil temperature that is used by the *primary* controller (TC) to establish the set point for the secondary loop controller. The *secondary* measurement is the fuel gas pressure, which is transmitted to the slave controller (PC). If a disturbance in supply pressure occurs, the pressure controller will act very quickly to hold the fuel gas pressure at its set point. The cascade control scheme provides improved performance, because the control valve will be adjusted as soon as the change in supply pressure is detected.

Alternatively, flow control rather than pressure control can be employed in the secondary loop to deal with discharge pressure variations. If the performance improvements for disturbances in oil flow rate or inlet temperature are not large enough, then feedforward control could be utilized for those disturbances (see Section 12.1).

The cascade control loop structure has two distinguishing features:

1. The output signal of the primary controller serves as the set point for the secondary controller.
2. The two feedback control loops are nested, with the secondary control loop (for the secondary controller) located inside the primary control loop (for the primary controller).

Thus there are two controlled variables, two sensors, and one manipulated variable, while the conventional control structure has one controlled variable, one sensor, and one manipulated variable.

The primary control loop can change the set point of the pressure control loop based on deviations of the hot oil temperature from its set point. Note that all variables in this configuration can be viewed as deviation variables. If the hot oil temperature is at its set point, the deviation variable for the pressure set point is also zero, which keeps the pressure at its desired steady-state value.

Figure 12.20 shows a second example of cascade control, a stirred chemical reactor where cooling water flows through the reactor jacket to regulate the reactor temperature. The reactor temperature is affected by changes in disturbance variables such as reactant feed temperature or feed composition. The simplest control strategy would handle such disturbances by adjusting a control valve on the cooling water inlet stream. However, an increase in the inlet cooling water temperature, an unmeasured disturbance, can cause unsatisfactory performance. The resulting increase in the reactor temperature, due to a reduction in heat removal rate, may occur slowly. If appreciable dynamic lags in heat transfer occur due to the jacket as well as in the reactor, the corrective action taken by the controller will be delayed. To avoid this disadvantage, a feedback controller for the jacket temperature, whose set point is determined by the reactor temperature controller, can be added to provide cascade control, as shown in Fig. 12.20. The control system measures the jacket temperature, compares it to a set point, and adjusts the cooling water makeup. The

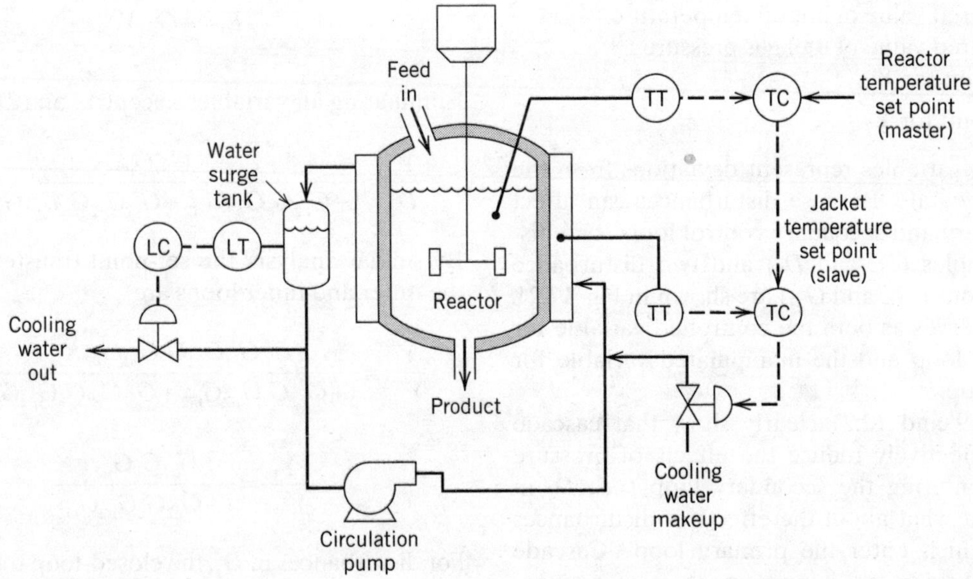

Figure 12.20 Cascade control of an exothermic chemical reactor.

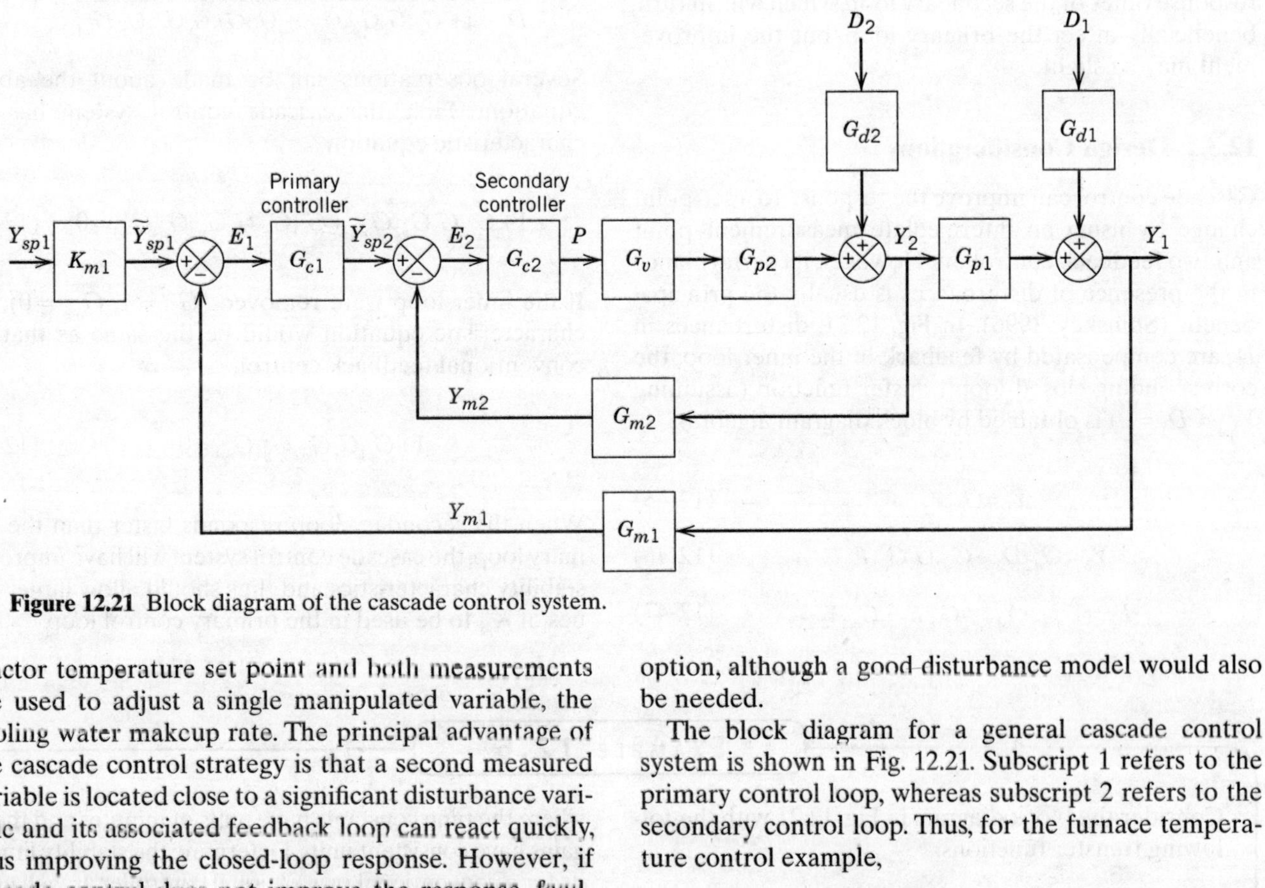

Figure 12.21 Block diagram of the cascade control system.

reactor temperature set point and both measurements are used to adjust a single manipulated variable, the cooling water makeup rate. The principal advantage of the cascade control strategy is that a second measured variable is located close to a significant disturbance variable and its associated feedback loop can react quickly, thus improving the closed-loop response. However, if cascade control does not improve the response, feedforward control should be the next strategy considered, with cooling water temperature as the measured disturbance variable. In this case, an inexpensive sensor (for temperature) makes feedforward control an attractive option, although a good disturbance model would also be needed.

The block diagram for a general cascade control system is shown in Fig. 12.21. Subscript 1 refers to the primary control loop, whereas subscript 2 refers to the secondary control loop. Thus, for the furnace temperature control example,

Y_1 = hot oil temperature
Y_2 = fuel gas pressure
D_1 = cold oil temperature (or cold oil flow rate)
D_2 = supply pressure of fuel gas

Y_{m1} = measured value of hot oil temperature
Y_{m2} = measured value of fuel gas pressure
Y_{sp1} = set point for Y_1
$\tilde{Y}_{sp2}$ = set point for Y_2

All of these variables represent deviations from the nominal steady state. Because disturbances can affect both the primary and secondary control loops, two disturbance variables (D_1 and D_2) and two disturbance transfer functions (G_{d1} and G_{d2}) are shown in Fig. 12.21. Note that Y_2 serves as both the controlled variable for the secondary loop and the manipulated variable for the primary loop.

Figures 12.19 and 12.21 clearly show that cascade control will effectively reduce the effects of pressure disturbances entering the secondary loop (i.e., D_2 in Fig. 12.21). But what about the effects of disturbances such as D_1, which enter the primary loop? Cascade control can provide an improvement over conventional feedback control when both controllers are well-tuned. The cascade arrangement will usually reduce the response times of the secondary loop, which will, in turn, beneficially affect the primary loop, but the improvement may be slight.

12.3.1 Design Considerations

Cascade control can improve the response to a set-point change by using an intermediate measurement point and two feedback controllers. However, its performance in the presence of disturbances is usually the principal benefit (Shinskey, 1996). In Fig. 12.21, disturbances in D_2 are compensated by feedback in the inner loop; the corresponding closed-loop transfer function (assuming $Y_{sp1} = D_1 = 0$) is obtained by block diagram algebra:

$$Y_1 = G_{p1} Y_2 \tag{12-45}$$

$$Y_2 = G_{d2} D_2 + G_{p2} G_v G_{c2} E_2 \tag{12-46}$$

$$E_2 = \tilde{Y}_{sp2} - Y_{m2} = G_{c1} E_1 - G_{m2} Y_2 \tag{12-47}$$

$$E_1 = -G_{m1} Y_1 \tag{12-48}$$

Eliminating all variables except Y_1 and D_2 gives

$$\frac{Y_1}{D_2} = \frac{G_{p1} G_{d2}}{1 + G_{c2} G_v G_{p2} G_{m2} + G_{c1} G_{c2} G_v G_{p2} G_{p1} G_{m1}} \tag{12-49}$$

By similar analysis, the set-point transfer functions for the outer and inner loops are

$$\frac{Y_1}{Y_{sp1}} = \frac{G_{c1} G_{c2} G_v G_{p1} G_{p2} K_{m1}}{1 + G_{c2} G_v G_{p2} G_{m2} + G_{c1} G_{c2} G_v G_{p2} G_{p1} G_{m1}} \tag{12-50}$$

$$\frac{Y_2}{\tilde{Y}_{sp2}} = \frac{G_{c2} G_v G_{p2}}{1 + G_{c2} G_v G_{p2} G_{m2}} \tag{12-51}$$

For disturbances in D_1, the closed-loop transfer function is

$$\frac{Y_1}{D_1} = \frac{G_{d1}\left(1 + G_{c2} G_v G_{p2} G_{m2}\right)}{1 + G_{c2} G_v G_{p2} G_{m2} + G_{c1} G_{c2} G_v G_{p2} G_{p1} G_{m1}} \tag{12-52}$$

Several observations can be made about the above equations. First, the cascade control system has the characteristic equation

$$1 + G_{c2} G_v G_{p2} G_{m2} + G_{c1} G_{c2} G_v G_{p2} G_{p1} G_{m1} = 0 \tag{12-53}$$

If the inner loop were removed ($G_{c2} = 1$, $G_{m2} = 0$), the characteristic equation would be the same as that for conventional feedback control,

$$1 + G_{c1} G_v G_{p2} G_{p1} G_{m1} = 0 \tag{12-54}$$

When the secondary loop responds faster than the primary loop, the cascade control system will have improved stability characteristics and thus should allow larger values of K_{c1} to be used in the primary control loop.

EXAMPLE 12.6

Consider the block diagram in Fig. 12.21 with the following transfer functions:

$$G_v = \frac{5}{s+1} \quad G_{p1} = \frac{4}{(4s+1)(2s+1)} \quad G_{p2} = 1$$

$$G_{d2} = 1 \quad G_{m1} = 0.05 \quad G_{m2} = 0.2 \quad G_{d1} = \frac{1}{3s+1}$$

where the time constants have units of minutes and the gains have consistent units. Determine the stability limits for a conventional proportional controller as well as for a cascade control system consisting of two proportional controllers. Assume $K_{c2} = 4$ for the secondary controller. Calculate the resulting offset for a unit step change in the secondary disturbance variable D_2.

Solution

For the cascade arrangement, first analyze the inner loop. Substituting into Eq. 12-51 gives

$$\frac{Y_2}{\widetilde{Y}_{sp2}}=\frac{4\left(\frac{5}{s+1}\right)(1)}{1+4\left(\frac{5}{s+1}\right)(0.2)}=\frac{20}{s+50}=\frac{4}{0.2s+1} \qquad (12\text{-}55)$$

From Eq. 12-55 the closed-loop time constant for the inner loop is 0.2 min. In contrast, the conventional feedback control system has a time constant of 1 min because in this case, $Y_2(s)/\widetilde{Y}_{sp2}(s)=G_v=5/(s+1)$. Thus, cascade control significantly speeds up the response of Y_2. Using a proportional controller in the primary loop ($G_{c1}=K_{c1}$), the rearranged characteristic equation becomes

$$1+(K_{c1})(4)\left(\frac{5}{s+1}\right)\left(\frac{4}{(4s+1)(2s+1)}\right)(0.05)+4\left(\frac{5}{s+1}\right)(0.2)=0 \qquad (12\text{-}56)$$

which reduces to

$$8s^3+46s^2+31s+5+4K_{c1}=0 \qquad (12\text{-}57)$$

By use of direct substitution (Chapter 9), the ultimate gain for marginal stability is $K_{c1,u}=43.3$.

For the conventional feedback system with proportional-only control, the characteristic equation in Eq. 12-54 reduces to

$$8s^3+14s^2+7s+1+K_{c1}=0 \qquad (12\text{-}58)$$

Direct substitution gives $K_{c1,u}=11.25$. Therefore, the cascade configuration has increased the stability margin by nearly a factor of four. Increasing K_{c2} will result in even larger values for $K_{c1,u}$. For this example, there is no theoretical upper limit for K_{c2}, except that large values will cause the valve to saturate for small set-point changes or disturbances.

The offset of Y_1 for a unit step change in D_2 can be obtained by setting $s=0$ in the right side of Eq. 12-49; equivalently, the Final Value Theorem of Chapter 3 can be applied for a unit step change in D_2 ($Y_{sp1}=0$):

$$e_1(t\rightarrow\infty)=y_{sp1}-y_1(t\rightarrow\infty)=\frac{-4}{5+4K_{c1}} \qquad (12\text{-}59)$$

By comparison, the offset for conventional control ($G_{m2}=0$, $G_{c2}=1$) is

$$e_1(t\rightarrow\infty)=\frac{-4}{1+K_{c1}} \qquad (12\text{-}60)$$

By comparing Eqs. 12-59 and 12-60, it is clear that for the same value of K_{c1}, the offset is much smaller (in absolute value) for cascade control.

For a cascade control system to function properly, the secondary control loop must respond faster than the primary loop. The secondary controller is normally a P or PI controller, depending on the amount of offset that would occur with proportional-only control. Note that small offsets in the secondary loop can be tolerated, because the primary loop will compensate for them. Derivative action is rarely used in the secondary loop. The primary controller is usually PI or PID.

For processes with higher-order dynamics and/or time delay, the model can first be approximated by a low-order model (see Chapter 6). The offset is checked to determine whether PI control is required for the secondary loop after K_{c2} is specified. The open-loop transfer function used for design of G_{c1} is

$$G_{OL}=\frac{G_{c1}K_{c2}G_vG_{p2}}{1+K_{c2}G_vG_{p2}G_{m2}}G_{p1}G_{m1} \qquad (12\text{-}61)$$

Figure 12.22 shows the closed-loop response for Example 12.6 and disturbance variable D_2. The cascade configuration has a PI controller in the primary loop and a proportional controller in the secondary loop. Figure 12.22 demonstrates that in this case the cascade control system is clearly superior to a conventional PI controller for a secondary loop disturbance. Figure 12.23 shows a similar comparison for a step change in the primary loop disturbance D_1 with a less dramatic improvement.

When a cascade control system is tuned after installation, the secondary controller should be tuned first with the primary controller in the manual mode. Then the primary controller is transferred to automatic, and it is tuned. If the secondary controller is retuned for some reason, usually the primary controller must also be retuned. Alternatively, Lee et al. (1998) have developed a tuning method based on Direct Synthesis where both loops are tuned simultaneously. When there are limits

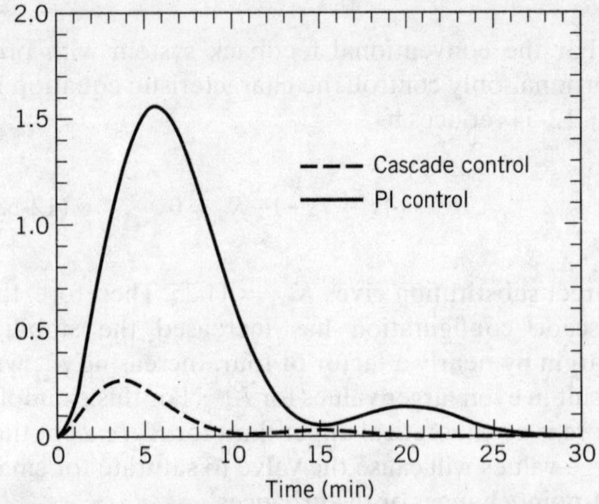

Figure 12.22 A comparison of D_2 unit step responses for Example 12.6 with and without cascade control.

on either controller output (saturation constraints), Brosilow and Joseph (2002) have recommended design modifications based on the Internal Model Control (IMC) approach. Usually many secondary loops in process industries are flow control loops. In such circumstances, Krishnaswamy et al. (1990) cautioned against use of a high gain value despite small time delay in the secondary loop. Because without significant benefit, this will result in high wear and tear of the control valve.

A commonly used form of cascade control involves a *valve positioner*, which addresses the problem of valve nonidealities. Valves may stick or exhibit dead zones or hysteresis, and so they may not achieve the same percentage stem position required for a given controller output. The valve positioner senses the valve stem position and uses an internal proportional controller in the inner loop of a cascade control system to attain the

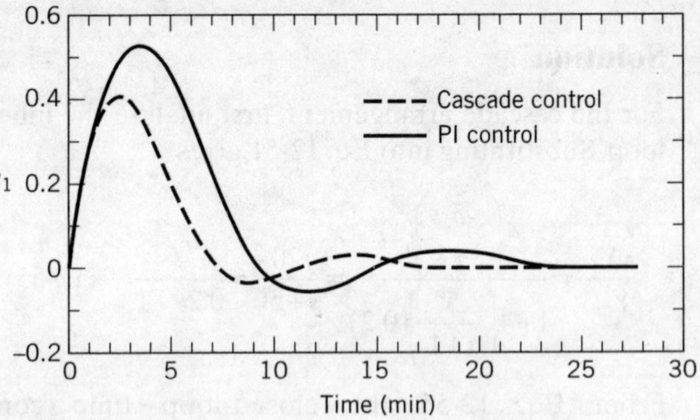

Figure 12.23 A comparison of D_1 step responses.

desired stem position. The set point is provided from the flow controller. Valve positioners are almost always beneficial when used in control loops. See Chapter 8 for more details on these devices.

12.4 TIME-DELAY COMPENSATION

In this section we present an advanced control technique, *time-delay compensation*, which deals with a problematic area in process control—namely, the occurrence of significant time delays. Time delays commonly occur in the process industries because of the presence of distance velocity lags, recycle loops, and the analysis time associated with composition measurement. The presence of time delays in a process reduces the stability and limits the performance of a conventional feedback control system. Consequently, the controller gain must be reduced below the value that could be used if no time delay were present, and the response of the closed-loop system will be sluggish compared to the case with no time delay.

EXAMPLE 12.7

Compare the set-point responses for a second-order process with a time delay ($\theta = 2$ min) and without the delay. The transfer function for the delay case is

$$G_p(s) = \frac{e^{-\theta s}}{(5s+1)(3s+1)} \tag{12-62}$$

Assume that $G_m = G_v = 1$, with time constants in minutes. Use the following PI controllers. For $\theta = 0$, $K_c = 3.02$ and $\tau_I = 6.5$ min, while for $\theta = 2$ min the controller gain must be reduced to meet stability requirements ($K_c = 1.23$, $\tau_I = 7.0$ min).

Solution

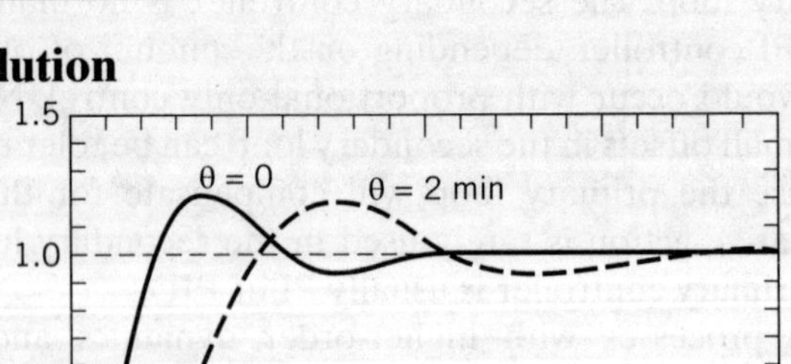

Figure 12.24 A comparison of closed-loop set-point changes for PI control.

The closed-loop responses are shown in Fig. 12.24. For $\theta = 2$ min, the resulting response is more sluggish. Clearly the closed-loop response for the time-delay case has deteriorated, with a 50% increase in response time (30 vs. 20 min). This response is much longer than might be expected from the size of the time delay.

In order to improve the performance of system containing time delays, special control strategies have been developed that provide effective time-delay compensation. The *Smith predictor* technique is the best-known strategy (Smith, 1957). A related method, the *analytical predictor*, has been developed specifically for digital control applications and is discussed in Chapter 13. Various investigators have found that the performance of a controller incorporating the Smith predictor for set-point changes is better than a conventional PI controller based on an integral-squared-error criterion. However, the Smith predictor performance may not be superior for all types of disturbances.

A block diagram of the Smith predictor controller is shown in Fig. 12.25, where $G_v = G_m = 1$ for simplicity. Here the process model $\widetilde{G}(s)$ is divided into two parts: the part without a time delay, $\widetilde{G}^*(s)$, and the time-delay term, $e^{-\tilde{\theta}s}$. Thus, the total transfer function model is $\widetilde{G}(s) = \widetilde{G}^*(s)e^{-\tilde{\theta}s}$.

Figure 12.25 shows an inner feedback loop, somewhat similar to that in cascade control. Assuming there is no model error $\left(\widetilde{G} = G\right)$, the inner loop has the effective transfer function

$$G' = \frac{P}{E} = \frac{G_c}{1 + G_c G^*\left(1 - e^{-\theta s}\right)} \tag{12-63}$$

where G^* is defined analogously to $\widetilde{G}^*$, that is $G = G^* e^{-\theta s}$. After some rearrangement, the closed-loop set-point transfer function is obtained:

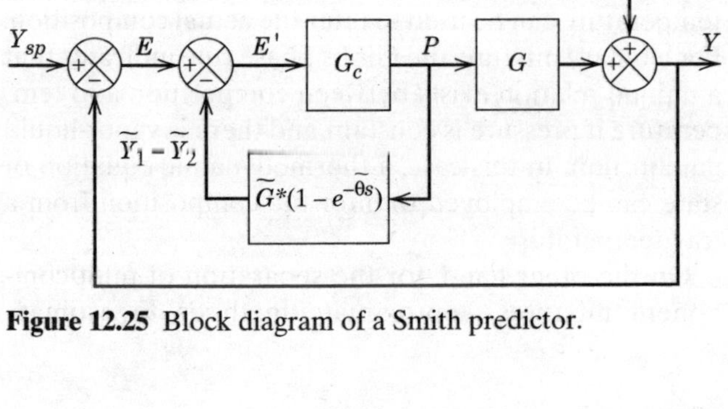

Figure 12.25 Block diagram of a Smith predictor.

$$\frac{Y}{Y_{sp}} = \frac{G_c G^* e^{-\theta s}}{1 + G_c G^*} \tag{12-64}$$

By contrast, for conventional feedback control

$$\frac{Y}{Y_{sp}} = \frac{G_c G^* e^{-\theta s}}{1 + G_c G^* e^{-\theta s}} \tag{12-65}$$

Comparison of Eqs. 12-64 and 12-65 indicates that the Smith predictor has the theoretical advantage of eliminating the time delay from the characteristic equation. Unfortunately, this advantage is lost if the process model is inaccurate due to lack of robustness (Åström and Hägglund, 2006). Kirtania and Choudhury (2012) and Morari and Zafiriou (1989) have also discussed the robustness aspects of the Smith predictor and have recommended that tuning be performed with other inputs besides step inputs.

Figure 12.26 shows the closed-loop responses for the Smith predictor (θ = 2 min) and PI control

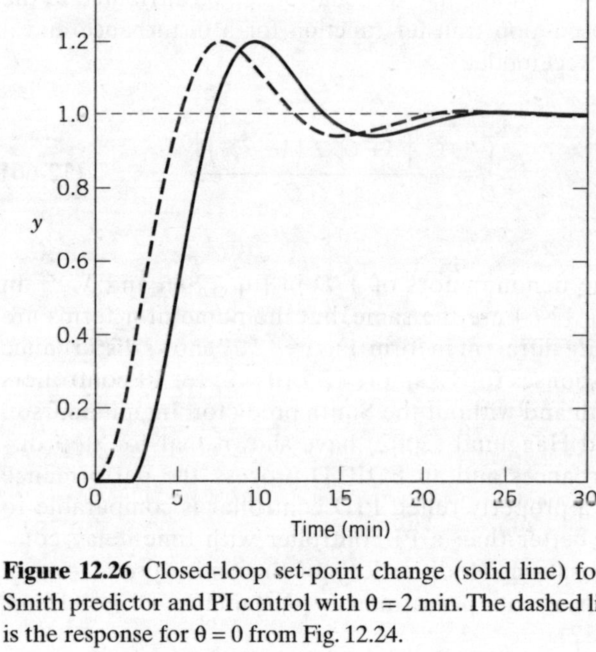

Figure 12.26 Closed-loop set-point change (solid line) for a Smith predictor and PI control with $\theta = 2$ min. The dashed line is the response for $\theta = 0$ from Fig. 12.24.

($\theta = 0$). The controller settings are the same as those developed in Example 12.7 for $\theta = 0$. A comparison of Fig. 12.24 (dashed line) and Fig. 12.26 shows the improvement in performance that can be obtained with the Smith predictor. Note that the responses in Fig. 12.26 for $\theta = 2$ and $\theta = 0$ are identical, except for the initial time delay. This closed-loop time delay results from the numerator delay term in Eq. 12-64. The process model $G^*(s)$ is second order and thus readily yields a stable closed-loop system for P-only control, but not necessarily for PI control (cf. Example 10.7).

One disadvantage of the Smith predictor approach is that it is model-based; that is, a dynamic model of the process is required. If the process dynamics change significantly, the predictive model will be inaccurate and the controller performance will deteriorate, perhaps to the point of closed-loop instability. For such processes, the controller should be tuned conservatively to accommodate possible model errors. Typically, if the time delay or process gain is not within ±30% of the actual value, the Smith predictor is inferior to a PID controller with no time-delay compensation. If the time delay varies significantly, it may be necessary to use some sort of adaptive controller to achieve satisfactory performance (see Section 12.8). Alternatively, Lee et al. (1999) and Åström and Hägglund (2006) have presented robust tuning procedures for PID controllers with Smith predictors.

The Smith predictor configuration generally is beneficial for handling disturbances. However, under certain conditions, a conventional PI controller can provide better regulatory control than the Smith predictor. This somewhat anomalous behavior can be attributed to the closed-loop transfer function for a disturbance and a perfect model:

$$\frac{Y}{D} = \frac{G_d\left[1 + G_c G^*\left(1 - e^{-\theta s}\right)\right]}{1 + G_c G^*} \tag{12-66}$$

The denominators of Y/D in Eq. 12-66 and Y/Y_{sp} in Eq. 12-64 are the same, but the numerator terms are quite different in form. Figure 12.27 shows disturbance responses for Example 12.7 ($\theta = 2$) for PI controllers with and without the Smith predictor. Ingimundarson and Hägglund (2002) have shown that for step disturbances and an FOPTD process, the performance of a properly tuned PID controller is comparable to or better than a PI controller with time-delay compensation. Modifications of the Smith predictor can improve its performance (Kirtania and Choudhury, 2012).

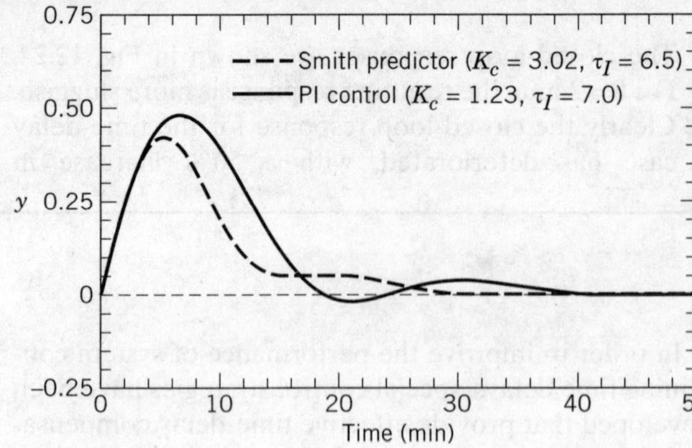

Figure 12.27 A comparison of disturbance changes for the Smith predictor and a conventional PI controller.

If the dominant time constant in G_d is relatively large, a controller designed for set-point changes by IMC will not be able to speed up the disturbance response, as demonstrated in Example 11.4.

Another control strategy for treating both disturbances and set-point changes is predictive control, which utilizes a prediction of the process behavior in the future based on the process and disturbance transfer functions, G and G_d. Further details are given in Chapters 15 and 19.

12.5 INFERENTIAL CONTROL

In some control applications, the process variable to be controlled cannot be conveniently measured on-line. For example, product composition measurement may require that a sample be sent to the plant analytical laboratory from time to time. In this situation, measurements of the controlled variable may not be available frequently enough or quickly enough to be used for feedback control.

One solution to this problem is to employ *inferential control*, where process measurements that can be obtained more rapidly are used with a mathematical model, sometimes called a *soft sensor*, to infer the value of the controlled variable. For example, if the overhead product stream in a distillation column cannot be analyzed on-line, sometimes measurement of a selected tray temperature can be used to infer the actual composition. For a binary mixture, the Gibbs phase rule indicates that a unique relation exists between composition and temperature if pressure is constant and there is vapor-liquid equilibrium. In this case, a thermodynamic equation of state can be employed to infer the composition from a tray temperature.

On the other hand, for the separation of multicomponent mixtures, approximate methods to estimate

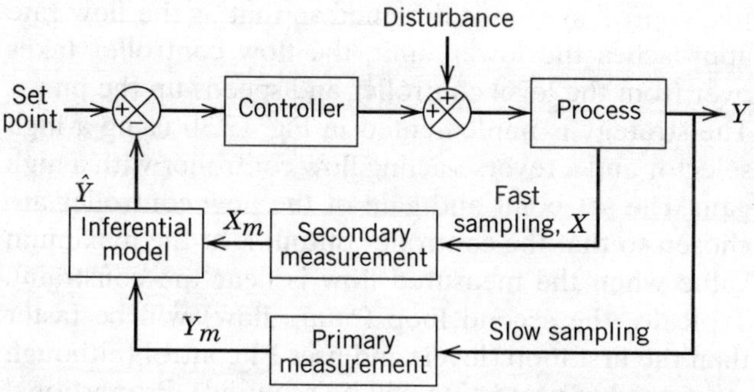

Figure 12.28 Block diagram of a soft sensor used in inferential control.

compositions must be used. Based on process models and plant data, simple algebraic correlations can be developed that relate the mole fraction of the heavy key component to several different tray temperatures (usually in the top half of the column above the feed tray). The overhead composition can then be inferred from the available temperature measurements and used in the control calculations. The parameters in the correlation may be updated, if necessary, as composition measurements become available. For example, if samples are sent to the plant's analytical laboratory once per hour, the correlation parameters can be adjusted so that the predicted values agree with the measured values. Figure 12.28 shows the general structure of an inferential controller. X_m is the secondary measurement, which is available on a nearly continuous basis (fast sampling), while Y_m is the primary measurement, which is obtained intermittently and less frequently (e.g., off-line laboratory sample analysis). Note that X_m and/or Y_m can be used to infer $\hat{Y}$, which is then compared to the set point. One type of nonlinear model that could be used as a soft sensor is a neural network (see Chapter 16). The inferential model is obtained by analyzing and fitting accumulated X and Y data. Dynamic linear or nonlinear models (called *observers*) can also be used for inferential control, as reviewed by Doyle (1998).

The concept of inferential control can be employed for other process operations, such as chemical reactors, where composition is normally the controlled variable. Selected temperature measurements can be used to estimate the outlet composition if it cannot be measured on-line. However, when inferential control does not perform satisfactorily, an incentive exists to introduce other on-line measurements for feedback control. Consequently, there is ongoing interest in the development of new sensors, such as novel process analyzers, which can be used on-line and whose response times are very short.

12.6 SELECTIVE CONTROL SYSTEMS

Most process control applications have an equal number of controlled variables and manipulated variables. However, if fewer manipulated variables than controlled variables are available, it is not possible to eliminate offset in all the controlled variables for arbitrary disturbances or set-point changes. This assertion is evident from a degrees-of-freedom analysis of a steady-state model. For control problems with fewer manipulated variables than controlled variables, selectors are employed for sharing the manipulated variables among the controlled variables.

12.6.1 Auctioneering control

An auctioneering control system is a practical solution for choosing the appropriate signal from among a number of available measurements. Selectors are used to improve the control system performance as well as to protect equipment from unsafe operating conditions. Selectors can be based on multiple measurement points, multiple final control elements, or multiple controllers, as discussed below. On instrumentation diagrams, the symbol (>) denotes a *high selector* and the symbol (<) a *low selector*.

For one type of selector, the output signal is the highest (or lowest) of two or more input signals. This approach is often referred to as *auctioneering* (Shinskey, 1996). For example, a high selector can be used to determine the *hotspot* temperature in a fixed-bed chemical reactor as shown in Fig. 12.29. In this reactor application, the output from the high selector is the input to the temperature controller. In an exothermic catalytic reaction, the process may "run away" due to disturbances or changes in the reactor, and immediate action should be taken to prevent a dangerous rise in temperature. Because a hotspot can potentially develop at one of several possible locations in the reactor, multiple (redundant) measurement points are employed. This approach helps identify when a temperature has become too high at some point in the bed.

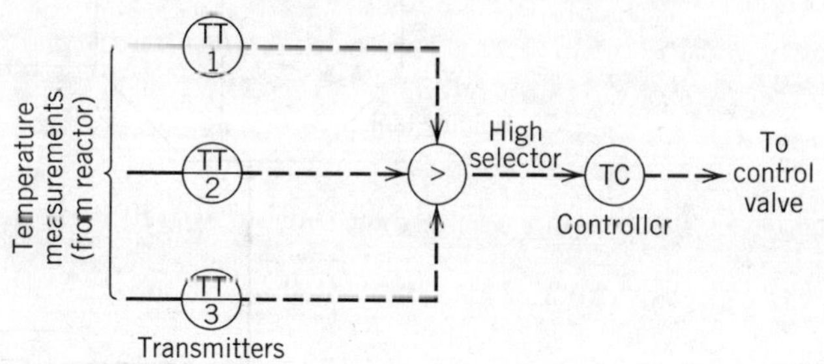

Figure 12.29 Control of a reactor hotspot temperature by using a high selector.

With a *median selector*, the selector output is the median of three or more input signals. These devices are useful for situations in which redundant sensors are used to measure a single process variable. By selecting the median value, reliability is improved, because a single sensor failure will not cause the loss of a meaningful feedback signal.

12.6.2 Override control

The use of high or low limits for process variables represents another type of selective control called an *override*, where a second controller can "override" or take over from the first controller. This is a less extreme action than an interlock, which is used for emergency shutdown of the process. The anti-reset windup feature in feedback controllers (cf. Chapter 7) is a type of override. Another example is a distillation column that has lower and upper limits on the heat input to the column reboiler. The minimum level ensures adequate liquid inventory on the trays, while the upper limit exists to prevent the onset of flooding (Buckley et al., 1985; Shinskey, 1996). Overrides are also often used in forced draft combustion control systems to prevent an imbalance between air flow and fuel flow, which could result in unsafe operating conditions (Bozzuto, 2009). See Smith and Corripio (2006) for other applications of overrides.

The selective control system shown in Fig. 12.30 is used to regulate the level and exit flow rate in a pumping system for a sand/water slurry. During normal operation, the level controller (LC) adjusts the slurry exit flow by changing the pump speed. A variable-speed pump is used rather than a control valve owing to the abrasive nature of the slurry. The slurry velocity in the exit line must be kept above a minimum value at all times to prevent the line from sanding up. Consequently, the selective control system is designed so that as the flow rate approaches the lower limit, the flow controller takes over from the level controller and speeds up the pump. The strategy is implemented in Fig. 12.30 using a high selector and a reverse-acting flow controller with a high gain. The set point and gain of the flow controller are chosen so that the controller output is at the maximum value when the measured flow is near the constraint. Typically, the second loop (pump flow) will be faster than the first loop (level) and uses PI control (although reset windup protection will be required). Proportional control could be employed in the slower loop (liquid level) because tight level control is not required.

One alternative arrangement to Fig. 12.30 would be to employ a single controller, using the level and flow transmitter signals as inputs to a high selector, with its output signal sent to the controller. The controller output would then adjust the pump speed. This scheme is simpler, because only one controller is needed. However, it suffers from an important operational disadvantage; namely, it may not be possible to tune the single controller to meet the needs of both the level and flow control loops. In general, these control loops and their transmitters will have very different dynamic characteristics. A second alternative would be to replace the flow transmitter and controller with a constant (override) signal to the high selector whose value corresponds to the minimum allowable flow rate. However, this scheme would be susceptible to changing pump characteristics.

12.6.3 Split-range control

In *split-range control* several manipulated variables are used to control a single controlled variable. The single controlled variable results in a single controller output.

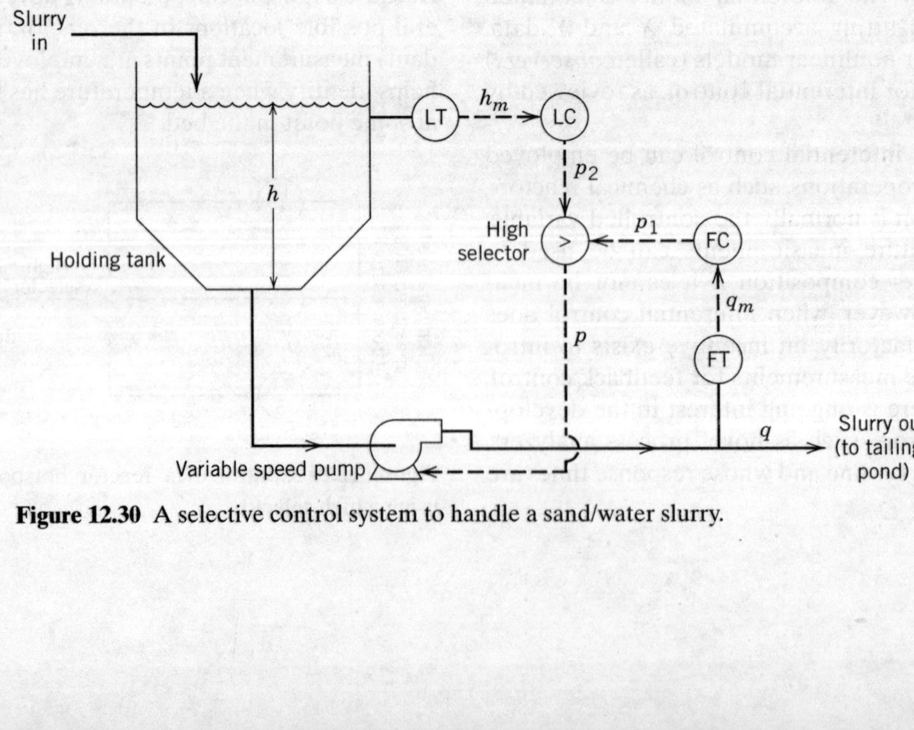

Figure 12.30 A selective control system to handle a sand/water slurry.

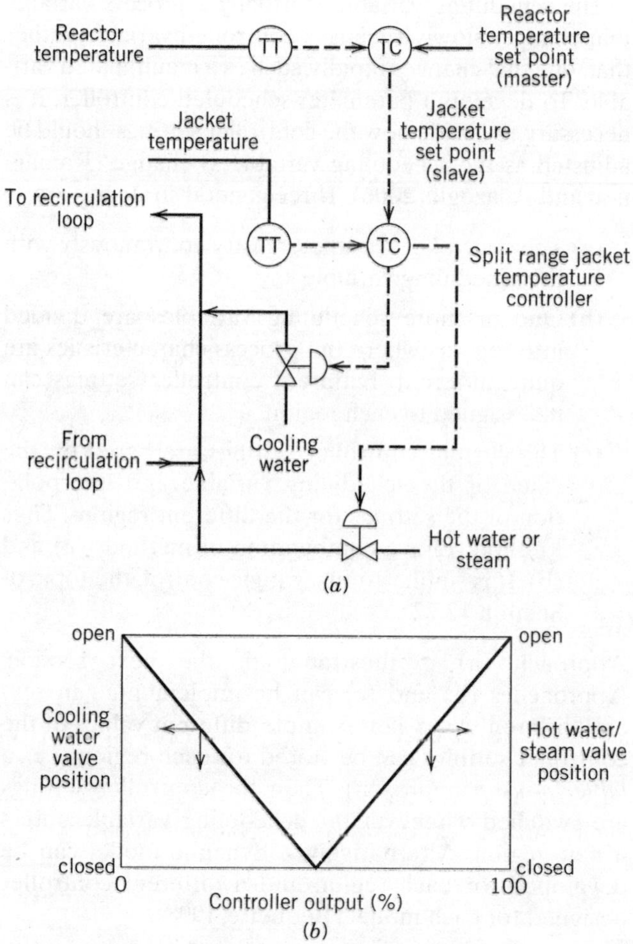

Figure 12.31 Split range control: (*a*) control loop configuration, (*b*) valve position–controller output relationship.

This output from the controller is split and each split signal is used to change each manipulated variable.

Typical examples include the adjustment of both inflow and outflow from a chemical reactor in order to control reactor pressure or the use of both acid and base to control pH in wastewater treatment. Another example is in reactor control, where both heating and cooling are used to maintain precise regulation of the reactor temperature. Figure 12.31 shows how the control loop in Fig. 12.20 can be modified to accommodate both heating and cooling using a single controller and two control valves. This split-range control is achieved using the controller input–output relationship shown in Fig. 12.31*b*.

An alternative solution for reactor temperature control can be applied when both heating and cooling are necessary, or when a median temperature is regulated. Although the physical processes are configured in the same way as in Fig. 12.31, the controller output is mapped to the valves so that both valves are always active; that is, cooling/heating medium is always flowing through the heat exchange equipment. However, it is not a requirement that both valves must be of the same size and characteristics. Note also that in this scheme, the heat removal capability is more or less linear with respect to the controller output, regardless of valve selection.

12.7 NONLINEAR CONTROL SYSTEMS

Most physical processes exhibit nonlinear behavior to some degree. However, linear control techniques such as conventional PID control are still very effective if (1) the nonlinearities are rather mild or (2) a highly nonlinear process operates over a narrow range of conditions. For some highly nonlinear processes, the second condition is not satisfied and as a result, linear control strategies may not be adequate. For these situations, nonlinear control strategies can provide significant improvements over PID control. In this section, we consider several traditional nonlinear control strategies that have been applied in industry. Newer model-based techniques are described by Henson and Seborg (1997) and Doyle et al. (2015).

Three types of nonlinear control strategies are essentially enhancements of single loop feedback control:

1. Nonlinear modifications of standard PID control algorithms
2. Nonlinear transformations of input or output variables
3. Controller parameter scheduling such as *gain scheduling*

Shinskey (1994) and Bequette (1998) have provided informative overviews of these methods and related techniques. Other enhanced single-loop control strategies considered earlier in this chapter, namely, inferential control, selectors, and adaptive control, can also be classified as nonlinear control strategies.

As one example of Method 1, standard PID control laws can be modified by making the controller gain a function of the control error. For example, the controller gain can be higher for larger errors and smaller for small errors by making the controller gain vary linearly with the absolute value of the error signal

$$K_c = K_{c0}\left(1 + a|e(t)|\right) \tag{12-67}$$

where K_{c0} and a are constants. The resulting controller is sometimes referred to as an *error-squared controller*, because the controller output is proportional to

$|e(t)|e(t)$. Error-squared controllers have been used for level control in surge vessels where it is desirable to take stronger action as the level approaches high or low limits. However, care should be exercised when the error signal is noisy (Shinskey, 1994).

The design objective for Method 2 is to make the closed-loop operation as linear as possible. If successful, this general approach allows the process to be controlled over a wider range of operating conditions and in a more predictable manner. One approach uses simple linear transformations of input or output variables. Common applications include using the logarithm of a product composition as the controlled variable for high-purity distillation columns or adjusting the ratio of feed flow rates in blending problems. The major limitation of this approach is that it is difficult to generalize because the appropriate variable transformations are application-specific.

In distillation column control, some success has been obtained in using logarithmic transformations to linearize the error signal so that the controller makes adjustments that are better scaled. For example, a transformed composition variable x_D^* has been used in industrial applications (Skogestad, 2007):

$$x_D^* = \log\frac{1-x_D}{1-x_{Dsp}} \tag{12-68}$$

where x_{Dsp} is the desired value of x_D. However, this approach may not work well in all cases. Another linearizing function can be used to treat the nonlinear behavior observed in flow systems. When there are pipe resistances in series with a control valve, a nonlinear gain results between stem position l and flow rate. In this case a nonlinear function, called a *valve characterizer*, can be used to transform the controller output p:

$$f(p) = \frac{p}{L+(1-l)\,p} \tag{12-69}$$

where L is a parameter used to fit the shape of the non-linearity. Shinskey (1994) has discussed this approach and related control strategies.

In Method 3, *controller parameter scheduling*, one or more controller settings are adjusted automatically based on the measured value of a scheduling variable. Adjustment of the controller gain, *gain scheduling*, is the most common method. The scheduling variable is usually the controlled variable or set point, but it could be the manipulated variable or some other measured variable. Usually, only the controller gain is adjusted, because many industrial processes exhibit variable steady-state gains but relatively constant dynamics (e.g., pH neutralization).

The scheduling variable is usually a process variable that changes slowly, such as a controlled variable, rather than one that changes rapidly, such as a manipulated variable. To develop a parameter-scheduled controller, it is necessary to decide how the controller settings should be adjusted as the scheduling variable(s) change (Romagnoli and Palazoglu, 2006). Three general strategies are:

(a) The controller parameters vary continuously with the scheduling variable.

(b) One or more scheduling variables are divided into regions where the process characteristics are quite different. Different controller settings can be assigned to each region.

(c) The current controller settings are based on the value of the scheduling variable and interpolation of the settings for the different regions. Thus Method (c) is a combination of methods (a) and (b). It is similar to fuzzy logic control, the topic of Section 12.7.2.

Approach (a) is illustrated in the next section. Approaches (b) and (c) can be implemented in several different ways. For example, different values of the controller settings can be stored for each region (i.e., a *table look-up* approach). Then the controller settings are switched whenever the scheduling variable enters a new region. Alternatively, a dynamic model can be developed for each region and a different controller designed for each model (Bequette, 1998).

12.7.1 Gain Scheduling

The most widely used type of controller parameter scheduling is *gain scheduling*. A simple version has a piecewise constant controller gain that varies with a single scheduling variable, the error signal e:

$$\begin{aligned} K_c &= K_{c1} \quad \text{for } e_1 \le e < e_2 \\ K_c &= K_{c2} \quad \text{for } e_2 \le e < e_3 \\ K_c &= K_{c3} \quad \text{for } e_3 \le e \le e_4 \end{aligned} \tag{12-70}$$

This gain-scheduling approach is shown in Fig. 12.32 and can easily be extended to more than three regions. A special case, the *error gap controller*, includes a dead band around $e = 0$. In this case, $K_c = 0$ for $e_2 \le e < e_3$, while $K_c \neq 0$ outside this region. Note that the nonlinear gain expression in Eq. 12-67 is another example of gain scheduling.

Next, we consider an example of Method 1 from the previous section. A relationship can be developed between the controller settings and the scheduling variable(s). The resulting strategy is sometimes called *programmed adaptation* (Lipták, 2006; Shinskey, 1996).

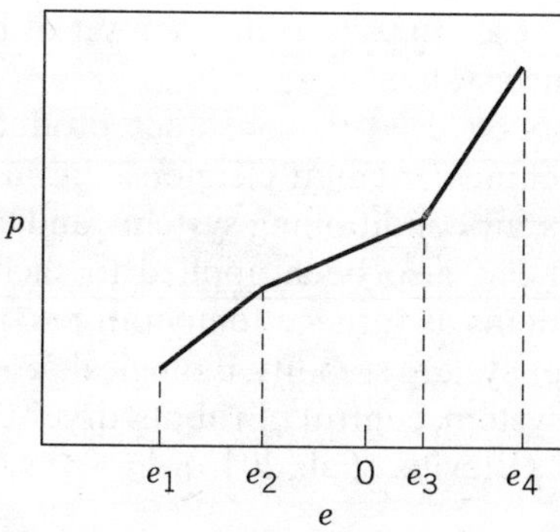

Figure 12.32 A gain-scheduled proportional controller with a controller gain that is piecewise constant.

Programmed adaptation is limited to applications where the process dynamics depend on known, measurable variables and the necessary controller adjustments are not too complicated. As an example of programmed adaptation, consider a once-through boiler (Lipták, 2006). Here, feedwater passes through a series of heated tube sections before emerging as superheated steam. The steam temperature must be accurately controlled by adjusting the flow rate of the hot gas that is used to heat the water. The feedwater flow rate has a significant effect on both the steady-state and dynamic behavior of the boiler. For example, Fig. 12.33 shows typical open-loop responses to a step change in controller output at two different feedwater flow rates, 50% and 100% of the maximum flow. Suppose that an empirical FOPTD model is chosen to approximate the process. The steady-state gain, time delay, and dominant time constant are all twice as large at 50% flow as the corresponding values are at 100% flow. Lipták's proposed solution to this control problem is to have the PID controller settings vary with w, the fraction of full-scale flow ($0 \leq w \leq 1$), in the following manner:

$$\begin{aligned} K_c &= w\overline{K}_c \\ \tau_I &= \overline{\tau}_I / w \\ \tau_D &= \overline{\tau}_D / w \end{aligned} \tag{12-71}$$

where $\overline{K}_c$, $\overline{\tau}_I$, and $\overline{\tau}_D$ are the controller settings for 100% flow. Note that this recommendation for programmed adaptation is qualitatively consistent with the controller tuning rules of Chapter 11. The recommended settings in Eq. 12-71 are based on the assumption that the effects of flow changes are linearly related to flow rate over the full range of operation.

In this example, step responses were available to categorize the process behavior for two different conditions. In other applications, such test data may not be available but there may be some knowledge of process

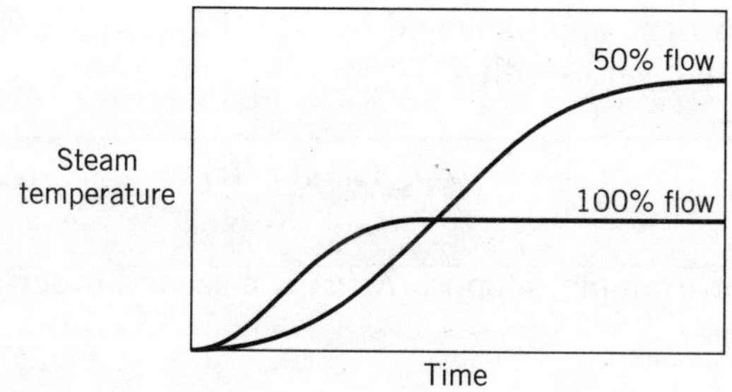

Figure 12.33 Open-loop step responses for a once-through boiler.

nonlinearities. For pH control problems involving a strong acid and/or a strong base, the pH curve can be very nonlinear, with gain variations over several orders of magnitude. If the process gain changes significantly with the operating conditions, an appropriate gain scheduling strategy is to keep the product of the controller and process gains constant, $K_c K_p = C$, where C is a specified constant. This strategy helps maintain the desired gain margin (see Chapter 10). This gain scheduling strategy could be implemented as follows. Suppose that pH is used as the scheduling variable and an empirical equation is available that relates K_p to pH. Then the current value of K_p could be calculated from the pH measurement and K_c could be determined as, $K_c = C/K_p$.

Another representative nonlinearity is illustrated by the step responses in Fig. 12.34, where the process gain K_p depends on the input signal u, $K_p(u)$. Note that the process gain doubles when the input size is increased from $u = 0.5$ to $u = 1.0$. Because the process dynamics are independent of the magnitude of u, the dynamic

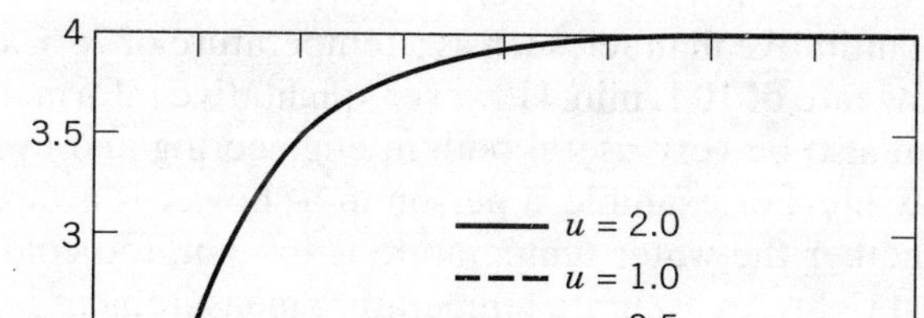

Figure 12.34 Step responses for a nonlinear model for different input magnitudes.

behavior is approximated by a FOPTD model with a gain that varies with u:

$$\tau \frac{dy}{dt} = -y + K_p(u)u(t-\theta) \quad (12\text{-}72)$$

For example, suppose $K_p(u)$ is a second-order polynomial,

$$K_p(u) = a_0 + a_1u + a_2u^2 \quad (12\text{-}73)$$

The model parameters in Eq. 12-73 can be obtained by regression of K_p values for steady-state values of u and y, using the parameter estimation methods of Chapter 16. Note that a transfer function with constant parameters, $Y(s)/U(s)$, cannot be derived for the dynamic model in Eqs. 12-72 and 12-73, because it is not a linear system. However, it is fairly straightforward to design a PID controller (with gain-scheduling) for this process. Suppose that the nominal gain is $K_{p0} = a_0$ for $u = 0$. Then the values of K_c, τ_I, and τ_D can be determined using any of the tuning rules in Chapter 11 for given values of K_{p0}, τ, and θ. As u varies, it is desirable to keep the product of $K_p(u)$ and $K_c(u)$ constant, in order to maintain a satisfactory gain margin. Thus, we specify that $K_c(u)K_p(u) = K_{c0}K_{p0}$, where K_{c0} is the nominal controller gain for K_{p0}. The controller gain for the current u can be calculated as, $K_c(u) = K_{c0}K_{p0}/K_p(u)$. A similar relationship can be developed for the case where K_p is a function of y, rather than u.

12.7.2 Fuzzy Logic Control

Engineers normally interpret physical variables in a quantitative manner, such as a temperature of 78°C or a flow rate of 10 L/min. However, qualitative information can also be very useful both in engineering and everyday life. For example, a person in a shower is aware of whether the water temperature is too hot, too cold, or just right. An accurate temperature measurement is not necessary. Also, such qualitative information can be used to good advantage for feedback control. For example, if the shower temperature is too cold and the flow rate is too low, one would increase the hot water flow rate. In the process industries, experienced plant operators sometimes take control actions based on qualitative information, such as the observed color or uniformity of a solid material when on-line measurements of such quality variables are not available.

Fuzzy logic control (FLC) is a feedback control technique that utilizes qualitative information and verbal or linguistic *rules* of the if–then form (Babuška and Verbruggen, 1996; Rhinehart et al., 1996; Passino and Yurkovich, 1998; Jantzen, 2007). To derive the control law, the FLC uses fuzzy set theory, a set of rules, and a fuzzy inference system.

FLC has been used in consumer products such as washing machines, vacuum cleaners, automobiles, battery chargers, air conditioning systems, and camera autofocusing. It has also been applied to such industrial control problems as furnace temperature control, wind energy, power system stability, biological processes, a jet engine fuel system, control of robots, dryer control, and level control (Blevins et al., 2013).

Fuzzy Sets

Consider a room temperature T to be a qualitative variable with three possible classifications: Hot, OK, or Cold. One possible classification scheme is

Hot: if $T > 24°C$
OK: if $18°C < T < 24°C$
Cold: if $T < 18°C$

However, these class boundaries are arbitrary and somewhat inappropriate. For example, do we really want to classify temperatures of 23.5°C as OK and 24.5°C as *Hot*? A more appropriate classification scheme would use a *fuzzy set*. A *fuzzy inference system* would determine which rule is appropriate for the fuzzy temperature classification.

12.8 ADAPTIVE CONTROL SYSTEMS

Process control problems inevitably require on-line tuning of the controller settings to achieve a satisfactory degree of control. If the process operating conditions or the environment changes significantly, the controller may then have to be retuned. If these changes occur frequently, then adaptive control techniques should be considered. An *adaptive control system* is one in which the controller parameters are adjusted automatically to compensate for changing process conditions. Many adaptive control techniques have been proposed for situations where the process changes are largely unknown or unpredictable, as contrasted with situations amenable to the gain-scheduling approach discussed in the previous section. In this section, we are concerned principally with automatic adjustment of feedback controller settings.

Examples of changing process conditions that may require controller retuning or adaptive control are

1. Changes in equipment characteristics (e.g., heat exchanger fouling, catalyst deactivation)
2. Unusual operational status, such as failures, startup, and shutdown, or batch operations
3. Large, frequent disturbances (feed composition, fuel quality, etc.)

4. Ambient variations (rain storms, daily cycles, etc.)
5. Changes in product specifications (grade changes) or product flow rates
6. Inherent nonlinear behavior (e.g., the dependence of chemical reaction rates on temperature)

In situations where the process changes can be anticipated or measured directly and the process is reasonably well understood, then the gain-scheduling approach (or programmed adaptation) discussed in the previous section can be employed. When the process changes cannot be measured or predicted, the adaptive control strategy must be implemented in a feedback manner, because there is little opportunity for a feedforward type of strategy such as programmed adaptation. Many such controllers are referred to as *self-tuning controllers* or *self-adaptive controllers* (Åström and Wittenmark, 1995; Lipták, 2006; Blevins et al., 2013).

In self-tuning control, the parameters in the process model are updated as new data are acquired (using on-line estimation methods), and the control calculations are based on the updated model. For example, the controller settings could be expressed as a function of the model parameters and the estimates of these parameters updated on-line as process input/output data are received. Self-tuning controllers generally are implemented as shown in Fig. 12.35 (Åström and Wittenmark, 1995; Åström and Hägglund, 2006; Corripio and Newell, 2015).

In Fig. 12.35, three sets of computations are employed: estimation of the model parameters, calculation of the controller settings, and calculation of the controller output in a feedback loop. Most real-time parameter estimation techniques require that an external forcing signal occasionally be introduced to allow accurate estimation of model parameters (Hang et al., 1993). Such a pertubation signal can be deliberately introduced through the set point or added to the controller output.

During each disturbance or set-point change, the process response is compared to the predicted model response, and then the model can be updated based on the prediction error. One approach that deals with models changing with varying operating conditions is *multiple model adaptive control* (Narendra et al., 1995; Schott and Bequette, 1997), where a set of models and corresponding controllers is employed. A weighting function is used to choose the combination of models that best matches the process input–output behavior. This technique has been used in a variety of applications, including drug infusion control (Schott and Bequette, 1997).

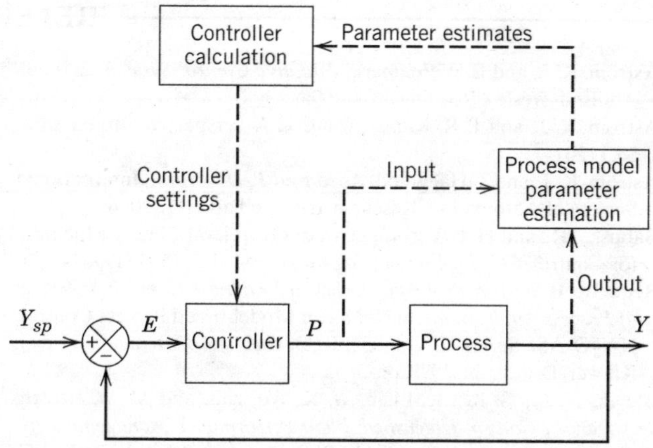

Figure 12.35 A block diagram for self-tuning control.

Two advantages of the self-tuning control approach are that the model in Fig. 12.35 is not restricted to low-order linear transfer functions or difference equations and the controller is not required to have a PID controller structure. The model structure can be multivariable in nature and even incorporate nonlinear modeling elements, such as artificial neural nets.

The subject of adaptive control continues to be of considerable interest. Many new and improved algorithms have been proposed since the early 1970s, and a number of commercial products introduced during the 1980s are in use today (Åström and Hägglund, 2006; Corripio and Newell, 2015). However, adaptive control has tended to be a niche application rather than being used pervasively in industrial applications. In addition, there is a concern that use of an adaptive controller can lead to unstable operations or unsafe operating conditions. In the future, it is expected that control products that use some form of adaptation will become more common as algorithms improve and process knowledge increases.

SUMMARY

In this chapter, we have presented a number of control strategies that offer the potential of enhanced performance over what can be achieved with conventional single-loop PID controllers. These techniques are especially attractive for difficult control problems, such as those characterized by unmeasured process variables and disturbances, long time delays, process constraints, changing operating conditions, and process nonlinearities and uncertainties. All of these characteristics can be treated by one or more of the advanced feedback control techniques discussed here: feed forward control, ratio control, cascade control, time-delay compensation, inferential control, selective control, nonlinear control, and adaptive control. Note that digital control (Chapter 13), multivariable control (Chapter 14), and model predictive control (Chapter 15) augment the more specialized methods treated here.

REFERENCES

Åström, K. J., and B. Wittenmark, *Adaptive Control Systems*, 2nd ed., Addison-Wesley, Reading, MA, 1995.

Åström, K. J., and P. R. Kumar, Control: A Perspective, *Automatica*, **50**,3 (2014).

Åström, K. J., and T. Hägglund, *Advanced PID Control*, International Society of Automation, Research Triangle Park, NC, 2006.

Babuška, R., and H. B. Verbruggen, An Overview of Fuzzy Modeling for Control, *Control Engineering Practice*, **4** (11), 1593 (1996).

Bequette, B. W., *Practical Approaches to Nonlinear Control: A Review of Process Applications*, in Nonlinear Model-based Process Control, NATO ASI Series E, Vol. **353**, R. Berber and C. Kravaris (Eds.), Kluwer, Dordrecht, 1998, pp. 3–32.

Blevins, T. L., G. K. McMillan, W. K. Wojsznis, and M. W. Brown, *Advanced Control Unleashed: Plant Performance Management for Optimum Benefit*, Appendix B, ISA, Research Triangle Park, NC, 2003.

Blevins, T., W. K. Wojsznis, and M. Nixon, *Advanced Control Foundation: Tools Techniques and Applications*, International Society of Automation, Research Triangle Park, NC, 2013.

Bozzuto, C. (Ed.), *Clean Combustion Technologies*, 5th ed., Allstom, Windsor, CT, 2009.

Brosilow, C., and B. Joseph, *Techniques of Model-Based Control*, Prentice Hall, Upper Saddle River, NJ, 2002.

Buckley, P. S., W. L. Luyben, and J. P. Shunta, *Design of Distillation Column Control Systems*, ISA, Research Triangle Park, NC, 1985.

Corripio, A., and M. Newell, *Tuning of Industrial Control Systems*, 3rd ed., ISA, Research Triangle Park, NC, 2015.

Daoutidis, P., and D. Bartusiak, Special Issue: CPC VIII, *Computers and Chemical Engineering*, **51**, 1–133 (2013).

Doyle, F. J. III, R. K. Pearson, and B. A. Ogunnaike, *Identification and Control Using Volterra Models*, Springer, New York, 2015.

Doyle, F. J., Nonlinear Inferential Control, *J. Process Control*, **8**, 339 (1998).

Guzmán, J. L., and T.Hägglund, Simple Tuning Rules for Feedforward Compensators, *J. Process Control*, **21**, 92 (2011).

Guzmán, J. L., T. Hägglund, M. Veronesi, and A. Visioli; Performance Indices for Feedforward Control, *J. Process Control*, **26**, 26 (2015).

Hägglund, T., The Blend Station—A New Ratio Control Structure, *Control Eng. Prac.*, **9**, 1215 (2001).

Hang, C. C., T. H. Lee, and W. K. Ho, *Adaptive Control*, ISA, Research Triangle Park, NC, 1993.

Hast, M., and T.Hägglund, Low-order Feedforward Controllers: Optimal Performance and Practical Considerations, *J. Process Control*, **24**, 1462 (2014).

Henson, M. A., and D. E. Seborg (Eds.), *Nonlinear Process Control*, Prentice-Hall, Upper Saddle River, NJ, 1997.

Henson, M. A., and T. A. Badgwell (Eds.), Chemical Process Control, **7**, *Comput. Chem. Eng.*, **30**, 1425 (2006).

Ingimundarson, I., and T. Hägglund, Performance Comparison Between PID and Dead-Time Compensating Controllers, *J. Process Control*, **12**, 887 (2002).

Jantzen, J., *Foundations of Fuzzy Control*, John Wiley and Sons, Hoboken, NJ, 2007.

Kirtania, K., and M. S. Choudhury, A Novel Dead Time Compensator for Stable Processes with Long Dead Times, *J. Process Control*, **22**, 612–625 (2012).

Krishnaswamy, P. R., G. P. Rangaiah, R. K. Jha, and P.B. Deshpande, When to use cascade control. *Ind. Eng. Chem. Res.* 29(10), 2163 (1990).

Lee D., M. Lee, S. Sung, and I. Lee, Robust PID Tuning for Smith Predictor in the Presence of Model Uncertainty, *J. Process Control*, **9**, 79 (1999).

Lee, Y., S. Park, and M. Lee, PID Controller Tuning to Obtain Desired Closed-Loop Responses for Cascade Control Systems, *IEC Research*, **37**, 1859 (1998).

Lipták, B. G., *Instrument Engineers Handbook*, 4th ed., *Process Control and Optimization*, CRC Press, Boca Raton, FL, 2006.

McMillan, G. K., *Tuning and Control Loop Performance*, 4th ed., Momentum Press, New York, 2015.

Morari, M., and E. Zafiriou, *Robust Process Control*, Prentice Hall, Englewood Cliffs, NJ, 1989.

Narendra, K. S., J. Balakrishnan, and M. K. Ciliz, Adaptation and Learning Having Multiple Models, Switching and Tuning, *IEEE Control Systems*, **15** (3), 37 (1995).

Passino, K. M., and S. Yurkovich, *Fuzzy Control*, Addison-Wesley, Reading, MA, 1998.

Rawlings, J. B., B. A. Ogunnaike, and J. W. Eaton (Eds.), *Chemical Process Control VI, AIChE Symp. Ser.*, **97**, No. 326, 2002.

Rhinehart, R. R., H. H. Li, and P. Murugan, Improve Process Control Using Fuzzy Logic, *Chem. Eng. Progress*, **92** (11), 60 (1996).

Riggs, J. B., Improve Distillation Column Control, *Chem. Eng. Prog.*, **94** (10), 31 (1998).

Romagnoli, J. A., and A. Palazoglu, *Introduction to Process Control*, Taylor and Francis, Boca Raton, FL, 2006.

Schott, K. D., and B. W. Bequette, *Multiple Model Adaptive Control*, in Multiple Model Approaches to Modeling and Control, R. Murray-Smith and T. Johansen (Eds.), Taylor and Francis, London, U.K., 1997, Chapter 11.

Shinskey, F. G., *Feedback Controllers for the Process Industries*, McGraw-Hill, New York, 1994.

Shinskey, F. G., *Process Control Systems*, 4th ed., McGraw-Hill, NewYork 1996.

Shinskey, F.G., *Process Control Systems: Application, Design, and Tuning*, 4th ed. McGraw-Hill, New York, 1996, Chapter 7.

Skogestad, S., The Dos and Don'ts of Distillation Columns Control, *Trans IChemE, Part A, Chem. Eng. Res. Des.*, **85**(A1), 13, 2007.

Smith, C. A. and A. Corripio, *Principles and Practice of Automatic Control Systems*, 3rd ed., John Wiley and Sons, Hoboken, NJ, 2006.

Smith, C. A., and A. B. Corripio, *Principles and Practice of Automatic Process Control*, 3rd ed., John Wiley and Sons, Hoboken, NJ, 2006.

Smith, O. J. M., Closer Control of Loops with Dead Time, *Chem. Eng. Prog.*, **53** (5), 217 (1957).

Visioli, A., A New Ratio Control Architecture, *Ind. Eng. Chem. Res.*, **44**, 4617 (2005a).

Visioli, A., Design and Tuning of a Ratio Controller, *Control Eng. Prac.*, **13**, 485 (2005b).

EXERCISES

12.1 It is desired to reduce the concentration of CO_2 in the flue gas from a coal-fired power plant, in order to reduce greenhouse gas emissions. The effluent flue gas is sent to an ammonia scrubber, where most of the CO_2 is absorbed in a liquid ammonia solution, as shown in Fig. E12.1. A feedforward control system will be used to control the CO_2 concentration in the flue gas stream leaving the scrubber Cco_2 which cannot be measured on-line. The flow rate of the ammonia solution entering the scrubber Q_A can be manipulated via a control valve. The inlet flue gas flow rate Q_F is a measured disturbance variable.

(a) Draw a block diagram of the feedforward control system. (It is *not* necessary to derive transfer functions.)

(b) Design a feedforward control system to reduce CO_2 emissions based on a *steady-state design* (Eq. 12-37).

Available Information:

(i) The flow sensor-transmitter and the control valve have negligible dynamics.

(ii) The flow sensor-transmitter has a steady-state gain of 0.08 mA/(L/min).

(iii) The control valve has a steady-state gain of 15 (L/min)/mA.

(iv) The following steady-state data are available for a series of changes in Q_A:

Q_A (L/min)	C_{CO_2} (ppm)
57	125
114	90
170	62

(v) The following steady-state data are available for a series of changes in Q_F,

Q_F (L/min)	C_{CO_2} (ppm)
100	75
200	96
300	122

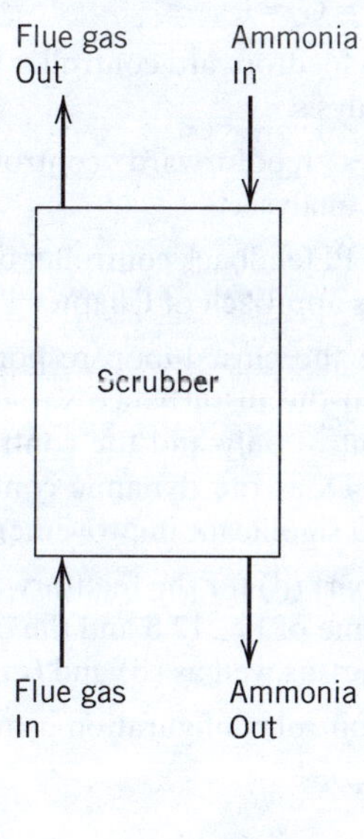

Figure E12.1

12.2 It is desired to control liquid level h_2 in the storage tank system shown in Fig. E12.2 by manipulating flow rate q_3 Disturbance variable q_1 can be measured. Use the information available to do the following:

(a) Draw a block diagram for a feedforward–feedback control system.

(b) Derive an ideal feedforward controller based on a steady-state analysis.

(c) Suppose that the flow-head relation for the hand valve is $q_2 = C\sqrt{h_1 - h_2}$. Does the ideal feedforward controller of part (b) change?

Available Information

(i) The two tanks have uniform cross-sectional areas, A_1 and A_2, respectively.

(ii) The valve on the exit line of Tank 1 acts as a linear resistance with a flow-head relation, $q_2 = (h_1 - h_2)/R$.

(iii) The transmitters and control valve are pneumatic instruments that have negligible dynamics.

(iv) The pump operates so that flow rate q_3 is independent of h_2 when the control valve stem position is maintained constant.

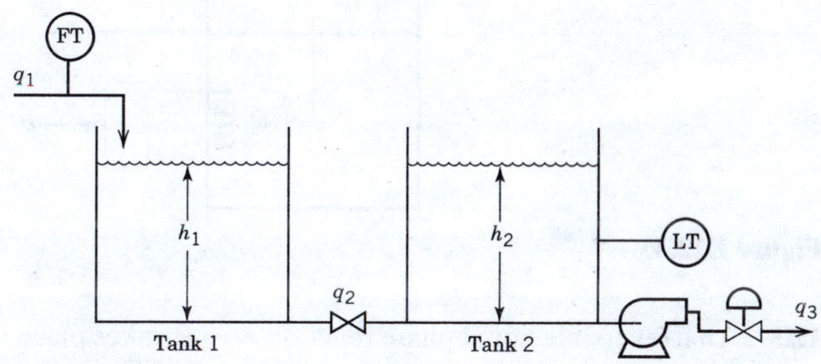

Figure E12.2

12.3 The closed-loop system in Fig. 12.8 has the following transfer functions:

$G_p(s) = 2/(s+1)$

$G_d(s) = 3/(s+1)(4s+1)$

$G_v = G_m = G_t = 1$

(a) Design a feedforward controller based on a steady-state design.

(b) Design a feedforward controller based on a dynamic analysis.

(c) Design a feedback controller based on the IMC approach of Chapter 11 and $\tau_c = 2$.

(d) Simulate the closed-loop response to a unit step change in the disturbance variable using feedforward control only and the controllers of parts (a) and (b).

(e) Repeat part (d) for the feedforward–feedback control scheme of Fig. 12.8 and the controllers of parts (a) and (c) as well as (b) and (c).

12.4 A feedforward control system is to be designed for the two-tank heating system shown in Fig.E12.4. The design objective is to regulate temperature T_4 despite variations in disturbance variables T_1 and w. The voltage signal to the heater p is the manipulated variable. Only T_1 and w are measured. Also, it can be assumed that the heater and transmitter dynamics are negligible and that the heat duty is linearly related to voltage signal p.

(a) Design a feedforward controller based on a steady-state energy balance. This control law should relate p to T_{1m} and w_m.

(b) Is dynamic compensation desirable? Justify your answer.

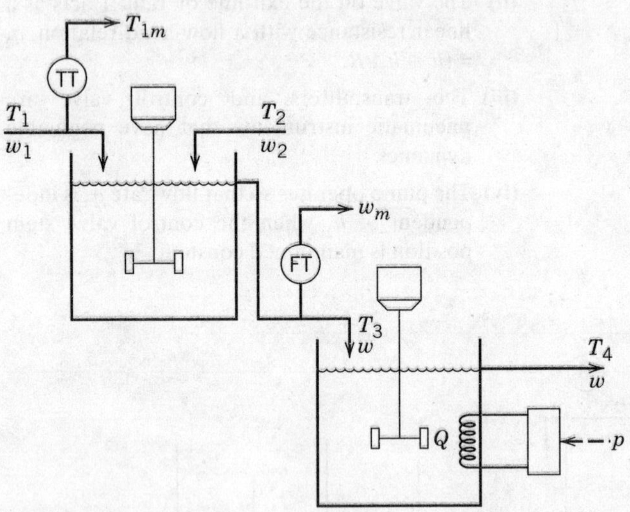

Figure E12.4

12.5 An irreversible liquid phase reaction A → B, takes place in a stirred-tank reactor, as shown in Fig. E12.5.

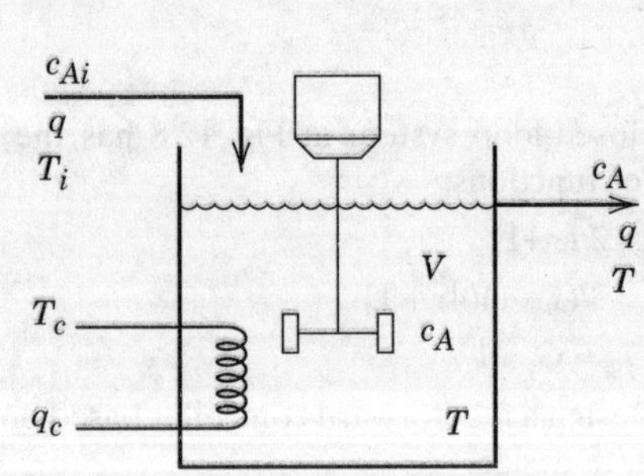

Figure E12.5

The reaction is zero order, and thus the rate of reaction does not depend on the concentrations of A or B. The rate of reaction, r (mols of A reacting/m^3 · s), is given by

$$r = k_0 e^{-E/RT}$$

where k_0 and E are constants. It has suggested that feedforward control be used to control reactor temperature T by adjusting the inlet coolant temperature, T_c. The chief disturbance variable is inlet temperature, T_i. Using the information given below, do the following:

(a) Design a feedforward controller based on steady-state considerations.

(b) Do you recommend that dynamic compensation in the form of a lead–lag unit be added to this feedforward controller? Briefly justify your answer.

Available Information

1. The volume of liquid V is held constant via an overflow line so that the volumetric flow rate q is the same for both inlet and outlet streams.
2. The variation in coolant temperature along the length of the cooling coil is negligible.
3. The rate of heat removal by the coolant can be expressed as $Q = UA(T - T_c)$ where U and A are constants.
4. The heat of reaction per mol A reacted ΔH, is constant.
5. The reactor can be assumed to be perfectly mixed.
6. The density of each liquid stream is the same.
7. The sensor/electronic transmitter combination for the feed temperature has negligible dynamics and a steady-state gain, K_t.
8. The electronic control valve has negligible dynamics and a steady-state gain K_v.

12.6 The distillation column in Fig. 12.5 has the following transfer function model:

$$\frac{Y'(s)}{D'(s)} = \frac{2e^{-20s}}{95s+1} \qquad \frac{Y'(s)}{F'(s)} = \frac{0.5e^{-30s}}{60s+1}$$

with $G_v = G_m = G_t = 1$.

(a) Design a feedforward controller based on a steady-state analysis.

(b) Design a feedforward controller based on a dynamic analysis.

(c) Design a PI feedback controller based on the Direct Synthesis approach of Chapter 11 with $\tau_c = 30$.

(d) Simulate the closed-loop response to a unit step change in the disturbance variable using feedforward control only and the controllers of parts (a) and (b). Does the dynamic controller of part (b) provide a significant improvement?

(e) Repeat part (d) for the feedforward–feedback control scheme of Fig. 12.8 and the controllers of parts (a) and (c), as well as (b) and (c).

(f) Which control configuration provides the best control?

12.7 A feedforward-only control system is to be designed for the stirred-tank heating system shown in Fig. E12.8. Exit temperature T will be controlled by adjusting coolant flow rate, q_c. The chief disturbance variable is the inlet temperature T_i which can be measured on-line. Design a feedforward-only control system based on a dynamic model of this process and the following assumptions:

1. The rate of heat transfer Q between the coolant and the liquid in the tank can be approximated by

$$Q = U(1+q_c)A(T-T_c)$$

where U, A, and the coolant temperature T_c are constant.

2. The tank is well mixed, and the physical properties of the liquid remain constant.
3. Heat losses to the ambient air can be approximated by the expression

$$Q_L = U_L A_L (T - T_a)$$

where T_a is the ambient temperature.

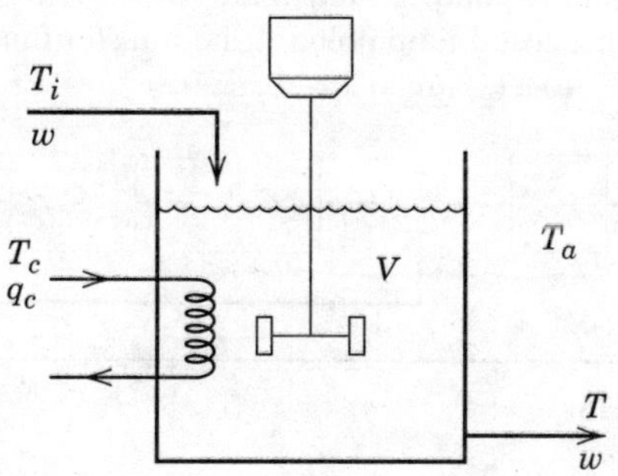

Figure E12.7

4. The control valve on the coolant line and the T_i sensor/transmitter (not shown in Fig. E12.8) exhibit linear behavior. The dynamics of both devices can be neglected, but there is a time delay θ associated with the T_i measurement due to the sensor location.

12.8 It is desired to design a feedforward control scheme in order to control the exit composition x_4 of the two-tank blending system shown in Fig. E12.8. Flow rate q_2 can be manipulated, while disturbance variables, q_5 and x_5, can be measured. Assume that controlled variable x_4 cannot be measured and that each process stream has the same density. Also, assume that the volume of liquid in each tank is kept constant by using an overflow line. The transmitters and control valve have negligible dynamics.

(a) Using the steady-state data given below, design an ideal feedforward control law based on steady-state considerations. State any additional assumptions that you make.

(b) Do you recommend that dynamic compensation be used in conjunction with this feedforward controller? Justify your answer.

Steady-State Data

Stream	*Flow (m³/min)*	*Mass Fraction*
1	7.19	0.000
2	3.78	0.990
3	9.08	0.167
4	12.87	0.409
5	1.89	0.800

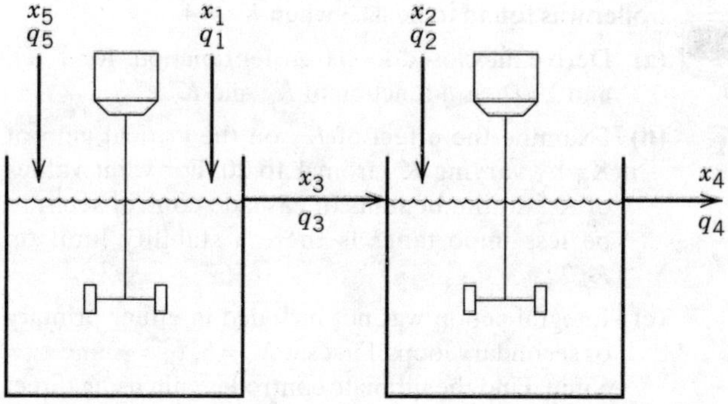

Figure E12.8

12.9 In ratio control, would the control loop gain for Method I (Fig. 12.15) be less variable if the ratio were defined as $R = d/u$ instead of $R = u/d$? Justify your answer.

12.10 Consider the ratio control scheme shown in Fig. 12.16. Each flow rate is measured using an orifice plate and a differential pressure (D/P) transmitter. The pneumatic output signals from the D/P transmitters are related to the flow rates by the expressions

$$d_m = d_{m0} + K_1 d^2$$

$$u_m = u_{m0} + K_2 u^2$$

Each transmitter output signal has a range of 20–100 kPa. The transmitter spans are denoted by S_d and S_u for the disturbance and manipulated flow rates, respectively. Derive an expression for the gain of the ratio station K_R in terms of S_d, S_u, and the desired ratio R_d.

12.11 Measurement devices and their dynamics influence the design of feedback controllers. Briefly indicate which of the two systems below would have its closed-loop performance enhanced significantly by application of cascade control (see Fig. 12.21 for notation and assume $G_{p2} = 1$ and $G_{d1} = G_{p1}$). Using the controller settings shown below, evaluate the effect of a unit step disturbance in D_2 on both systems A and B.

System A	*System B*
$G_v = 5$	$G_v = 5$
$G_{p1} = \dfrac{2}{10s+1}$	$G_{p1} = \dfrac{2}{10s+1}$
$G_{m2} = \dfrac{0.5}{0.5s+1}$	$G_{m2} = \dfrac{2}{5s+1}$

System A	System B
$G_{m1} = \dfrac{1}{5s+1}$	$G_{m1} = \dfrac{0.2}{5s+1}$
$K_{c1} = 0.5$	$K_{c1} = 2.5$
$\tau_{I1} = 15$	$\tau_{I1} = 15$
$K_{c2} = 1.0$	$K_{c2} = 0.25$

All time constants are in minutes.

12.12 In Example 12.6, the ultimate gain for the primary controller was found to be 43.3 when $K_{c2} = 4$.

(a) Derive the closed-loop transfer functions for Y_1/D_1 and Y_1/D_2 as a function of K_{c1} and K_{c2}.

(b) Examine the effect of K_{c2} on the critical gain of K_{c1} by varying K_{c2} from 1 to 20. For what values of K_{c2} do the benefits of cascade control seem to be less important? Is there a stability limit on K_{c2}?

(c) Integral action was not included in either primary or secondary loops. First set $K_{c2} = 5$, $\tau_{I1} = \infty$, and $\tau_{I2} =$ 5 min. Find the ultimate controller gain using direct substitution. Then repeat the stability calculation for $\tau_{I1} =$ 5 min and $\tau_{I2} = \infty$ and compare the two results. Is offset for Y_1 eliminated in both cases for step changes in D_1 or D_2?

12.13 Consider the cascade control system in Fig. E12.13

(a) Specify K_{c2} so that the gain margin $\geq$1.7 and phase margin $\geq$30° for the slave loop.

(b) Then specify K_{c1} and τ_I for the master loop using the Ziegler-Nichols tuning relation.

12.14 Consider the cascade control system in Fig. E12.14. Use IMC tuning rules for both the master and slave controllers. Design K_{c2} first, and then use that value to design G_{c1} (PI controller). The higher-order transfer function can be approximated first by a FOPTD model using a step test. Plot closed-loop responses for different values of the IMC closed-loop time constant for both outer loop and inner loop for a set point change.

12.15 A cascade control configuration is sometimes used to separate different objectives of feedback control. For instance, first an inner loop can be used to stabilize the process, and then an outer loop can be used to control the output to track set-point changes and reject disturbances. An example of a cascade control system is shown in Fig. E12.15. The process transfer function is unstable.

$$G_p(s) = \frac{s+1}{s-3}$$

(a) Determine the maximum range of K_{inner} values for which the inner loop will be stable.

(b) Now assume $K_{c2} = 6$ and G_{c1} is PI controller with gain K_{c1} and τ_I. Find values of K_{c1} and τ_I such that the closed-loop poles of the transfer function from Y_{sp} to $Y(s)$ are at $s = -0.5 \pm 0.2j$.

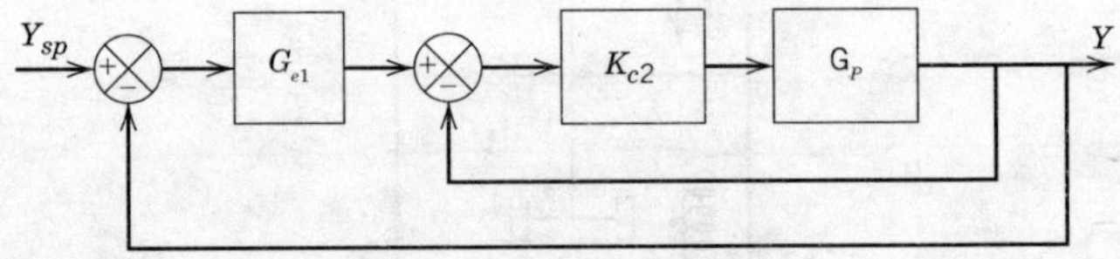

Figure E12.15

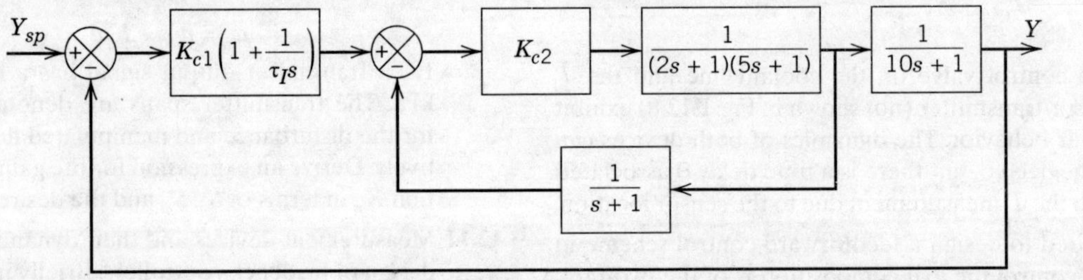

Figure E12.13

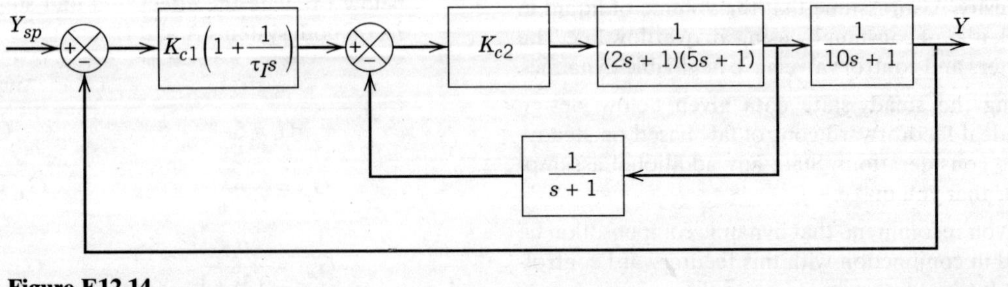

Figure E12.14

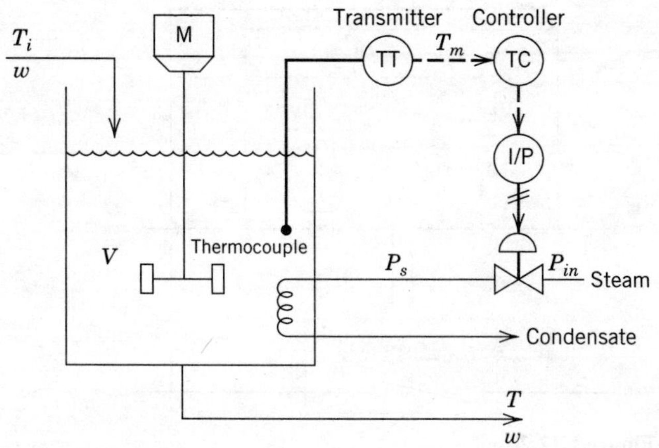

Figure E12.16

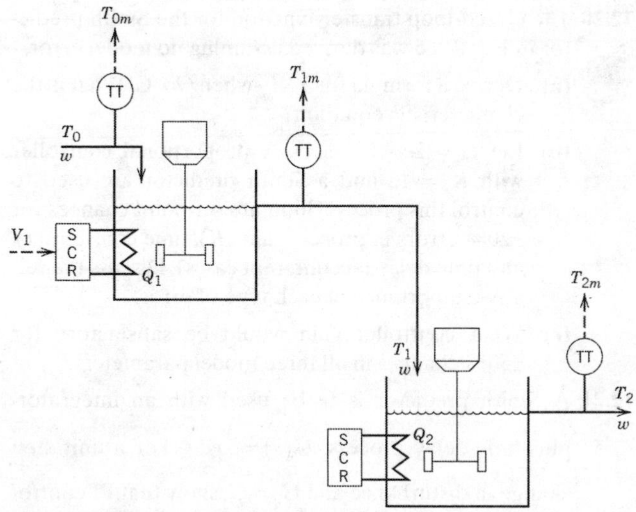

Figure E12.17

12.16 A water tank heating process shown in Fig. E12.17 controls the outlet temperature at a desired set point T_{SP} by manipulating the steam pressure P_S of steam sent to the tank heat exchanger. The steam delivery pressure P_{in} is subject to variations that affect P_S. Disturbances in the inlet stream temperature T_i are the primary disturbance to the process.

The transfer functions of the system in Fig. E12.17 are

$G_p = \dfrac{0.5}{(5s+1)(3s+1)}$	Process
$G_v = \dfrac{4}{s+1}$	Valve
$G_t = 0.5$	Temperature transmitter
$G_{d1} = \dfrac{2}{2s+1}$	Inlet stream temperature disturbance
$G_{d2} = \dfrac{0.4}{10s+1}$	Steam pressure disturbance

(a) Assuming that a PI controller with $\tau_I = 0.2$ min is used, find the range of K_c values for which the system is stable for a standard single control loop configuration.

(b) Suppose a cascade control strategy is implemented to handle variations in upstream steam pressure. How should the second loop be configured? Draw the block diagram for the process using this strategy and label the process variables.

(c) For the flowrate measurement, $G_m = 0.4$, and an inner loop proportional controller, compare the cascade controller closed-loop response with that of the standard PI controller ($K_{c1} = 2$ and $\tau_{I1} = 5$) for a unit step disturbance in steam pressure. For the cascade controller, use $K_{c1} = 2, \tau_{I1} = 5, K_{c2} = 2$.

12.17 Consider the stirred-tank heating system shown in Fig. E12.18. It is desired to control temperature T_2 by adjusting the heating rate Q_1 via voltage signal V_1 to the SCR. It has been suggested that measurements of T_1 and T_0, as well as of T_2, could provide improved control of T_2.

(a) Briefly describe how such a control system might operate, and sketch a schematic diagram. State any assumptions that you make.

(b) Indicate how you would classify your control scheme—for example, feedback, cascade, or feedforward. Briefly justify your answer.

(c) Draw a block diagram for the control system.

12.18 The closed loop transfer function for a feedback control system is given as $G_{CL} = \dfrac{G_p G_c}{1+G_p G_c}$. Take two instances of G_p : $\dfrac{1}{(5s+1)(s+1)(2s+1)}$ (without any time delay) and $\dfrac{e^{-4s}}{(5s+1)(s+1)(2s+1)}$ (with time delay).

Applying Ziegler-Nichols tuning technique, design PID controller for both these processes and plot the response for unit step change in set point. Discuss on the closed loop performance of the two processes.

12.19 Design a time-delay compensator (Smith predictor) for

$$G_p = \frac{e^{-\theta s}}{5s+1}$$

where $G_v = G_m = 1$ and $\theta = 1$. Show closed-loop responses for unit step set-point and disturbance changes ($G_d = G_p$). Use an IMC PI controller with $\tau_c = 3$ (Case A in Table 11.1).

12.20 The closed-loop transfer function for the Smith predictor in Eq. 12-66 was derived assuming no model error.

(a) Derive a formula for Y/Y_{sp} when $G \neq \widetilde{G}$. What is the characteristic equation?

(b) Let $G = 2e^{-2s}/(5s + 1)$. A proportional controller with $K_c = 15$ and a Smith predictor are used to control this process. Simulate set-point changes for ±20% errors in process gain (K), time constant (τ), and time delay (six different cases). Discuss the relative importance of each type of error.

(c) What controller gain would be satisfactory for ±50% changes in all three model parameters?

12.21 A Smith predictor is to be used with an integrator-plus-time-delay process, $G(s) = \dfrac{3}{s}e^{-5s}$. For a unit step change in disturbance and $G_d = G$, show that PI control will not eliminate offset even when the model is known perfectly. Use Eq. 12-66 as the starting point for your analysis.

12.22 In Chapter 11, we introduced the Direct Synthesis design method, in which the closed-loop set-point response is specified and the controller transfer function is calculated algebraically. For a FOPTD process and an IMC controller (see Chapter 11), show that setting $G_+ = e^{-\theta s}$ leads to a Smith predictor controller structure when $G \neq \widetilde{G}$.

12.23 The pressure of a reactor vessel can be adjusted by changing either the inlet or outlet gaseous flow rate. The outlet flow is kept fixed as long as the tank pressure remains between 7 kg/cm² and 14 kg/cm², and pressure changes are treated by manipulating the inlet flow control valve. However, if the pressure goes higher than these limits, the exit gas flow is then changed. Finally, if the pressure exceeds 14 kg/cm², a vent valve on the vessel is opened and transfers the gas to a storage vessel. Design a control scheme that meets the performance objectives. Draw a process instrumentation diagram for the resulting control system.

12.24 Selectors are normally used in combustion control systems to prevent unsafe situations from occurring. Figure E12.24 shows the typical configuration for high and low selectors are applied to air and fuel flow rates. The energy demand signal comes from the steam pressure controller. Discuss how the selectors operate in this control scheme when the steam pressure drops suddenly.

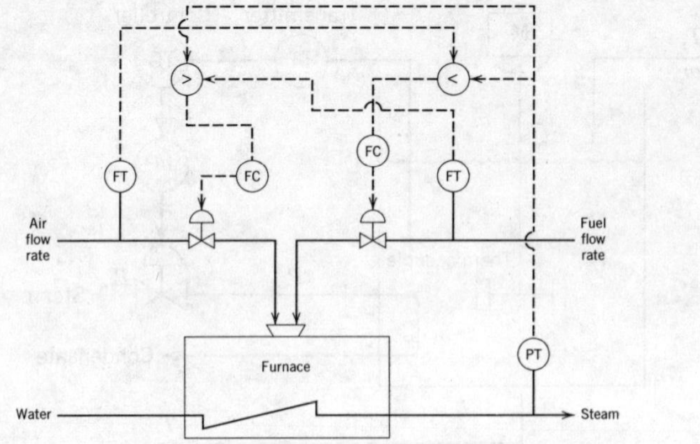

Figure E12.24

(a) Derive a control law that shows how to adjust K_c, τ_I, and τ_D based on variations in K_p, τ_1, and τ_2 and the desired closed-loop time τ_c.

(b) Suppose $\tau_1 = 3$, $\tau_2 = 5$, and $K_p = 1$. Calculate values of K_c, τ_I and τ_D to achieve $\tau_c = 1.5$. Show how the response deteriorates for changes in the following model parameters when the controller remains unchanged:

(i) $K_p = 2$

(ii) $K_p = 0.5$

(iii) $\tau_2 = 10$

(iv) $\tau_2 = 1$

12.25 A second-order process is controlled by a PID controller. The desired closed-loop servo transfer function is

$$\frac{Y}{Y_{sp}} = \frac{1}{\tau_c s + 1} = \frac{G_p G_c}{1 + G_p G_c} \qquad G_v = G_m = 1$$

and the process model is

$$G_p = \frac{K_p}{(\tau_1 s + 1)(\tau_2 s + 1)}$$

12.26 An instrumentation diagram for a fired heater control system is shown in Fig. E12.26. Identify advanced control strategies based on material from Chapter 12. Discuss the rationale for each advanced method.

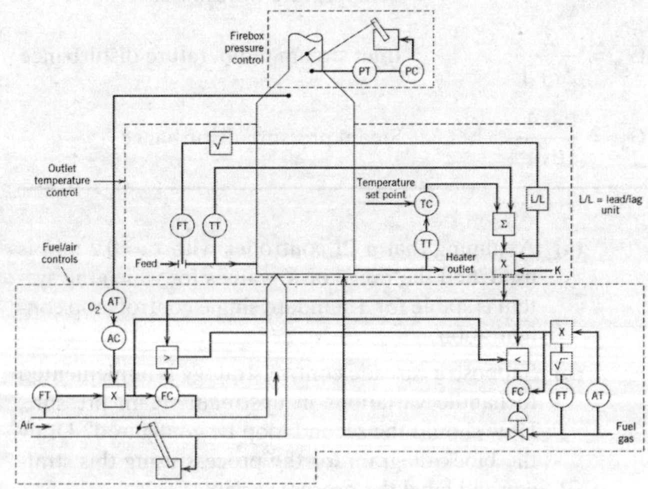

Figure E12.26

12.27 Figure E12.27 shows cascade temperature control of a polymerization reactor, which uses feed heat exchange to adjust the reactor temperature. Using the instrumentation diagram, explain how this cascade control system (both master and slave components) handles the following disturbances and describe what happens to the reactor temperature. Assume normal temperatures of coolant (22°C), polymerization feed (94°C), exchanger effluent (38°C), and reactor outlet (427°C).

(a) Feed temperature becomes too high.

(b) Feed flow rate becomes too high.

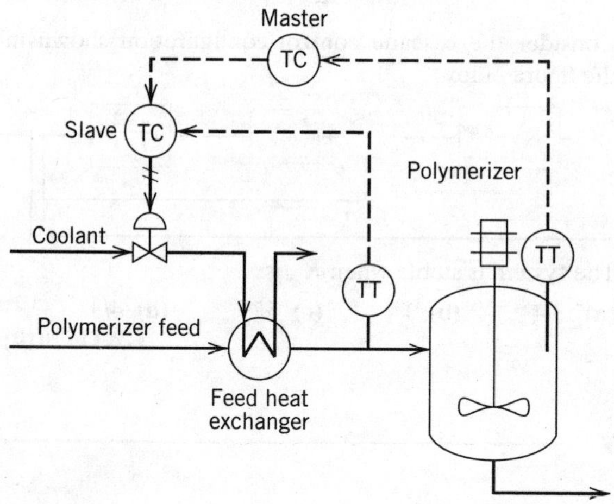

Figure E12.27

12.28 A ratio control scheme is to be used to maintain a desired air-fuel ratio of 2:1 fed to a furnace. Consider fuel flow rate is the disturbance variable

(a) Draw a schematic diagram for the ratio control scheme.

(b) Specify the appropriate gain for the ratio station KR.

Available information

(i) Span of the flow transmitters are 100 L/min for air and 50 L/min for fuel.

(ii) current to pressure transducer has a gain of 5.5 kPa/mA.

(iii) input and output to the ratio station are electronic signals in the range of 4 to 20 mA.

MULTIPLE CHOICE QUESTIONS

1. A control configuration used to control unmeasured outputs with the help of secondary measurements and process model is called

(a) feedback control scheme.

(b) feedforward control scheme.

(c) inferential control scheme.

(d) selective control scheme.

2. Ratio control is a special type of

(a) feedback control configuration.

(b) feedforward control configuration.

(c) inferential control configuration.

(d) selective control configuration.

3. Presence of dead time in the open loop transfer function:

(a) increases crossover frequency and the ultimate controller gain.

(b) decreases crossover frequency and the ultimate controller gain.

(c) increases crossover frequency and decreases the ultimate controller gain.

(d) decreases crossover frequency and increases the ultimate controller gain.

4. Which of the following is a non-linear control strategy?

(a) Cascade control

(b) Gain scheduling

(c) Ratio control

(d) Feed-forward control

5. If the primary loop time constant is less than the secondary loop time constant, the cascade control loop will

(a) make the closed loop response more sluggish as compared to feedback control loop.

(b) make the closed loop response faster as compared to feedback control loop.

(c) exhibit no better performance than the feedback control loop.

(d) None of the above.

6. Adaptive gain is used for controlling _______ processes.

(a) time delay

(b) integrating

(c) linear

(d) non-linear

7. The primary (master) controller in a cascade control system must always be tuned

(a) after the slave (secondary) controller is tuned.

(b) before the slave (secondary) controller is tuned.

(c) using the same parameters as that of the slave controller.

(d) Using the Ziegler-Nichols tuning method.

8. Which of the following is an example of a non-linear chemical process?

(a) Blending system

(b) Isothermal CSTR

(c) Non-isothermal CSTR

(d) None of the above

9. Given below is a simplified block diagram of a feed forward control system.

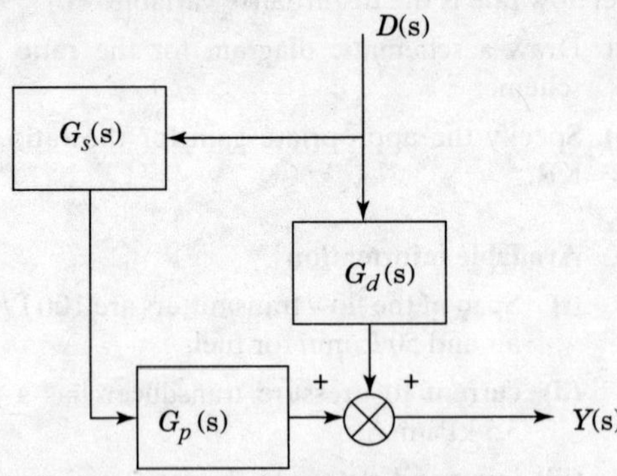

The transfer function of the process is $G_p(s) = \dfrac{5}{s+1}$ and the disturbance transfer function is $G_d(s) = \dfrac{1}{s^2 + 2s + 1}$. The transfer function of the PERFECT feed forward control $G_f(s)$ is

(a) $\dfrac{-5}{(s+1)^3}$ **(b)** $\dfrac{-5}{s+1}$ **(c)** $\dfrac{-1}{5(s+1)}$ **(d)** $-5(s+1)$

(GATE 2014)

10. Consider the cascade control configuration shown in the figure below.

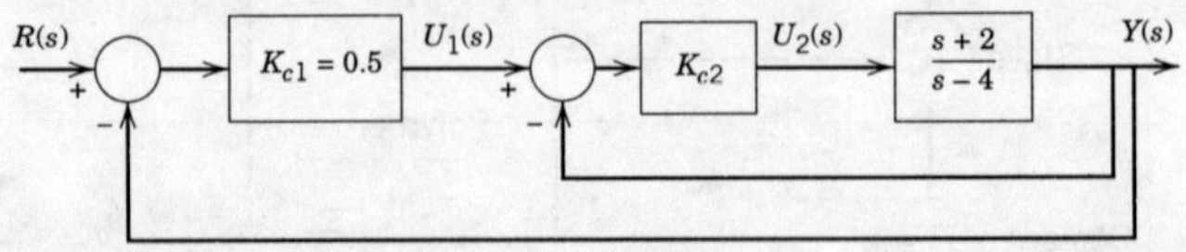

The system is stable when K_{c2} is:

(a) 3/4 **(b)** 1 **(c)** 5/4 **(d)** 4/3

(GATE 2010)

ANSWER KEY

1. (c)	**2.** (b)	**3.** (b)	**4.** (b)	**5.** (c)	**6.** (d)	**7.** (a)
8. (c)	**9.** (c)	**10.** (d)				

CHAPTER 13

Digital Sampling, Filtering, and Control

CHAPTER CONTENTS

All discussions on control systems so far are based on continuous-data systems or analog systems. Hardware elements used for process control have evolved during the last sixty to seventy years from pneumatic to electronic to microprocessor based control. During early days, pneumatic controllers were predominantly used where the input-output signals are air pressure in the range of 20 to 100 kPa(g). Subsequently during 1960s, electronic controllers came into existence in which input-output signals are in the range of 4 to 20 mA.

In the last 50 years, technological advancements in digital computing system coupled with large information (or data) storage capability have resulted in application of digital computer for process control. Unlike analog control systems, the inputs and outputs to a digital computer are discrete signals. The high computation speed ability of microprocessor or digital controllers have made it possible for implementation of advanced control algorithms in process industries. Therefore, a single digital computer can be used to perform computations for multiple control loops unlike analog system where one controller is required for each control loop. Further, digital components are more reliable and have better sensitivity to environmental and process parameter variations than analog components.

13.1 COMPONENTS OF DIGITAL COMPUTER CONTROL LOOP

The multiple advantages and benefits led to gradual shift from analog control system to digital control systems in the past 30 to 40 years. Almost all process industries today are under digital computer control. Components of analog control system implementing an electronic controller is shown in Figure 13.1.

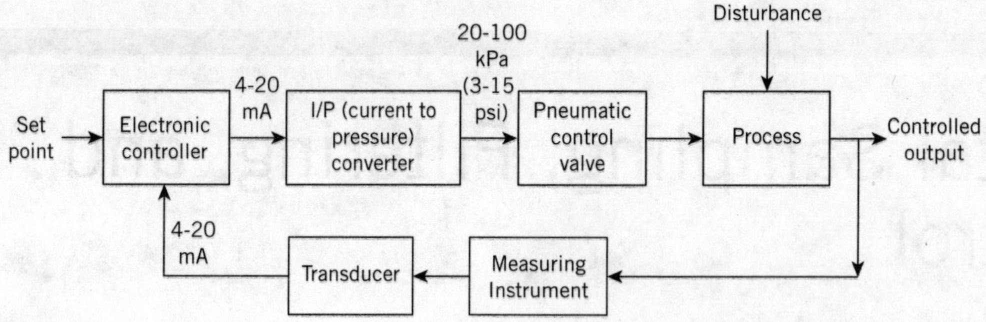

Figure 13.1 Simplified Block diagram of an analog control system with electronic controller.

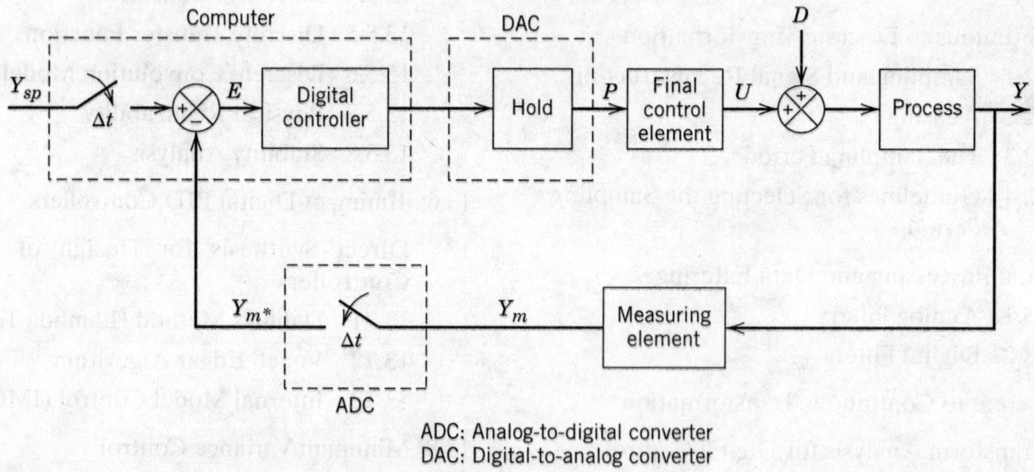

Figure 13.2 Simplified block diagram for computer control.

Replacement of pneumatic or electronic controller with digital computer results in different hardware requirements and controller design algorithms. Unlike analog controllers, the input and output signals are discrete. Therefore, there is a need to convert the continuous signal received from the measuring sensor to discontinuous form using ADC (Analog to Digital Conversion) cards. Similarly, the discrete outputs from the controller (computer) has to be converted to continuous signal using DAC (Digital to Analog Conversion) cards so that the final control element (a pneumatic control valve in most cases) can be operated uninterruptedly/continuously. Figure 13.2 is a typical schematic block diagram of a digital control system.

In this chapter we will discuss about the different components shown in Figure 13.2.

The specifications for a computer-based system to perform data acquisition and control must address several questions:

1. How often should data be acquired from each sensor? That is, what sampling rate should be employed?
2. Do the measurements contain a significant amount of noise? If so, can the data be conditioned (*filtered*) to reduce the effects of noise?
3. What digital control algorithm should be employed?

13.2 CONTINUOUS TO DISCRETE TRANSFORMATION

When a digital computer is used for control, continuous measurements are converted into digital form by an analog-to-digital converter (ADC) (see Chapter-19). This operation is necessary because the digital computer cannot directly process a continuous (analog) signal. In the computer controlled system, there is an interface device between the measuring instrument and the computer. This device prepares the information coming from the measuring instrument so that the information is transformed to suitable signal that is compatible with the computer.

13.2.1 Sampling and Signal Reconstruction

The continuous measurement signal received from the measuring instruments are sampled and then each analog value is assigned its corresponding digital value. The time interval between successive samples is referred to as the *sampling period* Δt, which corresponds to the sampling or *scan* rate, $f_s = 1/\Delta t$ (cycles/time), or, equivalently, the sampling frequency, $\omega_s = 2\pi/\Delta t$ (radians/time).

Figure 13.3 shows an idealized periodic sampling operation in which the sampled signal $y*(t)$ is a series of impulses that represents the measurements $y(0), y(1), y(2)....$ at the sampling instants $t = 0, \Delta t, 2\Delta t, \ldots$. The representation in Fig. 13.3, also referred to as *impulse*

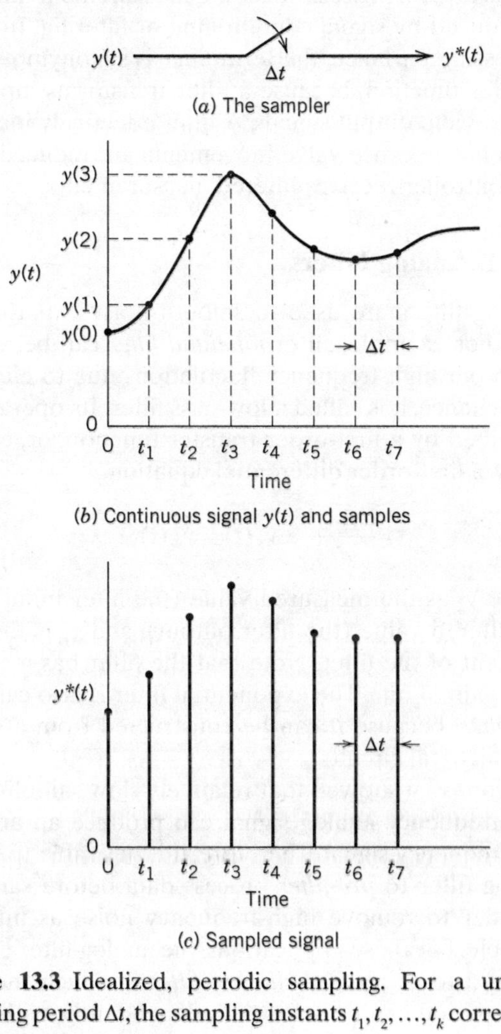

Figure 13.3 Idealized, periodic sampling. For a uniform sampling period Δt, the sampling instants $t_1, t_2, \ldots, t_k$ correspond to times $\Delta t, 2\Delta t, \ldots, k\Delta t$.

modulation (Franklin et al., 2005), is based on the assumption that the sampling operation occurs instantaneously.

Figure 13.2 shows the block diagram for a typical feedback control loop with a digital controller. Note that both continuous (analog) and sampled (digital) signals appear in the block diagram. The two samplers typically have the same sampling period and operate synchronously, which means that they acquire sampled signals at exactly the same time. However, *multirate sampling* is sometimes used, in which one sampler operates at a faster rate than the other. For example, we may wish to sample a process variable and filter the measurements quite frequently while performing the control calculations less often in order to avoid excessive wear in the actuator or control valve. The block diagram in Fig. 13.2 is symbolic in that the mathematical relations between the various signals (e.g., transfer functions) are not shown.

13.2.2 Aliasing

The sampling rate must be fast enough that significant process information is not lost, as illustrated in Fig. 13.4. Suppose that a sinusoidal signal is sampled at a rate of 4/3 samples per cycle (i.e., 4/3 samples per period). This sampling rate causes the reconstructed signal to appear as a sinusoid with a much longer period than the original signal, as shown in Fig. 13.4*b*. This phenomenon is known as *aliasing*. Note that if the original sinusoidal signal were sampled only twice per period, then a constant sampled signal would result, as shown in Fig. 13.4*d*. According to Shannon's Sampling Theorem (Franklin et al., 2005), a sinusoidal signal must be sampled *more* than twice each period to recover the original signal; that is, the sampling frequency must be at least twice the frequency of the sine wave.

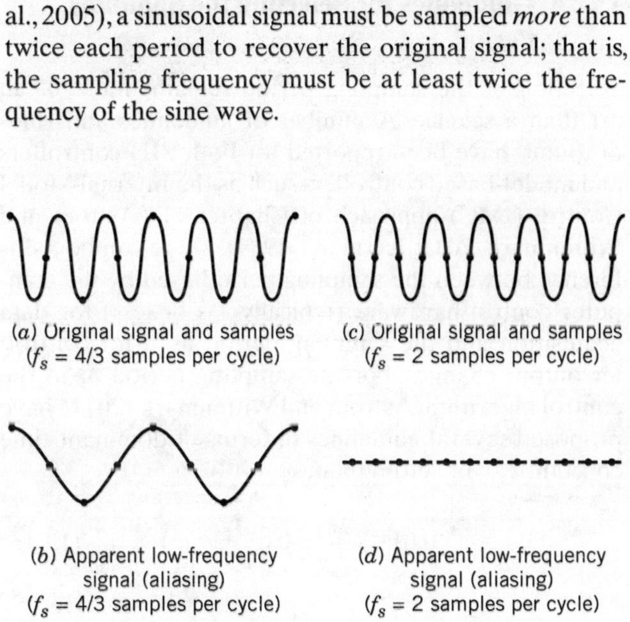

Figure 13.4 Aliasing error due to sampling too slowly.

Aliasing also occurs when a process variable that is *not* varying sinusoidally is sampled. In general, if a process measurement is sampled with a sampling frequency, ω_s, high-frequency components of the process variable with a frequency greater than $\omega_s/2$ appear as low-frequency components ($\omega < \omega_s/2$) in the sampled signal. Such low-frequency components can cause control problems if they appear in the same frequency range as the normal process variations (e.g., frequencies close to the critical frequency ω_c, as discussed in Chapter 10). Aliasing can be eliminated by using an *anti-aliasing* filter, as discussed in Section 13.3.

13.2.3 The Sampling Period

Sampling too slowly can reduce the effectiveness of a feedback control system, especially its ability to cope with disturbances. In an extreme case, if the sampling period is longer than the process response time, then a disturbance can affect the process, but the influence of the disturbance will disappear before the controller takes corrective action. In this situation, the control system cannot handle transient disturbances and is capable only of steady-state control. Thus, it is important to consider the process dynamics (including disturbance characteristics) in selecting the sampling period. Commercial digital controllers, typically employ a fixed scan rate less than or equal to 1 s but can vary the sampling period for control calculations. For $\Delta t \leq 1$ s, the performance of a digital controller closely approximates that for continuous (analog) control in normal process control applications.

13.2.4 Guidelines for Selecting the Sampling Period

Selection of the sampling period remains more of an art than a science. A number of guidelines and rules of thumb have been reported for both PID controllers and model-based controllers such as the Internal Model Control (IMC) approach of Chapter 11 (Åström and Wittenmark, 2011; Isermann, 1989). There can be a difference between the sampling period used by the computer control hardware (typically 1 s or less) for data acquisition and the sampling period used for controller output changes. For the sampling period Δt in the control algorithm, Åström and Wittenmark (2011) have proposed several guidelines in terms of dominant time constant τ_{dom} or settling time t_s.

$$0.01 \leq \frac{\Delta t}{\tau_{dom}} \leq 0.05 \qquad \text{(13-1a)}$$

$$\frac{t_s}{15} \leq \Delta t \leq \frac{t_s}{6} \qquad \text{(13-1b)}$$

In some cases, the process time delay can become a factor, and the sampling period must be reduced to speed up the response time for disturbances. Simulation using different sampling periods can be carried out to make the final selection. Again, we should mention that process data may be acquired at a higher rate than that indicated above.

13.3 SIGNAL PROCESSING AND DATA FILTERING

In process control applications, noise associated with measurements can arise from a number of sources: the measurement device, electrical equipment, or the process itself. The effects of electrically generated noise can be minimized by following established procedures concerning shielding of cables, grounding, and so forth (McConnell and Jernigan, 2005). Process-induced noise can arise from variations resulting from incomplete mixing, turbulence, and nonuniform multiphase flows. The effects of both process noise and measurement noise can be reduced by signal conditioning or filtering. In signal processing parlance, the term *filter* is synonymous with transfer function, because a filter transforms input signals to yield output signals. A filter effectively increases valve life, because valve movements are reduced when the controller receives filtered measurements.

13.3.1 Analog Filters

Analog filters are used to smooth noisy experimental data. For example, an *exponential filter* can be used to damp out high-frequency fluctuations due to electrical noise; hence, it is called a low-pass filter. Its operation is described by a first-order transfer function, or, equivalently, a first-order differential equation.

$$\tau_F \frac{dy_F(t)}{dt} + y_F(t) = y_m(t) \qquad \text{(13-2)}$$

where y_m is the measured value (the filter input), y_F is the filtered value (the filter output), and τ_F is the time constant of the filter. Note that the filter has a steady-state gain of one. The exponential filter is also called an *RC filter*, because it can be constructed from a simple RC electrical circuit.

Figure 13.4 showed that relatively slow sampling of a high-frequency analog signal can produce an artificial low-frequency signal. Therefore, it is desirable to use an analog filter to *pre-filter* process data before sampling in order to remove high-frequency noise as much as possible. For these applications, the analog filter is often referred to as an *anti-aliasing filter* in which the sampling period can be selected independently, with τ_F set to approximately $0.5\Delta t$. However, to treat slowly varying signals, digital filtering can also be used, as described in Section 13.3.2 (McConnell and Jernigan, 2005).

The filter time constant τ_F in Eq. 13-2 should be much smaller than the dominant time constant of the process τ_{dom} to avoid introducing a significant dynamic lag in the feedback control loop. For example, choosing $\tau_F < 0.1\tau_{dom}$ generally satisfies this requirement. On the other hand, if the noise amplitude is high, then a larger value of τ_F may be required to smooth the noisy measurements. The frequency range of the noise is another important consideration. Suppose that the lowest noise frequency expected is denoted by ω_N. Then τ_F should be selected so that $\omega_F < \omega_N$, where $\omega_F = 1/\tau_F$. For example, suppose we specify $\omega_F = 0.1\omega_N$, which corresponds to $\tau_F = 10/\omega_N$. Then noise at frequency ω_N will be attenuated by a factor of 10, according to Eq. 10-13 and the Bode diagram of Fig. 10.2. In summary, τ_F should be selected so that $1/\omega_N \leq \tau_F \leq 0.1\tau_{dom}$.

13.3.2 Digital Filters

In this section, we consider several widely used digital filters. A more comprehensive treatment of digital filtering and signal processing techniques is available elsewhere (Oppenheim and Shafer, 1999).

Exponential Filter

First we consider a digital version of the exponential filter in Eq. 13-2. Denote the samples of the measured variable as $y_m(k-1), y_m(k),\ldots$ and the corresponding filtered values as $y_F(k-1), y_F(k),\ldots$ where k refers to the current sampling instant. The derivative in Eq. 13-2 at time step k can be approximated by a first-order backward difference:

$$\frac{dy_F}{dt} \cong \frac{y_F(k) - y_F(k-1)}{\Delta t} \tag{13-3}$$

Substituting in Eq. 13-2 and replacing $y_F(t)$ by $y_F(k)$ and $y_m(t)$ by $y_m(k)$ yields

$$\tau_F \frac{y_F(k) - y_F(k-1)}{\Delta t} + y_F(k) = y_m(k) \tag{13-4}$$

Rearranging gives

$$y_F(k) = \frac{\Delta t}{\tau_F + \Delta t} y_m(k) + \frac{\tau_F}{\tau_F + \Delta t} y_F(k-1) \tag{13-5}$$

We define a dimensionless parameter

$$\alpha \triangleq \frac{\Delta t}{\tau_F + \Delta t} \tag{13-6a}$$

where $0 < \alpha \leq 1$. Then

$$1 - \alpha = 1 - \frac{\Delta t}{\tau_F + \Delta t} = \frac{\tau_F}{\tau_F + \Delta t} \tag{13-6b}$$

so that Eq. 13-5 can be written as

$$y_F(k) = \alpha y_m(k) + (1-\alpha) y_F(k-1) \tag{13-7}$$

Equation 13-7 indicates that the filtered measurement is a weighted sum of the current measurement $y_m(k)$ and the filtered value at the previous sampling instant $y_F(k-1)$. This operation is also called *single exponential smoothing* or the *EWMA filter*, or *exponentially weighted moving average*. Limiting cases for α are

$\alpha = 1$: No filtering (the filter output is the raw measurement $y_m(k)$).
$\alpha \to 0$: The measurement is ignored.

Equation 13-6a indicates that $\alpha = 1$ corresponds to a filter time constant of zero (no filtering).

Alternative expressions for α in Eq. 13-7 can be derived if the forward difference or other integration schemes for dy/dt are utilized (Franklin et al., 2005). Analytical integration of Eq. 13-2 to yield a difference equation can be performed for a piecewise constant input, leading to the result previously obtained in Chapter 16 (Eq. 16-24).

Double Exponential Filter

Another useful digital filter is the double exponential or second-order filter, which offers some advantages for dealing with signal drift and trending signals. The second-order filter is equivalent to two first-order filters in series, where the second filter input is the output signal $y_F(k)$ from the exponential filter in Eq. 13-7. The second filter (with output $\bar{y}_F(k)$ and filter constant γ) can be expressed as

$$\bar{y}_F(k) = \gamma y_F(k) + (1-\gamma)\bar{y}_F(k-1) \tag{13-8}$$

or

$$\bar{y}_F(k) = \gamma\alpha y_m(k) + \gamma(1-\alpha) y_F(k-1) + (1-\gamma)\bar{y}_F(k-1) \tag{13-9}$$

Writing Eq. 13-8 for the previous sampling instant gives

$$\bar{y}_F(k-1) = \gamma y_F(k-1) + (1-\gamma)\bar{y}_F(k-2) \tag{13-10}$$

Solving for $y_F(k-1)$,

$$y_F(k-1) = \frac{1}{\gamma}\bar{y}_F(k-1) - \frac{1-\gamma}{\gamma} y_F(k-2) \tag{13-11}$$

Substituting Eq. 13-11 into Eq. 13-9 and rearranging gives the following expression for the double exponential filter:

$$\begin{aligned}\bar{y}_F(k) &= \gamma\alpha y_m(k) + (2-\gamma-\alpha)\bar{y}_F(k-1)\\ &\quad -(1-\alpha)(1-\gamma)\bar{y}_F(k-2)\end{aligned} \tag{13-12}$$

A common simplification is to select $\gamma = \alpha$, yielding

$$\begin{aligned}\bar{y}_F(k) &= \alpha^2 y_m(k) + 2(1-\alpha)\bar{y}_F(k-1)\\ &\quad -(1-\alpha)^2\bar{y}_F(k-2)\end{aligned} \tag{13-13}$$

The advantage of the double exponential filter over the exponential filter of Eq. 13-7 is that it provides better filtering of high-frequency noise, especially if $\gamma = \alpha$. On the other hand, it is sometimes difficult to tune γ and α properly for a given application or data set.

The robustness of a PID controller is enhanced by use of a filter on the measured variable. Segovia et al. (2014) and Garpinger and Hägglund (2015) recommend a second-order critically damped filter that has one tuning parameter and exhibits amplitude attenuation at high frequency. This filter appears to be advantageous compared to the double exponential filter with two different time constants ($\alpha \neq \gamma$). Although the second-order filter is beneficial in some cases, the single exponential filter is more widely used in process control applications.

Moving-Average Filter

A *moving-average filter* averages a specified number of past data points, giving equal weight to each data point. It is usually less effective than the exponential filter, which gives more weight to the most recent data. The moving-average filter can be expressed mathematically as

$$y_F(k) = \frac{1}{N^*}\sum_{i=k-N^*+1}^{k} y_m(i) \tag{13-14}$$

where N^* is the number of past data points that are being averaged. Equation 13-14 also can be expressed in terms of the $k-1$ filtered value, $y_F(k-1)$:

$$y_F(k-1) = \frac{1}{N^*}\sum_{i=k-N^*}^{k-1} y_m(i) \tag{13-15}$$

Subtracting Eq. 13-15 from Eq. 13-14 gives the recursive form of the moving-average filter:

$$y_F(k) = y_F(k-1) + \frac{1}{N^*}(y_m(k) - y_m(k-N^*)) \tag{13-16}$$

The moving-average filter is a low-pass filter that eliminates high-frequency noise.

Noise-Spike Filter

If a noisy measurement changes suddenly by a large amount and then returns to the original value (or close to it) at the next sampling instant, a *noise spike* is said to occur. Figure 13.5 shows two noise spikes appearing in the experimental temperature data for a fluidized sand bath. In general, noise spikes can be caused by spurious electrical signals in the environment of the sensor. If noise spikes are not removed by filtering before the noisy measurement is sent to the controller, the controller will produce large, sudden changes in the manipulated variable.

Noise-spike filters (or *rate-of-change* filters) are used to limit how much the filtered output is permitted to change from one sampling instant to the next. If Δy denotes the maximum allowable change, the noise-spike filter can be written as

$$y_F(k) = \begin{cases} y_m(k) & \text{if } |y_m(k) - y(k-1)| \le \Delta y \\ y_F(k-1) - \Delta y & \text{if } y_F(k-1) - y_m(k) > \Delta y \\ y_F(k-1) + \Delta y & \text{if } y_m(k) - y_F(k-1) > \Delta y \end{cases} \tag{13-17}$$

If a large change in the measurement occurs, the filter replaces the measurement by the previous filter output plus (or minus) the maximum allowable change. This filter can also be used to detect instrument malfunctions such as a power failure, a break in a thermocouple or instrument line, or an ADC "glitch."

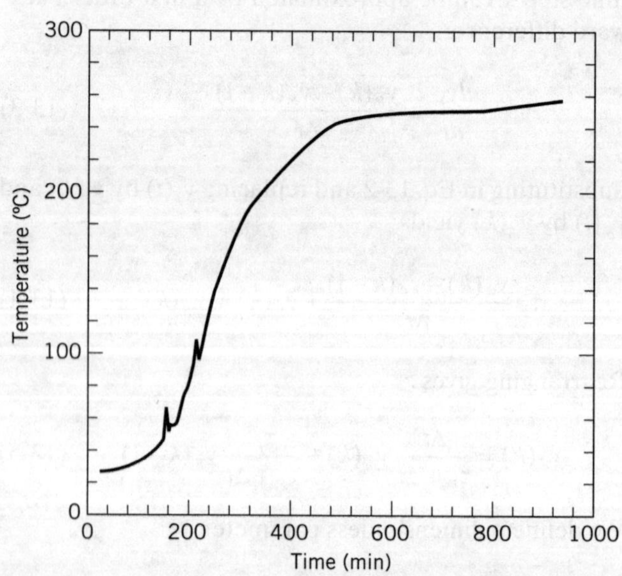

Figure 13.5 Temperature response data containing two noise spikes.

More complicated digital filters are available but have not been commonly used in process control applications. These include high-pass filters and band-pass filters (Isermann, 1989; Oppenheim and Shafer, 1999) as well as the Savitzky–Golay polynomial smoothing filter (Orfanidis, 1995; Xu et al., 2015), which can be used online.

EXAMPLE 13.1

To compare the performance of alternative filters, consider a square-wave signal with a frequency of $f = 0.33$ cycles/min and an amplitude 0.5 corrupted by

(i) High-frequency sinusoidal noise (amplitude = 0.25, $f_N = 9$ cycles/min)

(ii) Random (Gaussian) noise with zero mean and a variance of 0.01

Evaluate both analog and digital single exponential filters, as well as a moving-average filter, and assess the effect of sampling interval Δt.

Solution

(i) Sinusoidal Noise

Representative results for high-frequency sinusoidal noise are shown in Fig. 13.6. The square-wave with additive noise, the signal to be filtered, is shown in Fig. 13.6*a*, and the performance of two analog exponential filters with different filter time constants is shown in Fig.13.6*b*. Choosing a relatively large filter time constant ($\tau_F = 0.4$ min) results in a filtered signal that contains less noise but is more sluggish, compared to the response for $\tau_F = 0.1$ min.

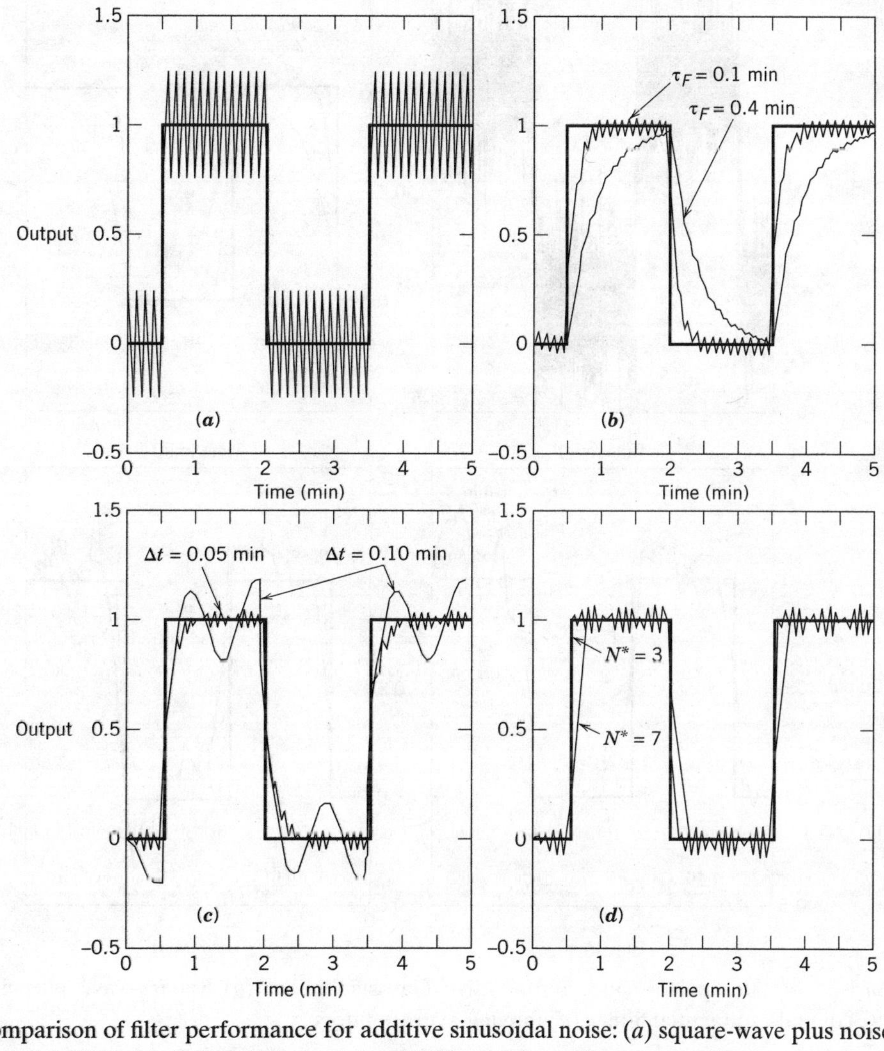

Figure 13.6 A comparison of filter performance for additive sinusoidal noise: (*a*) square-wave plus noise (original signal); (*b*) analog exponential filters; (*c*) digital single exponential filters; (*d*) moving-average filters.

The effect of sampling period Δt on digital filter performance is illustrated in Fig. 13.6*c*. A larger sampling interval ($\Delta t = 0.1$ min) results in serious aliasing, because $f_s = 1/\Delta t = 10$ cycles/min, which is less than $2f_N = 18$ cycles/min for the added sinusoidal noise. Reducing Δt by a factor of two results in much better performance. For each filter, a value of $\tau_F = 0.1$ min was chosen, because this value was satisfactory for the analog filter of Fig. 13.6*b*. The smaller value of α (0.33 for $\Delta t = 0.05$ min vs. 0.5 for $\Delta t = 0.1$ min) provides more filtering.

The performance of two moving-average filters ($N^* = 3$ and 7) with $\Delta t = 0.05$ min is shown in Fig. 13.6*d* Choosing $N^* = 7$ results in better filtering because this moving-average filter averages the sinusoidal noise over several cycles, while $N^* = 3$ gives a faster response but larger fluctuations.

(ii) Random Noise

The simulations illustrating the effects of this noise level are shown in Fig. 13.7. Figure 13.7*a* shows the unfiltered signal after Gaussian noise with zero mean and a variance of 0.01 was added to the square-wave signal. The analog exponential filters in Fig. 13.7*b* provide effective filtering and again show the tradeoff between degree of filtering and sluggish response that is inherent in the choice of τ_F. The digital filters in Fig. 13.7*c* and *d* are less effective, even though different values of Δt and N^* were considered. Some aliasing occurs owing to the high-frequency components of the random noise, which prevents the digital filter from performing as well as the analog filter.

In conclusion, both analog and digital filters can smooth noisy signals, providing that the filter design parameters (including sampling period) are carefully selected.

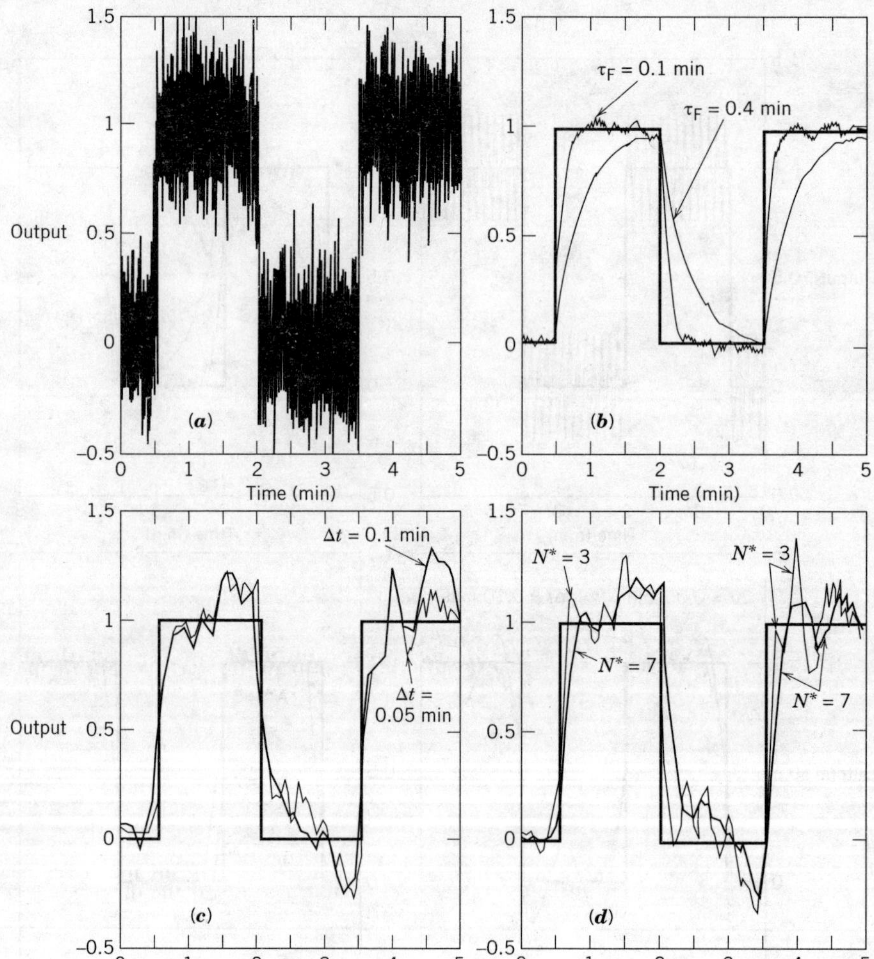

Figure 13.7 Comparison of filter performance for additive Gaussian noise: (*a*) Square-wave plus noise; (*b*) analog exponential filters; (*c*) digital exponential filters; (*d*) moving-average filters.

13.4 DISCRETE TO CONTINUOUS TRANSFORMATION

Most computer control systems require a device called a DAC (*digital-to-analog converter*), which changes a series of pulses (from the digital computer or controller) into a continuous signal. This signal is then transferred to a final control element such as a control valve. In process control, the final control element normally requires a continuous input signal rather than a pulsed input (although a stepping motor is one exception).

The DAC usually contains a *zero-order hold* (ZOH) to convert the digital (pulsed) signal from the controller into a staircase function $y_H(t)$, as shown in Fig. 13.8. Note that the output signal from the ZOH $y_H(t)$ is held constant for one sampling period until the next sample is received, which can be expressed as

$$y_H(t) = y(k-1) \quad \text{for } t_{k-1} \le t < t_k \tag{13-18}$$

Other types of hold devices can be employed for *signal reconstruction*; for example, a *first-order-hold* extrapolates the digital signal linearly during the time interval from t_{k-1} to t_k based on the change during the previous interval:

$$y_H(t) = y(k-1) + \left(\frac{t-t_{k-1}}{\Delta t}\right)[y(k-1) - y(k-2)] \quad \text{for } t_{k-1} \le t < t_k \tag{13-19}$$

Although second-order and other higher-order holds can be designed and implemented as special-purpose DACs

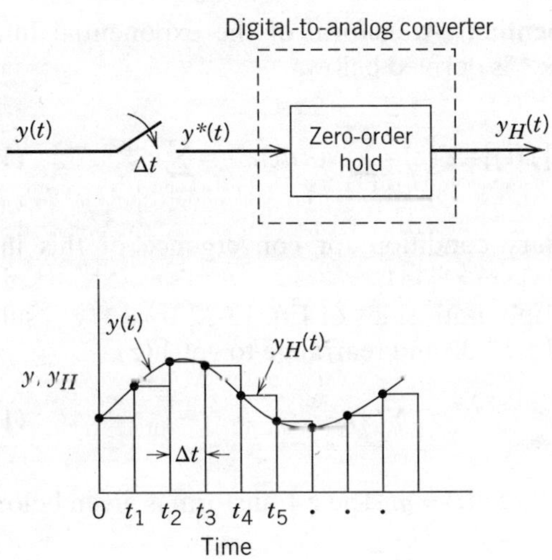

Figure 13.8 Digital-to-analog conversion using a zero-order hold.

(Ogata, 1995; Åström and Wittenmark, 2011; Franklin et al., 2005), these more complicated approaches do not offer significant advantages for most process control problems. Consequently, we will emphasize the ZOH, because it is the most widely used hold device for process control.

13.5 z-TRANSFORM ANALYSIS FOR DIGITAL CONTROL

In this section, we introduce the *z-transform* in order to analyze discrete-time systems. Once a continuous system is interfaced with a discrete system, such as shown in Fig. 13.3, it is necessary to analyze the behavior of the closed-loop system in discrete time. It is possible to simulate the discrete and continuous elements of the closed-loop control system using software such as Simulink; however, a simulation-based approach does not provide a rigorous basis to interpret or analyze discrete-time behavior. This analysis includes such items as process and controller discrete models; effect of poles, zeros, and system order on dynamic behavior; physical realizability; and stability of closed-loop systems. Key concepts for these topics are discussed below. More detailed analysis is available in Franklin et al. (2005), Phillips et al. (2014), and the first edition of this book (Seborg et al., 1989).

13.5.1 Definition

In the previous chapters continuous linear systems are discussed. These systems are represented by linear differential equations. The technique of Laplace transform offers a simplified method for solving linear differential equations. Therefore, the design and analysis of continuous control systems are based on transfer functions (*s*-domain). Similarly, the design and analysis of digital control systems is facilitated by the introduction of a discrete-time transform, namely, the *z*-transform. Consider the operation of the ideal, periodic sampler shown in Fig. 13.1. The sampler converts a continuous signal $y(t)$ into a discrete signal $y^*(t)$ at equally spaced intervals of time. Mathematically, it is convenient to consider impulse sampling, where $y^*(t)$ is the sampled signal formed by a sequence of impulses or Dirac delta functions based on the value of $y(t)$ at each sampling instant:

$$y^*(t) = \sum_{k=0}^{\infty} y(k\Delta t)\delta(t - k\Delta t) \tag{13-20}$$

Recall from Chapter 3 that $\delta(t - k\Delta t) = 1$ when $t = k\Delta t$, so an impulse is formed at each sampling instant with magnitude $y(k\Delta t)$.

Next, we derive the Laplace transform of Eq. 13-20, $Y^*(s)$. The value of $y(k\Delta t)$ is considered to be a constant in each term of the summation and thus is invariant when transformed. Since $\mathscr{L}[\delta(t)] = 1$, it follows that the Laplace transform of a delayed unit impulse is $\mathscr{L}[\delta(t - k\Delta t)] = e^{-k\Delta ts}$. Thus, the Laplace transform of (13-20) is given by

$$Y^*(s) = \sum_{k=0}^{\infty} y(k\Delta t)e^{-k\Delta ts} \quad (13\text{-}21)$$

Define the *z-transform variable* as

$$z \triangleq e^{s\Delta t} \quad (13\text{-}22)$$

Let $Y(z)$ denote the z-transform of $y^*(t)$,

$$Y(z) = Z[y^*(t)] = \sum_{k=0}^{\infty} y(k)z^{-k} \quad (13\text{-}23)$$

where the notation is simplified by using $y(k)$ to denote $y(k\Delta t)$.

13.5.2 Properties

The properties of Laplace transform are used to solve problems in the continuous time domain. Similarly, z-transforms possess certain useful properties which are applied to solve problems related to discrete time control system. Some useful properties are discussed below:

Linearity: Like Laplace transform, z-transform is a linear operation.

Let $y_1(t)$ and $y_2(t)$, are two functions in time domain and $Y_1(z)$ and $Y_2(z)$ are their respective z transforms. Then

$$Z[y_1(t) + y_2(t)] = Y_1(z) + Y_2(z) \quad (13\text{-}24)$$

$$Z[cy_1(t)] = cY_1(z) \quad (13\text{-}25)$$

where c is a constant.

Real translation theorem: Similar to Laplace transforms, z-transforms can express a transfer function for discrete time that corresponds to a difference equation. In order to derive some properties of z-transforms, Eq. 13-23 can be used to develop the *real translation theorem* as follows:

$$Z(y(t - i\Delta t)) = \sum_{k=0}^{\infty} y(k\Delta t - i\Delta t)z^{-k} \quad (13\text{-}26)$$

Substituting $j = k - i$ and because $y(j\Delta t) = 0$ for $j < 0$, then

$$Z(y(t - i\Delta t)) = z^{-i}\sum_{j=0}^{\infty} y(j\Delta t)z^{-j} = z^{-i}Y(z) \quad (13\text{-}27)$$

Complex translation theorem: If $Y(z)$is the z-transform of $y(t)$ then

$$Z = [e^{-at}y(t)] = Y(ze^{a\Delta t})$$

Initial value theorem and final value theorem: These theorems facilitate easy computation of initial and final values of a function from its z-transform.

The initial value theorem states that

$$\lim_{t\to 0} Y(t) = \lim_{z\to\infty} Y(z) \quad (13\text{-}28)$$

The final value theorem is

$$\lim_{t\to\infty} Y(t) = \lim_{z\to 1}(1 - z^{-1})Y(z) \quad (13\text{-}29)$$

It may be noted that as $z \to \infty$, all z^{-k} terms tend to zero for $k = 1, 2 \ldots$ in Eq. 13-28 and only the initial value corresponding to $k = 0$, remains. Similarly, in Eq. 13-29, as, $z \to 1$, all differences tend to zero and the only non-zero value left is $y(k)$ as $k \to \infty$

13.5.3 z-Transform of Some Basic Functions

Step: A unit step input has a value of 1 for all time; hence, at each sampling instant, $u(k\Delta t) = u(k) = 1$. Using Eq. 13-23, we find that the z-transform of a series of pulses of unit height is

$$U(z) = 1 + z^{-1} + z^{-2} + z^{-3} + \cdots \quad (13\text{-}30)$$

For $|z^{-1}| < 1$, $U(z)$ can be expressed in closed form as (Ogata, 1995)

$$U(z) = \frac{1}{1 - z^{-1}} \quad (13\text{-}31)$$

Exponential: z-transform of the exponential function $y(t) = e^{-at}$ is derived below:

$$Z[y(t)] = Y(z) = \sum_{k=0}^{\infty} y(k\Delta t)z^{-k} = \sum_{k=0}^{\infty} e^{-ak\Delta t}z^{-k} \quad (13\text{-}32)$$

Necessary condition for convergence of this infinite series is $|e^{-a\Delta t}z^{-1}| < 1$.

Multiply both sides of Eq. 13-32 by $e^{-a\Delta t}z^{-1}$, subtract from Eq. 13-32 and rearrange to get $Y(z)$:

$$Y(z) = \frac{z}{z - e^{-a\Delta t}} \quad (13\text{-}33)$$

Ramp: Let $y(t) = at$. The z-transform is given below:

$$\begin{aligned} Z[y(t)] = Y(z) &= \sum_{k=0}^{\infty} (ak\Delta t)z^{-k} \\ &= a[\Delta t z^{-1} + 2\Delta t z^{-2} + \ldots..] \end{aligned} \quad (13\text{-}34)$$

Multiply both sides of Eq. 13-34 by z^{-1} and subtract from Eq. 13-34 to get:

$$(1 - z^{-1})\, Y(z) = a[Tz^{-1} + 2\Delta t\, z^{-2} + \ldots]$$

Multiply again by z^{-1}:

$$(1 - z^{-1})^2\, y(z) = aTz^{-1}$$

Thus, $$Y(z) = \frac{aTz}{(z-1)^2} \tag{13-35}$$

z-transform values of other commonly used time domain functions are provided in Table 13.1.

Table 13.1 z-Transform of Commonly Used Functions

Function	z-Transform
$f(t) = \delta(t)$	1
$f(t) = S(t)$	$\dfrac{1}{1-z^{-1}}$
$f(t) = at$	$\dfrac{aTz^{-1}}{(1-z^{-1})^2}$
$f(t) = t^n$	$\lim\limits_{a\to 0}(-1)^n \dfrac{\partial^n}{\partial a^n}\dfrac{1}{1-e^{-aT}z^{-1}}$
$f(t) = e^{-at}$	$\dfrac{1}{1-e^{-aT}z^{-1}}$
$f(t) = te^{-at}$	$\dfrac{Te^{-aT}z^{-1}}{(1-e^{-aT}z^{-1})^2}$
$f(t) = \sin \omega t$	$\dfrac{z^{-1}\sin\omega T}{1-2z^{-1}\cos\omega T + z^{-2}}$
$f(t) = \cos \omega t$	$\dfrac{1-z^{-1}\cos\omega T}{1-2z^{-1}\cos\omega T + z^{-2}}$
$f(t) = 1 - e^{-at}$	$\dfrac{(1-e^{-aT})z^{-1}}{(1-z^{-1})(1-e^{-aT}z^{-1})}$
$f(t) = e^{-at}\sin \omega t$	$\dfrac{z^{-1}e^{-aT}\sin\omega T}{1-2z^{-1}e^{-aT}\cos\omega T + e^{-2aT}z^{-2}}$
$f(t) = e^{-at}\cos \omega t$	$\dfrac{1-z^{-1}e^{-aT}\cos\omega T}{1-2z^{-1}e^{-aT}\cos\omega T + e^{-2aT}z^{-2}}$

(In the table T is same as Δt, i.e., sampling period)

13.5.4 Inverse z-Transform

In Laplace transform analysis, inverse Laplace transform is conducted to get the response back in time domain. Similarly, inverse z-transform is performed on the z-domain function to get the values at sampling instants.

$$y(k\Delta t) = Z^{-1}[Y(z)] \tag{13-36}$$

Two popular methods for performing inverse-z transform are presented below:

Method of long division (The power series method)

The output z-transform can be calculated using long division, which is a power series expansion in terms of z^{-k}. The basic definition of z-transform as presented in Eq. 13-23 can also be written as given below:

$$Y(z) = y(0) + y(\Delta t)z^{-1} + y(2\Delta t)z^{-2} + \cdots \tag{13-37}$$

In general, the given $Y(z)$ can be expressed as a ratio of two polynomials of z^{-1}.

$$Y(z) = \frac{C(z^{-1})}{D(z^{-1})} = a_0 + a_1 z^{-1} + a_2 z^{-2} + \cdots \tag{13-38}$$

Comparison of Eqs. 13-37 and 13-38 gives us information about y at discrete time instants

$$y(0) = a_0;\ y(\Delta t) = a_1;\ y(2\Delta t) = a_2; \ldots \tag{13-39}$$

Partial fraction technique

Given $Y(z) = \dfrac{C(z^{-1})}{D(z^{-1})}$, expand $Y(z)$ as a series of fractions whose denominators contain low order polynomials in z^{-1}.

Let

$$Y(z) = \frac{C(z^{-1})}{D(z^{-1})} = \frac{C_1}{(1-p_1 z^{-1})} + \frac{C_2}{(1-p_2 z^{-1})} + \cdots + \frac{C_n}{(1-p_n z^{-1})} \tag{13-40}$$

$$Y_i(z) = \frac{c_i}{(1-p_i z^{-1})} \quad i = 1, 2, \ldots n \tag{13-41}$$

$$Z^{-1}[Y_i(z)] = Z^{-1}\left[\frac{c_i}{(1-p_i z^{-1})}\right] = C_i(p_i)^k \tag{13-42}$$

where k is the sampling instance

$$Z^{-1}[Y(z)] = \sum C_i(p_i)^k \tag{13-43}$$

Long division method is simpler than partial fraction technique. However, if k is very high, the long division activity to get same number of values may be cumbersome. On the other hand, the method of partial fraction gives general expression for $y(k\Delta t)$ which can be used for any value of k. Therefore, long division method is suitable when k is low and partial fraction method is the preferred technique when k is high.

13.5.5 Discrete Transfer Function

The response of a continuous process at discrete intervals of time $(y(k), k = 0, 1, 2, \ldots)$ to changes in the input at past intervals $(u(k), k = 0, 1, 2, \ldots)$ can be expressed using a difference equation. For the first-order difference equation,

$$y(k) + a_1 y(k-1) = b_1 u(k-1) \qquad (13\text{-}44)$$

the z-transform can be obtained using Eq. 13-27 for a general input $u(k)$:

$$Y(z) + a_1 z^{-1} Y(z) = b_1 z^{-1} U(z) \qquad (13\text{-}45)$$

Solving for $Y(z)$ in terms of $U(z)$,

$$Y(z) = \frac{b_1 z^{-1}}{1 + a_1 z^{-1}} U(z) = G(z)U(z) \qquad (13\text{-}46)$$

Eq. 13-46 defines the discrete transfer function $G(z)$ of the first-order difference equation, which is analogous to the transfer function obtained by applying Laplace transforms to a first-order linear differential equation. If the input $U(z)$ is known, then an expression for the output $Y(z)$ can be found by multiplying $G(z)$ times $U(z)$.

To calculate the response of a discrete transfer function, which corresponds to the response of the equivalent difference equation, we can use direct simulation of the difference equation based on the specified input. Alternatively, the output z-transform can be calculated using long division, which is a power series expansion in terms of z–k. We will illustrate this calculation in Examples 13.2 and 13.3.

EXAMPLE 13.2

Calculate the response of the first-order difference equation 13-44 for $a_1 = -0.368, b_1 = 1.264$, and $y(0) = 0$ using z-transforms and long division for $k = 0, 1, \ldots 5$. Compare the result with the unit step response for a first-order continuous-time system ($K = 20$, $\tau = 1$), where $a_1 = -e^{-\Delta t/\tau}$, $b_1 = K(1 - e^{-\Delta t/\tau})$, and $\Delta t = 1$.

Solution

Using Eq. 13-46, we find that the response for a unit step input $(U(z) = 1/(1 - z^{-1}))$ is

$$Y(z) = \frac{1.264z^{-1}}{1 - 0.368z^{-1}} \cdot \frac{1}{1 - z^{-1}} = \frac{1.264z^{-1}}{1 - 1.368z^{-1} + 0.368z^{-2}} \qquad (13\text{-}47)$$

Next long division is used to divide the denominator into the numerator. The order of the numerator and denominator polynomials starts with the lowest powers of z^{-k} for the division operation.

$$\begin{array}{r|l}
 & 1.264z^{-1} + 1.729z^{-2} + 1.900z^{-3} + \cdots \\
\hline
1 - 1.368z^{-1} + 0.368z^{-2} & 1.264z^{-1} \\
 & 1.264z^{-1} - 1.729z^{-2} + 0.465z^{-3} \\
 & \quad 1.729z^{-2} - 0.465z^{-3} \\
 & \quad 1.729z^{-2} - 2.365z^{-3} + 0.636z^{-4} \\
 & \qquad 1.900z^{-3} - 0.636z^{-4} \\
 & \qquad \text{(etc.)}
\end{array}$$

Because of space limitations, only the first three terms are shown above: $y(1) = 1.264$, $y(2) = 1.729$, and $y(3) = 1.900$. Continuing on, we calculate $y(4) = 1.963$ and $y(5) = 1.986$. Ultimately, $y(k)$ reaches its steady-state value of 2.0 (k large), which agrees with the fact that the process gain K is 2 and the input is a unit step change. The step response in continuous time is $y(t) = 2(1 - e^{-t})$, and the sampled values of the discrete-time response for $\Delta t = 1$ are the same ($k = 0, 1, 2, 3 \ldots$). Thus, the discretization is exact; that is, it is based on the analytical solution for a piecewise constant input.

The same answer could be obtained from simulating the first-order difference equation (13-44), with $u(k) = 1$ for $k \geq 0$; that is,

$$y(k) = 0.368\, y(k-1) + 1.264(1)$$

Starting with $y(0) = 0$, it is easy to generate recursively the values of $y(1) = 1.264$, $y(2) = 1.729$, and so on. Note that the steady-state value can be obtained in the above equation by setting $y(k) = y(k - 1) = y_{ss}$ and solving for y_{ss}. In this case $y_{ss} = 2.0$, as expected.

Solving using MATLAB

```
>> syms z;                        % creation of sym-
                                    bol z
>> Y(z) = (1.264*z^-              % creation       of
1)/(1-(1.368*z^-                    function Y(z)
1+ 0.368*z^-2);
>> iztrans(Y(z))                  % P e r f o r m i n g
                                    inverse z-trans-
                                    form
ans = 2 - 2*(46/125)^n
```

Note: The above is the general expression computed by MATLAB at any sampling instance n. Put different values of n to get the output.

For example, for $n = 2$:

```
>> 2 - 2*(46/125)^2
```

ans = 1.7292 (Check with the answer obtained above using long division method)

EXAMPLE 13.3

For the difference equation,

$$y(k) = 0.9744\,y(k-1) - 0.2231\,y(k-2) - 0.3225\,u(k-2) + 0.5712\,u(k-3) \quad (13\text{-}48)$$

derive its discrete transfer function and step response for $U(z) = \frac{1}{1-z^{-1}}$. Use long division to obtain $Y(z)$ for $k = 0$ to $k = 9$. Compare this result with simulating the original difference equation, Eq. 13-48, and a unit step change in $u(k)$ at $k = 0$. Assume $y(0) = 0$.

Solution

Taking the z-transform of Eq. 13-48,

$$Y(z) = 0.9744\,z^{-1}\,Y(z) - 0.2231\,z^{-2}\,Y(z) - 0.3225\,z^{-2}\,U(z) + 0.5712\,z^{-3}\,U(z) \quad (13\text{-}49)$$

Rearranging gives the discrete transfer function

$$G(z) = \frac{Y(z)}{U(z)} = \frac{-0.3225z^{-2} + 0.5712z^{-3}}{1 - 0.9744z^{-1} + 0.2231z^{-2}} \quad (13\text{-}50)$$

Note that the numerator of $G(z)$ has a common factor of z^{-2}, which indicates the presence of an apparent time delay of two sampling periods.

To determine the step response, set $U(z) = \frac{1}{1-z^{-1}}$ and multiply it by the transfer function to find the power series for $Y(z)$. Long division as done in Example 17.2 yields

$$Y(z) = -0.3225z^{-2} - 0.0655z^{-3} + 0.2568z^{-4} + 0.5136z^{-5} + 0.6918z^{-6} + 0.8082z^{-7} + 0.8820z^{-8} + 0.9277z^{-9} + \cdots \quad (13\text{-}51)$$

Note that $y(k) = 0$ for $k = 0$ and $k = 1$, and $y(2) = -0.3225$, which indicates a two-unit time delay in $G(z)$. After an initial transient period, it appears that $y(k)$ is steadily increasing and may approach a steady-state value. For a unit step change in $U(z)$, the steady-state value of the response $Y(z)$ can be found by determining the steady-state gain of $G(z)$. In analogy to continuous-time transforms, the steady-state gain can be found by setting $s = 0$ in $z = e^{s\Delta t}$, or $z = 1$. In this case, $G(z = 1) = 1$, so $y(k) = 1$ at steady state for a unit step change, which is apparent in Eq. 13-51.

The same result can be obtained using the original difference equation. A table (or spreadsheet) could be constructed to track the various terms of the difference equation for $k = 0, 1, 2, \ldots, 9$. The top row of the table is structured using the same terms as in the difference equation, and the step response is generated using spreadsheet software. We assume that the original system is at steady state at $k = 0$, so y and u terms corresponding to $k < 0$ are equal to 0. Using the difference equation, Eq. 13-48, the value of $y(k)$ can be obtained from the entries (cells) in the same row by performing the appropriate multiplications.

Next, consider a general higher-order difference equation given by

$$a_0 y(k) + a_1 y(k-1) + \cdots + a_m y(k-m) = b_0 u(k) + b_1 u(k-1) + \cdots + b_n u(k-n) \quad (13\text{-}52)$$

where $\{a_i\}$ and $\{b_i\}$ are sets of constant coefficients, n and m are positive integers, $u(k)$ is the input, and $y(k)$ is the output. Taking the z-transform of both sides of Eq. 13-52 gives

$$a_0 Y(z) + a_1 z^{-1} Y(z) + \cdots + a_m z^{-m} Y(z) = b_0 U(z) + b_1 z^{-1} U(z) + \cdots + b_n z^{-n} U(z) \quad (13\text{-}53)$$

Rearranging Eq. 13-53 gives the *discrete transfer function* form,

$$Y(z) = \frac{b_0 + b_1 z^{-1} + \cdots + b_n z^{-n}}{a_0 + a_1 z^{-1} + \cdots + a_m z^{-m}} U(z) \quad (13\text{-}54)$$

The ratio of polynomials in $G(z)$ can be derived by algebraic manipulations for any difference equation (Åström and Wittenmark, 2011). In this case,

$$G(z) = \frac{B(z)}{A(z)} = \frac{b_0 + b_1 z^{-1} + \cdots + b_n z^{-n}}{a_0 + a_1 z^{-1} + \cdots + a_m z^{-m}} \quad (13\text{-}55)$$

where $B(z)$ and $A(z)$ are polynomials in z^{-1}. For most processes, b_0 is zero, indicating that the input does not instantaneously affect the output; in Eq. 13-53, if $b_0 \neq 0$, then the input term would include $u(k)$ in the difference equation. In addition, the leading coefficient in the denominator can be set equal to unity by dividing both numerator and denominator by a_0. The steady-state gain of G in Eq. 13-55 can be found by setting $z = 1$ (Ogata, 1995).

The dynamic behavior of Eq. 13-55 can be characterized by its poles and zeros in analogy to continuous-time systems. To do this, we must first convert $G(z)$ to positive powers of z by multiplying Eq. 13-55 by z^n/z^m, leading to modified polynomials, $B'(z)/A'(z)$. The stability of $G(z)$ is determined by its poles, the roots of the characteristic equation given by $A'(z) = 0$; $A'(z)$ is called the characteristic polynomial. Note that z is a complex variable, because it is related to complex variable s by the definition in Eq. 13-22.

The *unit circle* in the complex z-plane is defined as a circle with unit radius, $|z| = 1$. The unit circle is the dividing line between the stable and unstable regions. Any pole that lies inside the unit circle is stable and thus provides a stable response to a bounded input. In contrast, a pole lying outside the unit circle is unstable (Åström and Wittenmark, 2011). The zeros of $G(z)$ are the roots of $B'(z) = 0$. In discrete time analysis, time delays are usually assumed for convenience to be an integer multiple N of the sampling period Δt. This time delay produces N roots of $A'(z)$ that are located at the origin and thus can be expressed as a factor, z^N. By definition, the order of $G(z)$ is $P + N$, where P is the number of poles.

Figure 13.9 shows the effect of pole location on the possible responses for a simple first-order transfer function, $G(z) = b_0/(1 - az^{-1})$, forced by an impulse input at $k = 0$. The corresponding continuous-time model responses are also shown. Poles 3 and 4 are inside the unit circle and thus are stable, while poles 1 and 6 are outside the unit circle and cause an unstable response. Poles 2 and 5 lie on the unit circle and are marginally stable. Negative poles such as 4–6 produce oscillatory

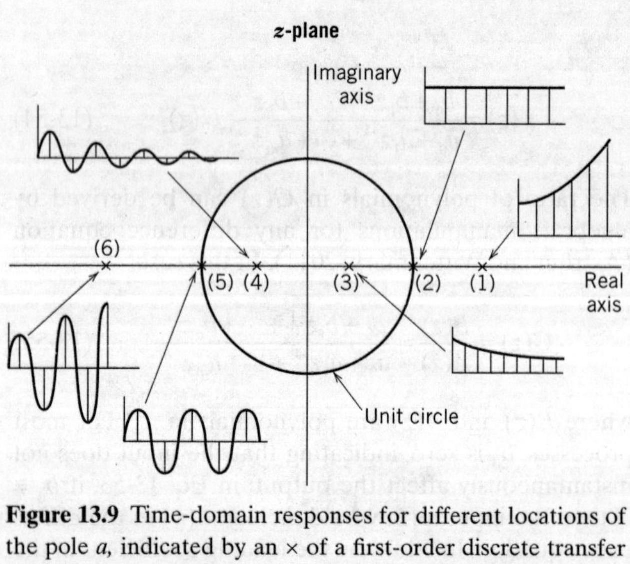

Figure 13.9 Time-domain responses for different locations of the pole a, indicated by an × of a first-order discrete transfer function and a pulse input at $k = 0$.

responses, even for a first-order discrete-time system, in contrast to continuous-time first-order systems.

Some important properties of sampled-data systems can be obtained from long division of their z-transforms. For example, for the first-order z-transform,

$$Y(z) = \frac{b_0}{1 - a_1 z^{-1}} \tag{13-56}$$

an equivalent sampled signal can be found by long division, resulting in the infinite series

$$\overline{Y}(z) = b_0(1 + a_1 z^{-1} + a_1^2 z^{-2} + \cdots + a_1^n z^{-n} + \cdots) \tag{13-57}$$

If $a_1 = b_0 = 1$, then $Y(z) = \frac{1}{1-z^{-1}}$ and $\overline{Y}(z) = \sum_{i=0}^{\infty} z^{-i}$.

Hence, the equivalent difference equation would be a summation of all previous values of $u(k)$.

13.5.6 Discrete Convolution Model Form

For the transfer function in Eq. 13-54, dividing the numerator by the denominator (starting with a_0 and determining the remainders) leads to a model equivalent to a discrete convolution model (see Section 16.4.1). Long division of Eq. 13-54, yields for the first three terms of the series $\sum_{k=0}^{\infty} c_k z^{-k}$

$$c_0 = \frac{b_0}{a_0} \tag{13-58}$$

$$c_1 = \frac{b_1}{a_0} - \frac{b_0 a_1}{a_0^2} \tag{13-59}$$

$$c_2 = b_2 - \frac{b_0 a_2}{a_0} - \frac{b_1 a_1}{a_0} + \frac{b_0 a_1^2}{a_0^2} \tag{13-60}$$

The transfer function is therefore

$$G(z) = c_0 + c_1 z^{-1} + c_2 z^{-2} + \cdots \tag{13-61}$$

which corresponds to the convolution model form,

$$y(k) = c_0 + c_1 u(k-1) + c_2 u(k-2) + \cdots \tag{13-62}$$

13.5.7 Physical Realizability

Chapter 4 addressed the notion of physical realizability for continuous-time transfer functions. An analogous condition can be stated for a difference equation or its transfer function—namely, that a discrete-time model cannot have an output signal that depends on future inputs. Otherwise, the model is not physically realizable.

Consider the ratio of polynomials given in Eq. 13-55. The discrete transfer function will be physically realizable as long as $a_0 \neq 0$, assuming that $G(z)$ has been reduced so that common factors in the numerator and denominator have been canceled. To show this property, examine Eq. 13-52. If $a_0 = 0$, the difference equation is

$$\begin{aligned} a_1 y(k-1) &= b_0 u(k) + b_1 u(k-1) + \cdots + b_k u(k-n) \\ &\quad - a_2 y(k-2) - \cdots - a_m y(k-m) \end{aligned} \tag{13-63}$$

or, by shifting the index from k to $k + 1$,

$$a_1 y(k) = b_0 u(k+1) + b_1 u(k) + \cdots \tag{13-64}$$

This equation requires a future input $u(k + 1)$ to influence the present value of the output $y(k)$, which is physically impossible (unrealizable). Physical realizability of models (process or controller) should be checked prior to their use for simulation or control.

Discrete Transfer Function of an Integral Controller

Now we derive a transfer function for a digital integral controller, where the output is $p(k)$ and the input is the error signal $e(k)$ (cf. Eq. 7-7). The integral of $e(t)$ in continuous time can be approximated by a summation in discrete time. By using a finite difference approximation to the integral

$$p(t) = \bar{p} + \frac{1}{\tau_I}\int_0^t e(t')dt' \approx \bar{p} + \frac{\Delta t}{\tau_I}\sum_{k=0}^{n} e(k) \tag{13-65}$$

Then

$$p(k) - \bar{p} = \frac{\Delta t}{\tau_I}\sum_{k=0}^{n} e(k) \tag{13-66}$$

Taking the z-transform

$$p(z) = \frac{\Delta t}{\tau_I}\left(\sum_{k=0}^{n} z^{-k}\right)E(z) \tag{13-67}$$

When n is large, the summation can be expressed in closed form as

$$p(z) = \frac{\Delta t}{(1-z^{-1})\tau_I}E(z) \tag{13-68a}$$

The continuous-time analog to Eq. 13-59 for an integral controller is

$$P(s) = \frac{E(s)}{\tau_I s} \tag{13-68b}$$

Comparing the above two expressions, we can observe a relationship between z^{-1} and s, which is

$$s \cong \frac{1-z^{-1}}{\Delta t} \tag{13-69}$$

This expression is known as the *backward-difference* (BD) approximation of s (equivalent to a first-order Taylor series), and it can be used to convert a continuous time expression (in s) into an approximate discrete-time expression in z^{-1} simply by direct substitution. The same relationship can be obtained by recognizing that $\mathcal{L}\left(\frac{de}{dt}\right) = sE(s)$. This expression can be compared with $\frac{de}{dt} \approx \frac{e(k) - e(k-1)}{\Delta t}$ for which the z-transform is $\frac{1-z^{-1}}{\Delta t}E(z)$. Therefore, we can approximate the Laplace transform of a continuous-time function such as a PID controller in discrete time by substituting the BD approximation for s given in Eq. 13-69, as shown in the next example.

EXAMPLE 13.4

Derive the discrete transfer function for the parallel form of a PID controller.

$$G_c(s) = K_c\left(1 + \frac{1}{\tau_I s} + \tau_D s\right) \tag{13-70}$$

using the BD substitution for S. Compare the result with the velocity form of the PID algorithm given in Eq. 7-27.

Solution

Substituting $s \cong (1 - z^{-1})/\Delta t$ into Eq. 13-70 gives

$$G_c(z) = \frac{K_c(a_0 + a_1 z^{-1} + a_2 z^{-2})}{1 - z^{-1}} \tag{13-71}$$

where $a_0 = 1 + \frac{\Delta t}{\tau_I} + \frac{\tau_D}{\Delta t}$, $a_1 = -\left(1 + \frac{2\tau_D}{\Delta t}\right)$, $a_2 = \frac{\tau_D}{\Delta t}$

Because $E(z)$ is the error signal (input) and $P(z)$ is the controller output, $P(z) = G_c(z)E(z)$. Multiplying both sides of Eq. 13-71 by $(1 - z^{-1})$ yields

$$(1-z^{-1})p(z) = K_c(a_0 + a_1 z^{-1} + a_2 z^{-2})E(z) \quad (13\text{-}72)$$

Converting the controller transfer function into difference equation form gives

$$\begin{aligned} p(k) - p(k-1) = K_c a_0 e(k) + K_c a_1 e(k-1) \\ + K_c a_2 e(k-2) \end{aligned} \quad (13\text{-}73)$$

Substituting for a_0, a_1, and a_2 and collecting terms with respect to the controller settings K_c, τ_I, and τ_D gives

$$\begin{aligned} p(k) - p(k-1) = K_c \Bigg[\Bigg(& e(k) - e(k-1)) + \frac{\Delta t}{\tau_I} e(k) \\ & + \frac{\tau_D}{\Delta t}(e(k) - 2e(k-1) + e(k-2) \Bigg) \Bigg] \end{aligned} \quad (13\text{-}74)$$

Note that this equation is identical to Eq. 7-27, which was derived using a finite-difference approximation in the time domain.

Ogata (1995) has listed more accurate formulas for algebraic substitution into a transfer function $G(s)$. Approximate substitution is a procedure that should always be used with care. When feasible, it is preferable to use exact conversion of the continuous process model into discrete time rather than finite difference approximations such as Eq. 13-69. Table 13.2 presents a conversion table of commonly used transfer functions $G(s)$ based on a ZOH (which yields a piecewise constant input u). This table is also consistent with the exact (analytical) conversion formulas for first- and second-order process models in Eqs. 16-23 through 16-30.

Table 13.2 Discrete Transfer Functions Obtained Using a Zero-Order Hold

Transfer Function $G(s)$	$G(z)$	
$\frac{K}{s}$	$\frac{b_1 z^{-1}}{1 + a_1 z^{-1}}$	$a_1 = -1$ $b_1 = K\Delta t$
$\frac{K}{s+r}$	$\frac{b_1 z^{-1}}{1 + a_1 z^{-1}}$	$a_1 = \exp(-r\Delta t)$ $b_1 = \frac{K}{r}[1 - \exp(-r\Delta t)]$
$\frac{K}{(s+r)+(s+p)}$	$\frac{b_1 z^{-1} + b_2 z^{-2}}{1 + a_1 z^{-1} + a_2 z^{-2}}$	$a_1 = -\exp(-r\Delta t) - \exp(-p\Delta t)$ $a_2 = \exp[-(r+p)\Delta t]$ $b_1 = [K/rp(r-p)][(r-p) - r\exp(-p\Delta t) + p\exp(-r\Delta t)]$ $b_2 = [K/rp(r-p)]\{(r-p) - \exp[-(r+p)\Delta t] + p\exp(-p\Delta t) - r\exp(-r\Delta t)\}$
$\frac{K}{s(s+r)}$	$\frac{b_1 z^{-1} + b_2 z^{-2}}{1 + a_1 z^{-1} + a_2 z^{-2}}$	$a_1 = -\{1 + \exp(-r\Delta t)\}$ $a_2 = \exp(-r\Delta t)$ $b_1 = -(K/r^2)[1 - r\Delta t - \exp(-r\Delta t)]$ $b_2 = (K/r^2)[1 - \exp(-r\Delta t) - r\Delta t\exp(-r\Delta t)]$
$\frac{K(s+v)}{(s+r)(s+p)}$	$\frac{b_1 z^{-1} + b_2 z^{-2}}{1 + a_1 z^{-1} + a_2 z^{-2}}$	$a_1 = -\{\exp(-p\Delta t + \exp(-r\Delta t)$ $a_2 = \exp[-(r+p)\Delta t]$ $b_1 = \frac{K}{p-r}\{\exp(-p\Delta t) - \exp(-r\Delta t) + (v/p)[1 - \exp(-p\Delta t)] - (v/r)[1 - \exp(-r\Delta t)]\}$ $b_2 = K\{(v/rp)\exp[-(r+p)\Delta t] + [(p-v)/p(r-p)]\exp(-r\Delta t) + [(v-r)/r(r-p)]\exp(-p\Delta t)\}$

Block Diagram Algebra

It is important to realize that in the block diagram of the closed-loop system, Fig. 13.3, the open-loop transfer function of the continuous components includes the product of the final control element (or valve), the process, and the measurement, or $G(s) = G_v(s) G_p(s)G_m(s)$. It is mathematically incorrect to find separate discrete-time versions $G_v(z)$, $G_p(z)$, and $G_m(z)$ and then multiply them together (Ogata, 1995; Seborg et al., 1989). Instead, the z-transform of $G(s)$ should be based on exact discretization of the Laplace transform (Åström and Wittenmark, 2011). In the case where $G_m = K_m$, this leads to the closed-loop expression (analogous to Chapter 11),

$$\frac{Y(z)}{Y_{sp}(z)} = \frac{G_c(z)G(z)}{1+G_c(z)G(z)} \qquad (13\text{-}75)$$

In Eq. 13-75, $G_c(z)$ is the discrete transfer function for the digital controller. A digital controller is inherently a discrete-time device, but with the ZOH, the discrete-time controller output is converted to a continuous signal that is sent to the final control element. So the individual elements of G are inherently continuous, but by conversion to discrete-time we compute their values at each sampling instant. The discrete closed-loop transfer function in Eq. 13-75 provides a framework to perform closed-loop analysis and controller design, as discussed in the next section. Additional material on closed-loop analysis for discrete-time systems is available elsewhere (Ogata, 1995; Seborg et al., 1989).

13.5.8 Stability Analysis

For sampled-data systems the stability of a transfer function can be tested by determining whether any roots of its characteristic polynomial lie outside the unit circle (see Fig.13.9). To apply this stability test, write the denominator of the transfer function in Eq. 13-55 in terms of positive powers of z:

$$\begin{aligned} A'(z) &= z^m A(z) \\ &= a_m z^m + a_{m-1} z^{m-1} + \cdots + a_1 z + a_0 = 0 \end{aligned} \qquad (13\text{-}76)$$

Note that if the time delay is an integer multiple of Δt, a polynomial in z will always result. Roots of characteristic polynomial $A'(z)$ can be found using a root-finding method. If any root lies outside the unit circle, then the discrete transfer function is unstable (see Fig. 13.9). This conclusion can be verified by simulating the transfer function response for a pulse input to see if the output $y(k)$ grows with respect to k.

Because of space limitation, we have deliberately not included rigorous coverage of stability of z-transform models, such as the bilinear transform or Jury's test. See Franklin et al. (2005), Phillips et al. (2014), and Ogata (1995) for more details on these techniques.

13.6 TUNING OF DIGITAL PID CONTROLLERS

In this section, the main emphasis is on the digital PID controller and how it should be tuned. Digital versions of the PID controller in the form of difference equations were previously presented in Eqs. 7-25 and 7-27 (position and velocity forms) by using the backward difference approximation. For small values of Δt (relative to the process response time), the finite difference approximations for integral and derivative control action discussed in Section 7.4 are reasonably accurate. Hence, suitable controller settings obtained for a continuous controller can also be utilized in a digital PID controller. As noted by Isermann (1989), the continuous and the discrete PID controllers will have essentially the same behavior as long as $\Delta t/\tau_I \leq 0.1$. Åström and Wittenmark (2011) have discussed the effect of sampling period for designing a wide range of digital controllers.

If Δt is not small, use of the ZOH in digital control systems requires a modification in the controller design procedure, because the sampler plus a ZOH introduces an effective time delay in the signal to the final control element. In this case, the dynamic effect of the sampler plus ZOH should be approximated by a time delay equal to one-half the sampling period (Franklin et al., 2005). Thus, it is a common practice in tuning digital PID controllers to add $\Delta t/2$ to the process time delay θ before using the controller tuning relations in Chapter 11. The BD version of a PID controller (Eq. 13-71) will provide greater stability margins in discrete time (vs. other finite difference schemes); see Franklin et al. (2005).

In earlier chapters, Simulink was used to simulate linear continuous-time control systems described by transfer function models. For digital control systems, Simulink can also be used to simulate open- and closed-loop responses of discrete-time systems. As shown in Fig. 13.2, a computer control system includes both continuous and discrete components. In order to carry out detailed analysis of such a hybrid system, it is necessary to convert all transfer functions to discrete time and then carry out analysis using z-transforms (Åström and Wittenmark, 2011; Franklin et al., 2005). On the other hand, simulation can be carried out with Simulink using the control system components in their native forms, either discrete or continuous. This approach facilitates tuning of digital controllers.

In this section, we show how to perform closed-loop simulations for various digital controllers. Although the controller is represented by a discrete transfer function, all other components of the control loop (models for the final control element, process, sensor, and disturbance) will normally be available as continuous transfer functions, which can be directly entered into a Simulink block diagram as functions of s. To introduce a step change in set point requires only selecting the icon in Simulink for the step input (no s or z transform is needed). A step change in the disturbance can be entered in a similar fashion. A ZOH icon can be placed after the digital controller to convert the series of controller pulses into a continuous signal to drive a final control element such as a control valve. However, in Simulink the user does not have to include the ZOH, because the software performs the calculations as if the ZOH were there. The user should also stipulate the sampling period in each discrete block.

Figure 13.10 shows the Simulink block diagram for a digital control system with

1. Two controller blocks that are multiplied ($G_c = G_{c1}G_{c2}$)
2. Zero-order hold
3. Continuous-time process G (time constants of 5 and 3, gain of 1)
4. Unit step change in set point and graphical output ("scope") for the process output, which is a continuous-time response

Other blocks can be added from the Simulink menu to explicitly include transfer functions for the final control element, disturbance, and sensor (as many as desired). Simulink can also be employed for open-loop simulation.

Simulink will be used in subsequent examples to demonstrate the performance of various types of digital feedback controllers, as was done in Chapter 11. The stability of closed-loop systems can also be checked using trial-and-error simulations of the block diagram to determine the maximum controller gain for stability.

Effect of Filter Selection on PID Controller Performance

Digital and analog filters are valuable for smoothing data and eliminating high-frequency noise, but they also affect control system performance. In particular, a filter in the feedback loop introduces additional phase lag. Consequently, it reduces the stability margin for a feedback controller, compared to the situation where no filter is present. Therefore, the controller may have to be retuned if the filter constant is changed. When derivative action is used, it is important to filter noisy signals before the derivative control calculations are performed. Because derivative action tends to amplify noise in the process measurement, filtering helps prevent controller saturation and wear in the final control elements. Some PID controllers include a derivative filter in the controller equation; see Chapter 7 for more details.

Wireless Feedback Control

Wireless feedback control is a promising new technology that is attractive for reducing wiring costs. To achieve the best closed-loop response, digital feedback control in such an environment should be executed 4–10 times faster than the process response time in order to minimize delays in sensing and data acquisition. In contrast to a wired transmitter, it is not practical to provide frequent sampling for a wireless measurement transmitter, because it quickly depletes the transmitter battery. With a slower sampling period, the closed-loop response to disturbances can become oscillatory. As an alternative, PID control can be reformulated to provide satisfactory control when a wireless measurement is not updated as frequently as a wired measurement. A PID implementation known as PID Plus (Blevins et al., 2015) automatically compensates for slow measurement sampling rates and possible dropped measurements, which yields performance comparable to that achieved using a wired transmitter. PID Plus is a specific example of *event-based control* (Ruiz et al., 2014), where a *send-on-delta* sampling strategy is used (Beschi et al., 2012).

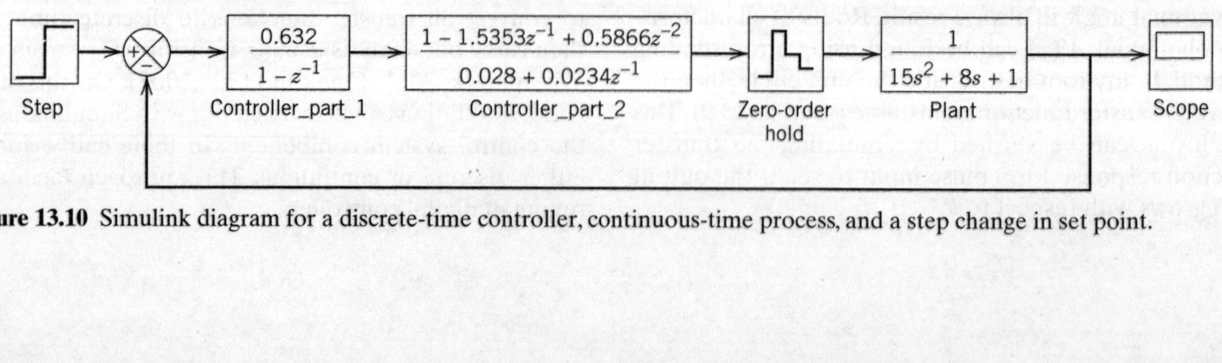

Figure 13.10 Simulink diagram for a discrete-time controller, continuous-time process, and a step change in set point.

13.7 DIRECT SYNTHESIS FOR DESIGN OF DIGITAL CONTROLLERS

In this section, the Direct Synthesis (DS) method presented in Section 11.2 is extended to the design of digital controllers. We begin with special cases that lead to a PID controller, and then show how other types of digital feedback controllers can be derived using the Direct Synthesis technique. Both G_c and G must be expressed as discrete-time in the closed-loop transfer function Eq. 13-75. In Direct Synthesis, the designer specifies the desired closed-loop transfer function $(Y/Y_{sp})_d$. The controller G_c that yields the desired performance is obtained from Eq. 13-75.

$$G_c = \frac{1}{G}\frac{(Y/Y_{sp})_d}{1-(Y/Y_{sp})_d} \tag{13-77}$$

Equation 13-77 is the model-based control law used for the Direct Synthesis design method in discrete time, analogous to the procedure discussed in Section 11.2 for continuous-time controllers.

Next, two Direct Synthesis algorithms for discrete-time application are considered: Dahlin's method and the Vogel-Edgar method. The discrete-time version of a related method considered in Chapter 11, Internal Model Control, is also presented.

13.7.1 Dahlin's Method (Lambda Tuning)

Dahlin's method (Dahlin, 1968) specifies that the desired closed-loop transfer function is an FOPTD model:

$$\left(\frac{Y}{Y_{sp}}\right)_d = \frac{e^{-hs}}{\tau_c s+1} \tag{13-78}$$

where τ_c is the desired closed-loop time constant (cf. Chapter 11) and h is the specified time delay. After selecting $h = \theta$ (the process time delay), Table 13.1 indicates that the discrete transfer function corresponding to Eq. 13-78 is

$$\left(\frac{Y}{Y_{sp}}\right)_d = \frac{(1-A)z^{-N-1}}{1-Az^{-1}} \tag{13-79}$$

where $A = e^{-\Delta t/\tau_c}$, $K = 1/\tau_c$, $r = 1/\tau_c$, $N = \theta/\Delta t$.

Substituting Eq. 13-79 into Eq. 13-77 yields the general form of G_c for Dahlin's method, which is denoted by G_{DC}:

$$G_{DC} = \frac{1}{G}\frac{(1-A)z^{-N-1}}{1-Az^{-1}-(1-A)z^{-N-1}} \tag{13-80}$$

As a special case, consider G to be the discrete version of a FOPTD transfer function with gain K and time constant τ.

$$G = \frac{K(1-a_1)}{1-a_1z^{-1}}z^{-N-1} \tag{13-81}$$

where $a_1 = e^{-\Delta t/\tau}$. Dahlin's controller is

$$G_{DC} = \frac{(1-A)}{1-Az^{-1}-(1-A)z^{-N-1}}\frac{1-a_1z^{-1}}{K(1-a_1)} \tag{13-82}$$

For all values of N, $(1 - z^{-1})$ is a factor of the denominator, indicating the presence of integral action. The result is consistent with steady-state gain calculations in Eq. 13-78 with $s = 0$ and in Eq. 13-79 with $z = 1$, which specify zero steady-state error for set-point changes $(Y/Y_{sp} = 1)$.

The desired time constant τ_c for the closed-loop system serves as a convenient tuning parameter, which is often referred to as *lambda tuning* (see Chapter 11). Small values of τ_c produce faster responses, while large values of τ_c give more sluggish control. This correlation is especially useful in situations where the model parameters, especially the time delay, are subject to error or are time-varying because of changes in the process. In an aggressively controlled process, an inaccurate time delay can cause poor control and an unstable response. By choosing a larger τ_c and having more conservative control action, the controller can better accommodate the inaccurate model. As $\tau_c \to 0$ (i.e., $A \to 0$), Dahlin's algorithm is equivalent to *minimal prototype control*, but such aggressive tuning is usually not desirable for process control applications, because it is quite sensitive to parameter changes.

One important feature of a Direct Synthesis method such as Dahlin's method is that the resulting controller contains the reciprocal of the process transfer function. This feature causes the poles of G to become zeros of G_c, while the zeros of G become poles of the controller, unless the poles and zeros of G are canceled by terms in $(Y/Y_{sp})_d$. The inversion of G in Eq. 13-77 can lead to operational difficulties, just as in the continuous-time case. If G contains a zero that lies outside the unit circle, then G_c will contain an unstable pole lying outside the unit circle. In this case, G_c is an *unstable* controller and produces an unbounded output sequence for a step change in set point. Although the product G_cG in Eq. 13-75 indicates that the unstable pole and zero will cancel, in practice there will always be some model error that prevents exact cancellation. Nevertheless, problems associated with unstable zeros can be successfully treated by judicious selection of $(Y/Y_{sp})_d$, as discussed below.

Digital controllers of the Direct Synthesis type share yet one more characteristic: namely, they contain time-delay compensation in the form of a Smith predictor (see Chapter 12). In Eq. 13-77, for G_c to be physically realizable, $(Y/Y_{sp})_d$ must also contain a term equivalent to $e^{-\theta s}$, which is z^{-N}, where $N = \theta/\Delta t$. In other words, if there is a term z^{-N} in the open-loop discrete transfer function, the closed-loop process cannot respond before $N\Delta t$ or θ units of time have passed. Using $(Y/Y_{sp})_d$ of this form in Eq. 13-79 yields a G_c containing the mathematical equivalence

EXAMPLE 13.5

A process is modeled in continuous time by a second-order plus-time-delay transfer function with $K = 1$, $\tau_1 = 5$, and $\tau_2 = 3$. For $\Delta t = 1$, the discrete-time equivalent (with ZOH) is

$$G = \frac{(b_1 + b_2 z^{-1}) z^{-N-1}}{1 + a_1 z^{-1} + a_2 z^{-2}} \tag{13-83}$$

where $a_1 = -1.5353$, $a_2 = 0.5866$, $b_1 = 0.0280$, $b_2 = 0.0234$, and $N = 0$ (cf. Eq. 16-26 to 16-30). For Dahlin's controller with $\tau_c = \Delta t = 1$, plot the response for a unit change in set point at $t = 5$ for $0 \le t \le 10$ using Simulink.

Solution

The Simulink diagram for the example was shown earlier in Fig. 13.10 Using Eq. 13-79, the desired closed-loop transfer function for $\theta = 0$ ($N = 0$) and $\tau_c = \Delta t$ is $(Y/Y_{sp})_d = 0.632z^{-1}/(1 - 0.368z^{-1})$. By applying Eq. 13-80 the formula for the controller is

$$G_{DC} = \frac{1 + a_1 z^{-1} + a_2 z^{-2}}{b_1 z^{-1} + b_2 z^{-2}} \frac{0.632z^{-1}}{1 - z^{-1}} \tag{13-84}$$

Substituting the numerical values for a_1, a_2, b_1, and b_2, the controller is

$$G_{DC} = \frac{1 - 1.5353z^{-1} + 0.5866z^{-2}}{0.0280 + 0.0234z^{-1}} \frac{0.632}{1 - z^{-1}} \tag{13-85}$$

When this controller is implemented, an undesirable characteristic appears, namely, *intersample ripple.* Figure 13.11*a* shows the response y and ZOH output p to a unit step change in set point at $t = 5$. Although the response does satisfy $y(k) = 1$ at each sampling instant ($\Delta t = 1$) for $t \ge 6$, the response is quite oscillatory; that is, inter-sample ripple occurs. This result is caused by the controller output cycling back and forth between positive and negative deviations from the steady-state value. The behavior, called *ringing*, of course is unacceptable for a control system.

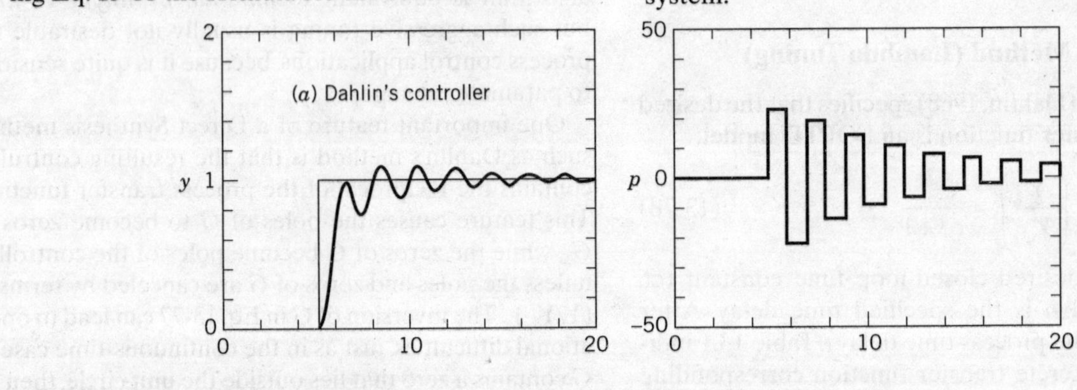

Figure 13.11 Comparison of (*a*) ringing and (*b*) nonringing Dahlin's controllers for a second-order process ($\tau_c = 1$), Example 13.6 (y = controlled variable, p = controller output after ZOH).

of time-delay compensation, because the time delay has been eliminated from the characteristic equation.

The controller ringing results from the presence of the term $(0.0280 + 0.0234z^{-1})$ in the denominator of Eq. 13-85 which corresponds to a controller pole at –0.836, quite close to the unit circle. This term, when transformed to the time domain, causes a change in sign at each sampling instant in the manipulated variable as can be shown using long division. Dahlin (1968) suggested that ringing can be eliminated by setting $z^{-1} = 1$ in the ringing term, in this case replacing $(0.0280 + 0.0234z^{-1})$ by a constant $(0.0280 + 0.0234 = 0.0514)$. Let the nonringing version of Dahlin's controller be denoted by $\overline{G}_{DC}$. Figure 13.11*b* shows $y(t)$ and $p(t)$ for this case, indicating that the ringing behavior has disappeared. Interestingly, the closed-loop response now exhibits an overshoot, which contradicts the original design criterion of a first-order approach to the set point (Eq. 13-68). Therefore, the closed-loop performance of Dahlin's controller modified for ringing is not always predictable. This lack of predictability represents a major disadvantage of this technique.

13.7.2 Vogel–Edgar Algorithm

For processes that can be described by a second-order plus-time-delay model (Eq. 13-83), Vogel and Edgar (1988) developed a controller that eliminates the ringing pole caused by inverting G. The desired closed-loop transfer function is similar to that for Dahlin's controller (cf. Eq. 13-79 and includes time-delay compensation (z^{-N-1} in the closed-loop transfer function):

$$\left(\frac{Y}{Y_{sp}}\right)_d = \frac{(1-A)}{1-Az^{-1}}\frac{b_1+b_2z^{-1}}{b_1+b_2}z^{-N-1} \tag{13-86}$$

However, the zeros of model G are also included as zeros of the closed-loop transfer function (in this case, divided by $b_1 + b_2$ to ensure that the closed-loop steady-state gain equals one). Although this choice may slow down the response somewhat, it makes the controller less sensitive to model errors and also eliminates the possibility of ringing. The Vogel–Edgar controller corresponding to Eq. 13-86 is

$$G_{VE} = \frac{(1+a_1z^{-1}+a_2z^{-2})(1-A)}{(b_1+b_2)(1-Az^{-1})-(1-A)(b_1+b_2z^{-1})z^{-N-1}} \tag{13-87}$$

Because of the form of $(Y/Y_{sp})_d$ in Eq. 13-86 this controller does not attempt to cancel the numerator terms of the process transfer function and thus does not include the potential ringing pole. Note that for $a_2 = b_2 = 0$ (a first-order process), Eq. 13-87 reverts to

EXAMPLE 13.6

For the same process model used in Example 13.6, plot the response and the controller output for a unit set-point change ($\tau_c = \Delta t = 1.0$) for the nonringing Dahlin's controller and compare it with the Direct Synthesis PID (BD conversion) and Vogel–Edgar approaches.

Solution

Applying Eq. 13-79 for $N = 0$, the controller transfer function for the nonringing version of Dahlin's algorithm is

$$\overline{G}_{DC} = \left(\frac{0.6321}{1-z^{-1}}\right)\left(\frac{1-1.5353z^{-1}+0.5866z^{-2}}{0.0514}\right) \tag{13-88}$$

For this controller the response $y(t)$ and the controller output $p(t)$ are shown in Fig. 13.12*a*. Next we derive a PID controller using the IMC approach presented in Chapter 11. Starting with the continuous-time second-order transfer function ($\tau_1 = 5, \tau_2 = 3, K = 1, \theta = 0$), τ_c in Table 11.1 is set equal to one to provide the best response for a set-point change. When the BD approximation $s \approx (1 - z^{-1})/\Delta t$ is substituted into the PID controller transfer function,

$$G_{BD} = 4.1111\frac{3.1486-5.0541z^{-1}+2.0270z^{-2}}{1.7272-2.4444z^{-1}+0.7222z^{-2}} \tag{13-89}$$

The closed-loop response is shown in Fig. 13.13*b*. When the BD-PID digital controller is designed for this system, there is no need to correct for ringing such as is required for Dahlin's controller. This is true for wide ranges of τ_c and Δt that have been investigated.

Figure 13.12 shows the closed-loop response for the Vogel–Edgar controller (G_{VE}) for the same second-order model. This controller is

$$G_{VE} = 0.6321\frac{1-1.5353z^{-1}+0.5866z^{-2}}{0.0514-0.0366z^{-1}-0.0148z^{-2}} \tag{13-90}$$

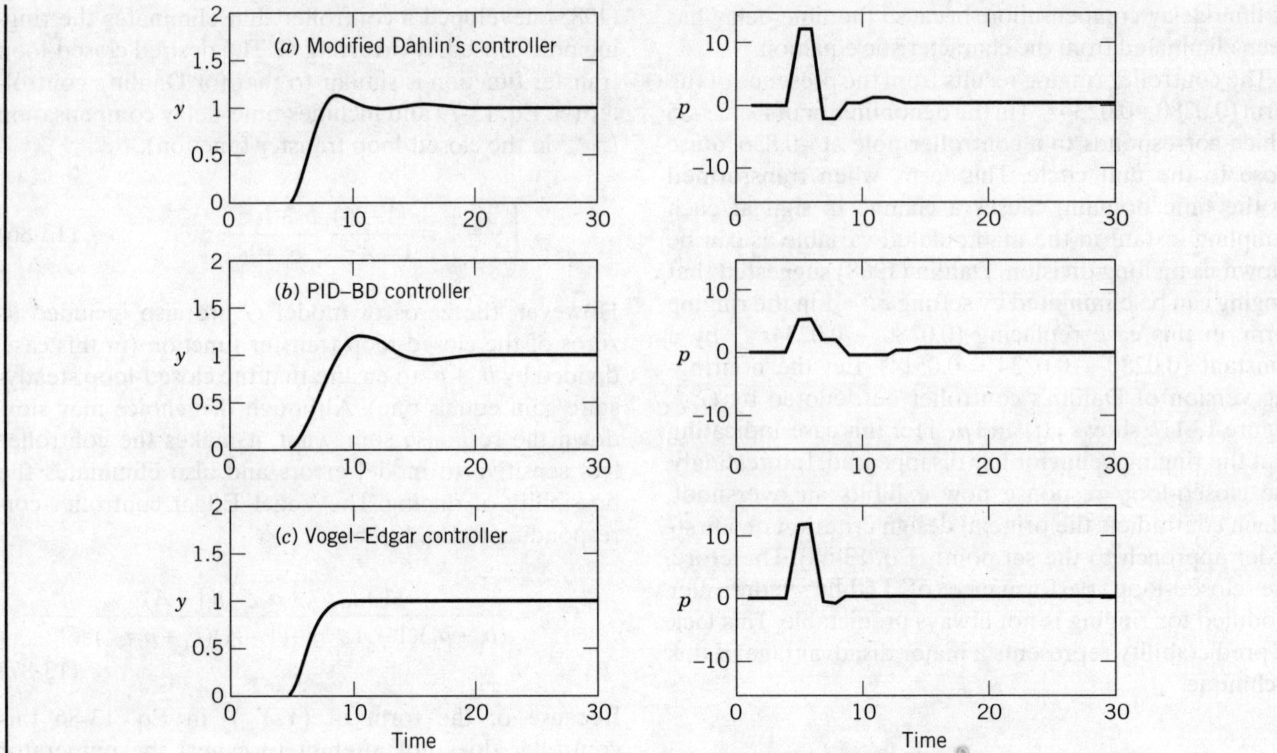

Figure 13.12 Comparison of closed-loop response for a second-order process in Example 13.7 using (*a*) nonringing Dahlin's controller ($\lambda = 1$), (*b*) backwards difference PID controller and (*c*) Vogel–Edgar controller (y = controlled variable, p = controller output after ZOH).

The tuning parameter A is selected to be 0.368 ($\tau_c = 1$). For this second-order system, the controlled variable response for G_{VE} is superior to $\bar{G}_{DC}$. If a time delay is added to the model, the comparative performance of G_{VE} and $\bar{G}_{DC}$ is still the same, because both controllers utilize the same form of the Smith predictor.

Note that the controller parameters in Eqs. 13-88 to 13-90 have been reported with four decimal points in order to avoid roundoff errors in the control calculations.

Dahlin's controller, Eq. 13-80 because a first-order process will not lead to a ringing controller.

Vogel and Edgar (1988) have shown that their controller satisfactorily handles first-order or second-order process models with positive zeros (inverse response) or negative zeros as well as simulated process and measurement noise. Many higher-order process models can be successfully controlled with G_{VE}. Neither G_{VE} nor $\bar{G}_{DC}$ are suitable for unstable process models, however. The robustness of the Vogel–Edgar controller is generally better than Dahlin's controller when model errors are present. G_{VE} can be used as an adaptive controller when the discrete-time model is updated on-line. For processes with zeros outside the unit circle, Dahlin's controller can become unstable, while the stability of the Vogel–Edgar controller is unaffected.

13.7.3 Internal Model Control (IMC)

The general design methodology of Internal Model Control presented in Section 11.2.2 for continuous-time systems can be extended to sampled-data systems (Garcia and Morari, 1982; Zafiriou and Morari, 1985). Figure 12.6 shows the block diagram used for IMC contrasted with that for conventional feedback control. Here the notation G_c^* is used instead of G_c for the controller transfer function because of the different block diagram structure and controller design methodology used with IMC.

The perfect IMC controller is simply the inverse of the process model.

$$G_c^*(z) = 1/\widetilde{G} \tag{13-91}$$

However, a perfect controller is usually not physically realizable, or it may be impractical because of model error. The two key steps involved in digital controller design are as follows (cf. Section 11.2.2).

1. The process model is factored as

$$\widetilde{G}(z) = \widetilde{G}_+(z)\widetilde{G}_-(z) \tag{13-92}$$

where $\widetilde{G}_+$ contains the time-delay term z^{-N-1}, zeroes that lie outside the unit circle, and zeroes that lie inside the unit circle near (−1,0). Also, $\widetilde{G}_+$ has a steady-state gain of unity.

2. The controller is obtained by inverting $\widetilde{G}_-$ (the invertible part of $\widetilde{G}$) and then multiplying by a first-order filter F to improve the robustness of the controller as well as to ensure the physical realizability of G_c^*:

$$G_c^*(z) = \frac{F(z)}{G\_(z)} \tag{13-93}$$

The filter F usually contains one or more tuning parameters. Zeros of $\widetilde{G}$ that lie outside the unit circle (the so-called nonminimum phase zeroes) would yield unstable controller poles if such terms were included in $\widetilde{G}_-$, instead of $\widetilde{G}_+$. Negative zeroes on the real axis near $z = -1$ will result in a ringing controller if they are inverted; hence, they are also included in $\widetilde{G}_+$.The closed-loop transfer function using the above design rules, assuming the process model is perfect, is

$$\frac{Y}{Y_{sp}} = \widetilde{G}_+(z)F(z) \tag{13-94}$$

The IMC design framework can yield Dahlin's controller and the Vogel–Edgar controller for appropriate choices of $F(z)$ and $\widetilde{G}_+(z)$. It can also readily be applied to higher-order systems, where Direct Synthesis is not as reliable. For details on treatment of process model zeroes and selection of the filter, see Zafiriou and Morari (1985).

The IMC block diagram in Fig. 11.6 can be expanded to include a block A^* in the feedback path as well as a disturbance transfer function G_d. The block A^* can be used to predict the effect of the disturbance on the error signal to the controller, and it can also provide time-delay compensation. This two-degree-of-freedom controller (see Chapter 11) is known as an *analytical predictor* (Kirtania and Choudhury, 2012; Wellons and Edgar, 1987).

EXAMPLE 13.7

A second-order process with transfer function

$$G(s) = \frac{-5e^{-4s}}{(3s+1)(5s+1)}; G_m = 1$$

is controlled with a digital IMC controller ($\Delta t = 0.2$). There is no modeling error. An IMC filter time constant of zero is used ($F(z) = 1$). A unit step change in load occurs at time $t = 10$, with a known disturbance transfer function

$$G_d = \frac{1}{7s+1}$$

Compute the IMC closed-loop response.

Solution

Using MATLAB c2d function with sampling time $\Delta t = 0.2$,

$$G(z) = \frac{(-0.006434z^{-1} - 0.00621z^{-2})}{1 - 1.896z^{-1} + 0.8988z^{-2}} z^{-20}$$

Since a zero is at −0.9652, it should be included in. $\widetilde{G}_+(z)$ Therefore,

$$\widetilde{G}_+(z) = \frac{(-0.006434z^{-1} - 0.00621z^{-2})}{(-0.006434 - 0.00621)} z^{-20}$$

$$= 0.5089z^{-21} + 0.4911z^{-22}$$

$$\widetilde{G}_-(z) = \frac{(-0.006434 - 0.00621)}{1 - 1.896z^{-1} + 0.8988z^{-2}}$$

For the filter $F(z) = 1$, analogous to Eq. 11-20, the IMC controller is

$$G_{c*}(z) = \widetilde{G}_-^{-1}(z)F(z) = \frac{1 - 1.896z^{-1} + 0.8988z^{-2}}{(-0.006434 - 0.00621)}$$

By using Simulink-MATLAB, the IMC response for a unit step in disturbance at $t = 10$ is shown in Fig. 13.13.

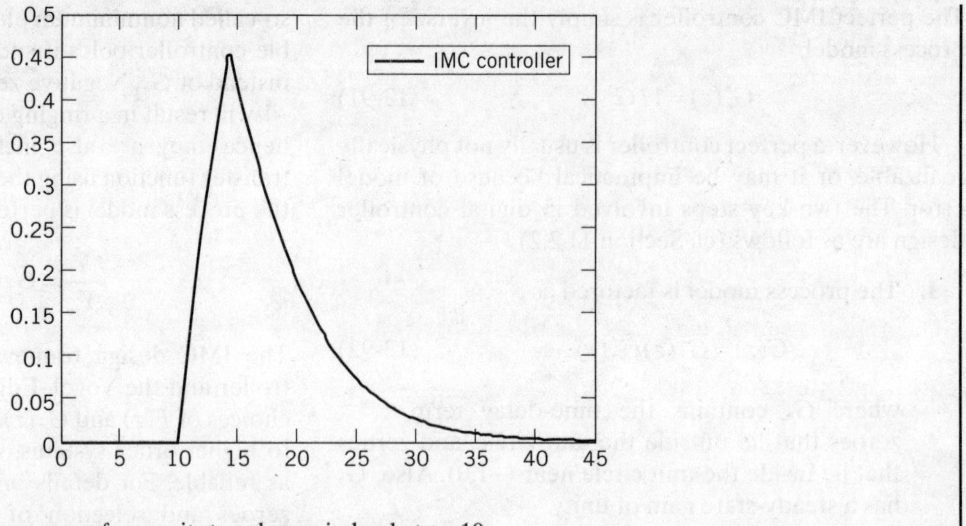

Figure 13.13 IMC closed-loop response for a unit step change in load at $t = 10$.

13.8 MINIMUM VARIANCE CONTROL

In this design method, the objective is to reduce the variability of the controlled variable y when the set point is constant and the process is subject to unknown, random disturbances. In statistical terms, the objective is to minimize the variance of y. This approach is especially relevant for processes where the disturbances are *stochastic* (that is, random) rather than *deterministic* (for example, steps or drifts). Sheet-making processes for producing paper and plastic film or sheets are common examples (Featherstone et al., 2000).

The *Minimum Variance Control* (MVC) design method generates the form of the feedback control law, as well as the values of the controller parameters. Like the Direct Synthesis and Internal Model Control design methods, the MVC method results in PI or PID controllers for simple transfer function models (MacGregor, 1988; Ogunnaike and Ray, 1994; Box and Luceño, 2009). Although MVC tends to be quite aggressive, the design method can be modified to be less aggressive (Bergh and MacGregor, 1987). Because Minimum Variance Control is a limiting case on actual controller performance, it provides a useful benchmark for monitoring control-loop performance (Harris and Seppala, 2002); see Chapter 17.

The starting point for the MVC design method is the following discrete transfer function model:

$$Y(z) = G(z)U(z) + D(z) \tag{13-95}$$

The disturbance $D(z)$ can be written as a zero, mean white (e.g., Gaussian) noise signal, $a(z)$, and a disturbance transfer function $G_d(z)$:

$$D(z) = G_d(z)a(z) \tag{13-96}$$

Previous discussions on accommodating disturbances focused on deterministic changes in the disturbance, such as step changes. Four alternative disturbance models are shown graphically in Fig. 13.14, These disturbances are persistent (as a result of the random component) but

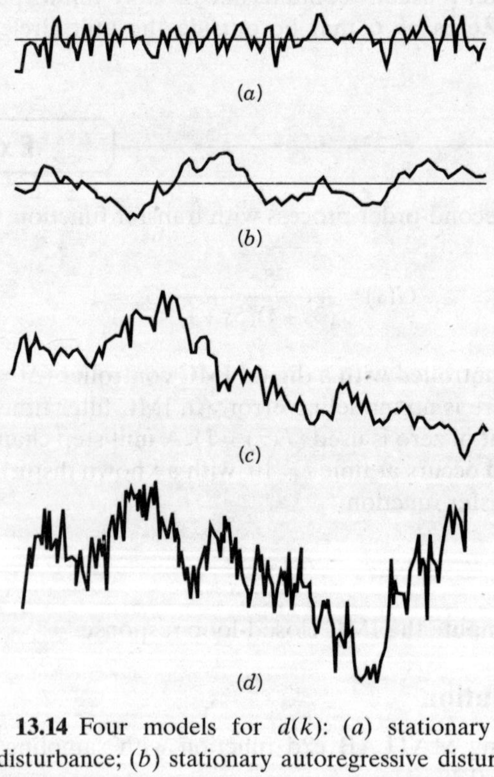

Figure 13.14 Four models for $d(k)$: (*a*) stationary white noise disturbance; (*b*) stationary autoregressive disturbance; (*c*) nonstationary disturbance (random walk); (*d*) integrated (nonstationary) moving-average disturbance (Reprinted with permission of John Wiley & Sons, Inc. and adapted from Box, G.E.P, and A. Luceño, Statistical Control by Monitoring and Adjustment, John Wiley and Sons, Hoboken, NJ, 2009).

may also exhibit features such as dynamics, drift, or trending. A typical process disturbance seldom will be random but will depend on past values of the disturbance. These models can be constructed by starting with an input $a(z)$ that is a white noise sequence. This input passes through a dynamic model such as a first-order transfer function or an integrating transfer function. The output $D(z)$ is an auto-correlated disturbance to the process.

Table 13.3 gives important time-series models (Box et al., 2015) that are commonly encountered in industrial process control, including statistical process control applications (see Chapter 17). Analogous to the four cases in Fig. 13.14, stationary disturbance models (a) and (b) have a fixed mean; that is, the sums of deviations above and below the line are equal to zero, but case (a) rarely occurs in industrial processes. Nonstationary disturbance models (c) and (d) do not have a fixed mean but are drifting in nature. Case (c), so-called *random walk behavior*, is often used to describe stock market index patterns. Case (b) is called an *autoregressive* (AR) model, while case (d) is called an *integrated moving-average* (IMA) model.

Table 13.3 Disturbance Models for Figure 13.14

Model
(a) $d(k) = a(k)$ (white noise)
(b) $d(k) = \phi d(k-1) + a(k)$ $(\phi \leq 1)$
(c) $d(k) = d(k-1) + a(k)$
(d) $d(k) = d(k-1) + a(k) + \psi a(k-1)$ $(\psi \leq 1)$

Suppose the IMA noise model is to be employed in a minimum variance controller for a process model with gain K that has no dynamics. It can be shown theoretically that for this simple case, the minimum variance controller has the same attributes as the IMC controller. The controller is the inverse of the process gain, and the IMC filter F is a first-order filter (Ogunnaike and Ray, 1994). A similar analysis can be performed for the case when the process model has first-order dynamics and the disturbance model is described by the IMA noise model. As shown by Ogunnaike and Ray (1994), these model assumptions lead to a PI controller. A Minimum Variance Controller can be obtained for higher-order process models, models with time delay, and different disturbance model structures (Bergh and MacGregor, 1987; Middleton and Goodwin, 1990). MVC can also be extended to problems where there is a cost associated with the control effort, as well as upper and lower bounds on the controller output. This approach requires that an objective function be defined, as in model predictive control (see Chapter 15).

SUMMARY

When a digital computer is used for process control, measurements of the process output are sampled and converted into digital form by an analog-to-digital converter (ADC). The sampling period Δt must be carefully selected. The choice of the sampling period should be based on the process dynamics, noise frequencies, signal-to-noise ratio, and the available computer control system.

Noisy measurements should be filtered before being sent to the controller. Analog filters are effective in removing high-frequency noise and avoiding aliasing. Digital filters are also widely used both for low-pass filters and other purposes such as the elimination of noise spikes. The choice of a filter and the filter parameters (e.g., τ_F) should be based on the process dynamics, the noise characteristics, and the sampling period. If a filter parameter is changed, it may be necessary to retune the controller, because the filter is a dynamic element in the feedback control loop.

We have introduced the z-transform technique and its properties, methods for obtaining discrete transfer function model of a system and methods for conducting inverse z-transform. Further different approaches for designing digital feedback controllers are also discussed. Digital controllers that emulate continuous-time PID controllers can include a number of special features to improve performance. Controllers based on Direct Synthesis or IMC can be tuned in continuous or discrete time, avoid ringing, eliminate offset, and provide a high level of performance for set-point changes. Minimum variance control can be very effective if a disturbance model is available.

REFERENCES

Åström, K. J., and B. Wittenmark, *Computer Controlled Systems: Theory and Design*, 3rd ed., Prentice Hall, Upper Saddle River, NJ, 2011.

Bergh, L. G., and J. F. MacGregor, Constrained Minimum Variance Controllers: Internal Model Structure and Robustness Properties, *Ind. Eng. Chem. Res.*, **26**, 1558 (1987).

Beschi, M., S. Dormido, J. Sanchez, and A. Visioli, Characterization of Symmetric Send-on-delta PI Controllers, *J. Process Control*, **22**, 1930 (2012).

Blevins, T., D. Chen, M. Nixon, and W. Wojsznis, *Wireless Control Foundation; Continuous and Discrete Control for the Process Industry*, ISA, Research Triangle Park, NC, 2015.

Box, G. E. P., G. M. Jenkins, and G. C. Reinsel, *Time Series Analysis, Forecasting, and Control*, 5th ed., Prentice Hall, Englewood Cliffs, NJ, 2015.

Box, G. E. P., and A. Luceño, *Statistical Control by Monitoring and Adjustment*, 2nd ed., John Wiley and Sons, Hoboken, NJ, 2009.

Dahlin, E. B., Designing and Tuning Digital Controllers, *Instrum. Control Systems*, **41** (6), 77 (1968).

Featherstone, A. P., J. G. VanAntwerp, and R. D. Braatz, *Identification and Control of Sheet and Film Processes*, Springer-Verlag, London, 2000.

Franklin, G. F., J. D. Powell, and M. L. Workman, *Digital Control of Dynamic Systems*, 3rd ed., Pearson, Washington D.C., 2005.

Garcia, C. E., and M. Morari, Internal Model Control, 1. A Unifying Review and Some New Results. *IEC Proc. Des.Dev.*, **21**, 308 (1982).

Garpinger, O., and T. Hagglund, Software-based Optimal PID Design with Robustness and Noise Sensitivity Constraints, *J. Process Control*, **33**, 90 (2015).

Harris, T. J., and C. T. Seppala, Recent Developments in Controller Performance Monitoring and Assessment Techniques, Chemical Process Control VI, *AIChE Symp.Ser.*, **98**, No. 326, 208 (2002).

Isermann, R., *Digital Control Systems*, 2nd ed., Springer-Verlag, New York, 1991.

Kirtania, K., and M. A. A. Choudhury, A Novel Dead Time Compensator for Stable Processes with Long Dead Times, *J. Process Control*, **22**, 612 (2012).

Ljung, L., *System Identification: Theory for the User*, 2nd ed., Prentice Hall, Upper Saddle River, NJ, 1999.

McConnell, E., and D. Jernigan, Data Acquisition, Sect. 18.3, p. 1938, in *The Electronics Handbook*, 2nd ed., J. C. Whitaker (Ed.), CRC Press, Boca Raton, FL, 2005.

Middleton, R. H., and G. C. Goodwin, *Digital Control and Estimation—A Unified Approach*, Prentice Hall, Englewood Cliffs, NJ, 1990.

Ogata, K., *Discrete-Time Control Systems*, 2nd ed., Prentice Hall, Englewood Cliffs, NJ, 1995.

Ogunnaike, B. A., and W. H. Ray, *Process Dynamics, Modeling, and Control*, Oxford University Press, New York, 1994.

Oppenheim, A. V., and R. W. Shafer, *Discrete Time Signal Processing*, 3rd ed., Prentice Hall, Englewood Cliffs, NJ, 2009.

Orfanidis, S., *Introduction to Signal Processing*, Prentice Hall, Englewood Cliffs, New Jersey, 1995.

Phillips, C. L., T. Nagle, and A. Chakraborty, *Digital Control System Analysis and Design*, 4th ed., Prentice Hall, Englewood Cliffs, New Jersey, 2014.

Ruiz, A., J. E. Jimenez, J. Sanchez, and S. Dormido, A Practical Tuning Methodology for Event-based PI Control, *J. Process Control*, **24**, 278 (2014).

Seborg, D. E., T. F. Edgar, and D. A. Mellichamp, *Process Dynamics and Control*, 1st ed., John Wiley and Sons, Hoboken, NJ,, 1989.

Segovia, V. R., T. Hagglund, and K. J. Astrom, Measurement Noise Filtering for Common PID Tuning Rules, *Contr. Eng. Prac.*, **32**, 43 (2014).

Vogel, E. G., and T. F. Edgar, An Adaptive Pole Placement Controller for Chemical Processes and Variable Dead Time, *Comp. Chem. Engr.*, **12**, 15 (1988).

Wellons, M. C., and T. F. Edgar, The Generalized Analytical Predictor, *Ind. Eng. Chem. Res.*, **26**, 1523 (1987).

Xu, S., B. Lu, T. F. Edgar, W. Wojsznis, T. Blevins, and M. Nixon, Data Cleaning in the Process Industries, *Rev. Chem. Eng.*, **31**(5), 453 (2015).

Zafiriou, E., and M. Morari, Digital Controllers for SISO Systems: A Review and a New Algorithm, *Int. J. Control*, **42**, 885 (1985).

EXERCISES

13.1 The mean arterial pressure P in a patient is subjected to a unit step change in feed flow rate F of a drug. Normalized response data are shown below. Previous experience has indicated that the transfer function,

$$\frac{P(s)}{F(s)} = \frac{4}{8s+1}$$

provides an accurate dynamic model. Filter these data using an exponential filter with two different values of α, 0.4 and 0.7. Graphically compare the noisy data, the filtered data, and the analytical solution for the transfer function model for a unit step input.

Time (min)	P	Time (min)	P
0	0	11	3.336
1	0.495	12	3.564
2	0.815	13	3.419
3	1.374	14	3.917
4	1.681	15	3.884
5	1.889	16	3.871
6	2.078	17	3.924
7	2.668	18	4.300
8	2.533	19	4.252
9	2.908	20	4.409
10	3.351		

13.2 A signal given by

$$y_m(t) = t^2 + 0.5\ \sin(t)$$

is to be filtered with an exponential digital filter over the interval $0 \le t \le 20$. Using three different values of α (0.85, 0.55, 0.25), determine the output of the filter at each sampling time. Do this for sampling periods of 1.0 and 0.1. Compare the three filters for each value of Δt.

13.3 The following product quality data y_m were obtained from a bioreactor, based on a photometric measurement evaluation of the product:

Time (min)	y_m (absorbance)
0	0
1	1.5
2	0.3
3	1.6
4	0.4
5	1.7
6	1.5
7	2.0
8	1.5

(a) Filter the data using an exponential filter with $\Delta t = 1$ min. Consider $\alpha = 0.2$ and $\alpha = 0.5$.

(b Use a moving-average filter with $N^* = 4$.

(c) Implement a noise-spike filter with $\Delta y = 0.5$.

(d) Plot the filtered data and the raw data for purposes of comparison.

13.4 The analog exponential filter in Eq. 13-2 is used to filter the output signal from a transfer function $G = 1/(2s+1)$. The input signal is $d(t) = 1 + 0.2 \sin t$. Compare the responses for no filtering ($\tau_F = 0$) and an exponential filter ($\tau_F = 3$ min) for $0 \le t \le 50$ min.

13.5 Find the response $y(k)$ for the difference equation

$$y(k) - 1.1y(k-1) + 0.3y(k-2) = u(k-1) + u(k-2)$$

Let $y(0) = y(1) = 0$, $u(0) = 1$, $u(k) = 0$ for $k \ge 1$. Perform direct integration using a spreadsheet. What is the steady-state value of y?

13.6 Derive expression for the transfer functions of

(i) zero-order hold

(ii) first-order hold

13.7 Determine the z-transform of $y(t)$ when:

(i) $y(t) = \dfrac{df(t)}{dt}$

(ii) $y(t) = \int_0^{t=k\Delta t} f(t)dt$

13.8 The dynamic behavior of a temperature sensor and transmitter can be described by the FOPTD transfer function

$$\frac{T'_m(s)}{T'(s)} = \frac{e^{-2s}}{8s+1}$$

where the time constant and time delay are in seconds and:

T = actual temperature (deviation)

T'_m = measured temperature (deviation)

The actual temperature changes as follows (t in s):

$$T = \begin{cases} 70\ °\text{C} & \text{for } t < 0 \\ 85\ °\text{C} & \text{for } 0 \le t < 100 \\ 70\ °\text{C} & \text{for } t \ge 10 \end{cases}$$

If samples of the measured temperature are automatically logged in a digital computer every 2 s beginning at $t = 0$, what is the maximum value of the logged temperature? Use simulation with a ZOH to find the answer.

13.9 Derive z-transform of the following function sampled at interval of $\Delta t = 0.5$

$$y(t) = u(t-1)[1 - e^{-(t-1)}]$$

13.10 z-transform of a discrete signal is given by

$$Y(z) = \frac{0.7z^{-1}}{1 - 1.4z^{-1} + 0.4z^{-2}}$$

Determine the value of $y^*(k\Delta t)$ at first four sampling instances (i.e., $k = 0,1,2$ and 3) using:

(i) Method of long division

(ii) Method of partial fraction

13.11 For a process given by

$$\frac{Y(z)}{U(z)} = \frac{2.7z^{-2} + 8.1z^{-3}}{1 - 0.5z^{-1} + 0.06z^{-2}}$$

(a) Calculate the response $y(k\Delta t)$ to a unit step change in u using integration of the difference equation.

(b) Check your answer in (a) by using simulation software, e.g., Simulink.

(c) What is the steady-state value of y?

13.12 The discrete-time transfer function of a process is given by

$$\frac{Y(z)}{U(z)} = \frac{5z^{-1} + 3z^{-2}}{1 - z^{-1} + 0.41z^{-2}}$$

(a) Convert this transfer function to an equivalent difference equation.

(b) Calculate the response $y(k)$ to a unit step change in u using integration of the difference equation.

(c) Check your answer in (a) by using simulation software.

(d) What is the steady-state value of y?

13.13 $x^*(k\Delta t)$ is a discrete signal coming from a digital element. The signal is passed through a zero-order hold and a first order lag process connected in series. Determine the output from the process as a function of sampling instance and show that the final value of output is equal to the gain of the first-order process.

13.14 To determine the effects of pole and zero locations, simulate the unit step responses of the discrete transfer functions shown below for the first six sampling instants, $k = 0$ to $k = 5$. What conclusions can you make concerning the effect of pole and zero locations?

(a) $\dfrac{1}{1 - z^{-1}}$

(b) $\dfrac{1}{1 + 0.7z^{-1}}$

(c) $\dfrac{1}{1 - 0.7z^{-1}}$

(d) $\dfrac{1}{(1 + 0.7z^{-1})(1 - 0.3z^{-1})}$

(e) $\dfrac{1 - 0.5z^{-1}}{(1 + 0.7z^{-1})(1 - 0.3z^{-1})}$

(f) $\dfrac{1 - 0.2z^{-1}}{(1 + 0.6z^{-1})(1 - 0.3z^{-1})}$

13.15 A temperature control loop includes a second-order overdamped process described by the discrete transfer function.

$$G(z) = \frac{(0.0826 + 0.0368z^{-1})z^{-1}}{(1 - 0894z^{-1})(1 - 0.295z^{-1})}$$

and a digital PI controller

$$G_c(z) = K_c\left(1 + \frac{1}{8(1 - z^{-1})}\right)$$

Find the maximum controller gain K_{cm} for stability by trial-and-error.

13.16 The block diagram of a digital control system is shown in Fig. E13.16. The sampling period is $\Delta t = 1$ min.

(a) Design the digital controller $G_c(z)$ so that the closed-loop system exhibits a first-order response to a unit step change in the set point (after an appropriate time delay).

(b) Will this controller eliminate offset after a step change in the set point? Justify your answer.

(c) Is the controller physically realizable? Justify your answer.

(d) Design a digital PID controller based on the ITAE (set-point) method in Chapter 11 and examine its performance for a step change in set point. Approximate the sampler and ZOH by a time delay of $\theta = \Delta t/2$.

13.17 The block diagram of a sampled-data control system is shown in Fig. E13.17. Design a Dahlin controller $G_c(z)$ that is physically realizable and based on a change in set point. The sampling period is $\Delta t = 1$ min. Calculate the closed-loop response when this controller is used and a unit step change in disturbance occurs.

13.18 A second-order system G with $K = 1$, $\tau_1 = 6$, and $\tau_2 = 4$ is to be controlled using the Vogel–Edgar controller with $\lambda = 5$ and $\Delta t = 1$. Assuming a step change in y_{sp}, calculate the controlled variable $y(k)$ for $k = 0, 1, \ldots, 25$, and plot $y(k)$ and the controller output $p(k)$.

13.19 For a process including control valve and sensor, $G(s) = 1.25e^{-5s}/(5s + 1)$, derive the equation for Dahlin's controller with Δt and $\lambda = 1$ and plot controller output $p(k)$ for a set-point change. Does ringing occur?

13.20 Compare the Dahlin and Vogel–Edgar controllers for $G(s) = 1/[(2s + 1)(s + 1)]$ and $\lambda = \Delta t = 1$. Does either controller ring? Derive the resulting difference equations for the closed-loop system $y(k)$ related to $y_{sp}(k)$. Does overshoot occur in either case?

13.21 Design a digital controller for the liquid level in the storage system shown in Fig. E13.21. Each tank is 0.76 m in diameter. The piping between the tanks acts as a linear resistance to flow with $R = 21.5$ min/m². The liquid level is sampled every 30 s. The digital controller also acts as a ZOH device for the signal sent to the control valve. The control valve and level transmitter have negligible dynamics. Their gains are $K_v = 0.0071$ m³/min/mA and $K_m = 26.23$ mA/m, respectively. The nominal value of q_1 is 0.014 m³/min.

(a) Derive Dahlin's control algorithm based on a step change in set point.

(b) Does the controller output exhibit any oscillation?

(c) For what values of λ is the controller physically realizable?

(d) If you were to tune this controller on-line, what value of λ would you use as an initial guess? Justify your answer.

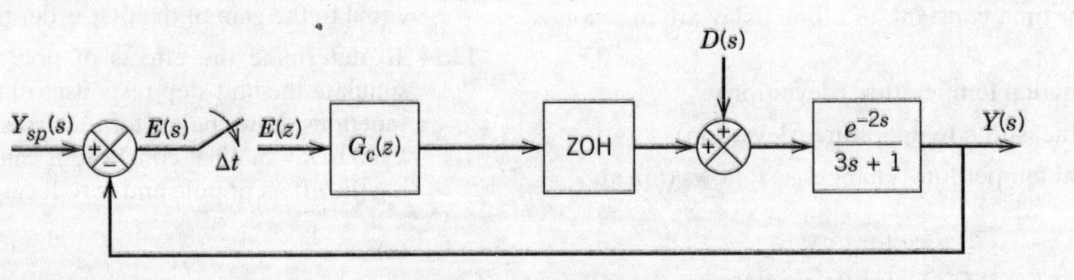

Figure E13.16

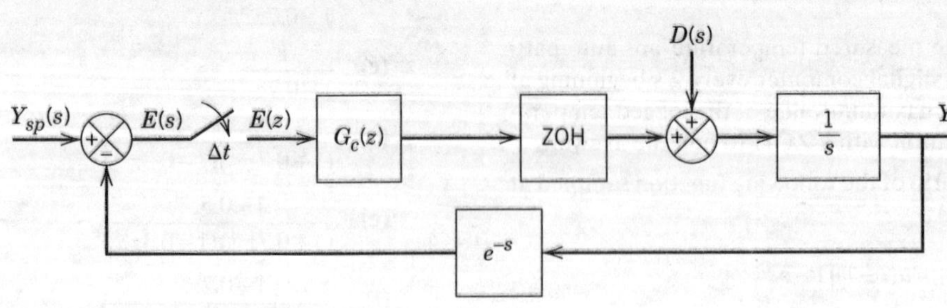

Figure E13.17

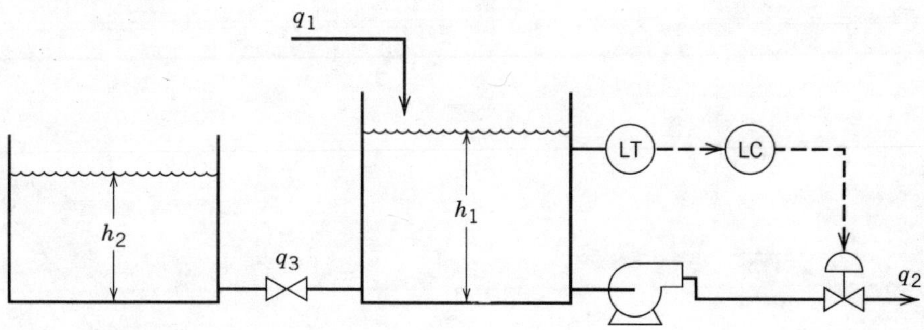

Figure E13.21

MULTIPLE CHOICE QUESTIONS

1. Which of the following action converts continuous signal to discrete signal?

(a) Sampling
(b) Holding
(c) Quantization
(d) Reconstruction

2. Which of the following action converts discrete signal to continuous signal?

(a) Hold
(b) Sampling
(c) Extrapolation
(d) Hold and extrapolation

3. z-transform and Laplace transform are related by

(a) $s = z$
(b) $s = \ln(z)$
(c) $s = (\ln z)/(\Delta t)$
(d) $s = (\Delta t)/\ln(z)$

4. Convolution of two sequences $x_1(n)$ and $x_2(n)$ are represented by

(a) $x_1(z)\, x_2(z)$
(b) $x_1(z) \times x_2(z)$
(c) $x_1(z) + x_2(z)$
(d) $x_1(z)/x_2(z)$

5. Final value theorem is used for

(a) unstable systems.
(b) stable systems.
(c) marginally stable systems.
(d) all types of systems.

6. Aliasing is caused when

(a) sampling frequency is equal to the input signal frequency.
(b) sampling frequency is more than the input signal frequency.
(c) sampling frequency is less than the input signal frequency.
(d) sampling frequency is equal to or more than the input signal frequency.

ANSWER KEY

1. (a) **2.** (d) **3.** (c) **4.** (b) **5.** (b) **6.** (c)

CHAPTER 14

Multiloop and Multivariable Control

CHAPTER CONTENTS

In previous chapters, we have emphasized control problems that have only one controlled variable and one manipulated variable. These problems are referred to as single-input, single-output (SISO), or *single-loop*, control problems. But in many practical control problems, typically a number of variables must be controlled, and a number of variables can be manipulated. These problems are referred to as multiple-input, multiple-output (MIMO) control problems. For almost all important processes, at least two variables must be controlled: product quality and throughput.

Several examples of processes with two controlled variables and two manipulated variables are shown in Fig. 14.1. These examples illustrate a characteristic feature of MIMO control problems, namely, the presence of *process interactions*; that is, each manipulated variable can affect both controlled variables. Consider the in-line blending system shown in Fig. 14.1*a*. Two streams containing species A and B, respectively, are to be blended to make a product stream with mass flow rate w and composition x, the mass fraction of A. Adjusting either manipulated flow rate, w_A or w_B, affects both w and x.

Similarly, for the distillation column in Fig. 14.1*b*, adjusting either reflux flow rate R or steam flow rate S will affect both distillate composition x_D and bottoms composition x_B. For the gas liquid separator in Fig. 14.1*c*, adjusting gas flow rate G will have a direct effect on pressure P and a slower, indirect effect on liquid level h. In contrast, adjusting the other manipulated variable L directly affects h but has only a relatively small and indirect effect on P.

When significant process interactions are present, the selection of the most effective control configuration may not be obvious. For example, in the blending problem, suppose that a conventional feedback control strategy, consisting of two PI controllers, is to be used. This control system, referred to as

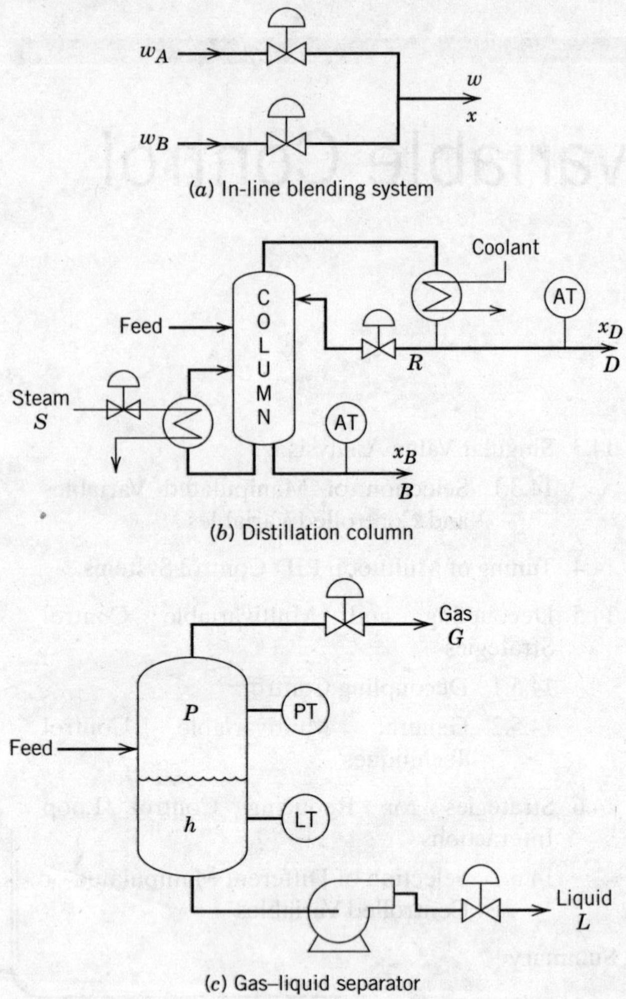

Figure 14.1 Physical examples of multivariable control problems.

a *multiloop control system* because it employs two single-loop feedback controllers, raises several questions. Should the composition controller adjust w_A and the flow controller adjust w_B, or *vice versa*? How can we predict which of these two multiloop control configurations will be more effective? Will control loop interactions generated by the process variables cause problems?

In the next section, we consider techniques for selecting an appropriate multiloop control configuration. If the process interactions are significant, even the best multiloop control system may not provide satisfactory control. In these situations there are incentives for considering *multivariable control strategies* such as decoupling control (Section 14.5) and model predictive control (Chapter 15). But first we examine the phenomenon of control loop interactions.

14.1 PROCESS INTERACTIONS AND CONTROL LOOP INTERACTIONS

A schematic representation of several SISO and MIMO control applications is shown in Fig. 14.2. For convenience, it is assumed that the number of manipulated variables is equal to the number of controlled variables. This allows pairing of a single controlled variable and a single manipulated variable via a feedback controller which is called a *square* system. This is preferred over a *non-square* system, where the number of manipulated variables is greater than or less than the number of controlled variables. General multivariable control strategies are not limited to single-loop feedback control and can treat any number of MVs and CVs (see Chapter 15).

MIMO control problems are inherently more complex than SISO control problems because process interactions occur between controlled and manipulated variables. In general, a change in a manipulated variable, say u_1, will affect all of the controlled variables $y_1, y_2, \ldots, y_n$. Because of the process interactions, the selection of the best single loop pairing of controlled and manipulated variables for a multiloop control scheme can be a difficult task. In particular, for a control problem with n controlled variables and n manipulated variables, there are $n!$ possible multiloop control configurations.

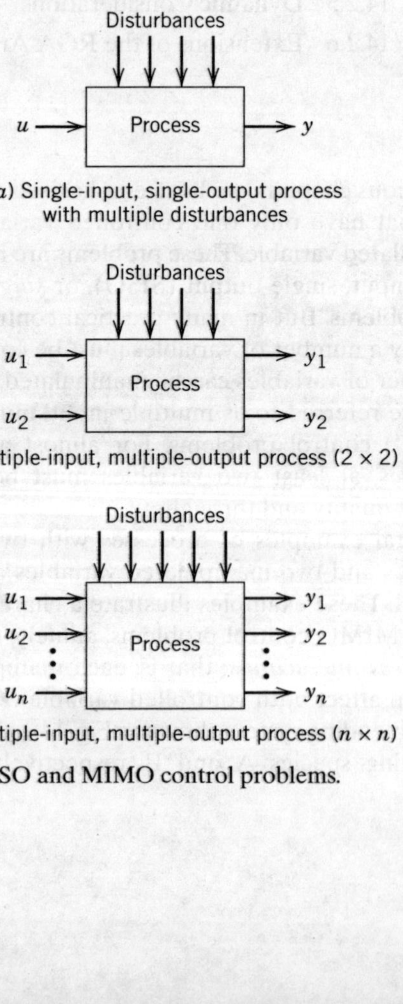

Figure 14.2 SISO and MIMO control problems.

14.1.1 Block Diagram Analysis

Consider the 2×2 control problem shown in Fig. 14.2*b*. Because there are two controlled variables and two manipulated variables, four process transfer functions are necessary to completely characterize the process dynamics:

$$\frac{Y_1(s)}{U_1(s)} = G_{p11}(s) \quad \frac{Y_1(s)}{U_2(s)} = G_{p12}(s)$$
$$\frac{Y_2(s)}{U_1(s)} = G_{p21}(s) \quad \frac{Y_2(s)}{U_2(s)} = G_{p22}(s) \tag{14-1}$$

The transfer functions in Eq. 14-1 can be used to determine the effect of a change in either U_1 or U_2 on Y_1 and Y_2. From the Principle of Superposition (Section 3.1), it follows that *simultaneous* changes in U_1 and U_2 have an additive effect on each controlled variable:

$$Y_1(s) = G_{p11}(s)U_1(s) + G_{p12}(s)U_2(s) \tag{14-2}$$

$$Y_2(s) = G_{p21}(s)U_1(s) + G_{p22}(s)U_2(s) \tag{14-3}$$

These input–output relations can also be expressed in vector-matrix notation as

$$\boldsymbol{Y}(s) = \boldsymbol{G}_p(s)\boldsymbol{U}(s) \tag{14-4}$$

where $\boldsymbol{Y}(s)$ and $\boldsymbol{U}(s)$ are vectors with two elements,

$$\boldsymbol{Y}(s) = \begin{bmatrix} Y_1(s) \\ Y_2(s) \end{bmatrix} \quad \boldsymbol{U}(s) = \begin{bmatrix} U_1(s) \\ U_2(s) \end{bmatrix} \tag{14-5}$$

and $\boldsymbol{G}_p(s)$ is the process transfer function matrix,

$$\boldsymbol{G}_p(s) = \begin{bmatrix} G_{p11}(s) & G_{p12}(s) \\ G_{p21}(s) & G_{p22}(s) \end{bmatrix} \tag{14-6}$$

For a system with N number of inputs and M number of outputs, the transfer function matrix is given below

$$G_p(s) = \begin{bmatrix} G_{p11}(s) & \cdots & G_{p1N}(s) \\ \vdots & \cdots & \vdots \\ G_{pM1}(s) & \cdots & G_{pMN}(s) \end{bmatrix} \tag{14-7}$$

Any output Y_i is represented as

$$Y_i(s) = G_{pi1}(s)U_1(s) + G_{pi2}(s)U_2(s) + \cdots + G_{piN}(s)U_N(s) = \sum_{j=1}^{N} G_{pij}(s) \tag{14-8}$$

The matrix notation in Eq. 14-4 provides a compact representation for problems larger than 2×2. Recall that a transfer function matrix for an MIMO system, a thermal mixing system, was derived in Section 6.4. The steady-state process transfer function matrix ($s = 0$) is called the process *gain matrix* and is denoted by $\boldsymbol{K}$.

Suppose that a conventional multiloop control scheme consisting of two feedback controllers is to be used. The two possible control configurations are shown in Fig. 14.3. In scheme (a), Y_1 is controlled by adjusting U_1, while Y_2 is controlled by adjusting U_2. Consequently, this configuration will be referred to as the 1-1/2-2 control scheme. The alternative strategy is to pair Y_1 with U_2 and Y_2 with U_1, the 1-2/2-1 control scheme shown in Fig. 14.3*b*. Note that these block diagrams have been simplified by omitting the transfer functions for the final control elements and the sensor-transmitters. Also, the disturbance variables have been omitted.

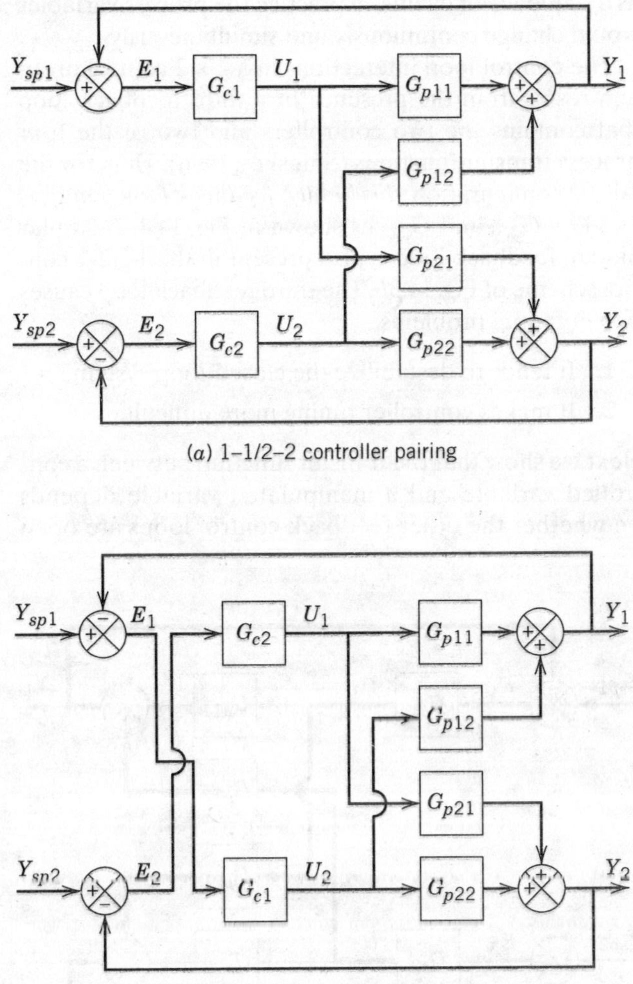

(*a*) 1–1/2–2 controller pairing

(*b*) 1–2/2–1 controller pairing

Figure 14.3 Block diagrams for 2×2 multiloop control schemes.

Figure 14.3 illustrates that process interactions can induce undesirable interactions between the control loops. For example, suppose that the 1-1/2-2 control scheme is used and a disturbance moves Y_1 away from its set point, Y_{sp1}. Then the following events occur:

1. The controller for loop 1 (G_{c1}) adjusts U_1 so as to force Y_1 back to the set point. However, U_1 also affects Y_2 via transfer function G_{p21}.
2. Since Y_2 has changed, the loop 2 controller (G_{c2}) adjusts U_2 to bring Y_2 back to its set point, Y_{2sp}. However, changing U_2 also affects Y_1 via transfer function G_{p12}.

These controller actions proceed simultaneously until a new steady state is reached. Note that the initial change in U_1 has two effects on Y_1: (1) a *direct* effect and (2) an *indirect* effect via the control loop interactions. Although it is instructive to view this dynamic behavior as a sequence of events, in practice the process variables would change continuously and simultaneously.

The control loop interactions in a 2 × 2 control problem result from the presence of a third feedback loop that contains the two controllers and two of the four process transfer functions (Shinskey, 1996). Thus, for the 1-1/2-2 configuration, *this hidden feedback loop* contains G_{c1}, G_{c2}, G_{p12}, and G_{p21}, as shown in Fig. 14.4. A similar hidden feedback loop is also present in the 1-2/2-1 control scheme of Fig. 14.3*b*. The third feedback loop causes two potential problems:

1. It tends to destabilize the closed-loop system.
2. It makes controller tuning more difficult.

Next we show that the transfer function between a controlled variable and a manipulated variable depends on whether the other feedback control loops are open or closed. Consider the control system in Fig. 14.3*a*. If the controller for the second loop G_{c2} is out of service or is placed in the manual mode with the controller output constant at its nominal value, then $U_2 = 0$. For this situation, the transfer function between Y_1 and U_1 is merely G_{p11}:

$$\frac{Y_1}{U_1} = G_{p11} \quad (Y_2 - U_2 \text{ loop open}) \tag{14-9}$$

If the lower loop is also closed, then the contributions to Y_1 from the two loops are added together:

$$Y_1 = G_{p11}U_1 + G_{p12}U_2 \tag{14-10}$$

However, if the second feedback controller is in the automatic mode with $Y_{2sp} = 0$, then, using block diagram algebra,

$$Y_2 = \frac{G_{p21}U_1}{1 + G_{c2}G_{p22}} \tag{14-11}$$

The signal to the first loop from the second loop is

$$G_{p21}U_2 = -G_{p12}G_{c2}Y_2 \tag{14-12}$$

If we substitute for $G_{p12}U_2$ in Eq. 14-10 using Eq. 14-12 and then substitute for Y_2 using Eq. 14-11, the overall closed-loop transfer function between Y_1 and U_1 is

$$\frac{Y_1}{U_1} = G_{p11} - \frac{G_{p12}G_{p21}G_{c2}}{1 + G_{c2}G_{p22}} \quad (Y_2 - U_2 \text{ loop closed}) \tag{14-13}$$

Thus, the transfer function between Y_1 and U_1 depends on the controller for the second loop G_{c2} via the interaction term. Similarly, transfer function Y_2/U_2 depends on G_{c1} when the first loop is closed.

$$\frac{Y_2}{U_2} = G_{p22} - \frac{G_{p21}G_{p12}G_{c1}}{1 + G_{c1}G_{p11}} \tag{14-14}$$

These results have important implications for controller tuning because they indicate that the two controllers should not be tuned independently.

The expression given above in Eq. 14-13, is sometimes called as effective transfer function between Y_1 and U_1, $G_{p11,\text{eff}}$. Similarly, Eq. 14-14 gives effective transfer function between Y_2 and U_2. Therefore, if the effect of input 1 on output 2 and input 2 on output 1 are not negligible (i.e., G_{p12} and G_{p21} are not close to zero), the effective transfer function between Y_1, U_1 and Y_2, U_2 should be used for controller design.

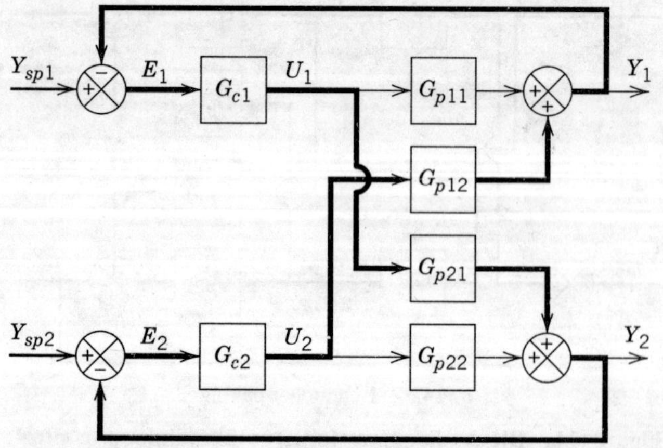

Figure 14.4 The hidden feedback control loop (in dark lines) for a 1-1/2-2 controller pairing.

EXAMPLE 14.1

Consider the following empirical model of a pilot-scale distillation column (Wood and Berry, 1973)

$$\begin{bmatrix} X_D(s) \\ X_B(s) \end{bmatrix} = \begin{bmatrix} \dfrac{12.8e^{-s}}{16.7s+1} & \dfrac{-18.9e^{-3s}}{21s+1} \\ \dfrac{6.6e^{-7s}}{10.9s+1} & \dfrac{-19.4e^{-3s}}{14.4s+1} \end{bmatrix} \begin{bmatrix} R(s) \\ S(s) \end{bmatrix} \quad (14\text{-}15)$$

where the notation is defined in Fig. 14.1*b*. Suppose that a multiloop control system consisting of two PI controllers is used. Compare the closed-loop set-point changes that result if the $X_D - R/X_B - S$ pairing is selected and

(a) A set-point change is made in each loop with the other loop in manual
(b) The set-point changes are made with both controllers in automatic

Assume that the controller settings are based on the ITAE tuning method for set-point changes in Chapter 11.

Solution

Table 14.1 shows the single-loop ITAE settings, and Fig. 14.5 shows simulation results for set-point changes for each controlled variable. The ITAE settings provide satisfactory set-point responses for either control loop when the other controller is in manual (solid line). However, when both controllers are in automatic, the control loop interactions produce very oscillatory responses especially in x_B (dashed line). McAvoy (1983) has discussed various approaches for improving the performance of the multiloop controllers. See Exercise 14.1 for a similar MIMO control problem where the loops also exhibit oscillations.

Table 14.1 Controller Settings for Example 14.1

Controller Pairing	K_c	τ_I (min)
$x_D - R$	0.604	16.37
$x_B - S$	− 0.127	14.46

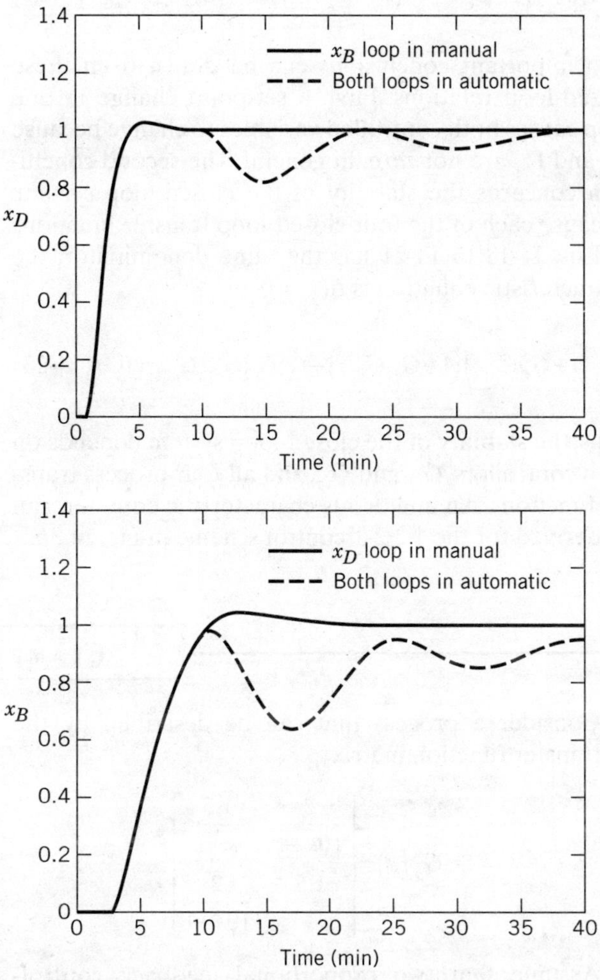

Figure 14.5 Set-point responses for Example 14.1 using ITAE tuning.

14.1.2 Closed-Loop Stability

To evaluate the effects of control loop interactions further, again consider the block diagram for the 1-1/2 2 control scheme in Fig. 14.3*a*. Using block diagram algebra (see Chapter 9), we can derive the following expressions relating controlled variables and set points:

$$Y_1 = \Gamma_{11} Y_{sp1} + \Gamma_{12} Y_{sp2} \quad (14\text{-}16)$$

$$Y_2 = \Gamma_{21} Y_{sp1} + \Gamma_{22} Y_{sp2} \quad (14\text{-}17)$$

where the closed-loop transfer functions are

$$\Gamma_{11} = \frac{G_{c1}G_{p11} + G_{c1}G_{c2}\left(G_{p11}G_{p22} - G_{p12}G_{p21}\right)}{\Delta(s)} \quad (14\text{-}18)$$

$$\Gamma_{12} = \frac{G_{c2}G_{p12}}{\Delta(s)} \quad (14\text{-}19)$$

$$\Gamma_{21} = \frac{G_{c1}G_{p21}}{\Delta(s)} \tag{14-20}$$

$$\Gamma_{22} = \frac{G_{c2}G_{p22} + G_{c1}G_{c2}\left(G_{p11}G_{p22} - G_{p12}G_{p21}\right)}{\Delta(s)} \tag{14-21}$$

and $\Delta(s)$ is defined as

$$\Delta(s) = \left(1+G_{c1}G_{p11}\right)\left(1+G_{c2}G_{p22}\right) - G_{c1}G_{c2}G_{p12}G_{p21} \tag{14-22}$$

Two important conclusions can be drawn from these closed-loop relations. First, a set-point change in one loop causes both controlled variables to change because Γ_{12} and Γ_{21} are not zero, in general. The second conclusion concerns the stability of the closed-loop system. Because each of the four closed-loop transfer functions in Eqs. 14-18 to 14-21 has the same denominator, the characteristic equation is $\Delta(s) = 0$, or

$$\left(1+G_{c1}G_{p11}\right)\left(1+G_{c2}G_{p22}\right) - G_{c1}G_{c2}G_{p12}G_{p21} = 0 \tag{14-23}$$

Thus, the stability of the closed-loop system depends on both controllers, G_{c1} and G_{c2}, and all four process transfer functions. An analogous characteristic equation can be derived for the 1-2/2-1 control scheme in Fig. 14.3*b*.

For the special case where either $G_{p12} = 0$ or $G_{p21} = 0$, the characteristic equation in Eq. 14-23 reduces to

$$\left(1+G_{c1}G_{p11}\right)\left(1+G_{c2}G_{p22}\right) = 0 \tag{14-24}$$

For this situation, the stability of the overall system merely depends on the stability of the two individual feedback control loops and their characteristic equations, which is expected.

$$1+G_{c1}G_{p11} = 0 \text{ and } 1+G_{c2}G_{p22} = 0 \tag{14-25}$$

Note that if either $G_{p12} = 0$ or $G_{p21} = 0$, the third (hidden) feedback control loop in Fig. 14.4 is broken. For example, if $G_{p12} = 0$, then the second control loop has no effect on Y_1, while the first control loop serves as a source of disturbances for the second loop via transfer function G_{p21}.

The above analysis has been based on the 1-1/2-2 control configuration in Fig. 14.3*a*. A similar analysis and conclusions can be derived for the 1-2/2-1 configuration (see Exercise 14.2). The results in Eqs. 14-16 to 14-25 can be extended to block diagrams that include the transfer functions for the sensors and control valves; see Exercise 14.3.

EXAMPLE 14.2

Consider a process that can be described by the transfer function matrix

$$G_p(s) = \begin{bmatrix} \dfrac{2}{10s+1} & \dfrac{1.5}{s+1} \\ \dfrac{1.5}{s+1} & \dfrac{2}{10s+1} \end{bmatrix}$$

Assume that two proportional feedback controllers are to be used so that $G_{c1} = K_{c1}$ and $G_{c2} = K_{c2}$. Determine the values of K_{c1} and K_{c2} that result in closed-loop stability for both the 1-1/2-2 and 1-2/2-1 configurations.

Solution

The characteristic equation for the closed-loop system (1-1/2-2 pairing) is obtained by substitution into Eq. 14-23 and collecting powers of s as follows (which can be obtained using a symbolic math toolbox):

$$a_4 s^4 + a_3 s^3 + a_2 s^2 + a_1 s + a_0 = 0 \tag{14-26}$$

where

$$\begin{aligned} a_4 &= 100 \\ a_3 &= 20K_{c1} + 20K_{c2} + 220 \\ a_2 &= 42K_{c1} + 42K_{c2} - 221K_{c1}K_{c2} + 141 \\ a_1 &= 24K_{c1} + 24K_{c2} - 37K_{c1}K_{c2} + 22 \\ a_0 &= 2K_{c1} + 2K_{c2} + 1.75K_{c1}K_{c2} + 1 \end{aligned}$$

Note that the characteristic equation in Eq. 14-26 is fourth-order, even though each individual transfer function in $\mathbf{G}_p(s)$ is first order.

The controller gains that result in a stable closed-loop system can be determined by applying direct substitution to the stability criterion (Chapter 9) for specified values of K_{c1} and K_{c2}. The resulting stability regions are shown in Fig. 14.6. If either K_{c1} or K_{c2} is close to zero, the other controller gain can be an arbitrarily large, positive value and still have a stable closed-loop system. This result is a consequence of having process transfer functions that are first order without time delay, which is an idealistic case. MIMO control systems normally have an upper bound for stability for both controller gains for all values of K_{c1}.

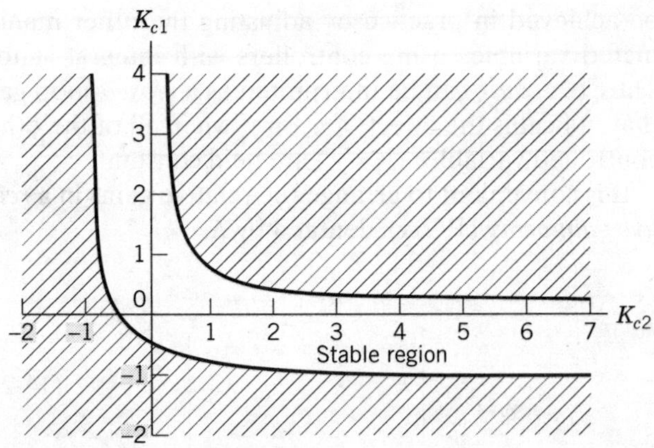

Figure 14.6 Stability region for Example 14.2 with 1-1/2-2 controller pairing.

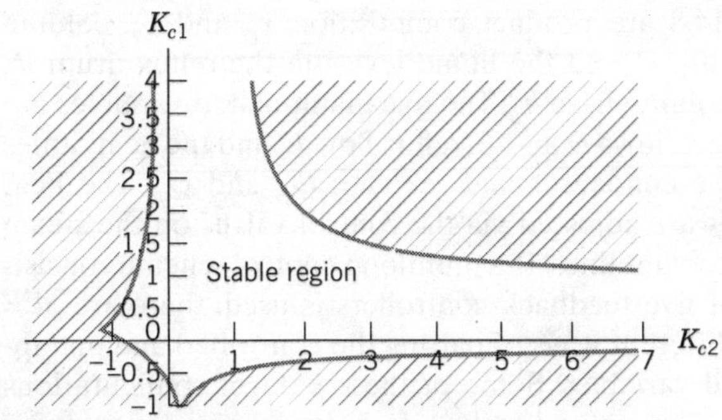

Figure 14.7 Stability region for Example 14.2 with 1-2/2-1 controller pairing.

A similar stability analysis can be performed for the 1-2/2-1 control configuration (see Fig. 14.7). A comparison of Figs. 14.6 and 14.7 indicates that the 1-2/2-1 control scheme results in a larger stability region because a wider range of controller gains can be used. For example, suppose that $K_{c1} = 2$. Then Fig. 14.6 indicates that the 1-1/2-2 configuration will be stable if $-0.8 < K_{c2} < 0.5$. By contrast, Fig. 14.7 shows that the corresponding stability limits for the 1-2/2-1 configuration are $-0.3 < K_{c2} < 2.0$.

This example illustrates that closed-loop stability depends on the control configuration as well as the numerical values of the controller settings. If PI control had been considered instead of proportional-only control, the stability analysis would have been much more complicated due to the larger number of controller settings and the higher-order characteristic equation.

14.2 PAIRING OF CONTROLLED AND MANIPULATED VARIABLES

In this section, we consider the important general problem of how the controlled variables and the manipulated variables should be paired in a multiloop control scheme. An incorrect pairing can result in poor control system performance and reduced stability margins, as was the case for the 1-1/2-2 pairing in Example 14.2. As an illustrative example, consider the distillation column shown in Fig. 14.8. A typical distillation column has five possible controlled variables and five manipulated variables (Shinskey, 1996). The controlled variables in

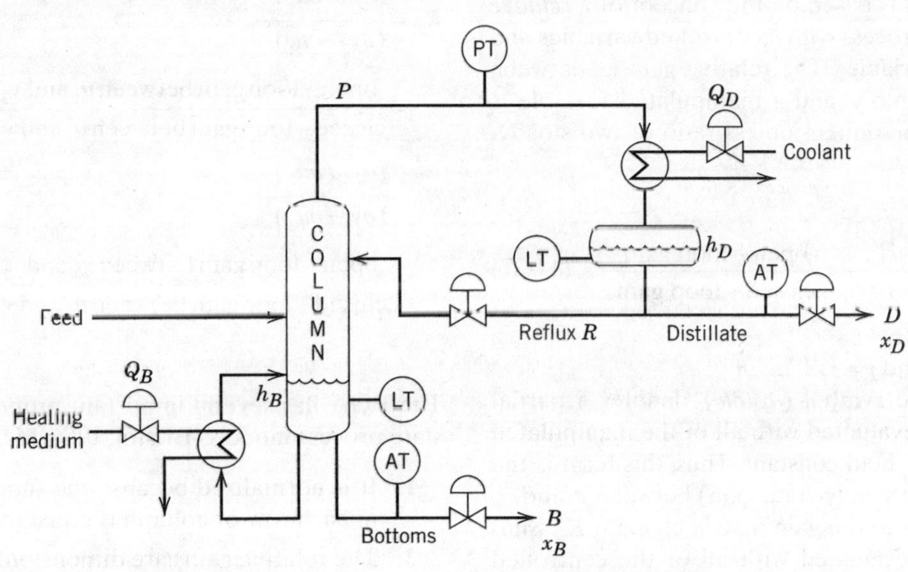

Figure 14.8 Controlled and manipulated variables for a typical distillation column.

Fig. 14.8 are product composition x_D and x_B, column pressure P, and the liquid levels in the reflux drum h_D and column base h_B. The five manipulated variables are product flows D and B, reflux flow R, and the heat duties for the condenser and reboiler, Q_D and Q_B. The heat duties are adjusted via the control valves on the steam and coolant lines. If a multiloop control scheme consisting of five feedback controllers is used, there are 5! = 120 different ways of pairing the controlled and manipulated variables. Some of these control configurations would be immediately rejected as being impractical or unworkable, for example, any scheme that attempts to control base level h_B by adjusting distillate flow D or condenser heat duty Q_D. However, there may be a number of alternative pairings that seem promising; the question then facing the control system designer is how to determine the most effective pairing.

Next, we consider a systematic approach for determining the best pairing of controlled and manipulated variables, the *relative gain array* method. An alternative approach based on singular value analysis is described later in this chapter.

14.2.1 Bristol's Relative Gain Array Method

Bristol (1966) developed a systematic approach for the analysis of multivariable process control problems. His approach requires only steady-state information (the process gain matrix $\boldsymbol{K}$) and provides two important items of information:

1. A measure of process interactions.
2. A recommendation concerning the most effective pairing of controlled and manipulated variables.

Bristol's approach is based on the concept of a *relative gain*. Consider a process with n controlled variables and n manipulated variables. The relative gain λ_{ij} between a controlled variable y_i and a manipulated variable u_j is defined to be the dimensionless ratio of two steady-state gains:

$$\lambda_{ij} \triangleq \frac{\left(\partial y_i / \partial u_j\right)_{u_{k,k\neq j}}}{\left(\partial y_i / \partial u_j\right)_{y_{k,k\neq i}}} = \frac{\text{open-loop gain}}{\text{closed-loop gain}} \qquad (14\text{-}27)$$

for $i = 1, 2, \dots, n$ and $j = 1, 2, \dots, n$.

In Eq. 14-27, the symbol $(\partial y_i/\partial u_j)_u$ denotes a partial derivative that is evaluated with all of the manipulated variables except u_j held constant. Thus, this term is the *open-loop gain* (or steady-state gain) between y_i and u_j, which corresponds to the gain matrix element K_{ij}. Similarly, $(\partial y_i/\partial u_j)_y$ is evaluated with all of the controlled variables except y_i held constant. This situation could be achieved in practice by adjusting the other manipulated variables using controllers with integral action. Thus, $(\partial y_i/\partial u_j)_y$ can be interpreted as a *closed*-loop gain that indicates the effect of u_j on y_i when all of the other controlled variables ($y_i \neq y_j$) are held constant.

It is convenient to arrange the relative gains in a *relative gain array* (RGA), denoted by $\boldsymbol{\Lambda}$:

$$\lambda = \begin{array}{c} \\ y_1 \\ y_2 \\ \dots \\ y_n \end{array} \begin{array}{c} \begin{array}{cccc} u_1 & u_2 & \dots & u_n \end{array} \\ \begin{bmatrix} \lambda_{11} & \lambda_{12} & \dots & \lambda_{1n} \\ \lambda_{21} & \lambda_{22} & \dots & \lambda_{2n} \\ \dots & \dots & \dots & \dots \\ \lambda_{n1} & \lambda_{n2} & \dots & \lambda_{nn} \end{bmatrix} \end{array} \qquad (14\text{-}28)$$

Equation 14-28 for a 2 × 2 system can be written as

$$\Lambda = \begin{vmatrix} \lambda_{11} & \lambda_{12} \\ \lambda_{21} & \lambda_{22} \end{vmatrix} \qquad (14\text{-}29)$$

where

$$\left.\begin{aligned}
\lambda_{11} &= \frac{(\partial y_1/\partial u_1)_{u_2}}{(\partial y_1/\partial u_1)_{y_2}} \\
&= \frac{\text{open-loop gain between } u_1 \text{ and } y_1 \text{ (constant } u_2)}{\text{closed-loop gain between } u_1 \text{ and } y_1 \text{ (constant } y_2)} \\
\lambda_{12} &= \frac{(\partial y_1/\partial u_2)_{u_1}}{(\partial y_1/\partial u_2)_{y_2}} \\
&= \frac{\text{open-loop gain between } u_2 \text{ and } y_1 \text{ (constant } u_1)}{\text{closed-loop gain between } u_2 \text{ and } y_1 \text{ (constant } y_2)} \\
\lambda_{21} &= \frac{(\partial y_2/\partial u_1)_{u_2}}{(\partial y_2/\partial u_1)_{y_1}} \\
&= \frac{\text{open-loop gain between } u_1 \text{ and } y_2 \text{ (constant } u_1)}{\text{closed-loop gain between } u_1 \text{ and } y_2 \text{ (constant } y_1)} \\
\lambda_{22} &= \frac{(\partial y_2/\partial u_2)_{u_1}}{(\partial y_2/\partial u_2)_{y_1}} \\
&= \frac{\text{open-loop gain between } u_2 \text{ and } y_2 \text{ (constant } u_1)}{\text{closed-loop gain between } u_2 \text{ and } y_2 \text{ (constant } y_1)}
\end{aligned}\right\} \qquad (14\text{-}30)$$

The RGA has several important properties for steady-state process models (Bristol, 1966; McAvoy, 1983):

1. It is normalized because the sum of the elements in each row or column is equal to one.
2. The relative gains are dimensionless and thus not affected by choice of units or scaling of variables.

3. The RGA is a measure of sensitivity to element uncertainty in the gain matrix $\boldsymbol{K}$. The gain matrix can become singular if a single element K_{ij} is changed to $K_{ij}(1 - 1/\lambda_{ij})$. Thus a large RGA element indicates that small changes in K_{ij} can markedly change the process control characteristics.

14.2.2 Calculation of the RGA

The relative gains can easily be calculated from either steady-state data or a process model. For example, consider a 2×2 process for which a steady-state model is available. Suppose that the model has been linearized and expressed in terms of deviation variables as follows:

$$y_1 = K_{11}u_1 + K_{12}u_2 \tag{14-31}$$

$$y_2 = K_{21}u_1 + K_{22}u_2 \tag{14-32}$$

where K_{ij} denotes the steady-state gain between y_i and u_j. This model can be expressed more compactly in matrix notation as

$$\boldsymbol{y} = \boldsymbol{K}\boldsymbol{u} \tag{14-33}$$

For stable processes, the steady-state (gain) model in Eq. 14-33 is related to the dynamic model in Eq. 14-4 by

$$\boldsymbol{K} = \boldsymbol{G}_p(0) = \lim_{s \to 0} \boldsymbol{G}_p(s) \tag{14-34}$$

Next, we consider how to calculate λ_{11}. It follows from Eq. 14-31 that

$$\left(\frac{\partial y_1}{\partial u_1}\right)_{u_2} = K_{11} \tag{14-35}$$

Before calculating $(\partial y_1 / \partial u_1)_{y_2}$ from Eq. 14-31, we first must eliminate u_2. This is done by solving Eq. 14-32 for u_2 and holding y_2 constant at its nominal set point value, $y_2 = 0$:

$$u_2 = \frac{K_{21}}{K_{22}} u_1 \tag{14-36}$$

Then substituting into Eq. 14-31 gives

$$y_1 = K_{11}\left(1 - \frac{K_{12}K_{11}}{K_{11}K_{22}}\right)u_1 \tag{14-37}$$

It follows that

$$\left(\frac{\partial y_1}{\partial u_1}\right)_{y_2} = K_{11}\left(1 - \frac{K_{12}K_{21}}{K_{11}K_{22}}\right) \tag{14-38}$$

Substituting Eqs. 14-35 and 14-38 into Eq. 14-27 gives an expression for relative gain λ_{11}:

$$\lambda_{11} = \frac{1}{1 - \dfrac{K_{12}K_{21}}{K_{11}K_{22}}} \tag{14-39}$$

Other relative gain values can be computed in the similar manner. The final relative gain array for a 2×2 system is:

$$\Lambda = \begin{bmatrix} \lambda_{11} & \lambda_{12} \\ \lambda_{21} & \lambda_{22} \end{bmatrix} = \begin{bmatrix} \dfrac{K_{11}K_{22}}{K_{11}K_{22} - K_{12}K_{21}} & \dfrac{-K_{12}K_{21}}{K_{11}K_{22} - K_{12}K_{21}} \\ \dfrac{-K_{21}K_{12}}{K_{11}K_{22} - K_{12}K_{21}} & \dfrac{K_{11}K_{22}}{K_{11}K_{22} - K_{12}K_{21}} \end{bmatrix} \tag{14-40}$$

$$\lambda_{12} = \lambda_{21} = 1 - \lambda_{11} \text{ and } \lambda_{22} = \lambda_{11} \tag{14-41}$$

It can be verified that the sum of relative gain values across any row or column equals to unity. This property also holds good for $N \times N$ system.

Thus, the RGA for a 2×2 system can be expressed as

$$\Lambda = \begin{bmatrix} \lambda & 1-\lambda \\ 1-\lambda & \lambda \end{bmatrix}$$

where the symbol λ is now used to denote λ_{11}. Note that the RGA for a 2×2 process is always symmetric. However, this will not necessarily be the case for a higher-dimension process ($n > 2$).

For higher-dimension processes, the RGA can be calculated from the expression

$$\Lambda = \boldsymbol{K} \otimes \mathbf{H} \tag{14-42}$$

where $\otimes$ denotes the Schur product (element by element multiplication):

$$\lambda_{ij} = K_{ij}H_{ij} \tag{14-43}$$

K_{ij} is the (i, j) element of $\boldsymbol{K}$ in Eq. 14-33, and H_{ij} is the (i, j) element of $\boldsymbol{H} = (\boldsymbol{K}^{b1})^T$; that is, H_{ij} is an element of the transpose of the matrix inverse of $\boldsymbol{K}$. Because computer software is readily available to perform matrix algebra, Eq. 14-43 can be easily evaluated. Note that Eq. 14-42 does not imply that $\Lambda = \boldsymbol{K}(\boldsymbol{K}^{b1})^T$.

14.2.3 Methods for Obtaining the Steady-State Gain Matrix

Equation 14-43 shows how the RGA can be calculated from a linearized steady-state process model with gains K_{ij}. The open-loop process gains can be obtained numerically from a simulation model or directly from experimental data. For a multivariable process, one input (u_1) can be changed in a stepwise fashion (Δu_1) while holding all other inputs constant. The responses for y_1, y_2, and so on are then observed. All loops are kept open during this test; that is, no feedback controllers are operational. Then step changes can be made in the other inputs, one at a time, and open-loop response data can be obtained for all the controlled variables. The steady-state gain depends only on the final value of each y_i, from which the change in y, Δy_i, can be calculated. Thus, the individual process gains are given by the formula:

$$K_{ij} = \frac{\Delta y_i}{\Delta u_i} \tag{14-44}$$

For example, K_{11} can be evaluated as $K_{11} = \Delta y_1/\Delta u_1$. This approach can be used whether the gains are obtained from a mathematical model (simulator) or from actual process data. Of course, in the latter case, usually more effort and cost are required to obtain the necessary information; hence, it is advantageous to use a model when one is available.

For experimental determination of K_{ij}, it is also desirable to perform several step tests for the same input, using different magnitudes and directions for the input change and then average the results. When using a simulator, it is easier to control the conditions of the simulated step change, and the results are less prone to error. However, it is not mandatory to have a dynamic simulator in order to perform the gain calculation. Alternatively, a steady-state simulator can be used by starting from a base case and then changing a single input to two new values, one higher ($+\Delta u_j$) and one lower ($-\Delta u_j$) than the base case, and then finding the corresponding changes in y for each input change. The perturbations in u_j should be chosen carefully so that the calculated gains (and hence the RGA) do not change significantly with the size of the perturbation (as it is increased or decreased) so that the steady-state model is in a linear region. See McAvoy (1983) for more details.

14.2.4 Measure of Process Interactions and Pairing Recommendations

Equation 14-37 can be used to interpret further the relative gain array of a 2×2 process. When y_2 is kept at its set point of zero under closed-loop control, rearranging Eq. 14-37 gives

$$\frac{y_1}{u_1} = K_{11} - \frac{K_{12}K_{21}}{K_{22}} = K_{11}\left[1 - \frac{K_{12}K_{21}}{K_{11}K_{22}}\right] \tag{14-45}$$

The term, $-K_{12}K_{21}/K_{11}K_{22}$, can be thought of as an interaction term that modifies the open-loop process gain K_{11} due to interaction from the other loop (G_{c2}). This effect can be positive or negative, depending on the ratio of the gains. Using Eq. 14-39 as the definition of $\lambda_{11} = \lambda$, we can then write

$$\frac{y_1}{u_1} = K_{11}\left[\frac{1}{\lambda}\right] = K_{11}\left[1 + \frac{1-\lambda}{\lambda}\right] \tag{14-46}$$

Thus λ can be interpreted as a divisor of the open-loop gain K_{11}, or the term $(1 - \lambda)/\lambda$ is a correction to K_{11}. Another way to view Eq. 14-46 is that closed-loop gain = open-loop gain/λ, which is a restatement of the definition in Eq. 14-27. For example, if $\lambda = 1$, there is no correction to the gain, and the open-loop gain is the same as the closed-loop gain. If λ is a large positive value, the closed-loop gain is much smaller than the open-loop gain. The practical implication of a large value of λ is that u_1 no longer has much influence on y_1, which could have important operational implications. Finally, if λ is negative, the closed-loop gain changes sign from K_{11}, which indicates serious difficulties in controller design if the first loop is closed.

RGA analysis of cases larger than 2×2 may not lead to clear conclusions, especially if one is evaluating an operating process that already has some combination of open and closed loops. It is important to understand in a multivariable process how other control loops affect the process gain of a given open loop. Sometimes it is not obvious that the pathological behavior of a particular input–output combination may be due to other controllers. Further troubleshooting analysis may lead to the conclusion that a different control strategy is required before a critical output variable can be successfully controlled (see Appendices I and J for a discussion of plantwide control).

For the 2×2 process in Eqs. 14-31 and 14-32, five pairing cases can be considered (recall that the single parameter λ defines the entire matrix $\mathbf{\Lambda}$ because the sum of the RGA elements is unity for each row and for each column):

1. **$\lambda = 1$.** In this situation, it follows from Eq. 14-27 that the open-loop and closed-loop gains between y_1 and u_1 are identical. In this ideal situation, opening or closing loop 2 has no effect on loop 1. It follows that y_1 should be paired with u_1 (i.e., a 1-1/2-2 configuration should be employed).

2. **$\lambda = 0$.** Equation 14-27 indicates that the open-loop gain between y_1 and u_1 is zero, and thus u_1 has no direct effect on y_1. Consequently, u_1 should be paired with y_2 rather than y_1 (i.e., the 1-2/2-1 configuration should be utilized).
3. **$0 < \lambda < 1$.** From Eq. 14-27, the closed-loop gain between y_1 and u_1 is larger than the open-loop gain. Within this range, the interaction between the two loops is largest when $\lambda = 0.5$, which indicates that the second term in Eq. 14-38 is equal to -1.
4. **$\lambda > 1$.** For this situation, closing the second loop reduces the gain between y_1 and u_1. Eq. 14-27 indicates that $(\partial y_1/\partial u_1)_{u_2} > (\partial y_1/\partial u_1)_{y_2}$. In other words, open-loop gain between y_1 and u_1 is greater than the closed-loop gain. Thus, the control loops interact. As λ increases, the degree of interaction increases and becomes most severe as $\lambda \to \infty$. When λ is very large, it is impossible to control both outputs independently (Skogestad and Postlethwaite, 2005).
5. **$\lambda < 0$.** When λ is negative, the open-loop and closed-loop gains between y_1 and u_1 have different signs. If the controller 1 gain is positive when loop 2 is open, then on closing loop 2, the controller gain has to be negative. Therefore, the controller gain sign has to be changed depending on the condition whether loop 2 is closed or open. Thus, opening or closing loop 2 has an adverse effect on the behavior of loop 1 such as oscillation. It follows that y_1 should not be paired with u_1. For $\lambda < 0$ the control loops interact by trying to "fight each other," and the closed-loop system may become unstable.

Based on these considerations, the RGA analysis for a 2×2 process leads to the conclusion that y_1 should be paired with u_1 only if $\lambda \geq 0.5$. Otherwise, y_1 should be paired with u_2, the reverse pairing. This reasoning can be extended to $n \times n$ processes and leads to Bristol's original recommendation for controller pairing:

Recommendation: *Pair the controlled and manipulated variables so that corresponding relative gains are positive and as close to one as possible.*

At this point, it is appropriate to make several remarks about the RGA approach.

1. The above recommendation is based solely on steady-state information. However, dynamic behavior should also be considered in choosing a controller pairing. In particular, closed-loop stability should be checked using a theorem that is presented in the next section.
2. If $\lambda = 0$ or $\lambda = 1$, the two control loops for a 2×2 process either do not interact at all or exhibit only a one-way interaction, based on this steady-state analysis. Furthermore, at least one of the four process gains must be zero, according to Eq. 14-39.
3. If a pairing of inputs and outputs in a 2×2 process corresponds to a negative relative gain, then the closed-loop system will exhibit instability either in the overall closed-loop system or in the loop with the negative relative gain by itself.
4. If the gain matrix of Eq. 14-34 has odd number of positive elements then all relative gains will lie between 0 and 1. On the other hand, if the number of positive elements in the gain matrix is even, then all relative gain values will lie outside the range 0~1 i.e., some values will be negative and other values will be greater than 1.

One property of interest in control loop design is called *loop integrity*. Skogestad and Postlethwaite (2005) have considered the case when the RGA element is negative for a given pairing, the process is stable, and the feedback controller contains integral action. If the control loop with the negative pairing is disabled owing to failure (or being taken out of service) or because of saturation of the controller output, the multiloop control system will become unstable. A related property is called *decentralized integral controllability* (DIC). When each loop contains integral action, the property of DIC means that the gain of each controller can be reduced to zero without the closed-loop system becoming unstable. Campo and Morari (1994) have developed conditions of λ_{ij} that ensure DIC; for the 2×2 case, for example, $\lambda_{11} > 0$ guarantees DIC.

A one-way interaction occurs when one loop affects the other loop but not vice versa. Suppose that $\boldsymbol{K}$ has the structure

$$\boldsymbol{K} = \begin{bmatrix} K_{11} & K_{12} \\ 0 & K_{22} \end{bmatrix}$$

Then loop 1 does not affect loop 2, because $K_{21} = 0$, and thus u_1 has no effect on y_2. However, loop 2 does affect loop 1 via u_2 if $K_{12} \neq 0$. This one-way interaction does not affect the closed-loop stability because $K_{21} = 0$; consequently, the characteristic equation in Eq. 14-23 reduces to the two equations in Eq. 14-25. Thus, for this one way interaction, loop 2 tends to act as a source of disturbances for loop 1.

To illustrate how the RGA can be used to determine controller pairing, we next consider several examples. Additional examples have been presented by McAvoy (1983).

EXAMPLE 14.3

Consider the in-line blending system of Fig. 14.1*a*. It is proposed that w and x be controlled using a conventional multiloop control scheme, with w_A and w_B as the manipulated variables. Derive an expression for the RGA and recommend the best controller pairing for the following conditions: $w = 4$ kg/min and $x = 0.4$.

Solution

Assuming perfect mixing, a process model can be derived from the following steady-state mass balances:

$$\text{Total mass:} \quad w = w_A + w_B \tag{14-47}$$

$$\text{Component A:} \quad xw = w_A \tag{14-48}$$

Substituting Eq. 14-47 into Eq. 14-48 and rearranging gives

$$x = \frac{w_A}{w_A + w_B} \tag{14-49}$$

The RGA for the blending system can be expressed as

$$\Lambda = \begin{array}{c} \\ w \\ x \end{array} \!\!\begin{array}{c} \begin{array}{cc} w_A & w_B \end{array} \\ \begin{bmatrix} \lambda & 1-\lambda \\ 1-\lambda & \lambda \end{bmatrix} \end{array}$$

Relative gain λ can be calculated from Eq. 14-39 after the four steady-state gains are calculated:

$$K_{11} = \left(\frac{\partial w}{\partial w_A}\right)_{w_B} = 1 \tag{14-50a}$$

$$K_{12} = \left(\frac{\partial w}{\partial w_B}\right)_{w_A} = 1 \tag{14-50b}$$

$$K_{21} = \left(\frac{\partial x}{\partial w_A}\right)_{w_B} = \frac{w_B}{(w_A + w_B)^2} = \frac{1-x}{w} \tag{14-50c}$$

$$K_{22} = \left(\frac{\partial x}{\partial w_B}\right)_{w_A} = \frac{-w_A}{(w_A + w_B)^2} = -\frac{x}{w} \tag{14-50d}$$

Substituting into Eq. 14-39 gives $\lambda = x$. Thus, the RGA is

$$\Lambda = \begin{array}{c} \\ w \\ x \end{array} \!\!\begin{array}{c} \begin{array}{cc} w_A & w_B \end{array} \\ \begin{bmatrix} x & 1-x \\ 1-x & x \end{bmatrix} \end{array}$$

Note that the recommended pairing depends on the desired product composition x. For $x = 0.4$, w should be paired with w_B and x with w_A. Because all four relative gains are close to 0.5, control loop interactions will be a serious problem. On the other hand, if $x = 0.9$, w should be paired with w_A and x with w_B. In this case, the control loop interactions will be small. Note that for both cases, total flow rate w is controlled by the larger component flow rate, w_A or w_B.

EXAMPLE 14.4

The relative gain array for a refinery distillation column associated with a hydrocracker discussed by Nisenfeld and Schultz (1971) is given by

$$\Lambda = \begin{array}{c} \\ y_1 \\ y_2 \\ y_3 \\ y_4 \end{array} \!\!\begin{array}{c} \begin{array}{cccc} u_1 & u_2 & u_3 & u_4 \end{array} \\ \begin{bmatrix} 0.931 & 0.150 & 0.080 & -0.164 \\ -0.011 & -0.429 & 0.286 & 1.154 \\ -0.135 & 3.314 & -0.270 & -1.910 \\ 0.215 & -2.030 & 0.900 & 1.919 \end{bmatrix} \end{array} \tag{14-51}$$

The four controlled variables are the compositions of the top and bottom product streams (y_1, y_2) and the two side streams (y_3, y_4). The manipulated variables are the four flow rates numbered from the top of the column; for example, the top flow rate is u_1. Find the recommended pairing using the RGA.

Solution

To determine the recommended controller pairs, we identify the positive relative gains that are closest to one in each row and column. From the rows, it is apparent that the recommended pairings are y_1-u_1, y_2-u_4, y_3-u_2, and y_4-u_3. Note that this pairing assigns u_2 to y_3 rather than to y_1, even though its relative gain of 3.314 is farther from one. This choice is required because pairing any other manipulated variable with y_3 corresponds to a negative relative gain, which is undesirable. The y_1-u_1 and y_4-u_3 relative gains are close to one, so these selections are straightforward.

The two previous examples have shown how the RGA can be calculated from steady-state gain information. For integrating processes such as the liquid storage system considered in Section 5.5, one or more steady-state gains do not exist. Consequently, the standard RGA analysis must be modified for such systems (Arkun and Downs, 1990). The RGA analysis proceeds in the usual manner, except that any controlled variable that is the output of an integrating element should be replaced by its rate of change. Thus, if a liquid level h is both a controlled variable and the output of an integrating element, then h will be replaced by dh/dt in the RGA analysis.

Useful information about the stability of a proposed multiloop control system can be obtained using a theorem originally reported by Niederlinski (1971) and later corrected by Grosdidier and Morari (1986). Like the RGA analysis, the theorem is based solely on steady-state information. It is assumed that the steady-state gain matrix $\boldsymbol{K}$ has been arranged so that the diagonal elements correspond to the proposed pairing; that is, it is assumed that y_1 is paired with u_1, y_2 and u_2, and so on. This arrangement can always be obtained by reordering the elements of the $\boldsymbol{y}$ and $\boldsymbol{u}$ vectors if necessary.

The following theorem is based on three assumptions:

1. Let $G_{pij}(s)$ denote the (i, j) element or process transfer function matrix, $\boldsymbol{G}_p(s)$. Each $G_{pij}(s)$ must be stable, rational, and proper; that is, the order of the denominator must be at least as great as the order of the numerator.
2. Each of the n feedback controllers in the multiloop control system contains integral action.
3. Each individual control loop is stable when any of the other $n - 1$ loops are opened.

Stability Theorem. Suppose that a multiloop control system is used with the pairing $y_1 - u_1, y_2 - u_2, \dots, y_n - u_n$. If the closed-loop system satisfies Assumptions 1–3, then the closed-loop system is unstable if

$$\frac{|\boldsymbol{K}|}{\prod_{i=1}^{n} K_{ii}} < 0 \qquad (14\text{-}52)$$

where $|\boldsymbol{K}|$ denotes the determinant of $\boldsymbol{K}$.

The term given in Eq. 14-52 is known as Niederlinski Index (NI). After pairing of manipulated and controlled variables is decided based on RGA analysis, NI is computed to evaluate the stability of the MIMO system. Note that this theorem provides a sufficient (but not necessary) condition for instability. Thus, if the inequality is satisfied, the closed-loop system will be unstable. However, if the inequality is not satisfied, the closed-loop system may or may not be unstable, depending on the numerical values of the controller settings. The inequality is also satisfied if the proposed pairing for a 2×2 system corresponds to a negative value of a relative gain. McAvoy (1983, p. 84) reports several examples where apparently reasonable RGA pairings result in unstable closed-loop systems. Thus, it is important to consider the process dynamics and also check to ensure that a proposed pairing does not satisfy the inequality in Eq. 14-52.

Assumption 1 requires that each $G_{pij}(s)$ be a rational function; hence, the theorem does not strictly apply to processes that contain time delays. Because time delays do not affect the steady-state matrix $\boldsymbol{K}$, the theorem still provides useful insight into the stability of such systems, even though the analysis is no longer rigorous.

Processes with poorly conditioned $\boldsymbol{K}$ matrices tend to require large changes in the manipulated variables in order to influence the controlled variables. This assertion can be justified as follows. Solving Eq. 14-33 for $\boldsymbol{u}$,

$$\boldsymbol{u} = \boldsymbol{K}^{-1}\boldsymbol{y} \qquad (14\text{-}53)$$

and substituting set point $\boldsymbol{y}_{sp}$ for $\boldsymbol{y}$ gives

$$\boldsymbol{u} = \boldsymbol{K}^{-1}\boldsymbol{y}_{sp} \qquad (14\text{-}54)$$

The inverse of $\boldsymbol{K}$ in Eq. 14-54 can be calculated from the standard formula,

$$\boldsymbol{K}^{-1} = \frac{\text{adjoint of } \boldsymbol{K}}{|\boldsymbol{K}|} \qquad (14\text{-}55)$$

The adjoint of $\boldsymbol{K}$ is formed from its cofactors (Strang, 2005).

If $|\boldsymbol{K}|$ is small ($\ll 1$), we conclude from Eqs. 14-54 and 14-55 that the required adjustments in $\boldsymbol{u}$ will be very large, resulting in excessive control actions. Small values of $|\boldsymbol{K}|$ also lead to large values of the relative gain array (cf. Section 14.2). For a 2 × 2 process, the relative gain array is characterized by a single parameter λ. The following expression for λ can be obtained by rearranging Eq. 14-39:

$$\lambda = \frac{K_{11}K_{22}}{K_{11}K_{22} - K_{12}K_{21}} = \frac{K_{11}K_{22}}{|\boldsymbol{K}|} \tag{14-56}$$

Thus, if $|\boldsymbol{K}|$ is small, λ becomes very large, and process interactions are extremely strong, leading to control difficulties.

14.2.5 Dynamic Considerations

An important disadvantage of the standard RGA approach is that it ignores process dynamics, which can be an important factor in the pairing decision. For example, if the transfer function between y_1 and u_1 contains a very large time delay or time constant (relative to the other transfer functions), y_1 will respond very slowly to changes in u_1. Thus, in this situation, a y_1-u_1 pairing is not desirable from a dynamic perspective (see Example 14.2 where the RGA favors the y_1-u_1 pairing). McAvoy (1983, p. 214) has noted that dynamic interactions tend to be more important for 2 × 2 processes when $\lambda > 1$ than when $0 < \lambda < 1$. However, dynamic considerations can still affect the pairing decision even when $0 < \lambda < 1$, as illustrated in the following example.

EXAMPLE 14.5

Consider the transfer function model of Example 14.2 but with a gain of −2 in $G_{p11}(s)$ and a time delay of unity in each transfer function:

$$\boldsymbol{G}_p(s) = \begin{bmatrix} \dfrac{-2e^{-s}}{10s+1} & \dfrac{1.5e^{-s}}{s+1} \\ \dfrac{1.5e^{-s}}{s+1} & \dfrac{2e^{-s}}{10s+1} \end{bmatrix} \tag{14-57}$$

Use the RGA approach to determine the recommended controller pairing based on steady-state considerations. Do dynamic considerations suggest the same pairing?

Solution

The corresponding steady-state gain matrix is

$$\boldsymbol{K} = \begin{bmatrix} -2 & 1.5 \\ 1.5 & 2 \end{bmatrix} \tag{14-58}$$

Using the formula in Eq. 14-39, we obtain $\lambda_{11} = 0.64$. Thus, the RGA analysis indicates that the 1-1/2-2 pairing should be used. However, the off-diagonal time constants in Eq. 14-57 are only one-tenth of the diagonal time constants. Thus, y_1 responds 10 times faster to u_2 than to u_1; similarly, y_2 responds 10 times faster to u_1 than to u_2. Consequently, the 1-2/2-1 pairing is favored based on dynamic considerations, and a conflict exists between steady-state and dynamic considerations that should be resolved by dynamic simulation studies.

Evaluation of RGA using MATLAB

```
>> K = [-2 1.5; 1.5 2]; % Creation of
   steady state gain matrix by apply-
   ing Eq. 14-34
>> H = (inv(K))'; % Computation of H
   matrix
>> lambda = K.*H; % Computation of RGA
   by element by element multiplication
   as given in equation 14-42
>> lambda
   lambda =
   0.6400        0.3600
   0.3600        0.6400
```

Extensions of the RGA to include process dynamics are discussed next.

14.2.6 Extensions of the RGA Analysis

Several researchers have suggested interaction measures that consider the process dynamics or frequency response as well as steady-state gains (Zhu et al., 1997; Skogestad and Postlethwaite, 2005). Although these newer methods are more complicated than the standard RGA approach, they offer additional insights concerning the closed-loop behavior of the system. The frequency domain interpretation of the RGA indicates how dynamics should be considered in the pairing of inputs and outputs. The frequency-dependent RGA, analogous to Eq. 14-42, is the Schur product (each variable is a function of $s = j\omega$):

$$\Lambda = G \otimes (G^{-1})^T$$
$$\Lambda(j\omega) = G(j\omega) \otimes (G^{-1}(j\omega))^T \qquad (14\text{-}59)$$

Skogestad and Postlethwaite (2005) recommend pairings for which the relative gains at the gain crossover and critical frequencies are close to one (see Appendix L). For this analysis, the input and output variables are reordered based on the recommended pairing, which yields an RGA that is diagonally dominant (diagonal terms have larger magnitudes than off-diagonal terms) and close in magnitude to the identity matrix (Grosdidier and Morari, 1986). Plants with large RGA elements around the critical frequency are inherently difficult to control because of sensitivity to errors in the model parameters or model mismatch, which makes design approaches such as decoupling unattractive (see Section 14.5).

Other papers have extended the RGA approach to consider the effect of model uncertainty and disturbances on multiloop control systems (Chen and Seborg, 2002). The *relative disturbance gain* (RDG) provides a measure of the change in the effect of a given disturbance caused by multiloop (decentralized) control. For the 2 × 2 case, the steady state RDG is

$$\beta_1 = \lambda_{11}\left(1 - \frac{K_{d2}K_{12}}{K_{d1}K_{22}}\right) \qquad (14\text{-}60)$$

where K_{di} is the gain of the disturbance variable d_i on y_i and β_1 is a dimensionless parameter. It is desirable to keep β_1 small, because small values indicate that the loop interactions actually reduce the effect of the disturbance. Skogestad and Postlethwaite (2005) have discussed the frequency dependence of RDG.

14.3 SINGULAR VALUE ANALYSIS

Singular value analysis (SVA) is a powerful analytical technique that can be used to solve several important control problems:

1. Selection of controlled, measured, and manipulated variables
2. Evaluation of the robustness of a proposed control strategy
3. Determination of the best multiloop control configuration

Singular value analysis and extensions such as singular value decomposition (SVD) also have many uses in numerical analysis and the design of multivariable control systems, which is beyond the scope of this book (Bjorck, 1996; Skogestad and Postlethwaite, 2005). In this section, we provide a brief introduction to SVA that is based on an analysis of steady-state gains from the process models.

Again, we consider the linear steady-state process model in Eq. 14-33.

$$y = Ku \qquad (14\text{-}61)$$

One desirable property of K is that the n linear equations in n unknowns represented by Eq. 14-61 be linearly independent. In contrast, if the equations are dependent, then not all of the n controlled variables can be independently regulated. This characteristic property of linear independence can be checked by several methods (Bjorck, 1996). For example, if the determinant of K is zero, the matrix is singular and the n equations in Eq. 14-61 are not linearly independent.

Another way to check for linear independence is to calculate one of the most important properties of a matrix: its eigenvalues. The eigenvalues of matrix K are the roots of the equation

$$|K - \alpha I| = 0 \qquad (14\text{-}62)$$

where $|K - \alpha I|$ denotes the determinant of matrix $K - \alpha I$, and I is the $n \times n$ identity matrix. The n eigenvalues of K will be denoted by $\alpha_1, \alpha_2, \ldots, \alpha_n$. If any of the eigenvalues are zero, K is a singular matrix, and difficulties will be encountered in controlling the process, as noted above. If one eigenvalue is very small compared to the others, then very large changes in one or more manipulated variables will be required to control the process, as will be shown at the end of this section.

Another important property of K is its *singular values*, $\sigma_1, \sigma_2, \ldots, \sigma_n$ (Roat et al., 1986; Bjorck, 1996). The singular values are nonnegative numbers that are defined as the positive square roots of the eigenvalues of the matrix product K^TK. The first r singular values are positive numbers, where r is the rank of K^TK. The remaining $n - r$ singular values are zero. Usually, the nonzero singular values are ordered, with σ_1 denoting the largest and σ_r the smallest.

The singular values arise from the decomposition of K (Laub, 2004):

$$K = W\Sigma V^T \qquad (14\text{-}63)$$

Where Σ is the diagonal matrix of singular values. W and V are unitary matrices such that

$$WW^T = I \qquad (14\text{-}64)$$

$$VV^T = I \qquad (14\text{-}65)$$

Note that for a unitary matrix, the transpose of W (or V) is also its inverse. The columns of W are referred to as the input singular vectors (and are orthonormal), and the columns of $\boldsymbol{V}$ are the output singular vectors (also orthonormal). $\boldsymbol{W}$, $\boldsymbol{V}$, and $\boldsymbol{\Sigma}$ can be easily calculated from computer software for matrix analysis.

The final matrix property of interest here is the *condition number* (CN). Assume that $\boldsymbol{K}$ is nonsingular. Then the condition number of $\boldsymbol{K}$ is a positive number defined as the ratio of the largest and smallest nonzero singular values:

$$CN = \frac{\sigma_1}{\sigma_r} \tag{14-66}$$

If $\boldsymbol{K}$ is singular, then it is ill-conditioned, and by convention, CN = ∞. The concept of a condition number can also be extended to nonsquare matrices (Bjorck, 1996).

One significant difference between the RGA and SVA is that the elements of the RGA are independent of scaling, whereas the singular values (and CN) depend on scaling or normalization of inputs and outputs. The usual SVA convention is to divide each u_i and y_i by its corresponding range. Thus, input u_i is scaled as

$$u_i^* = \frac{u_i}{u_i^{max} - u_i^{max}} \tag{14-67}$$

Where u_i is the scaled input. Skogestad and Postlethwaite (2005) discuss the notion of the *minimum condition number*, where all possible scalings are evaluated in order to find the minimum CN.

The condition number also provides useful information about the sensitivity of the matrix properties to variations in the elements of the matrices. This important topic, which is related to control system robustness, will be considered later in this section. But first we consider a simple example.

EXAMPLE 14.6

A 2 × 2 process has the following steady-state gain matrix:

$$\boldsymbol{K} = \begin{bmatrix} 1 & K_{12} \\ 10 & 1 \end{bmatrix} \tag{14-68}$$

Calculate the determinant, RGA, eigenvalues, and singular values of $\boldsymbol{K}$. Use $K_{12} = 0$ as the base case; then recalculate the matrix properties for a small change, $K_{12} = 0.1$.

Solution

By inspection, the determinant for the base case is $|\boldsymbol{K}| = 1$, and the RGA is $\boldsymbol{\Lambda} = \boldsymbol{I}$, so $\lambda_{11} = 1.0$ and pairing is straight-forward ($y_1 - u_1, y_2 - u_2$). The eigenvalues can be calculated as follows:

$$|\boldsymbol{K} - \alpha \boldsymbol{I}| = \begin{vmatrix} 1-\alpha & 0 \\ 10 & 1-\alpha \end{vmatrix} = 0 \tag{14-69}$$

Thus, $(1 - \alpha)^2 = 0$ and the eigenvalues are $\alpha_1 = \alpha_2 = 1$.

Now calculate the singular values, which arise from

$$\boldsymbol{K}^T\boldsymbol{K} = \begin{bmatrix} 1 & 10 \\ 0 & 1 \end{bmatrix}\begin{bmatrix} 1 & 0 \\ 10 & 1 \end{bmatrix} = \begin{bmatrix} 101 & 10 \\ 10 & 1 \end{bmatrix} \tag{14-70}$$

The eigenvalues of $\boldsymbol{K}^T\boldsymbol{K}$, denoted by α', can be calculated from $|\boldsymbol{K}^T\boldsymbol{K} - \alpha'\boldsymbol{I}| = 0$, which again yields a second-order polynomial:

$$(101 - \alpha')(1 - \alpha') - 100 = 0 \tag{14-71}$$

Solving Eq. 14-71 gives $\alpha_1' = 101.99$, and $\alpha_2' = 0.01$. The singular values of $\boldsymbol{K}$ are then

$$\sigma_1 = \sqrt{101.99} = 10.1 \tag{14-72}$$

$$\sigma_2 = \sqrt{0.01} = 0.1 \tag{14-73}$$

and the condition number is

$$CN = \frac{\sigma_1}{\sigma_2} = \frac{10.1}{0.1} = 101 \tag{14-74}$$

Thus, $\boldsymbol{K}$ is considered to be poorly conditioned because of the large CN value.

Now consider the case where $K_{12} = 0.1$, a small change from the base case. The determinant of $\boldsymbol{K}$ is zero, which indicates that $\boldsymbol{K}$ is singular and the RGA does not exist for this perturbation. The eigenvalues of $\boldsymbol{K}$ calculated from Eq. 14-69 are $\alpha_1 = 2$ and $\alpha_2 = 0$. The singular values of $\boldsymbol{K}$ are $\sigma_1 = 10.1$, $\sigma_2 = 0$, and the condition number is CN = ∞, because $\boldsymbol{K}$ is singular.

Condition number computation using MATLAB

```
>> K = [1 0; 10 1];
>>rga = K.*(inv(K))'

rga =

     1     0
     0     1
```

```
>>egn = eig(K) % calculating eigen
values of K

egn =

     1
     1

>>svsq = eig(K'*K) %eigen valu of
K'K

svsq =

    0.0098
  101.9902

>>sv = sqrt(svsq)

sv =

    0.0990
   10.0990

>> CN = sv(2)/sv(1)

CN =

  101.9902
```

This example shows that the original $\boldsymbol{K}$ matrix (with $K_{12} = 0$) is poorly conditioned and very sensitive to small variations in the K_{12} element. The large condition number (CN = 101) indicates the poor conditioning. In contrast, the value for the determinant ($|\boldsymbol{K}| = 1$) and the RGA give no indication of poor conditioning. The value of $\lambda = 1$ for $K_{12} = 0$ is quite misleading, because it suggests that the process model in Example 14.6 has no interactions and that a 1-1/2-2 controller pairing will be suitable. However, the large condition number of 101 for this case implies that the process is poorly conditioned and thus will be difficult to control with any controller pairing. The example demonstrates that the condition number is superior to the determinant in providing a more reliable measure of ill-conditioning and potential sensitivity problems.

14.3.1 Selection of Manipulated Variables and Controlled Variables

The SVA and RGA methods can be used as a way to screen subsets of the possible manipulated variables (MVs) and controlled variables (CVs) for a MIMO control system. Because these analyses are based on the steady-state gain matrix, it is recommended that promising combinations of MVs and CVs be identified and then investigated in more detail using simulation and dynamic analysis. The two steps shown below can be used to identify promising subsets of MVs and CVs, recognizing that for multiloop control the number of MVs should equal the number of CVs (a square system).

1. Arrange the singular values from largest to smallest ($\sigma_n, \sigma_{n-1}, \ldots, \sigma_1$); if $\sigma_i/\sigma_{i-1} > 10$ for some $i \geq 2$, then these singular values can be neglected, and at least one MV and one CV should be omitted, as discussed in step 2.
2. Generate alternative gain matrices by deleting one row and one column at a time and calculating the singular values and condition numbers. Elements of $\boldsymbol{W}$ and $\boldsymbol{V}$ can be used in some cases to guide the choice of which MV and CV should be removed (Skogestad, 1992). The most promising gain matrices have the smallest condition numbers. Then perform dynamic simulation to choose the best MV/CV set out of the remaining alternatives.

Skogestad and Postlethwaite (2005) have indicated that for nonsquare plants (more inputs than outputs or vice versa), the RGA can be used to eliminate some inputs or outputs. For this case, the RGA is also non-square, and elements in each row (or each column) do not necessarily sum to one. For more inputs than outputs, if all the elements in a column in the RGA are small ($\ll 1$), then the corresponding input can be deleted without much loss in performance. Similarly, for more outputs than inputs, if all elements in a row of the RGA are small, then that output cannot be controlled easily and other outputs should be selected. Chang and Yu (1990) have developed an RGA-based methodology for non square multivariable systems.

Roat et al. (1986) analyzed the choice of manipulated variables for a complex, four-component distillation column. The four components were propane, isobutane, n-butane, and isopentane. There were six possible manipulated variables, and ratios of these variables were also permissible. Table 14.2 shows the condition numbers for six control configurations that were evaluated for the column. Note that the last three strategies have approximately the same low CN. Subsequently, these three strategies were selected for further evaluation using dynamic simulation. Based on simulation results, the best control strategy in Table 14.2 was number 4.

Table 14.2 Condition Numbers for the Gain Matrices Relating Controlled Variables to Various Sets of Manipulated Variables for a Distillation Column (Roat et al., 1986)

Controlled Variables

x_D = Mole fraction of propane in distillate D

x_{64} = Mole fraction of isobutane in tray 64 sidedraw

x_{15} = Mole fraction of n-butane in tray 15 sidedraw

x_B = Mole fraction of isopentane in bottoms B

Possible Manipulated Variables

L = Reflux flow rate — B = Bottoms flow rate

D = Distillate flow rate — S_{64} = Sidedraw flow rate at tray 64

V = Steam flow rate — S_{15} = Sidedraw flow rate at tray 15

Strategy Number*	Manipulated Variables	Condition Number
1	$L/D, S_{64}, S_{15}, V$	9, 030
2	$V/L, S_{64}, S_{15}, V$	60, 100
3	$D/V, S_{64}, S_{15}, V$	116, 000
4	D, S_{64}, S_{15}, V	51.5
5	L, S_{64}, S_{15}, B	57.4
6	L, S_{64}, S_{15}, V	53.8

*In each control strategy, the first controlled variable is paired with the first manipulated variable, and so on. Thus, for Strategy 1, x_D is paired with L/D, and x_B is paired with V.

EXAMPLE 14.7

Determine the preferred multiloop control strategy for a process with the following steady-state gain matrix, which has been scaled by dividing the process variables by their maximum values:

$$\begin{bmatrix} y_1 \\ y_2 \\ y_3 \end{bmatrix} = \begin{bmatrix} 0.48 & 0.90 & -0.006 \\ 0.52 & 0.95 & 0.008 \\ 0.90 & -0.95 & 0.020 \end{bmatrix} \begin{bmatrix} u_1 \\ u_2 \\ u_3 \end{bmatrix} \quad (14\text{-}75)$$

Solution

The singular value analysis in Eqs. 14-63 through 14-65 yields

$$\mathbf{W} = \begin{bmatrix} 0.5714 & 0.3766 & 0.7292 \\ 0.6035 & 0.4093 & -0.6843 \\ -0.5561 & 0.8311 & 0.0066 \end{bmatrix} \quad (14\text{-}76)$$

$$\boldsymbol{\Sigma} = \begin{bmatrix} 1.618 & 0 & 0 \\ 0 & 1.143 & 0 \\ 0 & 0 & 0.0097 \end{bmatrix} \quad (14\text{-}77)$$

$$\mathbf{V} = \begin{bmatrix} 0.0541 & 0.9984 & 0.0151 \\ 0.9985 & -0.0540 & -0.0068 \\ -0.0060 & 0.0154 & -0.9999 \end{bmatrix} \quad (14\text{-}78)$$

$$\text{CN} = \frac{\sigma_1}{\sigma_3} = \frac{1.618}{0.0097} = 166.5$$

The RGA is as follows:

$$\Lambda = \begin{bmatrix} -2.4376 & 3.0241 & 0.4135 \\ 1.2211 & -0.7617 & 0.5407 \\ 2.2165 & -1.2623 & 0.0458 \end{bmatrix} \quad (14\text{-}79)$$

Note that a preliminary pairing based on the RGA would be y_1–u_2, y_2–u_3, y_3–u_1. However, two of the singular values (σ_1, σ_2) are of the same magnitude, but σ_3 is much smaller. The CN value suggests that only two output variables can be controlled effectively. If we eliminate one input variable and one output variable, the condition number, σ_1/σ_2, can be recalculated, as shown in Table 14.3.

Table 14.3 CN and λ for Different 2 × 2 Pairings, Example 14.7

Pairing Number	Controlled Variables	Manipulated Variables	CN	λ
1	y_1, y_2	u_1, u_2	184	39.0
2	y_1, y_2	u_1, u_3	72.0	0.552
3	y_1, y_2	u_2, u_3	133	0.558
4	y_1, y_3	u_2, u_1	1.51	0.640
5	y_1, y_3	u_1, u_3	69.4	0.640
6	y_1, y_3	u_2, u_3	139	1.463
7	y_2, y_3	u_2, u_1	1.45	0.634
8	y_2, y_3	u_1, u_3	338	3.25
9	y_2, y_3	u_2, u_3	67.9	0.714

In order to assess which two inputs should be used and which measured variables should be controlled, Table 14.3 shows nine pairings, along with CN and λ. Based on their having small condition numbers and acceptable values of λ, pairings 4 (y_1-u_2, y_3-u_1) and 7 (y_2-u_2, y_3-u_1) appear to be the most promising ones. In both cases, u_1 and u_2 are the preferred set of inputs, probably because the gain matrix has small entries in the column corresponding to u_3. Note also that λ is acceptable for pairings 5 and 9, but the CN is very high for each case, thus ruling them out. Pairing 4 is consistent with the original 3 × 3 RGA in Eq. 14-79, but pairing 7 is not. The final choice of either pairing 4 or 7 should be based on dynamic simulation of the closed-loop systems.

There may be considerable value in using the various measures discussed in this section (RGA and SVA) for plantwide control analysis, where the number of process variables can be very large. Screening approaches can identify possible control configurations, which reduces the number of dynamic simulation cases to a manageable number (McAvoy and Braatz, 2003).

This topic is currently an open research area; more details on plantwide control are provided in Appendices I and J.

14.4 TUNING OF MULTILOOP PID CONTROL SYSTEMS

Multiloop (decentralized) PID control systems are often used to control interacting multiple-input, multiple-output processes because they are easy to understand and require fewer parameters to tune than more general multivariable controllers. Another advantage of multiloop controllers is that *loop failure tolerance* of the resulting control system can be easily checked. Loop failure tolerance is important in practical applications, because some loops may be placed in manual mode, or the manipulated variables of some loops can be saturated at their limits so they cannot be changed to avoid instability.

We consider four types of tuning methods for multiloop PID control systems:

1. Detuning method (Luyben, 1986)
2. Sequential loop tuning method (Hovd and Skogestad, 1994)
3. Independent loop method (Grosdidier and Morari, 1987; Skogestad and Morari, 1989)
4. Relay auto-tuning (Shen and Yu, 1994)

In the detuning method, each controller of the multiloop control system is first designed, ignoring process interactions from the other loops. Then interactions are taken into account by detuning each controller until a performance criterion is met. Typically, controller settings are made more conservative; that is, the gains are decreased, and the integral times are increased in one or more loops. For example, in a 2 × 2 control problem, one could choose to detune the control loop for the less important controlled variable. The *biggest log-modulus tuning* (BLT) method proposed by Luyben (1986) is a well-known detuning method. Initially, the Ziegler–Nichols settings (Section 11.6) are determined for each control loop (K_{ZN}, τ_{ZN}). For PI controllers, the detuning is performed by adjusting a single parameter F that adjusts the controller gain and the integral time as follows.

$$K_c = \frac{K_{ZN}}{F} \qquad \tau_1 = F\tau_{ZN} \tag{14-80}$$

The detuning parameter F is increased from one until the biggest log modulus reaches a specified value. The biggest log modulus is a measure of how far the closed-loop system is from being unstable (Luyben, 1986).

In the sequential loop tuning method (Hovd and Skogestad, 1994), the controller for a selected input–output pair is tuned and this loop is closed. Then a second controller is tuned for a second pair while the first control loop remains closed, and so on. Because each controller can be tuned using SISO methods, it is simpler than the detuning method. A disadvantage is that the controller settings depend strongly on which loop is tuned first. Usually, the fastest loops are tuned first. In the independent loop method, each controller is designed based on the corresponding open-loop and closed-loop transfer functions, while satisfying inequality constraints on the process interactions (Grosdidier and Morari, 1987; Skogestad and Morari, 1989). Then the IMC approach is used to obtain PID controller settings for each loop, usually with a single tuning parameter for each loop.

Relay auto-tuning (Chapter 11) can also be used to tune multiloop control systems. The loops can be tuned in a sequential manner or simultaneously. Shen and Yu (1994) use relay auto-tuning of each single loop in succession. For a 2 × 2 system, they first put one loop in manual while tuning the second loop. Then with the first loop in automatic, they auto-tune the second loop. Then the first loop is tuned again with the second controller in automatic. This procedure is repeated until convergence occurs.

14.5 DECOUPLING AND MULTIVARIABLE CONTROL STRATEGIES

In this section, we discuss several strategies for reducing control-loop interactions.

14.5.1 Decoupling Control

One of the early approaches to multivariable control is *decoupling control*. By adding additional controllers called *decouplers* to a conventional multiloop configuration, the design objective to reducing control loop interactions can be realized. In principle, decoupling control schemes can reduce control loop interactions, and a set-point change for one controlled variable has little or no effect on the other controlled variables. In practice, these benefits may not be fully realized due to imperfect process models. A typical decoupler is based on a simple process model that can be either a steady-state or dynamic model.

One type of decoupling control system for a 2 × 2 process and a 1-1/2-2 control configuration is shown in Fig. 14.9. Note that four controllers are used: two conventional feedback controllers, G_{c1} and G_{c2}, plus two decouplers, T_{12} and T_{21}. The input signal to each decoupler is the output signal from a feedback controller. In Fig. 14.9, the transfer functions for the transmitters, disturbances, and final control elements have been omitted for the sake of simplicity. Skogestad and Postlethwaite (2005) have discussed the more general case where these transfer functions are included. The decouplers are designed to compensate for undesirable process interactions. For example, decoupler T_{21} can be designed so as to cancel Y_{21}, which arises from the undesirable process interaction between U_1 and Y_2.

This cancellation will occur at the Y_2 summer if the decoupler output U_{21} satisfies

$$G_{p21}U_{11} + G_{p22}U_{21} = 0 \tag{14-81}$$

Substituting for $U_{21} = T_{21}\,U_{11}$ and factoring gives

$$\left(G_{p21} + G_{p22}T_{21}\right)U_{11} = 0 \tag{14-82}$$

Note that $U_{11}(s) \neq 0$, because U_{11} is a controller output that is time dependent. Thus, to satisfy Eq. 14-82, it follows that

$$G_{p21} + G_{p22}T_{21} = 0 \tag{14-83}$$

Solving for T_{21} gives an expression for the ideal decoupler,

$$T_{21} = -\frac{G_{p21}}{G_{p22}} \tag{14-84}$$

Similarly, a design equation for T_{12} can be derived by imposing the requirement that U_{22} have no net effect on Y_1. Thus, the compensating signal U_{12} and the process interaction due to G_{p12} should cancel at the Y_1 summer. Analogous to Eq. 14-84, the ideal decoupler is given by

$$T_{12} = -\frac{G_{p12}}{G_{p11}} \tag{14-85}$$

The ideal decouplers in Eqs. 14-84 and 14-85 are very similar to the ideal feedforward controller in Eq. 12-18 with $G_t = G_v = 1$. In fact, one can interpret a decoupler as a type of feedforward controller with an input signal that is a manipulated variable rather than a disturbance variable. Recall from Chapter 12 that the ideal feed-forward controller may not be physically realizable. Similarly, ideal decouplers are not always physically realizable and they may suffer from model error; hence a steady-state model may be assumed for simplicity—for example,

$$T_{21} = -\frac{K_{p21}}{K_{p22}} \quad \text{and} \quad T_{12} = -\frac{K_{p12}}{K_{p11}} \tag{14-86}$$

This is called *static decoupling*. Another simplification is called *partial decoupling*, where only one decoupler is implemented. A disadvantage of static decoupling is that control loop interactions still exist during transient conditions; e.g., a set-point change for Y_1 will tend to upset Y_2.

If the dynamics of the two loops are similar, static decoupling can produce excellent transient responses (Åström and Hagglund, 2006). However, for most multivariable control problems, model predictive control (Chapter 15) is the preferred technique, and decoupling is rarely implemented in new control applications in industry.

14.5.2 General Multivariable Control Techniques

The term *multivariable control* refers generically to the class of control strategies in which each manipulated variable is adjusted based on the errors in all controlled variables, rather than the error in a single variable, as is the case for multiloop control. For example, a simple multivariable proportional control strategy for a 2×2 process could have four controller gains and the following form:

$$u_1(t) = K_{c11}e_1(t) + K_{c12}e_2(t) \tag{14-87}$$

$$u_2(t) = K_{c21}e_1(t) + K_{c22}e_2(t) \tag{14-88}$$

If $K_{c12} = K_{c21} = 0$, the multivariable control system reduces to a 1-1/2-2 multiloop control system, because each manipulated variable is adjusted based on a single error signal. Similarly, a 1-2/2-1 multiloop control system results if $K_{c11} = K_{c22} = 0$. Thus, multiloop control is a special case of the more general multivariable control. Note that the decoupling control scheme shown in Fig. 14.9 is also a multivariable control strategy, because each manipulated variable depends on both error signals.

Equations 14-87 and 14-88 illustrate multivariable proportional control for a 2×2 process. Multivariable control strategies can also be developed that include integral, derivative, and feedforward control action. The books by Goodwin et al. (2001) and Skogestad and Postlethwaite (2005) provide additional information. In this text, we emphasize the use of model predictive control as the method of choice for designing multivariable controllers, as discussed in Chapter 15.

14.6 STRATEGIES FOR REDUCING CONTROL LOOP INTERACTIONS

In Section 14.1, we described how process interactions between manipulated and controlled variables can result in undesirable control loop interactions. When control loop interactions are a problem, a number of alternative strategies are available:

1. Select different manipulated or controlled variables.
2. Re-tune one or more multiloop PID controllers, taking process interactions into account.
3. Consider a more general multivariable control method, such as model predictive control.

Strategy 1 is illustrated in the next section, while strategy 2 was considered in previous sections. Strategy 3 is considered in Chapter 15.

14.6.1 Selection of Different Manipulated or Controlled Variables

For some control problems, loop interactions can be significantly reduced by choosing alternative controlled and manipulated variables. For example, the new controlled or manipulated variable could be a simple function of the original variables such as a sum, difference, or ratio (Weber and Gaitonde, 1982; McAvoy, 1983; Waller and Finnerman, 1987). Industrial distillation columns have been controlled using simple, nonlinear functions of x_D and x_B as the controlled variables, rather than x_D and x_B. Very few guidelines for selecting functions of the original MVs and CVs have been published.

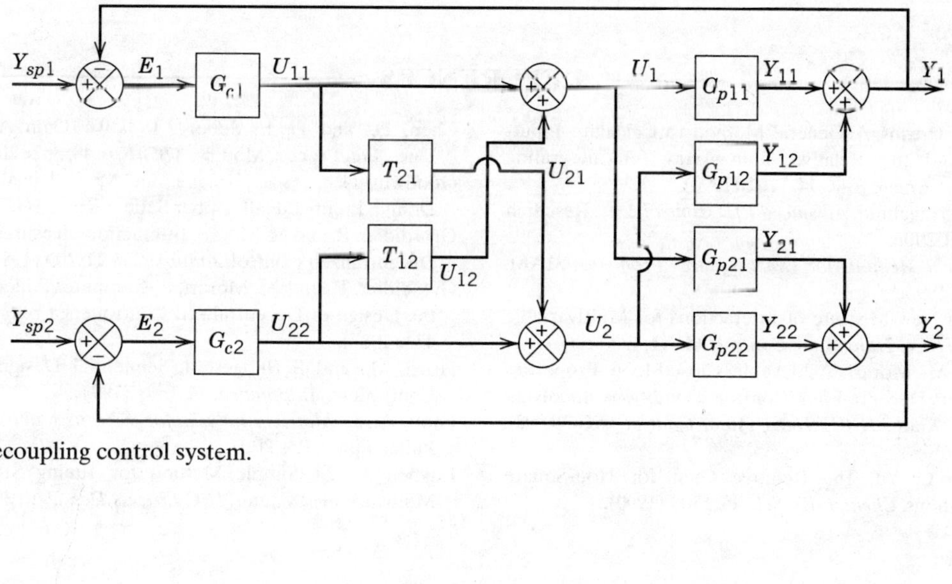

Figure 14.9 A decoupling control system.

EXAMPLE 14.8

For the blending system of Example 14.3, choose a new set of manipulated variables that will reduce control loop interactions by making $\lambda = 1$.

Solution

From the expression for the relative gain in Eq. 14-39, it is clear that $\lambda = 1$ if K_{12} and/or $K_{21} = 0$. Thus, we want to choose manipulated variables so that the steady-state gain matrix has a zero for at least one of the off-diagonal elements. Inspection of the process model in Eqs. 14-47 and 14-48 suggests that suitable choices for the manipulated variables are $u_1 = w_A + w_B$ and $u_2 = w_A$. Substitution into the process model gives an equivalent model in terms of the new variables:

$$w = u_1 \tag{14-89}$$

$$x = \frac{u_2}{u_1} \tag{14-90}$$

Linearizing Eq. 14-90 gives the gain matrix $\boldsymbol{K}$ in Eq. 14-33 where

$$\boldsymbol{K} = \begin{bmatrix} 1 & 0 \\ \dfrac{u_2}{u_1^2} & \dfrac{1}{u_1} \end{bmatrix} \tag{14-91}$$

vectors $\boldsymbol{y}$ and $\boldsymbol{u}$ are defined as $\boldsymbol{y} = [w, x]^T$ and $\boldsymbol{u} = [u_1, u_2]^T$. Because w depends on u_1 but not u_2, the only feasible controller pairing is w-u_1 and x-u_2. From Eqs. 14-91 and 14-39, it follows that $K_{12} = 0$ and $\lambda = 1$. Because $K_{21} \neq 0$, there will be a one-way interaction, with the w-u_1 control loop generating disturbances that affect the x-u_2 loop but not vice versa.

McAvoy (1983, p. 136) suggests alternative manipulated variables, namely, $u_1 = w_A + w_B$ and $u_2 = w_A/(w_A + w_B)$. This choice is motivated by the process model in Eqs. 14-47 and 14-48 and means that the controlled variables are identical to the manipulated variables! Thus, K is the identity matrix, $\lambda = 1$, and the two control loops do not interact at all. This situation is fortuitous, and also unusual, because it is seldom possible to choose manipulated variables that are, in fact, the controlled variables.

SUMMARY

In this chapter, we have considered control problems with multiple inputs (manipulated variables) and multiple outputs (controlled variables), with the main focus on using a set of single-loop controllers (multiloop control). Designing MIMO control systems is more difficult than SISO controller design because the multivariable process can cause undesirable control loop interactions for multiloop control.

If these control loop interactions are unacceptable, then one can either detune one or more of the control loops, choose a different set of manipulated or controlled variables, or employ multivariable control. Model-based multivariable control strategies such as model predictive control can provide significant improvements over conventional multiloop control, as discussed in Chapter 15.

REFERENCES

Arkun, Y., and J. J. Downs, A General Method to Calculate Input-Output Gains and the Relative Gain Array for Integrating Processes, *Comput. Chem. Eng.*, **14**, 1101 (1990).

Åström, K. J., and T. Hägglund, *Advanced PID Control*, ISA, Research Triangle Park, NC, 2006.

Bjorck, A., *Numerical Methods for Least Squares Problems*, SIAM, Philadelphia, PA, 1996.

Bristol, E. H., On a New Measure of Interactions for Multivariable Process Control, *IEEE Trans. Auto. Control*, **AC-11**, 133 (1966).

Campo, P. J., and M. Morari, Achievable Closed-loop Properties of Systems under Decentralized Control: Conditions Involving the Steady State Gain, *IEEE Trans. Auto. Control*, **AC-39**, 932 (1994).

Chang, J., and C. C. Yu, The Relative Gain for Non-Square Multivariable Systems, *Chem. Engr. Sci.*, **45**, 1309 (1990).

Chen, D., and D. E. Seborg, Relative Gain Array Analysis for Uncertain Process Models, *AIChE J.*, **48**, 302 (2002).

Goodwin, G. C., S. F. Graebe, and M. E. Salgado, *Control System Design*, Prentice Hall, Upper Saddle River, NJ, 2001.

Grosdidier, P., and M. Morari, Interaction Measures for Systems under Decentralized Control, *Automatica*, **22**, 309 (1986).

Grosdidier, P., and M. Morari, A Computer-Aided Methodology for the Design of Decentralized Controllers, *Comput. Chem. Eng.*, **11**, 423 (1987).

Hovd, M., and S. Skogestad, Sequential Design of Decentralized Controllers, *Automatica*, **30**, 1601 (1994).

Laub, A. J., *Matrix Analysis for Scientists and Engineers*, SIAM, Philadelphia, PA, 2004.

Luyben, W. L., Simple Method for Tuning SISO Controllers in Multivariable Systems, *IEC Process Des. Dev.*, **25**, 654 (1986).

McAvoy, T. J., *Interaction Analysis*, ISA, Research Triangle Park, NC, 1983.

McAvoy, T. J., and R. D. Braatz, Contollability of Processes with Large Singular Values, *Ind. Eng. Chem. Res.*, **42**, 6155 (2003).

Niederlinski, A., A Heuristic Approach to the Design of Linear Multivariable Interacting Control Systems, *Automatica*, **7**, 691 (1971).

Nisenfeld, A. E., and H. M. Schultz, *Interaction Analysis in Control System Design*, Paper 70-562, *Advances in Instrum.* **25**, Pt. 1, ISA, Pittsburgh, PA, 1971.

Roat, S. D., J. J. Downs, E. F. Vogel, and J. E. Doss, The Integration of Rigorous Dynamic Modeling and Control System Synthesis for Distillation Columns: An Industrial Approach, *Chemical Process Control CPC-III*, M. Morari and T. J. McAvoy (Eds.), CACHE-Elsevier, New York, 1986, p. 99.

Shen, S. H., and C.-C. Yu, Use of Relay-feedback Test for Automatic Tuning of Multivariable Systems, *AIChE J.*, **40**, 627 (1994).

Shinskey, F. G., *Process Control Systems*, 4th ed., McGraw-Hill, New York, 1996, Chapter 8.

Skogestad, S., Robust Control, *Practical Distillation Control*, W. L. Luyben (Ed.), Van Nostrand-Reinhold, New York, 1992, Chapter 14.

Skogestad, S., and M. Morari, Robust Performance of Decentralized Control Systems by Independent Design, *Automatica*, **25**, 119 (1989).

Skogestad, S., and I. Postlethwaite, *Multivariable Feedback Control*, 2nd ed., John Wiley and Sons, Hoboken, NJ, 2005.

Strang, G., *Linear Algebra and Its Applications*, 4th ed., Brooks-Cole, Independence, KY, 2005.

Wood, R. K., and M. W. Berry, Terminal Composition Control of a Binary Distillation Column, *Chem. Eng. Sci.*, **28**, 1707 (1973).

Zhu, Z., J. T. Lee, and T. F. Edgar, Steady State Structural Analysis and Interaction Characteristics in Multivariable Control Systems, *IEC Res.*, **36**, 3718 (1997).

EXERCISES

14.1 Luyben and Vinante (*Kem. Teollisuus*, **29**, 499 (1972)) developed a distillation column model relating temperatures on the 4th and 17th trays from the bottom of the column (T_4, T_{17}) to the reflux ratio R and the steam flow rate to the reboiler S:

$$\begin{bmatrix} T_{17}(s) \\ T_4(s) \end{bmatrix} = \begin{bmatrix} \dfrac{-2.16e^{-s}}{8.25s+1} & \dfrac{1.26e^{-0.3s}}{7.05s+1} \\ \dfrac{-2.75e^{-1.8s}}{8.25s+1} & \dfrac{4.28e^{-0.35s}}{9.0s+1} \end{bmatrix} \begin{bmatrix} R(s) \\ S(s) \end{bmatrix}$$

Compare the closed-loop set-point changes that result from the T_{17}-R/T_4-S pairing and the Ziegler–Nichols continuous cycling method in Chapter 11. Consider two cases:

(a) A set-point change is made in each loop with the other loop in manual.

(b) The set-point changes are made with both controllers in automatic.

14.2 Derive an expression for the characteristic equation for the 1-2/2-1 configuration in Fig. 14.3*b*. Simplify and interpret this equation for the special situation where either G_{p11} or G_{p22} is zero.

14.3 Derive equivalent closed-loop formulas to Eq. 14-11 through Eq. 14-13 for the case where there are sensor and valve transfer functions (G_{m1}, G_{m2}, G_{v1}, G_{v2}) for the outputs (y_1, y_2).

14.4 For the in-line blending system of Example 14.8, draw block diagrams for two multiloop control schemes:

(a) The standard scheme for $x = 0.4$

(b) The less interacting scheme where $u_1 = w_A + w_B$ and $u_2 = w_A$.

You may assume that each transmitter and controlvalve can be represented by a first-order transfer function and that PI controllers are utilized.

14.5 For the two-input-two-output process in Example 14.2, use Simulink to check the stability regions in Fig. 14.6 for the 1-1, 2-2 controller pairing. Set $K_{c1} = 1$, and try three values of K_{c2} (-1, 0, and +2) by simulating a unit set-point change in y_1 with both loops in automatic (proportional control only).

14.6 A conventional multiloop control scheme consisting of two PI controllers is to be used to control the product compositions x_D and x_B of the distillation column shown in Fig. 14.1*b*. The manipulated variables are the reflux flow rate R and the steam flow rate to the reboiler S. Experimental data for a number of steady-state conditions are summarized below. Use this information to do the following:

(a) Calculate the RGA and determine the recommended pairing between controlled and manipulated variables.

(b) Does this pairing seem appropriate from dynamic considerations? Justify your answer.

Table E14.6

Run	R (kg/min)	S (kg/min)	x_D	x_B
1	57	10	0.96	0.05
2	68	10	0.94	0.06
3	80	10	0.92	0.07
4	68	9	0.93	0.07
5	68	11	0.95	0.05

14.7 For the Wood-Berry distillation column model in Example 14.1:

(a) Which pairing of controlled and manipulated variables would you recommend based on steady-state considerations?

(b) Which pairing based on dynamic considerations? Justify your answers.

14.8 Two streams with flow rates w_A and w_B (Refer Figure 14-1a) and compositions x_A and x_B are mixed. The resulting flow rate and composition are w and x respectively. The desired values for w and x are 200 kg/hr and 0.65. If $x_A = 0.75$ and $x_B = 0.45$, determine the pairing of inputs and outputs using RGA analysis.

14.9 A binary distillation column has three tray temperature measurements (17th, 24th, 30th trays) that can be used as possible controlled variables. Controlling temperature is equivalent to controlling composition. Step testing gives the following steady-state input–output relationships (u_1 = steam pressure in reboiler; u_2 = reflux ratio):

$$T'_{17} = 1.5u_1 + 0.5u_2 \qquad (1)$$

$$T'_{24} = 2.0u_1 + 1.7u_2 \qquad (2)$$

$$T'_{30} = 3.4u_1 + 2.9u_2 \qquad (3)$$

All variables are deviation variables. Select the 2 × 2 control system that has the most desirable interactions, as determined by the RGA (note that there are three possible 2 × 2 control configurations). Explain why the combination of T_{24} and T_{30} is the least desirable controlled variable set, based on analyzing Eqs. (2) and (3) and the resulting determinant.

14.10 For the liquid storage system shown in Fig. E14.10, it is desired to control liquid levels h_1 and h_2 by adjusting volumetric flow rates q_1 and q_2. Flow rate q_6 is the major disturbance variable. The flow-head relations are given by

$$q_3 = C_{v1}\sqrt{h_1} \quad q_5 = C_{v2}\sqrt{h_2} \quad q_4 = K(h_1 - h_2)$$

where C_{v1}, C_{v2}, and K are constants.

(a) Derive an expression for the relative gain array for this system.

(b) Use the RGA to determine the recommended pairing of controlled and manipulated variables for the following conditions:

Parameter Values

$$K = 0.04 \text{ m}^3/\text{min} \cdot \text{m}$$
$$C_{v1} = 0.02 \text{ m}^3/\text{min} \cdot \text{m}^{0.5}$$
$$C_{v2} = 0.024 \text{ m}^3/\text{min} \cdot \text{m}^{0.5}$$
$$D_1 = D_2 = 1 \text{ m (tank diameters)}$$

Nominal Steady-State Values

$$\bar{h}_1 = 1.2\text{m} \quad \bar{h}_2 = 0.9\text{m}$$

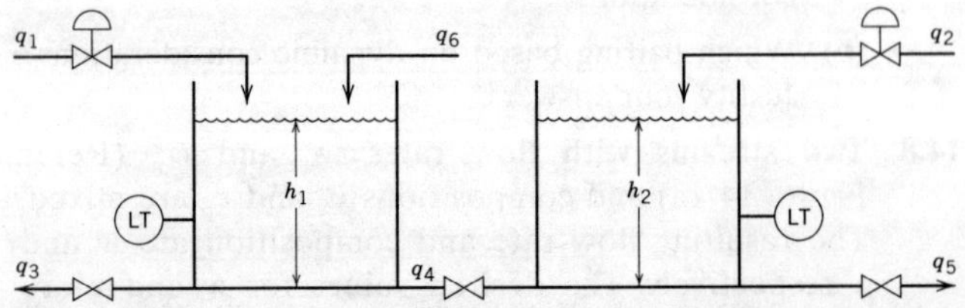

Figure E14.10

14.11 For the liquid-level storage system in Exercise 14.10:

(a) Derive a transfer function model of the form

$$Y(s) = G_p(s)U(s) + G_d(s)D(s)$$

where U represents the manipulated variables q_1 and q_2, D is the disturbance variable, and G_d is a 2 × 1 matrix of disturbance transfer functions.

(b) Draw a block diagram for a multiloop control system based on the following pairing: $h_1 - q_1/h_2 - q_2$. Do not attempt to derive transfer functions for the transmitters, control valves, or controllers.

14.12 For the flow-pressure process shown in Fig. E14.12, it is desired to control both pressure P_1 and flow rate F. The manipulated variables are the stem positions of the control valves, M_1 and M_2. For simplicity, assume that the flow-head relations for the two valves are given by

$$F = 18M_1(P_0 - P_1)$$
$$F = 25M_2(P_1 - P_2)$$

The nominal steady-state conditions are F = 4 m³/min, P_0 = 138 kPa, P_1 = 70 kPa, and P_2 = 35 kPa. Use the RGA approach to determine the best controller pairing.

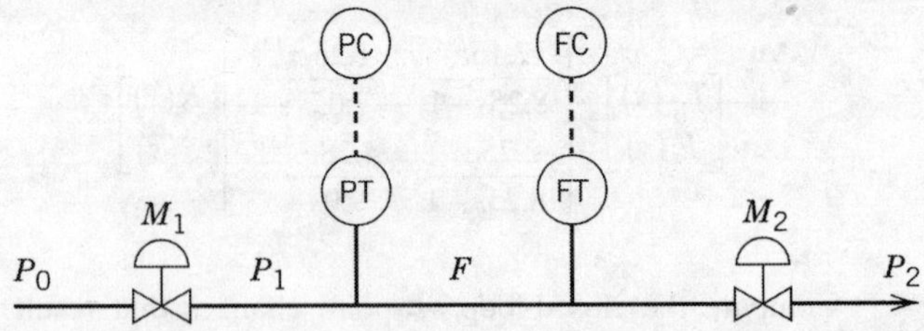

Figure E14.12

14.13 A blending system is shown in Fig. E14.13. Liquid level h and exit composition c_3 are to be controlled by adjusting flow rates q_1 and q_3. Based on the information below, do the following:

(a) Derive the process transfer function matrix, $\mathbf{G}_p(s)$.

(b) If a conventional multiloop control system is used, which controller pairing should be used? Justify your answer.

(c) Obtain expressions for the ideal decouplers $T_{21}(s)$ and $T_{12}(s)$ in the configuration of Fig. 14.9.

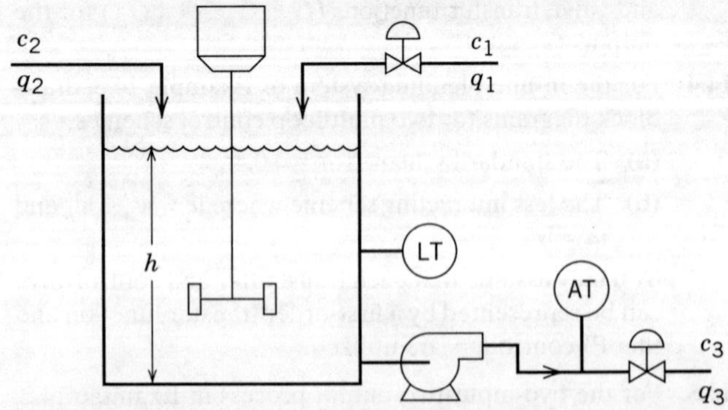

Figure E14.13

Available Information

(i) The tank is 1 m in diameter and is perfectly mixed.

(ii) Nominal steady-state values are

$$\bar{h} = 1 \text{ m} \qquad \bar{q}_3 = 0.6 \text{ m}^3/\text{min}$$
$$\bar{c}_1 = 15 \text{ mole/m}^3 \qquad \bar{c}_2 = 4 \text{ mole/m}^3$$
$$\bar{q}_1 = 0.3 \text{ m}^3/\text{min}$$

(iii) The density of each process stream remains constant at $\rho = 960$ kg/m³.

(iv) The primary disturbance variable is flow rate q_2.

(v) Inlet compositions c_1 and c_2 are constant.

(vi) The transmitter characteristics are approximated by the following transfer functions with time constants in minutes:

$$G_{m11}(s) = \frac{14}{0.1s+1}(\text{mA/m})$$
$$G_{m22}(s) = \frac{3}{0.2s+1}(\text{mA m}^3/\text{mole})$$

(vii) Each control valve has a gain of 4×10^{-3} m³/min mA and a time constant of 10 s.

14.14 A process to be controlled has two controlled variables Y_1 and Y_2, and three inputs that can be used as manipulated variables, U_1, U_2, and U_3. However, it is desired to use only two of these three manipulated variables in a conventional multiloop feedback control system. Transfer functions for the process are shown below. Which multiloop control configuration will result in the smallest amount of steady-state interaction between inputs and outputs? Justify your answer.

$$Y_1(s) = \frac{3}{3s+1}U_1(s) - \frac{0.5}{6s^2+9s+1}U_2(s) + \frac{1}{2s^2+5s+2}U_3(s)$$
$$Y_2(s) = -10U_1(s) + \frac{2}{3s+1}U_2(s) + \frac{4}{(1.5s+1)(2.5s+1)}U_3(s)$$

14.15 A stirred-tank heat exchanger with a bypass stream is shown in Fig. E14.15 with the available control valves. The possible manipulated variables are mass flow rate w_2, valve stem positions x_c and x_3, and f, the fraction of mass flow rate w_1 that bypasses the tank before being added to the exit stream. Using the information given here, do the following:

(a) Derive a dynamic model of the stirred-tank system. Define any additional symbols that you introduce.

(b) Determine the degrees of freedom for control that are available. Allocate the degrees of freedom by specifying manipulated variables and variables that are determined by the environment.

(c) Select controlled variables and briefly justify your choice.

(d) Suppose that only T_4 and h are to be controlled by using x_2 and f as the manipulated variables. (Valve stem positions, x_c and x_3, are held constant.) Derive an expression for the relative gain array for this control configuration.

(e) It has been proposed that x_2 be replaced by x_3 in the control problem of (d). Briefly analyze this proposal. (It is *not* necessary to perform another RGA analysis.)

Available Information

(i) The tank is perfectly mixed, and the temperature changes are relatively small so that constant physical properties can be assumed. Mass flow rates are denoted by w_1 to w_4 and temperatures by T_1 to T_4.

(ii) The exit flow rate w_3 depends on the pressures upstream and downstream of the control valve, and the valve stem position x_3. The following empirical relation is available where C_1 and C_2 are constants:

$$w_3 = x_3(C_1 h - C_2 f w_1)$$

(iii) The overall heat transfer coefficient for the cooling coil U depends on the velocity of the coolant in the line and hence on the valve stem position x_c, according to the relation below where C_3 is a constant:

$$U = C_3 x_c$$

(iv) The pump on the bypass line operates in the "flat part" of the pump curve so that the mass flow rate, fw_1, depends only on the control valve.

(v) The following process variables remain constant: T_1, w_1, T_2, and T_c.

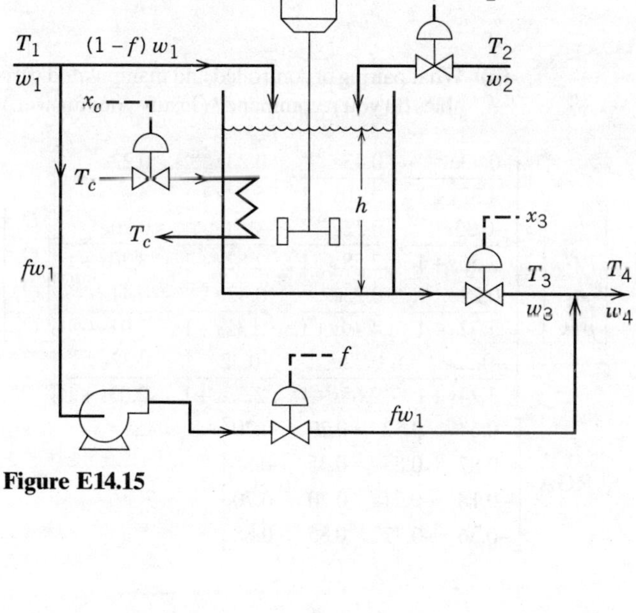

Figure E14.15

14.16 Water (F_1) is blended with a stream F_2 with 70% ethanol to make a whiskey product that is 40% ethanol. Assume $F_1 = 0.02$ m³/min and $F_2 = 0.025$ m³/min.

(a) Develop a steady-state material balance model for the blending operation. Find the linearized gains for the 2 × 2 transfer function model using Example 14.3 as a guide. F_1 and F_2 are manipulated variables, and the controlled variables are the outlet composition z and the total flow rate F.

(b) Derive the RGA in terms of F_1, F_2, and F.

(c) Determine the RGA and the preferred pairing for the controllers.

14.17 A schematic diagram for a pH neutralization process is shown in Fig. E14.17. The transfer function matrix and relative gain array are also shown.

(a) Suppose that a multiloop control system consisting of four PID controllers is to be designed. Recommend a pairing of controlled and manipulated variables. Briefly justify your recommendation based on steady-state, dynamic, and physical considerations.

(b) Suppose that only pH_2 and h_2 are to be controlled using Q_4 and Q_6 as the manipulated variables (Q_1 and Q_3 are held constant).

(i) What is the RGA for this 2×2 control problem?

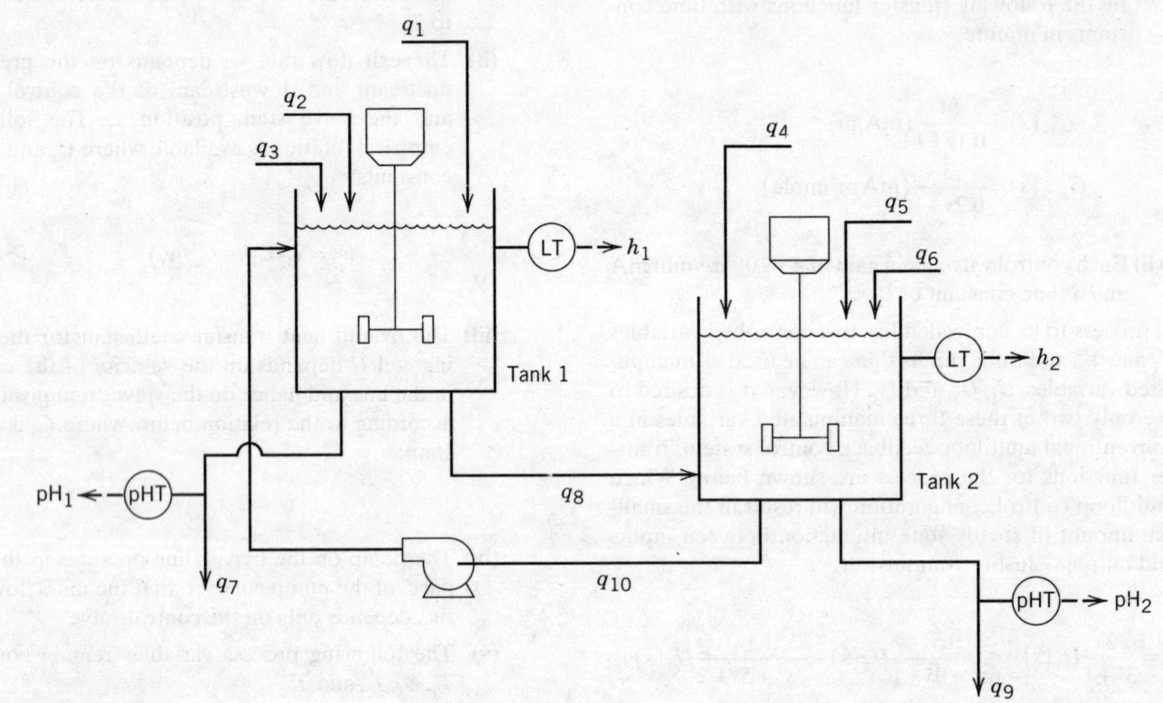

Figure E14.17

(ii) What pairing of controlled and manipulated variables do you recommend? (Justify your answer.)

$$\begin{bmatrix} h_1 \\ pH_1 \\ h_2 \\ pH_2 \end{bmatrix} = \begin{bmatrix} \dfrac{0.43e^{-0.8s}}{4.32s+1} & \dfrac{0.43e^{-0.1s}}{3.10s+1} & \dfrac{0.23e^{-1.0s}}{5.24s+1} & \dfrac{0.22e^{-0.5s}}{4.42s+1} \\ \dfrac{-0.33e^{-1.0s}}{2.56s+1} & \dfrac{0.32e^{-0.5s}}{2.58s+1} & \dfrac{-0.20e^{-1.8s}}{2.82s+1} & \dfrac{0.20e^{-0.8s}}{3.30s+1} \\ \dfrac{0.22e^{-1.1s}}{5.52s+1} & \dfrac{0.23e^{-0.3s}}{4.49s+1} & \dfrac{0.42e^{-0.4s}}{3.32s+1} & \dfrac{0.41e^{-0.1s}}{2.07s+1} \\ \dfrac{-0.22e^{-1.5s}}{3.24s+1} & \dfrac{0.22e^{-1.2s}}{2.65s+1} & \dfrac{-0.32e^{-0.8s}}{2.36s+1} & \dfrac{0.32e^{-0.4s}}{2.03s+1} \end{bmatrix} \begin{bmatrix} Q_1 \\ Q_3 \\ Q_4 \\ Q_6 \end{bmatrix}$$

$$\text{RGA} = \begin{bmatrix} 0.64 & 0.72 & -0.20 & -0.20 \\ 0.87 & 0.85 & -0.35 & -0.35 \\ -0.18 & -0.21 & 0.70 & 0.70 \\ -0.36 & -0.37 & 0.85 & 0.88 \end{bmatrix}$$

14.18 A combination of two drugs (hydrochlorothiazide and oxybutynin) is commonly used to regulate blood pressure in elderly patients. These two drugs mainly affect two physiological variables of the patient (blood pressure and urine production rate). Since the goal is to regulate both variables with these two drugs, interaction analysis has to be performed to design two SISO control loops. For the following model, there are two inputs and two outputs (Ogunnaike and Ray, *Process Dynamics, Modeling, and Control*, Oxford University Press, 1994, p. 771.):

$$\begin{bmatrix} y_1 \\ y_2 \end{bmatrix} = \begin{bmatrix} \dfrac{-0.04e^{-0.1s}}{0.11s+1} & \dfrac{0.0005e^{-0.15s}}{0.21s+1} \\ \dfrac{0.22}{0.12s+1} & \dfrac{-0.02}{0.21s+1} \end{bmatrix} \begin{bmatrix} u_1 \\ u_2 \end{bmatrix}$$

where

y_1 = normalized (dimensionless) blood pressure

y_2 = normalized urine production rate

u_1 = rate of hydrochlorothiazide ingestion

u_2 = rate of oxybutynin ingestion

(a) Calculate the relative gain array.

(b) What loop pairing would you suggest?

14.19 A rapid thermal processing system for microelectronics manufacturing uses three concentric lamp heater arrays to keep the wafer temperature uniform. The gain matrix for the system is

$$K = \begin{bmatrix} 3.38 & 2.50 & 0.953 \\ 3.20 & 2.38 & 0.986 \\ 3.13 & 2.33 & 1.054 \end{bmatrix}$$

The system experiences difficulties in controlling all three temperatures uniformly. Examine possible control difficulties using RGA and SVA analyses.

14.20 A 4×4 control system has the following gain matrix:

$$K = \begin{bmatrix} 9.82 & 4.49 & 1.40 & 4.56 \\ 8.99 & 5.47 & 1.68 & 5.24 \\ 4.59 & 5.80 & 2.31 & 7.33 \\ 2.41 & 4.33 & 2.82 & 8.92 \end{bmatrix}$$

Assess the potential difficulty of control using singular value analysis. Should any outputs or inputs be eliminated to achieve better control?

MULTIPLE CHOICE QUESTIONS

1. For a process with 4 controlled variables and 4 manipulated variables, how many multiloop control configurations are possible?

(a) 4 **(b)** 16 **(c)** 256 **(d)** 24

2. Which of the following methods can be used to determine the best pairing of controlled and manipulated variables in a multiple input-multiple output process?

(a) Routh array

(b) Relative gain array

(c) Ziegler-Nichols technique

(d) ITAE criterion

3. The relative gain array for a 2×2 system is given by $\Lambda = \begin{bmatrix} 0.7 & 0.3 \\ 0.3 & 0.7 \end{bmatrix}$. The recommended pairing will be

(a) y_1–u_1, y_2–u_2

(b) y_1–u_2, y_2–u_1

(c) loop interaction will be same for any pairing.

(d) Cannot be determined.

4. Which of the following cannot be a relative gain array for a 2×2 process?

(a) $\begin{bmatrix} 1 & 0 \\ 0 & 1 \end{bmatrix}$ **(b)** $\begin{bmatrix} 1.8 & -0.8 \\ -0.8 & 1.8 \end{bmatrix}$

(c) $\begin{bmatrix} 0.5 & 0.5 \\ 0.5 & 0.5 \end{bmatrix}$ **(d)** $\begin{bmatrix} 0.8 & 0.3 \\ 0.3 & 0.8 \end{bmatrix}$

5. A closed loop MIMO system is unstable if the value of Niederlinski index is

(a) <1 **(b)** >1 **(c)** <0 **(d)** >0

6. An $n \times n$ MIMO process is represented by the steady state model $y = Ku$. y and u are $n \times 1$ vector of outputs and inputs and K is the $n \times n$ matrix. The n equations will be linearly independent if, determinant value of K is

(a) $\neq 0$ **(b)** $= 0$ **(c)** $\neq 1$ **(d)** $= 1$

7. A 2×2 MIMO system is represented by the transfer function matrix $G = \begin{bmatrix} G_{p11} & G_{p12} \\ G_{p21} & G_{p22} \end{bmatrix}$. For 1-1/2-2 pairing, the ideal decoupler transfer functions to eliminate loop interaction are

(a) $-\dfrac{G_{p11}}{G_{p22}}, -\dfrac{G_{p22}}{G_{p11}}$ **(b)** $-\dfrac{G_{p11}}{G_{p21}}, -\dfrac{G_{p21}}{G_{p11}}$

(c) $-\dfrac{G_{p21}}{G_{p22}}, -\dfrac{G_{p12}}{G_{p11}}$ **(d)** $-\dfrac{G_{p11}}{G_{p12}}, -\dfrac{G_{p12}}{G_{p11}}$

8. The singular values of the steady state gain matrix K are

(a) squared eigen values of K.

(b) squared eigen values of $K^T K$.

(c) Positive square roots of eigen values of $K^T K$.

(d) Negative square roots of eigen values of $K^T K$.

9. The largest and smallest nonzero singular values of steady state gain matrix K are σ_{max} and σ_{min} respectively. The condition number of K is

(a) $\sigma_{max} \times \sigma_{min}$ **(b)** $\sigma_{max} + \sigma_{min}$

(c) $\sigma_{max} - \sigma_{min}$ **(d)** $\sigma_{max} / \sigma_{min}$

10. Condition number of a singular matrix is

(a) 0 **(b)** ∞ **(c)** 1 **(d)** –1

ANSWER KEY

1. (d) **2.** (b) **3.** (a) **4.** (d) **5.** (c)

6. (a) **7.** (c) **8.** (c) **9.** (d) **10.** (b)

CHAPTER 15

Model Predictive Control

CHAPTER CONTENTS

In this chapter we consider *model predictive control (MPC)*, an important advanced control technique for difficult multivariable control problems. The basic MPC concept can be summarized as follows. Suppose that we wish to control a multiple-input, multiple-output process while satisfying inequality constraints on the input and output variables. If a reasonably accurate dynamic model of the process is available, model and current measurements can be used to predict future values of the outputs. Then the appropriate changes in the input variables can be calculated based on both predictions and measurements. In essence, the changes in the individual input variables are coordinated after considering the input–output relationships represented by the process model. In MPC applications, the output variables are also referred to as *controlled variables* or *CVs*, while the input variables are also called *manipulated variables* or *MVs*. Measured disturbance variables are called *DVs* or *feedforward variables*. These terms will be used interchangeably in this chapter.

Model predictive control offers several important advantages: (1) the process model captures the dynamic and static interactions between input, output, and disturbance variables, (2) constraints on inputs and outputs are considered in a systematic manner, (3) the control calculations can be coordinated with the calculation of optimum set points, and (4) accurate model predictions can provide early warnings of potential problems. Clearly, the success of MPC (or any other model-based approach) depends on the accuracy of the process model. Inaccurate predictions can make matters worse, instead of better.

First-generation MPC systems were developed independently in the 1970s by two pioneering industrial research groups. Dynamic Matrix Control *(DMC)*, devised by Shell Oil (Cutler and Ramaker, 1980), and a related approach developed by ADERSA (Richalet et al., 1978) have quite similar capabilities. An adaptive MPC technique, Generalized Predictive Control (GPC), developed by Clarke et al. (1987) has also received considerable attention. Model predictive control has had a major impact on industrial practice. For example, an MPC survey by Qin and Badgwell (2003) reported that there were over 4500 applications worldwide by the end of 1999, primarily in oil refineries and petrochemical plants. In these industries, MPC has become the method of choice for difficult multivariable control problems that include inequality constraints.

In view of its remarkable success, MPC has been a popular subject for academic and industrial research. Major extensions of the early MPC methodology have been developed, and theoretical analysis has provided insight into the strengths and weaknesses of MPC. Informative reviews of MPC theory and practice are available in books (Camacho and Bordons, 2003; Maciejowski, 2002; Rossiter, 2003; Richalet and O'Donovan, 2009; Rawlings and Mayne, 2009); tutorials (Hokanson and Gerstle, 1992; Rawlings, 2000), and survey papers (Morari and Lee, 1999; Qin and Badgwell, 2003; Canney, 2003; Kano and Ogawa, 2010; Darby and Nikolaou, 2012; Qin and Badgwell, 2014).

15.1 OVERVIEW OF MODEL PREDICTIVE CONTROL

The overall objectives of an MPC controller have been summarized by Qin and Badgwell (2003):

1. Prevent violations of input and output constraints.
2. Drive some output variables to their optimal set points, while maintaining other outputs within specified ranges (see Section 15.4.2).
3. Prevent excessive movement of the input variables.
4. Control as many process variables as possible when a sensor or actuator is not available.

A block diagram of a model predictive control system is shown in Fig. 15.1. A process model is used to predict the current and future values of the output variables. The residuals, the differences between the actual and predicted outputs, serve as the feedback signal to a *Prediction* block. The predictions are used in two types of MPC calculations that are performed at each sampling instant: set-point calculations and control calculations. Inequality constraints on the input and output variables, such as upper and lower limits, can be included in either type of calculation. Note that the MPC configuration is similar to both the internal model control configuration in Chapter 11 and the Smith predictor configuration of Chapter 12, because the model acts in parallel with the process and the residual serves as a feedback signal. However, the coordination of the control and set-point calculations is a unique feature of MPC. Furthermore, MPC has had a much greater impact on industrial practice than IMC or Smith predictor, because it is more suitable for constrained MIMO control problems.

The set points for the control calculations, also called *targets*, are calculated from an economic optimization based on a steady-state model of the process, traditionally, a linear model. Typical optimization objectives include maximizing a profit function or minimizing a cost function, or maximizing a production rate. The optimum values of set points change frequently due to varying process conditions, especially changes in the inequality constraints (See Appendix F). The constraint changes are due to variations in process conditions, equipment, and instrumentation, as well as economic data such as prices and costs. In MPC, the set points are typically calculated each time the control calculations are performed, as discussed in Section 15.5.

The MPC calculations are based on current measurements and predictions of the future values of the outputs. The objective of the MPC control calculations is to determine a sequence of *control moves* (i.e., manipulated input changes) so that the predicted response moves to the set point in an optimal manner. The actual output y, predicted output $\hat{y}$, and manipulated input u for SISO control are shown in Fig. 15.2. At the current sampling instant, denoted by k, the MPC strategy calculates a set of M values of the input $\{u(k+i-1), i = 1, 2, \ldots, M\}$. The set consists of the current input $u(k)$ and $M-1$ future inputs. The input is held constant after the M control moves. The inputs are calculated so that a set of P predicted outputs $\hat{y}(k+i), i = 1, 2, \ldots, P\}$ reaches the set point in an optimal manner. The control calculations are based on optimizing an objective function. The number of predictions P is referred to as the *prediction horizon* while the number of calculated control moves M is called the *control horizon, or move horizon*.

In MPC strategy, a linear or quadratic objective function is optimized so as to achieve least prediction error with minimum change in control moves. The linear function is the sum of absolute values of predicted

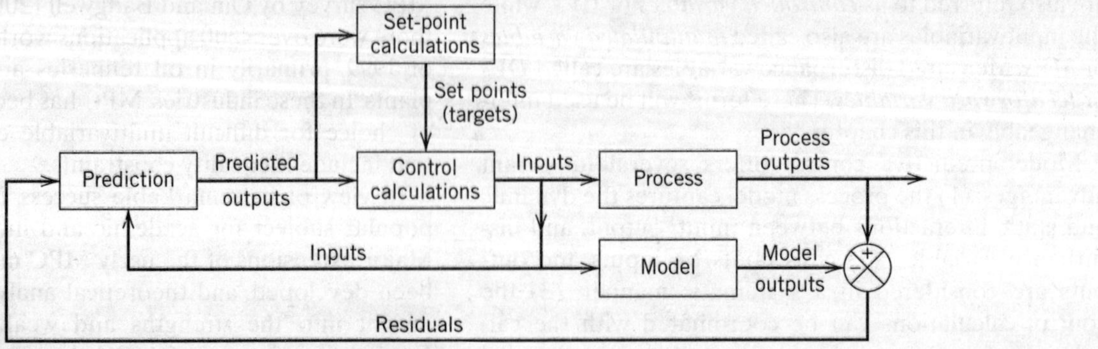

Figure 15.1 Block diagram for model predictive control.

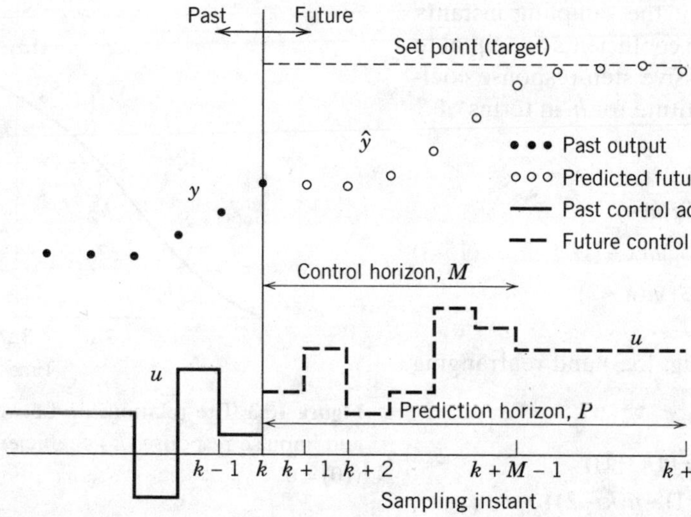

Figure 15.2 Basic concept for model predictive control.

errors and changes of control action. Quadratic function minimizes the sum of squared predicted errors and squares of control action changes.

A distinguishing feature of MPC is its *receding horizon approach*. Although a sequence of M control moves is calculated at each sampling instant, only the first move is actually implemented. Then a new sequence is calculated at the next sampling instant, after new measurements become available; again only the first control move is implemented. This procedure is repeated at each sampling instant. But why is an M-step control strategy calculated if only the first step is implemented? We will answer this question in Section 15.4.

15.2 PREDICTIONS FOR SISO MODELS

The MPC predictions are made using a dynamic model, typically a linear empirical model such as a multivariable version of the difference equation models or the related step response models that will be described in Section 15.2.1. Alternatively, transfer function or state-space models (Section 4.4) can be employed. For very nonlinear processes, it can be advantageous to predict future output values using a nonlinear dynamic model. Both physical models and empirical models, such as neural networks (Section 16.2), have been used in nonlinear MPC (Badgwell and Qin, 2001; White, 2008). Step-response models offer the advantage that they can represent stable processes with unusual dynamic behavior that cannot be accurately described by simple transfer function models. Their main disadvantage is the large number of model parameters. Although step-response models are not suitable for unstable processes, they can be modified to represent integrating processes, as shown in Section 15.2.4.

15.2.1 Impulse and Step Response Models

In chapter 13, the use of discrete time models in digital control was discussed. A continuous time function is converted to a discrete time function by approximating the derivative term with a difference term and the integral term by a summation term.

An important class of discrete-time models that find wide application for MPC design is the *finite impulse response (FIR)* or *convolution* model. This model can be written as

$$y(k+1) = y(0) + \sum_{i=1}^{N} h_i u[k-(i-1)] \qquad (15.1)$$

Integer N is selected so that $N\Delta t \geq t_s$, the settling time of the process. Note that an equivalent version of Eq. 15-1 can be written with $y(k)$ instead of $y(k+1)$ on the left-hand side by shifting the index backward one sampling period. Another kind of discrete-time models widely used in MPC is finite step response model or just the step response model. Step response models can be obtained by introducing step change in input and recording the output value with time as shown in Figure 15.3. Step response model can also be derived from the impulse response model Eq. 15-1. To illustrate this relationship, we consider a simple (finite) impulse response model where $N = 3$. Expanding the summation in Eq. 15-1 gives

$$y(k+1) = y(0) + h_1 u(k) + h_2 u(k-1) + h_3 u(k-2) \qquad (15\text{-}2)$$

The *step response coefficients* S_i are related to the impulse response coefficients h_i as shown in Fig. 15.3. By definition, the step response coefficients are simply the values of the response y at the sampling instants. Note that the impulse response coefficients are equal to the differences between successive step response coefficients, $h_i = S_i - S_{i-1}$. If we substitute for h_i in terms of S_i in Eq. 15-2, then

$$\begin{aligned} y(k+1) = y(0) &+ (S_1 - S_0)u(k) \\ &+ (S_2 - S_1)u(k-1) \\ &+ (S_3 - S_2)u(k-2) \end{aligned} \tag{15-3}$$

Recognizing that $S_0 = 0$ (see Fig. 15.3) and rearranging gives

$$\begin{aligned} y(k+1) = y(0) &+ S_1(u(k) - u(k-1)) \\ &+ S_2(u(k-1) - u(k-2)) \\ &+ S_3u(k-2) \end{aligned} \tag{15-4}$$

After defining $\Delta u(k) \triangleq u(k) - u(k-1)$, Eq. 15-4 becomes

$$\begin{aligned} y(k+1) = y(0) &+ S_1\Delta u(k) \\ &+ S_2\Delta u(k-1) + S_3u(k-2) \end{aligned} \tag{15-5}$$

Similarly, the step response model that corresponds to the full impulse response model in Eq. 15-1 is given by

$$\begin{aligned} y(k+1) = y(0) &+ \sum_{i=1}^{N-1} S_i\Delta u(k-i+1) \\ &+ S_N u(k-N+1) \end{aligned} \tag{15-6}$$

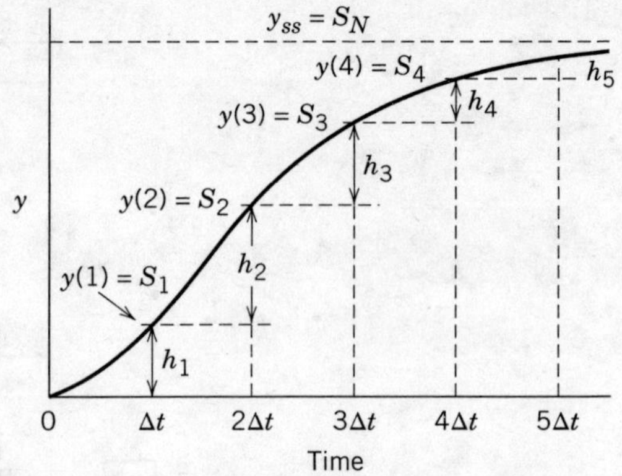

Figure 15.3 The relationship between the step response (S_i) and impulse response (h_i) coefficients for the situation where $y(0) = 0$.

The model parameters are the N step-response coefficients, S_1 to S_N. Typically, N is selected so that $30 \le N \le 120$. The initial value, $y(0)$, is denoted by y_0. For simplicity, we will assume that $y_0 = 0$. This derivation is left to the reader.

Note that Fig. 15.3 illustrates the case where there is no time delay in the process model. When a time delay is present, the initial step (or impulse) coefficients are zero. For example, if there is a time delay of d sampling periods, then $S_1, \ldots, S_d$ are zero in Eq. 15-6. Similarly, $h_i = 0$ for $0 \le i \le d$ in Eq. 15-1.

A discrete-time impulse or step response model can be developed from a transfer function model or a linear differential (or difference) equation model, as Example 15.1 illustrates.

EXAMPLE 15.1

Consider a first-order-plus-time-delay model:

$$\frac{Y(s)}{U(s)} = \frac{Ke^{-\theta s}}{\tau s + 1} \tag{15-7}$$

(a) Derive the equivalent step-response model by considering the analytical solution to a unit step change in the input.

(b) Calculate the step-response coefficients, $\{S_i\}$, for the following parameter values: $K = 5$, $\tau = 15$ min, $\theta = 3$ min, and a sampling period of $\Delta t = 1$ min. Also, calculate and plot the response $y(k)$ for $0 \le k \le 80$ after a step change in u from 0 to 3 occurs at $t = 2$ min.

Solution

(a) The step response for a first-order model without a time delay ($\theta = 0$) was derived in Chapter 5

$$y(t) = KM(1 - e^{-t/\tau}) \tag{5-16}$$

where M is the magnitude of the step change. The corresponding response for the model with a time delay is

$$\begin{aligned} y(t) &= 0 && \text{for } t \le \theta \\ y(t) &= KM(1 - e^{-(t-\theta)/\tau}) && \text{for } t > \theta \end{aligned} \tag{15-8}$$

The sampling instants are denoted by $t = i\Delta t$ where Δt is the sampling period and $i = 1, 2, \ldots$ Substituting $t = i\Delta t$ into Eq. 15-8 gives the response for $0 \le i \le 80$:

$$\left.\begin{aligned} &y(i\Delta t) = 0 && \text{for } i\Delta t \le \theta \\ &y(i\Delta t) = KM(1 - e^{-(i\Delta t - \theta)/\tau}) && \text{for } \Delta t > \theta \end{aligned}\right\} \quad (15\text{-}9)$$
where $i = 1, 2, \ldots, 80$

The number of step-response coefficients, N in Eq. 15-6, is specified to be $N = 80$ so that $N\Delta t$ is slightly larger than the process settling time of approximately $5\tau + \theta$. As indicated earlier in this section, the ith step-response coefficient is the value of the unit step response at the ith sampling instant. Thus, the step-response coefficients can be determined from Eq. 15-9 after setting $M = 1$:

$$\left.\begin{aligned} &S_i = 0 && \text{for } i\Delta t \le \theta \\ &S_i = K(1 - e^{-(i\Delta t - \theta)/\tau}) && \text{for } i\Delta t > \theta \end{aligned}\right\} \quad (15\text{-}10)$$
where $i = 1, 2, \ldots, 80$

Table 15.1 Step-Response Coefficients for Example 15.1

Sampling Instant	S_i	Sampling Instant	S_i	Sampling Instant	S_i
1	0	28	4.06	55	4.84
2	0	29	4.12	56	4.85
3	0	30	4.17	57	4.86
4	0.32	31	4.23	58	4.87
5	0.62	32	4.28	59	4.88
6	0.91	33	4.32	60	4.89
7	1.17	34	4.37	61	4.90
8	1.42	35	4.41	62	4.90
9	1.65	36	4.45	63	4.91
10	1.86	37	4.48	64	4.91
11	2.07	38	4.52	65	4.92
12	2.26	39	4.55	66	4.93
13	2.43	40	4.58	67	4.93
14	2.60	41	4.60	68	4.93
15	2.75	42	4.63	69	4.94
16	2.90	43	4.65	70	4.94
17	3.03	44	4.68	71	4.95
18	3.16	45	4.70	72	4.95
19	3.28	46	4.72	73	4.95
20	3.39	47	4.73	74	4.96
21	3.49	48	4.75	75	4.96
22	3.59	49	4.77	76	4.96
23	3.68	50	4.78	77	4.96
24	3.77	51	4.80	78	4.97
25	3.85	52	4.81	79	4.97
26	3.92	53	4.82	80	4.97
27	3.99	54	4.83		

(b) Substituting numerical values into Eq. 15-10 gives the step-response coefficients in Table 15.1. The step response $y(k)$ in Fig. 15.4 can be calculated either from Eqs. 15-6 and 15-10, or from Eq. 15-9. For $\theta = 2$ min, $S_1 = S_2 = 0$, and the step response is zero until $t > 2$ min. The new steady-state value is $y = 15$ because the steady-state gain in Eq. 15-7 is $K = 5$ and the magnitude of the step change is $M = 3$. Because $\Delta t = 1$ min and the step change occurs at $t = 3$ min, $\Delta u(k) = M = 3$ for $k = 3$ and $\Delta u(k) = 0$ for all other values of k. Recall that $\Delta u(k)$ is defined as $\Delta u(k) \triangleq u(k) - u(k-1)$.

For this example, the response $y(k)$ could be calculated analytically from Eq. 15-9 because a transfer function model was assumed. However, in many MPC applications, the transfer function model is not known, and thus the response must be calculated from the step-response model in Eq. 15-6

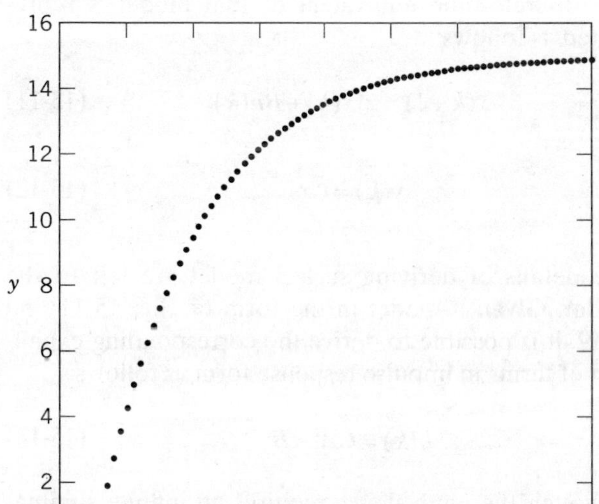

Figure 15.4 Step response for Example 15.1.

MATLAB APPLICATION

```
>> sys = tf([5],[15,1],'InputDelay',3); % Create a first order plus time delay model
>> t = 1:1:80; % Create a time vector from 1 to 80
>> sys_discrete = c2d(sys,1); % Convert the continuous system to discrete sytem with sampling time 1 and zero order hold
   [y,t] = step(sys_discrete,80); %Step response of the model till 80 minute (Check the values of y with the values given in Table 15.1).
>> step(sys,sys_discrete); % Plot step response for the continuous and discrete systems
```

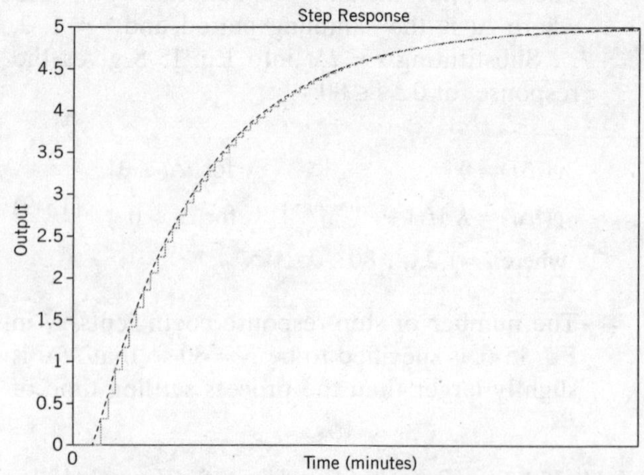

Figure 15.5 Comparison of the continuous and discrete FOPTD output for unit step change in input.

It may be noted that the above response in Figure 15.5 is for unit step change. For a step change of magnitude 3, the output will change by threefold.

This modeling approach leads to a step response model that is equivalent to the first-order difference with many parameters. Given the two alternatives, it is clear that the number of model parameters (*model parsimony*) is an issue in selecting the appropriate model.

One can also develop an impulse response model (and its equivalent step response model) from a linear state-space model (introduced in Eqs. 4-94 and 4-95). The discrete-time equivalent of that model is represented as follows:

$$x(k+1) = Ax(k) + Bu(k) \tag{15-11}$$

$$y(k) = Cx \tag{15-12}$$

The details of deriving such a model are left to the reader. Given a model in the form of Eqs. 15-11 and 15-12, it is possible to derive the corresponding expansion of terms in impulse response form as follows:

$$h(k) = CA^{k-1}B \tag{15-13}$$

Although the equivalence requires an infinite summation, that is, the so-called *Markov parameters* (Chen, 1984), a finite truncation will yield a reasonable approximation in practice.

When should a step response or impulse response model be selected? First, this type of model is useful when the actual model order or time delay is unknown, because this information is not required for step response models. The model parameters can be calculated directly from data using linear regression (see Chapter 16). Second, step or impulse response models are appropriate for processes that exhibit unusual dynamic behavior that cannot be described by standard low-order models.

Step response and impulse response models also have some disadvantages. These two kinds of models can be used in MPC calculation when the processes are open loop stable. Further, unlike a discrete transfer function model or a discrete state space model, step response and impulse response models are specified by using a large number of model parameters.

15.2.2 Prediction Using Step-Response Models

Next, we demonstrate how step-response models can be used to predict future outputs. Similar predictions can be made using other types of linear models such as transfer function or state-space models.

Model predictive control is based on predictions of future outputs over a prediction horizon, P. We now consider the calculation of these predictions. Let k denote the current sampling instant and $\hat{y}(k+1)$ denote the prediction of $y(k+1)$ that is made at time k. If $y_0 = 0$, this one-step-ahead prediction can be obtained from Eq. 15-6 by replacing $y(k+1)$ with $\hat{y}(k+1)$:

$$\hat{y}(k+1) = \sum_{i=1}^{N-1} S_i \Delta u[k-(i-1)] + S_N u[k-(N-1)] \tag{15-14}$$

Equation 15-14 can be expanded as

$$\hat{y}(k+1) = \underbrace{S_1\Delta u(k)}_{\text{Effect of current control action}} + \underbrace{\sum_{i=2}^{N-1} S_i\Delta u(k-i+1) + S_N u(k-N+1)}_{\text{Effect of past control actions}} \quad (15\text{-}15)$$

The first term on the right-hand side indicates the effect of the current manipulated input $u(k)$ because $\Delta u(k) = u(k) - u(k-1)$. The second term represents the effects of past inputs, $\{u(i), i < k\}$. An analogous expression for a two-step-ahead prediction can be derived in a similar manner. Substitute $k = k' + 1$ into Eq. 15-14:

$$\hat{y}(k'+2) = \sum_{i=1}^{N-1} S_i\Delta u(k'-i+2) + S_N u(k'-N+2) \quad (15\text{-}16)$$

Because Eq. 15-16 is valid for all positive values of k', without loss of generality, we can replace k' with k and then expand the right-hand side to identify the contributions relative to the current sampling instant, k:

$$\hat{y}(k+2) = \underbrace{S_1\Delta u(k+1)}_{\text{Effect of future control action}} + \underbrace{S_2\Delta u(k)}_{\text{Effect of current control action}} + \underbrace{\sum_{i=3}^{N-1} S_i\Delta u(k-i+2) + S_N u(k-N+2)}_{\text{Effect of past control actions}} \quad (15\text{-}17)$$

An analogous derivation provides an expression for a *j-step-ahead prediction* where j is an arbitrary positive integer:

$$\hat{y}(k+j) = \underbrace{\sum_{i=1}^{j} S_i\Delta u(k+j-i)}_{\text{Effect of current and future control actions}} + \underbrace{\sum_{i=j+1}^{N-1} S_i\Delta u(k+j-i) + S_N u(k+j-N)}_{\text{Effect of past control actions}} \quad (15\text{-}18)$$

The second and third terms on the right-hand side of Eq. 15-18 represent the predicted response when there are no current or future control actions, that is, the predicted response when $u(k+i) = u(k-1)$ for $i \geq 0$, or equivalently, $\Delta u(k+i) = 0$ for $i \geq 0$. Because this term accounts for past control actions, it is referred to as the *predicted unforced response* and is denoted by the symbol, $\hat{y}^o(k+j)$. Thus, we define $\hat{y}^o(k+j)$ as

$$\hat{y}^o(k+j) \triangleq \sum_{i=j+1}^{N-1} S_i\Delta u(k+j-i) + S_N u(k+j-N) \quad (15\text{-}19)$$

and write Eq. 15-18 as

$$\hat{y}(k+j) = \sum_{i=1}^{j} S_i\Delta u(k+j-i) + \hat{y}^o(k+j) \quad (15\text{-}20)$$

Examples 15.2 and 15.3 demonstrate that Eq. 15-20 can be used to derive a simple predictive control law based on a *single* prediction.

EXAMPLE 15.2

Derive a predictive control law that is based on the following concept. A single control move, $\Delta u(k)$, is calculated so that the J-step-ahead prediction is equal to the set point, that is, $\hat{y}(k+J) = y_{sp}$ where integer J is a tuning parameter. This sampling instant, $k + J$, is referred to as a *coincidence point*. Assume that u is held constant after the single control move, so that $\Delta u(k+i) = 0$ for $i > 0$.

Solution

In the proposed predictive control strategy, only a single prediction for J steps ahead is considered. Thus, we let $j = J$ in Eq. 15-20. Similarly, because we are only interested in calculating the current control move, $\Delta u(k)$, the future control moves in Eq. 15-20 are set equal to zero: $\Delta u(k+J-i) = 0$ for $i = 1, 2, \ldots, J-1$. Thus, Eq. 15-20 reduces to

$$\hat{y}(k+J) = S_J\Delta u(k) + \hat{y}^o(k+J) \quad (15\text{-}21)$$

Setting $\hat{y}(k+J) = y_{sp}$ and rearranging gives the desired predictive controller:

$$\Delta u(k) = \frac{y_{sp} - \hat{y}^o(k+J)}{S_J} \quad (15\text{-}22)$$

The predicted unforced response $\hat{y}^o(k+J)$ can be calculated from Eq. 15-19 with $j = J$.

The control law in Eq. 15-22 is based on a single prediction that is made for J steps in the future. Note that the control law can be interpreted as the inverse of the predictive model in Eq. 15-21.

EXAMPLE 15.3

Apply the predictive control law of Example 15.2 to a fifth-order process:

$$\frac{Y(s)}{U(s)} = \frac{1}{(5s+1)^5} \qquad (15\text{-}23)$$

Evaluate the effect of tuning parameter J on the set-point responses for values of $J = 3, 4, 6,$ and 8 and $\Delta t = 5$ min.

Solution

The y and u responses for a unit set-point change at $t = 0$ are shown in Figs. 15.6 and 15.7, respectively.

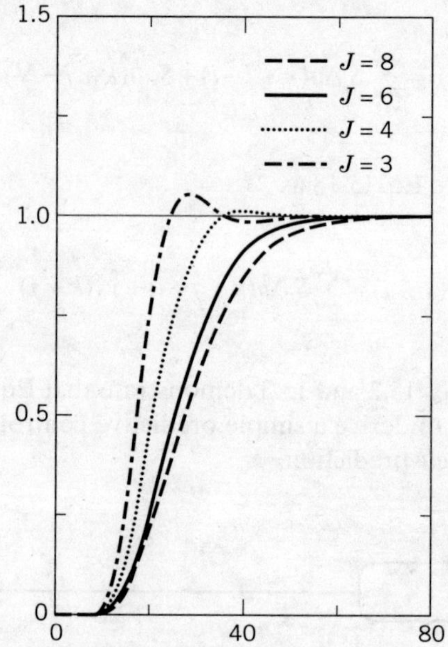

Figure 15.6 Set-point responses for Example 15.3 and different values of J.

As J increases, the y responses become more sluggish while the u responses become smoother. These trends occur because larger values of J allow the predictive controller more time before the J-step ahead prediction $\hat{y}(k+J)$ must equal the set point. Consequently, less strenuous control action is required. The Jth step-response coefficient S_J increases monotonically as J increases. Consequently, the input moves calculated from Eq. 15-22 tend to become smaller as S_J increases. (The u responses for $J = 4$ and 8 are omitted from Fig. 15.7.)

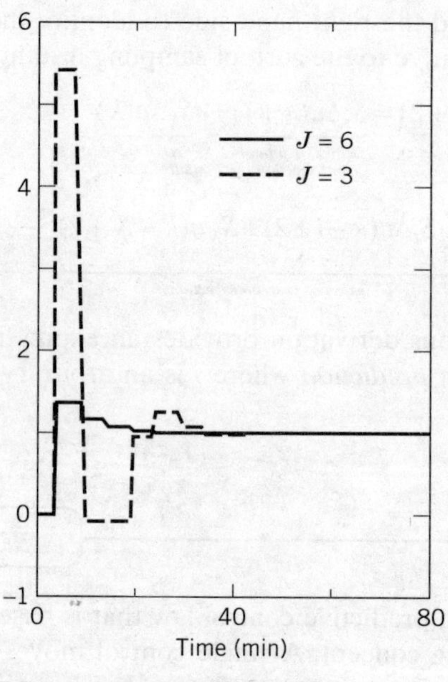

Figure 15.7 Input responses for Fig. 15.6.

The previous two examples have considered a simple predictive controller based on single prediction made J steps ahead. Now, we consider the more typical situation in which the MPC calculations are based on multiple predictions rather than on a single prediction. The notation is greatly simplified if vector-matrix notation is employed. Consequently, we define a vector of predicted responses for the next P sample instants as

$$\hat{\mathbf{Y}}(k+1) \triangleq col[\hat{y}(k+1), \hat{y}(k+2), \ldots, \hat{y}(k+P)] \qquad (15\text{-}24)$$

where col denotes a column vector. Similarly, a vector of predicted unforced responses from Eq. 15-19 is defined as

$$\hat{\mathbf{Y}}^o(k+1) \triangleq col[\hat{y}^o(k+1), \hat{y}^o(k+2), \ldots, \hat{y}^o(k+P)] \qquad (15\text{-}25)$$

Define $\Delta \mathbf{U}(k)$ to be a vector of control actions for the next M sampling instants:

$$\Delta \mathbf{U}(k) \triangleq col[\Delta u(k), \Delta u(k+1), \ldots, \Delta u(k+M-1)] \qquad (15\text{-}26)$$

The control horizon M and prediction horizon P are key design parameters, as discussed in Section 15.6. In general, $M \leq P$ and $P \leq N + M$.

The MPC control calculations are based on calculating $\Delta \boldsymbol{U}(k)$ so that the predicted outputs move optimally to the new set points. For the control calculations, the model predictions in Eq. 15-20 are conveniently written in vector-matrix notation as

$$\hat{\boldsymbol{Y}}(k+1) = \boldsymbol{S}\Delta \boldsymbol{U}(k) + \hat{\boldsymbol{Y}}^o(k+1) \quad (15\text{-}27)$$

where S is the $P \times M$ *dynamic matrix*:

$$S \triangleq \begin{bmatrix} S_1 & 0 & \cdots & 0 \\ S_2 & S_1 & 0 & \vdots \\ \vdots & \vdots & \ddots & 0 \\ S_M & S_{M-1} & \cdots & S_1 \\ S_{M+1} & S_M & \cdots & S_2 \\ \vdots & \vdots & \ddots & \vdots \\ S_P & S_{P-1} & \cdots & S_{P-M+1} \end{bmatrix} \quad (15\text{-}28)$$

Equations 15-27 and 15-28 can be derived from Eqs. 15-20 and 15-24 to 15-26.

15.2.3 Output Feedback and Bias Correction

The predictions in Eqs. 15-20 and 15-27 do not make use of the latest measurement $y(k)$. Consequently, the cumulative effects of model inaccuracy and unmeasured disturbances can lead to inaccurate predictions. However, prediction accuracy can be improved by utilizing the latest measurement in the predictions. This general strategy is referred to as *output* feedback (Qin and Badgwell, 2003). A typical approach is to add a *bias correction*, $b(k+j)$, to the prediction. The *corrected prediction*, $\tilde{y}(k+j)$, is defined as

$$\tilde{y}(k+j) \triangleq \hat{y}(k+j) + b(k+j) \quad (15\text{-}29)$$

We will refer to $\hat{y}(k+j)$ as the *uncorrected prediction*. In practice, the bias correction is often specified to be the difference between the latest measurement $y(k)$ and the corresponding predicted value, $\hat{y}(k)$:

$$b(k+j) = y(k) - \hat{y}(k) \quad (15\text{-}30)$$

The difference, $y(k) - \hat{y}(k)$, is also referred to as a *residual* or an *estimated disturbance* or *bias*. The block diagram for MPC in Fig. 15.1 includes the bias correction.

In Eq. 15-30, $\hat{y}(k)$ is a one-step ahead prediction made at the previous sampling instant, $k - 1$. Using Eq. 15-30 is equivalent to assuming that a process disturbance is added to the output and is constant for $j = 1, 2, \ldots, P$. Furthermore, the assumed value of the additive disturbance is the residual, $y(k) - \hat{y}(k)$.

Substituting Eq. 15-30 into Eq. 15-29 gives

$$\tilde{y}(k+j) \triangleq \hat{y}(k+j) + [y(k) - \hat{y}(k)] \quad (15\text{-}31)$$

In a similar fashion, adding the bias correction to the right side of Eq. 15-27 provides a vector of corrected predictions,

$$\tilde{\boldsymbol{Y}}(k+1) = \boldsymbol{S}\Delta \boldsymbol{U}(k) + \hat{\boldsymbol{Y}}^o(k+1) + [y(k) - \hat{y}(k)]\,\boldsymbol{1} \quad (15\text{-}32)$$

where $\boldsymbol{1}$ is a $\boldsymbol{P}$-dimensional column vector with each element having a value of one. Thus the same correction is made for all P predictions. Vector $\tilde{\boldsymbol{Y}}(k+1)$ is defined as

$$\tilde{\boldsymbol{Y}}(k+1) \triangleq col[\tilde{y}(k+1), \tilde{y}(k+2), \ldots, \tilde{y}(k+P)] \quad (15\text{-}33)$$

Incorporating output feedback as a bias correction has been widely applied, but it can result in excessively sluggish responses for certain classes of disturbances. Consequently, other types of output feedback and disturbance estimation methods have been proposed (Maciejowski, 2002; Qin and Badgwell, 2003; Rawlings and Mayne, 2009).

EXAMPLE 15.4

The benefits of using corrected predictions will be illustrated by a simple example, a first-order-plus-time-delay model:

$$\frac{Y(s)}{U(s)} = \frac{5e^{-2s}}{15s+1} \quad (15\text{-}34)$$

Assume that the disturbance transfer function is identical to the process transfer function, $G_d(s) = G_p(s)$. A unit change in u occurs at time $t = 2$ min, and a step disturbance, $d = 0.15$, occurs at $t = 8$ min. The sampling period is $\Delta t = 1$ min.

(a) Compare the process response $y(k)$ with the predictions that were made 15 steps earlier based on a step-response model with $N = 80$. Consider both the corrected prediction $\tilde{y}(k)$ and the uncorrected prediction $\hat{y}(k)$ over a time period, $0 \le k \le 90$.

(b) Repeat (a) for the situation where the step-response coefficients are calculated using an incorrect model:

$$\frac{Y(s)}{U(s)} = \frac{4e^{-2s}}{20s+1} \tag{15-35}$$

Solution

The output response to the step changes in u and d can be derived from Eq. 15-34 using the analytical techniques developed in Chapters 3 and 4. Because $\theta = 2$ min and the step in u begins at $t = 2$ min, $y(t)$ first starts to respond at $t = 5$ min. The disturbance at $t = 8$ min begins to affect y at $t = 11$ min. Thus, the response can be written as

$$\begin{aligned} y(t) &= 0 && \text{for } t \le 4\,\text{min} \\ y(t) &= 5(1)(1-e^{-(t-4)/15}) && \text{for } 4 < t \le 10\,\text{min} \\ y(t) &= 5(1)(1-e^{-(t-4)/15}) && \text{for } t > 10\,\text{min} \\ &\quad + 5(0.15)(1-e^{-(t-10)/15}) && \end{aligned} \tag{15-36}$$

The 15-step-ahead prediction, $\hat{y}(k+15)$, can be obtained using Eq. 15-20 with $j = 15$ and $N = 80$. The corrected prediction, $\tilde{y}(k+15)$, can be calculated from Eqs. 15-29 and 15-30 with $j = 15$. But in order to compare actual and predicted responses, it is more convenient to write these equations in an equivalent form:

$$\tilde{y}(k) \triangleq \hat{y}(k) + b(k) \tag{15-37}$$

$$b(k) = y(k-15) - \hat{y}(k-15) \tag{15-38}$$

(a) The actual and predicted responses are compared in Fig. 15.8. For convenience, the plots are shown as lines rather than as discrete points. After the step change in u at $t = 2$ min (or equivalently, at $k = 4$), the 15-step-ahead predictions are identical to the actual response until the step disturbance begins to affect $y(k)$ starting at $k = 11$. For $k > 10$, $\hat{y}(k) < y(k)$ because $\hat{y}(k)$ does not include the effect of the unknown disturbance. Note that $\tilde{y}(k) = \hat{y}(k)$ for $10 < k < 25$ because $b(k) = 0$ during this period. For $k > 25$, $b(k) \ne 0$ and $\tilde{y}(k)$ converges to $y(k)$. Thus, the corrected prediction $\tilde{y}(k)$ is more accurate than the uncorrected prediction, $\hat{y}(k)$.

(b) Figure 15.9 compares the actual and predicted responses for the case of the plant-model mismatch in Eqs. 15-34 and 15-35. The responses in Fig. 15.9 are similar to those in Fig. 15.8, but there are a few significant differences. Both of the predicted responses in Fig. 15.9 differ from the actual response for $t > 4$, as a result of the model inaccuracy. Figure 15.9 demonstrates that the corrected predictions are much more accurate than the uncorrected predictions even when a significant plant-model mismatch occurs. This improvement occurs because new information is used as soon as it becomes available.

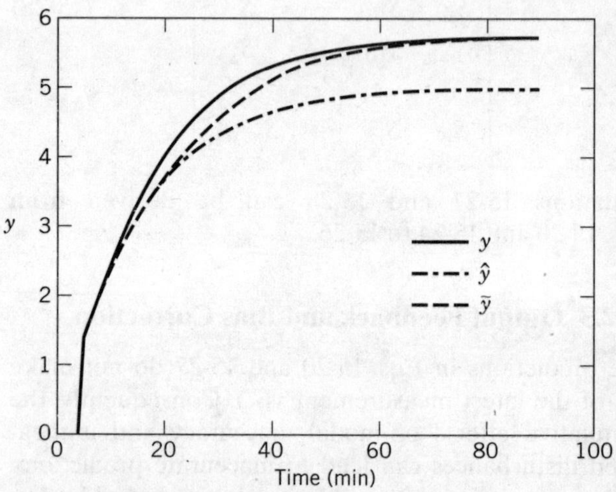

Figure 15.8 Comparison of actual (y), predicted ($\hat{y}$), and corrected ($\tilde{y}$) responses when the model is perfect.

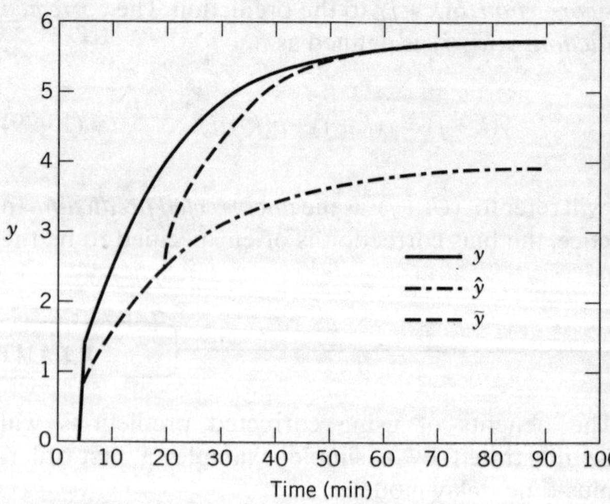

Figure 15.9 Comparison of actual and predicted responses for plant-model mismatch.

15.2.4 Extensions of the Basic MPC Model Formulation

We will now consider several extensions of the basic MPC problem formulation that are important for practical applications.

Integrating Processes

The standard step-response model in Eq. 15-14 is not appropriate for an integrating process because its step response is unbounded. However, because the output rate of change, $\Delta y(k) = y(k+1) - y(k)$, is bounded, a simple modification eliminates this problem. Replacing $\hat{y}(k+1)$ in Eq. 15-14 by $\Delta\hat{y}(k+1) = \hat{y}(k+1) - \hat{y}(k)$ provides an appropriate step-response model for integrating processes (Hokanson and Gerstle, 1992):

$$\Delta\hat{y}(k+1) = \sum_{i=1}^{N-1} S_i \Delta u[k-(i-1)] + S_N u[k-(N-1)] \tag{15-39}$$

or, equivalently,

$$\hat{y}(k+1) = \hat{y}(k) + \sum_{i=1}^{N-1} S_i \Delta u[k-(i-1)] + S_N u[k-(N-1)] \tag{15-40}$$

Although the bias correction approach of Eq. 15-30 is not valid for integrating processes, several modifications are available (Qin and Badgwell, 2003).

Known Disturbances

If a disturbance variable is known or can be measured, it can be included in the step-response model. Let d denote a measured disturbance and $\{S_i^d\}$ its step-response coefficients. Then the standard step-response model in Eq. 15-14 can be modified by adding a disturbance term,

$$\hat{y}(k+1) = \sum_{i=1}^{N-1} S_i \Delta u(k-i+1) + S_N u(k-N+1) + \sum_{i=1}^{N_d-1} S_i^d \Delta d(k-i+1) + S_N^d d(k-N_d+1) \tag{15-41}$$

where N_d is the number of step-response coefficients for the disturbance variable (in general, $N_d \neq N$). This same type of modification can be made to other step-response models such as Eq. 15-27 or 15-32. However, predictions made more than one step ahead require an assumption about future disturbances. If no other information is available, the usual assumption is that the future disturbances will be equal to the current disturbance: $d(k+j) = d(k)$ for $j = 1, 2, \ldots, P$. However, if a disturbance model is available, the prediction accuracy can improve.

15.3 PREDICTIONS FOR MIMO MODELS

The previous analysis for SISO systems can be generalized to MIMO systems by using the Principle of Superposition. For simplicity, we first consider a process control problem with two outputs, y_1 and y_2, and two inputs, u_1 and u_2. The predictive model consists of two equations and four individual step-response models, one for each input–output pair:

$$\hat{y}_1(k+1) = \sum_{i=1}^{N-1} S_{11,i}\Delta u_1(k-i+1) + S_{11,N}u_1(k-N+1) + \sum_{i=1}^{N-1} S_{12,i}\Delta u_2(k-i+1) + S_{12,N}u_2(k-N+1) \tag{15-42}$$

$$\hat{y}_2(k+1) = \sum_{i=1}^{N-1} S_{21,i}\Delta u_1(k-i+1) + S_{21,N}u_1(k-N+1) + \sum_{i=1}^{N-1} S_{22,i}\Delta u_2(k-i+1) + S_{22,N}u_2(k-N+1) \tag{15-43}$$

where $S_{12,i}$ denotes the ith step-response coefficient for the model that relates y_1 and u_2. The other step-response coefficients are defined in an analogous manner. This MIMO model is a straightforward generalization of the SISO model in Eq. 15-14. In general, a different model horizon can be specified for each input–output pair. For example, the upper limits for the summations in Eq. 15-43 can be specified as N_{21} and N_{22}, if y_2 has very different settling times for changes in u_1 and u_2.

Next, the analysis is generalized to MIMO models with arbitrary numbers of inputs and outputs. Suppose that there are r inputs and m outputs. In a typical MPC application, $r < 20$ and $m < 40$, but applications with much larger numbers of inputs and outputs have also been reported (Qin and Badgwell, 2003; Canney, 2003). It is useful to display the individual step-response models graphically as shown in Fig. 15.10 (Hokanson and Gerstle, 1992), where the output variables (or CVs) are arranged as the columns and the inputs and disturbances (the MVs and DVs) are arranged as the rows.

It is convenient to express MIMO step-response models in vector-matrix notation. Let the output vector be $\boldsymbol{y} = [y_1, y_2, \ldots, y_m]^T$ and the input vector be $\boldsymbol{u} = [u_1, u_2, \ldots, u_r]^T$ where superscript T denotes the transpose of a vector of matrix. In analogy with the derivation of Eq. 15-32 for SISO systems, the MIMO model for the corrected predictions can be expressed in dynamic matrix form:

$$\tilde{\boldsymbol{Y}}(\boldsymbol{k}+\boldsymbol{1}) = \boldsymbol{S}\Delta\boldsymbol{U}(k) + \hat{\boldsymbol{Y}}^o(k+1) + [y(k) - \hat{y}(k)]\boldsymbol{I} \tag{15-44}$$

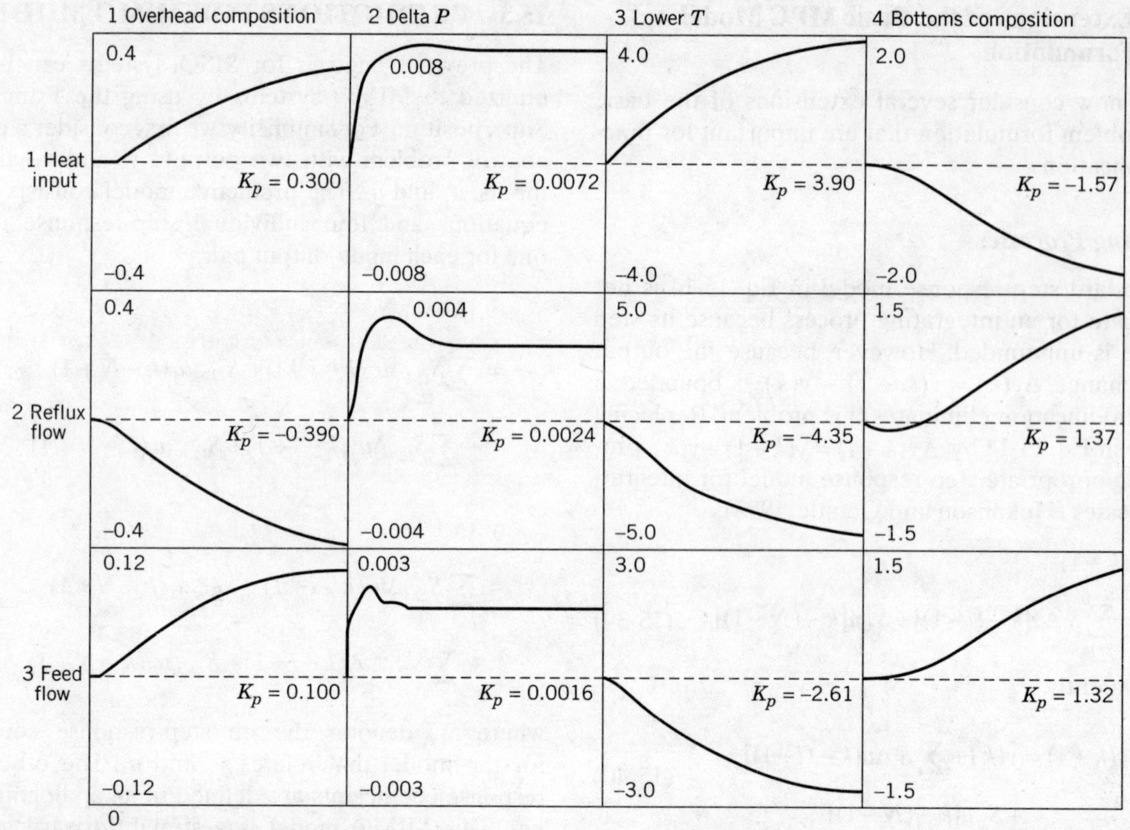

Figure 15.10 Individual step-response models for a distillation column with three inputs and four outputs. Each model represents the step response for 120 minutes (Hokanson and Gerstle, 1992).

where $\widetilde{Y}(k+1)$ is the mP-dimensional vector of corrected predictions over the prediction horizon P,

$$\widetilde{Y}(k+1) \triangleq col[\tilde{y}(k+1), \tilde{y}(k+2), \ldots, \tilde{y}(k+P)] \quad (15\text{-}45)$$

$\widetilde{Y}^o(k+1)$ is the mP-dimensional vector of predicted unforced responses,

$$\hat{Y}^o(k+1) \triangleq col\left[\hat{y}^o(k+1), \hat{y}^o(k+2), \ldots, \hat{y}^o(k+P)\right] \quad (15\text{-}46)$$

and $\Delta U(k)$ is the rM-dimensional vector of the next M control moves,

$$\Delta U(k) \triangleq col[\Delta u(k), \Delta u(k+1), \ldots, \Delta u(k+M-1)] \quad (15\text{-}47)$$

The $mP \times m$ matrix (in Eq. 15-44) is defined as

$$I \triangleq [\underbrace{I_m I_m \cdots I_m}_{P\ times}]^T \quad (15\text{-}48)$$

where I_m denotes the $m \times m$ identity matrix.

The dynamic matrix S is defined as

$$S \triangleq \begin{bmatrix} S_1 & 0 & \cdots & 0 \\ S_2 & S_1 & 0 & \vdots \\ \vdots & \vdots & \ddots & 0 \\ S_M & S_{M-1} & \cdots & S_1 \\ S_{M+1} & S_M & \cdots & S_2 \\ \vdots & \vdots & \ddots & \vdots \\ S_P & S_{P-1} & \cdots & S_{P-M+1} \end{bmatrix} \quad (15\text{-}49)$$

where S_i is the $m \times r$ matrix of step-response coefficients for the ith time step.

$$S_i \triangleq \begin{bmatrix} S_{11,i} & S_{12,i} & \cdots & S_{1r,i} \\ S_{21,i} & \cdots & \cdots & S_{2r,i} \\ \vdots & \vdots & \vdots & \vdots \\ S_{m1,i} & \cdots & \cdots & S_{mr,i} \end{bmatrix} \quad (15\text{-}50)$$

Note that the dynamic matrix in Eq. 15-49 for MIMO systems has the same structure as the one for SISO systems in Eq. 15-28.

The dimensions of the vectors and matrices in Eq. 15-44 are as follows. Both $\widetilde{Y}(k+1)$ and $\widetilde{Y}^o(k+1)$ are mP-dimensional vectors where m is the number of outputs and P is the prediction horizon. Also, $\Delta U(k)$ is an rM-dimensional vector where r is the number of manipulated inputs and M is the control horizon. Consequently, the dimensions of step-response matrix $\boldsymbol{S}$ are $mP \times rM$. The MIMO model in Eqs. 15-44 through 15-50 is the MIMO generalization of the SISO model in Eq. 15-32. It is also possible to write MIMO models in an alternative form, a generalization of Eqs. 15-42 and 15-43. An advantage of this alternative formulation is that the new dynamic matrix is partitioned into the individual SISO models, a convenient form for real-time predictions.

For stable models, the predicted unforced response, $\hat{\boldsymbol{Y}}^o(k+1)$ in Eq. 15-46, can be calculated from a recursive relation (Lundström et al., 1995) that is in the form of a discrete-time version of a state-space model:

$$\hat{\boldsymbol{Y}}^o(k+1) = \boldsymbol{M}\hat{\boldsymbol{Y}}^o(k) + \boldsymbol{S}^*\Delta \boldsymbol{u}(k) \tag{15-51}$$

where:

$$\hat{\boldsymbol{Y}}^o(k) = col\left[\hat{\boldsymbol{y}}^o(k), \hat{\boldsymbol{y}}^o(k+1), \ldots, \hat{\boldsymbol{y}}^o(k+P-1)\right] \tag{15-52}$$

$$\boldsymbol{M} \triangleq \begin{bmatrix} \boldsymbol{0} & \boldsymbol{I}_m & \boldsymbol{0} & \cdots & \boldsymbol{0} \\ \boldsymbol{0} & \boldsymbol{0} & \boldsymbol{I}_m & \ddots & \boldsymbol{0} \\ \vdots & \vdots & \vdots & \ddots & \boldsymbol{0} \\ \boldsymbol{0} & \boldsymbol{0} & \cdots & \boldsymbol{0} & \boldsymbol{I}_m \\ \boldsymbol{0} & \boldsymbol{0} & \cdots & \boldsymbol{0} & \boldsymbol{I}_m \end{bmatrix} \tag{15-53}$$

$$\boldsymbol{S}^* \triangleq \begin{bmatrix} \boldsymbol{S}_1 \\ \boldsymbol{S}_2 \\ \vdots \\ \boldsymbol{S}_{P-1} \\ \boldsymbol{S}_P \end{bmatrix} \tag{15-54}$$

where $\boldsymbol{M}$ is an $mP \times mP$ matrix and $\boldsymbol{S}^*$ is an $mP \times r$ matrix. The MIMO models in Eqs. 15-44 through 15-54 can be extended to include measured disturbances and integrating variables, in analogy to the SISO case in the previous section.

Most of the current MPC research, as well as industrial practice, is based on state-space models, because they provide an important theoretical advantage, namely, a unified framework for both linear and nonlinear control problems. They also facilitate state estimation (Yu et al., 1994), which is a powerful method for reconciling measurements and predicted variables. State-space models are also more convenient for theoretical analysis and facilitate a wider range of output feedback strategies (Rawlings, 2000; Maciejowski, 2002; Qin and Badgwell, 2003; Rawlings and Mayne, 2009; Qin and Badgwell, 2014). Both linear (Bonavita et al., 2003) and nonlinear (Poland et al., 2009) state-space models are also widely used in MPC applications in industry, with applications ranging from cement kilns (Stadler et al., 2011) to gas compressors (Cortinovis et al., 2015).

The architecture of the controller is similar in many respects to the developments in this chapter, except for the substitution of a state-space model in place of the step response model. For example, the prediction formulas in Eqs. 15-42 and 15-43 would be replaced by the appropriate version of Eqs. 15-11 and 15-12. Note that the state-space model is already in multivariable format and that the step response model can be viewed as a special case (the derivation is left to the reader).

15.4 MODEL PREDICTIVE CONTROL CALCULATIONS

The flowchart in Fig. 15.11 provides an overview of the MPC calculations. The seven steps are shown in the order they are performed at each control execution time. For simplicity, we assume that the control

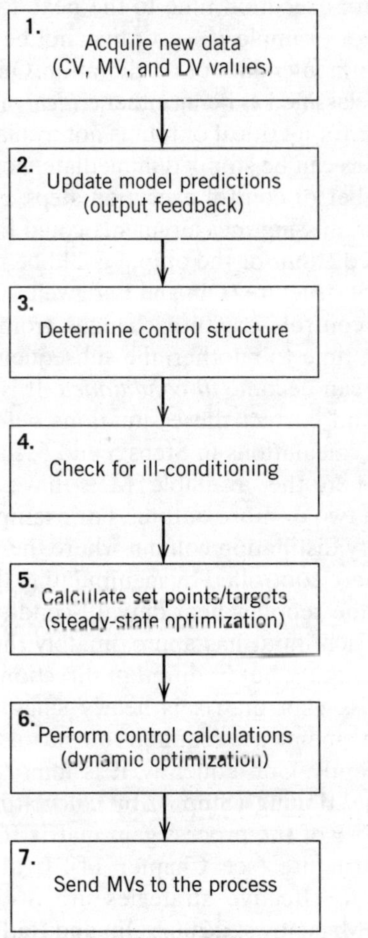

Figure 15.11 Flow chart for MPC calculations (modified from Qin and Badgwell (2003)).

execution times coincide with the measurement sampling instants.

In MPC applications, the calculated MV moves are usually implemented as set points for regulatory control loops at the Distributed Control System (DCS) level, such as flow control loops. If a DCS control loop has been disabled or placed in manual, the MV is no longer available for control. In this situation, the control degrees of freedom are reduced by one. Even though an MV is unavailable for control, it can serve as a disturbance variable if it is measured.

In Step 1 of the MPC calculations, new process data are acquired via the regulatory control system (DCS) that is interfaced to the process. Then new output predictions are calculated in Step 2 using the process model and the new data (e.g., see Eqs. 15-29 and 15-30).

Before each control execution, it is necessary to determine which outputs (CVs), inputs (MVs), and disturbance variables (DVs) are currently available for the MPC calculations. This Step 3 activity is referred to as determining the current control structure. The variables available for the control calculations can change from one control execution time to the next, for a variety of reasons. For example, a sensor may not be available due to routine maintenance or recalibration. Output variables are often classified as being either *critical* or *noncritical*. If the sensor for a critical output is not available, the MPC calculations can be stopped immediately or after a specified number of control execution steps. For a noncritical output, missing measurements could be replaced by model predictions or the output could be removed from the control structure (Qin and Badgwell, 2003).

If the control structure changes from one control execution time to another, the subsequent control calculations can become *ill-conditioned*. It is important to identify and correct these situations before executing the MPC calculations in Steps 5 and 6. Ill-conditioning occurs when the available MVs have very similar effects on two or more outputs. For example, consider a high-purity distillation column where the product compositions are controlled by manipulating the reflux flow rate and the reboiler heat duty. Ill-conditioning occurs because each input has approximately the same effect on both outputs, but in different directions. As a result, the process gain matrix is nearly singular, and large input movements are required to control these outputs independently. Consequently, it is important to check for ill-conditioning (Step 4) by calculating the *condition number* of the process gain matrix for the current control structure (see Chapter 14). If ill-conditioning is detected, effective strategies are available for its removal (Maciejowski, 2002; Qin and Badgwell, 2003).

In MPC applications, the major benefits result from determining the optimal operating conditions (set-point calculations) and from moving the process to these set points in an optimal manner based on the control calculations. Both types of calculations optimize a specified objective function while satisfying inequality constraints, such as upper and lower limits on the inputs or outputs. Set-point calculations are the subject of Section 15.5, while control calculations are considered in the next section.

The final step, Step 7 of Fig. 15.11, is to implement the calculated control actions, usually as set points to regulatory PID control loops at the DCS level.

15.4.1 Unconstrained MPC

This section considers the control calculations of Step 6 for the special case in which inequality constraints are not included in the problem formulation. In Section 15.4.2, the analysis is extended to the more typical situation where there are inequality constraints on $\boldsymbol{u}$, $\Delta\boldsymbol{u}$, and $\boldsymbol{y}$.

As noted earlier, the MPC control calculations are based on both current measurements and model predictions. The control objective is to calculate a set of control moves (MV changes) that make the corrected predictions as close to a reference trajectory as possible. Thus, an optimization approach is employed. For unconstrained linear control problems, an analytical expression for the MPC control law is available.

Reference Trajectory

In MPC applications, a reference trajectory can be used to make a gradual transition to the desired set point. The *reference trajectory* $\boldsymbol{y}_r$ can be specified in several different ways (Maciejowski, 2002; Qin and Badgwell, 2003; Rossiter, 2003). We briefly introduce several typical approaches.

Let the reference trajectory over the prediction horizon P be denoted as

$$\boldsymbol{Y}_r(k+1) \triangleq col[\boldsymbol{y}_r(k+1), \boldsymbol{y}_r(k+2), \ldots, \boldsymbol{y}_r(k+P)] \qquad (15\text{-}55)$$

where $\boldsymbol{Y}_r$ is an mP-dimensional vector. A reasonable approach is to specify the reference trajectory to be the *filtered set point*,

$$y_{i,r}(k+j) = (\alpha_i)^j y_{i,r}(k) + [1-(\alpha_i)^j] y_{i,sp}(k) \qquad (15\text{-}56)$$
$$\text{for } i = 1,2,\ldots,m \quad \text{and} \quad j = 1,2,\ldots,P$$

where $y_{i,r}$ is the ith element of $\boldsymbol{y}_r$, $\boldsymbol{y}_{sp}$ denotes the set point, and α_i is a filter constant, $0 < \alpha_i < 1$. For $j = 1$, Eq. 15-56 reduces to the set-point filtering expression for PID controllers that was considered in Chapter 11.

It is also equivalent to the exponential filter introduced in Chapter 13. Note that $y_r = y_{sp}$ for the limiting case of $\alpha_i = 0$. An alternative approach is to specify the reference trajectory for the ith output as an exponential trajectory from the current measurement $y_i(k)$ to the set point, $y_{i,sp}(k)$:

$$y_{i,r}(k+j) = (\alpha_i)^j y_i(k) + [1-(\alpha_i)^j] y_{i,sp}(k) \quad \text{for } i = 1,2,\ldots,m \text{ and } j = 1,2,\ldots,P \tag{15-57}$$

In some commercial MPC products, the desired reference trajectory for each output is specified indirectly by a *performance ratio* for the output. The performance ratio is defined to be the ratio of the desired closed-loop settling time to the open-loop settling time. Thus, small values of the performance ratios correspond to small values of α_i in Eq. 15-56 or 15-57.

Model Predictive Control Law (Dynamic Matrix Control)

The control calculations are based on minimizing the predicted deviations from the reference trajectory. Let k denote the current sampling instant. The *predicted error* vector, $\hat{\boldsymbol{E}}(k+1)$, is defined as

$$\hat{\boldsymbol{E}}(k+1) \triangleq \boldsymbol{Y}_r(k+1) - \widetilde{\boldsymbol{Y}}(k+1) \tag{15-58}$$

where $\widetilde{\boldsymbol{Y}}(k+1)$ was defined in Eq. 15-45. Similarly, $\hat{\boldsymbol{E}}^o(k+1)$ denotes the *predicted unforced error vector*,

$$\hat{\boldsymbol{E}}^o(k+1) \triangleq \boldsymbol{Y}_r(k+1) - \widetilde{\boldsymbol{Y}}^o(k+1) \tag{15-59}$$

where the *corrected prediction for the unforced case*, $\widetilde{\boldsymbol{Y}}^o(k+1)$, is defined as

$$\widetilde{\boldsymbol{Y}}^o(k+1) \triangleq \widetilde{\boldsymbol{Y}}^o(k+1) + \boldsymbol{I}[\boldsymbol{y}(k) - \hat{\boldsymbol{y}}(k)] \tag{15-60}$$

Thus, $\hat{\boldsymbol{E}}^o(k+1)$ represents the predicted deviations from the reference trajectory when no further control action is taken, that is, the predicted deviations when $\Delta\boldsymbol{u}(k+j) = \boldsymbol{0}$ for $j = 0, 1, \ldots, M-1$. Note that $\hat{\boldsymbol{E}}(k+1)$ and $\hat{\boldsymbol{E}}^o(k+1)$ are mP-dimensional vectors.

The general objective of the MPC control calculations is to determine $\Delta\boldsymbol{U}(k)$, the control moves for the next M time intervals,

$$\Delta\boldsymbol{U}(k) = col[\Delta\boldsymbol{u}(k), \Delta\boldsymbol{u}(k+1), \ldots, \Delta\boldsymbol{u}(k+M-1)] \tag{15-61}$$

The rM-dimensional vector $\Delta\boldsymbol{U}(k)$ is calculated so that an objective function (also called a *performance* index) is minimized. Typically, either a linear or a quadratic objective function is employed. For unconstrained MPC, the objective function is based on minimizing some (or all) of three types of deviations or errors (Qin and Badgwell, 2003):

1. The predicted errors over the predicted horizon, $\widetilde{\boldsymbol{E}}(k+1)$
2. The next M control moves, $\Delta\boldsymbol{U}(k)$
3. The deviations of $\boldsymbol{u}(k+i)$ from its desired steady-state value $\boldsymbol{u}_{sp}$ over the control horizon

For MPC based on linear process models, both linear and quadratic objective functions can be used (Maciejowski, 2002; Qin and Badgwell, 2003). The dynamic matrix control algorithm is based on quadratic objective function and step response model. To demonstrate the MPC control calculations, consider a quadratic objective function J based on the first two types of deviations:

$$\min_{\Delta\boldsymbol{U}(k)} J = \widehat{\boldsymbol{E}}(k+1)^T \boldsymbol{Q} \widehat{\boldsymbol{E}}(k+1) + \Delta\boldsymbol{U}(k)^T \boldsymbol{R} \Delta\boldsymbol{U}(k) \tag{15-62}$$

where $\boldsymbol{Q}$ is a positive-definite weighting matrix and $\boldsymbol{R}$ is a positive semi-definite matrix. Both are usually diagonal matrices with positive diagonal elements. The weighting matrices are used to weight the most important elements of $\hat{\boldsymbol{E}}(k+1)$ or $\Delta\boldsymbol{U}(k)$ as described in Section 15.6. If diagonal weighting matrices are specified, these elements are weighted individually.

The MPC control law utilizing dynamic matrix control algorithm, minimizes the value of objective function J of equation 15-62 subjected to the equality constraint of $\hat{E}(k+1)$ given by equation 15-58. The optimum value of control moves ΔU over a control horizon M, which minimizes the value of J can be calculated analytically.

$$\Delta\boldsymbol{U}(k) = (\boldsymbol{S}^T\boldsymbol{Q}\boldsymbol{S} + \boldsymbol{R})^{-1}\boldsymbol{S}^T\boldsymbol{Q}\hat{\boldsymbol{E}}^o(k+1) \tag{15-63}$$

This control law can be written in a more compact form,

$$\Delta\boldsymbol{U}(k) = \boldsymbol{K}_c \hat{\boldsymbol{E}}^o(k+1) \tag{15-64}$$

where the *controller gain matrix* $\boldsymbol{K}_c$ is defined to be

$$\boldsymbol{K}_c \triangleq (\boldsymbol{S}^T\boldsymbol{Q}\boldsymbol{S} + \boldsymbol{R})^{-1}\boldsymbol{S}^T\boldsymbol{Q} \tag{15-65}$$

Note that $\boldsymbol{K}_c$ is an $rM \times mP$ matrix that can be evaluated off-line rather than on-line provided that the dynamic matrix $\boldsymbol{S}$ and weighting matrices, $\boldsymbol{Q}$ and $\boldsymbol{R}$, are constant.

The MPC control law in Eq. 15-64 can be interpreted as a multivariable, proportional control law based on the predicted unforced error rather than the conventional control error (set point–measurement). The control law utilizes the latest measurement $\boldsymbol{y}(k)$ because it appears in the expressions for the corrected prediction $\tilde{\boldsymbol{y}}(k)$, and thus also in the predicted unforced error, $\tilde{\boldsymbol{E}}_o(k+1)$. Furthermore, the MPC control law in Eq. 15-64 implicitly contains integral control action because $\boldsymbol{u}$ tends to change until the unforced error $\hat{\boldsymbol{E}}_o$ becomes zero. Thus, offset is eliminated for set-point changes or sustained disturbances.

Although the MPC control law calculates a set of M input moves, $\Delta \boldsymbol{U}(k)$ only the first control move, $\Delta \boldsymbol{u}(k)$, is actually implemented. Then at the next sampling instant, new data are acquired and a new set of control moves is calculated. Once again, only the first control move is implemented. These activities are repeated at each sampling instant, and the strategy is referred to as a *receding horizon approach*. The first control move, $\Delta \boldsymbol{u}(k)$, can be calculated from Eqs. 15-61 and 15-64,

$$\mathrm{D}\boldsymbol{u}(k) = \boldsymbol{K}_{c1}\,\hat{\boldsymbol{E}}^{o}(k+1) \tag{15-66}$$

where matrix $\boldsymbol{K}_{c1}$ is defined to be the first r rows of $\boldsymbol{K}_c$. Thus, $\boldsymbol{K}_{c1}$ has dimensions of $r \times mP$.

It may seem strange to calculate an M-step control policy and then only implement the first move. The important advantage of this receding horizon approach is that new information in the form of the most recent measurement $\boldsymbol{y}(k)$ is utilized immediately instead of being ignored for the next M sampling instants. Otherwise, the multistep predictions and control moves would be based on old information and thus be adversely affected by unmeasured disturbances, as demonstrated in Example 15.4.

The calculation of $\boldsymbol{K}_c$ requires the inversion of an $rM \times rM$ matrix where r is the number of input variables and M is the control horizon. For large problems with many inputs, the required computational effort can be reduced by using *input blocking* (Maciejowski, 2002; Qin and Badgwell, 2003). In this approach, the inputs are not changed at every sampling instant. Instead, $\Delta \boldsymbol{u} = \boldsymbol{0}$ for "blocks" of sampling instants. Input blocking is illustrated in Fig. 15.12 where a single input is changed at each sampling instant for the first four sampling instants (k through $k+3$). Starting at $k+4$, u is blocked so that it changes every three sampling instants until the steady-state value is reached at $k+13$. The design parameters are the block length and the time at which blocking begins.

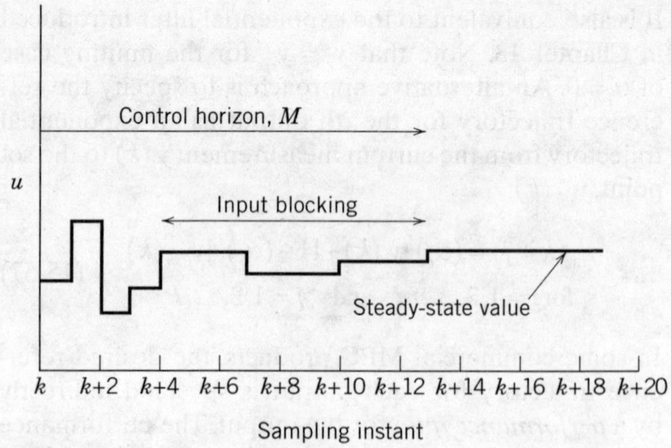

Figure 15.12 Input blocking.

15.4.2 MPC with Inequality Constraints

Inequality constraints on input and output variables are important characteristics for MPC applications. In fact, inequality constraints were a primary motivation for the early development of MPC. Input constraints occur as a result of physical limitations on plant equipment such as pumps, control valves, and heat exchangers. For example, a manipulated flow rate might have a lower limit of zero and an upper limit determined by the pump, control valve, and piping characteristics. The dynamics associated with large control valves impose rate-of-change limits on manipulated flow rates.

Constraints on output variables are a key component of the plant operating strategy. For example, a common distillation column control objective is to maximize the production rate while satisfying constraints on product quality and avoiding undesirable operating regimes such as flooding or weeping.

For processing plants different types of inequality and equality constraints often arise as given below:

1. *Production constraints.* Equipment throughput restrictions, storage limits, emissions restrictions, or market constraints (no additional product can be sold) are frequently encountered in manufacturing. These constraints have the form of $xi \leq ci$ or $gi = xi - ci \leq 0$.
2. *Raw material limitations.* Feedstock supplies are frequently limited owing to supplier capability or production levels of other plants within the same company.
3. *Safety restrictions.* Common examples are limitations on operating temperature and pressure.
4. *Product specifications.* Constraints placed on the physical properties or composition of the final product fall into this category. For blends of various liquid products in a refinery, it is commonly assumed that a blend property can be calculated by averaging pure component properties. Thus, a blend of Nc components with physical property

values ψ_k and volume fractions y_k (based on volumetric flow rates) has a calculated blend property of

$$\bar{\Psi} = \sum_{k=1}^{N_c} \Psi_k y_k \quad (15\text{-}67)$$

If there is an upper limit α on $\bar{\Psi}$, the resulting constraint is linear in y_k:

$$\sum_{k=1}^{N_c} \Psi_k y_k \leq \alpha \quad (15\text{-}68)$$

5. *Material and energy balances.* Although items 1–4 generally are considered to be inequality constraints, the steady-state material and energy balances are equality constraints.

The set of inequality constraints for $\boldsymbol{u}$ and $\boldsymbol{y}$ define an operating window for the process, as shown in Fig. 15.13

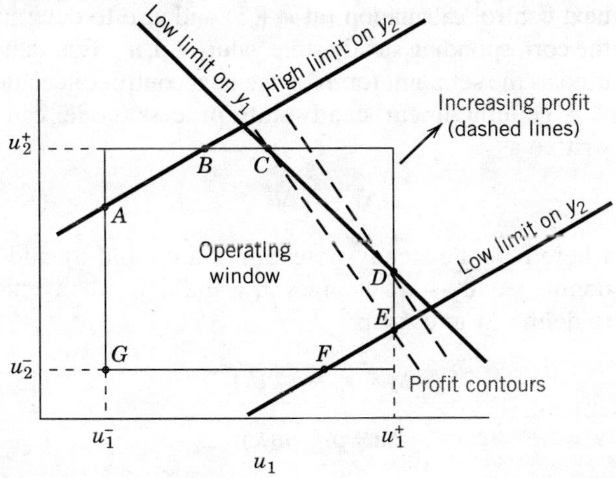

Figure 15.13 Operating window for a two input–two output optimization problem. The dashed lines are objective function contours, increasing from left to right. The maximum profit occurs where the profit line intersects the constraints at vertex *D*.

Inequality constraints can be included in the control calculations in many different ways (Maciejowski, 2002; Qin and Badgwell, 2003). It is convenient to make a distinction between *hard constraints* and *soft constraints*. As the name implies, a hard constraint cannot be violated at any time. By contrast, a soft constraint can be violated, but the amount of violation is penalized by a modification of the objective function, as described below. This approach allows small constraint violations to be tolerated for short periods of time.

For MPC, the inequality constraints for $\boldsymbol{u}$ and $\Delta\boldsymbol{u}$ are typically hard constraints specified as upper and lower limits:

$$\boldsymbol{u}^-(k) \leq \boldsymbol{u}(k+j) \leq \boldsymbol{u}^+(k) \quad j = 0, 1\ldots, M-1 \quad (15\text{-}69)$$

$$\Delta\boldsymbol{u}^-(k) \leq \Delta\boldsymbol{u}(k+j) \leq \Delta\boldsymbol{u}^+(k) \quad j = 0, 1\ldots, M-1 \quad (15\text{-}70)$$

The analogous hard constraints for the predicted outputs are:

$$\boldsymbol{y}^-(k+j) \leq \tilde{\boldsymbol{y}}(k+j) \leq \boldsymbol{y}^+(k+j) \quad j = 1, 2, \ldots, P \quad (15\text{-}71)$$

Unfortunately, hard output constraints can result in infeasible solutions for the optimization problem, especially for large disturbances. Consequently, output constraints are usually expressed as soft constraints involving *slack variables* $\boldsymbol{c}_j$ (Qin and Badgwell, 2003):

$$\boldsymbol{y}^-(k+j) - \boldsymbol{c}_j \leq \tilde{\boldsymbol{y}}(k+j) \leq \boldsymbol{y}^+(k+j) + \boldsymbol{c}_j j = 1, 2, \ldots, P \quad (15\text{-}72)$$

The numerical values of the slack variables can be determined during constrained optimization if the performance index in Eq. 15-62 is modified by adding a penalty term for the slack variables. Thus, an *mP*-dimensional vector of slack variables is defined as $\bar{\boldsymbol{C}} \triangleq col[\boldsymbol{c}_1, \boldsymbol{c}_2, \ldots, \boldsymbol{c}_P]$. The modified performance index is

$$\min_{\Delta U(k)} \boldsymbol{J} = \hat{\boldsymbol{E}}(k+1)^T \boldsymbol{Q} \hat{\boldsymbol{E}}(k+1) + \Delta\boldsymbol{U}(k)^T \boldsymbol{R} \Delta\boldsymbol{U}(k) + \bar{\boldsymbol{C}}^T \boldsymbol{T} \bar{\boldsymbol{C}} \quad (15\text{-}73)$$

where $\boldsymbol{T}$ is an $mP \times mP$ weighting matrix for the slack variables. Note that inequality constraints in Eqs. 15-71 and 15-72 are imposed on the corrected prediction $\tilde{\boldsymbol{y}}$, rather than the actual output $\boldsymbol{y}$, because future values of $\boldsymbol{y}$ are not available. Consequently, $\boldsymbol{y}$ may violate a constraint even though $\tilde{\boldsymbol{y}}$ does not, as a result of modeling errors. Slack variables can also be used to weight positive and negative errors, differently.

Range Control

An unusual feature of MPC applications is that many output variables do not have set points. For these outputs, the control objective is to keep them between upper and lower limits, an approach called *range control* (or *zone control*). The limits can vary with time, as shown in Eq. 15-71. The advantage of range control is that it creates additional degrees of freedom for the control calculations. Furthermore, many output variables such as the liquid level in a surge tank do not have to be regulated at a set point. In medical applications, there are examples of physiological variables that yield equivalent outcome performance for a range, as opposed to a single value. One well studied example is the use of zone Model Predictive Control for regulating glucose in Type 1 Diabetes (Grosman et al., 2010). Consequently, in many MPC applications, range control is the rule rather than the exception. Set points are only specified for output variables that must be kept close to a specified value (e.g., pH or a quality variable). Note that control to a set point can be considered to be a special case of range control that occurs when the upper and lower limits in Eq. 15-71 are equal.

The constraint limits in Eqs. 15-69 to 15-72 can vary with time as a result of changes in process equipment or instrumentation. However, it can also be beneficial to allow the limits to change in a specified manner over the control or prediction horizons. For example, in the *limit funnel* technique, the output limits in Eq. 15-71 or 15-72 gradually become closer together over the prediction horizon (Maciejowski, 2002; Qin and Badgwell, 2003).

The introduction of inequality constraints results in a constrained optimization problem that can be solved numerically using linear or quadratic programming techniques (Edgar et al., 2001). As an example, consider the addition of inequality constraints to the MPC design problem in the previous section. Suppose that it is desired to calculate the M-step control policy $\Delta U(k)$ that minimizes the quadratic objective function J in Eq. 15-54, while satisfying the constraints in Eqs. 15-69, 15-70, and 15-71. The output predictions are made using the step-response model in Eq. 15-44. This MPC design problem can be solved numerically using the quadratic programming technique given in Appendix F.

15.5 SET-POINT CALCULATIONS

As indicated in Section 15.1 and Fig. 15.11, the MPC calculations at each control execution time are typically performed in two steps. First, the optimum set points (or *targets*) for the control calculations are determined. Then, a set of M control moves are generated by the control calculations, and the first move is implemented. In practical applications, significant economic benefits result from both types of calculations, but the steady-state optimization is usually more important. In this section, the set-point calculations are described in more detail.

The MPC set points are calculated so that they maximize or minimize an economic objective function. The calculations are usually based on linear steady-state models and a simple objective function, typically a linear or quadratic function of the MVs and CVs. The linear model can be a linearized version of a complex nonlinear model or the steady-state version of the dynamic model that is used in the control calculations. Linear inequality constraints for the MVs and CVs are also included in the steady-state optimization. The set-point calculations are repeated at each sampling instant because the active constraints can change frequently due to disturbances, instrumentation, equipment availability, or varying process conditions.

Because the set-point calculations are repeated as often as every minute, the steady-state optimization problem must be solved quickly and reliably. If the optimization problem is based on a linear process model, linear inequality constraints, and either a linear or a quadratic cost function, the linear and quadratic programming techniques discussed in Appendix F can be employed.

15.5.1 Formulation of the Set-Point Optimization Problem

Next, we provide an overview of the set-point calculation problem. More detailed descriptions are available elsewhere (Sorensen and Cutler, 1998; Kassman et al., 2000; Rawlings and Mayne, 2009; Maciejowski, 2002).

Consider an MIMO process with r MVs and m CVs. Denote the current values of $\boldsymbol{u}$ and $\boldsymbol{y}$ as $\boldsymbol{u}(k)$ and $\boldsymbol{y}(k)$. The objective is to calculate the optimum set point $\boldsymbol{y}_{sp}$ for the next control calculation (at $k+1$) and also to determine the corresponding steady-state value of $\boldsymbol{u}$, $\boldsymbol{u}_{sp}$. This value is used as the set point for $\boldsymbol{u}$ for the next control calculation.

A general, linear steady-state process model can be written as

$$\Delta \boldsymbol{y} = \boldsymbol{K} \Delta \boldsymbol{u} \tag{15-74}$$

where $\boldsymbol{K}$ is the steady-state gain matrix and $\Delta \boldsymbol{y}$ and $\Delta \boldsymbol{u}$ denote steady-state changes in $\boldsymbol{y}$ and $\boldsymbol{u}$. It is convenient to define $\Delta \boldsymbol{y}$ and $\Delta \boldsymbol{u}$ as

$$\Delta \mathbf{y} \triangleq \boldsymbol{y}_{sp} - \boldsymbol{y}_{OL}(k) \tag{15-75}$$

$$\Delta \mathbf{u} \triangleq \boldsymbol{u}_{sp} - \boldsymbol{u}(k) \tag{15-76}$$

In Eq. 15-75 $\boldsymbol{y}_{OL}(k)$ represents the steady-state value of $\boldsymbol{y}$ that would result if $\boldsymbol{u}$ were held constant at its current value, $\boldsymbol{u}(k)$, until steady state was achieved. In general, $\boldsymbol{y}_{OL}(k) \neq \boldsymbol{y}(k)$ except for the ideal situation where the process is at steady state at time k. In order to incorporate output feedback, the steady-state model in Eq. 15-74 is modified as

$$\Delta \boldsymbol{y} = \boldsymbol{K} \Delta \boldsymbol{u} + [\boldsymbol{y}(k) - \hat{\boldsymbol{y}}(k)] \tag{15-77}$$

A representative formulation for the set-point optimization is to determine the optimum values, $\boldsymbol{u}_{sp}$ and $\boldsymbol{y}_{sp}$, that minimize a quadratic objective function,

$$\min_{\boldsymbol{u}_{sp},\ \boldsymbol{y}_{sp}} J_s = \boldsymbol{c}^T \boldsymbol{u}_{sp} + \boldsymbol{d}^T \boldsymbol{y}_{sp} + \boldsymbol{e}_y^T \boldsymbol{Q}_{sp} \boldsymbol{e}_y + \boldsymbol{e}^T \boldsymbol{R}_{sp} \boldsymbol{e} + \boldsymbol{C}^T \boldsymbol{T}_{sp} \boldsymbol{C} \tag{15-78}$$

subject to satisfying Eq. 15-74 and inequality constraints on the MVs and CVs:

$$\boldsymbol{u}^- \le \boldsymbol{u}_{sp} \le \boldsymbol{u}^+ \tag{15-79}$$

$$\Delta\boldsymbol{u}^- \le \Delta\boldsymbol{u}_{sp} \le \Delta\boldsymbol{u}^+ \tag{15-80}$$

$$\boldsymbol{y}^- - \boldsymbol{c} \le \boldsymbol{y}_{sp} \le \boldsymbol{y}^+ + \boldsymbol{c} \tag{15-81}$$

where

$$\boldsymbol{e}_y \triangleq \boldsymbol{y}_{sp} - \boldsymbol{y}_{\text{ref}} \tag{15-82}$$

$$\boldsymbol{e}_u \triangleq \boldsymbol{u}_{sp} - \boldsymbol{u}_{\text{ref}} \tag{15-83}$$

The $\boldsymbol{c}$ vector in Eq. 15-81 denotes the slack elements. In Eqs. 15-82 and 15-83, $\boldsymbol{y}_{\text{ref}}$ and $\boldsymbol{u}_{\text{ref}}$ are the desired steady-state values of $\boldsymbol{y}$ and $\boldsymbol{u}$ that are often determined by a higher-level optimization. The weighting factors in Eq. 15-78, $\boldsymbol{c}$, $\boldsymbol{d}$, $\boldsymbol{Q}_{sp}$, $\boldsymbol{R}_{sp}$, and $\boldsymbol{T}_{sp}$, are selected based on economic considerations. Although the weighting factors are constants in Eq. 15-78, in MPC applications they can vary with time to accommodate process changes or changes in economic conditions such as product prices or raw material costs. Similarly, it can be advantageous to allow the limits in Eqs. 15-79 to 15-81 ($\boldsymbol{u}^-$, $\boldsymbol{u}^+$, etc.) to vary from one execution time to the next, as discussed in Section 15.4. Fortunately, new values of weighting factors and constraint limits are easily accommodated, because the optimum set points are recalculated at each execution time.

It is important to make a distinction between $\boldsymbol{y}_{\text{ref}}$ and $\boldsymbol{u}_{\text{ref}}$, and $\boldsymbol{y}_{sp}$ and $\boldsymbol{u}_{sp}$. Both pairs represent desired values of $\boldsymbol{y}$ and $\boldsymbol{u}$, but they have different origins and are used in different ways. Reference values, $\boldsymbol{y}_{\text{ref}}$ and $\boldsymbol{u}_{\text{ref}}$, are often determined infrequently by a higher-level optimization. They are used as the desired values for the steady-state optimization of Step 5 of Fig. 15.11. By contrast, $\boldsymbol{y}_{sp}$ and $\boldsymbol{u}_{sp}$ are calculated at each MPC control execution time and serve as set points for the control calculations of Step 6.

We have emphasized that the goal of this steady-state optimization is to determine $\boldsymbol{y}_{sp}$ and $\boldsymbol{u}_{sp}$, the set points for the control calculations in Step 6 of Fig. 15.11. But why not use $\boldsymbol{y}_{\text{ref}}$ and $\boldsymbol{u}_{\text{ref}}$ for this purpose? The reason is that $\boldsymbol{y}_{\text{ref}}$ and $\boldsymbol{u}_{\text{ref}}$ are ideal values that may not be attainable for the current plant conditions and constraints, which could have changed since $\boldsymbol{y}_{\text{ref}}$ and $\boldsymbol{u}_{\text{ref}}$ were calculated. Thus, steady-state optimization (Step 5) is necessary to calculate $\boldsymbol{y}_{sp}$ and $\boldsymbol{u}_{sp}$, target values that more accurately reflect current conditions. In Eq. 15-78, $\boldsymbol{y}_{sp}$ and $\boldsymbol{u}_{sp}$ are shown as the independent values for the optimization. However, $\boldsymbol{y}_{sp}$ can be eliminated by substituting the steady-state model, $\boldsymbol{y}_{sp} = \boldsymbol{K}\boldsymbol{u}_{sp}$.

Next, we demonstrate that the objective function J_s is quite flexible, by showing how it is defined for three different types of applications.

Application 1: *Maximize operating profit.*
An economic model of a plant involves the costs of raw materials, values of products, and costs of production as functions of operating conditions, projected sales figures, and so on. An objective function is specified in terms of these quantities; in particular, *operating profit* over some specific period of time can be expressed as

$$P = \sum_s F_s V_s - \sum_r F_r C_r - OC \tag{15-84}$$

where

P = operating profit/time

$\sum_s F_s V_s$ = sum of product flow rates times respective product values

$\sum_r F_r C_r$ = sum of feed flow rate times respective unit cost

OC = operating costs/time

If the product, feedstock, and utility flow rates are manipulated or disturbance variables in the MPC control structure, they can be included in objective function J_s. In order to maximize the operating profit (OP), the objective function is specified to be $Js = -OP$, because minimizing J_s is equivalent to maximizing $-J_s$. The weighting matrices for the two quadratic terms, $\boldsymbol{Q}_{sp}$ and $\boldsymbol{R}_{sp}$, are set equal to zero.

Application 2: *Minimize deviations from the reference values.*
Suppose that the objective of the steady-state optimization is to calculate $\boldsymbol{y}_{sp}$ and $\boldsymbol{u}_{sp}$ so that they are as close as possible to the reference values, $\boldsymbol{y}_{\text{ref}}$ and $\boldsymbol{u}_{\text{ref}}$. This goal can be achieved by setting $\boldsymbol{c} = \boldsymbol{0}$ and $\boldsymbol{d} = \boldsymbol{0}$ in Eq. 15-78. Weighting matrices $\boldsymbol{Q}_{sp}$, $\boldsymbol{R}_{sp}$, and $\boldsymbol{T}_{sp}$ should be chosen according to the relative importance of the MVs, CVs, and constraint violations.

Application 3: *Maximize the production rate.*
Suppose that the chief control objective is to maximize a production rate while satisfying inequality constraints on the inputs and the outputs. Assume that the production rate can be adjusted via a flow control loop whose set point is denoted as u_{1sp} in the MPC control structure. Thus, the optimization objective is to maximize u_{1sp}, or equivalently, to minimize $-u_{1sp}$. Consequently, the performance index in Eq. 15-78 becomes $J_s = -u_{1sp}$. This expression can be derived by setting all of the weighting factors equal to zero except for c_1, the first element of $\boldsymbol{c}$. It is chosen to be $c_1 = -1$.

The set-point optimization problem can be summarized as follows. At each sampling instant, the optimum values of $\boldsymbol{u}$ and $\boldsymbol{y}$ for the next sampling instant ($\boldsymbol{u}_{sp}$ and $\boldsymbol{y}_{sp}$) are calculated by minimizing the cost function in Eq. 15-78, subject to satisfying the model equation

15-72 and the constraints in Eqs. 15-79 to 15-81. This optimization problem can be solved efficiently using the standard LP or QP techniques.

Infeasible calculations can occur if the calculations of Steps 5 and 6 are based on constrained optimization, because feasible solutions do not always exist (Edgar et al., 2001). Infeasible problems can result when the control degrees of freedom are reduced (e.g., control valve maintenance), large disturbances occur, or the inequality constraints are inappropriate for current conditions. For example, the allowable operating window in Fig. 19.6 could disappear for inappropriate choices of the y_1 and y_2 limits. Other modifications can be made to ensure that the optimization problem always has a feasible solution (Kassmann et al., 2000).

In view of the dramatic decreases in the ratio of computer cost to performance in recent years, it can be argued that physically based, nonlinear process models should be used in the set-point calculations, instead of approximate linear models. However, linear models are still widely used in MPC applications for three reasons: First, linear models are reasonably accurate for small changes in $\boldsymbol{u}$ and $\boldsymbol{d}$ and can easily be updated based on current data or a physically based model. Second, some model inaccuracy can be tolerated, because the calculations are repeated on a frequent basis and they include output feedback from the measurements. Third, the computational effort required for constrained, nonlinear optimization is still relatively large, but is decreasing.

15.6 SELECTION OF DESIGN AND TUNING PARAMETERS

A number of design parameters must be specified in order to design an MPC system. In this section, we consider key design issues and recommended values for the parameters. Several design parameters can also be used to tune the MPC controller. The effects of the MPC design parameters will be illustrated in two examples.

Sampling period Δt and model horizon N. The sampling period Δt and model horizon N (in Eq. 15-14) should be chosen so that $N\Delta t = t_s$ where t_s is the settling time for the open-loop response. This choice ensures that the model reflects the full effect of a change in an input variable over the time required to reach steady state. Typically, $30 \le N \le 120$. If the output variables respond on different time scales, a different value of N can be used for each output, as noted earlier. Also, different model horizons can be used for the MVs and DVs, as illustrated in Eq. 15-41.

Control M and prediction P horizons. As control horizon M increases, the MPC controller tends to become more aggressive and the required computational effort increases. However, the computational effort can be reduced by input blocking, as shown in Fig. 15.12. Some typical rules of thumb are $5 \le M \le 20$ and $N/3 < M < N/2$. A different value of M can be specified for each input.

The prediction horizon P is often selected to be $P = N + M$ so that the full effect of the last MV move is taken into account. Decreasing the value of P tends to make the controller more aggressive. A different value of P can be selected for each output if their settling times are quite different. An infinite prediction horizon can also be used and has significant theoretical advantages (Maciejowski, 2002; Rawlings and Mayne, 2009).

Weighting Matrices, Q and R

The output weighting matrix $\boldsymbol{Q}$ in Eq. 15-62 allows the output variables to be weighted according to their relative importance. Thus, an $mP \times mP$ diagonal $\boldsymbol{Q}$ matrix allows the output variables to be weighted individually, with the most important variables having the largest weights. For example, if a reactor temperature is considered more important than a liquid level, the temperature will be assigned a larger weighting factor. An important consideration in tuning with the $\boldsymbol{Q}$ matrix is that the output variables should be scaled to equivalent (e.g., dimensionless) ranges. That way one can prioritize, say, temperature versus concentrations without having to account for unit conversions. The same reasoning applies to the $\boldsymbol{R}$ matrix. The inverse of a diagonal weighting factor is sometimes referred to as an *equal concern factor* (Qin and Badgwell, 2003).

It can be advantageous to adjust the output weighting over the prediction horizon. For example, consider an SISO model with a time delay θ. Suppose that an input change Δu occurs at $k = 0$. Then $y(k) = 0$ until $k\Delta t > \theta$ due to the time delay. Consequently, it would be reasonable to set the corresponding elements of the $\boldsymbol{Q}$ matrix equal to zero, or, equivalently, to make the corresponding predictions zero. These approaches tend to make the control calculations better conditioned (see Section 15.4).

As a second example, the elements of $\boldsymbol{Q}$ that correspond to predicted errors early in the prediction horizon (e.g., at time $k + 1$) can be weighted more heavily than the predicted errors at the end of the horizon, $k + P$, or vice versa. The use of coincidence points is a special case of this strategy. Here, the corrected errors only have nonzero weights for a subset of the P sampling instants called coincidence points. The corrected errors at other times are given zero weighting. In Example 15.2 a simple predictive control strategy was derived based on a single coincidence point.

A time-varying $\boldsymbol{Q}$ matrix can also be used to implement soft constraints by real-time adjustment of $\boldsymbol{Q}$. For example, if an output variable approaches an upper or lower limit, the corresponding elements of $\boldsymbol{Q}$ would be temporarily increased.

In a similar fashion, $\boldsymbol{R}$ in Eq. 15-62 allows input MVs to be weighted according to their relative importance. This $rM \times rM$ matrix is referred to as the *input weighting matrix* or the *move suppression matrix*. It is usually chosen to be a diagonal matrix with the diagonal elements r_{ii}, referred to as *move suppression factors*. They provide convenient tuning parameters, because increasing the value of r_{ii} tends to make the MPC controller more conservative by reducing the magnitudes of the MV moves.

If a reference trajectory is employed, move suppression is not required, and thus $\boldsymbol{R}$ can be set equal to zero.

Reference Trajectory α_i

In MPC applications, the desired future output behavior can be specified in several different ways: as a set point, high and low limits, a reference trajectory, or a funnel (Qin and Badgwell, 2003). Both the reference trajectory and the funnel approaches have a tuning factor that can be used to adjust the desired speed of response for each output. Consider Eq. 15-56 or 15-57, for example. As α_i increases from zero to one, the desired reference trajectory becomes slower. Alternatively, the performance ratio concept can be used to specify the reference trajectories. As mentioned earlier, the performance ratio is defined to be the ratio of the desired closed-loop settling time to the open-loop settling time.

The influence of MPC design parameters is illustrated by a simple example.

EXAMPLE 15.5

A process has the transfer function,

$$\frac{Y(s)}{U(s)} = \frac{e^{-s}}{(10s+1)(5s+1)}$$

(a) Use Eq. 15-65 to calculate the controller gain matrix, $\boldsymbol{K}_c$, for $\boldsymbol{Q} = \boldsymbol{I}, \boldsymbol{R} = \boldsymbol{0}$ two cases:

(i) $P = 3, \quad M = 1$

(ii) $P = 4, \quad M = 2$

Assume that $N = 70$, $\Delta t = 1$ and that u is unconstrained for each case.

(b) Compare the set-point responses of two MPC controllers and a digital PID controller with $\Delta t =$ 0.5 and ITAE set-point tuning (Chapter 11): $K_c = 2.27$, $\tau_I = 16.6$ and $\tau_D = 1.49$. Compare both y and u responses.

(c) Repeat **(b)** for a unit step disturbance and a PID controller with ITAE disturbance tuning: $K_c =$ 3.52, $\tau_I = 6.98$, and $\tau_D = 1.73$.

Solution

(a) The step response coefficients are obtained by evaluating the step response at the sampling instants, $t = i\Delta t = i$ (because $\Delta t = 1$):

$$S_1 = 0$$

$$S_i = 1 - 2e^{-0.1(i-1)} + e^{-0.2^{(i-1)}} \quad \text{for } i = 2, 3, \ldots, 70$$

The controller matrix $\boldsymbol{K}_c$ for each case is shown in Table 15.2. Note that the dimensions of $\boldsymbol{K}$ are different for the two cases, because $\boldsymbol{K}_c$ has dimensions of $rM \times mP$, as noted earlier. For this SISO example, $r = m = 1$, and the values of M and P differ for the two cases.

Table 15.2 Feedback Matrices K_c for Example 15.5

For $P = 3$ and $M = 1$:	$K_c = [0 \quad 7.79 \quad 28.3]$
For $P = 4$ and $M = 2$:	$K_c = \begin{bmatrix} 0 & 33.1 & 48.8 & -13.4 \\ 0 & -71.4 & -97.4 & 57.3 \end{bmatrix}$

(b) The unit step response can be derived analytically using Lapace transforms:

$$y(t) = 0 \qquad \text{for } t \leq 1$$

$$y(t) = 1 - 2e^{-0.1(t-1)} + e^{-0.2(t-1)} \qquad \text{for } t > 1$$

Figure 15.12 compares the y and u responses for a unit set point change. The two MPC controllers provide superior output responses with very small settling times, but their initial MV changes are larger than those for the PID controller. (Note the expanded time scale for u.)

(c) For the step disturbance, the output responses for the MPC controllers in Fig. 15.15 have relatively small maximum deviations and are non-oscillatory. By comparison, the PID controller results in the largest maximum deviation and an oscillatory response. Of the two MPC controllers, the one designed using $P = 3$ and $M = 1$ provides a slightly more conservative response.

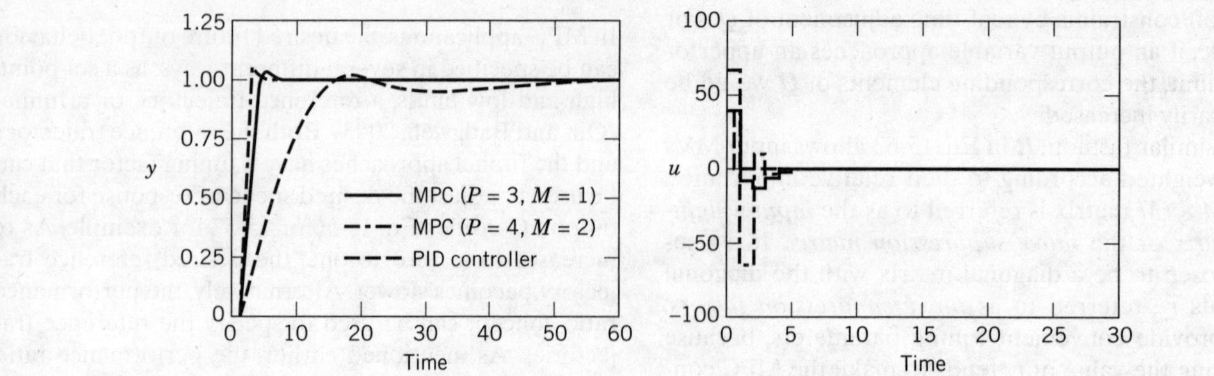

Figure 15.14 Set-point responses for Example 15.5.

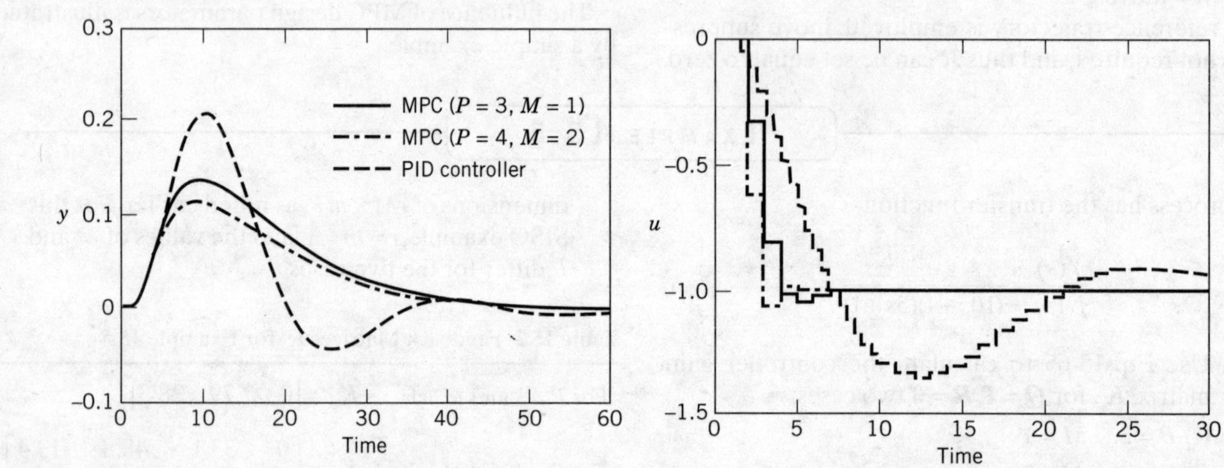

Figure 15.15 Disturbance responses for Example 15.5.

15.6.1 MPC Application: Distillation Column Model

In order to illustrate the effects of the MPC design parameters ($M, P, \boldsymbol{Q}$, and $\boldsymbol{R}$) for an MIMO problem, consider the Wood–Berry model that was introduced in Example 14.1:

$$\begin{bmatrix} X_D(s) \\ X_B(s) \end{bmatrix} = \begin{bmatrix} \dfrac{12.8e^{-s}}{16.7s+1} & \dfrac{-18.9e^{-3s}}{21s+1} \\ \dfrac{6.6e^{-7s}}{10.9s+1} & \dfrac{-19.4e^{-3s}}{14.4s+1} \end{bmatrix} \begin{bmatrix} R(s) \\ S(s) \end{bmatrix} + \begin{bmatrix} \dfrac{3.8e^{-8.1s}}{14.9s+1} \\ \dfrac{4.9e^{-3.4s}}{13.2s+1} \end{bmatrix} F(s) \tag{15-85}$$

The controlled variables (outputs) are the distillate and bottoms compositions (X_D and X_B); the manipulated variables (inputs) are the reflux flow rate and the steam flow rate to the reboiler (R and S); and feed flow rate F is an unmeasured disturbance variable.

Table 15.3 PI Controller Settings for the Wood–Berry model

Control Loop	K_c	τ_I (min)
$X_D - R$	0.85	7.21
$X_B - S$	−0.089	8.86

Next, we compare a variety of MPC controllers and a multiloop control system, based on simulations performed using the MATLAB Model Predictive Control Toolbox (Bemporad et al., 2009).[1] For each simulation the sampling period was $\Delta t = 1$ min, and saturation limits of ±0.15 were imposed on each input. Unconstrained MPC controllers were designed using Eq. 15-63, while the constrained MPC controllers were based on the input constraints in Eq. 15-69. Some constrained MPC

[1]The code for the Wood–Berry example is available in this MATLAB Toolbox. A modified version of the code is included with Exercise 15.9. A newer version of the MPC Toolbox is also available, but without this example (Bemporad et al., 2009).

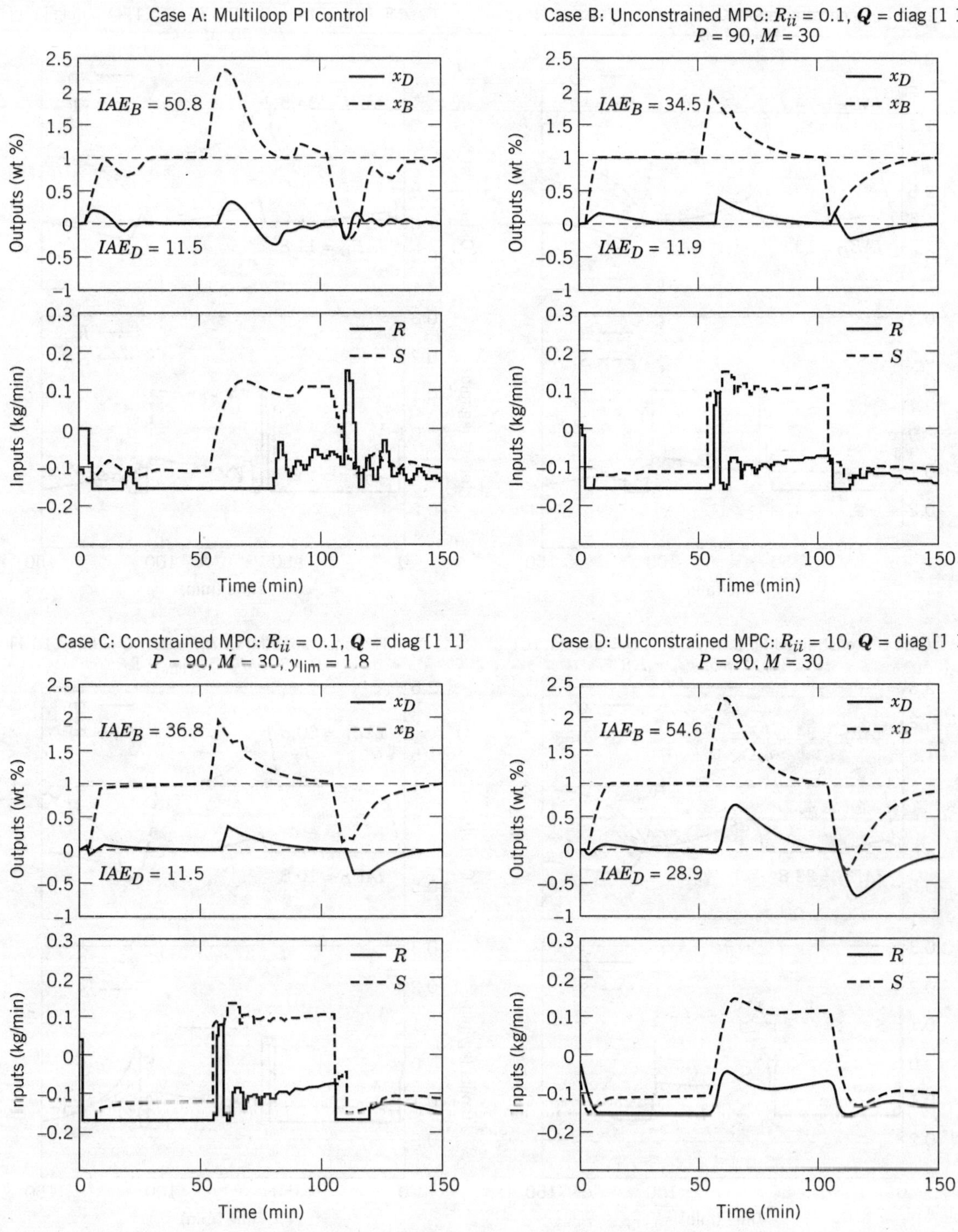

Figure 15.16 Comparison of multiloop PI control and MPC for the Wood–Berry model.

controllers were designed using an additional hard-output constraint of $|y_i| \leq 1.8$, while in practice this would be done with slack variables. In order to compare MPC and a standard multiloop control system, two PI controllers were simulated using the $X_D - R/X_B - S$ control configuration from Example 15.1 and the controller settings in Table 15.3 reported by Lee et al. (1998).

Figures 15.16 and 15.17 compare the performance of the MPC and multiloop control systems for a +1% set-point change in X_B at $t = 0$, followed by two feed flow rate disturbances: a +30% increase at t = 50 min and return to the original value at t = 100 min. The input and output variables are displayed as deviation variables. The numerical values of the integral of the absolute error

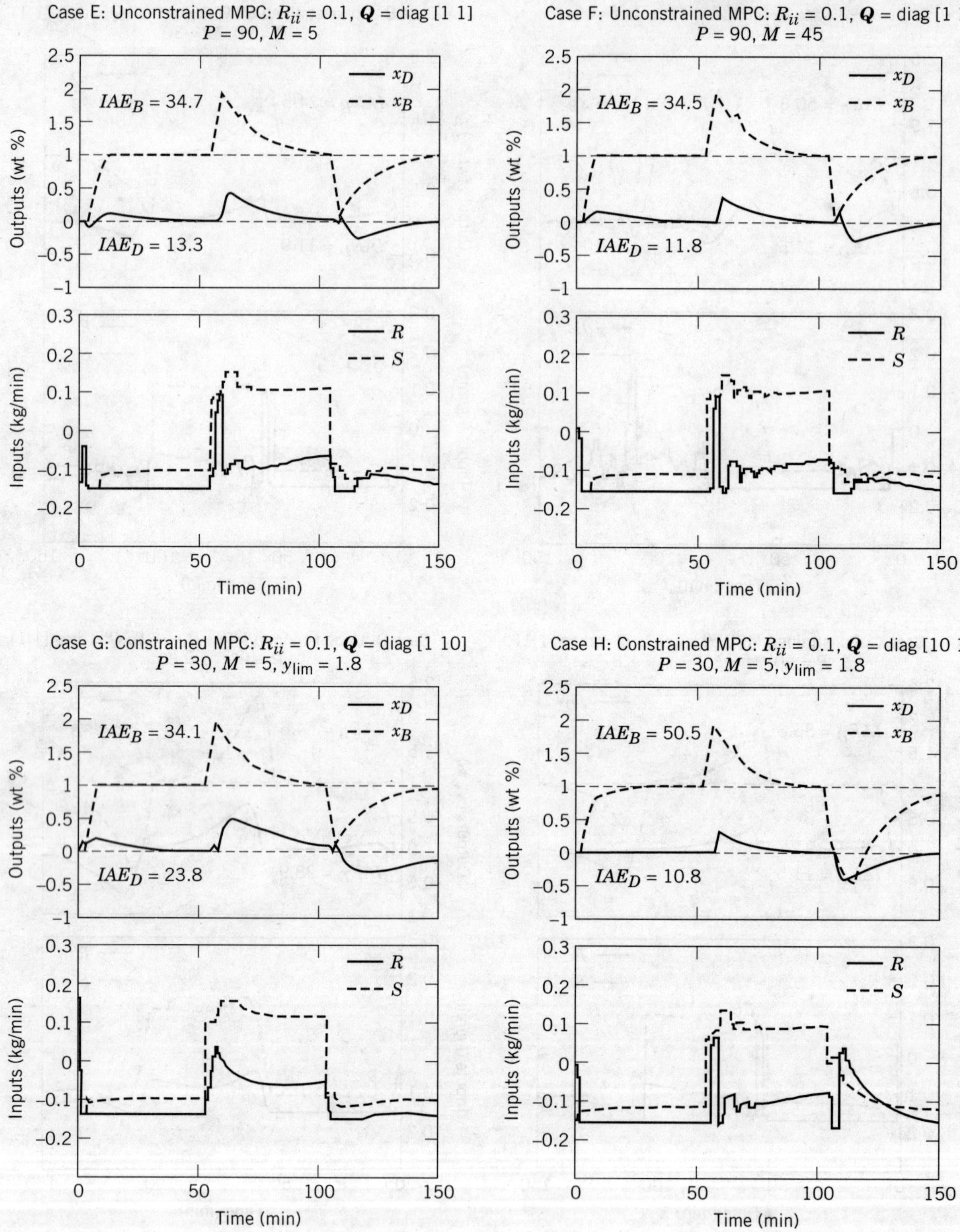

Figure 15.17 Effects of MPC design parameters for the Wood–Berry model.

(IAE) performance index (Chapter 11) are included for each output.

A comparison of Cases A and B in Fig. 15.16 indicates that unconstrained MPC is superior to the multiloop control system, because its output variables exhibit faster set-point responses, less oscillation, and smaller IAE values. In addition, the changes in the input variables are smoother for MPC. Case B is used as a "base case" for the comparisons in Figs. 15.16 and 15.17. Its MPC design parameters are shown in Fig. 15.16 and were selected according to the guidelines presented earlier.

Cases B and C in Fig. 15.16 provide a comparison of constrained and unconstrained MPC. These responses are very similar, with only small differences occurring, mainly for the second disturbance. This somewhat surprising result can be interpreted as follows. The responses for constrained and unconstrained MPC are very similar because the inputs are saturated much of the time for both controllers. When one input saturates, the MPC controller only has a single degree of freedom left, the other input. By contrast, for larger control problems (e.g., 10×10), constrained MPC will have many more degrees of freedom. For these larger problems, constrained MPC tends to provide improved control due to the extra degrees of freedom and its awareness of the constraints and process interactions.

The effect of a diagonal move suppression matrix $\boldsymbol{R}$ is apparent from a comparison of Cases B and D. When the diagonal elements, R_{ii}, are increased from 0.1 to 10, the MPC inputs become smoother and the output responses have larger deviations, higher IAE values, and longer settling times.

The effect of changing control horizon, M, is shown in Cases B, E, and F. The y responses and IAE values are quite similar for all three values of M: 5, 30, and 45. However, the u responses are smoother for $M = 5$.

Cases G and H demonstrate that improved control of a designated output variable can be achieved by adjusting the elements of the $\boldsymbol{Q}$ matrix in Eq. 15-62. For Case G, x_B is weighted 10 times more heavily than x_D, in contrast to Case H, where the reverse situation occurs. Control of the more heavily weighted output improves at the expense of the other output, as indicated by smaller maximum deviations, IAE values, and settling times. For Cases G and H, $P = 30$, and the results are similar to other cases where $P = 90$.

15.7 IMPLEMENTATION OF MPC

This section provides an overview of the activities that are involved in designing and implementing a model predictive control system. For a new MPC application, a cost/benefit analysis is usually performed prior to project approval. Then the steps involved in the implementation of MPC can be summarized as follows (Hokanson and Gerstle, 1992; Qin and Badgwell, 2003):

1. Initial controller design
2. Pretest activity
3. Plant tests
4. Model development
5. Control system design and simulation
6. Operator interface design and operator training
7. Installation and commissioning
8. Measuring results and monitoring performance

Step 1: Initial Controller Design

The first step in MPC design is to select the controlled, manipulated, and measured disturbance variables. These choices determine the structure of the MPC control system and should be based on process knowledge and control objectives. In typical applications the number of controlled variables is less than or equal to 40, and the number of manipulated (input) variables is less than or equal to 20. These preliminary selections are reviewed in Step 5 and revised, if necessary. The input and measured disturbance variables that are varied during the plant tests of Step 3 should be chosen carefully. For example, if it is decided to add a new input variable later during Step 5, additional plant tests would be required, a nontrivial task. By contrast, additional output variables can be added to the MPC control structure later, if necessary, provided that these measurements were recorded during the plant tests.

Step 2: Pretest Activity

During the pretest activity (*or pretest*, for short), the plant instrumentation is checked to ensure that it is working properly. Remedial action may be required for faulty sensors, sticking control valves, and the like. Also, a decision may be made to install sensors for some process variables that are not currently measured. The pretest also includes preliminary experimental tests to estimate the steady-state gains and approximate settling times for each input–output pair. This information is used to plan the full plant tests of Step 3.

As mentioned earlier, the results of the MPC control calculations are input moves that are implemented as set points for regulatory control loops. For example, if a cooling water flow rate is an MPC input variable, the MPC controller calculates the set point for the corresponding DCS control loop. Consequently, it is important to thoroughly check the performance of the DCS control system during the pretest, and to retune or reconfigure control loops if necessary.

These evaluation and maintenance activities are very important. If the basic instrumentation and DCS control system do not function properly, the MPC strategy will be ineffective, and the success of the MPC application will be jeopardized.

In the pretest experiments, each manipulated variable (MV) is *bumped* at least once by making a small step change. Steady-state gains and settling times are estimated from the step-response data using the techniques described in Chapter 16. Each measured disturbance variable (DV) should also be bumped, if possible. If not, the gains and settling times can be estimated from historical data for periods during which the disturbance variables changed significantly. During these bump tests, any existing DCS control loops for the output variables should be placed in manual. Thus, the pretest experiments are open-loop step tests (see Chapter 11). However, the MV and DV moves are usually implemented as set-point changes to the DCS loops for the DVs and MVs.

As part of the pretest, it is desirable to benchmark the performance of the existing control system for later comparison with MPC performance (Step 8). For example, the closed-loop responses for representative set-point changes and measured disturbances could be characterized using the performance criteria of Chapter 11. A baseline for the economic performance of the control system should also be established, although it is not always easy to do so.

Step 3: Plant Tests

The dynamic model for the MPC calculations is developed from data collected during special plant tests. The plant testing can be very time-consuming, typically requiring days, or even weeks, of around-the-clock experiments. The required test duration depends on the settling times of the outputs and the numbers of MVs and DVs. The excitation for the plant tests usually consists of changing an input variable or a disturbance variable (if possible) from one value to another, using either a series of step changes with different durations or the pseudorandom-binary sequence (PRBS) that will be introduced in Chapter 6. The plant test experiments are implemented in the same manner as the pretest experiments of Step 2.

It is traditional industrial practice to move each MV and DV individually. The magnitudes of the moves should be carefully chosen, because too small a move may result in the step responses being obscured by normal process fluctuations and measurement noise. On the other hand, too large a change may result in an output constraint violation or nonlinear process behavior that cannot be accurately described by a linear model.

The magnitude of the maximum allowable input changes can be estimated from knowledge of the output constraints and the estimated steady-state gains from the pretest. For example, suppose that $(\Delta u_j)_{max}$ denotes the maximum change that can be made in u_j without violating a constraint for y_i. It can be estimated from the expression,

$$(\Delta u_j)_{max} = \frac{(\Delta y_i)_{max}}{\hat{K}_{ij}} \tag{15-86}$$

where $(\Delta y_i)_{max}$ is the maximum allowable change in y_i and $\hat{K}_{ij}$ is the estimated steady-state gain between y_i and u_j. However, this steady-state analysis does not guarantee that each y_i satisfies its constraints during transient responses.

The duration of the longest step test is equal to t_{max}, the longest settling time that was observed during the pretest. Shorter step changes are also made, with the durations typically varying from $t_{max}/8$ to $t_{max}/2$. In order to ensure that sufficient data are obtained for model identification, each input variable is typically moved 8–15 times (Qin and Badgwell, 2003).

Some MPC vendors recommend a total plant testing period of $t_{test} = 6(r+p)t_{max}$ where r is the number of input variables and p is the number of measured disturbance variables. In principle, t_{test} can be reduced by making simultaneous changes to several input (or disturbance) variables rather than the traditional sequential ("one-at-a-time") approach. Also, it can be very difficult to identify poorly conditioned process models using the sequential approach. However, because of a number of practical considerations, input moves are traditionally made sequentially. In particular, simultaneous input moves tend to complicate the test management and make it more difficult to identify periods of abnormal operation by visual inspection of the test data. It is also more difficult to ensure that output constraints will not be violated. Because of similar practical considerations, step changes have been traditionally preferred over the pseudorandom binary sequence (PRBS).

Step 4: Model Development

The dynamic model is developed from the plant test data by selecting a model form (e.g., a step-response model) and then estimating the model parameters. However, first it is important to eliminate periods of test data during which plant upsets or other abnormal situations have occurred, such as control valve saturation or a DCS control loop having been placed in manual. Decision to omit portions of the test data is based on visual inspection of the data, knowledge of the process, and experience. Parameter estimation is usually based on least-squares estimation (Chapter 16).

As part of the model development step, the model accuracy should be characterized, because this information is useful for subsequent system design and tuning. The characterization can include confidence intervals for the model predictions and/or model parameters. The confidence intervals can be calculated using standard statistical techniques (Ljung, 1999).

Step 5: Control System Design and Simulation

The MPC design is based on the control and optimization objectives, process constraints, and the dynamic model of the process. The preliminary control system design from Step 1 is critically evaluated and modified, if necessary. Then the MPC design parameters in Section 15.6 are selected, including the sampling periods, weighting factors, and control and prediction horizons. Next, the closed-loop system is simulated using the identified process model and a wide variety of process conditions to evaluate control system performance. The MPC design parameters are adjusted, if necessary, to obtain satisfactory control system performance and robustness over the specified range of operating conditions.

Step 6: Operator Interface Design and Operator Training
Because plant operators play a key role in manufacturing plants, it is important that the MPC operator interface meet their needs. Operator training is also important, because MPC concepts such as predictive control, multivariable interactions, and constraint handling are very different from conventional regulatory control concepts. For a standard multiloop control system, each input is adjusted based on measurements of a single output. By contrast, in MPC each input depends on all of the outputs. Thus, understanding why the MPC system responds the way that it does, especially in unusual operating conditions, can be very challenging for both operators and engineers.

Step 7: Installation and Commissioning
After an MPC control system is installed, it is first evaluated in a "prediction mode." Model predictions are compared with measurements, but the process continues to be controlled by the existing control system (e.g., DCS). After the output predictions are judged to be satisfactory, the calculated MPC control moves are evaluated to see if they are reasonable. Finally, the MPC software is evaluated during closed-loop operation with the calculated control moves implemented as set points to the DCS control loops. The MPC design parameters are tuned, if necessary. The commissioning period typically requires some troubleshooting and can take as long as, or even longer than, the plant tests of Step 3.

Step 8: Measuring Results and Monitoring Performance
The evaluation of MPC system performance is not easy, and widely accepted metrics and monitoring strategies are not available. However, useful diagnostic information is provided by basic statistics, such as the means and standard deviations for both measured variables, and calculated quantities, such as control errors and model residuals. Another useful statistic is the relative amount of time that an input is saturated or a constraint is violated, expressed as a percentage of the total time the MPC system is in service. These types of routine monitoring activities are considered in more detail in Chapter 17.

In Chapter 11, we considered a number of classical metrics for characterizing control system performance, such as the IAE index, overshoot, settling time, and degree of oscillation. Though helpful, these metrics provide an incomplete picture of overall MPC performance. An important motivation for MPC is that it facilitates process operation closer to targets and limiting constraints. Thus, an evaluation of MPC performance should include measures of whether these objectives have been realized. If so, a noticeable improvement in process operation should be reflected in economically meaningful measures such as product quality, through-put, or energy costs. The various performance metrics should be calculated before and after the MPC system is installed.

MPC system performance should be monitored on a regular basis to ensure that performance does not degrade owing to changes in the process, instrumentation, or process conditions, including disturbances. If performance becomes significantly worse, retuning the controller or reidentifying all (or part of) the process model may be required. The development of MPC monitoring strategies is an active research area (Kozub, 2002; Jelali, 2006; McIntosh and Canney, 2008; Badwe et al., 2009).

SUMMARY

Model predictive control is an important model-based control strategy devised for large multiple-input, multiple-output control problems with inequality constraints on the inputs and/or outputs. This chapter has considered both the theoretical and practical aspects of MPC. Applications typically involve two types of calculations: (1) a steady-state optimization to determine the optimum set points for the control calculations, and (2) control calculations to determine the MV changes that will drive the process to the set points. The success of model-based control strategies such as MPC depends strongly on the availability of a reasonably accurate process model. Consequently, model development is the most critical step in applying MPC. As Rawlings (2000) has noted, "feedback can overcome some effects of poor models, but starting with a poor process model is akin to driving a car at night without headlights." Finally, the MPC design parameters should be chosen carefully, as illustrated by two simulation examples.

Model predictive control has had a major impact on industrial practice, with thousands of applications world-wide. MPC has become the method of choice for difficult control problems in the oil refining and petrochemical industries. However, it is not a panacea for all difficult control problems (Shinskey, 1994; Hugo, 2000; McIntosh and Canney, 2008). Furthermore, MPC has had much less impact on other process industries. Performance monitoring of MPC systems is an important topic of current research interest.

REFERENCES

Badgwell, T. A., and S. J. Qin, A Review of Nonlinear Model Predictive Control Applications, in *Nonlinear Predictive Control: Theory and Practice*, B. Kouvaritakis and M. Cannon (Eds.), Inst. Electrical Eng., London, 2001, Chapter 1.

Badgwell, T. A., and S. J. Qin, Model-Predictive Control in Practice, in *Encyclopedia of Systems and Control*, Springer-Verlag Ltd., London, 2014.

Badwe, A. S., R. D. Gudi, R. S. Patwardhan, S. L. Shah, and S. C. Patwardhan, Detection of Model-Plant Mismatch in MPC Applications, *J. Process Control*, **19**, 1305 (2009).

Bemporad, A., M. Morari, and N. L. Ricker, *Model Predictive Control Toolbox 3*, User's Guide, Mathworks, Inc., Natick, MA, 2009.

Bonavita, N., R. Martini, and T. Matsko, 2003, Improvement in the performance of online control applications via enhanced modeling techniques, *Proc. of the European Refining Technology Conference on Computing*, Milan, Italy (2003).

Camacho, E. F., and C. Bordons, *Model Predictive Control* 2nd ed., Springer-Verlag, New York, 2003.

Canney, W. M., The Future of Advanced Process Control Promises More Benefits and Sustained Value, *Oil & Gas Journal*, **101** (16), 48 (2003).

Clarke, D. W., C. Mohtadi, and P. S. Tufts, Generalized Predictive Control—Part I. *The Basic Algorithm, Automatica*, **23**, 137 (1987).

Cortinovis, A., H. J. Ferreau, D. Lewandowski, and M. Mercangöz, Experimental evaluation of MPC-based anti-surge and process control for electric driven centrifugal gas compressors, *J. Process Control*, **34**, 13 (2015).

Cutler, C. R., and B. L. Ramaker, Dynamic Matrix Control—A Computer Control Algorithm, *Proc. Joint Auto. Control Conf., Paper WP5-B*, San Francisco (1980).

Darby, M. L., and M. Nikolaou, MPC: Current Practice and Challenges, *Control Eng. Practice*, **20**, 328 (2012).

Edgar, T. F., D. M. Himmelblau, and L. Lasdon, *Optimization of Chemical Processes*, 2nd ed., McGraw-Hill, New York, 2001.

Forbes, M. G., R. S. Patwardhan, H. Hamadah, and R. B. Gopaluni, Model Predictive Control in Industry: Challenges and Opportunities, *Proc. IFAC ADCHEM Symposium*, Whistler, 532 (2015).

Grosman, B., E. Dassau, H. C. Zisser, L. Jovanovic, and F. J. Doyle III, Zone Model Predictive Control: A Strategy to Minimize Hyper-and Hypoglycemic Events, *J. Diabetes Sci. Tech.*, **4**, 961 (2010).

Hokanson, D. A., and J. G. Gerstle, Dynamic Matrix Control Multivariable Controllers, in *Practical Distillation Control*, W. L. Luyben (Ed.) Van Nostrand Reinhold, New York, 1992, p. 248.

Hugo, A., Limitations of Model Predictive Controllers, *Hydrocarbon Process.*, **79** (1), 83 (2000).

Jelali M., An Overview of Control Performance Assessment Technology and Industrial Applications, *Control Eng. Practice*, **14**, 441 (2006).

Kano, M., and M. Ogawa, The State of the Art in Chemical Process Control in Japan: Good Practice and Questionnaire Survey, *J. Process Control*, **20**, 969 (2010).

Kassman, D. E., T. A. Badgwell, and R. B. Hawkins, Robust Steady-State Target Calculations for Model Predictive Control, *AIChE J.*, **46**, 1007 (2000).

Kozub, D. J., Controller Performance Monitoring and Diagnosis. Industrial Perspective, *Preprints of the 15th Triennial World IFAC Congress*, Barcelona, Spain (July 2002).

Lee, J., W. Cho, and T. F. Edgar, Multiloop PI Controller Tuning for Interacting Multivariable Processes, *Comput. Chem. Eng.*, **22**, 1711 (1998).

Ljung, L., *System Identification*, 2nd ed., Prentice Hall, Upper Saddle River, NJ, 1999.

Lundström, P., J. H. Lee, M. Morari, and S. Skogestad, Limitations of Dynamic Matrix Control, *Comput. Chem. Eng.*, **19**, 409 (1995).

Maciejowski, J. M., *Predictive Control with Constraints*, Prentice Hall, Upper Saddle River, NJ, 2002.

Maurath, P. R., D. A. Mellichamp, and D. E. Seborg, Predictive Controller Design for SISO Systems, *IEC Research*, **27**, 956 (1988).

McIntosh, A. R., and W. M. Canney, The Dirty Secrets of Model Predictive Controller Sustained Value, *Proc. Internat. Sympos. on Advanced Control of Industrial Processes (ADCONIP 2008)*, Paper MoB1.4, Jasper, Alberta, Canada (May 2008).

Morari, M., and J. H. Lee, Model Predictive Control: Past, Present, and Future, *Comput. Chem. Eng.*, **23**, 667 (1999).

Morari, M., and N. L. Ricker, *Model Predictive Control Toolbox*, The Mathworks, Inc., Natick, MA, 1994.

Poland, J., A. J. Isaksson, and P. Aronsson, Building and Solving Non-linear Optimal Control and Estimation Problems, *Proc. Of the 7th Modelica Conference*, Como, Italy (2009).

Qin, S. J. and T. A. Badgwell, A Survey of Industrial Model Predictive Control Technology, *Control Eng. Practice*, **11**, 733 (2003).

Rawlings, J. B., Tutorial Overview of Model Predictive Control, *IEEE Control Systems*, **20**(3), 38 (2000).

Rawlings, J. B., and D. Q. Mayne, *Model Predictive Control: Theory and Design*, Nob Hill Publishing, Madison, Wisconsin, 2009.

Richalet, J., and D. O'Donovan, *Predictive Functional Control: Principles and Industrial Applications*, Springer-Verlag Ltd., London, 2009.

Richalet, J., A. Rault, J. L. Testud, and J. Papon, Model Predictive Heuristic Control: Applications to Industrial Processes, *Automatica*, **14**, 413 (1978).

Rossiter, J. A., *Model-Based Predictive Control: A Practical Approach*, CRC Press, Boca Raton, FL, 2003.

Shinskey, F. G., *Feedback Controllers for the Process Industries*, McGraw-Hill, New York, 1994.

Sorensen, R. C., and C. R. Cutler, LP Integrates Economics into Dynamic Matrix Control, *Hydrocarbon Process.*, **77**(9), 57 (1998).

Stadler, K., J. Poland, and E. Gallestey, Model predictive control of a rotary cement kiln. *Control Eng. Prac.*, **19**, 1 (2011).

White, D. C., Multivariable Control of a Chemical Plant Incinerator: An Industrial Case Study of Co-Linearity and Non-Linearity, *Proc. Internat. Sympos. on Advanced Control of Industrial Processes (ADCONIP 2008)*, Paper MoB1.1, Jasper, Alberta, Canada (May 2008).

Yu, Z. H., J. H. Lee, and M. Morari, State Estimation Based Model Predictive Control Applied to Shell Control Problem: A Case Study, *Chem. Eng. Sci.*, **49**, 285 (1994).

EXERCISES

15.1 For the transfer functions

$$G_P(s) = \frac{2e^{-2s}}{(10s+1)(5s+1)} \qquad G_v = G_m = 1$$

(a) Derive an analytical expression for the step response to a unit step change. Evaluate the step-response coefficients, $\{S_i\}$, for a sampling period of $\Delta t = 1$.

(b) What value of model horizon N should be specified in order to ensure that the step-response model covers a period of at least 99% of the open-loop settling time? (That is, we require that $N\Delta t \geq t_{99}$ where t_{99} is the 99% settling time.)

15.2 Control calculations for a control horizon of $M = 1$ can be performed either analytically or numerically. For the process model in Exercise 15.1, derive $\boldsymbol{K}_{c1}$ for $\Delta t = 1$, $N = 50$, and $P = 5$, using Eq. 15-65. Compare your answer with the analytical result reported by Maurath et al. (1988).

$$\boldsymbol{K}_{c1} = \frac{1}{\sum_{i=1}^{P} S_i^2}[S_1\, S_2\, S_3 \ldots S_P]$$

15.3 Consider the transfer function model of Exercise 15.1. For each of the four sets of design parameters shown below, design a model predictive controller. Then do the following:

(a) Compare the controllers for a unit step change in set point. Consider both the y and u responses.

(b) Repeat the comparison of (a) for a unit step change in disturbance, assuming that $G_d(s) = G(s)$.

(c) Which controller provides the best performance? Justify your answer.

Set No.	Δt	N	M	P	R
(i)	2	40	1	5	0
(ii)	2	40	20	20	0
(iii)	2	40	3	10	0.01
(iv)	2	40	3	10	0.1

15.4 For Exercise 15.1, suppose that a constraint is placed on the manipulated variable, $u_{k+j} \leq 0.2$ for $j = 1, 2, \ldots, M-1$. Let $\Delta t = 2$ and $N = 40$. Select values of M, P, and $\boldsymbol{R}$ so that these constraints are not violated after a unit step disturbance occurs.

15.5 For Exercise 15.1, consider two sets of design parameters and simulate unit step changes in both the disturbance and the set point. Assume that the disturbance model is identical to the process model. The design parameters are

(a) $M = 7 \quad P = 10 \quad \boldsymbol{R} = \boldsymbol{0}$

(b) $M = 3 \quad P = 10 \quad \boldsymbol{R} = \boldsymbol{0}$

Which controller provides the best control? Justify your answer.

15.6 A process (including sensor and control valve) can be modeled by the transfer function,

$$G(s) = \frac{2(1-9s)}{(15s+1)(3s+1)}$$

(a) Derive an analytical expression for the response to a unit step change in the input.

(b) Suppose that the maximum allowable value for the model horizon is $N = 30$. What value of the sampling period Δt should be specified to ensure that the step-response model covers a period of at least 99% of the open-loop settling time? (That is, we require that $N\Delta t \geq t_{99}$ where t_{99} is the 99% settling time.) Use the analytical solution and this value of Δt to obtain a step-response model in the form of Eq. 15-1.

15.7 Starting with discrete time equivalent of state space model (Eqs. 15-11 and 15-12), derive Eq. 15-13.

15.8 Starting with the quadratic objective function of Eq. 15-62, derive the expression for minimum number of controls moves given by Eq. 15-63 (Assume that the weighting matrix Q is unity).

15.9 Consider the unconstrained, SISO version of MPC in Eq. 19-65. Suppose that the controller is designed so that the control horizon is $M = 1$ and the weighting matrices are $\boldsymbol{Q} = \boldsymbol{I}$ and $\boldsymbol{R} = 1$. The prediction horizon P can be chosen arbitrarily. Demonstrate that the resulting MPC controller has a simple analytical form.

15.10 In Section 15.6.1, MPC was applied to the Wood–Berry distillation column model. A MATLAB program for this example and constrained MPC is shown in Table E15.8. The design parameters have the base case values (Case B in Fig. 15.14) except for $P = 10$ and $M = 5$. The input constraints are the saturation limits for each input (–0.15 and +0.15). Evaluate the effects of control horizon M and input weighting matrix $\boldsymbol{R}$ by simulating the set-point change and the first disturbance of Section 15.6.1 for the following parameter values:

(a) Control horizon, $M = 2$ and $M = 5$

(b) Input weighting matrix, $\boldsymbol{R} = 0.1\,\boldsymbol{I}$ and $\boldsymbol{R} = \boldsymbol{I}$

Consider plots of both inputs and outputs. Which choices of M and $\boldsymbol{R}$ provide the best control? Do any of these MPC controllers provide significantly better control than the controllers shown in Figs. 15.16 and 15.17? Justify your answer.

Table E15.10 MATLAB Program (Based on a program by Morari and Ricker (1994))

```
g11=poly2tfd(12.8,[16.7 1],0.1); % model
g21=poly2tfd(6.6,[10.9 1],0,7);
g12=poly2tfd(-18.9,[21.0 1],0,3);
g22=poly2tfd(-19.4,[14.4 1],0,3);
gd1=poly2tfd(3.8,[14.9 1],0,8.1);
gd2=poly2tfd(4.9,[13.2 1],0,3.4);
tfinal=120;          % Model horizon, N
delt=1;              % Sampling period
ny=2;                % Number of outputs
model=tfd2step(tfinal,delt,ny,g11,g21,g12,g22)
plant=model;         % No plant/model mismatch
dmodel=[ ]           % Default disturbance model
dplant=tfd2step(tfinal,delt,ny,gd1,gd2)
P=10; M=5;           % Horizons
ywt=[1 1]; uwt=[0.1 0.1]; % Q and R
tend=120;            % Final time for simulation
r=[0 1];             % Set-point change in XB
a=zeros([1,tend]);
for i=51:tend
  a(i)=0.3*2.45; % 30 % step in F at t=50 min.
end
dstep=[a'];
ulim=[-.15 -.15 .15 .15 1000 1000]; % u limits
ylim=[ ]; % No y limits
tfilter=[ ];
[y1,u1]=cmpc(plant,model,ywt,uwt,M,P,-
tend,r,ulim,ylim, tfilter,dplant,dmodel,dstep);
figure(1)
subplot(211)
plot(y1)
legend('XD','XB')
xlabel('Time (min)')
subplot(212)
stairs(u1)% Plot inputs as staircase functions
legend('R','S')
xlabel('Time (min)')
```

15.11 Design a model predictive controller for the process

$$G_P(s) = \frac{e^{-6s}}{10s+1} \qquad G_v = G_m = 1$$

Select a value of N based on 95% completion of the step response and $\Delta t = 2$. Simulate the closed-loop response for a set-point change using the following design parameters:

(a) $M = 1 \quad P = 7 \quad \boldsymbol{R} = \boldsymbol{0}$

(b) $M = 1 \quad P = 5 \quad \boldsymbol{R} = \boldsymbol{0}$

(c) $M = 4 \quad P = 30 \quad \boldsymbol{R} = \boldsymbol{0}$

15.12 PCM Consider the PCM furnace module of Appendix E with the following variables (HC denotes hydrocarbon):

CVs: HC exit temperature T_{HC} and oxygen exit concentration c_{O_2}

MVs: fuel gas flow rate F_{FG} and air flow rate F_A

DV: HC flow rate F_{HC}

Do the following, using the transfer function models given below:

(a) Design an MPC system using the following design parameters: Δt = 1 min, $\boldsymbol{Q}$ = diagonal [0.1, 1], $\boldsymbol{R}$ = diagonal [0.1, 0.1], $P = 20$, and M = 1.

(b) Repeat part (a) for the same design parameters, but where $\boldsymbol{R}$ = diagonal [0.5, 0.5].

(c) Simulate the two MPC controllers for a step change in the c_{o_2} set point to 1.0143 mol/m³ at $t = 10$ min.

(d) Repeat part (c) for a step change in F_{HC} at $t = 10$ min to 0.035 m³/min.

(e) Based on your results for parts (c) and (d), which MPC controller is superior? Justify your answer.

Process transfer function matrix:

$$\begin{array}{c|cc|} & F_{FG} & F_A \\ \hline T_{HC} & \dfrac{220e^{-2s}}{6.5s+1} & \dfrac{-13e^{-2s}}{6.2s+1} \\ c_{O_2} & \dfrac{-2.0e^{-4s}}{3.8s+1} & \dfrac{0.14e^{-4s}}{4.2s+1} \end{array}$$

15.13 PCM Consider the PCM distillation column module of Appendix E with the following variables:

CVs: Overhead Composition (x_D) and Bottom Composition (x_B)

MVs: Reflux Ratio (R) and Vapor Flow Rate (V)

DV: Column Feed Flow Rate (F)

Do the following, using the transfer function models given below:

(a) Design an MPC system using the following design parameters: Δt = 60 s, $\boldsymbol{Q}$ = diagonal [0.1, 0.1], $\boldsymbol{R}$ = diagonal [0.1, 1], $P = 40$, $M = 1$.

(b) Repeat part (a) for the same design parameters, but where $\boldsymbol{Q}$ = diagonal [0.5, 0.5].

(c) Simulate the two MPC controllers for a step change in the set point for Overhead Composition (x_D) from 0.85 to 0.80.

(d) Repeat part (c) for a step change in the set point for Bottom Composition (x_B) from 0.15 to 0.2.

(e) (Optional–disturbance test) Repeat part (c) for a step change in the Column Feed Flow Rate (F) from 0.025 to 0.03.

(f) Based on your results from parts (c)–(e), which MPC controller is superior? Justify your answer.

Process transfer function matrix (all times in minutes):

$$\begin{array}{c} \\ x_D \\ \\ x_B \end{array} \begin{vmatrix} \overset{R}{\dfrac{0.14e^{-0.94s}}{14s+1}} & \overset{V}{\dfrac{-9.0e^{-18.6s}}{20s+1}} \\ \dfrac{0.16e^{-3.7s}}{17s+1} & \dfrac{-14e^{-0.22s}}{10s+1} \end{vmatrix}$$

15.14 Repeat Exercise 15.13 for $\boldsymbol{R}$ = diagonal [0.1, 0.1] and (i) $M = 1$ and (ii) $M = 4$.

15.15 Repeat Exercise 15.14 for $\boldsymbol{Q}$ = diagonal [0.1 0.1] and (i) $M = 1$ and (ii) $M = 5$.

15.16 Prepare a comparative list of different MPC techniques used in industries.

MULTIPLE CHOICE QUESTIONS

1. Which of the following process models can be used for implementing model predictive control?

(a) Step response model

(b) Impulse response model

(c) State-space model

(d) All of these

2. Which of the following algorithm is used in MPC?

(a) Routh test

(b) Dynamic matrix control

(c) Ziegler-Nichols technique

(d) Cohen-Coon technique

3. If P is the prediction horizon and M is the control horizon, then

(a) $P > M$

(b) $P < M$

(c) $P = M$

(d) $P \leq M$

4. Which of the following process parameters is taken into consideration in choosing the value of model horizon N?

(a) Time constant

(b) Settling time

(c) Rise time

(d) Dead time

5. A distinguishing feature of MPC is

(a) elimination of process dead time.

(b) ensuring stable closed loop behavior.

(c) ability to handle constraints on process inputs and outputs.

(d) none of these.

6. In SISO MPC, the dimension of dynamic matrix S for a prediction horizon P and control horizon M is

(a) M × P

(b) P × M

(c) P × P

(d) M × M

7. Higher value of control horizon M,

(a) increases computational effort.

(b) decreases computational effort.

(c) has no effect on computation.

(d) leads to more aggressive controller.

ANSWER KEY

1. (d) **2.** (b) **3.** (a) **4.** (b) **5.** (c) **6.** (b) **7.** (a)

CHAPTER 16

Development of Empirical Models from Process Data

CHAPTER CONTENTS

Several modeling approaches are used in process control applications. Theoretical models based on the chemistry and physics of the process represent one alternative. However, the development of rigorous theoretical models may not be practical for complex processes if the model requires a large number of equations with a significant number of process variables and unknown parameters (e.g., chemical and physical properties). An alternative approach is to develop an empirical model directly from experimental data. Empirical models are sometimes referred to as *black box* models, because the process being modeled can be likened to an opaque box. Here the input and output variables (u and y, respectively) are known, but the inner workings of the box are not. (See Fig. 16.1, where vectors of time-varying variables $\boldsymbol{u}(t)$, $\boldsymbol{y}(t)$, and $\boldsymbol{d}(t)$ are shown.) The development of empirical steady-state and dynamic models is the subject of this chapter. This activity is referred to as *process* or *system identification* (Ljung and Glad, 1994; Ljung, 1999). In general, empirical dynamic models are simpler than theoretical models and offer the advantage that they can be solved in "real time." In other words, the computational time required for the model solution (e.g., transient response) is much shorter than the actual process response time. However, this may not be true for complex models with many variables and equations.

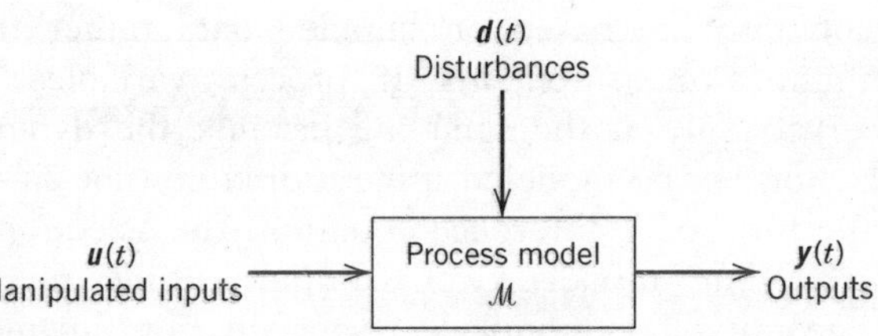

Figure 16.1 Input-output process model.

The key differences between process simulation and process identification can be summarized with the aid of Fig. 16.1. In simulation, the process model $\mathcal{M}$ is known, and we wish to generate the response $\boldsymbol{y}(t)$ for a specified input $\boldsymbol{u}(t)$ and a specified disturbance $\boldsymbol{d}(t)$. If $\mathcal{M}$ is a linear dynamic model and $\boldsymbol{u}(t)$ and $\boldsymbol{d}(t)$ are expressed analytically, $\boldsymbol{y}(t)$ can be derived using Laplace transforms (see Chapter 4). Alternatively, $\boldsymbol{y}(t)$ can be calculated numerically using software packages such as MATLAB (Ljung, 2007). If $\mathcal{M}$ is a nonlinear dynamic model, $\boldsymbol{y}(t)$ can be obtained by numerical integration (cf. Chapter 2) after $\boldsymbol{u}(t)$ and $\boldsymbol{d}(t)$ are specified. By contrast, in process identification the model $\mathcal{M}$ is determined from data for $\boldsymbol{u}(t)$, $\boldsymbol{y}(t)$, and $\boldsymbol{d}(t)$, if $\boldsymbol{d}$ can be measured. If the model structure is postulated but contains unknown model parameters, then the model parameters can be obtained using regression techniques. This parameter estimation can be done with commercially available software regardless of whether the process model is linear or nonlinear, or whether it is theoretically based or empirical in nature.

Steady-state empirical models can be used for instrument calibration, process optimization, and specific instances of process control. Single-input, single-output (SISO) models typically consist of simple polynomials relating an output to an input. Dynamic empirical models can be employed to understand process behavior during upset conditions. They are also used to design control systems and to analyze their performance. Empirical dynamic models typically are low-order differential equations or transfer function models (e.g., first- or second-order model, perhaps with a time delay), with unspecified model parameters to be determined from experimental data. However, in some situations, more complicated models are valuable in control system design, as discussed later in this chapter.

Discrete-time models are generally represented by difference equations rather than differential equations (see Chapter 13). Most process control tasks are implemented via digital computers, which are intrinsically discrete-time systems. In digital control, the continuous-time process variables are sampled at regular intervals (e.g., every 0.1 s); hence, the computer calculations are based on sampled data rather than continuous measurements. If process variables are observed only at the sampling instants, the dynamic behavior can be modeled using a discrete-time model in the form of a difference equation. The selection of discrete-time models over continuous-time models is becoming commonplace, especially for advanced control strategies.

Empirical process models can be developed for steady state as well as dynamic systems. Development of continuous time models using transient process information is discussed in Chapter 6 (See section 6.5). In this chapter several methods for determining steady state and dynamic empirical models for discrete-time model systems will be presented. We first consider general model-fitting techniques based on linear and nonlinear regression that can be used to calculate model parameters for any type of model. Methods for obtaining first-order and second-order linear dynamic models from step response data are presented in Chapter 6 (section 6.5). These methods yield models suitable for the design of control systems; however, the resulting models are usually accurate only for a narrow range of operating conditions close to the nominal steady state, where the process exhibits linear behavior. We also show the relationship between continuous-time and discrete-time models. Finally, we present several methods for developing linear discrete-time models for dynamic processes.

16.1 MODEL DEVELOPMENT USING LINEAR OR NONLINEAR REGRESSION

Before developing an empirical model for two variables, a single input u and a single output y, it is instructive first to plot the available data (e.g., y vs. u for steady-state data and y and u vs. time for transient response data). From these plots, it may be possible to visualize overall trends in the data and to select a reasonable form for the model. After the model form is selected, the unknown model parameters can be calculated and the model accuracy evaluated. This parameter calculation procedure is referred to as *parameter estimation* or regression (Ljung, 1999; Montgomery and Runger, 2013). These calculations are usually based on model fitting, that is, minimizing a measure of the differences between model predictions and data. However, the problem of fitting a model to a set of input-output data becomes complicated when the model relation is not simple or involves multiple inputs and outputs.

First, we consider steady-state models. Suppose that a set of steady-state input-output data is available and shown as circles in Fig. 16.2. Variable y represents a process output (e.g., a reactor yield), whereas u represents an input variable (e.g., an operating condition such as temperature). Although a straight-line model (Model 1) provides a reasonable fit, higher-order polynomial relations (Models 2 and 3) result in smaller errors between the data and the curve representing the empirical model. Models 2 and 3 provide better agreement with the data at the expense of greater complexity because more model parameters must be determined. Sometimes the model form may be known from theoretical considerations or past experience with the process.

In Fig. 16.2, if the actual process behavior is linear, the differences (or residuals) between Model 1 and the data could be due to process disturbances or measurement errors. In empirical modeling, it is preferable to choose the simplest model structure that yields a good fit of the data, providing that the model is physically reasonable. Note that in Fig. 16.2, if Model 3 is extrapolated beyond the data range, it would apparently yield significantly different model predictions than Model 1 or 2. The selection of the best model might require collecting

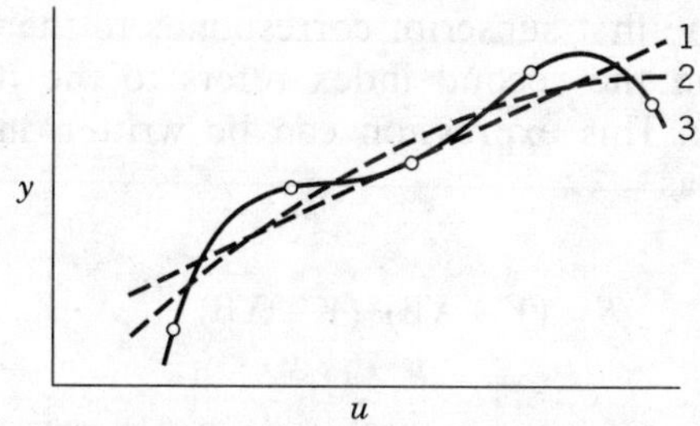

Figure 16.2 Three models for scattered data.

more data, perhaps outside the range shown in Fig. 16.2, which then could be used to *validate* each model.

16.1.1 Model Building Procedure

In this section, we present a systematic procedure for developing empirical dynamic models (Ljung, 1999). The procedure consists of the following steps:

1. Formulate the model objectives; that is, how will the model be used, and who will be the user?
2. Select the input and output variables for the model.
3. Evaluate available data and develop a plan to acquire additional data. A testing plan would specify the values of u or the form of $u(t)$, for example, a step change or some other input sequence (see Section 6.5).
4. Select the model structure and level of model complexity (e.g., steady-state vs. dynamic model, linear vs. nonlinear model).
5. Estimate the unknown model parameters using linear or nonlinear regression.
6. Using input and output data, evaluate model accuracy based on statistical considerations. It is desirable to use new data (if available) as well as the "old" data that were used to develop the model. If the model does not provide a satisfactory fit, return to Step 2 and try a different model. If possible, the model should be tested with new data (that is, *validation data*); if the model predictions agree with these data, the model is said to be validated.
7. For a dynamic model, nonstatistical criteria also can be used to evaluate a model, such as speed of response, shape of response, correct stability properties, and correct steady-state gain. The utility of a model for designing controllers is also important in process control, where an overly complex model can be a disadvantage. Thus *control-relevant models* are desirable (Rivera and Jun, 2000).

16.1.2 Linear Regression

Statistical analysis can be used to estimate unknown model parameters and to specify the uncertainty associated with the empirical model. It can also be used to compare several candidate models (Draper and Smith, 1998; Montgomery and Runger, 2013). For linear models, the *least-squares* approach is widely used to estimate model parameters. Consider the linear (or straight-line) model in Fig. 16.2 (Model 1) and let Y_i represent the data point where $\hat{y}_i$ is the model prediction for $u = u_i$. Then for the model, $y = \beta_1 + \beta_2 u + \varepsilon$, the individual data points can be expressed as

$$Y_i = \beta_1 + \beta_2 u_i + \varepsilon_i \tag{16-1}$$

where β_1 and β_2 are the model parameters to be estimated. ε_i is the random error for the particular data point.

The least-squares method is the standard approach for calculating the values of β_1 and β_2 that minimize the sum of the squares of the errors S for an arbitrary number of data points, N:

$$S = \sum_{i=1}^{N} \varepsilon_i^2 = \sum_{i=1}^{N} (Y_i - \beta_1 - \beta_2 u_i)^2 \tag{16-2}$$

In Eq. 16-2, note that the values of Y_i and u_i are known, while β_1 and β_2 are to be calculated so as to minimize S, the objective function. The optimal estimates of β_1 and β_2 calculated for a specific data set are designated as $\hat{\beta}_1$ and $\hat{\beta}_2$. The model predictions are given by the regression model:

$$\hat{y} = \hat{\beta}_1 + \hat{\beta}_2 u \tag{16-3}$$

and the *residuals* e_i are defined as

$$e_i \triangleq Y_i - \hat{y}_i \tag{16-4}$$

These least-squares estimates will minimize the least squares index in Eq. 16-2 if two ideal conditions are satisfied (Montgomery and Runger, 2013):

1. The model structure is correct.
2. The random errors ε_i are independent and normally distributed.

In practical problems, both ideal conditions are seldom realized. Nevertheless, the estimated least squares models can be satisfactory if the selected model structure is appropriate.

For a linear model and N data points, values of $\hat{\beta}_1$ and $\hat{\beta}_2$ that minimize (16-2) are obtained by first setting the derivatives of S with respect to β_1 and β_2 equal to zero. Because S is a quadratic function, this approach leads to two linear equations in two unknowns $\hat{\beta}_1$ and $\hat{\beta}_2$. The analytical solution (Edgar et al., 2001) is

$$\hat{\beta}_1 = \frac{S_{uu}S_y - S_{uy}S_u}{NS_{uu} - (S_u)^2} \tag{16-5}$$

$$\beta_{12} = \frac{NS_{uy} - S_uS_y}{NS_{uu} - (S_u)^2} \tag{16-6}$$

where

$$S_u \triangleq \sum_{i=1}^{N} u_i \qquad S_{uu} \triangleq \sum_{i=1}^{N} u_i^2$$

$$S_y \triangleq \sum_{i=1}^{N} Y_i \qquad S_{uy} \triangleq \sum_{i=1}^{N} u_iY_i$$

These calculations can be made using statistical packages or spreadsheets such as Excel.

This least-squares estimation approach (also called linear regression) can be extended to more general models with:

1. More than one input or output variable.
2. Functions of the input variables u, such as polynomials and exponentials, providing that the unknown parameters appear linearly.

A general nonlinear steady-state model which is linear in the parameters has the form

$$y = \sum_{j=1}^{p} \beta_j X_j + \varepsilon \tag{16-7}$$

The p unknown parameters (β_j) are to be estimated, and the X_j are the p specified functions of u. Note that the unknown parameters β_j appear linearly in the model, and a constant term can be included by setting $X_1 = 1$.

The sum of the squares analogous to (16-2) is

$$S = \sum_{i=1}^{N}\left(Y_i - \sum_{j=1}^{p}\beta_j X_{ij}\right)^2 \tag{16-8}$$

For X_{ij} the first subscript corresponds to the ith data point, and the second index refers to the jth function of u. This expression can be written in matrix notation as

$$S = (Y - X\beta)^T(Y - X\beta) \tag{16-9}$$

where the superscript T denotes the transpose of a vector or matrix and:

$$Y = \begin{bmatrix} Y_1 \\ \vdots \\ Y_N \end{bmatrix} \quad \beta = \begin{bmatrix} \beta_1 \\ \vdots \\ \beta_p \end{bmatrix} \quad X = \begin{bmatrix} X_{11} & X_{12} & \cdots & X_{1p} \\ X_{21} & X_{22} & \cdots & X_{2p} \\ \vdots & \vdots & & \vdots \\ X_{N1} & X_{N2} & \cdots & X_{Np} \end{bmatrix}$$

The least-squares estimate is given by Draper and Smith, 1998; Montgomery and Runger, 2003),

$$\hat{\beta} = (X^TX)^{-1}X^TY \tag{16-10}$$

providing that matrix X^TX is nonsingular so that its inverse exists. Note that the matrix X is comprised of functions of u; for example, if

$$y = \beta_1 + \beta_2 u + \beta_3 u^2 + \varepsilon, \text{ then } X_1 = 1, X_2 = u, \text{ and } X_3 = u^2$$

If the number of data points is equal to the number of model parameters, (i.e., $N = p$), Eq. 16-10 provides a unique solution to the parameter estimation problem, one that provides perfect agreement with the data points, as long as X^TX is nonsingular. For $N > p$ a *least-squares solution* results that minimizes the sum of the squared deviations between each of the data points and the model predictions.

The least-squares solution in Eq. 16-10 provides a *point estimate* for the unknown model parameters β_i but it does indicate how accurate the estimates are. The degree of accuracy is expressed by confidence intervals that have the form, $\hat{\beta}_i \pm \Delta\beta_i$. The $\Delta\beta_1$ are calculated from the (u, y) data for a specified *confidence level* (Draper and Smith, 1998).

Next we consider the development of a steady-state performance model, such as might be used in optimizing the operating conditions of an electrical power generator.

EXAMPLE 16.1

An experiment has been performed to determine the steady-state power delivered by a gas turbine-driven generator as a function of fuel flow rate. The following normalized data were obtained:

Fuel Flow Rate u_i	Power Generated Y_i
1.0	2.0
2.3	4.4
2.9	5.4
4.0	7.5
4.9	9.1
5.8	10.8
6.5	12.3
7.7	14.3
8.4	15.8
9.0	16.8

Compare the accuracies of best linear and quadratic models.

Solution

To solve for the linear model, Eqs. 16-5 and 16-6 could be applied directly. However, to illustrate the use of Eq. 16-10, first define the terms in the linear model: $X_1 = 1$ and $X_2 = u$. The following matrices are then obtained:

$$X^T = \begin{bmatrix} 1 & 1 & 1 & 1 & 1 & 1 & 1 & 1 & 1 & 1 \\ 1.0 & 2.3 & 2.9 & 4.0 & 4.9 & 5.8 & 6.5 & 7.7 & 8.4 & 9.0 \end{bmatrix}$$

$$Y^T = [2.0 \;\; 4.4 \;\; 5.4 \;\; 7.5 \;\; 9.1 \;\; 10.8 \;\; 12.3 \;\; 14.3 \;\; 15.8 \;\; 16.8]$$

$$X^T X = \begin{bmatrix} \sum_{i=1}^{10} (1)^2 & \sum_{i=1}^{10} u_i \\ \sum_{i=1}^{10} u_i & \sum_{i=1}^{10} u_i^2 \end{bmatrix}$$

$$X^T Y = \begin{bmatrix} \sum_{i=1}^{10} Y_i \\ \sum_{i=1}^{10} u_i Y_i \end{bmatrix}$$

Solving for β_1 and β_2 using Eq. 16-10 yields the same results given in Eqs. 16-5 and 16-6, with $\hat{\beta}_1 = 0.0785 \pm 0.0039$ and $\hat{\beta}_2 = 1.859 \pm 0.093$ (95% confidence limits are shown).

To determine how much the model accuracy can be improved by using a quadratic model, Eq. 16-10 is again applied, this time with $X_1 = 1$, $X_2 = u$, and $X_3 = u^2$. The estimated parameters and 95% confidence limits for this quadratic model are:

$$\hat{\beta}_1 = 0.1707 \pm 0.0085, \quad \hat{\beta}_2 = 1.811 \pm 0.096, \text{ and}$$

$$\hat{\beta}_3 = 0.0047 \pm 0.0002$$

The predicted values of $y(\hat{y})$ are compared with the measured values (actual data) in Table 16.1 for both the linear and quadratic models. It is evident from this comparison that the linear model is adequate and that little improvement results from the more complicated quadratic model.

Table 16.1 A Comparison of Model Predictions from Example 16.1 (SSE = Sum of squared errors)

u_i	y_i	Linear Model Prediction $\hat{y}_{1i} = \hat{\beta}_1 + \hat{\beta}_2 u_i$	Quadratic Model Prediction $\hat{y}_{2i} = \hat{\beta}_1 + \hat{\beta}_2 u_i + \hat{\beta}_3 u_i^2$
1.0	2.0	1.94	1.99
2.3	4.4	4.36	4.36
2.9	5.4	5.47	5.46
4.0	7.5	7.52	7.49
4.9	9.1	9.19	9.16
5.8	10.8	10.86	10.83
6.5	12.3	12.16	12.14
7.7	14.3	14.40	14.40
8.4	15.8	15.70	15.72
9.0	16.8	16.81	16.85
		SSE = 0.0613	SSE = 0.0540

Linear regression using MATLAB

```
>> u = [1;2.3;2.9;4  % Creation of input
.0;4.9;5.8;6.5;7.7;   vector
8.4;9.0];
>> y1 = [2;4.4;5.4;  % Creation of output
7.5;9.1;10.8;12.3;1   vector
4.3;15.8;16.8];
>>X=[ones(10,1),u];  % If the proposed lin-
                      ear equation contains
                      a constant term, a
                      column of ones is to
                      be appended at the
                      beginning of input
                      vector
```

```
>> B = X\y1                 % Getting the values
                            of regression coeffi-
                            cients
B =
   0.0785
   1.8593
>> ylin = X*B;              % Obtain predicted
                            values
```

Use of quadratic model

```
>> u2 = u.*u;               % get the values of u²
>> X = [ones(10,1),
u,u2];
>> B1 = X\y1
B1 =
   0.1707
   1.8113
   0.0047
>> ynlin = X*B1;
>> table(u,y1,ylin,
ynlin)
ans =
   10 × 4 table
```

u	y1	ylin	ynlin
1	2	1.9379	1.9867
2.3	4.4	4.355	4.3614
2.9	5.4	5.4706	5.4628
4	7.5	7.5158	7.4908
4.9	9.1	9.1892	9.1585
5.8	10.8	10.863	10.834
6.5	12.3	12.164	12.142
7.7	14.3	14.395	14.396
8.4	15.8	15.697	15.716
9	16.8	16.812	16.852

```
>> SSElin =                 % Accuracy of
sum((y1-ylin).^2)           linear model
SSElin =
   0.0613
>> SSEnlin =                % Accuracy of
sum((y1-ynlin).^2)          quadratic model
SSEnlin =
   0.0540
```

16.1.3 Nonlinear Regression

If the empirical model is nonlinear with respect to the model parameters, then nonlinear regression rather than linear regression must be used. For example, suppose that a reaction rate has an expression of the following form

$$r_A = kc_A^n \tag{16-11}$$

This is to be fit to experimental data, where r_A is the reaction rate of component A, c_A is the reactant concentration, and k and n are model parameters. This model is *linear* with respect to rate constant k but is *nonlinear* with respect to concentration c_A. In such cases, the least square regression can still be performed by first linearizing the non-linear equation as shown below

$$\ln(r_A) = \ln(k) + n\ln(c_A) \tag{16-12}$$

It may be noticed that while r_A and c_A are non-linearly related. However, the relation between $\ln(r_A)$ and $\ln(c_A)$ is linear with slope n and intercept $\ln(k)$. Therefore, instead of performing least square regression on the original variables, the transformed variable values are used.

A general nonlinear model can be written as

$$y = f(u_1,\ u_2,\ u_3,\ldots\beta_1,\beta_2,\beta_3\ldots) \tag{16-13}$$

where y is the model output, u_j are inputs, and β_j are the parameters to be estimated. In this case, the β_j do not appear linearly in the model. However, we can still define a sum of squares error criterion to be minimized by selecting parameters β_j:

$$\min_{\beta_j} S = \sum_{i=1}^{N} (Y_i - \hat{y}_i)^2 \tag{16-14}$$

where Y_i and $\hat{y}_i$ denote the ith output measurement and model prediction corresponding to the ith data point, respectively. The least-squares estimates are again denoted $\hat{\beta}_j$.

Consider the problem of estimating the time constants for first-order and overdamped second-order dynamic models based on the measured output response to a step input change of magnitude M. Analytical expressions for these step response were developed in Chapter 5.

Transfer Function	***Step Response***	
$\dfrac{Y(s)}{U(s)}=\dfrac{K}{\tau s+1}$	$y(t)=KM(1-e^{-t/\tau})$	(5-24)
$\dfrac{Y(s)}{U(s)}=\dfrac{K}{(\tau_1 s+1)(\tau_2 s+1)}$	$y(t)=KM\left(1-\dfrac{\tau_1 e^{-t/\tau_1}-\tau_2 e^{-t/\tau_2}}{\tau_1-\tau_2}\right)$	(5-62)

In the step response equations, t is the independent variable instead of the input u used earlier, and y is the dependent variable expressed in deviation form. Although steady-state gain K appears linearly in both response equations, the time constants are contained in a nonlinear manner, which means that linear regression cannot be used to estimate them.

Sometimes a variable transformation can be employed to transform a nonlinear model so that linear regression can be used (Montgomery and Runger, 2013). For example, if K is assumed to be known, the first-order step response can be rearranged:

$$\ln\left(1-\frac{y(t)}{KM}\right)=-\frac{t}{\tau} \tag{16-15}$$

Because $\ln(1 - y/KM)$ can be evaluated at each time t_i, this model is linear in the parameter $1/\tau$. Thus, this model has the standard linear form as Eq. 16-1, where the left-hand side of Eq. 16-15 is Y_i, $\beta_1 = 0$, and $u_i = t_i$.

The transformation in Eq. 16-15 leads to the *fraction incomplete response* method of determining first-order models discussed in the next section. However, for step responses of higher-order models, such as Eq. 5-62 the transformation approach is not feasible. For these calculations, we must use optimization software such as Excel Solver, as discussed in Section 16.1.2 to find the least-squares estimates of the time constants (Edgar et al., 2001).

An important assumption in simple linear regression is that the residuals ε_i in Eq.16.1 are independent and normally distributed. Further, before performing least square regression, it should be ensured that the dataset is free of outliers. Presence of a small number of outliers in a dataset will cause significant changes in the model parameters i.e., the slope and intercept values. Similarly, multiple linear regression cannot be performed if the matrix (X^TX) is singular (because the determinant value for a singular matrix is zero and therefore, matrix inverse cannot be computed). This means the independent variables must be uncorrelated (no independent variables can be expressed as a linear combination of other independent variables).

16.2 NEURAL NETWORK MODELS

Most industrial processes such as chemical reactors and separation systems exhibit nonlinear behavior. Unfortunately, many processes are so complex that significant engineering time and effort is required to develop and validate detailed theoretical dynamic models. As an alternative, an empirical nonlinear model can be obtained from experimental data. *Neural networks (NN)*, or artificial neural networks, are an important class of empirical nonlinear models. Neural networks have been used extensively in recent years to model a wide range of physical and chemical phenomena and to model other nonengineering situations such as stock market analysis, chess strategies, speech recognition, and medical diagnoses. Neural networks are attractive whenever it is necessary to model complex or little understood processes with large input–output data sets, as well as to replace models that are too complicated to solve in real time (Su and McAvoy, 1997; Himmelblau, 2008).

The exceptional computational abilities of the human brain have motivated the concept of an NN. The brain can perform certain types of computation, such as perception, pattern recognition, and motor control, much faster than existing digital computers (Haykin, 2009). The operation of the human brain is complex and nonlinear and involves massive parallel computation. Its computations are performed using structural constituents called *neurons* and the synaptic interconnections between them (i.e., a *neural network*). The development of artificial neural networks is an admittedly approximate attempt to mimic this biological neural network, in order to achieve some of its computational advantages.

A multilayer feedforward network, one of the most common NN structures, is shown in Fig. 16.3. The *neurons* (or *nodes*) are organized into layers (input, output, hidden); each neuron in the hidden layer is connected to the neurons in adjacent layers via connection weights. These weights are unknown parameters that are estimated based on the input/output data from the process

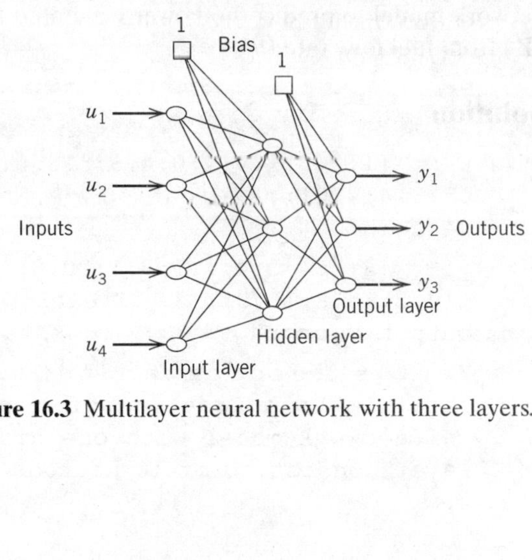

Figure 16.3 Multilayer neural network with three layers.

to be modeled. The number of unknown parameters can be quite large (e.g., 50–100), and powerful nonlinear programming algorithms are required to fit the parameters to the data using the least-squares objective function (Edgar et al., 2001). If enough neurons are utilized, an input–output process can be accurately modeled by a neural net model.

As shown in Fig. 16.4, at each neuron inputs are collected from other neurons or from bias terms, and their strength or magnitude is evaluated. These inputs are then summed and compared with a threshold level, and the appropriate output I_j is determined. The connection weight (W_{ij}) determines the relative importance of that input. The output from the neuron is a transformed value of the sum of weighted inputs to that neuron. This transformation is accomplished by the activation function associated with the neurons. Some common activation functions are: binary step, linear, sigmoid and hyperbolic tangent. Non-linear transformation using hyperbolic tangent activation function is shown in Figure 16.4.

The *training* of a neural network involves estimating the unknown parameters; this procedure generally utilizes

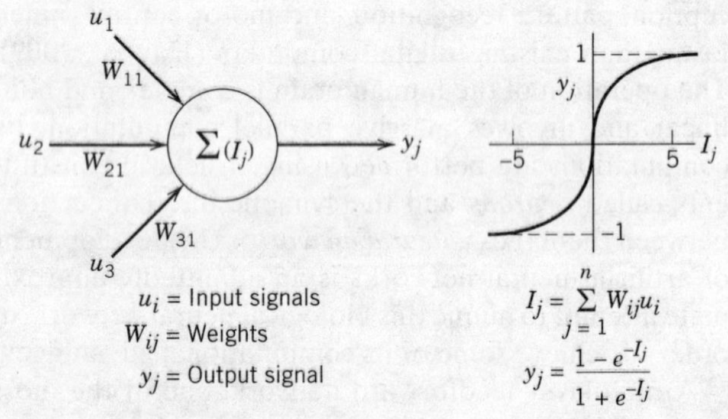

Figure 16.4 Signal diagram for a neuron.

normal operating data (often large data sets) taken in the operating region where the model is intended to be used. During the training phase, minimization of a performance function is undertaken. The performance functions reflect the average difference between model predicted output value and the actual output values. Different types of performance functions can be used such as mean absolute error, mean squared error, or root mean squared error. Some popular training algorithms for multilayer feed forward neural network models are: Gradient descent, conjugate gradient and Levenberg-Marquardt algorithm.

In most cases, carefully chosen model structure and training algorithm gives highly accurate prediction values for the data used for modeling. However, one should exercise caution before choosing such an accurate model, because this model may give highly erroneous value when simulated with unknown data (i.e., the data which were not used for training). This phenomenon is known as the problem of overfitting. Therefore, after the parameters are estimated (the network is trained), another large set of data can be used to validate the model performance. Sometimes the resulting NN model is not satisfactory, and changes in the model structure must be made, often by trial and error. Commercial software packages are available that make automatic empirical modeling of complex processes feasible.

Advanced applications of neural nets have been commercially implemented in the areas of fault detection and diagnosis, sensor errors, and dynamic modeling and control (Su and McAvoy, 1997). In some cases, neural nets have been used to determine controller settings in advanced control systems.

The following example demonstrates the design of a Neural network model using MATLAB:

EXAMPLE 16.2

Use the data of Example 16.1 and design a neural network model to predict the Power generated from (Y_i) from fuel flow rate (u_i)

Solution

```
>> u = [1;2.3;2.9;4.0;4.9;5.8;6.5;7.7;8.4;9.0]; % Creation of input vector

>> y1 = [2;4.4;5.4;7.5;9.1;10.8;12.3;14.3;15.8;16.8]; %% Creation of output vector

>> Ynet = feedforwardnet(10, 'trainlm'); % Creation of a neural network model with one hidden layer consisting of 10 neurons and to be trained using Levenberg-Marquardt algorithm

>> Ynet.performFcn = 'mae'; % Mean absolute error is the objective function to be minimized

>> Ynet.layers{1}.transferFcn = 'tansig'; % Hidden layer activation function is hyperbolic tangent

>> Ynet.layers{2}.transferFcn = 'purelin'; % Output layer activation function is linear

>> Ynet.divideFcn = ' '; % All samples of the supplied dataset will be used for training without any data division
```

```
>> view(Ynet) % This will display the
   created neural network model
>> Ynet.trainparam.goal = 0;
>> Ynet.trainparam.epochs = 500; %
   Training will stop when MAE value
   becomes 0 or 500 iterations are
   performed
```

After creating the neural network model and setting the model parameters and training parameters, we are now ready to train the network with our input-output dataset.

```
>> Ynet = train (Ynet,u,y1);
>> Ynetp = sim(Ynet,u); % Get pre-
   dicted values by the model
>> SSEYnet = sum(y1-Ynetp).^2 % Model
   accuracy testing
SSEYnet = 0
```

The neural network model fitted the input-output data with absolute accuracy.

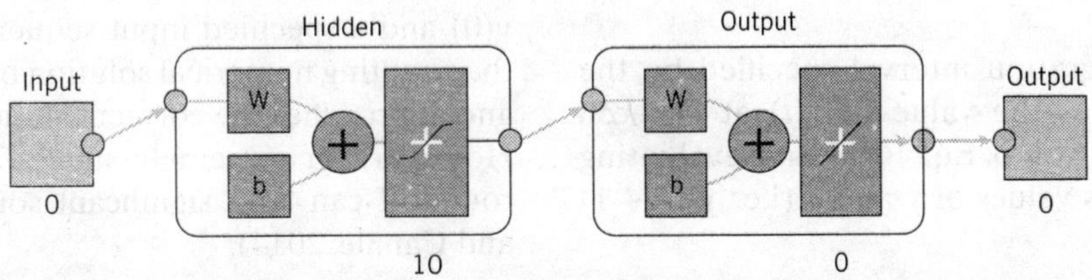

Figure 16.5 Neural network model as displayed in MATLAB with one hidden layer (hyperbolic tangent activation function) and output layer (linear activation function).

16.2.1 Soft Sensors

A common problem shared by many industrial processes is the inability to measure key process variables non-invasively and in real time, especially the compositions of process streams and product properties. The development of improved sensors, based on new techniques of analytical chemistry and modern electronic devices using fiber optics and semiconductors, has been an active area. As an alternative, the use of easily measured secondary variables to infer values of unmeasured process variables is now receiving great interest; the term soft *sensors* is often used to denote this approach. *Chemometrics* is a term related to soft sensors that describes how data from process analyzers (e.g., spectra) can be analyzed and modeled for use in process monitoring and control (Brown, 1998).

Soft sensors have become an attractive alternative to the high cost of accurate on-line measurements for applications where empirical models can accurately infer (i.e., predict) unmeasured variables (Kadlec et al., 2009). For example, the environmental regulatory agency in Texas permits NN models to be used for monitoring emissions from various process units such as power boilers. The NN models use measurements of selected input and output variables to predict pollutants at the parts per billion level (Martin, 1997). In materials manufacturing, the real-time detection of cracks, inclusions, porosity, dislocations, or defects in metallurgical or electronic materials would be highly desirable during processing, rather than after processing is completed and defective products are shipped. Use of virtual sensor models to predict quality control measures, such as the formation and location of defects, can greatly reduce the stringent requirements imposed on hardware-based sensors.

It may be noted that artificial neural network is not the only technique for soft sensor development. Multivariate statistical techniques based on principal component analysis and partial least square regression, soft computing methods such as neuro-fuzzy and support vector regression have also become quite popular among soft sensor researchers. With progress in computational speed, deep learning techniques such as convolution neural network, auto encoder and deep belief network are gaining more attention for process monitoring and control.

16.3 DEVELOPMENT OF DISCRETE-TIME DYNAMIC MODELS

A digital computer by its very nature deals internally with discrete-time data or numerical values of functions at equally spaced intervals determined by the sampling period. Thus, discrete-time models such as *difference equations* are widely used in computer control applications. One way a continuous-time dynamic model can be converted to discrete-time form is by employing a finite difference approximation (Chapra and Canale, 2014). Consider a nonlinear differential equation,

$$\frac{dy(t)}{dt} = f(y,u) \tag{16-16}$$

where y is the output variable and u is the input variable. This equation can be numerically integrated (although with some error) by introducing a finite difference approximation for the derivative. For example, the first-order, backward difference approximation to the derivative at $t = k\Delta t$ is

$$\frac{dy(t)}{dt} \cong \frac{y(k) - y(k-1)}{\Delta t} \tag{16-17}$$

where Δt is the integration interval specified by the user and $y(k)$ denotes the value of $y(t)$ at $t = k\Delta t$. Substituting Eq. 16-16 into Eq. 16-17 and evaluating $f(y, u)$ at the previous values of y and u (i.e., $y(k-1)$ and $u(k-1)$) gives

$$\frac{y(k - y(k-1))}{\Delta t} \cong f(y(k-1), u(k-1)) \tag{16-18}$$

or

$$y(k) = y(k-1) + \Delta t f(y(k-1), u(k-1)) \tag{16-19}$$

Equation 16-19 is a first-order difference equation that can be used to predict $y(k)$ based on information at the previous time step $(k - 1)$. This type of expression is called a *recurrence relation*. It can be used to numerically integrate Eq. 16-18 by successively calculating $y(k)$ for $k = 1,2,3, \ldots$ starting from a known initial condition $y(0)$ and a specified input sequence, $\{u(k)\}$. In general, the resulting numerical solution becomes more accurate and approaches the correct solution $y(t)$ as Δt decreases. However, for extremely small values of Δt, computer roundoff can be a significant source of error (Chapra and Canale, 2014).

EXAMPLE 16.3

For the first-order differential equation,

$$\tau \frac{dy(t)}{dt} + y(t) = Ku(t) \tag{16-20}$$

derive a recursive relation for $y(k)$ using a first-order backwards difference for $dy(t)/dt$.

Solution

The corresponding difference equation after approximating the first derivative is

$$\frac{\tau(y(k) - y(k-1))}{\Delta t} + y(k-1) = Ku(k-1) \tag{16-21}$$

Rearranging gives

$$y(k) = \left(1 - \frac{\Delta t}{\tau}\right) y(k-1) + \frac{K\Delta t}{\tau} u(k-1) \tag{16-22}$$

The new value $y(k)$ is a weighted sum of the previous value $y(k - 1)$ and the previous input $u(k - 1)$. Equation 16-22 can also be derived directly from Eq. 16-19.

As shown in numerical analysis textbooks, the accuracy of Eq. 16-22 is influenced by the integration interval. However, discrete-time models involving no approximation errors can be derived for any linear differential equation under the assumption of a piecewise constant input signal, that is, the input variable u is held constant over Δt. Next, we develop discrete-time modeling methods that introduce no integration error for piecewise constant inputs, regardless of the size of Δt. Such models are important in analyzing computer-controlled processes where the process inputs are piecewise constant.

16.3.1 Exact Discrete-Time Models

For a process described by a linear differential equation, the corresponding discrete-time model can be derived from the analytical solution for a piecewise constant input. This analytical approach eliminates the discretization error inherent in finite-difference approximations. Consider a first-order model in Eq. 16-20 with previous output $y[(k - 1)\Delta t]$ and a piecewise constant input $u(t) = u[(k - 1)\Delta t]$ over the time interval, $(k-1)\Delta t \le t < k\Delta t$. The analytical solution to Eq. 16-20 at $t = k\Delta t$ is

$$y(k\Delta t) = (1 - e^{-\Delta t/\tau})Ku[(k-1)\Delta t] + e^{-\Delta t/\tau}y[(k-1)\Delta t] \qquad (16\text{-}23)$$

Equation 16-23 can be written more compactly as

$$y(k) = e^{-\Delta t/\tau}y(k-1) + K(1 - e^{-\Delta t/\tau})u(k-1) \qquad (16\text{-}24)$$

Equation 16-24 is the exact solution to Eq. 16-20 at the sampling instants provided that $u(t)$ is piecewise constant for each sampling interval of length Δt. Note that the continuous output $y(t)$ is not necessarily constant between sampling instants, but Eq. 16-24 provides an exact solution for $y(t)$ at the sampling instants, $k = 1, 2, 3, \ldots$

In general, when a linear differential equation of order p is converted to discrete time, a linear difference equation of order p results. For example, consider the second-order model:

$$G(s) = \frac{Y(s)}{U(s)} = \frac{K(\tau_a s + 1)}{(\tau_1 s + 1)(\tau_2 s + 1)} \qquad (16\text{-}25)$$

The analytical solution for a piecewise constant input provides the corresponding difference equation, which is also referred to as an autogressive model with external (or exogenous) input, or *ARX model* (Ljung, 1999):

$$y(k) = a_1 y(k-1) + a_2 y(k-2) + b_1 u(k-1) + b_2 u(k-2) \qquad (16\text{-}26)$$

Where

$$a_1 = e^{-\Delta t/\tau_1} + e^{-\Delta t/\tau_2} \qquad (16\text{-}27)$$

$$a_2 = -e^{-\Delta t/\tau_1} e^{-\Delta t/\tau_2} \qquad (16\text{-}28)$$

$$b_1 = K\left(1 + \frac{\tau_a - \tau_1}{\tau_1 - \tau_2}e^{-\Delta t/\tau_1} + \frac{\tau_2 - \tau_a}{\tau_1 - \tau_2}e^{-\Delta t/\tau_2}\right) \qquad (16\text{-}29)$$

$$b_2 = K\left(e^{-\Delta t(1/\tau_1 + 1/\tau_2)} + \frac{\tau_a - \tau_1}{\tau_1 - \tau_2}e^{-\Delta t/\tau_2} + \frac{\tau_2 - \tau_a}{\tau_1 - \tau_2}e^{-\Delta t/\tau_1}\right) \qquad (16\text{-}30)$$

In Eq. 16-26 the new value of y depends on the values of y and u at the two previous sampling instants; hence, it is a second-order difference equation. If $\tau_2 = \tau_a = 0$ in Eqs. 16-26 through 16-30. the first-order difference equation in Eq. 16-23 results.

The steady-state gain of the second-order difference equation model can be found by considering steady-state conditions. Let $\bar{u}$ and $\bar{y}$ denote the new steady-state values after a step change in u. Substituting these values into Eq. 16-26 gives

$$\bar{y} = a_1\bar{y} + a_2\bar{y} + b_1\bar{u} + b_2\bar{u} \qquad (16\text{-}31)$$

Because y and u are deviation variables, the steady-state gain is simply $\bar{y}/\bar{u}$, the steady-state change in y divided by the steady state change in u. Rearranging Eq. 16-31

$$\text{Gain} = \frac{\bar{y}}{\bar{u}} = \frac{b_1 + b_2}{1 - a_1 - a_2} \qquad (16\text{-}32)$$

Substitution of Eqs. 16-27 through 16-30 into Eq. 16-32 gives K, the steady-state gain for the transfer function model in Eq. 16-25.

Higher-order linear differential equations can be converted to a discrete-time difference equation model using a state-space analysis (Åstörm and Wittenmark, 1997).

16.4 IDENTIFYING DISCRETE-TIME MODELS FROM EXPERIMENTAL DATA

If a linear discrete-time model is desired, one approach is to fit a continuous-time model to experimental data (cf. Section 6.5) and then to convert it to discrete-time form using the above approach. A more attractive approach is to estimate parameters in a discrete-time model directly from input–output data based on linear regression. This approach is an example of *system identification* (Ljung, 1999) As a specific example, consider the second-order difference equation in Eq. 16-26. It can be used to predict $y(k)$ from data available at times, $(k-1)\Delta t$ and $(k-2)\Delta t$. In developing a discrete-time model, model parameters a_1, a_2, b_1, and b_2 are considered to be unknown. They are estimated by applying linear regression to minimize the error criterion in Eq. 16-8 after defining

$$\beta^T = [a_1\ a_2\ b_1\ b_2],\ U_1 = y(k-1),\ U_2 = y(k-2),$$
$$U_3 = u(k-1), \text{ and } U_4 = u(k-2)$$

EXAMPLE 16.4

Consider the step response data $y(k)$ in Table 16.2 which were obtained from Example 6.5 and Fig. 6.16 for $\Delta t = 1$. At $t = 0$ a unit step change in u occurs, but the first output change is not observed until the next sampling instant.

Estimate the model parameters in Eq. 16-26 from the input–output data. Compare this model with the models obtained in Example 6.5 using nonlinear regression. For $k < 0$, assume $y(k) = 0$ and $u(k) = 0$, that is, the system is at rest.

Table 16.2 Step Response Data

k	$y(k)$
1	0.058
2	0.217
3	0.360
4	0.488
5	0.600
6	0.692
7	0.772
8	0.833
9	0.888
10	0.925

Solution

For linear regression, there are four independent variables, $y(k-1)$, $y(k-2)$, $u(k-1)$, $u(k-2)$, one dependent variable $y(k)$, and four unknown parameters (a_1, a_2, b_1, b_2). We structure the data for regression as shown in Table 16.3 and solve using Excel.

Table 16.3 Data Regression for Example 16.4

$$Y = \begin{bmatrix} 0.058 \\ 0.217 \\ 0.360 \\ 0.488 \\ 0.600 \\ 0.692 \\ 0.722 \\ 0.833 \\ 0.888 \\ 0.925 \end{bmatrix} \qquad U = \begin{bmatrix} 0 & 0 & 1 & 0 \\ 0.058 & 0 & 1 & 1 \\ 0.217 & 0.058 & 1 & 1 \\ 0.360 & 0.217 & 1 & 1 \\ 0.488 & 0.360 & 1 & 1 \\ 0.600 & 0.488 & 1 & 1 \\ 0.692 & 0.600 & 1 & 1 \\ 0.772 & 0.692 & 1 & 1 \\ 0.833 & 0.772 & 1 & 1 \\ 0.888 & 0.833 & 1 & 1 \end{bmatrix}$$

Table 16.4 compares the estimated parameters obtained by the two approaches. The results labeled nonlinear regression were obtained by fitting a continuous-time model (overdamped second order with time constants τ_1 and τ_2 and gain K) to the data using nonlinear regression. The continuous-time model was then converted to the corresponding discrete-time model using Eqs. 16-26 to 16-30.

Table 16.4 Comparison of Estimated Model Parameters for Example 16.4

	Linear Regression	Nonlinear Regression
a_1	0.975	0.984
a_2	−0.112	−0.122
b_1	0.058	0.058
b_2	0.102	0.101
K	1.168	1.159

The parameters obtained from linear regression in Table 16.4 are slightly different from those for nonlinear regression. This result occurs because for linear regression, four parameters were estimated; with nonlinear regression, three parameters were estimated. The estimated gain for linear regression, $K = 1.168$, is about 1% higher than the value obtained from nonlinear regression.

Table 16.5 compares the simulated responses for the two empirical models. Linear regression gives slightly better predictions, because it fits more parameters. However, in this particular example, it is difficult to distinguish graphically among the three model step responses.

Table 16.5 Comparison of Simulated Responses for Various Difference Equation Models*

n	y	$\hat{y}_L$	$\hat{y}_N$
1	0.058	0.058	0.058
2	0.217	0.217	0.216
3	0.360	0.365	0.366
4	0.488	0.487	0.487
5	0.600	0.595	0.596
6	0.692	0.690	0.690
7	0.772	0.768	0.767
8	0.833	0.835	0.835
9	0.888	0.886	0.885
10	0.925	0.933	0.932

*y, experimental data; $\hat{y}_L$, linear regression; $\hat{y}_N$, nonlinear regression

Example 16.4 has shown how a second-order difference equation model can be fit to data directly. The linear regression approach can also be used for higher-order models, provided that the parameters still appear linearly in the model. It is important to note that the estimated parameter values depend on the sampling period Δt for the data collection.

An advantage of the regression approach is that it is not necessary to make a step change in u in order to estimate model parameters. Other types of discrete-time models are the finite impulse response (FIR) model and the step response model, which are considered in Chapter 15 because of their close association with model predictive control.

16.4.1 Process Identification of More Complicated Models

In this section, we briefly consider three classes of more complicated process models: MIMO (multiple input, multiple output) models, stochastic models, and nonlinear discrete-time models.

MIMO process modeling is inherently more complicated than SISO modeling. For linear systems, the Principle of Superposition holds, which allows MIMO models to be developed through a series of single step tests for each input, while holding the other inputs constant. For a process with three inputs ($\boldsymbol{u}$) and three outputs ($\boldsymbol{y}$), we can introduce a step change in u_1, and record the responses for y_1, y_2, and y_3. The three transfer functions involving u_1, namely

$$\frac{Y_1}{U_1} = G_{11}, \quad \frac{Y_2}{U_1} = G_{21}, \quad \frac{Y_3}{U_1} = G_{31}$$

can be obtained using the techniques described in Section 6.5. In a similar fashion, step changes in U_2 and U_3 can be introduced in order to determine the other six G_{ij}. Alternatively, discrete-time models can be developed for each output, as discussed earlier in this section, using linear regression techniques. See Chapter 15 for a discussion of how such models are developed and used in model predictive controller calculations.

It is also possible to use PRBS forcing (Fig. 6.18(b)) to obtain MIMO models. To generate a multi-input PRBS signal, it is desirable that the input changes be independent. One way to accomplish this is to implement "shifted" or delayed versions of a single PRBS signal in each input. This means that if a single-input PRBS test requires 20 hours, a three-input test is designed to take three times as long. Hokanson and Gerstle (1992) suggest that about 20 total moves in each independent variable should be made.

Linear discrete-time models can also be developed that include the effects of unmeasured stochastic disturbances. For example, separate process models and disturbance models, also called *noise models*, can be obtained from data (Ljung, 1999). In this case the error term ε in Eq. 16-1 is not white noise but *colored noise* that is autocorrelated. In other words, there are underlying disturbance dynamics that causes ε to depend on previous values.

A variety of nonlinear discrete-time models have also been used in process control (Pearson, 1999). They include the neural net models discussed in Section 16.2 as well as nonlinear models obtained by adding nonlinear terms to the linear models of the previous section.

SUMMARY

When theoretical models are not available or are very complicated, empirical process models provide a viable alternative. In these situations, a model that is sufficiently accurate for control system design can often be obtained from experimental input/output data. Discrete-time models in the form of linear difference equations are frequently used in process control. These models can be readily obtained by least-squares fitting of experimental response data.

REFERENCES

Åström, K. J., and B. Wittenmark, *Computer-Controlled Systems: Theory and Design*, 3rd ed., Prentice-Hall, Englewood Cliffs, NJ, 1997.

Brown, S. D., Information and Data Handling in Chemistry and Chemical Engineering: The State of the Field from the Perspective of Chemometrics, *Computers and Chem. Engr.*, **23**, 203 (1998).

Chapra, S. C., and R. P. Canale, *Numerical Methods for Engineers*, 7th ed, McGraw-Hill, New York, 2014.

Draper, N. R., and H. Smith, *Applied Regression Analysis*, 3rd ed., John Wiley and Sons, Hoboken, NJ, 1998.

Edgar, T. F., D. M. Himmelblau, and L. S. Lasdon, *Optimization of Chemical Processes*, 2nd ed., McGraw-Hill, New York, 2001.

Godfrey, K. (Ed.), *Perturbation Signals for System Identification*, Prentice-Hall, Englewood Cliffs, NJ (1993).

Haykin, S. S., *Neural Networks and Learning Machines*, 3rd ed., Prentice-Hall, Upper Saddle River, NJ, 2009.

Himmelblau, D. M., Accounts of Experience in the Application of Artificial Neural Networks in Chemical Engineering, *Ind. Eng. Chem. Res.*, **47**, 5782 (2008).

Hokanson, D. A., and J. G. Gerstle, Dynamic Matrix Control Multivariable Controllers, in *Practical Distillation Control*, W. L. Luyben (Ed.), Van Nostrand Reinhold, New York (1992), p. 248.

Kadlec, P., B. Gabrys, and S. Strandt, Data-driven Soft Sensors in the Process Industries, *Computers and Chem. Engr.*, **33**, 795 (2009).

Ljung, L., *System Identification: Theory for the User*, 2nd ed., Prentice-Hall, Upper Saddle River, NJ, 1999.

Ljung, L., *System Identification Toolbox for Use with Matlab. Version 7.* MathWorks, Inc, Natick, MA, 2007.

Ljung, L., and T. Glad, *Modeling of Dynamic Systems*, Prentice-Hall, Upper Saddle River, NJ, 1994.

Martin, G., Consider Soft Sensors, *Chem. Engr. Prog.*, **93**(7), 66 (1997).

Montgomery, D. C., and G. C. Runger, *Applied Statistics and Probability for Engineers*, 6th ed., John Wiley and Sons, Hoboken, NJ (2013).

Pearson, R. K., *Discrete-Time Dynamic Models*, Oxford University Press, New York, 1999.

Rangaiah, G. P., and P. R. Krishnaswamy, Estimating Second-Order plus Dead Time Model Parameters, *Ind. Eng. Chem. Res.*, **33**, 1867 (1994).

Rangaiah, G. P., and P. R. Krishnaswamy, Estimating Second-Order Dead Time Parameters from Underdamped Process Transients, *Chem. Engr. Sci.*, **51**, 1149 (1996).

Rivera, D. E., and K. S. Jun, An Integrated Identification and Control Design Methodology for Multivariable Process System Applications, *IEEE Control Systems Magazine*, **20**(3), 25 (2000).

Smith, C. L., *Digital Computer Process Control*, Intext, Scranton, PA, 1972.

Su, H. T., and T. J. McAvoy, Artificial Neural Networks for Nonlinear Process Identification and Control, Chapter 7 in *Nonlinear Process Control*, M. A. Henson and D. E. Seborg (Eds.), Prentice Hall, Upper Saddle River, NJ, 1997.

Sundaresan, K. R., and R. R. Krishnaswamy, Estimation of Time Delay, Time Constant Parameters in Time, Frequency, and Laplace Domains, *Can. J. Chem. Eng.*, **56**, 257 (1978).

EXERCISES

16.1 Equation 16.1 is a linear relation between input u and output y with slope β_2 and intercept β_1. Let a linear approximation for an input x_i and output Y_i is of the form $Y_i = \beta x_i$. Following the method given in 16.1, derive least square formula for β.

16.2 The following data in Table E16.2 were obtained in a control valve experiment consisting of flow rates through the valve at different stem position of a linear control valve.

Table E16.2

Stem position (% of valve opening)	0	10	20	30	50	70	85	100
Flow rate	0	9	22	31	52	73	82	98

(a) Determine the linear equation relating flow and valve opening.

(b) Determine the valve opening required for a desired flow of 70.

16.3 The input (temperature) – output (mV) for a thermocouple is of the form: $V = \beta_0 + \beta_1 T$. Experimentally determined values of V and T are given in Table E16.3

(a) Determine the value of regression parameters and evaluate model accuracy by computing the sum of squared errors.

(b) Repeat calculations when one value of V is incorrectly measured as 2.719 instead of 1.719 (row 3 of Table E16.3). Comment on the result.

Table E16.3

T	37	50	73	95	112	137	193
V	0.591	0.778	1.118	1.453	1.719	2.118	3.049
V with outlier	0.591	0.778	1.118	1.453	2.719	2.118	3.049

16.4 Consider the first-order differential equation

$$5\frac{dy}{dt} + y(t) = 6u(t) \qquad y(0) = 3$$

where $u(t)$ is piecewise constant and has the following values:

$$u(0)=0 \qquad u(3)=3$$
$$u(1)=1 \qquad u(4)=0$$
$$u(2)=2 \qquad u(t)=0 \qquad \text{for } t > 4$$

Derive a difference equation for this ordinary equation using $\Delta t = 1$ and

(a) Exact discretization

(b) Finite difference approximation

Compare the integrated results for $0 \le t \le 10$. Examine whether $\Delta t = 0.1$ improves the accuracy of finite difference model.

16.5 The following data were collected from a cell concentration sensor measuring absorbance in a biochemical stream. The input u is the flow rate deviation (in dimensionless units) and the sensor output y is given in volts. The flow rate (input) is piecewise constant between sampling instants. The process is not at steady state initially, so y can change even though $u = 0$.

Table E16.5

Time (s)	u	y
0	0	3.000
1	3	2.456
2	2	5.274
3	1	6.493
4	0	6.404

(Continued)

Table E16.5 (Continued)

Time (s)	u	y
5	0	5.243
6	0	4.293
7	0	3.514
8	0	2.877
9	0	2.356
10	0	1.929

Fit a first-order model, $y(k) = a_1y(k-1) + b_1u(k-1)$, to the data using the least-squares approach. Plot the model response and the actual data. Can you also find a first-order continuous transfer function $G(s)$ to fit the data?

16.6 Obtain a first order discrete-time model from the step response data in Table E16.5. Compare your results with the first order graphical method for step response data, fitting the gain and time constant. Plot the two simulated responses for comparison with the observed data.

16.7 Unit step response data are given in Table E16.7 for a process with gain $K = 2$. Fit the data to a first-order model with no time delay. Next use linear regression to fit a first-order discrete-time equation to the data with $\Delta t = 1$. Then plot the predicted value of $y(t)$ for both models and compare with the data in Table E16.7. Calculate the gain of the discrete-time model and compare with the known value of $K = 1$. Why are the two values different?

Table E16.7

Time	$y(t)$
0	0.0
1	0.05
2	0.22
3	0.37
4	0.47
5	0.58
6	0.68
7	0.77
8	0.82
9	0.87
10	0.90

[1] Book web site: www.wiley.com/college/seborg

16.8 Data for a person with type 1 diabetes are available as both MATLAB and Excel data files on the book website.[1] Glucose measurements (y) were recorded every five minutes using a wearable sensor that measures subcutaneous glucose concentration. The insulin infusion rate (u) from a wearable subcutaneous insulin pump was also recorded every five minutes. The data files consist of experimental data for two step changes in the insulin infusion rate. The data are reported as deviations from the initial values that are considered to be the nominal steady-state values. It is proposed that the relationship between the glucose concentration y and the insulin infusion rate u can be described by a discrete-time, dynamic model of the form:

$$y(k) = a_1y(k-1) + a_2y(k-2) + b_1u(k-1) + b_2u(k-2)$$

(a) Use the least squares approach to estimate the model parameters from the *basal1* dataset. This data will be referred to as the *calibration data.* Graphically compare the model response and this data.

(b) In order to assess the accuracy of the model from part (a), calculate the model response $\hat{y}$ to the u step changes in the *validation data* (*basal2*). Then graphically compare the model response $\hat{y}$ *with* the validation data y.

(c) Repeat Steps (a) and (b) using an alternative transfer function model:

$$\frac{Y(s)}{U(s)} = \frac{K}{\tau s + 1}$$

Estimate the model parameters using graphical techniques and the *basal1* dataset. Then compare the model and experimental response data for both datasets.

(d) Which model is superior? Justify your answer by considering the least squares index for the one-step-ahead prediction errors,

$$S = \sum_{k=1}^{N}[y(k) - \hat{y}(k)]^2$$

where N is the number of data points.

MULTIPLE CHOICE QUESTIONS

1. The model $y = mx^2$ is to be fit to the data given below

x	1	2	3
y	2	$\sqrt{5}$	$\sqrt{8}$

Using linear regression, the value (rounded off to the second decimal place) of m is

(a) 5.32
(b) 2.57
(c) 3
(d) 0.387

2. What is the significance of least squares method in linear regression?

(a) To minimize the distance between the data points.
(b) To determine the least problematic regression line.
(c) To determine the optimum values of the intercept and slope that provide the smallest value of the sum of squared residuals.
(d) To determine the optimum values of the intercept and slope that provide the smallest value of the sum of residuals.

3. In a simple linear regression model (One independent variable), If the input variable is changed by 1 unit, what will be the change in the output variable?

(a) By 1 unit
(b) No change
(c) By intercept
(d) By its slope

4. If the correlation coefficient is 0.8, the percentage of variation in the output (response)variable explained by the variation in the input (explanatory) variable is

(a) 0.80% **(b)** 80%
(c) 0.64% **(d)** 64%

5. If the coefficient of determination (R^2) is a positive value, then the regression equation

(a) must have a positive slope.
(b) must have a negative slope.
(c) could have either a positive or a negative slope.
(d) must have a zero y-intercept.

6. Artificial neural network models can be used for

(a) function approximation.
(b) pattern recognition.
(c) regression.
(d) all of these.

7. The process of updating weight in neural network model is known as

(a) activation.
(b) learning.
(c) synchronization.
(d) none of these.

8. The output signal when a signal $x(n) = (0,2,4,6)$ is processed through a 'Delay' system is

(a) (2,4,6,0)
(b) (0,4,6,2)
(c) (6,0,4,2)
(d) (6,4,0,2)

9. Which of the following is a dynamic discrete time model?

(a) $y(k) = ax(k) + bx^3(k)$
(b) $y(k) = x(k)/(x^3(k) + b)$
(c) $y(k) = ax(k) + bx(k-1)$
(d) $y(k) = (ax(k) + bx^3(k))^2$

ANSWER KEY

1. (b) **2.** (c) **3.** (d) **4.** (d) **5.** (c) **6.** (d) **7.** (b) **8.** (a) **9.** (c)

CHAPTER 17

Process Monitoring

CHAPTER CONTENTS

In industrial plants, large numbers of process variables must be maintained within specified limits in order for the plant to operate properly. Excursions of key variables beyond these limits can have significant consequences for plant safety, the environment, product quality, and plant profitability. Earlier chapters have indicated that industrial plants rely on feedback and feedforward control to keep process variables at or near their set points. A related activity, *process monitoring*, also plays a key role in ensuring that the plant performance satisfies the operating objectives. In this chapter, we introduce standard monitoring techniques as well as newer strategies that have gained industrial acceptance in recent years. In addition to process monitoring, the related problem of monitoring the performance of the control system itself is also considered.

The general objectives of process monitoring are:

1. ***Routine Monitoring.*** Ensure that process variables are within specified limits.
2. ***Detection and Diagnosis.*** Detect abnormal process operation and diagnose the root cause.
3. ***Preventive Monitoring.*** Detect abnormal situations early enough that corrective action can be taken before the process is seriously upset.

Abnormal process operation can occur for a variety of reasons, including equipment problems (heat exchanger fouling), instrumentation malfunctions (sticking control valves, inaccurate sensors), and unusual disturbances (reduced catalyst activity, slowly drifting feed composition). Severe abnormal situations can have serious consequences, even forcing a plant shutdown. It has been estimated that improved handling of abnormal situations could result in savings of \$10 billion each year to the U.S. petrochemical industry (ASM, 2015). Thus, process monitoring and abnormal situation management are important activities.

The traditional approach for process monitoring is to compare measurements against specified limits. This *limit checking* technique is a standard feature of computer control systems and is widely used to validate measurements of process variables such as flow rate, temperature, pressure, and liquid level. Process variables are measured quite frequently with sampling periods that typically are much smaller than the process settling time (see Chapter 13). However, for most industrial plants, many important *quality variables* cannot be measured on-line. Instead, samples of the product are taken on an infrequent basis (e.g., hourly or daily) and sent to the quality control laboratory for analysis. Due to the infrequent measurements, standard feedback control methods like PID control cannot be applied. Consequently, statistical process control techniques are implemented to ensure that the product quality meets the specifications.

The terms *statistical process control* (*SPC*) and *statistical quality control* (*SQC*) refer to a collection of statistically-based techniques that rely on *quality control charts* to monitor product quality. These terms tend to be used on an interchangeable basis. However, the term SPC is sometimes used to refer to a broader set of statistical techniques that are employed to improve process performance as well as product quality. In this chapter, we emphasize the classical SPC techniques that are based on quality control charts (also called *control charts*). The simplest control chart, a *Shewhart chart*, merely consists of measurements plotted vs. sample number, and control limits that indicate the upper and lower limits for normal process operation.

The major objective in SPC is to use process data and statistical techniques to determine whether the process operation is normal or abnormal. The SPC methodology is based on the fundamental assumption that normal process operation can be characterized by random variations about a mean value. If this situation exists, the process is said to be *in a state of statistical control* (or *in control*), and the control chart measurements tend to be normally distributed about the mean value. By contrast, frequent control chart violations would indicate abnormal process behavior or an *out-of-control* situation. Then, a search would be initiated to attempt to identify the root cause of the abnormal behavior. The root cause is referred to as the *assignable cause* or the *special cause* in the SPC literature, while the normal process variability is referred to as *common cause* or *chance cause*. From an engineering perspective, SPC is more of a monitoring technique than a control technique because no automatic corrective action is taken after an abnormal situation is detected. A brief comparison of conventional feedback control and SPC is presented in Section 17.2.4.

The basic SPC concepts and control chart methodology were introduced by Shewhart (1931). The current widespread interest in SPC techniques began in the 1950s when they were successfully applied first in Japan and then in North America, Europe, and the rest of the world. Control chart methodologies are now widely used in discrete-parts manufacturing and in some sectors of the process industries, especially for the production of semiconductors, synthetic fibers, polymers, and specialty chemicals. SPC techniques are also widely used for product quality control and for monitoring control system performance (Shunta, 1995). The basic SPC methodology is described in introductory statistics texts (Montgomery and Runger, 2013) and books on SPC (Ryan 2011; Montgomery, 2012).

SPC techniques played a key role in the renewed industrial emphasis on product quality that is sometimes referred to as the *Quality Revolution*. During the 1980s, Deming (1986) had a major impact on industrial management in North America by convincing corporations that quality should be a top corporate priority. He argued that the failure of a company to produce quality products was largely a failure in management rather than a shortcoming of the plant equipment or employees. His success led to the establishment of many process and quality improvement programs, including the *Six Sigma* methodology that is considered in Section 17.3.

In this chapter, we first introduce traditional process monitoring techniques (Section 17.1) that are based on limit checking of measurements and process performance calculations. In Section 17.2, the theoretical basis of SPC monitoring techniques and the most widely used control charts are considered. We also introduce *process capability indices* and compare SPC with standard automatic feedback control. Traditional SPC monitoring techniques consider only a single measured variable at a time, a *univariate* approach. But when the measured variables are highly correlated, improved monitoring can be achieved by applying the *multivariate* techniques that are introduced in Section 17.4. In addition to monitoring process performance, it can be very beneficial to assess control system performance. This topic is considered in Section 17.5.

Monitoring strategies have been proposed based on process models, neural networks, and expert systems (Davis et al., 2000; Chiang et al., 2001). However, these topics are beyond the scope of this book.

17.1 TRADITIONAL MONITORING TECHNIQUES

In this section, we consider two relatively simple but very effective process monitoring techniques: limit checking and performance calculations.

17.1.1 Limit Checking

Process measurements should be checked to ensure that they are between specified limits, a procedure referred to as *limit checking.* The most common types of measurement limits are:

1. High and low limits
2. High limit for the absolute value of the rate of change
3. Low limit for the sample variance

The limits are specified based on safety and environmental considerations, operating objectives, and equipment limitations. For example, the high limit on a reactor temperature could be set based on metallurgical limits or the onset of undesirable side reactions. The low limit for a slurry flow rate could be selected to avoid having solid material settle and plug the line. Sometimes a second set of limits serves as *warning limits.* For example, in a liquid storage system, when the level drops to 15% (the low limit), a low-priority alarm signal could be sent to the operator. But when the level decreases to 5% (the low-low limit), a high-priority alarm would be generated for this more serious situation. Similarly, in order to avoid having the tank overflow, a high limit of 85% and a high-high limit of 95% level could be specified. The high-high and low-low limits are also referred to as *action limits.*

In practice, there are physical limitations on how much a measurement can change between consecutive sampling instants. For example, we might conclude that a temperature in a process vessel cannot change by more than 2°C from one sampling instant to the next, based on knowledge of the energy balance and the process dynamics. This *rate-of-change limit* can be used to detect an abnormal situation such as a noise spike or a sensor failure. (Noise-spike filters were considered in Chapter 13.)

A set of process measurements inevitably exhibits some variability, even for "steady-state operation." This variability occurs as a result of measurement noise, turbulent flow near a sensor, and other process disturbances. However, if the amount of variability becomes unusually low, it could indicate an abnormal situation such as a "dead sensor" or a sticking control valve. Consequently, it is common practice to monitor a measure of variability such as the variance or standard deviation of a set of measurements. For example, the variability of a set of n measurements can be characterized by the *sample standard deviation*, s, or the *sample variance*, s^2,

$$s^2 \triangleq \frac{1}{n-1}\sum_{i=1}^{n}(x_i - \bar{x})^2 \tag{17-1}$$

where x_i denotes the ith measurement and $\bar{x}$ is the sample mean:

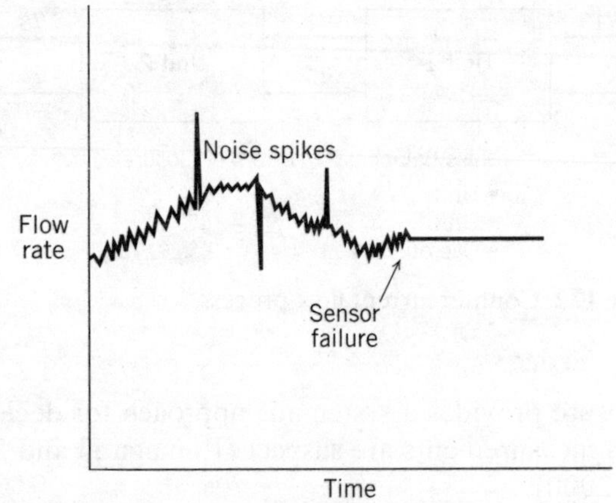

Figure 17.1 Flow rate measurement.

$$\bar{x} \triangleq \frac{1}{n}\sum_{i=1}^{n} x_i \tag{17-2}$$

For a set of data, $\bar{x}$ indicates the average value, while s and s^2 provide measures of the spread of the data. Either s or s^2 can be monitored to ensure that it is above a threshold that is specified based on process operating experience.

The flow rate data in Fig. 17.1 includes three noise spikes and a sensor failure. The rate of change limit would detect the noise spikes, while an abnormally low sample variance would identify the failed sensor. After a limit check violation occurs, an alarm signal can be sent to the plant operator in a number of different ways. A relatively minor alarm might merely be "logged" in a computer file. A more important alarm could be displayed as a flashing message on a computer terminal and require operator acknowledgment. A critical alarm could result in an audible sound or a flashing warning light in the control room.

17.1.2 Performance Calculations

A variety of performance calculations can be made to determine whether the process and instrumentation are working properly. In particular, steady-state mass and energy balances are calculated using data that are averaged over a period of time (for example, one hour). The percent error of closure for a total mass balance can be defined as

$$\%\ error\ of\ closure \triangleq \frac{rate\ in - rate\ out}{rate\ in} \times 100\% \tag{17-3}$$

A large error of closure may be caused by an equipment problem (e.g., a pipeline leak) or a sensor problem. *Data reconciliation* based on a statistical analysis of the errors

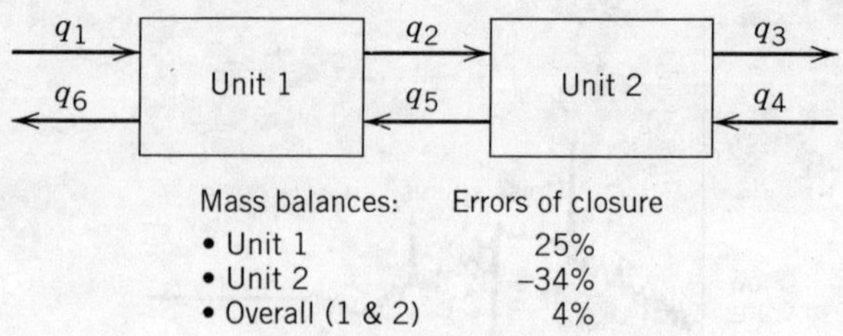

Figure 17.2 Countercurrent flow process.

of closure provides a systematic approach for deciding which measurements are suspect (Romagnoli and Sanchez, 2000).

Both redundant measurements and conservation equations can be used to good advantage. A process consisting of two units in a countercurrent flow configuration is shown in Fig. 17.2. Three steady-state mass balances can be written, one for each unit plus an overall balance around both units. Although the three balances are not independent, they provide useful information for monitoring purposes. Figure 17.2 indicates that the error of closure is small for the overall balance but large for each individual balance. This situation suggests that the flow rate sensor for one of the two interconnecting streams, q_2 or q_5, may be faulty.

Process performance calculations also are very useful for diagnostic and monitoring purposes. For example, the thermal efficiency of a refrigeration unit or the selectivity of a chemical reactor could be calculated on a regular basis. A significant decrease from the normal value could indicate a process change or faulty measurement.

17.2 QUALITY CONTROL CHARTS

Industrial processes inevitably exhibit some variability in their manufactured products regardless of how well the processes are designed and operated. In statistical process control, an important distinction is made between normal (random) variability and abnormal (non-random) variability. Random variability is caused by the cumulative effects of a number of largely unavoidable phenomena such as electrical measurement noise, turbulence, and random fluctuations in feedstock or catalyst preparation. The random variability can be interpreted as a type of "background noise" for the manufacturing operation. Nonrandom variability can result from process changes (e.g., heat exchanger fouling, loss of catalyst activity), faulty instrumentation, or human error. As mentioned earlier, the source of this abnormal variability is referred to as a *special cause* or an *assignable cause*.

17.2.1 Normal Distribution

Because the *normal distribution* plays a central role in SPC, we briefly review its important characteristics. The normal distribution is also known as the *Gaussian distribution.*

Suppose that a random variable x has a normal distribution with a mean μ and a variance σ^2 denoted by $N(\mu, \sigma^2)$. The probability that x has a value between two arbitrary constants, a and b, is given by:

$$P(a < x < b) = \int_a^b f(x)dx \tag{17-4}$$

where $P(\cdot)$ denotes the probability that x lies within the indicated range and $f(x)$ is the *probability density function* for the normal distribution:

$$f(x) = \frac{1}{\sigma\sqrt{2\pi}} \exp\left[-\frac{(x-\mu)^2}{2\sigma^2}\right] \tag{17-5}$$

The following probability statements are valid for the normal distribution (Montgomery and Runger, 2013):

$$\begin{aligned} P(\mu - \sigma < x < \mu + \sigma) &= 0.6827 \\ P(\mu - 2\sigma < x < \mu + 2\sigma) &= 0.9545 \\ P(\mu - 3\sigma < x < \mu + 3\sigma) &= 0.9973 \end{aligned} \tag{17-6}$$

A graphical interpretation of these expressions is shown in Fig. 17.3 where each probability corresponds to an area under the $f(x)$ curve. Equation 17-6 and Fig. 17.3 demonstrate that if a random variable x is normally distributed, there is a very high probability (0.9973) that a measurement lies within 3σ of the mean μ. This important result provides the theoretical basis for widely used SPC techniques. Similar probability statements can be formulated based on statistical tables for the normal distribution. For the sake of generality, the tables are expressed in terms of the *standard normal distribution,* $N(0, 1)$, and the *standard normal variable,* $z \triangleq (x - \mu)/\sigma$.

It is important to distinguish between the theoretical mean μ and the sample mean $\bar{x}$. If measurements of a process variable are normally distributed, $N(\mu, \sigma^2)$, the sample mean is also normally distributed. Of course, for any particular sample, $\bar{x}$ is not necessarily equal to μ.

17.2.2 The $\bar{x}$ Control Chart

In statistical process control, Control Charts (or Quality Control Charts) are used to determine whether the process operation is normal or abnormal. The widely used $\bar{x}$ control chart is introduced in the following example. This type of control chart is often referred to as a

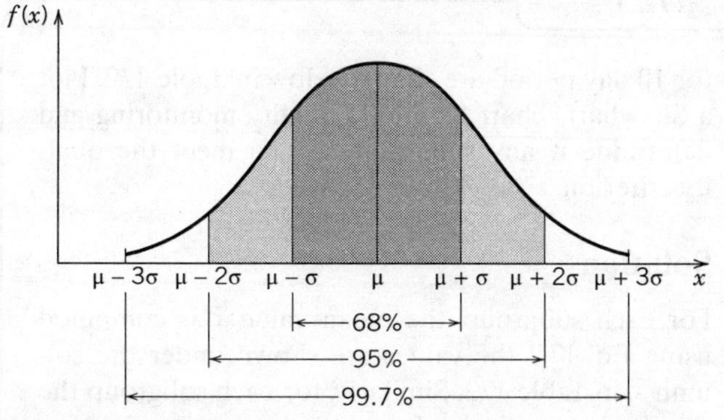

Figure 17.3 Probabilities associated with the normal distribution. From Montgomery and Runger (2013).

Table 17.1 Control Chart Constants

	Estimation of σ	s Chart	
n	c_4	B_3	B_4
2	0.7979	0	3.267
3	0.8862	0	2.568
4	0.9213	0	2.266
5	0.9400	0	2.089
6	0.9515	0.030	1.970
7	0.9594	0.118	1.882
8	0.9650	0.185	1.815
9	0.9693	0.239	1.761
10	0.9727	0.284	1.716
15	0.9823	0.428	1.572
20	0.9869	0.510	1.490
25	0.9896	0.565	1.435

Source: Adapted from Ryan (2000).

Shewhart Chart, in honor of the pioneering statistician, Walter Shewhart, who first developed it in the 1920s.

The first step in devising a control chart is to select a set of representative data for a period of time when the process operation is believed to be normal, rather than abnormal. Suppose that these test data consist of N subgroups that have been collected on a regular basis (for example, hourly or daily) and that each subgroup consists of n randomly selected items. Let x_{ij} denote the jth measurement in the ith subgroup. Then, the subgroup sample means can be calculated:

$$\bar{x}_i \triangleq \frac{1}{n}\sum_{j=1}^{n} x_{ij} \quad (i = 1, 2, \ldots, N) \tag{17-7}$$

The *grand mean* $\bar{\bar{x}}$ is defined to be the average of the subgroup means:

$$\bar{\bar{x}} \triangleq \frac{1}{N}\sum_{i=1}^{N} \bar{x}_i \tag{17-8}$$

The general expressions for the control limits are

$$UCL \triangleq T + c\hat{\sigma}_{\bar{x}} \tag{17-9}$$

$$LCL \triangleq T - c\hat{\sigma}_{\bar{x}} \tag{17-10}$$

where $\hat{\sigma}_{\bar{x}}$ is an estimate of the standard deviation for $\bar{\bar{x}}$ and c is a positive integer; typically, $c = 3$. The choice of $c = 3$ and Eq. 17-6 imply that the measurements will lie within the control chart limits 99.73% of the time, for normal process operation. The target T is usually specified to be either $\bar{\bar{x}}$ or the desired value of $\bar{\bar{x}}$.

The estimated standard deviation $\hat{\sigma}_{\bar{x}}$ can be calculated from the subgroups in the test data by two methods: (1) the standard deviation approach and (2) the range approach (Montgomery and Runger, 2013). By definition, the *range* R is the difference between the maximum and minimum values. Historically, the R approach has been emphasized, because R is easier to calculate than s, an advantage for hand calculations. However, the standard deviation approach is now preferred because it uses all of the data, instead of only two points in each subgroup. It also has the advantage of being less sensitive to *outliers* (i.e., bad data points). However, for small values of n, the two approaches tend to produce similar control limits (Ryan, 2011). Consequently, we will only consider the standard deviation approach.

The average sample standard deviation $\bar{s}$ for the N subgroups is

$$\bar{s} \triangleq \frac{1}{N}\sum_{i=1}^{N} s_i \tag{17-11}$$

where the standard deviation for the ith subgroup is

$$s_i \triangleq \sqrt{\frac{1}{n-1}\sum_{j=1}^{n}(x_{ij} - \bar{x}_i)^2} \tag{17-12}$$

If the x data are normally distributed, then $\hat{\sigma}_{\bar{x}}$ is related to $\bar{s}$ by

$$\hat{\sigma}_{\bar{x}} = \frac{1}{c_4\sqrt{n}}\bar{s} \tag{17-13}$$

where c_4 is a constant that depends on n (Montgomery and Runger, 2013) and is tabulated in Table 17.1.

Example 17.1

A manufacturing plant produces 10,000 plastic bottles per day. Because the product is inexpensive and the plant operation is normally satisfactory, it is not economically feasible to inspect every bottle. Instead, a sample of 3 bottles is randomly selected and inspected each day. The inspection includes measuring the toughness x of each bottle in the subgroup. Values for 10 day period are shown below in Table 17.2. Plot a Shewhart chart for bottle quality monitoring and determine if any subgroup fails to meet the quality criterion.

Table 17.2 Product Quality Data for Example 17.1

Sub Group No	x Data			$\overline{x}$	s_i
1	100.8	108	104	104.27	3.61
2	83	85	82	83.33	1.53
3	98	89.6	96.8	94.80	4.54
4	97.6	94.4	98	96.67	1.97
5	108	91.2	94.4	97.87	8.92
6	100.8	101.6	89.6	97.33	6.71
7	89	82	83	84.67	3.79
8	89.6	101.6	101.2	97.47	6.82
9	86	94.8	93.2	91.33	4.69
10	105.2	99	107	103.73	4.20
				$\overline{\overline{x}} = 95.15$	$\overline{s} = 4.68$

Solution

For each subgroup, the mean value $\overline{x}$ is computed using Eq. 17-7 the values are shown under the column $\overline{x}$ in Table 17.2. Similarly, for each subgroup the standard deviation values are computed using Eq. 17-12 and the values are shown in the last column.

The values of grand mean and average sample standard deviation are computed using Eq. 17-8 and 17-11 respectively. These two values are shown in the last row of Table 17.2.

Target is specified as the grand mean value. Next task is to compute the upper and lower control limits.

Value of c_4 for $n = 3$ is 0.8862 (Table 17.1).

Estimate of standard deviation is computed using Eq. 17-13: $\hat{\sigma}_{\overline{x}} = 3.05$

Using $T = 95.15, c = 3$ and $\hat{\sigma}_{\overline{x}} = 3.05$ in Eqs. 17-9 and 17-10, the values are:

$$\text{UCL} = 104.29, \text{LCL} = 86.01.$$

Figure 17.4 shows the mean values of each subgroup, UCL, LCL and target:

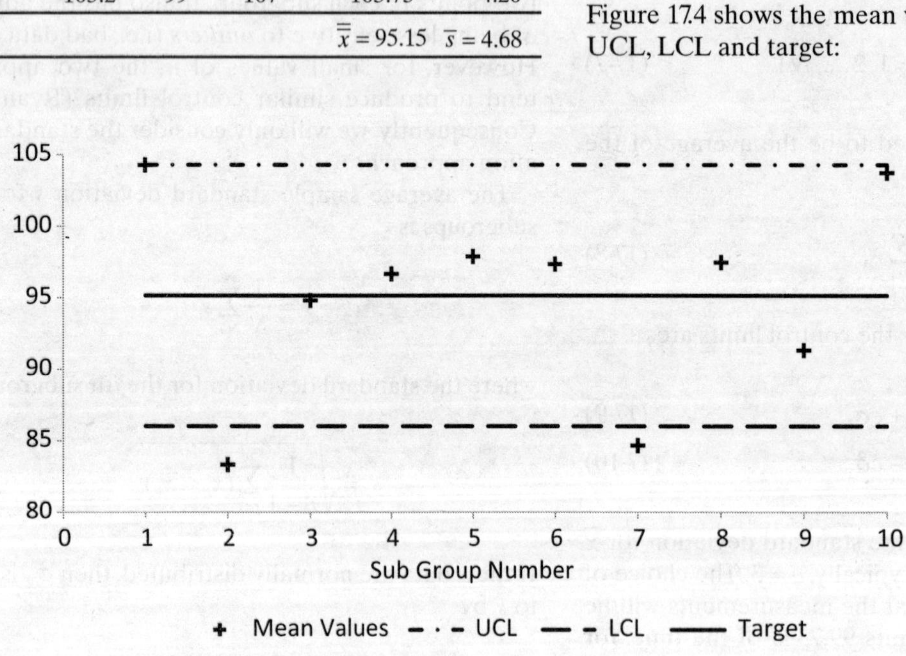

Figure 17.4 The $\overline{x}$ Control Chart for Example 17.1.

The chart indicates that subgroup number 2 and 7 may contain defective samples since the values lie outside the control limit values.

MATLAB Application

Initially, all values are stored in a matrix x of dimension 10×3. Rows correspond to number of subgroups and columns, number of elements in each subgroup.

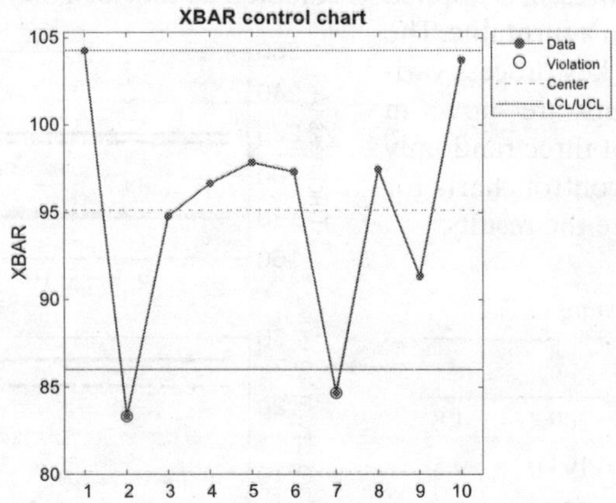

The following command generated the $\bar{x}$ control chart:

```
>> controlchart(x)
```

The concept of a *rational subgroup* plays a key role in the development of quality control charts. The basic idea is that a subgroup should be specified so that it reflects typical process variability but not assignable causes. Thus, it is desirable to select a subgroup so that a special cause can be detected by a comparison of subgroups, but it will have little effect within a subgroup (Montgomery, 2012). For example, suppose that a small chemical plant includes six batch reactors and that a product quality measurement for each reactor is made every hour. If the monitoring objective is to determine whether overall production is satisfactory, then the individual reactor measurements could be pooled to provide a subgroup size of $n = 6$ and a sampling period of $\Delta t = 1$ h. On the other hand, if the objective is to monitor the performance of individual reactors, the product quality data for each reactor could be plotted on an hourly basis ($n = 1$) or averaged over an eight-hour shift ($n = 8$ and $\Delta t = 8$ h). When only a single measurement is made at each sampling instant, the subgroup size is $n = 1$ and the control chart is referred to as an *individuals chart*.

17.2.3 The *s* Control Chart

In addition to monitoring average process performance, it is also advantageous to monitor process variability. The variability within a subgroup can be characterized by its range, standard deviation, or sample variance. Control charts can be developed for all three statistics, but our discussion will be limited to the control chart for the standard deviation, the *s control chart*.

The centerline for the s chart is s, which is the average standard deviation for the test set of data. The control limits are

$$UCL = B_4\bar{s} \tag{17-14}$$

$$LCL = B_3\bar{s} \tag{17-15}$$

Constants B_3 and B_4 depend on the subgroup size n, as shown in Table 17.1.

The control chart limits for the $\bar{x}$ and s charts in Eqs. 17-9 to 17-15 have been based on the assumption that the x data are normally distributed.

When individual measurements are plotted ($n = 1$), the standard deviation for the subgroup does not exist. In this situation, the *moving range* (MR) of two successive measurements can be employed to provide a measure of variability. The moving range is defined as the absolute value of the difference between successive measurements. Thus, for the kth sampling instant, $MR(k) = |x(k) - x(k-1)|$. The $\bar{x}$ and s control charts are also applicable when the sample size n varies from one sample to the next.

Example 17.2 illustrates the construction of $\bar{x}$ and s control charts.

EXAMPLE 17.2

In semiconductor processing, the photolithography process is used to transfer the circuit design to silicon wafers. In the first step of the process, a specified amount of a polymer solution, *photoresist*, is applied to a wafer as it spins at high speed on a turntable. The resulting photoresist thickness x is a key process variable. Thickness data for 25 subgroups are shown in Table 17.3. Each subgroup consists of three randomly selected wafers. Construct $\bar{x}$ and s control charts for these test data and critically evaluate the results.

Table 17.3 Thickness Data (in Å) for Example 17.2

No.	x Data			$\bar{x}$	s
1	209.6	207.6	211.1	209.4	1.8
2	183.5	193.1	202.4	193.0	9.5
3	190.1	206.8	201.6	199.5	8.6
4	206.9	189.3	204.1	200.1	9.4
5	260.0	209.0	212.2	227.1	28.6
6	193.9	178.8	214.5	195.7	17.9
7	206.9	202.8	189.7	199.8	9.0
8	200.2	192.7	202.1	198.3	5.0
9	210.6	192.3	205.9	202.9	9.5
10	186.6	201.5	197.4	195.2	7.7
11	204.8	196.6	225.0	208.8	14.6
12	183.7	209.7	208.6	200.6	14.7
13	185.6	198.9	191.5	192.0	6.7
14	202.9	210.1	208.1	207.1	3.7
15	198.6	195.2	150.0	181.3	27.1
16	188.7	200.7	207.6	199.0	9.6
17	197.1	204.0	182.9	194.6	10.8
18	194.2	211.2	215.4	206.9	11.2
19	191.0	206.2	183.9	193.7	11.4
20	202.5	197.1	211.1	203.6	7.0
21	185.1	186.3	188.9	186.8	1.9
22	203.1	193.1	203.9	200.0	6.0
23	179.7	203.3	209.7	197.6	15.8
24	205.3	190.0	208.2	201.2	9.8
25	203.4	202.9	200.4	202.2	1.61

Solution

The following sample statistics can be calculated from the data in Table 17.3: $\bar{\bar{x}} = 199.8$ Å, $\bar{s} = 10.4$ Å. For $n = 3$ the required constants from Table 17.1 are $c_4 = 0.8862$, $B_3 = 0$, and $B_4 = 2.568$. Then the $\bar{x}$ and s control limits can be calculated from Eqs. 17-9 to 17-15. The traditional value of $c = 3$ is selected for Eqs. 17-9 and 17-10. The resulting control limits are labeled as the "original limits" in Fig. 17.5.

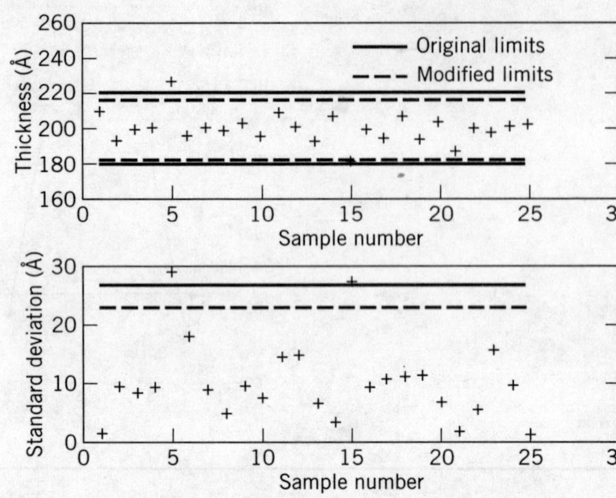

Figure 17.5 The $\bar{x}$ and s control charts for Example 17.2.

Figure 17.5 indicates that sample #5 lies beyond the UCL for both the $\bar{x}$ and s control charts, while sample #15 is very close to a control limit on each chart. Thus, the question arises whether these two samples are "outliers" that should be omitted from the analysis. Table 17.3 indicates that sample #5 includes a very large value (260.0), while sample #15 includes a very small value (150.0). However, unusually large or small numerical values by themselves do not justify discarding samples; further investigation is required.

Suppose that a more detailed evaluation has discovered a specific reason as to why measurements #5 and #15 should be discarded (e.g., faulty sensor, data misreported, etc.). In this situation, these two samples should be removed and the control limits should be recalculated based on the remaining 23 samples. These modified control limits are tabulated below as well as in Fig. 17.5.

	Original Limits	Modified Limits (omit samples #5 and #15)
$\bar{x}$ Chart Control Limits		
UCL	220.1	216.7
LCL	179.6	182.2
s Chart Control Limits		
UCL	26.6	22.7
LCL	0	0

17.2.4 Theoretical Basis for Quality Control Charts

The traditional SPC methodology is based on the assumption that the natural variability for "in control" conditions can be characterized by random variations around a constant average value,

$$x(k) = x^* + e(k) \tag{17-16}$$

where $x(k)$ is the measurement at time k, x^* is the true (but unknown) value, and $e(k)$ is an additive random error. Traditional control charts are based on the following assumptions:

1. Each additive error, $\{e(k),\ k = 1, 2,...\}$, is a zero-mean, random variable that has the same normal distribution, $N(0, \sigma^2)$.
2. The additive errors are statistically independent and thus uncorrelated. Consequently, $e(k)$ does not depend on $e(j)$ for $j \neq k$.
3. The true value x^* is constant.
4. The subgroup size n is the same for all of the subgroups.

The second assumption is referred to as *independent and identically distributed* (*IID*).

Consider an ideal individuals control chart for x with x^* as its target and "3σ control limits":

$$UCL \triangleq x^* + 3\sigma \tag{17-17}$$

$$LCL \triangleq x^* + 3\sigma \tag{17-18}$$

These control limits are a special case of Eqs. 17-9 and 17-10 for the idealized situation where σ is known, $c = 3$, and the subgroup size is $n = 1$. The typical choice of $c = 3$ can be justified as follows. Because x is $N(0, \sigma^2)$, the probability p that a measurement lies outside the 3σ control limits can be calculated from Eq. 17-6: $p = 1 - 0.9973 = 0.0027$. Thus on average, approximately three out of every 1,000 measurements will be outside of the 3σ limits. The average number of samples before a chart violation occurs is referred to as the *average run length* (ARL). For the normal ("in control") process operation,

$$\text{ARL} \triangleq \frac{1}{p} = \frac{1}{0.0027} = 370 \tag{17-19}$$

Thus, a Shewhart chart with 3σ control limits will have an average of one control chart violation every 370 samples, even when the process is *in a state of control*.

This theoretical analysis justifies the use of 3σ limits for $\bar{x}$ and other control charts. However, other values of c are sometimes used. For example, 2σ warning limits can be displayed on the control chart in addition to the 3σ control limits. Although the 2σ warning limits provide an early indication of a process change, they have a very low average run length value of ARL = 22. In general, larger values of c result in wider chart limits and larger ARL values. Wider chart limits mean that process changes will not be detected as quickly as they would be for smaller c values. Thus, the choice of c involves a classical engineering compromise between early detection of process changes (low value of c) and reducing the frequency of false alarms (high value of c).

Standard SPC techniques are based on the four assumptions listed above. However, because these assumptions are not always valid for industrial processes, standard techniques can give misleading results. In particular, the implications of violating the normally distributed and IID assumptions have received considerable theoretical analysis (Ryan, 2011). Although modified SPC techniques have been developed for these nonideal situations, commercial SPC software is usually based on these assumptions.

Industrial plant measurements are not normally distributed. However, for large subgroup sizes ($n > 25$), $\bar{x}$ is approximately normally distributed even if x is not, according to the famous *Central Limit Theorem* of statistics (Montgomery and Runger, 2013). Fortunately, modest deviations from "normality" can be tolerated. In addition, the standard SPC techniques can be modified so that they are applicable to certain classes of nonnormal data (Jacobs, 1990).

In industrial applications, the control chart data are often *serially correlated*, because the current measurement is related to previous measurements. For example, the flow rate data in Fig. 17.1 are serially correlated. Standard control charts such as the $\bar{x}$ and s charts can provide misleading results if the data are serially correlated. But if the degree of correlation is known, the control limits can be adjusted accordingly (Montgomery, 2012). Serially correlated data also can be modeled using time-series analysis, as described in Section 13.8.

17.2.5 Pattern Tests and the Western Electric Rules

We have considered how abnormal process behavior can be detected by comparing individual measurements with the $\bar{x}$ and s control chart limits. However, the pattern of measurements can also provide useful information. For example, if 10 consecutive measurements are all increasing, then it is very unlikely that the process is *in a state of control.*

A wide variety of pattern tests (also called *zone rules*) can be developed based on the IID and normal distribution assumptions and the properties of the normal distribution. For example, the following excerpts from the *Western Electric Rules* (Montgomery and Runger, 2013) indicate that the process is *out of control* if one or more of the following conditions occur:

1. One data point is outside the 3σ control limits.
2. Two out of three consecutive data points are beyond a 2σ limit.
3. Four out of five consecutive data points are beyond a 1σ limit and on one side of the centerline.
4. Eight consecutive points are on one side of the centerline.

Note that the first condition corresponds to the familiar Shewhart chart limits of Eqs. 17-9 and 17-10 with $c = 3$. Additional pattern tests are concerned with other types of nonrandom behavior (Montgomery, 2012). Pattern tests can be used to augment Shewhart charts. This combination enables out-of-control behavior to be detected earlier, but the false alarm rate is higher than that for a Shewhart chart alone.

17.2.6 CUSUM and EWMA Control Charts

Although Shewhart charts with 3σ limits can quickly detect large process changes, they are ineffective for small, sustained process changes (e.g., changes in μ smaller than 1.5σ). Two alternative control charts have been developed to detect small changes: the CUSUM and EWMA control charts. They also can detect large process changes (e.g., 3σ shifts), but detection is usually somewhat slower than for Shewhart charts. Because the CUSUM and EWMA control charts can effectively detect both large and small process shifts, they provide viable alternatives to the widely used Shewhart charts. Consequently, they will now be considered. The *cumulative sum* (CUSUM) is defined to be a running summation of the deviations of the plotted variable from its target. If the sample mean is plotted, the cumulative sum, $C(k)$, is

$$C(k) = \sum_{j=1}^{k} (\bar{x}(j) - T) \quad (17\text{-}20)$$

where T is the target for $\bar{x}$. During normal process operation, $C(k)$ fluctuates around zero. But if a process change causes a small shift in $\bar{x}$, $C(k)$ will drift either upward or downward.

The CUSUM control chart was originally developed using a graphical approach based on *V-masks* (Montgomery, 2012). However, for computer calculations, it is more convenient to use an equivalent algebraic version that consists of two recursive equations,

$$C^+(k) = \max[0, \bar{x}(k) - (T + K) + C^+(k-1)] \quad (17\text{-}21)$$

$$C^-(k) = \max[0, (T - K) - \bar{x}(k) + C^-(k-1)] \quad (17\text{-}22)$$

where C^+ and C^- denote the sums for the high and low directions and K is a constant, the *slack parameter*. The CUSUM calculations are initialized by setting $C^+(0) = C^-(0) = 0$. A deviation from the target that is larger than K increases either C^+ or C^-. A control limit violation occurs when either C^+ or C^- exceeds a specified control limit (or *threshold*), H. After a limit violation occurs, that sum is reset to zero or to a specified value.

The selection of the threshold H can be based on considerations of average run length. Suppose that we want to detect whether the sample mean $\bar{x}$ has shifted from the target by a small amount, δ. The slack parameter K is usually specified as $K = 0.5\,\delta$. For the ideal situation where the normally distributed and IID assumptions are valid, ARL values have been tabulated for specified values of δ, K, and H (Ryan, 2011; Montgomery, 2012).

Table 17.4 summarizes ARL values for two values of H and different values of δ. (The values of δ are usually expressed as multiples of $\hat{\sigma}_{\bar{x}}$.) The ARL values indicate the average number of samples before a change of δ is detected. Thus, the ARL values for $\delta = 0$ indicate the average time between "false alarms," that is, the average time between successive CUSUM alarms when no shift in $\bar{x}$ has occurred. Ideally, we would like the ARL value to be very large for $\delta = 0$, and small for $\delta \neq 0$. Table 17.4 shows that as the magnitude of the shift δ increases, ARL decreases, and thus the CUSUM control chart detects the change faster. Increasing the value of H from 4σ to 5σ increases all of the ARL values and thus provides a more conservative approach.

CUSUM control charts also are constructed for measures of variability such as the range or standard deviation (Ryan, 2011; Montgomery, 2012).

EWMA Control Chart

Information about past measurements can also be included in the control chart calculations by exponentially

Table 17.4 Average Run Lengths for CUSUM Control Charts

Shift from Target (in multiples of $\hat{\sigma}_{\bar{x}}$)	ARL for $H = 4\hat{\sigma}_{\bar{x}}$	ARL for $H = 5\hat{\sigma}_{\bar{x}}$
0	168.0	465.0
0.25	74.2	139.0
0.50	26.6	38.0
0.75	13.3	17.0
1.00	8.38	10.4
2.00	3.34	4.01
3.00	2.19	2.57

Source: Adapted from Ryan (2000).

weighting the data. This strategy provides the basis for the *exponentially weighted moving-average* (EWMA) *control chart*. Let $\bar{x}$ denote the sample mean of the measured variable and z denote the EWMA of $\bar{x}$. A recursive equation is used to calculate $z(k)$,

$$z(k) = \lambda\bar{x}(k) + (1 - \lambda)z(k - 1) \tag{17-23}$$

where λ is a constant, $0 \le \lambda \le 1$. Note that Eq.17-23 has the same form as the first-order (or exponential) filter that was introduced in Chapter 13. The EWMA control chart consists of a plot of $z(k)$ vs. k, as well as a target and upper and lower control limits. Note that the EWMA control chart reduces to the Shewhart chart for $\lambda = 1$. The EWMA calculations are initialized by setting $z(0) = T$.

If the $\bar{x}$ measurements satisfy the IID condition, the EWMA control limits can be derived. The theoretical 3σ limits are given by

$$T \pm 3\hat{\sigma}_{\bar{x}}\sqrt{\frac{\lambda}{2-\lambda}} \tag{17-24}$$

where $\hat{\sigma}_{\bar{x}}$ is determined from a set of test data taken when the process is *in a state of control* (Montgomery, 2012). The target T is selected to be either the desired value of $\bar{x}$ or the grand mean for the test data, $\bar{\bar{x}}$. Time-varying control limits can also be derived that provide narrower limits for the first few samples, for applications where early detection is important. Tables of ARL values have been developed for the EWMA method, similar to Table 17.4 for the CUSUM method (Ryan, 2011).

The EWMA performance can be adjusted by specifying λ. For example, $\lambda = 0.25$ is a reasonable choice, because it results in an ARL of 493 for no mean shift ($\delta = 0$) and an ARL of 11 for a mean shift of $\hat{\sigma}_{\bar{x}}$ ($\delta = 1$). EWMA control charts can also be constructed for measures of variability such as the range and standard deviation.

EXAMPLE 17.3

In order to compare Shewhart, CUSUM, and EWMA control charts, consider simulated data for the tensile strength of a phenolic resin. It is assumed that the tensile strength x is normally distributed with a mean of $\mu = 70$ MPa and a standard deviation of $\sigma = 3$ MPa. A single measurement is available at each sampling instant. A constant ($\delta = 0.5\sigma = 1.5$) was added to $x(k)$ for $k \ge 10$ in order to evaluate each chart's ability to detect a small process shift. The CUSUM chart was designed using $K = 0.5\sigma$ and $H = 5\sigma$, while the EWMA parameter was specified as $\lambda = 0.25$.

The relative performance of the Shewhart, CUSUM, and EWMA control charts is compared in Fig. 17.6. The Shewhart chart fails to detect the 0.5σ shift in x. However, both the CUSUM and EWMA charts quickly detect this change, because limit violations occur about 10 samples after the shift occurs (at $k = 20$ and $k = 21$, respectively). The mean shift can also be detected by applying the Western Electric Rules in the previous section.

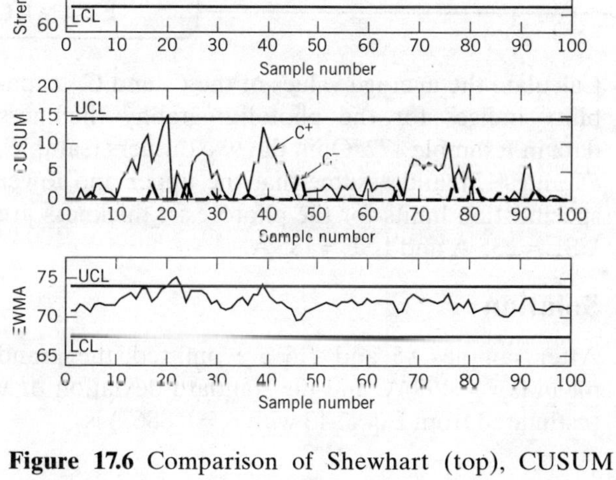

Figure 17.6 Comparison of Shewhart (top), CUSUM (middle), and EWMA (bottom) control charts for Example 17.3.

17.3 EXTENSIONS OF STATISTICAL PROCESS CONTROL

Now that the basic quality control charts have been presented, we consider several other important topics in statistical process control.

17.3.1 Process Capability Indices

Process capability indices (or *process capability ratios*) provide a measure of whether an "in control" process is meeting its product specifications. Suppose that a quality variable x must have a volume between an *upper specification limit* (*USL*) and a *lower specification limit* (*LSL*) in order for product to satisfy customer requirements. The C_p *capability index* is defined as

$$C_p \triangleq \frac{USL - LSL}{6\sigma} \tag{17-25}$$

where σ is the standard deviation of x. Suppose that $C_p = 1$ and x is normally distributed. Based on Eq. 17-6, we would expect that 99.73% of the measurements satisfy the specification limits. If $C_p > 1$, the product specifications are satisfied; for $C_p < 1$, they are not.

A second capability index C_{pk} is based on average process performance ($\bar{x}$), as well as process variability (σ). It is defined as

$$C_{pk} \triangleq \frac{\min\,[\bar{x} - LSL, USL - \bar{x}]}{3\sigma} \tag{17-26}$$

Although both C_p and C_{pk} are used, we consider C_{pk} to be superior to C_p for the following reason. If $\bar{x} = T$, the process is said to be "centered" and $C_{pk} = C_p$. But for $\bar{x} \neq T$, Cp does not change, even though the process performance is worse, while C_{pk} decreases. For this reason, C_{pk} is preferred.

If the standard deviation σ is not known, it is replaced by an estimate $\hat{\sigma}$ in Eqs. 17-25 and 17-26. For situations where there is only a single specification limit, either USL or LSL, the definitions of C_p and C_{pk} can be modified accordingly (Ryan, 2011).

In practical applications, a common objective is to have a capability index of 2.0, while a value greater than 1.5 is considered to be acceptable (Shunta, 1995). If the C_{pk} value is too low, it can be improved by making a change that either reduces process variability or causes $\bar{x}$ to move closer to the target. These improvements can be achieved in a number of ways, including better process control, better process maintenance, reduced variability in raw materials, improved operator training, and changes in process operating conditions.

Three important points should be noted concerning the C_p and C_{pk} capability indices:

1. The data used in the calculations do *not* have to be normally distributed.
2. The specification limits, USL and LSL, and the control limits, UCL and LCL, are *not* related. The specification limits denote the desired process performance, while the control limits represent actual performance during normal operation when the process is *in control.*

EXAMPLE 17.4

Calculate the average values of the C_p and C_{pk} capability indices for the photolithography thickness data in Example 17.2. Omit the two outliers (samples #5 and #15), and assume that the upper and lower specification limits for the photoresist thickness are USL = 235 Å and LSL = 185 Å.

Solution

After samples #5 and #15 are omitted, the grand mean is $\bar{\bar{x}} = 199$ Å, and the standard deviation of $\bar{x}$ (estimated from Eq. 17-13 with $c_4 = 0.8862$) is

$$\hat{\sigma}_{\bar{x}} = \frac{s}{c_4\sqrt{n}} = \frac{8.83}{0.8862\sqrt{3}} = 5.75\ \text{Å}$$

From Eqs. 17-25 and 17-26,

$$C_p = \frac{235 - 185}{6(5.75)} = 1.45$$

$$C_{pk} = \frac{\min[199.5 - 185, 235 - 199.5]}{3(5.75)} = 0.84$$

Note that C_{pk} is much smaller than the C_p, because $\bar{\bar{x}}$ is closer to the LSL than the USL.

3. The numerical values of the C_p and C_{pk} capability indices in Eqs 17-25 and 17-26 are only meaningful when the process is *in a state of control*. However, other *process performance indices* are available to characterize process performance when the process is not *in a state of control*. They can be used to evaluate the incentives for improved process control (Shunta, 1995).

17.3.2 Six Sigma Approach

Product quality specifications continue to become more stringent as a result of market demands and intense worldwide competition. Meeting quality requirements is especially difficult for products that consist of a very large number of components and for manufacturing processes that consist of hundreds of individual steps. For example, the production of a microelectronics device typically requires 100–300 batch processing steps. Suppose that there are 200 steps, and that each one must meet a quality specification in order for the final product to function properly. If each step, is independent of the others and has a 99% success rate, the overall yield of satisfactory product is $(0.99)^{200} = 0.134$, or only 13.4%. This low yield is clearly unsatisfactory. Similarly, even when a processing step meets 3σ specifications (99.73% success rate), it will still result in an average of 2,700 "defects" for every million produced. Furthermore, the overall yield for this 200-step process is still only 58.2%.

These examples demonstrate that for complicated products or processes, 3σ quality is no longer adequate, and there is no place for failure. These considerations and economic pressures have motivated the development of the *six sigma approach* (Pande et al., 2014). The statistical motivation for this approach is based on the properties of the normal distribution. Suppose that a product quality variable x is normally distributed, $N(\mu, \sigma^2)$. As indicated on the left portion of Fig. 17.7, if the product specifications are $\mu \pm 6\sigma$, the product will meet the specifications 99.999998% of the time. Thus, on average, there will only be two defective products for every billion produced. Now suppose that the process operation changes so that the mean value is shifted from $\bar{x} = \mu$ to either $\bar{x} = \mu + 1.5\sigma$ or $\bar{x} = \mu - 1.5\sigma$, as shown on the right side of Fig. 17.7. Then the product specifications will still be satisfied 99.99966% of the time, which corresponds to 3.4 defective products per million produced.

In summary, if the variability of a manufacturing operation is so small that the product specification limits are equal to $\mu \pm 6\sigma$, then the limits can be satisfied even if the mean value of x shifts by as much as 1.5σ. This very desirable situation of near perfect product quality is referred to as *six sigma quality*.

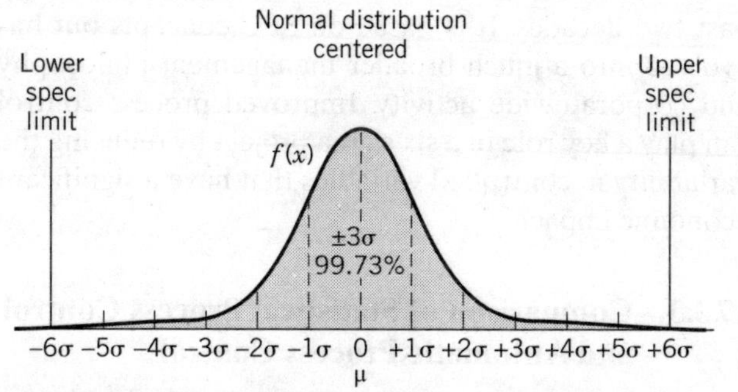

Spec limit	Percent	Defective ppm
±1σ	68.27	317,300
±2σ	95.45	45,500
±3σ	99.73	2,700
±4σ	99.9937	63
±5σ	99.999943	0.57
±6σ	99.9999998	0.002

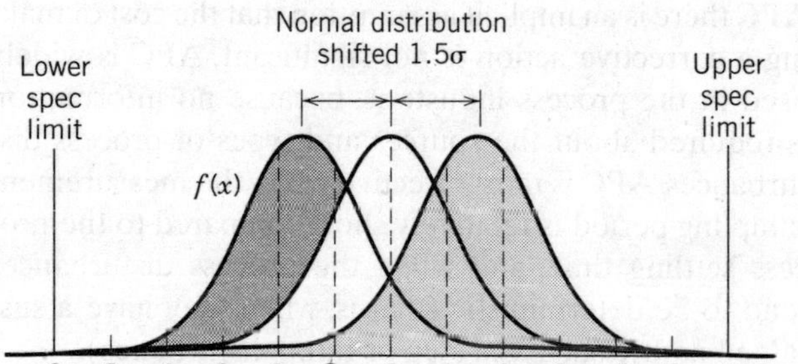

Spec limit	Percent	Defective ppm
±1σ	30.23	697,700
±2σ	69.13	308,700
±3σ	93.32	66,810
±4σ	99.3790	6,210
±5σ	99.97670	233
±6σ	99.999660	3.4

Figure 17.7 The Six Sigma Concept (Montgomery and Runger, 2013). Top: No shift in the mean. Bottom: 1.5σ shift.

The six sigma approach was pioneered by the Motorola and General Electric companies in the early 1980s as a strategy for achieving both six sigma quality and continuous improvement. Since then, other large corporations have adopted companywide programs that apply the six sigma approach to all of their business operations, both manufacturing and nonmanufacturing. Thus, although the six sigma approach is "data-driven" and based on statistical techniques, it has evolved into a broader management philosophy that has been implemented successfully by many large corporations. Six sigma programs have also had a significant financial impact. Large corporations have reported savings of billions of dollars that were attributed to successful six sigma programs.

In summary, the six sigma approach based on statistical monitoring techniques has had a major impact on both manufacturing and business practice during the past two decades. It is based on SPC concepts but has evolved into a much broader management philosophy and corporatewide activity. Improved process control can play a key role in a six sigma project by reducing the variability in controlled variables that have a significant economic impact.

17.3.3 Comparison of Statistical Process Control and Automatic Process Control

Statistical process control and *automatic process control* (APC) are complementary techniques that were developed for different types of problems. As indicated in earlier chapters, APC takes corrective action when a controlled variable deviates from the set point. The corrective action tends to change at each sampling instant. Thus, for APC there is an implicit assumption that the cost of making a corrective action is not significant. APC is widely used in the process industries, because no information is required about the sources and types of process disturbances. APC is most effective when the measurement sampling period is relatively short compared to the process settling time, and when the process disturbances tend to be deterministic (that is, when they have a sustained nature such as a step or ramp disturbance).

In statistical process control, the objective is to decide whether the process is behaving normally, and to identify a special cause when it is not. In contrast to APC, no corrective action is taken when the measurements are within the control chart limits. This philosophy is appropriate when there is a significant cost associated with taking a corrective action, such as when shutting down a process unit or taking an instrument out of service for maintenance. From an engineering perspective, SPC is viewed as a *monitoring*, rather than a *control*, strategy. It is very effective when the normal process operation can be characterized by random fluctuations around a mean value. SPC is an appropriate choice for monitoring problems where the sampling period is long compared to the process settling time and the process disturbances tend to be random rather than deterministic. SPC has been widely used for quality control in both discrete-parts manufacturing and the process industries.

In summary, SPC and APC should be regarded as complementary rather than competitive techniques. They were developed for different types of situations and have been successfully used in the process industries. Furthermore, a combination of the two methods can be very effective. For example, in model-based control such as model predictive control (Chapter 15), APC can be used for feedback control, while SPC is used to monitor the model residuals, the differences between the model predictions and the actual values.

17.4 MULTIVARIATE STATISTICAL TECHNIQUES

In Chapter 14, we have emphasized that many important control problems are multivariable in nature because more than one process variable must be controlled and more than one variable can be manipulated. Similarly, for common SPC monitoring problems, two or more quality variables are important, and they can be highly correlated. For example, ten or more quality variables are typically measured for synthetic fibers (MacGregor, 1996). For these situations, multivariable SPC techniques can offer significant advantages over the single-variable methods discussed in Section 17.2. In the statistics literature, these techniques are referred to as *multivariate methods*, while the standard Shewhart and CUSUM control charts are examples of *univariate methods*. The advantage of a multivariate monitoring approach is illustrated in Example 17.5.

Figure 17.9 provides a general comparison of univariate and multivariate SPC techniques (Alt et al., 1998). When two variables, x_1 and x_2, are monitored individually, the two sets of control limits define a rectangular region, as shown in Fig. 17.9. In analogy with Example 17.5, the multivariate control limits define the dark, ellipsoidal region that represents *in-control* behavior. Figure 17.9 demonstrates that the application of univariate SPC techniques to correlated multivariate data can result in two types of misclassification: false alarms and out-of-control conditions that are not detected. The latter type of misclassification occurred at sample #8 for the two Shewhart charts in Fig. 17.8(a) and 17.8(b).

In the next section, we consider some well-known multivariate monitoring techniques.

17.4.1 Hotelling's T^2 Statistic

Suppose that it is desired to use SPC techniques to monitor p variables, which are correlated and normally distributed. Let $\boldsymbol{x}$ denote the column vector of these p variables, $\boldsymbol{x} = col[x_1, x_2, \ldots, x_p]$. At each sampling instant, a subgroup of n measurements is made for each variable. The subgroup sample means for the kth sampling instant can be expressed as a column vector: $\bar{\boldsymbol{x}}(k) = col\,[\bar{x}_1(k), \bar{x}_2(k), \ldots, \bar{x}_p(k)]$. Multivariate control charts are traditionally based on *Hotelling's T^2 statistic* (Montgomery, 2012).

$$T^2(k) \triangleq n[\bar{\boldsymbol{x}}(k) - \bar{\bar{\boldsymbol{x}}}]^T \boldsymbol{S}^{-1} [\bar{\boldsymbol{x}}(k) - \bar{\bar{\boldsymbol{x}}}] \qquad (17\text{-}27)$$

where $T^2(k)$ denotes the value of the T^2 statistic at the kth sampling instant. The vector of grand means $\bar{\bar{\boldsymbol{x}}}$ and the

EXAMPLE 17.5

The effluent stream from a wastewater treatment process is monitored to make sure that two process variables, the biological oxidation demand (BOD) and the solids content, meet specifications. Representative data are shown in Table 17.5. Shewhart charts for the sample means are shown in parts (*a*) and (*b*) of Fig. 17.8. These univariate control charts indicate that the process appears to be in-control because no chart violations occur for either variable. However, the bivariate control chart in Fig. 17.8*c* indicates that the two variables are highly correlated, because the solids content tends to be large when the BOD is large, and vice versa. When the two variables are considered together, their joint confidence limit (e.g., at the 99% confidence level) is an ellipse, as shown in Fig. 17.8.[1] Sample #8 lies well beyond the 99% limit, indicating an out-of-control condition. By contrast, this sample lies within the Shewhart control chart limits for both individual variables.

This example has demonstrated that univariate SPC techniques such as Shewhart charts can fail to detect abnormal process behavior when the process variables are highly correlated. By contrast, the abnormal situation was readily apparent from the multivariate analysis.

Table 17.5 Wastewater Treatment Data

Sample Number	BOD (mg/L)	Solids (mg/L)
1	17.7	1380
2	23.6	1458
3	13.2	1322
4	25.2	1448
5	13.1	1334
6	27.8	1485
7	29.8	1503
8	9.0	1540
9	14.3	1341
10	26.0	1448
11	23.2	1426
12	22.8	1417
13	20.4	1384
14	17.5	1380
15	18.4	1396
16	16.8	1345
17	13.8	1349
18	19.4	1398
19	24.7	1426
20	16.8	1361
21	14.9	1347
22	27.6	1476
23	26.1	1454
24	20.0	1393
25	22.9	1427
26	22.4	1431
27	19.6	1405
28	31.5	1521
29	19.9	1409
30	20.3	1392

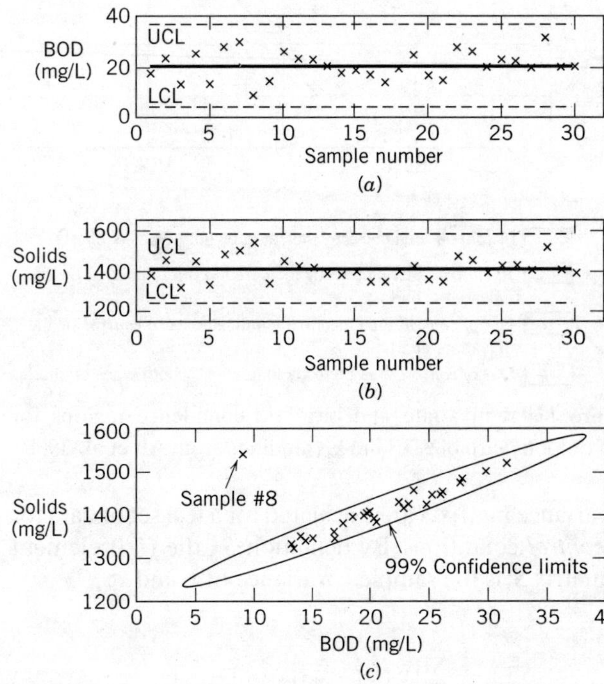

Figure 17.8 Confidence regions for Example 17.5. Univariate in (a) and (b), bivariate in (c).

[1] If two random variables are correlated and normally distributed, the confidence limit is in the form of an ellipse and can be calculated from the well-known *F* distribution (Montgomery and Runger, 2013).

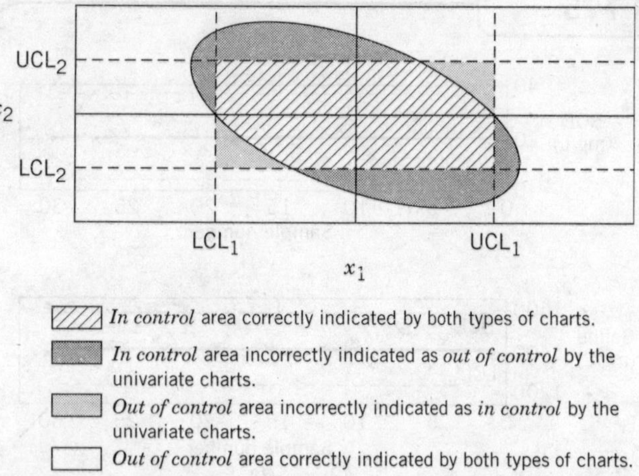

Figure 17.9 Univariate and bivariate confidence regions for two random variables, x_1 and x_2 (modified from Alt et al., 1998).

covariance matrix $\mathbf{S}$ are calculated for a test set of data for *in-control* conditions. By definition S_{ij}, the (i, j)-element of matrix $\mathbf{S}$, is the sample covariance of x_i and x_j:

$$S_{ij} \triangleq \frac{1}{N}\sum_{k=1}^{N}[\bar{x}_i(k) - \bar{\bar{x}}_i][\bar{x}_j(k) - \bar{\bar{x}}_j] \tag{17-28}$$

In Eq. 17-28 N is the number of subgroups, and $\bar{\bar{x}}_i$ denotes the mean for $\bar{x}_i$.

Note that T^2 is a scalar, even though the other quantities in Eq. 17-27 are vectors and matrices. The inverse of the sample covariance matrix, S^{-1}, scales the p variables and accounts for correlation among them.

A multivariate process is considered to be out-of-control at the kth sampling instant if $T^2(k)$ exceeds an upper control limit (UCL). (There is no target or lower control limit.) The UCL values are tabulated in statistics books and depend on the number of variables p and the subgroup size n. The T^2 control chart consists of a plot of $T^2(k)$ vs. k and an UCL. Thus, the T^2 control chart is the multivariate generalization of the $\bar{x}$ chart introduced in Section 17.2.2. Multivariate generalizations of the CUSUM and EWMA charts are also available (Montgomery, 2012).

17.4.2 Principal Component Analysis and Partial Least Squares

Multivariate monitoring based on Hotelling's T^2 statistic can be effective if the data are not highly correlated and the number of variables p is not large (for example, $p < 10$). For highly correlated data, the $\mathbf{S}$ matrix is poorly conditioned and the T^2 approach becomes problematic. Fortunately, alternative multivariate monitoring techniques have been developed that are very effective for monitoring problems with large numbers of variables and highly correlated data. The *Principal Component Analysis* (*PCA*) and *Partial Least Squares* (*PLS*) methods have received the most attention in the process control community. Both techniques can be used to monitor process variables (e.g., temperature, level, pressure, and flow measurements) as well as product quality variables. These methods can provide useful diagnostic information after a chart violation has been detected.

EXAMPLE 17.6

Construct a T^2 control chart for the wastewater treatment problem of Example 17.5. The 99% control chart limit is $T^2 = 11.63$. Is the number of T^2 control chart violations consistent with the results of Example 17.5?

Solution

The T^2 control chart is shown in Fig. 17.10. All of the T^2 values lie below the 99% confidence limit except for sample #8. This result is consistent with the bivariate control chart in Fig. 17.8c.

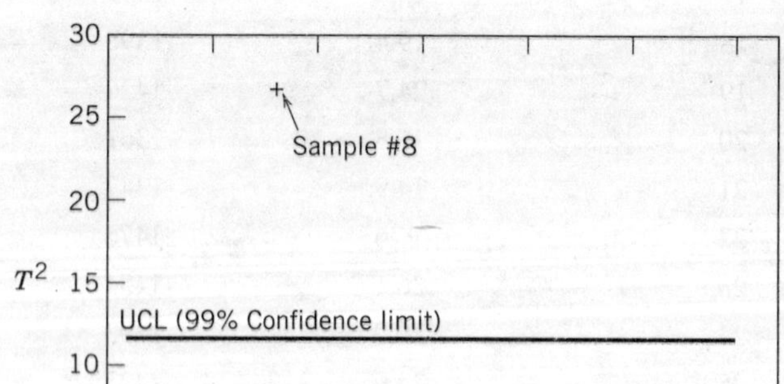

Figure 17.10 T^2 control chart for Example 17.5.

17.4.3 Process Monitoring Using Principal Component Analysis

The technique of PCA is widely used for dimensionality reduction, feature extraction, outlier detection, image analysis and process monitoring. Application of PCA for the purpose of process monitoring is briefly explained below.

The initial step involves application of orthogonal transformation to convert a set of correlated variables into a set of uncorrelated variables. These uncorrelated variables are called as principal components and are linear combinations of the original variables. Let $X \in \mathbb{R}^{n \times m}$ be the $n \times m$ process data matrix with n samples and m process variables. Compute the $m \times m$ covariance matrix S as given below:

$$S = \frac{1}{n-1} X^T X \tag{17-29}$$

It may be noted that in most cases the values in process data matrix are normalised so that the PCA is not influenced by the magnitude of the variables. The normalization is performed by subtracting mean value and dividing by standard deviation for a particular variable.

The eigen vectors and corresponding eigen values are computed for the covariance matrix S, The eigen vector with the highest eigen value is the first principal component and like this principal components are chosen in the order of decreasing eigen values.

Although for an $m \times m$ matrix, there will be m eigen vectors (or m principal components), in case of correlated dataset, p number of principal components ($p < m$) account for most of the variance of the original data and hence $m–p$ number of principal components can be ignored with little loss of information.

The p principal components chosen to constitute the P matrix of dimension $m \times p$. The score matrix is obtained as below:

$$T = XP \tag{17-30}$$

The projections of T back into the original space gives the predicted values.

$$\hat{X} = TP^T \tag{17-31}$$

The difference between X and $\hat{X}$ is the residual matrix, E.

$$E = X - \hat{X} \tag{17-32}$$

Popular methods for choosing the number of principal components are cumulative percentage variance, SCREE plot analysis and PRESS (Prediction sum of squares) statistic.

The subspaces covered by T and E are known as principal component space and residual space, respectively. For process monitoring, Hotelling T^2 statistic in the principal component subspace and squared prediction error in the residual subspace are computed for each sample. The values are compared with a threshold value (cut-off value). The threshold values are computed as below (Qin, 2003).

For Hotelling T^2 statistic,

$$J_{th,T^2} = \frac{p(n^2-1)}{n(n-p)} F_\alpha(p, n-p) = \chi^2_{p;\alpha} \tag{17-33}$$

$F_\alpha(.)$ is the critical value in the F-distribution having p and $(n-p)$ as the degrees of freedom with $100\alpha\%$ as the confidence value.

For squared prediction error,

$$J_{th,SPE} = \theta_1 \left(\frac{c_\alpha \sqrt{2\theta_2 h_0^2}}{\theta_1} + 1 + \frac{\theta_2 h_0 (h_0 - 1)}{\theta_1^2} \right)^{1/h_0} = \delta_\alpha^2 \tag{17-34}$$

$$\theta_i = \sum_{j=p+1}^{n} \sigma_j^{2i} \quad \text{and} \quad h_0 = 1 - \frac{2\theta_1\theta_3}{3\theta_2^2} \tag{17-35}$$

c_α is the normal deviate which is associated with $(1-\alpha)$ percentile and J_{th} is the threshold value.

Higher than threshold value of fault detection index for a particular sample indicates a faulty sample.

17.4.4 Recent Advances

Though PCA has been the most widely used technique for process monitoring so far, there are certain disadvantages associated with the traditional PCA technique. PCA is applicable to linear data with Gaussian distribution. Therefore, in order to account for process non-linearity, a modified version of PCA known as kernel PCA has become quite popular. Similarly, for non-Gaussian data, another unsupervised process monitoring technique i.e., independent component analysis (ICA) is commonly used.

Besides this some other machine learning techniques that find application in process monitoring are Gaussian mixture modeling, support vector data description and k-means clustering. In the last few years, there is a growing trend of application of deep learning techniques such as convolution neural network, auto encoder and deep belief network for industrial process monitoring.

17.5 CONTROL PERFORMANCE MONITORING

In order to achieve the desired process operation, the control system must function properly. As indicated in Chapter 11, industrial surveys have reported that many control loops perform poorly and even increase variability in comparison with manual control. Contributing factors include poor controller tuning and control valves that are incorrectly sized or tend to stick due to excessive frictional forces. In large processing plants, each plant operator is typically responsible for 200 to 1000 loops. Thus, there are strong incentives for automated *control performance monitoring* (*CPM*). The overall objectives of CPM are (1) to determine whether the control system is performing in a satisfactory manner and (2) to diagnose the cause of any unsatisfactory performance.

17.5.1 Basic Information for Control Performance Monitoring

In order to monitor the performance of a single standard PI or PID control loop, the basic information in Table 17.6 should be available.

Service factors should be calculated for key components of the control loop such as the sensor and final control element. Low service factors and/or frequent maintenance suggest chronic problems that require attention. The fraction of time that the controller is in the automatic mode is a key metric. A low value indicates that the loop is frequently in the manual mode and thus requires attention. Service factors for computer hardware and software should also be recorded.

Simple statistical measures such as the sample mean and standard deviation can indicate whether the controlled variable is achieving its target and how much control effort is required. An unusually small standard deviation for a measurement could result from a faulty sensor with a constant output signal, as noted in Section 17.1. By contrast, an unusually large standard deviation could be caused by equipment degradation or even failure, for example, inadequate mixing caused by a faulty vessel agitator. Raw measurement data is usually of low quality and should be "cleaned" for subsequent analysis such as process monitoring (Xu et al., 2015).

Table 17.6 Basic Data for Control Loop Monitoring

- Service factors (time in use/total time period)
- Mean and standard deviation for the control error (set point − measurement)
- Mean and standard deviation for the controller output
- Alarm summaries
- Operator logbooks and maintenance records

A high alarm rate can be indicative of poor control system performance. Operator logbooks and maintenance records are valuable sources of information, especially if this information has been captured in a computer database.

17.5.2 Control Performance Monitoring Techniques

Chapters 5 and 11 introduced traditional control loop performance criteria such as rise time, settling time, overshoot, offset, degree of oscillation, and integral error criteria. CPM methods have been developed based on these and other criteria, and commercial CPM software is available. Comprehensive reviews of CPM techniques and industrial applications have been reported by Jelali (2006) and Shardt et al. (2012).

If a process model is available, then process monitoring techniques based on monitoring the model residuals can be employed (Chiang et al., 2001; Davis et al., 2000; Cinar et al., 2007). But in order to be effective, model-based monitoring requires a reasonably accurate process model that is not always available. Simple CPM methods have also been developed that do not require a process model. Control loops that are excessively oscillatory or very sluggish can be identified using correlation or frequency response techniques (Hägglund, 1999; Seborg and Miao, 1998, Tangirala et al., 2005), or by evaluating standard deviations (Rhinehart, 1995; Shunta, 1995). A common problem, control valve stiction, can be detected from routine operating data (Choudhury et al., 2008; Bacci di Capaci and Scali, 2015).

Control system performance can be assessed by comparison with a *benchmark*. For example, historical data representing periods of satisfactory control can be used as a benchmark. Alternatively, the benchmark could be an ideal control system performance, such as *minimum variance control*. As the name implies, a minimum variance controller minimizes the variance of the controlled variable when unmeasured, random disturbances occur. This ideal performance limit can be estimated from closed-loop operating data; then the ratio of minimum variance to the actual variance is used as the measure of control system performance. This statistically based approach has been commercialized, and many successful industrial applications have been reported (Kozub, 1997; Desborough and Miller, 2002; Harris and Seppala, 2002; Hoo et al., 2003; Paulonis and Cox, 2003).

Additional information on statistically-based CPM is available in survey articles (Piovoso and Hoo, 2002; Kourti, 2005; Shardt et al., 2012) and books (Box et al., 2009; Huang and Shah, 1999; Cinar et al., 2007). Extensions to MIMO control problems (e.g., MPC) have also been reported (Qin and Yu, 2007; Cinar et al., 2007; Badwe et al., 2010; Shardt et al., 2012).

SUMMARY

Process monitoring is essential to ensure that plants operate safely and economically while meeting environmental standards. In recent years, control system performance monitoring has also been recognized as a key component of the overall monitoring activity. Process variables are monitored by making simple limit and performance calculations. Statistical process control (SPC) techniques based on control charts are monitoring techniques widely used for product quality control and other applications where the sampling periods are long relative to process settling times. In particular, Shewhart control charts are used to detect large shifts in mean process behavior, while CUSUM and EWMA control charts are better at detecting small, sustained changes. Multivariate monitoring techniques such as PCA and PLS can offer significant improvements over these traditional univariate methods when the measured variables are highly correlated. In order to overcome the limitations of traditional PCA and PLS techniques, performances of other machine learning algorithms are being investigated from time to time and results are reported in reputed research journals. SPC and advanced process control (APC) are complementary techniques that can be used together to good advantage. Control performance monitoring techniques have been developed and commercialized, especially methods based on on-line, statistical analysis of operating data.

REFERENCES

Abnormal Situation Management (ASM) Consortium, *http://www.asmconsortium.com*, 2015.

Alt, F. B., N. D. Smith, and K. Jain, Multivariate Quality Control, in *Handbook of Statistical Methods for Scientists and Engineers*, 2nd ed., H. M. Wadsworth (Ed.), McGraw-Hill, New York, 1998, Chapter 21.

Bacci di Capaci, R., and C. Scali, Review on Valve Stiction, Part I: From Modeling to Smart Diagnosis. *Processes*, **3**(2), 422–451 (2015), www.mdpi.com/journal/processes.

Badwe, A. S., R. S. Patwardhan, S. L. Shah, S. C. Patwardhan, and R. D. Gudi, Quantifying the Impact of Model-Plant Mismatch on Controller Performance. *J. Process Control*, **20**, 408 (2010).

Box, G. E. P., A. Luceño, and M. C. Paniagua-Quiñones, *Statistical Control by Monitoring and Feedback*, 2nd ed., John Wiley, and Sons, Hoboken, NJ, 2009.

Chiang, L. H., E. L., Russell, and R. D. Braatz, *Fault Detection and Diagnosis in Industrial Systems*, Springer, New York, 2001.

Choudhury, M. A. A. S., M. Jain, and S.L. Shah, Stiction–Definition, Modelling, Detection and Quantification, *J. Process Control*, **18**, 232 (2008).

Cinar, A. A. Palazoglu, and F. Kayihan, *Chemical Process Performance Evaluation*, CRC Press, Boca Raton, FL, 2007.

Davis, J. F., M. J. Piovoso, K. A. Hoo, and B. R. Bakshi, Process Data Analysis and Interpretation, *Adv. Chem. Eng.*, **25**, Academic Press, New York (2000).

Deming, W. E., *Out of the Crisis*, MIT Center for Advanced Engineering Study, Cambridge, MA, 1986.

Desborough, L., and R. Miller, Increasing Customer Value of Industrial Control Performance Monitoring–Honeywell's Experience, *Chemical Process Control, CPC-VI*, J. B. Rawlings, B. A. Ogunnaike, and J. Eaton (Eds.), *AIChE Symposium Series*, **98**, 169 (2002).

Hägglund, T., Automatic Detection of Sluggish Control Loops, *Control Eng. Prac.*, **7**, 1505 (1999).

Harris, T. J., and C. T. Seppala, Recent Developments in Controller Performance Monitoring and Assessment Techniques, *Chemical Process Control, CPC-VI*, J. B. Rawlings, B. A. Ogunnaike, and J. Eaton (Eds.), *AIChE Symposium Series*, **98**, 208 (2002).

Hoo, K. A., M. J. Piovoso, P. D. Schnelle, and D. A. Rowan, Process and Controller Performance Monitoring: Overview with Industrial Applications, *Int. J. Adapt. Control Signal Processing*, **17**, 635 (2003).

Huang, B., and S. L. Shah, *Performance Assessment of Control Loops: Theory and Applications*, Springer-Verlag, New York (1999).

Jackson, J. E. *A User's Guide to Principal Components*, Wiley-Interscience, New York, 1991.

Jacobs, D. C., Watch Out for Nonnormal Distributions, *Chem. Eng. Prog.*, **86**, (11), 19 (1990).

Jelali, M., An Overview of Performance Assessment Technology and Industrial Applications, *Control Eng. Prac.*, **14**, 441 (2006).

Kourti, T., Process Analysis and Abnormal Situation Detection: From Theory to Practice, *IEEE Control Systems*, **22** (5), 10 (2002).

Kourti, T., Application of Latent Variable Methods to Process Control and Statistical Process Control in Industry, *Int. I. Adap. Control Signal Process*, **19**, 213 (2005).

Kozub, D. J., Monitoring and Diagnosis of Chemical Processes with Automated Process Control in *Chemical Process Control, CPC-V*, J. C. Kantor, C. E. Garcia, and B. Carnahan (Eds.), *AIChE Symposium Series*, **93** (No. 316), 83 (1997).

MacGregor, J. F., Using On-line Process Data to Improve Quality, *ASQC Statistics Division Newsletter*, **16** (2), 6 (1996).

Montgomery, D. C., *Introduction to Statistical Quality Control*, 7th ed., John Wiley and Sons, Hoboken, NJ, 2012.

Montgomery, D. C., and G. C. Runger, *Applied Statistics and Probability for Engineers*, 6th ed., John Wiley and Sons, Hoboken, NJ, 2013.

Pande, P. S., R. P. Neuman, and R. R. Cavanagh, *The Six Sigma Way*, 2nd ed., McGraw-Hill, New York, 2014.

Paulonis, M. A., and J. W. Cox, A Practical Approach for Large-Scale Controller Performance Assessment, Diagnosis, and Improvement, *J. Process Control*, **13**, 155 (2003).

Piovoso, M. J., and K. A. Hoo, Multivariate Statistics for Process Control, *IEEE Control Syst.*, **22** (5), 8 (2002).

Qin, S. J., Statistical process monitoring: basics and beyond, *J. Chemom.* **17** (2003) 480–502. https://doi.org/10.1002/cem.800.

Qin, S. J., and J. Yu, Recent Developments in Multivariable Controller Performance Monitoring, *J. Process Control*, **17**, 221 (2007).

Rhinehart, R. R., A Watchdog for Controller Performance Monitoring, *Proc. Amer. Control Conf.*, 2239 (1995).

Romagnoli, J., and M. C. Sanchez, *Data Processing and Reconciliation in Chemical Process Operations*, Academic Press, San Diego, CA, 1999.

Ryan, T. P., *Statistical Methods for Quality Improvement*, 3rd ed., John Wiley and Sons, Hoboken, NJ, 2011.

Seborg. D. E., and T. Miao, Automatic Detection of Excessively Oscillatory Feedback Control Loops, U.S. patent # 5.719.788, awarded February 17, 1998.

Shardt, Y., Y. Zhao, F. Qi, K. Lee, X. Yu, B. Huang, and S. Shah, Determining The State of a Process Control System: Current Trends and Future Challenges, *Can. J. Chem. Eng.*, **90**, 217 (2012).

Shewhart, W. A., *Economic Control of Quality*, Van Nostrand, New York, 1931.

Shunta, J. P., *Achieving World Class Manufacturing Through Process Control*, Prentice Hall PTR, Englewood Cliffs, NJ, 1995.

Tangirala, A. K., S. L. Shah, and N. F. Thornhill, PSCMAP: A New Tool for Plant-Wide Oscillation Detection, *J. Process Control*, **15**, 931 (2005).

Xu, S., M. Baldea, T. F. Edgar, W. Wojsznis, T. Blevin, and M. Nixon, An Integrated Data Cleaning Scheme for Quality Improvement of Process Models. *Reviews Chem. Engr.*, **31**(5), 453 (2015).

EXERCISES

17.1 A standard signal range for electronic instrumentation is 4–20 mA. For purposes of monitoring instruments using limit checks, would it be preferable to have an instrument range of 0–20 mA? Justify your answer.

17.2 An analyzer measures the pH of a process stream every 15 minutes. During normal process operation, the mean and standard deviation for the pH measurement are $\overline{x} = 5.5$ and $s = 0.06$, respectively. When the process is operating normally, what is the probability that a pH measurement will exceed 5.74?

17.3 In a computer control system, the high and low warning limits for a critical temperature measurement are set at the "2-sigma limits," $\overline{T} \pm 2\hat{\sigma}_T$, where $\overline{T}$ is the nominal temperature and $\hat{\sigma}_T$ is the estimated standard deviation. If the process operation is normal and the temperature is measured every second, how many "false alarms" (that is, measurements that exceed the warning limits) would you expect to occur in a day.

17.4 In a manufacturing process, the impurity level of the product is measured on a daily basis. When the process is operating normally, the impurity level is approximately normally distributed with a mean value of 0.800%. The laboratory measurements for a period of eight consecutive days are shown below. Make a Cusum analysis to assess if there is any process abnormality. Take slack parameter as half of estimated standard deviation and threshold as four times estimated standard deviation. (Estimate of standard deviation can be obtained by dividing average moving range with 1.128)

Table E17.4

Day	Impurity	Day	Impurity
1	0.821	5	0.794
2	0.789	6	0.825
3	0.836	7	0.815
4	0.804	8	0.781

17.5 Quality of paper is assessed by a parameter called GSM (grams per square meter). Hourly GSM values from three paper machines are reported below for an 8 hour shift:

Table E17.5

Sampling hour	GSM values		
1	72.00	76.00	74.00
2	68.00	70.00	73.00
3	75.00	72.00	78.00
4	65.00	62.00	63.00
5	77.00	82.00	66.00
6	64.00	66.00	74.00
7	60.00	62.00	61.00
8	77.00	74.00	72.00

Using $\overline{x}$ method identify if there is any faulty sample.

17.6 For the problem number 17.5, determine process fault using *s* control chart.

17.7 For the BOD data of Example 17.5, develop CUSUM and EWMA charts. Do these charts indicate an "abnormal situation"? Justify your answer. For the CUSUM chart, use $K = 0.5s$ and $H = 5s$ where s is the sample standard deviation. For the EWMA chart, use $\lambda = 0.25$.

17.8 Calculate the average values of the C_p and C_{pk} capability indices for the BOD data of Example 17.5, assuming that LSL = 5 mg/L and USL = 35 mg/L. Do these values of the indices indicate that the process performance is satisfactory?

17.9 Consider the wastewater treatment problem of Examples 17.5 and 17.6 and five new pairs of measurements shown below. Calculate the value of Hotelling's T^2 statistic for each pair using the information for Example 17.6, and plot the data on a T^2 chart. Based on the number of chart violations for the new data, does it appear that the current process behavior is normal or abnormal?

Sample Number	BOD (mg/L)	Solids (mg/L)
1	18.1	1281
2	36.8	1430
3	16.0	1510
4	28.2	1343
5	31.0	1550

Note: The required covariance matrix **S** in Eq. 3 can be calculated using either the *cov* command in MATLAB or the *covar* command in EXCEL.

17.10 An SPC chart is to be designed for a key process variable, a chemical composition, which is also a controlled variable. Because the measurements are very noisy, they must be filtered before being sent to a PI controller. The question arises whether the variable plotted on the SPC chart should be the filtered value or the raw measurement. Are both alternatives viable? If so, which one do you recommend? (Briefly justify your answers.)

17.11 In order to improve the reliability of a critical control loop, it is proposed that redundant sensors be used. Suppose that three independent sensors are employed and each sensor works properly 95% of the time.

(a) What is the probability that all three sensors are functioning properly?

(b) What is the probability that none of the sensors are functioning properly?

(c) It is proposed that the average of the three measurements be used for feedback control. Briefly critique this strategy.

Hint: See Appendix F for a review of basic probability concepts.

17.12 Outline the limitations associated with traditional PCA technique which algorithms can be used in cases where standard PCA results are unsatisfactory?

MULTIPLE CHOICE QUESTIONS

1. Upper control limit for a $\overline{X}$ chart is expressed by

(a) $\overline{\overline{X}} + c\hat{\sigma}_{\bar{x}}$ **(b)** $\overline{\overline{X}} - c\hat{\sigma}_{\bar{x}}$

(c) $\overline{X} + c\hat{\sigma}_{\bar{x}}$ **(d)** $\overline{X} - c\hat{\sigma}_{\bar{x}}$

2. Which of the following is the most appropriate reason explaining better performance of CUSUM chart than Shewhart chart?

(a) Each value plotted on the CUSUM chart contains information of a single object.

(b) Each value plotted on the CUSUM chart contains information of more than one object.

(c) Each value plotted on Shewhart control chart is variable.

(d) None of these.

3. If the average value increases, how will the cumulative sum change?

(a) Cumulative sum will not change.

(b) Cumulative sum decreases.

(c) Cumulative sum increases.

(d) Cumulative sum may increase or decrease, difficult to predict.

4. Which of these is the correct expression for EWMA?

(a) $z_i = \lambda x_i + (1 - \lambda)\, z_{i-1}$

(b) $z_i = \lambda x_i - (1 - \lambda)\, z_{i-1}$

(c) $z_i = \lambda x_i + (1 - \lambda)\, z_{i+1}$

(d) $z_i = \lambda x_i + (1 + \lambda)\, z_{i-1}$

5. Which of the following is a supervised technique in process monitoring?

(a) Principal component analysis

(b) Partial least square

(c) Independent component analysis

(d) Support vector data description

6. Which of the following is an unsupervised technique in process monitoring?

(a) Principal component analysis

(b) Partial least square

(c) Artificial neural network

(d) Support vector regression

7. Which of the following techniques can be applied to non-linear data sets?

(a) Principal component analysis

(b) Independent component analysis

(c) Dynamic principal component analysis

(d) Kernel principal component analysis

8. Which is a multivariate process monitoring technique?

(a) Shewhart chart

(b) Hotelling's T^2 statistic

(b) Cusum chart

(d) EWMA chart

9. The EWMA chart reduces to Shewhart chart for

(a) $\lambda = 1$ **(b)** $\lambda = 0$

(c) $\lambda = -1$ **(d)** $\lambda = \infty$

10. PCA is conducted on a process data matrix of 100 observations and 10 variables. The number of eigen vectors of the covariance matrix will be

(a) 100 **(b)** 10

(c) 110 **(d)** 90

11. A product quality parameter in an industry is monitored by Shewhart chart. The size of sample is 5. If the average sample standard deviation $\overline{s}$ is 0.15 and the mean of process is 0.95, which of the following will be an estimate of standard deviation of grand mean?

(a) 0.084 **(b)** 0.0713

(c) 0.1578 **(d)** 14.025

12. In the above question, the value of 3 sigma upper and lower control limits will be

(a) 1.1639 and 0.7361

(b) 1.202 and 0.698

(c) 1.4234 and 0.4766

(d) 43.025 and –41.125

13. Composition of product samples from a process as measured in the laboratory follows a normal distribution. On a particular day, if 95% of the samples have a composition varying from 0.9 to 0.96, what is the mean and standard deviation of the composition values?

(a) 0.93 and 0.06

(b) 0.93 and 0.03

(c) 0.93 and 0.015

(d) 0.95 and 0.06

ANSWER KEY

1. (a) **2.** (b) **3.** (c) **4.** (a) **5.** (b) **6.** (a) **7.** (d)

8. (b) **9.** (a) **10.** (b) **11.** (b) **12.** (a) **13.** (c)

CHAPTER 18

Batch Process Control

CHAPTER CONTENTS

Batch processing is an alternative to continuous processing. In batch processing, a sequence of one or more steps, either in a single vessel or in multiple vessels, is performed in a defined order, yielding a specific quantity of an intermediate or finished product. Because the volume of product is normally small, large production runs are achieved by repeating the process steps on a predetermined schedule. In batch processing, the production amounts are usually smaller than for continuous processing; hence, it is usually not economically feasible to dedicate processing equipment to the manufacture of a single product. Instead, batch processing units are organized so that a range of products (from a few to possibly hundreds) can be manufactured with a given set of process equipment.

Batch processing can be complicated by having multiple stages, multiple products made in the same equipment, or parallel processing lines, and process steps that often operate at non steady state (i.e., are time varying). In addition, many batch processes in industry are only semi-automated. The automation system must integrate with many manual operations, such as those required to prepare equipment in moving from one batch step to the next. The key challenge for batch plants is to consistently manufacture each product to meet specifications while maximizing the utilization of available equipment. Benefits include reduced inventories and shortened response times to make a specialty product compared to continuous processing plants. Typically, it is not possible to use blending of multiple batches in order to obtain the desired product quality, so product quality specifications must be satisfied by each batch.

Batch processing is widely used to manufacture specialty chemicals, metals, electronic materials, ceramics, polymers, food, biochemicals and pharmaceuticals, multiphase materials/blends, coatings, and composites, an extremely broad range of processes and products. The unit operations in batch processing are also quite diverse, and some are analogous to operations for continuous processing.

As one example of batch processing, batch distillation is used in the production of many chemicals and pharmaceuticals. A batch column or still can be used to separate products with different purity specifications. Compared to continuous distillation, it is easier to tailor product specification on a batch-to-batch basis, giving a flexible, easily operated separation unit with low capital cost (Mujtaba, 2004; Diwekar, 1995). The general arrangement of a typical batch still and the important controlled and manipulated variables are shown in Fig.18.1. As is typical of batch processes, a sequence of steps must be carried out; each step involves the opening and closing of different valves at specified times and in a specific order. After charging the kettle at the base of the column, the feed flow is stopped and heat is applied at the reboiler. Using cooling at the condenser, the reflux flow rate can be manipulated so that the column reaches a certain overhead distillate composition x_{Dsp} (the set point) prior to product withdrawal. At this point, distillate is withdrawn into a product receiver at a flow rate D in order to meet a product specification. At selected times, the product receiver may be switched to make multiple products with different purity specification in various tanks. At the end of the

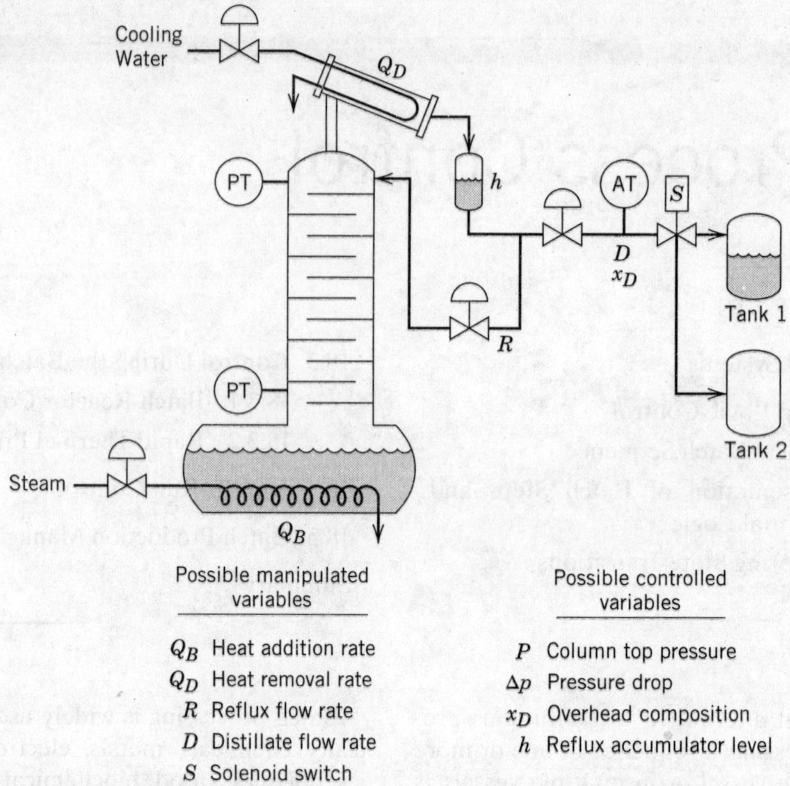

Figure 18.1 Batch distillation schematic.

last product withdrawal, the column is shut down, the remaining bottoms residue and receiver holdup(s) are pumped to storage, and the column is readied for the next batch.

Figure 18.2 shows the time profile of ethanol composition for an ethanol–water batch fractionation at a constant distillate rate. Notice that the overhead ethanol mole fraction (the main product) remains nearly constant for the first 45 minutes of the run, while the bottoms composition undergoes a gradual decline.

In an effort to increase the safety, efficiency, and affordability of medicines, the FDA has recently proposed a new framework for the regulation of pharmaceutical development, manufacturing, and quality assurance. The primary focus of the initiative is to reduce variability through a better understanding of processes and on-line decision making than can be obtained by the traditional approach. PAT (Process Analytical Technology) has become an acronym in the pharma industry for designing, analyzing, and controlling manufacturing through measurements (i.e., during processing) of critical quality and performance attributes of raw and in-process materials and processes, with the goal of ensuring final product quality (Buckbee and Alford, 2007). Process variations that could possibly contribute to patient risk are determined through modeling and timely measurements of critical quality attributes, which are then addressed by process control.

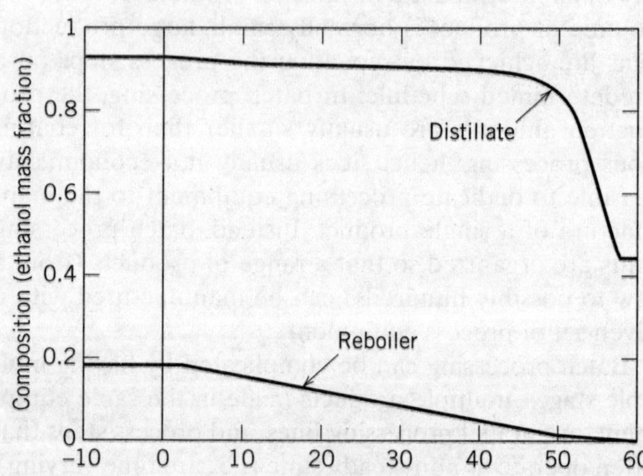

Figure 18.2 The time variation of distillate and bottom ethanol compositions during batch fractionation of ethanol and water for constant distillate flow.

This chapter provides an introduction to batch process control. First we introduce the operational practices and control system design for batch plants, which differ markedly from continuous plants. In batch processing, there is a much greater emphasis on production scheduling of batch equipment; this procedure is critical to matching available production equipment and raw materials with the demands for a range of specialty products, each having different specifications. Batch control systems, in contrast to continuous process control, involve *binary logic* and *discrete event analysis* applied to the sequencing of different process steps in the same vessel, usually requiring the application of *programmable logic controllers* (PLCs) or a Distributed Control System (DCS). Note that many batch processes can be controlled entirely with a DCS because it can duplicate essentially all of the capabilities of PLCs (see Chapter 19). One exception is high-speed discrete product processing, which requires multiple input/output (I/O) turnarounds per second. Feedback controllers are utilized in order to handle set-point changes and disturbances, but they may require certain enhancements to treat the wide operating ranges, because there is no steady-state operating point. While the lack of a steady-state operating point is often the case with batch processes, some batch steps (e.g., use of centrifuges to separate cell mass from broth in fermentation) operate at steady state.

In several sections below, we highlight the use of batch process control in semiconductor manufacturing, where individual wafers or groups of wafers are repetitively processed through a variety of unit operations such as etching and lithography. The practice of *run-to-run control*, a form of supervisory control in which operating conditions or trajectories are changed only between runs (batches) and not during a batch, is also described.

18.1 BATCH CONTROL SYSTEMS

Batch control systems operate at various levels:

1. Batch sequencing and logic control
2. Control during the batch
3. Run-to-run control
4. Batch production management

The ANSI/ISA 95 Standard provides an international standard for defining the interface between enterprise and control systems for batch processing, based on four levels. In the standard, almost all control is done in levels 1 and 2 and contained in the PLC and DCS environment. Control during the batch would be a level 2 function. In the standard, Level 3 corresponds to Manufacturing Operations Management and includes activities like batch reports and manufacturing tickets. Level 4 is business management and includes enterprise resource planning (ERP), bill of materials, purchasing, inventory management, etc. (ISA, 2012).

Figure 18.3 shows the interconnections of the different types of control used in a typical batch process. Run-to-run control is a type of supervisory control that is part of the production management block. In contrast to continuous processing, the focus of control shifts from regulation to set-point changes, and sequencing of batches and equipment takes on a more important role.

Batch control systems must be very versatile to be able to handle pulse inputs and discrete I/O as well as analog signals for sensors and actuators. Functional control activities are summarized as follows:

1. ***Batch sequencing and logic control:*** The sequence of control steps that follow a recipe involves, for example: mixing of ingredients, heating, waiting for a reaction to complete, cooling, and discharging the resulting product. Transfer of materials to and from batch tanks or reactors includes metering of materials as they are charged (as specified by the recipe), as well as transfer of materials at the completion of the process operation. In addition to discrete logic for the control steps, logic is needed for safety interlocks to protect personnel, equipment, and the environment from unsafe conditions. Process interlocks ensure that process operations can only occur in the correct logical sequence.

2. ***Control during the batch:*** Feedback control of flow rate, temperature, pressure, composition, and level, including advanced control strategies, falls in this category, which is also called "*within-the-batch*" *control* (Bonvin, 1998). In complex applications, this requires specification of an operating trajectory for the batch (for example, temperature or flow rate as a function of time). In simpler cases, it involves tracking of set points of the controlled variables, which includes ramping the controlled variables up and down and/or holding them constant for a prescribed

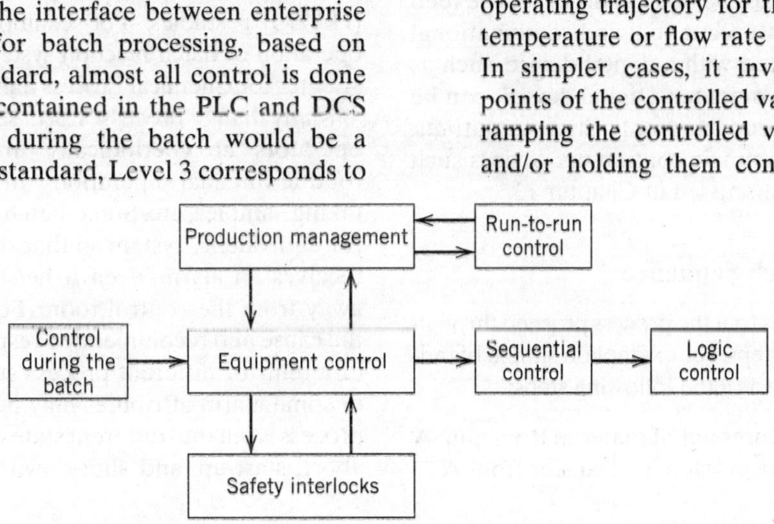

Figure 18.3 Overview of a batch control system.

period of time. Detection of when the batch operations should be terminated (*end points*) may be performed by inferential measurements of product quality, if direct measurement is not feasible (see Chapter 12).

3. ***Run-to-run control:*** Also called batch-to-batch control, this supervisory function is based on offline product quality measurements at the end of a run. Operating conditions and profiles for the batch are adjusted between runs to improve the product quality using tools such as optimization.
4. ***Batch production management:*** This activity entails advising the plant operator of process status and how to interact with the recipes and the sequential, regulatory, and discrete controls. Complete information (recipes) is maintained for manufacturing each product grade, including the names and amounts of ingredients, process variable set points, ramp rates, processing times, and sampling procedures. Other database information includes batches produced on a shift, daily, or weekly basis, as well as material and energy balances It is especially important to include the batch lot number on all historical records. Scheduling of process units is based on availability of raw materials and equipment and customer demand.

18.2 SEQUENTIAL AND LOGIC CONTROL

Sequential logic is used to ensure that the batch process undergoes the proper sequence of events, because the order of steps is important. Sequential logic must not be confused with combinational logic, which depends only on instantaneous values for the variables. This type of logic is especially suitable for interlocks or for permissive actions; for example, the reactor discharge valve must be closed, or the vent must be open in order for the feed valve to be opened. Both sequential and combinational logic can be implemented with a digital device, such as a microprocessor, or a computer. Digital devices can be intrinsically discrete (producing only discrete outputs, such as integers) or can mimic continuous devices such as a PID controller, as discussed in Chapter 13.

18.2.1 A Typical Batch Sequence

Batch processing requires that the process proceed through the proper sequence of steps. For example, a simple blending sequence might consist of the following steps:

1. Transfer specified amount of material from tank A to tank R. The process step is "Transfer from A."
2. Transfer specified amount of material from tank B to tank R. The process step is "Transfer from B."
3. Agitate for a specified period of time after the feeds are added. The process step is "Agitate."
4. Discharge the product to storage tank C. The process step is "Transfer from R."

A more detailed example of the sequence of operations for a batch mixing tank is described later (see Example 18.1).

For each process step, the various discrete-acting devices are expected to be in a specified *device state*,[1] usually a binary value (0 or 1). Then, for process step "Transfer from A," the device states might be as follows:

1. Tank A discharge valve: open
2. Tank R inlet valve: open
3. Tank A transfer pump: running
4. Tank R agitator: off
5. Tank R cooling valve: closed

Sequential logic is coupled to device states. For example, device state 0 could be a valve closed, agitator off, and so on, while device state 1 could be the valve open or the agitator on. Basically, the sequential logic determines when the process should proceed from the current set of operating conditions to the next. Sequential logic must encompass both normal and abnormal process operations, such as equipment failures.

When failure occurs, the simplest action is to stop or hold at the current operating state in response to any abnormal condition, and let the process operator determine the cause of the problem. However, some failures lead to hazardous conditions that require immediate action; waiting for the operator to decide what to do is not acceptable. The appropriate response to such situations is best determined in conjunction with process hazards and operability (HAZOP) studies. For example, guidelines for safe operation of batch reaction systems have been published (Center for Chemical Process Safety, 1999).

Many batch processes are semi-automated because operators are periodically in the field performing routine manual operations in preparing equipment, taking samples, etc. Some batch plants require a paging (or equivalent) system so that the appropriate operator receives an alarm, even if he/she is on the plant floor away from the control room. For batch processes, both the cause and recommended response to an alarm can be different for different process steps. Further, the values of some alarm attributes may need to change when the process is put into different states, such as run, pause, hold, abort, start-up, and shut-down (ISA, 2012). Normally,

[1] Not to be confused with state variables discussed in Section 4.4.

the automated batch recipe must prompt the operator using a display of interlocks for the next operation to start and then wait for manual acknowledgment from the operator. Batch processes are more prone to human error due to the operator component; see Smith (2015b) for a discussion of batch operational safety issues.

Regarding data collection and storage, it is common practice that at the end of a batch production run a batch report is created. In highly regulated industries, such a report normally includes information on any abnormal events that occurred during the batch run (involving any critical process parameters) as well as the disposition of the batch lot resulting from those abnormal situations (Alford, 2006).

18.2.2 Representation of Batch Steps and Sequential Logic

There are several ways to depict the sequential logic in batch operation, which is a prerequisite to incorporating binary logic into the computer control system. Two process-oriented representations are considered: the *information flow diagram* and the *sequential function chart*. There can be used to develop digital logic diagrams including *ladder logic diagrams* and *binary logic diagrams*.

To create an information flow chart, a complete list of steps for a batch process must be documented and displayed. From this representation, it is straightforward to prepare a sequential function chart. Figure 18.4 shows the flowchart symbols that indicate the point of decisions, processing operations, input–output structure, and the sequence they should follow, which will be illustrated later in Example 18.1

A sequential function chart (SFC) describes graphically the sequential behavior of a control program (Parshall and Lamb, 2000). More sophisticated than the information flow chart, it is derived from two earlier approaches, Petri Nets and Grafcet (David, 1995; Desrochers and Al-Jaar, 1995). SFCs consist of steps that alternate between action blocks and transitions. Each step corresponds to a state of the batch process. A transition is associated with a condition that, when true, activates the next step and deactivates the previous step. Steps are linked with action blocks that perform a specific control action. SFC and Grafcet are standard languages established by the International Electrotechnical Commission (IEC) and are supported by an association of vendors and users called PLCopen; see *www.plcopen.org*.

Figure 18.5 gives a simple illustration of the SFC notation. The steps are denoted as rectangles (a double rectangle is the initial step), and the transition symbol is a small horizontal bar on the line linking control steps. A double bar is used for branching, and it can precede a transition when two or more paths can be followed. Similarly, a double bar indicates where two or more parallel paths join together into a single path. In Fig.18.5, both steps 5 and 6 must be completed before moving to step 7. The active steps are shown with a black dot in the box.

Ladder logic and binary logic diagrams provide alternative graphical formats for representing logical functions and can be analyzed using *truth tables* (Platt and Gilbert, 1995). In binary logic, the main operations are AND, OR, NAND (not AND), and NOR (not OR). When two input variables A and B are "ANDed" together ($= A \cdot B$), the output is 1 if and only if both inputs A and B are 1. When two inputs A and B are ORed together, the output $(A + B)$ is 1 if either A or B is 1. The NOT operation changes the input A to the

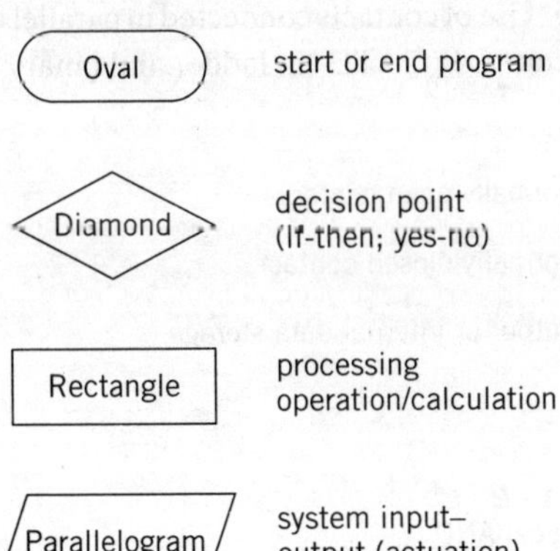

Figure 18.4 Flowchart symbols and their definitions.

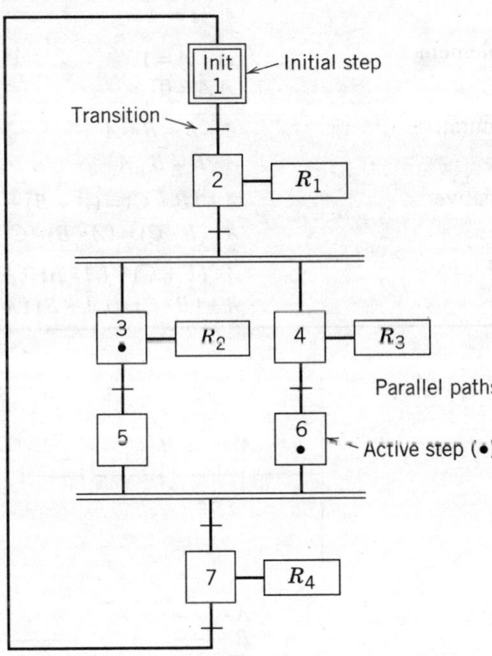

Figure 18.5 A generic sequential function chart (active steps are 3 and 6). R_1 through R_4 denote specified actions.

complementary binary value $\bar{A}$. Table 18.1 gives the *truth table* for several standard operations with four combinations of binary variables A and B.

Mathematical operations on binary logic are also known as Boolean algebra. This algebra is widely used for construction and analysis of logic gates. Some properties of Boolean algebra are given in Table 18.2.

A ladder diagram contains two vertical uprights, which are the power source (on the left) and neutral (on the right). A number of horizontal rungs indicate various paths between the two uprights, which can contain logical switches (normally open or closed) and an output. Very few symbols are required to construct a ladder diagram (Petruzella, 2010; Johnson, 2005). Two or more switches (also called contacts) on the same rung form an AND gate. Contacts on two or more parallel branches of a rung form an OR gate, as discussed below. Two vertical bars are used to depict a normally open contact, while a slash across the bars indicates a normally closed contact.

Table 18.1 Truth Table for AND, OR, NOT Binary Logic Operations

A	B	$A \times B$ (AND)	$A + B$ (OR)	$\bar{A}$ (NOT)	$\bar{B}$ (NOT)
0	0	0	0	1	1
0	1	0	1	1	0
1	0	0	1	0	1
1	1	1	1	0	0

Table 18.2 Properties of Boolean Algebra

Rule	Boolean Expression
Annulment	$A + 1 = 1$
	$A \cdot 0 = 0$
Identity	$A \cdot 1 = A$
	$A + 0 = A$
Idempotent	$A + A = A$
	$A \cdot A = A$
Complement	$A + \bar{A} = 1$
	$A\bar{A} = 0$
Commutative	$A + B = B + A$
	$A \cdot B = B \cdot A$
Associative	$A + (B + C) = (A + B) + C$
	$A \cdot (B \cdot C) = (A \cdot B) \cdot C$
Distributive	$A \cdot (B + C) = (A \cdot B) + (A \cdot C)$
	$A + (B \cdot C) = (A + B)\,(A + C)$

By clever construction of parallel and series relay circuits, designers can implement sophisticated logical statements for a sequence of logical steps (Platt and Gilbert, 2005). As an example, a set of three relays wired in series can be used to implement the three-input AND condition shown in Fig. 18.6*a*. The output is actuated only when all three input relays (A, B, and C) are actuated.

Similarly, a set of relays wired in parallel can be used to implement the OR condition shown in Fig. 18.6*b*. Here actuation of any one or more of these input relays will cause the output to be actuated. To illustrate NOR or NAND gates in ladder logic, slashed contacts are drawn. In binary logic diagrams, a small circle is appended to the OR or AND symbol. These contacts are then normally closed rather than normally open as is the case shown in Fig. 18.6. The application of the Boolean logic rules of Table 18.2 and the simplification of binary logic diagram is illustrated in Example 18.1.

Sequencing operations typically require the use of *latching relays* (which hold a state indefinitely once actuated, much like a solid-state flip-flop) and *delay*

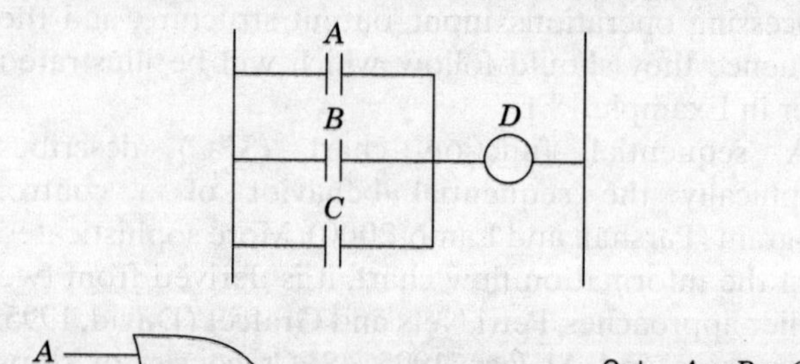

Figure 18.6*b* Use of contacts connected in parallel to implement OR logic as an "OR GATE" in ladder and binary logic.

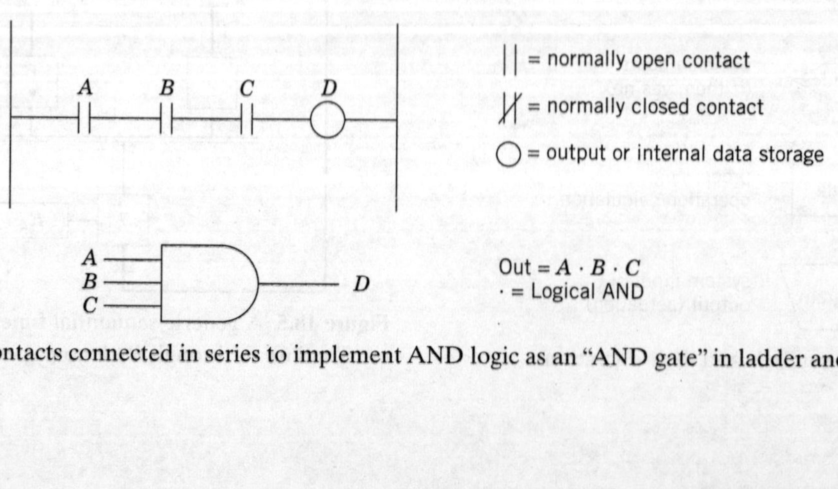

Figure 18.6*a* Use of contacts connected in series to implement AND logic as an "AND gate" in ladder and binary logic.

relays (which delay a preprogrammed time interval before operating, after actuation). Latching relays are shown as parallel connected rungs, as shown in Example 18.2. From a process point of view, the sequential function chart is preferred to relay ladder logic and binary logic because it clearly shows the sequence of steps and also indicates concurrency, that is, when some subsystems are partially independent. The input–output structure and behavior are more clearly delineated for these subsystems by SFC. SFCs are also useful for interfacing with expert system software for supervisory control, monitoring, and diagnosis (Arzen, 1994). It is possible to convert binary ladder logic diagrams into SFCs using computer algorithms (Falcione and Krogh, 1993).

EXAMPLE 18.1

Simplify the following digital circuit shown in Figure 18.7, evaluate Y and draw the simplified binary logic diagram and ladder logic diagram.

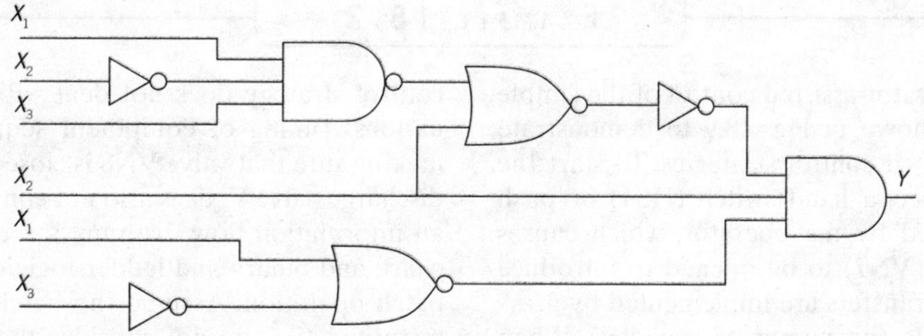

Figure 18.7 The Boolean circuit

Solution

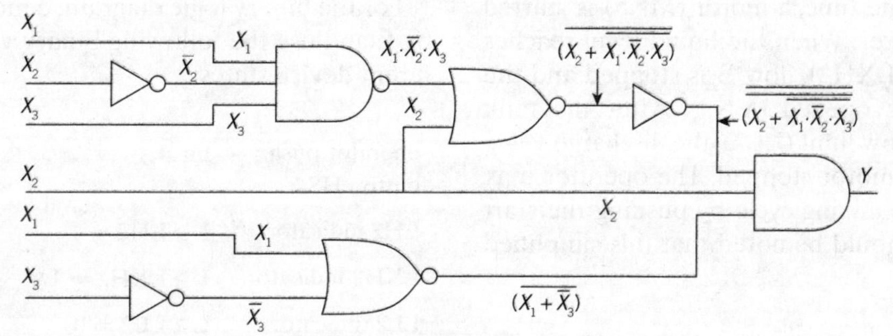

The outputs from a particular logic gate is shown above

$$Y = \overline{\overline{(X_2 + \overline{X_1 \cdot \bar{X}_2 \cdot X_3})}} \cdot X_2 \cdot \overline{(X_1 + \bar{X}_3)}$$

Apply De-Morgan rule and double negative rule ($\bar{\bar{A}} = A$):

$$Y = (\bar{X}_1 + X_2 + \bar{X}_3 + X_2) \cdot X_2 \cdot (\bar{X}_1 \cdot X_3)$$

$$= \bar{X}_1 \cdot X_2 \cdot \bar{X}_1 \cdot X_3 + X_2 \cdot X_2 \cdot \bar{X}_1 \cdot X_3 + \bar{X}_3 \cdot X_2 \cdot \bar{X}_1 \cdot X_3 + X_2 \cdot X_2 \cdot \bar{X}_1 \cdot X_3$$

$$= X_2 \cdot \bar{X}_1 \cdot X_3 + X_2 \cdot \bar{X}_1 \cdot X_3 + 0 + X_2 \cdot \bar{X}_1 \cdot X_3$$

(The third term is zero as per annulment and complement rule).

$$= X_2 \cdot \bar{X}_1 \cdot X_3$$

The simplified binary logic and ladder logic diagrams are presented below:

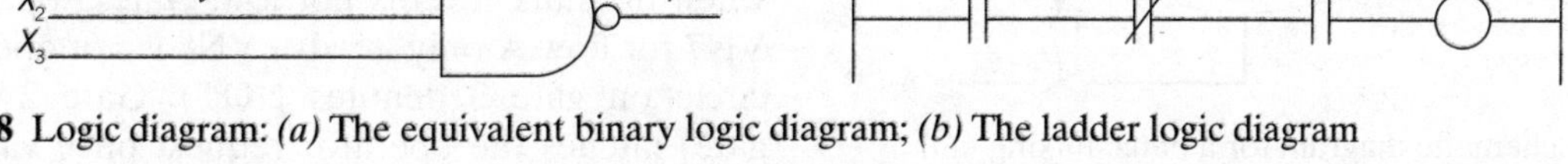

Figure 18.8 Logic diagram: *(a)* The equivalent binary logic diagram; *(b)* The ladder logic diagram

Programmable Logic Controllers

Programmable Logic Controllers (*PLCs*) are widely used in batch process control to execute the desired binary logic operations and to implement the desired sequencing. The inputs to the PLC are a set of relay contacts representing various device states (e.g., limit switches indicate whether a valve is fully open or fully closed). Various operator inputs (for example, start/stop buttons) are also provided. The PLC output signals energize (actuate) a set of relays that turn pumps on or off, actuate lights on a display panel, operate solenoid or motor-driven valves, and so on. PLCs are discussed in more detail in Chapter 19, Section 19.3.1; see also Hughes (2005) and Petruzella (2010). PLCs can easily implement all variations of PID control. Consequently, it is relatively easy to program a single PLC to deal with most requirements of batch processing. However, it would be difficult for a PLC to optimize batch cycle operations or to implement inferential control of a reactor or separator product composition, functions that are easier to implement by integrating the PLC with a general-purpose computer or a distributed control system.

EXAMPLE 18.2

Consider the operator-assisted control of the simple mixing process shown in Fig. 18.9 to demonstrate representative batch control strategies. To start the operation sequence, a hand switch (HS4) or push button is activated by the operator, which causes a solenoid valve (VN7) to be opened to introduce liquid A. Liquid transfers are implemented by gravity in this example (no pumps are needed). When the liquid level in the tank reaches an intermediate value (LH2), flow A is stopped and flow B is started (VN8). At the same time, a motor (MN5) is started to turn on the mixer. When the liquid level reaches a specified level (LXH2), flow B is stopped and the discharge valve is opened (VN9). After the tank level reaches the low limit (LL2), the discharge valve is closed and the motor stopped. The operator may now start another mixing cycle by pushing the start button again. It should be noted that this simplified control strategy does not deal with emergency situations. Timing of equipment sequencing, such as making sure that valve VN8 is closed before opening discharge valve VN9, is also not considered. Develop an information flow diagram, a sequential function chart, and binary and ladder logic diagrams for this batch operation. Assume the batch proceeds uninterrupted (i.e., do not consider the case where the operator could accidentally activate the hand switch after the batch sequence starts).

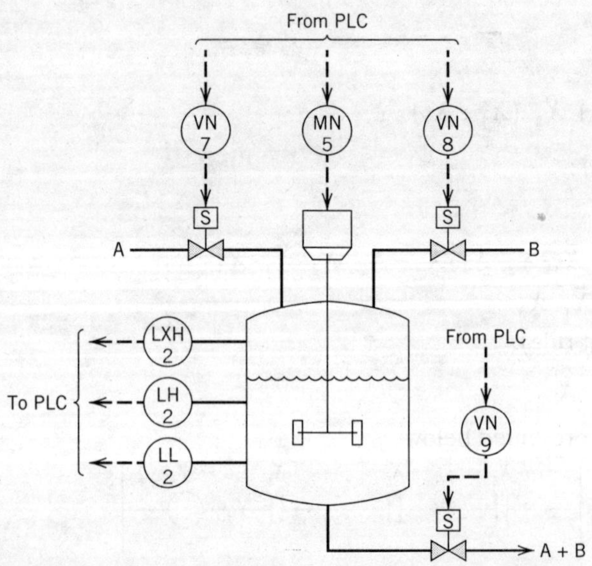

Figure 18.9 Schematic diagram for a batch mixing tank.

For the binary logic diagram, denote the tank level by L and use the following binary values for the different device states:

operator push button HS4:	on $\Rightarrow 1$	off $\Rightarrow 0$
LH2 indicator:	$L \geq LH2 \Rightarrow 1$	$L < LH2 \Rightarrow 0$
LXH2 indicator:	$L \geq LXH2 \Rightarrow 1$	$L < LXH2 \Rightarrow 0$
LL2 indicator:	$L \geq LL2 \Rightarrow 0$	$L < LL2 \Rightarrow 1$
Valves VN7, VN8, VN9:	open $\Rightarrow 1$	closed $\Rightarrow 0$
Mixer MN5:	motor on $\Rightarrow 1$	off $\Rightarrow 0$

Solution

Figures 18.10 and 18.11 show the series of events on the information flow diagram and the sequential function chart. For implementation via hardware (or software) interlocks, the binary logic diagram is shown in Fig. 18.12. Figure 18.6 defines the symbols used in this logic diagram. Gate 1 (an AND gate) ensures that the process will not start, if requested, when the tank level is not low. Gate 3 opens valve VN7 for flow A only if valve VN8 is not open (small circle on gate 3 denotes NOT). Gate 2 (an OR gate) latches the operator request once valve VN7

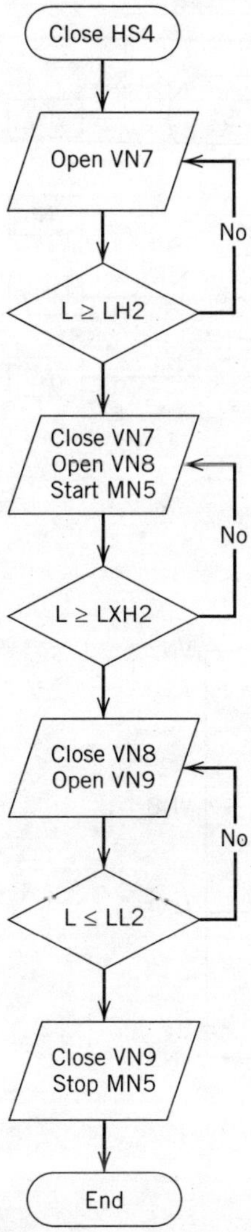

Figure 18.10 Information flow diagram for control of the mixing tank.

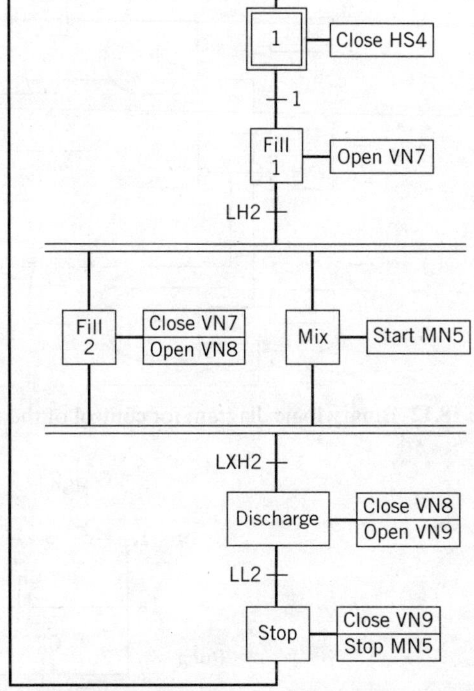

Figure 18.11 Sequential function chart (SFC) for control of the mixing tank.

is opened so that the operator may release the push button. Gates 4 and 7 start flow B and the mixer motor when the intermediate liquid level is reached. This start signal (AND gate 3) terminates flow A. At the high tank liquid level, gate 6 opens the discharge valve. Gate 7 is used to prevent VN8 from opening during the discharge cycle. Gate 8 starts the mixer MN5 at the same time that valve VN8 is opened and ensures that the mixer continues to operate when VN8 is closed and VN9 is opened. The high-level signal LH2 is fed into gate 4 to stop flow B and the mixer motor. Gate 5 holds the discharge signal until the tank is drained.

Figure 18.13 presents the ladder logic diagram for the same mixing process. An example of a normally open contact is C1 (no slash), and an example of a normally closed contact is CR8. The operator-actuated push button HS4 and contact C1 form an AND gate equivalent to gate 1 in Fig. 18.12. The junction connecting rungs 1 and 1a is equivalent to the output of the OR gate 2 in Fig. 18.12. Contact C1 on rung 1 is normally open unless the tank level is low (condition LL2). Contact relay CR8 is usually closed unless relay CR8 on rung 2 is energized. When the level is at (or below) LL2, contact C1 is closed. Then if push button HS4 is engaged, rung 1 becomes energized. Because contact CR8 is normally closed, this action energizes CR7, which results in starting flow A. At the same time, because CR7 is energized, it closes the normally open CR7 on rung 1a. As soon as CR7 on rung 1a is closed, the operator can release HS4, but rung 1/1a will stay energized (or latched) to continue introducing flow A. As long as the level is not at or above LH2, C2 on rung 2 is open. Although CR9 and C3 on rung 2 are closed, rung 2 is not energized because of C2. Thus, contact CR8 remains "not energized," and the mixer motor and flow B valve are not active.

This all changes when the level reaches LH2. First, C2 on rung 2 closes. Because all three contacts on rung 2 are now closed, rung 2 and CR8 are energized. This activates solenoid valve control VN8. At the same time, as soon as CR8 on rung 2 is energized, it opens CR8 on rung 1 and closes CR8 on rung 3. This deenergizes rungs 1 and 1a and CR7, causing two

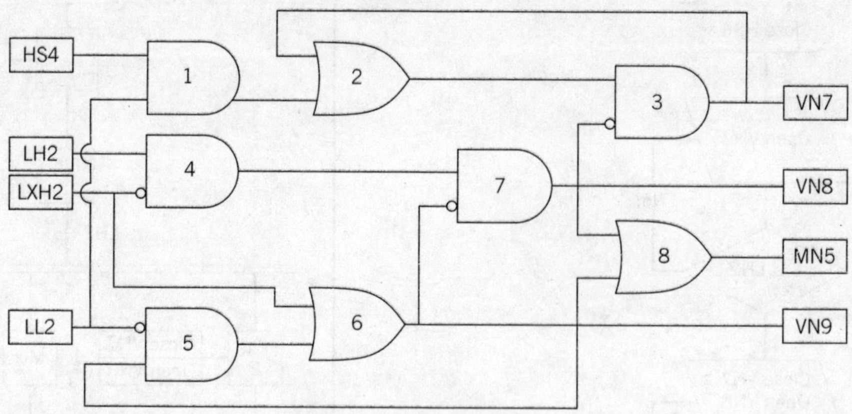

Figure 18.12 Binary logic diagram for control of the mixing tank.

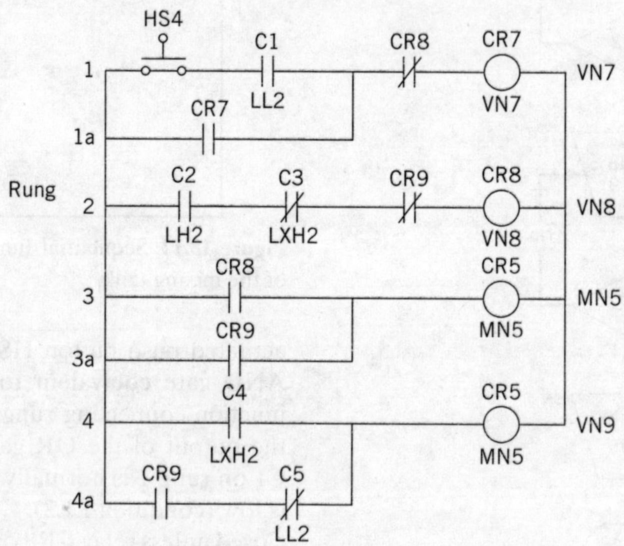

Figure 18.13 Ladder logic diagram for control of the mixing tank.

things to happen. First, CR7 on rung 1a is opened; that is, rungs 1 and 1a are "tripped." More importantly, because CR7 is deenergized, the solenoid valve for flow A is also deenergized, closing the valve and stopping flow A.

In rung 3, CR5 is energized, turning on the mixer motor control MN5. Rung 3a assures that CR5 remains energized after the contact CR8 is deenergized. The junction connecting rungs 3 and 3a is equivalent to the output of the OR gate 8 in Fig. 18.12. For rung 4, as soon as the liquid level reaches LXH2, C4 closes energizing the discharge solenoid valve VN9. The discharge should not stop until the level drops below LXH2, so rung 4a is designed to latch LXH2 (CR9). This strategy allows discharge to continue until the level reaches LL2 (which is tripped by normally closed contact C5).

18.2.3 Monitoring State Transitions

An automated batch facility utilizes sequential logic and discrete devices to change the process state subject to the interlocks. Discrete devices such as two-position valves can be driven to one of two possible states (open/closed). Such devices can be equipped with limit switches that indicate the state of the device. The discrete-device driver is the software routine that generates the output to a discrete device such as a valve and that also monitors process measurements to ascertain that the discrete device actually attains the desired state.

Valves do not instantly change states; rather, they require *travel times* associated with the change from one state to another. Thus, the processing logic within the discrete device driver must provide for a user-specified transition time for each field device. The transition between states can be implemented as:

1. **Drive and wait.** Further actions are delayed until the device attains its assigned state.
2. **Drive and proceed.** Further actions are initiated while the device is in the transition state.

Although two-state devices are most common, the need occasionally arises for devices with three or more states. For example, an agitator may be on high speed, low speed, or off.

Batch control software packages permit the control computer to:

1. Generate the necessary commands to drive each device to its proper state.
2. Monitor the status of each device to determine when all devices have attained their proper states.
3. Continue to monitor the state of each device to ensure that the devices are in their proper states.

If any discrete device does not reach its target state, failure logic is initiated.

18.3 CONTROL DURING THE BATCH

Control during a batch or "within-the-batch" control (Bonvin, 1998) is different from the sequential and logic control discussed above because it is concerned with an *operating trajectory*, that is, how the manipulated and controlled variables change as a function of time (vs. a sequence of on–off device states). Tracking of the set point (which may be a function of time) is challenging for this type of control, because there is no steady-state operating point and wide operating ranges may be encountered due to frequent start-up and shutdown or to large changes in load during the batch. For example, the number of cells in aerobic fermentation at the end of a batch is several orders of magnitude greater than the beginning of the batch, resulting in a wide range of aeration and agitation requirements during the batch.

Bonvin (1998) and Juba and Hamer (1986) have discussed the operational challenges for dynamic control during a batch and provide the following observations:

1. ***Time-varying characteristics.*** Sometimes there is no steady-state operating point, and the transition in the controlled variable may be large compared to typical excursions for continuous systems. Thus, if a standard linear transfer function model is used, the gain and time constants may be time-varying. Batch characteristics can change from run to run, and even the process chemistry may change over a period of months due to changes in the product specifications.
2. ***Nonlinear behavior.*** Because of the potentially wide range of operation, linearized models may be inaccurate and inadequate for controller design. For example, batch chemical reaction rates may have a nonlinear dependence on temperature and concentration, and a nonlinear relationship may exist between heat transferred from a reactor and the flow rate of the cooling medium. Bioprocess unit operations often exhibit nonlinear behavior (Buckbee and Alford, 2007).
3. ***Model inaccuracies.*** Often, mechanistic or fundamental models are not available for batch processes, thus limiting the ability to design and tune controllers *a priori*.
4. ***Sensors.*** Often on-line sensors are not available or are inaccurate due to the wide operating ranges; hence, infrequent samples are analyzed by the plant laboratory. The inability to measure a process variable in real time reduces the safety margin for a process, potentially leading to an undesirable operating condition, for example, a runaway reaction (Center for Chemical Process Safety, 1999). Virtual (inferential) or soft sensors can be used to estimate variables in real time for which direct sensors are not available (see Chapter 17).
5. ***Constrained operation.*** This is a consequence of the wide operating ranges, which makes operating against constraints more likely.
6. ***Unmeasured disturbances.*** Operator error (e.g., wrong feed tank chosen), fouling of vessel walls and heat transfer surfaces, and raw material impurities are sources of major disturbances.
7. ***Irreversible behavior.*** It is often impossible to reverse the effects of history-dependent evolution in product properties such as molecular weight distribution in a polymer or crystal size distribution in a pharmaceutical product. In the semiconductor industry, once a semiconductor wafer is made, it is difficult to modify its electrical properties by further processing or rework.

On the other hand, a batch process has several advantages over a continuous process in meeting product quality requirements.

1. The batch duration can be adjusted in order to meet quality specifications.

2. Because a batch process is repetitive in nature, it offers the possibility of making improvements on a run-to-run basis (see Section 18.4).
3. Batch processes tend to be fairly slow so that improved operating conditions can be computed in real time.

Many of these advantages and disadvantages of batch processes are discussed in the next section, which has reactor control as the focus.

18.3.1 Batch Reactor Control

Figure 18.14 shows a schematic diagram of a batch reactor and its control system. Batch reactors are designed to operate primarily in an unsteady-state manner and are exemplary of the seven control challenges for batch processes that were cited earlier. Many batch reactors exhibit nonlinear behavior due to the coupling of reaction kinetics and reactor temperature while operating over a wide temperature range. Exothermic reactions produce heat that must be removed by a cooling system. Figure 18.14 shows the recommended control system for an exothermic batch reactor. The circulating pump for the coolant loop is essential to minimize the time delay and keep it constant; without it, the time delay varies inversely with cooling load. Because heating is also required to raise the temperature to reaction conditions, the valves are operated in split range. The heating valve opens when the controller output is between 50 and 100%, and the cooling valve opens for the 0–50% range. Sometimes cascade control of the cooling water temperature is utilized (see Chapter 12).

Figure 18.15 shows a typical *batch reactor cycle* consisting of (1) charging each of three reactants sequentially, (2) a heat-up operation, (3) reaction, (4) a cool-down sequence, and (5) discharge of the final

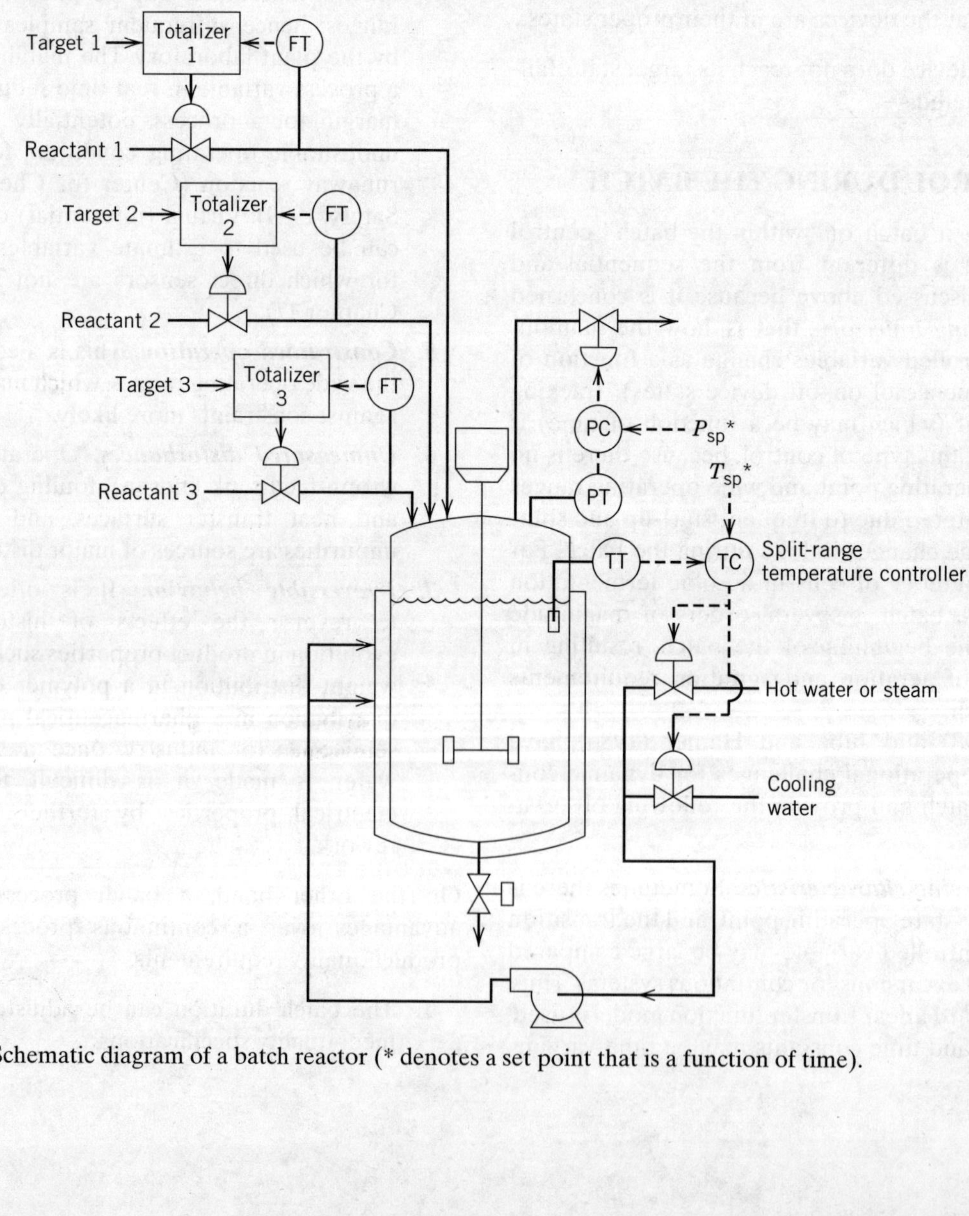

Figure 18.14 Schematic diagram of a batch reactor (* denotes a set point that is a function of time).

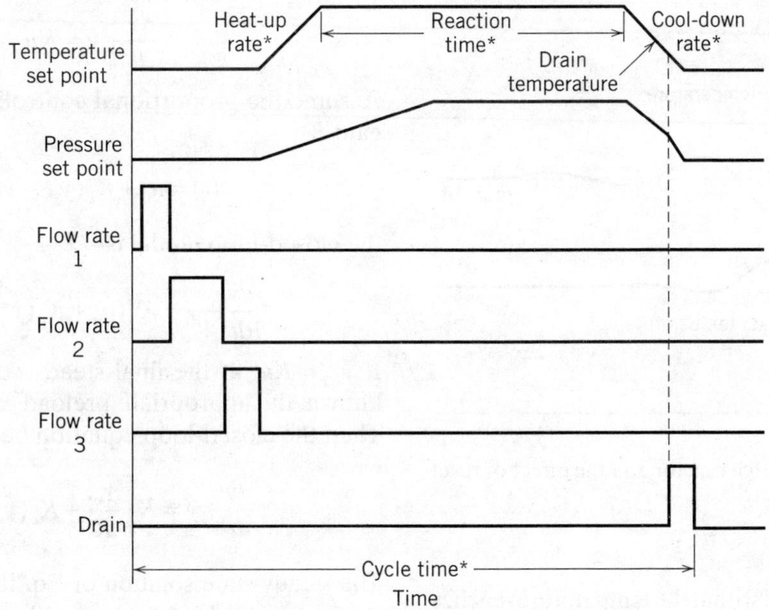

Figure 18.15 Set points and flow sequences of a batch reactor cycle (* denotes externally set function of time).

product mix for separation and subsequent processing. In implementing batch process control, there are several important differences compared with control of continuous processes. The start-up of a batch process can be carried out by operators with all controllers placed in the manual mode. In charging materials to the unit, *totalizers* (see Fig. 18.14) are often used to determine the end point of a charge, that is, the total amount of material that has been transferred to the reactor. Thus, the ability to control flow rate accurately is not as important as the ability to measure and integrate flow rate accurately. If a weight mechanism is used, such as a load cell for the batch reactor, flow rate measurement is not particularly important. For the reaction period, it is important to determine when to terminate the batch reaction (the end-point composition). If the reactor conversion cannot be measured directly in real time, end-point composition can be predicted by some type of indirect (inferential) method, such as measuring off-gas evolution or agitator drive power consumption (torque requirements) in conjunction with mass balances.

Design of PID Controllers for Batch Reactors

Temperature control of exothermic batch reactors is especially challenging, because the process is inherently nonlinear and does not operate at a steady state. In some cases, the reaction can run away if the temperature becomes too high. In addition, the rather large changes in set points during batch start-up can cause controller saturation. Sometimes the set point is ramped instead of using a step change in order to reduce the possibility of controller saturation. Standard PID control may be satisfactory for reactor temperature, although enhanced single-loop controllers can be much more effective (see Chapter 12). Customized profiles of the temperature set point vs. time are often employed to obtain the maximum yield or selectivity in a reactor.

Reset windup is a common problem encountered in batch process control. In a batch reactor, the integral term (the summation in Eq. 7-26) can increase (windup) as a result of the large error between the temperature and its set point. Once the reaction speeds up, reset windup can cause the temperature to exhibit a large overshoot that may be detrimental to product quality control (Smith, 2014).

Figure 18.16 depicts the typical behavior of manipulated and controlled variables in a batch reactor with split range cooling/heating. Note that the manipulated variable (jacket temperature) is initially saturated at its maximum value because of the large error between the set point and the controlled variable (reactor temperature). Placing a limit on the integral term is a common way of implementing anti-reset windup (see Chapter 7). It acts to keep the controller from saturating by placing upper or lower limits on the summation term.

Alternatively, Shinskey (1996) advocates using a bias term (or *preload*) u_0 in the control law to prevent reset windup:

$$u(t) = u_0 + K_c\left[e + \frac{1}{\tau_I}\int_0^t e(t^*)\,dt^*\right] \tag{18-1}$$

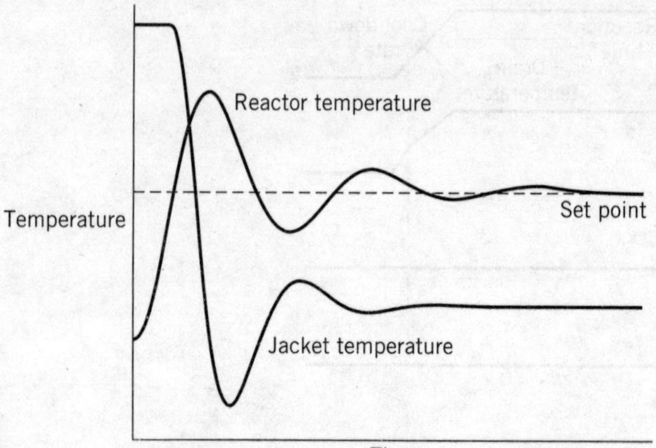

Figure 18.16 Start-up of a batch reactor and the effect of reset windup.

The preload can be selected so that the temperature reaches the set point more rapidly and allows a lower value of K_c to be used in the controller. Shinskey (1996) recommends that the preload setting be adjusted to be slightly lower than the steady-state value of the controller output once the controller reaches the set point. Figure 18.17 shows typical set-point changes with different values of u_0 (Hawkins and Fisher, 2006). How to determine the correct preload is discussed later in this section.

For some batch reactors, it is advantageous to reach the steady-state temperature (at the end of the heat-up period in Fig. 18.15) as rapidly as possible. A combination of an on–off controller and a PID controller can achieve this goal (Lipták, 2007; Bonvin, 1998). This *dual mode approach* seems to be well suited for exothermic reactions where the batch has to be heated to the desired reaction temperature quickly and then cooled using the split range control configuration shown in Fig. 18.14. Maximum heating can be applied until just before the set point is reached; then a switch is made to maximum cooling (Bonvin, 1998). Alternatively, once the controlled variable approaches within 5% of the set point, the on–off controller can be switched off and the PID controller activated to bring the controlled variable smoothly to the desired value.

Figure 18.17 shows the effect of preload u_0 on the typical responses of batch reactor temperature, which is a nonlinear response. u_0 should be ideally selected to achieve zero offset without using integral action. The desired value of u_0 can be developed by experimental testing or from the process model. It is instructive to analyze a linear dynamic model to see how preload eliminates offset with a proportional controller (see Chapter 11). For a first-order model with gain $K(=K_v K_p K_m)$ and time constant τ,

$$\tau\frac{dy}{dt}+y=Ku \tag{18-2}$$

Assume the proportional controller has preload u_0 and gain K_c:

$$u(t)=u_0+K_c(y_{sp}-y(t)) \tag{18-3}$$

the closed-loop model is

$$\tau\frac{dy}{dt}+y=K[(u_0+K_c(y_{sp}-y))] \tag{18-4}$$

If $y_{sp}=Ku_0$ at the final steady state and the gain K is known, the appropriate preload expression is $u_0=y_{sp}/K$. Then the closed-loop equation becomes

$$\tau\frac{dy}{dt}+y=K\left[\frac{y_{sp}}{K}+K_c(y_{sp}-y)\right] \tag{18-5}$$

The steady-state solution of Eq. 18-5 yields $y_{sp}=y$. Thus, there is no steady-state error with proportional control (K_c can be any value) as long as the process gain is known. K_c can be adjusted in order to tune the speed of the closed-loop response. In theory, it is possible to eliminate offset without using integral action, but in practice integral action should be included to deal with model inaccuracy and unanticipated disturbances. For nonlinear processes, the preload will not match the required controller output ("100% preload" in Fig. 18.17), thus an adjustment in u_0 must be made ("correct preload" in Fig. 18.17). This controller will normally be superior to the standard PI controller for a batch reactor shown in Fig. 18.17.

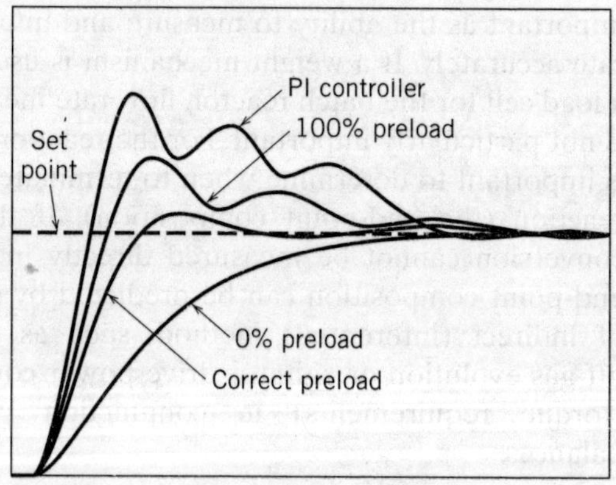

Figure 18.17 Transient responses for batch switch with preload.

Advanced Batch Reactor Control

When the process nonlinearities are significant during the batch transition or start-up, a standard PID controller may not be satisfactory. As discussed in Section 12.8, a gain scheduling or multimodel control approach can be used to deal with excessive nonlinearity. If transfer function models are available for the starting and ending points of the batch trajectory, and model-based PID controllers are available (cf. Chapter 11), then the controller settings can be switched at some point during the set-point change. On the other hand, an adaptive control strategy can be employed, as discussed in Chapter 12. Huzmezan et al. (2002) applied an adaptive control strategy to both a PVC reactor and an ethoxylated fatty acid reactor. In both cases, the variability of the reactor temperatures was reduced by 60% or more.

Juba and Hamer (1986) have described the advantages of using model-based controllers to address the challenges of control of composition or yield when there are highly exothermic reactions (and the potential for a runaway reaction). Typically, three process characteristics must be determined:

1. How steam and/or cooling water affects the reactor temperature
2. How reactor temperature affects reaction chemistry and reaction rates
3. How reaction rate affects heat generation

Simplified nonlinear relationships can be developed based on operating data. When the development of a detailed reaction kinetics model (model 2) is not feasible, a *heat-release-rate estimator* using material balance information or data-driven empirical models can be employed, in effect combining models 2 and 3. It is especially important to understand the sensitivity of the resulting controller to variations in reaction chemistry in order to protect against unsafe conditions.

18.3.2 Rapid Thermal Processing

Semiconductor devices are manufactured in a series of physical and/or chemical batch unit operations. An integrated circuit or semiconductor consists of several layers of carefully patterned thin films, each chemically altered to achieve desired electrical characteristics (Edgar et al., 2000). From 30 to 300 separate steps (such as deposition, etching, and lithography) over a total duration of two months (*cycle time*) are typically required to construct a set of circuits (devices). The wafers range in size up to 450 mm in diameter and are 400–700 μm thick.

Rapid thermal processing (*RTP*) is a unit operation employed in semiconductor manufacturing for thin-film deposition, such as nitridation or oxidation, and for annealing. RTP provides high ramp rates in wafer temperature that lead to short thermal processing times, thus increasing wafer throughput. In RTP, the wafer temperatures have specified ramp rates and steady-state values. It is imperative that the wafer temperature be controlled precisely to the specified ramp rate and steady-state temperature in order to meet process specifications. The temperature trajectory can be divided into three regions: the ramp, steady state, and cool-down periods shown in Fig. 18.18. The overall duration of the three steps is less than several minutes. In the ramp region, the ramp rates can vary between 25 and 200 °C/s.

Steady-state temperatures depend on the RTP process (e.g., oxidation, nitridation, annealing, etc.). Cool-down can be either a controlled ramp or an uncontrolled cooling of the process.

Feedback control is the preferred method for controlling the wafer temperature trajectory. The design of feedback controllers for RTP is a challenging problem because of the nonlinear nature of the radiative and other heat transfer phenomena, the fast process dynamics, and the additional constraints that are placed on the wafer temperature response. The wafer temperature is controlled by adjusting the power output from an array of lamp heaters above and below the wafer to track the set-point trajectory shown in Fig. 18.18. The three key requirements for control are as follows:

1. ***Ramp error.*** It is important that the wafer temperature follow the set-point trajectory as closely as possible. Therefore, one measure of controller performance is the ramp error, defined as the difference between the set point and wafer temperature at any given time during the ramp. It should be less than 10 °C during the ramp.

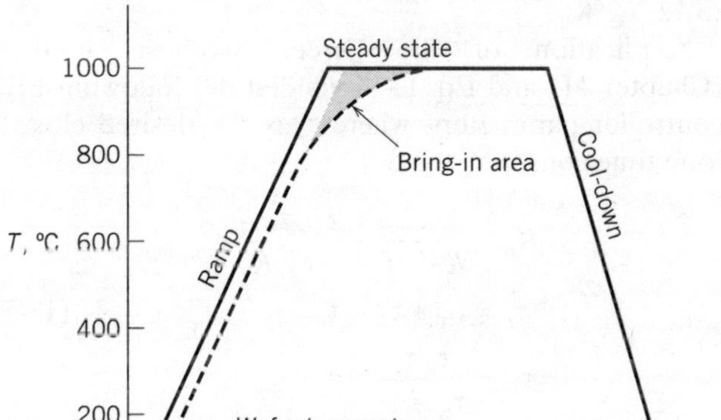

Figure 18.18 RTP set-point change. The solid line is the temperature set point, and the dashed line is the wafer temperature. The wafer temperature has been shifted to the right, for clarity. Consequently, the ramp error exceeds 10° C.

2. ***Bring-in.*** As mentioned earlier, it is important that the wafer temperature reach steady state as rapidly as possible. Bring-in is a criterion that indicates controller performance during the transition from ramp to steady state (i.e., where the trajectory "turns the corner"). Bring-in is defined as the enclosed area between the desired set-point trajectory (shifted in time, as shown in Fig. 18.18) and the set-point temperature at steady state; bring-in should be minimized. The shifting of the set-point signal takes into account the dynamic error that normally occurs.
3. ***Overshoot.*** The corner of the set-point trajectory should be turned without overshoot.

Balakrishnan and Edgar (2000) evaluated gain-scheduled control of a commercial RTP reactor. They determined that a PID controller based on a semi-empirical model of the heating process provided effective temperature control of the reactor. Derivation of a fundamental heat transfer model based on an unsteady-state energy balance yielded an approximate second-order transfer function with wafer heating time constant τ_W and heating lamp time constant τ_L ($\tau_L \ll \tau_W$).

$$G(s) = \frac{K}{(\tau_W s+1)(\tau_L s+1)} \tag{18-6}$$

The controlled variable is the measured wafer temperature and the manipulated variable is the percent power to the heating lamps above and below the wafer. For a series of step tests, the time constants varied with temperature between 650 and 1000 °C ($3 \le \tau_W \le 8$ s; $0.5 \le \tau_L \le 0.7$ s), and the gain K varied from 4 to 12 °C/%.

Application of the Direct Synthesis method (Chapter 11) and Eq. 11-14 yielded the following PID controller parameters, where τ_c is the desired closed-loop time constant:

$$K_c = \frac{1}{K}\frac{\tau_W+\tau_L}{\tau_c} \cong \frac{\tau_W}{\tau_c K}$$

$$\tau_I = \tau_W + \tau_L \cong \tau_W \tag{18-7}$$

$$\tau_D = \frac{\tau_W \tau_L}{\tau_W+\tau_L} \cong \tau_L$$

A bias term for the controller was calculated from a physical heat transfer model. It was necessary to include this preload in order to avoid overshoot and to satisfy the ramp error and bring-in requirements. The control algorithm was programmed with gain-scheduled PID parameters determined from the linear models (see Chapter 12). Because the model gain and time constants varied with temperature, seven gain-scheduled regions were used between room temperature and 1100 °C. Experiments were performed for final steady-state temperatures ranging from 750 to 1050 °C and a ramp rate of 50 °C/s. The closed-loop experimental response for 950 °C is shown in Fig. 18.19.

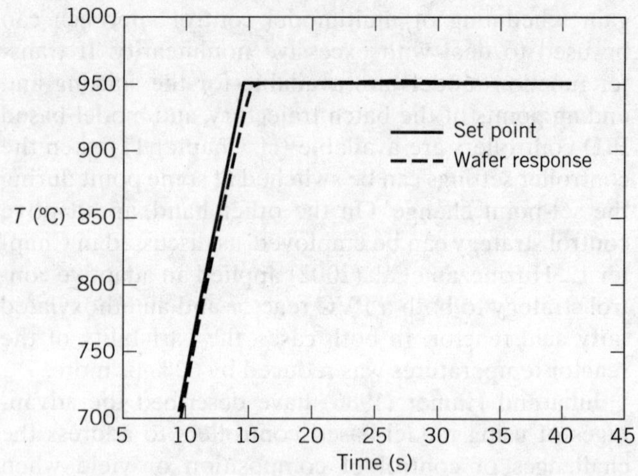

Figure 18.19 Gain-scheduled PID controller results for RTP temperature cycle and a steady-state temperature of 950 °C.

18.4 RUN-TO-RUN CONTROL

Recipe modifications from one run to the next are common in many batch processes. Typical examples are modifying the reaction time, feed stoichiometry, or reactor temperature. When such modifications are done at the beginning of a run (rather than during a run), the control strategy is called *run-to-run (RtR) control.* Run-to-run control is frequently motivated by the lack of on-line measurements of the product quality during a batch run. In batch chemical production, on-line measurements are often not available during the run, but the product can be analyzed by laboratory samples at the end of the run. The process engineer must specify a recipe that contains the values of the inputs (which may be time-varying) that will meet the product requirements. The task of the run-to-run controller is to adjust the recipe after each run to reduce variability in the output product from the stated specifications.

In semiconductor manufacturing, the goal is to control qualities such as film thickness or electrical properties that are difficult, if not impossible, to measure in real-time in the process environment. Most semiconductor products must be transferred from the processing chamber to a metrology tool (measuring device) before an accurate measurement of the controlled variable can be taken. The scope of run-to-run control applications in the semiconductor industry is significant.

Batch run-to-run control can be viewed as implementing a series of set-point changes to the underlying batch process controllers at the end of each run. By analyzing the results of previous batches, the run-to-run controller adjusts the batch recipe in order to reduce quality variations. Thus, run-to-run control is equivalent to controlling a sequence of the controlled variable at times k, $k + 1$, $k + 2$, ... analogous to a standard control problem.

Significant challenges in some industries limit making process control changes (including run-to-run changes). Regulatory hurdles may prevent modifications to computer software in pharmaceutical, biotech, and medical device manufacturing. Changing control procedures requires formal documenting and operator retraining, which may require process revalidation. Desired changes may not be allowable if they involve setpoints falling outside the range of values critical to product quality that are contained in the company's application to the U.S. Food and Drug Administration (FDA).

Run-to-run control is particularly useful to compensate for processes where the controlled variable drifts over time. For example, in a chemical vapor deposition process or in a batch chemical reactor, the reactor walls may become fouled due to byproduct deposition. This slow drift in the reactor chamber condition requires occasional changes to the batch recipe in order to ensure that the controlled variables remain on-target. Eventually, the reactor chamber must be cleaned to remove the wall deposits, effectively causing a step disturbance to the process outputs when the inputs are held constant. Just as the RtR controller compensates for the drifting process, it can also return the process to target after a step disturbance (Edgar et al., 2000; Moyne et al., 2001).

Because RtR controllers generally are model-based controllers, the availability of a process model (fundamental or empirical) is of great utility. However, if a dynamic model is not easily obtained due to the complexity of the process, run-to-run control can still be carried out. The majority of models used in RtR control for semiconductor applications are steady-state models (Moyne et al., 2001). These pure gain models assume that the process drift is slow. The RtR controller is typically an integral controller plus a bias term. When the output measurement is noisy, it is useful to employ an observer or filter to estimate the actual process output. In this case, controllers can be designed using the techniques presented in Sections 13.7 and 13.8.

Use of Optimization in Batch Control

A batch trajectory may be changed on a run-to-run basis in order to optimize product yield or selectivity while satisfying process constraints. The best set-point profile can be obtained theoretically using optimal control techniques (Bonvin et al., 2002). An alternative approach uses parameterization of the manipulated variable as a function of the batch time t, for example, $u(t) = a_0 + a_1 t + a_2 t^2$. This type of control law is not based on feedback from the available sensors; parameters a_0, a_1, and a_2 would be adjusted after each batch based on the product quality measurements at the end of the run. This approach is beneficial when unmeasured slow disturbances are encountered, namely, those that do not change much from run to run. Faster-acting disturbances would need to be managed with a feedback control system, as discussed in Section 18.3.

The minimal information needed in carrying out this type of RtR control is a static model relating the manipulated variable to the quality variables at the end of a batch. It can be as simple as a steady-state (constant) gain relationship, or as complicated as a nonlinear model that includes the effects of different initial conditions and the batch time. In contrast, a time-dependent profile for the manipulated variable during the batch can be adjusted from run-to-run to meet the end-of-the-batch quality requirements as well as operating constraints, for example, upper and lower bounds on the manipulated variables (Bonvin et al., 2002). Other variations of RtR controllers in different applications have been reported by Zafiriou et al. (1995) in rapid thermal processing, Clarke-Pringle and MacGregor (1998) in polymerization, and Zisser et al. (2005) for medical drug dosing.

18.5 BATCH PRODUCTION MANAGEMENT

Batch process equipment must be properly configured in a plant, process, or unit operation in order to be operated and maintained in a reasonable manner. A flowsheet for a general batch plant is shown in Fig. 18.20. The Instrument Society of America (ISA) SP-88 standard deals with the terminology involved in batch control (Parshall and Lamb, 2000; Strothman, 1995).

Figure 18.21 shows the hierarchy of activities that take place in a batch processing system (Hawkins and Fisher, 2006; Smith, 2014). At the highest level, *procedures* identify how the products are made, that is, the actions to be performed (and their order) as well as the associated control requirements for these actions. *Operations* are equivalent to unit operations in continuous processing

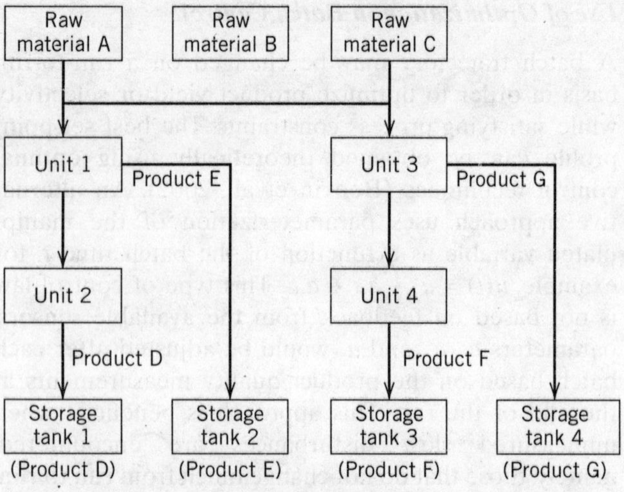

Figure 18.20 Flowsheet for a multiproduct batch plant.

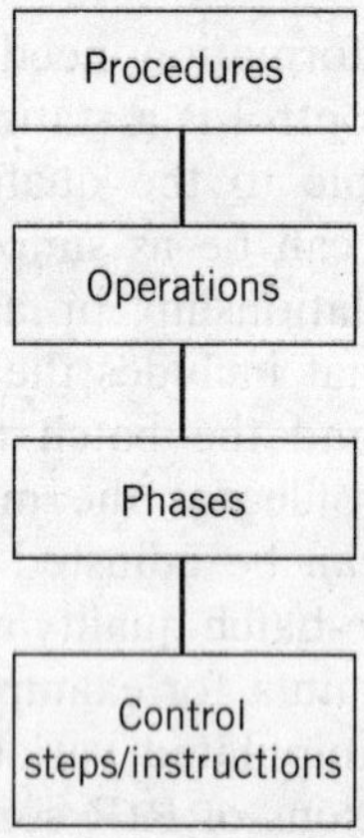

Figure 18.21 Hierarchy of activities in implementing a batch recipe.

and include such steps as charging, reacting, separating, and discharging. Within each operation are logical points called *phases*, where processing can be interrupted by operator or computer interaction. Examples of different phases include the sequential addition of ingredients, heating a batch to a prescribed temperature, mixing, and so. *Control steps* involve direct commands to final control elements, specified by individual control instructions in software. As an example, for {*operation = charge reactant*} and {*phase = add ingredient B*}, the control steps would be (1) open the B supply valve, (2) total the flow of B over a period of time until the prescribed amount has been added, and (3) close the B supply valve. Such sequential control operations were discussed in Section 18.2.

The term *recipe* has a range of definitions in batch processing, but in general a recipe is a procedure with the set of data, operations, and control steps required to manufacture a particular grade of product. A *formula* is the list of recipe parameters, which includes the raw materials, processing parameters, and product outputs. A recipe *procedure* has operations for both normal and abnormal conditions. Each *operation* contains resource requests for certain ingredients (and their amounts). The operations in the recipe can adjust set points and turn equipment on and off. The complete production run for a specific recipe is called a *campaign* (multiple batches).

Cyclical batch processing is the simplest category, where the same logic is used with no product recipe or formula, for example, in catalyst regeneration (Smith, 2015a). In *multigrade batch processing*, the instructions remain the same from batch to batch, but the formula can be changed to yield modest variations in the product. For example, in emulsion polymerization, different grades of polymers are manufactured by changing the formula. In *flexible batch processing*, both the formula (recipe parameters) and the processing instructions can change from batch to batch. The recipe for each product must specify both the raw materials required and how conditions within the reactor are to be sequenced in order to make the desired product.

Many batch plants, especially those used to manufacture pharmaceuticals, are certified by the *International Standards Organization* (*ISO*). ISO 9000 (and related ISO standards 9001–9004) state that every manufactured product should have an established, documented procedure, and the manufacturer should be able to document that the procedure was followed. Companies must pass periodic audits to main ISO 9000 status. Both ISO 9000 and the FDA require that only a certified recipe be used. Thus, if the operation of a batch becomes "abnormal," performing any unusual corrective action to bring it back within the normal limits is not an option. In addition, if a slight change in the recipe apparently produces superior batches, the improvement cannot be implemented unless the entire recipe is recertified. The FDA typically requires product and raw materials tracking, so that product abnormalities can be traced back to their sources.

Batch Scheduling and Planning

For recipe management, each batch is tracked as it moves through the production stages, which may involve sequential processing operations on various pieces of equipment. As the batch proceeds from one piece of equipment to the next, recipe management is responsible for ensuring that the proper type of process equipment is used, the specific equipment is not currently being used by another batch, and materials are charged to the correct batch. The complexity in such operations demands that a computer control system be utilized to minimize operator errors and off-specification batches.

A production run typically consists of a sequence of a specified number of batches using the same raw materials and making the same product to satisfy customer demand; the accumulated batches are called a *lot*. When a production run is scheduled, the necessary equipment items are assigned and the necessary raw materials are allocated to the production run, As the individual batches proceed, the consumption of raw materials must be monitored for consistency with the original allocation of raw materials to the production run. because parallel trains of equipment mat be involved. A typical scheduling and planning scenario is shown in Table 18.3. Various optimization techniques can be employed to solve the problem, ranging from linear programming to mixed-integer nonlinear programming (Pekny and Reklaitis, 1998).

When several products are similar in nature, they require the same processing steps and hence pass through the same series of processing units; often the batches are produced sequentially. Because of different processing time requirements, the total time required to produce a set of batches (also called the *makespan* or cycle time) depends on the sequence in which they are produced. To maximize plant productivity, that is. the maximum amounts of each product for the fixed capital investment, the batches should be produced in a sequence that minimizes the makespan. The plant schedule corresponding to such a sequence can be represented graphically in the form of a Gantt chart. The Gantt chart provides a timetable of plant operations showing which products are produced by which units and at what times.

Consider four products ($p1, p2, p3, p4$) that are to be produced as a series of batches in a multiproduct plant

Table 18.3 Characteristics of Batch Scheduling and Planning

Determine	Given
What?	**Product requirements**
Product amounts: lot sizes, batch sizes	Time horizon, demands, starting and ending inventories
When?	**Operational steps**
Timing of specific operations, run lengths	Precedence order, resource utilization
Where?	**Production facilities**
Sites, units, equipment	Types, capacities
How?	**Resource limitations**
Resource types and amounts	Types, amounts, rates

Source: Pekny and Rekaitis (1998).

consisting of three batch reactors in series shown in Fig. 18.22. The processing times for each batch reactor and each product are given in Table 18.4. Suppose that no intermediate storage is available between the processing units. If a product finishes its processing on unit k and unit $k + 1$ is not free because it is still processing a previous product, then the completed product must be kept in unit k until unit $k + 1$ becomes free. As an example, product $p1$ must be held in unit 1 until unit 2 finishes processing $p3$. When a product finishes processing in unit 3. it is sent immediately to product storage. Assume that the times required to transfer products from one unit to another are negligible compared with the processing times.

Optimization can be used to determine the time sequence for producing one batch of each of the four products so as to minimize the total production time. The schedule corresponding to this production sequence is conveniently displayed in the form of a Gantt chart in Fig. 18.23, which shows the status of the units at different times. For instance, unit 1 is processing pl during [0, 3.5]. When $p1$ leaves unit 1 at $t = 3.5$ h, it starts processing $p3$. It processes $p3$ during [3.5, 7]. However, it is unable to discharge $p3$ to unit 2, because unit 2 is still processing $p1$. So unit 1 holds $p3$ during [7, 7.8]. When unit 2 discharges $p3$ to unit 3 at 16.5 h,

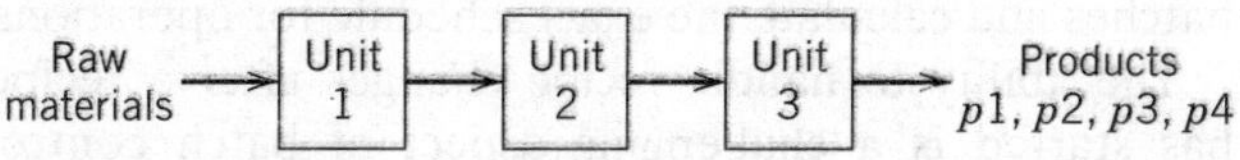

Figure 18.22 Multiproduct plant.

Table 18.4 Processing Times (h) of Products

	Products			
Units	$p1$	$p2$	$p3$	$p4$
1	3.5	4.0	3.5	12.0
2	4.3	5.5	7.5	3.5
3	8.7	3.5	6.0	8.0

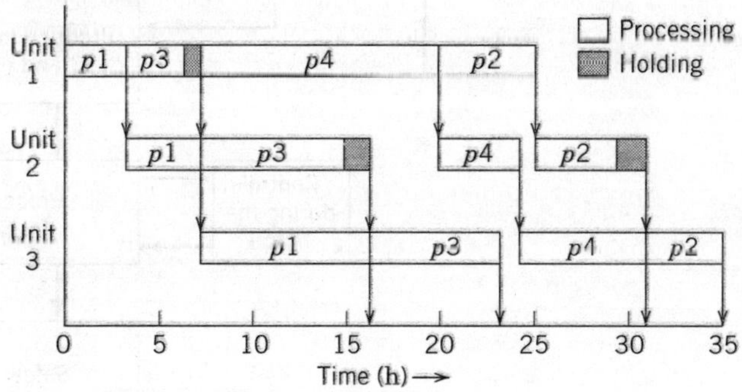

Figure 18.23 Gantt chart for the optimal multiproduct plant schedule.

unit 1 is still processing $p4$; therefore, unit 2 remains idle during [16.5, 19.8]. it is common in batch plants to have units blocked due to busy downstream units, or units waiting for upstream units to finish. This happens because the processing times vary from unit to unit and from product to product, reducing the time utilization of units in a batch plant. The finished batches of $p1$, $p3$, $p4$, and $p2$ are completed at times 16.5, 23.3, 31.3, and 34.8 h. The minimum makespan is 34.8 h.

Many different kinds of planning and scheduling software systems are used in batch processing. Figure 18.24 gives an expanded view of batch scheduling and recipe management, along with the different types of control involved. In the top half of Fig.18.24 Enterprise Resource Planning (ERP) software provides the following information to the operator console: production planning, equipment scheduling, recipe management. selection of resources and rates, and lot sizing (Harjunkoski et al., 2014; Wassick, 2009). Typically, the activities are structured hierarchically, with higher-level tasks carried out infrequently to determine the operating conditions and set points that must be addressed by regulatory control as well as by interlocks at the equipment level. Within the scope of a scheduling area (defined as a few units or machines that make a group of products), and a time horizon of hours to day, much greater detail is needed to sequence batches and calculate the exact schedule for operations.

The ability to handle recipe changes after a recipe has started is a challenging aspect of batch control systems. Many times it is desirable to change the grade of the batch to meet product demand, or to change the resources used by the batch after the batch has started In other cases, the grade is unknown until near the end of the batch, or off-spec laboratory analysis necessitates a change in grade. Because every batch of product is not always good, special-purpose control recipes are needed to fix, rework. blend. or dispose of bad batches. if that is allowable. It is important to be able to respond to unusual situations by creating special-purpose recipes and still meet the demand. This procedure is referred to as *reactive scheduling*.

Batch durations can vary greatly in the length of time required. Some batch processes may involve repeating a step or even recycling a batch to an earlier step if the results of a given step are unsatisfactory, for example, in semiconductor manufacturing or bioprocessing. If that step does not result in satisfactory yield and/or purity, the batch is not rejected and thrown away, but is rerouted to an appropriate batch step and the processing tried again or "reworked." In batch sterilization and "clean in place" operations of equipment, the batch is not sent on to the next step if the results of the step are not acceptable; rather, the step is repeated.

When ample storage capacity is available, the normal practice has been to build up large inventories of raw materials and ignore the inventory carrying cost. However, improved scheduling can be employed to minimize inventory costs, which implies that *supply chain management* techniques may be necessary to implement the schedule (Pekny and Reklaitis, 1998; Grossmann, 2008).

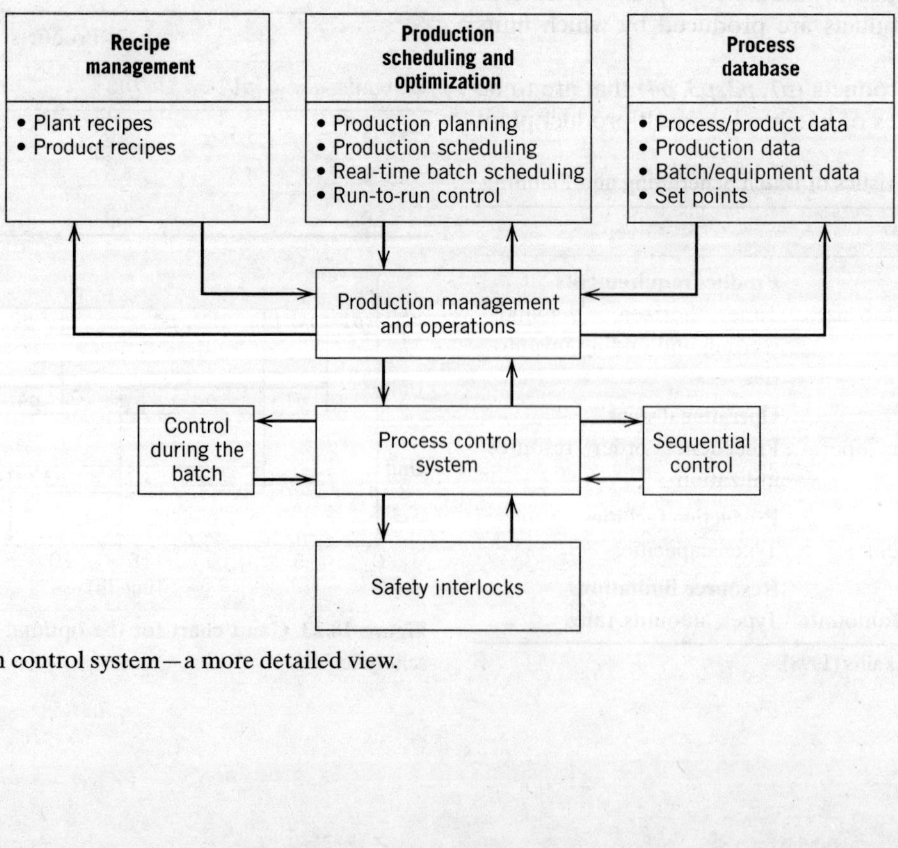

Figure 18.24 Batch control system—a more detailed view.

Movable storage tanks and processing equipment are used in a number of plants. In *flexible manufacturing plants*, flexhoses, manned vehicles, and automated guided vehicles (AGVs) are used to move material between the different groups of equipment (Piana and Engell, 2011), instead of having permanently installed connections. Even most nonflexible chemical plants have a great number of flexhoses; with an array of pipe headers, cleaned out flexhoses can connect virtually any two pieces of equipment. Some pharmaceutical plants have taken the flexhose concept much further by eliminating all permanent piping. Junctions are made and broken as needed. Although this setup is not economical for large plants with long distances to traverse, it is practical for small plants that make a large number of products. A third approach is to have fixed piping and movable units. The batch reactors and separators are moved when transfers need to be made. If a batch reactor needs to operate for a few hours with no transfers occurring, it can be moved off to the side and left unconnected.

SUMMARY

This chapter has surveyed the broad field of batch process control and emphasized topics and techniques that are unique to batch processing, for example, sequential logic and batch scheduling, Binary logic and ladder logic diagrams, sequential function charts, and Gantt charts are specialized tools that are introduced and applied in this chapter. Batch process present significant challenges for the design of feedback control systems, especially because of the process nonlinearity that is normally encountered. Various methods to design controllers for batch proccess have been presented. In the future, the chemical industry will rely more heavily on specialty products via batch processing. Many electronic materials and pharmaceutical products are already manufactured by batch processing. Thus, it is important to understand the key concepts in this developing area of process control.

REFERENCES

Alford, J. S., Bioprocess Control: Advances and Challenges, *Comput. Chem. Eng.*, **30**, 1464 (2006).

Arzen, K. E., Grafcet for Intelligent Supervisory Control Applications, *Automatica*, 1513 (1994).

Balakrishnan, K. S., and T. F. Edgar, Model-Based Control of Rapid Thermal Processing, *J. Thin Solid Films*, **365**, 322 (2000).

Bonvin, D., Optimal Operation of Batch Reactors—A Personal View, *J. Process Control*, **8**, 355 (1998).

Bonvin, D., B. Srinivasan, and D. Ruppen, Dynamic Optimization in the Chemical Industry, Chemical Process Control-CPC VI, J. B. Rawlings, B. A. Ogunnaike, and J. B. Eaton (Eds.), *AIChE Symp. Ser.*, **97**, No. 326, 255 (2002).

Buckbee, G., and J. Alford, *Automation Applications in Biopharmaceuticals*, ISA, Research Triangle Park, NC, 2008.

Center for Chemical Process Safety, *Guidelines for Process Safety in Batch Reaction Systems*, AIChE, New York, 1999.

Clarke-Pringle, T. L., and J. F. MacGregor, Optimization of Molecular-Weight Distribution Using Batch to Batch Adjustments, *Ind. Eng. Chem. Res.*, **37**, 3660 (1998).

David, R., Grafcet: A Powerful Tool for Specification of Logic Controllers, *IEEE Trans. Control System Technology*, **3**, 253 (1995).

Desrochers, A. A., and R. Y. Al-Jaar, *Application of Petri Nets in Manufacturing Systems*, IEEE Press, Piscataway, NJ, 1995.

Diwekar, U., *Batch Distillation Simulation, Optimal Design and Control*, Taylor and Francis, Washington, DC, 1995.

Edgar, T. F., S. W. Butler, W. J. Campbell, C. Pfeiffer, C. Bode, S. B. Hwang, K. S. Balakrishnan, and J. Hahn, Automatic Control in Microelectronics Manufacturing: Practices, Challenges and Possibilities, *Automatica*, **36**, 1567 (2000).

Falcione, A., and B. H. Krogh, Design Recovery for Relay Ladder Logic, *IEEE Control Systems Magazine*, **13**(4), 90 (1993).

Grossmann, I. E., Design of Responsive Supply Chains under Demand Uncertainty, *AIChE J.*, **32**, 3090 (2008).

Harjunkoksi, I., C. Maravelias, P. Bongers, P. Castro, S. Engell, I. Grossmann, J. Hooker, C. Mendez, G. Sand, and J. Wassick, Scope of Industrial Applications of Production Scheduling Models and Solution Methods, *Comput. Chem. Eng.*, **62**, 161 (2014).

Hawkins, W. M. J., and T. G., Fisher, *Batch Control Systems: Design, Application, and Implementation* 2nd ed., ISA, Research Triangle Park, NC, 2006.

Hughes, T. A., *Programmable Controllers*, 4th ed., ISA, Research Triangle Park, NC, 2005.

Huzmezan, M., B. Gough, and S. Kovac, Advanced Control of Batch Reactor Temperature, *Proc. Amer. Control Conf.*, 1156 (2002).

Johnson, C. D., *Process Control Instrumentation Technology*, 8th ed., Prentice Hall, Upper Saddle River, NJ, 2005.

Juba, M. R., and J. W. Hamer, Progress and Challenges in Batch Process Control, *Chemical Process Control—CPC IV*, M. Morari, and T. J. McAvoy (Eds.) CACHE-Elsevier, Amsterdam, 1986, p. 139.

Lipták, B. G. (Ed.), *Process Control and Optimization, Instrument Engineers' Handbook*, vol. 8, CRC Press, Boca Ratón, FL, 2007.

Moyne, J., E. del Castillo, and A. M. Hurwitz (Eds.), *Run to Run Control in Semiconductor Manufacturing*, CRC Press, Boca Raton, FL, 2001.

Mujtaba, I. M., *Batch Distillations: Design and Operations*, Imperial College Press, London, 2004.

Parshall, J., and L. Lamb, *Applying S88: Batch Control from a User's Perspective,* ISA, Research Triangle Park, NC, 2000; see also ANSI-ISA-88.01–1995.

Pekny, J. F., and G. V. Reklaitis, Towards the Convergence of Theory and Practice: A Technology Guide for Scheduling/Planning Methodology, in Foundations of Computer-Aided Process Operations (FOCAPO) *AIChE Symp. Ser.*, **94**, 91, (1998).

Petruzella, F. *Programmable Logic Controllers*, 4th ed., McGraw-Hill, New York, 2010.

Piana, S., and S. Engell, Hybrid Evolutionary Optimization of the Operation of Pipeless Plants, *J. Heuristics*, **16**, 311 (2011).

Platt, G., and R. A. Gilbert, Binary Logic Diagrams, in *Instrument Engineers' Handbook*, 3rd ed., B. G. Lipták (Ed.), CRC Press, Boca Raton, FL, 2005.

Rosenof, H. P., and A. Ghosh, *Batch Process Automation*, Van Nostrand-Reinhold, New York, 1987.

Shinskey, F. G., *Process Control Systems*, 4th ed., McGraw-Hill, New York, 1996.

Smith, C. L., Automate Your Batch Process-Part 1, *Chem. Eng. Prog.*, **111**, 28 (August, 2015a).

Smith, C. L., Automate Your Batch Process-Part 2, *Chem. Eng. Prog.*, **111**, 37 (September, 2015b).

Smith, C. L., *Control of Batch Processes*, John Wiley and Sons, Hoboken, NJ, 2014.

Strothman, J., Batch Standards Group Agrees on Terminology, *InTech*, **42**(8), 31 (1995).

Wassick, J., Enterprise-wide Optimization in an Integrated Chemical Complex, *Comput. Chem. Eng.*, **33**, 1950 (2009).

Webb, J. W., and R. A. Reis, *Programmable Logic Controllers*, 5th ed., Prentice Hall, Upper Saddle River, NJ, 2002.

Zafiriou, E., R. Adomaitis, and G. Gattu, An Approach to Run-to-run Control for Rapid Thermal Processing, *Proc. Amer. Cont. Conf.*, 1286 (1995).

Zisser, H., L. Jovanović, F. Doyle III, P. Ospina, and C. Owens, Run-to-Run Control of Meal Related Insulin Dosing, *Diabetes Technol. Ther.*, **7**, 48 (2005).

EXERCISES

18.1 Consider the microwave oven as an example of a discrete state process. The process variables that can be on or off include fan, light, timer, rotating base, microwave generator, and door switch. The process steps include opening the oven door/placing food inside, closing the door, setting the timer, heating up food, and cooking completed. Prepare a table for each process step that shows the process variable status (on or off). What interlocks between variables are important safety issues?

18.2 A pump motor operates by a push button that once actuated keeps the pump on until the operator pushes a stop button. Also there is an emergency stop if the pump overheats ($T > T_H$). Draw a logic diagram using AND/OR symbols. Also draw the equivalent ladder logic diagram.

18.3 A truth table for a set of inputs is shown below where A and B are inputs and Y is an output:

A	B	Y
0	0	1
1	0	1
0	1	0
1	1	1

Construct binary operations (AND, OR, NOT) in series or parallel that yield Y from A and B.

18.4 For the following binary logic gate, express the output Y in terms of the two inputs X_1 and X_2. Also construct the truth table for various binary values of X_1 and X_2.

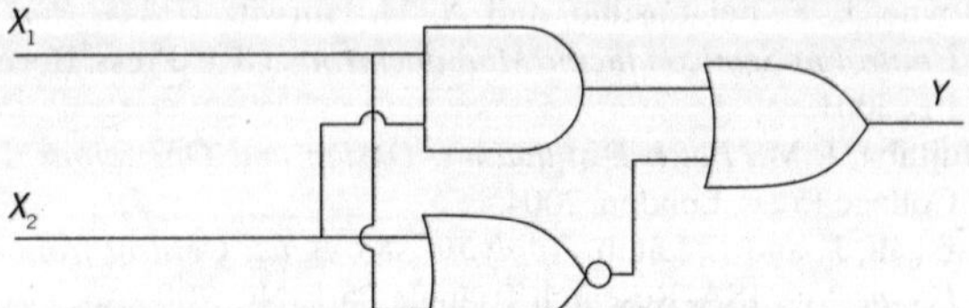

18.5 Simplify the following Boolean expression:

$$(X_1 + X_2)(X_1X_3 + X_1\bar{X}_3) + X_1X_2 + X_2$$

18.6 A batch operation is used to heat a liquid to a specified temperature. There is a start button, a stop button, inlet/outlet valves, and limit sensors for low tank level (LL) and high tank level (LH). Flow is performed by gravity transfer. The process steps are

(a) Push the start button to start the process.

(b) Fill the tank up to LH by opening an inlet valve with the exit line closed.

(c) Heat the liquid to the temperature set point while stirring.

(d) Turn off the stirrer and empty the tank down to LL by opening the exit valve (but the inlet valve must be closed). Draw an information flow diagram, sequential function chart, and ladder logic diagram.

18.7 Consider a process that consists of a liquid chemical tank with two level indicators, a heater, inlet and outlet pumps, and two valves. Assume that the following sequence of operations are to be performed:

(a) Start the sequence by pressing button S.

(b) Fill the tank with liquid by opening valve V_1 and turning on pump P_1 until the upper level L_1 is reached.

(c) Heat the liquid until the temperature is greater than T_H. The heating can start as soon as the liquid is above level L_0.

(d) Empty the liquid by opening the valve V_2 and turning on pump P_2 until the lower level L_0 is reached.

(e) Close the valves and go to step (a) and wait for a new sequence to start.

Draw an information flow diagram, sequence function chart, and ladder logic diagram.

18.8 A furnace has two heating chambers for simultaneous heating of two samples in the two chambers. But when only one sample is available for heating, the empty chamber also gets heated unnecessarily. To avoid overheating of the empty chamber a heat removal system (a fan) is available. Develop the ladder logic diagram for this control problem.

18.9 A two-tank filling system is shown in Fig. E18.9. Both tanks are used in a similar way. Tank 1 is considered to be empty when the level is less than L_1 and is considered to be full when the level is greater than L_2. Initially, both tanks are empty. If push button S is pressed, both tanks are filled by opening valves V_1 and V_2. When a tank is full (e.g., tank 1), filling stops by closing valve V_1, and its contents start to be used (by opening valve W_1). When tank 1 is empty, valve W_1 is closed. Filling may only start up again when both tanks are empty, and if button S is pressed. Draw an information flow diagram, a sequential function chart, and ladder logic diagram for the system. Use the notation $V_1 = 1$ to denote that valve V_1 is open.

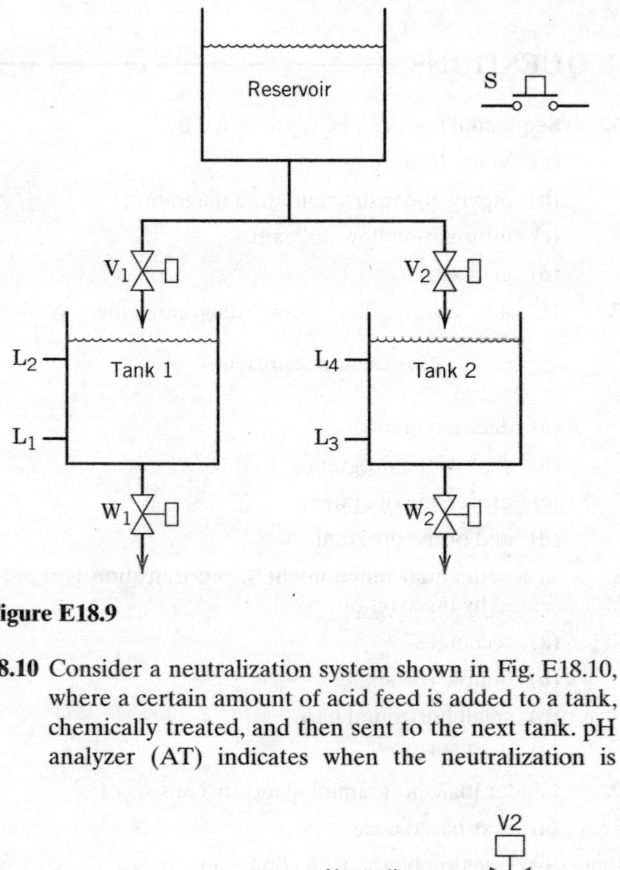

Figure E18.9

18.10 Consider a neutralization system shown in Fig. E18.10, where a certain amount of acid feed is added to a tank, chemically treated, and then sent to the next tank. pH analyzer (AT) indicates when the neutralization is complete. Level switches LS1, LS2, and LS3 are activated when the level in the tank is at or above a given level. The neutralization process proceeds with the following steps:

(a) Initially, all the valves are closed, the mixer is off, and the neutralization tank is empty.

(b) When the start button (not shown) is pressed, V1 opens and LS2 is activated. These actions fill the tank with the solution to be neutralized.

(c) When the solution level rises above LS2, start mixer M. When the level drops below LS1, stop the mixer.

(d) Whenever the pH of the solution is too low, open V2 to add neutralizer.

(e) If the tank becomes full before the acid feed is neutralized, indicated by the activation of LS3, close V2 to stop the inflow of neutralizer. Next, open V4 to reduce the liquid level to the point indicated by LS2; this solution will be reprocessed later. Then close V4 and proceed with step (d) again.

(f) When the pH of the solution is correct, close V2 and open V3 to drain the tank. When the tank is empty, indicated by the deactivation of LS1, close V3 and proceed with step (a). Draw information flow and ladder logic diagrams and a sequential function chart.

18.11 Consider the tank car unloading of a hazardous chemical, as shown in Fig. 5.6. A pump is used to empty the tank car and fill the storage tank. Then the chemical is transferred to a chemical reactor. Describe the discrete steps necessary to carry out such an operation and discuss any safety issues associated with the transfer of the hazardous chemical; that is, what instrumentation and alarms would you recommend installing?

18.12 Suppose a batch reactor such as the one in Fig. 18.14 has a gas ingredient added to the liquid feed. As long as the reaction is proceeding normally, the gas is absorbed in the liquid (where it reacts), keeping the pressure low. However, if the reaction slows or the gas feed is greater

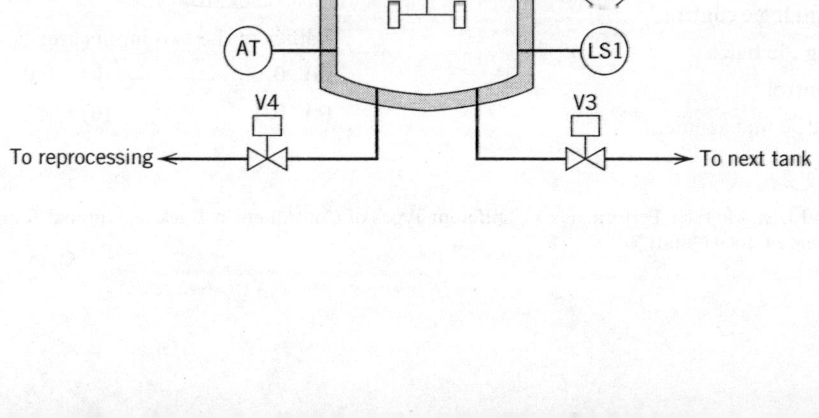

Figure E18.10

than can be absorbed, the pressure will start to rise. The pressure rise can be compensated by an increase in liquid feed, but this may cause the cooling capacity to be exceeded. Describe a solution to this problem using overrides (see Chapter 12).

18.13 Consider the batch reactor system simulated by Aziz et al.[2] The two reactions, $A + B \rightarrow C$ and $A + C \rightarrow D$, are carried out in a jacketed batch reactor, where C is the desired product and D is a waste product. The manipulated variable is the temperature of the coolant in the cooling jacket. There are two inequality constraints: upper and lower bounds on the coolant temperature and an upper limit on the maximum reactor temperature. Using the model parameters specified by Aziz et al., evaluate the following control strategies for a set-point change from 20 °C to 92 °C.

(a) PID controller

(b) Batch unit

(c) Batch unit with preload

(d) Dual-mode controller

18.14 In Chapter 17, different process monitoring strategies are discussed. Like automatic process control, statistical process control is more challenging for batch processes than continuous processes. Prepare a list of challenges and difficulties associated with monitoring of batch processes.

MULTIPLE CHOICE QUESTIONS

1. In comparison to a continuous process, the flexibility of operation for a batch process is

(a) higher.

(b) lower.

(c) both have same flexibility.

(d) flexibility does not depend on type of operation.

2. Which of the following is usually not produced in batch operation?

(a) Specialty chemical

(b) Polymer

(c) Pharmaceutical

(d) fertilizer

3. Which of the following is a distinguishing feature of most batch processes in comparison to continuous processes?

(a) Steady state operating point

(b) Unsteady state operating point

(c) Static operating point

(d) Time invariant operating point

4. DCS stands for

(a) Distributed control system.

(b) Digital control system.

(c) Discrete control system.

(d) Distributed cascade system.

5. Feedback control mechanism can be implemented for which of the following level in batch process control?

(a) Sequencing and logic control.

(b) Control during the batch.

(c) Run to run control.

(d) Batch production management.

6. Sequential logic can be represented by

(a) Venn diagram.

(b) piping and instrumentation diagram.

(c) information flow diagram.

(d) pi chart.

7. In an information flow diagram, the symbol

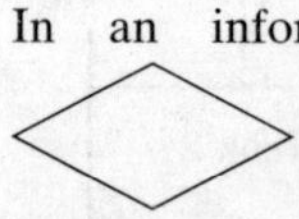

is used to represent

(a) decision point.

(b) start of the program.

(c) processing operation.

(d) end of the program.

8. In a sequential function chart, the transition is represented by the symbol

(a) rectangle.

(b) double rectangle.

(b) small horizontal bar.

(d) double bar.

9. Ladder logic programming mostly consists of

(a) text-based code.

(b) function blocks with connecting lines.

(c) coils and virtual relay contacts.

(d) none of these.

10. The output of the following binary logic gate is 1.

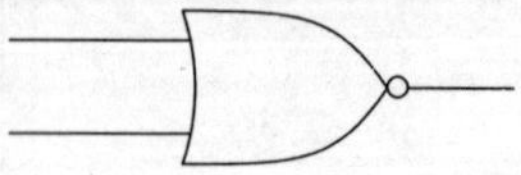

Values of the two inputs are:

(a) 0, 0 **(b)** 1, 0

(c) 0, 1 **(d)** 1, 1

[2] N. Aziz, M. A. Hussain, and I. M. Mujtaba, Performance of Different Types of Controllers in Tracking Optimal Temperature Profiles in Batch Reactors, *Comput. Chem. Eng*, **24**, 1069 (2000).

ANSWER KEY

1. (a)	**2.** (d)	**3.** (b)	**4.** (a)	**5.** (b)
6. (c)	**7.** (a)	**8.** (b)	**9.** (c)	**10.** (a)

CHAPTER 19

Digital Process Control Systems: Hardware and Software

CHAPTER CONTENTS

Process control implemented by computers has undergone extensive changes in both concepts and equipment during the past 50 years. The feasibility of digital computer control in the chemical process industries was first investigated in the mid-1950s. During that period, studies were performed to identify chemical processes that were suitable for process monitoring and control by computers. These efforts culminated in several successful applications, the first ones being a Texaco refinery and a Monsanto chemical plant (both on the Gulf Coast) using mainframe computers. The first commercial systems were slow in execution and massive in size compared with the computers available today. They also had very limited capacity. For example, a typical first-generation process control computer had 32K RAM and disk storage of 1 MB.

The functionalities of these early control systems were limited by capabilities of the existing computers rather than the process characteristics. These limitations, coupled with inadequate operator training and an unfriendly user interface, led to designs that were difficult to operate, maintain, and expand. In addition, many systems had customized specifications, making them extremely expensive. Although valuable experience was gained in systems design and implementation, the lack of financial success hindered the infusion of digital system applications into the process industries until about 1970, when inexpensive microprocessors became available commercially (Lipták, 2005).

During the past 40 years, developments in microelectronics and software technologies have led to the widespread application of computer control systems. Digital control systems have largely replaced traditional analog instrument panels, allowing computers to control process equipment while monitoring process conditions. Technological advancements, such as VLSI (very large-scale integrated) circuitry, object-oriented programming techniques, and distributed configurations have improved system reliability and maintainability while reducing manufacturing and implementation cost. This cost reduction has allowed small-scale applications in new areas, for example, microprocessors in single-loop controllers and smart instruments (Thompson and Shaw, 2016). Programmable logic controllers have also gained a strong foothold in the process industries.

Increased demand for digital control systems created a new industry, consisting of systems engineering and service organizations. Manufacturing companies moved toward enterprise-wide computer networks by interfacing process control computers with business computer networks. These networks permit all computers to use the same databases in planning and scheduling (see Appendix F), and they also allow access to operator station information from locations outside the plant.

In the following sections, we provide an overview of the hardware and software used for process control. The distributed control system configuration is described first, followed by data acquisition for different signal types. Digital hardware is then considered, and concluding with a description of control system software organization and architectures.

19.1 DISTRIBUTED DIGITAL CONTROL SYSTEMS

The revolutionary development in microelectronics and telecommunications hastened the evolution of distributed computer networks. In the 1970s, first-generation *distributed control systems* (DCS) replaced the single-mainframe design used previously in process control with a number of identical minicomputers that operated independently of each other. Removable media such as magnetic tapes were used for information transmission. Networking allowed these computers to share resources and/or information.

Computers physically located in different plant areas, which control nearby processes, are said to be *geographically distributed*. More than one computer may share the control of one or more processes. When the control functions are distributed over more than one computer or device, the system is said to be *logically distributed*. Process control networks tend to be distributed both geographically and logically, the extent of which depends on execution priority and complexity. Applications often utilize a variety of digital devices, such as workstations in a distributed control system, personal computers (PC), single-loop controllers (SLC), programmable logic controllers (PLC), and embedded microcomputers.

During the 1980s, the standard distributed digital control network topology was the star configuration, where individual satellite nodes communicated with each other via an arbitrator node. The arbitrator was often the main computer of that system, located in or

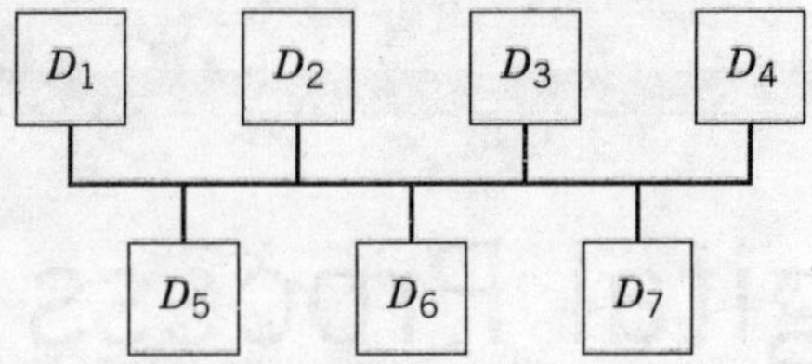

Figure 19.1 Global bus architecture for digital process control with different devices D_i.

near the central control room. This computer supported the operator interface and a number of other functions not normally implemented in the satellite computers, which were located in processing areas. One inherent flaw of this scheme is that the operator supervisory and control capability was lost when the main computer failed, even though the satellites continued to function (Lipták, 2005).

Currently open system designs with global bus architecture and local area networks (LANs) are being used for computer control, as shown in Fig. 19.1. Unlike earlier networks, which were normally isolated, the LANs are often connected to other networks via gateway devices. The traditional host computer functions are divided functionally and are implemented in separate autonomous computers, which share the same data bus. When more than one operator interface node is installed, the operator interface to the process can be maintained even when several operator stations fail. A DCS for process control is fundamentally the same as for other real-time distributed systems used in business data centers or server facilities (Kopetz, 2011; Lewis et al., 2006), although specialized hardware such as data acquisition equipment is required.

19.2 ANALOG AND DIGITAL SIGNALS AND DATA TRANSFER

Field instrumentation is essential for process control and monitoring. For digital computers to monitor and control processes, they must be able to acquire data from these instruments and implement control based on the sensed information. Special devices are required to convert analog signals to and from digital form. Although analog signals have traditionally been used to transmit data within the plant, the availability of digital architectures such as Fieldbus and Profibus (Mehta and Reddy, 2016) is changing this situation.

19.2.1 Analog Signal Representation

Analog signals occur in the physical world as continuous time-varying signals that can have any value in a

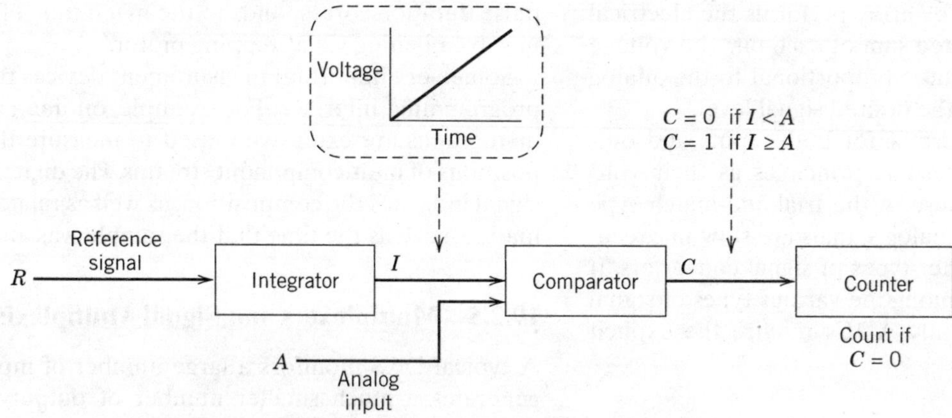

Figure 19.2 A conceptual diagram of a voltage A/D converter.

specified range. In contrast, discrete signals are limited to a defined set of values. To interface digital computers with measurements from field instruments, electrical signals must be converted to a form acceptable to digital computers, and vice versa. Analog-to-digital and digital-to-analog conversion is performed by simple devices called A/D and D/A converters[1] (ADCs and DACs). Analog electrical signals are in either voltage or current form. As a result of the transmitter standards discussed in Chapter 8, standard interfaces are available for every measured and manipulated variable.

19.2.2 Binary Signals and Converters

Binary signals carry two-valued information (0 or 1), which is used to represent the operating state of two-state devices, for example, whether a motor or pump is off or running. Similarly, binary signals may be used to start or stop such devices. Because binary values are expressed in bits (elements that are 0 or 1), they are packed in clusters of a certain length (bytes = 8 bits) or according to computer word sizes. Often, a light-emitting diode (LED) is attached to an instrument for state indication.

Practically all binary signals use a zero voltage (AC or DC) to represent *logical zero*. Different voltage levels, for example, 5, 15, and 24 VDC and 24 and 110 VAC, are used to represent *logical one*. Other commonly used types of binary signals include pulse trains, which are described below. More details are available elsewhere (Patterson and Hennessey, 2013; Johnson, 2005).

19.2.3 Analog Signals and Converters

Because digital computers are not capable of storing data with infinite precision, measurements must be *quantized*. Similarly, the control actions calculated from these process measurements are quantized according to the computer precision. Thus, a fixed number of bits is used to represent the digitized version of an analog measurement. A n-bit storage can be used to represent 2^n different numbers. Most process control-oriented ADCs and DACs utilize a 12-bit unsigned integer representation. Thus, there are 2^{12} or 4096 quantization levels for each process variable. This resolution is better than 0.025%, which is lower than typical noise levels in electrical signals. For high-precision applications, up to 24-bit representation is used.

To digitize an analog input, the unknown process signal is compared with a known signal (Johnson, 2005). Figure 19.2 illustrates a simple voltage ADC, which includes an integrator, a comparator, and a counter. The unknown signal is used as an input of the comparator, which compares it with a trial signal generated by the integrator. In this example, the trial signal is a ramp voltage. Assuming 12-bit representation, the ramp voltage is increased by 1/4096 of the nominal voltage span of the unknown signal, at a preset frequency determined by a quartz crystal. The counter is incremented at the same frequency. The comparator signals the counter to freeze its content when the ramp voltage equals or exceeds the unknown voltage. The time it takes the ramp voltage to equal or exceed the unknown signal is proportional to the magnitude of the unknown signal.

At the beginning of the next sampling interval, the value stored in the counter is transferred to another register for processing. The counter is zeroed, and the comparator output is reset. The ramp voltage is then returned to the lower bound of the nominal voltage range, and the process described above is repeated. Other A/D conversion methods exist, but these methods use different techniques to generate the trial signals.

D/A converters are based on a different principle that involves arrays of resistors. To convert a digital value, the bits of its digital representation are fed into the resistor

[1]Pronounced "A-to-D" and "D-to-A."

array simultaneously. The array performs the electrical equivalence of a weighted sum of each bit. The voltage level of the array output is proportional to the analog value and amplified to the desired signal level.

Current signal converters, for both inputs and outputs, operate using the same principles as their voltage counterparts. Because of the trial-and-match type operations, ADCs for analog signals are slow in execution compared with other types of signal converters. It should be noted that among the various types of signal converters discussed, only ADCs require the explicit use of microprocessors.

19.2.4 Pulse Trains

A *pulse train* is a special type of binary signal that is used to convey analog information. This can be accomplished by measuring the frequency of the pulses (usually for inputs), while on-time ratio over a period (the fraction that the period of the pulse is equal to 1) is used for outputs. This is also called *duty cycle* or absolute-on-time (Johnson, 2005). Although process control computers are quite capable of handling low-frequency pulse signals, this is rarely done unless only a few signals are involved. To process high-frequency pulses for a large number of signals, special pulse-counting ADCs and pulse-generating DACs are used. A value indicating the pulse frequency is required to process a pulse output. The low and high instrument limits represent 0% and 100%, respectively, on-time of the pulses.

A pulse input consisting of a train of pulses can be digitized by using a pulse counter, which measures the pulse frequency and converts pulse frequency to a digital representation. The computer maintains an accumulator for the pulse counts; its output after a period of time is proportional to the pulse frequency (Johnson, 2005). For example, a turbine flow meter utilizes a pulse counter to measure the rate of fluid flow. In one full revolution, a fixed amount of fluid flows through the meter, and a single pulse is generated. By determining the pulse frequency, the fluid flow rate can be calculated. Pulse outputs are normally used to manipulate two-state devices to control process variables. For example, suppose a heater is equipped with a constant wattage power supply. Temperature can be controlled by limiting power consumed by the heater, which can be accomplished by turning the heater on and off periodically while regulating the percent on-time. The higher the on-off frequency, the smoother the maintained temperature. For pulse duration outputs (PDOs), the duration of a pulse is proportional to the incremental control applied to an analog device. For example, the pulse duration corresponds to the magnitude of change in valve opening via a stepping motor.

Some very important measurement devices require a programming interface. For example, on-line gas chromatographs are extensively used to measure the compositions of multicomponent streams. The digital output signal indicates the composition, as well as related information, such as the time that the sample was analyzed.

19.2.5 Multiplexers and Signal Multiplexing

A typical DCS monitors a large number of inputs and generates a much smaller number of outputs. Multiplexer is a combinational circuit with multiple inputs and single output. For 2^n inputs, there are n selection lines in a multiplexer. It is used to send multiple signals over a common transmission line at different times or speed so that the channel is utilized with maximum efficiency. The multiplexing (MUX) and data retrieval are synchronized by a computer and are applicable to high-level signals that are measured in volts. For low-level input signals, such as millivolts from thermocouples and strain gauges, low-level multiplexing must be performed. Single channel I/O cards simplify marshalling, reduce maintenance costs, and eliminate the need for marshalling cabinets. These MUXs are electromechanical in nature. Alternatively, amplifiers can be used to boost low-level signals in order to employ high-level MUXs directly (Johnson, 2005). Although hardware costs have dropped, the use of MUXs to reduce the number of ADCs still merits consideration in certain cases.

19.3 MICROPROCESSORS AND DIGITAL HARDWARE IN PROCESS CONTROL

Digital systems employed for process control increase in size, scope, and cost according to the following hierarchy:

1. Single-loop controllers
2. Programmable logic controllers
3. Personal computer controllers
4. Distributed control system

These categories are discussed in four subsections below (Sections 19.3.1 through 19.3.4). Even at the lowest level (SLCs), miniaturization of the integrated circuits permits up to 16 control loops to be incorporated into special-purpose microprocessors. All four types of control hardware systems include redundant hardware for failure protection. They can operate under extreme environmental conditions such as high temperature and can withstand vibrations and shocks. They are often enclosed in special cabinets when sited in explosive or corrosive atmospheres. Nitrogen purge gas is used to maintain a slight positive pressure inside the cabinets and isolate the systems from the hazardous environment or airborne contaminants.

19.3.1 Single-Loop Controllers

The single-loop controller (SLC) is the digital equivalent of analog single-loop controllers. It is a self-contained microprocessor-based unit that can be rack-mounted. Although the basic three-mode (PID) controller function is the same as its analog counterpart, the processor-based SLC allows the operator to select a control strategy from a predefined set of control functions, such as PID, on/off, lead/lag, adder/subtractor, multiply/divider, filter functions, signal selector, peak detector, and analog track. SLCs feature auto/manual transfer switching, multi-set point, self-diagnosis, gain scheduling, and perhaps also time sequencing. Many manufacturers produce single processor units that handle cascade control or multiple loops, typically 4, 8, or 16 loops per unit, and incorporate self-tuning or auto-tuning PID control algorithms. Although designed to operate independently, single-loop controllers have digital communications capability similar to that for a distributed control system (DCS), as discussed in Section 19.3.4.

19.3.2 Programmable Logic Controllers

Programmable logic controllers (PLCs) are simple digital devices that are widely used to control sequential and batch processes (see Chapter 18). Although PLCs were originally designated to replace electromechanical relays, they now have additional functions that are usually associated with microprocessors. For example, PLCs can implement PID control and other mathematical operations via specialized software (Hughes, 2005; Petruzella, 2010).

PLCs can be utilized as standalone devices or in conjunction with digital computer control systems. Hughes (2005) and Lipták (2005) have summarized the general characteristics of PLCs:

1. ***Inputs/Outputs (I/O).*** Up to several thousand discrete (binary) inputs and outputs can be accommodated. Large PLCs have several hundred analog inputs and outputs for data logging and/or continuous PID control.
2. ***Logic handling capability.*** All PLCs are designed to handle binary logic operations efficiently. Because the logical functions are stored in main memory, one measure of a PLC's capability is its memory scan rate. Another measure is the average time required to scan each step in a logic or ladder diagram (see Chapter 18). Thousands of steps can be processed by a single unit. Most PLCs also handle sequential logic and are equipped with an internal timing capability to delay an action by a prescribed amount of time, to execute an action at a prescribed time, and so on.
3. ***Continuous control capability.*** PLCs with analog I/O capability usually include PID control algorithms to handle up to several hundred control loops. More elaborate PLCs incorporate virtually all of the commonly used control functions covered in Chapters 11 and 12, including PID, on/off, integral action only, ratio and cascade control, low-or high-signal select, lead–lag elements, and so forth. Such PLCs are quite efficient, because internal logic signals are available to switch controller functions.
4. ***Operator communication.*** Older PLCs provide virtually no operator interface other than simple signal lamps to indicate the states of discrete inputs and outputs. Newer models often are networked to serve as one component of a DCS control system, with operator I/O provided by a separate component in the network.
5. ***PLC programming.*** A distinction is made between *configurable* and *programmable* PLCs. The term *configurable* implies that logical operations (performed on inputs to yield a desired output) are located in PLC memory, perhaps in the form of *ladder diagrams* by selecting from a PLC menu or by direct interrogation of the PLC. Usually, the logical operations are put into PLC memory in the form of a higher-level programming language. Most control engineers prefer the simplicity of configuring the PLC to the alternative of programming it. However, some batch applications, particularly those involving complex sequencing, are best handled by a programmable approach, perhaps through a higher-level, computer control system.

19.3.3 Personal Computer Controllers

Because of their high performance, low cost, and ease of use, personal computers (PCs) are a popular platform for process control. When configured to perform scan, control, alarm, and data acquisition (*SCADA*) functions, and when combined with a spreadsheet or database management application, the PC controller can be a low-cost, basic alternative to the DCS.

In order to use a PC for real-time control, it must be interfaced to the process instrumentation. The I/O interface can be located on a board in an expansion slot, or the PC can be connected to an external I/O module using a standard communication port on the PC (e.g., RS-232, RS-422, or IEEE-488). The controller card/module supports 16- or 32-bit microprocessors. Standardization and the high-volume PC market has resulted in a large selection of hardware and software tools for PC controllers (McConnell and Jernigan, 1996; Auslander and Ridgely, 2002).

In comparison with PLCs, PCs have the advantages of lower purchase cost, graphics output, large memory, large selection of software products (including databases and development tools), more programming options (use of C or Java vs. ladder logic), richer operating systems, and open networking. PLCs have the following advantages: lower maintenance cost, operating system and hardware optimized for control, fast boot times, ruggedness, low mean time between failures, longer support for product models, and self-contained units. PC-based control systems are predicted to continue to grow at a much faster rate than PLCs and DCSs during the next decade.

Process control systems should also be *scalable*, which means that the size of the control and instrumentation system is easily expanded by simply adding more devices. This feature is possible because of the availability of open systems (i.e., "plug-and-play" between devices), smaller size, lower cost, greater flexibility, and more off-the-shelf hardware and software in digital control systems. A typical system includes personal computers, an operating system, object-oriented database technology, modular field-mounted controllers, and plug-and-play integration of both system and intelligent field devices. New devices are automatically recognized and configured with the system. Advanced control algorithms can be executed at the PC level (Lipták, 2005).

19.3.4 Distributed Control System

Figure 19.3 depicts a representative distributed control system. The DCS system consists of many commonly used DCS components, including MUXs, single-loop and multiple-loop controllers, PLCs, and smart devices. A system includes some or all of the following components (Lipták, 2005):

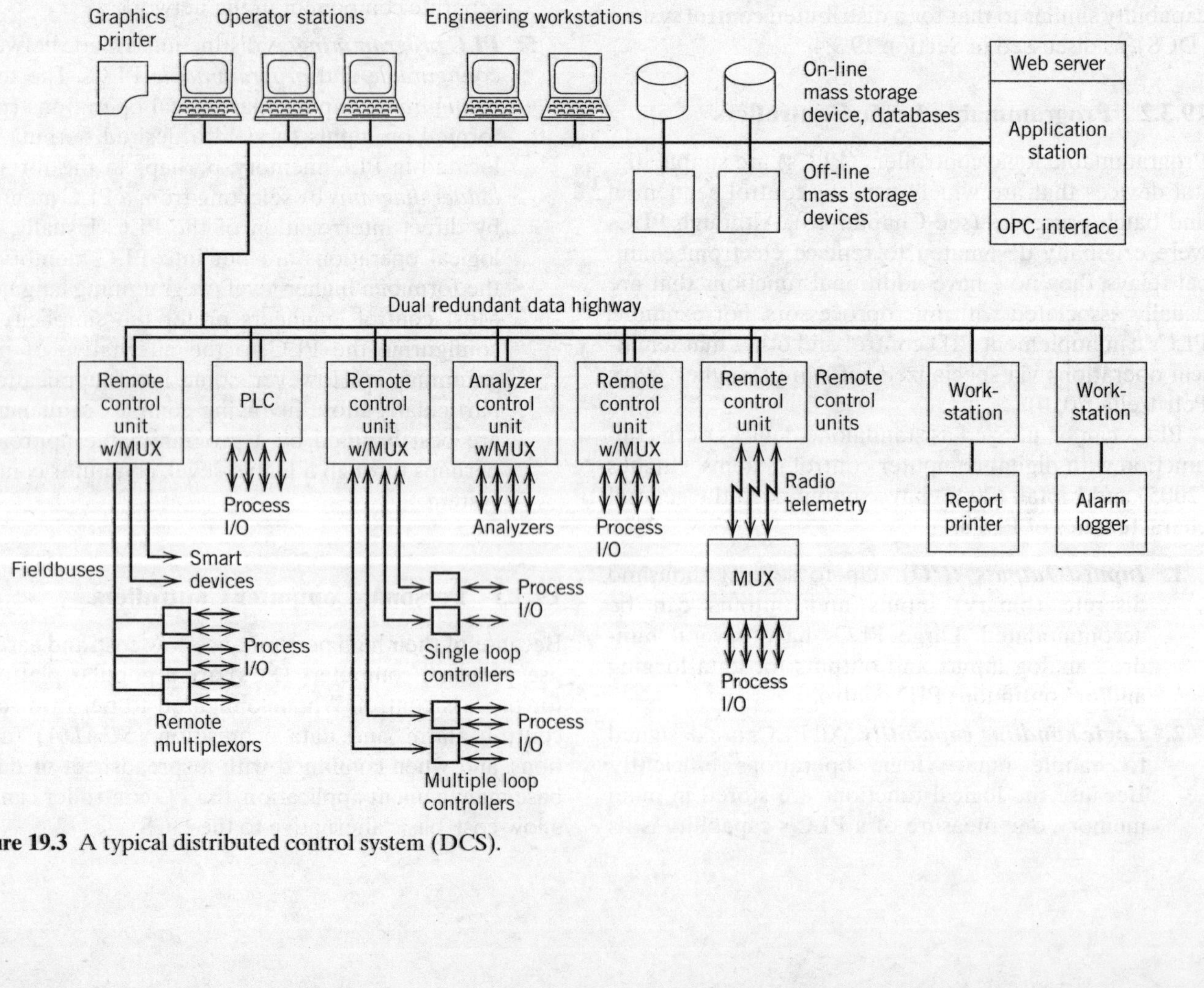

Figure 19.3 A typical distributed control system (DCS).

1. ***Control Network.*** The control network is the communication link between the individual components of a network. Coaxial cable and, more recently, fiber-optic cable have often been used, in competition with ethernet protocols. A redundant pair of cables (dual redundant highway) is normally supplied to reduce the possibility of link failure.
2. ***Workstations.*** Workstations are the most powerful computers in the system, capable of performing functions not normally available in other units. A workstation acts both as an arbitrator unit to route internodal communications and the database server. An operator interface is supported, and various peripheral devices are coordinated through the workstations. Computationally intensive tasks, such as real-time optimization (Appendix F) or model predictive control (Chapter 15), are implemented in a workstation.
3. ***Real-Time Clocks.*** Process control systems must respond to events in a timely manner and should have the capability of real-time control. Some DCSs are connected to atomic clock signals to maintain accuracy.
4. ***Operator Stations.*** Operator stations typically consist of color graphics monitors with special keyboards to perform dedicated functions. Operators supervise and control processes from these workstations. Operator stations may be connected directly to printers for alarm logging, printing reports, or process graphics.
5. ***Engineering Workstations.*** These are similar to operator stations but can also be used as programming terminals, that is, used to develop system software. This arrangement reduces compatibility problems between the development and application environments for the system software. Typically, users may also develop their own application programs on the engineering workstations.
6. ***Remote Control Units (RCUs).*** These components are used to implement basic control functions such as PID control. Some RCUs may be configured to acquire or supply set points to single-loop controllers. Radio telemetry systems (and/or wireless HART technology) or fiber optic cable for higher bandwidth may be installed to communicate with MUX units located at great distances.
7. ***Application Stations.*** These separate computers run application software such as databases, spreadsheets, financial software, and simulation software via an *OPC interface*. OPC is an acronym for object linking and embedding for process control, a software architecture based on standard interfaces. These stations can be used for e-mail and as web-servers, for remote diagnosis, configuration, and even for operation of devices that have an IP (Internet protocol) address. Applications stations can communicate with the main database contained in on-line mass storage systems (see Section 19.4).
8. ***Mass Storage Devices.*** Typically, hard disk drives are used to store active data, including on-line and historical databases and nonmemory resident programs. Memory resident programs are also stored to allow loading at system start-up.
9. ***Fieldbuses/Smart Devices.*** An increasing number of field-mounted devices are available that support digital communication of the process I/O in addition to, or in place of, the traditional 4–20 mA current signal. These devices have greater functionality, resulting in reduced setup time, improved control, combined functionality of separate devices, and control-valve diagnostic capabilities. Digital communication also allows the control system to become completely distributed where, for example, a PID control algorithm could reside in a valve positioner or in a sensor/transmitter. See Section 19.4.3 for more details.

19.4 SOFTWARE ORGANIZATION

In distributed control systems, computers and other components from a number of vendors may be part of the network. Consequently, software compatibility and portability is a major concern. Portable software and virtual (cloud-based) storage is used to ensure consistent computer performance, avoid duplicating development efforts, reduce IT management overhead, and provide improved cybersecurity. Object-oriented programming techniques are employed to minimize customization for different computers and applications. Internet connections allow monitoring of system health and maintenance.

For a DCS to function properly, a concerted effort of many software tasks is required (Kopetz, 2011; Lipták, 2005). The core of each network node must be a reliable real-time multitasking operating system that is divided functionally into different tasks—that is, communication between DCS nodes, data acquisition and control, operator interface, process control software, system utility libraries, and report generation. All these tasks are interdependent and share process data stored in a database. Because all network nodes must possess communication capability while maintaining a local database, these tasks may be distributed in different nodes.

19.4.1 Distributed Database and the Database Manager

A database is a centralized repository for data storage that reduces data redundancy at different network nodes. Multiple databases can be accessed through the network, although some local databases may not be accessible. Central database server systems are set up based on equipment storage capabilities and cost. Detailed discussions of database and transaction processing may be found in Lewis et al. (2006) and Garcia-Molina et al. (2008).

The database manager is a set of system utility programs that acts as the gatekeeper to the various databases on the network. All functions that retrieve or modify data must first submit a request to the manager. Information required to access the database includes the tag name of the database entity (often referred to as a point), the attributes to be accessed, and the values if they are to be modified. The database manager maintains the integrity of the databases by executing a request only when it is not processing other conflicting requests. Although a number of tasks may simultaneously read the same data, simultaneous read/write of the same data item is not permitted.

19.4.2 Internodal Communications

In order for a group of computers to become a network, intercomputer communication is required. The introduction of standardized communication protocols has decreased capital cost. Most current DCS network protocol designs are based on the ISO-OSI[2] seven-layer model with physical, data link, network, transport, session, presentation, and application layers (Thompson and Shaw, 2016).

An effort in standardizing communication protocols for plant automation was initiated by General Motors in the early 1980s. This work culminated in the Manufacturing Automation Protocol (MAP), which adopted the ISO-OSI standards as its basis. MAP specifies a broadband backbone local area network (LAN) that incorporates a selection of existing standard protocols suitable for discrete component manufacturing. MAP was intended to address the integration of DCSs used in process control. Subsequently, TCP/IP (transmission control protocol/Internet protocol) was adopted for communication between nodes that have different operating systems. The most widely used LAN technology at levels two and three is *Ethernet*.

Communication programs also act as links to the database manager. When data are requested from a remote node, the database manager transfers the request to the remote node database manager via the communication programs. The remote node communication programs then relay the request to the resident database manager and return the requested data. The remote database access and the existence of communications equipment and software are transparent to the user.

19.4.3 Digital Field Communications and Fieldbus

Microprocessor-based equipment, such as smart instruments and single-loop controllers with digital communications capability, are now used extensively in process plants. A *fieldbus*, which is a low-cost protocol, is necessary to perform efficient communication between the DCS and devices that may be obtained from different vendors. Figure 19.4 illustrates a LAN-based DCS with fieldbuses and smart devices connected to a data highway.

Presently, there are several regional and industry-based fieldbus standards, including the French standard (FIP), the German standard (Profibus), and proprietary standards by DCS vendors, generally in the United States, led by the Fieldbus Foundation, a not-for-profit corporation (Mehta and Reddy, 2016; Thomesse, 1999). International standards organizations have adopted all of these fieldbus standards rather than a single unifying standard. However, there will likely be further developments in fieldbus standards in the future. A benefit of standardizing the fieldbus is that it has encouraged third-party traditional equipment manufacturers to enter the smart equipment market, resulting in increased competition and improved equipment quality.

Several manufacturers have made available *fieldbus controllers* that reside in the final control element or measurement transmitter. A suitable communications modem is present in the device to interface with a proprietary PC-based, or hybrid analog/digital bus network. At the present time, fieldbus controllers are single-loop controllers containing 8- and 16-bit microprocessors that support the basic PID control algorithm as well as other functionalities. Case studies in implementing such digital systems have shown significant reductions in cost of installation (mostly cabling and connections) vs. traditional analog field communication.

[2]Abbreviated from *International Organization for Standardization-Open System Interconnection.*

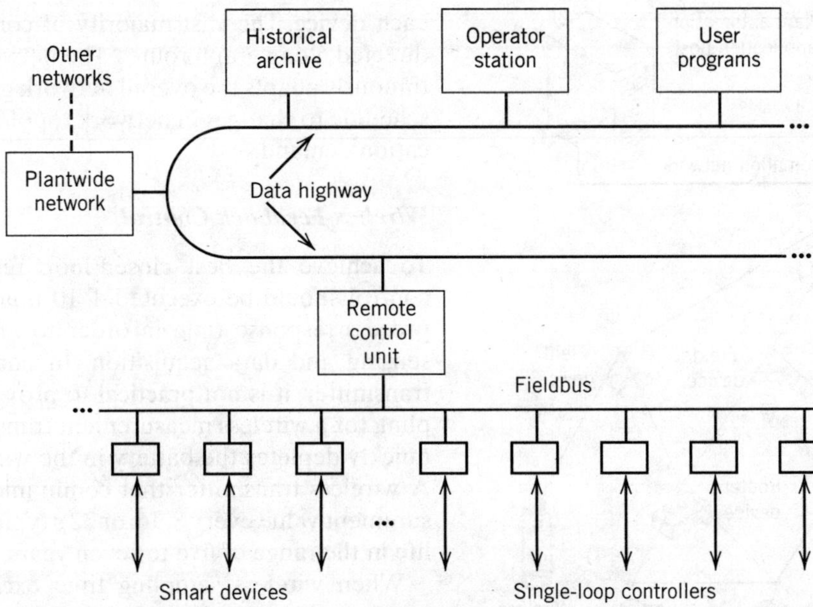

Figure 19.4 A DCS using a broadband (high-bandwidth) data highway and fieldbus connected to a single remote control unit that operates smart devices and single-loop controllers.

An example of a hybrid analog/digital protocol that is open (not proprietary) and used by several vendors is the *HART* (Highway Addressable Remote Transducer) protocol. Digital communications utilize the same two wires to provide the 4–20 mA process control signal without disrupting the actual process signal. This is done by superimposing a frequency-dependent sinusoid ranging from –0.5 to +0.5 mA to represent a digital signal.

A general movement has also begun in the direction of using the high-speed ethernet standard (100 Mbit/s or higher), allowing data transfer by TCP/IP that is used pervasively in computer networking. This allows any smart device to communicate directly with others in the network or to be queried by the operator regarding its status and settings. However, considerable changes in the ethernet standard will be required to make it suitable for process control applications, which provides a more challenging environment than corporate data networks.

Because the HART protocol is widely used due to its similarity to the traditional 4- to 20-mA field signaling, it represents a safe, controlled transition to wireless field communications as an alternative to fieldbus. The HART protocol is principally a master/slave protocol, which means that a field device (slave) speaks only when requested by a master device. An optional communication mode, "burst mode," allows a HART slave device to continuously broadcast updates without stimulus requests from the master device, which is an important attribute for wireless data transmission.

19.4.4 Wireless Technology in the Process Industries

Since 2010, the commercial application of wireless field devices in the process industries has grown significantly. Almost all of these field devices are based on the IEC 62591 WirelessHART international standard, which evolved from the 4–20 mA based protocol for wired technology. The advantage of a wireless field network is the potentially reduced cost versus a wired installation. Hurdles for wireless transmissions include security from non-network sources, transmission reliability in the plant environment, limited bus speed, battery life, and the resistance of the process industry to change. Both point-to-point and mesh architectures have been commercialized at the device level. *Mesh architectures* utilize the other transmitting devices in the area to receive and then pass on any data transmission, thus rerouting communications around sources of interference. Multiple frequencies within the radio band are utilized to transmit data. Figure 19.5 illustrates the basic network device types for a mesh architecture. The path of an RF signal from a device to the gateway can approach 230 m but is lower when obstructions are present. To obtain a clean line of sight, the antenna for the device should be mounted above obstructions when the angle of the terrain change is less than 5°.

The power modules used in wireless devices are designed to be periodically replaced, and field replacements can be performed without the need to remove the transmitter from the process. Lithium thionyl chloride

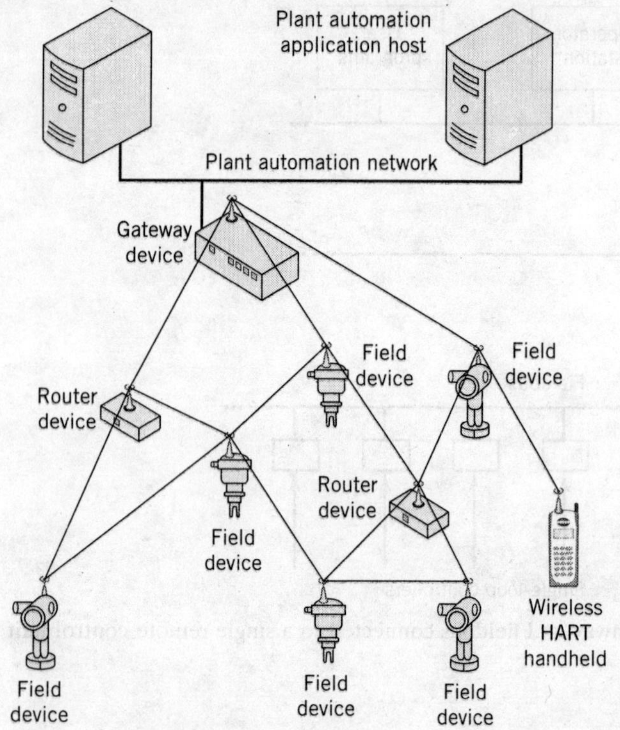

Figure 19.5 Example of a WirelessHART network (Mesh Architecture).

battery cells are used by many wireless devices to provide high energy density, long shelf life, and a wide working temperature range. WirelessHART adapters use power drawn from the wired transmitter current loop and can be installed on a traditional wired transmitter for wireless access to the measured values of devices used for monitoring or control.

A *Field Communicator* is used to configure HART devices. WirelessHART devices are preconfigured using the same tools and methods used on wired HART devices. However, WirelessHART devices require a Network ID to join the correct wireless network, because a large installation can span multiple process areas. Once the device has identified nearby neighbors, it uses the Join Key to encrypt the join messages exchanged with the network manager. The network manager, in turn, authenticates the device and incorporates the device into the network. Several protocols including Modbus, OPC, and HART-IP are supported. Diagnostics in wireless products include information about the device and the equipment that is being monitored.

WirelessHART communications are scheduled with precise time synchronization, using an approach referred to as Time Division Multiple Access (TDMA). Scheduling is performed by a centralized network manager that uses overall network routing information, in combination with communication requirements of each device. The vast majority of communications are directed along graph routes. The network manager continuously adapts the overall network graph and network schedule to changes in network topology and communication demand.

Wireless Feedback Control

To achieve the best closed-loop response, feedback control should be executed 4–10 times faster than the process response time in order to minimize delays in sensing and data acquisition. In contrast to a wired transmitter, it is not practical to provide frequent sampling for a wireless measurement transmitter, because it quickly depletes the battery in the wireless transmitter. A wireless transmitter that communicates a new measurement value every 8, 16, or 32 s typically has a battery life in the range of five to seven years.

When wireless sampling time exceeds the process response time, the closed-loop response to disturbances can become oscillatory. For applications such as temperature control and level control that are characterized by slow process dynamics, it is possible to use wireless transmitter update rates that are four times faster than the process response time and still achieve a satisfactory battery life. In faster loops, such update rates are impractical. However, PID control can be reformulated to provide satisfactory control when a measurement is not updated frequently. A PID implementation known as PIDPlus (Blevins et al., 2015) automatically compensates for slow measurement sampling rates and the possibility of a dropped measurement value.

When the PIDPlus algorithm is used with a wireless transmitter in a control application, the performance can be comparable to that achieved using a wired transmitter. An example of PIDPlus using a wireless transmitter compared to a standard PI controller with a wired measurement value is shown below (Blevins et al., 2015), based on simulation results for a disturbance change with a first-order-plus-time-delay process.

Communications/ Controller	Number of Communications	Integral Absolute Error (IAE)
Wired Periodic/ Standard PID	692	123
Wireless/PID Plus	25	159

In many common applications such as flow or pressure control of a liquid or gas stream, the PIDPlus can implement control using a slower sampling rate such as 8 or 16 s.

Wireless Final Control Elements

Recently, manufacturers have developed and introduced wireless actuators for on/off valves. Future technological developments should lead to wireless throttling valves with a wireless transmitter to implement closed-loop control. The feedback loop may be modified to minimize the number of changes made in the valve position when control is implemented using a wireless valve and a wireless transmitter. To minimize the power consumed by the valve positioner, the calculated PID output is transmitted to the wireless valve only if the error from target exceeds some threshold. In simulation tests, the number of changes in valve position using this strategy was reduced by a factor of 70 vs. that for 0.1-s loop execution, which cut total valve movement by over 50%. Minimizing the number of valve position changes had no impact on loop stability with less than a 50% increase in integral absolute error (Blevins et al., 2015).

Wireless Control Foundations by Blevins et al. (2015) details the recent technical innovations that address feedback control using wireless measurements and final control elements. Wireless measurements may also be utilized in model predictive control (MPC). This reference and associated website includes simulations that explore key concepts associated with wireless control.

19.4.5 Data Acquisition

The data acquisition software is utilized to coordinate signal converters and MUXs discussed in Section 19.2. Process data are preprocessed before being transferred to databases for storage and retrieval. Alarm condition screening is performed on process data on a periodic basis. A number of data fields and parameters are required for data acquisition and utilization in process control. A *tag name* is an alphanumeric string that uniquely identifies a process I/O point. Most commercial systems use some numeric sequences to associate database points to signal converters and MUXs. Process system and smart devices can frequently monitor the quality of each point and direct it to appropriate operator and control strategies. Lists of tag names and parameters are stored in EEPROM or Flash ROM to prevent loss due to system failure.

Most DCSs provide a pair of alarm bits associated with the instrument limits. For an instrument output signal, the limits prevent transmitting a value that is outside of the specified ranges. If an input value is outside the limits, an alarm action is taken (see Appendix E).

19.4.6 Process Control Languages

Originally, software for process control utilized high-level programming languages such as FORTRAN and BASIC. Some companies have incorporated libraries of software routines for these languages, but others have developed specialty languages characterized by natural language statements. The most widely adopted user-friendly approach is the fill-in-the-forms or table-driven *process control languages* (PCLs). Typical PCLs include function block diagrams, ladder logic, and programmable logic. The core of these languages is a number of basic function blocks or software modules, such as analog in, digital in, analog out, digital out, PID, summer and splitter. Using a module is analogous to calling a subroutine in conventional Fortran or C programs.

In general, each module contains one or more inputs and an output. The programming involves connecting outputs of function blocks to inputs of other blocks via the graphical user interface. Some modules may require additional parameters to direct module execution. Users are required to fill in templates to indicate the sources of input values, the destinations of output values, and the parameters for forms/tables prepared for the modules. The source and destination blanks may specify process I/O channels and tag names when appropriate. To connect modules, some systems require filling in the tag names of modules originating or receiving data. A completed control strategy resembles a data flow diagram such as the one shown in Fig. 19.6.

Many DCSs allow users to write custom code (much as with BASIC) and attach it to data points, so that the code is executed each time the point is scanned. The use of custom code allows many tasks to be performed that cannot be carried out by standard blocks.

All process control languages contain PID control blocks of different forms (Fig. 19.7, also see Chapter 7). Other categories of function blocks include

1. ***Logical operators.*** AND, OR, and exclusive OR (XOR) functions.
2. ***Calculations.*** Algebraic operations such as addition, multiplication, square root extraction, or special function evaluation.
3. ***Selectors.*** Min and max functions, transferring data in a selected input to the output or the input to a selected output.
4. ***Comparators.*** Comparison of two analog values and transmission of a binary signal to indicate whether one analog value exceeds the other.
5. ***Timers.*** Delayed activation of the output for a programmed duration after activation by the input signal.
6. ***Process Dynamics.*** Emulation of a first-order process lag (or lead) and time delay.

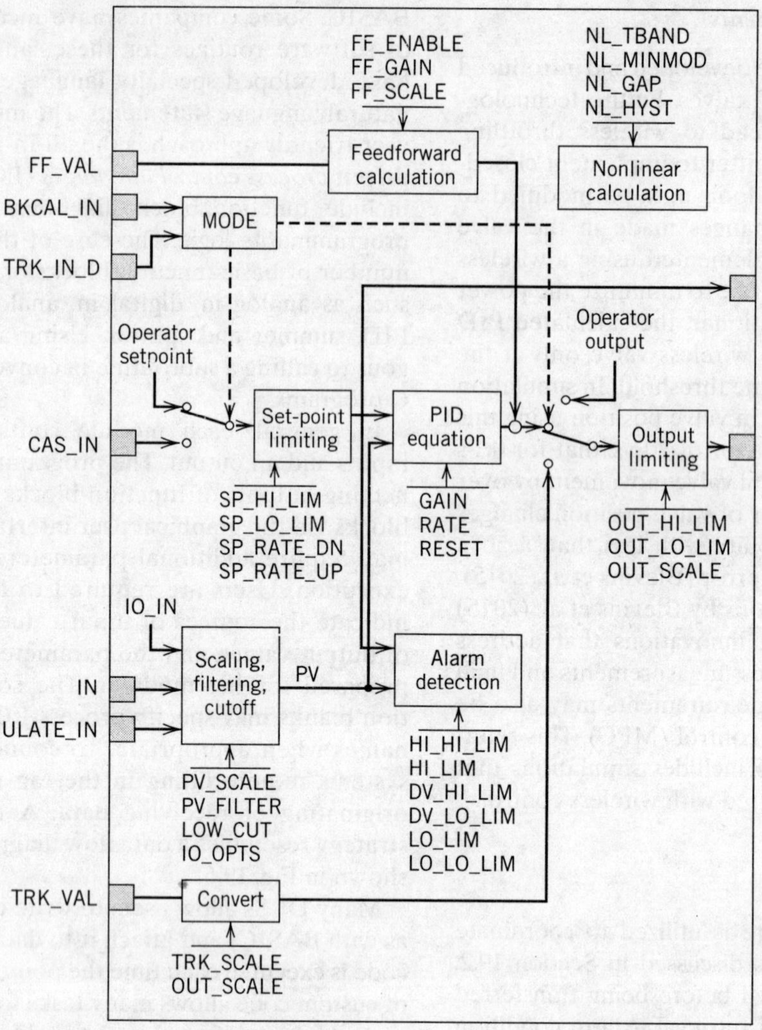

Figure 19.6 Function block representation (Courtesy Fisher-Rosemount Systems).

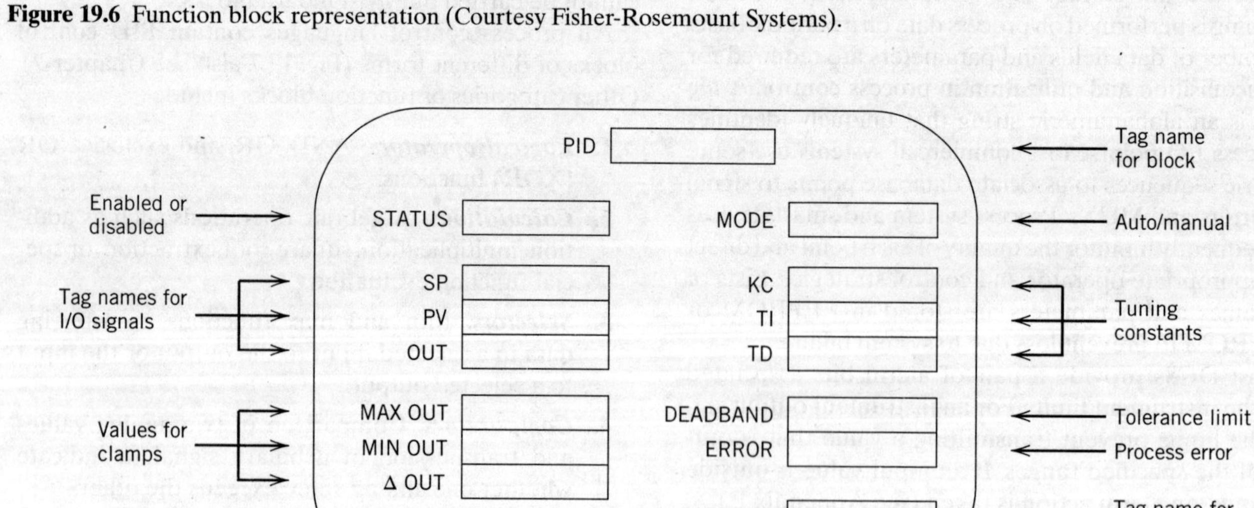

Figure 19.7 A display template for PID blocks.

Figure 19.8*a* shows the process diagram of a mixing process under analog ratio control. A flow controller (FC) is used to maintain ingredient A at the desired amount. An analog calculator (FY) computes the amount of ingredient B to be maintained (by a second FC), based on the desired amount of A and the ratio between the two ingredients. All hardware components beyond the process equipment can be replaced by PCL modules, as illustrated in Fig. 19.8*b*. The fill-in-the-forms process control languages owe their success, at least partly, to their resemblance to process schematics and control strategy diagrams.

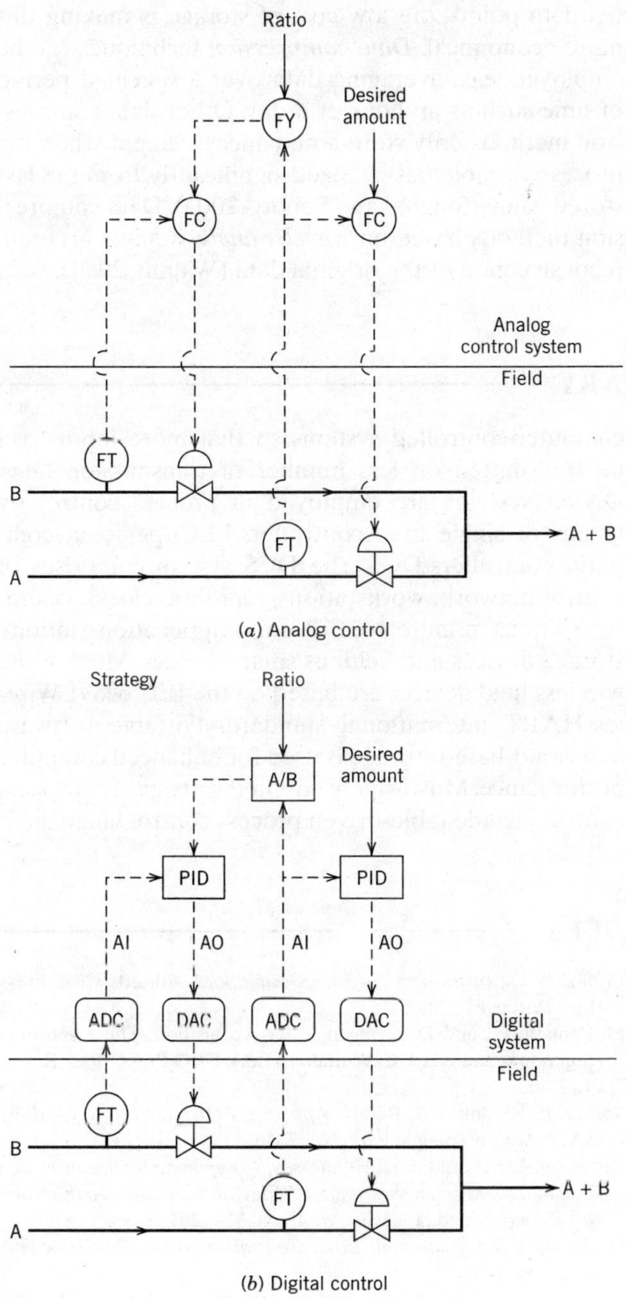

Figure 19.8 Analog and digital control of a mixing process.

19.4.7 Operator–Machine Interface

Most DCS tasks execute in a manner that is transparent to the operators or engineers. Most of the interface functions are integrated in the operator control stations equipped with graphics monitors. Through monitor displays, the operators observe the process operations and their status and issue commands via associated peripheral devices. Operator stations support some graphics building/generation capability, allowing system users to construct process graphics as needed. Most DCSs display color-coded device symbols to indicate device status and targets dynamically. There are many initiatives to increase operator effectiveness and safety through improved displays and color contrast, especially for actionable items.

The process displays serve as gateways to the databases and constantly poll the databases to retrieve process information for updates. A system may contain a function key to retrieve the active alarms log display directly, because fast response is especially critical under alarm conditions. Displays of how a process variable changes over a time horizon can be used to compare the magnitude of several data points dynamically and depict real-time trends to monitor process variations over time. The live trends show the values stored in the databases at the time the data are requested.

System event monitoring and operator action logging are important functions maintained by the operator–machine interface software (see Appendix E).

Practically all systems display active alarms in reverse chronological order. A list entry is removed when the alarm condition is resolved. Most systems provide visual and audible alarms that are activated when an alarm condition is detected. To deactivate, the operators must acknowledge all alarms. Alarms not acknowledged by the operators are typically displayed as blinking messages. Many systems maintain system alarms, system events, and operator action log.

Smart devices such as tablets and smart phones along with wireless networks and web technologies have enabled new methods for operators, engineers, and other personnel to interact with the process plant. The ability for an operator to move around the plant with a smart device and have pertinent process information available on the device has created what is often called the *mobile worker*. These wireless devices might be connected directly to the data highway or through the web server, which are both shown in Fig. 19.3. Security requirements may necessitate that such devices are "read-only." The ability to sense the location of a smart device either through GPS or near-field wireless communication can provide systems with the ability to automatically present the information that is most pertinent.

19.4.8 Data Historians

The use of on-line databases was briefly discussed in Section 19.4.1. Because on-line databases have limited capacity, the oldest data points are periodically transferred to a historical database. The data stored in a historical database are not normally accessed directly by other subsystems for process control and monitoring. These databases tend to be set up as relational databases, similar to corporate databases (Garcia-Molina et al., 2008; Silberschatz et al., 2005). Periodic reports and long-term trends are generated based on the historical (or archived) data. The reports are often used for long-term planning and system performance evaluations such as statistical process control. The trends may be used to detect process drifts or to compare process variations at different times.

Large industrial plants can have as many as 50,000 measured variables. Sampling periods for many process variables range from seconds to a few minutes. All the acquired data can be stored for relatively short periods of time (for example, weeks to months). While it is often not feasible to store years of historical data as individual data points, the low cost of storage is making this more economical. *Data compression* techniques can be employed, e.g., averaging data over a specified period of time such as an hour or a day. Other data compression methods only store a new measurement when the process variable has changed significantly from the last stored value (Singhal and Seborg, 2003). Data compression methods based on *wavelet analysis* allow accurate reconstruction of the original data (Walnut, 2004).

SUMMARY

In the past 40 years, technological developments in microelectronics and VLSI coupled with advancements in software technologies such as object-oriented programming have shifted the process control application from analog platform to digital platform. First generation distributed control systems were used in 1970s followed by distributed digital control network in the 80s. Presently, computer control systems are implemented through open system designs with global bus architecture and LANs. ADC is used as an interface between field instrument and digital computers. Similarly, DAC is used as an interface between digital computer and the final control element. Multiplexing is used in the computer-controlled systems so that more inputs can be transmitted on less number of transmission lines. Digital systems are employed in process control by means of single loop controllers/PLCs/personal computer controllers/DCS. The DCS system comprises of control network, workstations, real-time clocks, operator stations, remote control units, application stations, storage devices and fieldbus/smart devices. Most of the wireless field devices are based on the IEC 62591 WirelessHART international standard. Portable software and cloud-based storage is used for enhanced computer performance. Most widely adopted software for process control include table-driven process control languages.

REFERENCES

Auslander, D. M., and J. R. Ridgely, *Design and Implementation of Real-Time Software for the Control of Mechanical Systems*, Prentice Hall, Upper Saddle River, NJ, 2002.

Blevins, T., D. Chen, M. Nixon, and W. Wojsznis, *Wireless Control Foundation; Continuous and Discrete Control for the Process Industry*, ISA, Research Triangle Park, NC, 2015.

Garcia-Molina, H., J. D. Ullman, and J. D. Widom, *Database Systems: The Complete Book*, 2nd ed., Prentice Hall, Upper Saddle River, NJ, 2008.

Hughes, T. A., *Programmable Controllers*, 2nd ed., ISA, Research Triangle Park, NC, 2005.

Johnson, C. D., *Process Control Instrumentation Technology*, 8th ed.,Prentice Hall, Upper Saddle River, NJ, 2005.

Kopetz, H., *Real-Time Systems: Design Principles for Distributed Embedded Applications*, 2nd ed. Springer, Berlin, 2011.

Lewis, P. M., A. Bernstein, and M. Kifer, *Databases and Transaction Processing: An Application-Oriented Approach*, 2nd ed., Addison-Wesley, New York, 2006.

Lipták, B. G., *Instrument Engineers Handbook*, 4th ed., CRC Press, Boca Raton, FL, 2005.

McConnell, E., and D. Jernigan, Data Acquisition, *The Electronics Handbook*, 2nd ed., J. C. Whitaker (Ed.), CRC Press, Boca Raton, FL, 2005.

Mehta, B. R., and Y. J. Reddy, *Applying FOUNDATION Fieldbus*, ISA, Research Triangle Park, NC, 2016.

Patterson, D. A., and J. L. Hennessey, *Computer Organization and Design: The Morgan Kaufmann Series in Computer Architecture and Design*, 5th ed., Elsevier, Waltham, MA, 2013.

Petruzella, F., *Programmable Logic Controllers*, 4th ed., McGraw-Hill, New York, 2010.

Silberschatz, A., H. Korth, and S. Sudarshan, *Database Systems Concepts*, 5th ed., McGraw-Hill, New York, 2005.

Singhal, A., and D. E. Seborg, Data Compression Issues with Pattern Matching in Historical Data, *Proc. Amer. Control Conf.*, 3696 (2003).

Thomesse, J. P., Fieldbuses and Interoperability, *Control Engr. Practice*, **7**, 81 (1999).

Thompson, L. M., and T. Shaw, *Industrial Data Communications*, 5th ed. ISA, Research Triangle Park, NC, 2016.

Walnut, D. F., *An Introduction to Wavelet Analysis*, Birkhanser, Boston, MA, 2004.

EXERCISES

19.1 A 5 bit digital to analog converter has output in the form of current. The output is 10 mA for a digital input of 10100. For the same DAC:

(i) What will be the output current if the digital input is 11011?

(ii) If an output current of 12 mA is required, what will be the digital input?

19.2 A temperature sensor is installed in a digital temperature control loop which has a range of 250 to 550 K. The sensor is connected to a transducer which linearly converts the temperature in the range of 4 to 20 mA. The current value is digitized by an 8 bit ADC. Determine the resolution of the ADC.

19.3 One type of A/D converter is shown in Figure 19.2 (Dual slope A/D converter) and its working principle is explained in 19.2.3. Prepare a list of a few other type of A/D converters. Also mention the advantages and disadvantages of each type.

19.4 A 4×1 multiplexer with two selectors is shown in the figure below. Determine the output Y and construct the binary logic diagram.

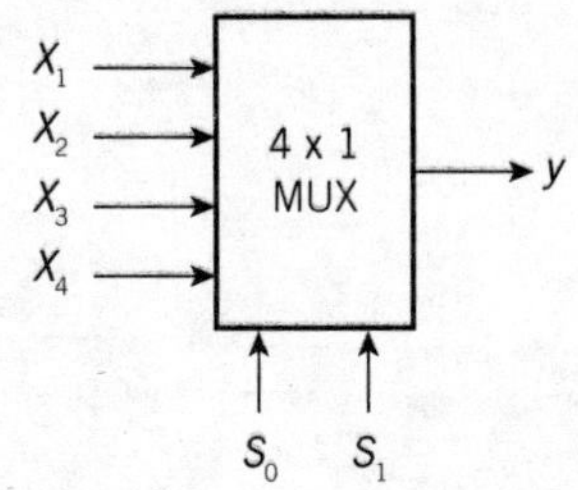

19.5 With the help of a schematic diagram, explain the functions of different components of a distributed control system.

MULTIPLE CHOICE QUESTIONS

1. A 5 bit DAC produces an output of 1 V corresponding to a digital input of 01010. What is the largest output value possible from this DAC?

(a) 5 V **(b)** 3.1 V **(c)** 6.7 V **(d)** 8.5 V

2. A 8 bit digital to analog converter produces an output in the range of 0 to 9 V. If an output of 3.56 V is required, what is the digital input?

(a) 11000011 **(b)** 1000011
(c) 01100101 **(d)** 10101010

3. Which of the following A/D converters is the fastest?

(a) Dual slope integrator A/D converter
(b) Successive approximation A/D converter
(c) Delta-sigma A/D converter
(d) Flash A/D converter

4. In computer control systems, ADC stands for

(a) Advanced Digital Control.
(b) Analog to Digital Control.
(c) Advanced Digital Converter.
(d) Analog to Digital Converter.

5. A computer utilizing 8-bit unsigned integer representation can represent _______ different numbers.

(a) 16 **(b)** 256 **(c)** 32 **(d)** 64

6. An 8-bit signed integer has a range of

(a) –128 to 127 **(b)** 0 to 256
(c) –127 to 128 **(d)** –64 to 63

7. A multiplexer has 3 selection lines. Number of inputs to the multiplexer are

(a) 3 **(b)** 6 **(c)** 8 **(d)** 9

8. HART stands for

(a) High Analog Residual Transformer.
(b) Highway Addressable Remote Transducer.
(c) High Addressable Remote Terminal.
(d) None of these.

9. HART protocol is widely used for

(a) digital field communications.
(b) database management.
(c) internodal communication.
(d) data acquisition.

ANSWER KEY

1. (b) **2.** (c) **3.** (d) **4.** (d) **5.** (b) **6.** (a) **7.** (c)
8. (b) **9.** (a)

APPENDIX A

Review of Thermodynamic Concepts for Conservation Equations

APPENDIX CONTENTS

The general energy balances in Eqs. 2-10 and 2-11 provide a useful starting point for the development of dynamic models. However, expressions for U_{int} and $\hat{H}$ (or $\tilde{H}$) are required and can be derived from thermodynamic principles. In this appendix, we review fundamental thermodynamic concepts, first for single components and then for multicomponent mixtures. Additional background information is available in thermodynamics textbooks such as Sandler (2006).

A.1 SINGLE-COMPONENT SYSTEMS

Consider a fluid or a solid that consists of a single component such as water or silicon. The enthalpy per unit mass, $\hat{H}$, depends on temperature and pressure. With a slight abuse of standard mathematical notation, we can write

$$\hat{H} = \hat{H}(T, P) \tag{A-1}$$

For differential changes in T and P,

$$d\hat{H} = \left(\frac{\partial \hat{H}}{\partial T}\right)_P dT + \left(\frac{\partial \hat{H}}{\partial P}\right)_T dP \tag{A-2}$$

By definition, the heat capacity at constant pressure, C_p, is defined as

$$Cp \triangleq \left(\frac{\partial \hat{H}}{\partial T}\right)_P \tag{A-3}$$

Substituting Eq. A-3 into Eq. A-2 gives

$$d\hat{H} = C_p dT + \left(\frac{\partial \hat{H}}{\partial P}\right)_T dP \tag{A-4}$$

For liquids and solids, $(\partial \hat{H}/\partial P)_T \approx 0$ and $\hat{U}_{\text{int}} \approx \hat{H}$. Consequently, we can write

$$d\hat{U}_{\text{int}} \approx d\hat{H} = C_p dT \textit{ (for liquids and solids)} \tag{A-5}$$

and

$$\frac{d\hat{U}_{\text{int}}}{dt} \approx \frac{d\hat{H}}{dt} = C_p \frac{dT}{dt} \textit{ (for liquids and solids)} \tag{A-6}$$

The total internal energy of the system, U_{int}, can be expressed in terms of the internal energy per unit mass, $\hat{U}_{\text{int}}$, multiplied by the mass in the system, ρV,

$$U_{\text{int}} = \rho V \hat{U}_{\text{int}} \tag{A-7}$$

where ρ and V are the density and volume of the liquid, respectively. Differentiating Eq. A-7 with respect to time gives

$$\frac{dU_{\text{int}}}{dt} = \frac{d(\rho V \hat{U}_{\text{int}})}{dt} \tag{A-8}$$

Suppose that ρ and V are constant. Then substituting Eq. A-6 into Eq. A-8 gives

$$\frac{dU_{\text{int}}}{dt} = \rho V \frac{d\hat{U}_{\text{int}}}{dt} = \rho V C_p \frac{dT}{dt}$$

$$\textit{(for liquids with constant } \rho \textit{ and } V) \tag{A-9}$$

For some modeling activities, it is more convenient to express U_{int} in terms of molar quantities,

$$U_{\text{int}} = n\tilde{U}_{\text{int}} \tag{A-10}$$

where n is the total number of moles. Then equations analogous to Eqs. A-8 and A-9 can be derived. Equations A-8 and A-9 provide general expressions for the accumulation term in the energy balance of Eq. 2-10.

For ideal gases, $\hat{H}$ and $\hat{U}_{\text{int}}$ are functions only of temperature, and the following relationships hold:

$$d\hat{H} = C_p dT \ (\textit{for ideal gases}) \tag{A-11}$$

$$\hat{H} = \hat{U}_{\text{int}} + RT \ (\textit{for ideal gases}) \tag{A-12}$$

For nonideal (*real*) gases, $\hat{H}$ and $\hat{U}_{\text{int}}$ depend on pressure, as well as temperature, as shown in Eq. A-4. Numerical values of $\hat{H}$ and $\hat{U}_{\text{int}}$ can be obtained from tables of thermodynamic data or relations.

Consider a liquid or ideal gas at a temperature T. Integrating Eq. A-11 from a reference temperature T_{ref} to T provides an expression for the difference between $\hat{H}$ and $\hat{H}_{\text{ref}}$, the value of $\hat{H}$ at T_{ref}:

$$\hat{H} - \hat{H}_{\text{ref}} = C(T - T_{\text{ref}}) \tag{A-13}$$

In Eq. A-13, C is the mean heat capacity over the temperature range from T_{ref} to T. Without loss of generality, we assume that $\hat{H}_{\text{ref}} = 0$.

The value of T_{ref} for enthalpy calculations can be selected arbitrarily. For example, the triple point of water is used as the reference point for the steam tables, while 25 C is a typical choice for physical property tables. For process control calculations, it is often convenient to set $T_{\text{ref}} = 0$ or to choose T_{ref} to be an inlet temperature or an initial temperature.

A.2 MULTICOMPONENT SYSTEMS

A key issue for multicomponent systems is: how are the properties of the mixture related to pure component properties? Consider a system that consists of k components. Because the enthalpy depends on composition as well as temperature and pressure, the enthalpy per unit mole of the system, $\tilde{H}$, can be expressed as

$$\tilde{H} = \tilde{H}(T, P, \tilde{x}) \tag{A-14}$$

where $\tilde{x}$ denotes chemical composition. In general,

$$\tilde{H}(T, P, \tilde{x}) = \sum_{i=1}^{k} x_i \bar{H}_i(T, P, \tilde{x}) \tag{A-15}$$

where $\tilde{x}_i$ is the mole fraction of component i and $\bar{H}_i$ is the *partial molar enthalpy of component i*:

$$\bar{H}_i(T, P, \tilde{x}) \triangleq \left(\frac{\partial(n\tilde{H}(T, P, \tilde{x}))}{\partial n_i} \right)_{T, P, n_j \neq n_i} \tag{A-16}$$

In Eq. A-16, n_i is the number of moles of component i and n is the total number of moles, $n \triangleq \sum n_i$.

An important simplification occurs if the mixture can be considered to be an *ideal solution*. For an ideal solution, $\bar{H}_i(T, P, \tilde{x}) = \tilde{H}_i(T, P)$, where $\tilde{H}_i(T, P)$, is the molar enthalpy of pure component i. The mixture can be analyzed as a set of individual components, and Eq. A-15 can be written as

$$\tilde{H}(T, P, x) = \sum_{i=1}^{k} \tilde{x}_i \tilde{H}_i(T, P) \ (\textit{for ideal solutions}) \tag{A-17}$$

Similarly, it can be shown that the enthalpy per unit mass of an ideal solution, $\hat{H}(T, P, x)$, can be expressed as

$$\hat{H}(T, P, x) = \sum_{i=1}^{k} x_i \hat{H}_i(T, P) \ (\textit{for ideal solutions}) \tag{A-18}$$

where x denotes the composition in mass units and x_i is the mass fraction of component i.

Equations A-17 and A-18 are very useful in developing dynamic models from the general energy balances in Eqs. 2-10 and 2-11. Similar expressions can be derived for $\hat{U}_{\text{int}}$ and $\tilde{U}_{\text{int}}$. Then the total internal energy of the system, U_{int}, can be expressed in terms of $\hat{U}_{\text{int}}$ and $\tilde{U}_{\text{int}}$ according to Eqs. A-7 and A-10 where ρ and V are now the density and volume of the mixture, respectively, and n is the total number of moles.

REFERENCE

Sandler, S. I., *Chemical, Biochemical, and Engineering Thermodynamics*, 4th ed., John Wiley and Sons, Hoboken, NJ, 2006.

APPENDIX B

Control Simulation Software

APPENDIX CONTENTS

MATLAB is a general-purpose software package for mathematical computations, analysis, and visualization available from The Mathworks (2010). This Appendix introduces the basic functionality of the MATLAB software and shows how to solve simple algebraic equations and ordinary differential equations (ODEs). Vector and matrix manipulations are considered first, and then solving simple linear algebraic equations is demonstrated using MATLAB. The basics of functions and scripts are presented next, and use of the ODE integration function, *ode45*, is described. Subsequent sections introduce the graphical modeling tools, Simulink and Lab VIEW, and their usage for computing responses for open-loop and closed-loop block diagrams. For more details on MATLAB usage, see Palma (2010), Bequette (1998), Doyle III et al. (2000), and Ogata (2007).

B.1 MATLAB OPERATIONS AND EQUATION SOLVING

In MATLAB statements, square brackets denote vectors and matrices. Elements in a row vector are separated by commas or spaces. For example, the row vector $\boldsymbol{\upsilon} = (1, 2, 3)$ can be represented by

$$\upsilon = [1\ 2\ 3] \text{ or } \upsilon = [1, 2, 3]$$

The elements of a column vector are separated by semicolons. Thus, the column vector, $\boldsymbol{w} = col\,[4, 5, 6]$ is represented as $w = col\,[4; 5; 6]$. A matrix $\boldsymbol{M}$,

$$\boldsymbol{M} = \begin{bmatrix} 3 & 7 & 9 \\ 2 & 6 & 8 \\ 1 & 0 & 4 \end{bmatrix}$$

has the MATLAB representation:

$$\text{M} = [3\ 7\ 9; 2\ 6\ 8; 1\ 0\ 4]$$

Note that MATLAB variables cannot be **boldface** or *italicized*. Similarly, subscripts, superscripts, and other accent marks are not allowed. Also, MATLAB is case-sensitive.

B.1.1 Matrix Operations

The transpose of a matrix M is calculated using the command M'. The inverse of a matrix M is calculated as inv(M). In MATLAB, the multiplication of matrices **A** and **B** is denoted by $A * B$, while their addition and subtraction are denoted by $A + B$ and $A - B$, respectively. Commands for element-by-element multiplication and division are also available. For more functions and help on any MATLAB operation, type *help function_name* or *doc function_name*.

EXAMPLE B.1

Consider the following matrices:

$$A=\begin{bmatrix}1 & -3\\ 0 & 2\end{bmatrix} \quad B=\begin{bmatrix}5 & 2\\ -1 & 4\end{bmatrix}$$

$$C=\begin{bmatrix}1 & 2\\ -2 & -4\end{bmatrix} \quad D=\begin{bmatrix}2 & 3\\ 1 & 4\end{bmatrix}$$

Calculate the following:

(a) AB **(b)** AB^T **(c)** A^{-1} **(d)** DCD^T

(e) C^{-1} **(f)** $(ADA^T)^{-1}$ **(g)** $BC - D^{-1}$

Solution

(a) $\begin{bmatrix}8 & -10\\ -2 & 8\end{bmatrix}$ **(b)** $\begin{bmatrix}-1 & -13\\ 4 & 8\end{bmatrix}$ **(c)** $\begin{bmatrix}1 & 1.5\\ 0 & 0.5\end{bmatrix}$

(d) $\begin{bmatrix}-32 & -36\\ -56 & -63\end{bmatrix}$ **(e)** $\begin{bmatrix}\text{Inf} & \text{Inf}\\ \text{Inf} & \text{Inf}\end{bmatrix}$

C is a singular matrix (not invertible).

(f) $\begin{bmatrix}0.8 & 0.9\\ 1.1 & 1.3\end{bmatrix}$ **(g)** $\begin{bmatrix}-0.2 & 2.6\\ -8.8 & -18.4\end{bmatrix}$

Other matrix operations in MATLAB include:

- Eigenvalues and eigenvectors: *eig*
- Singular value decomposition: *svd*
- Pseudoinverse: *pinv*

B.1.2 Solution of Algebraic Linear or Nonlinear Equations

The solution to a set of linear algebraic equations, $\boldsymbol{Mx} = \boldsymbol{b}$, is given by $\boldsymbol{x} = \boldsymbol{M}^{-1}\boldsymbol{b}$. The MATLAB solution can be written as either x = inv(M)*b or x = M\b, where the backslash operator (\) is used as a shortcut for the solution. The solution to a set of nonlinear algebraic equations can be obtained using the MATLAB routine *fsolve.*

EXAMPLE B.2

Solve the equation $\boldsymbol{Mx} = \boldsymbol{b}$ for $\boldsymbol{x}$ using the values of $\boldsymbol{A}$, $\boldsymbol{B}$, and $\boldsymbol{D}$ from Example B.1.

(a) $\boldsymbol{M} = \boldsymbol{A}, b = [1; 2]$ **(b)** $\boldsymbol{M} = \boldsymbol{B}, b = [1; 2]$

(c) $\boldsymbol{M} = \boldsymbol{ADA^T}, b = [5; 1]$.

Solution

(a) $\boldsymbol{x} = \begin{bmatrix}4\\ 1\end{bmatrix}$ **(b)** $\boldsymbol{x} = \begin{bmatrix}0\\ 0.5\end{bmatrix}$ **(c)** $\boldsymbol{x} = \begin{bmatrix}4.9\\ 5.8\end{bmatrix}$

B.1.3 m-files

A MATLAB code, or *m-file*, is a collection of commands that are executed sequentially. Commands can be mathematical operations, function calls, flow control statements, and calls to the functions and scripts described in Section B.1.4. m-files are written using the MATLAB editor and have names such as *myfile.m*. They are executed from the MATLAB command window by typing the name of the m-file (without the .m). Saving an m-file will avoid many hours of retyping the same commands.

B.1.4 Functions and Scripts

There are two types of m-files, *functions* and *scripts*. A MATLAB function has variables that can be passed into and out of the function. Any other variables except for the global ones used inside the function are not saved in memory when the function is finished. Scripts, on the other hand, save all their variables in the MATLAB workspace. Functions and scripts have names like *myfunction.m*. The first line of a function must contain a function declaration, using the following format:

```
function [output1, output2,...]
= myfunction(input1, input2,...)
```

Commented lines immediately following the function declaration comprise the help file for the function. To obtain information on any function, simply type *help function*. Some MATLAB functions that are useful for process control include

- Unit step response of a transfer function: *step*
- Transfer function matrix derived from a state-space model: *ss2tf*
- State-space model derived from a transfer function matrix: *tf2ss*
- Transfer function multiplication: *series*
- Roots of the characteristic equation: *roots*
- Polynomial fitting of input–output data: *polyfit*
- Minimization of a multivariable function: *fminunc*
- Frequency response of a linear, time-invariant system: *bode*

B.1.5 Solving a System of Differential Equations

MATLAB has several built-in functions for solving systems of differential equations. The basic use of the standard integration algorithm, *ode45*, is described in this section. First, a function containing the differential equations to be integrated must be created. This function must have at least two arguments, *t* and the state vector, *y* (see Section 4.4). The function returns a column vector containing the derivatives evaluated at the current time. The commands inside the function calculate these derivatives. Additional arguments for the function are optional and can be used to pass parameter values from the script that calls the function.

Once the differential equation function is written and saved, a script (i.e., an m-file) containing the call to the integrator must be written. Here, parameter values, initial conditions, and options are specified, and the integration routine is called with the following command:

```
[t, y] = ode45(@myfunction, [ti tf], y0,
         options, P1, P2, P3,...);
```

where *myfunction* is the function containing the differential equations as described above, *ti* and *tf* are the initial and final integration times, and *y*0 is the vector of initial conditions. *Options* is a parameter vector for *ode45*. More information is available in the help files. Empty brackets [ ] can be used in place of the options argument. P1, P2,... are additional parameter values that are passed to *myfunction*.

B.1.6 Plots

It is easy to display results in MATLAB graphically. The *plot* function is used to create simple plots. The command syntax is

```
plot(x1, y1, format1, x2, y2, format2,...)
```

*x*1 and *x*2 are independent variables (usually time), and *y*1 and *y*2 are dependent variables. The format1 and format2 arguments are short combinations of characters containing the plot-formatting commands. For example, a blue solid line is "b-" (include the single quotes), a red dashed line is "r--", and a green dotted line is "g:" More formats can be viewed by typing *help plot*. Axis labels, title, and legend can be created using *xlabel*, *ylabel*, *title*, and *legend* commands. These and other properties can also be edited directly on the figure by selecting the arrow icon and double-clicking on an object contained in the figure.

Additional plot commands in MATLAB are *loglog* for log–log plots, and *semilogx* and *semilogy* for semilog plots, such as the Bode plots used in Chapter 10.

B.1.7 MATLAB Toolboxes

For advanced techniques in modeling, identification, and control, MATLAB has a variety of additional *toolboxes* that are licensed individually. Relevant toolboxes for process control include control system, fuzzy logic, system identification, model predictive control, neural networks, optimization, partial differential equations, robust control, and statistics.

B.2 COMPUTER SIMULATION WITH SIMULINK

Simulink, a companion software package to MATLAB, is an excellent interactive environment for simulation and analysis of control systems. Simulink enables the rapid creation of block diagrams based on transfer functions, followed by simulation for a given input signal. To facilitate model definitions, Simulink has a block diagram window in which blocks are created from the Simulink library browser and edited primarily by implementing drag-and-drop commands using a mouse. Blocks can be configured as additive transfer functions (see Fig. 4.1) or as multiplicative transfer functions (see Fig. 4.2), simply by connecting the output of one block to the input of another block. The coefficients of descending powers of *s* of the numerator and denominator polynomials in each block are entered as vectors. Time delays (called *transport delays* in Simulink) can be inserted in series with blocks for rational transfer functions. Input signals, called *sources*, include step, sinusoidal, and random inputs, but not the impulse function.[1] Clicking on the input block allows the user to specify the time when the

[1] To obtain the unit impulse response of a single transfer function, use the function impulse from the MATLAB command window.

input changes from an initial value of zero, and, for a step input, its initial and final values.

Consider a dynamic system consisting of a single output Y and two inputs U and D:

$$Y(s) = G_p(s)\,U(s) + G_d(s)D(s) \tag{B-1}$$

where

$$G_p(s) = \frac{2e^{-5s}}{50s^2 + 15s + 1} \quad \textit{(process transfer function)}$$

$$G_d(s) = \frac{0.3e^{-5s}}{15s + 1} \quad \textit{(disturbance transfer function)}$$

Figure B.1 shows the Simulink diagram for Eq. B-1 (transport delay 1 = 5 for both models). To generate a transient response, the simulation menu is selected to allow parameters for the simulation to be specified (start time, stop time, integration routine, maximum integration step size). Numerical values of time t are entered into the input-output data set via a *clock* block. After the simulation has been completed, the resulting data can be plotted (see Fig. B.2), manipulated, and analyzed from the MATLAB command window.

To simulate a closed-loop system, the procedure is somewhat more involved than for an open-loop system. Changing the previous example somewhat, start with Fig. B.1, but let $G_d = G_p$. Click on the connection between the U block and the *Process TF* block and delete it. Rename the U block, *Ysp*. This block will be used to produce a step change in the set point.

Place a copy of the *Sum* block to the right of *Ysp*. Double click above the *Sum* icon and label it *Sum1*. Open its dialog box and change the + + sequence to + –. The *top* left input will have a + located to the right of it, while the bottom input will have a – located above it. Connect the output of *Ysp* to the left input of *Sum1*. Also, connect the output from *Sum* to the bottom input of *Sum1*. This can be done by clicking on the bottom input of *Sum1* and dragging the arrow to the line following the output of *Sum*. The output of *Sum1* is the error between the set point *Ysp* and the controlled variable *Y*. This leads to the block diagram in Fig. B.3.

To insert the controller, right-click the *Simulink Extras* block. Click on the *Additional Linear* block; then select the *PID Controller* and drag it to the right of the *Sum1* block. Connect the output of *Sum1* to the input of *PID controller* and the output of *PID controller* to the input of *Process TF*. Double-click on *PID controller* and use the following controller settings: $K_c = 1.65$, $\tau_I = 7.12$, $\tau_D = 1.80$. Note that Simulink PID controller settings are entered in the expanded form (see Eq. 8-16) as *P*, *I*, and *D* where $P = K_c$, $I = K_c/\tau_I$, and $D = K_c/\tau_D$. Thus, the numerical values of *P*, *I*, and *D* should reflect these definitions. The model developed above represents the closed-loop system, as shown in Fig. B.4. Text can be

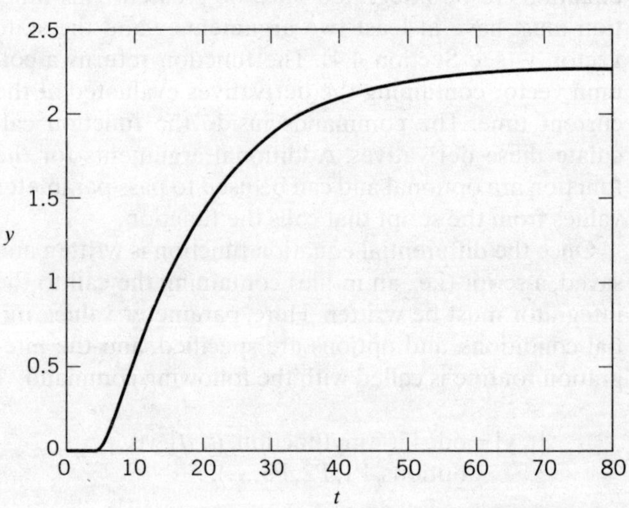

Figure B.2 Response for simultaneous unit step changes at $t = 0$ in U and D from the Simulink diagram in Fig. B.1.

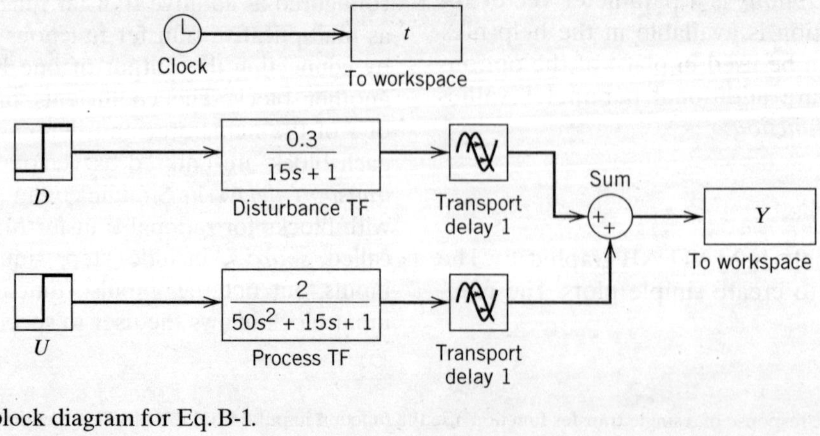

Figure B.1 Simulink block diagram for Eq. B-1.

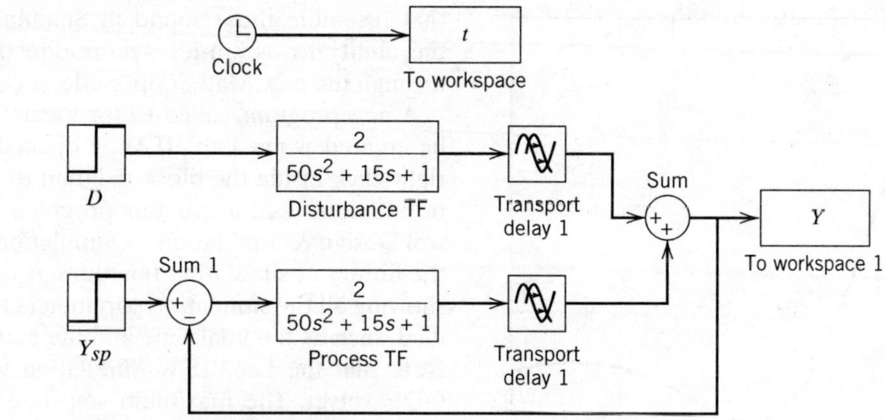

Figure B.3 Partially completed closed-loop diagram.

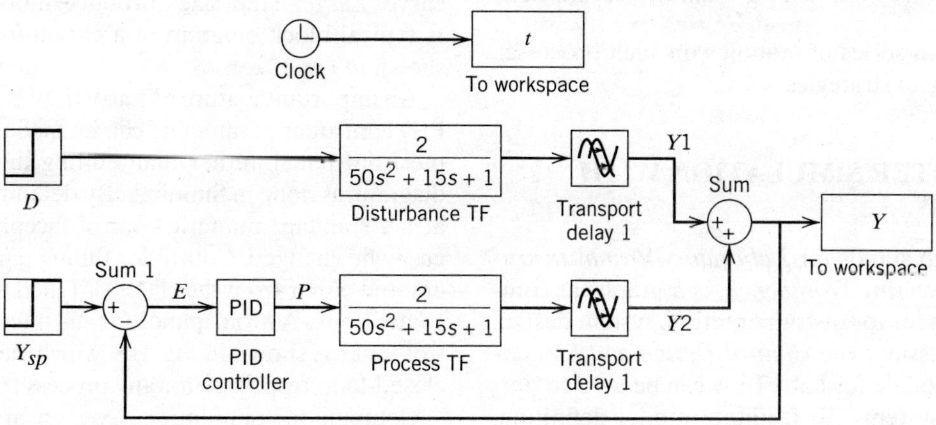

Figure B.4 Closed-loop diagram.

added to the block diagram simply by double-clicking on a point in the diagram and typing the desired words (see Fig. B.4).

Now the closed-loop response of the system can be simulated. Starting with the set-point response, click on block D and set the *Final value* to 0 so that no step disturbance will occur. Create a step in the set point by clicking on *Ysp* and setting the *Final value* to 1. In the *Simulation Parameters* menu, change the stop time to 50. Start the simulation by selecting *Start* from the *Simulation* menu. Because D (the disturbance) has been disabled, the resulting Y and t variables in the workspace will be for the unit set-point response, as shown in Fig. B.5.

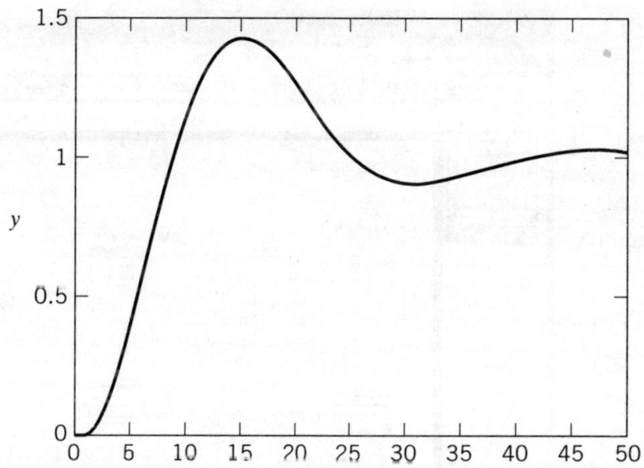

Figure B.5 Unit set-point response for the closed-loop system in Fig. B.4 with $P = 1.65$, $I = 0.23$, $D = 2.97$.

Now simulate the unit response to a unit step disturbance. Double-click on *Ysp* and set *Final value* to 0. Double-click on D and set *Final value* to 1. Again, select *Start* from the *Simulation* menu to begin the simulation. Type *plot (t, Y)* to view the response. Figure B.6 shows the resulting disturbance response plot after modifying some of the labels. Simulink can be used to simulate the effects of different control strategies with realistic multivariable process models such as a distillation column or a furnace. See Appendix E and Doyle III

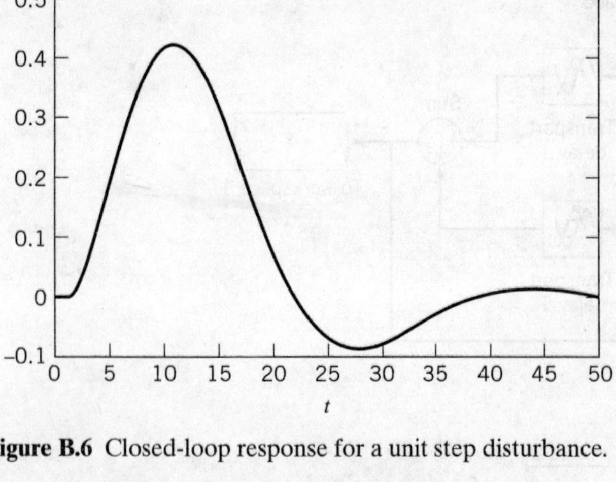

Figure B.6 Closed-loop response for a unit step disturbance.

et al. (2000) for a series of modules on such processes and various control strategies.

B.3 COMPUTER SIMULATION WITH LabVIEW

LabVIEW, which stands for *Laboratory Virtual Instrumentation Engineering Workbench*, is a graphical computing environment for instrumentation, system design, and signal processing. The *Control Design and Simulation (CDSim)* module for LabVIEW can be used to simulate dynamic systems. To facilitate model definition, CDSim has functions in the LabVIEW environment that resemble those found in Simulink. There is also the ability to use m-file syntax directly in LabVIEW through the new MathScript node.

A new program, called *VI* for *V*irtual *I*nstrument, can be created when LabVIEW is opened. Then one can right-click inside the block diagram to view the palette of functions used in creating programs. Select the Control Design & Simulation → Simulation palette to view the library of simulation functions. A dialog box opens, showing all the simulation parameters that can be modified, such as the final time and the maximum step size. Note that the LabVIEW Simulation loop includes an ODE solver. The maximum step size is used in LabVIEW for numerically integrating the ODE. A typical linear dynamic system is easy to integrate numerically, so a maximum step size of 1 usually result in a smooth curve. Larger step sizes produce more jagged curves. A typical block diagram of a closed-loop simulation is shown in Fig. B.7.

An important feature of LabVIEW is interactivity. The PID controller parameters can be made interactive from the front panel, rather than editing them on the block diagram, as done in Simulink. By default, Lab VIEW creates a standard numeric control faceplate, but this can easily be changed. Controller tuning parameters may be entered either from the slider or typed in to the numeric control box. A front panel for an interactive PID control tuner is shown in Fig. B.8, which can depict open or closed-loop responses for any process transfer function.

Companion simulation courseware for selected examples in this book using National Instruments VIs

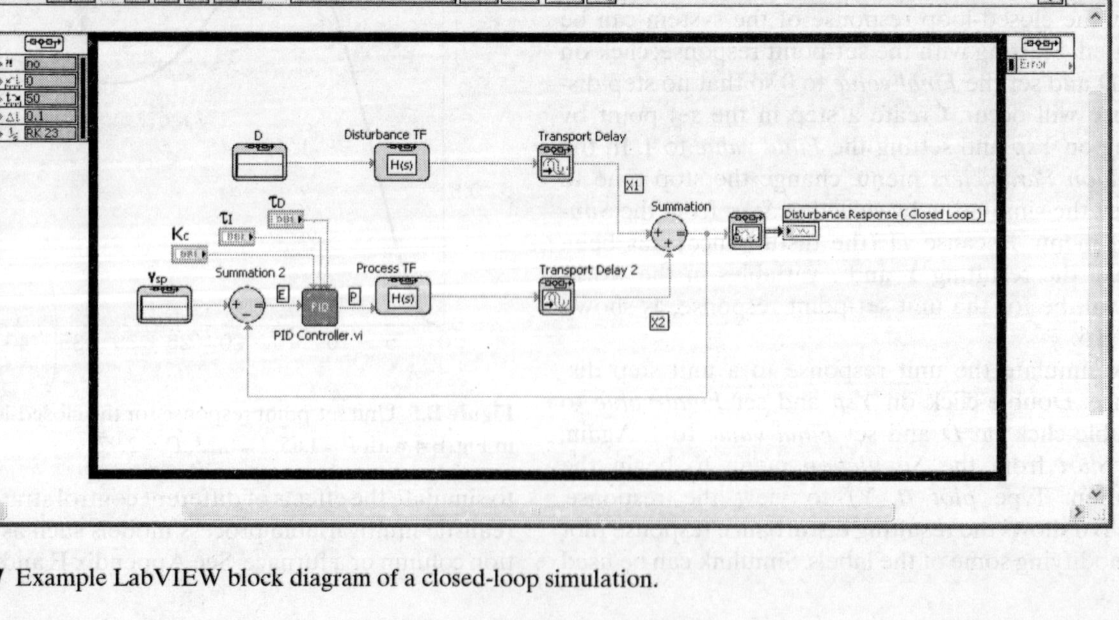

Figure B.7 Example LabVIEW block diagram of a closed-loop simulation.

Process PID Control Tuner

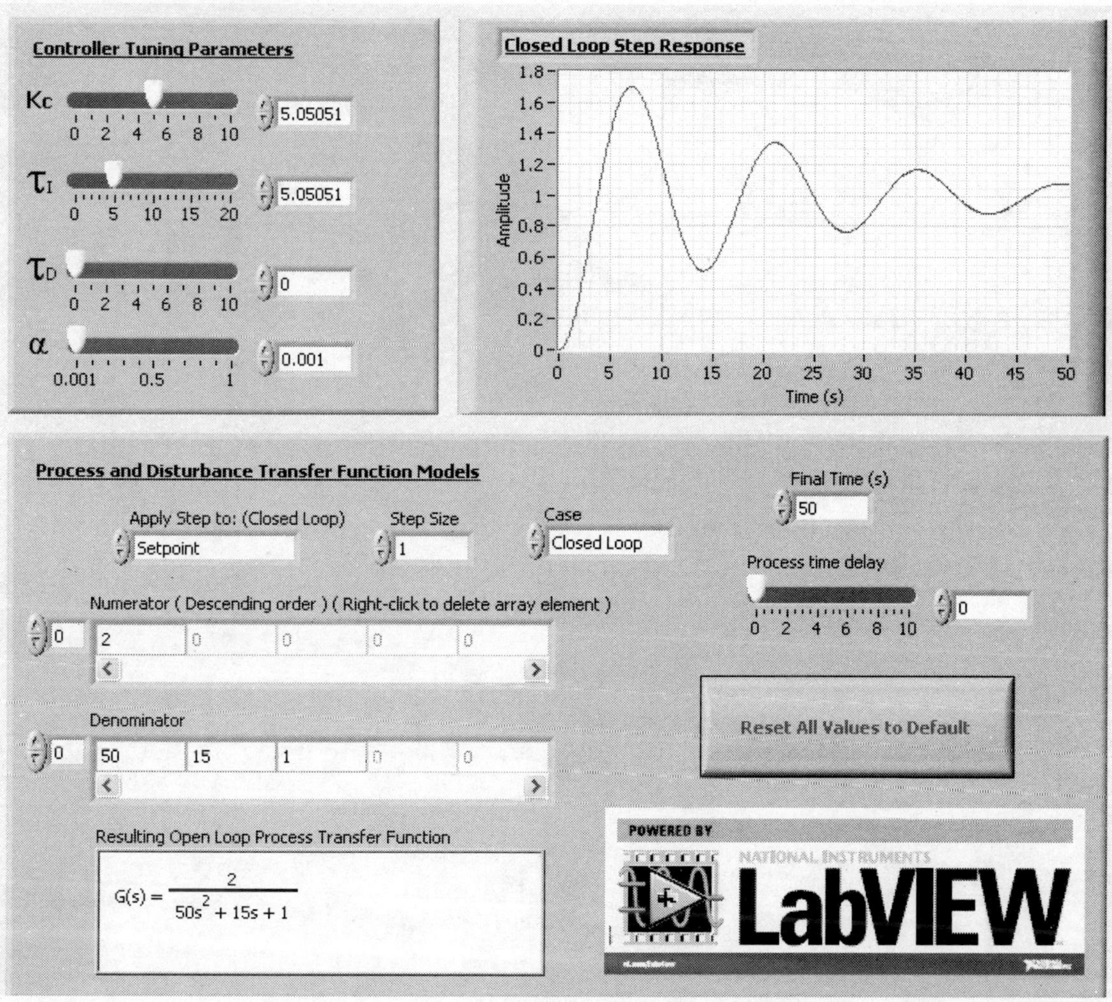

Figure B.8 The front panel of the Virtual Instruments (VI) for the interactive PID control tuner (www.che.utexas.edu/course/che360/Process_Tuner.html).

is found on the book's student companion website at www.wiley.com/college/seborg.

REFERENCES

Bequette, B. W., *Process Dynamics: Modeling, Analysis, and Simulation*, Prentice Hall, Upper Saddle River, NJ, 1998.

Doyle III, F. J., E. P. Gatzke, and R. J. Parker, *Process Control Modules: A Software Laboratory for Control Design*, Prentice Hall, Upper Saddle River, NJ, 2000.

Ogata, K., *MATLAB for Control Engineers*, Prentice Hall, England Cliffs, NJ, 2007.

Palm, W. J., *Introduction to MATLAB for Engineers*, 3rd ed., McGraw-Hill, New York, 2010.

The Mathworks, www.mathworks.com, 2010.

APPENDIX C

Process Control Modules

APPENDIX CONTENTS

C.1 INTRODUCTION

The Process Control Modules (PCM), originally developed at the University of Delaware, have been designed to address the key engineering educational challenge of realistic problem solving within the constraints of a typical lecture course in process dynamics and control (Doyle III et al., 1998; Doyle III, 2001). A listing of the modules appears in Table C.1. These modules have been updated and adapted by Dr. Eyal Dassau at the University of California, Santa Barbara, to be used in conjunction with the third edition of *Process Dynamics and Control*. The primary objectives in creating these MATLAB® modules were to develop the following:

- Realistic computer simulation case studies, based on physical properties that exhibited nonlinear, high order dynamic behavior in a rapid simulation environment
- A convenient graphical interface for students that allowed real-time interaction with the evolving virtual experiment

Table C.1 Organization of Process Control Modules (PCM)

Module	Modes						
Furnace	Operator Interface	PID	Feedforward	Multivariable	MPC		
Distillation Column	Operator Interface	PID	Feedforward	Multivariable	Decoupling	MPC	
Bioreactor	Operator Interface	PID	Feedforward	Multivariable	Decoupling		
Four Tanks	Operator Interface	PID	Feedforward	Multivariable	MPC		
Fermentor	Operator Interface	PID					
Diabetes	Operator Interface	PID	MPC				
First and Second Order Systems	First Order System	Second Order System	System Identification #1	System Identification #2			
Discrete	Aliasing	Model ID	PID-Furnace	PID-Column	PID-Four Tanks	IMC-Furnace	IMC-Column

- A set of challenging exercises that reinforce the conventional lecture material through active learning and problem-based methods

C.2 MODULE ORGANIZATION

Eight distinct chemical and biological process applications, which range from simple single input-single output (SISO) processes to more complex 2 × 2 control loops, are formulated with a modular approach. The progression of the modules follows a typical undergraduate process dynamics and control course, starting with low-order dynamic system analysis and continuing through multivariable controller synthesis.

C.3 HARDWARE AND SOFTWARE REQUIREMENTS

The Process Control Modules are a set of MATLAB/Simulink routines that require either a full license or the Student Version of MATLAB and Simulink. The current version of the modules has been tested with version 2011b of MATLAB and Simulink. The minimum recommended system configuration is a Windows 7 PC with 1 GB RAM.

C.4 INSTALLATION

The Process Control Modules (PCM) software can be downloaded from www.wiley.com/college/seborg onto the user's computer. Then double-click on the PCM file, and follow the instructions on the installer to install the software. Note that MATLAB should be installed in order to use these modules. During the installation, users can create a shortcut icon to the software on their desktop (recommended).

C.5 RUNNING THE SOFTWARE

There are two ways to execute the software: the first is to double-click the PCM button on the desktop, which launches MATLAB and the PCM interface (Fig. C.1), and the other way is to open MATLAB manually and to call the PCM software by pointing to the PCM installation folder and typing "PCM," followed by the Enter key.

The website for this textbook contains a more detailed tutorial on PCM, including case studies for the furnace and distillation column modules.

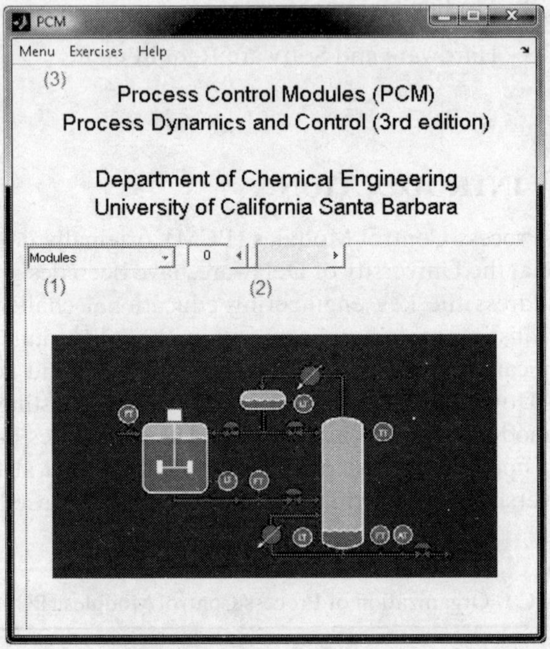

Figure C.1 PCM main interface.

REFERENCES

Doyle III, F. J., E. P. Gatzke, and R. S. Parker, Practical Case Studies for Undergraduate Process Dynamics and Control Using Process Control Modules, *Comp. Appls. Eng. Educ.*, 6, 181 (1998).

Doyle III, F. J. *Process Control Modules: A Software Laboratory for Control Design,* Prentice Hall PTR, Upper Saddle River, NJ, 2000.

APPENDIX D

Review of Basic Concepts From Probability and Statistics

APPENDIX CONTENTS

D.1 PROBABILITY CONCEPTS

The term *probability* is used to quantify the likely outcome of a random event. For example, if a fair coin is flipped, the probability of a head is 0.5 and the probability of a tail is 0.5. Let $P(A)$ denote that probability that a random event A occurs. Then $P(A)$ is a number in the interval $0 \le P(A) \le 1$, such that the larger $P(A)$ is, the more likely it is that A occurs. Let A' denote the *complement* of A, that is, the event that A does not occur. Then,

$$P(A') = 1 - P(A) \tag{D-1}$$

Now consider two events, A and B, with probabilities $P(A)$ and $P(B)$, respectively. The probability that one or both events occurs $(A \cup B)$ can be expressed as

$$P(A \cup B) = P(A) + P(B) - P(A \cap B) \tag{D-2}$$

If A and B are *mutually exclusive*, this means that if one event occurs, the other cannot; consequently, their intersection is the null set $A \cap B = \phi$. Then $P(A \cap B) = 0$ and Eq. D-2 becomes

$$P(A \cup B) = P(A)+P(B) \quad \text{(for mutually exclusive events)} \tag{D-3}$$

Analogous expressions are available for the union of more than two events (Montgomery and Runger, 2013).

If A and B are *independent*, then the probability that both occur is

$$P(A \cap B) = P(A)\,P(B) \quad \text{(for independent events)} \tag{D-4}$$

Similarly, the probability that n independent events, E_1, E_2, … , E_n, occur is

$$P(E_1 \cap E_2 \cap \ldots \cap E_n) = P(E_1)\,P(E_2)\,\ldots\,P(E_n) \quad \text{(for independent events)} \tag{D-5}$$

These probability concepts are illustrated in two examples.

EXAMPLE D.1

A semiconductor processing operation consists of five independent batch steps where the probability of each step having its desired outcome is 0.95. What is the probability that the desired end product is actually produced?

Solution

In order to make the product, each individual step must be successful. Because the steps are independent, the probability of a success, $P(S)$, can be calculated from Eq. D-5:

$$P(S) = (0.95)^5 = 0.77$$

EXAMPLE D.2

In order to increase the reliability of a process, a critical process variable is measured on-line using two sensors. Sensor A is available 95% of the time, while Sensor B is available 90% of the time. Suppose that the two sensors operate independently and that their periods of unavailability occur randomly. What is the probability that neither sensor is available at any arbitrarily selected time?

Solution

Let A denote the event that Sensor A is not available and B denote the event that Sensor B is not available. The event that neither Sensor is available can be expressed as $(A \cup B)'$. Then from Eqs. D-1 and D-2,

$$P(A \cup B)' = 1 - P(A \cup B)$$

$$P(A \cup B)' = 1 - [P(A) + P(B) - P(A \cap B)]$$

$$P(A \cup B)' = 1 - [0.95 + 0.90 - (0.95)(0.90)] = 0.005$$

D.2 MEANS AND VARIANCES

Next, we consider two important statistical concepts, *means* and *variances*, and how they can be used to characterize both probability distributions and experimental data.

D.2.1 Means and Variances for Probability Distributions

In Section D.1, we considered the probability of one or more *events* occurring. The same probability concepts are also applicable for *random variables* such as temperatures or chemical compositions. For example, the product composition of a process could exhibit random fluctuations for several reasons, including feed disturbances and measurement errors. A temperature measurement could exhibit random variations due to turbulence near the sensor. Probability analysis can provide useful characterizations of such random phenomena.

Consider a continuous random variable, X, with an assumed probability distribution, $f(x)$, such as a Gaussian distribution. The probability that X has a numerical value in an interval $[a, b]$ is given by (Montgomery and Runger, 2013),

$$P(a \le X \le b) = \int_a^b f(x)dx \tag{D-6}$$

where x denotes a numerical value of random variable, X. By definition, the *expected value* of X, μ_X, is defined as

$$\mu_x = E(X) = \int_{-\infty}^{\infty} xf(x)dx \tag{D-7}$$

The expected value is also called the *population mean* or *average*. It is an average over the expected range of values, weighted according to how likely each value is.

The *population variance* of X, σ_X^2, indicates the variability of X around its population mean. It is defined as

$$\sigma_X^2 = E[(X - u_X)^2] = \int_{-\infty}^{\infty} (x - \mu_X)^2 f(x)dx \tag{D-8}$$

The positive square root of the variance is the *population standard deviation*, σ_X.

These calculations are illustrated in Example D.3.

EXAMPLE D.3

A mass fraction of an impurity X varies randomly between 0.3 and 0.5 with a uniform probability distribution:

$$f(x) = \frac{1}{0.2}$$

Determine its population mean and population standard deviation.

Solution

Substituting $f(x)$ into Eq. D-7 gives

$$\mu_X = \int_{-\infty}^{\infty} x f(x)dx = \int_{0.3}^{0.5} x\left(\frac{1}{0.2}\right)dx$$

$$\mu_X = \left(\frac{1}{0.2}\right)\left(\frac{1}{2}x^2\right)\Bigg|_{0.3}^{0.5} = 0.4$$

Thus μ_X is the midpoint of the [0.3, 0.5] interval for X. To determine σ_X, substitute $f(x)$ into Eq. D-8:

$$\sigma_X^2 = \int_{-\infty}^{\infty} (x-u_X)^2 f(x)dx = \int_{0.3}^{0.5} (x-0.4)^2\left(\frac{1}{0.2}\right)dx$$

$$\sigma_X^2 = \left(\frac{1}{0.2}\right)\left(\frac{1}{3}(x-0.4)^3\right)\Bigg|_{0.3}^{0.5} = 0.00333$$

$$\sigma_X = 0.0577$$

D.2.2 Means and Variances for Experimental Data

A set of experimental data can be characterized by its *sample mean* and *sample variance* (or simply, its *mean* and *variance*). Consider a set of N measurements, $\{x_1, x_2, \ldots, x_N\}$. Its mean, $\bar{x}$, and variance, s^2, are defined as (Montgomery and Runger, 2013)

$$\bar{x} \triangleq \frac{1}{N}\sum_{i=1}^{N} x_i \quad \text{(D-9)}$$

$$s^2 \triangleq \frac{1}{N-1}\sum_{i=1}^{N} (x_i - x)^2 \quad \text{(D-10)}$$

The standard deviation s is the positive square root of the variance.

The mean is the average of the data set while the variance and standard deviation characterize the variability in the data.

D.3 STANDARD NORMAL DISTRIBUTION

The normal (or Gaussian) probability distribution plays a central role in both the theory and application of statistics. It was introduced in Section 17.2.1. For probability calculations, it is convenient to use the *standard normal distribution,* $N(0, 1)$, which has a mean of zero and a variance of one. Suppose that a random variable X is normally distributed with a mean μ_X and variance σ_X^2. Then, the corresponding *standard normal variable* Z is

$$Z = \frac{X - \mu_X}{\sigma_X} \quad \text{(D-11)}$$

Statistics book contains tables of the *cumulative standard normal distribution,* $\phi(z)$.

By definition, $\phi(z)$ is the probability that Z is less than a specified numerical value, z (Ogunnaike, 2010; Montgomery and Runger, 2013):

$$\phi(z) = P(Z \le z) \quad \text{(D-12)}$$

D.4 ERROR ANALYSIS

In engineering calculations, it can be important to determine how uncertainties in independent variables (or inputs) lead to even larger uncertainties in dependent variables (or outputs). This analysis is referred to as *error analysis*. Due to the uncertainties associated with input variables, they are considered to be random variables. The uncertainties can be attributed to imperfect measurements or uncertainties in unmeasured input variables. Error analysis is based on the statistical concepts of means and variances, considered in the previous section.

As an important example of error analysis, consider a linear combination of p variables,

$$Y = \sum_{i=1}^{p} c_i X_i \quad \text{(D-13)}$$

where X_i is an independent random variable with expected value μ_i and variance σ_i^2. Then, Y has the following mean and variance (Montgomery and Runger, 2013):

$$\mu_Y = \sum_{i=1}^{P} c_i \mu_i \tag{D-14}$$

$$\sigma_Y^2 = \sum_{i=1}^{P} c_i^2 \sigma_i^2 \tag{D-15}$$

Equations D-14 and D-15 show how the variability of the individual X_i variables determines the variability of their linear combination, Y.

EXAMPLE D.4

Experimental tests are to be performed to determine whether a new catalyst A is superior to the current catalyst B, based on their yields for a chemical reaction. Denote the yields by X_A and X_B and their standard deviations by $\sigma_A = 3\%$ and $\sigma_B = 2\%$. What is the standard deviation for the difference in yields, $X_A - X_B$?

Solution

Let $Y = X_A - X_B$, an expression in the form of Eq. D-13 with $c_A = 1$ and $c_B = -1$. Thus Eq. D-15 becomes

$$\sigma_Y^2 = \sigma_A^2 + \sigma_B^2$$

and

$$\sigma_Y = \sqrt{\sigma_A^2 + \sigma_B^2} = \sqrt{(3\%)^2 + (2\%)^2} = 3.6\%$$

Thus, the standard deviation of the difference is larger than the individual standard deviations.

REFERENCES

Montgomery, D. C., and G. C. Runger, *Applied Statistics and Probability for Engineers*, 6th ed., John Wiley and Sons, Hoboken, NJ, 2013.

Ogunnaike, B. A., Random Phenomena: *Fundamentals of Probability and Statistics for Engineers*, CRC Press, Boca Raton, FL, 2010.

APPENDIX E

Process Safety and Process Control

APPENDIX CONTENTS

The chemical and other process industries have played a major role in developing the goods and services that are key components of modern technological society. They include inexpensive food and energy, synthetic materials, pharmaceuticals and medical devices. But there is a downside in this development: industrial plants are complex systems that deal with hazardous materials and create potentially dangerous situations. Although the overall safety record of the process industries has been quite good, unfortunate accidents can occur, resulting in the loss of life and property (Mannan, 2012; Kletz, 2009; Venkatasubramanian, 2011). In fact, the accident and loss statistics for the chemicals and allied products industries are among the best of the manufacturing sectors (Crowl and Louvar, 2002).

Process safety has been a primary concern of the process industries for decades. But in recent years, safety issues have received increased attention as a result of increased public awareness of potential risks, stricter legal requirements, and the increased complexity of modern industrial plants. Chemical engineers have a special role to perform in assuring process safety. As Turton et al. (2012) have noted, "As the professional with the best knowledge of the risks of a chemical processing operation, the chemical engineer has a responsibility to communicate those risks to employers, employees, clients, and the general public." Furthermore, the American Institute of Chemical Engineers (AIChE) Code of Ethics states that the first responsibility of chemical engineers is to "hold paramount the safety, health, and welfare of the public in performance of their professional duties."

An extensive literature on process safety is available in both the public domain and proprietary industrial documents. Professional societies such as AIChE and the International Society of Automation (ISA) have played a leading role in developing safety standards and reference materials. For example, the AIChE Center for Chemical Process Safety (CCPS) has published a large number of books and reports (e.g., AIChE, 1993, 2001, 2007, 2016) and a journal, *Process Safety Progress*. Valuable information is also available in handbooks (Perry and Green, 2008; Mannan, 2012), textbooks (Crowl and Louvar, 2002; Banerjee, 2003), and review articles (Venkatasubramanian, 2011; Leveson and Stephanopoulos, 2014; Mannan et al., 2015).

Process safety is considered at various stages during the development and operation of a process:

1. An initial safety analysis is performed during the preliminary process design.
2. A very thorough safety review is conducted during the final stage of the process design, using techniques such as *hazard and operability* (HAZOP)

studies, failure mode and effect analysis, and fault tree analysis (AIChE, 1993; Kletz, 2001; Crowl and Louvar, 2002).

3. After plant operation begins, HAZOP studies are conducted on a periodic basis in order to identify and eliminate potential hazards.
4. Many companies require that any proposed plant change or change in operating conditions requires formal approval via a *Management-of-Change* (MOC) process that considers the potential impact of the change on the safety, environment, and health of the workers and the nearby communities. Furthermore, facilities that process significant quantities of hazardous materials must comply with government regulations from agencies such as the Environmental Protection Agency (EPA) and the Occupational Safety and Health Administration (OSHA).
5. After a serious industrial plant accident or incident in the United States, a thorough investigation is conducted by an independent government agency, the Chemical Safety Board (CSB), to determine the root cause of the incident, assess responsibility, and suggest safety improvements. The subsequent accident report is made available on an Internet site: www.chemsafety.gov.

The process control system, instrumentation, and alarms play critical roles in ensuring plant safety. In fact, it can be argued that process safety is really a problem in process control, a broader version of the same underlying themes and objectives (Venkatasubramanian, 2011). But if control systems do not function properly, they can be a contributing factor or even a root cause of a serious incident (Kletz, 2009). In this chapter, we provide an overview of the influence of process dynamics and control on process safety.

E.1 LAYERS OF PROTECTION

In modern industrial plants, process safety relies on the principle of multiple layers of protection (AIChE, 1993, 2001; ISA, 2004). A typical configuration is shown in Fig. E.1. Each layer of protection consists of a grouping of equipment and/or human actions. The layers of protection are shown in the order of activation that occurs as a plant incident develops, with the most effective layers used first. The basic concept is that an incident should be handled at the lowest possible layer. In the interior of the diagram, the process design itself provides the first level of protection. The next two layers consist of the *basic process control system* (BPCS), augmented with two levels of alarms and operator supervision or intervention. An alarm indicates that a measurement has exceeded its specified limits and may require either operator intervention or an automated response.

The fourth layer consists of a *safety instrumented system* (SIS) and/or an *emergency shutdown* (ESD) *system*. The SIS, formerly referred to as a *safety interlock system*, automatically takes corrective action when the process and BPCS layers are unable to handle an emergency. For example, the SIS could automatically turn off the reactant and catalyst pumps for a chemical reactor after a high temperature alarm occurs. The SIS is described in Section E.1.4.

In the fifth layer of protection, passive relief devices, such as rupture disks and relief valves, provide physical protection by preventing excessive pressures from being generated within process vessels and pipelines. If over-pressurization occurs, the relief valve or rupture disk opens and the fluid is vented to an appropriate location, such as a combustion flare stack, incinerator, scrubber, waste treatment facility, or the atmosphere. Such passive devices operate independently of the SIS system.

Finally, dikes are located around process units and storage tanks to contain liquid spills. Emergency response plans are used to address extreme situations, inform the nearby community, and implement evacuation plans, if necessary.

The functioning of the multiple-layer protection system can be summarized as follows (AIChE, 1993): "Most failures in well-designed and operated chemical processes are contained by the first one or two protection layers. The middle levels guard against major releases and the outermost layers provide mitigation response to very unlikely major events. For a major hazard potential, even more layers may be necessary."

It is evident from Fig. E.1 that process control and automation play important roles in ensuring process safety. In particular, many of the protective layers in Fig. E.1 involve instrumentation and control equipment. Furthermore, the process and instrument dynamics are key considerations in safety analysis. For example, after a major incident develops, how much time elapses before the sensors detect the new conditions and corrective action is taken? If the incident remains undetected, how long will it take for an emergency situation to result?

Next, we consider the role of process control and instrumentation in the protection layers of Fig. E.1.

E.1.1 The Role of the Basic Process Control System

The BPCS consists of feedback and feedforward control loops that regulate process variables such as temperatures, flow rates, liquid levels, and pressures. Although

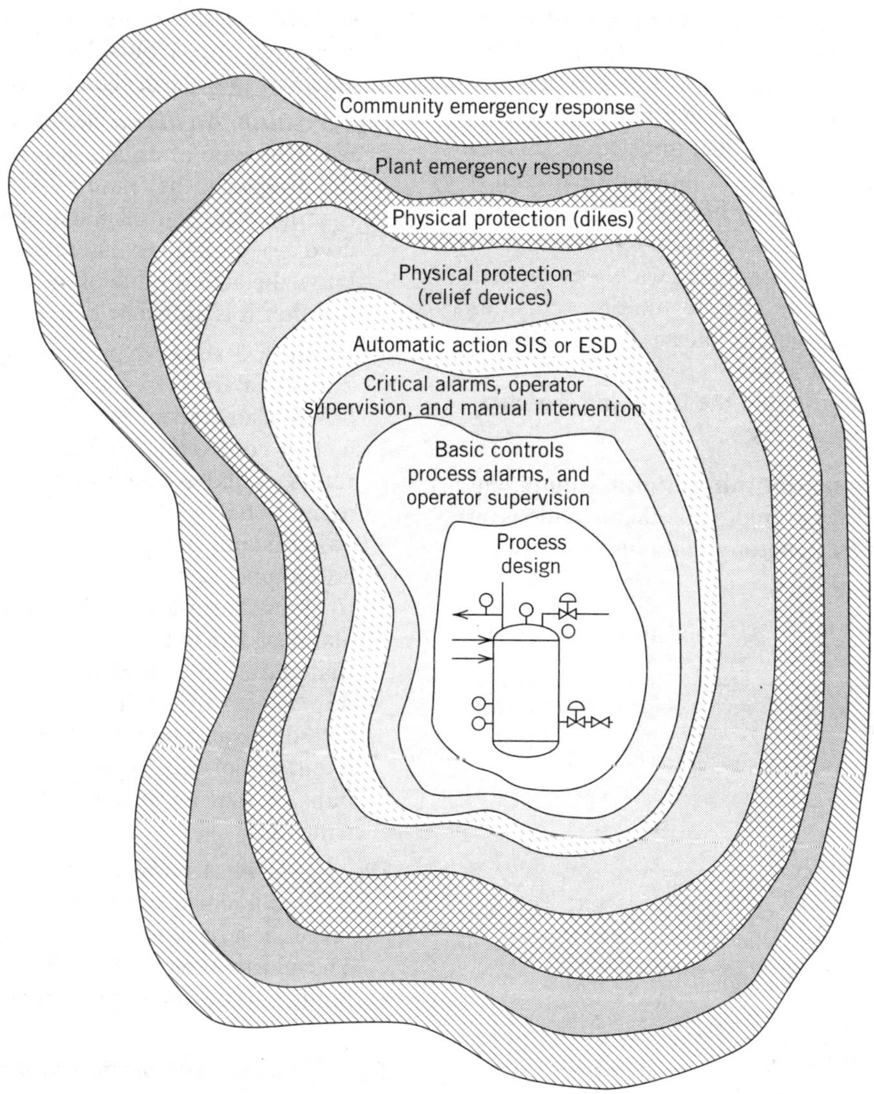

Figure E.1 Typical layers of protection in a modern chemical plant (AIChE, 1993).

Note: Protection layers for a typical process are shown in the order of activation expected as a hazardous condition is approached
ESD = Emergency shutdown
SIS = Safety Interlock system

the BPCS typically provides satisfactory control during routine process operation, it may not do so during abnormal conditions. For example, if a controller output or manipulated variable saturates, the controlled variable may exceed allowable limits. Similarly, the failure or malfunction of a component in the feedback loop, such as a sensor, control valve, or transmission line, could cause the process operation to enter an unacceptable region. In this case, the operation of the process is transferred to the SIS system. Typical component failure rates are shown in Table E.1, expressed as the average number of faults per year.

E.1.2 Process Alarms

The second and third layers of protection rely on process alarms to inform operators of abnormal situations. A block diagram for an alarm system is shown in Fig. E.2. An alarm is generated automatically when a measured variable exceeds a specified high or low limit. The logic block is programmed to take appropriate corrective action when one or more alarm switches are triggered. After an alarm occurs, the logic block activates an annunciator, either a visual display or an audible sound such as a horn or bell. For example, if a reactor temperature exceeds a high alarm limit, a light might flash on a computer screen, with the color indicating the alarm priority (e.g., yellow for a less serious situation, red for a critical situation). An alarm continues until it is acknowledged by an operator action, such as pressing a button or a key on a computer keyboard. If the alarm indicates a potentially hazardous situation, an automated corrective action is initiated by the SIS. Two types of high and low alarm limits are widely employed. *Warning limits* are used to denote minor excursions from nominal values, whereas alarm limits indicate larger, more serious excursions.

Connell (1996) has proposed the following classification system for process alarms:

Type 1 Alarm: Equipment status alarm. Equipment status indicates, for example, whether a pump is on or off, or whether a motor is running or stopped.

Table E.1 Failure Rates for Selected Components (Mannan, 2012)

Instrument	Failure Frequency (faults per year)
Control valve	0.60
Valve positioner	0.44
Current/pressure transducer	0.49
Pressure measurement	1.41
Flow measurement (fluids)	
Orifice plate and DP transmitter	1.73
Magnetic flowmeter	2.18
Temperature measurement	
Thermocouple	0.52
Mercury-in-steel thermometer	0.027
Controller (electronic)	0.29
Flow switch	1.12
Pressure switch	0.34
Alarm indicator lamp	0.044
Gas-liquid chromatograph	30.6

Type 2 Alarm: Abnormal measurement alarm. If a measurement is outside specified limits, a high alarm or low alarm signal is triggered.

Type 3 Alarm: An alarm switch without its own sensor. This type of an alarm is directly activated by the process, rather than by a sensor signal; thus it is utilized in lieu of a sensor. Type 3 alarms are used for situations where it is not necessary to know the actual value of the process variable, only whether it is above or below a specified limit.

Figure E.3 shows typical configurations for Type 2 and 3 alarms. In the Type 2 alarm system, the flow sensor/transmitter (FT) signal is sent to both a flow controller (FC) and a flow switch (FSL refers to "flow-switch-low"). When the measurement is below the specified low limit, the flow switch sends a signal to an alarm that activates an annunciator in the control room (FAL refers to "flow-alarm-low"). By contrast, for the Type 3 alarm system in Fig. E.3b, the flow switch is self-actuated and thus does not require a signal from a flow sensor/transmitter. Type 3 alarms are preferred because they still function when the sensor is out of service. Type 3 alarms are also used to indicate that an automatic shutdown system has tripped.

Type 4 Alarm: An alarm switch with its own sensor. A Type 4 alarm system has its own sensor, which serves as a backup in case the regular sensor fails. The alarm sensor also allows sensor drifts and failures to be detected more readily than a switch does.

Type 5 Alarm: Automatic shutdown or startup system. These important and widely used systems are described in the next section on Safety Instrumented Systems.

It is tempting to specify tight alarm limits for a large number of process variables. But this temptation should be resisted, because an excessive number of unnecessary alarms can result. Furthermore, too many alarms can be as detrimental as too few, for several reasons. First, fre-

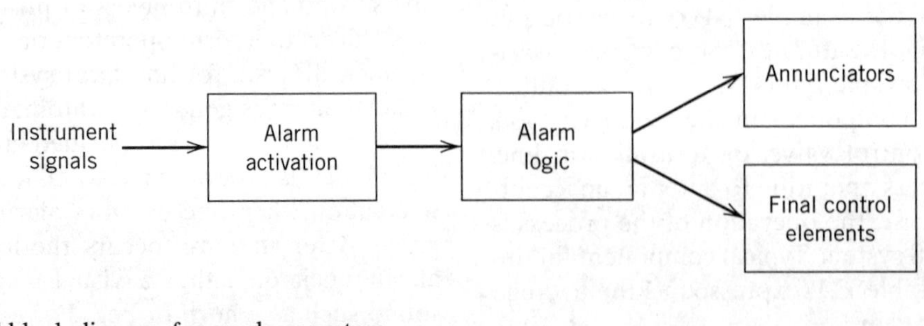

Figure E.2 A general block diagram for an alarm system.

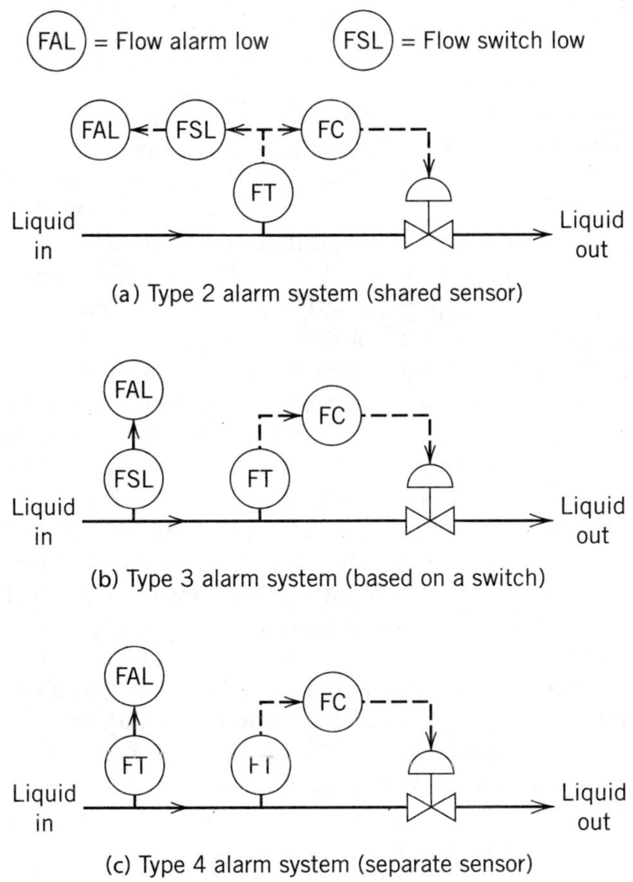

Figure E.3 Alternative flow alarm configurations.

quent nuisance alarms tend to make the plant operators less responsive to important alarms. For example, if a tank is being filled during a plant startup, low-level alarms could occur repeatedly but be of no value to the plant operator. Second, in an actual emergency, a large number of unimportant alarms can obscure the root cause of the problem. Third, the relationships between alarms need to be considered. Thus, the design of an appropriate alarm management system is a challenging task.

E.1.3 Safety Instrumented System (SIS)

The SIS in Figure E.1 serves as an emergency backup system for the BPCS. The SIS starts automatically when a critical process variable exceeds specified alarm limits that define its allowable operating range. Its initiation results in a drastic action, such as starting or stopping a pump or shutting down a process unit. Consequently, it is used only as a last resort to prevent injury to people or equipment. The term *Safety Interlock System* was previously used, but the newer term *Safety Instrumented System* is now preferred because it is more general.

The proper design of a SIS is of critical importance because it can prevent dangerous situations and potential disasters. By contrast, SIS failures have been responsible for major industrial disasters including the worst industrial accident that has ever occurred. A runaway chemical reaction caused the release of deadly methyl isocyanate gas in a pesticide plant (Bhopal, India) in 1984. This accident resulted in the deaths of an estimated 8000–16,000 people and several hundred thousand injuries.

Fortunately, SIS design is facilitated by the wealth of information, experience, and guidelines that are available from professional and government organizations such as AIChE, ISA, and OSHA (see References).

It is very important that the SIS function independently of the BPCS; otherwise, emergency protection will be unavailable during periods when the BPCS is not operating (e.g., as a result of a malfunction or power failure). Thus, it is recommended that the SIS be physically separated from the BPCS and have its own sensors and actuators (AIChE, 1993).

Sometimes redundant sensors and actuators are utilized. For example, *triply redundant*[1] sensors are used for critical measurements, with SIS actions based on the median of the three measurements. This strategy prevents a single sensor failure from crippling SIS operation. The SIS also has a separate set of alarms so that the operator can be notified when the SIS initiates an action (e.g., turning on an emergency cooling pump), even if the BPCS is not operational.

As an alternative approach, redundant computer control systems can be employed, with each system having the BPCS, SIS, and ESD functions. This approach provides greater security but tends to be more complex and expensive.

E.1.4 Interlocks and Emergency Shutdown Systems

The SIS operation is designed to provide automatic responses after alarms indicate potentially hazardous situations. The objective is to have the process reach a safe condition. The automatic responses are implemented via interlocks and automatic shutdown and startup systems. Distinctions are sometimes made between safety interlocks and process interlocks; the latter are used for less critical situations to provide protection against minor equipment damage and undesirable process conditions, such as the production of off-specification product.

Two simple interlock systems are shown in Fig. E.4. For the liquid storage system, the liquid level must stay

[1] Arguably, this strategy should be referred to as doubly redundant, rather than triply redundant, because three sensors are used.

above a minimum value in order to avoid pump damage such as cavitation. If the level drops below the specified limit, the low-level switch (LSL) triggers both an alarm (LAL) and a solenoid switch (S) (or solenoid) that turns the pump off. For the gas storage system in Fig. E.4, the solenoid valve is normally closed. But if the pressure of the hydrocarbon gas in the storage tank exceeds a specified limit, the high pressure switch (PSH) activates an alarm (PAH) and causes the solenoid valve to open fully, thus reducing the pressure in the tank. For interlock and other safety systems, a solenoid switch can be replaced by a sensor transmitter if the measurement is required.

Another common interlock configuration is to locate a solenoid switch between a controller and a control valve. When an alarm is actuated, the solenoid trips and causes the air pressure in the pneumatic control valve to be vented; consequently, the control valve reverts to either its fail-open or fail-close position. Interlocks have traditionally been implemented as hard-wired systems that are independent of the control hardware. But, for most applications, software implementation of the interlock logic via a digital computer or a programmable logic controller is a viable alternative. *Programmable logic controllers (PLCs)* used for batch processes are considered in Chapters 18 and 19.

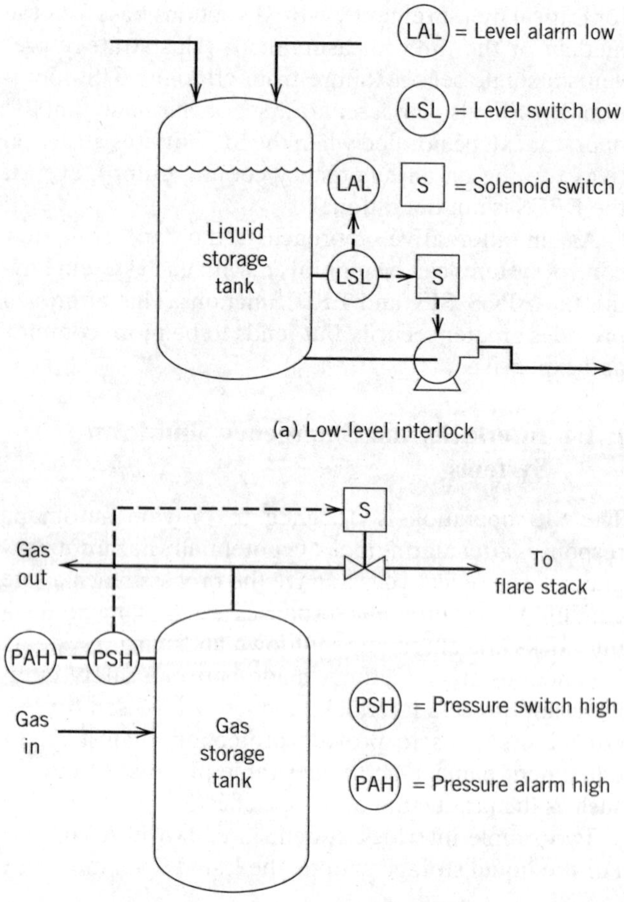

Figure E.4 Two interlocks configurations.

If a potential emergency situation is very serious, the Emergency Shutdown System automatically shuts down or starts up equipment. For example, a pump would be turned off (or *tripped*) if it overheats or loses lubricant pressure. Similarly, if an exothermic chemical reaction starts to run away, it may be possible to add a quench material that stops the reaction quickly. For some emergency situations, the appropriate response is an automatic startup of equipment, rather than an automatic shutdown. For example, a backup generator or a cooling water pump could be started if the regular unit shuts down unexpectedly.

Although the ESD function is essential for safe process operation, unnecessary plant shutdowns and startups should be avoided because they result in loss of production and generate off-specification product during the subsequent plant startup. Also, emergency shutdowns and startups for a process unit involve risks and may activate additional safety systems that shutdown other process units. Such nuisance shutdowns can create additional hazards. The use of redundant sensors can reduce unnecessary shutdowns.

E.2 ALARM MANAGEMENT

As industrial processes have become more complex and integrated, the topics of alarm management and abnormal situation management have become increasingly important, both from safety and economic considerations. For example, it has been estimated that preventable abnormal situations have an annual impact of over \$20 billion on the operations of the U.S.-based petrochemical industry because of production loss, equipment damage, litigation etc. (ASM, 2015). Alarm management and the occurrence of an excessive number of alarms during an abnormal situation (an *alarm flood*) have often been cited as contributing factors in investigations of major industrial plant accidents. In this section, we consider key elements of effective alarm management and response to abnormal situations.

Before computer-control systems were available, a process alarm was connected to a light box and an audible alarm on a panel board. When an alarm occurred, the light box flashed and/or a horn sounded, thus attracting the operator's attention. The operator then acknowledged the alarm by pressing a button.

Because panel board space was quite limited, only key process variables could be alarmed. But the introduction of modern computer-control systems, beginning in the 1970s, drastically changed this situation. It became feasible to alarm virtually any measured variable and to display the alarms on a computer monitor. As a result, engineers are able to alarm very large numbers of process variables, which can inadvertently generate many spurious or nuisance alarms, an *alarm overload*. For example, during the Three Mile Island accident in 1979, the nuclear power plant operators were overwhelmed with information, much of which was irrelevant or incorrect. Fortunately, no one was injured in the accident (Nuclear Regulatory Commission, 2015).

Representative alarm experience for the process industries is compared with recommended guidelines in Table E.2. A standing (or *stale*) alarm is one that remains in an alarm state for an extended period of time, such as 24 h. Unfortunately, the actual alarm rates in Table E.2 exceed the corresponding guidelines by large margins. Note the *average* alarm rates per 10-min interval in Table E.2 correspond to less than one alarm per minute. These alarm rates could be dealt with by an experienced operator, particularly if many alarms were low priority. On the hand, the *peak* alarm rates per 10-min interval (180-350 alarms) are *alarm floods* that would tend to overwhelm even experienced operators, especially if the alarm floods persisted for long periods of time (e.g., several hours).

Alarm floods can distract or confuse operators, making it difficult for them to determine which alarms are most important for the current situation. Dire consequences, such as incorrect decisions, lost production, and unsafe process operation, can result. Consequently, alarm management and abnormal situation management have become increasingly important issues in recent years. Alarm requirements are the subject of government regulations (OSHA, 2000).

E.2.1 Alarm Guidelines

The design and maintenance of alarm management systems has been the subject of numerous articles, standards, and books (AIChE, 2007; Hollifield and Habibi, 2007, 2011; ASM, 2015). Some important guidelines are shown in Table E.3.

Table E.3 Guidelines for Alarm Design and Management

1. Each alarm should have two important characteristics; it should result from an abnormal situation and require a specific operator response.
2. Alarm systems and displays should be designed with the user (the plant operator) in mind.
3. Each alarm should have a priority level to indicate its level of importance. Typically, two to four priority levels are employed. For example, the four levels could be designated as critical, emergency, high, and low (Hollifield and Habibi, 2007).
4. Protective alarms related to process safety or alarms that require an immediate response should be assigned the highest priority.
5. An alarm should continue until it is acknowledged.
6. Operators should respond to every alarm, regardless of priority, but alarm overload should be avoided.
7. Alarm suppression should only be allowed when there is a legitimate reason (e.g., an instrument is out of service). Automated reminders should be employed so that it is not possible for an operator to suppress an alarm and then forget about it (i.e., alarm storing is preferable to alarm suppression).
8. Each alarm should be logged with a time/date stamp and a record of the corresponding operator actions. Suppressed alarms should be included in the log.
9. Alarm histories should be reviewed on a regular basis in order to reduce standing alarms and nuisance alarms.
10. Alarm limit changes should be carefully considered and well-documented. Changes for critical protective alarms should be approved via an Mangagement of Change order.
11. State-based alarming can greatly improve alarm management by eliminating nuisance alarms. It involves automatic adjustment of alarm limits to accommodate different process conditions, such as startups, shutdowns, changes in feedstocks or production rates (Hollifield and Habibi, 2007).

These alarm guidelines are illustrated by two common situations.

Table E.2 Alarm Rates Per Plant Operator in the Process Industries[†]

	EEMUA Guideline	Oil and Gas	Petrochemical	Power	Other
Average alarms per day	144	1200	1500	2000	900
Average standing alarms	9	50	100	65	35
Peak alarms (per 10-min interval)	10	220	180	350	180
Average alarms (per 10-min interval)	1	6	9	8	5
Alarm distribution % (low/medium/high)	80/15/5	25/40/35	25/40/35	25/40/35	25/40/35

[†] Engineering Equipment and Materials Users' Association (2007).

Tank-Filling Operation

A tank is filled using a pump on an inlet line. In order to avoid nuisance alarms, the low-level alarm should be suppressed (or the alarm limit changed) during this operation. Similarly, the low-flow alarm should be suppressed (or its limit changed) when the pump is turned off. Automatic adjustment of the alarm limits for routine tank filling is an example of state-based alarming (Guideline 11). It also satisfies Guideline 1 by not generating an alarm for a routine and known situation.

Regular and Backup Pumps

Consider a situation where there is a regular pump and a spare pump (Hollifield and Habibi, 2007). During typical operations, there could be zero, one, or two pumps running, with the latter situation occurring during a routine switch from one pump to the other. Alarms can be used to indicate the status of each pump. If the alarm strategy is to have an alarm activate when a pump is not running, one of the two pumps is always in an alarm state.

An alternative approach is to allow the operator to specify the number of pumps that are supposed to be running. Then digital logic could be used to activate an alarm only when the actual number of operating pumps differs from the desired number. This approach conforms with Guideline 1, because nuisance alarms are avoided during routine changes in process operations.

EXAMPLE E.1

Consider the liquid storage system shown in Fig. E.5. A high-level alarm is used to prevent tank overflow. After a high alarm sounds, the operator has 10 minutes to respond before the ESD system turns off the pumps on the inlet streams. The tank is 180 cm in diameter, 240 cm tall, and half full at the nominal conditions. The density of the liquid is constant. The nominal values and ranges of the flow rates are as follows:

Flow rate	Nominal Value (L/min)	Range (L/min)
q_1	280	220–340
q_2	140	110–170
q_3	420	360–480

(a) Assuming that the pumps are operating properly, determine an appropriate set-point value for the high-level alarm, that is, the numerical value of liquid level h that triggers the alarm.

(b) What additional safety features are required to handle unanticipated pump shutdowns?

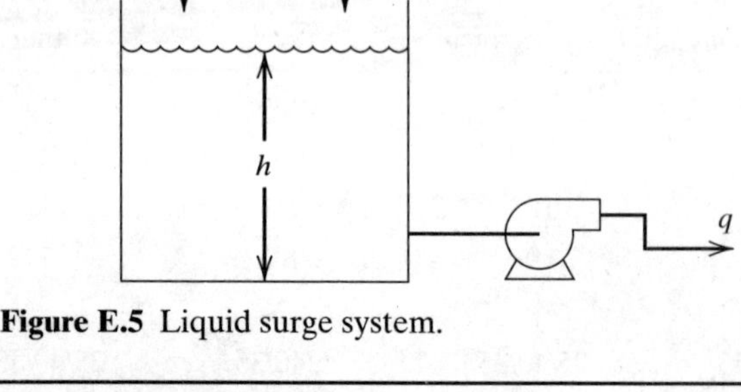

Figure E.5 Liquid surge system.

Solution

(a) For constant liquid density, the mass balance for the liquid in the tank is

$$\rho A \frac{dh}{dt} = \rho(q_1 + q_2 - q) \tag{E-1}$$

In order to specify the high alarm limit, consider the worst-case conditions that result in the largest rate of change, dh/dt. This situation occurs when the inlet flow rates are at their maximum values and the exit flow rate is a minimum. Substituting q_1 = 340 L/min, q_2 = 170 L/min, and q = 420 L/min gives

$$\frac{dh}{dt} = \frac{140 \text{ L/min}}{\pi(0.9\,\text{m})^2} = 0.055 \text{ m/min} \tag{E-2}$$

Consequently, the maximum increase in liquid level that can occur during any 10-min period is

$$\Delta h_{max} = (10 \text{ min})(0.055 \text{ m/min}) = 0.55 \text{ m} \tag{E-3}$$

Thus, the alarm limit should be at least 55 cm below the tank height. As a safety margin, choose 50 cm and set the high alarm limit at 100 cm.

(b) A pump failure could result in tank overflow or a tank level so low that the exit pump could be damaged. Thus, high and low level alarms should be connected to interlocks on the inlet and exit pumps, respectively.

E.2.2 Alarm Rationalization

Because processes and process conditions change over time, it is necessary to review alarm system performance on a periodic basis. The review should be based on the alarm history and a number of key performance metrics. Some of the most important alarm metrics are the following (Jofriet, 2005):

(a) Frequency of alarm activations (e.g., alarms per day)

(b) Priority distribution (e.g., numbers of high, medium, and low alarms)

(c) Alarm performance during upset conditions (e.g., the numbers of alarms activated during the first 10 min and the subsequent 10-min periods)

(d) Identification of bad actors—individual alarm points that generate a large fraction of the total alarms

These metrics are illustrated in Table E.2 for a set of industrial plant data (Hollifield and Habibi, 2007). A recommended alarm priority distribution is shown in Table E.4 with the Table E.2 guidelines in parentheses.

A critical analysis of the alarm metrics can result in significant improvements in alarm management. In particular, it can identify potentially dangerous alarm floods where operators are overwhelmed by a large number of alarms in a small time interval. It can also delineate bad actors. An industrial example of an alarm flood is shown in Fig. E.6; identification of the associated bad actors is shown in Fig. E.7. Analysis of bad actor alarm data can identify process problems or alarm limits that are inappropriate (e.g., too aggressive). Figure E.7 indicates that only four alarms generate over 40% of the total number of alarms.

Table E.4 Recommended Alarm Event Priority Distribution†

Alarm Priority	% of Total Alarms
Emergency	3–7% (5%)
High	15–25% (15%)
Low	70–80% (80%)

† Reprinted by permission of ISA—The International Society of Automation. © Copyright 2007, ISA—all rights reserved. Hollifield and Habibi, 2007.

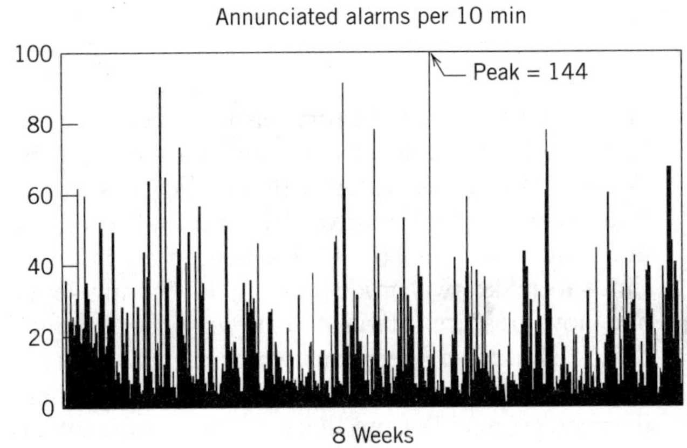

Figure E.6 Example of annunciated graphs for one operator station over a period of 8 weeks (Reprinted by permission of ISA—The International Society of Automation. © Copyright 2007, ISA—all rights reserved. Hollifield and Habibi, 2007).

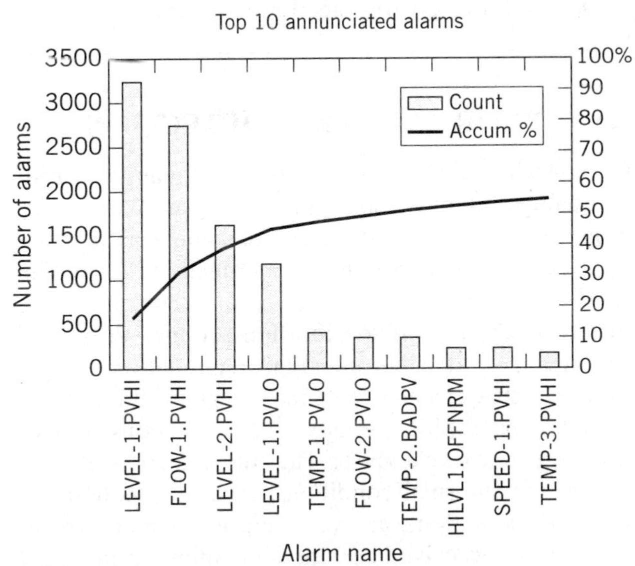

Figure E.7 An example of the 10 most frequent alarms (bad actors) over an 8-week period (Reprinted by permission of ISA—The International Society of Automation. © Copyright 2007, ISA—all rights reserved. Hollifield and Habibi, 2007).

EXAMPLE E.2

A review of an alarm history has identified that the bad actors include level, pressure, and temperature alarms that are associated with a liquid-phase, continuous, stirred-tank reactor. The chemical reactions are exothermic, and the CSTR is used at different times to make two products, A or B. The low-level and low-pressure alarm violations occur mainly during shutdown operations, whereas high temperature alarms for the jacket cooling water occur primarily when product B is produced. Devise a strategy for reducing these bad actor alarms.

Solution

Because these nuisance alarms tend to occur for different known-process conditions, a state-based alarming strategy is warranted (Guideline 11 of Table E.3). In particular, different alarm levels are required for the three reactor conditions. For example, the following alarm limits could be employed with variables expressed in %.

Condition	Level (high)	Level (low)	Pressure (high)	Cooling water temperature (high)
Product A	90%	10%	85%	75%
Product B	90%	10%	85%	85%
Shutdown/ empty	5%	0%	10%	30%

Production of B involves a more exothermic reaction and thus tends to result in a higher cooling water temperature. When the reactor is shut down and evacuated, the level alarm settings are reduced to very low values. In particular, if reactor is not completely evacuated, the alarm sounds if the liquid level exceeds 5%. The state-based alarm settings eliminate many of the nuisance alarms.

E.3 ABNORMAL EVENT DETECTION

The overall goal of process control is to maintain product quality under safe operating conditions. When operating conditions exceed acceptable limits as a result of external causes, equipment malfunctions, or human error, unacceptable situations can occur, including unsafe conditions, off-specification product, decreased production, and environmental degradation. Major process excursions may even require plant shutdown to prevent catastrophic events, such as explosions, fires, or discharge of toxic chemicals. Thus the detection and resolution of abnormal conditions, especially potentially disastrous situations, are very important plant activities that have received considerable attention in recent years, especially from government organizations (e.g., OSHA, 1996), engineering societies (AIChE, 2007), and industrial organizations (EEMUA, 1997; ASM, 2015).

In this section, we consider the early detection of abnormal conditions, an activity referred to as *abnormal event detection* (AED). It is important to note the crucial role of plant operating personnel, especially plant operators, in AED. Their process knowledge and experience can provide both early detection of abnormal situations and remedial action to either return the plant to normal levels of operation or to shut down the process. However, an operator's response depends on many factors: the number of alarms and the frequency of occurrence of abnormal conditions; how information is presented to the operator; the complexity of the plant; and the operator's intelligence, training, experience, and reaction to stress. Consequently, computational tools that assist plant personnel are crucial to the success of operating complex manufacturing plants. These computational tools can be embedded in the process control system. In this section, we consider three general approaches for AED.

E.3.1 Fault Detection Based on Sensor and Signal Analysis

Analysis of past and current values of measured variables can provide valuable diagnostic information for the detection of abnormal events. As described in Section E.1, on-line measurements are routinely checked to ensure that they are between specified high and low limits and to ensure that the rate of change is consistent with the physical process. Also, a simple calculation can identify a common type of sensor malfunction or fault, a *dead (or frozen) sensor*.

Suppose that the sample variance of a measured variable, x, is calculated from n consecutive measurements with a constant sample period (e.g., $\Delta t = 1$ min). The sample variance s^2 is defined as (see Appendix D):

$$s^2 = \frac{1}{n-1}\sum_{i=1}^{n} x_i^2 \tag{E-4}$$

Based on the expected process variability, measurement noise, and past experience, a lower limit for s^2, s^2_{min}, can be specified. For subsequent on-line monitoring, if $s^2 < s^2_{min}$, there is reason to be believe that the measurement may be essentially constant, due to a dead sensor.

Other simple signal analyses can be used to detect common problems, such as sticking control valves (Choudhury et al., 2008) and oscillating control loops (Babji et al., 2012). For AED applications that involve many process variables, multivariable statistical techniques, such as Principal Component Analysis (PCA) and Partial Least Squares (PLS), can be very effective (see Chapter 17).

E.3.2 Model-Based Methods

Both steady-state and dynamic process models can provide useful information for AED. The model can be either a physical model (Chapter 2) or an empirical model (Chapter 16). For example, equipment performance calculations based on mass and energy balances can be used to calculate thermal efficiencies for energy-intensive processes such as furnaces. An unusually low value could indicate a potentially hazardous situation, such as a burner malfunction.

If a chemical composition or a product quality variable cannot be measured on-line, it may be possible to predict it from measured process variables, such as flow rates, temperatures, and pressures. When the predictions are used for feedback control, this strategy is referred to as inferential control (see Chapter 12). On-line prediction of future values of controlled variables based on a dynamic model is a key element of a widely used, advanced control strategy: model predictive control (see Chapter 15).

Periodic evaluations of steady-state conservation equations provide a powerful tool for abnormal event detection. The error of closure, E_c, of a steady-state mass or component balance for a continuous process can be defined as

$$E_c = Rate\ In - Rate\ Out \tag{E-5}$$

An unusually large error of closure (in absolute value) suggests an abnormal event, for example, a malfunctioning sensor or an equipment problem, such as heat exchanger fouling, as illustrated by Example E.3.

EXAMPLE E.3

A feed mixture of two components is separated in a distillation column, as shown in Fig. E.8 where F, D, and B are molar flow rates and z, y, and x are mole fractions of the more volatile component. Assume that the three flow rates are constant: $F = 4$ mol/min, $D = B = 2$ mol/min. During normal operation, each composition measurement contains random errors. Based on previous experience, it can be assumed that the population means μ and population standard deviations σ (see Appendix E for definitions) have the following values:

	z	y	x
μ	0.500	0.800	0.200
σ	0.010	0.020	0.005

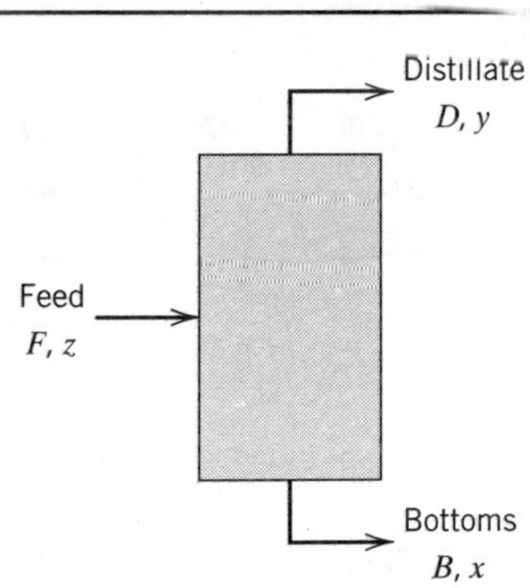

Figure E.8 Schematic diagram of a distillation column.

At steady state, the measured compositions were

$$\bar{z} = 0.485, \quad \bar{y} = 0.285, \quad \bar{x} = 0.205$$

Is there statistically significant evidence to support the assertion that one or more of the composition sensors are not working properly?

Solution

The error of closure for the steady-state component balance is

$$E_c = Fz - Dy - Bx$$

Substituting the measured values gives

$$E_c = 4(0.485) - 2(0.825) - 2(0.205) = -0.120 \text{ mol/min}$$

As shown in Appendix F, the standard deviation of E_c is

$$\sigma_{E_c} = \sqrt{F^2\sigma_z^2 + D^2\sigma_y^2 + B^2\sigma_x^2}$$

$$\sigma_{E_c} = \sqrt{(4)^2(0.01)^2 + (2)^2(0.02)^2 + (2)^2(0.005)^2}$$

$$= 0.0574 \text{ mol/min}$$

Because the absolute value of the calculated value of E_c is more than two standard deviations away from the expected value of zero, there is strong statistical evidence that an abnormal event has occurred. For example, if E_c is normally distributed, with a mean of $\mu_{Ec} = 0$ and a standard deviation of $\sigma_{Ec} = 0.0574$, the corresponding Z value (see Appendix E) is

$$Z = \frac{E_c - \mu_{E_c}}{\sigma_{E_c}} = \frac{-0.120 - 0}{0.0574} = -2.09$$

The probability that an abnormal event has occurred is the probability that $|Z| > 2.09$. From the table for the standard normal probability distribution (Montgomery and Runger, 2013), this probability is 0.963. Thus, there is statistically significant evidence to include that an abnormal event has occurred.

This error analysis can be extended to the situation where the flow measurements also contain random errors.

E.3.3 Knowledge-Based Methods

In the previous section, quantitative models were used to detect abnormal events. As an alternative, the AED assessment can utilize *qualitative* information based on process knowledge and past experience. This general strategy is referred to as a *knowledge-based approach.* It relies on specific methods, such as causal analysis, fault-tree diagrams, fuzzy logic, logical reasoning, and expert systems (Chiang et al., 2001; Dash and Venkatasubramanian, 2003).

To illustrate the use of qualitative knowledge in AED, consider the liquid storage system in Fig. E.5. Using qualitative concepts such as low, normal, and high, an AED would be indicated if the following logical IF-THEN statement is true based on measurements of q_1, q_2, q_3, and dh/dt:

IF ("q_1 is normal or low") AND ("q_2 is normal or low") AND ("q is normal or high") AND ("dh/dt is high"), THEN ("an AED has occurred"). This type of analysis requires thresholds for each qualitative concept and process variable.

For this simple example, a similar analysis could be based on a quantitative approach, namely, a dynamic model based on an unsteady-state mass balance. The level could be changing as a result of process changes or a sensor failure. However, for more complicated processes, a reasonably accurate dynamic model may not be available, and thus a qualitative approach can be used to good advantage. The diagnosis of the abnormal event could lead to a subsequent root cause analysis, where the source of the abnormality is identified and appropriate corrective actions are taken.

E.4 RISK ASSESSMENT

Risk assessment considers the consequences of potential hazards, faults, failures, and accident scenarios. In particular, it provides a quantitative assessment of risk in contrast to other approaches (e.g., a HAZOP study) that provide qualitative assessments. Because industrial processes are complex and interconnected, risk assessment determines overall failure probabilities from those of individual components. For example, the *failure rate* of a typical temperature control loop depends on the failure rates of the individual components: sensor/transmitter, controller, I/P transducer (if required), and the control valve. Like HAZOP, risk analysis should consider the possibility of human mistakes, as well as equipment and instrumentation problems or failures.

In this section, we consider relevant reliability concepts based on probability theory and statistics. Appendix F reviews concepts from probability and statistics that are needed for this analysis.

E.4.1 Reliability Concepts

A typical failure rate curve for process equipment and other hardware has the "bathtub shape" shown in Fig. E.9. For much of its lifetime, the failure rate is approximately constant, with a value denoted by μ. In the subsequent analysis, a constant failure rate is assumed for each component; however, extensions can be made to include time-varying failure rates.

The probability that the component does not fail during the interval, $(0, t)$, is defined to be its *reliability*, *R*.

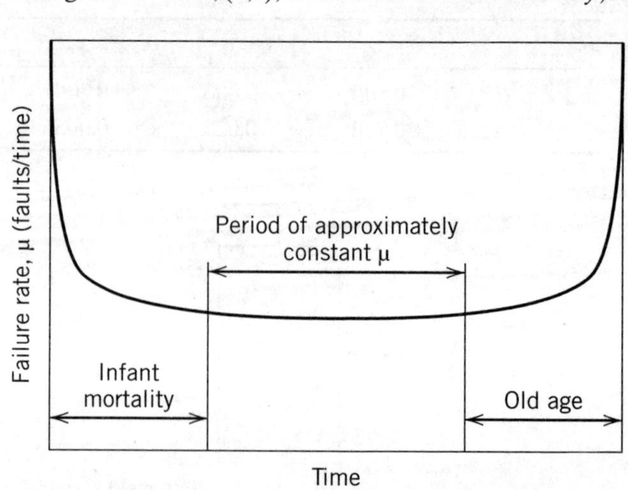

Figure E.9 The "bathtub curve," showing a typical failure rate plot for hardware and instruments.

For a component with a constant failure rate R is given by the exponential probability distribution (Crowl and Louvar, 2002),

$$R = e^{-\mu t} \tag{E-6}$$

and the corresponding *failure probability* P is

$$P = 1 - R \tag{E-7}$$

Both P and R are bounded by [0, 1]. The expected *mean time between failures* (MTBF) is given by

$$MTBF = \frac{1}{\mu} \tag{E-8}$$

E.4.2 Overall Failure Rates

We consider overall failure rates for two common situations.

Components in series: Consider a system of n independent components, each with an individual failure rate, μ_i. Suppose that all of the components must operate properly in order for the overall system to function. The reliability of the overall system R is given by

$$R = \prod_{i=1}^{n} R_i \tag{E-9}$$

where R_i is the reliability of the ith component. Then the overall failure probability P is

$$P = 1 - R = 1 - \prod_{i=1}^{n}(1 - P_i) \tag{E-10}$$

where $P_i = 1 - R_i$.

Components in parallel (redundant components): Improved system reliability can be achieved by using redundant components. For example, in Section E.1, redundant sensors and control loops were used in SIS and ESD systems. Suppose that n independent sensors are available and only one needs to be operational in order for the feedback control loop to operate properly. The probability that all m sensors fail is

$$P = \prod_{i=1}^{n} P_i \tag{E-11}$$

and the reliability of the overall system is

$$R = 1 - P = 1 - \prod_{i=1}^{n}(1 - R_i) \tag{E-12}$$

The calculation of overall failure rates from individual failure rates is illustrated in the following example.

EXAMPLE E.4

A flow control loop consists of a differential pressure flow sensor/transmitter, a digital PI controller, an I/P transducer, a control valve, and a valve positioner. Determine the reliability, failure rate, and MTBF for this control loop.

Solution

Using the failure rate data in Table E.1, the calculated values are

Component	Failure rate, μ (failures/yr)	Reliability (per yr) $R = e^{-\mu t}$	Failure probability $P = 1 - R$
DP flowmeter[†]	1.73	0.18	0.82
Digital controller	0.05	0.95	0.05
I/P transducer	0.49	0.61	0.39
Control valve	0.60	0.55	0.45
Valve positioner	0.44	0.64	0.36

[†] Assumed value (not in Table E.1).

All of the components must function properly in order for the control loop to operate. Thus, the components can be considered to be in a series configuration. The overall reliability of components in series is the probability that no failures occur. It can be calculated using Eq. E-9:

$$R = \prod_{i=1}^{5} R_i = (0.18)(0.95)(0.61)(0.55)(0.64) = 0.037$$

The overall failure probability per year is

$$P = 1 - R = 1 - 0.037 = 0.96$$

The overall failure rate is calculated from Eq. E-6:

$$0.037 = e^{-\mu}$$

$$\mu = -\ln(0.037) = 3.3 \text{ failures/yr}$$

The mean time between failures is

$$\text{MTBF} = \frac{1}{\mu} = 0.3 \text{ yr}$$

E.4.3 Fault and Event Tree Analysis

In order to evaluate the reliability and failure rates for complex processes, it is necessary to consider carefully the available physical information about causes and effects of possible failure modes. This information can be incorporated into the risk assessment using two types of analyses based on logic diagrams: *fault tree analysis* (FTA) and *event tree analysis* (ETA). Fault trees display all of the component failures that can lead to a very serious situation, such as an accident or an explosion, and the subsequent chain of events. Event trees are similar to fault trees but focus on a single initiating event, such as a component failure, and then evaluate the consequences, classified according to how serious they are. Both methods should consider human mistakes by plant personnel, as well as equipment and instrumentation problems or failures. Both FTA and ETA are used to analyze proposed designs for new processes and to diagnose the reliability of existing or retrofit processes (Crowl and Louvar, 2002; Banerjee, 2003).

For fault tree analysis, the starting point is to specify an undesirable serious situation, called the *top effect*, and then to consider all possible causes that could produce it. For example, the specified top effect could be the over-pressurization of a chemical reactor. Possible causes could include a reduction or loss of coolant, excess catalyst, an ineffective pressure control loop, etc. Each possible cause is analyzed further to determine why it occurred. For example, the pressure control loop problem could be to the result of a sensor or control valve malfunction. Thus, FTA is a *top-down approach* that generates a tree of causal relations, starting with the specified top event and working backward. Standard logic concepts, such as AND and OR, are used in the logic diagrams.

Event tree analysis is a similar technique, but its starting point is a single cause, rather than a single outcome. For the previous chemical reactor example, the specified starting point might be a blockage in the coolant piping or a sensor failure in the pressure control loop. Logic diagrams are used to illustrate how the initial root cause propagates through different levels in the event tree and produces one or more undesirable outcomes. Thus ETA is called a *bottom-up approach*.

After an FTA or ETA diagram is generated, a probability or probability distribution is specified for each step in the tree, and a risk assessment analysis is made. Although the FTA and ETA can be time-consuming and involve a degree of subjectivity, they can provide considerable insight into reliability, provided that there is considerable physical insight and/or experience available. Also, fault and event trees can be devised with different levels of detail (Deshotels and Dejmek, 1995).

SUMMARY

Process safety is both a paramount concern in manufacturing plants and a primary issue in control system design. In this chapter, it has been shown that automation and process control play key roles in ensuring safe process operation. The primary strategy for process safety is based on the concept of layers of protection that is illustrated in Fig. E.1. Alarm management and alarm resolution are essential components of a plant safety system. The selection of the variables to be alarmed and their alarm limits are based on hazard identification and risk assessment. Abnormal event detection and risk assessment can play key roles in enhancing plant safety.

Finally, as Rinard (1990) has poignantly noted, “The regulatory control system affects the size of your paycheck; the safety control system affects whether or not you will be around to collect it.”

REFERENCES

AIChE Center for Chemical Process Safety, *Guidelines for Safe and Reliable Instrumented Protective Systems*, AIChE, NY, 2007.

AIChE Center for Chemical Process Safety, *Guidelines for Safe Automation of Chemical Processes*, AIChE, NY, 1993.

AIChE Center for Chemical Process Safety, *Layer of Protection Analysis: Simplified Process Risk Assessment*, AIChE, NY, 2001.

AIChE Center for Chemical Process Safety, *Introduction to Process Safety for Undergraduates and Engineers*, John Wiley and Sons, Hoboken, NJ, 2016.

Abnormal Situation Management (ASM) Consortium, http://www.asmconsortium.com, 2015.

Babji S., U. Nallasivam, and R. Rengaswamy, Root Cause Analysis of Linear Closed-Loop Oscillatory Chemical Process Systems, *Ind. Eng. Chem. Res.*, **51** 13712 (2012).

Banerjee, S., *Industrial Hazards and Plant Safety*, Taylor and Francis, New York, 2003.

Chiang, L. H., E. L. Russell, and R. D. Bratz, *Fault Detection and Diagnosis in Industrial Systems*, Springer-Verlag, London, U.K., 2001.

Choudhury, M. A. A. S., M. Jain, and S. L. Shah, Stiction—Definition, Modelling, Detection and Quantification, *J. Process Contr.*, **18**, 232 (2008).

Connell, B., *Process Instrumentation Applications Manual*, McGraw-Hill, New York, 1996.

Crowl, D. A., and J. F. Louvar, *Chemical Process Safety: Fundamentals with Applications*, 2nd ed., Prentice Hall, Englewood Cliffs, NJ, 2002.

Dash, S., and V. Venkatasubramanian, Integrated Framework for Abnormal Event Management and Process Hazard Analysis, *AIChE J.*, **49**, 124 (2003).

Deshotels, R., and M. Dejmek, Choosing the Level of Detail for Hazard Identification, *Process Saf. Prog.*, **14**, 218 (1995).

Engineering Equipment & Materials Users' Association (EEMUA), *Alarm Systems—A Guide to Design, Management and Procurement*, 2nd ed., Publication 191, London, 2007.

Hollifield, B. R., and E. Habibi, *Alarm Management: Seven Effective Methods for Optimum Performance*, ISA, Research Triangle Park, NC, 2007.

Hollifield, B. R., and E. Habibi, *Alarm Management: A Comprehensive Guide*, ISA, Research Triangle Park, NC, 2011.

International Society of Automation (ISA), *ANSI/ISA-84.00.01-2004: Application of Safety Instrumented Systems for the Process Industries*, Research Triangle Park, NC (2004).

Jofriet, P., Alarm Management, *Chem. Eng.*, **112** (2), 36 (2005).

Kletz, T. A., *HAZOP and HAZAN: Identifying and Assessing Process Industry Hazards*, 4th ed., IChemE, Rugby, Warwickshire, England, 2001.

Kletz, T. A., What Went Wrong? Case Histories of Process Plant Disasters and How They Could Have Been Avoided, 5th ed., Elsevier, New York, 2009.

Leveson N. G., and G. Stephanopoulos, A System-Theoretic, Control-Inspired View and Approach to Process Safety, *AIChE J.*, **60**,2 (2014).

Mannan, S. (Ed.), *Lees' Loss Prevention in the Process Industries: Hazard Identification, Assessment, and Control, Vol. 1*, 4th ed., Butterworth-Heinemann Elsevier, New York, 2012.

Mannan, M. S., S. Sachdeva, H. Chen, O. Reyes-Valdes, Y. Liu, and D. M. Laboureur, Trends and Challenges in Process Safety, *AIChE J.*, **61**, 3558 (2015)

McGuire, C., J. Lo, and E. Mathiason, Selecting Sensors for Safety Instrumented Systems, *Chem. Eng. Progr.*, **111** (7), 19 (2015).

Montgomery, D. C., and G. C. Runger, *Applied Statistics and Probability for Engineers*, 6th ed., John Wiley and Sons, Hoboken, NJ, 2013.

Nuclear Regulatory Commission, Backgrounder on the Three Mile Island Accident, http://www.nrc.gov/reading-rm/doc-collections/fact-sheets/3mile-isle.html, 2015.

Occupational Safety and Health Administration (OSHA), *Process Safety Management*, Publication 3132, OSHA, U.S. Dept. of Labor, Washington D.C. 2000.

Perry, R. H., and D. W. Green (Eds.), *Chemical Engineers' Handbook*, 8th ed., Section 23, Process Safety, McGraw-Hill, New York, 2008.

Rinard, I., Discussion, *Chem. Eng. Educ.*, **24**, Spring Issue, 76 (1990).

Turton, R., R. C. Bailie, W. B. Whiting, and J. A. Shaeiwitz, *Analysis, Synthesis and Design of Chemical Processes*, 4th ed., Prentice-Hall PTR, Upper Saddle River, NJ, 2012.

Venkatasubramanian, V., Systemic Failures: Challenges and Opportunities in Risk Management in Complex Systems, *AIChE J.*, **57**,2 (2011).

EXERCISES

E.1 Air samples from a process area are continuously drawn through a 6 mm diameter tube to an analytical instrument that is located 60 m away. The tubing has an outside diameter of 6.35 mm and a wall thickness of 0.762 mm. The flow rate through the transfer line is 10 cm³/s for ambient conditions of 20°C and 1 atm. The pressure drop in the transfer line is negligible. Because chlorine gas is used in the process, a leak can poison workers in the area. It takes the analyzer 10 s to respond after chlorine first reaches it. Determine the amount of time that is required to detect a chlorine leak in the processing area. State any assumptions that you make. Would this amount of time be acceptable if the hazardous gas were carbon monoxide, instead of chlorine?

(Adapted from: *Student Problems for Safety, Health, and Loss Prevention in Chemical Processes*, AIChE Center for Chemical Process Safety, NY, 1990).

E.2 A cylindrical storage tank is 2 m tall and 1 m in diameter. A low-level alarm is activated when the liquid level decreases to 0.25 m. Suppose that the tank is initially half full when a slow leak develops near the bottom of the tank due to corrosion. The relationship between the liquid level h and the leakage flow rate q is

$$q = C\sqrt{h}$$

where q has units of m³/min, h is in m and $C = 0.065$ m$^{2.5}$/min.

(a) If the leak begins at 5 AM, when will the alarm sound?

(b) How much liquid has leaked out when the alarm sounds?

E.3 The two-phase feed stream for the gas-liquid separator (or flash drum) shown in Fig. EE.3 consists of a hydrocarbon mixture. Because the pressure in the vessel is significantly lower than the feed pressure, part of the liquid feed flashes to form a gas phase. The hydrocarbons are flammable and somewhat hazardous. Discuss the process safety issues and propose an alarm/SIS strategy.

E.4 The loss of the coolant to a process vessel can produce an unacceptably high pressure in the vessel. As a result, a pressure relief valve is used to reduce the pressure by releasing the vapor mixture to the atmosphere. But if the mixture is toxic or flammable, the release can be hazardous. For the distillation column in Fig. EE.4,

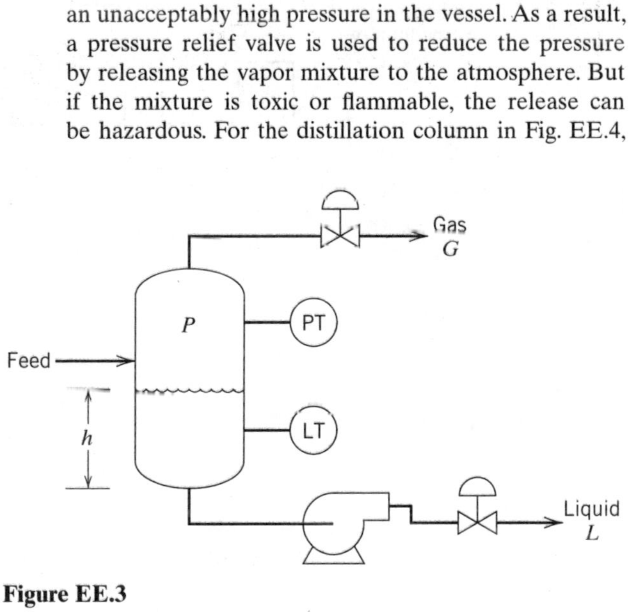

Figure EE.3

which operates at above ambient temperature, propose an alarm/SIS system that will reduce the number of releases to the environment, even though the occasional loss of coolant flow to the condenser is unavoidable. (*Note:* The pressure relief valve at the top of the column is not shown in Fig. EE.4.)

E.5 The probability of a particular type of sensor functioning properly is 0.95. Consequently, a triply redundant sensor system has been proposed for a critical measurement. Thus, three independent sensors will be installed, and the median of the three measurements will be used for the alarms and control calculations. What is the probability that at least two of the sensors will be working at any time?

E.6 Consider the liquid storage tank with a low-level interlock, as shown in Fig. EE.4. Suppose that an independent low-level alarm is added, with its set-point value above the value for the low-level switch. If both the low-level alarm and the low-level interlock system fail simultaneously, the pump could be seriously damaged. What is the probability that this occurs? What is the mean time between failures?

Failure rates (faults per year):

Solenoid switch: $\mu_S = 0.01$

Level switch: $\mu_{LS} = 0.45$

Level alarm: $\mu_A = 0.3$

E.7 For the reliability analysis of the flow control loop in Example E.4, the DP flowmeter is the least reliable component. Suppose that a second, identical flowmeter is used in a backup mode so that it could be automatically and immediately employed if the first flowmeter failed. How much would the overall system reliability improve by adding the second sensor?

E.8 Repeat Exercise E.7 for the case where the flowmeters are triply redundant; that is, there are three identical flowmeters with two in the backup mode. How much would the overall system reliability improve?

E.9 Using the failure rate data in Table E.1, evaluate the reliability and mean time between failures for the high-pressure interlock in Fig. E.4. Assume that the failure rate for the solenoid switch and valve is $\mu = 0.42$ faults per year.

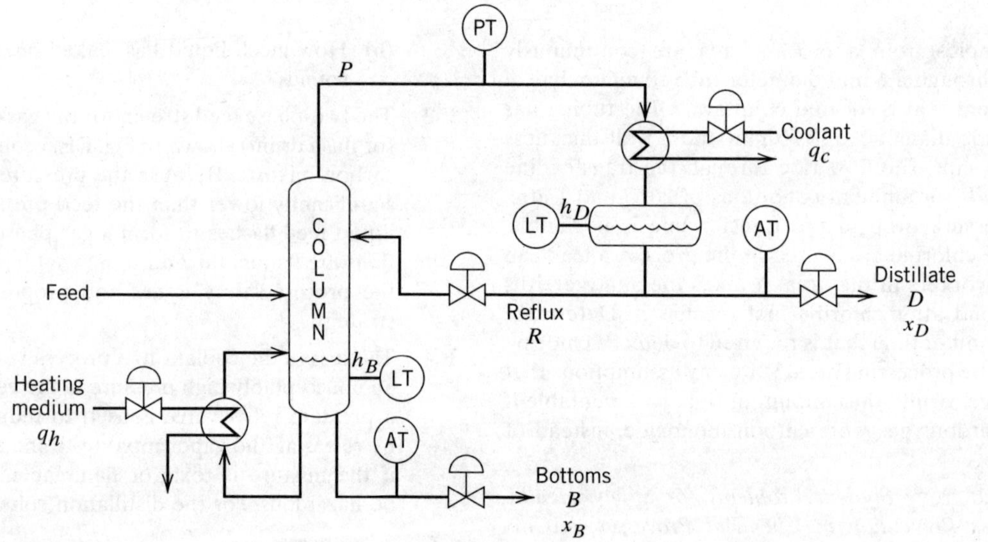

Figure EE.4

APPENDIX F

Real-Time Optimization

APPENDIX CONTENTS

Previous chapters have considered the development of process models and the design of controllers from an unsteady-state (dynamic) point of view. Such an approach focuses on obtaining reasonable closed-loop responses for set-point changes and disturbances. Up to this point, we have only peripherally mentioned how set points should be specified for the process. The on-line calculation of optimal set points, also called *real-time optimization* (RTO), allows the profits from the process to be maximized (or costs to be minimized) while satisfying operating constraints. The appropriate optimization techniques are implemented in the computer control system. Steady-state models are normally used, rather than dynamic models, because the process is intended to be operated at steady state except when the set point is changed.

This chapter first discusses basic RTO concepts and then describe typical applications to process control. Guidelines for determining when RTO can be advantageous are also presented. Subsequently, set-point selection is formulated as an optimization problem, involving economic information and a steady-state process model. Optimization techniques that are used in the process industries are briefly described. For more information, see textbooks on optimization methodology (Ravindran et al., 2006; Biegler, 2010; Edgar et al., 2001).

Figure F.1 is a detailed version of Fig. 1.7, which shows the five levels in the process control hierarchy where various optimization, control, monitoring, and data acquisition activities are employed. The relative position of each block in Fig. F.1 is intended to be conceptual, because there can be overlap in the functions carried out, and often several levels may utilize the same computing platform. The relative time scale for each level's activity is also shown. Process data (flows, temperatures, pressures, compositions, etc.) as well as *enterprise data*, consisting of commercial and financial information, are used with the methodologies shown to make decisions in a timely fashion. The highest level (planning and scheduling) sets production goals to meet supply and logistics constraints and addresses time-varying capacity and manpower utilization decisions. This *enterprise resource planning* (ERP) and the *supply chain management* in level 5 refer to the links in a web of relationships involving retailing (sales), distribution, transportation, and manufacturing (Grossman, 2005). Planning and scheduling usually operate over relatively long time scales and tend to be decoupled from the rest of the activities in lower levels (Geddes and Kubera, 2000). For example, Shobrys and White (2002) indicate that all of the refineries owned by an oil company are usually included in a comprehensive planning and scheduling model. This model can be optimized to obtain target levels and prices for inter-refinery transfers, crude oil and product allocations to each refinery, production targets, inventory targets, optimal operating conditions, stream allocations, and blends for each refinery (Alattas et al., 2011).

In Level 4, RTO provides optimal set points for each unit in a plant, which is also called *supervisory control*.

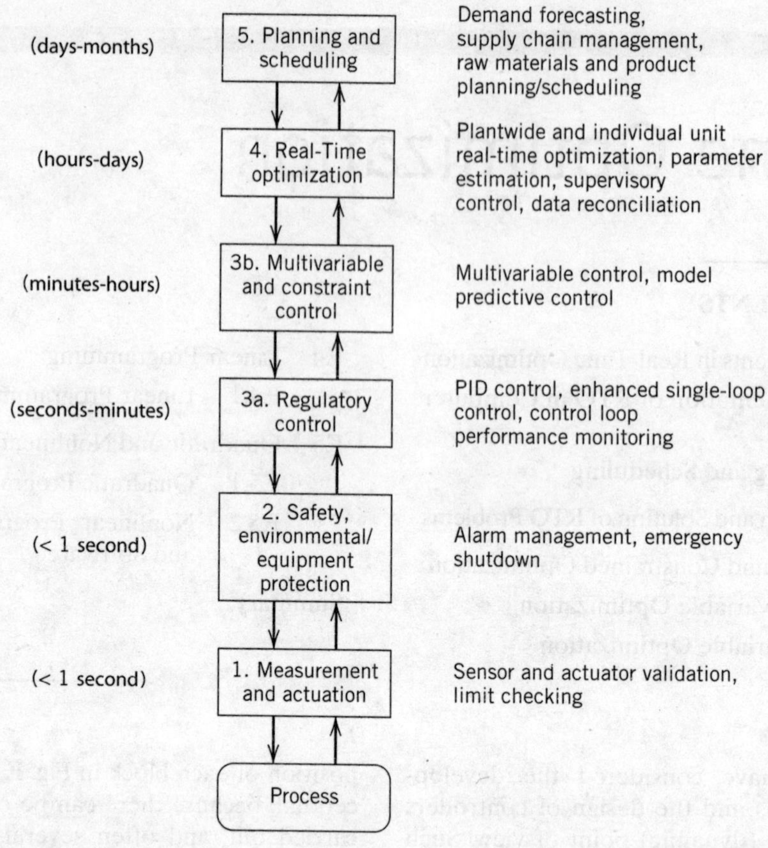

Figure F.1 The five levels of process control and optimization in manufacturing. Time scales are shown for each level.

For multivariable control or processes with active constraints, set-point changes are performed in Level 3b (e.g., model predictive control discussed in Chapter 15). For single-loop or multiloop control the regulatory control is performed at Level 3a. Level 2 (safety and environmental/equipment protection) includes activities such as alarm management and emergency shutdowns. Although software implements the tasks shown, there is also a separate hardwired safety system for the plant, as discussed in Appendix E. Level 1 (process measurement and actuation) provides data acquisition and on-line analysis and actuation functions, including some sensor validation. Ideally, there is bidirectional communication between levels, with higher levels setting goals for lower levels and the lower levels communicating constraints and performance information to the higher levels. The time scale for decision-making at the highest level (planning and scheduling) may be of the order of months, while at lower levels (for example, regulatory control), decisions affecting the process can be made frequently (e.g., in fractions of a second). The main focus of this chapter is on Level 4.

Plant profitability can be enhanced by performing optimization of operating conditions on a regular basis. In a large plant, the improved profits attained with RTO can be substantial (White, 2010). Optimal operating points can sometimes change markedly from day to day, or even during the course of one day. For example, the price of delivered electrical power can vary by a factor of five from highest to lowest price (due to time-of-day pricing by electrical utilities). Other changes that require periodic optimization of operating conditions include variations in the quality and cost of feedstocks, processing and storage limits, and product demands. With recent advances in digital hardware and optimization software, RTO can be easily incorporated into computer control systems. The scale at which industrial RTO can be implemented is impressive. Problems with over 100,000 variables and equality/inequality constraints are routinely solved (Biegler, 2010).

F.1 BASIC REQUIREMENTS IN REAL-TIME OPTIMIZATION

The steady-state model used in RTO typically is obtained either from fundamental knowledge of the plant or from experimental data. It utilizes the plant operating conditions for each unit, such as temperature, pressure, and feed flow rates to predict properties such as product yields (or distributions), production rates, and measurable product characteristics (e.g., purity, viscosity, and molecular weight). The economic model involves the costs of raw materials, values of products, and costs of production as functions of operating conditions, projected sales figures, and so on. An objective function is specified in terms of these quantities; in particular, operating profit over some specific period of time can be expressed as functions of product flow rate, feed flow rate and operating cost as given in Eq. 15-84.

Both the operating and economic models typically will include constraints on

1. ***Operating conditions:*** Process variables must be within certain limits due to valve ranges (0% to 100% open), equipment limitations (pressures or temperatures), and environmental restrictions (e.g., emissions regulations).
2. ***Feed and production rates:*** A feed pump has a maximum capacity; sales are limited by market projections.
3. ***Storage and warehousing capacities:*** Storage tank capacity cannot be exceeded during periods of low demand.
4. ***Product impurities:*** A salable product cannot contain more than the maximum amount of a specified contaminant or impurity.
5. ***Environmental regulations:*** Emission limits for key pollutants must be met and sustainability objectives satisfied.

Process operating situations that are relevant to maximizing operating profits include

1. ***Sales limited by production.*** In this type of market, sales can be increased by increasing production. This can be achieved by optimizing operating conditions and production schedules.
2. ***Sales limited by market.*** This situation is susceptible to optimization only if improvements in efficiency at current production rates can be obtained. An increase in thermal efficiency, for example, usually leads to a reduction in manufacturing costs (e.g., utilities or feedstocks).
3. ***Large throughput.*** Units with large production rates (or throughputs) offer great potential for increased profits. Small savings in product costs per unit throughput or incremental improvements in yield, plus large production rates, can result in major increases in profits.
4. ***High raw material or energy consumption.*** These are major cost factors in a typical plant and thus offer potential savings. For example, the optimal allocation of fuel supplies and steam in a plant can reduce costs by minimizing fuel consumption.
5. ***Product quality better than specification.*** If the product quality is significantly better than the customer requirements, it can cause excessive production costs and wasted capacity. By operating closer to the customer requirement (e.g., impurity level), cost savings can be obtained, but this strategy also requires lower process variability (see Fig. 1.9).
6. ***Losses of valuable or hazardous components through waste streams.*** The chemical analysis of plant waste streams, both to air and water, will indicate whether valuable materials are being lost or if undesirable pollutants are being emitted. Adjustment of air/fuel ratios in furnaces to minimize unburned hydrocarbon losses and to reduce nitrogen-oxide emissions is one such example.

Timmons et al. (2000) have discussed opportunities for the application of on-line optimization or supervisory control in refinery operations. Three general types of optimization problems commonly encountered in industrial process operations are discussed next.

Operating Conditions

Common examples include optimizing distillation column reflux ratio and reactor temperature. Consider the RTO of a fluidized catalytic cracker (FCC) (Latour, 1979). The FCC reaction temperature largely determines the conversion of a light gas oil feedstock to lighter (i.e., more volatile) components. The product distribution (gasoline, middle distillate, fuel oil, light gases) changes as the degree of conversion is increased. Accurate process models of the product distribution as a function of FCC operating conditions and catalyst type are required for real-time optimization. Feedstock composition, downstream unit capacities (e.g., distillation columns), individual product prices, product demand, feed preheat, gas oil recycle, and utilities requirements must be considered in optimizing an FCC unit. The large throughput of the FCC implies that a small improvement in yield translates to a significant increase in profits. Biegler et al. (1997) have discussed an RTO case study on a hydrocracker and fractionation plant.

Olefins plants in which ethylene is the main product are another application where RTO has had a significant impact (Darby and White, 1998; Starks and Arrieta, 2007). A full plant model can have as many as 1500 submodels, based on the development of fundamental chemical engineering relations for all unit operations involved, that is, furnaces, distillation columns, mixers, compressors, and heat exchangers (Georgiou et al.,

1997). Although standard approaches are used for developing separation and heat exchange models, the furnace models are quite elaborate and are typically proprietary. Key optimization variables are conversion, feed rate, and steam/oil ratio, subject to feedstock availability and equipment constraints with improved profits as large as millions of dollars per year (Wang et al., 2014).

Allocation

Allocation problems involve the optimal distribution of a limited resource among several parallel (alternative) process units. Typical examples include (Marlin and Hrymak, 1997):

Steam Generators. Optimum load distribution among several boilers of varying size and efficiency.

Refrigeration Units. Optimum distribution of a fixed refrigeration capacity among several low-temperature condensers associated with distillation columns.

Parallel Distillation Columns. Minimization of "off-spec" products and utilities consumption while maximizing overall capacity.

F.1.1 Implementation of RTO in Computer Control

The RTO system performs all data transfer and optimization calculations and sends set-point information to the controllers. It should perform all tasks without unduly upsetting plant operations. Several steps are necessary for implementation of RTO, including data gathering and validation (or reconciliation), determination of the plant steady state, updating of model parameters (if necessary) to match current operations, calculation of the new (optimized) set points, and implementation of these set points.

To determine whether a process unit is at steady state, the RTO system monitors key plant measurements (e.g., compositions, product rates, flow rates, etc.) and checks if the plant operating conditions are close enough to steady state to start the RTO sequence. Only when all of the key measurements are within the allowable tolerances is the plant considered to be at steady state and the optimization calculations are started; see Kelly and Hedengren (2013) for a statistical technique that determines the existence of steady-state conditions. The optimization software screens the measurements for unreasonable data (*gross error detection*). Data validity checking automatically adjusts the model updating procedure to reflect the presence of bad data or equipment that has been taken out of service. *Data reconciliation* based on satisfying material and energy balances can be carried out using separate optimization software (Narasimhan and Jordache, 2000). Data validation and reconciliation is a critical part of any RTO activity. If measurement errors resulting from poor instrument calibration are not considered, the data reconciliation step or subsequent parameter estimation step will not provide meaningful answers (Soderstrom et al., 2000).

The optimization software can update model parameters to match current plant data, using regression techniques. Typical model parameters include exchanger heat transfer coefficients, reactor performance parameters, and furnace efficiencies. The parameters appear in material and energy balances for each unit in the plant as well as constitutive equations for physical properties. Parameter updating compensates for plant changes and degradation of process equipment, although there is a loss of performance when the model parameters are uncertain or the plant data contain noise (Perkins, 1998). Considerable plant knowledge and experience is required in deciding which parameters to update and which data to use for the updates.

After completion of the parameter estimation, the current plant constraints, the control status data, and the economic values for feeds, products, utilities, and other operating costs are collected. Planning and scheduling software updates the economic values on a regular basis. The optimization software then calculates the optimum set points. The steady-state condition of the plant is rechecked after the optimization calculation. If the individual processes are confirmed to still be at the same steady state, then new set points are transferred to the computer control system for implementation. Subsequently, the process control computer repeats the steady-state detection calculations, restarting the cycle. If the new optimum set points are not statistically different from the previous ones, no changes are made (Marlin and Hrymak, 1997).

The combination of RTO and regulatory control can be viewed as analogous to cascade control. As shown in Fig. F.2, the outer RTO loop will operate more slowly than the inner loop, and a poor design of this interaction results in inferior performance. The dynamic controller (or layer 3) handles the transition between the steady-state model used in RTO and the actual dynamic operation of the process. If the RTO model and dynamic model have very different gains, the resulting combination can perform poorly (Marlin and Hrymak, 1997).

F.1.2 Planning and Scheduling

The hierarchical levels shown in Fig. F.1 are designed to reduce the complexity in operational decision-making. As discussed by Darby et al. (2011), *planning* and *scheduling* are carried out less frequently than RTO

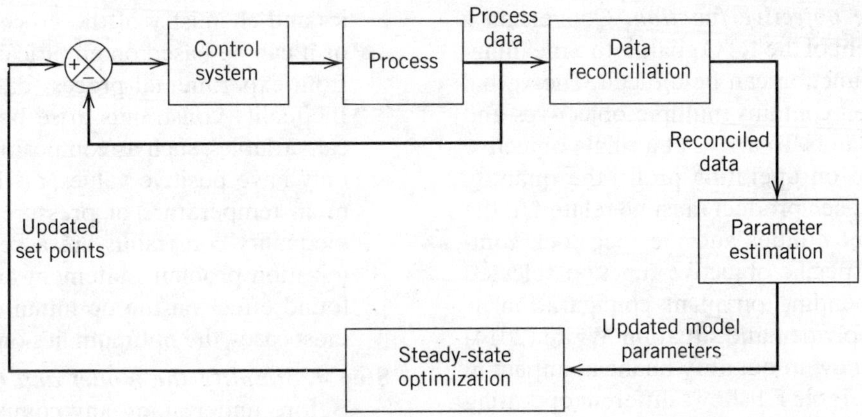

Figure F.2 A block diagram for RTO and regulatory feedback control.

and reside at higher levels in Fig. F.1. Planning deals with "what and how" based on economic forecasting. It determines which feedstocks and the amounts to purchase and which products to make and the amounts of each product. Scheduling focuses on the timing of actions that must be taken to execute the plan (the "when"). It addresses when specific feeds are to be delivered and when operating modes are to be changed. Both planning and scheduling are considered to be part of *supply chain management*. For an oil production and refining company, the supply chain goes from the wellhead and includes the trading and scheduling of crude oils and other feedstocks, the trading and scheduling of products from a refinery, transfer of products from the refinery to the terminal, and terminal loading/storage. *Enterprise-wide optimization* is used to maximize profitability by managing the operations of supply, manufacturing, and distribution activities in the supply chain, thereby reducing costs and inventories (Grossmann, 2005).

Examples of scheduling problems encountered in continuous plants include catalyst regeneration, furnace decoking, and heat exchanger cleaning, which deal with the tradeoff between operating efficiency and lost production due to maintenance. Planning problems normally entail optimization of plant operations over a period of months. This approach is commonly used in refinery optimization (Alattas et al., 2011). In batch processing, optimal scheduling is crucial to match equipment to product demands and to minimize cycle times (see Chapter 18). In a batch campaign, several batches of product may be produced using the same recipe. In order to optimize the production process, the engineer needs to determine the recipe that satisfies product quality requirements, the production rates to fulfill the product demand, the availability of raw material inventories, product storage availability, and the run schedule (Harjunkoski et al., 2014).

Recently, there has been considerable research on integrating two or more of the five levels in Fig. F.1, which has mostly focused on the merging of planning and scheduling levels. Applying powerful optimization tools can address combined planning and scheduling problems with a large number of decision variables and constraints. There is also interest in solving problems that encompass both scheduling and control, where dynamic considerations come into play (Baldea and Harjunkoski, 2014). Illustrative applications of enterprise-wide optimization have been presented by Wassick (2009), Grossmann (2012), and Wang et al. (2014).

F.2 THE FORMULATION AND SOLUTION OF RTO PROBLEMS

Once a process has been selected for RTO, an appropriate problem statement must be formulated and then solved. As mentioned earlier, the optimization of set points requires

1. The economic model, an objective function to be maximized or minimized, that includes costs and product values
2. The operating model, which includes a steady-state process model and all constraints on the process variables

Edgar et al. (2001) have listed six steps that should be used in solving any practical optimization problem. A summary of the procedure with comments relevant to RTO is given below.

Step 1. ***Identify the process variables.*** The important input and output variables for the process must be identified. These variables are employed in the objective function and the process model (see Steps 2 and 3).

Step 2. ***Select the objective function.*** Converting a verbal statement of the RTO goals into a meaningful objective function can be difficult. The verbal statement often contains multiple objectives and implied constraints. To arrive at a single objective function based on operating profit, the quantity and quality of each product must be related to the consumption of utilities and the feedstock composition. The specific objective function selected may vary depending on plant configuration as well as the supply/demand situation. White (2014) has discussed how to quantify financial impact in a typical plant. Table F.1 shows different operating objectives that may arise for a fluidized catalytic cracker.

Step 3. ***Develop the process model and constraints.*** Steady-state process models are formulated, and operating limits for the process variables are identified. The process model can be based on the physics and chemistry of the process (see Chapter 2), or it can be based on empirical relations obtained from experimental process data (see Chapter 7). Inequality constraints arise because many physical variables, such as composition or pressure, can only have positive values, or there may be maximum temperature or pressure restrictions. These inequality constraints are a key part of the optimization problem statement and can have a profound effect on the optimum operating point. In most cases, the optimum lies on a constraint.

Table F.1 Alternative Operating Objectives for a Fluidized Catalytic Cracker

1. Maximize gasoline yield subject to a specified feed rate.
2. Minimize feed rate subject to required gasoline production.
3. Maximize conversion to light products subject to load and compressor/regenerator constraints.
4. Optimize yields subject to fixed feed conditions.
5. Maximize gasoline production with specified cycle oil production.
6. Maximize feed with fixed product distribution.
7. Maximize FCC gasoline plus olefins for alkylate.

Step 4. ***Simplify the model and objective function.*** Before undertaking any computation, the mathematical statement developed in steps 1-3 may be simplified to be compatible with the most effective solution techniques. A nonlinear objective function and nonlinear constraints can be linearized in order to use a fast, reliable optimization method such as linear programming.

Step 5. ***Compute the optimum.*** This step involves choosing an optimization technique and calculating the optimum set points. Over the past 20 years, much progress has been made in developing efficient and robust numerical methods for optimization calculations (Biegler, 2010; Griva et al., 2008; Nocedal and Wright, 2006).

Step 6. ***Perform sensitivity studies.*** It is useful to know which parameters in an optimization problem are the most important in determining the optimum. By varying model and cost parameters individually and recalculating the optimum, the most sensitive parameters can be identified.

Example F.1 illustrates the six steps.

EXAMPLE F.1

A section of a chemical plant makes two specialty products (E, F) from two raw materials (A, B) that are in limited supply. Each product is formed in a separate process as shown in Fig. F.3. Raw materials A and B do not have to be totally consumed. The reactions involving A and B are as follows:

$$\text{Process 1:} \quad A + B \rightarrow E$$
$$\text{Process 2:} \quad A + 2B \rightarrow F$$

The processing cost includes the costs of utilities and supplies. Labor and other costs are Rs 14000/day for process 1 and Rs 24500/day for process 2. These costs occur even if the production of E or F is zero. Formulate the objective function as the total operating profit per day. List the equality and inequality constraints (Steps 1, 2, and 3).

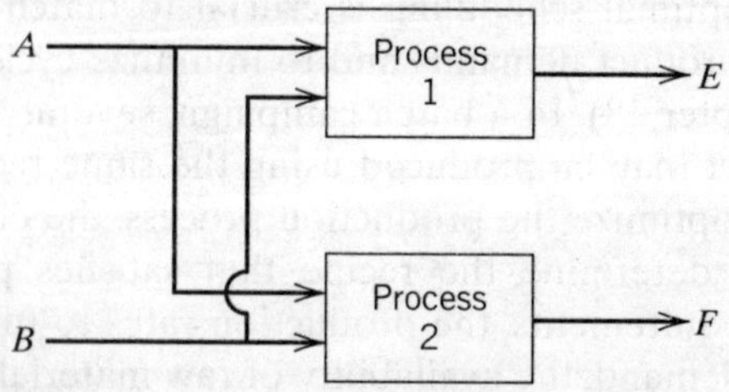

Figure F.3 A flow diagram of a chemical plant (Example F.1).

Available Information

Raw Material	Maximum Available (kg/day)	Cost (Rs/kg)
A	18,000	10
B	14,000	14

Process	Product	Reactant Requirements (kg) per kg Product	Processing Cost	Selling Price of Product	Maximum Production Level (kg/day)
1	E	2/3 A, 1/3 B	Rs 10/kg E	Rs 28/kg E	14,000
2	F	1/2 A, 1/2 B	Rs 4/kg F	Rs 23/kg F	14,000

Solution

The optimization problem is formulated using the first three steps delineated above.

Step 1. The relevant process variables are the mass flow rates of reactants and products (see Fig. F.3):

x_1 = kg/day A consumed
x_2 = kg/day B consumed
x_3 = kg/day E produced
x_4 = kg/day F produced

Step 2. In order to use Eq. F.1 to compute the operating product per day, we need to specify product sales income, feedstock costs, and operating costs:

$$\text{Sales income (kg/day)} = \sum_s F_s V_s = 28x_3 + 23x_4 \tag{F-1}$$

$$\text{Feedstock costs (kg/day)} = \sum_r F_r C_r = 10x_1 + 14x_2 \tag{F-2}$$

$$\text{Operating costs (kg/day)} = \text{OC} = 10x_3 + 4x_4 + 14{,}000 + 24{,}500 \tag{F-3}$$

Substituting into Eq. F-1 yields the daily profit:

$$\begin{aligned} P &= 28x_3 + 23x_4 - 10x_1 - 14x_2 \\ &\quad - 10x_3 - 4x_4 - 14{,}000 - 24{,}500 \\ &= 18x_3 + 19x_4 - 10x_1 - 14x_2 \\ &\quad - 14{,}000 - 24{,}500 \end{aligned} \tag{F-4}$$

Step 3. Not all variables in this problem are unconstrained. First consider the material balance equations, obtained from the reactant requirements, which in this case comprise the process operating model:

$$x_1 = 0.667x_3 + 0.5x_4 \tag{F-5a}$$
$$x_2 = 0.333x_3 + 0.5x_4 \tag{F-5b}$$

The limits on the feedstocks and production levels are

$$0 \le x_1 \le 18{,}000 \tag{F-6a}$$
$$0 \le x_2 \le 14{,}000 \tag{F-6b}$$
$$0 < x_3 \le 14{,}000 \tag{F-6c}$$
$$0 \le x_4 \le 14{,}000 \tag{F-6d}$$

Equations F-4 through F-6 constitute the optimization problem to be solved. Because the variables appear linearly in both the objective function and constraints, this formulation is referred to as a *linear programming problem*, which is discussed in Section F.4.

F.3 UNCONSTRAINED AND CONSTRAINED OPTIMIZATION

Unconstrained optimization refers to the situation where there are no inequality constraints and all equality constraints can be eliminated by variable substitution in the objective function. First we consider single-variable optimization, followed by optimization problems with multiple variables. Because optimization techniques are iterative in nature, we focus mainly on efficient methods that can be applied on-line. Most RTO applications are multivariable problems, which are considerably more challenging than single-variable problems.

F.3.1 Single-Variable Optimization

Some RTO problems involve determining the value of a single independent variable that maximizes (or minimizes) an objective function. Examples of single-variable optimization problems include optimizing the reflux ratio in a distillation column or the air/fuel ratio in a furnace. Optimization methods for single-variable problems are typically based on the assumption that the objective function $f(x)$ is *unimodal* with respect to x over the region of the search. In other words, a single maximum (or minimum) occurs in this region. To use these methods, it is necessary to specify upper and lower bounds for x^{opt}, the optimum value of x, by evaluating $f(x)$ for trial values of x within these bounds and observing where $f(x)$ is a maximum (or minimum). The values of x nearest this apparent optimum are specified to be the region of the search. This region is also referred to as the *interval of uncertainty* or *bracket*, and is used to initiate the formal optimization procedure.

Efficient single-variable (or *one-dimensional*) optimization methods include Newton and quasi-Newton methods and polynomial approximation (Edgar et al., 2001). The second category includes quadratic interpolation, which utilizes three points in the interval of uncertainty to fit a quadratic polynomial to $f(x)$ over this interval. Let x_a, x_b, and x_c denote three values of x in the interval of uncertainty and f_a, f_b and f_c denote the corresponding values of $f(x)$. Then a quadratic polynomial, $\hat{f}(x) = a_0 + a_1 x + a_2 x^2$, can be fit exactly to these data to provide a local approximation to $f(x)$. The resulting equation for $\breve{f}(x)$ can be differentiated, set equal to zero, and solved for its optimum value, which is denoted by x^*. The expression for x^* is

$$x^* = \frac{1}{2}\frac{(x_b^2 - x_c^2)f_a + (x_c^2 - x_a^2)f_b + (x_a^2 - x_b^2)f_c}{(x_b - x_c)f_a + (x_c - x_a)f_b + (x_a - x_b)f_c} \tag{F-7}$$

After one iteration, x^* usually is not equal to x^{opt}, because the true function $f(x)$ is not necessarily quadratic. However, $f(x^*)$ will be an improvement. By saving the best two of the three previous points and finding the actual objective function $f(x^*)$, the search can be continued until convergence is indicated.

EXAMPLE F.2

A free radical reaction involving nitration of decane is carried out in two sequential reactor stages, each of which operates like a continuous stirred-tank reactor (CSTR). Decane and nitrate (as nitric acid) in varying amounts (D_i and N_i, respectively) are added to each reactor stage, as shown in Fig. F-4. The reaction of nitrate with decane is very fast and forms the following products by successive nitration: DNO_3, $D(NO_3)_2$, $D(NO_3)_3$, $D(NO_3)_4$, and so on. The desired product is DNO_3, whereas dinitrate, trinitrate, and so on, are undesirable products.

The flow rates of D_1 and D_2 are selected to satisfy temperature requirements in the reactors, while N_1 and N_2 are optimized to maximize the amount of DNO_3 produced from stage 2, subject to satisfying an overall level of nitration. In this case, we stipulate that $(N_1 + N_2)/(D_1 + D_2) = 0.4$. There is an excess of D in each stage, and $D_1 = D_2 = 0.5$ mol/s. A steady-state reactor model has been developed to maximize selectivity. Define $r_1 \triangleq N_1/D_1$ and $r_2 \triangleq N_2/(D_1 + D_2)$. The amount of DNO_3 leaving stage 2 (as mol/s in F_2) is given by

$$f_{DNO_3} = \frac{r_1 D_1}{(1+r_1)^2(1+r_2)} + \frac{r_2 D_2}{(1+r_1)(1+r_2)^2} \tag{F-8}$$

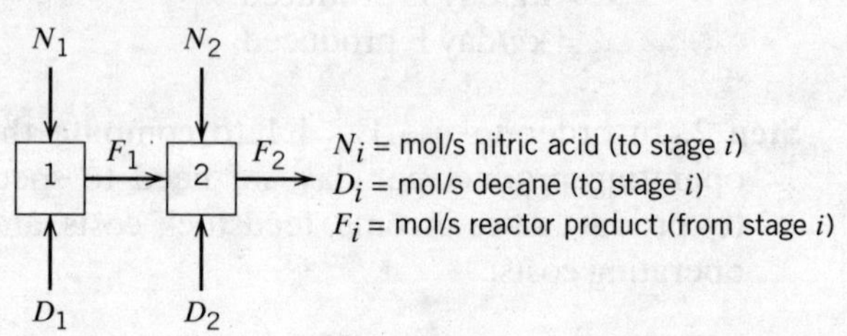

Figure F.4 A schematic diagram of a two-stage nitration reactor.

This equation can be derived from the steady-state equations for a continuous stirred reactor with the assumption that all reaction rate constants are equal.

Formulate a one-dimensional search problem in r_1 that will permit the optimum values of r_1 and r_2 to be found. Employ quadratic interpolation using an initial interval of$0 \le r_1 \le 0.8$. Use enough iterations so that the final value of f_{DNO_3} is within ±0.0001 of the maximum.

Solution

The six steps described earlier are used to formulate the optimization problem.

Step 1. ***Identify the process variables.*** The process variables to be optimized are N_1 and N_2, the nitric acid molar flow rates for each stage. Because D_1 and D_2 are specified, we can just as well use r_1 and r_2, because the conversion model is stated in terms of r_1 and r_2.

Step 2. ***Select the objective function.*** The objective is to maximize production of DNO_3 that can be made into useful products, while the other nitrates are unwanted byproducts with a value of zero. The objective function f is given in Eq. F-8 which is proportional to the profit function, as in Eq. 15-84.

Step 3. ***Develop models for the process and constraints.*** The values of N_1 and N_2 are constrained by the overall nitration level:

$$\frac{N_1 + N_2}{D_1 + D_2} = 0.4 \tag{F-9}$$

which can be expressed in terms of r_1 and r_2 as

$$\frac{r_1 + D_1 + r_2 + D_1 + r_2 + D_2}{D_1 + D_2} = 0.4 \tag{F-10}$$

Inequality constraints on r_1 and r_2 do exist, namely, $r_1 \geq 0$ and $r_2 \geq 0$ — because all N_i and D_i are positive. These constraints can be ignored except when the search method incorrectly leads to negative values of r_1 or r_2.

Step 4. ***Simplify the model.*** Given $D_1 = D_2 = 0.5$, then, from Eq. F-10,

$$r_2 = 0.4 - 0.5r_1 \tag{F-11}$$

Because r_1 and r_2 are nonnegative, Eq. F-11 implies that $r_1 \leq 0.8$ and $r_2 \leq 0.4$. After variable substitution, there is only one independent variable (r_1) in the objective function in Eq. F-8.

Step 5. ***Compute the optimum.*** Because r_1 lies between 0 and 0.8 (the interval of uncertainty), select the three interior points for the search to be r_1 = 0.2, 0.4, and 0.6. The corresponding values of r_2 are 0.3, 0.2, and 0.1. Table F.2 shows the numerical results for three iterations, along with objective function values. After the first iteration, the worst point (r_1 = 0.2) is discarded and the new point (r = 0.4536) is added. After the second iteration, the point with the lowest value of $f(r_1 = 0.6)$ is discarded. The tolerance on the objective function change is satisfied after only three iterations, with the value of r_1 that maximizes f_{DNO_3} computed to be r_1^{opt} = 0.4439. The converted mononitrate is 0.1348 mol/s from stage 2; the remainder of the nitrate is consumed to make higher molecular weight byproducts.

Step 6. ***Perform sensitivity studies.*** Based on the results in Table F.2, the yield is not significantly different from the optimum as long as $0.4 \leq r_1 \leq 0.6$. Practically speaking, this situation is beneficial, because it allows a reasonable range of decane flows to achieve temperature control. If either D_1 or D_2 changes by more than 10%, one should recalculate the optimum. There also might be a need to reoptimize r_1 and r_2 if ambient conditions change (e.g., summer vs. winter operation). Even a 1% change in yield can be economically significant if production rates and the selling price of the product are sufficiently high.

Table F.2 Search Iterations for Example F.2 (Quadratic Interpolation)

Iteration	x_a	f_a	x_b	f_b	x_c	f_c	x^*
1	0.2	0.1273	0.4	0.1346	0.6	0.1324	0.4536
2	0.4	0.1346	0.6	0.1324	0.4536	0.1348	0.4439
3	0.4	0.1346	0.4536	0.1348	0.4439	0.1348	(not needed)

$r_1^{opt} = 0.4439$

If the function to be optimized is not unimodal, then some care should be taken in applying the quadratic interpolation method. Selecting multiple starting points for the initial scanning before quadratic interpolation is initiated ensures that an appropriate search region has been selected. For a single variable search, scanning the region of search is a fairly simple and fast procedure, but evaluating the presence of multiple optima can become problematic for multivariable optimization problems.

F.3.2 Multivariable Optimization

In multivariable optimization problems, there is no guarantee that a given optimization technique will find the optimum point in a reasonable amount of computer time. The optimization of a general nonlinear multivariable objective function, $f(x) = f(x_1, x_2, ..., x_{N_V})$, requires that efficient and robust numerical techniques be employed. Efficiency is important, because the solution requires an iterative approach. Trial-and-error solutions or enumeration of many combinations of process variables are usually out of the question for problems with more than two or three variables.

The difficulty of optimizing multivariable functions often is resolved by treating the problem as a series of single-variable (or one-dimensional) searches. From a given starting point, a search direction is specified, and then the optimum point along that direction is determined by a one-dimensional search. Then a new search direction is determined, followed by another one-dimensional search in that direction. In choosing an algorithm to determine the search direction, we can draw upon extensive numerical experience with various optimization methods (Nocedal and Wright, 2006; Griva et al., 2008; Biegler, 2010).

Multivariable RTO of nonlinear objective functions using function derivatives is recommended with more than two variables. In particular, the conjugate gradient and quasi-Newton methods (Griva et al., 2008; Edgar et al., 2001) are extremely effective in solving such problems. Applications of multivariable RTO have experienced rapid growth as a result of advances in computer hardware and software. We consider such methods in more detail in Section F.5.

An important application of unconstrained optimization algorithms is to update parameters in steady-state models from the available data. Usually, only a few model parameters are estimated on-line, and then RTO is based on the updated model. Guidelines for parameter estimation have been provided by Marlin and Hrymak (1997) and Forbes et al. (1994).

Most practical multivariable problems include constraints, which must be treated using linear or nonlinear programming. A problem statement with a nonlinear objective function (profit) and nonlinear constraints is called a *nonlinear programming problem:*

$$\text{maximize } f(x_1, x_2, ..., x_{N_v}) \tag{F-12}$$

$$\text{subject to: } h_i(x_1, x_2, ..., x_{N_v}) = 0 \ (i = 1, ..., N_E) \tag{F-13}$$

$$g_i(x_1, x_2, ..., x_{N_v}) \leq 0 \ (i = 1, ..., N_I) \tag{F-14}$$

In this case, there are N_V process variables, N_E equality constraints and N_I inequality constraints.

Skogestad (2000) and Perkins (1998) have discussed the interplay of constraints and the selection of the optimal operating conditions. Skogestad identified three different cases for RTO that are illustrated in Fig. F.5. In each case, a single variable x is used to maximize a profit function, $f(x)$.

(a) ***Constrained optimum:*** The optimum value of the profit is obtained when $x = x_a$. Implementation of an active constraint is straightforward; for example, it is easy to close a valve.

(b) ***Unconstrained flat optimum:*** In this case, the profit is insensitive to the value of x, and small process changes or disturbances do not affect profitability very much.

(c) ***Unconstrained sharp optimum:*** A more difficult problem for implementation occurs when the profit is sensitive to the value of x. If possible, we may want to select a different input variable for which the corresponding optimum is flatter, so that the operating range can be wider without reducing the profit very much.

In some cases, an actual process variable (such as yield) can be the objective function, and no process model is required. Instead, the process variables are varied systematically to find the best value of the objective function from the specific data set, sometimes involving design of experiments as discussed by Myers and Montgomery (2009). In this way, improvements in the objective function can be obtained gradually. Usually, only a few vari-

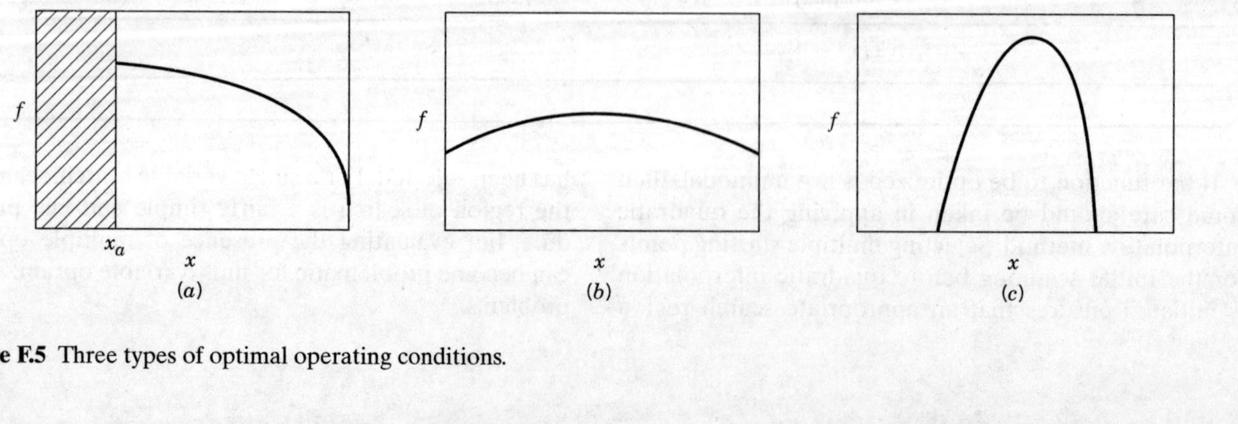

Figure F.5 Three types of optimal operating conditions.

ables can be optimized in this way, and it is limited to batch operations. Methods used in industrial batch process applications include EVOP (evolutionary operation) and response surface analysis (Edwards and Jutan, 1997; Box and Draper, 1998; Myers and Montgomery, 2009).

F.4 LINEAR PROGRAMMING

An important class of constrained optimization problems has a linear objective function and linear constraints. The solution of these problems is highly structured and can be obtained rapidly via *linear programming* (LP). This powerful approach is widely used in RTO applications because it is guaranteed to reach the optimum assuming a feasible solution exists.

F.4.1 Linear Programming Concepts

The number of independent variables in a constrained optimization problem can be found by a procedure analogous to the degrees of freedom analysis in Chapter 2. For simplicity, suppose that there are no constraints. If there are N_V process variables (which includes process inputs and outputs) and the process model consists of N_E independent equations, then the number of independent variables is $N_F = N_V - N_E$. This means N_F set points can be specified independently to maximize (or minimize) the objective function. The corresponding values of the remaining $(N_V - N_F)$ variables can be calculated from the process model. However, the presence of inequality constraints that can become active changes the situation, because the N_F set points cannot be selected arbitrarily. They must satisfy all of the equality and inequality constraints.

The standard linear programming (LP) problem can be stated as follows:

$$\text{minimize } f = \sum_{r=1}^{N_V} c_i x_i \qquad \text{(F-15)}$$

subject to

$$x_i \geq 0 \qquad i = 1, 2, \ldots, N_V$$

$$\sum_{j=1}^{N_V} a_{ij} x_j \geq b_i \qquad i = 1, 2, \ldots, N_I \qquad \text{(F 16)}$$

$$\sum_{j=1}^{N_V} \tilde{a}_{ij} x_j \geq d_i \qquad i = 1, 2, \ldots, N_E \qquad \text{(F-17)}$$

The LP solution can be obtained by a method called the Simplex algorithm (Edgar et al., 2001; Griva et al., 2008). The Simplex algorithm using software such as CPLEX can quickly solve virtually any number of inequality constraints and any number of variables in the objective function. Maximization problems can be converted to the minimization form of Eq. F-15, by multiplying the objective function by −1. Because there are a limited number of intersections of constraint boundaries where the optimum must occur, the amount of computer time required to search for the optimum is reduced considerably compared to more general nonlinear optimization problems. Hence, many nonlinear optimization problems (even those with nonlinear constraints) are often linearized so that the LP algorithm can be employed. This procedure allows optimization problems with over one million variables to be solved.

In the 1980s, a major change in optimization software occurred when linear programming solvers and then nonlinear programming solvers were interfaced to spreadsheet software for desktop computers. The spreadsheet has become a popular user interface for entering and manipulating numeric data. Spreadsheet software increasingly incorporates analytic tools that are accessible from the spreadsheet interface and permit access to external databases. For example, Microsoft Excel incorporates an optimization-based routine called Solver that operates on the values and formulas of a spreadsheet model. Current versions (2010 and later) include LP and NLP solvers and mixed integer programming (MIP) capability for both linear and nonlinear problems. The user specifies a set of cell addresses to be independently adjusted (the decision variables), a set of formula cells whose values are to be constrained (the constraints), and a formula cell designated as the optimization objective.

In the process industries, the Simplex algorithm has been applied to a wide range of problems, such as the optimization of a total plant utility system. A general steam utility configuration, typically involving as many as 100 variables and 100 constraints, can be easily optimized using linear programming (Hirata et al., 2004; Edgar et al., 2001; Marlin, 2000). The process variables can be updated on an hourly basis because steam demands in process units can change. In addition, it may be economical to generate more electricity locally during times of peak demand, due to variable time-of-day electricity pricing by utilities. Larger LP problems are routinely solved in refineries, numbering in the thousands of variables and spanning several months of operations (Alattas et al., 2011).

F.5 QUADRATIC AND NONLINEAR PROGRAMMING

The most general optimization problem occurs when both the objective function and constraints are nonlinear, a case referred to as *nonlinear programming* (NLP), which is stated mathematically in Eqs. (F-12 to F-14).

The leading constrained optimization methods discussed by Nocedal and Wright (2006), Griva et al. (2008), and Biegler (2010) include:

1. Quadratic programming (QP)
2. Generalized reduced gradient (GRG)
3. Successive quadratic programming (SQP)
4. Successive linear programming (SLP)

F.5.1 Quadratic Programming

In quadratic programming (QP), the objective function is quadratic and the constraints are linear. Although the solution is iterative, it can be obtained quickly. A quadratic programming problem minimizes a quadratic function of n variables subject to m linear inequality or equality constraints. A convex QP is the simplest form of a nonlinear programming problem with inequality constraints. A number of practical optimization problems are naturally posed as a QP problem, such as constrained least squares and some model predictive control problems.

In compact notation, the quadratic programming problem is

$$\text{Minimize } f(x) = c^T x + \frac{1}{2} x^T Q x \tag{F-18}$$

$$\text{Subject to } \begin{array}{l} Ax = b \\ x \geq 0 \end{array} \tag{F-19}$$

where c is a vector $(n \times 1)$, $\boldsymbol{A}$ is an $m \times n$ matrix, and $\boldsymbol{Q}$ is a symmetric $n \times n$ matrix.

Computer codes for quadratic programming allow arbitrary upper and lower bounds on $\boldsymbol{x}$; here we assume $\boldsymbol{x} \geq \boldsymbol{0}$ for simplicity. QP software finds a solution by using LP operations to minimize the sum of constraint violations. Because LP algorithms are employed as part of the QP calculations, most commercial LP software also contains QP solvers. Quadratic programming is often used in computing model predictive controller calculations (see Chapter 15).

F.5.2 Nonlinear Programming Algorithms and Software

A variety of methods have been used to solve nonlinear programming problems in the field of chemical engineering design and operations. One popular NLP algorithm called the *generalized reduced gradient (GRG)* algorithm uses iterative linearization and is used in the Excel Solver. The CONOPT software package uses a reduced gradient algorithm that works well for large-scale problems and nonlinear constraints (Alattas et al., 2011). CONOPT and GRG work best for problems where the number of degrees of freedom is small (the number of constraints is nearly equal to the number of variables). Successive quadratic programming (SQP) solves a sequence of quadratic programs that approach the solution of the original NLP by linearizing the constraints and using a quadratic approximation to the objective function. It is used in MATLAB's constrained optimizer (fmincon). The search procedure generally uses some variation of Newton's method, a second-order method that approximates the Hessian matrix using first derivatives (Biegler, 2010; Edgar et al., 2001). MINOS and NPSOL, SQP-based software packages that were originally developed in the 1980s are suitable for NLPs that are nearly linear. For large-scale NLPs, IPOPT is an interior point line search with barrier functions that is very effective (Biegler, 2010). Successive linear programming (SLP) is used less often for solving RTO problems and requires linear approximations of both the objective function and constraints. It sometimes exhibits poor convergence to optima that are not located at constraint intersections.

Software libraries such as GAMS (General Algebraic Modeling System) or NAG (Numerical Algorithms Group) offer one or more NLP algorithms, but rarely are all algorithms available from a single source. Web sources that serve as comprehensive repositories of optimization packages include OPTI Toolbox (http://www.i2c2.aut.ac.nz/Wiki/OPTI/), NEOS Server (http://www.neos-server.org/neos/) (Czyzyk et al., 2015), and COIN-OR (http://www.coin-or.org/) (Lougee- Heimer, 2003; Nocedal and Wright, 2006). No single NLP algorithm is best for every problem, so several solvers should be tested on a given application. Other optimization software packages are designed to solve mixed-integer problems, with both discrete and continuous variables (MILP, MINLP) and also the so-called Global Optimization problems that can converge to a nonconvex optimum. Impressive speedups (~10^5) in MIP solution have occurred during the past 20 years. See Biegler (2010) for more details.

EXAMPLE F.3

Consider the problem of minimizing fuel costs in a boiler-house. The boiler-house contains two turbine generators, each of which can be simultaneously operated with two fuels: fuel oil and medium Btu gas (MBG); see Fig. F.6 The MBG is produced as a waste off-gas from another part of the plant, and it must be flared if it cannot be used on site. The goal of the RTO scheme is to find the optimum flow rates of fuel oil and MBG and provide 50MW of power at all times, so that steady-state operations can be maintained while minimizing costs. It is desirable to use as much of the MBG as possible (which has zero cost) while minimizing consumption of expensive fuel oil. The two turbine generators (G_1, G_2) have different operating characteristics; the efficiency of G_1 is higher than that of G_2.

Data collected on the fuel requirements for the two generators yield the following empirical relations:

$$P_1 = 4.5x_1 + 0.1x_1^2 + 4.0x_2 + 0.06x_2^2 \qquad \text{(F-20)}$$

$$P_2 = 4.0x_3 + 0.05x_3^2 + 3.5x_4 + 0.02x_4^2 \qquad \text{(F-21)}$$

where

P_1 = power output (MW) from G_1
P_2 = power output (MW) from G_2
x_1 = fuel oil to G_1 (tons/h)
x_2 = MBG to G_1 (fuel units/h)
x_3 = fuel oil to G_2 (tons/h)
x_4 = MBG to G_2 (fuel units/h).

The total amount of MBG available is 5 fuel units/h. Each generator is also constrained by minimum and maximum power outputs: generator 1 output must lie between 18 and 30 MW, while generator 2 can operate between 14 and 25 MW.

Formulate the optimization problem by applying the methodology described in Section F.2. Then solve for the optimum operating conditions ($x_1, x_2, x_3, x_4, P_1, P_2$) using the Excel Solver.

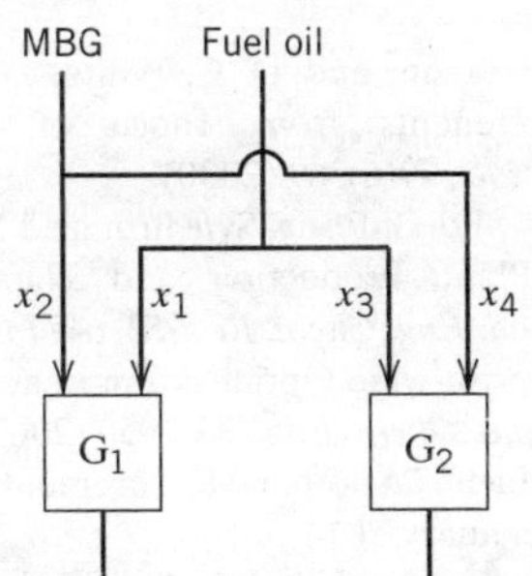

Figure F.6 The allocation of two fuels in a boilerhouse with two turbine generators (G_1, G_2).

Solution

Step 1. *Identify the variables.* Use x_1 through x_4 as the four process variables. Variables P_1 and P_2 are dependent because of the equality constraints (see Steps 3 and 4).

Step 2. *Select the objective function.* The way to minimize the cost of operation is to minimize the amount of fuel oil consumed. This implies that we should use as much MBG as possible, because it has zero cost. The objective function can be stated in terms of variables defined above; that is, we wish to minimize

$$f = x_1 + x_3 \qquad \text{(F-22)}$$

Step 3. *Specify process model and constraints.* The constraints given in the problem statement are as follows:

(1) Power relation

$$P_1 = 4.5x_1 + 0.1x_1^2 + 4.0x_2 + 0.06x_2^2 \qquad \text{(F-23)}$$

$$P_2 = 4.0x_3 + 0.5x_3^2 + 3.5x_4 + 0.02x_4^2 \qquad \text{(F-24)}$$

(2) Power range $18 \le P_1 \le 30$ (F-25)

$14 \le P_2 \le 25$ (F-26)

(3) Total power $50 = P_1 + P_2$ (F-27)

(4) MBG supply $5 = x_2 + x_4$ (F-28)

Note that all variables defined above are nonnegative.

Step 4. *Simplify the model and objective function.* This step is not need because this is a fairly small NLP problem.

Step 5. *Compute the optimum.* Using Excel Solver, the optimum is $f = 6.39$, $x_1 = 5.89$, and $x_3 = 0.50$, meaning that 5.89 tons/h of fuel oil are delivered to generator G_1, while 0.50 tons/h are used in G_2. G_2 utilizes all of the MBG (x_4), while G_1 uses none ($x_2 = 0$).

Step 6. *Perform a sensitivity analysis.* Many operating strategies may be satisfactory, though not optimal, for the above problem. In fact, the Excel Solver will converge to a local minimum if a starting point of $x_1 = 2, x_2 = x_3 = 4, x_4 = 1$ is used ($f = 6.54$). Parameters in the original constraint equations can change as plant operating conditions vary. For example, suppose the total power requirement is changed to 55 MW; as an exercise, determine whether any of the active constraints change for the increased power requirement.

SUMMARY

Although the economic benefits from feedback control are not always readily quantifiable, RTO offers a direct method of maximizing the steady-state profitability of a process or group of processes. The optimization of the set points is performed as frequently as necessary, depending on changes in operating conditions or constraints. It is important to formulate the optimization problem carefully; a methodology for formulation and solution of optimization problems is presented in this chapter. A wide range of optimization techniques can be used, depending on (1) the number of variables, (2) the nature of the equality and inequality constraints, and (3) the nature of the objective function. Because we have presented only introductory concepts in optimization here, the reader is advised to consult other comprehensive references on optimization such as Edgar et al. (2001) before choosing a particular method for RTO.

REFERENCES

Alattas, A. M., I. E. Grossmann, and I. Palou-Rivera, Integration of Nonlinear Crude Distillation Unit Models in Refinery Planning Optimization, *Ind. Eng. Chem. Res.*, **50**, 6860 (2011).

Bailey, J. K., A. N. Hrymak, S. S. Treiba, and R. B. Hawkins, Nonlinear Optimization of a Hydrocracker Fractionation Plant, *Comput. Chem. Eng.*, **17**, 123 (1993).

Baldea, M., and I. Harjunkoski, Integrated Production Scheduling and Process Control: A Systematic Review, *Comput. Chem. Eng.*, **71**, 377 (2014).

Biegler, L. T., *Nonlinear Programming: Concepts, Algorithms, and Applications to Chemical Processes,* MPS-SIAM Series on Optimization, Philadelphia, PA, 2010.

Box, G. E. P., and N. R. Draper, *Evolutionary Operation: A Statistical Method for Process Improvement,* John Wiley and Sons, Hoboken, NJ, 1998.

Czyzyk, J., M. P. Mesnier, J. J. More, E. Dolan, and W. Gropp, NEOS Server: State-of-the-Art Solvers for Numerical Optimization, University of Wisconsin-Madison, http://www.neos-server.org/neos/ (2015).

Darby, M. L., M. Nikolaou, J. Jones, and D. Nicholson, RTO: An Overview and Assessment of Current Practice, *J. Process Control,* **21**, 874 (2011).

Darby, M. L., and D. C. White, On-line Optimization of Complex Process Units, *Chem. Engr. Prog.*, **84**(10), 51 (1998).

Edgar, T. F., D. M. Himmelblau, and L. S. Lasdon, *Optimization of Chemical Processes*, 2nd ed., McGraw-Hill, New York, 2001.

Edwards, I. M., and A. Jutan, Optimization and Control Using Response Surface Methods, *Comput. Chem. Eng.*, **21**, 441 (1997).

Forbes, F., T. Marlin, and J. F. MacGregor, Model Selection Criteria for Economics-Based Optimizing Control. *Comput. Chem. Eng.*, **18**, 497 (1994).

Geddes, D., and T. Kubera, Integration of Planning and Real-Time Optimization in Olefins Productions, *Comput. Chem. Eng.*, **24**, 1645 (2000).

Georgiou, A., P. Taylor, R. Galloway, L. Casey, and A. Sapre. Plantwide Closed-Loop Real Time Optimization and Advanced Control of Ethylene Plant – (CLRTO) Improves Plant Profitability and Operability, *Proc. NPRA Computer Conference*, New Orleans, LA, November 1997.

Griva, I., S. G. Nash, and A. Sofer, *Linear and Nonlinear Optimization*, 2nd ed., SIAM, Philadelphia, PA, 2008.

Grossmann, I., Advances in Mathematical Programming Models for Enterprise-wide Optimization, *Comput. Chem. Eng.*, **47**, 2 (2012).

Grossmann, I., Enterprise-wide Optimization: A New Frontier in Process Systems Engineering, *AIChE J.*, **51**, 1846 (2005).

Harjunkoksi, I., C. Maravelias, P. Bongers, P. Castro, S. Engell, I. Grossmann, J. Hooker, C. Mendez, G. Sand, and J. Wassick, Scope of Industrial Applications of Production Scheduling Models and Solution Methods, *Comput. Chem. Eng.*, **62**, 161 (2014).

Hirata, K., H. Sakamoto, L. O'Young, K. Cheung, and C. Hui, Multisite Utility Integration—An Industrial Case Study, *Comput. Chem. Eng.*, **28**, 139 (2004).

Kelly, J. D., and J. D. Hedengren, A Steady-state Detection (SSD) Algorithm to Detect Non-stationary Drifts in Processes, *J. Process Control*, **23**, 326 (2013).

Lougee-Heimer, R., The Common Optimization INterface for Operations Research, *IBM J.* Res. Dev., **47**, 57 (2003) http://www.coin-or.org/.

Marlin, T. E., *Process Control*, 2nd ed., McGraw-Hill, New York, 2000.

Marlin, T. E., and A. N. Hrymak, Real-Time Operations Optimization of Continuous Processes, in *Chemical Process Control V, AIChE Symp. Ser.* **93**, No. 316, 156 (1997).

Myers, R. H., and D. C. Montgomery, *Response Surface Methodology: Process and Product Optimization Using Designed Experiments*, 3rd ed., John Wiley and Sons, Hoboken, NJ, 2009.

Narasimhan, S., and C. Jordache, *Data Reconciliation and Gross Error Detection*, Gulf Publishing, Houston, TX, 2000.

Nocedal, J., and S. J. Wright, *Numerical Optimization*, 2nd ed., Springer, New York, 2006.

Perkins, J. D., Plant-wide Optimization: Opportunities and Challenges, *Foundations of Computer-Aided Process Operations*, J. F. Pekny and G. E. Blau (Eds.), *AIChE Symp. Ser.*, **94**, No. 320, 15 (1998).

Shobrys, D. E., and D. C. White, Planning, Scheduling, and Control Systems: Why They Cannot Work Together, *Comput. Chem. Eng.*, **26**, 149 (2002).

Skogestad, S., Self-optimizing Control: The Missing Link Between Steady-State Optimization and Control. *Comput. Chem. Eng.*, **24**, 569 (2000).

Soderstrom, T. A., T. F. Edgar, L. P. Russo, and R. E. Young, Industrial Application of a Large-Scale Dynamic Data Reconciliation Strategy, *Ind. Eng. Chem. Res.*, **39**, 1683 (2000).

Starks, D. M., and E. Arrieta, Maintaining AC & O Applications, Sustaining the Gain, AIChE Spring Meeting Houston, TX, March, 2007.

Timmons, C., J. Jackson, and D. C. White, Distinguishing On-line Optimization Benefits from Those of Advanced Controls, *Hydrocarbon Proc.*, **79**(6), 69 (2000).

Wang, Z., Y. Feng, and G. Rong, Synchronized Scheduling Approach of Ethylene Plant Production and Naphtha Oil Inventory Management, *Ind. Eng. Chem. Res.*, **53**, 6477 (2014).

Wassick, J., Enterprise-wide Optimization in an Integrated Chemical Complex, *Comput. Chem. Eng.*, **33**, 1950 (2009).

White, D., How Much is Another Measurement Worth?, *Chem. Engr. Prog.*, **100**, 50 (January, 2014).

White, D. C., Save Energy Through Automation, *Chem. Eng. Prog.*, **106**, 26 (January, 2010).

EXERCISES

F.1 A laboratory filtration study has been carried out at constant rate. The filtration time (t_f in hours) required to build up a specific cake thickness has been correlated as

$$t_f = 5.3\, x_i e^{-3.6xi+2.7}$$

where x_i = mass fraction solids in the cake. Find the value of x_i that maximizes t_f using quadratic interpolation.

F.2 The thermal efficiency of a natural gas boiler versus air/fuel ratio is plotted in Fig. EF.2. Using physical arguments, explain why a maximum occurs.

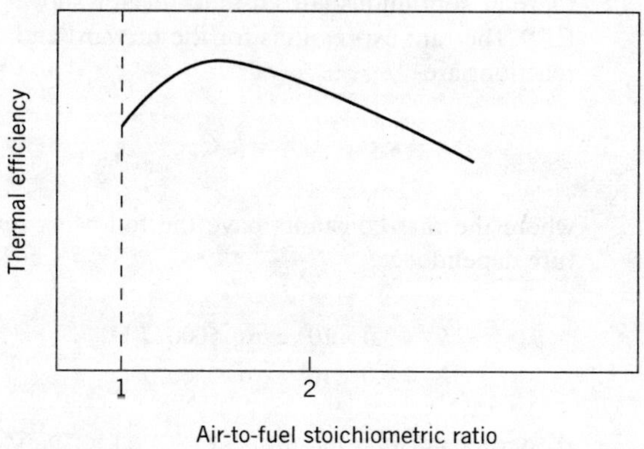

Figure EF.2

F.3. A plasma etcher has a yield of good semiconductor chips that is influenced by pressure X_1 and gas flow rate X_2. Both X_1 and X_2 are scaled variables ($0 \le X_i \le 2$). A model has been developed based on operating data as follows:

$$Y = -0.1X_1^4 + 0.2X_2X_1^2 - 0.09X_2^2 - 0.11X_1^2 + 0.15X_1 + 0.5$$

Use Excel to maximize yield Y, using starting points of (1,1) and (0,0).

F.4. Reconciliation of inaccurate process measurements is an important problem in process control that can be solved using optimization techniques. The flow rates of streams B and C have been measured three times during the current shift (shown in Fig. EF.4). Some errors in the measurement devices exist. Assuming steady-state operation (w_A = constant), find the optimal value of w_A (flow rate in kg/h) that minimizes the sum of the squares of the errors for the material balance, $w_A + w_C = w_B$ using an analytical solution.

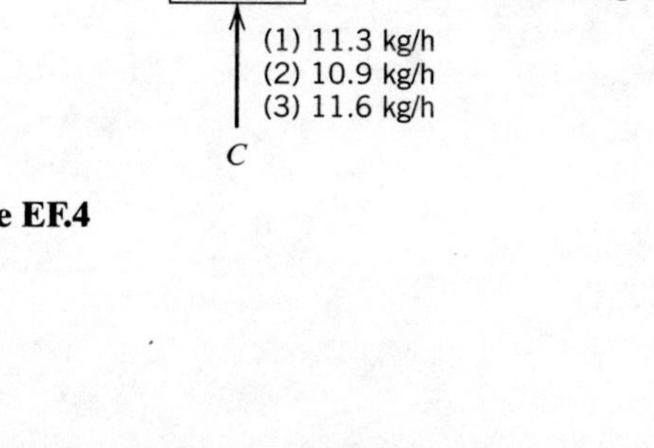

Figure EF.4

F.5 A batch reactor converts component A into B, which in turn decomposes into C:

$$\mathrm{A} \xrightarrow{k_1} \mathrm{B} \xrightarrow{k_2} C$$

where $k_1 = k_{10}e^{-E_1/RT}$ and $k_2 = k_{20}e^{-E_2/RT}$.

The concentrations of A and B are denoted by x_1 and x_2, respectively. The reactor model is

$$\frac{dx_1}{dt} = -k_{10}x_1e^{-E_1/RT}$$

$$\frac{dx_2}{dt} = k_{10}x_1e^{-E_1/RT} - k_{20}x_2e^{-E_2/RT}$$

Thus, the ultimate values of x_1 and x_2 depend on the reactor temperature as a function of time. For

$$k_{10} = 1.335\times10^{10}\,\text{min}^{-1},\quad k_{20} = 1.149\times10^{17}\,\text{min}^{-1}$$
$$E_1 = 75{,}000\,\text{J/g mol},\quad E_2 = 125{,}000\,\text{J/g mol}$$
$$R = 8.31\,\text{J/(g mol K)}\quad x_{10} = 0.7\,\text{mol/L},\ x_{20} = 0$$

Find the constant temperature that maximizes the amount of B, for $0 \le t \le 6$ min.

F.6 Optimization methods can be used to fit equations to data. Parameter estimation involves the calculation of unknown parameters that minimize the squared error between data and the proposed mathematical model. The step response of an overdamped second-order dynamic process can be described using the equation

$$\frac{y(t)}{K} = \left(1 - \frac{\tau_1 e^{-t/\tau_1} - \tau_2 e^{-t/\tau_2}}{\tau_1 - \tau_2}\right)$$

where τ_1 and τ_2 are process time constants and K is the process gain. The following normalized data have been obtained from a unit step test (K = 1.0):

time, t	0	1	2	3	4	5
y_i	0.0	0.0583	0.217	0.360	0.488	0.600
t	6	7	8	9	10	11
y_i	0.692	0.772	0.833	0.888	0.925	0.942

Use Excel with a starting point (1,0) to find values of τ_1 and τ_2 that minimize the sum of squares of the errors.

F.7 Supervisory control often involves the optimization of set points in order to maximize profit. Can the same results be achieved by optimizing PID controller tuning (K_c, τ_I, τ_D), in order to maximize profits? Are regulatory (feedback) control and supervisory control complementary?

F.8 A dynamic model of a continuous-flow, biological chemostat has the form

$$\dot{X} = 0.063C - Dx$$
$$\dot{C} = 0.9S[X - C] - 0.7C - DC$$
$$\dot{S} = -0.9S[X - C] + D[10 - S]$$

where X is the biomass concentration, S is the substrate concentration, and C is a metabolic intermediate concentration. The dilution rate, D, is an independent variable, which is defined to be the flow rate divided by the chemostat volume.

Determine the value of D, which maximizes the steady-state production rate of biomass, f, given by

$$f = DX$$

F-9 A reversible chemical reaction, A ⇄ B, occurs in the isothermal continuous stirred-tank reactor shown in Fig. EF.9. The rate expressions for the forward and reverse reactions are

$$r_1 = k_1 C_A$$
$$r_2 = k_2 C_B$$

For the information given below, use a numerical search procedure to determine the value of F_B (L/h) that maximizes the production rate of C_B (i.e., the amount of C_B that leaves the reactor, mol B/h). The allowable values of F_B are $0 \le F_B \le 200$ L/h.

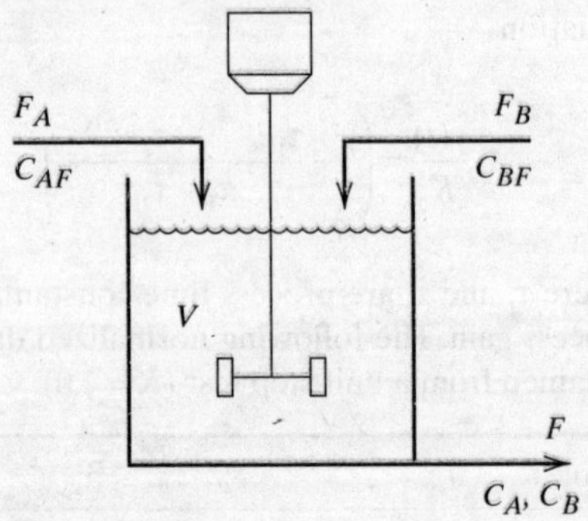

Figure EF.9

Available Information

(i) The reactor is perfectly mixed.

(ii) The volume of liquid, V, is maintained constant using an overflow line (not shown in the diagram).

(iii) The following parameters are kept constant at the indicated numerical values:

$$V = 200\text{L} \qquad F_A = 150\,\text{L/h}$$
$$C_{AF} = 0.3\,\text{mol A/L} \qquad C_{BF} = 0.3150\,\text{mol B/L}$$
$$k_1 = 2\text{h}^{-1} \qquad k_2 = 1.5\text{h}^{-1}$$

F.10 A reversible chemical reaction, A ⇄ B, occurs in the isothermal continuous stirred-tank reactor shown in Fig. EF.9. The rate expressions for the forward and reverse reactions are

$$r_1 = k_1 C_A \qquad r_2 = k_2 C_B$$

where the rate constants have the following temperature dependence:

$$k_1 = 3.0 \times 10^6 \exp(-5000/T)$$
$$k_2 = 6.0 \times 10^6 \exp(-5500/T)$$

Each rate constant has units of h^{-1}, and T is in K.

Use the MATLAB Optimization Toolbox or Excel to determine the optimum values of temperature T (K) and flow rate F_B (L/h) that maximize the steady-state production rate of component B. The allowable values are $0 \le F_B \le 200$ and $300 \le T \le 500$.

Available Information

(i) The reactor is perfectly mixed.

(ii) The volume of liquid, V, is maintained constant using an overflow line (not shown in Fig. EF.9).

(iii) The following parameters are kept constant at the indicated numerical values:

$$V = 200\text{ L} \qquad F_A = 150\,\text{L/h}$$
$$C_{AF} = 0.3\text{ mol A/L} \qquad C_{BF} = 0.3\,\text{mol B/L}$$

APPENDIX G

Biosystems Control Design

APPENDIX CONTENTS

Previous chapters have introduced the concepts of process dynamics and strategies for process control, emphasizing traditional applications from the petrochemical industries, such as chemical reactors and distillation columns. In this chapter, we introduce the application fields of bioprocessing and biomedical devices, and elucidate the characteristics that these processes share with traditional chemical processes. Differences will also be highlighted, including the nature of uncertainty in biological processes, as well as the safety considerations in medical closed-loop systems. Control system design for three bioprocessing operations is described: crystallization, fermentation, and granulation. Finally, a number of problems in controlled drug delivery are reviewed, and control strategies are demonstrated in the areas of diabetes and blood pressure regulation. Biological applications are expanded in Appendix H, with a discussion of control systems opportunities, including applications to systems biology.

G.1 PROCESS MODELING AND CONTROL IN PHARMACEUTICAL OPERATIONS

A typical flowsheet in the pharmaceutical industry contains many of the same categories of operations as occur in a traditional petrochemical processing plant: reactors to generate products from raw materials, purification steps to extract desired products from the by-products and unreacted feed materials, and downstream processing associated with the final formulation of the product. Pharmaceutical processes are unique in several respects: (1) the main reactions involve biological materials, such as cells and tissues from more complex organisms, and (2) most of the products are formulated in solid form, which requires a unique set of bulk solids processing steps to purify and formulate the desired end product (e.g., a medicinal tablet). Consequently, the upstream processing involves sterilization and fermentation, and the manipulated inputs for the reactor often include *inducers* to activate the expression of particular genes in microbes in the reactor (gene expression is covered in more detail in Appendix H). The downstream section of the flowsheet includes crystallization or chromatographic purification, to extract a high-purity product with desirable properties (e.g., chirality). Subsequent steps may involve solids handling and processing to produce final particulates with desirable properties, including dissolution attributes and tableting capability. These processes include mixing, classification, milling, grinding, crushing, granulation (agglomeration), tableting, coating, molding, and extrusion. Each of these operations has its own challenges and unique dynamic characteristics.

In the following sections, we consider three of the main processing steps in the pharmaceutical flowsheet: fermentation, crystallization, and granulation. Several of these processes appear in other industries as well (e.g., food, mining, semiconductor, and specialty chemical), so the process control methods described find broad application in industry. It is important to note that the industry has a new emphasis on process systems engineering methods, driven by changes in FDA regulations (see PAT discussion in Chapter 18). The pharmaceutical industry is placing increasing emphasis on integrated manufacturing, linking the control of synthesis, purification, and final (drug)product formulation (Mascia et al., 2013). Furthermore, the push for sustainable manufacturing has identified unique challenges and opportunities for this sector (Sheldon, 2007).

G.1.1 Bioreactors

Fermentation reactors are widely used in the pharmaceutical industries to make an array of important compounds, including penicillin, insulin, and human growth hormone. In recent years, genetic engineering has further expanded the portfolio of useful products that can be synthesized using fermentation methods (Buckland, 1984; Lim, 1991; Schügerl, 2001). Despite the importance of this unit operation, the state of pharmaceutical fermentation operations is often characterized as more art than science, as with the winemaking industry (Fleet, 1993; Alford, 2006). In beermaking, there are similar traditions of "art", although modern facilities employ a sophisticated array of unit operations (e.g., cooking, washing, lautering, brewing, fermentation) with the corresponding control challenges. Since 1990, there has been a focused effort to develop more sophisticated control architectures for fermentation operation, driven by the availability of new technologies for monitoring the quality of the contents of the fermentor (see, e.g., Van Impe et al., 1998; Boudreau and McMillan, 2007). Key measurements and their associated controlled variables include dissolved oxygen (status of aerobic metabolic process), pH (optimal growth conditions), temperature (metabolic heat generation), and offgas concentration (indicator of culture activity). A schematic of a fermentation process is given in Figure G.1.

In a general sense, fermentation involves the generation of cell mass (product) from a substrate according to a simple reaction:

$$aC_\alpha H_\beta O_\gamma + bO_2 + cNH_3 \rightarrow C_\delta H_\varepsilon O_\zeta N_\eta + dCO_2 + eH_2O \qquad \text{(G-1)}$$

where $C_\alpha H_\beta O_\gamma$ is the substrate (reactant) and $C_\delta H_\varepsilon O_\zeta N_\eta$ is the cell mass (product). For example, in beer making, the substrate is glucose (derived from the wort from grains), and the products are the alcohol and carbon dioxide gas, both of which contribute to the quality of the final product. This apparently straightforward reaction is complicated by the fact that it does not obey simple mass action kinetics; instead, the

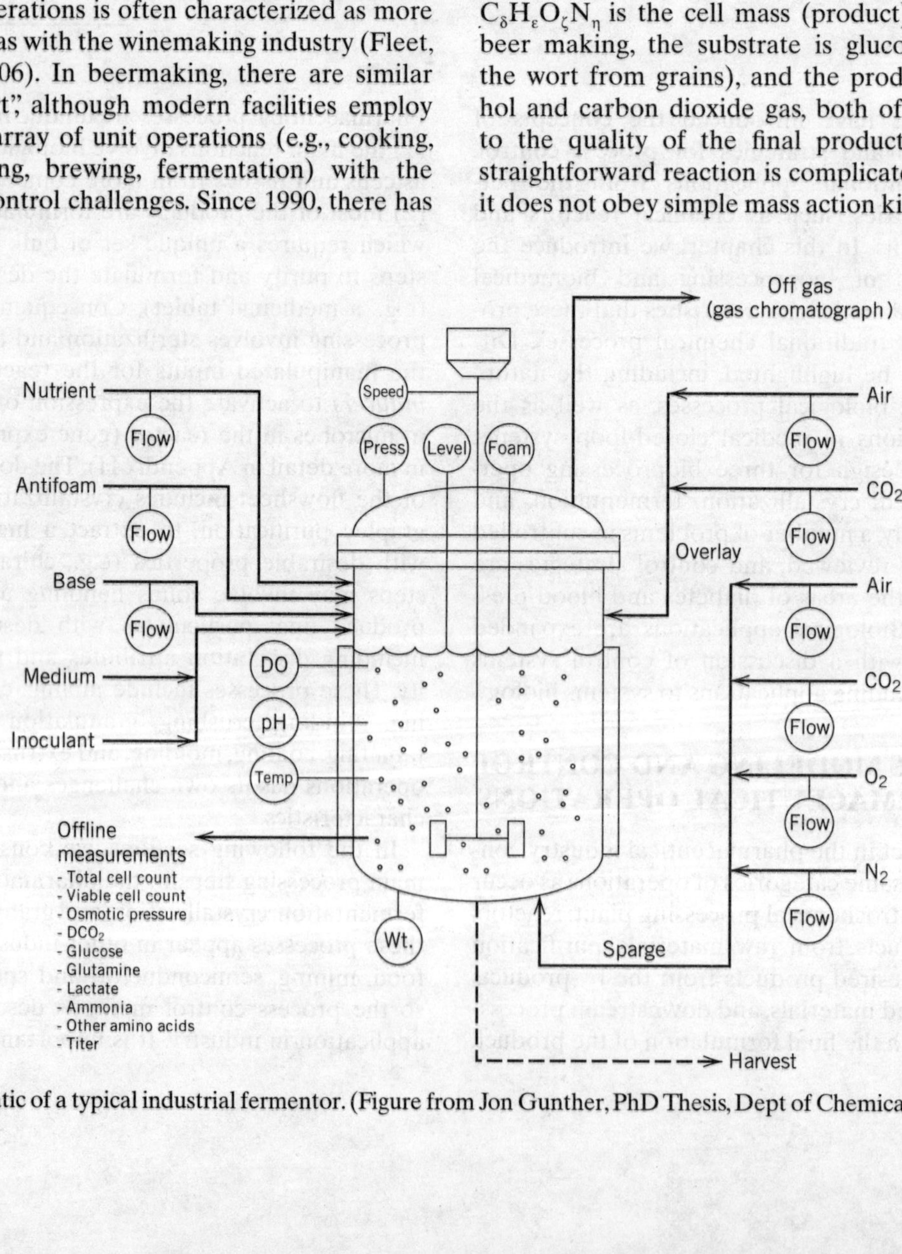

Figure G.1 Schematic of a typical industrial fermentor. (Figure from Jon Gunther, PhD Thesis, Dept of Chemical Eng., UCSB, 2008).

complex biochemistry underlying the reaction gives rise to unusual nonlinear rate expressions that characterize the enzymatic processes.

A simple dynamic model of a fed-batch bioreactor was given in Chapter 2, Eqs. 2-88 to 2-91. This model can be converted into mass balances on individual components as follows:

$$\frac{dX}{dt}=\mu(S)X-\frac{F}{V}X \tag{G-2}$$

$$\frac{dP}{dt}=Y_{P/X}\,\mu(S)X-\frac{F}{V}P \tag{G-3}$$

$$\frac{dS}{dt}=\frac{F}{V}(S_f-S)-\mu(S)\frac{X}{Y_{X/S}} \tag{G-4}$$

$$\frac{dV}{dt}=F \tag{G-5}$$

where the material balance in Eq. G-2 details the conservation of biomass X, Eq. G-3 describes the production of metabolites by the cells (biomass), Eq. G-4 details the conservation of substrate S, and Eq. G-5 is the overall material balance. The ratio F/V is often denoted as the dilution rate, D. The constants that appear in this equation include the feed concentration of the substrate S_f, the yield of cell mass from substrate $Y_{X/S}$, and the product yield coefficient $Y_{P/X}$. The rate of the biochemical reaction, $\mu(S)$, typically utilizes Monod kinetics given by the saturating function:

$$\mu(S)=\frac{\mu_m S}{K_m+S} \tag{G-6}$$

where μ_m is the maximum specific growth rate (limiting value of the rate), and K_m is the substrate saturation constant. Control of this simple reactor involves manipulating the influx of substrate (via the dilution rate, D) to achieve an optimal level of production. More sophisticated control inputs are also possible, including inducers that stimulate the transcription of key genes in the microorganisms, leading to the synthesis of enzymes that maximize product yield.

One of the primary challenges to controlling these reactors in industry is the difficulty in measuring the status of the microorganisms in the fermentor. Specialized sensors (Mandenius, 1994) include enzyme electrodes (e.g., to measure glucose, lactate), calorimetric analyzers (e.g., to measure penicillin), specific gravity (e.g., beer fermentation), and immunosensors (e.g., to measure antigens). However, it remains an open challenge to develop *in situ* sensors that can monitor a variety of metabolites within the microorganisms in real time.

EXAMPLE G.1

Consider a fermentor, operated at constant volume, in which a single, rate-limiting substrate promotes biomass growth and product formation. Under the assumption of constant yield, one can derive the following material balances that describe the concentrations in the fermentor (Henson and Seborg, 1991):

$$\dot{X}=-DX+\mu(S,P)X \tag{G-7}$$

$$\dot{S}=D(S_f-S)\frac{1}{Y_{X/S}}\mu(S,P)X \tag{G-8}$$

$$\dot{P}=-DP+[\alpha\mu(S,P)+\beta]X \tag{G-9}$$

For this reactor, the growth term has a more complex shape than the simple Monod expression presented earlier, because both substrate and product can inhibit growth:

$$\mu(S,P)=\frac{\mu_m\left(1-\frac{P}{P_m}\right)S}{K_m+S+S^2/K_i} \tag{G-10}$$

The variables X, S, and P are the biomass, substrate, and product concentrations, respectively; D is the manipulated variable (dilution rate); S_f is the feed substrate concentration, and the remaining variables are fixed constants (yield parameters). Take the following values for the fixed parameters:

Table G.1 Parameter values and units for fermentor in Example G.1

$Y_{X/S}$	0.4 g/g	α	2.2 g/g
β	0.2 h^{-1}	μ_m	0.48 h^{-1}
P_m	50 g/L	K_m	1.2 g/L
K_i	22 g/L	S_f	20 g/L

(a) Assume a nominal operating point is $D = 0.202$ h^{-1} (dilution rate). The corresponding steady-state or equilibrium values of X, S, and P are 6.0, 5.0, and 19.14 g/L, respectively. Calculate the linearized model at this operating point, and determine the poles, zeros, and steady-state gain.

(b) Simulate the biomass X response to ±10% relative changes in dilution rate.

(c) Next, change the nominal dilution rate to $D = 0.0389\ \text{h}^{-1}$. The corresponding equilibrium values of X, S, and P are 6.0, 5.0, and 44.05 g/L, respectively. (Does anything look unusual here?). Recalculate the linearized model at this operating point, as well as the poles, zeros, and steady-state gain.

(d) Simulate the biomass response to ±10% relative changes in dilution rate.

(e) Comment on the extreme differences in behavior of the fermentor at these two operating points. What does this indicate about this nonlinear system? What are the implications for control design?

Solution

(a) Using the approach described in Section 4.3, a linearization of the nonlinear model (Eqs. G-7 through G-10) is performed. From the resulting model, one can derive a transfer function model with three poles and two zeros. The stable poles are calculated as the complex conjugate pair, −0.1469 ± 0.0694j, and the stable real pole at −0.2020. The zeros are both real and have the values: −0.1631, −0.2020. Finally, the steady-state process gain is −39.54.

(b) The simulated responses are depicted in Fig. G.2.

(c) As in part (a), the model is linearized, now at the new operating point. The poles are calculated as the complex conjugate pair, −0.0632 ± 0.0852j, and the stable real pole at −0.0389. The zeros are both real and have the values: −0.1630, −0.0389. Notice that one of the zeros is now nonminimum phase. Finally, the steady-state process gain is 86.90, indicating that the sign of the gain has been reversed.

(d) The simulated responses are depicted in Fig. G.3.

(e) In the first case ($D = 0.202\ \text{h}^{-1}$), the process gain was negative and the zeros were both negative. In the second case ($D = 0.0389\ \text{h}^{-1}$), the process gain was positive and one of the zeros becomes nonminimum phase, exhibiting inverse response. This suggests that the fermentor exhibits a dramatic nonlinearity, in which the gain can change sign and process zeros can change from negative to positive (indeed, a plot of the steady-state relationship between the dilution rate and the biomass for this fermentor reveals a parabolic shape). This suggests that operation across this gain change requires a nonlinear controller or an adaptive control scheme.

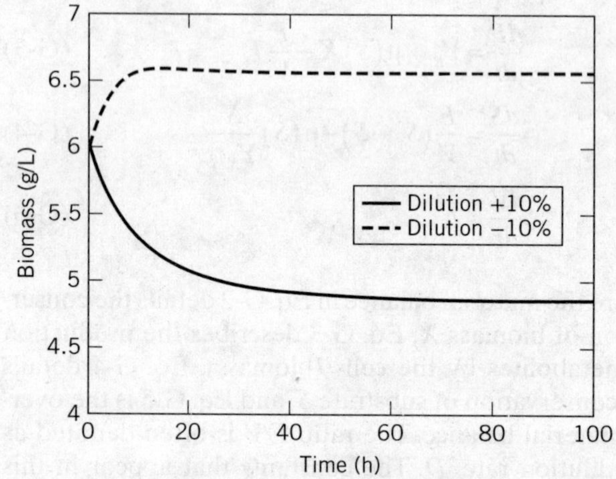

Figure G.2 Step response of fermentor model to symmetric changes in D of magnitude 10% from the nominal value of $D = 0.202\ \text{h}^{-1}$.

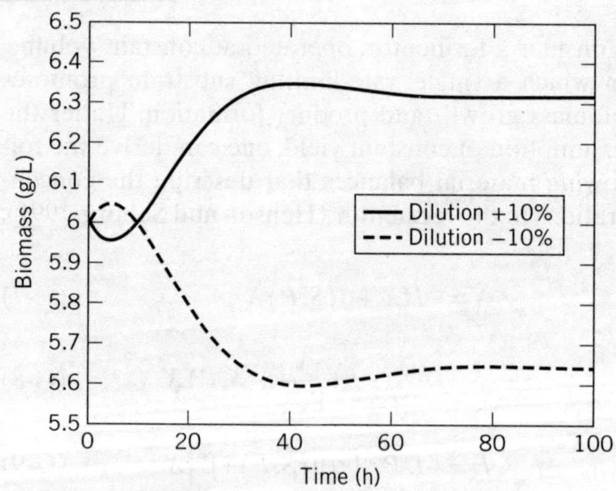

Figure G.3 Step response of fermentor model to symmetric changes in dilution of magnitude 10% from the nominal value of $D = 0.0389\ \text{h}^{-1}$.

G.1.2 Crystallizers

The operation of crystallization allows the separation of one phase (in this case, the solid from a solution) so that the product has desirable properties. The solid product that results from crystallization is a highly ordered solid structure, which may have other desirable attributes, including morphology (e.g., shape), that are of direct benefit to the value of the final product.

In the pharmaceutical industry, crystal size and shape may facilitate downstream solids processing and/or may be directly related to the final drug formulation, such as bioavailability, shelf life, toxicity, and drug dissolution (Fujiwara et al., 2005; Nagy and Braatz, 2012). Crystallization also finds application in the food industry to improve taste, as well as shelf life, for a diverse range of products (Larsen et al., 2006).

In order to explain the process control strategies employed in the operation of an industrial crystallizer, it is important to review briefly the concept of supersaturation and its relevance to crystallization (Larsen et al., 2006). Saturation refers to the property of phase equilibrium, in this case the equilibrium between the liquid and the dissolved solid (i.e., the solubility of the solid in the liquid). The state of supersaturation refers to the condition in which the liquid solution contains more solid than the amount that corresponds to the solubility (equilibrium), and the system exists in a so-called metastable state. Crystal formation can be induced by changing the operating conditions, such as temperature, so that the supersaturation state cannot be sustained, and a crystal is nucleated, or created, from the solution. As the dissolved component moves from the solution to the solid crystal phase, the concentration is, of course, lowered. Once a crystal is formed, it continues to grow as a function of the operating temperature and the concentration in the solution. In effect, the operation of crystallization involves the manipulation of this supersaturation state, trading off the formation of new crystals against the growth of existing ones.

An industrial crystallizer is operated typically in a batch mode, so that the management of the supersaturation state is accomplished over the course of the batch cycle time. The available manipulated inputs are the cooling jacket and steam flow rates for temperature management, and the inflow of antisolvent (a component that lowers the solvation capability of the liquid) and solvent, to regulate the concentration of the solution (Zhou et al., 2006). A typical crystallization flowsheet is depicted in Fig. G.4. Measurement of temperature is straightforward, and there are an increasing number of sophisticated instruments available for measurement of the crystal properties. These include turbidity sensors (to detect the presence of solids), laser scattering instruments (to extract the distribution of crystal sizes in the unit), and spectroscopic instruments, e.g., attenuated total reflectance-Fourier transformed infrared (ATR-FTIR), for measuring solution concentrations (Fujiwara et al., 2005; Larsen et al., 2006; Nagy and Braatz, 2012). More recently, a variety of imaging techniques have been used to measure crystal-shape properties (morphology), such as width and length. As mentioned earlier, these size and shape properties are major determinants for the resulting utility of the product (e.g., drug solubility), and it may be desirable to produce crystals with very uniform properties (i.e., narrow size distribution). Model-based control strategies, for example, optimization-based methods, are being employed to control crystal shape, including polymorphic forms (Nagy and Braatz, 2012).

G.1.3 Granulation

Granulation is a widely used process in which small particles agglomerate into larger granules. In wet granulation processes, the coagulation of particles is improved by the addition of a *binder liquid*, sprayed over an agitated powder in a tumbling drum or pan. The particles are wetted by the binder and a nucleate. The resulting binder-coated granules then collide and stick to form larger granules. These granules can also compact and consolidate as the binder liquid is brought to the surface of the aggregates by stirring in the granulator. Particles

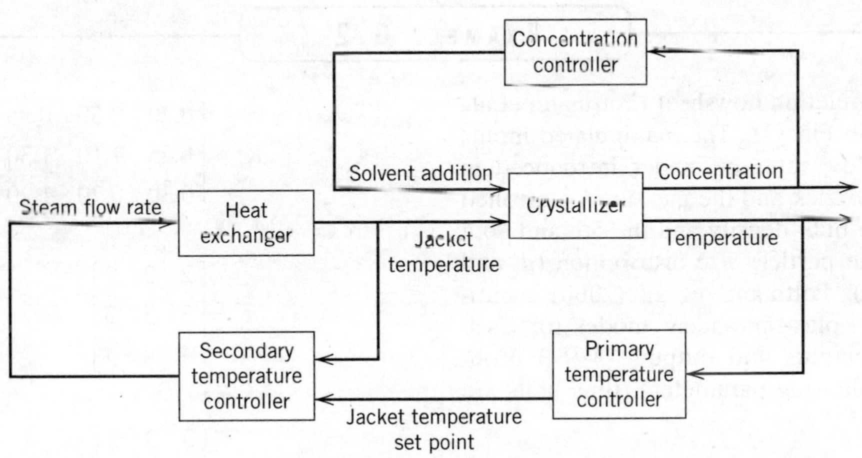

Figure G.4 Flowsheet of a typical industrial batch crystallizer, showing concentration and temperature controllers, including cascade control for temperature.

can also break because of collisions with the other particles or the granulator walls during mixing. Thus, the main phenomena in granulation processes are granule wetting and nucleation, consolidation and growth, and aggregation and breakage (Mort et al., 2001).

Granulation plays a key role in producing particles with special characteristics, such as time-release attributes (e.g., fertilizer, pharmaceutical tablet). However, in practice, inefficient operation, with very small yields and large recycle ratios (typically 4:1, recycle: product), often occurs. This inefficiency is due to the difficulty in designing and controlling granulation circuits that allow maintenance of specified size ranges for the granules.

As with crystallization, the key challenges for controlling a granulator are to produce particles with desirable attributes, to simplify downstream processing, and to realize end-product properties. In the pharmaceutical industry, granulation is usually accomplished in batch reactors, owing to the relatively small amount of material throughput. The key particle process that must be regulated is the agglomeration of smaller particles into larger particles. Manipulated inputs include binder spray addition (and/or viscosity), particle flow rate, temperature, mixing conditions (including shear), recycle of oversize (crushed) and undersize (fines) particles, and changing the rate of agitation (mixing) in the vessel (Pottmann et al., 2000; Mort et al., 2001; Faure et al., 2001). Some applications also incorporate heating, which introduces temperature control considerations. The measurements currently available are the torque on the agitator (which yields an inference of the load in the vessel and its size and moisture content) and, in more recent installations, measurements of particle size (possibly as a distribution). When a particle size distribution (PSD) is measured (e.g., by imaging methods or laser scattering), it is typically consolidated into one or more scalar measures of the distribution (e.g., the mean size, or d_x, the size of the particle in the *xth* percentile of the distribution (d_5, d_{90}, etc.)).

Process control strategies that utilize these measurements are finding application in wet granulation processes, that is, pharmaceutical industry (Faure et al., 2001), fluid bed granulation (Burggraeve, et al., 2011), and continuous drum granulation processes (Glaser et al., 2009). A typical continuous granulation flowsheet is depicted in Figure G.5. Batch granulation operation can benefit from traditional monitoring and quality approaches, as detailed in Chapters 17 and 18 (Burggraeve et al., 2011).

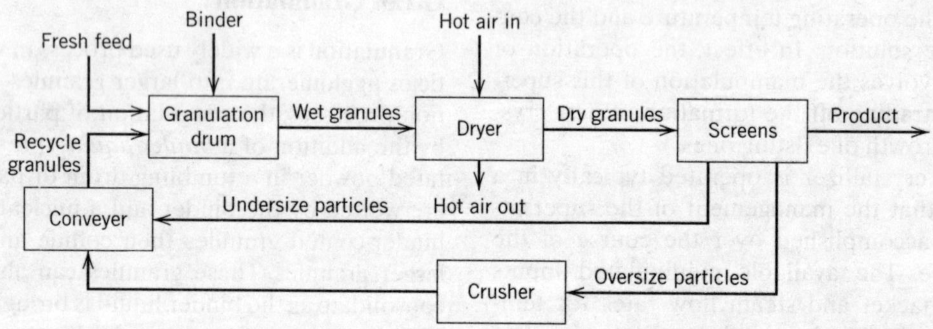

Figure G.5 Process flowsheet for granulation circuit with recycle.

EXAMPLE G.2

A simplified granulation flowsheet (Pottmann et al., 2000) is shown in Fig. G.6. The manipulated inputs are the liquid flow rates of binder introduced in three different nozzles, and the measured controlled variables are the bulk density and the 5th and 90th percentiles of the particle size distribution (d_5 and d_{90}, respectively). Pottmann et al. (2000) identified a first-order-plus-time-delay model for each combination of inputs and outputs (3 × 3 problem) with the following parameters (time units are dimensionless):

$$G_{ij}(S) = \frac{K_{ij}}{\tau_{ij}s+1}e^{-\theta_{ij}s} \tag{G-11}$$

$$K_{ij} = \begin{bmatrix} 0.20 & 0.58 & 0.35 \\ 0.25 & 1.10 & 1.30 \\ 0.30 & 0.70 & 1.20 \end{bmatrix} \tag{G-12}$$

$$\tau_{ij} = \begin{bmatrix} 2 & 2 & 2 \\ 3 & 3 & 3 \\ 4 & 4 & 4 \end{bmatrix} \tag{G-13}$$

$$\theta_{ij} = \begin{bmatrix} 3 & 3 & 3 \\ 3 & 3 & 3 \\ 3 & 3 & 3 \end{bmatrix} \tag{G-14}$$

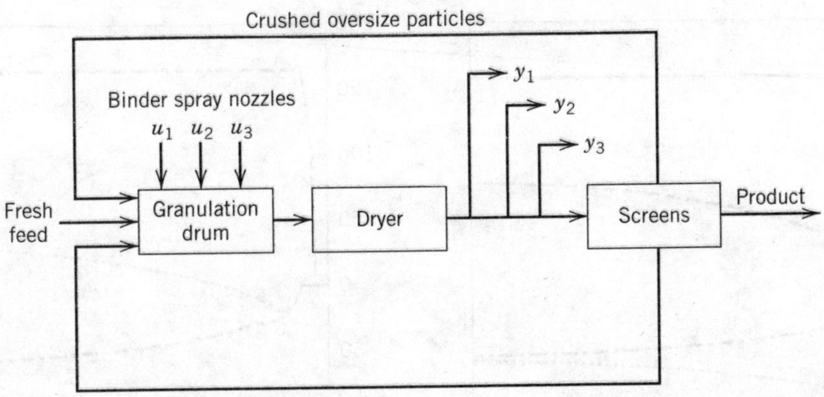

Figure G.6 Simplified process flowsheet for granulator example. Here u_1, u_2, and u_3 are, respectively, nozzles 1, 2, and 3, and y_1, y_2, and y_3 are, respectively, bulk density, d_5, and d_{90}.

The units for both the manipulated inputs and the measurements are dimensionless, and the nominal conditions (on which deviation variables are based) are 180 for all three nozzles, 40 for bulk density, 400 for d_5, and 1600 for d_{90}.

(a) Using the relative gain array (RGA), determine the most effective pairings between the inputs and the outputs.

(b) Design three PI + Smith predictor controllers using the IMC design method (see Table 11.1, and assume that a value of $\tau_c = 5$ is employed). Keep in mind that the IMC/PI tuning is for the delay-free part of the plant with a Smith predictor. For the Smith Predictor design, use the method detailed in Chapter 12 and keep in mind that you need to select one delay value for each input–output pair.

(c) Simulate the system response for the three PI + Smith predictor controllers using a step set-point change of [10 0 0]. Be sure to enforce the constraints on the manipulated variables (lower bound of 105; upper bound of 345). Repeat the simulation for a step set-point change of [50 0 0].

Solution

(a) The relative gain array is calculated as detailed in Section 14.2 from the process gains:

$$RGA = K \otimes \left(K^{-1}\right)^T = \begin{bmatrix} 1.0256 & 0.6529 & -0.6785 \\ -1.4103 & 1.8574 & 0.5528 \\ 1.3846 & -1.5103 & 1.1257 \end{bmatrix} \quad \text{(G-15)}$$

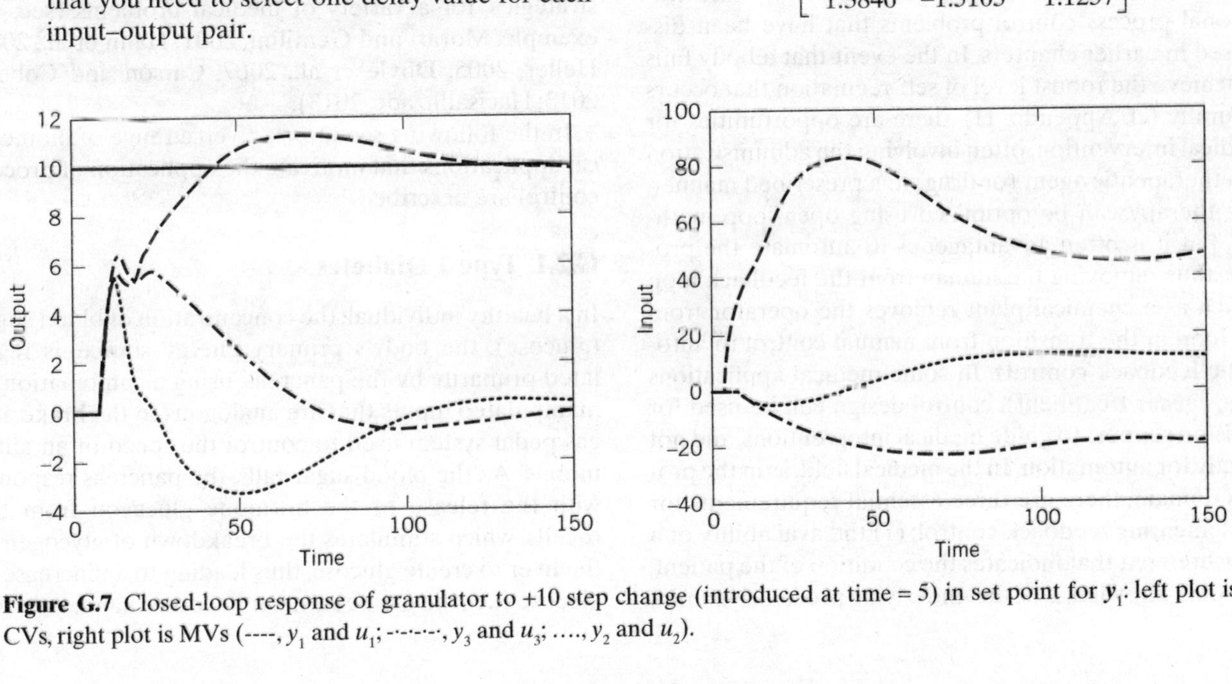

Figure G.7 Closed-loop response of granulator to +10 step change (introduced at time = 5) in set point for y_1: left plot is CVs, right plot is MVs (----, y_1 and u_1; -·-·-·-, y_3 and u_3; …., y_2 and u_2).

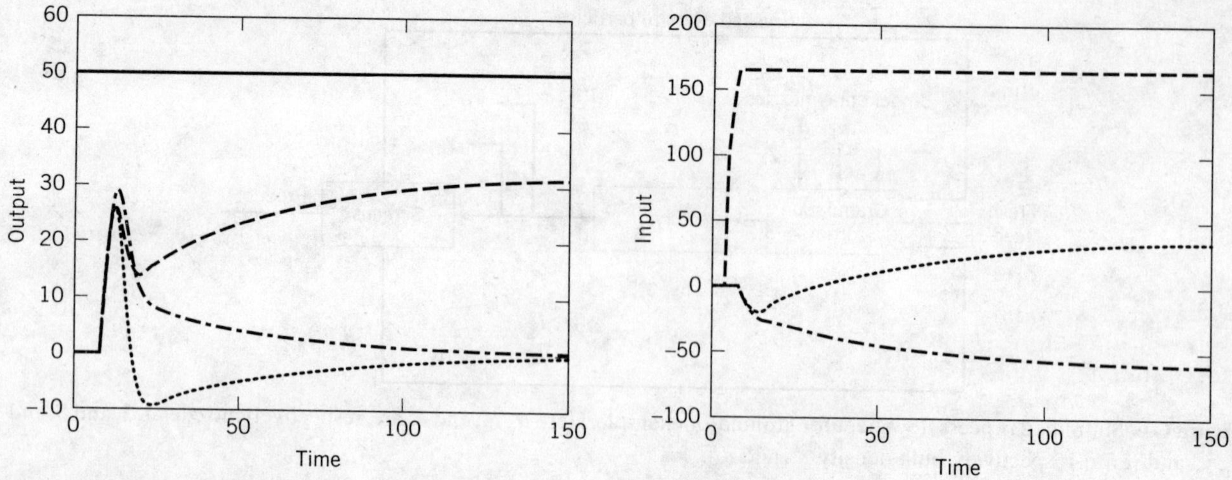

Figure G.8 Closed-loop response of granulator to +50 step change (introduced at time = 5) in set point for y_1, with constraints enforced on the inputs. The left plot is CVs, right plot is MVs (----, y_1 and u_1; - - - -, y_3 and u_3; …., y_2 and u_2).

Hence, the diagonal pairing of the controllers is recommended (1-1/2-2/3-3), because there are no negative values, and two of the three loops are paired on *RGA* values very close to 1.

(b) The individual controllers are given calculated from Table 11.1, row A:

Loop 1: $K_c = (2/5)/.2 = 2.0; \tau_i = 2$

Loop 2: $K_c = (3/5)/1.1 = 0.545; \tau_i = 3$

Loop 3: $K_c = (4/5)/1.2 = 0.667; \tau_i = 4$

(c) The simulation results are shown in Figs. G.7 and G.8. Note that enforcing the constraints for the second case leads to an unattainable set point for the first output.

G.2 PROCESS MODELING AND CONTROL FOR DRUG DELIVERY

The human body is a remarkably complex biochemical process, and it shares many attributes with more traditional process control problems that have been discussed in earlier chapters. In the event that a body fails to achieve the robust level of self-regulation that occurs naturally (cf. Appendix H), there are opportunities for medical intervention, often involving the administration of a therapeutic agent (or drug) in a prescribed manner. The therapy can be optimized using open-loop methods, but it is often advantageous to automate the process, thus removing the human from the feedback loop (much as a chemical plant removes the operator from the loop in the transition from manual control to automatic feedback control). In some medical applications (e.g., cancer treatment), control design can be used for decision support to guide medical interventions, and not strictly for automation. In the medical field, as in the process domain, there are three essential requirements for implementing feedback control: (1) the availability of a measurement that indicates the condition of the patient, (2) some knowledge of the underlying process dynamics (e.g., the effect of a drug on a patient's response), and (3) a suitable manipulated variable (e.g., drug or medication). Since 1990, there have been dramatic advances in sensor technology, as well as modeling and control strategies, for a variety of medical problems (see, for example, Morari and Gentilini, 2001; Hahn et al., 2002; Heller, 2005; Doyle et al., 2007; Carson and Cobelli, 2013; Hacisalihzade, 2013).

In the following sections, a diverse range of biomedical applications that motivate the application of process control are described.

G.2.1 Type 1 Diabetes

In a healthy individual, the concentration of blood sugar (glucose), the body's primary energy source, is regulated primarily by the pancreas, using a combination of manipulated inputs that are analogous to the brake and gas pedal system used to control the speed of an automobile. As the blood sugar falls, the pancreas responds with the release of the hormone glucagon from the α-cells, which stimulates the breakdown of glycogen in the liver to create glucose, thus leading to an increase in glucose (i.e., the gas pedal). On the other hand, as blood glucose rises, the pancreatic β-cells release the hormone insulin that stimulates the uptake of glucose by muscle and fat tissue (Carson and Cobelli, 2013), and, consequently, the blood glucose level is decreased (i.e., the brake).

Type 1 diabetes mellitus is a disease characterized by failure of the pancreatic β-cells. In contrast, the primary manifestation of Type 2 diabetes is an inability, or resistance, of the cells to respond to insulin. The only treatment for Type 1 diabetes consists of exogenous insulin injections, traditionally administered in an open-loop manner by the patient. The insufficient secretion of insulin by the pancreas results in large excursions of blood glucose outside of the target range of approximately 80-120 mg/dL, leading to brief, or often sustained, periods of hyperglycemia (elevated glucose levels). Intensive insulin therapy can often have the unintended consequence of overdosing, which can then lead to hypoglycemia (low glucose levels). The consequences of such inadequate glucose regulation include an increased risk for retinopathy, nephropathy, and peripheral vascular disease (DCCT, 1993; Jovanovič, 2000; Zisser at al., 2005).

As illustrated in Fig. G.9, a feedback controller can be used to regulate blood glucose using an insulin pump (widely available on the market today). There are preliminary clinical trials testing the efficacy of PID controllers for this delivery (Steil et al., 2006). The ADA has published guidelines (American Diabetes Association, 2014) recommending the following target zones for a blood sample drawn from a vein (a whole-blood sample):

- 70 mg/dL to 130 mg/dL before meals
- Less than 180 mg/dL 1 to 2 hours after meals

As indicated in Chapter 18 batch processes can benefit from recipe modifications in between consecutive batches or cycles, using a run-to-run (RtR) strategy. Run-to-run control strategies have also been developed for diabetes control, by considering glucose data for a meal response or an entire day to be the batch of interest. The similarities between the diabetic patient and the batch reactor recipe that motivate the application of this technique are the following:

1. The recipe (24-h cycle) for a human patient consists of a repeated meal protocol (typically three meals), with some variation on meal type, timing, and duration.
2. There is not an accurate dynamic model available to describe the detailed glucose response of each individual to the meal profile.
3. There are selected measurements available that might be used to characterize the quality of the glucose response for a 24-h day, including maximum and minimum glucose values.

Using currently available glucose meters, the blood sampling is very sparse, typically about six to eight measurements per day; hence, the overall quality (i.e., glycemic regulation) has to be inferred from these infrequent samples. The results of a subsequent clinical trial (Zisser et al., 2005) demonstrated that a large fraction of the patients responded favorably to this type of control. Current technology that is available for patients allows the continuous measurement of (subcutaneous) blood sugar with a 5-min sampling period (Harvey et al., 2010). This has enabled a wide array of closed-loop strategies for closed-loop control of insulin pumps, leading to the so-called *artificial pancreas*. From 2004 to 2014, there have been over 40 publications that report closed-loop clinical testing of an artificial pancreas, using PID, MPC, and fuzzy logic-based algorithms (Doyle III et al., 2014).

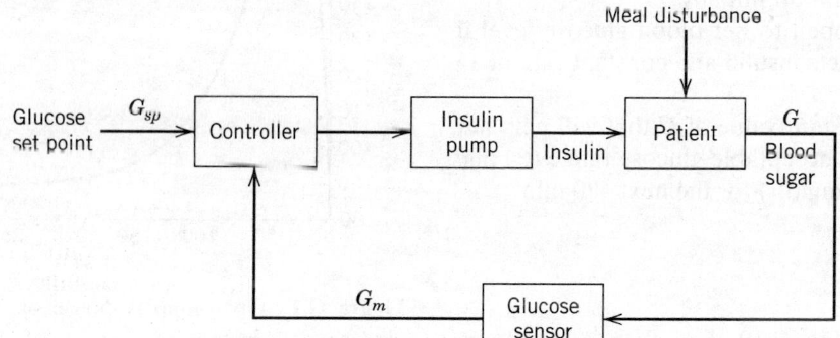

Figure G.9 Block diagram for artificial β-cell, illustrating the meal as the most common disturbance. G denotes the blood sugar of the patient, G_m is the output of the glucose sensor, and G_{sp} is the glucose set point.

EXAMPLE G.3

A patient with Type I diabetes needs an automated scheme to maintain her glucose within an acceptable range, widened here to allow less conservative control (54 mg/dL < G < 144 mg/dL). She has just eaten a large meal (a disturbance) that you estimate will introduce glucose into her bloodstream according to $D(t) = 9.0e^{-0.05t}$, where t is in minutes and $D(t)$ is in mg/dL-min. She has a subcutaneous insulin pump that can release insulin up to 115 mU/min (mU = 10^{-3} Units of insulin). The "U" is a standard convention used to denote the strength of an insulin solution. The flow rate of insulin is the manipulated variable.

A simple model of her blood glucose level is given by Bequette (2002):

$$\frac{dG}{dt} = -p_1 G - X(G + G_{Basal}) + D \qquad \text{(G-16)}$$

$$\frac{dX}{dt} = -p_2 X + p_3 I \qquad \text{(G-17)}$$

$$\frac{dI}{dt} = -n(I + I_{Basal}) + \frac{U}{V_1} \qquad \text{(G-18)}$$

where the constants are defined as follows: p_1 = 0.028735 [min^{-1}], p_2 = 0.028344 [min^{-1}], p_3 = 5.035E-5 [min^{-1}], V_1 = 12 [L], and n = 0.0926 [min^{-1}]. G, X, and I are values for glucose concentration (deviation) in the blood (mg/dL), insulin concentration (deviation) at the active site (mU/L), and blood insulin concentration, expressed in deviation variables. Basal values refer to the initial or baseline values for G and I (G_{Basal} = 81 mg/dL and I_{Basal} = 15 mU/L). D is the rate of glucose release into the blood (mg/dL-min) as the disturbance. U is the flow rate of insulin (mU/min) as the manipulated variable.

(a) What will happen to her blood glucose level if the pump is shut off initially?

(b) What will happen to her blood glucose level if the pump injects insulin at a constant rate of 15 mU/min?

(c) Is there a constant value of U that will help her stay within an acceptable glucose range (54 mg/dL < G < 170 mg/dL) for the next 400 min?

Solution

(a) As shown in Fig. G.10, the patient's blood glucose will rise in a ramplike fashion if the insulin pump fails (i.e., shuts off). This can also occur as a result of a catheter occlusion (blockage) with the insulin pump.

(b) In this case (Fig. G.11), the patient's blood sugar peaks, at slightly over 175 mg/dL, and takes 4 h to converge back to a steady-state glucose value of approximately 90 mg/dL.

(c) A setting of 25 mU/min yields the response in Fig. G.12, which might be deemed too aggressive by many doctors, because of the low post-meal glucose values, motivating a more advanced (i.e., closed-loop) approach to glucose management.

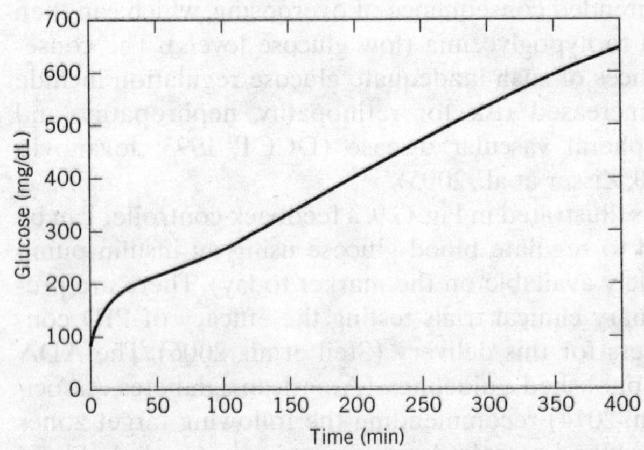

Figure G.10 Open-loop response of patient's blood glucose when the insulin pump is turned off.

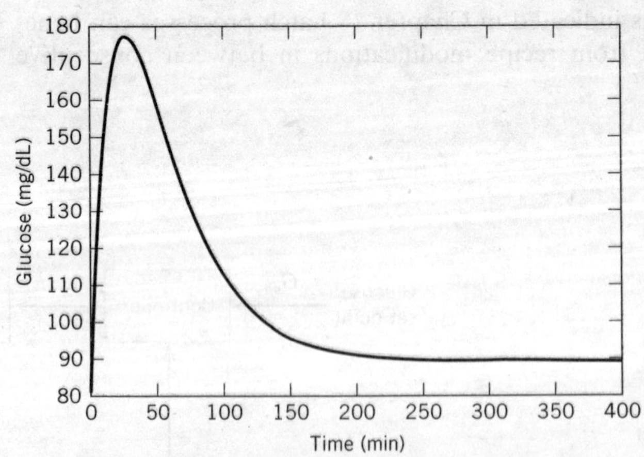

Figure G.11 Open-loop response of the patient's blood glucose to a constant infusion rate of 15 mU/min from her insulin pump.

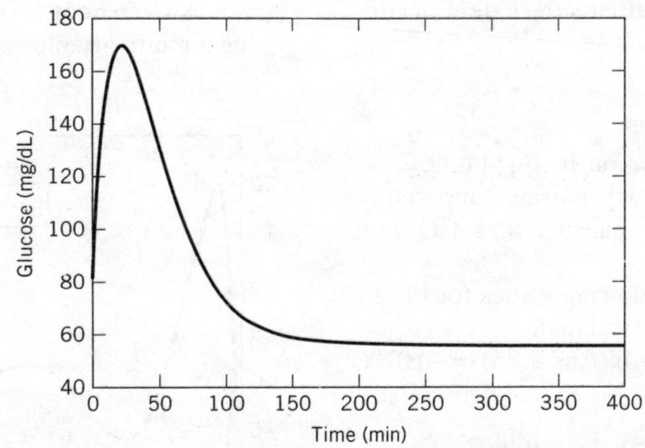

Figure G.12 Open-loop response of patient's blood glucose to a constant infusion rate of 25 mU/min from her insulin pump.

G.2.2 Blood Pressure Regulation

In both the operating room and postoperative care contexts, closed loop control of blood pressure and related variables (such as cardiac output and depth of anesthesia) have been studied for a number of years (e.g., Rao et al., 1999; Dumont and Ansermino, 2013), and human clinical trials have proved the efficacy of the approach (Bailey and Haddad, 2005; Araki and Furutani, 2005; Sarabadani et al., 2013). The postoperative application was handled typically by the administration of sodium nitroprusside (SNP) by a nurse via a continuous intravenous (IV) pump. SNP is a vasodilator that achieves blood pressure reduction by relaxing the muscles controlling the vascular resistance to flow through blood vessels. The current technology for both sensors and infusion pumps is facilitating the design of completely automated control strategies.

The context of the operating room is more complicated, with many critical variables that must be monitored. But an advantage of this setting is that nonportable sensors can be employed that would be too cumbersome or impractical for ambulatory applications. The measured variables include mean arterial pressure (MAP), cardiac output (CO), and depth of anesthesia (DOA). The DOA has been the subject of intense research activity over the last decade, and sensors are available to determine the depth of anesthesia through correlations. These sensors are inferential (see Chapter 12), in that they do not directly measure the medical state of anesthesia, which is characterized by such patient responses as hypnosis, amnesia, analgesia, and muscle relaxation (Araki and Furutani, 2005); rather, they measure the state of electrical activity in the patient's brain. One of the more promising methods is the bispectral index, derived from signal analysis of an electroencephalograph (EEG) (Bailey and Haddad, 2005). A variety of manipulated inputs are also available, resulting in an intrinsically multivariable control problem. Some candidate manipulated variables include vasoactive drugs, such as dopamine and SNP, as well as anesthetics (isoflurane, propofol, etc.).

EXAMPLE G.4

Consider the following model for predicting the influence of two drugs: SNP, [μg/kg min]) and dopamine (DPM, [μg/kg-min]), on two medical variables (mean arterial pressure MAP, [mmHg]) and cardiac output CO, [L/(kg-min)]), where time is measured in minutes (Bequette, 2007):

$$\begin{bmatrix} MAP \\ CO \end{bmatrix} = \begin{bmatrix} \dfrac{-6e^{-0.75s}}{0.67s+1} & \dfrac{3e^{-s}}{2.0s+1} \\ \dfrac{12e^{-0.75s}}{0.67s+1} & \dfrac{5e^{-s}}{5.0s+1} \end{bmatrix} \begin{bmatrix} SNP \\ DPM \end{bmatrix} \tag{G-19}$$

(a) Calculate the RGA for this problem and propose the appropriate control-loop pairing.

(b) Consider the pairing, SNP-MAP and DPM-CO, as is typically used in practice. Design a pair of PI controllers for this process, using the IMC tuning rules (Table 11.1) and choosing a value of τ_c for each controller that is equal to the corresponding open-loop time constant for that subsystem.

(c) Simulate the closed-loop response to a 10 mmHg change in the MAP set point, while holding CO constant. Discuss the extent of control-loop interactions.

Solution

(a) Using the RGA calculation in Eq. 14-39, λ_{11} = 0.4545; therefore, the loop pairings apparently should be the 1-2/2-1 pairing, SNP-CO and DPM-MAP.

(b) From Table 11.1, the following values for the PI controller settings are calculated:

Loop 1: $K_c = -(0.67)/(6^*(0.67+.75)) = -0.0786$

$\tau_i = 0.67$

Loop 2: $K_c = (5)/(5^*(5+1)) = 0.1667$

$\tau_i = 5$

(c) The simulated response for the MAP set point change is depicted in Fig. G.13, where there is a modest undershoot in the MAP response; however, the interacting nature of the process leads to a large excursion in CO. The control tuning constant τ_c could be refined to trade-off speed versus overshoot and interaction, or by designing a multivariable controller, such as MPC.

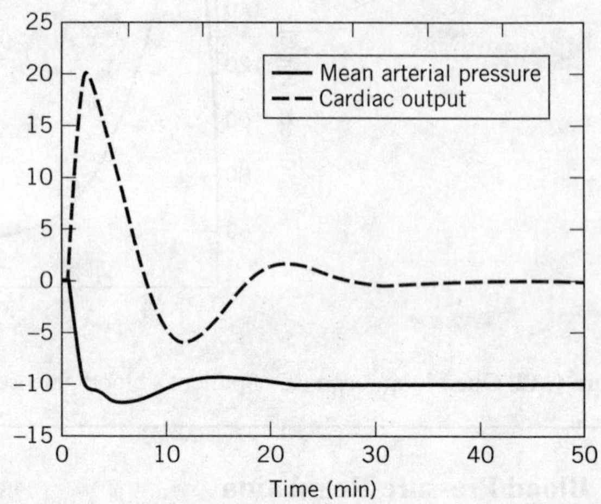

Figure G.13 Closed-loop response of patient's mean arterial blood pressure and cardiac output to a 10 mmHg change in the MAP set point.

G.2.3 Cancer Treatment

Cancer treatment has changed dramatically over the past decade, in large part enabled by advances in imaging technology. Surgery has been the classical method for attacking cancerous tumors, and more recently X-ray radiation has been employed. An unfortunate side effect in both cases is that healthy tissue can be compromised by inappropriate surgery or delivery of radiation, respectively. Chemotherapy is often be used, alone or in conjunction with surgery or X-ray treatment, and has the advantage that undetected metastases (cancer cells that have circulated through the bloodstream) can be attacked with this method. Thermal therapies (radiofrequency, microwave, or laser techniques) have also been demonstrated to be effective, with similar requirements on targeting the energy to the localized region of the tumor (Dodd et al., 2000). In thermal and radiation treatment, feedback control is finding application to the optimized delivery of the treatment (radiation, heat) to the targeted area (Salomir et al., 2000; Davison and Hwang, 2003; Ledzewicz and Schattler, 2007; Moonen, 2007). In one feedback-based therapy (Salomir et al., 2000), the heat source power was adjusted based on the deviation of temperature from a target at a particular location in the body, including an integral term, very similar to a PID controller. The desired response was that the temperature should rise quickly to the target without overshoot or oscillations.

Parker (2007) describes a strategy for "model-informed" treatment design for delivery of a chemotherapeutic agent. Using a combination of pharmacokinetic and pharmacodynamic models, predictions can be made about the patient's response (e.g., tumor volume) to the manipulated variables, which in this case could include the drug dosage level and schedule for drug administration. This strategy can be implemented by specifying the time horizon over which the patient's response is monitored and by calculating the optimal drug delivery protocol using the RTO methods of Appendix F.

More recent developments include chemotherapy using antiangiogenic agents, which deprive the tumor from developing blood cells required for growth (Ledzewicz and Schattler, 2007). In this application, information about the state of the tumor (e.g., the tumor volume, derived from MRI data) is used to control the rate of dosing of the antiangiogenic therapy. More recently, model predictive control designs have been proposed for chemotherapeutic protocols, as an example of a decision support tool, as contrasted with an automation tool (Florian et al., 2008).

G.2.4 Controlled Treatment for HIV/AIDS

To address the global problem of HIV/AIDS, a number of mathematical models and control algorithms have been proposed to help design better treatments for the disease.

The drug categories that have been considered include reverse transcriptase inhibitors and protease inhibitors, which affect reproduction of the virus via transcription and production of the virus from infected cells, respectively. The most effective strategies to date have involved a so-called cocktail of multiple drugs, thus attacking the disease in a vector direction (i.e., multiple, simultaneous targets). Measurements are problematic, consisting of relatively slow techniques based on off-line sampling of blood. However, the slow progression of the disease does not warrant real-time measurements, and thus feedback can still be accomplished on this slow time scale.

In their simplest form, mathematical models have been developed that describe the interactions of healthy CD4+ T cells, infected CD4+ T cells, and free viruses in the form of three coupled ordinary differential equations (Craig and Xia, 2005). Such a model can be the basis of simple model-based feedback strategies for control and can also be extended to generate more complex models suitable for a model predictive control strategy (Zurakowski et al., 2004), including the design of optimal treatment protocols (Pannocchia et al., 2012).

G.2.5 Cardiac-Assist Devices

Cardiac-assist devices are mechanical pumps that provide cardiac output at an appropriate pressure, to allow normal circulation of blood through the patient's body, subject to the changing demands for cardiac output as a function of the patient's state (e.g., level of exercise, emotion, posture, etc.). The ideal device would mimic the body's own mechanisms for maintaining cardiac output at target levels; however, currently available devices are rather primitive in terms of automation, requiring the patient to adjust the set point (Boston et al., 2000). The first such implantable device received approval by the FDA over a decade ago. Research continues on the design of control algorithms for implantable rotary blood pumps, including PID, fuzzy logic, adaptive control, and optimization-based methods (AlOmari et al., 2013).

One of the more interesting aspects of the control design problem for ventricular-assist devices is the placement of the sensors and actuators: there are the issues of susceptibility to infection, as well as anatomical placement (Paden et al., 2000).

G.2.6 Additional Medical Opportunities for Process Control

There are many other challenges in drug therapy, in which an optimized delivery regimen could be calculated using principles of process control and process optimization, e.g., the modeling and control of the anticoagulant drug, heparin (McAvoy, 2007). Another medical application is the treatment of acute neuropatients with brain hypothermia, to lower the intracranial pressure (ICP). A mathematical model can be developed to relate temperature effects with blood flow. The model can then be used to create an automated closed-loop controller (Gaohua and Kimura, 2006) to adjust the coolant temperature (e.g., by using cold-water circulating blankets) in an effort to regulate the ICP. Control design opportunities have been identified for hemodialysis (Javed et al., 2012), mechanical ventilation for respiration (Li and Haddad, 2013), leukemia treatment and vaccination (Laurino et al., 2013) and Parkinsons treatment (Hacisalihzade, 2013).

SUMMARY

Biological and biomedical processes share a great deal in common with the process applications considered in preceding chapters. The latter applications have a characteristic time constant, often exhibit time delays associated with measurements, and typically are multivariable in nature. In contrast, the types of uncertainties in bioprocesses are quite different, owing to the complex nature of biological regulation (see Appendix H). In addition, there are multiple safety and regulatory issues that are unique to medical closed-loop systems. Process control strategies for several key unit operations in the bioprocess industries (fermentation, crystallization, and granulation) have been described, along with biomedical applications, such as drug delivery for diabetes and blood pressure control and other examples. In Appendix H we focus on the molecular scale of biological systems and consider the feedback mechanisms inherent in naturally occurring biophysical networks.

REFERENCES

Alford, J., *Bioprocess Control: Advantages and Challenges*, Proc. Conf. Chemical Process Control, Lake Louise, Alberta, Canada (2006).

AlOmari, A. H., A. V. Savkin, M. Stevens, D. G. Mason, D. L. Timms, R. F. Salamonsen, and N. H. Lovell, Developments in Control Systems for Rotary Left Ventricular Assist Devices for Heart Failure Patients: A Review, *Physiol. Meas.*, **34**, R1 (2013).

American Diabetes Association, Standards of Medical Care in Diabetes—2014, *Diabetes Care*, **37**, S14 (2014).

Araki, M., and E. Furutani, Computer Control of Physiological States of Patients under and after Surgical Operation, *Annu. Rev. Control*, **29**, 229 (2005).

Bequette, B. W., *Process Control*, Prentice-Hall, Upper Saddle River, NJ, 2002.

Bequette, B. W., Modeling and Control of Drug Infusion in Critical Care, *J. Process Control*, **17**, 582 (2007).

Boston, J., M. Simaan, J. Antaki, and Y. Yu, Control Issues in Rotary Heart Assist Devices, *Proc. American Control Conf.*, **3473** (2000).

Boudreau, M. A., and G. K. McMillan, *New Directions in Bioprocess Modeling and Control*, ISA, NC, 2007.

Burggraevea, A., T. Van Den Kerkhofb, M. Hellingsb, J. P. Remonc, C. Vervaetc, and T. De Beer, Batch Statistical Process Control of a Fluid Bed Granulation Process using In-line Spatial Filter Velocimetry and Product Temperature Measurements, *Eur. J. Pharm. Sci.*, **42**, 584 (2011).

Buckland, B. C., The Translation of Scale in Fermentation Processes: The Impact of Computer Process Control, *Nature Biotechnology*, **2**, 875 (1984).

Carson, E., and C. Cobelli, *Modeling Methodology for Physiology and Medicine*, Elsevier, London, 2013.

Craig, I., and X. Xia, Can HIV/AIDS Be Controlled?, *IEEE Control Sys. Mag.*, **25**, 80-83 (2005).

Davison, D. E., and E.-S. Hwang, Automating Radiotherapy Cancer Treatment: Use of Multirate Observer-Based Control, *Proc. American Control Conf.*, **1194** (2003).

DCCT–The Diabetes Control and Complications Trial Research Group, The Effect of Intensive Treatment of Diabetes on the Development and Progression of Long-Term Complications in Insulin-Dependent Diabetes Mellitus, *New Engl. J. Med.*, **329**, 977 (1993).

Dodd, G. D., M. C. Soulen, R. A. Kane, T. Livraghi, W. R. Lees, Y. Yamashita, A. R. Gillams, O. I. Karahan, and H. Rhim, Minimally Invasive Treatment of Malignant Hepatic Tumors: At the Threshold of a Major Breakthrough, *Radiographics*, **20**, 9 (2000).

Doyle III, F. J., L. M. Huyett, J. B. Lee, H. C. Zisser, and E. Dassau, Closed-Loop Artificial Pancreas Systems: Engineering the Algorithms, *Diabetes Care*, **37**, 1191 (2014).

Doyle, F., L. Jovanovič, D. Seborg, R. S. Parker, B. W Bequette, A. M. Jeffrey, X. Xia, I. K. Craig, and T. J. McAvoy, A Tutorial of Biomedical Process Control, *J. Process Control*, **17**, 571 (2007).

Doyle, F., L. Jovanovič, and D. E. Seborg, Glucose Control Strategies for Treating Type 1 Diabetes Mellitus, *J. Process Control*, **17**, 572 (2007).

Dumont, G., and J. M. Ansermino, Closed-Loop Control of Anesthesia: A Primer for Anesthesiologists, *Anesthesia & Analgesia*, **117**, 1130 (2013).

Faurea, A., P. Yorkb, and R. C. Rowe, Process Control and Scale-up of Pharmaceutical Wet Granulation Processes: A Review, *Eur. J. Pharm. Biopharm.*, **52**, 269 (2001).

Fleet, G. H. (Ed.), *Wine Microbiology and Biotechnology*, Taylor & Francis, London, U.K., 1993.

Florian, J. A., J. L. Eiseman, and R. S. Parker. Nonlinear Model Predictive Control for Dosing Daily Anticancer Agents: A Tamoxifen Treatment of Breast Cancer Case Study, *Comput. Biol. Med.*, **38**, 339 (2008).

Fujiwara, M., Z. K. Nagy, J. W. Chew, and R. D. Braatz, First-Principles and Direct Design Approaches for the Control of Pharmaceutical Crystallization, *J. Process Control*, **15**, 493 (2005).

Gaohua, L., and H. Kimura, A Mathematical Model of Intracranial Pressure Dynamics for Brain Hypothermia Treatment, *J. Theor. Biol.*, **238**, 882 (2006).

Gatzke, E. P., and F. J. Doyle III, Model Predictive Control of a Granulation System Using Soft Output Constraints and Prioritized Control Objectives, *Powder Tech.*, **121**, 149 (2001).

Glaser, T., C. F. W. Sanders, F. Y. Wang, I. T. Cameron, R. Ramachandran, J. D. Litster, J. M. H. Poon, C. D. Immanuel, and F. J. Doyle III, Model Predictive Control of Drum Granulation, *J. Process Contr.*, **19**, 615 (2009).

Hacisalihzade, S. S., *Biomedical Applications of Control Engineering*, Springer, 2013.

Hahn, J., T. Edison, and T. F. Edgar, Adaptive IMC Control for Drug Infusion for Biological Systems, *Control Eng. Practice*, **10**, 45 (2002).

Harvey, R., Y. Wang, B. Grosman, M. W. Percival, W. Bevier, D. A. Finan, H. Zisser, D. E. Seborg, L. Jovanovic, F. J. Doyle III, and E. Dassau, Quest for the Artificial Pancreas, *IEEE Eng. Med. Biol. Mag.*, **March/April**, **53** (2010).

Heller, A., Integrated Medical Feedback Systems for Drug Delivery, *AIChE J.*, **51**, 1054 (2005).

Henson, M. A., and D. E. Seborg, An Internal Model Control Strategy for Nonlinear Systems, *AIChE J.*, **37**, 1065 (1991).

Javed, F., A. V. Savkin, G. S. H. Chan, J. D. Mackie, and N. H. Lovell, Recent Advances in the Monitoring and Control of Haemodynamic Variables during Haemodialysis: A Review, *Physiol. Meas.*, **33**, R1 (2012).

Jovanovič, L., The Role of Continuous Glucose Monitoring in Gestational Diabetes Mellitus, *Diabetes Technol. Ther.*, **2**, S67 (2000).

Larsen, P. A., D. B. Patience, and J. B. Rawlings, Industrial Crystallization Process Control, *IEEE Control Systems*, **26**, 70 (2006).

Laurino, M., M. Stano, M. Betta, G. Pannocchia, and A. Landi, Combining Pharmacological Therapy and Vaccination in Chronic Myeloid Leukemia via Model Predictive Control, *Proc. IEEE EMBS Conf.*, 3925 (2013).

Ledzewicz, U., and H. Schattler, Antiangiogenic Therapy in Cancer Treatment as an Optimal Control Problem, *SIAM J. Control Optim.*, **46**, 1052 (2007).

Li, H., and W. H. Haddad, Model Predictive Control for a Multicomponent Respiratory, *IEEE Trans. Control Sys. Tech.*, **21**, 1988 (2013).

Mahadevan, R., S. K. Agrawal, and F. J. Doyle III, Differential Flatness Based Nonlinear Predictive Control of Fed-Batch Bioreactors, *Control Eng. Prac.*, **9**, 889 (2001).

Mandenius, C.-F., Process Control in the Bioindustry, *Chemistry Today*, **19**, 19-22 (1994).

Mascia, S., P. L. Heider, H. Zhang, R. Lakerveld, B. Benyahia, P. I. Barton, R. D. Braatz, C. L. Cooney, J. M. B. Evans, T. F. Jamison, K. F. Jensen, A. S. Myerson, and B. L. Trout, End-to-End Continuous Manufacturing of Pharmaceuticals: Integrated Synthesis, Purification, and Final Dosage Formation, *Angew. Chem.*, **52**, 12359 (2013).

McAvoy, T. J., Modeling and Control of the Anticoagulant Drug Heparin, *J. Process Control*, **17**, 590 (2007).

Moonen, C. T. W., Spatio-Temporal Control of Gene Expression and Cancer Treatment Using Magnetic Resonance Imaging-Guided Focused Ultrasound, *Clin. Cancer Res.*, **13**, 3482 (2007).

Morari, M., and A. Gentilini, Challenges and Opportunities in Process Control: Biomedical Processes, *AIChE J.*, **47**, 2140 (2001).

Mort, P. R., S. W. Capeci, and J. W. Holder, Control of Agglomeration Attributes in a Continuous Binder-Agglomeration Process, *Powder Tech.*, **117**, 173 (2001).

Nagy, Z. K., and R. D. Braatz, Advances and New Directions in Crystallization Control, *Annu. Rev. Chem. Biomol. Eng.*, **3**, 55 (2012).

O. de Sa, D., Applied Technology and Instrumentation for Process Control, CRC Press, New York, 2004.

Paden, B., J. Ghosh, and J. Antaki, Control System Architectures for Mechanical Cardiac Assist Devices, *Proc. American Control Conf.*, 3478 (2000).

Pannocchia, G., Laurino, M., and A. Landi, A Model Predictive Control Strategy Toward Optimal Structured Treatment Interruptions in Anti-HIV Therapy, *IEEE Trans. Biomed. Eng.*, **57**, 1040 (2012).

Parker, R. S., Modeling for Anti-Cancer Chemotherapy Design, *J. Process Control*, **17**, 576 (2007).

Pottmann, M., B. A. Ogunnaike, A. A. Adetayo, and B. J. Ennis, Model-Based Control of a Granulation System, *Powder Tech.*, **108**, 192 (2000).

Rao, R. R., J. W. Huang, B. W. Bequette, H. Kaufman, and R. J. Roy, Control of a Nonsquare Drug Infusion System: A Simulation Study, *Biotechnol. Prog.*, **15**, 556 (1999).

Rohani, S., M. Haeri, and H. C. Wood, Modeling and Control of a Continuous Crystallization Process Part 1. Linear and Non-Linear Modeling, *Comput. Chem. Eng.*, **23**, 263 (1999).

Salomir, R., F. C. Vimeux, J. A. de Zwart, N. Grenier, and C. T. W. Moonen, Hyperthermia by MR-Guided Focused Ultrasound: Accurate Temperature Control Based on Fast MRI and a Physical Model of Local Energy Deposition and Heat Conduction, *Magnetic Resonance in Medicine*, **43**, 342 (2000).

Sarabadani, A., S. Bernasconi, V. Klamroth-Marganska, S. Nussbaumer, and R. Riener, Model-free Predictive Control of Human Heart Rate and Blood Pressure, *Proc. IEEE Intl. Conf. Bionformatics and Bioengineering*, **1**, (2013).

Schügerl, K., Progress in Monitoring, Modeling and Control of Bioprocesses during the Last 20 Years, *J. Biotechnol.*, **85**, 149 (2001).

Sheldon, R. A., The E Factor: Fifteen Years On, *Green Chem.*, **9**, 1273 (2007).

Steil, G. M., K. Rebrin, C. Darwin, F. Hariri, and M. F. Saad, Feasibility of Automating Insulin Delivery for the Treatment of Type 1 Diabetes, *Diabetes*, **55**, 3344 (2006).

Van Impe, J. F., P. A. Vanrolleghem, and D. M. Iserentant, *Advanced Instrumentation, Data Interpretation, and Control of Biotechnological Processes*, Kluwer, Dordrecht, 1998.

Zisser, H., L. Jovanovič, F. Doyle III, P. Ospina, and C. Owens, Run-to-Run Control of Meal-Related Insulin Dosing, *Diabetes Technol. Ther.*, **7**, 48 (2005).

Zhou, G. X., M. Fujiwara, X. Y. Woo, E. Rusli, H.-H. Tung, C. Starbuck, O. Davidson, Z. Ge, and R. D. Braatz, Direct Design of Pharmaceutical Antisolvent Crystallization through Concentration Control, *Crystal Growth & Design*, **6**, 892 (2006).

Zurakowski, R., M. J. Messina, S. E. Tuna, and A. R. Teel, HIV Treatment Scheduling Via Robust Nonlinear Model Predictive Control, *Proc. Asian Control Conf.*, **25** (2004).

EXERCISES

G.1 Consider the fermentor problem in Example G.1.

(a) Design an IMC controller for the first operating point (dilution = 0.202 h^{-1}), and simulate the response to both a +0.5 [g/L] and a –0.5 [g/L] change in the biomass concentration set point. Then, simulate the response to both a +1 [g/L] and a –1 [g/L] change in the biomass concentration set point.

(b) Simulate the response of a –12.5% step change in the maximum growth rate (μ_m). How well does the controller perform?

(c) Comment on the observed nonlinearity in the system.

(d) Discuss how the controller design would change if there were a requirement to operate at the lower dilution operating point. What do you need to consider in this case?

G.2 Consider the granulation model that was given in Example G.2.

(a) Design an MPC controller, using the nominal process model. Initially consider a control horizon of $M = 2$ and a prediction horizon of $P = 40$ (with a sampling period of $\Delta t = 1$). Use equal weights on the manipulated inputs and penalize the two percentile outputs equally, but use a larger weight on the bulk density y_1.

(b) Consider the effect of a plant-model mismatch. Use the problem statement for control design, but assume that the actual process is characterized by the following parameters:

$$\tilde{K}_{i,j} = \begin{bmatrix} 0.10 & 0.90 & 0.15 \\ 0.25 & 1.10 & 1.30 \\ 0.50 & 0.80 & 1.00 \end{bmatrix}$$

$$\tilde{\tau}_{i,j} = \begin{bmatrix} 1 & 2 & 1.5 \\ 3 & 3 & 3 \\ 3 & 3 & 3 \end{bmatrix}$$

$$\tilde{\theta}_{i,j} = \begin{bmatrix} 2 & 2 & 4 \\ 2 & 3 & 4 \\ 2 & 3 & 4 \end{bmatrix}$$

(c) These models are in deviation variables, but the actual steady-state flow rates for the nozzles are 175, 175, and 245, respectively. The steady-state outputs are 40, 400, and 1620, respectively. Nozzle flow rates are limited to values between 100 and 340, and it is desired to keep the 5th percentile (y_2) above 350 and the 90th percentile y_3 below 1650. Simulate the response of the controller to the following changes:

(i) Step change in bulk density from 40 to 90.

(ii) Simultaneous change in the 5th percentile from 400 to 375 and 90th percentile from 1620 to 1630.

Comment on the performance of your controller (and retune as necessary).

G.3 Gaohua and Kimura (2006) derived an empirical patient model for the manipulation of ambient temperature u (°C) to influence the patient's brain intracranial pressure (ICP) y (mmHg). The medical data support the following empirical values for a first-order-plus-time-delay model to describe the effect of cooling temperature (°C) on the ICP (mmHg) in time units of hours:

$$G(s) = \frac{4.7}{9.6s+1} e^{-s}$$

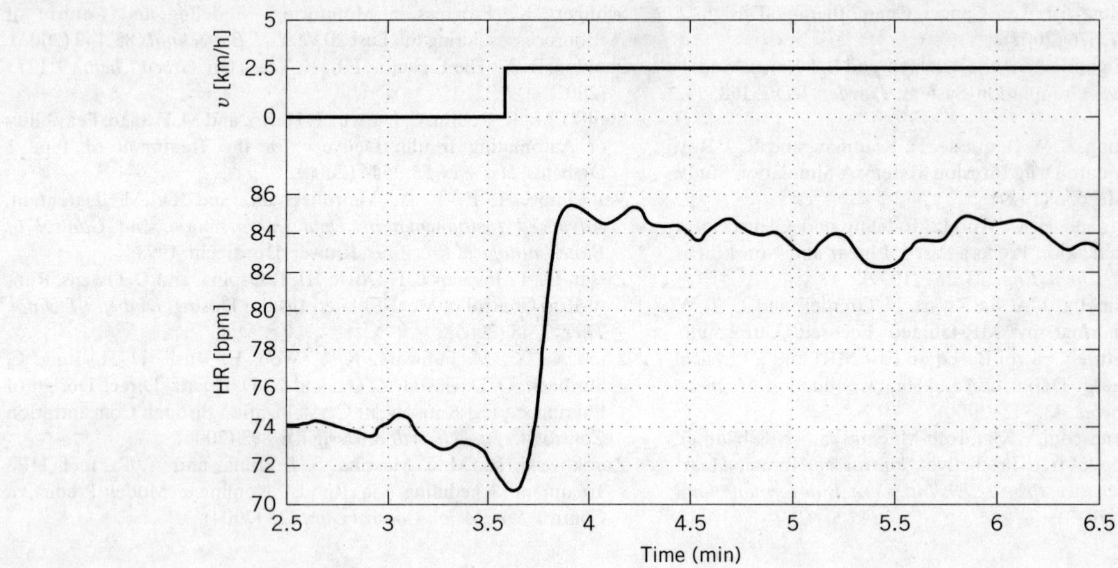

Figure EG.4

The nominal values for the process variables are: ICP = 20 mmHg; ambient temperature = 30 °C.

(a) Using the IMC tuning rules, derive an appropriate PI controller for this medical experiment. (*Hint*: begin with a value τ_c = 1.0 h). What does that value of τ_c mean?

(b) Simulate the response of a 10-mmHg reduction in ICP. What is the overshoot? What is the minimum value of the temperature? What is the settling time?

(c) Comment on whether this is a reasonable controller design for a biomedical application. How might you improve the design?

G.4 In a rehabilitation training experiment for a neurological patient, a step change in treadmill speed of +2.5 km/h was made. The patient heart rate response HR is given in Fig. EG.4.

(a) Derive an appropriate first-order-plus-time-delay model for the patient dynamics.

(b) The doctors wish to control the patient's heart rate to a nearly constant value by adjusting treadmill speed. Using the IMC tuning rules, design a suitable PI controller for this patient. Simulate the response of the controller for a step change in the HR of +10 bpm. Calculate the settling time, overshoot, and rise time for the controller. Do these values seem reasonable for a medical application?

(c) How would you improve the procedure for fitting the patient's initial dynamics?

G.5 A crystallizer is used to separate a pharmaceutical product from the fermentation extract. The three manipulated variables are the fines dissolution rate u_1, the crystallizer temperature u_2, and the flow rate in the overflow u_3. The nominal values of these three inputs are 2.25×10^{-6} m³/s, 310 K, and 1.5×10^{-6} m³/s, respectively. The three variables to be controlled are the crystal size distribution, as calculated by the fines suspension density y_1; the crystal purity, as calculated by supersaturation conditions y_2; and the product rate y_3. The nominal values of these three inputs are 0.55 K, 11.23 K, and 0.12 kg/kg H_2O, respectively. These variables have multiple interactions, and the following model has been identified from experimental data for a continuous crystallizer, where time is measured in s (Rohani et al., 1999):

$$\begin{bmatrix} y_1 \\ y_2 \\ y_3 \end{bmatrix} = \begin{bmatrix} \dfrac{72{,}600}{s+0.2692} & \dfrac{0.025082(s-20.0)(s-10.4)}{s^2+10.11s+96.57} & \dfrac{125{,}000(s-1.25)}{s+0.39} \\ \dfrac{568{,}000}{s+2.11} & \dfrac{-0.15095}{s+0.1338} & \dfrac{-1{,}830{,}000(s+0.089)}{s+0.43} \\ \dfrac{-1870}{s+0.21} & \dfrac{-0.0071}{s+0.235} & \dfrac{16{,}875}{s+0.2696} \end{bmatrix} \begin{bmatrix} u_1 \\ u_2 \\ u_3 \end{bmatrix}$$

(a) Calculate the RGA, and determine the appropriate pairings for SISO feedback control. Comment on the role of dynamics in your decision.

(b) Using the IMC tuning rules, design three PI controllers for this process.

(c) Simulate the process response to a step set point, separately, in each of the controlled outputs [use a magnitude of +10% (relative) change]. Next, simulate the response of the system to a simultaneous pair of step changes (again, use a magnitude of +10% relative) in each of the second (purity) and third (product rate) controlled outputs. Try to tune the controller to improve the transient response to the simultaneous step changes.

G.6 Consider the diabetic patient in Example G.3. Your goal is to design an automated device to administer insulin infusion in response to meal disturbances.

(a) Considering only the insulin-glucose dynamics, calculate an approximate second-order patient model by fitting the responses (changes in insulin) obtained from simulations of the equations given in the example.

(b) Using the IMC tuning rules, design a PID controller for this process.

(c) Simulate the closed-loop system response to a step set point change in blood glucose of –20 mg/dl. Try to tune the controller to improve the transient response.

(d) Simulate the closed-loop system response to the meal disturbance described in Example G.3. Is the controller able to maintain the safety boundaries for blood glucose (54 mg/dL $< G <$ 144 mg/dL)?

(e) In practice, sensors are available for measuring blood glucose sample from the subcutaneous tissue (the layer of fat under the skin, as opposed to directly from the blood stream). Assuming that such a procedure introduces a pure time delay, repeat the simulation from part (d) with a 10-min sensor delay. How has the performance changed? What is the maximum time delay that the closed-loop design will tolerate before it becomes unstable?

G.7 Recall the artificial pancreas control problem from Section G.2. In order to regulate blood glucose BG, you have two possible manipulated variables to choose from: pumping insulin F_I and pumping glucagon F_G. As you already learned, insulin can be a life-saving drug, but an overdose of insulin can send the body into a coma. Glucagon operates by converting glycogen in the liver to glucose. Their corresponding transfer functions are given below (time is in minutes):

$$G_{p1}(s) = \frac{BG(s)}{F_I(s)} = \frac{-1.5e^{-30s}}{(20s+1)(25s+1)}$$

$$G_{p2}(s) = \frac{BG(s)}{F_G(s)} = \frac{3.4e^{-15s}}{15s+1}$$

In addition, you have the following information relating the BG to a meal (disturbance) D:

$$G_d(s) = \frac{BG(s)}{D(s)} = \frac{5e^{-10s}}{30s+1}$$

(a) Using principles you have learned about processes in general, which manipulated variable would you select (insulin pump or glucagon pump) for the design of a feedback control device? Be very specific.

(b) For the insulin pump, design an appropriate PID controller for the system, using IMC tuning rules (assuming $\tau_c = 20$ min).

(c) Derive the theoretically optimal feedforward controller for this device, assuming you use insulin as the MV. Is this controller realizable? (explain why or why not) If not, what might you do to implement an appropriate controller?

(d) For the insulin pump controller you designed above, you are now asked to design a control system for the *pump subsystem only*. Of particular concern is the situation where the pump is not delivering insulin because of an occluded tube. Draw a block diagram for the loop consisting of the pump and a flow controller. Include a type 4 alarm system for safety (See Appendix E).

G.8 Data from an actual clinical trial for the closed-loop artificial pancreas is summarized in the following cumulative histograms, where the percentage of blood glucose measurements below a particular value is plotted against the blood glucose value. The three curves denote: (i) actual blood glucose BG measurements collected from finger stick measurements (solid), (ii) one continuous glucose measurement sensor CGM2 (dashed), and (iii) a (redundant) second glucose measurement sensor CGM2 (dotted).

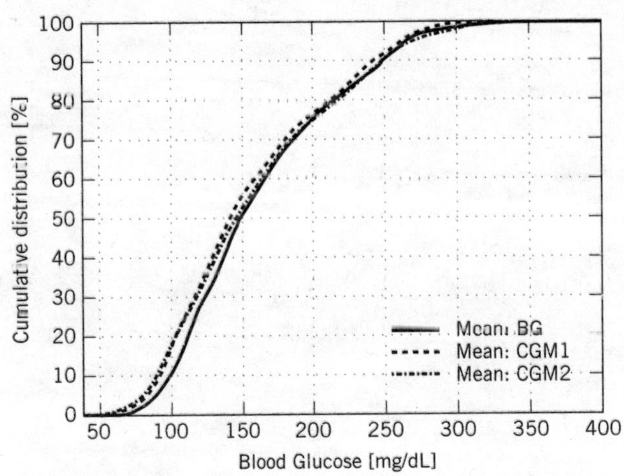

Figure EG.8

From a medical *safety* perspective, one would like to keep the blood glucose measurements in the range of 70–180 mg/dL.

(a) For each of the three measurements (two CGMs and actual BG), compute the fraction of the data that lies in the range 70–180 mg/dL. Comment on the absolute and relative safety of an algorithm that uses these measurements.

(b) Repeat the calculations for the more stringent performance range of 80–140 mg/dL.

APPENDIX H

Dynamics and Control of Biological Systems

APPENDIX CONTENTS

Previous chapters have emphasized the design of controllers for chemical process systems, as well as for biomedical systems (Appendix G). In this chapter, we consider the analysis of intrinsically closed-loop systems that exist in biological circuits, from gene level through cellular level. There is no external controller to be synthesized; rather, the tools that were developed in the first half of this textbook are applied to the analysis of networks that exploit principles of feedback and feedforward control. These biophysical networks display the same rich character as those encountered in process systems engineering: multivariable interactions, complex dynamics, and nonlinear behavior. Examples are drawn from gene regulatory networks, as well as from protein signal transduction networks, with an emphasis on the role of feedback. A glossary of key technical terms is provided at the end of the chapter.

H.1 SYSTEMS BIOLOGY

Biophysical networks are remarkably diverse, cover a wide spectrum of scales, and are characterized by a range of complex behaviors. These networks have attracted a great deal of attention at the level of gene regulation, where dozens of input connections may characterize the regulatory domain of a single gene in a eukaryote, as well as at the protein level, where hundreds to thousands of interactions have been mapped in protein interactome diagrams that illustrate the potential coupling of pairs of proteins (Campbell and Heyer, 2007; Barabasi, 2004). However, these networks also exist at higher levels, including the coupling of individual cells via signaling molecules, the coupling of organs via endocrine signaling, and, ultimately, the coupling of organisms in ecosystems. The biochemical notion of signaling is discussed in Section H.3. To elucidate the mechanisms employed by these networks, biological experimentation and intuition by themselves are insufficient. Instead, investigators characterize dynamics via mathematical models and apply control principles, with the goal of guiding further experimentation to better understand the biological network (Kitano, 2002). Increased understanding can facilitate drug discovery and therapeutic treatments.

A simple example that illustrates the roles of feedback and feedforward control in nature is the heat shock response exhibited by simple bacteria (El-Samad et al., 2006), as illustrated in Fig. H.1. When the organism experiences an increase in temperature, it leads to the misfolding of protein, which disrupts a number of metabolic processes. One of the immediate effects of a heat disturbance is the feedforward activation of a component, σ^{32}, which turns on the transcription process for a pair of genes (FtsH and DnaK) that facilitates the repair mechanism for a misfolded protein. In particular, the FtsH gene is a protease, which catalyzes the destruction of the improperly folded protein. In parallel, and independently, the protein product of one of those genes (DNAK) monitors the state of protein misfolding. It binds to σ^{32} and releases σ^{32} when misfolded protein is detected, leading to feedback activation of DnaK transcription.

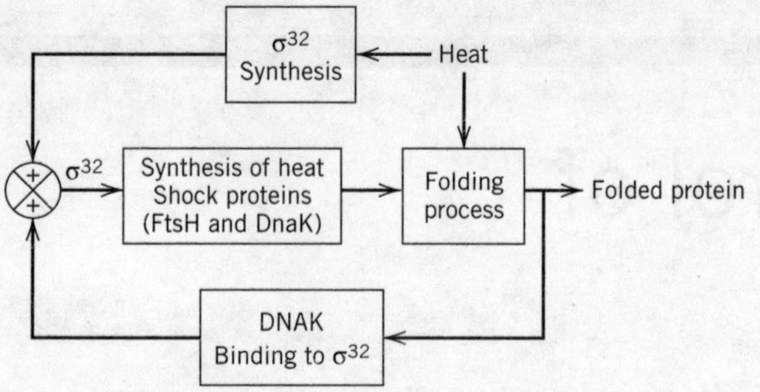

Figure H.1 Feedback and feedforward control loops that regulate heat shock in bacteria (modified from El-Samad, et al., 2006) (positive feedback is common in biological systems).

A second example of networked biological control is the circadian clock, which coordinates daily physiological behaviors of most organisms. The word *circadian* comes from the Latin for "approximately one day," and the circadian clock is vital to regulation of metabolic processes in everything from simple fungi to humans. The mammalian circadian master clock resides in the hypothalamus region of the brain (Reppert and Weaver, 2002). It is a network of multiple autonomous noisy oscillators, which communicate via neuropeptides to synchronize and form a coherent oscillator (Herzog et al., 2004; Liu et al., 2007). At the core of the clock is a gene regulatory network, in which approximately six classes of genes are regulated through an elegant array of time-delayed negative feedback circuits (see Fig. H.2, which illustrates two of those six gene classes). The activity states of the proteins in this network are modulated (activated/inactivated) through a series of chemical reactions, including phosphorylation and dimerization. These networks exist at the subcellular level. Above this layer is the signaling that leads to a synchronized response from the population of thousands of clock neurons in the brain. Ultimately, this coherent oscillator then coordinates the timing of daily behaviors, such as the sleep/wake cycle. An interesting property of the clock is that, under conditions of constant darkness, the clock free-runs, with a period of approximately 24 h (i.e., *circa*), such that its internal time, or phase, drifts away from that of its environment. However, in the presence of an entraining cue (i.e., forcing signal, such as the rising and setting of the sun), the clock locks on to the period of that cue (Boulos et al., 2002; Dunlap et al., 2004; Daan and Pittendrigh, 1976). This gives rise to a precise 24-h period for the oscillations in protein concentrations for the feedback circuit in Fig. H.2.

The *Central Dogma* tenet that most students learn in high school biology is a good starting point to understand these complex networks. Information in the cell is encoded in the DNA, and that information is expressed by the gene to produce messenger RNA. The mRNA is translated into a protein, which is one of the key building blocks of cells and which plays a critical role in cellular regulation. This form of the Central Dogma suggests a serial process, or a feedforward process, in which the genetic code influences the outcome (protein level and protein function). In some of the early publicity surrounding the Human Genome project, this type of logic was pervasive, and there was an understanding in some circles that the "parts list" (genetic code) would illuminate the cause of diseases. An engineer immediately recognizes the flaw in this logic: by analogy, if one were provided with the raw materials list for an aircraft (sheet metal, nuts, bolts, rivets, etc.), it would be an impossible leap to conclude anything about the principles of aerodynamics. Critical missing elements are the manner in which the parts are arranged into a network and, more important, how the components are controlled (or regulated). The same reasoning applies

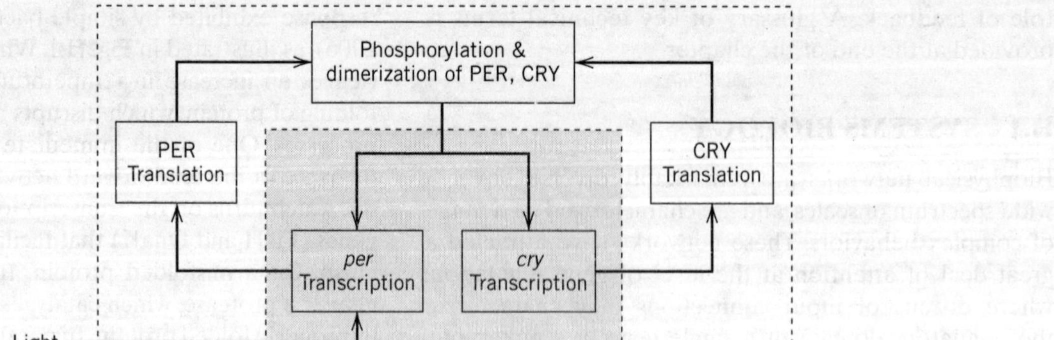

Figure H.2 The gene regulatory circuit responsible for mammalian circadian rhythms (by convention, italics and lowercase refer to genes, uppercase refers to proteins).

equally to biological networks as well, and this notion of the systems perspective has driven current research in systems biology.

According to the Central Dogma tenet, the additional layers of control and regulation that are mentioned in the preceding paragraph can be incorporated schematically. Feedback control plays a key role in (i) regulation of the transcription event; (ii) processing of the RNA, including its stability and potential silencing via RNA interference; (iii) regulation of the ribosomal machinery that accomplishes translation; and (iv) modulation of the activity state of protein, through, for example, degradation, conformation changes, and phosphorylation. There are additional layers of control not indicated in this block diagram that include microRNA modules for feedback and feedforward (Tsang et al., 2007; Inui et al., 2010), as well as the influence of the external or environmental factors through epigenetic control (Delcuve et al., 2009).

Systems biology holds great promise to revolutionize the practice of medicine, enabling a far more predictive and preventative capability (Hood et al., 2004). As scientists and engineers begin to understand the complex networks of genes and proteins that are regulated through feedback and feedforward control, it is possible to develop novel therapies through systematic modification of these closed-loop systems. These modification sites are referred to as targets, and they are opportunities for the design of drugs by the pharmaceutical sector. A drug may target a particular gene, or a protein, or an activity state of a protein (e.g., phosphorylated form), suggesting that there are multiple intervention points in the Central Dogma process. In control terminology, they are potential manipulated variables to restore a healthy state to the network. Likewise, medical scientists and engineers are looking for markers that reveal the pattern of a disease in the signature of the network response. Again, they are understood in control terms as novel sensors that form the basis of an inferential strategy to monitor the status of an unmeasurable disease state. Just as process control engineers test the efficacy of their control system designs through simulation, systems biologists evaluate these new drug targets through extensive simulations of patient populations.

H.2 GENE REGULATORY CONTROL

As described in the previous section, genes are regulated through complex feedback control networks. These networks exhibit a remarkable degree of robustness, because the transcription of critical genes is reliable and consistent, even in the face of disturbances from both within the cell and external to the organism. One of the very compelling features of gene regulatory networks is the recurring use of circuit elements that occur in engineering networks. It has been shown that groups of two to four genes exhibit recurring connection topologies, so-called motifs, which have direct analogs in digital electronic circuits (several examples are illustrated in Fig. H.3). Thus, nature employs these fundamental building blocks in constructing a wide array of gene regulatory networks.

There are a couple of technical terms associated with gene regulatory networks that require explanation. A gene is a portion of the DNA sequence of an organism, which has two primary subregions that are relevant for feedback control: (i) the regulatory or noncoding region can be considered as the input for transcription feedback, and (ii) the coding region determines the products of the expression process, in other words, the output of transcription. The noncoding region can be further divided into discrete regions of separate regulation, called promoters, to which transcription factors bind, leading to activation or inhibition of the expression of the gene (the transcription process). The combination of transcription factors and promoter regions are the controller for the gene transcription process.

There are three dominant network motifs found in *Escherichia coli* (Shen-Orr et al., 2002): (i) a feedforward loop, in which one transcription factor regulates another factor, and, in turn, the pair jointly regulates a

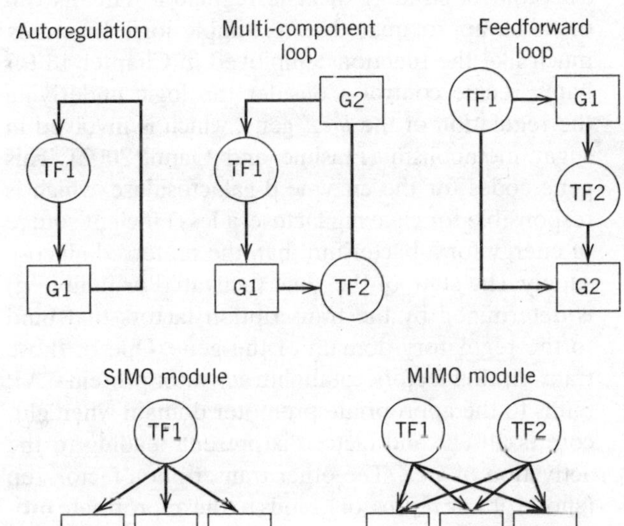

Figure H.3 Examples of circuit motifs in yeast (adapted from Lee et al., 2002). The rectangles denote promoter regions on a gene (G1, G2, etc.), and the circles are transcription factors (TF1, TF2, etc.).

third transcript factor; (ii) a single-input multiple-output (SIMO) block architecture; and (iii) a multiple-input multiple-output (MIMO) block architecture, referred to as a densely overlapping regulon by biologists.

A completely different organism, *Saccharomyces cerevisiae*, has six closely related network motifs (Lee et al., 2002): (i) an autoregulatory motif, in which a regulator binds to the promoter region of its own gene; (ii) a feed-forward loop; (iii) a multicomponent loop, consisting of a feedback closed-loop with two or more transcription factors; (iv) a regulator chain, consisting of a cascade of serial transcription factor interactions; (v) a single-input multiple-output (SIMO) module; and (vi) a multiple-input multiple-output (MIMO) module. These motifs are illustrated in Fig. H.3.

In effect, these studies prove that, in both eukaryotic and prokaryotic systems, cell function is controlled by complex networks of control loops, which are cascading and interconnected with other (transcriptional) control loops. The complex networks that underlie biological regulation appear to be constructed of elementary systems components, not unlike a digital circuit. This lends credibility to the notion that analysis tools from process control are relevant in systems biology.

Some of the analogies between process control concepts and biological control concepts are summarized in Table H.1, at the level of gene transcription. Keep in mind that there are many levels of analysis in biological circuits, and one can draw comparisons to engineering circuits at each of these levels.

Table H.1 Analogies between Process Control Concepts and Gene Transcription Control Concepts

Process Control Concept	Biological Control Analog
Sensor	Concentration of a protein
Set point	Implicit: equilibrium concentration of protein
Controller	Transcription factors
Final control element	Transcription apparatus; ribosomal machinery for protein translation
Process	Cellular homeostasis

EXAMPLE H.1

The control strategy of gene regulatory circuits can often be approximated using simple logic functions, much like the functions employed in Chapter 18 for batch recipe control. Consider the logic underlying the regulation of the *lacZ* gene, which is involved in sugar metabolism (Ptashne and Gann, 2002). This gene codes for the enzyme β-galactosidase, which is responsible for cleaving lactose, a less efficient source of energy for a bacterium than the preferred glucose supply. The state of the gene (activated or inhibited) is determined by the transcription factors that bind to the regulatory domain of the gene. One of those transcription factors, catabolite activator protein CAP, binds to the appropriate promoter domain when glucose is absent and lactose is present, leading to the activation of *lacZ*. The other transcription factor, rep (short for Lac repressor), binds to the appropriate promoter domain in the absence of lactose. Once bound, rep inhibits the expression of the gene. If neither rep nor CAP is present, you may assume that only a very small (basal) rate of gene expression occurs.

(a) Develop a logic table for the permutations in outcome (transcription of gene *lacZ*) as a function of the two input signals, CAP and rep.

(b) Write a simple logic rule for the expression of the *lacZ* gene as a function of the presence of lactose and glucose (ignore the basal state).

Solution

(a) The logic table is given in Table H.2.

Table H.2 Logic Table for Activity State of Gene *lacZ* as a Function of Input Signals CAP and rep

CAP	rep	*lacZ* state
+	–	off
+	+	basal
–	+	activated
–	–	off

(b) A simple rule for the expression logic is given as:

$$lacZ = \text{lactose AND (NOT(glucose))}$$

because the gene (and its enzyme product) are only required when the primary sugar source (glucose) is not present and the secondary source (lactose) is present.

H.2.1 Circadian Clock Network

Recall from the previous section that the circadian clock orchestrates a number of important metabolic processes in an organism. It does this by regulating the concentration of key proteins in a cycle manner, with a period of (approximately) 24 h. Consider a simplified model of the *Drosophila melanogaster* circadian clock involving two key genes: the period gene (denoted *per*) and the timeless gene (denoted *tim*). Those genes are transcribed into mRNA, exported from the nucleus, and translated into their respective proteins (denoted in Fig. H.4 by the uppercase convention as PER and TIM). The protein monomers form a dimer, and the dimers of both PER and TIM combine to form a heteromeric complex that reenters the nucleus and suppresses the rate of transcription of the two genes via negative feedback. The kinetic mechanisms for the phosphorylation events are assumed to be Michaelis-Menten form, and the kinetic mechanism for gene regulation (inhibition) follows a Hill mechanism (with a Hill coefficient of 2).

For the assumptions made by Tyson et al. (1999), the two genes can be lumped together, as well as their corresponding proteins and the nuclear and cytoplasmic forms of the dimer. Finally, assuming rapid equilibrium between the monomer and dimer, a second-order set of balances can be developed for the mRNA state M and the protein state P. The resulting pair of differential

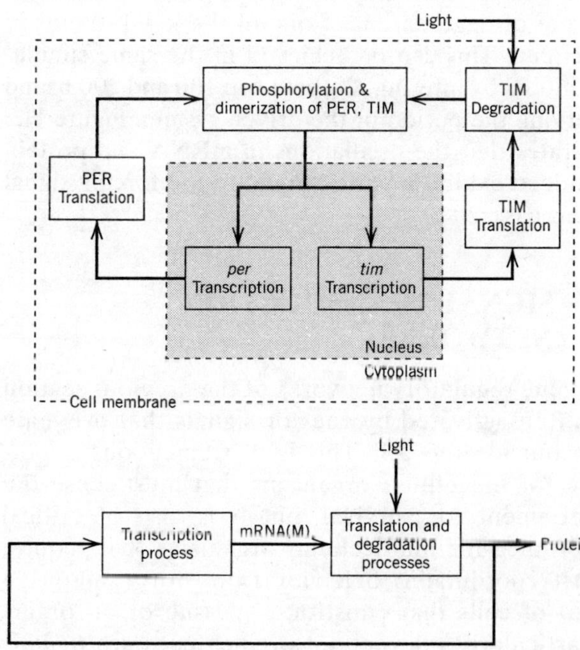

Figure H.4 Schematic of negative feedback control of *Drosophila* circadian clock (adapted from Tyson et al., 1999): detailed system (top), and simplified model (bottom).

equations captures the dynamics of the feedback-controlled circuit:

$$\frac{dM}{dt} = \frac{v_m}{1+\left(P(1-q)/2P_{crit}\right)^2} - k_m M \qquad \text{(H-1)}$$

$$\frac{dP}{dt} = v_p M \frac{k_{p1} P q + k_{p2} P}{J_p + P} - k_{p3} P \qquad \text{(H-2)}$$

An additional algebraic relationship introduces a more complex dependence of the transcription rate on the protein concentration P:

$$q = \frac{2}{1+\sqrt{1+8K_{eq}P}} \qquad \text{(H-3)}$$

The model parameters and their definitions are a result of the work of Tyson et al. (1999) and are summarized in Table H.3.

Using a computer package, such as Simulink/MATLAB, the gene regulatory circuit using these defined parameters can be simulated with initial values of M and P equal to [2.0; 2.0]. A 100-h simulation is shown in Fig. H.5; the period can be calculated from either the mRNA (M) or the protein (P) trajectory (e.g., time between peaks) and is 24.2 h (i.e., approximately 24 h or "circadian").

Table H.3 Parameter Values for Circadian Clock Circuit in Figure H.4 (C_m Denotes Transcript Concentration and C_p Denotes Protein Concentration).

Parameter	Value	Units	Description
v_m	1	$C_m h^{-1}$	Maximum rate of mRNA synthesis
k_m	0.1	h^{-1}	First-order constant for mRNA degradation
v_p	0.5	$C_p C_m h^{-1}$	Rate constant for translation of mRNA
k_{p1}	10	$C_p h^{-1}$	V_{max} for monomer phosphorylation
k_{p2}	0.03	$C_p h^{-1}$	V_{max} for dimer phosphorylation
k_{p3}	0.1	h^{-1}	First-order rate constant for proteolysis
k_{eq}	200	C_p^{-1}	Equilibrium constant for dimerization
P_{crit}	0.1	C_p	Dimer concentration at half-maximum transcription rate
J_p	0.05	C_p	Michaelis constant for protein kinase

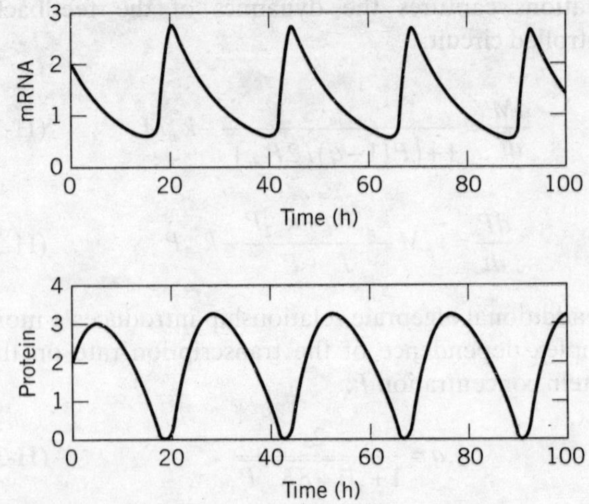

Figure H.5 Simulation of the circadian clock model.

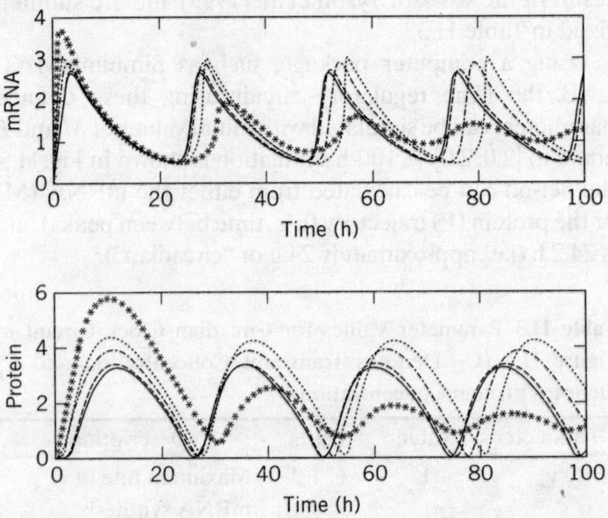

Figure H.6 Simulation of circadian clock model for varying values of v_m [1.0 (solid), 1.1 (dashed), 1.5 (dash-dot), 2.0 (dotted), 4.0 (asterisk)].

A common property of biological closed-loop circuits is that they exhibit remarkable robustness to disturbances and fluctuations in operating conditions. For example, the clock should maintain a nearly 24-h period, even though the organism is exposed to temperature changes, which affect the rates of biochemical reactions. The model circadian clock can be simulated by perturbing values of the kinetic constants. The same clock simulation is evaluated for the following values of the parameter μ_m: [1.0; 1.1; 1.5; 2.0; 4.0]. The period of the clock lengthens as μ_m is increased, as shown in Fig. H.6. The period increases as follows: [24.2; 24.5 25.5; 26.4] corresponding to the first four values of μ_m. At the extreme value of 4.0, oscillations are no longer observed, and the system settles to a stable equilibrium. The stability of the oscillations is quite remarkable for such large perturbations in μ_m (over 100%).

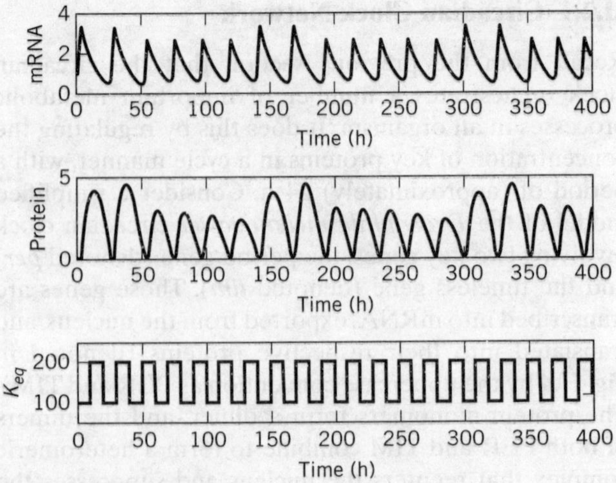

Figure H.7 Simulation of circadian clock model for entraining signal with period of 20 h.

Another important feature of the circadian clock is its ability to entrain (i.e., track) an external signal (sunlight), so that the period of the oscillations of mRNA and Protein match exactly the period of the external signal. In this manner, the organism's clock is reset to a period of precisely 24 h. Tyson et al. (1999) show that this can be simulated in the present model by switching the value of K_{eq} to emulate dark-light cycles (i.e., using a square wave with even intervals of light and dark and a 24-h period). In the fly, sunlight appears to modulate the rate of degradation of one of the key proteins in the circuit. This can be achieved in the same simulation model by altering K_{eq}, between 100 and 200 h, and observing the period of the driven system. Figure H.7 illustrates that the oscillations in mRNA and protein do indeed exhibit a period equal to the forcing signal (in this case, 20 h).

H.3 SIGNAL TRANSDUCTION NETWORKS

The gene regulatory networks of the previous section are often activated by cues or signals that originate from outside the cell. This is of tremendous importance for unicellular organisms that must sense the environment for survival, but it is also of critical importance for multicellular organisms that require robust coordinated behavior from, for example, a group of cells that constitute a tissue or an organ. A particularly relevant set of such cues are *ligands* (from the Latin "to bind"), which are molecules that bind to proteins that typically span the surface membrane of a cell. These ligands, called receptors, induce particular responses within the cell, depending on the conditions. They include a number of interesting stimulus-response mechanisms (Lauffenburger and Linderman, 1993):

- Growth factors → cell division
- Necrosis factor → programmed cell death (apoptosis)
- Chemoattractant → chemotaxis
- Insulin → glucose uptake
- Neurotransmitter → secretion by nerve cell
- Extracellular matrix (ECM) protein → adhesion

Once the ligand binds to the receptor, it initiates a series of biochemical reactions that induce a short-term response (e.g., phosphorylation state of an intermediate protein) and/or a longer-term response as a result of a regulated gene response. These networks respond relatively rapidly, exhibiting dynamics with characteristic time scales of seconds to minutes. A cell is often presented with multiple, competing cues, and it processes that information in rich signal transduction networks, to result in the appropriate cellular fate, depending on the context.

In this section, we highlight several signal transduction cascades, to illustrate the rich processing dynamics manifested by these networks.

H.3.1 Chemotaxis

The process of *chemotaxis* is the directed motion of a cell or cellular organism toward a chemical source, typically a food molecule. This mechanism is also invoked in the response to a detected toxin (i.e., motion away from that source) and is involved in more complex processes, such as development. The process is initiated by the detection of a ligand (e.g., a food molecule) at the cell surface, which invokes a signal transduction cascade and results in the alteration of the motor apparatus responsible for moving the cell.

A simplified version of the biochemical pathway that underlies chemotaxis in *E. coli* is shown in Fig. H.8. The binding of an attractant molecule (ligand) to the receptor complex CheW-CheA (denoted as W-A) induces the phosphorylation of protein CheY (Y), and the phosphorylated form (Yp) invokes a tumbling motion from the bacteria's flagella. This tumbling motion allows the organism to reorient and search the surrounding space; otherwise, the organism proceeds in a straight run. The ability of CheW-CheA to phosphorylate CheY depends on the methylation state of that complex, which is fine-tuned by the proteins CheR (R) and the phosphorylated form of CheB (Bp), as illustrated in the figure. Feedback is evident in Fig. H.8, because CheB phosphorylation is mediated by the CheW-CheA complex.

The signal transduction system that mediates chemotaxis exhibits a type of adaptation in which the response to a persistent stimulus is reset to the pre-stimulus value, thereby enabling an enhanced sensitivity. Several mechanistic explanations can be postulated for this robust behavior, including the following: (i) precise fine-tuning of several parameters to yield a consistent (robust) response under varied conditions, or (ii) inherent regulation that yielded this robust behavior. Utilizing

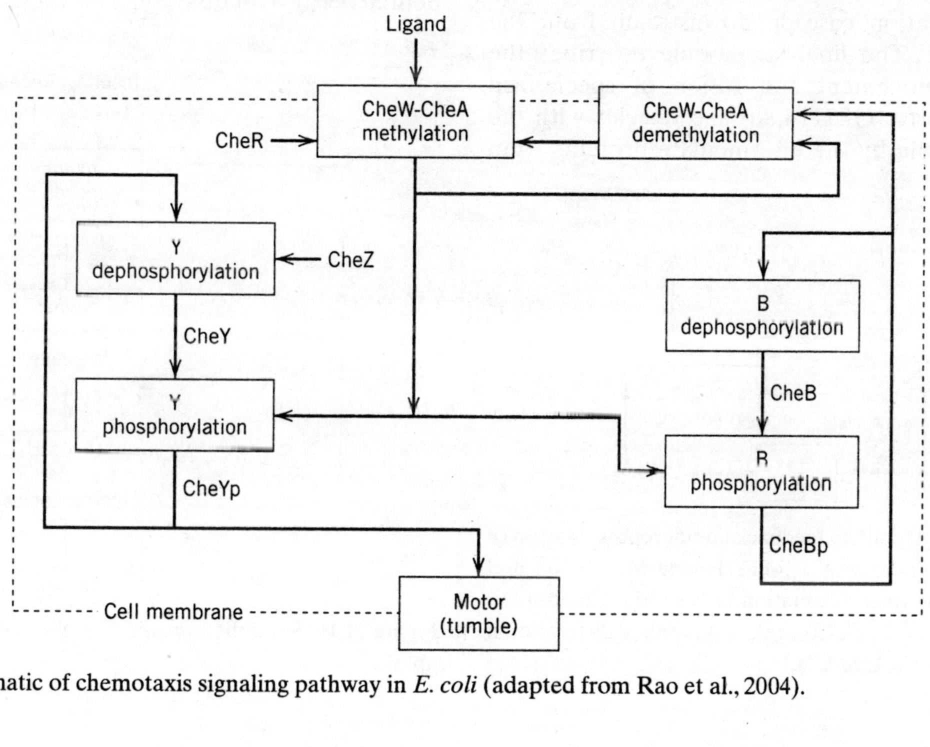

Figure H.8 Schematic of chemotaxis signaling pathway in *E. coli* (adapted from Rao et al., 2004).

process control principles, it has been demonstrated that the regulatory system exploits integral feedback control to achieve the robust level of adaptation exhibited in chemotaxis (Yi et al., 2000). The chemotaxis network can be reduced to the simple block diagram in Fig. H.9, in which u denotes the chemoattractant, y denotes the receptor activity, and $-x$ denotes the methylation level of the receptors. It is left as an exercise to show that this circuit ensures that perfect adaptation is achieved (i.e., the receptor activity always resets to zero asymptotically).

This understanding suggests that many seemingly complex biological networks may employ redundancy and other structural motifs or modules to achieve relatively simple overall system behavior.

H.3.2 Insulin-Mediated Glucose Uptake

Muscle, liver, and fat cells in the human body take up glucose as an energy source in response to, among other signals, the hormone insulin, which is secreted by the pancreas. As discussed in Appendix G, the release of insulin is regulated in a feedback manner by the blood glucose level. In Type 2 diabetes, the insulin signal transduction network is impaired such that insulin does not lead to glucose uptake in these cells. A simplified model of the insulin signaling network can be decomposed into three submodules, as shown in Fig. H.10. The first submodule describes insulin receptor dynamics: insulin binds to insulin receptor, causing subsequent receptor autophosphorylation. The receptor can also be recycled, introducing additional dynamics in the network. The second submodule describes the phosphorylation cascade downstream from the insulin receptor. The final submodule describes the activation of movement and fusion of specialized glucose transporter GLUT4 storage vesicles with the plasma membrane by the intermediate proteins from the second module. These GLUT4 transporters allow glucose molecules to enter the cell. Each of the three modules contains submodules that consist of layers of feedback.

H.3.3 Simple Phosphorylation Transduction Cascade

In signal transduction, a receptor signal is processed in a cascaded pathway, to yield a cellular response. For the example considered here, the processing consists of a sequence of kinase- and phosphatase-catalyzed reaction steps, consisting of phosphorylation and dephosphorylation, respectively. The key performance attributes of such a system are (i) the speed at which a signal arrives to the destination, (ii) the duration of the signal, and (iii) the strength of the signal. Under conditions of weak activation (low degree of phosphorylation), the individual steps in the signal transduction cascade can be modeled as a set of linear ODEs (Heinrich et al., 2002):

$$\frac{dX_i}{dt} = \alpha_i X_{i-1} - \beta_i X_i \tag{H-4}$$

where α_i is a pseudo first-order rate constant for phosphorylation, β_i is the rate constant for dephosphorylation, and X_i is the phosphorylated form of the kinase (i). Assume that the cascade consists of four stages (levels of phosphorylation), that the corresponding rate constants are equal for all stages ($\alpha_i = \alpha$; $\beta_i = \beta$), and that the receptor inactivation is approximated as an exponential decay with time constant $1/\lambda$ (see Fig. H.11). The

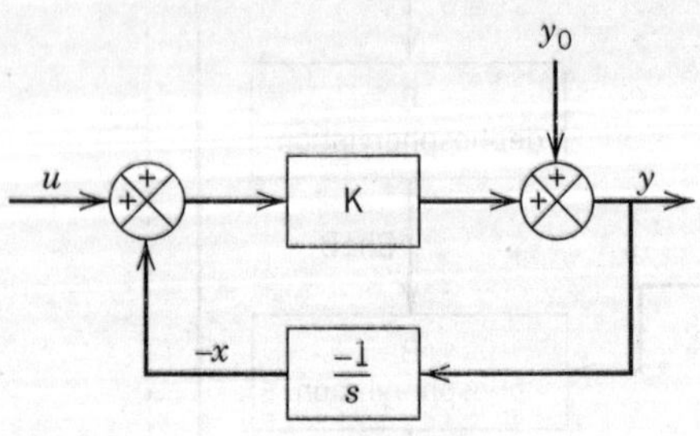

Figure H.9 Integral control feedback circuit representation of chemotaxis (Adapted from Yi, T.M., Y. Huang, M. I. Simon, and J. Doyle, Robust Perfect Adaptation in Bacterial Chemotaxis through Integral Feedback Control, "Copyright 2000 National Academy of Sciences, U.S.A.").

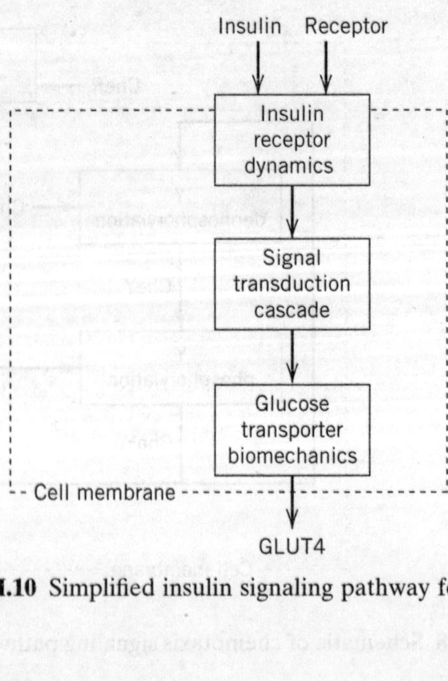

Figure H.10 Simplified insulin signaling pathway for glucose uptake.

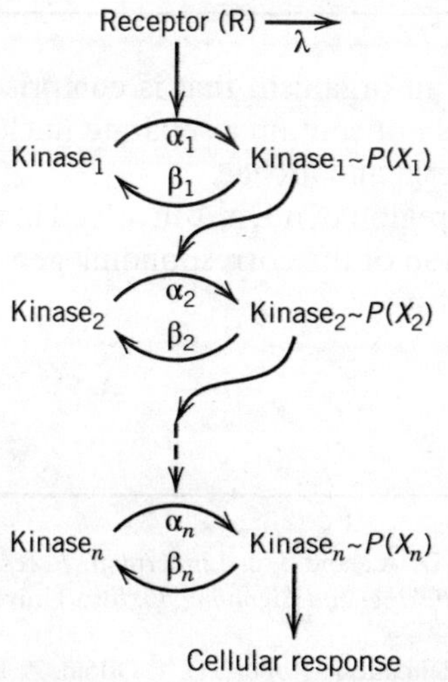

Figure H.11 Schematic of fourth-order signal transduction cascade for Example H.3, combined with first-order receptor activation (adapted from Heinrich et al., 2002).

resulting cellular response can be written in the Laplace domain as

$$Y(s) - \left(\frac{s}{\frac{1}{\lambda}s+1}\right)\left(\frac{\alpha^4}{(s+\beta)^4}\right)R(s) \tag{H-5}$$

where $R(s)$ is the receptor input and $Y(s)$ is the cellular response.

If the signaling T_{sig} time is defined as the average time to activate a kinase, a suitable expression in the time domain for this quantity is

$$T_{sig} = \frac{\int_0^\infty ty(t)dt}{\int_0^\infty y(t)dt} \tag{H-6}$$

where $y(t)$ is the unit step response ($R(s) = 1/s$ in Eq. H-5). It is possible to derive the analytical expression for the signaling time for this network. Recalling a few properties of Laplace transforms (see Chapter 3):

$$\mathscr{L}(tf(t)) = -\frac{d}{ds}F(s) \tag{H-7}$$

and

$$\mathscr{L}\left(\int_0^\infty f(t)dt\right) = F(s=0) \tag{H-8}$$

Then the following expression for the signal time can be derived:

$$T_{sig} = \frac{\left(-\frac{d}{ds}Y(s)\right)_{s=0}}{Y(0)} = \frac{\left[\lambda\alpha^4(s+\lambda)^{-2}(s+\beta)^{-4} + 4\lambda\alpha^4(s+\lambda)^{-1}(s+\beta)^{-5}\right]_{s=0}}{\alpha^4/\beta^4} \tag{H-9}$$

which simplifies to

$$T_{sig} = \frac{1}{\lambda} + \frac{4}{\beta} \tag{H-10}$$

Notice that the average time through the network (i.e., the signaling time) is not dependent on the rate of phosphorylation (α).

SUMMARY

In this chapter, a number of biological circuit diagrams have been introduced that illustrate the rich array of dynamics and feedback control that exist in all living organisms. Two particular biological processes were considered: the regulation of gene transcription and the protein signal transduction that characterizes cellular stimulus-response mechanisms. The recurring motifs of feedback and feedforward control motivated the application of process control analysis to these problems, to shed light on both the healthy functioning state as well as to promote the investigation of therapies for cases where the natural circuit is impaired (i.e., a disease state).

Some of the most recent work in the field has utilized light as a manipulated variables to perturb protein signaling pathways (optogenetics). These novel tools allow scientists to engineer feedback regulation into natural circuits, resulting in synthetic biology circuits (Toettcher et al., 2011; Chen et al., 2013). The field of synthetic biology holds great promise for the design of robust control circuits using biological building blocks (Endy, 2005; Cameron et al., 2014).

The rapidly developing field of systems biology continues to make great advancements in the area of medical problems, and the increased understanding of the biological circuits underlying diseases will likely lead to novel therapeutic strategies, as well to the discovery of new drugs. More information is available in more specialized books, including those of Klipp et al. (2005), Palsson (2006), Alon (2007), Liu and Lauffenburger (2010), Young and Michelson (2011), and Dubitzky et al. (2013).

GLOSSARY

Eukaryote: an organism that is comprised of cells (or possibly a single cell, as in yeast) that are divided into substructures by membranes, notably containing a nucleus. Examples include animals, plants, and fungi.

Kinase: an enzyme that catalyzes the transfer of phosphate group to a substrate, leading to phosphorylation of that substrate.

Prokaryote: an organism that is comprised of a single cell that does not contain a separate nucleus. Examples include bacteria and archae.

Promoter: a region of a DNA involved in the regulation of transcription of the corresponding gene.

REFERENCES

Alon, U., *An Introduction to Systems Biology: Design Principles of Biological Circuits*, Chapman & Hall/CRC, New York, 2007.

Alberts, B., D. Bray, A. Johnson, J. Lewis, M. Raff, K. Roberts, and P. Walters, *Essential Cell Biology*, Garland Pub., Inc., New York, 1998.

Barabasi, A. L., Network Biology: Understanding the Cell's Functional Organization, *Nature Rev. Genetics*, **5**, 101 (2004).

Boulos, Z., M. M. Macchi, M. P. Sturchler, K. T. Stewart, G. C. Brainard, A. Suhner, G. Wallace, and R. Steffen, Light Visor Treatment for Jet Lag after Westward Travel across Six Time Zones, *Aviat. Space Environ. Med.*, **73**, 953 (2002).

Cameron, D. E., C. J. Bashor, and J. J. Collins, A Brief History of Synthetic Biology, *Nat. Rev. Microbiol.*, **12**, 381, 2014.

Campbell, A. M., and L. J. Heyer, *Discovering Genomics, Proteomics, & Bioinformatics*, Benjamin Cummings, San Francisco, CA, 2007.

Chen, S., P. Harrigan, B. Heineike, J. Stewart-Ornstein, and H. El-Samad, Building Robust Functionality in Synthetic Circuits using Engineered Feedback Regulation, *Curr. Opin. Biotech.*, **24**, 790, 2013.

Cluve, G. P., M. Rastegar, and J. R. Davie, Epigenetic Control, *J. Cell. Physiol.*, **219**, 243, 2009.

Daan, S., and C. S. Pittendrigh, A Functional Analysis of Circadian Pacemakers in Nocturnal Rodents. II. The Variability of Phase Response Curves, *J. Comp. Physiol.*, **106**, 253 (1976).

Dubitzky, W., O. Wolkenhauer, K.-H. Cho, and H. Yokota, *Encyclopedia of Systems Biology*, Springer, 2013.

Dunlap, J. C., J. J. Loros, and P. J. DeCoursey (Eds.), *Chronobiology: Biological Timekeeping*, Sinauer Associates, Inc., Sunderland, MA, 2004.

El-Samad, H., S. Prajna, A. Papachristodoulou, J. Doyle, and M. Khammash, Advanced Methods and Algorithms for Biological Networks Analysis, *Proc. IEEE*, **94**, 832 (2006).

Endy, D., Foundation for Engineering Biology, *Nature*, **438**, 449, 2005.

Heinrich, R., B. G. Neel, and T. A. Rapoport, Mathematical Models of Protein Kinase Signal Transduction, *Mol. Cell*, **9**, 957 (2002).

Herzog, E. D., S. J. Aton, R. Numano, Y. Sakaki, and H. Tei, Temporal Precision in the Mammalian Circadian System: A Reliable Clock from Less Reliable Neurons, *J. Biol. Rhythms*, **19**, 35 (2004).

Hood, L., J. R. Heath, M. E. Phelps, and B. Lin, Systems Biology and New Technologies Enable Predictive and Preventative Medicine, *Science*, **306**, 640 (2004).

Inui, M., G. Martello, and S. Piccolo, MicroRNA Control of Signal Transduction, *Nature Revs. Mol. Cell Biol.*, **11**, 252, 2010.

Kitano, H., Systems Biology: A Brief Overview, *Science*, **295**, 1662 (2002).

Klipp, E., R. Herwig, A. Kowald, C. Wierling, and H. Lehrach, *Systems Biology in Practice*, Wiley-VCH, Weinheim, 2005.

Lauffenburger, D. A., and J. J. Linderman, *Receptors: Models for Binding, Trafficking, and Signaling*, Oxford University Press, New York, 1993.

Lee, T. I., N. J. Rinaldi, F. Robert, D. T. Odom, Z. Bar-Joseph, G. K. Gerber, N. M. Hannett, C. T. Harbison, C. M. Thompson, I. Simon, J. Zeitlinger, E. G. Jennings, H. L. Murray, D. B. Gordon, B. Ren, J. J. Wyrick, J. B. Tagne, T. L. Volkert, E. Fraenkel, D. K. Gifford and R. A. Young, Transcriptional Regulatory Networks in *Saccharomyces cerevisiae*. *Science*, **298**, 799 (2002).

Liu, A. C., D. K. Welsh, C. H. Ko, H. G. Tran, E. E. Zhang, A. A. Priest, E. D. Buhr, O. Singer, K. Meeker, I. M. Verma, F. J. Doyle III, Takahashi, J. S., and S. A. Kay, Intercellular Coupling Confers Robustness against Mutations in the SCN Circadian Clock Network, *Cell*, **129**, 605 (2007).

Liu, E., and D. Lauffenburger, *Systems Biomedicine*, Academic Press, 2010.

National Research Council, *Network Science*, National Academies Press, Washington DC, 2005.

Palsson, B., *Systems Biology: Properties of Reconstructed Networks*, Cambridge University Press, New York, 2006.

Ptashne, M., and A. Gann. *Genes and Signals*, Cold Spring Harbor Laboratory Press, Cold Spring Harbor, New York, 2002.

Rao, C. V., J. R. Kirby, and A. P. Arkin, Design and Diversity in Bacterial Chemotaxis: A Comparative Study in *Escherichia coli* and *Bacillus subtilis*, *PLoS Biology*, **2**, 239 (2004).

Reppert, S. M., and D. R. Weaver, Coordination of Circadian Timing in Mammals, *Nature*, **418**, 935 (2002).

Sedaghat, A. R., A. Sherman, and M. J. Quon, A Mathematical Model of Metabolic Insulin Signaling Pathways, *Amer. J. Physio-End. Met.*, **283**, E1084 (2002).

Shen-Orr, S. S., R. Milo, S. Mangan, and U. Alon, Network Motifs in the Transcriptional Regulation Network of *Escherichia coli*, *Nature Genetics*, **31**, 64 (2002).

Toettcher, J. E., D. Gong, W. A. Lim, and O. D. Weiner, Light-based Feedback for Controlling Intracellular Signaling Dynamics, *Nat. Met.*, **8**, 837, 2011.

Tsang, J., J. Zhu, and A. van Oudenaarden, MicroRNA-mediated Feedback and Feedforward Loops are Recurrent Network Motifs in Mammals, *Mol. Cell*, **26**, 753, 2007.

Tyson, J. J., C. I. Hong, C. D. Thron, and B. Novak, A Simple Model of Circadian Rhythms Based on Dimerization and Proteolysis of PER and TIM, *Biophys. J.*, **77**, 2411 (1999).

Yi, T. M., Y. Huang, M. I. Simon, and J. Doyle, Robust Perfect Adaptation in Bacterial Chemotaxis through Integral Feedback Control, *Proc. Nat. Acad. Sci. USA*, **97**, 4649 (2000).

Young, D. L., and S. Michelson, *Systems Biology in Drug Discovery and Development*, John Wiley and Sons, 2011.

EXERCISES

H.1 In this exercise, treat the components as simple (reactive) chemical species and perform the appropriate (dynamic) material balance. Assume that a messenger RNA *mRNA* is produced by a constant (basal) expression rate from a particular gene. In addition, assume that the *mRNA* degrades according to a first-order decay rate.

(a) Write the equation for the dynamics of the mRNA concentration as a function of the expression rate G_0 and the decay rate constant k_d^{mRNA}.

(b) Assume that each *mRNA* molecule is translated to form *p* copies of a protein product, *P*. Furthermore, the protein is subject to first-order degradation, with a decay rate constant k_p^{mRNA}. Write the equation for the dynamics of the protein concentration.

(c) Assume that the system has been operating for some time at a constant gene expression rate G_0, and then the expression rate changes instantaneously to a value G_1. Derive an analytical expression for the transient responses for *mRNA* and *P*.

H.2 Consider the block diagram in Fig. EH.2 of the multiple feedback loops involved in the Central Dogma, namely, genetic regulation C_1, translational regulation C_2, and enzyme inhibition C_3. Assume that the processes P_1, P_2, and P_3 obey first-order dynamics, with corresponding gains and time constants K_i, τ_i.

(i) Derive the transfer function from the external input (u) to the output (y) for each of the three cases shown in Fig. EH.2 (a), (b), and (c).

(ii) Assume that the feedback mechanisms operate via proportional control with corresponding controller gains K_{ci}. Derive the closed-loop transfer function from the external input u to the output y in block diagram (b).

(iii) Consider a simplified biological circuit in which only genetic regulation is active (C_1). Derive the closed-loop transfer function and comment on the key differences between this transfer function and the one from part (b).

(iv) Give several reasons why the natural feedback architecture with all three controllers operating is more effective than the control architecture in part (c).

H.3 As a specific biological example for Exercise H.2 and Fig. EH.2(b),* the synthesis of tryptophan can be described by the following set of material balances:

$$\frac{d}{dt}[O_R] = k_1[O_t]C_1[T] - k_{d1}[O_R] - \mu[O_R]$$

$$\frac{d}{dt}[mRNA] = k_2[O_R]C_2[T] - k_{d2}[mRNA] - \mu[mRNA]$$

$$\frac{d}{dt}[E] = k_3[mRNA] - \mu[E]$$

$$\frac{d}{dt}[T] = k_4C_3[T][E] - g\frac{[T]}{[T]+K_g} - \mu[T]$$

*The authors acknowledge Profs. Bhartiya, Venkatesh, and Gayen for their help with formulating this problem.

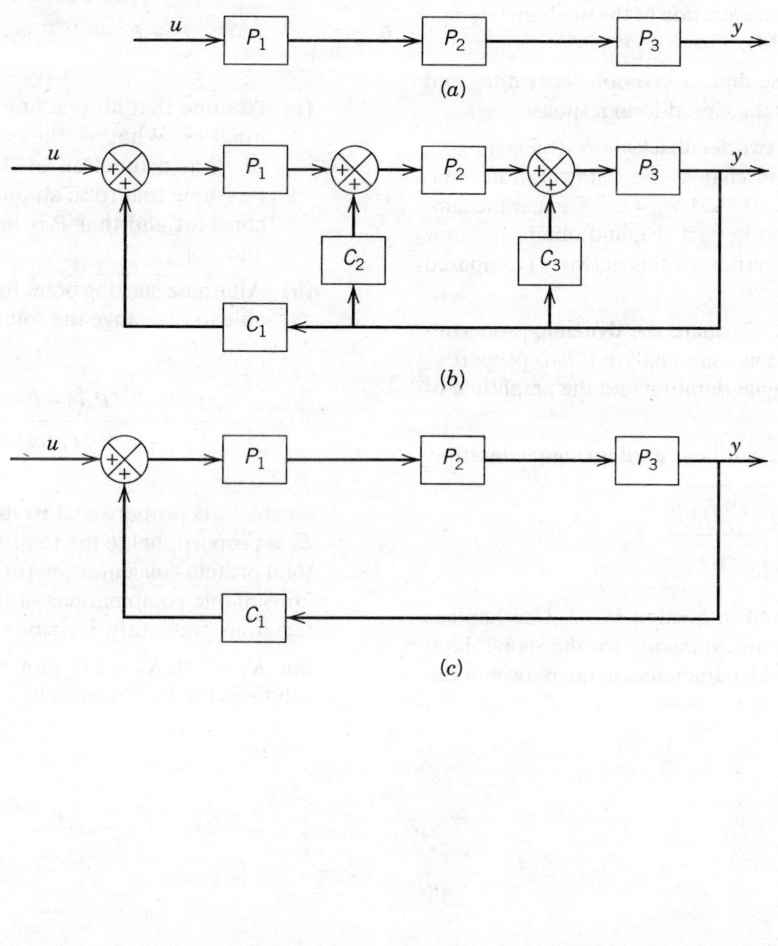

Figure EH.2

where k_1, k_2, k_3, and k_4 represent kinetic rate constants for the synthesis of free operator, mRNA transcription, translation, and tryptophan synthesis, respectively. Parameters O_t, μ, k_{d1}, and k_{d2} refer to total operator site concentration, specific growth rate of *E. coli*, degradation rate constants of free operator O_R, and mRNA, respectively. E and T represent concentrations of enzyme anthranilate synthase and tryptophan, respectively, in the cell. K_g and g are the half saturation constant and kinetic constant for the uptake of tryptophan for protein synthesis in the cell. Model parameter values are as follows: $k_1 = 50\ \text{min}^{-1}$; $k_2 = 15\ \text{min}^{-1}$; $k_3 = 90\ \text{min}^{-1}$; $k_4 = 59\ \text{min}^{-1}$; $O_t = 3.32$ nM; $k_{d1} = 0.5\ \text{min}^{-1}$; $k_{d2} = 15\ \text{min}^{-1}$; $\mu = 0.01\ \text{min}^{-1}$; $g = 25\ \mu\text{M}\cdot\text{min}^{-1}$; $K_g = 0.2\ \mu$M. Here, controllers $C_1(T)$, $C_2(T)$, and $C_3(T)$ represent repression, attenuation, and inhibition, respectively, by tryptophan and are modeled by a particular form of Michaelis-Menten kinetics (the Hill equation) as follows:

$$C_1(T)=\frac{K_{i,1}^{\eta H}}{K_{i,1}^{\eta H}+T^{\eta H}}, C_2(T)=\frac{K_{i,2}^{1.72}}{K_{i,2}^{1.72}+T^{1.72}}, C_3(T)=\frac{K_{i,3}^{1.2}}{K_{i,3}^{1.2}+T^{1.2}}$$

$K_{i,1}$, $K_{i,2}$, and $K_{i,3}$ represent the half-saturation constants, with values $K_{i,1} = 3.53\ \mu$M; $K_{i,2} = 0.04\ \mu$M; $K_{i,3} = 810\ \mu$M, whereas sensitivity of genetic regulation to tryptophan concentration, $\eta_H = 1.92$.

(a) Draw a block diagram, using one block for each of the four states. Comment on the similarities between this diagram and schematic (b) in Fig. EH.2.

(b) Simulate the response of the system to a step change in the concentration of the medium (change g from 25 to 0 μM).

(c) Calculate the rise time, overshoot, decay ratio, and settling time for the closed-loop response.

(d) Omit the inner two feedback loops (by setting C_2 and C_3 to 0) and change the following rate constants: $K_{P1} = 8 \times 10^{-8}\ \mu$M; $\eta_H = 0.5$. Repeat the simulation described in part (b), and obtain the new closed-loop properties for this network (compared to part (c)).

H.4 Consider Section H.3.3, where the dynamic properties of a signal transduction were analyzed. Two properties of interest are the signal duration and the amplitude of the signal.

(a) The following definition is used for signal duration:

$$T_{dur} = \sqrt{\frac{\int_{t=0}^{\infty} t^2 y(t)dt}{\int_{t=0}^{\infty} y(t)dt} - T_{sig}^2}$$

where T_{sig} was defined in Section H.3.3. Use Laplace transforms to derive an expression for the signal duration as a function of the parameters in the phosphorylation cascade.

(b) Define the signal amplitude as:

$$A = \frac{\int_{t=0}^{\infty} y(t)dt}{2T_{dur}}$$

Use Laplace transforms to derive an expression for the signal amplitude as a function of the parameters in the phosphorylation cascade.

H.5 Consider the simplified version of the chemotaxis circuit in Fig. H.9.

(a) Derive the conditions for the process gain K that ensure that the receptor activity is always reset to zero and even for the case of a persistent ligand signal.

(b) Show that the closed-loop transfer function from the ligand to the receptor activity is equivalent to a first-order transfer function with numerator dynamics.

(c) Comment on the biological relevance of the result in part (b), particularly for a ligand signal that is fluctuating.

H.6 An interesting motif in biological circuits is a switch, in which the system can change from (effectively) one binary state to another. An analysis of a continuous reaction network reveals a rise to a switchlike response (also referred to as ultrasensitivity). Consider interconversion of a protein from its native state P to an activated form P^*, catalyzed by the enzymes E_1 and E_2:

$$P + E_1 \leftrightarrow PE_1 \leftrightarrow P^* + E_1$$
$$P^* + E_2 \leftrightarrow P^*E_2 \leftrightarrow P + E_2$$

(a) Assume that all reaction steps obey mass-action kinetics. What is the steady-state dependence of P^* as a function of the concentration of E_1? (Assume that total amount of E_1 E_2, and P are all constant and that P is in excess compared to E_1 and E_2).

(b) Alternate starting point for problem: you should be able to rearrange the solution as follows

$$\frac{V_1}{V_2} = \frac{P^*/P_T(1-P^*/P_T+K_1)}{(1-P^*/P_T)(P^*/P_T+K_2)}$$

where V_1 is proportional to the total E_1 in the system, E_2 is proportional to the total V_2 in the system, P_T is the total protein concentration (in all forms), and K_1 and K_2 are suitable combinations of the rate constants for the reactions previously described.

For $K_1 = 1.0$, $K_2 = 1.0$, plot the steady-state locus of solutions for P^*/P_T versus V_1/V_2.

(c) Assume that the two enzymes operate in a saturated regime, that is, the reactions follow zero-order kinetics with respect to the enzymes. Use the expression from part (b) to plot the steady-state locus for this extreme situation (i.e., $K_1 = 0, K_2 = 0$).

(d) Comment on the difference in shape of the gain functions in parts (b) and (c). Based on the initial problem description, explain how biology can produce switchlike behavior in this system.

H.7 Recall the Central Dogma for biology, which is depicted in simplified form in the block diagram Fig. EH.7. In this case, it is assumed that nuclear transport can be ignored. Let X denote an environmental perturbation to the system.

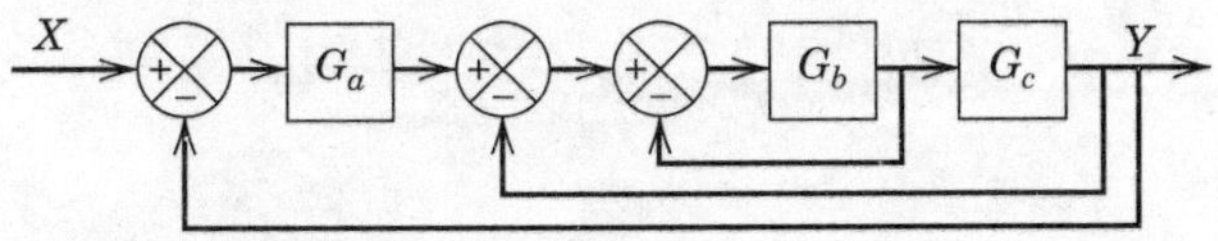

Figure EH.7

(a) Explain briefly the biological process denoted by G_a, G_b, G_c, and the variable Y.

(b) Find a general expression for $Y(s)/X(s)$, that is, the overall closed-loop transfer function. It should be simplified to contain only a numerator expression and a denominator expression without additional fractions.

(c) Consider now the specific values for the biological processes: $G_a = 5$, $G_b = 0.5/s$, and $G_c = 3/s$. Evaluate your expression from part (b) for this case and simplify accordingly. Is the closed-loop stable?

H.8 Consider the simple block diagram for the feedback control of circadian rhythms (i.e., 24 h) in Fig. H.4. To simplify things considerably, assume that transcription obeys a first-order process with gain = 5 [mRNA/Protein] and time constant = 1.0 h. Furthermore, assume that translation and degradation act together like a simple delay, with unknown magnitude.

(a) Calculate the frequency response characteristics for all the elements in this feedback loop. Be careful with your time units.

(b) In order to generate sustained circadian rhythms, how much time delay is required in the translation/degradation step? (requires a numerical answer)

(c) At this oscillation condition, what is the gain of the translation/degradation process?

APPENDIX I

Introduction to Plantwide Control

APPENDIX CONTENTS

Previous chapters have generally concentrated on the analysis and design of simple control systems, from single loops (such as reactor temperature control) to single processing units with multiple loops (e.g., distillation column control). However, most industrial applications involve larger problems with multiple processing units that interact with each other. The subject of *plantwide control* deals with unit-to-unit interactions through the choice of measured and manipulated variables in each unit and the selection of a control strategy—namely, how to pair controlled and manipulated variables in individual loops, where to use multiloop controllers (Chapter 14), where to use multivariable controllers such as MPC (Chapter 15), and so on.

For a new plant, the problem of designing the control system can be quite difficult as a consequence of unit-to-unit interactions. Thus, understanding the potential sources of these interactions and finding ways in which they can be substantially mitigated are important to achieve effective plant operations. In this chapter, we introduce several key concepts in plantwide control; Appendix J deals specifically with how to develop a control system design for a new plant.

Most continuous processing plants contain many units, such as reactors, furnaces, heat exchangers, and distillation columns. The goal of process design is to minimize capital costs while operating with optimum utilization of materials and energy. Unfortunately, achieving lower plant capital costs and higher processing efficiencies inevitably makes the individual units interact more with each other and thus makes them harder to control (see Chapter 14). The process control engineer deals with these unit-to-unit interactions by designing a control system that counteracts disturbances before they propagate from their source to other units.

A typical plantwide control system will consist of many single-loop controllers as well as multivariable controllers such as model predictive control (Chapter 15). A key characteristic of many plantwide control systems is the very large number of process variables, involving

1. Thousands of measurements
2. Hundreds to thousands of manipulated variables
3. Hundreds of disturbance variables

Fortunately, a plant with a large number of processing units can be analyzed as smaller clusters of units—for example, a gas treatment plant and a separations plant that interact very little with each other. Then, with even simple steady-state and dynamic process models, it is possible to develop a design using the standard analytical methods we developed in Chapter 14 (RGA, SVA, etc.) for multivariable control problems. In the absence of process models, one must resort to heuristic (rule-of-thumb) approaches. Although these approaches generally are based on prior experience, they also incorporate an understanding of the fundamental physics and chemistry that apply to all plants. In this chapter, several case studies are used to introduce important plantwide concepts. In the final chapter (Appendix J), we present a general strategy for designing plantwide control systems.

I.1 PLANTWIDE CONTROL ISSUES

One of the most basic issues in plantwide control is flow/inventory control. If a train of continuous processing units (reactors, columns, etc.) is considered, where should the production rate be controlled? It can be controlled at the exit of the line (e.g., a series of unit operations (as in Fig. I.1*a*)), at the beginning of the line (Fig. I.1*b*), or at any point in between. In these figures, the sensors/transmitters have been omitted for clarity. It might seem logical to use a feed flow rate into each unit to control the inventory (level) in that unit as illustrated in the *downstream method* of Fig. I.1*a*. However, as discussed below, adjusting each unit's effluent flow rate may be an easier way to control inventories if the flow rates of multiple streams into a unit are ratioed (see the *upstream method* in Fig. I.1*b*).

The objectives for any of these methods are (1) to maintain the production rate of the line (or the production rates, if there is more than one product), (2) to regulate the liquid level in each unit, and (3) to reduce the effect of disturbances as much as possible. Note that the downstream method has the advantage of fixing the actual product rate, but any flow disturbance to a single unit will tend to propagate successively to upstream units as manipulated flow rates are changed to deal with the disturbance. Also, in this case each additional stream into a unit may have to be regulated at a fixed ratio to one of the streams (the primary manipulated variable) if composition upsets as well as inventory disturbances are to be avoided.

The upstream method has the disadvantage that production rate is established via flow of materials into the first unit. Thus, flow or level disturbances will propagate downstream, eventually affecting the plant's production rate of the desired product, the flow rate from the final unit. Such a situation can be quite undesirable. Consider a bank of extruders or fiber-spinning machines that utilize the product of a continuous polymerization line. An increase in flow rate to the final polymerization unit causes its level to be increased. The resulting increased residence time can lead to increased degradation of the polymer as a result of extended high-temperature processing. In such a situation, excess product may have to be recycled back to an earlier unit and reprocessed, or even sent to "waste." If the flow rate to the final polymerization unit is reduced, one or more extruders/spinning machines may have to be shut down for a period of time to maintain a reasonably constant level in the final unit.

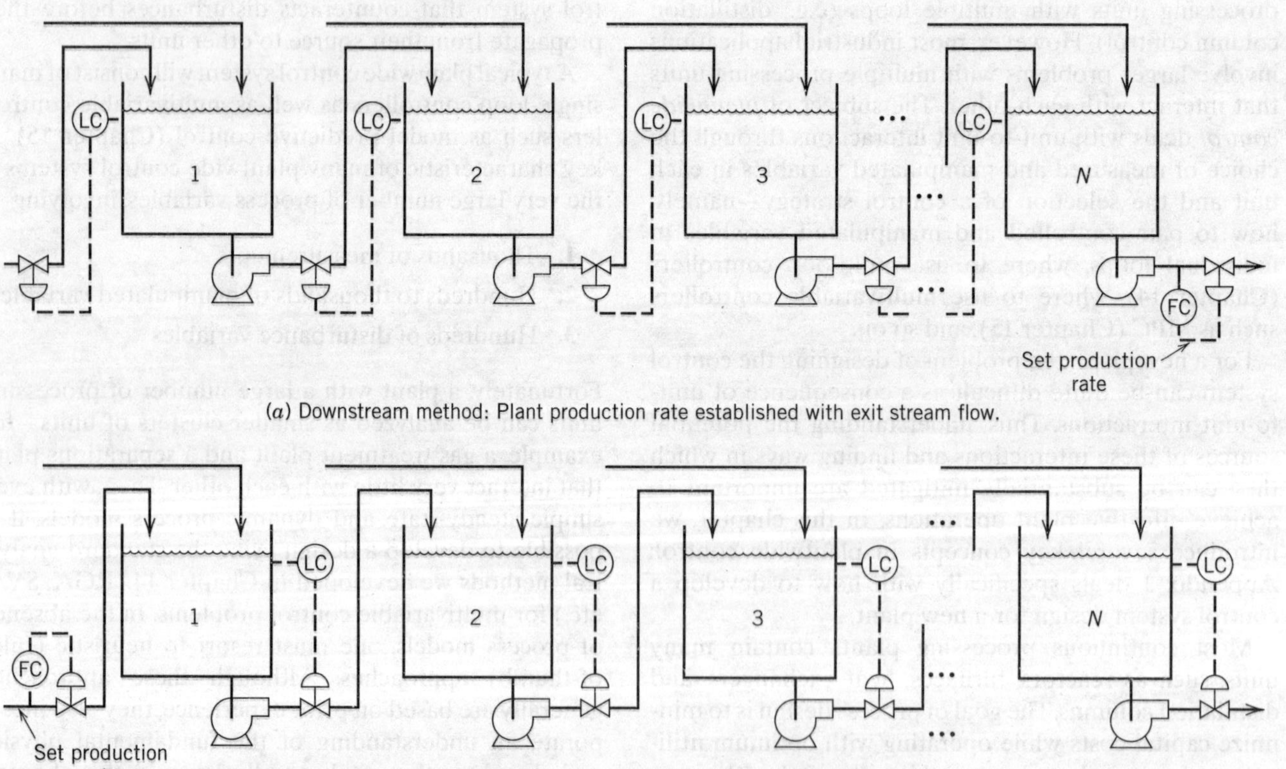

(*a*) Downstream method: Plant production rate established with exit stream flow.

(*b*) Upstream method: Plant production rate established with inlet stream flow.

Figure I.1 Train of continuous processing units.

Modern processing plants cannot be operated in this manner.

When continuous processing methods first achieved widespread usage in industry, disturbance propagation was reduced by placing surge vessels between key processing units. This arrangement allowed separate control systems to be used for each unit. In Fig. I.2 a reactor and distillation column are separated by a surge vessel. The surge tank prevents flow disturbances from the reactor from upsetting the column, and also prevents short-term production rate changes for the column from propagating back to the reactor. Note that the level in a surge vessel either is not controlled unless it reaches the high or low alarm position. Alternatively, it can be loosely controlled by averaging level control (see Chapter 11). The net effect is to dampen flow disturbances by allowing the level to "float" between low and high limits.

Modern plants are designed to avoid the extra capital and operating costs of surge tanks, related piping, and space in the operations area. Thus, extraneous vessels, whose only function is to make the plant easier to operate, are normally avoided.

In the continuing search for lower plant operating costs, two other process design techniques often are employed that make plants more difficult to control. One of these techniques is *heat integration*, in which the overhead vapor from one distillation column provides the energy for vaporizing liquid in the reboiler of another column, typically in the same separation train. Recapturing energy in this manner is a major concern in the design of modern processing plants. However, in obtaining the increased energy efficiency available through heat integration, designers must pay close attention to the more complicated plant that results, as noted below and, in more detail, in Section I.3.

Figure I.3 illustrates another commonly employed process design technique, *material recycle*. Here two reactors are connected in series, followed by a flash unit whose vapor product is recycled back to the first reactor. Unreacted reactants concentrated in the vapor stream are recycled to increase the reaction conversion or yield.

Although heat integration and material recycle can significantly reduce plant capital and operating costs, these techniques inevitably increase the amount of interaction among operating units and reduce the control degrees of freedom (see Chapter 2). Nevertheless, appropriate control strategies can deal with such undesirable consequences.

Plantwide control is concerned with designing control systems for large numbers of individual process units that may be highly interacting. Several additional issues arise from these interactions, which further distinguish plantwide control from the control of single units. A hypothetical plant consisting of a reactor and

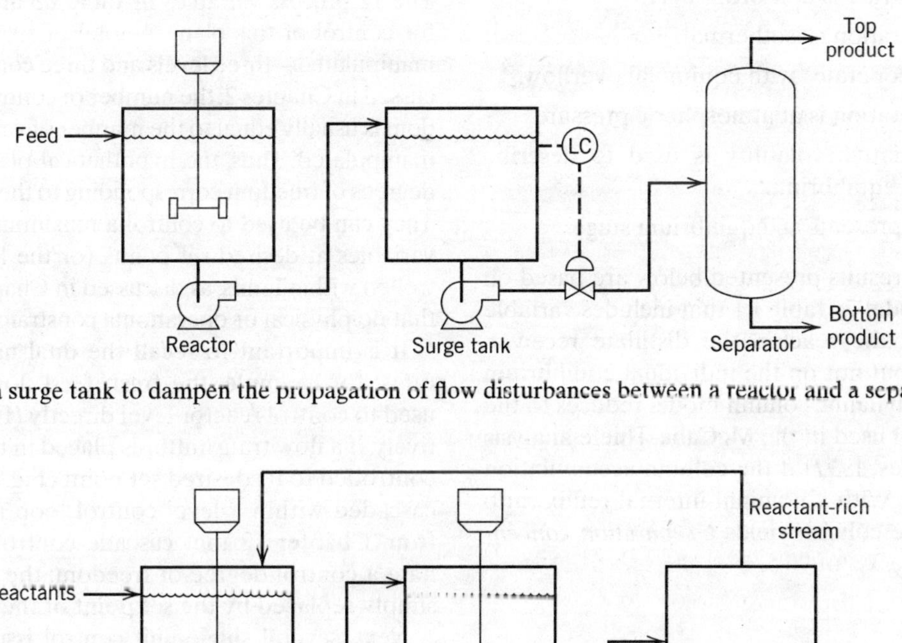

Figure I.2 Use of a surge tank to dampen the propagation of flow disturbances between a reactor and a separator.

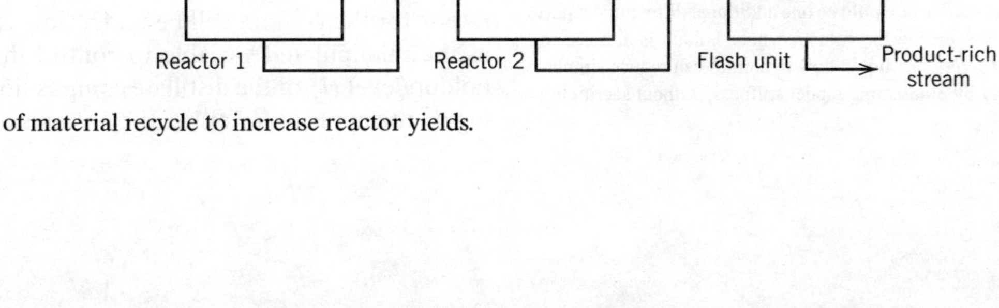

Figure I.3 Use of material recycle to increase reactor yields.

separation unit provides the basis for useful analytical and simulation results that are presented in Section I.3.

I.2 HYPOTHETICAL PLANT FOR PLANTWIDE CONTROL STUDIES

This type of plant has been considered by Papadourakis et al. (1987), and Luyben (1993). Though conceptually simple, the use of recycle considerably complicates steady-state and transient operations.

I.2.1 Reactor/Distillation Column Plant

Figure I.4 illustrates a simple generic plant, an isothermal reactor coupled with a distillation column. A mixture of two species, mainly A but also some B, is fed to a reactor where the reaction A → B takes place isothermally. The binary distillation column has 20 stages and produces two product streams: an overhead (distillate) stream rich in A and a bottoms stream rich in the desired product B. The A-rich distillate is recycled to the reactor to increase the conversion of A to B.

Table I.1 provides the dynamic model for the two process units. Parameter values for the individual process units and the nominal operating conditions of Luyben (1993) and Wu and Yu (1996) are shown in Table I.2. A number of simplifications are used here:

1. The reaction rate is first-order in A.
2. Reactor operation is isothermal.
3. The column operates with equimolal overflow.
4. Column operation is at atmospheric pressure.
5. Constant relative volatility is used to describe vapor/liquid equilibrium.
6. Each tray represents an equilibrium stage.

The simulation results presented below are based on the 26th-order model in Table I.1 that includes variable liquid holdups in the reactor, the distillate receiver, and the reboiler, but not on the individual equilibrium stages.[1] Thus, the dynamic column model reduces to the steady-state model used in the McCabe-Thiele analysis (Seader and Henley, 1997) if the column accumulation terms are all zero. With a nominal internal reflux ratio (V/D) of 1.455, the column yields a *separation concentration ratio*, $S \triangleq \bar{x}_D/\bar{x}_B$, of 90.5.

[1]Including variable liquid flow holdup for each stage would increase the model order from 26 to 46. There would be one additional differential equation for each of the 20 stages in the column whose holdup is allowed to vary. Elimination of the very fast liquid flow dynamics can reduce simulation times considerably by eliminating model stiffness, without sacrificing accuracy.

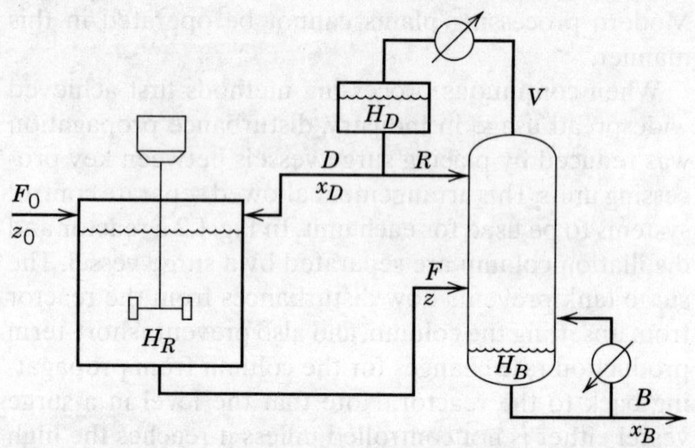

Figure I.4 Reactor/distillation column plant for plantwide control studies.

The control objective is to maintain the composition of B in the product stream x_B at the nominal value given in Table I.2, despite disturbances in the fresh feed composition z_0 and the feed flow rate F_0. We assume initially that the production rate is established either upstream or downstream of the plant. Later, we discuss ways of accommodating that objective using alternative plant control structures.

I.2.2 Degrees of Freedom Analysis

The 12 process variables in Table I.3 are now considered for control of this plant. A total of six flow rates can be manipulated—three levels and three compositions. As discussed in Chapter 2, the number of control degrees of freedom is usually equal to the number of variables that can be manipulated. Thus, the hypothetical plant has six control degrees of freedom corresponding to the six control valves. They can be used to control a maximum of six measured variables at desired set points (or the levels can be controlled within limits, as discussed in Chapter 15), assuming that no physical or operational constraints are violated.

It is important to recall the dual nature of the flow rates: for example, the fresh feed flow rate F_0 can be used to control reactor level directly (Fig. I.5*a*). Alternatively, if a flow transmitter is placed in this line, F_0 can be controlled to its desired set point (Fig. I.5*b*), or it can be cascaded within a level control loop (Fig. I.5*c*). Recall from Chapter 12 that cascade control does not eliminate a control degree of freedom; the flow rate itself is simply replaced by the set point of the flow controller.

Next, several single-unit control issues for this plant will be considered—for example, whether the reflux flow rate R for the column will be under flow control or used as the manipulated variable to control the reflux drum holdup/level H_D or the distillate composition x_D. Depending on the application, either the bottoms composition x_B

Table I.1 Dynamic Model for Reactor/Distillation Column Plant (Symbol definitions and values provided in Table I.2)

Reactor	
General Information:	Reaction: $A \rightarrow B$ Reaction rate expression is first-order in reactant A. $r_A = -k_R H_R z$
Reactor Model:	$\frac{dH_R}{dt} = F_0 + D - F$ (= 0 for perfect reactor level control) $\frac{d(H_R z)}{dt} = F_0 z_0 + D x_D - F z + r_A$
Column	
General Information:	Saturated liquid feed is to 12th stage (of 20) numbered from the top down. Equimolal overflow is assumed. A is the more volatile component; assume equilibrium holds for each stage: $y_i = \frac{\alpha x_i}{1 + x_i}$
Column Model:	
Reflux drum:	$\frac{dH_D}{dt} = V - R - D$ (= 0 for perfect level control) $\frac{d(H_D x_D)}{dt} = V y_{20} - R x_D - D x_D$
Stage *i* above feed:	$H_S \frac{dx_i}{dt} = L(x_{i+1} - x_i) V(y_{i-1} - y_i)$ for $13 \le i \le 19$ where $L = R$
Feed stage:	$H_S \frac{dx_{12}}{dt} = (L x_{13} - L' x_{12}) + V(y_{11} - y_{12}) + Fz$ where $L' = L + F$
Stage *j* below feed:	$H_S \frac{dx_j}{dt} = L'(x_{j+1} - x_j) + V(y_{j-1} - y_j)$ for $1 \le j \le 11$
Reboiler:	$\frac{dH_B}{dt} = L' - V - B$ (= 0 for perfect reboiler level control) $\frac{d(H_B x_B)}{dt} = L' x_1 - V x_B - B x_B$

can be controlled (Luyben, 1993), or both x_D and x_B can be explicitly controlled to their set points (Luyben, 1994). Several alternative control configurations can be used to accomplish the latter (two-point composition control). In the *material balance configuration*, H_D is controlled by manipulating D, and H_B is controlled by adjusting B. This choice leaves R and V to control, respectively, x_D and x_B. By contrast, in the *energy balance configuration*, the two manipulated variables at the top of the column are switched. Thus, H_D is controlled by R, and x_D is controlled by D. In addition, the control loop pairings at the bottom are switched (Shinskey, 1996).

In order to analyze either column control configuration, we assume for simplicity that the result is perfect control at the desired steady state. In other words, the levels and compositions will be held at the nominal values in Table I.2, while F and z vary. The steady-state material and component balances for the column are

$$\bar{F} = \bar{D} + \bar{B} \tag{I-1}$$

$$F\bar{z} = \bar{D}\bar{x}_D + \bar{B}x_B \tag{I-2}$$

Equations I-1 and I-2 indicate that fixing the values of x_D and x_B (via perfect control) determines the *steady-state* flow rates $\bar{D}$ and $\bar{B}$ for any values of $\bar{F}$ and $\bar{z}$. Here $\bar{x}_D$ and $\bar{x}_B$ denote the steady-state values of x_D and x_B (0.95 and 0.0105, respectively).

Table I.2 Parameter Values and Steady-State Conditions for the Reactor/Distillation Column Recycle Process (adapted from Wu and Yu (1996) with conversion from lb-mole to kmole)

Reactor	
Fresh feed, F_0	230 kmol/h
Fresh feed composition, z_0	0.9 mole fraction A
Reactor holdup, H_R	1200 kmol
Recycle flow rate, D	250 kmol/h
Recycle composition, x_D	0.95 mole fraction A
Reactor residence time, $H_R/(F_0 + D)$	2.5 h
Specific reaction rate, k_R	0.33 h^{-1}
Distillation Column	
Column feed rate, F	480 kmol/h
Column feed composition, z	0.5 mole fraction A
Distillate flow rate, D	250 kmol/h
Relux flow rate, R	550 kmol/h
Reflux ratio, R/D	2.20
Bottoms flow rate, B	230 kmol/h
Vapor boilup, V	800 kmol/h
Number of equilibrium stages	20
Feed stage	12
Distillate composition, x_D	0.95 mole fraction A
Bottoms composition, x_B	0.0105 mole fraction A
Relative volatility, α	2
Bottoms holdup, H_B	137.5 kmol
Reflux drum holdup, H_D	92.5 kmol
Individual stage holdup, H_S	11.75 kmol

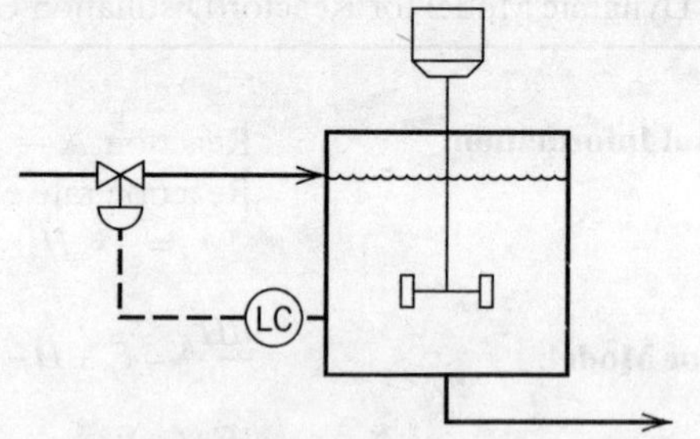

(*a*) Reactor feed flow rate controls reactor level

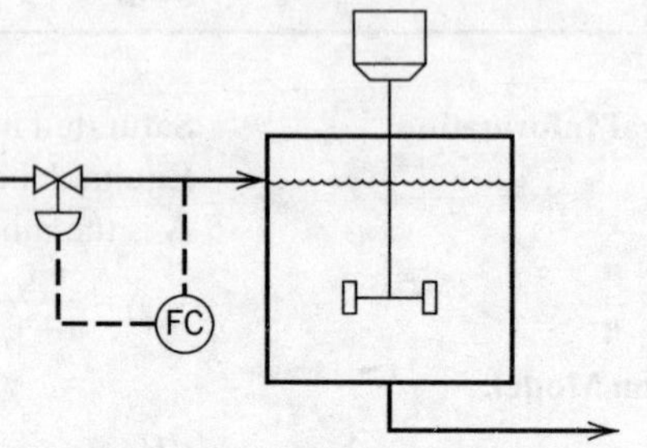

(*b*) Flow control of reactor feed rate

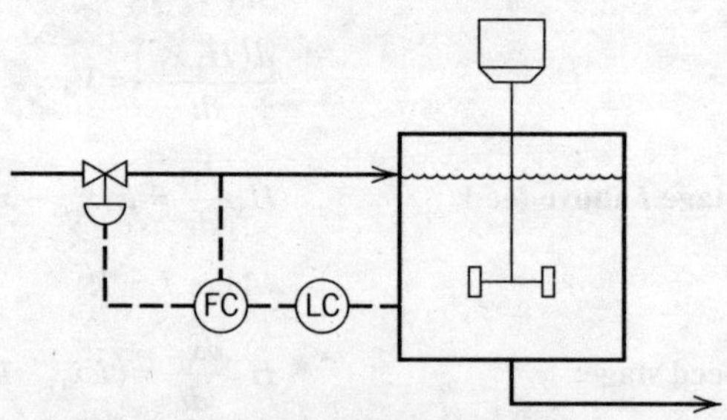

(*c*) Cascade control of reactor level via secondary controller for feed flow rate

Figure I.5 Multiple uses of a flow variable.

Assume that a two-point composition control system has been designed using the material balance configuration. Note that whether a material balance or energy balance column control structure is chosen does not restrict the discussion of plantwide issues below in any way. The column control structure can consist of four single-loop controllers:

Controlled Variable	Manipulated Variable
H_D	D
x_D	R
H_B	B
x_B	V

In this analysis, column pressure control has been disregarded, as would be the case, for example, if the column overhead is vented to another vessel at atmospheric pressure. When pressure control must be considered, the flow rate of cooling water to the condenser will be a logical manipulated variable, and an energy balance around the condenser/reflux drum must be added to the model. The number of single-loop controllers would then be five.

If control of reflux drum and bottoms holdups and product compositions is perfect, we can consider the plant as represented in Fig. I.6. Here D and B can vary, because the two flow rates are manipulated variables; hence, they vary with the column feed flow rate and feed composition whenever the plant is disturbed in order to control x_D and x_B at their set-point values.

Table I.3 Process Variables in the Reactor/Distillation Column Plant Identified as Important for Control

F_0^*	Reactor feed flow rate
z_0	Reactor feed composition
H_R	Reactor level (proportional to the holdup)
F^*	Column feed flow rate (saturated liquid)
z	Column feed composition
H_D	Distillate reflux drum level
R^*	Reflux flow rate
D^*	Distillate (recycle) flow rate
H_B	Bottoms level
B^*	Bottoms (product) flow rate
V^*	Reboiler (column) vapor flow rate
x_D	Distillate composition
x_B	Bottoms (product) composition

*Denotes a stream flow rate that can be measured and adjusted by a control valve.

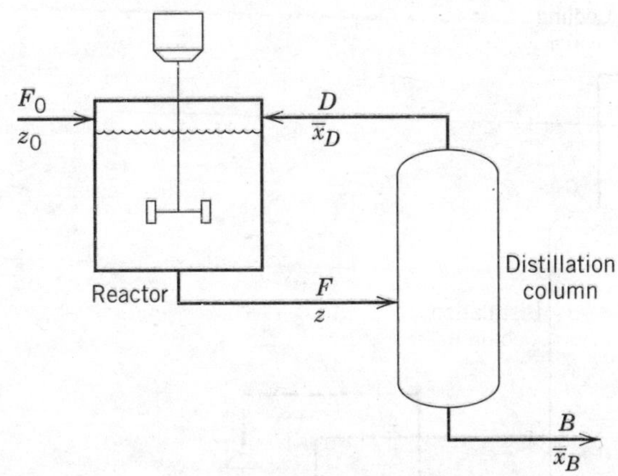

Figure I.6 Schematic diagram of reactor/distillation column plant with perfect control of all three levels and both column product compositions.

In summary, the column controllers have utilized four control degrees of freedom, and eight variables have been removed from the list in Table I.3, leaving

F_0	Fresh feed rate
H_R	Reactor level
F	Column feed rate
z	Column feed composition
z_0	Fresh feed composition

Next, we assume that the primary control objective is to maintain plant operation as much as possible at the set points, despite fresh feed flow rate or composition changes. Thus, F_0 (assuming it is not chosen as a manipulated variable) and z_0 are the disturbance variables. At this point in the analysis, a true plantwide control problem is encountered. Two control degrees of freedom remain, and either F_0 or F could be manipulated to control reactor level. However, it is not obvious whether one choice or the other is better in some sense. In principle, the remaining flow rate (control degree of freedom) could be used to control z or B; for example, if F is available, F_0 is used as a manipulated variable for reactor-level control.

In summary, if we choose to deal with this plantwide control situation by using a multiloop strategy, the consequences of material feedback need to be considered before making any more controller pairing decisions. The reactor/distillation column recycle system, with its two remaining control degrees of freedom, is fairly simple. Nevertheless, it provides several general results about plantwide control strategies.

I.3 INTERNAL FEEDBACK OF MATERIAL AND ENERGY

Processes that include recycle systems have an important design requirement—namely, that there must be an exit path for every chemical species. For example, inert components can be kept at reasonable levels by "bleeding off" a small portion of the recycle stream. However, Luyben (1994) discussed a subtle problem with recycle systems, the *snowball effect*, which is characterized by a large sensitivity of one or more of the variables in a recycle loop to small changes in a disturbance variable. This problem arises from both a small reactor holdup and a particular control structure.

In particular, if changes in fresh feed composition "load the reactor excessively"—that is, beyond its ability to provide the required conversion—then the separator/recycle system will be called on to make up the deficiency. For the case where extra reactor capacity is available through an increase in the reactor level, the particular choice of level/flow control structure within the recycle loop can radically affect plant gains (sensitivities). In the following, we assume that the reactor design is fixed and its level is set at less than full capacity. The question to be considered is how alternative designs of the level and flow loops mitigate the effect of fresh feed flow rate or composition disturbances.

I.3.1 Steady-State Behavior: The Snowball Effect

Because the snowball effect is a steady-state phenomenon, it can be analyzed by considering a steady-state model. We first consider two alternatives for controlling reactor level H_R (Luyben, 1994). For Alternative 1 in Fig. I.7*a*, H_R is controlled by manipulating the column feed rate F (i.e., the reactor effluent rate). For Alternative 2 in Fig. I.7*b*, H_R is allowed to "float" while F is held constant. This strategy is possible because, in theory, the reactor level in this structure is self-regulating (Larsson et al., 2003). For the moment, we assume that the plant production rate is established either upstream or downstream of the plant and analyze these two simple cases to see what insight can be obtained. Later, we consider the implications of setting production rate within the plant.

Alternative 1 (Fig. I.7a)

The key feature in this alternative is that H_R is held constant by manipulating F, with implications for operation of the column. To examine the steady state sensitivities of key variables within the recycle loop (F, z, and D) with respect to the disturbance variables (F_0 and z_0), consider the steady-state version of the dynamic model in Table I.1.

Reactor

$$\bar{F}_0 + \bar{D} = \bar{F} \tag{I-3}$$

$$\bar{F}_0\bar{z}_0 + \bar{D}\bar{x}_D = \bar{F}\bar{z} + k_R H_R \bar{z} \tag{I-4}$$

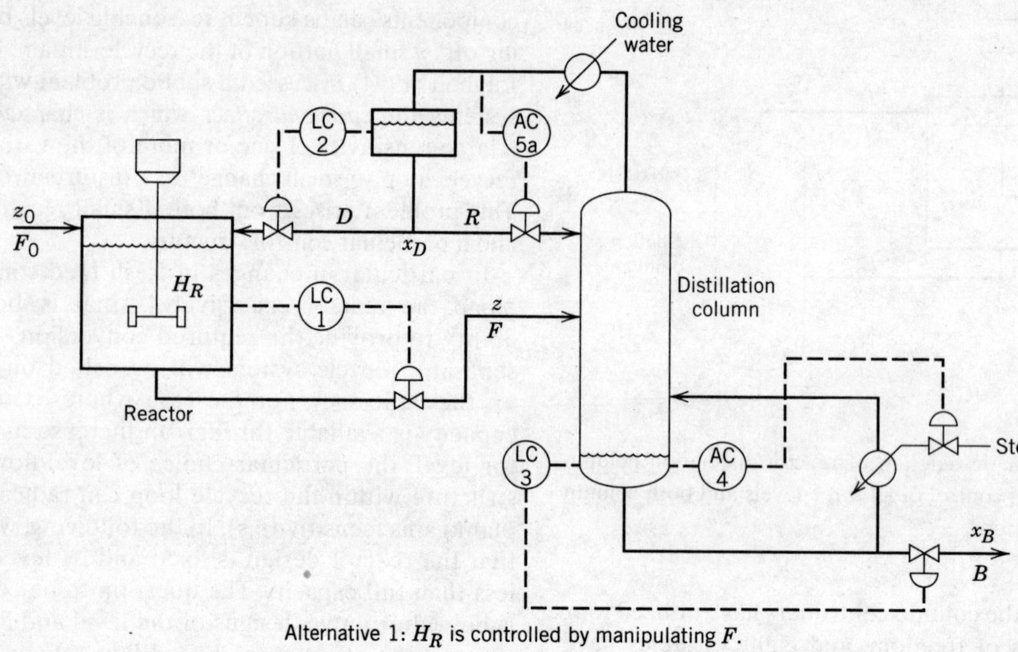

Figure I.7*a* Alternative control structures for the reactor/distillation column plant.

Column. The column equations were developed in the previous section:

$$\bar{F}_0 = \bar{D} + \bar{B} \tag{I-1}$$

$$\bar{F}\bar{z} = \bar{D}\bar{x}_D + \bar{B}\bar{x}_B \tag{I-2}$$

Combining Eqs. I-1 and I-3 (or, equivalently, by writing an overall balance around both units),

$$\bar{B} = \bar{F}_0 \tag{I-5}$$

Similarly, from Eqs. I-2 and I-4,

$$\bar{F}_0\bar{z}_0 = \bar{B}\bar{x}_B + k_R\bar{H}_R\bar{z} \tag{I-6}$$

To simplify the sensitivity analysis, consider the following limiting case:

$$\bar{x}_D \approx 1 \tag{I-7}$$

$$\bar{x}_B \approx 0 \tag{I-8}$$

Now, substitute the approximations of Eqs. I-7 and I-8 into Eq. I-2 to obtain

$$\bar{D} \approx \bar{F}\bar{z} \tag{I-9}$$

Similarly, from Eq. I-6,

$$\bar{F}_0\bar{z}_0 \approx k_R\bar{H}_R\bar{z} \tag{I-10}$$

Finally, by manipulating Eqs. I-1, I-3, I-5, I-9, and I-10, the desired expressions for $\bar{z}$, $\bar{D}$, and $\bar{F}$ can be obtained in terms of the reactor fresh feed variables, $\bar{F}_0$ and $\bar{z}_0$.

$$\bar{z} = \frac{\bar{F}_0\bar{z}_0}{k_R\bar{H}_R} \tag{I-11}$$

$$\bar{D} = \frac{(\bar{F}_0)^2\bar{z}_0}{k_R\bar{H}_R - \bar{F}_0\bar{z}_0} \tag{I-12}$$

$$\bar{F} = \frac{\bar{F}_0 k_R\bar{H}_R}{k_R\bar{H}_R - \bar{F}_0\bar{z}_0} \tag{I-13}$$

Equation I-12 indicates that any change in $\bar{F}_0$ or $\bar{z}_0$ will be considerably amplified in $\bar{D}$ because of the presence of the difference between two terms in the denominator, which is significantly increased or decreased by small changes in feed conditions. In a similar fashion, $\bar{D}$ and $\bar{F}$ are sensitive to changes in $\bar{F}_0$ and $\bar{z}_0$. High sensitivity to a disturbance is termed the *snowball effect* by analogy to a snowball, which grows larger as it rolls downhill.

An important point should be emphasized here—namely, that the snowball effect in D and F, while resulting from a particular control structure, is a steady-state phenomenon. In that sense, it is similar to the RGA, which is also a measure of steady-state sensitivities. Luyben (1994) suggested an alternative control method that was intended to reduce the snowball effect in D and F. We investigate a variation of his proposed method next.

Alternative 2 (Fig. I.7b)

In this alternative, F is held constant via a flow controller while H_R is allowed to float. Note that allowing

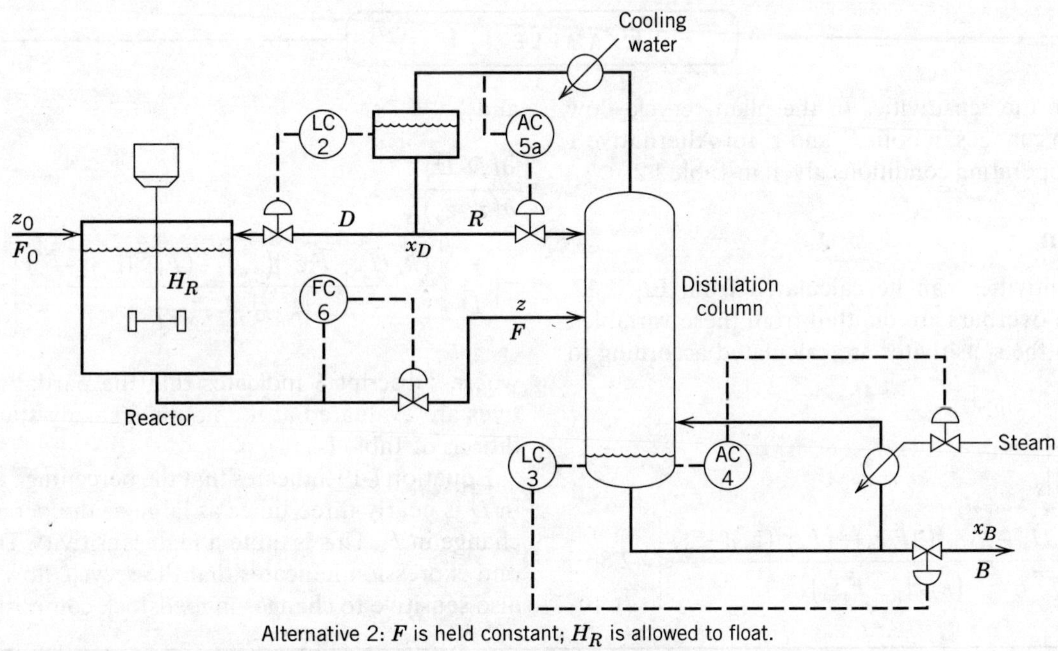

Figure I.7*b* Alternative control structures for the reactor/distillation column plant.

the reactor level (holdup) to vary as disturbance variables F_0 and z_0 change still allows z to change. Luyben (1994) originally proposed controlling H_R with F_0. Larsson et al. (2003) recognized this structure to be self-regulating because H_R adjusts as required to match changes in F. Thus, there is no need to manipulate F_0.

Because F is held constant instead of H_R, as in Alternative 1, we derive approximate expressions for the key recycle loop variables at steady-state ($\bar{D}$, $\bar{z}$, and $\bar{H}_R$) in terms of the disturbances ($\bar{F}_0$ and $\bar{z}_0$). Rearranging Eq. I-3 yields

$$\bar{D} = \bar{F} - \bar{F}_0 \tag{I-14}$$

From Eqs. I-9 and I-14

$$\bar{z} = \frac{\bar{F} - \bar{F}_0}{\bar{F}} \tag{I-15}$$

Substituting Eq. I-15 into Eq. I-10 yields

$$\bar{H}_R = \frac{\bar{F}\bar{F}_0\bar{z}_0}{k_R(\bar{F} - \bar{F}_0)} \tag{I-16}$$

Rearrangement of Eq. I-16 yields

$$\bar{H}_R = \frac{\bar{z}_0}{k_R\left(\dfrac{1}{\bar{F}_0} - \dfrac{1}{\bar{F}}\right)} \tag{I-17}$$

Equation I-14 shows clearly that Alternative 2 does not produce a snowball effect in distillate flow rate, because $\bar{D}$ is simply a linear function of $\bar{F}_0$. However, $\bar{H}_R$ now changes in a manner that is proportional to $\bar{z}_0$ and, as is shown below in the examples, is even more strongly related to $\bar{F}_0$. Larsson et al. (2003) showed that the reactor level is intrinsically self-regulating for Alternative 2, a feature that is evaluated in Exercise I.4. In considering Alternative 2, note that a level controller may be incorporated for safety reasons, even if not specifically required—for example, to prevent tank overflow.

Using the equations derived above, we can evaluate and compare quantitatively the sensitivities of key recycle loop variables to sustained changes in either input, z_0 or F_0, for each of the two alternative control structures. Recall that the sensitivity, or gain on a fractional basis, of any output variable y_i *at a specified steady state* $(\bar{x}, \bar{y})$ to a sustained change in an input variable x_j is given by the expression (see Appendix L.6):

$$\left.\frac{\partial(y_i/\bar{y})}{\partial(x_j/\bar{x})}\right|_S = \left.\frac{\partial y_i}{\partial x_j}\right|_S \left(\frac{\bar{x}}{\bar{y}}\right) \tag{I-18}$$

where subscript S indicates that the partial derivatives are evaluated at steady state $(\bar{x}, \bar{y})$.

EXAMPLE I.1

Calculate the sensitivities of the plant recycle flow rate D to changes in both F_0 and z_0 for Alternative 1 and the operating conditions given in Table I.2.

Solution

The sensitivities can be calculated from Eq. I-12. First, the overbars are omitted from these variables, and then the sensitivites are calculated according to Eq. I-18.

$$\left.\frac{\partial(D/\bar{D})}{\partial(F_0/\bar{F}_0)}\right|_S = \frac{\bar{F}_0}{\bar{D}}\left[\frac{(k_R\bar{H}_R-\bar{F}_0\bar{z}_0)(2\bar{F}_0\bar{z}_0)-(\bar{F}_0)^2(\bar{z}_0)(-\bar{z}_0)}{(k_R\bar{H}_R-\bar{F}_0\bar{z}_0)^2}\right]=2.86 \tag{I-19}$$

and

$$\left.\frac{\partial(D/\bar{D})}{\partial(z_0/\bar{z}_0)}\right|_S = \left(\frac{\bar{z}_0}{\bar{D}}\right)\left[\frac{(k_R\bar{H}_R-\bar{F}_0\bar{z}_0)(\bar{F}_0)^2-(\bar{F}_0)^2(\bar{z}_0)(-\bar{F}_0)}{(k_R\bar{H}_R-\bar{F}_0\bar{z}_0)^2}\right]=1.92 \tag{I-20}$$

where subscript S indicates that the partial derivatives are evaluated at the nominal steady-state conditions of Table I.2.

Equation I-19 indicates that the percentage change in D is nearly three times as large as the percentage change in F_0. This is quite a high sensitivity. The second expression indicates that the recycle flow rate is also sensitive to changes in feedstock composition.

EXAMPLE I.2

Repeat Example I.1, analyzing sensitivities for Alternative 2.

Solution

The relative sensitivities for D are obtained from Eq. I-14 using the method in Example I.1:

$$\left.\frac{\partial(D/\bar{D})}{\partial(F_0/\bar{F}_0)}\right|_S = \frac{\bar{F}_0}{\bar{D}}=-0.92 \tag{I-21}$$

and

$$\left.\frac{\partial(D/\bar{D})}{\partial(z_0/\bar{z}_0)}\right|_S = 0 \tag{I-22}$$

In the latter case, the sensitivity is zero, because D in Eq. I-14 is not a function of z_0.

Thus, the sensitivity of D to F_0 is less than one-third of the value for Alternative 1. With respect to sensitivity to z_0, the Alternative 2 control system completely eliminates the dependence of D on z_0.

Because H_R is allowed to float, we should analyze its sensitivities to F_0 and z_0. From Eq. I-16,

$$\left.\frac{\partial(H_R/\bar{H}_R)}{\partial(F_0/\bar{F}_0)}\right|_S = \frac{\bar{F}_0}{H_R}\frac{k_R(\bar{F}-\bar{F}_0)-\bar{F}\bar{F}_0\bar{z}_0(-k_R)}{[k_R(\bar{F}-\bar{F}_0)]^2}=4.8 \tag{I-23}$$

and

$$\left.\frac{\partial(H_R/\bar{H}_R)}{\partial(z_0/\bar{z}_0)}\right|_S = \frac{\bar{z}_0}{\bar{H}_R}\frac{\bar{F}\bar{F}_0}{k_R(\bar{F}-\bar{F}_0)}=1 \tag{I-24}$$

Although the sensitivity of H_R to changes in z_0 is satisfactory, it is quite large with respect to F_0. Thus, Alternative 2 does not eliminate the snowball effect; it simply shifts it from D and F to H_R. In typical industrial practice, changing the reactor level over a relatively wide range would be undesirable; it normally is held reasonably constant.

Luyben (1994) has investigated these and similar relations for more complex reaction kinetics over a wide range of the disturbances (F_0 and z_0). The snowball effect is not an artifact of the simplifying assumptions employed (e.g., perfect composition control in the column). It appears to be a general effect in recycle systems that can arise from inadequate reactor holdup or a particular choice of the plant inventory/flow control structure. However, before attempting to generalize, we look at two other control structures and their sensitivity characteristics.

Other Level/Flow Control Structures

Wu and Yu (1996) identified the major disadvantage associated with the Alternative 2 control structure discussed in Example I.2—namely, that it eliminates snowballing in D but introduces the same effect in H_R, which becomes sensitive to z_0 and F_0. With this objection in mind, they proposed two control structures which they referred to as "balanced" in the sense that feed disturbances are intended to be distributed to both units to smooth out the effects on any particular unit. Their configurations (designated here as Alternatives 3 and 4) include the following features:

Alternative 3 (Fig. I.8a). H_R is controlled by manipulating D; however, the set point of the H_R controller is manipulated to control reactor composition z (cascade control). Thus, H_R floats, but only as required to control z.

Alternative 4 (Fig. I.8b). H_R is controlled by manipulating D, but the H_R set point is manipulated to control distillate composition x_D. Again, H_R floats, but only as required to control x_D.

In both of these alternatives, Wu and Yu (1996) proposed that the ratio of F/F_0 be maintained constant by ratio control, a type of feedforward control. In summary, the key features are as follows:

1. D is used to control reactor level in the secondary loop of a cascade controller. H_R, in turn, floats to control one of the recycle loop compositions (z in Alternative 3 or x_D in Alternative 4) by adjusting the set point of the level controller.
2. Unlike Alternatives 1 and 2, disturbance F_0 is now measured and used as input to a ratio controller regulating F; thus, variations in F_0 are reflected directly in changes in F, loading the column somewhat.
3. Variations in F_0 and z_0 will cause changes in both H_R and D if a composition somewhere within the recycle loop is forced to remain constant. Luyben (1994) had suggested that a flow rate be specified (fixed) somewhere within each recycle loop. The more complicated Wu and Yu proposal is to specify a composition within the loop while fixing the ratio of the recycle loop flow rate to the reactor feed rate.

Figure I.8 illustrates the control configurations for Alternatives 3 and 4. Both utilize R to control H_D. Table I.4 provides a detailed comparison of all four alternative control configurations. Note that the only features common to all alternatives are the two loops controlling H_B and x_B.

Wu and Yu (1996) performed an extensive steady-state analysis of these control structures using a 2 × 2

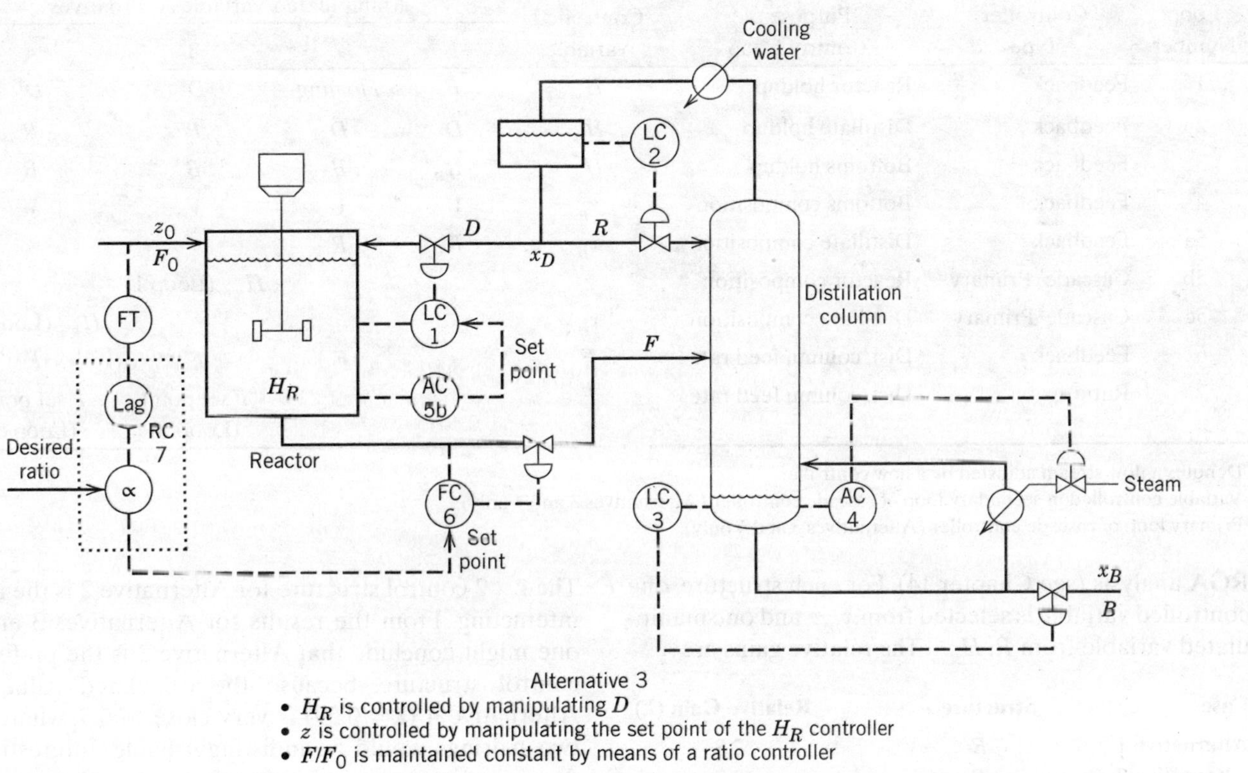

Figure I.8a Additional control structure for the reactor/distillation column plant.

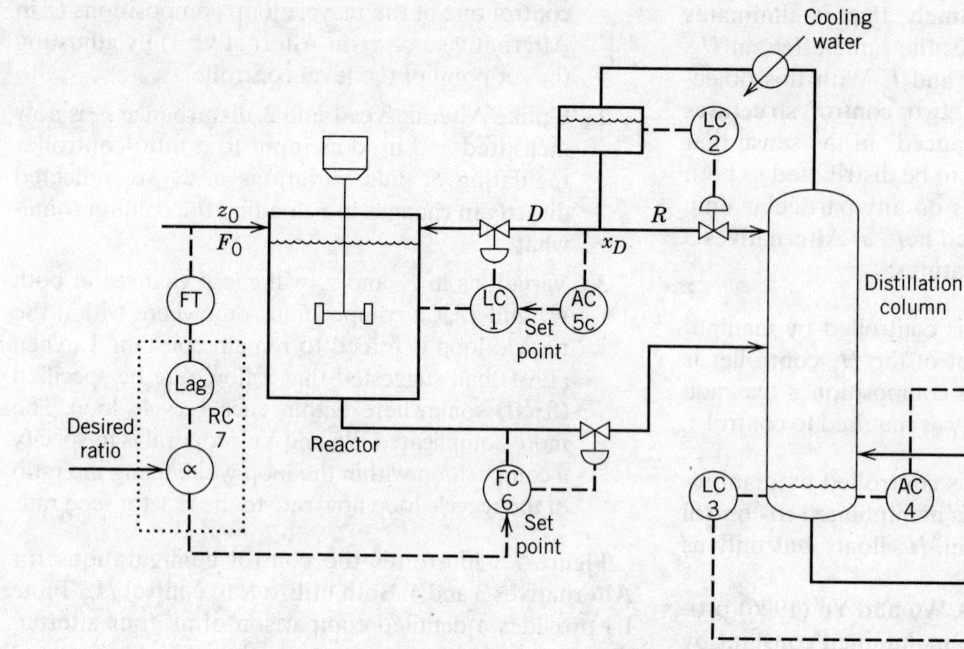

Alternative 4
- H_R is controlled by manipulating D
- x_D is controlled by manipulating the set point of the H_R controller
- F/F_0 is maintained constant by means of a ratio controller

Figure I.8*b* Additional control structure for the reactor/distillation column plant.

Table I.4 A Comparison of Alternative Control Strategies for the Reactor for the Reactor/Distillation Column Plant

Loop Number	Controller Type	Purpose of Control Loop	Controlled Variable	Manipulated Variable Alternatives			
				1	2	3	4
1	Feedback	Reactor holdup	H_R	F	*Floating*	$D^†$	$D^†$
2	Feedback	Distillate holdup	H_D	D	D	R	R
3	Feedback	Bottoms holdup	H_B	B	B	B	B
4	Feedback	Bottoms composition	x_B	V	V	V	V
5a	Feedback	Distillate composition	x_D	R	R		
5b	Cascade[‡] Primary	Reactor composition	z			$H_{R,sp}$ (Loop 1)	
5c	Cascade[‡] Primary	Distillate composition	x_D				$H_{R,sp}$ (Loop 1)
6	Feedback	Dist. column feed rate	F		F^*	F^*	F^*
7	Ratio	Dist. column feed rate	F			F set point (Loop 6)	F set point (Loop 6)

*Denotes a flow stream adjusted by a flow controller.
†Variable controlled in secondary loop of cascade controller (Alternatives 3 and 4 only).
‡Primary loop of cascade controller (Alternatives 3 and 4 only).

RGA analysis (see Chapter 14). For each structure, one controlled variable is selected from x_D, z and one manipulated variable from R, $H_{R,sp}$. The relative gains are

Case	Structure	Relative Gain (λ)
Alternative 1	$x_D - R/x_B - V$	2.8
Alternative 2	$x_D - R/x_B - V$	12.2
Alternative 3	$z - H_{R,sp}/x_B - V$	0.78
Alternative 4	$x_D - H_{R,sp}/x_B - V$	0.59

The 2×2 control structure for Alternative 2 is the most interacting. From the results for Alternatives 3 and 4, one might conclude that Alternative 3 is the preferred control structure, because the calculated value for Alternative 4 ($\lambda = 0.59$) is very close to 0.5, where the two pairings would be indistinguishable. Interestingly, dynamic simulation of these four control configurations led to the recommendation of Alternative 4 by Wu and Yu (1996). It exhibited the best closed-loop responses—that is, less interaction between the x_D/H_R primary control loop and other loops—than did Alternative 3 with its corresponding z/H_R loop. When compared with Alternatives 1 and 2, Alternative 4 provided better control of key product composition output x_B and handled larger disturbance changes without violating process constraints. This last point is important: transferring disturbances to more than one plant unit reduces the possibility of intermediate variables' violating a constraint, with the accompanying loss of controllability.

Figures I.9*a* and I.9*b* compare Alternatives 1, 2, and 4, showing the response of several intermediate plant variables (F, D, and H_R) to step changes in F_0 and z_0, respectively. Note that the responses of Alternative 4 in Fig. I.9*a* for feed flow changes lie between Alternatives 1 and 2 as expected; however, the Alternative 4 responses to feed composition changes in Fig. I.9*b* closely resemble those of Alternative 2.

The following generalizations can be made from this case study:

1. Wu and Yu (1996) recommend controlling one composition measurement somewhere in each recycle loop (x_D or z) to accomplish the desired balancing; however, how to couple that composition to key variables in other units, such as H_R, is not clear. Also, they ratio F to F_0 in order to hold the recycle loop flows at constant ratios to F_0. With these changes, Alternative 4 mitigates flow rate changes to reactor level much better than Alternative 2, as expected, but that is not the case for feed composition changes.

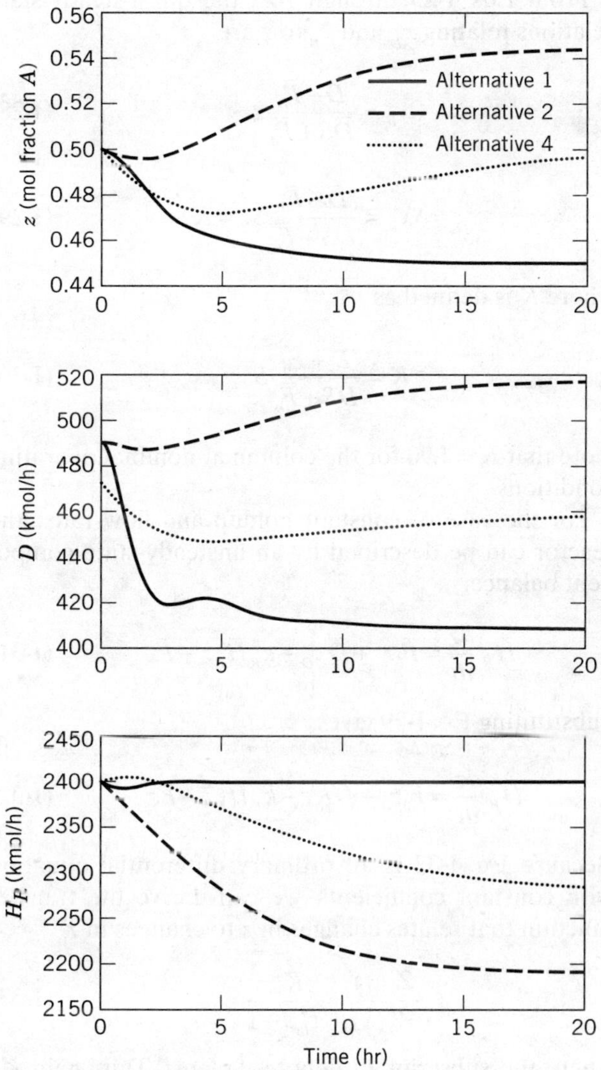

Figure I.9*a* Disturbance response of the reactor/distillation column plant using three alternative flow/level control structures (–10% change in F_0).

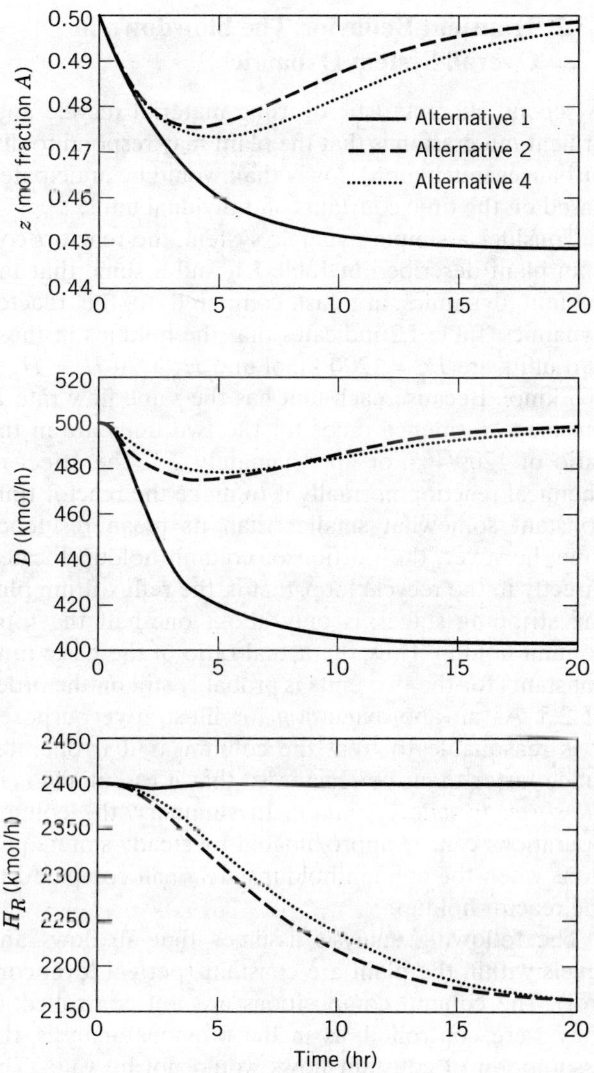

Figure I.9*b* Disturbance response of the reactor/distillation column plant using three alternative flow/level control structures (–10% change in z_0).

2. In general, the best choice of how to "allocate" anticipated disturbances to one or more units in a plant is an unresolved problem.
3. Although all design tools (both steady-state and dynamic) can be important in deciding among alternative control structures, determining the "best" structure should involve a rigorous dynamic simulation of the entire plant without using the sort of simplifying assumptions made in this chapter. Of course, final evaluation of the chosen method should be based on plant tests.

This last point, the need to consider process dynamics, is well illustrated by a discussion of how recycling material within a plant can drastically affect its overall dynamics. This topic is considered in the next section.

I.3.2 Transient Behavior: The Slowdown in Overall System Dynamics

A second characteristic of using material recycle and/or heat integration is that the plant may respond to disturbances much more slowly than would be anticipated based on the time constants of individual units.

Consider a simple dynamic system, the reactor/column plant described in Table I.1, and assume that the column dynamics are fast compared to the reactor dynamics. Table I.2 indicates that the holdups in these two units are $H_R = 1200$ kmol and $H_B + 20\,H_S + H_D = 465$ kmol. Because each unit has the same flow rate F, the mean residence times for the two units are in the ratio of 1200/465, or approximately 2.5. The effect of chemical reaction normally is to make the reactor time constant somewhat smaller than its mean residence time; however, the portion of column holdup located directly in the recycle loop, that is, the reflux drum plus the stripping stages, is only about one-half the total column holdup. Thus, the actual ratio of the basic time constants for the two units is probably still on the order of 2.5. As an approximation for illustrative purposes, it is reasonable to treat the column as if it operates much faster than the reactor, so that it essentially is in *quasi-steady-state operation*. In summary, the column operations can be approximated by steady-state equations when the column holdups are small compared to the reactor holdup.

The following analysis assumes that all flows and levels within the plant are constant (perfect level control). The column compositions are not controlled. If they were controlled, as in the previous analysis, the assumption of constant flows would not be valid. The only plant disturbance is the feed composition z_0. With the assumption of quasi-steady-state operation for the column, a simple linear dynamic model for compositions can be developed.

A steady-state model of the column can be derived using the two balances around the entire column (Eqs. I-1 and I-2) obtained earlier. Substitution of Eq. I-5 yields

$$\bar{F} = \bar{D} + \bar{F}_0 \tag{I-25}$$

$$\bar{F}\bar{z} = \bar{D}\bar{x}_D + \bar{F}_0\bar{x}_B \tag{I-26}$$

A third equation is obtained from the definition of the steady-state separation concentration ratio (see Section I.2.1) at the nominal operating conditions:

$$S = \frac{\bar{x}_D}{\bar{x}_B} \tag{I-27}$$

For this example, $S = 0.95/0.0105 = 90.5$.

From Eqs. I-25 through I-27, the quasi-steady-state relations relating x_D and x_B to z are

$$x_B \approx \frac{\bar{D}+\bar{F}_0}{\bar{D}S+\bar{F}_0}\,z \tag{I-28}$$

$$x_D \approx Sx_B = \frac{\bar{D}+\bar{F}_0}{\bar{D}S+\bar{F}_0}\,Sz = Kz \tag{I-29}$$

where K is defined as

$$K \triangleq \frac{\bar{D}+\bar{F}_0}{\bar{D}S+\bar{F}_0}\,S \tag{I-30}$$

Note that $K = 1.90$ for the column at nominal operating conditions.

For the case of constant holdup and flow rates, the reactor can be described by an unsteady-state component balance:

$$\bar{H}_R\frac{dz}{dt} = \bar{F}_0 z_0 + \bar{D}x_D - k_R\bar{H}_R\bar{z} - \bar{F}z \tag{I-31}$$

Substituting Eq. I-29 gives

$$\bar{H}_R\frac{dz}{dt} = \bar{F}_0 z_0 + \bar{D}Kz - k_R\bar{H}_R\bar{z} - \bar{F}z \tag{I-32}$$

Because Eq. I-32 is an ordinary differential equation with constant coefficients, we can derive the transfer function that relates changes in z to changes in z_0

$$\frac{Z'(s)}{Z'_0(s)} = \frac{K_{Pl}}{\tau_{Pl}s+1} \tag{I-33}$$

where the subscript Pl denotes "plant." Thus, gain K_{Pl} and time constant τ_{Pl} represent the entire plant (reactor, column, and recycle) and are defined as follows:

$$K_{Pl} \triangleq \frac{\bar{F}_0}{\bar{F}+k_R\bar{H}_R-\bar{D}K} \tag{I-34}$$

$$\tau_{Pl} \triangleq \frac{\bar{H}_R}{\bar{F}+k_R\bar{H}_R-\bar{D}K} \tag{I-35}$$

Substituting Eq. I-25 into Eq. I-35 yields

$$\tau_{Pl} = \frac{\bar{H}_R}{\bar{F}_0+k_R\bar{H}_R+\bar{D}(1-K)} \tag{I-36}$$

Note that the plant time constant without recycle $(\bar{D}=0)$ reduces to the reactor time constant

$$\tau_{pl}^0 = \frac{\bar{H}_R}{\bar{F}_0+k_R\bar{H}_R} \tag{I-37}$$

This result is obtained if there is no separation of A and B in the column ($K = 1$), regardless of the magnitude of the recycle flow rate! The effect of having a recycle stream that is richer in reactant than the product stream ($K > 1$) is to slow down the operation of the two-unit plant, because

$$\tau_{Pl} \geq \tau_{pl}^0 \tag{I-38}$$

as a result of the negative $\bar{D}(1-K)$ term in the denominator of Eq. I-36.

This analysis can be performed for the same assumptions that were used in investigating the snowball effect; that is, x_D and x_B are perfectly controlled. In this case, the slowdown effect is even more pronounced, but a simple expression for the plant time constant is not obtained. This exercise is left for the reader.

It is interesting to generalize the results of this example:

1. Any multi-unit plant with a recycle stream from a separation unit is likely to exhibit slower dynamics. Just as negative feedback normally speeds up the process response, the positive feedback of material in this recycle stream slows down the response.
2. The process response becomes slower as either the degree of separation or the recycle flow rate increases.

I.3.3 Propagation and Recirculation of Disturbances

A third major effect often encountered with recycle and heat integration is the propagation of disturbances from unit to unit, and their recirculation around internal process flow paths. To understand why this plantwide control issue differs so substantially from single-unit issues, it is worthwhile to review briefly the objectives of single-unit regulation.

In Chapter 13 we indicated that one desirable effect of using feedback control to attenuate disturbances in a process unit is to transfer these variations to a utility stream. For example, if a reactor temperature is disturbed, the cooling water flow rate will be changed by the temperature controller so as to reduce the effect of the disturbance. Even so, some variation in reactor temperature inevitably will remain, and this will propagate to downstream units as a disturbance.

EXAMPLE I.3

Find the time constant of the reactor/steady state column model for the operating conditions given in Table I.2 with $K = 1.90$. Determine how much the plant dynamics are slowed by the effect of material recycle.

Solution

From Eq. I-36,

$$\tau_{Pl} = \frac{1200}{230+(0.33)(1200)+(250)(1.0-19)} = 3.0 \text{ h} \tag{I-39}$$

Without recycle, from Eq. I-37,

$$\tau_{pl}^0 = \frac{1200}{230+(0.33)(1200)} = 1.92 \text{ h} \tag{I-40}$$

Thus, the approximate effect of recycle on this plant is to increase the time constant by

$$\frac{\tau_{Pl}}{\tau_{pl}^0}-1 = \frac{3.0}{1.92}-1 = 0.56 \text{ or } 56\% \tag{I-41}$$

This result means that a change in z_0 will take 56% longer to work its way through the system with recycle than it would without recycle. Kapoor and McAvoy (1987) provide a more general analysis of how internal recycle affects the time constants of a distillation column.

A minor side effect of these actions is that the supply header temperature itself will change slightly as cooling water demand is raised/lowered by actions of a reactor temperature controller. Although utility supply systems are built with their own internal controllers, and these are designed to attempt to regulate the utility outputs in the face of process disturbances, it is not possible to attenuate utility disturbances entirely. These propagate directly throughout the plant.

In older plants, surge tanks were used to damp flow variations between units. Material holdup in a surge tank can also serve as a thermal capacitance and thus reduce effluent temperature variations; only reduced flow and temperature variations propagate to downstream units. In today's more highly integrated plants, containing material recycle and/or heat integration but little surge capacity, unattenuated disturbances propagate directly to downstream units, even to adjacent (coupled) units and to upstream units.

I.4 INTERACTION OF PLANT DESIGN AND CONTROL SYSTEM DESIGN

In the past, when continuous processing plants were designed to be much less interacting than now, it was possible to complete the plant design before considering control system design. After the proposed plant's flowsheet and equipment specifications were completed, process control engineers were responsible for specifying instruments and controllers. By providing an excess of measurements (instruments) and control valves, plus a feedback controller for every important process variable, the control system designer was reasonably sure that the new plant could be started up and controlled. Continuous processing plants designed or retrofitted today no longer can utilize a sequential design process in which plant design is followed by control system design (Keller and Bryan, 2000), nor can designers specify redundant equipment, except for safety purposes.

Without careful attention to design, highly integrated plants may have too few control degrees of freedom, which makes them difficult to start up and operate safely. For example, in designing the heat exchanger and related equipment for heat integration the heating and cooling loads first must be approximately balanced by the process designer. Then the designer must establish whether the approach temperatures are satisfactory to meet the heat transfer requirements with a reasonably sized heat exchanger; in this step, adjustment of column operating pressures may be required (Seider et al., 2003). Because the energy supply capability in one unit usually will not balance the demand in another unit exactly, a "trim exchanger" (small heat exchanger sized to make up the difference in heating/cooling capability) generally has to be provided in the steady-state design.

Note that introducing a heat integration scheme also causes two control degrees of freedom to be "lost": the cooling water flow rate control valve that would have been located in the Column 1 condenser, plus the steam control valve that would have been used in the Column 2 reboiler. If process control engineers are not involved in the plant design process from the beginning, the critical process dynamic and control evaluations may be omitted that would provide such information and an opportunity to resolve any problems (Keller and Bryan, 2000). In short, a suitably sized trim unit must be available to make up for any steady-state heating/cooling deficiency plus lost control degrees of freedom necessary for normal operations. It also can assist in start-up and shutdown operations.

The control system designer must determine whether a proposed plant design will be controllable and operable (Fisher et al., 1988b; Downs and Ogunnaike, 1995). For example, highly integrated distillation columns can cause problems in a number of ways:

1. One or both column products cannot be controlled at the desired set point(s).
2. Disturbances in the Column 1 overhead cannot be prevented from propagating to Column 2.
3. The "lost degrees of freedom" from plant integration need to be restored by the addition of one or two trim heat exchangers operated and controlled using plant utility supplies.
4. The plant cannot be started up easily because of the need to have Column 1 "hot" before Column 2 can be brought into service.

This chapter has provided a brief overview of process integration issues and possible solutions. For a much broader discussion of the topic of heat integration, the reader should consult Douglas (1988) or Seider et al. (2008). For a more extensive development of control system design issues, including a number of simulation case studies, Luyben (2002) should be consulted.

SUMMARY

For new process designs, the control system designer may have little precise knowledge of how to control the proposed plant. Because the plant design may never be replicated, there will be little incentive to spend thousands of hours designing and optimizing the control system structure, as would be done, for example, for a new airplane design. Whether or not the final control structure will be successful depends to a large extent on the knowledge, skill, and intuition of the control system design team. The plant initially can be considered as a collection of reasonably well-understood processing units, but it can operate quite differently than would be expected from knowledge only of its individual units considered separately.

In this chapter, we have introduced some basic plantwide issues that are fundamentally based on multi-unit interactions. These topics have included steady-state issues (sensitivities), dynamic issues (settling times of integrated plants vs. individual units), and the propagation of disturbances from unit to unit in highly integrated plants that involve recycle of material and of energy.

In the next appendix (Appendix J), we present a systematic procedure that can guide the development of plantwide control system designs. The goal is to design a viable control system structure for a new processing plant that has a high probability of working satisfactorily when actually installed.

REFERENCES

Buckley, P. S., *Techniques of Process Control*, John Wiley and Sons, Hoboken, NJ, 1964.

Douglas, J. M., *Conceptual Design of Chemical Processes*, McGraw-Hill, New York, 1988.

Downs, J. J. and B. Ogunnaike, Design for Control and Operability: An Industrial Perspective, *AIChE Symp. Ser.*, **91**, L. T. Biegler and M. F. Doherty (Eds.), 115 (1995).

Fisher, W. R., M. F. Doherty, and J. M. Douglas, The Interface Between Design and Control: 1. Process Controllability, *Ind. Eng. Chem. Res.*, **27**, 597 (1988a).

Fisher, W. R., M. F. Doherty, and J. M. Douglas, The Interface Between Design and Control: 2. Process Operability, *Ind. Eng. Chem. Res.*, **27**, 606 (1988b).

Kapoor, N. and T. J. McAvoy, An Analytical Approach to Approximate Dynamic Modeling of Distillation Towers, *Ind. Eng. Chem. Res.*, **26**, 2473 (1987).

Keller, G. E. and P. F. Bryan, Process Engineering: Moving in New Directions, *Chem. Eng. Prog.*, **96**, 41 (2000).

Larsson, T., M. M. S. Govatsmark, S. Skogestad, and C.-C. Yu, Control Structure Selection for Reactor, Separator and Recycle Processes, *Ind. Eng. Chem. Res.*, **42**, 1225 (2003).

Luyben, W. L., Dynamics and Control of Recycle Systems. 1. Simple Open-loop and Closed-loop Systems, *Ind. Eng. Chem. Res.*, **32**, 466 (1993).

Luyben, W. L., Snowball Effects in Reactor/Separator Processes with Recycle, *Ind. Eng. Chem. Res.*, **33**, 299 (1994).

Luyben, W. L., *Plantwide Dynamic Simulators in Chemical Processing*, Marcel Dekker, New York, 2002.

Papadourakis, A., M. F. Doherty, and J. M. Douglas, Relative Gain Array for Units with Recycle, *Ind. Eng. Chem. Res.*, **26**, 1259 (1987).

Rinard, I. and J. J. Downs, Plantwide Control: A Review and Critique, Presented at the AIChE Spring National Meeting, Houston (1992).

Seader, J. D. and E. J. Henley, *Separation Process Principles*, John Wiley and Sons, Hoboken, NJ, 1997.

Seider, W. D., J. D. Seader, D. R. Lewin, and S. Widagdo, *Process Design Principles: Synthesis, Analysis, and Design*, 3rd ed. John Wiley and Sons, Hoboken, NJ, 2008.

Shinskey, F. G., *Process Control Systems*, 4th ed., McGraw-Hill, New York, 1996.

Wu, K.-L., and C.-C. Yu, Reactor/Separator Processes with Recycle-1. Candidate Control Structure for Operability, *Computers Chem. Engng.*, **20**, 1291 (1996).

EXERCISES

I.1 Figure EI.1 illustrates two CSTRs in a chemical manufacturing plant. Reactants A and B must be fed to the first stirred-tank at a constant molar ratio. Reactant C is introduced to the second stirred-tank at a constant molar ratio to reactant A. Five control valves are available for purposes of controlling the plant production rate and concentrations. Flow rates shown in the figure are in mass units. Reactor volumes are constant.

The assumed reaction kinetics are:

$$A + B \rightarrow D$$

$$C + D \rightarrow E$$

If each reaction goes to completion in its respective reactor, how can you control the plant production rate of the desired product E using each of the five valves? Specify how you would use ratio controllers to maintain the desired stoichiometry in each case, and explain the advantages and disadvantages relative to the other possible locations.

Note: *In the following exercises, a Simulink model is used to approximate the reactor and distillation column units discussed in this chapter. Information is given in Appendix K.1.*

I.2 In this exercise, you will evaluate the individual units at the nominal steady state for purposes of understanding how the plant would operate *without recycle.* Use Simulink to simulate the full differential equation model given in Table I.1. Then, for purposes of this problem only, "tear" the recycle stream to the reactor—that is, disconnect the distillate line and replace it with a constant stream to the reactor that is set at the recycle stream's nominal conditions of flow rate and concentration.

(a) Using a material balance control configuration and any of the techniques discussed in Chapter 11, find P or PI controller settings that will regulate the

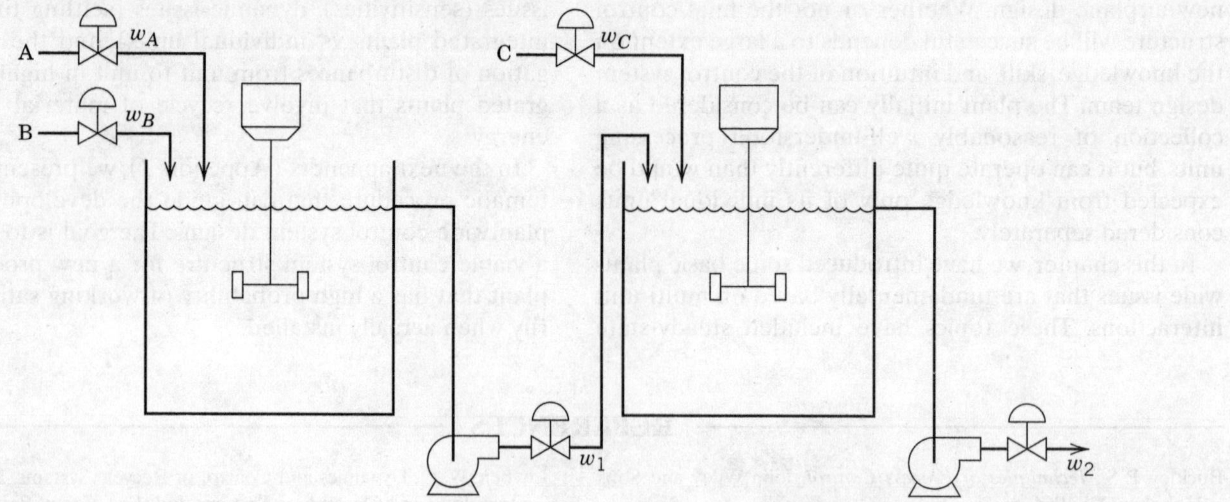

Figure II.1

liquid levels in the reboiler and the reflux drum with little overshoot.

(b) Complete the column control structure by finding PI controllers that will satisfactorily maintain the distillate and bottoms composition, again with little overshoot. Test your column level/composition control system by making small step changes in the column feed flow rate and composition.

(c) In a similar manner, develop a P or PI controller for reactor level using F as the manipulated variable. (Note that level controller settings obtained using F_0 for the manipulated variable will be identical to those using F.) Again, test your reactor level control system by making small step changes in the feed flow rate and composition.

For each of the following exercises, either work Exercise I.2 first or use controller settings similar to those provided with the parameters and Simulink model of the two-component plus recycle process in Appendix K.1.

I.3 Starting with a Simulink model of the recycle process, implement a reactor level controller using F as the manipulated variable. Confirm via simulation that control scheme Alternative 1 works effectively for a step change in F_0. If necessary, detune any of the controllers to keep oscillations to a minimum.

I.4 Starting with a Simulink model of the recycle process,

(a) Place a tightly tuned flow controller on F. Confirm, via a step change in F_0, that Alternative 2 is self-regulating; that is, that the level in the reactor automatically seeks a suitable steady-state value if the reactor feed flow rate is subjected to a sustained change.

(b) Show that this level is identically equal to the value given by Eq. I-16.

I.5 Luyben's original proposal (1994) for the Alternative 2 control structure incorporated a reactor level controller using F_0 as the manipulated variable. However, the level controller prevents specifying the plant's production rate by a flow controller on F_0, as can be done with Alternative 1. To deal with the problem that arises when F_0 is allocated for level control, Luyben proposed that the steady-state relation given by Eq. I-17 be rewritten to provide a type of feedforward control based on measurements of z_0 and F_0. F_{sp}, the set point of the reactor effluent flow controller, can be used to approximate F. The set point for the reactor level controller would then be

$$H_{R,sp}(t) = \frac{z_0(t)}{k_R\left(\frac{1}{F_0(t)} - \frac{1}{F_{sp}}\right)}$$

In the following steps, you are to evaluate Luyben's proposed alternative. If necessary, detune any of the controllers to keep oscillations to a minimum.

(a) Determine how Luyben's proposed alternative structure responds to a step change in z_0.

(b) Implement Luyben's proposed feedforward controller—that is the equation above—and implement a similar step change in z_0.

(c) What can you say about the speed of response of this controlled plant with and without the feedforward controller?

(d) What are the advantages and disadvantages of Luyben's proposal?

I.6 Implement Alternatives 3 and 4 using the Simulink model. This will require tuning a cascade loop to control composition in each case.

(a) Evaluate the response of these controlled plants for a step change in F_0 without using the ratio controller proposed by Wu and Yu.

(b) Repeat (a) with a controller used to ratio the column feed flow rate to the reactor feed flow rate. How do the responses in (a) and (b) compare?

I.7 Evaluate any two of the four alternatives we have looked at in this chapter, (i.e., Exercises I.3, I.4*a*, and/or I.6).

(a) Compare the responses of each control structure to step changes in reactor feed flow rate.

(b) Compare the sensitivities of each alternative to changes in this variable.

I.8 The recycle plant discussed in this chapter utilizes a composition-only model; that is, thermal effects are neglected. Appendix K contains equations and parameters that can be used to model temperature effects in the reactor. Implement the cooling coil equations for the reactor and, using the cooling water flow rate as manipulated variable, design a PI controller that will control reactor temperature. Test your plant's response to a step change in reactor feed flow rate using any of the control structure alternatives discussed in this chapter. What can you conclude regarding the effect of a well-tuned controller for reactor temperature on the responses of the other system variables such as F and D?

APPENDIX J

Plantwide Control System Design

APPENDIX CONTENTS

In this chapter, we describe a hierarchical design procedure that can be used to develop multiloop and multivariable measurement and control strategies for plantwide control systems. The procedure assists the engineer in determining how to choose the best controlled, manipulated, and measured variables in the plant, when to use advanced control techniques such as MPC, and how to select appropriate multiloop control structures with minimum interactions among the coupled processes in the plant. The proposed design procedure is based on the hierarchy of process control activities described in Chapter 1, the control system design guidelines discussed in Chapter 2, RGA and SVA multivariable methods presented in Chapter 14 the model predictive control approach of Chapter 15, plantwide control concepts of Appendix I, and designers' experience. It is important to realize that the design of plantwide control systems is an art as well as a science. Typically, more than one design will be satisfactory; thus, there is no single solution to the design problem. Furthermore, a design procedure generally involves iteration of individual steps until a satisfactory design results. Thus, the application of a systematic design procedure, such as the hierarchical approach of this chapter, produces preliminary designs that are subject to further exploration and refinement. Simulation methods should be employed to examine alternative control configurations while exploring the effect of controller tuning on the response of key process variables. The hierarchical procedure recommended in this chapter is illustrated by a case study.

The goal is a plantwide control system design that is no more complicated or expensive than necessary and that, when built, can be operated easily by typical plant operators. Ultimately, the only definitive way of validating a selected plantwide control system design is by plant tests and by the operating plant's performance.

J.1 PROCEDURES FOR THE DESIGN OF PLANTWIDE CONTROL SYSTEMS

The design of a plantwide control system consists of four major steps:

1. The overall specifications for the plant and its control system are stated.
2. The control system structure is developed. This step includes selecting controlled, measured, and

manipulated variables; choosing multiloop or multivariable control; deciding how to control production rate, product quality, and inventories; and handling operating constraints. Decomposition of the plantwide control problem into smaller problems for the purpose of analysis may also be employed here.

3. Design is followed by a detailed specification of all instrumentation/hardware and software, cost estimation, evaluation of alternatives, and the ordering and installation of equipment.
4. Following design and construction of the plant, plant tests, including startups, operation at design conditions, and shutdowns, are carried out prior to commissioning of the plant.

This chapter is concerned with the first two steps, beginning with the plant control system design specifications.

In principle, a comprehensive *top-down* formulation could be used to develop the required plantwide control design. We assume that general requirements for the plant, such as product specifications and production rates, have been established at the plant, division, or corporate level. The specifications for plant operating conditions have been developed by the plant design group working in collaboration with product development and process control specialists. Starting with the above specifications plus knowledge of the potential measured, manipulated, and controlled variables, optimization methods could be employed to develop the control system design based on a comprehensive dynamic model of the plant. Unfortunately, such an approach is impractical because of the large number of process variables involved in modern processing plants. On the other hand, some aspects of a top-down design approach may be quite useful as part of a realistic design procedure.

The traditional design procedure used for industrial control systems has been a *bottom-up*, unit-by-unit approach. Even though it incorporates systematic methods to develop the control structure, this approach also relies on heuristic design methods and rules of thumb developed from previous designs and the experience of both the process and control system design groups.

Control system design has been organized into a logical, sequential, process-oriented methodology by numerous control researchers—for example, Buckley (1964), Downs (1992), Luyben et al. (1998), Larsson and Skogestad (2000), and Stephanopoulos and Ng (2000). However, it would be naive to assume that any proposed design procedure can generate a suitable plantwide control system design after one pass through a set of sequential steps. At virtually every step of a given procedure, alternative choices will present themselves, each leading to a different final design alternative. The knowledge, skill, intuition, and persistence of the plantwide control design team will always be key elements in the design process. In addition, the control design specifications and models used for design will exhibit uncertainty, which means that multiple designs may need to be developed (Skogestad, 2002).

An effective way to make the large number of decisions is to organize the procedures in a generally hierarchical manner. Thus, detailed studies should not be undertaken until important general questions have been answered. Skogestad (2002) has developed a design procedure based on the intrinsically hierarchical nature of plantwide control systems while incorporating the best aspects of top-down and bottom-up design approaches. As shown in Fig. J.1, the most critical control tasks deal with the safety system (Appendix E) and with regulating the integrating response modes usually associated with liquid levels (holdups in the vessels). Thus, the basic objective at Levels 1, 2, and 3 is to provide

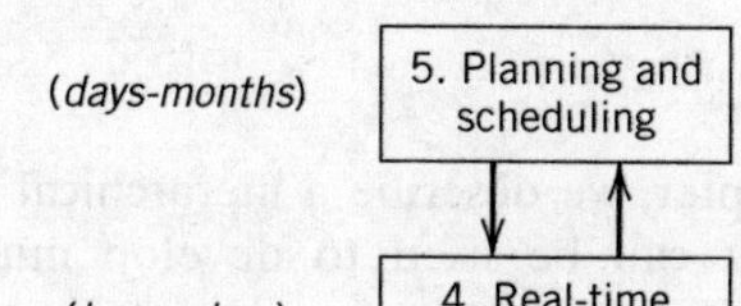

Figure J.1 Hierarchy of process control activities.

safe, stable control of the plant. Level 4 is concerned with economic optimization of plant operating conditions, and this step is usually decoupled from the control system operation.

Missing from many control system design methodologies, even hierarchical ones, is the important role that *decomposition* and *decentralization* play in a plantwide design approach. Procedures that lead to decomposition of the overall design into smaller subproblems can be advantageous. Even highly integrated plants do not require a multivariable approach linking all of the controlled variables with all of the manipulated variables. The extent to which a plantwide control system can be decentralized into smaller control systems designed to work at the process unit level invariably determines how easily the control system can be designed, tuned, and understood by plant operators. Decentralized control system designs generally are more robust when operating conditions change and are more tolerant to individual component failures.

J.2 A SYSTEMATIC PROCEDURE FOR PLANTWIDE CONTROL SYSTEM DESIGN

Table J.l provides the key steps in a systematic procedure recommended here for design of plantwide control structures. It is based on the combined top-down/bottom-up approach of Larsson and Skogestad (2000) and Skogestad (2002) and the hierarchical organization that generally matches Fig. J.1. The proposed systematic plantwide control design approach consists of the four major steps shown in Table J.1.

J.2.1 Control System Design Objectives

Plant operating/control objectives must be established at the outset of the design process. Two categories of information must be provided: (1) plant production and control objectives and (2) process constraints (Step I).

Step I. Specify the control system design objectives.

A. State the plant production, economic, and control objectives, including composition and production rates of all products.

B. Identify process constraints that must be satisfied, including safety, environmental, and quality restrictions.

Table J.1 Recommended Procedure to Design a Plantwide Control System

I. Specify the control system design objectives.
- A. State the plant production, economic, and control objectives, including composition and production rates of all products.
- B. Identify process constraints that must be satisfied, including safety, environmental, and quality restrictions.

II. Perform a top-down analysis.
- A. Identify the process variables, control degrees of freedom, control structure, and options for decomposition.
- B. Establish the overall control structure (in conceptual form).

III. Develop a bottom-up design.
- A. Develop a strategy for regulatory control.
- B. Examine the potential of applying advanced control strategies.
- C. Evaluate the economic benefits of real-time optimization.

IV. Validate the proposed control structure.
- A. Perform a final control degrees of freedom analysis. Check the allocation of the N_{FC} degrees of freedom.
- B. Check control of individual process units.
- C. Check the effect of constraints and disturbances on manipulated and controlled variables.
- D. Simulate control system performance for a wide range of conditions.

In this chapter we use box outlines to summarize the tasks in each step. A full case study and references to related work are provided to clarify the detailed procedures.

J.2.2 Top-Down Analysis

The top-down analysis identifies both the scope and complexity of a plantwide control design project and its control structure. (See Step II for an outline of individual tasks.) Among the conceptual issues considered at this point in the design are where to control the key production and quality measurements, how the overall plant might be divided into smaller subsystems (*decomposition*) to simplify control system design, and where variable coupling or constraint handling may justify, or even require, the use of multivariable control. For example, it is important to identify certain subsystems whose control system designs cannot be developed separately because the processes are so closely coupled, such as in heat integration.

Step II. Perform a top-down analysis.

A. Identify the process variables, control degrees of freedom, control structure, and options for decomposition.
1. Identify the potential controlled variables.
2. Determine how the CVs can be measured or inferred, and identify other process variables to be measured.
3. Select the potential manipulated variables.
4. Perform a preliminary control degrees of freedom analysis (compare the numbers of potential manipulated and controlled variables).
5. Identify the source and nature of the significant disturbances that must be mitigated.
6. Perform a structural analysis based on a steady-state model, select the final controlled and manipulated variables, and evaluate the possibilities for decomposition of the control problem.

B. Establish the overall control structure (in conceptual form).
1. Identify where the production rate of each product will be measured and controlled.
2. Identify how quality will be measured for each product, and how quality will be controlled.
3. Determine how each recycle loop throughput/composition will be controlled.
4. Specify how the constraints will be satisfied.
5. Determine how major disturbances will be handled.
6. Analyze the energy management scheme, and indicate conceptually how it will be controlled.

Note that the number of control degrees of freedom can be influenced by constraints imposed during the control system design process. Once the scope of the design problem has been determined and guidance is available to begin the control system design task, it is much easier to develop a preliminary (conceptual) control structure. An important goal at this level is to utilize structural analysis techniques (SVA, RGA) subject to the availability of a steady-state and/or a dynamic model of the plant. As part of the conceptual design of the plant, one should attempt to identify the most effective measured and manipulated variables and identify any highly decoupled or highly interacting process units that will need special attention.

Normally, careful consideration of the process design itself will indicate how the control system of an entire facility (for example, a refinery) might be decomposed to control systems for its individual sections—for example, the gas treatment section or the separations section. In addition, a top-down analysis generally will provide further clues as to how the overall control problem can be reduced to a set of smaller problems. From this discussion, it should be clear why recycle, heat integration, and constraint handling systems are best dealt with conceptually before decomposition decisions are made.

After completing the top-down design step, the designer should have an excellent overview of the plantwide control system design task in terms of subsystems of processes rather than as many single-unit control systems. For example, a train of distillation columns coupled via heat integration is probably best considered as a single subsystem for purposes of control. Of course, several single-loop controllers may have already been identified in the top-down sequence to deal with production rate and quality variables, and a structural analysis may point to certain specific pairings of the CVs and MVs that will be worth considering. However, a detailed design of the control system is properly treated in the bottom-up procedure considered next.

J.2.3 Bottom-Up Design

Once a conceptual control structure has been developed and the plant has been decomposed into subsystems, the control design procedure reverts to a traditional bottom-up approach. However, there are good reasons to treat the different control activities in a multilevel hierarchy, as shown in Fig. J.1. The first task in Step III is to identify the essential controllers, those that are absolutely required. The safety and regulatory levels in Fig. J.1 enable safe and stable operation of the plant. The advanced control functions are handled at Level 3 and keep the controlled variables close to their optimum set points through standard methods such as cascade, ratio, feedforward, and multivariable control. Level 4 in Fig. J.1 considers the real-time optimization of the process operations. The purpose of control at this level is to choose operating conditions that meet overall objectives in an economically optimum fashion.

Step III. Develop a bottom-up design.

A. Develop a strategy for regulatory control.
1. Specify how the control system will respond to unsafe or abnormal operating conditions and deal with constraints.
2. Identify control loops to regulate production rates and inventories.
3. Identify control loops that will mitigate major disturbances.

B. Examine the potential of applying advanced control strategies.
1. Evaluate the use of enhanced single-loop control strategies, including feedforward, ratio, cascade, and selective control schemes.
2. Employ MIMO control for highly interactive processes.

C. Evaluate the economic benefits of real-time optimization.

J.2.4 Validation of the Plantwide System Design

Finally, the design of the plantwide control system needs to be checked carefully and validated. At this point, a series of checks should be performed to ensure that the plantwide control structure is complete, is internally consistent, and functions appropriately, as shown in Step IV.

> **Step IV. Validate the proposed control structure.**
>
> **A.** Perform a final degrees of freedom analysis. Check the allocation of the N_{FC} degrees of freedom.
> **B.** Check control of individual process units.
> **C.** Check the effect of constraints and disturbances on manipulated and controlled variables.
> **D.** Simulate control system performance for a wide range of conditions.

After Steps I–IV are completed, a number of other tasks must be finished to complete the control system design. They include detailed specification and costing of instrumentation and control equipment, purchase, installation, and checkout. Then the control system must be evaluated during actual plant operation. The final step is to certify that the plant and control system meet safety, environmental, production, and quality requirements (the *commissioning* step).

J.3 CASE STUDY: THE REACTOR/FLASH UNIT PLANT

We now apply the principles from the previous two sections to a specific case study—a reactor/flash unit plant with recycle similar to the plants discussed in Section J.2 and by Robinson et al. (2001). The plant consists of a reactor, flash unit, and recycle (surge) tank as shown in Fig. J.2. The reactor produces a product C from two feed streams consisting of pure A and B, which contains a small amount of *D*. The reaction is

$$A + B \rightarrow C$$

A single-stage flash unit separates unreacted A and product C (liquid phase) from reactant B and an impurity D (vapor phase). A small portion of the vapor stream is purged to keep the composition of D from building up to a point where the reaction would be reduced significantly. Figure J.2 indicates that the recycle tank is intended to operate at a high enough pressure to recondense B and D for introduction back into the CSTR in the liquid phase; a condensor in the recycle line is used for this purpose. It is assumed that a compressor is not required.

The reactor is fitted with a cooling coil for temperature control. A heat exchanger (preheater) is provided to heat the feed stream to the flash unit to ensure that

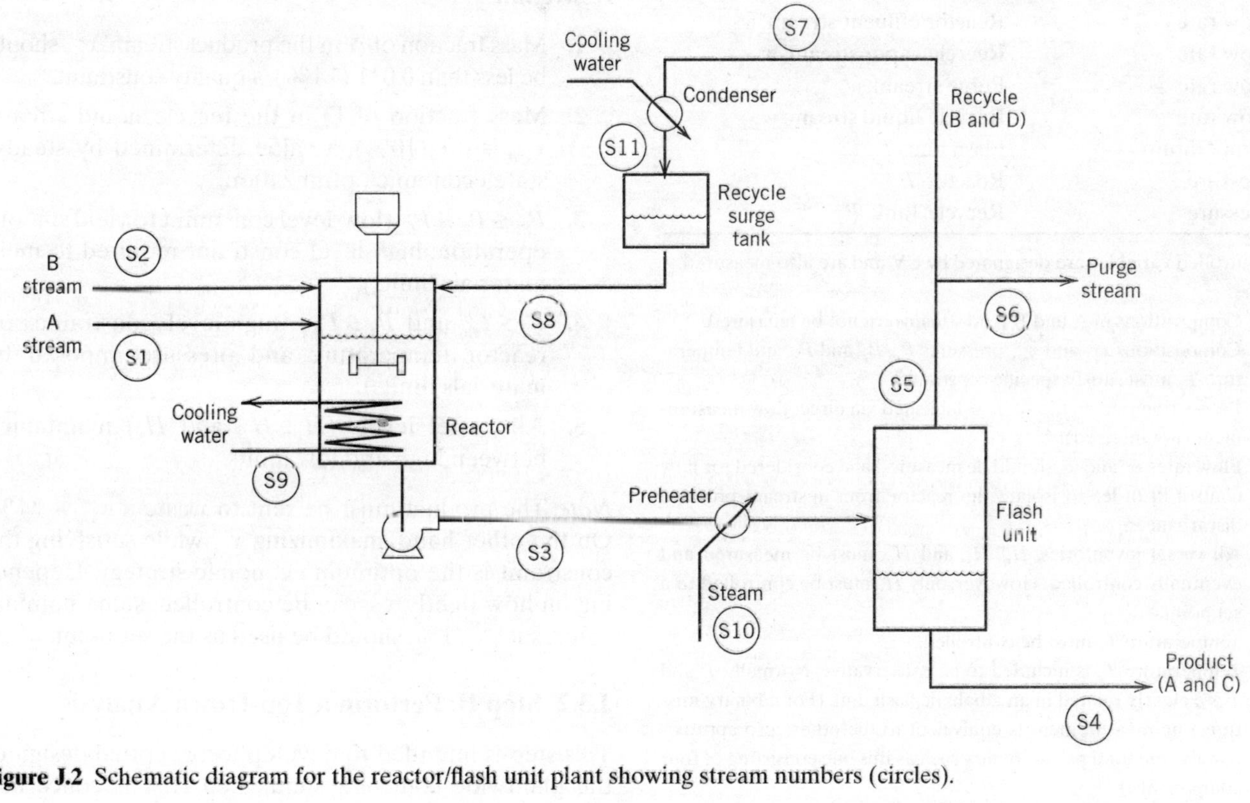

Figure J.2 Schematic diagram for the reactor/flash unit plant showing stream numbers (circles).

the feed enthalpy is sufficient to provide a complete separation of B and D (vapor) from A and C (liquid). Several dynamic models of the primary process units in this plant are presented in Appendix K.2. For simplicity, the flash unit is modeled as a splitter rather than by a more complex flash model.

We now discuss in detail each step in the design procedure presented in the previous section. Table J.2 lists the controlled and manipulated variables.

Table J.2 Potential Controlled and Measured Variables for the Reactor/Flash Unit Plant[a]

Controlled and Measured Variables	Location/Symbol
Composition, CV	Product stream, x_{4A}
Composition	Reactor effluent, x_{3A}
Composition, CV	Recycle stream, x_{8D}
Flow rate, CV	A feed stream to reactor, w_1
Flow rate, CV	B feed stream to reactor, w_2
Flow rate, CV	Product stream, w_4
Temperature, CV	Reactor, T_R
Temperature, CV	Flash unit feed stream, T_{FF}
Temperature, CV	Recycle tank (condenser exit temperature), T_C
Liquid level, CV	Reactor, H_R
Liquid level, CV	Flash unit, H_F
Liquid level, CV	Recycle tank, H_T
Pressure, CV	Flash unit, P_F
Flow rate	Reactor effluent stream, w_3
Flow rate	Recycle vapor stream, w_7
Flow rate	Purge stream, w_6
Flow rate	Recycle liquid stream, w_8
Temperature	Flash unit, T_F
Pressure	Reactor, P_R
Pressure	Recycle Tank, P_T

[a] Controlled variables are designated by CV and are also measured.

Notes:

i. Compositions in A and B feed streams cannot be measured.

ii. Compositions x_{4A} and x_{8D}, pressures P_R, P_F, and P_T, and temperature T_R must satisfy specific constraints.

iii. Production rate w_4 has to be established via direct flow measurement (not inferred).

iv. Flow rates w_1 and w_2 should be measured and considered for flow control in order to isolate the reactor from upstream pressure disturbances.

v. All vessel inventories, H_R, H_F, and H_T, must be measured and eventually controlled. However, only H_R must be controlled to a set point.

vi. Temperature T_R must be controlled.

vii. Temperature T_F is included to be conservative. Normally, P and T are closely related in an adiabatic flash unit. (For a binary mixture, one measurement is equivalent to the other; also approximately true for a pseudobinary such as this one consisting of four components.)

J.3.1 Step I: Specify the Control System Design Objectives

A. State the plant production and control objectives, including composition and production rates of all products plus economic objectives.

We assume that plant management and the design group have already developed product quality and production rate specifications, nominal operating conditions, and operating constraints for the plant. The control objectives are determined so as to meet customer requirements and anticipated sales figures, to reflect plant raw material and operating costs, and to satisfy materials of construction and environmental limitations:

1. The product should contain approximately 99% C; the remaining impurity is A.
2. The desired production rate w_4 to the downstream unit should meet the following specifications: Nominal value ± 1% on long-term basis (days); Nominal value ± 3% on short-term basis (hours).
3. The reactor should be operated with approximately constant conversion as production rate varies within expected limits. Because a suitable value of conversion will depend on the production rate, no specific requirement can be provided. The nominal reactor temperature T_R is specified.

B. Identify process constraints that must be satisfied, including safety, environmental, and quality restrictions.

1. Mass fraction of A in the product stream, x_{4A}, should be less than 0.011 (1.1%), a quality constraint.
2. Mass fraction of D in the recycle liquid stream, x_{8D}, is 0.1 (10%), a value determined by steady-state economic optimization.
3. $P_F^L \le P_F \le P_F^H$ (low-level constraint to yield smooth operation: high-level constraint required to meet materials limits).
4. $T_R \le T_R^H$ and $P_R \le P_R^H$ (high-level constraints on reactor temperature and pressure imposed by materials limits).
5. All vessel levels (H_R, H_F, and H_S) maintained between high and low limits.

Note: The product must be sent to waste if $x_{4A} > 1.1\%$. On the other hand, maximizing x_{4A} while satisfying the constraint is the optimum economic strategy. Depending on how tightly x_{4A} can be controlled, some nominal value, such as 1%, should be used as the set point.

J.3.2 Step II: Perform a Top-Down Analysis

This step is intended to develop a conceptual design of the plantwide control system. Step II.A is concerned primarily with analysis; in Step II.B, the overall control structure is established in a conceptual form.

A. Identify the process variables, control degrees of freedom, control structure, and options for decomposition.

A.1. Identify the potential controlled variables. The schematic flow diagram in Fig. J.3 shows the most important measurement locations and the process variables. The operating objectives clearly require that two key variables be controlled, x_{4A} and w_4. Composition x_{4A} has to be tightly controlled, but w_4 only needs to be tightly controlled on a long-term basis. Whether composition x_{3A} must be controlled explicitly is unclear at this point. However, we assume that it can be maintained approximately constant without using a specific composition controller. Furthermore, the composition of D in the recycle stream, x_{8D} (and consequently x_{8B}), also will likely need to be controlled. If x_{8D} is controlled, operation of the reactor/flash plant probably will not be unduly subject to disturbances (snowball effect). However, these issues need to be checked carefully at the end of the design using simulation.

Table J.2 lists the controlled variables for the plant developed using the specific arguments given above and the general guidelines given in Section 13.2. At this point, each process measurement could potentially be used as a control variable.

A.2. Determine how the controlled variables can be measured or inferred and identify other process variables to be measured. The schematic flow diagram in Fig. J.3 also shows the locations of the most important sensors/transmitters. In addition to measurements for the controlled variables, actual plants are routinely provided with many additional, but less important, measurements. Measurements such as cooling water inlet and outlet temperatures on the reactor cooling coil and heat exchanger steam supply pressure are required to give the operators a clear picture of the process behavior and its environment. Such information is particularly important during plant start-up, shutdown, and periods when the plant is upset.

The measured variables listed in Table J.2 are based on the problem description, the control system specifications (Step I.A), and the general guidelines of Section 13.2. Included are several potentially useful measurements that are not needed in the control system design.

A.3. Select the potential manipulated variables. Unless a stream is "wild" and cannot be manipulated (such as an exit stream from an upstream unit) or cannot be manipulated independently, its flow rate will be adjusted via a control valve. An example of the latter restriction would be a valve in Stream 5 (Fig. J.2), which cannot be used to manipulate flow rate independently if control valves are installed in both streams 6 and 7.

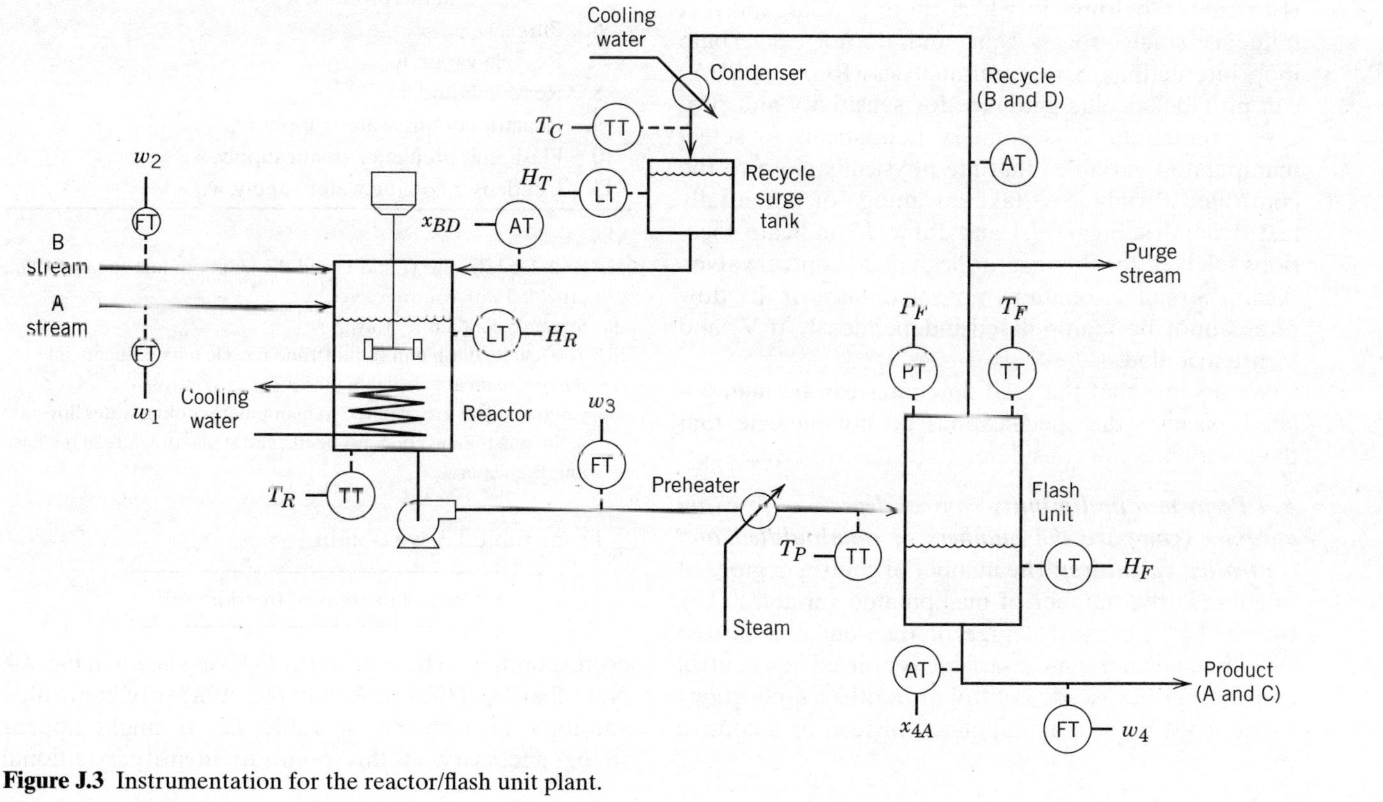

Figure J.3 Instrumentation for the reactor/flash unit plant.

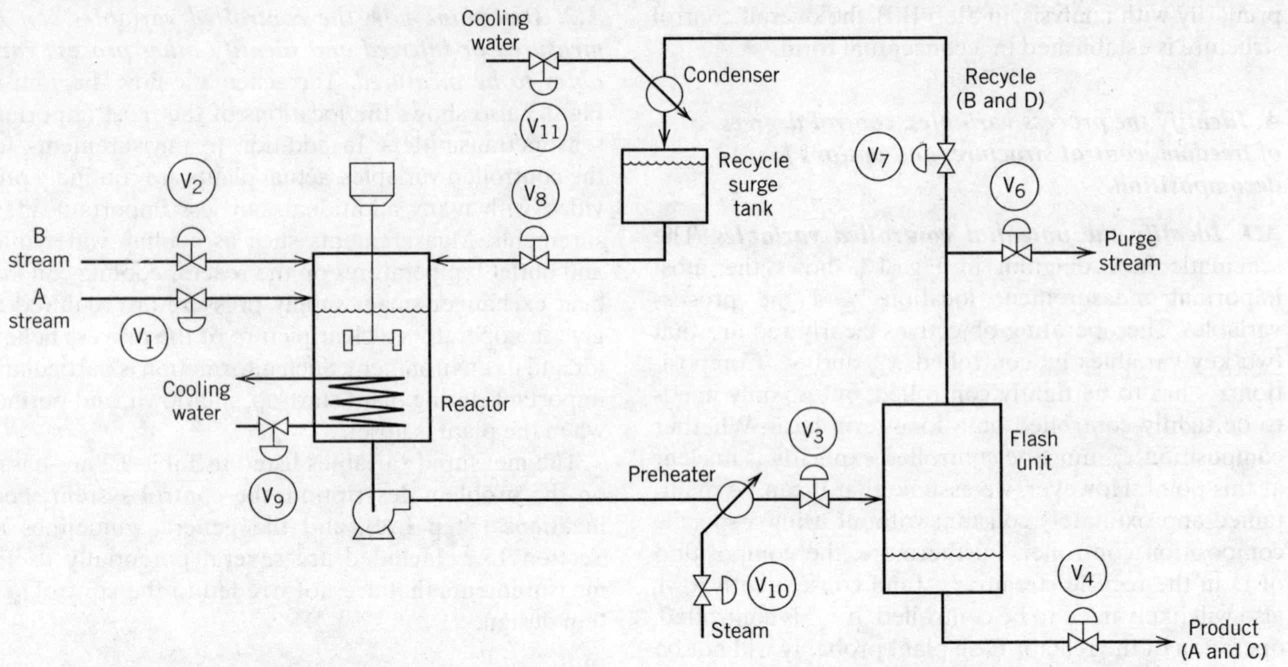

Figure J.4 Control valves for reactor/flash unit plant.

General guidelines for selecting manipulated variables are given in Section J.2. All of the manipulated variables in the case study are adjusted by control valves. In general, we try to select manipulated variables that have the most direct influence on the controlled variables—that is, largest sensitivity (gain) and fastest dynamic effects. The primary requirement is to enable pairings in which there is a large, direct influence (high process gain) and, do not exacerbate loop interactions. Structural analysis (RGA or SVA) can provide specific guidance for sensitivity and process interactions. A secondary heuristic is to select manipulated variables that are physically close to the controlled variables to take advantage of potentially fast dynamics. Figure J.4 and Table J.3 indicate locations selected for the reactor/flash plant control valves. Again, Stream 5 contains no valve, because its flow rate cannot be manipulated independently if V_6 and V_7 are installed.

We assume that the feed flow rates can be manipulated, because the specifications do not indicate that these variables are considered to be disturbances.

***A.4 Perform a preliminary control degrees of freedom analysis* (*compare the numbers of manipulated and controlled variables*).** The number of control degrees of freedom is the number of manipulated variables (10). Recall that a control degree of freedom is allocated each time a manipulated variable is utilized in a control loop, except in cascade control or in other applications where a set point is manipulated instead of a control valve.

Table J.3 Manipulated Variables (and Associated Valves) of the Reactor/Flash Unit Plant

Stream Number/MV	Valve
1. Reactor A feed, w_1	V_1
2. Reactor B feed, w_2	V_2
3. Reactor effluent, w_3	V_3
4. Flash unit liquid product, w_4	V_4
6. Purge, w_6	V_6
7. Recycle vapor, w_7	V_7
8. Recycle liquid, w_8	V_8
9. Reactor cooling water supply, w_9	V_9
10. Flash unit preheater steam supply, w_{10}	V_{10}
11. Condenser cooling water supply, w_{11}	V_{11}

Notes:

i. Both feed streams (1 and 2) and the plant product stream (4) are provided with control valves.

ii. Stream 5 contains no valve.

iii. The reactor, flash unit (liquid), and recycle tank effluents, and the purge stream are available for inventory control.

iv. Control valves are required to manipulate cooling water flow rate in the reactor and condenser coils, and steam pressure in the flash unit preheater.

From Table J.3, we obtain

Control degrees of freedom = 10

corresponding to the ten control valves shown in Fig. J.4. Note that N_{FC} (10) is less than the number of controlled variables (12) shown in Table J.2. It might appear to be necessary at this point to identify additional manipulated variables or to omit some of the controlled variables. However, if certain variables do not have to be independently controlled, it is possible to handle this situation by using advanced control methods (cascade control) or partial control (Kothare et al., 2000). This feature will be illustrated in the bottom-up design.

A.5. Identify the source and nature of the significant disturbances that must be mitigated. There are four primary sources of disturbances, three from within the plant itself or its immediate environment: composition variations in the feed streams and temperature or pressure variations in the cooling water and steam utility streams. The fourth disturbance is caused by planned changes in production rate:

1. x_D stream feed; random variation.
2. T_W (temperature of cooling water supply to reactor and to recycle condenser); diurnal (24-hour) cycle.
3. P_S (pressure of steam supply to flash unit preheater); relatively slowly varying supply pressure as other units load the steam supply header. We discuss in Step III.A.3 what to do in the event this disturbance turns out to be more difficult to handle.
4. Operator-implemented changes in desired production rate w_4.

A.6. Perform structural analysis based on a steady-state model and evaluate the possibilities for decomposition of the control problem. To simply this step, we assume that the pressure and temperature control loops are essentially decoupled from the plant holdups (integrating modes), the compositions, and the liquid flows. If this assumption is approximately valid, we can analyze a core plant model ("core model") that comprises the reactor, flash unit, and recycle tank–all assumed to operate isothermally and isobarically (see Fig. J.5). Thus, the approximate plant model consists only of material balances but includes the key flows, levels, and compositions. This type of approach, in which temperatures and pressures are assumed to remain constant at their

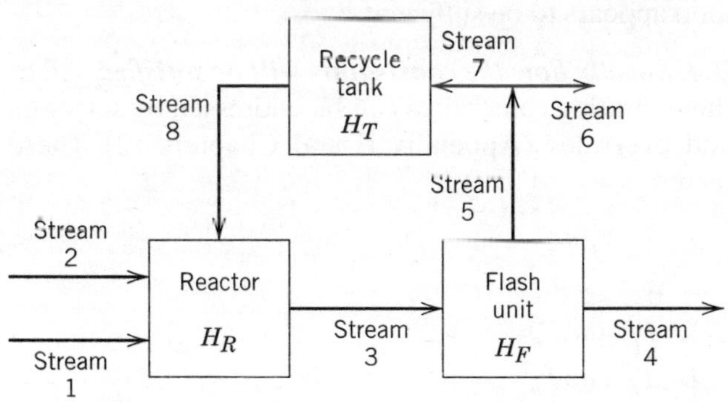

Figure J.5 Process flow diagram for the core model of the plant: the core model consists of reactor, flash unit, and recycle tank, all operated isothermally and isobarically.

nominal values, was employed by Robinson et al. (2001) in their analysis of a similar plant.

The resulting core model (see Appendix K.2) contains six controlled variables:

Production rate, w_4
Composition of A in the product stream, x_{4A}
Reactor holdup, H_R
Flash unit holdup, H_F
Recycle tank holdup, H_T
Composition of D in the recycle stream, x_{8D}

where the flow rates and holdups are in mass units and the compositions are mass fractions. Six manipulated variables (all flow rates established by control valves shown in parentheses) are available:

A feed flow rate w_1 (V_1)
B feed flow rate w_2 (V_2)
Reactor exit flow rate w_3 (V_3)
Flash unit liquid flow rate w_4 (V_4)
Purge flow rate w_6 (V_6)
Recycle flow rate w_8 (V_8)

At this point, one could develop a 6×6 RGA that would provide guidance on how the plant might be decomposed for multivariable control and how variables might be paired in a subsequent bottom-up (detailed) design. First, we recognize that the most direct way of controlling the plant production rate w_4 is to use V_4. However, making that choice leads to a problem discussed in Appendix I regarding the design of flow/level controllers for vessels in series. If V_4 is used to control w_4, then only V_3 can be used to control flash unit holdup H_F. Furthermore, there is no easy way to control the reactor holdup H_R, because use of any reactor inlet valve (V_1, V_2, or V_8) to adjust the reactor level can change the molar ratios of reactants. Of course, that problem could be mitigated by ratioing all three valves, but normally this approach is undesirable. Thus, we conclude that it is better to control H_R by V_3 and H_F by V_4, and to control the production rate in an indirect manner.

These preliminary decisions leave four controlled variables and four manipulated variables that can be analyzed using a 4×4 relative gain array. For the core plant model (mass balance equations only) and values of the operating parameters given in Appendix K, the steady-state gain matrix is

Steady-State Gain Matrix

$$
K = \begin{array}{c|cccc}
 & w_1 & w_2 & w_6 & w_8 \\
\hline
w_4 & 1.93 & 2.34\times10^{-2} & 0 & 6.29\times10^{-3} \\
x_{8D} & 8.46\times10^{-4} & -7.97\times10^{-4} & 0 & 5.72\times10^{-6} \\
x_{4A} & 2.51\times10^{-5} & -1.18\times10^{-5} & 0 & -3.17\times10^{-6} \\
H_T & -0.93 & 0.977 & -1 & -6.29\times10^{-3}
\end{array}
\quad (J.1)
$$

Note that w_6 only affects H_T. The elements in the H_T row (K_{4j}) consist of rate-of-change coefficients instead of gains, because it is an integrating variable. Woolverton (1980) and Arkun and Downs (1990) showed that, in order to calculate the RGA, the rate-of-change coefficients for an integrating variable can be treated just as if they were gains. Using their approach, we can obtain the RGA:

Relative Gain Array

$$\Lambda = \begin{array}{c|cccc} & w_1 & w_2 & w_6 & w_8 \\ \hline w_4 & 0.975 & 0.013 & 0 & 0.012 \\ x_{8D} & 0 & 0.974 & 0 & 0.026 \\ x_{4A} & 0.025 & 0.013 & 0 & 0.962 \\ H_T & 0 & 0 & 1 & 0 \end{array} \tag{J-2}$$

From the RGA, it is clear that the core plant model is not very interacting; however, it gives little insight into potential decomposition of the full plant. Thus, a control system design developed with a multiloop approach based on a simplified model should be tested using simulation, and eventually with the actual plant, to see how well the simplifying assumptions hold. Note that the RGA and similar analytical methods are intended to be used for initial screening.

Because the degree of interaction is low, there appears to be no compelling reason to employ a multivariable control methodology such as MPC in dealing with the core plant. Thus, it is possible to decompose down to the individual unit, and except for cascade and ratio control applications discussed below, even to the single-loop level.

Those potential pairings exhibiting relative gain elements approximately equal to one serve to guide the detailed bottom-up design that follows. For this simple plant model, with its straightforward reaction kinetics and separator modeled by a splitter rather than a flash model, enough information is already provided at this point to design the control system structure. However, we continue with application of the recommended design procedure to illustrate its application.

B. Establish the overall control structure in conceptual form.

B.1. Identify where the production rate of each product will be measured and controlled. The production rate w_4 is measured by means of a flow transmitter placed directly in the product line, rather than inferred from a measurement elsewhere in the plant, as sometimes is required. Because the desired variability of the production rate is small (±1%), measuring a related flow rate further upstream (e.g., w_3) could introduce too much variability if the flash unit level controller manipulates product stream flow rate w_4. However, with only two units in the downstream path of this plant (reactor and flash unit), we have already discussed why it is reasonable to manipulate the production rate at an upstream location using a variable that directly influences this flow rate. In principle, either w_1 or w_2 could be manipulated for this purpose, because both reactants are required to make product C. However, that is true only as long as A and B compositions in the reactor are near the stoichiometric ratio. We know that A is the limiting reactant. Thus, that is why the RGA indicates that only flow rate w_1 has a meaningful effect. Valve V_1 is allocated for this purpose.

Initially, we assume that flow rate w_1 will be maintained using a flow controller whose set point is adjusted manually to hold w_4 within the desired limits. However, what type of control loop to use or how its set point is to be adjusted is uncertain until we develop the detailed bottom-up design. Following the introduction of several additional considerations in the bottom-up design phase, these details can be developed.

> **Result:** Valve V_1 is allocated for control of production rate.
> Remaining control degrees of freedom = 10 − 1 = 9

B.2. Identify how quality will be measured for each product and how quality will be controlled. Composition x_{4A} is a key quality variable, because it is strictly limited to less than 1.1%. Because the RGA recommendations are unambiguous (Eq. J-2), the recycle stream valve V_8 (flow rate w_8) is chosen as the manipulated variable.

> **Result:** Valve V_8 is used to control x_A.
> Remaining control degrees of freedom = 9 − 1 = 8

Although one of the secondary control objectives is to keep the reactor exit composition x_{3A} reasonably constant, control of this intermediate variable does not appear difficult enough to require a separate feedback controller.

B.3. Determine how each recycle loop throughput/composition will be controlled. Because this plant does not appear to be sensitive to disturbances leading to effects such as snowballing, controlling x_{8D} in the recycle loop appears to be sufficient.

B.4. Specify how the constraints will be satisfied. All of the operating constraints can be addressed by selectors and overrides (Appendix E and Chapters 12). These include

1. x_{4A}
2. x_{8D}
3. P_F^H and P_F^L
4. T_R^H and P_R^H
5. High and low levels in all three vessels.

Note that Constraint 2 on x_{8D} has been specified by plant designers in advance. If x_{8D} should be changed in response to operating and economic conditions, it could be determined on-line via real-time optimization (Appendix F).

There is one implicit condition, not part of the design specifications, that needs to be considered in the bottom-up design. Because the flash unit separates a pseudobinary mixture (A/C and B/D), its temperature and pressure cannot be independently specified. Thus, either T_F or P_F, but not both, can be controlled.

B.5. Determine how major disturbances will be handled. The effect of temperature changes in the cooling water supply on reactor temperature or condenser exit temperature presumably can be attenuated by their temperature control loops without upsetting the composition and flow loops. The same is true for the effect of pressure changes in the steam supply header on flash unit temperature (or pressure). All three of these loops will have to be tested carefully after design is complete, through simulations and/or plant tests, to determine whether the assumptions underlying the decision to design the actual plant using a reduced model (the core model) are correct.

Variations in the remaining environmental disturbance x_{2D} require an explicit feedback controller to maintain x_{8D} near its desired value. The purge stream w_6 (the only path by which D leaves the plant) could be manipulated to control x_{8D}. However, RGA results clearly indicate that approach will only work if *all* control loops are closed (see Exercise J.4); but an effective manipulated variable (w_2) is available to deal with this disturbance without imposing such a severe restriction.

B.6. Analyze the energy management scheme and indicate conceptually how it will be controlled. A potential source of energy savings for this process appears to be the energy liberated in the exothermic reaction. That "lost energy" could provide some of the energy required to heat the flash unit feed to flash conditions. However, it is impossible to control the reactor temperature at a desired value and simultaneously heat up its exit stream, the flash unit feed stream, to a higher value. In other words, this energy is *thermodynamically unavailable* (Sandler, 2006).

The flash unit vapor condenser might supply some part of the flash preheater energy requirements, but recovering it with the small approach temperature differences that are available would require too large a heat exchanger. Thus, it appears that the potential for energy savings through heat integration is small, and plant heating/cooling utilities will have to be used.

J.3.3 Step III: Develop a Bottom-Up Design

A. Develop a strategy for regulatory control.

The primary tool for regulatory control is the SISO control loop. However, many applications benefit from the use of (and may even require) advanced methods such as selective control techniques and multivariable control (Chapters 12 and 14).

A.1. Specify how the control system will treat unsafe or abnormal operating conditions and deal with constraints. First identify those variables that potentially pose a safety or environmental hazard (cf. Chapter 13):

P_R *subject to high-pressure limit* P_R^H: Pressure-operated overrides (see Chapter 12) should be placed on the reactor level controller and on any flow controllers that feed or empty the reactor. To ensure that this hard constraint is never violated, a quick opening valve and rupture disk should be installed in a separate line leading to a flare or blow-down tank.

T_R *subject to high-temperature limit* T_R^H: A temperature-operated override should be placed on the reactor temperature controller and a fail-open valve used to manipulate the cooling water flow rate.

P_F *subject to high-pressure limit* P_F^H: Pressure-operated overrides should be placed on the flash unit level controller and on the recycle and purge stream controllers. Again, the contents should be released to a flare or blow-down tank if the hard constraint is reached.

Then identify those variables whose constraints might compromise satisfactory operation of the plant:

P_F *subject to a low limit* P_F^L: Place an override on the purge stream flow controller to make sure that the recycle tank is not upset by a low-flow situation in the recycle line.

H_R, H_F, H_T *subject to high- and low-level limits*: Place an override on the level controller for each vessel to open or close the pertinent control valve in the event that an "overflow" or "running dry" state is approached.

A.2. Identify control loops to regulate production rates and inventories. First, we specify control loops for the holdups in the major vessels. This step deals with the plant's integrating modes discussed in the top-down analysis (Step II.A.6). The RGA results (Eq. J-2) indicate that the recycle tank level is controlled best by manipulating the purge stream valve rather than the recycle stream valve:

Results: Reactor exit stream valve (V_3) is used to control H_R.
Liquid product stream valve (V_4) is used to control H_F.
Purge stream valve (V_6) is used to control H_T.

Remaining control degrees of freedom = 8 − 3 = 5

Note that *averaging control* can be used for H_F and H_T where tight level control is not required to smooth out the effect of disturbances, but not for H_R.

A.3 Identify control loops that will mitigate major disturbances. Variations in x_{2D} will produce deviations in x_{8D} from its desired nominal operating value. The RGA analysis has identified the B feed stream valve (V_2) as the most effective actuator.

Recall that plant designers included the purge stream in order to remove D from the plant. However, the RGA results in Eq. J-2 indicate that manipulating V_6 is not an effective way to control x_{8D}. Instead, V_2 is chosen, based on the RGA analysis:

Result: V_2 is used to control x_{8D}.

Remaining control degrees of freedom = 5 – 1 = 4

Upstream pressure variations in the two reactor feed streams (w_1 and w_2) can be attenuated by using a flow controller in each line. However, x_{8D} only needs to be controlled approximately at the desired value of 10%; thus, a flow controller for w_2 appears to be an unnecessary complication. Using a flow controller on w_1 implies that its set point will be adjusted to maintain production rate w_4. Note that an additional control degree of freedom is not required, because control valve V_1 was already allocated in the top-down analysis (Step II.B.1) to adjust production rate.

Result: Flow controller manipulates V_1 to control w_1. Set point of w_1 controller is adjusted to set production rate w_4.

Remaining control degrees of freedom are unchanged = 4 – 0 = 4

Because energy management is simple for this plant, disturbances in energy balances presumably can be handled by single-loop controllers. It is assumed that disturbances to reactor temperature (caused by a varying reaction rate or cooling water temperature changes) can be mitigated by using a reactor temperature control loop. The same is true for the effects of pressure changes at the steam supply header on flash unit pressure or temperature. Because there are explicit constraints on pressure but not on temperature, pressure is chosen. This item is considered more fully in Step IV.D, where the accommodation of constraints is discussed in detail.

Similarly, the effects of temperature variations in the cooling water supply on operation of the condenser can be mitigated by use of a temperature control loop. If disturbances are particularly large in a utility supply, a cascade secondary controller can be employed to control the temperature or pressure of the utility stream at the point it leaves the process, with a primary controller used to maintain the process temperature (Chapter 12). Cascade control, which is applied in Step III.B, is not used here for reasons of simplicity.

Result: V_9 is used to control T_R.
V_{10} is used to control flash unit feed temperature T_{FF}.
V_{11} is used to control condenser exit temperature T_C.

Remaining control degrees of freedom = 4 – 3 = 1

One major disturbance remains: the variation in P_F caused by changes in w_6. P_F can be controlled by manipulating valve V_7 in the recycle vapor line.

Result: V_7 is used to control P_F.

Remaining control degrees of freedom = 1 – 1 = 0

Note that some designers would choose to operate V_7 fully open and let P_F "float" in order to save pumping costs associated with the pump in the reactor effluent line. We assume here that pressure control is necessary to maintain flash unit pressure constant. Disturbance sensitivity is assumed not to be an issue for this plant, so any need to control a flow rate or composition variable within the recycle loop will be satisfied by controlling x_{8D}.

At this point in the analysis, there are no remaining control degrees of freedom. However, the design of the plantwide control system is by no means complete: advanced control methods that adjust the set points of already specified feedback controllers can be used to make the plant operate better. The objective here is to structure the control system in ways that avoid the need for operator intervention except when absolutely necessary.

B. Examine the potential of applying advanced control strategies.

Advanced control is intended to provide improved performance over traditional single-loop control. It includes such techniques as multivariable, cascade, feedforward (including ratio), and inferential control.

As noted already, overrides (Chapter 12) can be particularly helpful in dealing with variable constraints.

Cascaded flow controllers can reduce the effect of any upstream pressure variations or changes in control valve characteristics resulting from nonlinearities or from fouling, as noted in Step II.A.3. Ratio control between w_1 and w_2 can maintain the desired stoichiometric ratio of reactants approximately constant, despite changes in production rate or feed composition. Finally, cascade control can help deal with disturbances introduced by intentional changes in production rate w_4, as is discussed next.

B.1. Evaluate the use of advanced single-loop control strategies, including feedforward, ratio, cascade, and selective control schemes. In reviewing the plant processing and control objectives, a variable that needs further attention is the production rate w_4. One way of automating w_4 is to measure it and use a cascade controller to adjust the set point of the w_1 controller. An additional degree of freedom is not required to implement the cascade (master) controller, because the set point of the w_1 controller is available. A slow change in the w_{1sp} should meet the long-term production rate requirements and not interfere on a short-term basis with the H_F control loop.

> **Result:** A cascade controller for w_4 is employed to adjust the set point of the w_1 flow controller. Its set point, w_{4sp}, is used to set the desired production rate.
>
> Remaining control degrees of freedom = 0 (unchanged)

It is also desirable to speed up the adjustment of w_2 so that the ratio of B to A remains approximately at its correct (stoichiometric) value. A ratio controller (Section 12.2) whose internal ratio is adjusted by a primary composition controller is used to control x_{8D}.

> **Result:** Controller for x_{8D} adjusts the B to A ratio controller set point. The ratio controller manipulates V_2.
>
> Remaining control degrees of freedom = 0 (unchanged)

B.2. Employ multivariable control for highly interactive processes. So far we have assumed that a multiloop control approach will be sufficient and that multivariable control will not be necessary. One way to help ensure that this assumption will eventually be validated is to design the individual control loops so they interact as little as possible by careful selection of controlled variables and their pairing with manipulated variables. For example, adjusting the value of w_1 and the ratio of w_2/w_1 (in order to control w_4 and x_{8D}, respectively), instead of directly controlling the two flow rates individually, is one way of physically decoupling the two control loops (see Chapter 14).

Only a dynamic simulation of the controlled plant can determine whether the multiloop control strategy works satisfactorily. If the proposed loops interact too much or fail to achieve the desired control objectives, a more powerful multivariable approach such as MPC may be required.

C. Evaluate the economic benefit of real-time optimization.

The major process variables that are candidates for real-time optimization are T_R and H_R (to optimize operation of the reaction process), P_F (to optimize the separation process), and x_{8D} (here assumed to be constant at 10%). A steady-state process model must be available to carry out such calculations; see Appendix F for more details.

J.3.4 Step IV: Validate the Proposed Control Structure

A. Perform a final control degrees of freedom analysis. Check the allocation of the NFC degrees of freedom.

Because we have kept track of the control degrees of freedom as individual control loops have been proposed (control valves have been allocated), it is clear that we have not attempted to use too many degrees of freedom. Nor have any possibilities been neglected to obtain better control through the use of additional control loops that utilize already allocated degrees of freedom.

B. Check control of individual process units.

The next step is to make sure that no physically unrealizable control schemes have been proposed—for example, to attempt to control all of the component concentrations in a stream plus its total flow rate. Even with controllers in place, there must be some way for each species to leave the plant. For example, when the purge line is closed, there is no way for component D to leave the recycle path; thus, constraint handling methods associated with the control system cannot close V_6 for a significant period of time.

Finally, if steady-state simulation software is available, this is a good place to check anticipated concentrations and flows throughout the plant with the controllers implemented. Failure of the simulator to converge to the design operating conditions may be an indication that something is fundamentally wrong.

In designing the plantwide control system, we have essentially dealt with control of individual process units, with one exception. P_F must be maintained high enough to provide a sufficient pressure drop across V_7. If not, the recycle vapor stream valve may have to be operated fully open, and P_F controlled by adjusting V_{11}. If V_7 is always open, one control degree of freedom will be lost.

C. Check the effect of constraints and disturbances on manipulated and controlled variables.

All of the design constraints have been addressed by the proposed feedback control loops:

1. x_{4A} *constraint*: Manipulate w_1.
2. x_{8D} *constraint*: Manipulate w_2 by means of the w_2:w_1 ratio.
3. P_F^H *and* P_F^L: Manipulate w_7 with an override probably required on the flash unit feed temperature to handle the situation where V_7 either opens or closes fully.
4. T_R^H: Manipulate w_9.
5. T_R^H: Provide an override on w_6.

In the top-down design phase, we identified an additional, implicit constraint to be addressed that is not part of the design specifications. Specifically, the flash unit processes a pseudobinary mixture (A/C and B/D), and thus, its temperature and pressure cannot be specified independently. This design issue has been handled by controlling the flash unit feed temperature (flash unit preheater exit temperature) T_{FF} rather than T_F itself.

D. Simulate control system performance for a wide range of conditions.

If a dynamic process simulator is available, it should be used to evaluate the proposed plantwide control strategy and to determine recommended initial controller settings. It also should be used to evaluate the assumptions behind the core model analysis—namely, that the pressure and thermal control loops can be considered to be substantially decoupled from the flow/level/composition loops.

We present closed-loop simulation results for the core model (the reduced holdup form) and controller settings given in Appendix K.2, showing its responses to two important process disturbances. For simplicity, the reactor holdup (level) is assumed to be controlled perfectly. In addition, holdup in the flash unit is assumed to be zero, because it normally will be quite small compared to the reactor holdup. The base case control structure consists of 4 of the 12 control loops given in Table J.4: composition loop 7 ($x_{4A} - w_8$), modified composition loop 12 ($x_{8D} - w_2$—that is, manipulating V_2 directly), level loop 10 ($H_T - w_6$), and flow loop 11 ($w_4 - w_1$, primary loop only). Note that for the core model, flow rates are manipulated directly so the secondary controller for w_1 (flow loop 1) is not required.

Figure J.6 shows how three key controlled variables, w_4, x_{4A}, and x_{8D}, react to a +0.03 change in x_{2D}. In this simulation, the B-to-A ratio controller (ratio loop 2) was not implemented, because it does not affect the responses when there is no change in w_1. Note that all three controlled variables exhibit only small deviations from their set points as a result of the tight controller tuning that can be used in these loops. Recall that these controller pairings were chosen because of the direct influence of the manipulated variable in each loop and the relatively low degree of process interactions indicated by the RGA in Eq. J-2.

Figure J.7 shows the responses for x_{4A} and x_{8D} to a production rate change in w_4. A set-point change of +100 kg/h was made to the w_4 controller, first without ratio control of w_2. When ratio control is not used, larger deviations in x_{8D} occur as a result of the induced changes in A feed flow rate (w_1), with no corresponding immediate change in B feed flow rate w_2.

Table J.4 Proposed Control System Structure (Control Loops) for the Reactor/Flash Unit Plant

Loop Number	Controller Type	Controlled Variable	Manipulated Variable/Valve
1	Cascade (Secondary)	A stream flow rate, w_1	A feed stream, V_1
2	Ratio	B stream flow rate, w_2	B feed stream, V_2
3	Feedback	Reactor temperature, T_R	Cooling water, V_9
4	Feedback	Reactor level, H_R	Reactor effluent, V_3
5	Feedback	Flash unit feed temperature, T_{FF}	Steam supply, V_{10}
6	Feedback	Flash unit liquid level, H_F	Plant product, V_4
7	Feedback	A composition in product, x_{4A}	Recycle liquid stream V_8
8	Feedback	Flash unit pressure, P_F	Recycle vapor stream, V_7
9	Feedback	Condenser exit temperature, T_C	Cooling water, V_{11}
10	Feedback	Recycle surge tank level, H_T	Purge line, V_6
11	Cascade (Primary)	Plant production rate, w_4	Set point for w_4 (FC 1)
12	Feedback	D composition in recycle, x_{8D}	Ratio w_2:w_1 (RC 2)

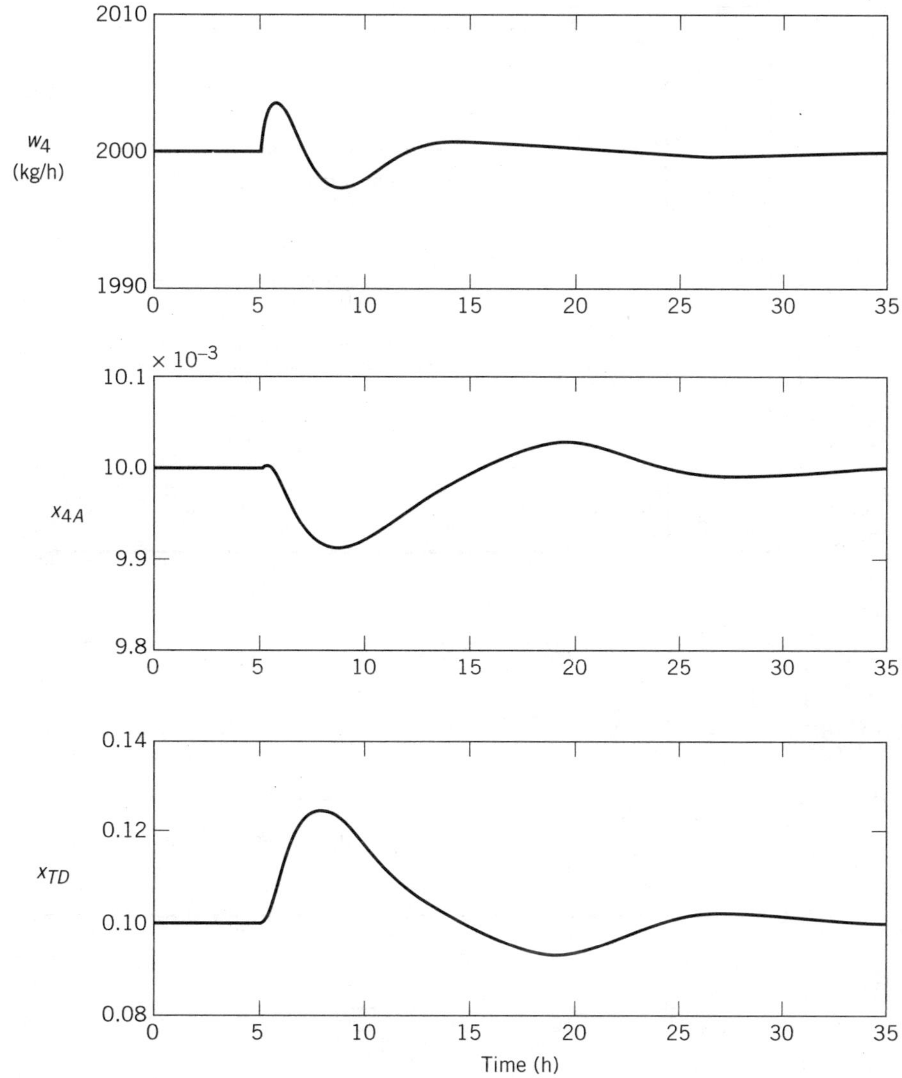

Figure J.6 Closed-loop responses of the reactor/flash unit core model: +0.03 change in x_{2D}, without ratio control. (Controller settings are in Table K.7.)

A form of ratio control was implemented in the second test by including ratio loop 2 $(w_2 - w_1)$ with an initial desired ratio of 1.09, while retaining the four base case controllers (loops, 7, 10, 11, and 12 in Table J.4). The addition of ratio control results in essentially no deviation in x_{8D}, and the beneficial effect of maintaining the B to A ratio during production rate changes is seen to affect only an unimportant recycle stream variable. Control of production rate w_4 and product quality x_{4A} is not significantly improved.

J.3.5 Summary and Interpretation of Control Structure for the Reactor/Flash Unit Plant

The proposed plantwide control system determined with the guidance of the 4 × 4 RGA-recommended pairings is summarized in Fig. J.8 and Table J.4. This case study represents one hypothetical plant and may give a misleading picture as to how the recommended control system design procedures lead to a particular structure. In general, design procedures are iterative, and thus they can lead to many alternative designs.

An example can be given of just how much the plantwide control system design changes if a slightly different set of assumptions is made. What if the short-term operating constraint on production rate was tighter than specified above (±1% instead of ±3%), or the plant involved more than just three process units? In such a situation, the decision to control the production rate by a cascade loop that extends back to the A feed stream flow rate (w_1) may not be practical. The intervening dynamic lags within the master loop might then

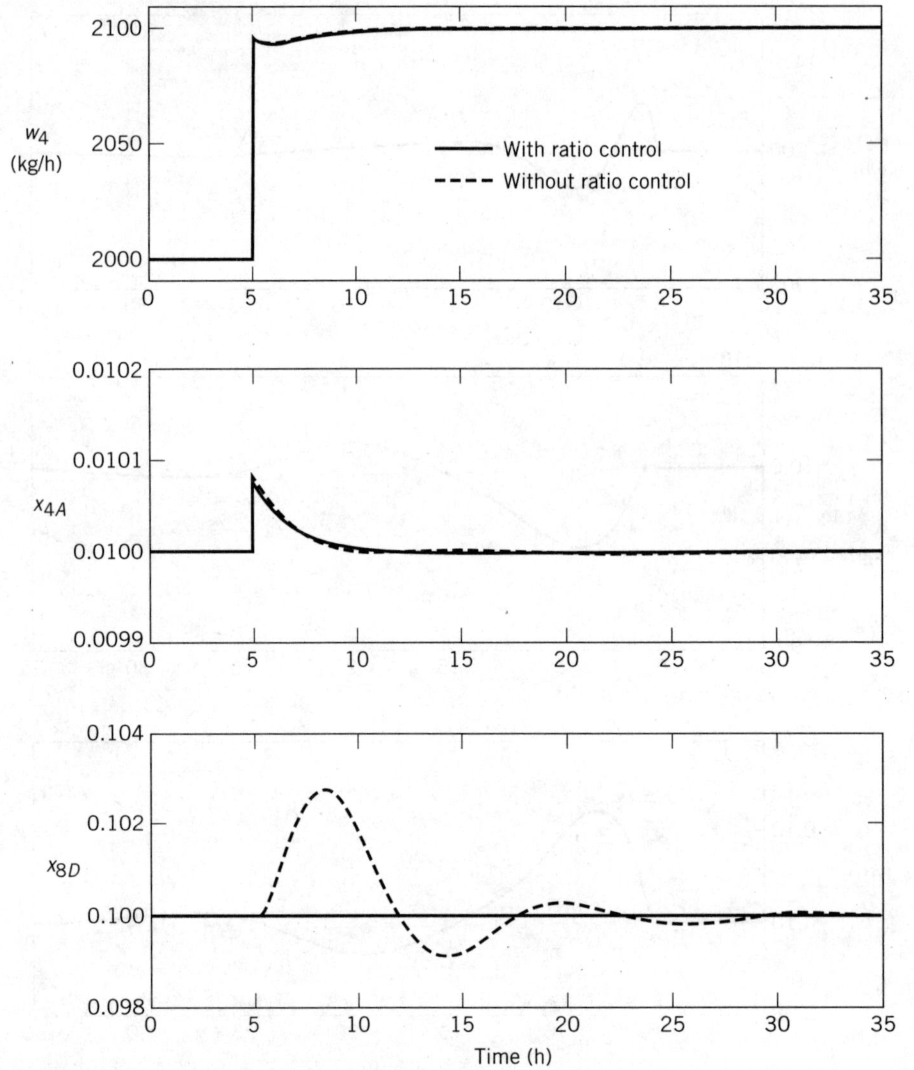

Figure J.7 Closed-loop responses of the reactor/flash unit core model: +100 kg/h step change in w_{4sp}, with and without ratio control. (Controller settings are in Table K.8.).

preclude its holding the required long-term tolerances. In this case, one alternative would be to control the production rate directly (via a flow controller on w_4) and to employ "upstream" control of H_R and H_F with the related complexities.

A control structure obtained using the hierarchical procedures in the previous section normally can be expected to work reasonably well. However, the only valid test of that conjecture is actually to perform simulations or plant tests after individual controllers have been tuned. In that way, one can determine just how well the controlled system deals with disturbances, production rate changes, and so on. For our purposes, we have focused initially on the core process units in the plant (reactor, flash unit, and recycle tank) to determine how well a design likely would work if it were developed using heuristics, strongly guided by simplified structural analysis. Other credible alternatives are possible. Which of the many alternatives are acceptable and which one is "best" in some sense can only be explored via simulation of the full plant model. An extended design including simulation of the full plant is left for the reader.

In making these comparisons, we developed the core model, a level/flow/composition model that neglects the effect of thermal (temperature) and pressure dynamics. For this plant, with only one recycle stream and no heat integration, the assumption is that the temperature and pressure control loops are largely isolated and noninteracting. This assumption has to be tested for accuracy via simulation.

A number of plant-scale control studies have been published. Luyben (2002) has presented a series of case

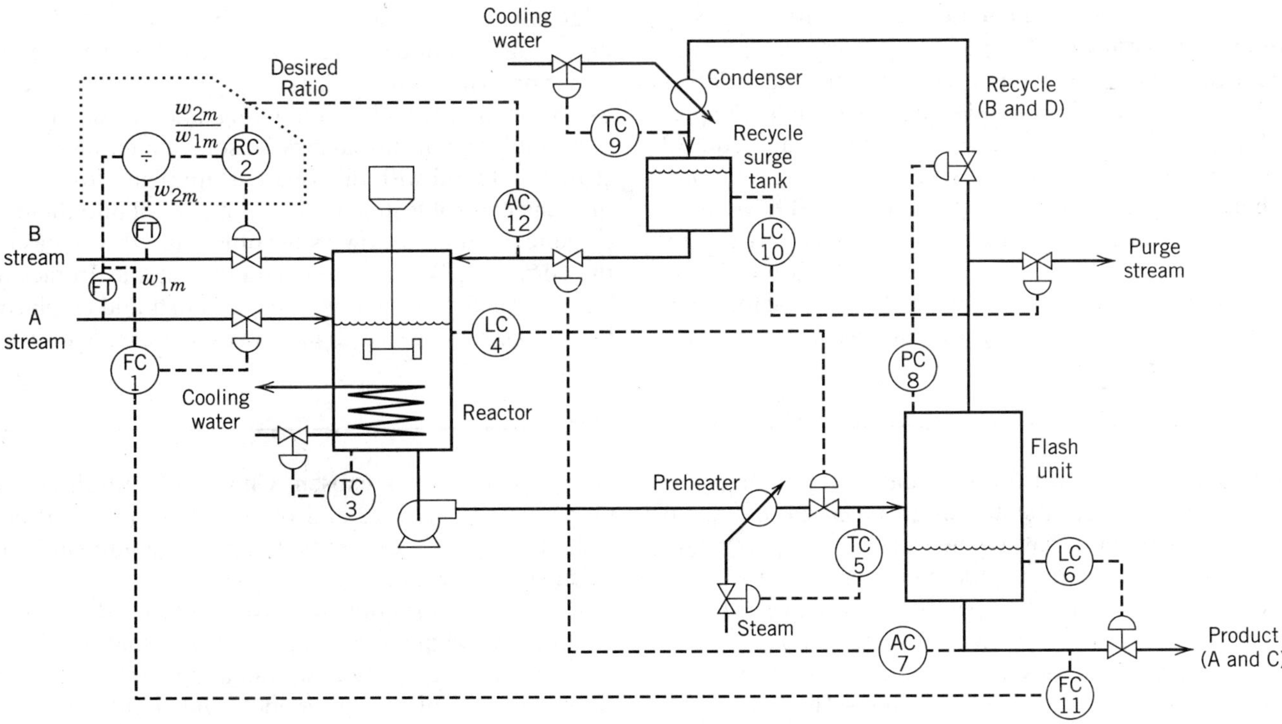

Figure J.8 Proposed decentralized (multiple-loop) control system specification for the full reactor/flash unit plant.

studies using dynamic simulation. Downs and Vogel (1993) documented a Tennessee Eastman Company challenge problem that has received considerable subsequent attention from control researchers. Larsson and Skogestad (2000) cite many efforts to deal with the Tennessee Eastman problem and support the conclusions that alternative control system designs, while sharing certain common features, are almost always highly idiosyncratic. Different plant control engineers or researchers will propose very different control system structures depending on the specific background they bring to the task, the specific design methodology employed, and the simplifying assumptions made.

J.4 EFFECT OF CONTROL STRUCTURE ON CLOSED-LOOP PERFORMANCE

In developing the control structure of the reactor/flash plant, an RGA analysis of the 4 × 4 core plant was used for guidance. In the final design steps of the case study, we introduced ratio control to maintain the feed stream flow rates at a B:A ratio, R, that is adjusted by the D-composition controller.

The RGA analysis was performed again, first replacing w_2 by R, the adjusted ratio. Linearizing the core model leads to slightly improved RGA values; in particular, the interaction measure relating w_4 to changes in w_1 becomes slightly higher (0.988 vs. 0.975). Improvement was seen in the dynamic responses of x_{8D}, whose oscillations were seen to be eliminated in Fig. J.7 with ratio control, compared to the responses without ratio control and to those in Fig. J.6.

Based on these results, it appears that the reactor may operate better if the ratio of combined B (feed plus recycle) to the A feed were controlled instead of just the feed stream ratio. To analyze this case, we let R^* be the ratio of combined B-stream flow rate ($w_2 + w_8$) to the A-stream flow rate (w_1). Recalculating the relative gain array yields a surprising result; two of the recommended control loops now become highly interacting (λ values of ~3.6) as given in Eq. J-3. This undesired result can be confirmed by simulation.

Relative Gain Array Using a "Combined B" Ratio

$$\Lambda = \begin{array}{c|cccc} & w_1 & R^* & w_6 & w_8 \\ \hline w_4 & 0.998 & 0.035 & 0 & -0.034 \\ x_{8D} & 0 & -2.66 & 0 & 3.66 \\ x_{4A} & 0.002 & 3.62 & 0 & -2.63 \\ H_T & 0 & 0 & 1 & 0 \end{array} \qquad \text{(J-3)}$$

Apparently, an important point has been overlooked—namely, that the recycle stream consists of the same flow of B + D that leaves the reactor. Because this material simply recirculates and does not participate in the reaction process under steady-state operating conditions, it should be ignored in applying feedforward control of the ratio. Often, a recombination of variables can lead to a less coupled system, as shown in Chapter 14 and, again, with the addition of simple ratio control between the B and A feed streams in the case study. That is not the situation for the combined B flow rate.

This type of control structure issue also can arise quite naturally through a particular physical feature of the steady-state plant design. Suppose that the reactor is not piped as shown before (Fig. J.2), with a separate recycle inlet port in the reactor, but with the two B streams piped together and entering the reactor through a common port. This design might be chosen to reduce reactor fabrication costs by eliminating an unneeded port, a potentially significant savings for some materials of construction. Faced with this steady-state plant design feature, the control system designer might fall into the trap of believing that a flow transmitter placed to measure the combined B flow rate would be suitable. In fact, as was shown above, the flow transmitter should be placed before the recycle stream inlet so as to measure only w_2.

SUMMARY

It is both exciting and yet intimidating for an engineer to be given a steady-state design proposal and some general ideas about how the new plant is to operate and then be asked to specify the complex system of controllers, safety interlocks, operator interfaces, hardware, and software that comprise a modern control system. In this chapter, we have presented a general procedure for designing plantwide control systems. The steps in the proposed hierarchical design procedure provided here are by no means either unique or complete. However, it is important to use an organized approach in the design of a plantwide control strategy, regardless of which design procedure is chosen. Choosing controlled and manipulated variables and pairing them in an ad hoc fashion without a coherent design procedure can lead to serious problems.

Throughout this book, we have emphasized that both qualitative and quantitative process information should be utilized in designing and evaluating control systems. Intelligent use of process models and simulation tools is required to develop a successful design. In addition, difficulties faced by the control system designer often can be mitigated or even eliminated by timely communication with the process design group.

REFERENCES

Arkun, Y., and J. J. Downs, A General Method to Calculate Input-Output Gains and the Relative Gain Array for Integrating Processes, *Comput. Chem. Engrg.*, **14**, 1101 (1990).

Buckley, P. S., *Techniques of Process Control*, John Wiley and Sons, Hoboken, NJ, 1964.

Downs, J. J., Distillation Control Design in a Plantwide Control Environment, in *Practical Distillation Column Control*, W. L. Luyben (Ed.), John Wiley and Sons, Hoboken, NJ, 1992, Chapter 20.

Downs, J. J., and E. F. Vogel, A Plantwide Industrial Process Control Problem, *Comput. Chem. Engrg.*, **17**, 245 (1993).

Kothare, M. V., R. Shinnar, I. Rinard, and M. Morari, On Defining the Partial Control Problem: Concepts and Examples, *AIChE J.*, **46**, 2456 (2000).

Larsson, T., and S. Skogestad, Plantwide Control: A Review and a New Design Procedure, *Model. Ident. and Control*, **21**, 209 (2000).

Luyben, W. L., *Plantwide Dynamic Simulators in Chemical Processing and Control*, Marcel Dekker, New York, 2002.

Luyben, W. L., B. D. Tyreus, and M. L. Luyben, *Plantwide Process Control*, McGraw-Hill, New York, 1998.

Robinson, D., R. Chen, T. J. McAvoy, and P. D. Schnelle, An Optimal Control Based Approach to Designing Plantwide Control System Architectures, *J. Process Control*, **11**, 223 (2001).

Sandler, S. I., *Chemical and Engineering Thermodynamics*, 4th ed., John Wiley and Sons, Hoboken, NJ, 2006.

Skogestad, S., Plantwide Control: Toward a Systematic Procedure, ESCAPE 12 Symposium, The Hague, Netherlands (June 2002).

Stephanopoulos, G., and C. Ng, Perspectives on the Synthesis of Plantwide Control Structures, *J. Process Control*, **10**, 97 (2000).

Woolverton, P. F., How to Use Relative Gain Analysis in Systems with Integrating Variables, *InTech*, **27**(9), 63 (1980).

EXERCISES

J.1 In Appendix K.2 three versions of the core model of the reactor/flash unit plant are developed. One is a "full-composition model" (Eqs. K-9 through K-31) that provides the relations needed to calculate every stream variable and every vessel holdup in the plant design. The second model (Eqs. K-33 through K-40) is a reduced-composition model, obtained from the full model by elimination of all variables and equations not needed to implement the control loops in this chapter. Thus only the necessary manipulated and disturbance variables, the dependent variables in the differential equations (predominantly reactor and recycle tank compositions), and the controlled (output) variables remain in the second model. The third model (Eqs. K-47 through K-55) is a reduced version of the original model equations in which component mass holdups have been used instead of vessel concentrations as the dependent variables.

Note that all the numerical and simulation results in this chapter were obtained using the third model.

(a) Provide a degrees of freedom analysis for each of the three models. Identify all variables, and list all of the required equations by number. Specify the parameters required for each model.

(b) What are the advantages and disadvantages of each model form?

(c) Implement the full concentration model and either of the reduced models using Simulink. Investigate the dynamic nature of the recycle plant and compare the responses of the two uncontrolled plants using changes in one or more disturbances.

J.2 The gain matrix in Eq. J-1 for the plant in this chapter was obtained using analytical methods (Mathematica: see www.mathematica.com) with the reduced holdup model. An alternative way to evaluate the plant interactions is to find the gain matrix and the relative gain array (Chapter 14) using a Simulink model by making small step changes in each input, and then determining the steady-state output changes in order to estimate the gains. Evaluate the relative gain array of this plant using the alternative approach along with a simulation of the full concentration model. Compare your results with the analytical results given in Eq. J-2.

J.3 Using the control loops in Table K.7 and K.8 with the full-composition model created in Exercise J.1, evaluate the use of ratio control in Loop 2 of Table J.4. In particular, indicate why ratio control should improve the plant performance for production rate (w_4 set point) changes, but not for disturbance (x_{2D}) changes. Your responses should be similar to those in Figures J.6 and J.7.

J.4 A purge stream often is included in recycle plants, such as the reactor/flash plant discussed in this chapter, to keep the concentration of a contaminant from building up within the plant. Thus, one might conclude that the best way to control the concentration of the contaminant below some acceptable level would be to manipulate the purge stream flow rate—i.e., w_6. Examine whether such an approach will work well here, using the following approach:

(a) Modify the Simulink program of the reactor/flash unit core model so that it includes an x_{8D}-w_6 loop and an H_T-R loop along with the remaining two loops summarized in Table I.8. Tune each of the new controllers so as to obtain a low level of interaction with the other control loops.

(b) What features in the RGA analysis given in Eq. J-2 or in the discussion in the first two paragraphs of Section J.4 explain the response results for your alternative control structure, when compared to those shown in Figure J.7?

(c) Can you give a physical explanation for the observed process gain between x_{8D} and w_6?

(d) How would you expect the plant to respond if one of the control loops was inoperable—e.g., as a result of a sensor failure? Remove the x_{4A}-w_8 control loop and compare the structure presented in this problem with the RGA suggested pairing for a +5% step change in production rate.

(e) How well do these two control structures (with all loops closed) handle a larger set-point change (10%) in the production rate, w_4? In particular, what modifications (if any) must be made to handle a set-point change of this magnitude? Can a 20% change in w_4 be accommodated? Why, or why not?

J.5 Using a Simulink representation of the reactor/flash plant, add an additional feedback control loop for reactor level H_R (assumed in this chapter to be perfectly controlled) by the following steps:

(a) Modify the gain matrix (Eq. J-1) to incorporate both a new manipulated variable w_3 and a new controlled variable H_R. What does the RGA indicate about the pairing of controlled and manipulated variables for this situation?

(b) Use RGA-recommended pairings, or any others that are appropriate, to control the plant (including H_R). Can you achieve essentially the same responses as shown in Figure J.7 while controlling reactor level with a PI controller? Using P-only control of level? Explain your results, and discuss whether it is important to control reactor level exactly at the set point.

J.6 If the flash unit in the example plant operates as an ideal splitter but with a non-negligible liquid holdup (e.g., 1000 kg), what would be the effect on the response of composition loop 12 in Table K.8 for a change in x_{2D}? On the response of composition loop 7? Simulate the modified plant and give logical arguments why one would or would not expect a difference.

J.7 Design an MPC controller for the reactor/flash unit plant and test it using a simulation of the linearized model of this plant. For purposes of this exercise, first design and implement a PI controller for the reactor level using the reactor outflow rate w_3 as manipulated variable (see Exercise J.5). Also, include a flash unit holdup of 500 kg and implement a PI controller for the liquid level with the flash unit outflow rate w_4 as manipulated variable (see Exercise J.6).

The following manipulated and controlled variables are to be used in the 4×4 MPC:

Manipulated Variables	Controlled Variables
A feed stream flow (w_1)	Production rate (w_4)
B feed stream flow (w_2)	Composition of A in the product stream (w_{4A})
Purge stream flow (w_6)	Recycle tank holdup (H_T)
Recycle line flow (w_8)	Composition of D in recycle stream (x_{8D})

Your design must meet the following control objectives:

(i) The product should contain approximately 99% C; the remaining impurity is A.

(ii) The desired production rate of product w_4 to the downstream unit should meet the following specifications: nominal value ±1% on long-term basis (days), nominal value ±3% on short-term basis (hours).

(iii) The reactor should be operated with approximately constant conversion (unspecified) as production rate varies within expected limits. The nominal reactor temperature T_R is fixed.

(iv) Important quality constraint: mass fraction of A in the product stream (x_{4A}) must be less than 0.011.

(v) For the purposes of this design, the only manipulated variable constraint is that flows are required to be positive.

Once the controller has been designed, evaluate its performance for the following set-point and disturbance sequences (each one separately):

(a) Disturbance response to a +0.03 change in x_{2D}.

(b) Set-point response to a 5% change in the production rate (w_4).

(c) Same as (a), but with the rate constant k increased by 20%.

(d) Set-point response to a 20% change in the production rate (w_4).

Hints: Use the MATLAB MPC Toolbox, if desired, for this exercise. Two commands are used to produce a linear model of the plant in the representation needed for controller design. First, the *dlinmod* command obtains a state-space representation (**A, B, C, D**). To use this command, be sure that the Simulink diagram is drawn so that the process manipulated inputs and disturbances correspond to "in ports" on the top level of the Simulink flow sheet; similarly, the outputs must correspond to "out ports." Then the *ss2mod* command produces a model in MPC *mod* format, specifying inputs that are manipulated variables, measured disturbances, and unmeasured disturbances. The *scmpc* command simulates control of the linearized plant with the MPC controller.

APPENDIX K

Dynamic Models and Parameters Used for Plantwide Control Chapters

APPENDIX CONTENTS

K.1 ENERGY BALANCE AND PARAMETERS FOR THE REACTOR/DISTILLATION COLUMN MODEL (APPENDIX I)

Energy Balance for the Reactor

Assumptions made in developing the balance:

1. Cooling temperature T_C is constant.
2. There is no vapor phase in the reactor.
3. There are no heat losses to the environment.
4. All streams and reactor contents have the constant heat capacity C_p.

$$\frac{dT_R}{dt} = \frac{1}{C_p H_R} \times \left[F_0 C_p (T_0 - T_R) + D C_p (T_D - T_R) - UA(T_R - T_C) - H_R \lambda k_R z\right] \tag{K-1}$$

Parameters in Table K.1 and controller settings in Table K.2 were used in the simulations shown in Figs. I9a and I.9b.

K.2 CORE REACTOR/FLASH UNIT MODEL AND PARAMETERS (APPENDIX J)

K.2.1 Simulation Models

In developing a simulation model that corresponds to the core elements of the reactor/flash unit plant, we first state the general modeling assumptions, then develop the model equations in several forms. The first model is based on the methods of Chapter 2. The second model is a simplified version. The third is developed specifically to use with a symbolic equation manipulator such as Mathematica for calculating the RGA. An evaluation of the three models, including comparison of their degrees of freedom, is considered in Exercise J.1. Figure J.5 illustrates the process units and stream numbering conventions.

Modeling Assumptions:

1. Mixing is perfect in each vessel.
2. Volume V_R is controlled by w_3 and assumed to be constant.
3. Temperature in each vessel is constant.

Table K.1 Parameters Used in the Energy Balance

Parameter	Symbol	Value	Units
Reactor feed rate	F_0	230	kmol/h
Feed temperature	T_0	21	°C
Distillate (recycle) rate	D	250	kmol/h
Distillate temperature	T_D	53	°C
Reactor holdup	H_R	1200	kmol
Reactor composition	z	0.5	mass fraction
Reactor temperature	T_R	69	°C
Cooling coil temperature	T_C	58	°C
Heat capacity	C_P	3.14	kJ/mol°C
Overall heat transfer coefficient	U	855	W/m²°C
Area for heat exchange	A	337	m²
Specific reaction rate	k_R	0.33	h⁻¹
Heat of reaction	λ	–69,780	kJ/mol

Parameters in Tables K.1 and K.2 reported by Wu and Yu (1996).

4. Volume V_F is controlled and assumed to be constant. It is small compared to Volume V_R.
5. Volume V_T varies to provide surge capacity between the flash unit and reactor.
6. Density ρ is constant.
7. Stream 1 is pure A ($x_{1A} = 1$).
8. Stream 2 is a mixture of B and D.
9. The holdup in the piping is negligible; materials move from vessel to vessel in zero time.

Material Balances:

Reactor:

Ordinarily, reaction rate is expressed in terms of reactant concentrations—for example, in units of kg-mol/m³:

$$r_c = k'\left[c_{R,A}\right]\left[c_{R,B}\right]$$

where subscript R denotes reactor. This expression can be converted to compositions expressed in mass fraction by rewriting the molar concentration terms as follows:

$$r_C = k'\left[\frac{\rho x_{R,A}}{MW_A}\right]\left[\frac{\rho x_{R,B}}{MW_B}\right] \tag{K-2}$$

where x denotes mass fraction; the mass density ρ, with units of kg/m³, is assumed to be constant; and MW_A and MW_B are the molecular weights for A and B, respectively.

$$r_C = \rho\left(\frac{\rho k'}{MW_A MW_B}\right)x_{R,A}x_{R,B} \tag{K-3}$$

$$= \rho\, k\, x_{R,A}x_{R,B} \tag{K-4}$$

where k has units of time⁻¹. Then, define a dimensionless mass ratio,

$$\alpha \triangleq \frac{MW_A}{MW_C} \tag{K-5}$$

Table K.2 Controller Settings for the Reactor/Distillation Column Closed-Loop Responses (Figures G.9a and G.9b)

Alternative	Control Loop	CV – MV	K_c	τ_I, min
1	Composition	$x_B - V$	-1.2×10^4 kmol/h	57.6
	Composition	$x_D - R$	3.41×10^3 kmol/h	76.7
	Level	$H_R - F$	-2.26 hr⁻¹	21.2
2	Composition	$x_B - V$	-8.5×10^3 kmol/h	76.7
	Composition	$x_D - R$	2.66×10^3 kmol/h	86.4
	Level	Not used		
4	Composition	$x_B - V$	-5.25×10^3 kmol/h	37.6
	Level	$H_R - D$	1.44×10^3 h⁻¹	45.7
	Composition	$x_D - H_{R,sp}$	6.1 kmol	

K_c normally is dimensionless. For simulation purposes, the controller is assumed to contain the gains of both the measurement transmitter and the control valve.

to represent the relationship between the masses of kg-mol of component A that reacts to yield a kgmol of C; thus $1-\alpha$ represents the mass of B used in the same reaction:

$$rA = -\alpha r_C \tag{K-6}$$

and

$$r_B = -(1-\alpha)r_C \tag{K-7}$$

where the dimensions are mass (of component i)/time.

The mass holdup in the reactor is controlled and constant. With constant ρ, the overall balance is

$$\rho\frac{dV_R}{dt} = w_1 + w_2 + w_8 - w_3 = 0 \tag{K-8}$$

yielding

$$w_3 = w_1 + w_2 + w_8 \tag{K-9}$$

The balance for each component in the reactor can be written as follows (recall subscript T denotes the recycle tank):

$$\frac{d(\rho V_R x_{R,A})}{dt} = w_1 - w_3 x_{R,A} - \rho V_R \alpha k x_{R,A} x_{R,B} \tag{K-10}$$

$$\frac{d(\rho V_R x_{R,B})}{dt} = w_2 x_{2B} + w_8 x_{T,B} - w_3 x_{R,B} - \rho V_R (1-\alpha) k x_{R,A} x_{R,B} \tag{K-11}$$

$$\frac{d(\rho V_R x_{R,C})}{dt} = -w_3 x_{R,C} + \rho V_R k x_{R,A} x_{R,B} \tag{K-12}$$

$$\frac{d(\rho V_R x_{R,D})}{dt} = w_2 x_{2D} + w_8 x_{T,D} - w_3 x_{R,D} \tag{K-13}$$

Because V_R is constant, only three of these balances are independent. We eliminate the C balance, Eq. K-12, which contains no inputs, using in its place:

$$x_{R,C} = 1 - (x_{R,A} + x_{R,B} + x_{R,D}) \tag{K-14}$$

Flash Unit (modeled as a splitter):

The flash unit liquid volume V_F is controlled, hence assumed to be constant. It also is negligible ($V_F \approx 0$) for modeling purposes. Therefore,

$$w_5 = w_3 - w_4 = w_1 + w_2 + w_8 - w_4 \tag{K-15}$$

In a hypothetical splitter, only A and C leave in the liquid phase while B and D leave as vapor. Thus a mass balance on the A and C components in the flash unit yields

$$w_4 = w_3(x_{R,A} + x_{R,C}) = (w_1 + w_2 + w_8)(x_{R,A} + x_{R,C}) \tag{K-16}$$

The exiting liquid-phase compositions (including critical quality measurement x_{4A}) are

$$x_{4A} = \frac{x_{R,A}}{x_{R,A} + x_{R,C}} \tag{K-17}$$

$$x_{4C} = \frac{x_{R,C}}{x_{R,A} + x_{R,C}} = 1 - x_{4A} \tag{K-18}$$

Similarly, the exiting vapor-phase flow rate and compositions are obtained from a mass balance on the vapor-phase components:

$$w_5 = w_3(x_{R,B} + x_{R,D}) = (w_1 + w_2 + w_8)(x_{R,B} + x_{R,D}) \tag{K-19}$$

$$x_{7B} = x_{6B} = x_{5B} = \frac{x_{R,B}}{x_{R,B} + x_{R,D}} = 1 - x_{7D} \tag{K-20}$$

$$x_{7D} = x_{6D} = x_{5D} = \frac{x_{R,D}}{x_{R,B} + x_{R,D}} \tag{K-21}$$

Note that substituting K-16 and K-19 into the overall mass balance around the flash unit (K-15) satisfies it exactly, as it must.

Purge stream node mass balance:

$$w_7 = (w_1 + w_2 + w_8)(x_{R,B} + x_{R,D}) - w_6 \tag{K-22}$$

Recycle tank:

The overall mass balance is given by

$$\rho\frac{dV_T}{dt} = w_7 - w_8 \tag{K-23}$$

and the component balances by

$$\rho\frac{d(V_T x_{T,B})}{dt} = w_7 x_{7B} - w_8 x_{T,B} \tag{K-24}$$

$$\rho\frac{d(V_T x_{T,D})}{dt} = w_7 x_{7D} - w_8 x_{T,D} \tag{K-25}$$

Only two of these relations are independent. Because we are primarily interested in the D component in the recycle, we choose Eqs. K-23 and K-25. Using the methods of Chapter 2, the derivative in Eq. K-23 can be expanded as

$$\rho\frac{dV_T}{dt}x_{T,D} + \rho V_T\frac{dx_{T,D}}{dt} = w_7 x_{7D} - w_8 x_{T,D} \tag{K-26}$$

$$(w_7 - w_8)x_{T,D} + \rho V_T\frac{dx_{T,D}}{dt} = w_7 x_{7D} - w_8 x_{T,D} \tag{K-27}$$

so that

$$\rho V_T\frac{dx_{T,D}}{dt} = w_7(x_{7D} - x_{T,D}) \tag{K-28}$$

The D composition of Stream 8 is provided by the output relation

$$x_{8D} = x_{T,D} \tag{K-29}$$

and the holdup of the recycle tank by

$$H_T = \rho V_T \tag{K-30}$$

Feed stream concentration relations:

Stream 1 is pure A. Therefore,

$$x_{1A} = 1 \quad \text{(K-31)}$$

Because Stream 2 consists only of reactant B and contaminant D,

$$x_{2B} = 1 - x_{2D} \quad \text{(K-32)}$$

K.2.2 Control Degrees of Freedom (CDOF) Analysis

Additional Assumptions for Control Structure Analysis:

10. V_R is assumed to be controlled by V3, so V_R and w_3 are removed from further consideration in control structure analysis.
11. V_F is assumed to be controlled by V4, so V_F and w_4 also can be eliminated from further consideration in control structure analysis.
12. Manipulated inputs are w_1, w_2, w_6, and w_8.
13. Controlled outputs are w_4, x_{8D}, x_{4A}, and H_T.
14. Primary disturbance variable is x_{2D}.

$$N_{CF} = 4 \text{ (maximum number of independently controlled variables)} \quad \text{(K-33)}$$

This formulation of the model leads to the following set of CVs and MVs used in Appendix J.

Controlled Variables	Manipulated Variables
w_4	w_1
x_{4A}	w_2
x_{8D}	w_6
H_T	w_8

K.2.3 The Simplified "Composition Model"

By eliminating extraneous variables, the model can be simplified to a set of equations containing only the desired four input and four output variables used in the control simulations of Appendix H, yielding the following:

$$\frac{dx_{R,A}}{dt} = \frac{1}{\rho V_R}[w_1 - (w_1 + w_2 + w_8)x_{R,A} - \rho V_R \alpha k x_{R,A} x_{R,B}] \quad \text{(K-34)}$$

$$\frac{dx_{R,B}}{dt} = \frac{1}{\rho V_R}[w_2(1 - x_{2D}) + w_8(1 - x_{T,D}) - (w_1 + w_2 + w_8)x_{R,B} - \rho V_R(1-\alpha)k x_{R,A} x_{R,B}] \quad \text{(K-35)}$$

$$\frac{dx_{R,D}}{dt} = \frac{1}{\rho V_R}[w_2 x_{2D} + w_8 x_{T,D} - (w_1 + w_2 + w_8)x_{R,D}] \quad \text{(K-36)}$$

$$\frac{dV_T}{dt} = \frac{1}{\rho}[(w_1 + w_2 + w_8)(x_{R,B} + x_{R,D}) - w_6 - w_8] \quad \text{(K-37)}$$

$$\frac{dx_{T,D}}{dt} = \left[\frac{(w_1 + w_2 + w_8)(x_{R,B} + x_{R,D}) - w_6}{\rho V_T}\right] \times \left[\frac{x_{R,D}}{x_{R,B} + x_{R,D}} - x_{T,D}\right] \quad \text{(K-38)}$$

$$w_4 = (w_1 + w_2 + w_8)(1 - x_{R,B} - x_{R,D}) \quad \text{(K-39)}$$

$$x_{4A} = \frac{x_{R,A}}{1 - x_{R,B} - x_{R,D}} \quad \text{(K-40)}$$

$$x_{8D} = x_{T,D} \quad \text{(K-41)}$$

$$H_T = \rho V_T \quad \text{(K-42)}$$

K.2.4 The Simplified "Holdup Model" Used for Analysis and Simulation

In developing the component balances, there are alternative ways of defining the state variables—i.e., the dependent variables in the accumulation terms (Chapter 4). Another alternative is to use the species mass holdups in each vessel as the state variables and then to find the compositions from the mass holdups, as in Equations K-43–K-48, below. The compositions in the reactor are given by

$$x_{R,A} = \frac{H_{R,A}}{H_R} \quad \text{(K-43)}$$

$$x_{R,B} = \frac{H_{R,B}}{H_R} \quad \text{(K-44)}$$

$$x_{R,C} = \frac{H_{R,C}}{H_R} \quad \text{(K-45)}$$

$$x_{R,D} = \frac{H_{R,D}}{H_R} \quad \text{(K-46)}$$

and, in the recycle tank, by

$$x_{T,B} = \frac{H_{T,B}}{H_T} \quad \text{(K-47)}$$

$$x_{T,D} = \frac{H_{T,D}}{H_T} \quad \text{(K-48)}$$

Making this change of variables in the previous model equations prior to expanding K-24 and K-25, we obtain

$$\frac{dH_{R,A}}{dt} = w_1 - (w_1 + w_2 + w_8)\frac{H_{R,A}}{H_R} - \frac{\alpha k}{H_R} H_{R,A} H_{R,B} \tag{K-49}$$

$$\frac{dH_{R,B}}{dt} = w_2(1 - x_{2D}) + w_8 \frac{H_{T,B}}{H_{T,B} + H_{T,D}} - (w_1 + w_2 + w_8)\frac{H_{R,B}}{H_R} - \frac{(1-\alpha)k}{H_R} H_{R,A} H_{R,B} \tag{K-50}$$

$$\frac{dH_{R,D}}{dt} = w_2 x_{2D} + w_8 \frac{H_{T,D}}{H_{T,B} + H_{T,D}} - (w_1 + w_2 + w_8)\frac{H_{R,D}}{H_R} \tag{K-51}$$

$$\frac{dH_{T,B}}{dt} = (w_1 + w_2 + w_8)\frac{H_{R,B}}{H_R} - w_6 \frac{H_{R,B}}{H_{R,B} + H_{R,D}} - w_8 \frac{H_{T,B}}{H_{T,B} + H_{T,D}} \tag{K-52}$$

$$\frac{dH_{T,D}}{dt} = (w_1 + w_2 + w_8)\frac{H_{R,D}}{H_R} - w_6 \frac{H_{R,D}}{H_{R,B} + H_{R,D}} - w_8 \frac{H_{T,D}}{H_{T,B} + H_{T,D}} \tag{K-53}$$

with the four outputs that remain to be controlled given by

$$w_4 = (w_1 + w_2 + w_8)\left(\frac{H_R - H_{R,B} - H_{R,D}}{H_R}\right) \tag{K-54}$$

$$x_{4A} = \frac{H_{R,A}}{H_R - H_{R,B} - H_{R,D}} \tag{K-55}$$

$$x_{0D} = \frac{H_{T,D}}{H_t} \tag{K-56}$$

$$H_t = H_{T,B} + H_{T,D} \tag{K-57}$$

Note that the holdup model, while complicated slightly by the quotient terms, has an advantage in that Eq. K-53 is symmetric with K-52 and thus is easier to manipulate symbolically than with Eqs. K-37 and, particularly, K-38. The model also has one less parameter.

K.2.5 Plant Parameters and Steady-State Operating Values

Table K.3 gives the parameters used in this case study. In addition, values of the input variables (both manipulated and disturbance variables) are specified in Table K.4.

Table K.3 Parameters of the Core Reactor/Flash Unit Model

Parameter	Value
H_R (kg)	3000
k (h^{-1})	330
α	0.5
ρ (kg/m^3)	1000

Table K.4 Initial Values of Inputs for the Core Reactor/Flash Unit Model

Parameter	Value
w_1 (kg/h)	1010
w_2 (kg/h)	1100
w_6 (kg/h)	110
w_8 (kg/h)	890
x_{2D}	0.01

Table K.5 Steady-State Values of the Core Reactor/Flash Unit Model

Variable	Steady-State Value
$H_{R,A}$ (kg)	20
$H_{R,B}$ (kg)	900
$H_{R,D}$ (kg)	100
$H_{T,B}$ (kg)	450
$H_{T,D}$ (kg)	50
w_4 (kg/h)	2000
x_{4A}	0.01
x_{8D}	0.10
H_T (kg)	500

These parameters and constant model inputs result in the steady-state in Table K.5. Note that the first five variables listed are the state variables in the mass holdup formulation of the model; the last four are the output variables:

The resulting steady-state flow rates are then found as shown in Table K.6.

K.2.6 RGA Analysis

The model equations can now be linearized about the nominal steady state. The gain matrix (Eq. J-1) is then obtained, and the RGA (Eq. J-2) is calculated. Note that in linearizing these relations, the steady state is defined by the values of the steady-state holdups. Therefore, the gain matrix and the RGA may differ slightly from those obtained using the previous model in terms of the recycle tank total holdup and the process concentrations.

Table K.6 Steady-State Flow Rates of Stream Variables

	Stream Number, S							
Variable	1	2	3	4	5	6	7	8
w_{SA}	1010	0	20	20	0	0	0	0
w_{SB}	0	1089	900	0	900	99	801	801
w_{SC}	0	0	1980	1980	0	0	0	0
w_{SD}	0	11	100	0	100	11	89	89
w_S	1010	1100	3000	2000	1000	110	890	890

Table K.7 Controller Settings for Figure J-6

Control Loop	CV – MV	Value of K_c	Dimensions of K_c
Composition Loop 7	$x_{4A} - w_8$	-1.6×10^6	kg/h
Level Loop 10	$H_T - W_6$	-5.0	h^{-1}
Flow Loop 11	$w_4 - w_1$	2.5	dimensionless
Primary controller only			
Composition Loop 12			
Modified to manipulate w_2 (V2) directly	$x_{8D} - w_2$	-6.3×10^3	kg/h

Table K.8 Controller Settings for Figure J.7

Control Loop	CV – MV	Value of K_c	Dimensions of K_c
Composition Loop 7	$x_{4A} - w_8$	-1.6×10^6	kg/h
Level Loop 10	$H_T - w_6$	-5.0	h^{-1}
Flow Loop 11	$w_4 - w_1$	2.5	dimensionless
Primary controller only			
Composition Loop 12	$x_{8D} - R$	-6.3	dimensionless
Manipulate w_2:w_1 ratio			

Notes:

1. All controller gains include measurement/transmitter and valve gains. The values were obtained by sequential trial- and-error tuning to obtain minimum-overshoot responses for step changes in set point.
2. All controllers are proportional-integral with $\tau_I = 1$ h.
3. The initial value of R (w_2:w_1 ratio) is 1.09.

K.2.7 Closed-Loop Control

Two sets of closed-loop simulations were made using the holdup version of the core reactor/flash unit model; the results are plotted in Figures J.6 and J.7. For each case, the four control loops listed in Table J.4 were implemented as shown in Tables K.7 and K.8, respectively.

REFERENCE

Wu, K.-L., and C. C. Yu, Reactor/Separator with Recycle-1. Candidate Control Structure for Operability, *Computers Chem. Engng.*, **20**, 1291 (1996).

APPENDIX L

Additional Closed-Loop Frequency Response Material

APPENDIX CONTENTS

Frequency response concepts and techniques play an important role in control system design and analysis. In particular, they are very useful for stability analysis, control system design, and robustness analysis. Historically, frequency response techniques provided the conceptual framework for early control theory and important applications in the field of communications (MacFarlane, 1979).

This chapter begins by presenting useful background information in Section L.1. The Bode and Nyquist stability criteria in Sections L.2 and L.3 are generally applicable and provide exact results for systems with time delays. These stability criteria also provide measures of *relative stability*, how close a system is to becoming unstable.

Two useful metrics for relative stability, the *gain* and *phase margins*, are introduced in Section L.4. The frequency response of closed-loop systems is considered in Section L.5, followed by an introduction to *robustness analysis* in Section L.6. This last topic addresses the important question of the sensitivity of a control system to process variations and to uncertainty in the process model used to design the control system.

L.1 CLOSED-LOOP BEHAVIOR

In Chapter 11, we observed that control system design involves tradeoffs between conflicting objectives such as performance and robustness. Now we consider these design objectives in more detail and introduce another issue, the effect of measurement noise. In general, a feedback control system should provide the following desirable characteristics (see Chapter 11):

1. Closed-loop stability
2. Good disturbance rejection (without excessive control action)
3. Fast set-point tracking (without excessive control action)
4. A satisfactory degree of robustness to process variations and model uncertainty
5. Low sensitivity to measurement noise

The block diagram of a general feedback control system is shown in Fig. L.1. It contains three external input signals: set point Y_{SP}, disturbance D, and additive measurement noise, N. The noisy output Y_n is the sum of the noise N and the noise-free output Y. The following analysis illustrates the fundamental limitations and engineering tradeoffs that are inherent in achieving these characteristics.

The dynamic behavior of the closed-loop system in Fig. L.1 can be described by the following set of equations where G is defined as $G \triangleq G_v G_p G_m$:

$$Y = \frac{G_d}{1+G_cG}D - \frac{G_cG}{1+G_cG}N + \frac{K_mG_cG_vG_p}{1+G_cG}Y_{sp} \quad \text{(L-1)}$$

$$E = -\frac{G_dG_m}{1+G_cG}D - \frac{G_m}{1+G_cG}N + \frac{K_m}{1+G_cG}Y_{sp} \quad \text{(L-2)}$$

$$U = -\frac{G_dG_mG_cG_v}{1+G_cG}D - \frac{G_mG_cG_v}{1+G_cG}N + \frac{K_mG_cG_v}{1+G_cG}Y_{sp} \quad \text{(L-3)}$$

These equations can be derived easily using the block diagram algebra of Chapter 9. They illustrate how the three external inputs (D, N, and Y_{sp}) affect three output variables: the actual output Y, the error E, and the controller output U. The nine transfer functions in Eqs. L-1 to L-3 completely characterize the closed-loop performance of the control system. Because each transfer function has the same denominator, there is a single characteristic equation, $1 + G_cG = 0$. Consequently, all nine closed-loop transfer functions have identical stability characteristics.

We consider two examples to illustrate the insight provided by these equations. Example L.1 demonstrates that the control system design should never include the cancellation of a pole by a zero when both are located in the unstable region (i.e., to the right of the imaginary axis). This undesirable situation is referred to as an *unstable pole-zero cancellation*.

EXAMPLE L.1

Consider the feedback system in Fig. L.1 and the following transfer functions:

$$G_p = G_d = \frac{-0.5}{1-2s}, G_v = G_m = 1$$

Suppose that controller G_c is designed to cancel the unstable pole in G_p, as in the IMC design method:

$$G_c = \frac{-3(1-2s)}{s+1}$$

Evaluate closed-loop stability and characterize the output response for a sustained disturbance.

Solution

The characteristic equation, $1 + G_cG = 0$, becomes

$$1 + \frac{3(1-2s)}{s+1}\frac{0.5}{1-2s} = 0$$

or

$$s + 2.5 = 0$$

In view of the single root at $s = -2.5$, it appears that the closed-loop system is stable. However, if we consider Eq. K-1 for $N = Y_{sp} = 0$,

$$Y = \frac{G_d}{1+G_cG}D = \frac{-0.5(s+1)}{(1-2s)(s+2.5)}D$$

This transfer function has an unstable pole at $s = +0.5$. Thus, the output response to a disturbance is unstable. Furthermore, other transfer functions in Eqs. L-1 to L-3 also have unstable poles. This apparent contradiction occurs because the characteristic equation does not include all of the information, namely, the unstable pole-zero cancellation.

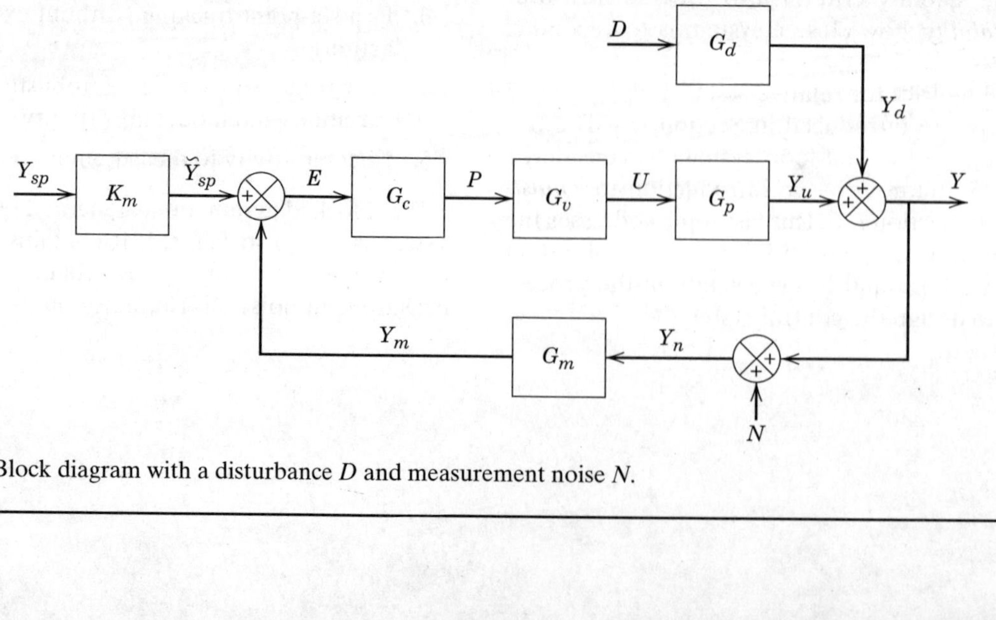

Figure L.1 Block diagram with a disturbance D and measurement noise N.

This example has demonstrated that even an exact cancellation of an unstable pole leads to instability. Consequently, an unstable pole should *never* be canceled with a right-half plane zero. In contrast, open-loop unstable systems can be stabilized with feedback control, as was demonstrated in Example 9.9.

The potential problem of instability arising from an unstable pole-zero cancellation motivates the concept of *internal stability*. A closed-loop system is said to be *internally stable* if all of the closed-loop transfer functions in Eqs. L-1 to L-3 are stable (Goodwin et al., 2001). This definition is equivalent to requiring that all signals in the feedback loop are bounded (Morari and Zafiriou, 1989). For the rest of this book, we assume that no unstable pole-zero cancellations occur in the formation of G_{OL}.

The next example demonstrates that design tradeoffs inherent in specifying G_c can be inferred from Eqs. L-1 to L-3.

In the next section, we consider one of the most important and useful frequency response results, the Bode stability criterion.

L.2 BODE STABILITY CRITERION

The Bode stability criterion has two important advantages in comparison with the Routh stability criterion of Chapter 9:

1. It provides exact results for processes with time delays, while the Routh stability criterion provides only approximate results due to the polynomial approximation that must be substituted for the time delay.
2. The Bode stability criterion provides a measure of the *relative stability* rather than merely a yes or no answer to the question, "Is the closed-loop system stable?"

Before considering the basis for the Bode stability criterion, it is useful to review the General Stability Criterion of Section 9.1:

A feedback control system is stable if and only if all roots of the characteristic equation lie to the left of the imaginary axis in the complex plane.

Thus, the imaginary axis divides the complex plane into stable and unstable regions. Recall that the characteristic equation was defined in Chapter 9 as

$$1+G_{OL}(s)=0 \tag{L-4}$$

where the open-loop transfer function is $G_{OL}(s) = G_c(s) G_v(s)G_p(s)G_m(s)$.

The root locus diagrams of Fig. L.2 and Section 9.5 show how the roots of the characteristic equation change as controller gain K_c changes. By definition, the roots of the characteristic equation are the numerical values of the complex variable s that satisfy Eq. L-4. Thus, each point on the root locus satisfies Eq. L-5, which is a rearrangement of Eq. L-4:

$$G_{OL}(s)=-1 \tag{L-5}$$

EXAMPLE L.2

Suppose that $G_d = G_p$, $G_m = K_m$ and that G_c is designed so that the closed-loop system is stable and $|GG_c| \gg 1$ over the frequency range of interest. Evaluate this control system design strategy for set-point changes, disturbances, and measurement noise. Also consider the behavior of the manipulated variable, U.

Solution

Because $|GG_c| \gg 1$,

$$\frac{1}{1+G_cG} \approx 0$$

The first expression and Eq. L-1 suggest that the output response to disturbances will be very good, because $Y/D \approx 0$. Next, we consider set-point responses. From Eq. L-1,

$$\frac{Y}{Y_{sp}}=\frac{K_mG_cG_vG_p}{1+G_cG}$$

Because $G_m = K_m$, $G = G_vG_pK_m$ and the above equation can be written as

$$\frac{Y}{Y_{sp}}=\frac{G_cG}{1+G_cG}$$

for $|GG_c| \gg 1$,

$$\frac{Y}{Y_{sp}} \approx 1$$

Thus, ideal (instantaneous) set-point tracking would occur.

Choosing G_c so that $|GG_c| \gg 1$ also has an undesirable consequence. The output Y becomes sensitive to noise, because $Y \approx -N$ (see the noise term in Eq. L-1). Thus, a design tradeoff is required between the set-point responses and sensitivity to noise.

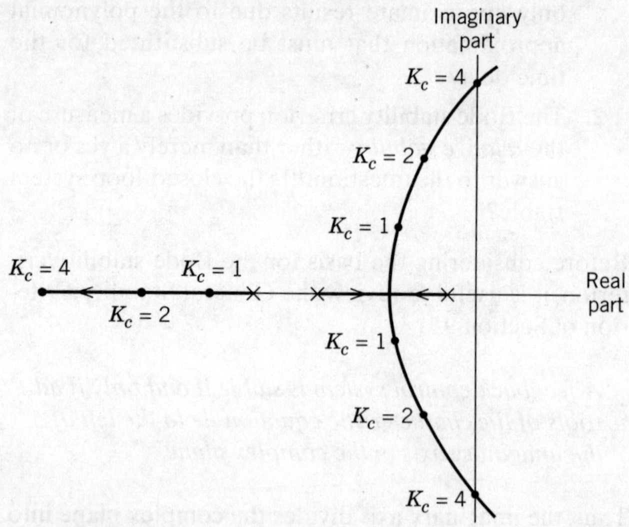

Figure L.2 Root locus diagram for a process with three poles and no zeroes.

The corresponding magnitude and argument are

$$|G_{OL}(j\omega)| = 1 \text{ and } \angle G_{OL}(j\omega) = -180° \quad \text{(L-6)}$$

In general, the *i*th root of the characteristic equation can be expressed as a complex number, $r_i = a_i \pm b_i j$ Note that complex roots occur as complex conjugate pairs. When a pair is located on the imaginary axis, the real part is zero ($a_i = 0$) and the closed-loop system is at the stability limit. As indicated in Chapter 9, this condition is referred to as *marginal stability* or *conditional stability*. When the closed-loop system is marginally stable and $b_i \neq 0$, the closed-loop response exhibits a sustained oscillation after a set-point change or a disturbance. Thus, the amplitude neither increases nor decreases. However, if K_c is increased slightly, the closed-loop system becomes unstable, because the complex roots on the imaginary axis move into the unstable region.[1]

For a marginally stable system with $b_i \neq 0$, the frequency of the sustained oscillation, ω_c, is given by $\omega_c = b_i$. This oscillatory behavior is caused by the pair of roots on the imaginary axis at $s = \pm\omega_c j$ (see Chapter 3). Substituting this expression for *s* into Eq. L-6 gives the following expressions for a conditionally stable system:

$$\text{AR}_{OL}(\omega_c) = |G_{OL}(j\omega_c)| = 1 \quad \text{(L-7)}$$

$$\phi_{OL}(\omega_c) = \angle G_{OL}(j\omega_c) = -180° \quad \text{(L-8)}$$

for some particular value of $\omega_c > 0$. Equations L-7 and L-8 provide the basis for both the Bode stability criterion and the Nyquist stability criterion of Section L.3.

Before stating the Bode stability criterion, we need to introduce two important definitions:

[1] For $K_c < 0$, instability occurs if K_c becomes more negative, that is, if $|K_c|$ becomes larger. In the subsequent analysis, assume that K_c is positive, but the results are also valid for $K_c < 0$ if K_c is replaced by $|K_c|$.

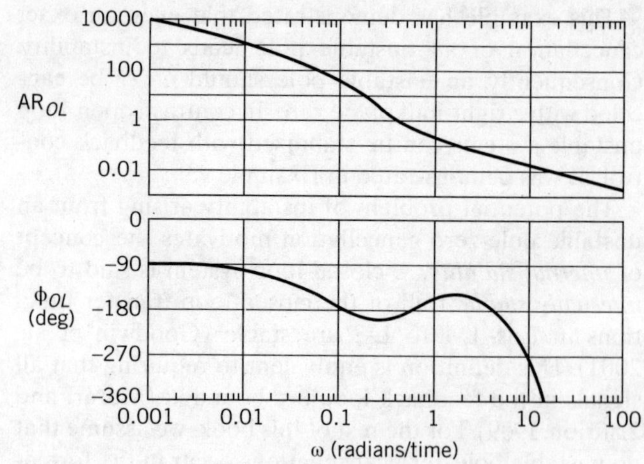

Figure L.3 Bode plot exhibiting multiple critical frequencies.

1. A *critical frequency* ω_c is defined to be a value of ω for which $\varphi_{OL}(\omega) = -180°$. This frequency is also referred to as a *phase crossover frequency*.
2. A *gain crossover frequency* ω_g is defined to be a value of ω for which $\text{AR}_{OL}(\omega) = 1$.

For a marginally stable system, $\omega_c = \omega_g$.

For many control problems, there is only a single ω_c and a single ω_g. But multiple values can occur, as shown in Fig. L.3 for ω_c. In this somewhat unusual situation, the closed-loop system is stable for two different ranges of the controller gain (Luyben and Luyben, 1997). Consequently, increasing the absolute value of K_c can actually improve the stability of the closed-loop system for certain ranges of K_c.

Next we state one of the most important results of frequency response analysis, the *Bode stability criterion*. It allows the stability of a closed-loop system to be determined from the open-loop transfer function.

Bode Stability Criterion. *Consider an open-loop transfer function $G_{OL} = G_c G_v G_p G_m$ that is strictly proper (more poles than zeros) and has no poles located on or to the right of the imaginary axis, with the possible exception of a single pole at the origin. Assume that the open-loop frequency response has only a single critical frequency ω_c and a single gain crossover frequency ω_g. Then the closed-loop system is stable if $AR_{OL}(\omega_c) < 1$. Otherwise, it is unstable.*

Some of the important properties of the Bode stability criterion are that

1. It provides a necessary and sufficient condition for closed-loop stability based on the properties of the open-loop transfer function.
2. Unlike the Routh stability criterion of Chapter 9, the Bode stability criterion is applicable to systems that contain time delays.

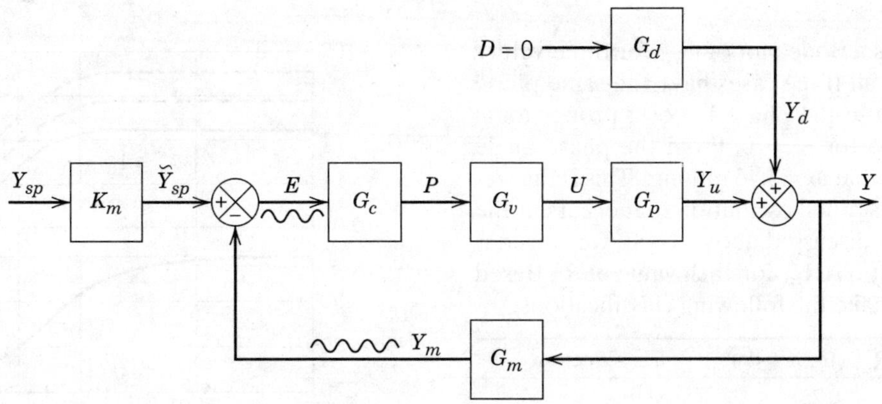

Figure L.4 Sustained oscillation in a feedback control system.

3. The Bode stability criterion is very useful for a wide variety of process control problems. However, for any $G_{OL}(s)$ that does not satisfy the required conditions, the Nyquist stability criterion of Section L.3 can be applied.
4. For systems with multiple ω_c or ω_g, the Bode stability criterion has been modified by Hahn et al. (2001) to provide a sufficient condition for stability.

In order to gain physical insight into why a sustained oscillation occurs at the stability limit, consider the analogy of an adult pushing a child on a swing. The child swings in the same arc so long as the adult pushes at the right time and with the right amount of force. Thus, the desired "sustained oscillation" places requirements on both timing (i.e., *phase*) and applied force (that is, *amplitude*). By contrast, if either the force or the timing is not correct, the desired swinging motion ceases, as the child will quickly exclaim. A similar requirement occurs when a person bounces a ball.

To further illustrate why feedback control can produce sustained oscillations, consider the following "thought experiment" for the feedback control system in Fig. L.4. Assume that the open-loop system is stable and that no disturbances occur ($D = 0$). Suppose that the set point is varied sinusoidally at the critical frequency, $y_{sp}(t) = A \sin(\omega_c t)$, for a long period, of time. Assume that during this period the measured output, y_m, is disconnected so that the feedback loop is broken before the comparator. After the initial transient dies out, y_m will oscillate at the excitation frequency ω_c because the response of a linear system to a sinusoidal input is a sinusoidal output at the same frequency (see Section 10.2). Suppose that two events occur simultaneously: (i) the set point is set to zero, and (ii) y_m is reconnected. If the feedback control system is *marginally stable*, the controlled variable *y* will then exhibit a sustained sinusoidal oscillation with amplitude *A* and frequency ω_c.

To analyze why this special type of oscillation occurs only when $\omega = \omega_c$, note that the sinusoidal signal *E* in Fig. L.4 passes through transfer functions G_c, G_v, G_p, and G_m before returning to the comparator. In order to have a sustained oscillation after the feedback loop is reconnected, signal Y_m must have the same amplitude as *E* and −180° phase shift relative to *E*. Note that the comparator also provides −180° phase shift due to its negative sign. Consequently, after Y_m passes through the comparator, it is in phase with *E* and has the same amplitude, *A*. Thus, the closed-loop system oscillates indefinitely after the feedback loop is closed because the conditions in Eqs. L-7 and L-8 are satisfied. But what happens if K_c is increased by a small amount? Then, $AR_{OL}(\omega_c)$ is greater than one, the oscillations grow, and the closed-loop system becomes unstable. In contrast, if K_c is reduced by a small amount, the oscillation is "damped" and eventually dies out.

EXAMPLE L.3

A process has the third-order transfer function (time constant in minutes)

$$G_p(s) = \frac{2}{(0.5s+1)^3}$$

Also, $G_v = 0.1$ and $G_m = 10$. For a proportional controller, evaluate the stability of the closed-loop control system using the Bode stability criterion and three values of K_c: 1, 4, and 20.

Solution

For this example,

$$G_{OL} = G_c G_v G_p G_m = (K_c)(0.1)\frac{2}{(0.5s+1)^3}(10) = \frac{2K_c}{(0.5s+1)^3}$$

Figure L.5 shows a Bode plot of G_{OL} for three values of K_c. Note that all three cases have the same phase angle plot, because the phase lag of a proportional controller is zero for $K_c > 0$. From the phase angle plot, we observe that $\omega_c = 3.46$ rad/min. This is the frequency of the sustained oscillation that occurs at the stability limit, as discussed above. Next, we consider the amplitude ratio AR_{OL} for each value of K_c. Based on Fig. L.5, we make the following classifications:

K_c	AR_{OL} (for $\omega = \omega_c$)	Classification
1	0.25	Stable
4	1	Marginally stable
20	5	Unstable

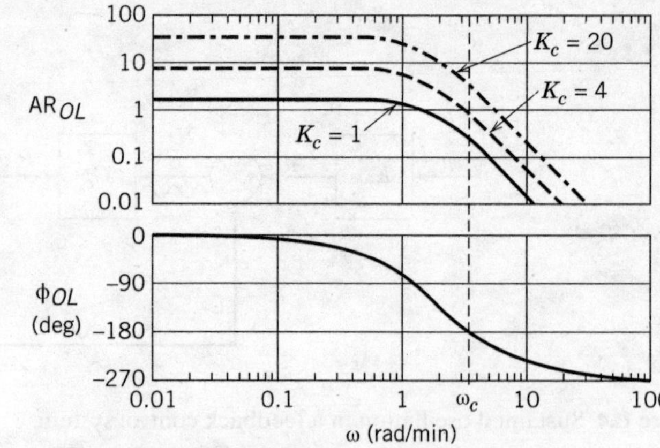

Figure L.5 Bode plots for $G_{OL} = 2K_c/(0.5s+1)^3$.

In Section 12.5.1, the concept of the ultimate gain was introduced. For proportional-only control, the ultimate gain K_{cu} was defined to be the largest value of K_c that results in a stable closed-loop system. The value of K_{cu} can be determined graphically from a Bode plot for transfer function $G = G_v G_p G_m$. For proportional-only control, $G_{OL} = K_c G$. Because a proportional controller has zero phase lag, ω_c is determined solely by G. Also,

$$AR_{OL}(\omega) = K_c AR_G(\omega) \tag{L-9}$$

where AR_G denotes the amplitude ratio of G. At the stability limit, $\omega = \omega_c$, $AR_{ol}(\omega_c) = 1$, and $K_c = K_{cu}$. Substituting these expressions into (L-9) and solving for K_{cu} gives an important result:

$$K_{cu} = \frac{1}{AR_G(\omega_c)} \tag{L-10}$$

The stability limit for K_c can also be calculated for PI and PID controllers, as demonstrated by Example L.4.

EXAMPLE L.4

Consider PI control of an overdamped second-order process (time constants in minutes),

$$G_p(s) = \frac{5}{(s+1)(0.5s+1)}$$

$$G_m = G_v = 1$$

Determine the value of K_{cu}. Use a Bode plot to show that controller settings of $K_c = 0.4$ and $\tau_i = 0.2$ min produce an unstable closed-loop system. Find K_{cm}, the maximum value of K_c that can be used with $\tau_i = 0.2$ min and still have closed-loop stability. Show that $\tau_i = 1$ min results in a stable closed-loop system for all positive values of K_c.

Solution

In order to determine K_{cu}, we let $G_c = K_c$. The open-loop transfer function is $G_{OL} = K_c G$ where $G = G_v G_p G_m$. Because a proportional controller does not introduce any phase lag, G and G_{OL} have identical phase angles. Consequently, the critical frequency can be determined graphically from the phase angle plot for G. However, curve A in Fig. L.6 indicates that ω_c does not exist, because φ_{OL} is always greater than −180°. As a result, K_{cu} does not exist, and thus K_c does not have a stability limit.

Conversely, the addition of integral control action can produce closed-loop instability. Curve B in Fig. L.6 indicates that an unstable closed-loop system occurs for $G_c(s) = 0.4(1 + 1/0.2s)$, because $AR_{OL} > 1$ when $\varphi_{OL} = -180°$.

To find K_{cm} for $\tau_i = 0.2$ min, we note that ω_c depends on τ_i, but not on K_c, because K_c has no effect on φ_{OL}. For curve B in Fig. L.6, $\omega_c = 2.2$ rad/min and the corresponding amplitude ratio is $AR_{OL} = 1.38$. To find K_{cm}, multiply the current value of K_c by a factor, 1/1.38. Thus, $K_{cm} = 0.4/1.38 = 0.29$.

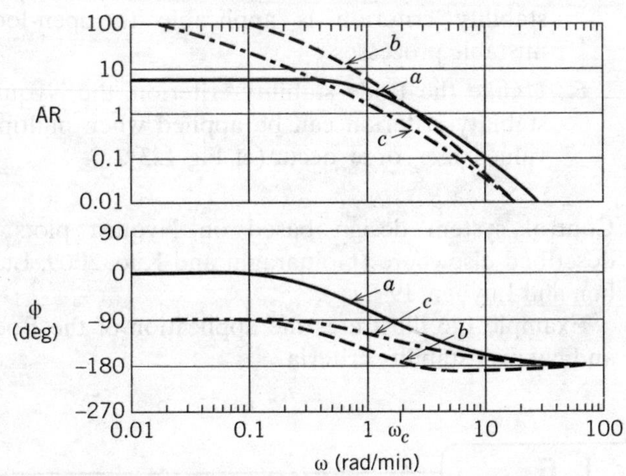

Figure L.6 Bode plots for Example L.4:

When τ_i is increased to 1 min, curve *C* in Fig. L.6 results. Because curve *C* does not have a critical frequency, the closed-loop system is stable for all positive values of K_c.

Curve A: $G_p(s)$

Curve B: $G_{OL}(s)$; $G_c(s) = 0.4\left(1+\dfrac{1}{0.2s}\right)$

Curve C: $G_{OL}(s)$; $G_c(s) = 0.4\left(1+\dfrac{1}{s}\right)$

EXAMPLE L.5

Find the critical frequency for the following process and PID controller, assuming $G_c = G_m = 1$.

$$G_p(s) = \frac{e^{-0.3s}}{(9s+1)(11s+1)} \qquad G_c(s) = 20\left(1+\frac{1}{2.5s}+s\right)$$

Solution

Figure L.3 shows the open-loop amplitude ratio and phase angle plots for G_{OL}. Note that the phase angle crosses −180° at three points. Because there is more than one value of ω_c, the Bode stability criterion cannot be applied. However, the Nyquist stability criterion presented in Section L.3 can be used to determine stability.

L.3 NYQUIST STABILITY CRITERION

The Nyquist stability criterion is similar to the Bode criterion in that it determines closed-loop stability from the open-loop frequency response characteristics. Both criteria provide convenient measures of relative stability, the *gain and phase margins*, which will be introduced in Section L.4. As the name implies, the Nyquist stability criterion is based on the Nyquist plot for $G_{OL}(s)$, a polar plot of its frequency response characteristics (see Chapter 10). The Nyquist stability criterion does not have the same restrictions as the Bode stability criterion, because it is applicable to open-loop unstable systems and to systems with multiple values of ω_c or ω_g. The Nyquist stability criterion is the most powerful stability test that is available for linear systems described by transfer function models.

The Nyquist stability criterion is based on two concepts from complex variable theory, *contour mapping* and the *Principle of the Argument*. We briefly review these concepts in Appendix M. More detailed descriptions are available elsewhere (Brown and Churchill, 2008; Franklin et al., 2014). We now present one of the most important results of frequency domain analysis:

Nyquist Stability Criterion. *Consider an open-loop transfer function $G_{OL}(s)$ that is proper and has no unstable pole-zero cancellations. Let N be the number of times that the Nyquist plot for $G_{OL}(s)$ encircles the (−1, 0) point in the clockwise direction. Also let P denote the number of poles of $G_{OL}(s)$ that lie to the right of the imaginary axis. Then, Z = N + P where Z is the number of roots (or zeros) of the characteristic equation that lie to the right of the imaginary axis. The closed-loop system is stable if and only if Z = 0.*

Some important properties of the Nyquist stability criterion are

1. It provides a necessary and sufficient condition for closed-loop stability based on the open-loop transfer function.
2. The reason that the (−1, 0) point is so important can be deduced from the characteristic equation, $1 + G_{OL}(s) = 0$. This equation can also be written as $G_{OL}(s) = -1$, which implies that $AR_{OL} = 1$ and

$\varphi_{OL} = -180°$, as noted earlier. This point is referred to as the *critical point*.

3. Most process control problems are open-loop stable. For these situations, $P = 0$, and thus $Z = N$. Consequently, the closed-loop system is unstable if the Nyquist plot for $G_{OL}(s)$ encircles the critical point, one or more times.
4. A negative value of N indicates that the critical point is encircled in the opposite direction (counterclockwise). This situation implies that each countercurrent encirclement can stabilize one unstable pole of the open-loop system.
5. Unlike the Bode stability criterion, the Nyquist stability criterion is applicable to open-loop unstable processes.
6. Unlike the Bode stability criterion, the Nyquist stability criterion can be applied when multiple values of ω_c or ω_g occur (cf. Fig. L.3).

Control system design based on Nyquist plots is described elsewhere (Golnaraghi and Kuo, 2009; Luyben and Luyben, 1997).

Example L.6 illustrates the application of the Bode and Nyquist stability criteria.

EXAMPLE L.6

Evaluate the stability of the closed-loop system in Fig. L.1 for

$$G_p(s) = \frac{4e^{-s}}{5s+1}$$

The time constant and time delay have units of minutes, and

$$G_v = 2, G_m = 0.25, G_c = K_c$$

Obtain ω_c and K_{cu} from a Bode plot. Let $K_c = 1.5K_{cu}$ and draw the Nyquist plot for the resulting open-loop system.

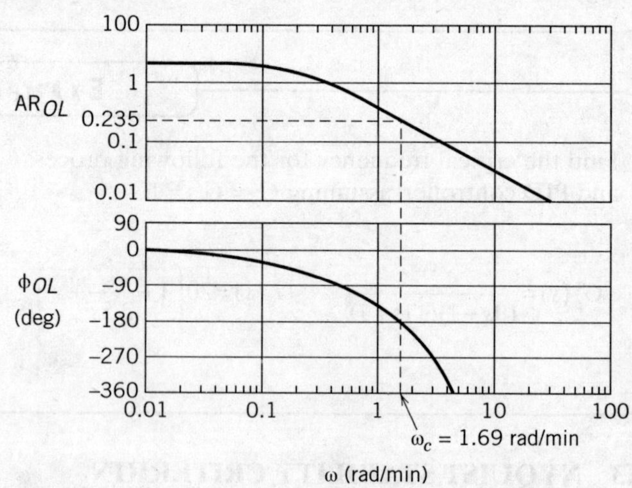

Figure L.7 Bode plots for Example L.6, $K_c = 1$.

Solution

The Bode plot for G_{OL} and $K_c = 1$ is shown in Fig. L.7. For $\omega_c = 1.69$ rad/min, $\varphi_{OL} = -180°$, and $AR_{OL} = 0.235$. For $K_c = 1$, $AR_{OL} = AR_G$ and K_{cu} can be calculated from Eq. L-10. Thus, $K_{cu} = 1/0.235 = 4.25$. Setting $K_c = 1.5K_{cu}$ gives $K_c = 6.38$.

The Nyquist plot for $K_c = 6.38$ is shown in Fig. L.8 (low-frequency data for $0 < \omega < 0.4$ have been omitted). Note that the −1 point is encircled once. Applying the Nyquist stability criterion gives $N = 1$, $P = 0$, and $Z = 1$. Thus, the larger value of K_c causes the closed-loop system to become unstable. Only values of K_c less than K_{cu} result in a stable closed-loop system.

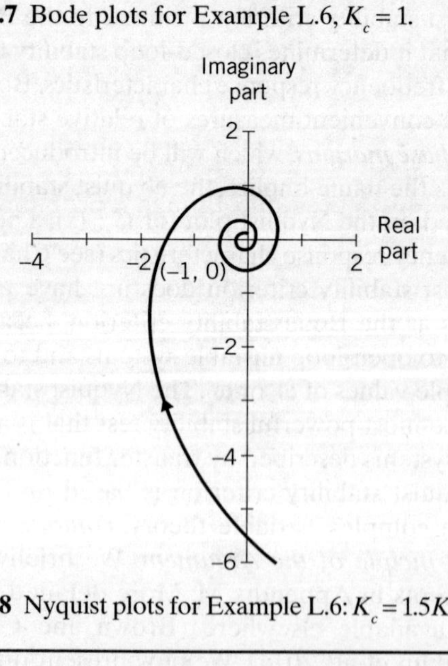

Figure L.8 Nyquist plots for Example L.6: $K_c = 1.5K_{cu} = 6.38$.

At this point, it is appropriate to summarize the relative advantages and disadvantages of the Bode and Nyquist plots. The Bode plot provides more information than the Nyquist plot, because the frequency is shown explicitly. In addition, it facilitates analysis over a wide range of frequencies due to its logarithmic frequency scale. Another advantage of the Bode plot is that it allows the open-loop frequency response characteristic to be graphically constructed from the characteristics for the individual transfer functions, G_c, G_v, G_p, and G_m, as shown in Chapter 10. The chief advantage of the Nyquist plot is that the Nyquist stability criterion is more widely applicable than the Bode stability criterion.

L.4 GAIN AND PHASE MARGINS

If the process changes or the controller is poorly tuned, the closed-loop system can become unstable. Thus, it is useful to have quantitative measures of *relative stability* that indicate how close the system is to becoming unstable. The concepts of *gain margin (GM)* and *phase margin (PM)* provide useful metrics for relative stability.

Let AR_c be the value of the open-loop amplitude ratio at the critical frequency ω_c. Gain margin GM is defined as

$$GM \triangleq \frac{1}{AR_c} \tag{L-11}$$

According to the Bode stability criterion, AR_c must be less than one for closed-loop stability. An equivalent stability requirement is that $GM > 1$. The gain margin provides a measure of relative stability, because it indicates how much any gain in the feedback loop can increase before instability occurs. For example, if $GM = 2.1$, either process gain K_p or controller gain K_c could be doubled and the closed-loop system would still be stable, though it probably would be very oscillatory.

Next, we consider the phase margin. In Fig. L.9, φ_g denotes the phase angle at the gain-crossover frequency ω_g where $AR_{ol} = 1$. Phase margin PM is defined as

$$PM \triangleq 180 + \phi_g \tag{L-12}$$

The phase margin also provides a measure of relative stability. In particular, it indicates how much additional time delay can be included in the feedback loop before instability will occur. Denote the additional time delay

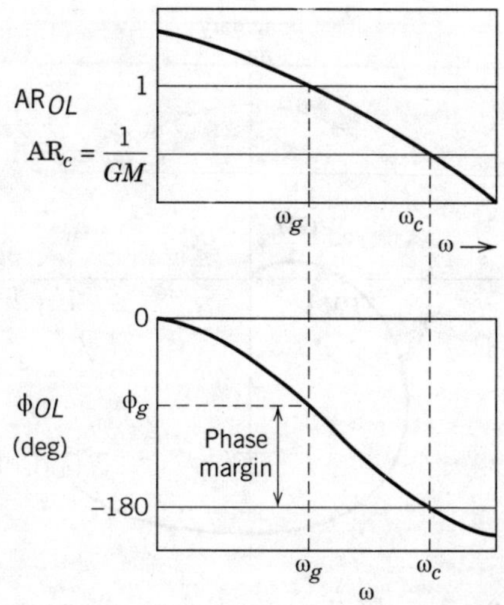

Figure L.9 Gain and phase margins on a Bode plot.

as Δ_{max}. For a time delay of Δ_{max}, the phase angle is $-\Delta_{max}\omega$ (see Section 14.3.5). Thus, Δ_{max} can be calculated from the expression

$$PM = \Delta\theta_{max}\,\omega_g \left(\frac{180^\circ}{\pi}\right) \tag{L-13}$$

or

$$\Delta\theta_{max} = \left(\frac{PM}{\omega_g}\right)\left(\frac{\pi}{180^\circ}\right) \tag{L-14}$$

where the $(\pi/180^\circ)$ factor converts PM from degrees to radians. Graphical representations of the gain and phase margins in Bode and Nyquist plots are shown in Figs. L.9 and L.10.

The specification of phase and gain margins requires a compromise between performance and robustness. In general, large values of GM and PM correspond to sluggish closed-loop responses, while smaller values result in less sluggish, more oscillatory responses. The choices for GM and PM should also reflect model accuracy and the expected process variability.

Guideline. *In general, a well-tuned controller should have a gain margin between 1.7 and 4.0 and a phase margin between 30° and 45°.*

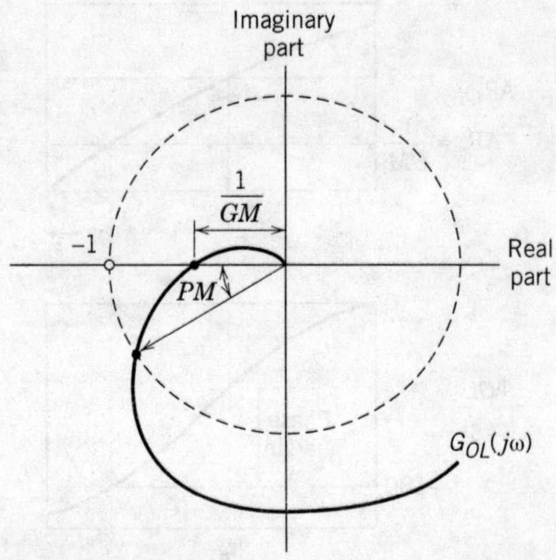

Figure L.10 Gain and phase margins on a Nyquist plot.

Recognize that these ranges are approximate, and that it may not be possible to choose PI or PID controller settings that result in specified *GM* and *PM* values. Tan et al. (1999) have developed graphical procedures for designing PI and PID controllers that satisfy *GM* and *PM* specifications. The *GM* and *PM* concepts are easily evaluated when the open-loop system does not have multiple values of ω_c or ω_g. However, for systems with multiple ω_g, gain margins can be determined from Nyquist plots (Doyle et al., 2009).

Although the gain and phase margins provide useful metrics for robustness, they can give misleading results for unusual situations. For example, the Nyquist plot for a stable open-loop process in Fig. L.12 exhibits large *GM* and *PM* values,

EXAMPLE L.7

For the FOPTD model of Example L.6, calculate PID controller settings for the two tuning relations in Table 12.6:

(a) Ziegler–Nichols
(b) Tyreus–Luyben

Assume that the two PID controllers are implemented in the parallel form with a derivative filter ($\alpha = 0.1$) in Table 8.1. Plot the open-loop Bode diagram and determine the gain and phase margins for each controller.

For the Tyreus–Luyben settings, determine the maximum increase in the time delay $\Delta\theta_{max}$ that can occur while still maintaining closed-loop stability.

Solution

From Example L.6, the ultimate gain is $K_{cu} = 4.25$ and the ultimate period is $P_u = 2\pi/1.69 = 3.72$ min. Therefore, the PID controllers have the following settings:

Controller Settings	K_c	τ_I (min)	τ_D (min)
Ziegler–Nichols	2.55	1.86	0.46
Tyreus–Luyben	1.91	8.18	0.59

The open-loop transfer function is

$$G_{OL} = G_c G_v G_p G_m = G_c \frac{2e^{-s}}{5s+1}$$

Figure L.11 shows the frequency response of G_{OL} for the two controllers. The gain and phase

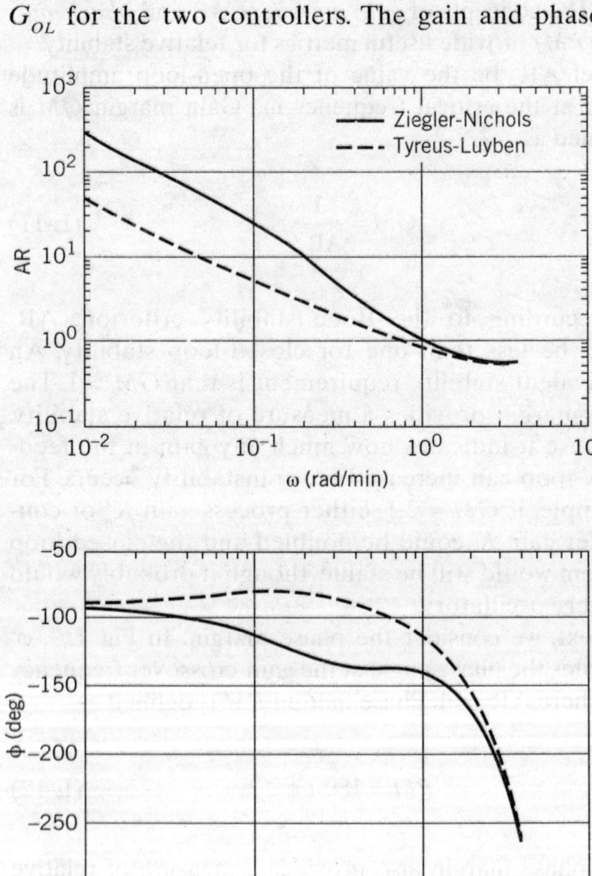

Figure L.11 Comparison of G_{OL} Bode plots for Example L.7

margins can be determined by inspection of the Bode diagram or by using the MATLAB command *margin*.

Controller	*GM*	*PM*	ω_c (rad/min)
Ziegler–Nichols	1.6	40°	2.29
Tyreus–Luyben	1.8	76°	2.51

The Tyreus–Luyben controller settings are more conservative due to the larger gain and phase margins.

The value of $\Delta\theta_{max}$ is calculated from Eq. L-14 and the information in the above table:

$$\Delta\theta_{max} = \frac{(76°)(\pi \text{ rad})}{(0.79 \text{ rad/min})(180°)} = 1.7 \text{ min}$$

Thus, time delay θ can increase by as much as 70% and still maintain closed-loop stability.

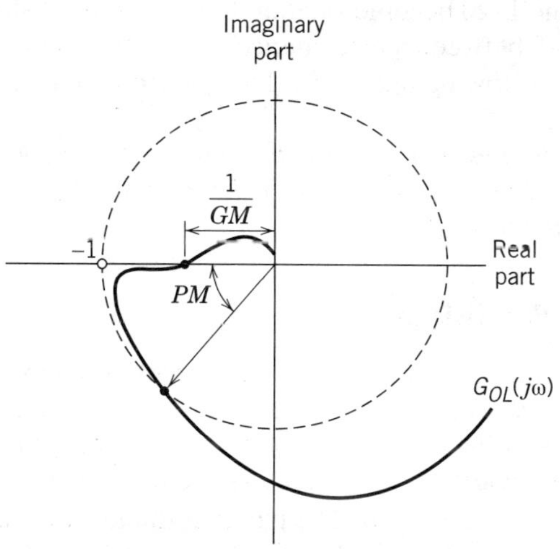

Figure L.12 Nyquist plot where the gain and phase margins are misleading.

but the Nyquist curve passes very close to the critical point. Thus, the closed-loop system is not very robust and a small process perturbation could cause instability. This potential shortcoming of the gain and phase margins can be avoided by considering the sensitivity functions that are introduced in the next section.

The open-loop transfer function and its frequency response characteristics play a key role in control system design and analysis. In the *loop-shaping* approach, the controller is designed so that the open-loop transfer function has the desired characteristics. These techniques are described in books on advanced control theory (Doyle et al., 2009; Skogestad and Postlethwaite, 2005).

L.5 CLOSED-LOOP FREQUENCY RESPONSE AND SENSITIVITY FUNCTIONS

The previous sections have demonstrated that open-loop frequency response characteristics play a key role in the analysis of closed-loop stability and robustness. However, the closed-loop frequency response characteristics also provide important information, as discussed in this section.

L.5.1 Sensitivity Functions

The following analysis is based on the block diagram in Fig. L.1. Define G as $G \triangleq G_v G_p G_m$ and assume that $G_m = K_m$ and $G_d = 1$. Two important concepts are now defined:

$$S \triangleq \frac{1}{1+G_c G} \quad \text{sensitivity function} \tag{L-15a}$$

$$T \triangleq \frac{G_c G}{1+G_c G} \quad \text{Complementary sensitivity function} \tag{L-15b}$$

Comparing Fig. L.1 and Eq. L-15 indicates that S is the closed-loop transfer function for disturbances (Y/D), while T is the closed-loop transfer function for set-point changes (Y/Y_{sp}). It is easy to show that

$$S + T = 1 \tag{L-16}$$

As will be shown in Section L.6, S and T provide measures of how sensitive the closed-loop system is to changes in the process.

Let $|S(j\omega)|$ and $|T(j\omega)|$ denote the amplitude ratios of S and T, respectively. The maximum values of the amplitude ratios provide useful measures of robustness. They also serve as control system design criteria, as discussed below. Define M_S to be the maximum value of $|S(j\omega)|$ for all frequencies:

$$M_s \triangleq \max_{\omega} |S(j\omega)| \tag{L-17}$$

The maximum value M_S also has a geometrical interpretation (Äström and Hägglund, 2006). Let $G_{OL}(s) = G_c(s)G(s)$. Then M_S is the inverse of the shortest distance from the Nyquist plot for $G_{OL}(s)$to the critical point. Thus, as M_S decreases, the closed-loop system becomes more robust.

The second robustness measure is M_T, the maximum value of $|T(j\omega)|$:

$$M_T \triangleq \max_{\omega} |T(j\omega)| \tag{L-18}$$

M_T is also referred to as the *resonant peak*. Typical amplitude ratio plots for S and T are shown in Fig. L.13. If the feedback controller includes integral action, offset is eliminated for set-point changes or sustained disturbances. Thus, at low frequencies, $|T(j\omega)| \rightarrow 1$ and $|S(j\omega)| \rightarrow 0$, as shown in Fig. L.13.

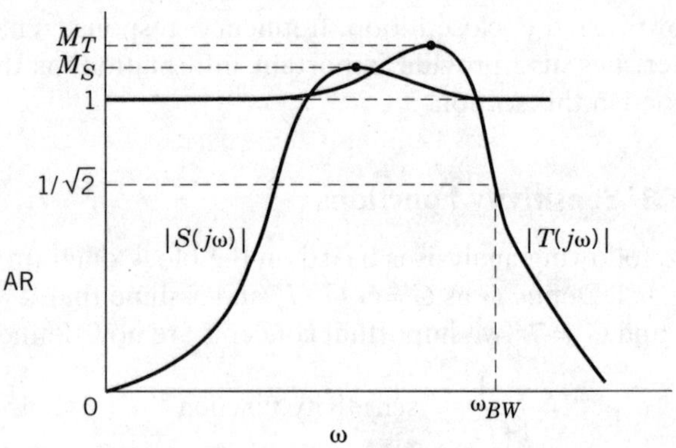

Figure L.13 Typical S and T magnitude plots (Modified from Maciejowski (1989)).

It is easy to prove that M_S and M_T are related to the gain and phase margins of Section L.4 (Skogestad and Postlethwaite, 2005):

$$GM \geq \frac{M_s}{M_s - 1}, \quad PM \geq 2\sin^{-1}\left(\frac{1}{2M_s}\right) \tag{L-19}$$

$$GM \geq 1 + \frac{1}{M_T}, \quad PM \geq 2\sin^{-1}\left(\frac{1}{2M_T}\right) \tag{L-20}$$

Designing a controller to have a specified value of M_T or M_S establishes lower bounds for *GM* and *PM*. For example, when $M_T = 1.25$, Eq. L-20 indicates that $GM \geq 1.8$ and $PM \geq 47°$, numerical values that satisfy the guidelines given in Section L.4. Equations L-19 and L-20 imply that better performance (corresponding to a larger value of M_T) is achieved at the expense of robustness, because the lower bounds for *GM* and *PM* in Eqs. L-19 and L-20 become smaller. In general, a satisfactory tradeoff between performance and robustness occurs for the following range of values (Äström et al., 1998):

Guidelines. *For a satisfactory control system, M_T should be in the range of 1.0-1.5, and M_S should be in the range of 1.2-2.0.*

L.5.2 Bandwidth

In this section, we introduce an important concept, the *bandwidth*. A typical amplitude ratio plot for T and the corresponding set-point response are shown in Fig. L.14. The bandwidth ω_{BW} is defined as the frequency at which $|T(j\omega)| = 1/\sqrt{2} = 0.707$. The bandwidth indicates the frequency range for which satisfactory set-point tracking occurs. In particular, ω_{BW} is the maximum frequency for a sinusoidal set point to be attenuated by no more than a factor of 0.707. The bandwidth is also related to speed of response. In general, the bandwidth is (approximately) inversely proportional to the closed-loop settling time.

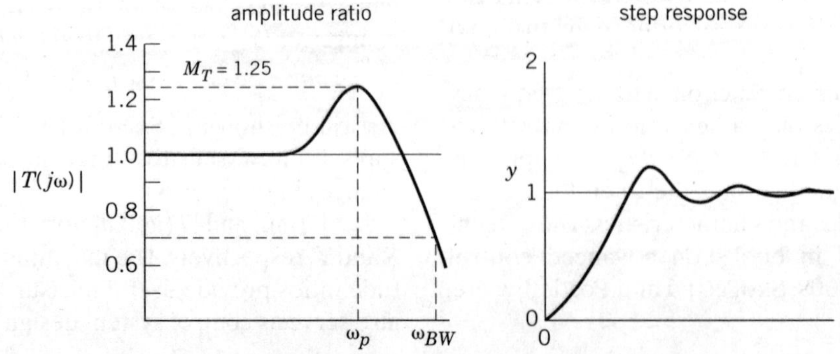

Figure L.14 Typical closed-loop amplitude ratio $|T(/ro)|$ and set-point response.

An alternative definition of the bandwidth is sometimes used. The bandwidth can also be defined as the frequency for which $|S(j\omega)| = 0.707$ when $|S(j\omega)|$ crosses 0.707 from below, as shown in Fig. L.13 (Skogestad and Postlethwaite, 2005). Fortunately, the two bandwidth definitions typically result in similar numerical values. Furthermore, the values of ω_{BW} and the gain crossover frequency ω_g are usually very close.

The curves in Fig. L.14 are similar to the corresponding plots for an underdamped second-order system (see Sections 5.2 and 10.3.3). For Fig.10.3, a peak amplitude ratio of 1.25 corresponds to a damping coefficient of $\zeta = 0.5$ for an underdamped second-order system. This analogy provides support for the previous guideline that M_T should have a value between 1.0 and 1.5.

L.5.3 Closed-Loop Performance Criteria

Ideally, a feedback controller should satisfy the following criteria:

1. In order to eliminate offset, $|T(j\omega)| \rightarrow 1$ as $\omega \rightarrow 0$.
2. $|T(j\omega)|$ should be maintained at unity up to as high as frequency as possible. This condition ensures a rapid approach to the new steady state during a set-point change.
3. As indicated in the guidelines, M_T should be selected so that $1.0 < M_T < 1.5$.
4. The bandwidth ω_{BW} and the frequency ω_T for M_T should be as large as possible. Large values result in fast closed-loop responses.

Satisfying these criteria typically requires a compromise. For example, the requirement that $M_T < 1.5$ means that the controller gain K_c cannot be too large. However, smaller values of K_c result in smaller values of ω_{BW} and ω_T.

Next, we consider the desired closed-loop frequency response characteristics for disturbances. Ideally, we would like to have the closed-loop amplitude ratio $|S(j\omega)|$ be zero for all frequencies. However, this ideal situation is physically impossible for feedback control, and thus a more realistic goal is to minimize $|S(j\omega)|$ over as wide a frequency range as possible. According to the guideline, the controller should be designed so that $1.2 < M_S < 2.0$.

L.5.4 Nichols Chart

The closed-loop frequency response can be calculated analytically from the open-loop frequency response. Again, consider Fig. L.1, and assume that the sensor dynamics are negligible so that $G_m(s) = K_m$. From Fig. L.1 and Eq. L-15b, it follows that the closed-transfer function for set-point changes Y/Y_{sp} is a function of G_{ol},

$$\frac{Y}{Y_{sp}} = T = \frac{G_{OL}}{1+G_{OL}} \tag{L-21}$$

where $G_{OL} \triangleq G_c G_v G_p G_m$. Denote the open-loop frequency response characteristics by AR_{OL} and φ_{OL}, and the corresponding closed-loop quantities, AR_{CL} and φ_{CL}. Equation L-21 and the analytical techniques of Chapter 10 can be employed to derive analytical expressions for AR_{CL} and φ_{CL} as functions of AR_{OL} and φ_{OL} (Luyben and Luyben, 1997).

The *Nichols chart* in Fig. L.15 provides a graphical display of the closed-loop frequency response characteristics for set-point changes when $G_m(s) = K_m$. Contours of

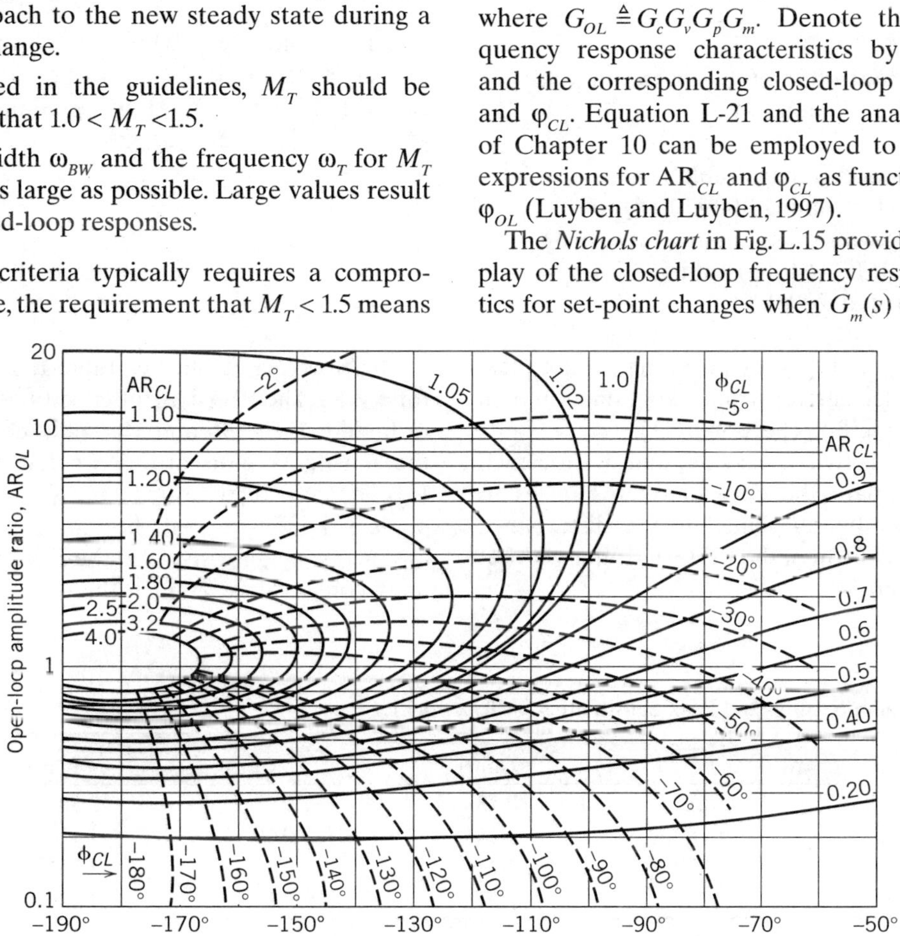

Figure L.15 A Nichols chart. [The closed-loop amplitude ratio AR_{CL} (___) and phase angle φ_{CL} (- - -) are shown as families of curves.]

constant AR_{CL} and φ_{CL} are shown on a plot of AR_{ol} vs. φ_{OL}. In a typical Nichols chart application, AR_{ol} and φ_{OL} are calculated from $G_{OL}(s)$ and plotted on the Nichols chart as a series of points. Then AR_{CL} and φ_{CL} are obtained by interpolation. For example, if $AR_{OL} = 1$ and $\varphi_{OL} = -100°$ at a certain frequency, then interpolation of Fig. L.15 gives $AR_{cl} = 0.76$ and $\varphi_{CL} = -50°$ for the same frequency. The Nichols chart can be generated in the *MATLAB Control Toolbox®* by a single command, *nichols*.

The Nichols chart served as a useful graphical technique for designing control systems prior to 1960, but it has become less important now that control software is widely available. Control system designs based on Nichols charts, including applications where $G_m(s) \neq K_m$, are described elsewhere (Golnaraghi and Kuo, 2009; Franklin et al., 2014).

The following example illustrates PID controller design for an M_S robustness constraint.

EXAMPLE L.8

Consider a fourth-order process with a wide range of time constants that have units in min (Åstrom et al., 1998):

$$G = G_v G_p G_m = \frac{1}{(s+1)(0.2s+1)(0.04s+1)(0.008s+1)} \quad \text{(L-22)}$$

Calculate PID controller settings based on following tuning relations in Chapter 11:

(a) Ziegler–Nichols tuning (Table 11.4)
(b) Tyreus–Luyben tuning (Table 10.5)
(c) IMC tuning with $\tau_c = 0.25$ min (Table 11.1)
(d) Simplified IMC (SIMC) tuning with $\tau_c = 0.25$ min (Skogestad, 2003)

For parts (c) and (d), use a second-order-plus-time-delay model derived from Skogestad's model approximation method (Section 6.3.1).

Determine sensitivity peaks M_S and M_T for each controller. Compare the closed-loop responses to step changes in the set-point and the disturbance using the parallel form of the PID controller without a derivative filter:

$$\tilde{G}_c(s) = K_c\left[1 + \frac{1}{\tau_I s} + \tau_D s\right] \quad \text{(L-23)}$$

Assume that $G_d(S) = G(S)$.

Solution

The Bode plot for $G(s)$ is not shown but indicates that $\omega_c = 11.18$ rad/min, $P_u = 2\pi/\omega_c = 0.562$ min, and $K_{cu} = 1/AR_c = 30.2$. Applying Skogestad's model approximation procedure gives:

$$\tilde{G}(s) = \frac{e^{-0.028s}}{(s+1)(0.22s+1)} \quad \text{(L-24)}$$

The SIMC controller in Table L.1 was calculated for $\tau_i < 8\theta$. The PID controller settings are compared in Table L.1. The M_S and M_T values determined from the sensitivity plots are also summarized in Table L.1. The controller settings and sensitivity values in Table L.1 indicate that the Z-N controller settings are the most aggressive and the IMC settings are the most conservative. In fact, the sensitivity values for

Table L.1 Controller Settings and Peak Sensitivities for Example L.8

Controller	K_c	τ_i (min)	τ_D (min)	M_S	M_T
Ziegler–Nichols	18.1	0.28	0.070	2.38	2.41
Tyreus–Luyben	13.60	1.25	0.089	1.45	1.23
IMC	4.39	1.22	0.180	1.13	1.00
Simplified IMC	21.8	1.22	0.180	1.58	1.16

Z-N controller are unacceptably large, according to the guidelines.

The closed-loop responses for a unit set-point change at $t = 0$ and a step disturbance ($d = 5$) at $t =$ 4 min are shown in Fig. L.16. The ZN controller provides the worst set-point response but the best disturbance response. The IMC controller is quite sluggish, because the approximate SOPTD model has a relatively small time delay. Similar results were obtained for the FOPTD model in Example 11.4. The SIMC and T-L controllers provide the best overall performance of these four controllers. However, a PID controller with improved performance for this example (not shown) can be obtained using a Direct Synthesis approach based on disturbance rejection (Chen and Seborg, 2002).

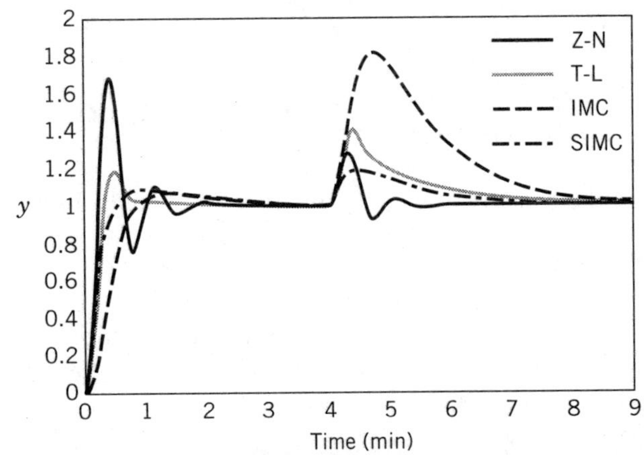

Figure L.16 Closed-loop responses for Example 10.8. A set-point change occurs at $t = 0$ and a step disturbance at $t = 4$ min.

Example L.8 has demonstrated that a variety of PID controllers can have approximately the same M_S value but different performance characteristics. Thus, although the guidelines for M_S and M_T provide useful limits concerning controller robustness, controller performance should also be considered. Robustness metrics such as M_S, M_T, GM, and PM should be evaluated in conjunction with controller design methods, especially the model-based techniques of Chapter 11.

L.6 ROBUSTNESS ANALYSIS

In order for a control system to function properly, it should not be unduly sensitive to small changes in the process or to inaccuracies in the process model, if a model is used to design the control system. A control system that satisfies this requirement is said to be *robust* or *insensitive*. It is very important to consider robustness as well as performance in control system design. First, we explain why the S and T transfer functions in Eq. L-15 are referred to as "sensitivity functions."

L.6.1 Sensitivity Analysis

In general, the term *sensitivity* refers to the effect that a change in one transfer function (or variable) has on another transfer function (or variable). Suppose that G changes from a nominal value G_0 to an arbitrary new value, $G_0 + dG$. This differential change dG causes T to change from its nominal value T_0 to a new value, $T_0 + dT$. Thus, we are interested in the ratio of these changes, dT/dG, as well as in the ratio of the relative changes:

$$\frac{dT/T}{dG/G} \triangleq \text{sensitivity} \tag{L-25}$$

The sensitivity is of particular interest because it is dimensionless and independent of the units of G and T. It is evaluated for a specific condition such as G_0 and T_0. We can write the sensitivity in an equivalent form:

$$\frac{dT/T}{dG/G} = \left(\frac{dT}{dG}\right)\frac{G}{T} \tag{L-26}$$

The derivative in Eq. L-26 can be evaluated after substituting the definition of T in Eq. L-15b:

$$\frac{dT}{dG} = G_c S^2 \tag{L-27}$$

Substitute Eq. L-27 into Eq. L-26. Then substituting the definition of S in Eq. L-15a and rearranging gives the desired result:

$$\frac{dT/T}{dG/G} = \frac{1}{1+G_cG} = S \tag{L-28}$$

Equation L-28indicates that the sensitivity is equal to S. For this reason, S is referred to as the *sensitivity function*. In view of the important relationship in Eq. L-16, T is called the *complementary sensitivity function* (Joshi et al., 2014).

L.6.2 Effect of Feedback Control on Sensitivity

Next, we show that feedback reduces sensitivity by comparing the relative sensitivities for open-loop control and closed-loop control. By definition, open-loop con trol occurs when the feedback control loop in Fig. L.1 is disconnected from the comparator. For this condition:

$$\left(\frac{Y}{Y_{sp}}\right)_{OL} = T_{OL} \triangleq G_cG \tag{L-29}$$

Substituting T_{OL} for T in Eq. L-25 and noting that $dT_{OL}/dG = G_c$ gives:

$$\frac{dT_{OL}/T_{OL}}{DG/G} = \left(\frac{dT_{OL}}{dG}\right)\frac{G}{T_{OL}} = G_c\frac{G}{G_cG} = 1 \tag{L-30}$$

Thus, the sensitivity is unity for open-loop control and is equal to S for closed-loop control, as indicated by Eq. L-28. Equation L-15a indicates that $|S| < 1$ if $|G_cG_p| > 1$, which usually occurs over the frequency range of interest. Thus, we have identified one of the most important properties of feedback control: *Feedback control makes process performance less sensitive to changes in the process.*

As indicated in the previous section, it would be desirable to make S very small at all frequencies. But this is not possible owing to the following integral constraint:

The Bode Sensitivity Integral **(Goodwin et al., 2001).** Consider a stable feedback control system with an open-loop transfer function, $G_{OL}(s) = G_c(s)G(s)e^{-\theta s}$. Assume that $G(s)$ has no poles to the right of the imaginary axis. Then,

$$\int_0^\infty \ln|S(j\omega)|d\omega = 0 \tag{L-31}$$

Thus, if S is small for a particular frequency range, it has to be large for other frequencies. This phenomenon is sometimes referred to as the "waterbed effect," because pushing down on one part of a waterbed causes another part to rise. Similar integral constraints are available for specific situations, including systems with right-half plane poles and zeros (Skogestad and Postlethwaite, 2005).

L.6.3 Robust Stability

The robustness of a control system can be analyzed theoretically if the degree of process variability (or model uncertainty) can be characterized. The uncertainty can be structured (e.g., parameter variations) or unstructured (e.g., variations in the transfer function or frequency response). In this section, we consider the important concept of *robust stability*. The related concept of *robust performance* is described elsewhere (Morari and Zafiriou, 1989; Skogestad and Postlethwaite, 2005).

In robustness analysis, the unstructured uncertainty is often expressed as bounds on a nominal process model. For example, suppose that $\tilde{G}$ is the process model used for control system design, and that G is the transfer function for the actual (but unknown) process. They can be related by an unstructured model uncertainty, $\Delta(s)$:

$$G(s) = [1 + \Delta(s)]\tilde{G}(s) \tag{L-32}$$

Rearranging Eq. L-32 indicates that $\Delta(s)$ can also be interpreted as the relative model error:

$$\Delta(s) = \frac{G(s) - \tilde{G}(s)}{\tilde{G}(s)} \tag{L-33}$$

Robustness analysis is typically based on a magnitude bound for the uncertainty

$$|\Delta(j\omega)| \le \ell(\omega) \tag{L-34}$$

where $\ell(\omega)$ is a known function of frequency. The multiplicative uncertainty description in Eq. L-32 provides the basis for an important *robust stability* result. The following theorem guarantees closed-loop stability for a specified degree of model uncertainty.

Robust Stability Theorem. Consider the feedback control system in Fig. L.1 and a specific controller G_c. Assume that

(i) The closed-loop system is stable for the nominal closed-loop system that consists of G_c and the nominal process model, $\tilde{G}$.

(ii) The magnitude of the unstructured uncertainty $\Delta(s)$ in Eq. L-33 is bounded by $\ell(\omega)$ in Eq. L-34. Also assume that each $G(s)$ in Eq. L-32 that is generated by this uncertainty structure has the same number of right-half-plane poles.

Then the closed-loop system is robustly stable for all G that satisfy Eqs. L-33 and L-34, if and only if

$$|\tilde{T}(j\omega)| < \frac{1}{\ell(\omega)} \quad \text{for all } \omega > 0 \tag{L-35}$$

or, equivalently,

$$|\tilde{T}(j\omega)|\ell(\omega) < 1 \quad \text{for all } \omega > 0 \tag{L-36}$$

where $\tilde{T}$ denotes the nominal complementary sensitivity function, $\tilde{T} = G_c\tilde{G}/(1 + G_c\tilde{G})$.

The proof for this theorem is available in textbooks on advanced control such as Bélanger (2005). Some important consequences of this theorem are (Skogestad and Postlethwaite, 2005) that

1. The theorem has a graphical interpretation, as shown in Fig. L.17. Robust stability is guaranteed if the uncertainty region at each frequency ro,a circle with radius $|G_{OL}(j\omega)|\ell(\omega)$, does not include

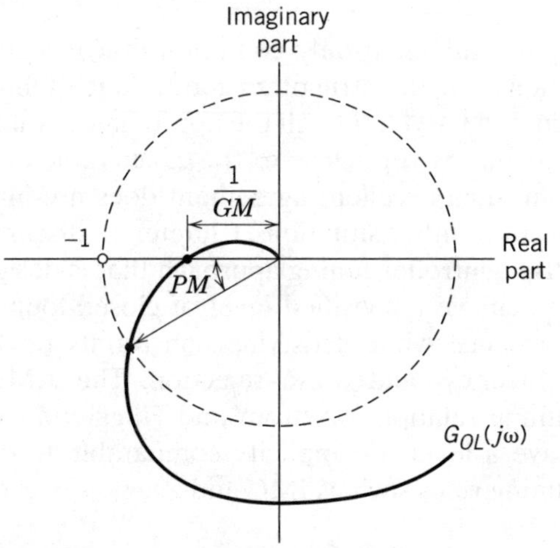

Figure L.17 Graphical interpretation of the Robust Stability Theorem.

the critical point at (−1, 0). Figure L.17 shows the uncertainty region for one frequency; the circles for the other frequencies have been omitted. In many practical problems, not all of the perturbations allowed by Eq. L-34 are possible. For example, upper and lower limits on model parameters such as a gain or time constant produce uncertainty regions that are not circular, as they appear in Fig. L.17. For these situations, the robust stability theorem provides sufficient (but not necessary) conditions. Consequently, the results tend to be conservative, as will be demonstrated in Example L.9.

2. In general, when the uncertainty bound $\ell(\omega)$ is large, the controller design must be more conservative so that $|\tilde{T}(j\omega)|$ will be small enough to satisfy Eqs. L-35 and L-36.

In summary, if the conditions of this theorem are satisfied, robust stability is guaranteed for the assumed process uncertainty description in Eqs. L-32 to L-34. The theorem also allows determination of how much modeling error can be tolerated for a particular controller design, as is demonstrated by the

EXAMPLE L.9

Consider the nominal process model:

$$\tilde{G}(s)=\frac{-1.6(1-0.5s)}{s(3s+1)}$$

The PID controller settings for the IMC method and $\tau_c = 1.25$ are: $K_c = -1.22$, $\tau_i = 6$ min, $\tau_d = 1.5$ min and derivative filter $\alpha = 0.1$. This value of τ_c was chosen to provide a peak sensitivity value of 1.96 (Chen and Seborg, 2002).

(a) Suppose that a small measurement time constant τ_m was neglected in developing the nominal model. Thus, the actual process transfer function G is given by

$$G(s)=\frac{\tilde{G}(s)}{\tau_m s+1}$$

Use the robust stability theorem to determine the largest value of τ_m for which robust stability can be guaranteed.

(b) Repeat part (a) using an exact stability analysis.

(c) Do the answers of part (a) and (b) agree? If not, explain which estimate is more accurate and why the discrepancy occurs.

Solution

(a) The complementary sensitivity function for the nominal model and the PID controller is

$$\tilde{T}=\frac{G_c\tilde{G}}{1+G_cG}$$

$$=\frac{\left(\dfrac{-1.6(1-0.5s)}{s(3s)+1}\right)(-1.22)\left(1+\dfrac{1}{6s}+\dfrac{1.5s}{(0.1)1.5s+1}\right)}{1+\left(\dfrac{-1.6(1-0.5s)}{s(3s+1)}\right)(-1.22)\left(1+\dfrac{1}{6s}+\dfrac{1.5s}{(0.1)1.5s+1}\right)}$$

Simplifying gives

$$\tilde{T}=\frac{-9.66s^3+13.3s^2+11.0s+1.95}{2.75s^4+9.21s^3+19.3s^2+11.0s+1.95}$$

The model uncertainty expression is given by Eq. L-33:

$$\Delta(s)=\frac{G(s)-\tilde{G}(s)}{\tilde{G}(s)}=\frac{-\tau_m s}{\tau_m s+1} \tag{L-37}$$

The bound $\ell(\omega)$ in Eq. L-34 can be set equal to the magnitude of $\Delta(s)$:

$$\ell(\omega) = |\Delta(j\omega)| = \frac{\tau_m \omega}{\sqrt{\tau_m^2 \omega^2 + 1}} \qquad \text{(L-38)}$$

The robust stability theorem can be used to determine the maximum value of τ_m for which robust stability can be guaranteed. For example, a plot of $|\tilde{T}(j\omega)|\,\ell(\omega)$ vs. ω can be constructed for a small value of τ_m, and the condition in Eq. L-35 can be checked to confirm that it is satisfied. Then τ_m is gradually increased until Eq. L-35 is no longer valid. Representative results for three values of τ_m are shown in Fig. L.18. By additional trial and error, the maximum value of τ_m = 1.88 min.

(b) The characteristic equation is given by $1 + G_c G = 0$. Substituting gives

$$1+\left(\frac{-1.6(1-0.5s)}{s(3s+1)}\right)(-1.22)\left(1+\frac{1}{6s}+\frac{1.5s}{(0.1)1.5s+1}\right)\times\left(\frac{1}{\tau_m s+1}\right)=0$$

Because the characteristic equation is a polynomial in s, the maximum value is found to be τ_m = 1.93 using direct substitution (see Chapter 9).

(c) For this example, the robust stability theorem provides a slightly conservative estimate due to the sufficient-only condition that is a consequence of the structured model uncertainty in Eq. L-33. Even though the approximate value of 1.88 min is very close to the exact value of 1.93 min, such excellent agreement does not always occur in other situations. Chapter 11 discusses a PID controller tuning approach that is designed to achieve a specified level of closed-loop performance while satisfying constraints on both robustness and noise rejection. The AMIGO tuning relations (Äström and Hägglund, 2006) have a level of simplicity comparable to other tuning rules such as IMC and Ziegler–Nichols.

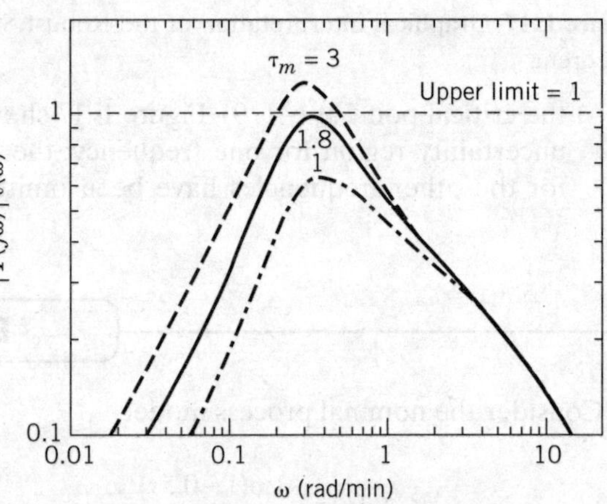

Figure L.18 Robustness results for Example L.9.

following example. An application of robust control theory to a pH control example is presented by Joshi et al. (2014).

Chapter 11 discusses a PID controller tuning approach that is designed to achieve a specified level of closed-loop performance while satisfying constraints on both robustness and noise rejection. The AMIGO tuning relations (Äström and Hägglund, 2006) have a level of simplicity comparable to other tuning rules such as IMC and Ziegler–Nichols.

SUMMARY

Frequency response techniques are powerful tools for the design and analysis of feedback control systems. The Bode and Nyquist stability criteria provide exact stability results for a wide variety of control problems, including processes with time delays. They also provide convenient measures of relative stability, such as gain and phase margins. Closed-loop frequency response concepts such as sensitivity functions and bandwidth can be used to characterize closed-loop performance.

Control system design involves tradeoffs between control system performance and robustness. Modern control systems are typically designed using a model-based technique such as the ones described in Chapter 11. The robustness of a control system can then be analyzed using the concepts of the sensitivity and complementary sensitivity functions that were introduced in this chapter. The robust stability theorem can be used to ensure that the control system can accommodate a specified degree of model uncertainty, or to determine how much uncertainty can be tolerated before the closed-loop system becomes unstable. Frequency response analysis is greatly facilitated by software tools that generate the required graphical and numerical results.

REFERENCES

Åström, K. J., and T Hägglund, *Advanced PID Control,* ISA, Research Triangle Park, NC, 2006.

Åström, K. J., H. Panagopoulos, and T. Hägglund, Design of PI Controllers Based on Non-Convex Optimization, *Automatica*, 34, 585 (1998).

Bélanger, P. R., *Control Engineering: A Modern Approach*, Oxford University Press London, 2005.

Brown, J.W., and R.V. Churchill, *Complex Variables and Applications*, 8th ed., McGraw-Hill, New York, 2008.

Chen, D., and D.E. Seborg, PI/PID Controller Design Based on Direct Synthesis and Disturbance Rejection, *Ind. Eng. Chem. Res.*, 41, 4807 (2002).

Doyle, J. C., B. A. Francis, and A. R. Tannenbaum, *Feedback Control Theory*, Dover, New York, 2009.

Franklin, G. F., J. D. Powell, and A. Emami-Naeini, *Feedback Control of Dynamic Systems*, 7th ed., Prentice Hall, Upper Saddle River, NJ, 2014.

Golnaraghi, F., and B. C. Kuo, *Automatic Control Systems*, 8th ed., John Wiley and Sons, Hoboken, NJ, 2009.

Goodwin, G. C., S. F. Graebe, and M. E. Salgado, *Control System Design*, Prentice Hall, Upper Saddle River, NJ, 2001.

Hahn, J., T. Edison, and T. F. Edgar, A Note on Stability Analysis Using Bode Plots, *Chem. Eng. Educ.* 35(3), 208 (2001).

Joshi, R., K. Tsakalis, J. W. McArthur, and S. Dash, Account for Uncertainty with Robust Control Design, *Chem. Engr. Prog.*, 110, 31 (November, 2014); 46 (December, 2014).

Luyben, W. L., and M. L. Luyben, *Essentials of Process Control*, McGraw-Hill, New York, 1997, Chapter 11.

MacFarlane, A. G. J., The Development of Frequency Response Methods in Automatic Control, *IEEE Trans. Auto. Control*, AC-24, 250 (1979).

Maciejowski, J. M., *Multivariable Feedback Design*, Addison-Wesley, New York, 1989.

Morari, M., and E. Zafiriou, *Robust Process Control*, Prentice Hall, Englewood Cliffs, NJ, 1989.

Skogestad, S., Simple Analytic Rules for Model Reduction and PID Controller Tuning, *J. Process Control*, 13, 291 (2003).

Skogestad, S., and I. Postlethwaite, *Multivariable Feedback Design: Analysis and Design*, 2nd ed., John Wiley and Sons, Hoboken, NJ, 2005.

Tan, K. K., Q.-G. Wang, C. C. Hang, and T. Hägglund, *Advances in PID Control*, Springer, New York, 1999.

APPENDIX M

Contour Mapping and the Principle of the Argument

APPENDIX CONTENT

The concept of contour mapping is illustrated in Fig. M.1. A closed contour C_s in the complex s-plane is mapped by a transfer function $H(s)$ into another closed contour C_H in the complex H-plane. For each point on C_s, there is a corresponding point on C_H. For example, three arbitrarily chosen points s_1, s_2, and s_3 on the C_s contour map into points $H(s_1)$, $H(s_2)$, and $H(s_3)$ on the C_H contour. Suppose that C_s is traversed in the clockwise direction, starting at s_1 and continuing along C_s to s_2 and s_3, before eventually returning to s_1. Then C_H will also be traversed, starting at $H(s_1)$ and continuing to $H(s_2)$ and $H(s_3)$ before eventually returning to the starting point. In Fig. M.1, a clockwise traverse of C_s results in a clockwise traverse of C_H. However, this is not always the case; a counterclockwise traverse of C_H could result, depending on the particular $H(s)$ that is considered.

The concept of contour encirclement plays a key role in Nyquist stability theory. A contour is said to make a *clockwise encirclement* of a point if the point is always to the right of the contour as the contour is traversed in the clockwise direction. Thus, a single traverse of either C_H or C_s in Fig. M.1 results in a clockwise encirclement of the origin. The number of encirclements by C_H is related to the poles and zeroes of $H(s)$ that are located inside of C_s, by a well-known result from complex variable theory (Brown and Churchill, 2014; Franklin et al., 2014).

Principle of the Argument. *Consider a transfer function $H(s)$ and a closed contour C_s in the complex s-plane that is traversed in the clockwise (positive) direction. Assume that C_s does not pass through any poles or zeroes of $H(s)$. Let N be the number of clockwise (positive) encirclements of the origin by contour C_H in the complex H-plane.*

Define P and Z to be the numbers of poles and zeroes of $H(s)$, respectively, that are encircled by C_s in the clockwise direction. Then $N = Z - P$.

Note that N is negative when $P > Z$. For this situation, the C_H contour encircles the origin in the *counterclockwise* (or *negative*) direction. Next, we show that the Nyquist Stability Criterion is based on a direct application of the Principle of the Argument.

M.1 DEVELOPMENT OF THE NYQUIST STABILITY CRITERION

According to the General Stability Criterion of Chapter 9, a feedback control system is stable if and only if all roots of the characteristic equation lie to the left of the imaginary axis. This condition motivates the following choices for function $H(s)$ and contour C_s:

1. Let $H(s) = 1 + G_{OL}(s)$, where $G_{OL}(s)$ is the open loop transfer function, $G_{OL}(s) = G_c(s)G_v(s)G_p(s)G_m(s)$. Assume that $G_{OL}(s)$ is *strictly proper* (more poles than zeros) and does not contain any unstable pole-zero cancellations.
2. Contour C_s is chosen to be the boundary of the open right-half-plane (RHP). Assume that it is traversed in the clockwise (positive) direction.

This choice of C_s creates a dilemma–how do we evaluate $H(s)$ on the boundary of an infinite region? This problem is solved by choosing C_s to be the *Nyquist contour* shown in Fig. M.2. The Nyquist contour consists of the imaginary axis and a semicircle with radius, $R \rightarrow \infty$. Because $G_{OL}(s)$ is strictly, $G_{OL}(s) \rightarrow 0$ as $R \rightarrow \infty$ and the semicircular arc of the Nyquist contour maps into

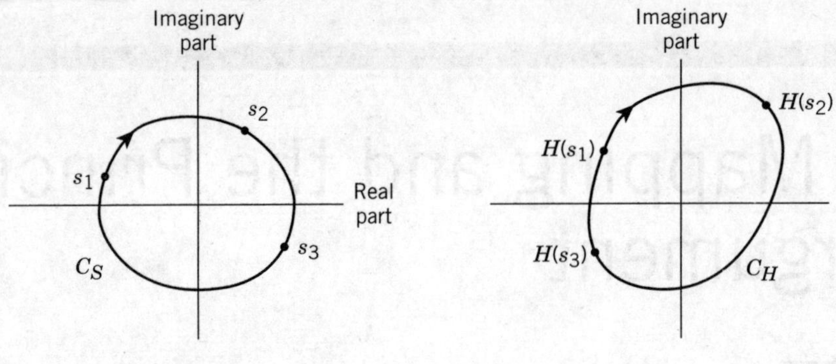

(*a*) *s*-plane (*b*) *H*(*s*)-plane

Figure M.1 Contour mapping for a transfer function, $H(s)$.

the origin of the H-plane. Thus the imaginary axis is the only portion of the Nyquist contour that needs to be considered. In other words, we only have to evaluate $G_{OL}(s)$ for $s = j\omega$ and $-\infty < \omega < \infty$.

In order to apply the Principle of the Argument, we need to determine P, the number of poles of $H(s) = 1 + G_{OL}(s)$ that lie in the RHP. Fortunately, this is easy because P is equal to the number of poles of $G_{OL}(s)$ that lie in the RHP. To prove this, suppose that $G_{OL}(s)$ has the general form,

$$G_{OL}(s) = \frac{A(s)e^{-\theta s}}{B(s)} \tag{M-1}$$

where $A(s)$ and $B(s)$ are polynomials in s and $G_{OL}(s)$ does not contain any unstable pole-zero cancellations. Then,

$$H(s) = 1 + G_{OL}(s) = 1 + \frac{A(s)e^{-\theta s}}{B(s)} = \frac{B(s) + A(s)e^{-\theta s}}{B(s)} \tag{M-2}$$

Because $H(s)$ and $G_{OL}(s)$ have the same denominator, they have the same number of RHP poles.

Recall that $H(s)$ was defined as $H(s) = 1 + G_{OL}(s)$. Thus, the C_H and $C_{G_{OL}}$ contours have the same shape but the C_H contour is shifted to the left by -1, relative to the $C_{G_{OL}}$ contour. Consequently, encirclements of the origin by C_H are identical to encirclements of the -1 point by $C_{G_{OL}}$. As a result, it is more convenient to express the Nyquist Stability Criterion in terms of $G_{OL}(s)$ rather than $H(s)$.

One more issue needs to be addressed, namely, the condition that C_s contour cannot pass through any pole or zero of $G_{OL}(s)$. Open-loop transfer functions often have a pole at the origin owing to an integrating element or integral control action. This difficulty is avoided by using the modified Nyquist contour in Fig. M.2, where $\varepsilon \ll 1$. A similar modification is available for the unusual situation where $G_{OL}(s)$ has a pair of complex poles on the imaginary axis. These modifications are described elsewhere (Golnaraghi and Kuo, 2009; Franklin et al., 2014). Although conceptually important, we do not have to be overly concerned with these modifications, because they are typically incorporated in software for control applications.

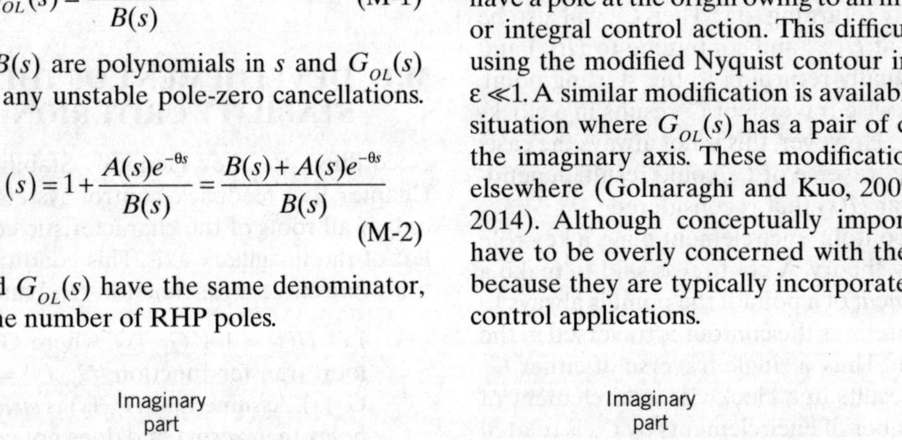

(*a*) Original Nyquist Contour (*b*) Modified Nyquist Contour (for the pole at the origin)

Figure M.2 Original and modified Nyquist contours.

Several slightly different versions of the Nyquist Stability Criterion are available. The version in Appendix L is based on two mildly restrictive assumptions:

A1. The open-loop transfer function $G_{OL}(s)$ is proper; that is, the number of zeros does not exceed the number of poles. Also, $G_{OL}(s)$ has no unstable pole-zero cancellations.

A2. The modified Nyquist contour C_s circumvents any open-loop poles that lie on the imaginary axis. This contour is traversed in the clockwise (positive) direction.

Assumption A1 is not very restrictive for process control problems. For practical control problems, $G_{OL}(s)$ will be proper, and an unstable pole-zero cancellation would be both unusual and avoidable. Similarly, Assumption A2 is not restrictive, because an analogous Nyquist Stability Criterion is available for traverses in the counterclockwise direction (Golnaraghi and Kuo, 2009).

REFERENCES

Brown, J. W., and R. V. Churchill, *Complex Variables and Application*, 9th ed., McGraw-Hill, New York, 2014.

Franklin, G. F., J. D. Powell, and A. Emami-Naeini, *Feedback Control of Dynamic Systems*, 7th ed., Pearson Prentice Hall, Upper Saddle River, NJ, 2014.

Golnaraghi, M. F., and B. C. Kuo, *Automatic Control Systems*, 9th ed., John Wiley and Sons, Hoboken, NJ, 2009.

APPENDIX N

Partial Fraction Expansions for Repeated and Complex Factors

APPENDIX CONTENTS

In this appendix, the partial fraction expansions (PFEs) of Laplace transform expressions are illustrated for two special cases: repeated factors and complex factors.

N.1 REPEATED FACTORS

In order to illustrate the PFE for repeated factors, consider the following example.

EXAMPLE N.1

Evaluate the α_i coefficients for

$$Y(s) = \frac{s+1}{s(s+2)^2} = \frac{\alpha_1}{s+2} + \frac{\alpha_2}{s(s+2)^2} + \frac{\alpha_3}{s} \tag{N-1}$$

Solution

Note that the PFE in Eq. N-1 contains an $(s + 2)$ term, in addition to the $(s + 2)^2$ term. But why is the $(s + 2)$ term needed? The answer is that omitting it would prevent Eq. N-1 from having an unique set of α_i that would satisfy Eq. N-1 for all values of s.

The Heaviside rule cannot be used to find α_i because multiplication of Eq. N-1 by $s = -2$ causes the middle term on the right side to be unbounded, rather than zero as desired. But the Heaviside expansion method can still be employed to find the other two coefficients (α_2 and α_3). Thus multiplying Eq. N-1 by $(s + 2)^2$ and letting $s = -2$ yields

$$\alpha_2 = \left.\frac{s+1}{s}\right|_{s=-2} = \frac{1}{2} \tag{N-2}$$

Multiplying Eq. N-1 by s and letting $s = 0$ gives

$$\alpha_3 = \left.\frac{s+1}{s(s+2)^2}\right|_{s=0} = \frac{1}{4} \tag{N-3}$$

Then solve for α_1 by specifying an arbitrary value of s in Eq. N-1. For example, substituting $s = -1$ into Eq. N-1 gives

$$0 = \alpha_1 + \alpha_2 - \alpha_3 \tag{N-4}$$

$$\alpha_1 = \alpha_2 + \alpha_3 = -\frac{1}{4} \tag{N-5}$$

An alternative approach for finding α_1 is to use differentiation. First, Eq. N-1 is multiplied by $s(s + 2)^2$,

$$s+1 = \alpha_1(s+2)s + \alpha_2 s + \alpha_3(s+2)^2 \tag{N-6}$$

Then Eq. N-6 is differentiated twice with respect to s,

$$0 = 2\alpha_1 + 2\alpha_3 \Rightarrow \alpha_1 = -\alpha_3 = -\frac{1}{4} \tag{N-7}$$

Note that differentiation gives the same results as the first approach in Eqs. N-3 through N-5.

N.1.1 A General Method for Partial Fraction Expansion with Repeated Factors

The differentiation approach illustrated above can be used as the basis of a more general method for repeated factors. Consider an expression of the form

$$Y(s) = \frac{N(s)}{D(s)} \tag{N-8}$$

If the denominator polynomial $D(s)$ contains a repeated factor $(s+b)^r$, multiply Eq. N-8 by this factor:

$$Q(s)=\frac{N(s)}{D(s)}(s+b)^r=(s+b)^{r-1}\alpha_1+(s+b)^{r-2}\alpha_2+\dots$$
$$+\alpha_r+(s+b)^r\times[\text{other partial fraction}] \qquad \text{(N-9)}$$

where $N(s)$ and $D(s)$ are polynomials in s.

Setting $s=-b$ will generate α_r directly. Differentiating $Q(s)$ with respect to s and then setting $s=-b$ generates α_{r-1}. Successive differentiation of a total of $r-1$ times will generate all α_i, $i=1,2,\dots,r$. Thus, the general expression is

$$\alpha_{r-i}=\frac{1}{i!}\frac{d^{(i)}Q(s)}{ds^{(i)}}\bigg|_{s=-b}\; i=0,\dots,r-1 \qquad \text{(N-10)}$$

For $i=0$ in Eq. N-10, the factorial symbol 0! is defined to be one and the zeroth-order derivative of $Q(s)$ is defined to be $Q(s)$itself.

N.2 COMPLEX FACTORS

An important special case for PFE occurs when $Y(s)$ in Eq. N-8 is factored and has a factor of the form

$$\frac{c_1s+c_0}{s^2+d_{1s}+d_0} \qquad \text{(N-11)}$$

with

$$\frac{d_1^2}{4}<d_0 \qquad \text{(N-12)}$$

Then the denominator of Eq. N-11 has complex roots (cf. the quadratic formula) and $Y(s)$ is said to have *complex factors.*

N.2.1 A General Method Partial Fraction Expansion with Complex Factors

The general procedure for performing a PFE when there is a complex factor can be summarized as follows.

First, convert Eq. N-11 to the standard form on the right side of Eq. N-13 by *completing the square:*

$$\frac{c_1s+c_0}{s^2+d_1s+d_0}=\frac{a_1(s+b)+a_2}{(s+b)^2+\omega^2} \qquad \text{(N-13)}$$

Next, determine the values of four parameters: a_1, a_2, b, and ω^2. Parameters b and ω^2 can easily be obtained by equating the denominators of both sides of Eq. N-13 and expanding the term on the left side:

$$s^2+d_1s+d_0=s^2+2bs+b^2+\omega^2 \qquad \text{(N-14)}$$

Equating coefficients for like powers of s gives expressions for b and ω^2 in terms of the known d_1 and d_2:

$$s^1: d_1=2b\Rightarrow \boxed{b=\frac{d_1}{2}}$$

$$s^0: d_0=b^2+\omega^2\Rightarrow\omega^2=d_0-b^2\Rightarrow\boxed{\omega^2=d_0-\frac{d_1^2}{4}} \qquad \text{(N-15)}$$

To find a_1 and a_2, multiply both sides of Eq. N-13 by $(s^2+d_1s+d_0)[(s+b)^2+\omega^2]$:

$$[(s+b)^2+\omega^2](c_1s+c_0)=(s^2+d_1s+d_0)[a_1(s+b)+a_2] \qquad \text{(N-16)}$$

Then substitute two arbitrary value of s into Eq. N-16 to generate two equations and solve simultaneously for a_1 and a_2.

Table 3.1 indicates that the inverse Laplace transform of the right side of Eq. N-13 is

$$a_1e^{-bt}\cos\omega t+\frac{a2}{\omega}e^{-bt}\sin\omega t \qquad \text{(N-17)}$$

The following example illustrates how the PFE needs to be modified for complex roots.

EXAMPLE N.2

Determine the PFE and the inverse Laplace transform, $y(t)$, for

$$Y(s)=\frac{s+2}{s^2+s+1} \qquad \text{(N-18)}$$

Solution

First, we complete the square in the denominator:

$$Y(s)=\frac{s+2}{(s+0.5)^2-0.25+1}=\frac{(s+0.5)+1.5}{(s+0.5)^2+\left(\frac{\sqrt{3}}{2}\right)^2} \qquad \text{(N-19)}$$

Divide $Y(s)$ into two terms,

$$Y(s) = \frac{s+0.5}{(s+0.5)^2 + \left(\frac{\sqrt{3}}{2}\right)^2} + \frac{1.5}{(s+0.5)^2 + \left(\frac{\sqrt{3}}{2}\right)^2} \quad \text{(N-20)}$$

To determine $y(t)$, invert each term separately. In Table 3.1, $\frac{s+b}{(s+b)^2+\omega^2}$ transforms to $e^{-bt}\cos \omega t$, while $\frac{\omega}{(s+b)^2+\omega^2}$ transforms to $e^{-bt}\sin \omega t$. Therefore, the corresponding time-domain expression is

$$y(t) = e^{-0.5t}\cos\frac{\sqrt{3}}{2}t + \sqrt{3}e^{-0.5t}\sin\frac{\sqrt{3}}{2}t \quad \text{(N-21)}$$

This example has demonstrated the general result that when $Y(s)$ has complex factors, $y(t)$ exhibits oscillatory behavior.

In the next example, $Y(s)$ has both repeated and complex factors.

EXAMPLE N.3

Find the inverse Laplace transform of

$$Y(s) = \frac{s+1}{s^2(s^2+4s+5)} \quad \text{(N-22)}$$

Solution

The denominator of Eq. N-22 has two repeated roots at $s = 0$. The roots of $s^2 + 4s + 5$ are complex because the inequality in Eq. N-12 is satisfied. The appropriate partial fraction for Eq. N-22 is

$$Y(s) = \frac{s+1}{s^2(s^2+4s+5)} = \frac{\alpha_1}{s} + \frac{\alpha_2}{s^2} + \frac{\alpha_3 s + \alpha_4}{s^2+4s+5} \quad \text{(N-23)}$$

Multiply both sides of Eq. 3-73 by $s^2(s^2 + 4s + 5)$ and collect terms:

$$s+1 = (\alpha_1+\alpha_3)s^3 + (4\alpha_1+\alpha_2+\alpha_4)s^2 + (5\alpha_1+4\alpha_2)s + 5\alpha_2 \quad \text{(N-24)}$$

Equate coefficients of like powers of s:

$$s^3: \alpha_1 + \alpha_3 = 0 \quad \text{(N-25a)}$$

$$s^2: 4\alpha_1 + \alpha_2 + \alpha_4 = 0 \quad \text{(N-25b)}$$

$$s^1: 5\alpha_1 + 4\alpha_2 = 1 \quad \text{(N-25c)}$$

$$s^0: 5\alpha_2 = 1 \quad \text{(N-25d)}$$

Solving simultaneously gives $\alpha_1 = 0.04$, $\alpha_2 = 0.2$, $\alpha_3 = 0.04$, and $\alpha_4 = -0.36$.

Before using Table 3.1, the denominator term $(s^2 + 4s + 5)$ must be converted to the standard form by completing the square. In order to match the expressions in Table 3.1, the PFE for the last term in Eq. N-23 must be expanded as

$$\frac{-0.04s-0.36}{(s+2)^2+1} = \frac{-0.04(s+2)}{(s+2)^2+1} + \frac{-0.28}{(s+2)^2+1} \quad \text{(N-26)}$$

This procedure yields the desired expression:

$$y(t) = 0.04 + 0.2t - 0.04e^{-2t}\cos t - 0.28e^{-2t}\sin t$$

It is clear from this example that the Laplace transform solution for complex or repeated roots can be quite cumbersome when $Y(s)$ has a high-order denominator. For this situation, using Symbolic Math software techniques may be a more efficient way to obtain a solution (see Section 3.6).

INDEX

D

E

F